QUATERNARY GEOLOGY OF THE UNITED STATES

INQUA 2003 Field Guide Volume

Edited by

Don J. Easterbrook
Dept. of Geology
Western Washington University
Bellingham, WA 98225

XVI INQUA Congress

Desert Research Institute
2215 Raggio Parkway
Reno, Nevada 89512-1095

Published by The Desert Research Institute
2215 Raggio Parkway, Reno, Nevada, USA 89512-1095

Printed in U.S.A.

Cover: Cirques and glaciated peaks in the North Cascades Washington. Photo by D.A. Rahm, courtesy of EPIC.

Contents

Foreword iv

1 ***Marine terraces, sea level history and Quaternary tectonics of the San Andreas fault on the coast of California*** 1
Muhs, D.R., Prentice, C., and Merritts, D.J.

2 ***Quaternary geology of the western United States*** 19
Easterbrook, D. J., Pierce, K., Gosse, J., Gillespie, A., Evenson, E., and Hamblin, K.

3 ***Quaternary stratigraphy, geomorphology, soils, and alpine archaeology in an alpine-to-plains transect, Colorado Front Range*** 81
Dethier, D.P., Benedict, J.B., Birkeland, P.W., Caine, N., Davis, P.T., Madole, R.F., Patterson, P.E., Price, A.B., Schildgen, T.F., and Shroba, R. R.

4 ***Soil geomorphological studies in the Mojave Desert: Impacts of Quaternary tectonics, climate, and rock type on soils, landscapes, and plant_community ecology*** 105
Eppes, M. C., McDonald, E. V., and McFadden, L. D., 2003,

5 ***The Palouse loess and the Channeled Scabland: A paired Ice–Age geologic system*** 123
Gaylord D. R., Busacca, A. J., and Sweeney, M. R.

6 ***Contrasting glacial landscapes created by ice lobes of the southern Laurentide Ice Sheet*** 135
Patterson, C.J., Hansel, A. K., Mickelson, D. M., Quade, D. J., Bettis E. A.III, Colgan, P. M., McKay, E. D., and Stumpf, A. J.

7 ***Pliocene to Holocene lakes in the western Great Basin: New perspectives on paleoclimate, landscape dynamics, tectonics, and paleodistribution of aquatic species*** 155
Reheis, M., Redwine, J., Adams, K., Stine, S., Parker, K., Negrini, R., Burke, R., Kurth, G., McGeehin, J., Paces, J., Phillips, F., Sarna–Wojcicki, A., and Smoot, J.

8 ***Cenozoic eolian sedimentary systems of the USA midcontinent*** 195
Bettis E.A. III, Mason J.P., Swinehart, J.B., Miao, X., Hanson, P. R., Goble, R.J., Loope, D.B., Jacobs, P.M., and Roberts, H.M.

9 ***Margin of Laurentide ice to the Atlantic Coastal Plain: Miocene-Pleistocene landscape evolution in the Central Appalachians*** 219
Braun, D.D., Pazzaglia, F.J., and Potter, N., Jr.

10 ***Glacial geology of upper Cook Inlet, Matanuska Glacier and Denali highway, Alaska*** 245
Larson, G.J., Evenson, E.B., Lawson, D.E., Ensminger, S.L., Baker, G., and Alley, R.B.

11 ***Cordilleran Ice Sheet glaciation of the Puget Lowland and Columbia Plateau and alpine glaciation of the North Cascade Range, Washington*** 265
Easterbrook, D.J.

12 ***Mountain glaciations of the Sierra Nevada*** 287
Clark, D., Gillespie, A.R., Clark, M., Burke, R.

13 ***Quaternary geology and ecology of the greater Yellowstone area*** 313
Pierce, K.L., Despain, D.,Whitlock, C., Cannon, K.P., Meyer, G.A., and Morgan, L.A.

14 ***The Rocky Mountain glacial model: The Wind River Range, Wyoming*** 345
Dahms, D.E., Hall, R.D., Shroba, R.R., Sorenson, C.J., Lynch, E.A., Applegarth, M.T., and Dillon, J.

15 ***Quaternary and geomorphic processes and landforms along a traverse across northern New England*** 365
Davis, P.T., Bierman, P.R., Brigham-Grette, J., Fitzgerald, D.M., Fowler, B.K., Retelle, M.J., Ridge, J.C., Thompson W.B., and Weddle, T.K.

16 ***Geo-ubiquity: Chemical, hydrological and physical evolution of coastal terraces near Santa Cruz, California*** 399
Schulz, M., W. Harden, J., and Weber, G.E.

17 ***Pleistocene and Holocene geomorphology of Marble and Grand Canyons, canyon cutting to adaptive management*** 407
Pederson, J.L., Schmidt, J.C., and Anders, M.D.

Foreword

Much of the landscape in the United States was shaped by climatic events during the Quaternary, especially the erosion and deposition in the Pleistocene. The wealth of Quaternary features found in the U.S. includes continental ice sheet glaciation, alpine glaciation, marine shorelines, marine deposits, faulting, tectonic uplift, effects of isostatic rebound, pluvial lakes, large scale eolian deposits, the world's largest floods from bursting of glacial lake Missoula, the world's largest volcanic eruptions, the world's largest geothermal area, and much more.

This volume contains field guides for 17 field excursions in the U.S. It consists of contributions from 97 authors across the country from Alaska and the west coast to New England, including much new, previously unpublished information. Each field guide includes specific sites with interpretations of the features to be seen and discussions of critical issues.

I would like to thank all of the authors, field trip organizers, and leaders for the thousands of hours of work that went into making this volume. I would especially like to thank Heather Sutphin at the Geological Society of America for her efforts in designing and laying out the material and for her patience in dealing with all of the details from such a large number of authors. We all hope that you will enjoy using the field guides and discover many new any interesting features of the Quaternary geology of the U.S.

Don J. Easterbrook

Marine terraces, sea level history and Quaternary tectonics of the San Andreas fault on the coast of California

Authors and Leaders: **Daniel R. Muhs,** *U.S. Geological Survey, MS 980, Box 25046, Federal Center, Denver, Colorado 80225*
Carol Prentice, *U.S. Geological Survey, MS 977, 345 Middlefield Road, Menlo Park, California 94025*
Dorothy J. Merritts, *Department of Geosciences, P.O. Box 3003, Franklin and Marshall College, Lancaster, Pennsylvania 17604-3003*

TECTONIC SETTING OF NORTHERN CALIFORNIA

The interaction of the Pacific and North American plates has dominated the tectonics of what is now coastal California since the early Miocene (Atwater, 1970). The major plate-boundary structures belong to the San Andreas fault system, a set of northwest-trending, right-lateral, strike-slip faults. In the San Francisco Bay area, the most important faults of this system are the San Andreas, Hayward, Calaveras, and San Gregorio (Fig. 1). The San Andreas fault continues north of San Francisco at least as far as Shelter Cove (Brown, 1995; Prentice et al., 1999; Merritts et al., 2000), but other faults within the San Andreas system accommodate a significant percentage of the plate motion north and east of San Francisco (Argus and Gordon, 2001; Savage et al., 1998; Kelson et al., 1992). This field trip will focus primarily on the San Andreas fault, but also includes one stop along the San Gregorio fault.

Although the Pacific-North American plate boundary is a dominantly right-lateral, transform system, slight variations in plate motion vectors give rise to a small amount of fault-normal compression. This component of compression produces the ongoing uplift of the Coast Ranges that has resulted in the emergence of Quaternary marine terraces along most of the California coast. These marine terraces provide a tool for understanding rates of tectonic uplift and deformation, as well as data that allow a better understanding of climatically driven, eustatic sea level change during the Quaternary.

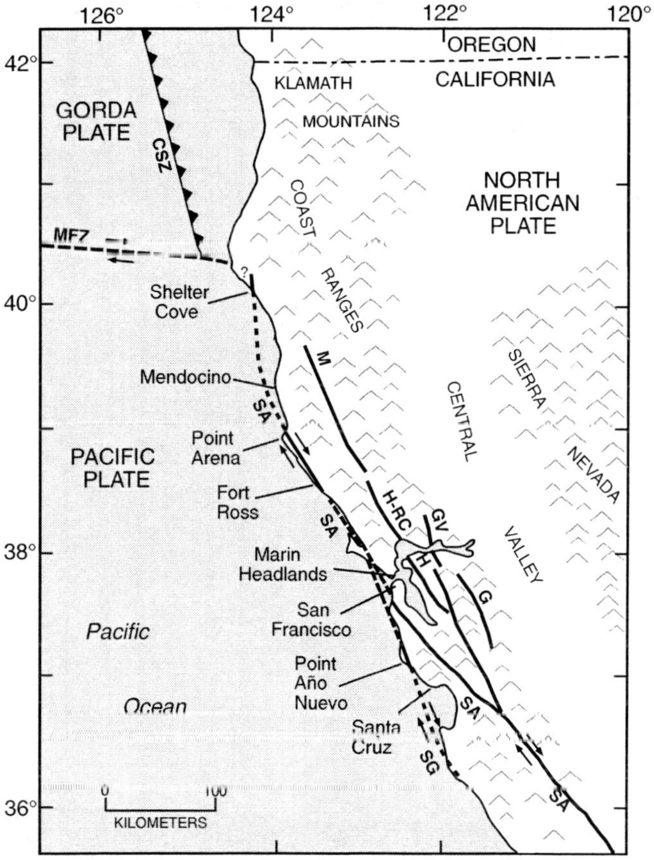

Figure 1. Map of coastal northern California showing major faults and other tectonic features and selected field trip stops. Fault and tectonic features from Merritts (1996) and Wakabayashi (1999). Abbreviations for faults and other structural features: SA, San Andreas; SG, San Gregorio; H, Hayward; H-RC, Healdsburg-Rodgers Creek; GV, Green Valley; G, Greenville; M, Maacama; MFZ, Mendocino fracture zone; CSZ, Cascadia subduction zone.

MARINE TERRACES AND THE NATURE OF THE SEA LEVEL RECORD

The Quaternary geologic record of high sea-stands takes the form of emergent reefs (on tropical shores) or marine terraces (on high-energy, mid-latitude coasts), on either tectonically stable or rising coasts. In contrast to tropical coral reefs, which are constructional landforms, marine terraces are emergent, wave-cut platforms veneered with thin, sometimes fossiliferous, marine sand and gravel. On rising crustal blocks, such as most of coastal California, interglacial high sea-stands leave a stair-step flight of marine terraces (Fig. 2).

Marine terraces have been studied in central California for more than a century, since the pioneering study of Lawson (1893). Alexander (1953), studying the region just south of Santa Cruz, first recognized that California marine terraces are the result of Quaternary sea level high-stands superimposed on a tectonically rising crustal block.

Bradley and Griggs (1976) mapped six prominent terraces along ~40 km of coastline between Santa Cruz and Point Año Nuevo (Fig. 3). The lowest of these, the Santa Cruz terrace, appears

Muhs, D.R., Prentice, C., Merritts, D.J., 2003, Marine terraces, sea level history and Quaternary tectonics of the San Andreas fault on the coast of California: *in* Easterbrook, D., ed., Quaternary Geology of the United States, INQUA 2003 Field Guide Volume, Desert Research Institute, Reno, NV, p. 1–18.

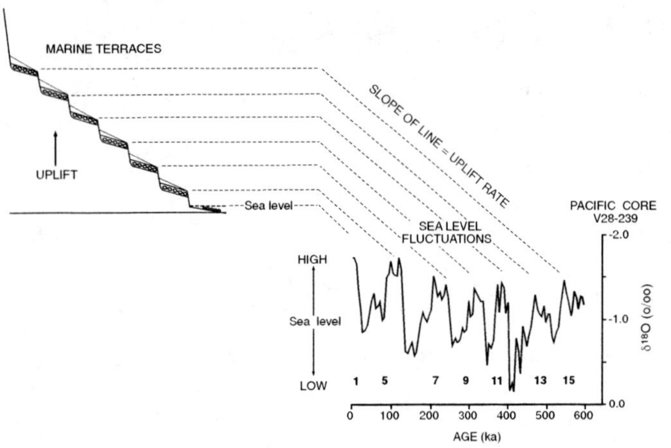

Figure 2. Relation of a flight of marine terraces to the long-term sea level record, as shown in the oxygen isotope variations of deep-sea foraminifera. Example of a marine terrace flight on an uplifting coast is hypothetical; sea level fluctuations from deep-sea foraminifera are from equatorial Pacific core V28-239 (Shackleton and Opdyke, 1976). Concept modified from one presented originally by Horsfield (1975).

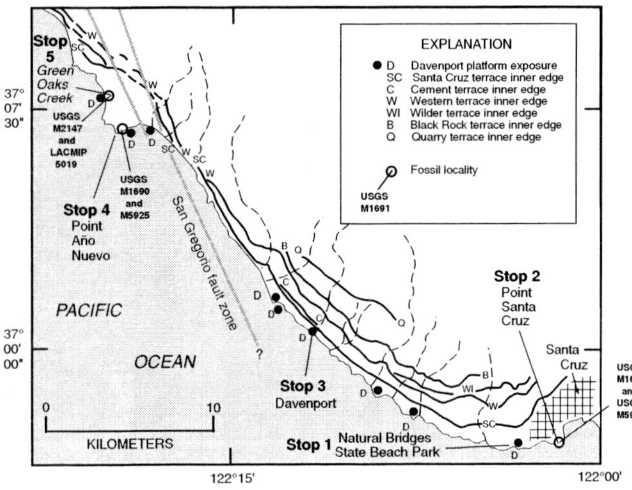

Figure 3. Map of a portion of the central coast of California, from Santa Cruz to just north of Point Año Nuevo, showing marine terrace inner edges (solid black lines, dashed where uncertain), fossil localities, and location of the San Gregorio fault zone (solid gray lines; dashed where uncertain). Marine terrace inner edges redrawn from Bradley and Griggs (1976) and Weber et al. (1979); location of the San Gregorio fault zone from Weber et al. (1979) and Weber (1990).

as a single, relatively flat surface, but Bradley and Griggs (1976) demonstrated it is actually a complex of three platforms, from oldest to youngest the Greyhound, Highway 1, and Davenport platforms. Marine and nonmarine deposits have "smoothed" the surface topographically into a single, broad landform. Most workers agree that the Santa Cruz terrace complex probably represents the last interglacial period, but correlation to specific sea level stands has been a matter of debate (Bradley and Griggs, 1976; Kennedy et al., 1982; Weber, 1990; Lajoie et al., 1991; Anderson and Menking, 1994; Perg et al., 2001, 2002; Brown and Bourles, 2002; Muhs et al., 2002c). Soils on terraces in the Santa Cruz area have been studied by Schulz et al. (2003) in a complimentary field trip (this volume).

DAY 1: San Francisco International Airport to Santa Cruz; field trip stops are from Santa Cruz to Half Moon Bay, CA

STOP 1. NATURAL BRIDGES STATE BEACH

At this first stop, some of the best examples of modern coastal landforms in California may be seen at low tide. The bedrock exposed in a well-developed, wave-cut platform along the coast at Natural Bridges State Beach is the Santa Cruz mudstone (Miocene). Although this rock is well cemented, it has thin bedding and is characterized by numerous fractures, which, as pointed out by Bradley and Griggs (1976), can result in easy erosion by waves. The platform is a planar, erosion surface on bedrock that formed by wave erosion through hydraulic impact, abrasion and quarrying, and gravitational collapse by undercutting. Wave erosion is strongest in winter, coastal California's rainy season, when storms develop in the eastern Pacific Ocean. Griggs and Johnson (1979) showed that the long-term average sea cliff retreat rate in some parts of Santa Cruz County is ~30 cm/yr. However, much cliff retreat takes place episodically during strong winter storms. Accelerated sea-cliff erosion took place during the winters of 1982-1983 and 1997-1998, when large storms related to El Niño conditions dominated the eastern Pacific Ocean (Storlazzi and Griggs, 2000).

STOP 2. DAVENPORT TERRACE AT POINT SANTA CRUZ

Much of the city of Santa Cruz, California is built on the Santa Cruz terrace complex (Fig. 3). Looking to the north from many vantage points within the city, the inner edge of the Santa Cruz terrace complex and the sea cliff rising to the next-highest ("Western") terrace can be seen. Exposures of the Davenport platform are visible in the modern sea cliff, near Point Santa Cruz. The platform, cut into sandstone, is ~6 m above sea level and is overlain by thin (~20 cm), fossiliferous, marine, terrace deposits containing paired valves of *Saxidomus gigantea* and *Protothaca staminea*. At this locality, a single *Balanophyllia elegans* coral, collected and archived by Hoskins (1957), gave a U-series age of 71,500 ± 400 yrs. (Muhs et al., 2002c).

STOP 3. TERRACE OVERLOOK AT DAVENPORT

This brief stop provides excellent views of the seaward extent of the Santa Cruz terrace complex. In addition, many of the higher terraces (Cement, Western, Blackrock, and Quarry) mapped by Bradley and Griggs (1976) are visible here.

STOP 4. TERRACE DEPOSITS, FOSSILS, AND FAULTS AT POINT ANO NUEVO

Año Nuevo State Reserve is home to the world's largest, mainland, elephant seal population, as well as Steller sea lions and harbor seals. The hills above the Reserve are part of Big Basin Redwoods State Park, California's first state park. The park was established for the occurrence here of the beautiful old-growth coastal redwood (*Sequoia sempervirens*), a tree found only in a coastal strip from Santa Cruz County, California, to just over the Oregon border.

Near Point Año Nuevo (Figs. 3, 4), fossiliferous, Davenport-platform, terrace deposits are exposed in the modern sea cliff. Note also the presence of both bivalves and gastropods in the Tertiary bedrock here. Quaternary marine deposits that can be observed at this locality are ~0.5 m thick and contain the bivalve *Saxidomus gigantea* in growth position. The platform has also been bored by pholads, some still in their holes in the platform. The overlying alluvium, which is ~10 m thick, has a well-developed soil (A/E/Bt/C profile) in its upper part. The terrace platform in this area is displaced vertically by at least two northwesterly-striking faults (the Frijoles and Coastways faults) (Fig. 4) that are part of the San Gregorio fault zone (Weber, 1990; Weber et al., 1979). On the uplifted crustal block west of the Frijoles fault, the Davenport platform is ~8-10 m above sea level and can be seen from the modern beach. A sag pond has formed on the downthrown side of the fault (Fig. 4) and is visible from the trail at the top of the terrace. At or near this locality, Hoskins (1957) collected and archived a single individual *Balanophyllia elegans* coral, which gave a U-series age of 75,800 ± 400 yrs. (Muhs et al., 2002c).

STOP 5. TERRACE AND FOSSILS AT GREEN OAKS CREEK

(Note that this is State Reserve property, so permission is needed for access to this outcrop.)

Figure 4. Aerial photograph of the Point Año Nuevo-Green Oaks Creek area, showing geology (from the present authors), geomorphic features, fossil localities, and faults (from Weber et al., 1979 and Weber, 1990). Qal, alluvium; Qmt, marine terrace deposits; Qes, eolian sand.

One of the richest, marine-terrace, fossil beds in central California is found on the Davenport platform near the mouth of Green Oaks Creek, just north of Point Año Nuevo (Figs. 3, 4). The Davenport platform is ~5 m above sea level and has a fossiliferous basal gravel ~0.5 m thick, overlain by ~4 m of marine (?) sand, capped by ~2 m of eolian sand. A paleosol separates the eolian sand from the underlying marine sediments. The Davenport platform here is bored extensively by pholadid bivalves. Corals are abundant in the terrace deposits at Green Oaks Creek and were collected by Muhs et al. (2002c). U-series ages of 11 individual corals (*Balanophyllia elegans*) from Green Oaks Creek range from 75,800 ± 800 to 84,000 ± 600 yrs. B.P. This age span includes the ~76,000 yrs. B.P. age of the coral from Point Año Nuevo (STOP 1-4) and is close to the ~72,000 yrs. B.P. age of the coral from Point Santa Cruz (STOP 1-2). All fossil corals analyzed from Green Oaks Creek have back-calculated initial $^{234}U/^{238}U$ values that are in close agreement with those measured for modern seawater (Fig. 5). Therefore, the corals have probably experienced closed-system conditions with respect to U and its daughters, and ages are considered to be reliable.

The coral ages from Green Oaks Creek, Point Año Nuevo, and Point Santa Cruz all indicate that the Davenport platform dates to the ~80,000 yr. B.P. high-stand of the sea. This high-stand is also recorded as marine oxygen isotope substage (OIS) 5a (Martinson et al., 1987), emergent reef terraces on New Guinea (Bloom et al., 1974) and Barbados (Mesolella et al., 1969; Ku et al., 1990; Edwards et al., 1997), and marine deposits on Bermuda (Muhs et al., 2002b). The multiple coral ages from the Green Oaks Creek locality, with their evidence of closed-system histories, give a high degree of confidence that this sea-level high-stand had a duration from ~84,000 to at least 77,000 yrs. B.P., similar to that recorded on Bermuda (Fig. 5).

Coral ages from Green Oaks Creek indicate some differences between the ~80,000-yr.-B.P., high sea stand and the present interglacial high sea stand. The present, relatively high, sea level is the result of melting of the Laurentide and Fennoscandian ice sheets at the end of the last glacial period. Melting of these ice sheets is thought to be the result of relatively high summer insolation in the Northern Hemisphere around 11,000 yrs. B.P. However, sea level did not reach near-present elevations until about 7,000 to 5,000 yrs. B.P., a sea-level lag of 4,000 to 6,000 yrs. In contrast, U-series ages of corals from the Davenport terrace at Green Oaks Creek indicate that sea level must have been relatively high at the time of, or prior to, a previous period of high summer insolation in the Northern Hemisphere (around 82,000 yrs. B.P.; Fig. 5).

The U-series ages for the Davenport terrace differ from the inferred age of this terrace based on cosmogenic isotope dating of higher terraces, reported by Perg et al. (2001). The age of the Davenport terrace, based on Perg et al.'s (2001) data, must be younger than ~65,000 yrs. B.P. Ages reported by these workers may reflect the ages of alluvium that overlies the marine deposits. Our own observations suggest that the terrestrial sedimentary cover in this area is usually much thicker than the marine cover (for example, at Point Año Nuevo), and that marine sediments are rarely, if ever, exposed at the ground surface. One of the implications of the study of Perg et al. (2001) is that uplift rates along this part of the California coast are relatively high, on the order of 1 m/ka. The U-series ages presented here indicate much lower rates of uplift, around 0.20-0.35 m/ka, which are consistent with rates found elsewhere along the coast from Baja California to Oregon (Muhs et al., 1992).

FAUNAS OF THE MARINE TERRACE DEPOSITS IN CENTRAL CALIFORNIA

Species that are most important in paleoclimatic interpretation of marine terrace faunas are those with modern geographic

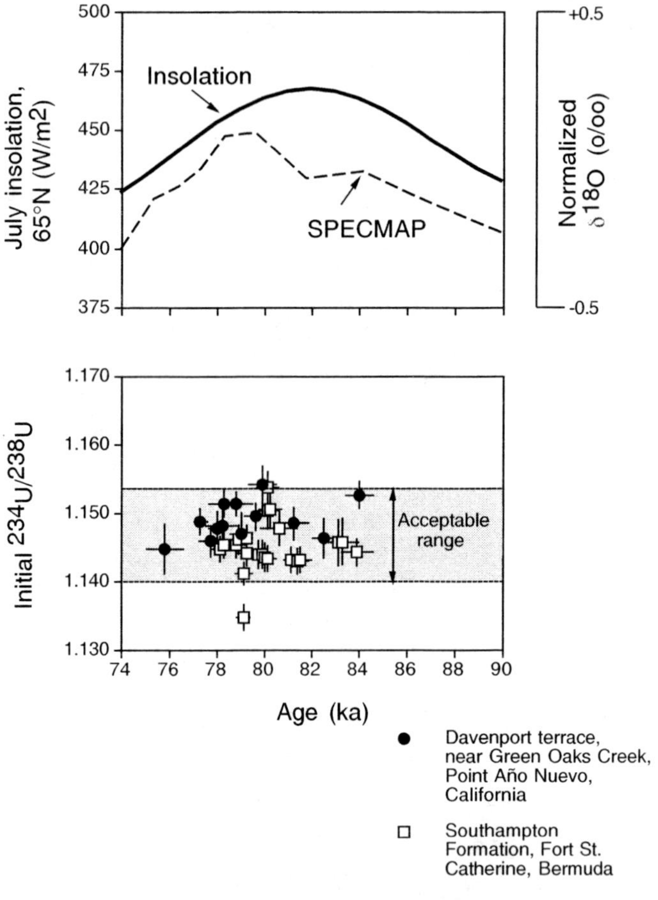

Figure 5. Plot showing U-series ages and initial $^{234}U/^{238}U$ activity values of solitary corals from the Davenport terrace at Green Oaks Creek (near Point Año Nuevo), California compared to similar data for colonial corals from the Southampton Formation on Bermuda (Bermuda data from Muhs et al., 2002b). Error bars represent ± 2 sigma analytical uncertainties. "Acceptable range" is defined as the range of initial $^{234}U/^{238}U$ activity values that would indicate probable closed-system conditions with respect to U-series isotopes over the history of the fossil. Also shown are normalized oxygen isotope values from the deep-sea record of SPECMAP (Martinson et al., 1987) and July insolation at the top of the atmosphere at 65°N (data from Berger and Loutre, 1991) for the same time period.

distributions that extend only to the south ("extralimital southern") or only to the north ("extralimital northern") of a fossil locality. Paleoclimatic inferences can also be made from species that are not strictly extralimital, but whose range endpoints are at or near a given fossil locality. These are referred to herein as "northward ranging" or "southward ranging" species. Our discussion here is based on the faunas, mostly mollusks, reported from the ~80,000-yr-old Davenport terrace by Addicott (1966). We have updated the modern geographic ranges of the species because information now available is better than it was in the 1960s (Fig. 6). Deposits of the Davenport terrace at Green Oaks Creek (USGS loc. M2147) contain three extralimital northern species plus three northward-ranging species whose modern southern ranges terminate at or near Point Año Nuevo. At Point Año Nuevo (USGS loc. M1690), seven extralimital northern species but no extralimital southern or southward-ranging species are found. The Davenport terrace fauna from Point Santa Cruz (USGS loc. M1691) contains three extralimital northern species and five northward-ranging species. As with the other central California localities, it lacks any extralimital southern or southward-ranging species. Based on the combined data from Point Santa Cruz, Point Año Nuevo, and Green Oaks Creek, all localities with extensive cool-water elements in their faunas, we conclude that central California had cooler-than-modern waters at ~80,000 yrs. B.P.

In contrast, during the peak of the last interglacial period at ~120,000 yrs. B.P. (OIS 5e), waters around much of North America, including California, were warmer than present (Muhs et al., 2002a, 2002b). Either the Highway 1 or Greyhound terraces probably date to this high sea stand, but fossils have not been found in the deposits of these terraces. However, an emergent marine deposit called the Millerton Formation, exposed around Tomales Bay north of San Francisco, also probably dates to this high sea-stand, based on both amino acid ratios in mollusks (Kennedy et al., 1982) and thermoluminescence dating of sediments (Grove and Niemi, 1999). The fauna from the Millerton Formation (Johnson, 1962) contains no northern species but has abundant extralimital southern and southward-ranging species (Fig. 6). Other climate proxies agree with the marine faunal data. Alkenone, radiolarian, and pollen data from cores off northern California and southern Oregon indicate cooler-than-present water at ~80,000 yrs. B.P. and warmer-than-present water at ~120,000 yrs. B.P. (Heusser et al., 2000; Herbert et al., 2001; Pisias et al., 2001). Cooler-than-modern waters are probably best

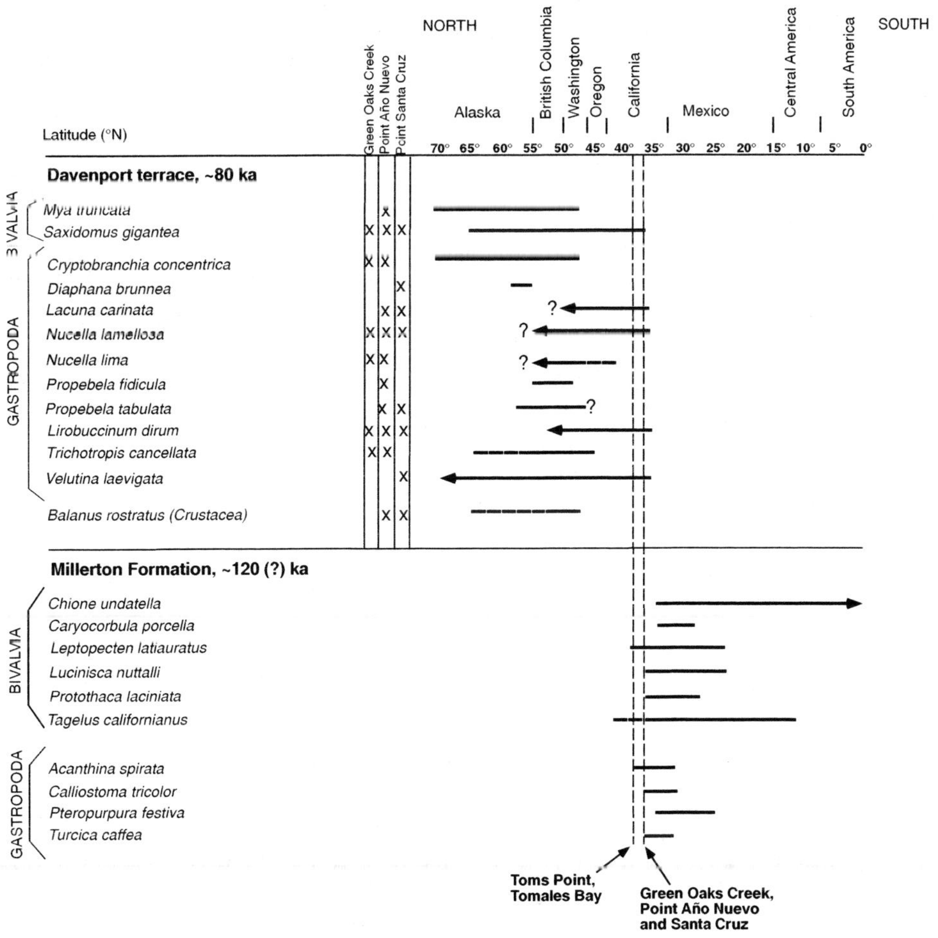

Figure 6. Modern geographic ranges of extralimital and northward or southward-ranging fossil mollusks found in ~80,000 yr B.P. marine terrace deposits at Green Oaks Creek, Point Año Nuevo and Santa Cruz, California and the ~120,000 yr B.P. Millerton Formation at Toms Point, Tomales Bay, California. Fossil data for the Millerton Formation are from Johnson (1962); Davenport terrace fossil data are from Addicott (1966). Modern species names and geographic ranges updated by the authors from Abbott (1974), Abbott and Haderlie (1980), O'Clair and O'Clair (1998), and Coan et al. (2000).

explained by a stronger California Current (subarctic water from the north), whereas warmer-than-modern waters are probably the result of a stronger Davidson Countercurrent (southern water). Mark Twain once said that the coldest winter he ever spent was a summer in San Francisco: during the ~80,000 yrs. B.P. high sea-stand, it would have been even colder!

End of DAY 1: Overnight in Half Moon Bay, CA

DAY 2: HALF MOON BAY TO GUALALA, CA

STOP 6. OVERLOOK OF SAN ANDREAS FAULT FROM VISTA POINT EAST OF HIGHWAY 280

The "rift valley" of the San Andreas fault zone forms a prominent geomorphic feature visible (on a clear day!) from the STOP 6 vista point. The San Francisco earthquake of April 18th, 1906, is the most-recent, large earthquake to rupture the entire northern San Andreas fault (SAF), including the section we see from this vista point. Surface rupture associated with this earthquake was reported along about 435 km of the fault, from near San Juan Bautista to several km NW of Shelter Cove, about 120 km northwest of Point Arena (Fig. 1) (Lawson, 1908; Prentice, 1999). Today, we will stop at several localities to view the active traces of the SAF and to discuss the current understanding of the seismic hazard this fault represents to the San Francisco Bay area.

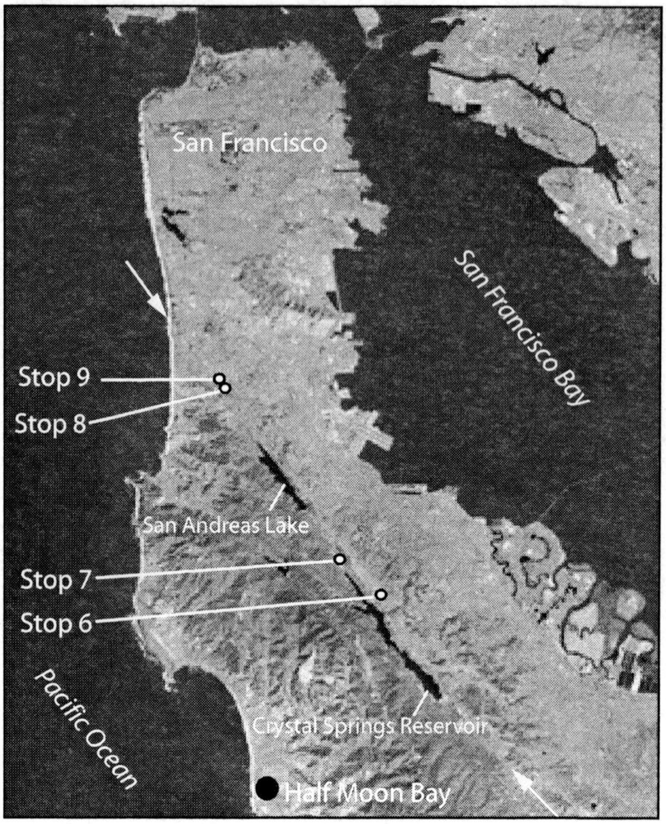

Figure 7. Landsat image showing locations of stops 6-9 on the San Francisco peninsula. White arrows indicate the San Andreas fault. Landsat image rendered by Michael Rymer, USGS.

Crystal Springs Reservoir and San Andreas Lake (Fig. 7) are artificial reservoirs situated in the fault zone, and the active trace of the SAF runs through both. The two dams that impound these reservoirs are San Andreas dam, initially constructed in 1868, and Crystal Springs dam, initially constructed in 1888. Both reservoirs survived the 1906 earthquake, despite their close proximity to the SAF. The brick waste weir tunnel below San Andreas dam was displaced right-laterally 2.7 m (9 ft). The dam embankment did not fail because shearing along the SAF was confined to the bedrock ridge that forms the eastern abutment of the dam (Hall, 1984). The water distribution system, however, was severely damaged (Lawson, 1908; Schussler, 1906), which contributed to the great damage caused by fire in San Francisco after the earthquake.

Due to its proximity to the San Francisco urban area, the San Francisco Peninsula section of the fault is one of the most potentially hazardous segments along the entire length of the SAF. However, it remains one of the most poorly studied in terms of slip rate and earthquake recurrence. Probabilistic earthquake hazard assessments depend largely on recurrence and slip-rate data gathered from analyses of geologic information about prehistoric fault behavior. Recent subsurface investigations at sites along the SAF north of the Golden Gate (Prentice, 1989; Prentice et al., 1991; Niemi and Hall, 1992; Baldwin et al., 2000; Noller et al., 1993; Prentice et al., 2001) have provided some data on pre-1906 earthquakes and the Holocene slip rate, but few successful paleoseismic sites on the San Francisco Peninsula have been developed. Paleoseismic work at the Filoli site, situated at the southern end of Crystal Springs reservoir (Fig. 7), has yielded a late Holocene slip-rate estimate of 17 ± 4 mm/yr (Hall et al., 1999), similar to slip-rate estimates for the section of the fault north of San Francisco (≤23 ± 2 mm/yr (Prentice, 1989); 24 ± 3 mm/yr (Niemi and Hall, 1992); 18 ± 3 mm/yr (Prentice et al., 2001)). Although these studies have added significantly to the understanding of the recent behavior of the northern SAF, many questions remain, and more work is needed to understand this critical segment of the SAF in the San Francisco area.

STOP 7. OFFSET CYPRESS TREES, FENCE, AND STREAM CHANNEL

(Note that this locality is within the San Francisco watershed, and special permission is needed for access).

The best-surviving cultural features that were offset during the 1906 earthquake on the San Francisco Peninsula are a fence and a row of cypress trees near Crystal Springs Reservoir (Fig. 7). Here, right-lateral displacement of 2.7 m (9 ft) is still clearly visible. This locality was photographed in 1906, and is shown in plate 61B of Lawson (1908) (reproduced here as Fig. 8). Note that the displacement is distributed over a zone sev-

Figure 8. Photograph taken after the 1906 earthquake in the vicinity of stop 7. Offset row of cypress trees (CT) and adjacent fence remain visible here today. Fence is offset 2.7m (9 ft.). Photograph appears as Plate 61B in Lawson, 1908. Glass plate negative is archived with the papers of Andrew Lawson in the Bancroft Library, University of California, Berkeley, call number 1957.007.12, Series 2. Courtesy of the Bancroft Library, University of California, Berkeley.

eral meters wide. Note also the excellent preservation of delicate geomorphic features showing the nature and exact location of the surface rupture (seen in Fig. 8 to the right of the person standing in the middle ground of the photograph).

North of the offset fence, the 1906 trace of the SAF lies in a trough and is marked by a small closed depression. The fault trace can be followed to the northwest where prior fault movements have offset an incised stream channel between 52 and 87 m (170 to 285 ft.). This offset represents the sum of 19 to 32 seismic events comparable in slip to that of 1906, suggesting that the active trace has not changed location significantly for thousands of years (Hall, 1984).

STOP 8. URBAN SAN ANDREAS FAULT, CORNER OF WESTBOROUGH BOULEVARD AND FLEETWOOD DRIVE, SAN BRUNO

The urban development in this area (Fig. 9) occurred prior to enactment of the California law that prohibits such construction directly on top of an active fault (Alquist-Priolo Earthquake Fault Zoning Act, 1972). The fault was once well expressed through this area (Fig. 9), but construction has entirely obliterated the geomorphic features, such as sag ponds and shutter ridges, that once marked the active trace. Using photographs taken of the rupture after the earthquake in 1906 and analysis of pre-development aerial photographs, a detailed GIS of the best estimate of precisely where the fault broke in 1906 in this area is being developed by the USGS (Prentice and Hall, 1996). At this

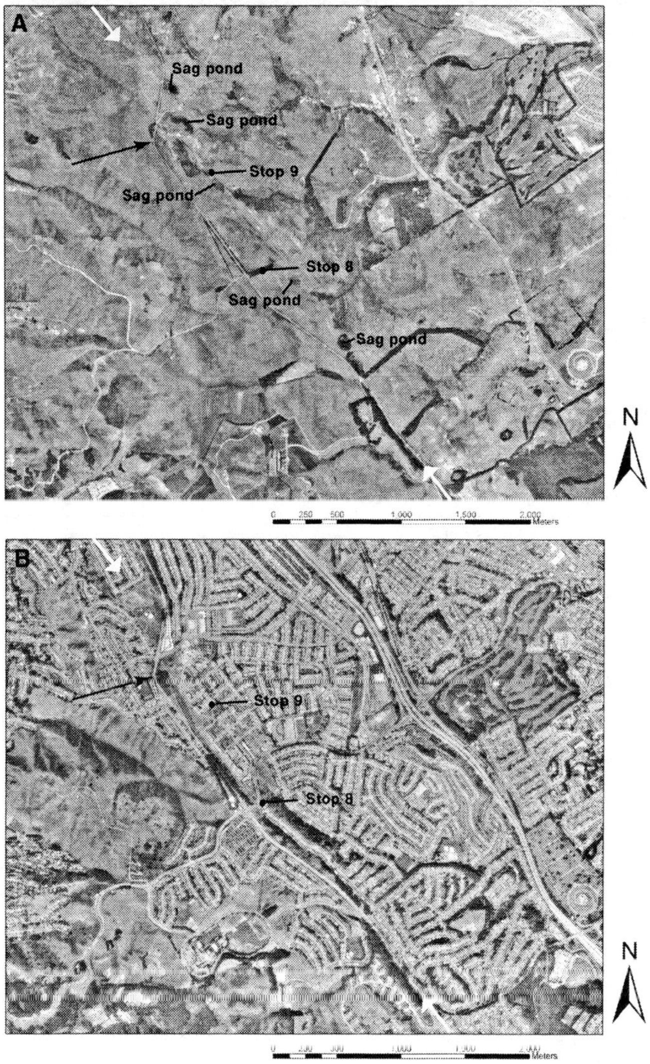

Figure 9. A. Part of a DOQ (photo taken in 1946) showing locations of stops 8 and 9 and vicinity prior to construction of housing developments. Note geomorphic features indicative of active faulting, such as sag ponds and shutter ridges, that mark the trace of the San Andreas fault. B. Black arrow points to curve in Skyline Boulevard, also pointed out in 9A. White arrows indicate the San Andreas fault. Part of a DOQ (photo taken in 1993) showing same area as 9A, illustrating the pre-Alquist-Priolo Act development that has obscured the location of the San Andreas fault. The 1906 surface rupture appears to run through many of these buildings.

location, a fence offset in 1906 was still visible in 1956 when M.G. Bonilla of the USGS took the picture shown in Figure 10A, but was gone by 1962 when he took the photograph shown in Figure 10B. A photograph taken in 1906 (Fig. 10C) at this location shows the offset by three fault strands, a main trace with just under 2 m of right-lateral offset and two smaller, parallel breaks. Our mapping indicates that a number of the houses visible from this location are located directly on top of the 1906 rupture trace.

Figure 10. A. Photograph of offset fence taken from stop 8 by M.G. Bonilla of the USGS in 1956. Line shows location of 1906 surface rupture, which offsets fence. B. Photograph taken at same location by M.G. Bonilla in 1962 illustrating impact of housing development. White arrow points to same building visible in A and B. C. Photograph taken of same fence by H.O. Wood in 1906. Note three 1906 rupture traces. Glass plate negative is archived with the papers of Andrew Lawson in the Bancroft Library, University of California, Berkeley, call number 1957.007.110, Series 1. Courtesy of the Bancroft Library, University of California, Berkeley.

STOP 9. URBAN SAN ANDREAS FAULT, MYRNA LANE, SOUTH SAN FRANCISCO

This development (Fig. 9) was constructed prior to the Alquist-Priolo Act. The developer in this instance apparently tried to keep homes off the fault by keeping a green belt and a road over the area where they believed the fault came through. However, many of these structures appear to be directly on top of the 1906 fault rupture trace, according to our mapping.

STOP 10. SLIP RATE OF THE SAN ANDREAS FAULT AT MILL GULCH

The Gualala block consists of the area west of the San Andreas fault from south of Fort Ross to north of Point Arena (Fig. 11). Our next stop is just north of where the San Andreas fault comes onshore (Fig. 12). Mill Gulch, a deeply incised stream that is offset 80-100 m across the San Andreas fault, (Fig. 13) is visible from STOP 10. An abandoned channel of Mill Gulch is visible on Fig. 13, north of STOP 10. We excavated trenches in the abandoned channel and collected charcoal samples from pre- and post-abandonment sediments to estimate the age of the 80-100 m offset. Radiocarbon analyses of these samples provide minimum ages that range from 4290–4520 to 4890-5290 cal yrs. B.P., and a single maximum age of 5040–5320 cal yrs. B.P. These data suggest a slip rate of 19 ± 4 mm/yr (best estimate of 18 ± 3 mm/yr) (Prentice et al., 2000, 2001).

STOP 11. OFFSET FENCE AND TERRACE OVERLOOK

From this location (Fig. 12), the view toward the ocean on a clear day shows the flight of Pleistocene marine terraces present along the coast of the Gualala block. A mapping and detailed survey measurement of these terraces is underway to improve understanding of coastal uplift in the region.

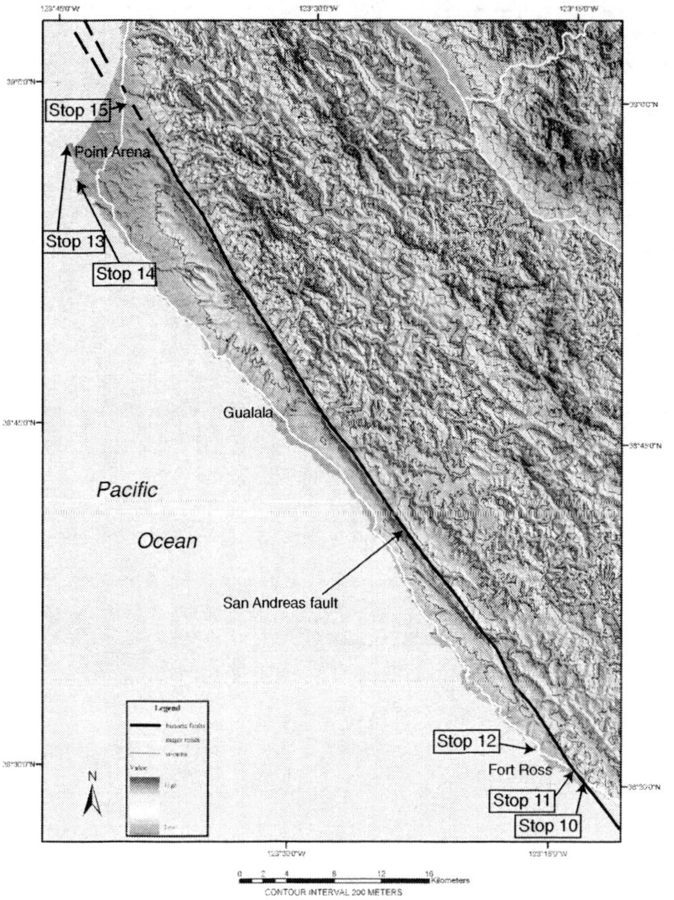

Figure 11. Map of Gualala block showing locations of Stops 10-15. Hillshade base produced from DEM's by Christopher J. Crosby, USGS.

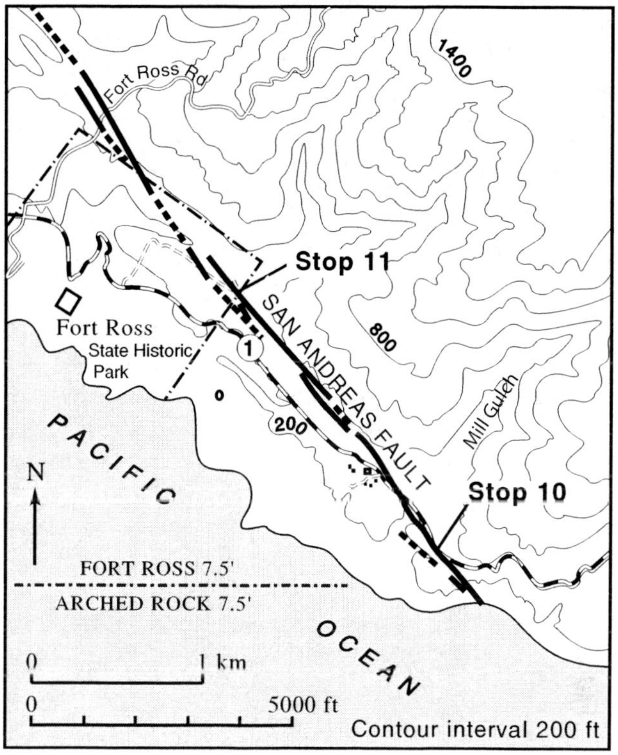

Figure 12. Map showing locations of Stops 10 and 11. Modified from parts of Fort Ross and Arched rock USGS 7.5-minute quadrangle topographic maps.

At this location, an old picket fence, offset by the 1906 earthquake, is still visible. Although many of the fence posts have fallen, enough remain standing in 2003 to see the offset. The survey of this fence shown in Lawson (1908, p. 64) shows that the total offset of 3.7 m was distributed over a distance of 126 m, with only 2.3 m occurring where the main fault crosses the fence (Fig. 14). Historical research shows that Esper Larsen did the original survey between January 20 and February 3, 1907, about a year after the earthquake (Letters, 20 January 1907, and 3 February, 1907). Resurvey of this fence in 1985 shows virtually no change since the 1907 survey. This indicates that no measurable afterslip occurred on this fault between February 1907 and September 1985, consistent with resurveys of other offset fences along the San Francisco peninsula (R.E. Wallace, USGS, personal communication, 1994).

A few meters south of the fence, the old Russian road leading south from Fort Ross crosses the fault at a low angle. Careful observation shows that this road is displaced about the same amount as the fence suggesting that only one fault movement (1906) has offset this road in the time since the Russians established Fort Ross (1812). Geomorphic features typical of active faults are well displayed in the area south of the fence.

STOP 12. VIEW OF MARINE TERRACE SHORELINE ANGLE EXPOSED IN SEA CLIFF:

This is an outstanding (and rare) example of an exposure of a shoreline angle (Fig.12). The elevation of this feature is 26 m above mean sea level. This is the lowest marine terrace in this area, and our mapping suggests it is correlative with the Point Arena terrace, which is dated to the ~80,000 yrs. B.P. (OIS 5a) sea-level high-stand (Muhs et al., 1994, 2002c).

End of DAY 2: Overnight in Gualala, CA

DAY 3. GUALALA TO POINT ARENA TO MENDOCINO, CA; RETURN TO GUALALA AT END OF DAY

STOP 13. POINT ARENA MARINE TERRACE AGES AND FAUNA

Well-preserved marine terraces are present along much of the coastline of northern California. In the vicinity of Point Arena (Fig. 11), Mendocino County, ~175 km north of San Francisco,

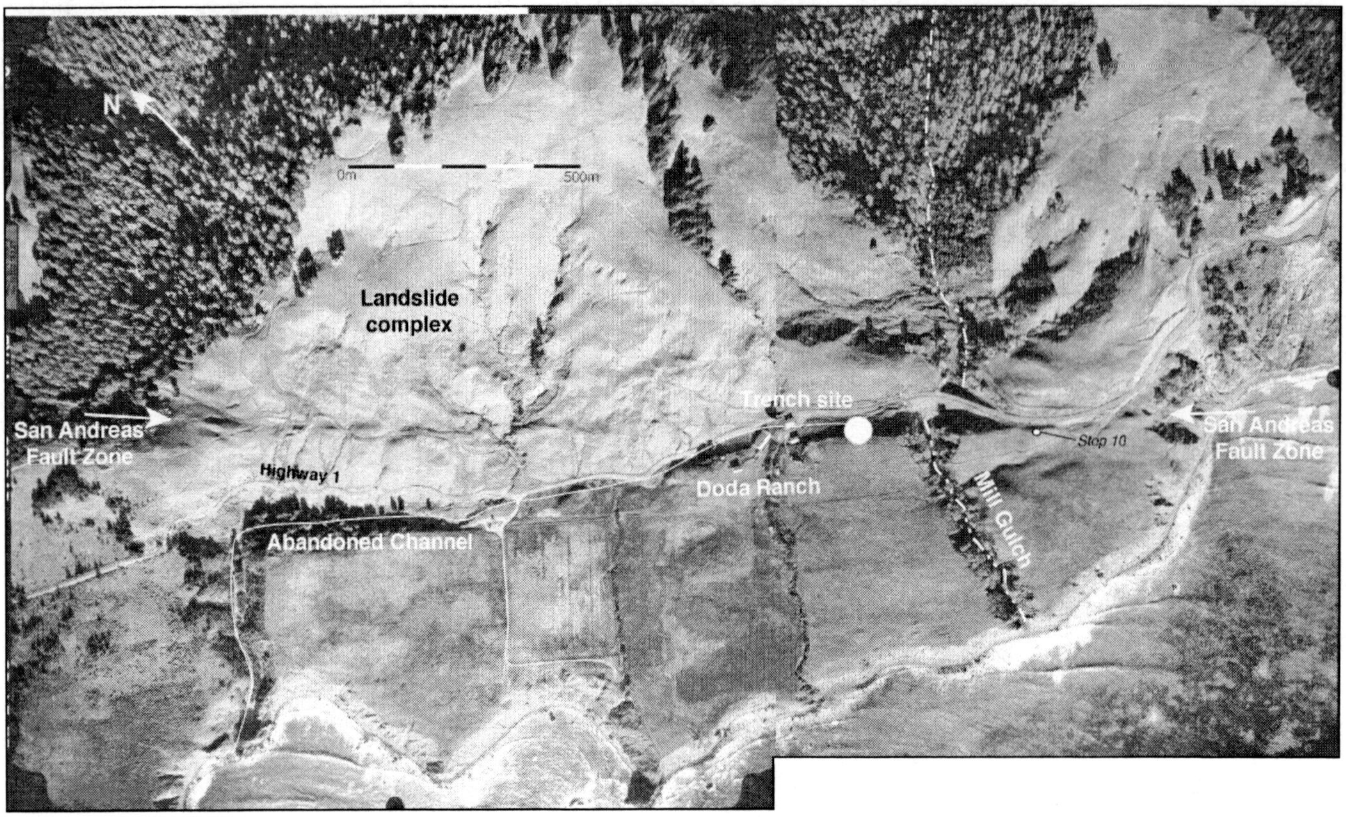

Figure 13. Mosaic of aerial photographs, showing offset of Mill Gulch, abandoned channel and trench site. Study at this site provides an estimate of the late Holocene slip rate for the San Andreas fault of 18 ± 3 mm/yr (Prentice et al., 2001).

marine terraces were mapped by Prentice (1989) and, over a more limited area, by the present authors (Fig. 15). The three lowest terraces, informally designated Qt_1, Qt_2, and Qt_3 in Fig. 15, have inner edge elevations of ~19-23, ~38-42 and ~56-64 m, respectively. The lowest terrace is called the "Point Arena terrace." This terrace achieved fame in 1992 when Mel Gibson landed an airplane on it at the end of the movie "Forever Young."

The Point Arena, or Qt_1 terrace is well expressed geomorphically, and the terrace platform has a sharp contact with the overlying marine sediments. On sea-cliff exposures, the platform is typically riddled with pholad-bored holes and has a variable elevation, ranging from ~5 to ~17 m. Terrace deposits are well stratified, and horizontal beds of sand and gravel vary in thickness from ~2 to 7 m. Well-developed soils with A/E/Bw/C or A/E/Bs/C profiles have formed in the upper part of the marine terrace deposits. These soils are interesting in that many of them have a mollic epipedon, developed under the modern coastal prairie vegetation, yet have well-developed E and sometimes Bs horizons, suggesting a former forest cover.

Fossils from the Point Arena terrace deposits (map unit Qt_1 on Fig. 15) at Point Arena were studied by Kennedy (1978), Kennedy et al. (1982), Kennedy and Armentrout (1989), and Muhs et al. (1990, 1994) for both their paleozoogeographic aspects and age estimates. Alpha- spectrometric U-series analyses of marine terrace corals from Point Arena are ~76,000 yrs. B.P. and ~88,000 yrs. B.P. (Muhs et al., 1990, 1994). Fieldwork conducted in August 2001, revealed that the fossil localities no longer exist. A new mass-spectrometric U-series analysis of a single *Balanophyllia* coral from the lowest terrace at Point Arena (collected before the fossil localities disappeared) has an apparent age of 83,000 ± 800 yrs. B.P. (Muhs et al., 2002c) and is therefore broadly consistent with the earlier ages.

The marine terrace fauna at Point Arena also contains extralimital northern species and lacks any extralimital southern or southward-ranging species. The fauna contains the extralimital northern mollusk *Mya truncata* (Kennedy, 1978; Muhs et al., 1990), which presently ranges only from Barrow, Alaska to Neah Bay, Washington (Coan et al., 2000). A particularly dramatic example of an extralimital northern species is the bivalve *Penitella hopkinsi*, whose modern distribution is limited to the Gulf of Alaska, but is found in the ~80,000-yrs.-B.P.-deposits at Point Arena (Kennedy and Armentrout, 1989). Thus, as with central California, the faunal data indicate cooler-than-modern marine paleotemperatures off the northern California coast at ~80,000 yrs. B.P.

STOP 14. LORAN STATION ON THE POINT ARENA TERRACE:

(Note: This is private property. Permission is needed for access.)

At this location (Fig. 15), the Point Arena terrace is well expressed geomorphically. Sea cliff exposures reveal two Quaternary thrust faults that have been active since the time this terrace formed (Prentice, 1989; Prentice et al., 1991). The first of these

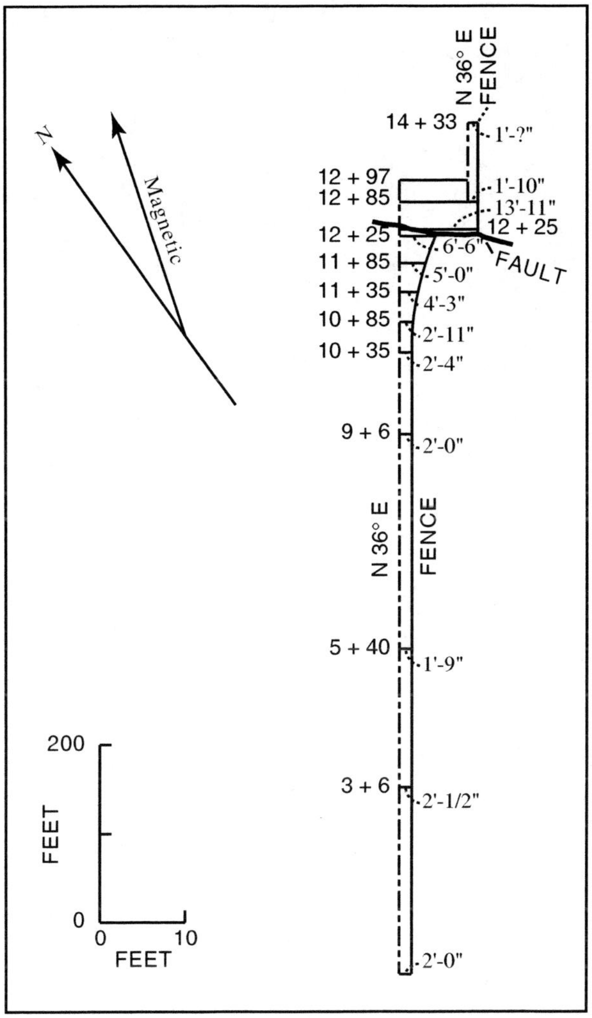

Figure 14. Offset fence at Stop 12. From Lawson, 1908. Note that slip is distributed over a broad zone.

exposures is illustrated in Fig. 16. Miocene bedrock has been thrust over deposits that overlie the ~80,000 yrs. B.P. wave-cut platform. Several other exposures in this vicinity show similar relationships. Two nearby sea cliff exposures reveal a thrust fault that soles into a bedding plane within the Miocene bedrock. Faulted terrace deposits are also well exposed in a nearby sink-hole. These relations indicate Quaternary compression along the plate boundary in this area.

STOP 15. SAN ANDREAS FAULT AND OFFSET TERRACE RISER VIEW POINT

(Note: This is private property. Permission is needed for access.)

From this location (Fig. 17), we look across the SAF to two marine terrace risers (old sea cliffs) that are truncated on the northeast side of the SAF and see a prominent marine terrace riser on the southwest side of the fault (Fig. 18). Because right-lateral strike slip has occurred across the SAF since the time these terraces formed, the riser southwest of the fault at this location cannot correlate to either of the risers northeast of the fault. The only potential correlative to the risers on the northeast side of the fault is located southwest of the SAF and right-laterally offset 1.3 to 1.8 km (Fig 18, not visible from STOP 15). The prominent riser at this location on the southwest side of the fault at STOP 15 must also have a correlative northeast of the fault. The risers on the northeast side of the fault visible from STOP 15 are both unlikely candidates because right-lateral movement on the SAF would have displaced the correlative feature to the SE. The best candidate for a correlative is located northeast of the fault, right-laterally offset 2.3 to 3.2 km (not visible from this location, and not shown on Fig. 18). Estimates of the ages of these features (~80,000 yr, ~100,000 yr, and ~120,000 yr or substages 5a, 5c, and 5e) suggest average late Pleistocene slip rates of 16 to 24 mm/yr (Prentice, 1989; Prentice et al., 2000), consistent with slip rates estimated from study of Holocene features.

STOP 16. MENDOCINO HEADLANDS

The town of Mendocino (Figs. 1, 19) was founded during the California gold rush in 1852 when San Franciscans came north to recover salvage from the Chinese sailing ship Frolic, which was laden with luxurious goods for the rapidly growing city of San Francisco. Pomo Indians apparently found the wreckage first, but the San Franciscans returned to the city with tales of immense red-wood forests, and, soon after, the first successful sawmill was established at the tip of the point. Although the mill now is gone, the New England-style town is a haven for artists and tourists.

Our multiple surveys (both real-time, differentially corrected GPS and total station) of flights of broad marine terraces between Mendocino Headlands and the town of Westport, about 30 km to the north (Fig. 19), give consistent altitudes of the shoreline angle/inner edges of the four lowest terraces: 20-24 m, 38-42 m, 58-63 m, and 83-89 m (Fig. 20). At Mendocino Headlands, the 22-m terrace is well exposed, revealing the wave-cut bedrock platform and as much as several meters of stratified near-shore marine deposits.

Mendocino Terrace Inner-Edge Elevation (m)	Possible Age By Correlation (ka)
20-24	~80 (OIS 5a)
38-42	~120 (OIS 5e)
58-63	~200 (OIS 7)
83-89	~300 (OIS 9)

The correlations given above are tentative, but are based on the following lines of reasoning. The ~22-m terrace at Mendocino has a similar elevation to that of the dated terrace at Point Arena (STOP 13). The ~42-m terrace is correlated with the ~120,000-yrs.-B.P.-high-sea stand (OIS 5e) because the elevation difference between this terrace and the ~22-m-high terrace is similar to that seen for terraces of these ages elsewhere in California (Muhs et

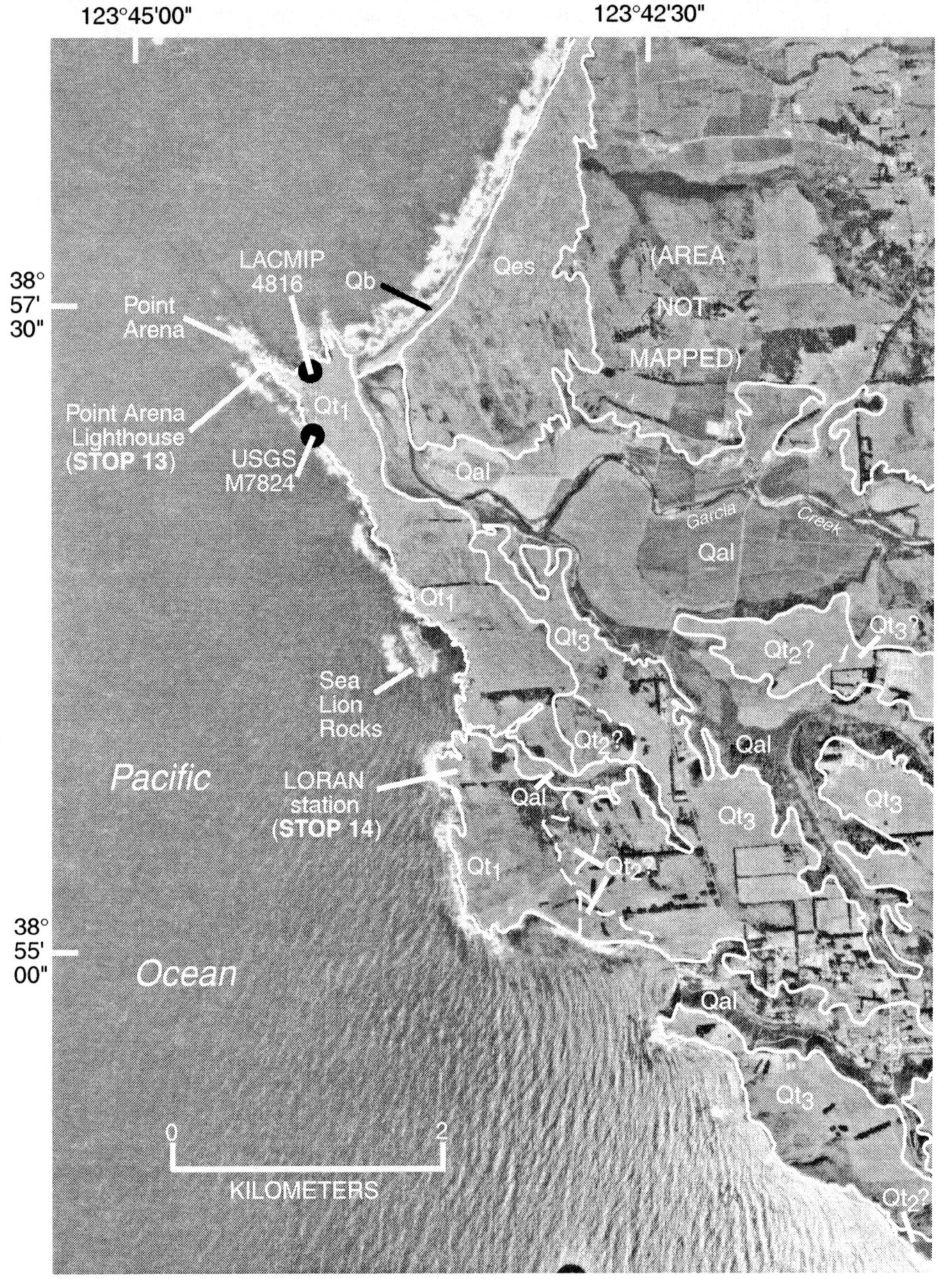

Figure 15. Aerial photograph and Quaternary geologic map of the area around Point Arena, California and fossil localities studied. White lines are contacts; dashed where uncertain. Units: Qb, beach deposits; Qal, alluvium; Qes, eolian sand; Qt_1, Qt_2, Qt_3, deposits of the 1st, 2nd, and 3rd marine terraces, respectively. Geologic mapping by the authors based on interpretation of aerial photographs and reconnaissance field checking.

al., 1994, 2002a). The ~60-m terrace is correlated with the penultimate interglacial period, when sea level was near or above present at ~200,000 yrs. B.P. (OIS 7; see Muhs et al., 2002b). Finally, the ~86-m terrace is correlated with the long interglacial at ~310,000-330,000 yrs. B.P. (OIS 9) when sea level was also near or above present (Stirling et al., 2001). If these correlations are correct, then the rate of uplift for this part of the coastline is ~0.3 m/ka (Figure 20). Note that these terrace correlations differ slightly from those reported by Merritts and Bull (1989) for this segment of coastline. Between Mendocino and the San Andreas fault near Alder Creek, correlation of these four low terraces is more difficult than it is to the north, between Mendocino and Westport. We have completed seven GPS surveys of terraces along this coastal stretch, and, apparently, terraces rise to the south from Mendocino to Alder Creek, where the San Andreas fault crosses the coastline and extends offshore. Planned work in 2003 will help resolve the terrace correlations in this area.

STOP 17. POINT CABRILLO LIGHTHOUSE

Point Cabrillo Preserve was protected as of 1992 when the Coastal Conservancy acquired it. Its 300 acres are prime habitat

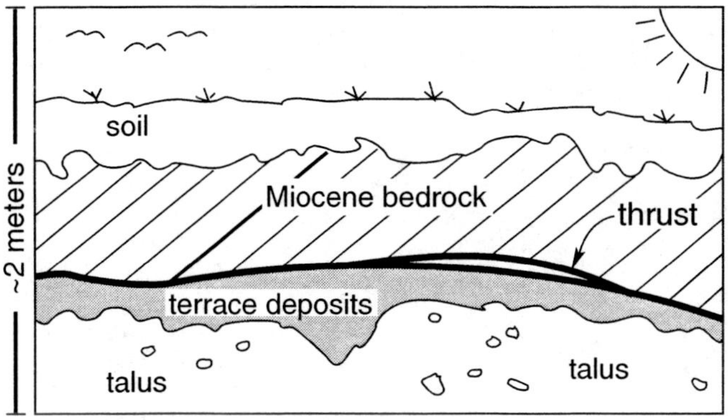

Figure 16. Line drawing made from a photograph showing Miocene bedrock thrust over Pleistocene marine terrace deposits exposed in the sea-cliff at Stop 14. Modified from Prentice (1989) and Prentice et al. (1991).

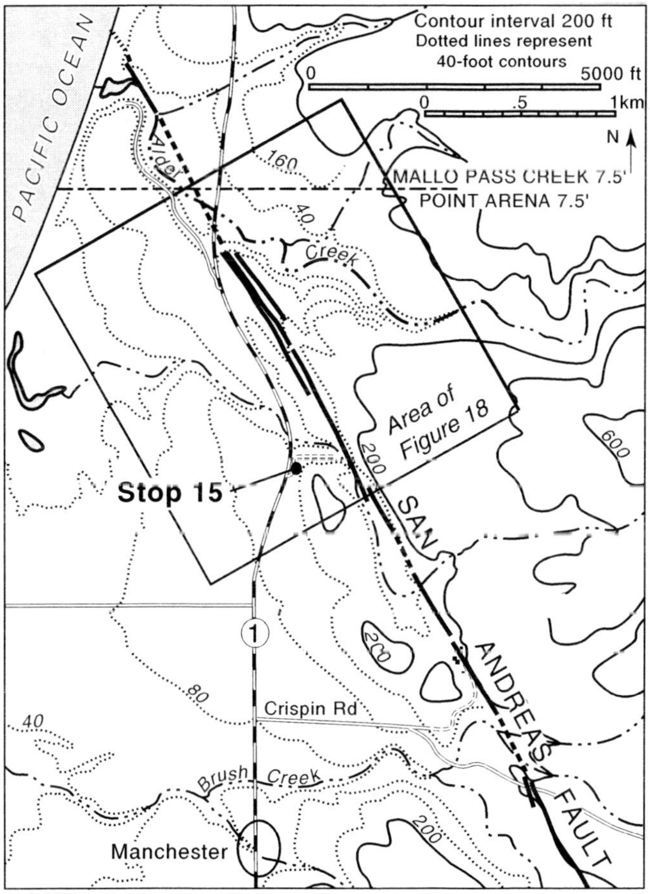

Figure 17. Map showing the location of Stop 15. Modified from parts of the Point Arena and Mallo Pass Creek USGS 7.5-minute quadrangle topographic maps. From Prentice et al. (1991).

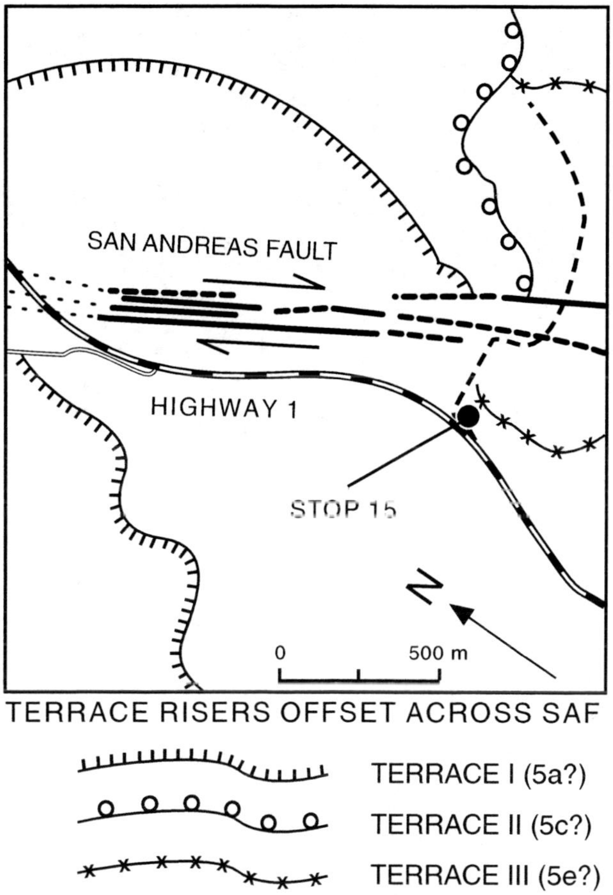

Figure 18. Map showing marine terrace risers offset by the SAF visible at stop 15. Location shown on Fig. 17. Modified from Prentice (1989) and Prentice et al. (1991).

for raptors and other birds, deer, wildflowers, and river otters. The lighthouse, still operating today, was established in 1909 and is open for tours. We will hike seaward across the three lowest marine terraces (going down in elevation from the 60-m to the 42-m to the 22-m terraces) until we reach the sea cliff. In the wall of the sea cliff, a low platform notch at about 10 m elevation lies below an unusual weathered zone that resembles marine-gravel cover sediments from a distance. Up close, however, the orange-colored zone is actually weathered bedrock of the next higher (22-m) terrace platform. The weathered zone generally has a uniform thickness of 2-3 m. In some places, small springs of groundwater seep from the base of the zone, suggesting that the weathering is associated with a fluctuating groundwater table. Because the weathered bedrock has little resistance to wave attack, it recedes more quickly than the underlying bedrock, resulting in a morphology that resembles a shoreline angle and paleo-sea-cliff. The same feature occurs at numerous sites along the coast between Mendocino and Westport and is consistently at an altitude of about 10 m. This feature was considered to represent a marine terrace shoreline angle at a height of 10 m and was correlated to the 80,000-yrs. -B.P. sea level high-stand by Merritts and Bull (1989). However, our subsequent work suggests that the 20-24-m-high shoreline angle, and not the 10-m-high feature, was cut by the 80,000 yrs. B.P. high-stand. Kennedy (1978) and Kennedy et al. (1982) sampled mollusk shells from a low terrace at Laguna Point in MacKerricher State Park, midway between Mendocino and

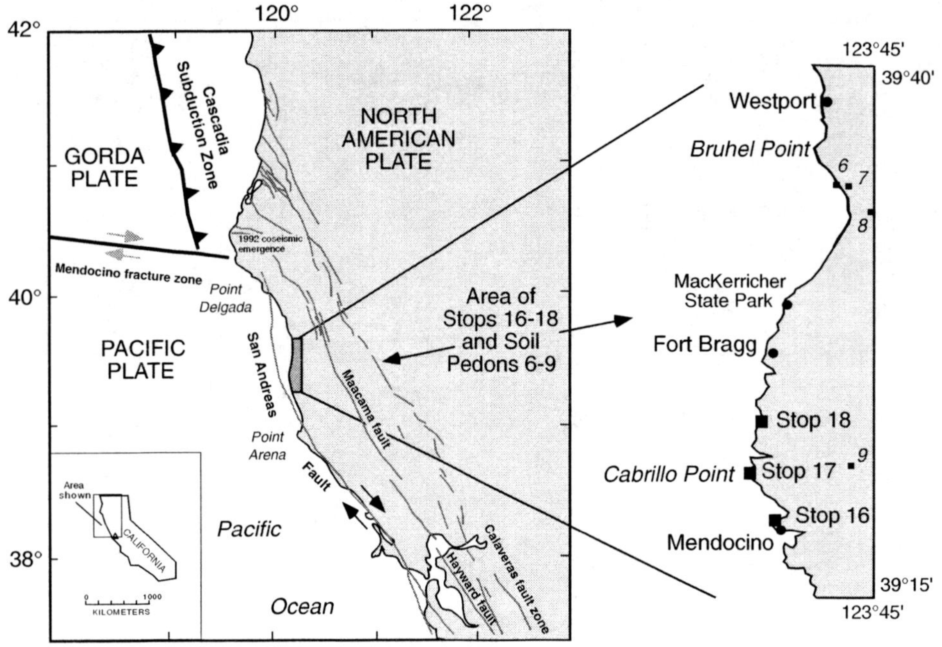

Figure 19. Broad marine terraces form a staircase topography along the coast of northern California between the towns of Mendocino and Westport. Locations of Stops 16, 17, and 18 are shown. [From Figure 1 of Merritts et al., 1991].

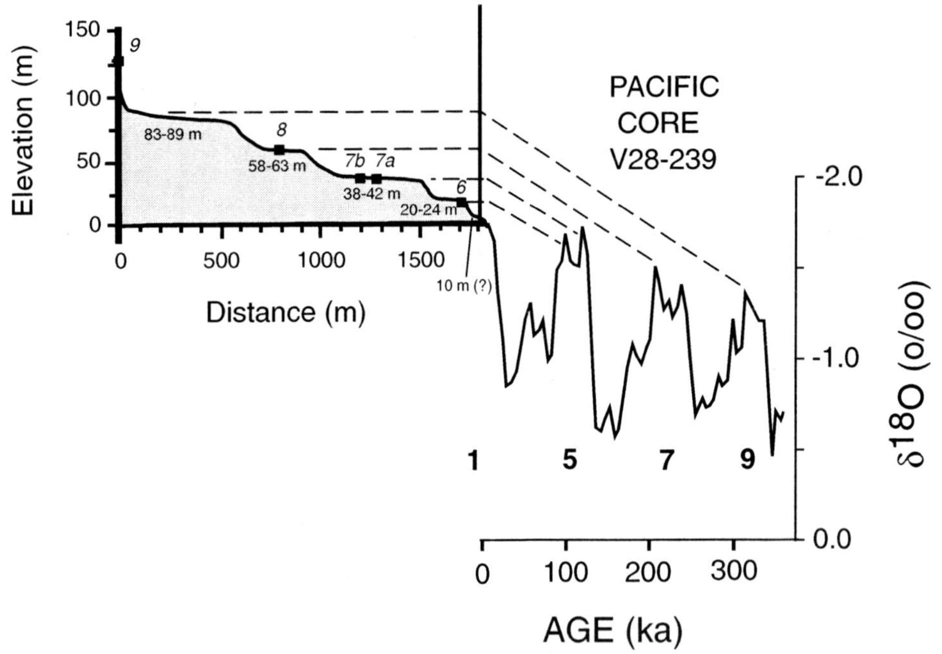

Figure 20. Elevations of the 4 lowest emergent marine terraces, plus possible 10-m notch, along the Mendocino-Cabrillo coast. Multiple surveyed transects completed with both a total geodetic station and a Trimble Pathfinder GPS unit with real-time differential correction (±1 m altitude) give consistent results for the terrace elevations. Terraces are tentatively correlated with four sea-level high-stands, recorded as dated terraces elsewhere (see text) and in the marine oxygen isotope record, shown here. Sea level fluctuations shown are from deep-sea foraminifera in equatorial Pacific core V28-239 (Shackleton and Opdyke, 1976), as in Fig. 2. These correlations result in an inferred long-term average rate of uplift of ~0.3 m/ka.

Westport (see Figure 19). Amino acid ratios in the shells plus the cool-water aspect of the fauna allowed Kennedy et al. (1982) to correlate the terrace with either the ~80,000 or ~100,000 yrs. B.P. high sea-stands. Because the elevation of the terrace sampled by Kennedy (1978) and Kennedy et al. (1982) is not reported, we are unsure if it is the 10-m-high platform notch or the 20-24-m-high terrace that we correlate with the ~80,000 yrs. B.P. high sea-stand. At Cabrillo Point, we will examine the possible paleo-sea cliff at 10 m and discuss whether or not it might be a lower terrace associated with an interstadial sea-level high-stand that post-dates the 80,000 yrs. B.P. high-stand.

STOP 18. JUGHANDLE ECOLOGICAL STAIRCASE

The headland area around the mouth of Jughandle Creek is part of the Jughandle State Reserve that includes the famous

Figure 21. Aerial view looking east and south along the Point Cabrillo to Mendocino coastline. Dashed lines indicate inner edges (IE) of tread surfaces of two lowest terraces. Road in the distance, just west of the forested area, is California HWY 1. [From Figure 2 of Merritts et al., 1991].

Jughandle Ecological Staircase Trail. This ~5-mile (~8 km) trail traverses the three lowest marine terraces of the flight that is nearly continuous from Mendocino Headlands (STOP 16) and Cabrillo Point (STOP 17) to the town of Westport (Merritts and Bull, 1989). We will begin the trail on the lowest terrace, then hike across the second terrace and end at the outer edge of the third and oldest terrace. If our age estimates are correct (see STOP 16), the three terraces are ~80,000, ~120,000 and ~200,000 yrs. B.P. Each older terrace has a better-developed soil profile, and hence, a different suite of biota. The two lowest terraces are covered with coastal prairie bunchgrasses and scrubs, and Bishop pines are found along the walls of ravines incised into the terraces. The third (and higher) terrace(s) has a unique botanical habitat called the Pygmy Forest, which includes mature Mendocino cypress, Bolander and Bishop pines, and redwood trees over 100 years old that are less than 1-2 m high. The flat terrain of the terrace landform, combined with sandy (marine nearshore and eolian) parent material and poor drainage at the platform-sand interface has resulted in extremely acidic, nutrient-deficient soils.

Gently sloping marine terraces in northern California provide excellent conditions for minimizing variations in non-temporal, soil-forming factors and for preserving the chronological record of soil properties that vary as a function of time. Jenny et al. (1969) recognized the value of the elevated northern California marine terraces for soil chronosequence studies, noting especially the *"remarkable and fortunate mineralogical uniformity in the soil parent materials [that] are either weathering greywacke sandstone or sandstone-derived [Franciscan Assemblage] beach materials and dunes"* (p. 62). Climatic conditions also are similar along much of the northern California coast between Mendocino and Westport, with cool, foggy, and dry summers; mean annual temperatures of 12-14° C; mean monthly temperatures that vary less than 6° C throughout the year; mean annual precipitation of 100-102 cm; and more than 90% of mean annual precipitation as rain during mild winters.

In places where the soil profiles include a silica-and- iron-rich hardpan (illuvial horizon), pockets of Pygmy Forest habitat are most well developed. At these locations, dark, organic-rich, upper A horizons rest on a leached, light-colored, thixotropic (i.e., can become liquid when shaken) horizon. Below the eluvial horizon are illuvial (B) horizons that are rich in clay, very sticky, hard when dry, buff-to-orange colored, mottled, and abundant in iron concretions and iron and manganese streaks (Figure 22). Maximum clay

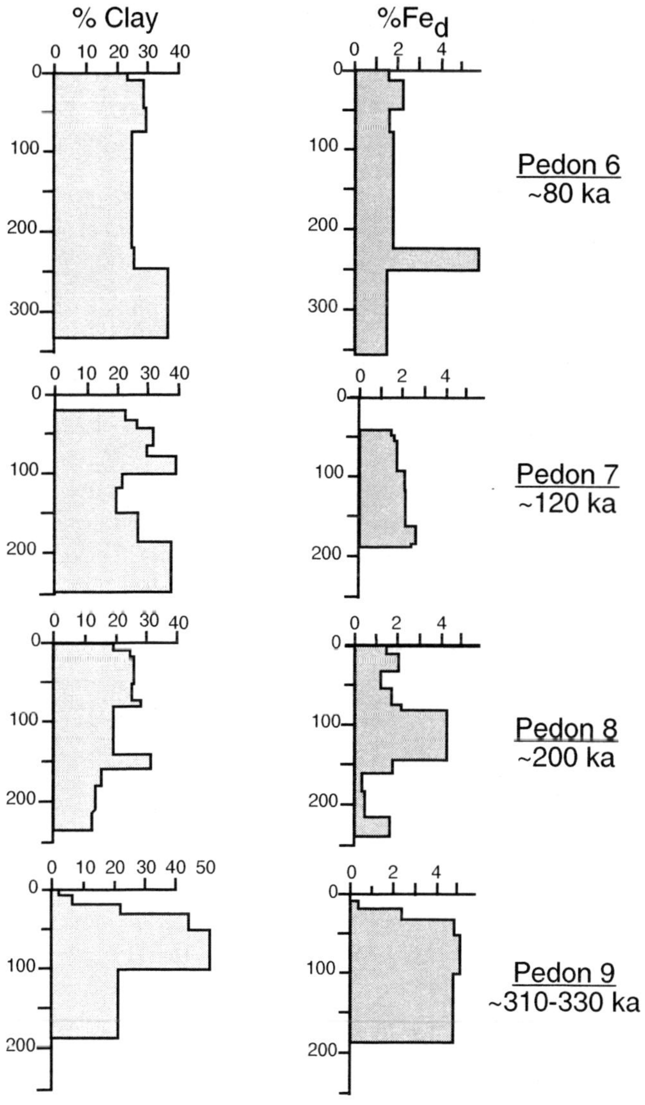

Figure 22. Depth profiles of five pedons from the Mendocino to Westport area. Pedon site numbers and inferred terrace ages are indicated on the right. Pedon 6 is on the 20-24-m terrace, inferred to be ~80,000 yr old. Pedon 7 is on the 38-42 m terrace, inferred to be ~120,000 yr old. Pedon 8 is on the 58-63 m terrace, inferred to be ~200,000 yr B.P. Pedon 9 is on the 83-89 m terrace, inferred to be ~310,000-330,000 yr old. The soil properties diagrammed are percent clay and Fe_d. [From Figure 8 in Merritts et al., 1991].

content in the B-horizons varies from 30 to 52%, maximum iron from 2.6 to 5.9 %, and maximum B-horizon thickness from 152 to 178 cm on the second and third terraces (Merritts et al., 1991, 1992).

End of DAY 3: Overnight in Gualala, CA

DAY 4: End of trip—thank you for joining us! Return to San Francisco International Airport

ACKNOWLEDGMENTS

We dedicate this field trip to the pioneers of marine terrace geology of the central and northern California coast: Warren Addicott, Charles S. Alexander, William C. Bradley, Hans Jenny and Andrew C. Lawson. Muhs' work was supported by the USGS Earth Surface Dynamics Program and is a contribution to the LITE (Last Interglacial: Timing and Environment) project. Prentice's work was supported by the USGS National Earthquake Hazards Reduction Program. Merritts' work was supported by the Keck Geology Consortium (awards in 1995-96 and 1999-2000) and the National Science Foundation (grants EAR-8405360 and EAR-9418682). We thank John Sisto (Point Arena lighthouse curator) and Jim Riley (Mendocino College) for granting access that enabled us to map the deposits at Point Arena, and Gary Strachan (Supervising Ranger at Año Nuevo State Reserve) for allowing access to exposures near Green Oaks Creek and Point Año Nuevo. Many thanks go to Josh Been (USGS, Denver) for help in organizing and carrying out the trip and Kathleen Simmons (USGS, Denver) who dated all of the corals. Thanks also go to Ken Ludwig (Berkeley Geochronology Center) who helped collect the corals and George Kennedy (Brian F. Smith and Associates) for helpful discussions of fossil mollusks. Oliver Chadwick (University of California, Santa Barbara) and David Hendricks (University of Arizona) contributed significantly to the field and laboratory work for the soil analyses in the Mendocino-Westport area. William Bull (University of Arizona) assisted with the interpretation of marine terrace ages at Mendocino and Cabrillo Point. Jorie Schulz and Marith Reheis (USGS) provided helpful reviews of an earlier version of the manuscript, and we thank Don Easterbrook for careful editing.

REFERENCES

Abbott, D.F., and Haderlie, E.C., 1980, Prosobranchia: Marine snails: *in* Morris, R.H., Abbott, D.P., and Haderlie, E.C., eds., Intertidal invertebrates of California, Stanford University Press, Stanford, CA, p. 230-307.

Abbott, R.T., 1974, American seashells: The marine mollusca of the Atlantic and Pacific coasts of North America (second edition): Van Nostrand Reinhold Co., New York, 663 p.

Addicott, W.O., 1966, Late Pleistocene marine paleoecology and zoogeography in central California: U.S. Geological Survey Professional Paper 523-C, p. C1-C21.

Alexander, C.S., 1953, The marine and stream terraces of the Capitola- Watsonville area: University of California Publications in Geography, v. 10, p. 1-44.

Anderson, R.S., and Menking, K.M., 1994, The Quaternary marine terraces of Santa Cruz, California: Evidence for coseismic uplift on two faults: Geological Society of America Bulletin, v. 106, p. 649-664.

Argus, D.F., and Gordon, R.G., 2001, Present tectonic motion across the Coast Ranges and San Andreas fault system in central California: Geological Society of America Bulletin, v. 113, p.1580-1592.

Atwater, T., 1970, Implications of plate tectonics for the Cenozoic tectonic evolution of western North America: Geological Society of America Bulletin, v. 81, p. 3513-3536.

Baldwin, J.N., Knudsen, K.L., Lee, A., and Prentice, C.S., and Gross, R., 2000, Preliminary estimate of coseismic displacement of the penultimate earthquake on the northern San Andreas fault, Pt. Arena, California: *in* Bokelmann, G. and Kovach, R.L., eds., Tectonic problems of the San Andreas fault system, Stanford University Publications, p. 355-368.

Berger, A., and Loutre, M. F., 1991, Insolation values for the climate of the last 10 million years: Quaternary Science Reviews, v. 10, p. 297-317.

Bloom, A.L., Broecker, W.S., Chappell, J.M.A., Matthews, R.K., and Mesolella, K.J., 1974, Quaternary sea level fluctuations on a tectonic coast: New $^{230}Th/^{234}U$ dates from the Huon Peninsula, New Guinea: Quaternary Research, v. 4, p. 185-205.

Bradley, W.C., and Griggs, G.B., 1976, Form, genesis, and deformation of central California wave-cut platforms: Geological Society of America Bulletin, v. 87, p. 433-449.

Brown, E.T., and Bourles, D.L., 2002, Use of a new ^{10}Be and ^{26}Al inventory method to date marine terraces, Santa Cruz, California, USA: Comment: Geology, v. 30, p. 1147-1148.

Brown, R.D., 1995, 1906 surface faulting on the San Andreas fault near Point Delgada, California: Bulletin of the Seismological Society of America, v. 85, p. 100-110.

Coan, E.V., Scott, P.V., and Bernard, F.R., 2000, Bivalve seashells of western North America: Marine bivalve mollusks from Arctic Alaska to Baja California: Santa Barbara Museum of Natural History Monographs, no. 2, 764 p.

Edwards, R.L., Cheng, H., Murrell, M.T., and Goldstein, S.J., 1997, Protactinium-231 dating of carbonates by thermal ionization mass spectrometry: Implications for Quaternary climate change: Science, v. 276, p. 782-786.

Griggs, G.B., and Johnson, R.E., 1979, Coastline erosion Santa Cruz County: California Geology, v. 32, p. 67-76.

Grove, K., and Niemi, T.M., 1999, The San Andreas fault zone near Point Reyes: Late Quaternary deposition, deformation, and paleoseismology: *in* Wagner, D.L. and Graham, S.A., eds., Geologic field trips in northern California, California Department of Conservation, Division of Mines and Geology, Special Publication 119, p. 176-187.

Hall, N. T., 1984, Holocene history of the San Andreas fault between Crystal Springs Reservoir and San Andreas Dam, San Mateo County, California: Seismological Society of America Bulletin, v. 7, p. 281-299.

Hall, N.T., Wright, R.H., and Clahan, K.B., 1999, Paleoseismic studies of the San Francisco peninsula segment of the San Andreas fault zone near Woodside, California: Journal of Geophysical Research, v. 104, p. 23,215-23,236.

Herbert, T.D., Schuffert, J.D., Andreasen, D., Heusser, L., Lyle, M., Mix, A., Ravelo, A.C., Stott, L.D., and Herguera, J.C., 2001, Collapse of the California current during glacial maxima linked to climate change on land: Science, v. 293, p. 71-76.

Heusser, L.E., Lyle, M., and Mix, A., 2000, Vegetation and climate of the northwest coast of North America during the last 500 k.y.: High-resolution pollen evidence from the northern California margin: *in* Lyle, M., Richter, C., and Moore, T.C., Jr., eds., Proceedings of the Ocean Drilling Program, Scientific Results, v. 167, p. 217-226.

Horsfield, W.T., 1975, Quaternary vertical movements in the Greater Antilles: Geological Society of America Bulletin, v. 86, p. 933-938.

Hoskins, C.W., 1957. Paleoecology and correlation of the lowest emergent California marine terrace, from San Clemente to Halfmoon Bay: Ph.D. thesis, Stanford University, 226 p.

Jenny, H., Arkley, R.J., and Schultz, A.M., 1969, The pygmy forest-podsol ecosystem and its dune associates of the Mendocino coast: Madroño, v. 20, p. 60-74.

Johnson, R.G., 1962, Mode of formation of marine fossil assemblages of the Pleistocene Millerton Formation of California: Geological Society of America Bulletin, v. 73, p. 113-130.

Kelson, K.I., Lettis, W.R., Lisowski, M., 1992, Distribution of geologic slip and creep along faults in the San Francisco Bay region: California Division of Mines and Geology Special Publication, v. 113, p. 31-38.

Kennedy, G.L., 1978, Pleistocene paleoecology, zoogeography and geochronology of marine invertebrate faunas of the Pacific Northwest Coast (San Francisco Bay to Puget Sound): Ph.D. thesis, University of California-Davis, 824 p.

Kennedy, G.L., and Armentrout, J.M., 1989, A new species of chimney-building *Penitella* from the Gulf of Alaska (Bivalvia: Pholadidae): The Veliger, v. 32, p. 320-325.

Kennedy, G.L., Lajoie, K.R., and Wehmiller, J.F., 1982, Aminostratigraphy and faunal correlations of late Quaternary marine terraces, Pacific Coast, USA: Nature, v. 299, p. 545-547.

Ku, T.-L., Ivanovich, M., and Luo, S., 1990, U-series dating of last interglacial high sea stands: Barbados revisited: Quaternary Research, v. 33, p. 129-147.

Lajoie, K.R., Ponti, D.J., Powell, C.L., II, Mathieson, S.A., and Sarna-Wojcicki, A.M., 1991, Emergent marine strandlines and associated sediments, coastal California; A record of Quaternary sea-level fluctuations, vertical tectonic movements, climatic changes, and coastal processes: *in* Morrison, R.B., ed., Quaternary nonglacial geology; conterminous U.S., Geological Society of America, Boulder, Colorado, The Geology of North America, v. K-2, p. 190-203.

Lawson, A.C., 1893, The post-Pliocene diastrophism of the coast of southern California: University of California Department of Geological Sciences Bulletin, v. 1, p. 115-160.

Lawson, A.C., 1908, The California earthquake of April 18, 1906, Report of the State Earthquake Investigation Commission: Carnegie Institute of Washington Publication 87, 254 p.

Martinson, D. G., Pisias, N. G., Hays, J. D., Imbrie, J., Moore, T. C., Jr., and Shackleton, N. J., 1987, Age dating and the orbital theory of the ice ages: Development of a high-resolution 0 to 300,000-year chronostratigraphy: Quaternary Research, v. 27, p. 1-29.

Merritts, D.J., 1996, The Mendocino triple junction: Active faults, episodic coastal emergence, and rapid uplift: Journal of Geophysical Research, v. 101, p. 6051-6070.

Merritts, D., and Bull, W.B., 1989, Interpreting Quaternary uplift rates at the Mendocino triple junction, northern California, from uplifted marine terraces: Geology, v. 17, p. 1020-1024.

Merritts, D.J., Chadwick, O.A., and Hendricks, D.M., 1991, Rates and processes of soil evolution on uplifted marine terraces, northern California: Geoderma, v. 51, p. 241-275.

Merritts, D.J., Bodin, P., Beutner, E.C., Prentice, C.S., and Muller, J.R., 2000, Active surface deformation in the Mendocino triple junction process zone: *in* Bokelman G., and Kovach, R.L., eds., Proceedings of the 3rd conference on tectonic problems of the San Andreas fault system: Stanford University Publications in Geological Sciences, v. 21, p. 128-143.

Merritts, D.J., Chadwick, O.A., Hendricks, D.M., Brimhall, G.H., and Lewis, C.J., 1992, The mass balance of soil evolution on late Quaternary marine terraces, northern California: Geological Society of America Bulletin, v. 104, p. 1456-1470.

Mesolella, K.J., Matthews, R.K., Broecker, W.S., and Thurber, D.L., 1969, The astronomical theory of climatic change: Barbados data: Journal of Geology, v. 77, p. 250-274.

Muhs, D. R., Kennedy, G. L., and Rockwell, T. K., 1994, Uranium-series ages of marine terrace corals from the Pacific coast of North America and implications for last-interglacial sea level history: Quaternary Research, v. 42, p. 72-87.

Muhs, D.R., Rockwell, T.K., and Kennedy, G.L., 1992, Late Quaternary uplift rates of marine terraces on the Pacific coast of North America, southern Oregon to Baja California Sur: Quaternary International, v 15/16, p. 121-133.

Muhs, D.R., Simmons, K.R., and Steinke, B., 2002b, Timing and warmth of the last interglacial period: New U-series evidence from Hawaii and Bermuda and a new fossil compilation for North America: Quaternary Science Reviews, v. 21, p. 1355-1383.

Muhs, D.R., Kelsey, H.M., Miller, G.H., Kennedy, G.L., Whelan, J.F., and McInelly, G.W., 1990, Age estimates and uplift rates for late Pleistocene marine terraces: Southern Oregon portion of the Cascadia forearc: Journal of Geophysical Research, v. 95, p. 6685-6698.

Muhs, D.R., Simmons, K.R., Kennedy, G.L., and Rockwell, T.K., 2002a, The last interglacial period on the Pacific Coast of North America: Timing and paleoclimate: Geological Society of America Bulletin, v. 114, p. 569-592.

Muhs, D.R., Simmons, K.R., Kennedy, G.L., Ludwig, K.R., and Groves, L.T., 2002c, A cool eastern Pacific Ocean at the close of the last interglacial period, ~80,000 yr BP: Geological Society of America Abstracts with Programs, v. 34, no. 6, p. 130.

Niemi, T.M. and Hall, N.T., 1992, Late Holocene slip rate and recurrence of great earthquakes on the San Andreas Fault in northern California: Geology, v. 20, no. 3, p. 195-198.

Noller, J.S., Kelson, K.I., Lettis, W.R., Wickens, K.A., Simpson, G.D., Lightfoot, K., Wake, T., 1993, Preliminary characterizations of Holocene activity on the San Andreas fault based on offset archaeologic sites, Ft. Ross State Historic Park, California: NEHRP Final Technical Report, 16 p.

O'Clair, R.M., and O'Clair, C.E., 1998, Southeast Alaska's rocky shores: Animals: Auke Bay, Alaska, Plant Press, 561 p.

Perg, L.A., Anderson, R.S., and Finkel, R.C., 2001, Use of a new ^{10}Be and ^{26}Al inventory method to date marine terraces, Santa Cruz, California, USA: Geology, v. 29, p. 879-882.

Perg, L.A., Anderson, R.S., and Finkel, R.C., 2002, Use of a new ^{10}Be and ^{26}Al inventory method to date marine terraces, Santa Cruz, California, USA: Reply: Geology, v. 30, p. 1148.

Pisias, N.G., Mix, A.C., and Heusser, L., 2001, Millennial scale climate variability of the northeast Pacific Ocean and northwest North America based on radiolaria and pollen: Quaternary Science Reviews, v. 20, p. 1561-1576.

Prentice, C.S., 1989, Earthquake geology of the northern San Andreas fault near Point Arena, California: Ph.D. thesis, California Institute of Technology, 252 p.

Prentice, C.S., 1999, San Andreas Fault: The 1906 Earthquake and subsequent evolution of ideas: *in* Moores, E.M., Sloan, D., and Stout, D., eds., Classic Cordilleran Concepts: A view from California: Geological Society of America Special Paper 338, p. 70-85.

Prentice, C.S. and Hall, N.T., 1996, Using historical research to locate the 1906 rupture of the San Andreas fault on the San Francisco Peninsula: Geological Society of America Abstracts with Programs, v. 28, p. 102.

Prentice, C.S., Langridge, R., and Merritts, D J., 2000, Paleoseismic and Quaternary tectonic studies of the San Andreas fault from Shelter Cove to Fort Ross: *in* Bokelmann, G. and Kovach, R.L., eds., Proceedings of the Third Conference on Tectonic Problems of the San Andreas System, Stanford University Publications, p. 349-350.

Prentice, C.S., Niemi, T.M., and Hall, N.T., 1991, Quaternary tectonics of the northern San Andreas fault, San Francisco Peninsula, Point Reyes, and Point Arena, California: *in* Sloan, D., and Wagner, D.L., eds., Geologic Excursions in Northern California: San Francisco to the Sierra Nevada, California Department of Conservation, Division of Mines and Geology, Special Publication 109, p. 25-34.

Prentice, C.S., Merritts, D.J., Bodin, P., Beutner, E., Schill, A., and Muller, J., 1999, The northern San Andreas fault near Shelter Cove, California: Geological Society of America Bulletin, v. 111, no. 4, p. 512-523.

Prentice, C. S., Prescott, W.H., Langridge, R., And Dawson, T., 2001, New geologic and geodetic slip rate estimates on the North Coast San Andreas Fault: Approaching agreement?: Seismological Research Letters, v. 72, p. 282.

Savage, J.C., Simpson, R.W, Murray, M.H., 1998, Strain accumulation rates in the San Francisco Bay area, 1972-1989: Journal of Geophysical Research, v. 103, p. 18,039-18,051.

Schulz, M., Harden, J.W., and Weber, G.E., 2003, Geo-ubiquity: Chemical, hydrological and physical evolution of coastal terraces near Santa Cruz, California, this volume.

Schussler, H., 1906, The water supply of San Francisco, California, before, dur ing, and after the earthquake of April 18, 1906: Marthin B. Brown Press, New York, 103 p.

Shackleton, N.J., and Opdyke, N.D., 1976, Oxygen-isotope and paleomagnetic stratigraphy of Pacific core V28-239 late Pliocene to latest Pleistocene: Geological Society of America Memoir 145, p. 449-464.

Stirling, C.H., Esat, T.M., Lambeck, K., McCulloch, M.T., Blake, S.G., Lee, D.C., and Halliday, A.N., 2001, Orbital forcing of the marine isotope stage 9 interglacial: Science, v. 291, p. 290-293.

Storlazzi, C.D., and Griggs, G.B., 2000, Influence of El Niño-Southern Oscillation (ENSO) events on the evolution of central California's shoreline: Geological Society of America Bulletin, v. 112, p. 236-249.

Wakabayashi, J., 1999, The Franciscan Complex, San Francisco Bay area: A record of subduction complex processes: *in* Wagner, D.L. and Graham, S.A., eds., Geologic Field Trips in Northern California, California Department of Conservation, Division of Mines and Geology Special Publication 119, p. 1-21.

Weber, G.E., 1990, Late Pleistocene slip rates on the San Gregorio fault zone at Point Año Nuevo, San Mateo County, California: *in* Greene, H.G., Weber, G.E., and Wright, T.L., eds., Geology and tectonics of the central California coast region, San Francisco to Monterey: Bakersfield, California: Pacific Section of the American Association of Petroleum Geologists, p. 193-203.

Weber, G.E., Lajoie, K.R., and Griggs, G.B., 1979, Coastal tectonics and coastal geologic hazards in Santa Cruz and San Mateo Counties, California: Geological Society of America Cordilleran Section, Field Trip Guide, 187 p.

Quaternary geology of the western United States

Authors and Leaders:

Don J. Easterbrook, *Dept. of Geology, Western Washington University, Bellingham, WA*
Kenneth Pierce, *U.S. Geological Survey, Bozeman, MT*
John Gosse, *Dalhousie University, Halifax, Nova Scotia, Canada*
Alan Gillespie, *University of Washington, Seattle, WA*
Ed Evenson, *Lehigh University, Bethlehem, PA*
Ken Hamblin, *Brigham Young University, Provo, Utah*

Organizer: **Don J. Easterbrook,** *Dept. of Geology, Western Washington University, Bellingham, WA 98225*

INTRODUCTION

A rich record of Quaternary features is to be found in the western U.S., including continental glaciation, alpine glaciation, glaciomarine deposits, isostatic rebound, giant floods from glacial Lake Missoula, the world's largest known volcanic eruptions, calderas, the world's largest geothermal area, faulting, pluvial lakes, and much more. This field trip includes classic examples of all of these features, spanning the time from early Pleistocene to the present day.

The trip is divided into segments (Fig. 1), led by experts in specific areas, as reflected by multiple authors of the guidebook. Contents of the guidebook are intended not only for use during the 2003 INQUA field trip but also to be useable by anyone who wishes to visit these areas on their own.

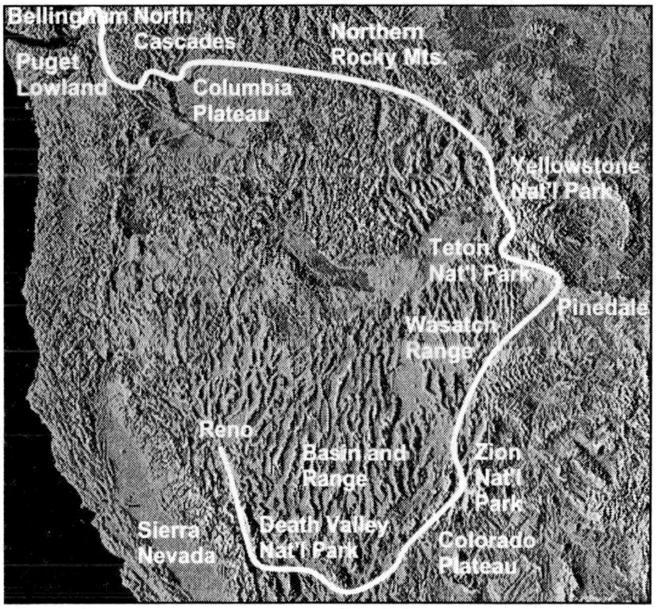

Figure 1. Field trip route map.

THE CORDILLERAN ICE SHEET AND NORTH CASCADE ALPINE GLACIATION (Don J. Easterbrook)

The Puget Lowland

The Puget Lowland (Fig. 2) is mantled with deposits of Pleistocene glaciations recording repeated advances and retreats of the Cordilleran Ice Sheet from British Columbia. Deposits of till, outwash sand and gravel, glaciomarine drift, and related deposits are well exposed in hundreds of miles of sea cliffs adjacent to inland waterways.

DAY 1. LATE PLEISTOCENE GLACIATIONS; DISINTEGRATION OF THE CORDILLERAN ICE SHEET

Early Pleistocene deposits are known only from the southern Puget Lowland. North of Seattle, Quaternary sediments greatly thicken and deposits older than 300 ka are deeply buried. For data on the early Pleistocene chronology see Easterbrook et al. (1981), Westgate et al. (1987), and Easterbrook (1994).

Drifts of three, late Pleistocene glaciations, occur throughout the central Puget Lowland, each separated by nonglacial sediments. The Double Bluff, Possession, and Vashon Drifts represent deposits of the Cordilleran Ice Sheet in the lowland. The interglacial Whidbey Fm. and Olympia nonglacial sediments that occur between the drifts were deposited on floodplains and in lakes and swamps. Each of the three glacial units typically consists of a package of deposits that represent characteristic patterns of advance and retreat of the ice sheet. The advance phase of each glaciation typically consists of (1) deposition of outwash sand and gravel ahead of the advancing ice, (2) scouring by overriding of the glacier, (3) deposition of till and related drift on the previously scoured surface, and (4) deposition of recessional outwash in the southern Puget Lowland and glaciomarine drift in the central and northern Puget Lowland during retreat of the ice (Easterbrook, 1992).

STOP 1–1. DOUBLE BLUFF DRIFT AND WHIDBEY FM. TYPE SECTIONS AT DOUBLE BLUFF

Enroute to STOP 1–1, travel north from Seattle on I–5, take the ferry from Mukilteo to Clinton on southern Whidbey Island, then drive northwest to Freeland. Turn south from

Easterbrook, D. J., Pierce, K., Gosse, J., Gillespie, A., Evenson, E., and Hamblin, K., 2003, Quaternary geology of the western United States: *in* Easterbrook, D.J., ed., Quaternary Geology of the United States, INQUA 2003 Field Guide Volume, Desert Research Institute, Reno, NV, p. 19–79.

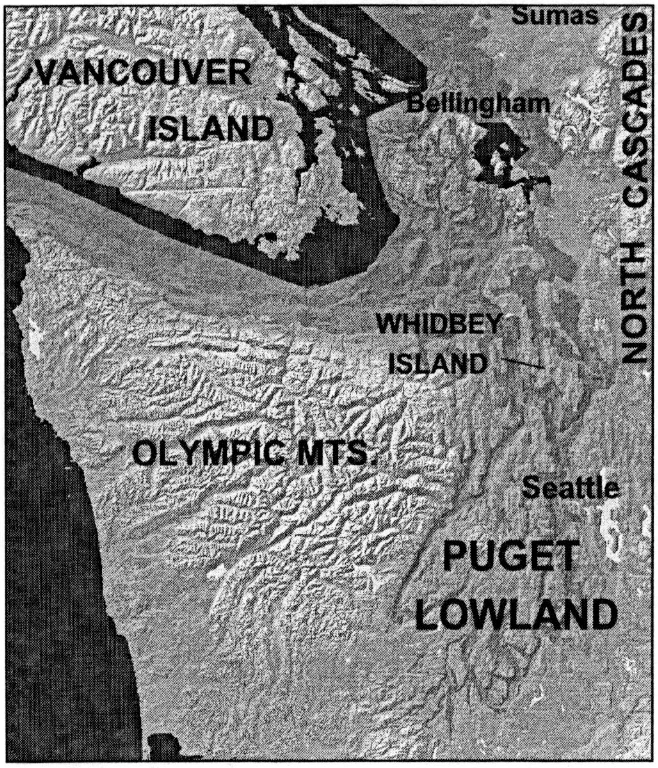

Figure 2. The Puget Lowland and adjacent areas. (Modified from digital image by PRISM, University of Washington)

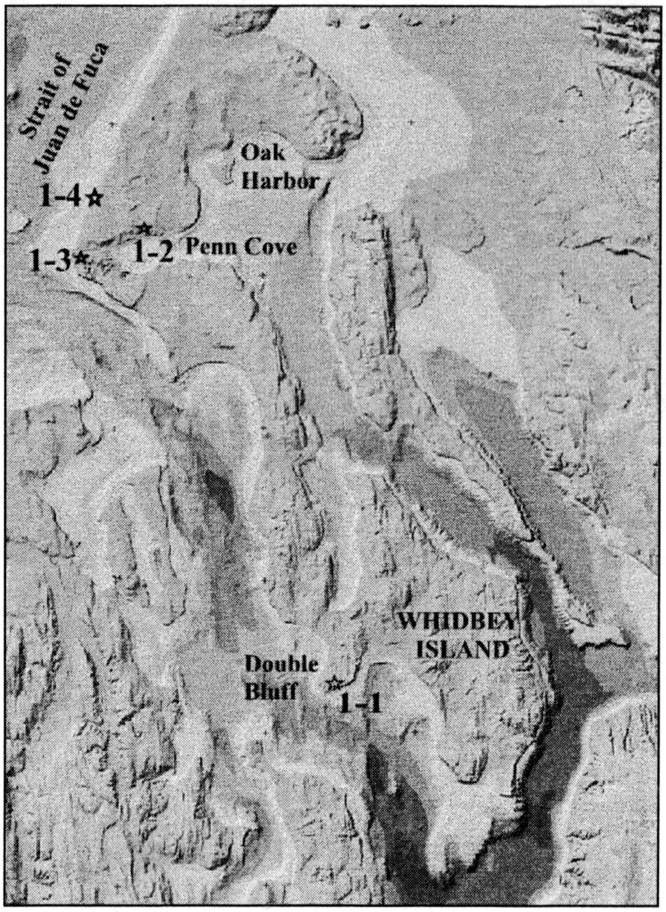

Figure 3. Whidbey Island showing field trip stops. (Modified from digital image by PRISM, University of Washingon.

Freeland to Double Bluff. Walk the beach from Useless Bay to Double Bluff (about 1.5 km; 1 mi.).

The sea cliffs between Useless Bay and Double Bluff (Fig. 3) have been designated the type localities for the Double Bluff Glaciation and the Whidbey Interglaciation (Easterbrook et al., 1967). The purpose of this stop is to examine the stratigraphic relationships of these two deposits.

Double Bluff Glaciation. Sea cliff exposures at Double Bluff consist of glacial outwash, till, glaciomarine drift, and glaciolacustrine sediments, overlain by peat–bearing silt and sand of the Whidbey Fm. (Fig. 4). (Easterbrook, 1968). Elsewhere, Double Bluff drift is rarely exposed, but crops out in a few sea cliffs at or near sea level.

Thermoluminesence (TL) dating of clay beneath Double Bluff Drift has yielded ages of 289 ± 74 ka and 291 ± 86 ka. A TL date of 177 ± 38 ka was obtained from Double Bluff glaciomarine drift, and a date of 320 ± 100 ka was obtained from clay associated with Double Bluff till (Berger and Easterbrook, 1993). Amino–acid ages from mollusk shells in Double Bluff glaciomarine drift range from 111 to 178 ka at the type locality and from 150 to 250 ka elsewhere in the central Puget Lowland (Easterbrook et al., 1982; Blunt et al., 1987).

Whidbey Interglaciation. The type section of the Whidbey Fm. consists of silt and sand, interbedded with, clay, peat, and widely scattered gravel exposed in the ~2–km–long sea cliffs between Double Bluff and Useless Bay (Fig. 3) (Easterbrook et al., 1967). Possession Drift, which overlies the Whidbey Fm. elsewhere, is absent here and younger sediments unconformably overlie the Whidbey (Easterbrook, 1968). Along the sea cliffs in the eastern part of the section, Whidbey sediments are overlain by Olympia lacustrine silt, ^{14}C–dated at 23,600 ± 275 and 22,210 ± 530 ^{14}C yrs. B.P. (Stoffel, 1980), Vashon advance outwash, and Vashon till (Easterbrook, 1968). Everson fossiliferous glaciomarine drift is exposed at a low swale just east of the SE point of Double Bluff.

Hansen and Mackin (1949) suggested that the Whidbey sediments originated as floodplain deposits formed from "very slow aggradation by meandering streams flanked by floodplain lakes and swamps" in which silty clay and peat were deposited. Pollen studies suggest that the Whidbey was an interglacial period with a climate somewhat warmer than the present (Hansen and Mackin, 1949; Hansen and Easterbrook, 1974; Heusser and Heusser, 1981).

More than 20 ^{14}C ages beyond the range of ^{14}C dating have been obtained from peat in the Whidbey Fm. Amino–acid ages of 97 ± 35, 96 ± 35 ka, and 107 ± 9 ka have been obtained from shells in Whidbey marine sediments to the north (Easterbrook, 1968; Blunt et al., 1987).

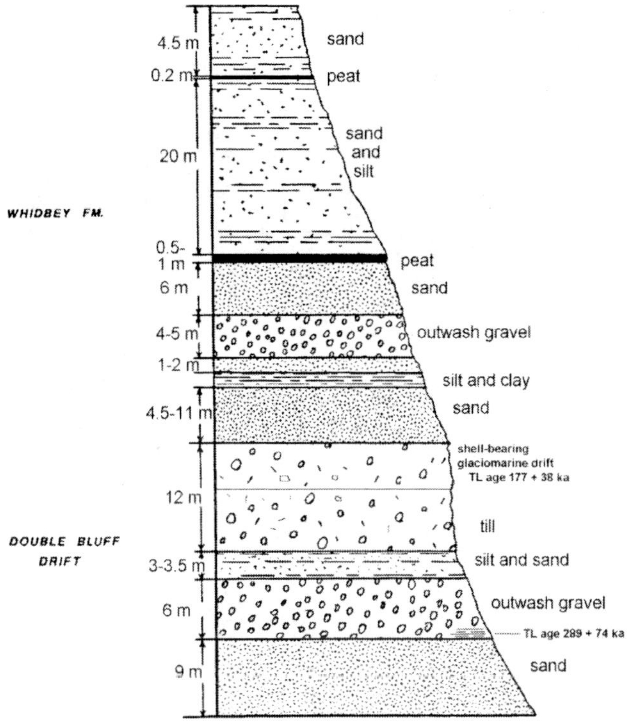

Figure 4. Stratigraphic section at Double Bluff. (Easterbrook, 1994)

TL ages measured on clay from Whidbey fluvial sediments are shown below. (Berger and Easterbrook, 1993; Easterbrook, 1994).

TL Age (ka)	Locality
102 ± 38	Lagoon Point, Whidbey Island
106 ± 17	Blowers Bluff, Whidbey Island
142 ± 10	Northern West Beach, Whidbey Island
151 ± 43	Point Wilson, NW of Port Townsend

Remarkable deformation structures (Fig. 5) occur in the silt and fine sand (Bretz, 1913; Easterbrook, 1968, 1992). Clastic dikes and swirling mass of silt and fine sand resembling bows of ribbon abound, especially near the east end of the exposure near Useless Bay. The structures are clearly not glacial in origin because many of them are truncated by overlying beds within the Whidbey Fm. They are distinctly intraformational and penecontemporaneous with deposition. Johnson et al. (1996, 2002) have proposed that they represent seismic shaking from a fault that passes under the area.

Mammoth tusks, bones, and teeth commonly weather out of the Whidbey Fm. and accumulate on the beach between Double Bluff and Useless Bay at low tide.

On route to STOP 1–2, return to Freeland and drive north on Highway 525 (which becomes Highway 20 northward) to the north side of Penn Cove (Fig. 3). Turn right off Highway 20 at San de Fuca onto Penn Cove Road and continue to Penn Cove Park. Walk along the beach from here to Blowers Bluff.

Figure 5. Deformation structures in the Whidbey Fm. West of Useless Bay.

STOP 1–2. POSSESSION DRIFT OVER WHIDBEY SILT/SAND; EVERSON GMD, BLOWERS BLUFF.

Possession Glaciation. Massive, cliff–making, Possession till and glaciomarine drift overlie peat–bearing sand and silt of the Whidbey Fm. at mid–cliff and are overlain by Vashon advance outwash, till, and fossiliferous Everson gmd. (Fig. 6) (Easterbrook, 1968). Whidbey Fm. sand and silt at the base of the section contains many granual–size pumice clasts. Electron microprobe data on the pumice is given in Easterbrook (1994).

Massive, poorly sorted, till–like Possession gmd that overlies till at mid–cliff contains scattered *Nuculana* shells. Amino–acid analyses of the shells suggest a mean age of 80 ± 22 ka

Figure 6. Possession till and glaciomarine drift overlying Whidbey Fm. sand and silt and overlain by Vashon Drift and Everson gmd at Blowers Bluff.

(Blunt et al., 1987; Easterbrook and Rutter, 1981, 1982). Vashon advance outwash, Vashon till, and Everson gmd overlie the Possession Drift.

Olympia Interglaciation. Nonglacial floodplain and lacustrine silt, clay, and peat, were deposited in the lowland between the Possession and Fraser glaciations during the Olympia Interglaciation (Armstrong et al., 1965). Radiocarbon dates from Olympia sediments just east of Blowers Bluff range from 22–28 ka ^{14}C yrs. B.P. (Hansen and Easterbrook, 1974; Easterbrook, 1976a, b 1986).

Fraser Glaciation. The late Wisconsin Fraser Glaciation consists of several stades; (1) the Evans Creek Stade, an early alpine phase; (2) the Vashon Stade, the maximum advance of the Cordilleran Ice Sheet (CIS); (3) the Everson Interstade, an interval of glaciomarine deposition during deglaclation of the lowland; and (4) the Sumas Stade, consisting of several short readvances of the ice sheet in the Fraser Lowland to the north before complete deglaciation (Easterbrook, 1963, 1969, 1992; Armstrong et al., 1965; Kovanen and Easterbrook, 2002a; Kovanen, 2002).

Vashon Stade. The last major advance of the CIS is defined as the Vashon Stade (Armstrong et al., 1965). It began with advance of the CIS into the lowland and ended with the beginning of marine and glaciomarine conditions of the Everson Interstade. Meltwater streams deposited large volumes of outwash sand and gravel ahead of the advancing ice sheet that was later overridden by the Vashon glacier and covered with till now exposed at or near the top of most of the sea cliffs in the lowland.

Timing of the advance is limited by ^{14}C dates from sediments beneath Vashon till. These dates indicate that Vashon ice advanced southward across the Canadian border sometime after 17–18 ka ^{14}C yrs. B.P. and reached the Seattle area soon after 14.5 ka ^{14}C yrs. B.P. (Easterbrook, 1992; Porter and Swanson, 1998). However, an anomalous mixture of ^{14}C dates near the Canadian border ranging in age from 16–20,000 ^{14}C yrs. B.P. were used by Clague et al. (1988) and Porter and Swanson (1998) to suggest that the Cordilleran Ice Sheet (CIS) did not cross the Canadian border until after 16,000 ^{14}C yrs B.P. However, these sediments are not demonstrably overlain by Vashon till, the 16 ka dates are mixed in the same horizon as 20 ka dates, and are anomalously young because the CIS was depositing an outwash plain in front of its advancing margin ~115 km to the south by 18,000 ± 400 ^{14}C yrs B.P. and along the Strait of Juan de Fuca 17,350 to 18,265 ^{14}C yrs B.P. (Blunt et al., 1987; Easterbrook, 1992) Advance Vashon outwash was deposited ~150 km south of the Canadian border at 16,510 ± 320, 15,450 ± 450, and 15,350 ± 210 ^{14}C yrs B.P. (Deeter, 1979), and in Seattle at ~14.5–16,000 ^{14}C yrs B.P. (Mullineaux et al., 1965). If the CIS didn't cross the border until after 16,000 ^{14}C yrs B.P., it could hardly have been depositing outwash far to the south 2000 years before then.

Newly developed imagery from DEM and LIDAR data show clearly defined elongate drumloidal hills and flutes over much of the Puget Lowland that indicate directions of Vashon ice flow (Fig. 7).

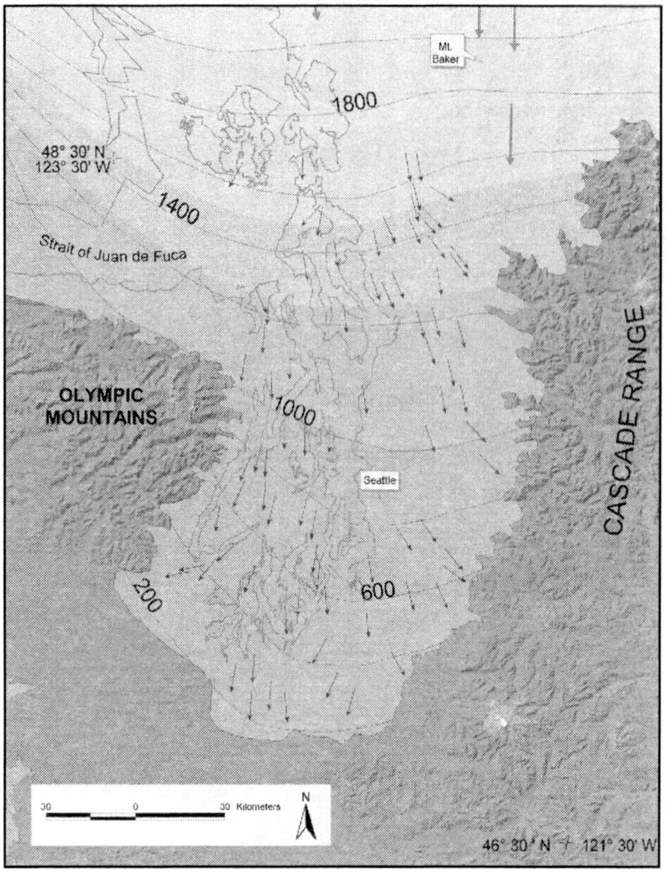

Figure 7. The Cordilleran Ice Sheet at its late Wisconsin maximum. (Shaded digital topographic model image by D.J. Kovanen and M. Price)

At its maximum, Vashon ice was more than 1,830 m. (6,000 ft) thick near Bellingham, about 915 m. (3,000 ft) thick near Seattle, and about 550 m (1,800 ft) thick southeast of Tacoma (Fig 7).

Return to Highway 20 and drive west. Turn right on the Libby Road to and continue to West Beach.

STOP 1–3. DEGLACIATION, EVERSON GLACIOMARINE DRIFT CHRONOLOGY, WEST BEACH

Everson Interstade. About 14.5 ka ^{14}C yrs. B.P., only a few hundred years after it reached its maximum, the CIS, began large–scale recession and downwasting. The Puget lobe backwasted northward from its southern terminus past the Seattle area by about 14 ka ^{14}C yrs. B.P. (Rigg and Gould, 1957; Leopold et al., 1982; Porter and Swanson, 1998). Rapid thinning of Vashon ice after the terminus had receded nearly to the Strait of Juan de Fuca, floated the remaining ice as marine water flooded the lowland, causing rapid disintegration of the remaining CIS. During this interval, Everson glaciomarine drift was deposited from south of Whidbey Island to the Canadian border.

The glaciomarine sediments consist largely of poorly sorted diamictons deposited when melting of floating ice released rock debris that accumulated on the sea floor. Unbroken, delicate, articulated marine shells, some in growth positions, and diseminated foraminifera indicate that gmd represents *in situ* deposition (Easterbrook, 1963, 1992). More than 150 ^{14}C dates from Washington and British Columbia fix the age of the Everson glaciomarine drift at 11,500 to 13,500 ^{14}C yrs. B.P., (Fig. 8) (Easterbrook, 1992, 2003), making it a valuable stratigraphic marker over the central and northern Puget Lowland.

Sea cliffs just north of the park at West Beach consist of fossiliferous Everson gmd, dated here at 12,535 ± 300 ^{14}C yrs. B.P. (uncorrected) (Easterbrook, 1968). Applying the new marine reservoir correction for this region (Kovanen and Easterbrook, 2002b) gives a corrected age of 11,400 ^{14}C yrs. B.P.

The first recognition and dating of glaciomarine drift on Whidbey and nearby islands in the 1960s were supported by 11 ^{14}C dates (Easterbrook, 1966a,b, 1968, 1969), and in the intervening years, 140 additional ^{14}C dates were obtained (Fig. 8). The inescapable conclusions from Figure 8 are:

(1). The distribution of dates indicates that glaciomarine conditions were essentially contemporaneous over the entire central and northern Puget Lowland,

(2). Because ^{14}C dates are the same in Canada as in the southernmost part of the lowland, the ice must have disintegrated rapidly, rather than by backwasting and calving.

(3) The age of final emergence was 11,500 ^{14}C yrs. B.P. for the entire area.

The topography in the Partridge Point area is deeply indented with kettles (Fig. 9), many >30 m deep. Just south of West Beach Park, wave erosion has breached a 30–m–deep kettle, exposing the Partridge Point gravel that makes up the kettled topography. Collapse structures are common around the margins of the kettles. The gravel typically has strong foreset bedding with dips up to 20–25° and contains pumice clasts and many small shell fragments (Easterbrook, 1966a, 1968, Carlstad, 1992).

The top of the Partridge Point gravel, where not indented by kettles, consists of a flat surface about 55–60 m above present sea level. Relict meltwater channels can be observed on this surface. Everson gmd caps the Partridge Point gravel in the bluffs just south of the park (Easterbrook, 1968).

Drive east to Oak Harbor and continue north to the north end the island. Turn right and drive east to I–5. Turn left and drive north on I–5 toward Bellingham.

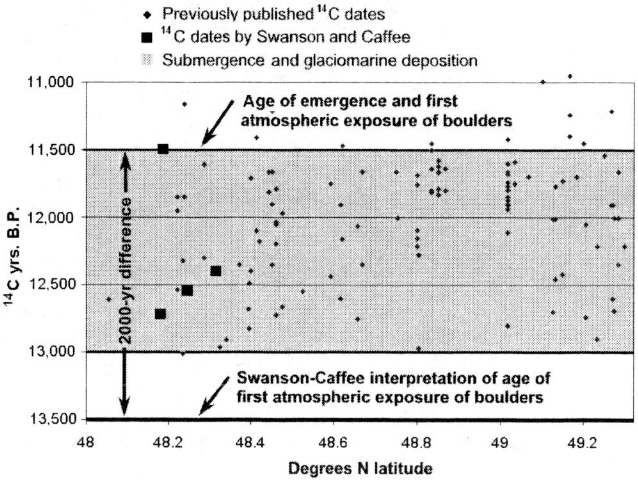

Figure 8. Chronology of Everson glaciomarine drift deposition. (Easterbrook, 2003)

Figure 9. 30m–deep kettles in west–dipping deltaic gravel at Partridge Pt. (Photo by D.A. Rahm, courtesy of EPIC)

STOP 1–4. SUMAS OUTWASH CHANNEL INCISED INTO EVERSON GMD ACROSS SHORELINES

Sumas Stade. Soon after 11,500 ^{14}C yrs. B.P., the Puget Lowland emerged from the sea and the CIS readvanced over its own outwash several times during the Sumas Stade, which spans the Younger Dryas and part of the Allerød (Easterbrook, 1963, 1992; Easterbrook and Kovanen, 1998; Kovanen and Easterbrook, 2002a; Kovanen, 2002). We will see the field evidence and chronology of these on DAY 2.

At the Samish valley, gmd extends to elevations of ~65 m, and shorelines cut on the gmd are deeply incised by the Samish River (Fig. 10). A limiting age for similar strandlines south of here is 11,700 ± 110 ^{14}C yrs. B.P. from basal peat at 47 m (Siegfried, 1978). The Samish River is badly underfit. Its valley was obviously made by a much larger stream than the present Samish

Figure 10. Sumas outwash channel cut into Everson glaciomarine drift and across late Everson shorelines. (Modified from shaded digital topographic model image by D.J. Kovanen)

River. A prominent terrace within the valley terminates at a marine delta about 30 m (100 ft.) above present sea level (Easterbrook, 1979, 1994). The outwash channel and terraces were made by meltwater from glaciers in the North Cascades to the east during the Sumas Stade, C^{14}–dated elsewhere at about 10–11,400 ^{14}C yrs. B.P. The flood of the valley was graded to a sea level near that of the present. The significance of this is that it shows that relative sea level had dropped to the present level by the end of the Pleistocene, rather than being spread throughout the Holocene.

Arrive in Bellingham. End of DAY 1.

DAY 2 EVERSON RELATIVE SEA LEVEL CHANGES AND SUMAS READVANCES

Everson Interstade

Everson glaciomarine drift (gmd) was deposited from floating ice in the Fraser Lowland (Fig. 11) at elevations up to 180 m (600 ft) above present sea level and extended more than 10 km (6 mi.) up the Nooksack valley into the North Cascades (Easterbrook, 1963, 1992, 1994; Kovanen and Easterbrook, 2001).

The Everson Stade in the Fraser Lowland consists of two fossiliferous glaciomarine drifts, the Kulshan (lower) and Bellingham (upper) separated by subaerial, fluvial, Deming sand with crossbedding, stream channels, *in situ* tree stumps, and basal peat lacking pollen from any salt–tolerant plants. The presence of non–marine Deming sand between the Bellingham and Kulshan gmds demonstrates a remarkable fluctuation of relative sea level in this area over short periods of time. During deposition of the Kulshan gmd, relative sea level stood at ~90 m (300 ft) above present sea level, but, as shown by beach deposits in Deming sand at Bellingham Bay, dropped to within 8–12 m (25–65 ft.) of present sea level during deposition of the Deming sand. Relative sea level then rose again to more than 180 m (600 ft.) during deposition of Bellingham gmd, and finally dropped to, or somewhat below, present sea level by the beginning of the Sumas Stade 11,400 ^{14}C yrs. B.P. (Easterbrook, 1963, 1992; Kovanen and Easterbrook, 2002a). As we saw at the Samish River valley yesterday and as we will see today at Tenmile and Squalicum Creek valleys, Sumas outwash channels that cut deeply into Bellingham gmd were graded to a sea level close to that of the present, indicating that the 180 m (600 ft.) of post–Bellingham–gmd emergence took place rapidly.

Only a complicated combination of isostatic, eustatic, and tectonic events can explain the ups and downs of relative sea level during about 1500 years at the end of the Pleistocene (Easterbrook 1963, 1992). The rapid changes in relative sea level cannot be explained by any combination of only isostatic and eustatic changes because of submergences that occurred during times of isostatic uplift. A strong local tectonic influence is required to adequately explain these late Pleistocene relative sea level changes, but the rates of emergence and submergence appear to be exceedingly rapid and the processes involved are not yet clearly understood.

Sumas Stade

The Sumas Stade began with emergence of the area from the Everson sea and readvance of the Cordilleran Ice Sheet (Armstrong et al., 1965). Details of the physical evidence for multiple advances of Sumas ice may be found in Easterbrook (1994), Easterbrook and Kovanen, (1998), Kovanen and Easterbrook, (2002a), and Kovanen, (2002)

The paleogeography of the ice margin and timing of ice advance and retreat during the Sumas Stade is bracketed by 70 radiocarbon dates, which are secured by morphologic and stratigraphic evidence. Four topographically distinct phases of the Sumas readvances have been documented (Fig. 12).

Leave WWU campus, drive east on Garden St. and turn left (north) on Holly St. Holly St. becomes Marine Drive to the north. Turn left on secondary road north of the old cement plant. Walk down the trail to the beach and proceed northward up the beach to sea cliff exposures.

STOP 2–1. BELLINGHAM GLACIOMARINE DRIFT OVERLYING BEACH DEPOSITS OF DEMING SAND, BELLINGHAM BAY

The purpose of this stop is to examine the Kulshan–Deming–Bellingham sequence at Bellingham Bay and establish relative sea level during deposition of the Deming sand. Sea cliffs expose Bellingham gmd lying on beach deposits of the Deming sand (Fig. 13), which in turn rest on Kulshan gmd (Easterbrook, 1963; Weber, 2001). Bellingham gmd consists of stony silt/clay containing marine fossils and can be traced continuously at the surface to the Cascade foothills to the east where it extends to ele-

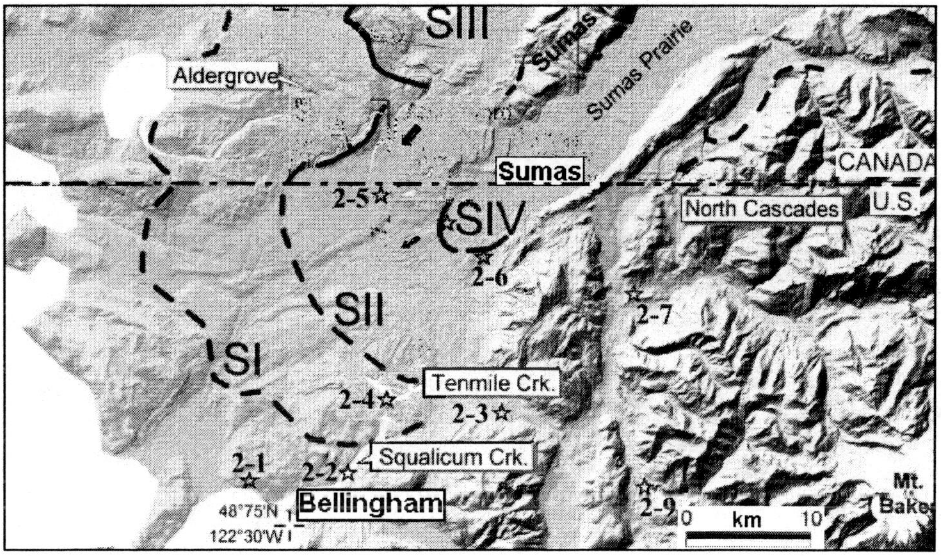

Figure 11. Location of stops in the Fraser Lowland and position of ice margins of the four Sumas phases. (Modified from Kovanen and Easterbrook, 2002a)

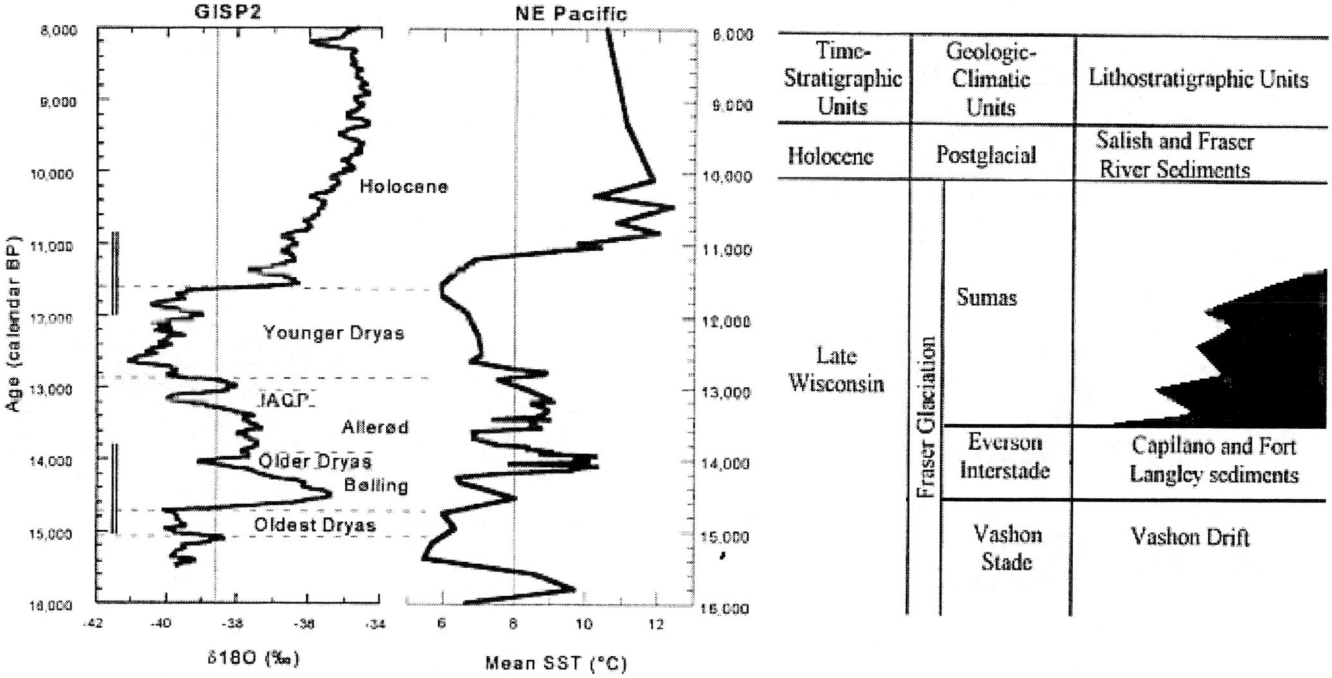

Figure 12. Correlation of the chronology of Sumas phases with the GISP2 ice core and sea surface paleotemperatures in the northeast Pacific. (Kovanen and Easterbrook, 2002a; Kienast and McKay, 2001)

vations up to 180 m (600 ft.). Six ^{14}C dates from wood and marine shells in the gmd just north of Bellingham Bay range from 11,580 to 11,830 ^{14}C yrs. B.P. and marine fossils in beach deposits of the Deming sand here were dated at 11,685 ± 85 and 11,760 ± 85 (Weber, 2001; Kovanen and Easterbrook, 2002a). Fossiliferous Deming beach deposits range in elevation from 7.5 m (25 ft.) to 20 m (65 ft.) and establish sea level limits for the Deming sand. Evidence for the beach origin of these deposits is documented in Easterbrook (1963) and Weber (2001). Marine shells from Kulshan gmd, exposed beneath the Deming sand in bluffs just above

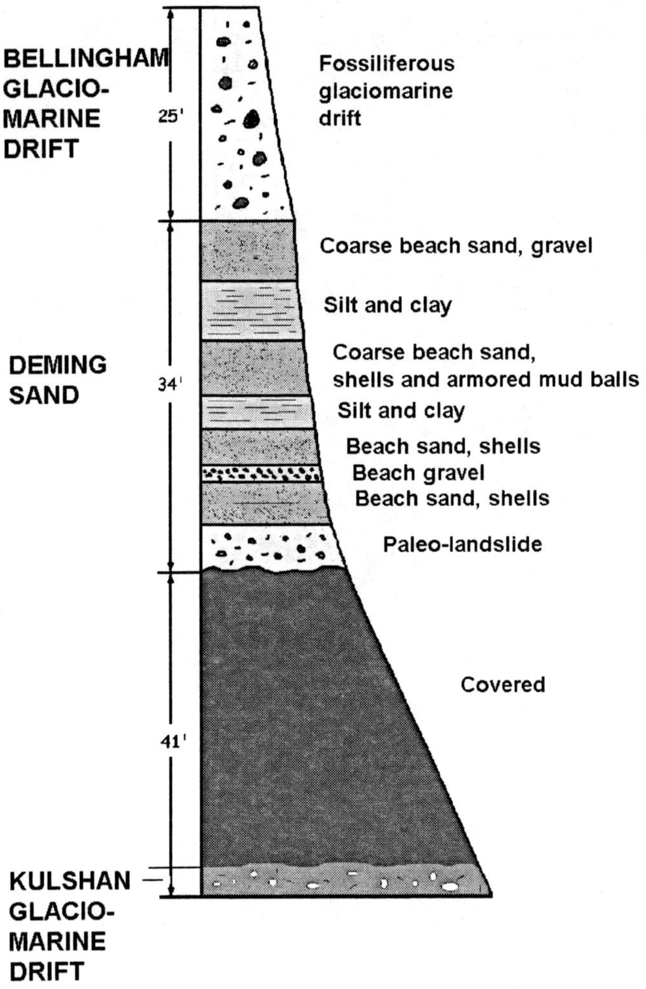

Figure 13. Bellingham glaciomarine drift overlying fossiliferous Deming beach sand at Bellingham Bay.

the modern beach and in the wave platform seaward from the bluffs have been dated at 12,210 ± 82 ^{14}C yrs. B.P.

The significance of the Kulshan–Deming–Bellingham sequence is that it replicates the same sequence exposed at the Everson type section (STOP 2–2) and establishes the position of sea level for the Deming sand between the two gmds.

Drive east on Sunset Drive (Mt. Baker Highway). Turn left and drive north on Hannegan Road. Squalicum outwash channel (Sumas I)

STOP 2–2. SQUALICUM OUTWASH CHANNEL

The Squalicum outwash channel (Fig. 11) is deeply incised into Bellingham gmd and now occupied only by a small underfit stream. It was cut by meltwater that originated from a brief stand of ice during the earliest phase of the Sumas Stade (Sumas I). The channel is not directly dated but incises Bellingham gmd and is thus post–11,500 ^{14}C yrs. B.P., and it is older than Sumas II (11,400 ^{14}C yrs. B.P.). No well–defined moraine occurs at the head of the channel—it just ends abruptly at the position of the glacier terminus.

Turn right on Van Wyck Road, cross the Squalicum outwash channel, and continue east on Mt. Baker Highway.

STOP 2–3. EVERSON INTERSTADE TYPE LOCALITY

The type locality of the Everson Interstade is a large bluff along the Nooksack River, kept constantly exposed by erosion along the outside of a meander against the valley side. The base of the section consists of ~5 m (15 ft.) of stratified, crossbedded, pebbly sand, interbedded with thin silt beds (Fig. 14). A twig and a thin, *in situ*, organic mat a few mm thick interbedded with silty–clay were dated at 12,070 ± 80 and 12, 185 ± 80 ^{14}C yrs. B.P., providing a lower limiting age for the Kulshan gmd and indicating that this area was above sea level before it was submerged to deposit Kulshan gmd.

The basal sand is overlain by Kulshan gmd, consisting of 9–12 m (30–40 ft.) of poorly sorted pebbly silt and clay containing widely scattered marine shells. No ^{14}C dates have been obtained directly from the Kulshan here, but shells in the same unit 6 km (3.7 mi.) to the east were dated at 11,870 ± 380 years. The top of the Kulshan is about 75 m (250 ft.) above sea level here. Adding another 30 m for probable water depth gives a Kulshan sea level about 100 m (350 ft.) higher than the present.

The Kulshan gmd is overlain by about 10 m (33 ft.) of Deming sand (Fig. 14), which consists mainly of fluvial crossbedded sand with some interbedded silt. Prominent channeling may be seen in bluff exposures. *In situ* rooted stumps in a peat bed about 20–30 cm (8–12 in.) thick at the contact between the Kulshan gmd and Deming sand give incontrovertible proof that the Deming is nonmarine (Easterbrook, 1992; Kovanen and Easterbrook, 2002a). Additional evidence of the nonmarine origin of the Deming sand has been documented in Easterbrook (1963 and 1992).

As seen at STOP 2–1, beach deposits in Deming sand show that sea level at this time was only about 7–20 m (25–65 ft.) above present sea level, compared to ~100–200 m (350–700 ft.) during the overlying and underlying gmd deposition. ^{14}C ages from the Deming at Bellingham Bay are 11,685 ± 85 and 11, 760 ± 85 ^{14}C yrs. B.P., and the rooted stump at the Everson type locality is 11,810 ±60 ^{14}C yrs. B.P. This means that the Deming sand at the Everson type locality must have been deposited ~60 m (200 ft.) above sea level at that time.

Deming sand is overlain by ~20 m of massive, poorly sorted, Bellingham gmd. Marine fossils, mostly *Nuculana* and *Clinocardium*, are more abundant than in the Kulshan gmd, but are widely scattered. An allocthanous piece of wood in the middle portion of the Bellingham gmd was dated at 11,800 ± 400 years (Easterbrook, 1963). Outcrops of Bellingham gmd extend to elevations of 180–210 m (600–690 ft.) above present sea level, and adding another 30 m for probable water depth gives a relative sea level of 210–240 m (700–800 ft.) above present sea level, a relative submergence of >200 m (>660 ft.) from Deming to Bellingham time. A similar, contemporaneous 'sandwich' of fluvial sediment between two gmds in British Columbia demonstrates that this is a regional, rather than local, feature. Ten ^{14}C dates from the Bradner pit in B.C. range from 11,600 to 11,800 ^{14}C yrs. B.P., almost identical to the ages from the Everson type locality and Bellingham Bay.

Post–Bellingham gmd relative sea level dropped abruptly about 11,500 ^{14}C yrs. B.P. Sumas outwash, graded to a sea level close to that of the present, was deposited on Bellingham gmd and outwash channels were incised into the gmd, indicating yet another abrupt relative sea level change of about 200 m. Remarkable as they may seem, these ups and downs of relative sea level (the "*yo–yo*" effect) appear to be irrefutable–the physical evidence for them can be seen in the single riverbank exposure at the Everson type locality and identical ^{14}C dates from identical contemporaneous sediments across the Fraser Lowland. No correlation of strata, extrapolation of data, or abstract interpretation is required. However, the *rates* of emergence and submergence must have been so high as to stretch one's imagination. Others have argued that because of this, what we see in ^{14}C–dated outcrops must be impossible, but as Ed Evenson once pointed out, if something happened, then it must be possible!

Another significant point demonstrated by the stratigraphic sequence at the Everson type section is that deposition of both the Kulshan and Bellingham gmds was preceded by non–marine fluvial sedimentation so neither gmd could possibly have resulted from a retreating, calving Vashon ice sheet.

Return to Mt. Baker Highway and drive west to the Smith Road intersection. Turn right and continue to Starry Road. Turn right and drive 1.7 km (1 mi.) to the Kulshan Golf Club on the right. Pull into the parking lot.

STOP 2–4. OVERLOOK OF TENMILE CREEK SUMAS II OUTWASH CHANNEL.

The purpose of this stop is to see a deep, Sumas, meltwater channel incised into Bellingham gmd (Fig. 15), indicating that emergence closely followed deposition of the Bellingham gmd and providing a lower limiting age for the Sumas II phase. The Sumas outwash channel is incised ~33m (100 ft.), exposing Bellingham gmd underlying Deming sand in the valley walls. Basal peat from a 6–9 m–(20–30 ft.)–thick bog that floors the channel was dated at 11,080 ± 100 and 11,113 ± 88 ^{14}C yrs. B.P. These dates correlate well with the age of a prominent Sumas II moraine in southwestern British Columbia that we will see at the next stop. A basal bog date of 10,400 ± 85 ^{14}C yrs. B.P. was obtained from Fazon Lake, a large kettle just to the north.

Drive west on Axton Road, turn right on Hannegan Road. Cross peat–filled Sumas IV outwash channels, continue through Lynden to near the Canadian border.

Figure 14. Type section of the Everson glaciomarine drift.

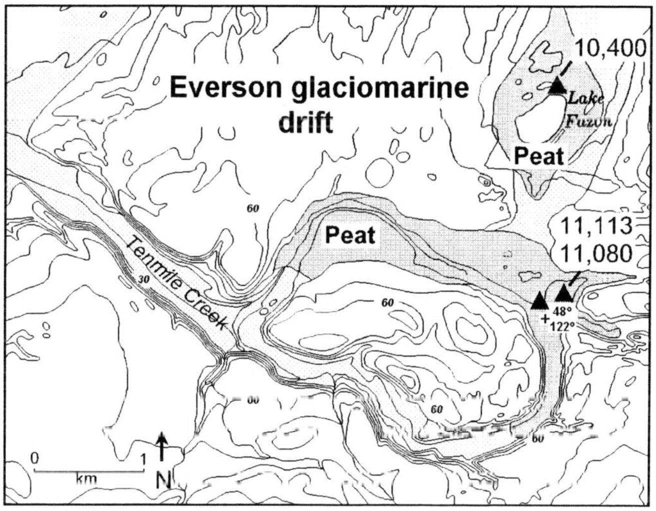

Fig. 15. Sumas II outwash channel incised ~30 m in Bellingham gmd and filled with peat. (Kovanen and Easterbrook, 2002a)

STOP 2–5. SUMAS II MORAINE, SUMAS III OUTWASH PLAIN, ^{14}C LIMITING DATES

Just north of the Canadian border, a prominent, elongate, Sumas II moraine (Fig.16A) rests on Everson gmd from which many ^{14}C dates on wood and shells have been obtained, limiting the Sumas II advance at younger than 11,600. ^{14}C yrs. B.P. A ^{14}C date of 11,413 ± 75 ^{14}C yrs. B.P. from basal peat in an outwash channel from the moraine (Fig. 16B) indicates that the glacier had retreated back from the end moraine by 11.4 ka, cutting off meltwater flow in this channel. Thus, the age of Sumas II moraine must be between 11.6 and 11.4 ka ^{14}C yrs. B.P.

A date of 10,980 ± 250 ^{14}C yrs. B.P. from basal peat in a channel *behind* the moraine (Fig. 16B) indicates that the glacier had receded 20 km to the Sumas III moraines by this time,

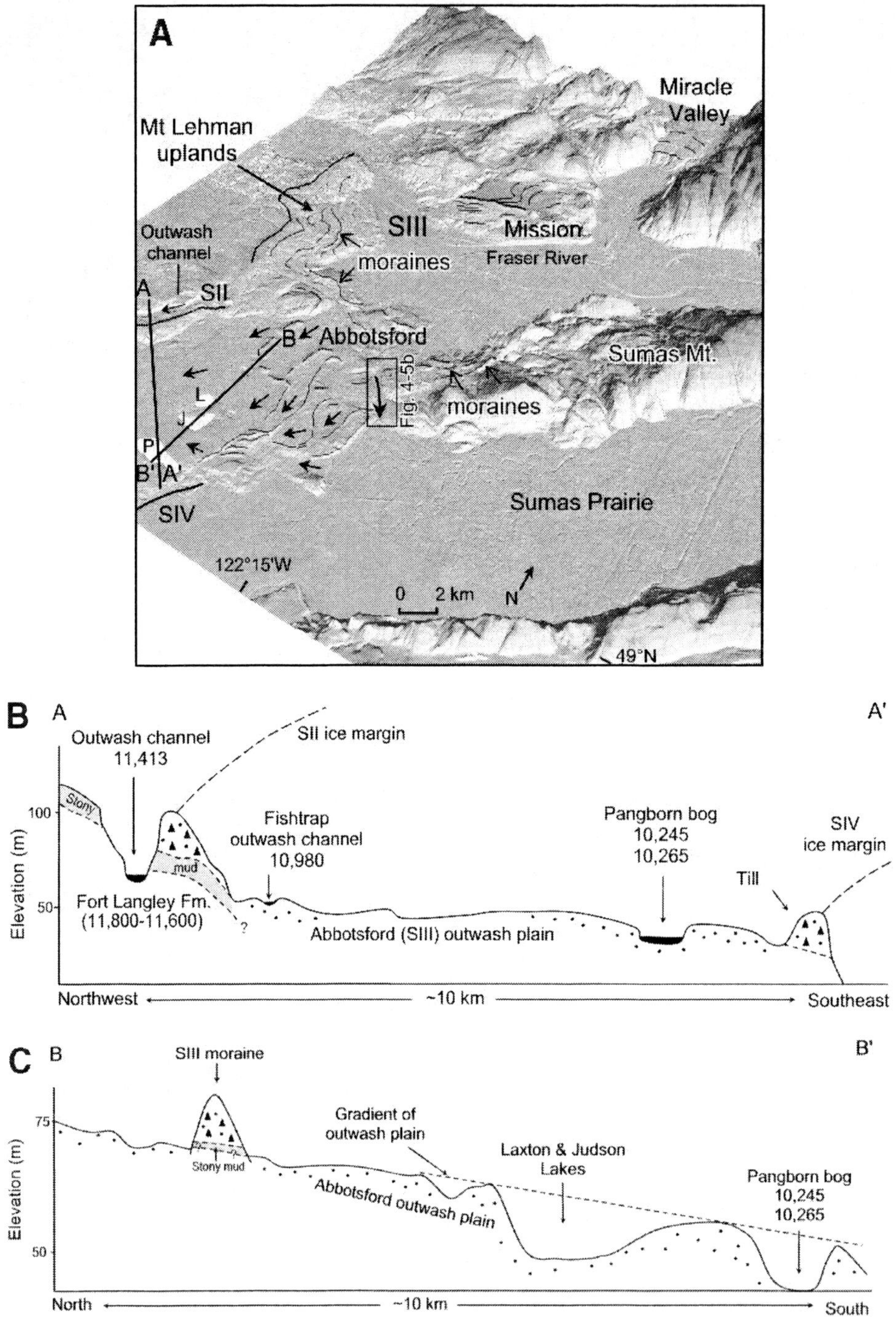

Figure 16. Sumas II and III moraines. (Kovanen and Easterbrook, 2002a). A. Sumas II and III moraines and Abbotsford outwash plain. B. East–west geologic cross–section from the Sumas II moraine to the Sumas IV moraine. C. North–south geologic cross–section from the Sumas III moraines to the Abbotsford outwash plain.

depositing the broad Abbotsford outwash plain that we are presently on. The significance of the position of the Abbotsford outwash plain behind the Sumas II moraine is that the CIS must have retreated to the position of the Sumas III moraines (Fig. 16C) and built them in the Younger Dryas.

Drive east on Halverstick Road past Pangborn bog Clearbrook (Sumas IV moraine across Sumas III outwash channel). Turn right off Halverstick Road onto road to Clearbrook, then to Van Buren Road. Turn left on Frick Road and take road across the front of the Sumas IV moraine.

STOP 2–6. SUMAS IV MORAINE AND OUTWASH TERRACE

The Sumas IV moraine (Fig. 11) consists of a hummocky, linear ridge that extends across the Sumas valley at the head of an outwash terrace that slopes southward away from the moraine. The moraine and outwash terrace have been breached in the central part of the Sumas Valley, but a remnant occurs on the west side of the valley, filling the distal end of a Sumas III meltwater channel (Fig. 17), showing that the Sumas IV glacier readvanced across earlier outwash. This final phase of the Sumas represents a late Younger Dryas readvance of the CIS.

Continue east on Telegraph Road–Easterbrook Road to Sumas. Turn right on road to Columbia Valley, drive to Kendall.

STOP 2–7. KENDALL MORAINE AND OUTWASH TERRACE

By about 12 ka ^{14}C yrs. B.P., the CIS had disappeared from the North Cascades and the lower Nooksack valley was submerged below sea level during deposition of Everson gmd. Shortly thereafter, long valley glaciers heading on Mt. Baker and fed by tributary glaciers from Mt. Shuksan and the Twin Sisters Range occupied the Nooksack drainage (Figs. 18 and 19) (Kovanen and Easterbrook, 2001). The purpose of this stop is to examine the Kendall moraine and outwash train deposited by a post–CIS, Cascade, alpine glacier ~45 km downvalley from its sources on Mt. Shuksan and Mt. Baker. The Kendall moraine forms a ridge several hundred meters long whose axis is transverse to the valley (Fig. 20). It is composed of ~12 m (40 ft.) of till containing abundant, glacially–faceted, striated boulders and cobbles derived from the upper North Fork valley, which has a U–shaped cross–profile at right angles to the direction of flow of the CIS.

The number of well–faceted and striated cobbles and boulders in the till is striking, probably because the North Fork glacier was so long. The lithology of cobble and boulders in the Kendall moraine is dominated by Chuckanut sandstone (54%), Mt. Baker andesite (12%), and graywacke/greenstone (14%) (presumably from the Jurassic Nooksack Group). These lithologies demonstrate that the source of ice was from the upper North Fork valley to the east, rather than from the CIS to the north. The Mt. Baker andesite was probably contributed via tributary valleys on the south side of the upper North Fork. Chuckanut sandstone was derived from the valley sides immediately upvalley, and the graywacke/ greenstone came from the valley sides.

Outwash from the Kendall moraine and the Maple Falls moraine (just upvalley) merge with the Sumas outwash terrace in the Columbia valley (Fig. 20). A few km downvalley, the outwash terrace rests on gmd ^{14}C–dated at 11,910 ± 80 ^{14}C yrs. B.P. Distinct charcoal layers interbedded with sandy terrace outwash gravel gave dates of 10,788 ± 77 and 10,603 ± 69 ^{14}C yrs. B.P. (Fig. 20), indicating a Younger Dryas age for the outwash (Kovanen and Easterbrook, 2001).

Time and weather permitting, continue east on Mt. Baker Highway to Heather Meadows.

STOP 2–8 (OPTIONAL) ARTISTS POINT AND HEATHER MEADOWS, SCENIC VIEWS OF MT. BAKER, MT. SHUKSAN, AND NORTH CASCADES; CORDILLERAN ICE SHEET GROOVES; HOLOCENE CIRQUE GLACIATION; MAZAMA ASH

Table Mt. is an andesite flow that has become topographically inverted by erosion. Erratics of Paleozoic greenstone on the summit of Table Mt. at an elevation of 1725 m (5700 ft.) indicate that the CIS flowed over Mt. Herman just to the north and

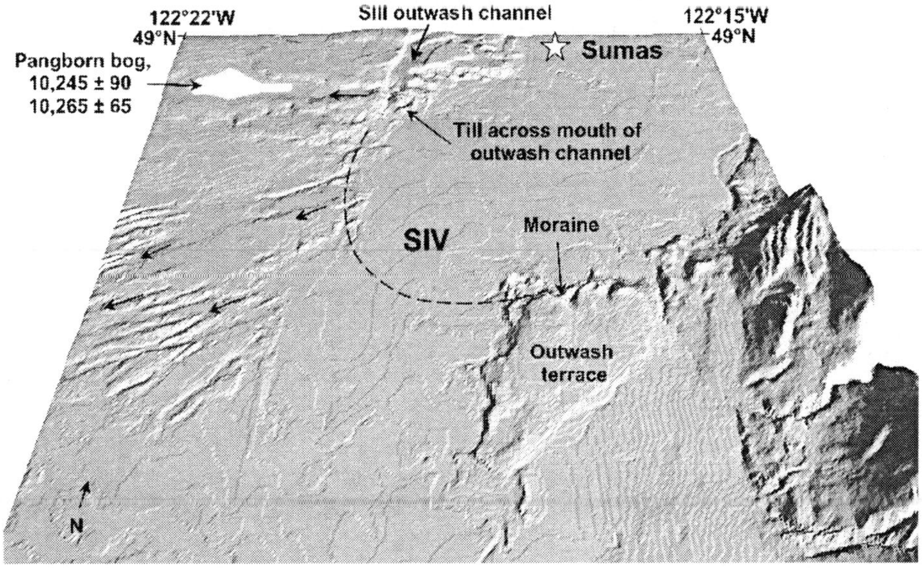

Figure 17. Sumas IV moraine and outwash terrace. (Kovanen and Easterbrook, 2002).

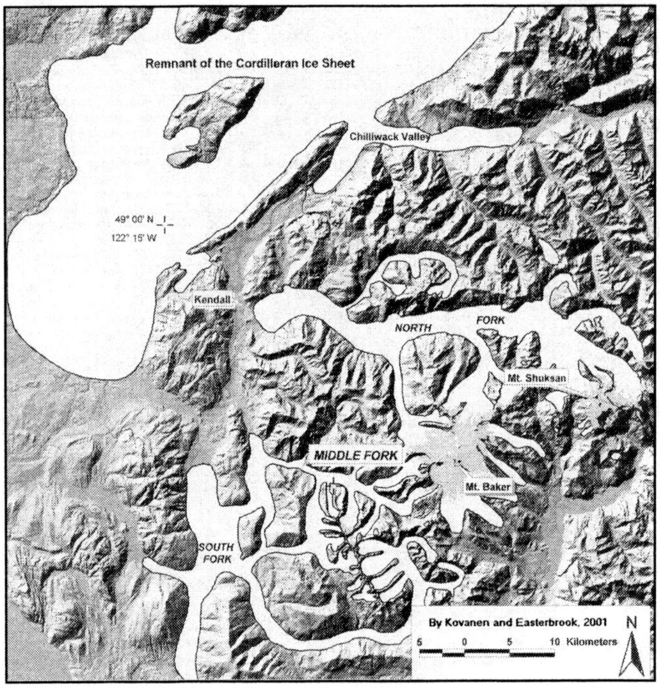

Figure 18. Reconstruction of Nooksack Alpine Glacier System. (Kovanen and Easterbrook, 2001)

deposited greenstone erratics on Table Mt. Adding a few hundred feet for ice thickness means that the surface of the CIS here must have been more than 1800 m (6000 ft.). From the parking lot, walk eastward along the trail to Artists Point and note the beautifully developed grooves and striations carved into the andesite by the CIS. Significantly, all of the grooves are oriented north–south, indicating N–S ice flow here. Thus, the CIS did not flow westward from here down the Nooksack drainage toward the lowland—it flowed almost due south down the axis of the range (Kovanen and Easterbrook, 2002a).

Return to Picture Lake at Heather Meadows. Picture Lake is held in by a moraine deposited by a cirque glacier. Mazama ash is exposed in a road cut beside the lake.

Return west on Mt. Baker Highway to Welcome. Turn left onto Mosquito Lake Road, cross Nooksack River, stop at overview of Deep Kettle bog.

STOP 2–9. DEEP KETTLE BOG, MOSQUITO LAKE KETTLE–KAME COMPLEX AND LATERAL MORAINE

The purpose of this stop is to examine evidence for the Cascade alpine ice origin of a post–CIS kettle–kame complex and lateral moraines. The kettle–kame complex is composed of poorly-to-moderately-sorted sand and gravel containing many glacially faceted and striated boulders whose lithologies are dominated by Mt. Baker andesite, (17.4%), dunite from the Twin Sisters Range (16.6%), and Chuckanut sandstone (45.2%), demonstrating that the glacier that transported them must have originated in the headwaters of the Nooksack Middle Fork (Kovanen and Easterbrook, 1996, 2001).

Rootlets in gravel at the base of a 20–m core from Deep Kettle bog (Fig. 21) were dated at 12,356 ± 115 ^{14}C yrs. B.P. and wood lying on the basal gravel was dated at 12,165 ± 115 ^{14}C yrs. B.P. This organic material could not have grown at the bottom of a 20–m deep lake, so it had to be growing in rock debris on stagnant ice before the ice melted out to make the kettle.

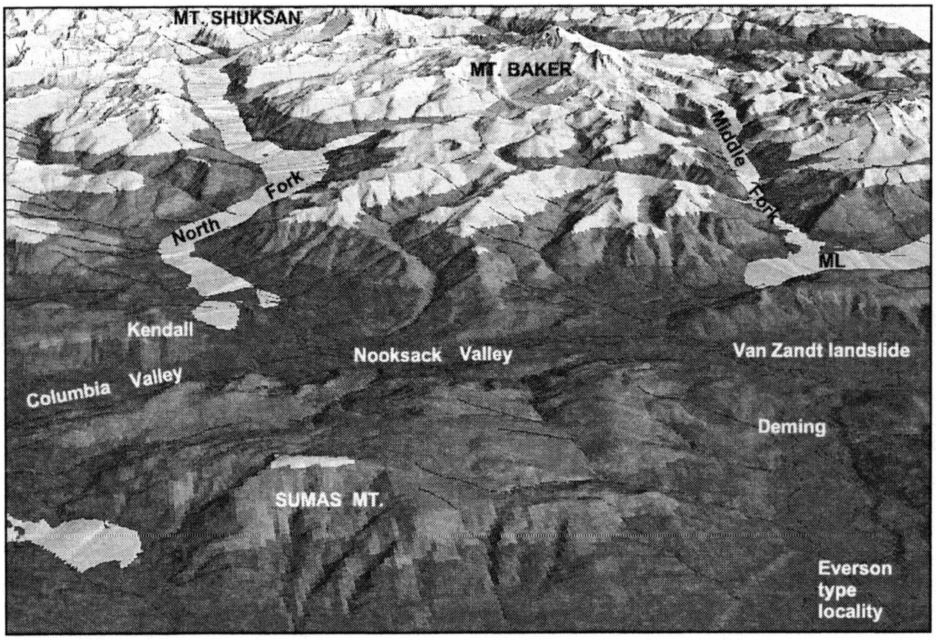

Figure 19. Reconstruction of post–CIS alpine glaciers in the North Cascades. (3D shaded digital topographic model image by D.J. Kovanen).

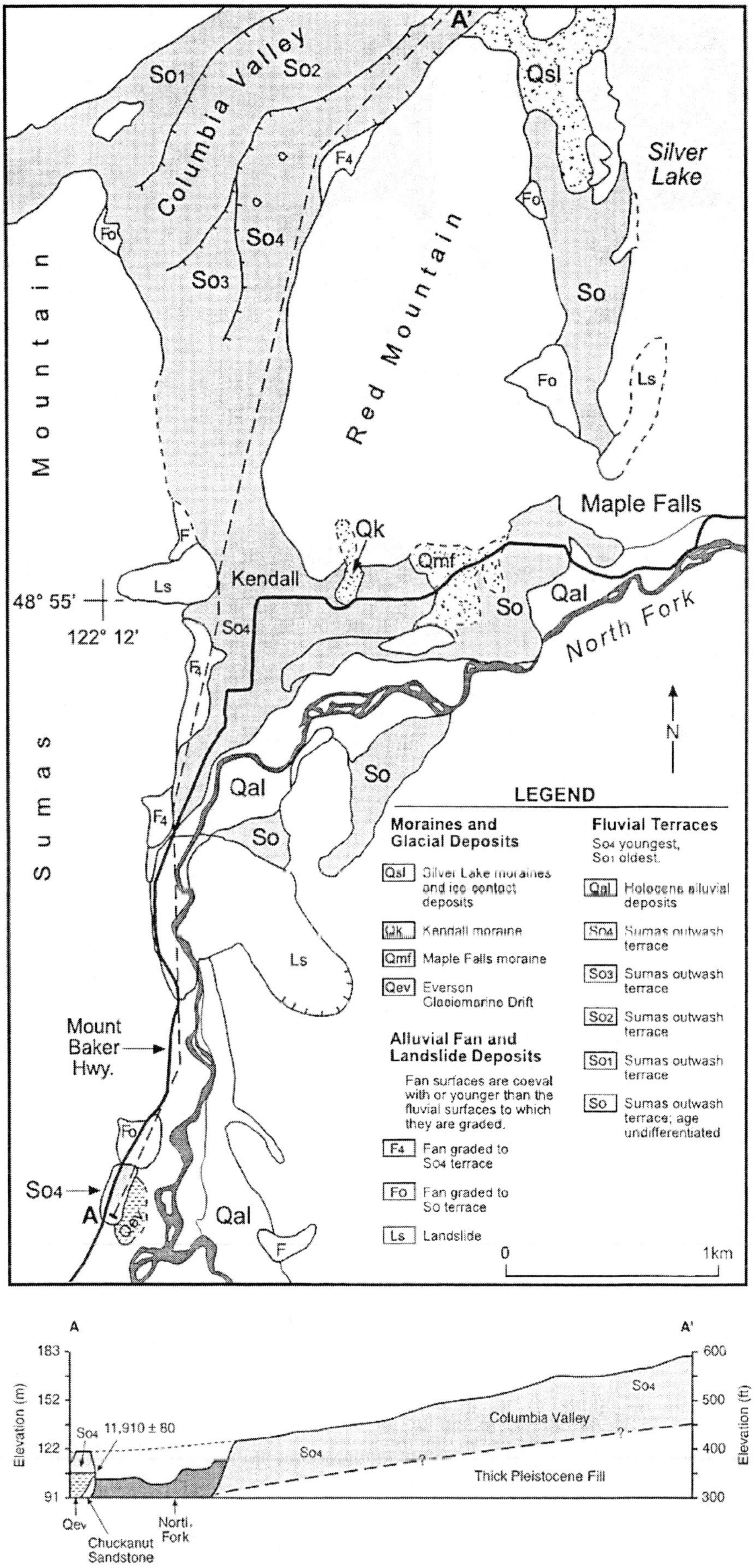

Figure 20. The Kendall moraine and outwash terrace. (Kovanen and Easterbrook, 2001).

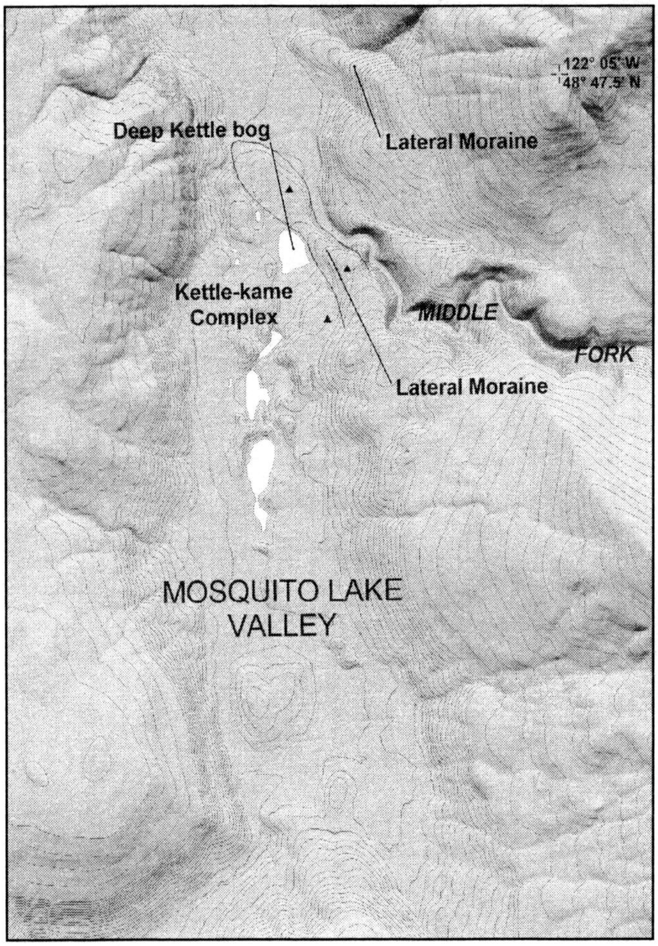

Figure 21. Ice–contact drift, Deep Kettle bog, and lateral moraines, Nooksack Middle Fork. (Kovanen and Easterbrook, 2001)

Following the stagnation that produced the kettle–kame complex, ice in the Middle Fork readvanced and deposited a lateral moraine that drapes across the kettle–kame complex (Fig. 21). The clast lithology of the moraine is dominated by glacially faceted and striated Mt. Baker andesite and Twin Sisters dunite, proving that it was deposited by an alpine glacier from the upper Middle Fork. The alpine glacier could not have been fed by Cordilleran ice because of the lithology of the drift and, as shown by gmd in the Nooksack drainage, the area to the north free of Cordilleran ice by then. The age of the moraine must be younger than the 12,000 year–old kettle–kame complex because it overlies it, both morphologically and stratigraphically.

Till making up the lateral moraine is exposed in a shallow landslide, giving stratigraphic, as well as morphologic, evidence of the relationship of the lateral moraine to the kettle–kame complex. The lithology of the till is dominated by glacially faceted and striated Mt. Baker andesite and Twin Sisters dunite from the upper drainage. The lateral moraine indicates that the Middle Fork alpine glacier readvanced over its previously–deposited kettle–kame complex. Upvalley, logs in a lateral moraine have been dated at 10,680 ± 70 and 10,500 ± 70 ^{14}C yrs. B.P. (Kovanen and Easterbrook, 2001).

Continue west to I–5, turn right, return to Bellingham.

DAY 3. GLACIATION OF THE NORTH CASCADES

Depart Bellingham, drive 58 mi. (97 km) south on I–5 to Everett. Exit I–5 onto Highway 2, drive east to Monroe, turn right (south) to Snoqualmie falls.

ICE–MARGINAL FEATURES OF THE VASHON GLACIER.

As the CIS moved southward down the Puget Lowland, its eastern margin was banked against the Cascade Range and the glacier extended up into major Cascade valleys, blocking drainage and impounding ice–marginal lakes into which large deltas were built (Cary and Carlston, 1937; Mackin, 1941; Booth, 1987). In the Snoqualmie valley near North Bend, the ice built a high, flat–topped, ice–marginal embankment that extended across the Snoqualmie River valley. Deltaic sand and gravel were built into the ice–dammed lake near the glacier terminus while lacustrine silt and clay were deposited upvalley. Wood from rhythmically–bedded (varved?) lake sediments upvalley was ^{14}C dated at 13,570 ± 130 (UW–35) years. As the ice retreated to the north and west, the newly–opened, ice–free valleys held ice–marginal lakes that became successively lower as new outlets were uncovered (Mackin, 1941; Booth, 1986).

STOP 3–1. TOKUL CREEK DELTA AND SNOQUALMIE FALLS.

As the Puget lobe retreated northward, it still blocked northward–draining valleys and an ice–marginal lake formed in the Snoqualmie valley. Tokul Creek, a tributary of the Snoqualmie River, built a delta into the lake, filling nearly all of the valley. As the glacier terminus retreated northward, new, lower outlets were uncovered and the lake eventually drained. When the Snoqualmie River re–established its course, the delta sediments forced the river to the west side of the valley at the distal edge of delta. When the Snoqualmie cut down through these sediments, it encountered the bedrock of the west valley wall, and, once notched into the rock, couldn't get out. Downstream, the river quickly eroded the valley–filling sediments, but where the river was notched into the bedrock, it could not erode its bed as rapidly, and Snoqualmie falls were born. As the falls retreat upvalley, they will eventually erode through the bedrock spur and into the valley fill upstream. When this happens, the river will quickly cut down through the sediments and bring an end to the falls.

Continue through North Bend, turn left on I–90 and drive 22 mi. (37 km) to Snoqualmie Pass. Exit I–90 to the ski area.

STOP 3–2. LATEST PLEISTOCENE ALPINE MORAINES

Multiple, as–yet–undated, post–LGM moraines occur at Snoqualmie Pass in the North Cascades. Half a dozen, closely–spaced moraines at Snoqualmie Pass, well upvalley from the LGM, record several periods of glacial retreat and stillstand. No numerical ages have been published for any of these moraines, although a ^{14}C date of 11,050 ± 50 was obtained from wood in gravel on till between two of the moraines (Porter, 1976a).

Rejoin I–90 and drive to Exit 85 to Blewett Pass. Contine 9 mi. (15 km) to Swauk Prairie.

STOP 3–3. SWAUK PRAIRIE MORAINE

Long valley glaciers extended down several tributaries of the Yakima River. At this stop we will examine the oldest moraines of this system.

Drive 37 mi. (62 km) to the junction with Highway 2. Turn left and continue 6 mi. (10 km) to Leavenworth. At the west end of town, turn left up Icicle Creek road and continue 6 mi. (10 km) to Rat Creek. Pull into the campground on the left.

STOP 3–4. YOUNGER DRYAS (?) MORAINE AT RAT CREEK

A double moraine in Icicle Creek (Page, 1939) from an alpine glacier in Rat Creek near Leavenworth has been correlated with the Sumas Stade of the CIS, which has been shown to include the YD. However, no numerical dates have yet been published for these moraines and the correlation is based on the position of the moraines upvalley from LGM moraines.

Return to Leavenworth, turn right on Highway 2 and drive 26 mi. (43 km) to Wenatchee. Exit to the right.

DAY–4. GLACIAL LANDFORMS OF THE OKANOGAN LOBE OF THE CORDILLERAN ICE SHEET; MISSOULA FLOOD FEATURES

The Cordilleran Ice Sheet extended down both sides of the North Cascades, topping the crest of the Cascades and connecting the Okanogan lobe with the Puget Lowland ice across the north Cascades. At the late Wisconsin maximum, the Okanogan lobe reached about 50 km south of the Columbia River onto the Columbia Plateau where it built a massive end moraine and left a virtual museum of glacial landforms beautifully preserved in the arid climate. Drumlins and flutes show the radial movement of ice, and eskers, kames, kame terraces, and recessional moraines were left during ice recession. The Okanogan lobe played an important role in directing giant floods from the bursting of ice dams and draining of glacial Lake Missoula. Glacial landforms are intimately associated with Missoula flood features.

From Wenatchee, drive north on Highway 2 for 50 miles to Jameson Lake junction. Turn left and drive 2 – mi. to Moses Coulee moraine

STOP 4–1. WITHROW MORAINE AND OUTWASH TERRACE DRAPED ACROSS A GIANT MISSOULA FLOOD BAR, MOSES COULEE

Moses Coulee was scoured out by a giant flood from Lake Missoula prior to the last glaciation, leaving a large, dry gorge, over 1.7 km (1 mi.) wide and 180 km (600 ft.) deep. During the late Wisconsin glaciation, the Okanogan lobe built the Withrow moraine which drapes into Moses Coulee and across an earlier giant flood bar (Fig. 22).

Return to Highway 2, turn right and return to Farmer. Turn right and drive 9 mi. to the Withrow moraine.

STOP 4–2. THE WITHROW MORAINE

The Withrow moraine (Figs. 22, 23), is a hummocky moraine, 1–2 km wide and 30–70 m high, extending across the Waterville Plateau. Large basalt erratics commonly dot the surface. The Columbia Plateau here slopes gently northward so meltwater streams could not flow directly away from the ice terminus. They flowed parallel to the ice margin, then down Moses to the Columbia River (Easterbrook and Rahm, 1970).

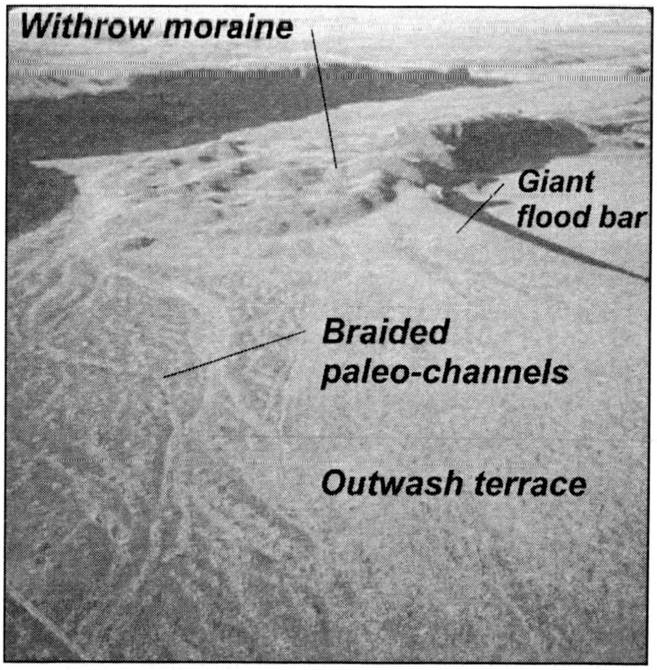

Figure 22. Withrow moraine and outwash terrace draped over a giant Missoula flood bar in Moses Coulee. (Photo by D.A. Rahm, courtesy of EPIC)

Dutch Henry Draw, which we crossed just before this stop is one such channel.

Drive 12 mi. to Mansfield. Cross recessional moraines. Continue for 3 mi. to huge erratic (Fig. 24), then 1 mi. to large erratic. Continue for 4.3 mi. to Barnes Butte kame complex. Drive 4.8 mi. past several eskers to Sims Corner.

Figure 23. The Withrow moraine. (Photo by D.A. Rahm, courtesy of EPIC)

Figure 24. Large basalt erratic resting on till east of Mansfield, WA.

STOP 4–3. ESKERS AND DISTAL FANS, SIMS CORNER

As the Okanogan lobe retreated northward from the Withrow moraine, it halted a number of times to deposit many eskers, kames, kettles, kame terraces, and related ice–contact deposits. Eskers that terminate at ice–marginal fans (Fig. 25) and low morainal ridges (Fig. 26) mark temporary halts in the recession of the glacier terminus. Three eskers near Sims Corner end abruptly at ice–marginal fans which still retain the traces of abandoned braided channels on their surface. The proximal side of the fans mark the former ice margin.

STOP 4–4. GRAND COULEE, DRY FALLS, MISSOULA FLOOD

Grand Coulee was described by Bretz (1969) as the "deepest, longest, and steepest in gradient. Its system possesses the largest number of abandoned cataracts, the widest one, a record of the highest, the greatest cascade, and the largest number of distributary canyons. It is unique in that, at its initiation, it found no preglacial valley head but crossed 20 miles or so of the northern divide of the plateau before encountering such a valley, and it trenched that divide nearly 1000 feet deep".

Dry Falls (Fig. 27) is about three miles across and is one of several falls excavated by spectacular quarrying of jointed basalt during catastrophic floods spilling down Grand Coulee when the Okanogan Lobe blocked flow down the Columbia River near the site of Grand Coulee Dam and forced it to spill over into Grand Coulee. Much of the large size and depth of Grand Coulee is due to waterfall recession.

Drive 14 mi. to junction with Highway 2 and Dry Falls.

OPTIONAL ROUTE UP GRAND COULEE TO ELECTRIC CITY

The drive up Grand Coulee from Dry Falls to its head at the Columbia River gives a striking impression of the immensity of the Missoula floods that carved the coulee.

Drive north on Highway 155 for 20 mi. to Steamboat Rock.

STOP 4–5. STEAMBOAT ROCK IN GRAND COULEE

Steamboat Rock is an island of basalt in upper Grand Coulee. Its surface is crossed by a prominent channel cut in the basalt and is mantled with glacial deposits, including large erratics. The surface gives clues concerning the pre–coulee history of the area.

Continue north on Highway 155 for 7 mi. to Electric City. Take Highway 174 east for ~20 mi. to junction with Highway 2. Turn left and continuefor 65 mi. to Spokane then 75 mi. to Kellog, Idaho and 12 mi. to Wallace, Idaho. Overnight in Wallace, Idaho.

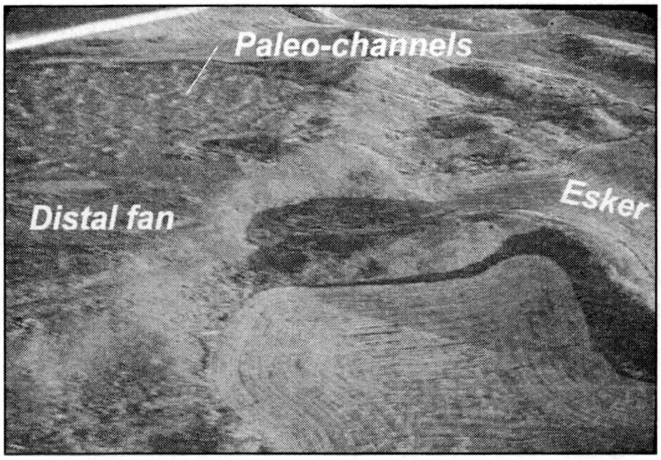

Figure 25. Esker terminating in a distal fan near Sims Corner.

Figure 26. Recessional moraine at the distal edge of a field of eskers near Sims Corner.

Figure 27. Dry falls. (Photo by D.A. Rahm, courtesy of EPIC)

DAY 5. WALLACE TO BOZEMAN, GLACIAL LAKE MISSOULA

Today we will drive through the northern Rocky Mts. to Bozeman, MT. On the way, we will pass through the area of Glacial Lake Missoula, made by damming of the Clark Fork by the Cordilleran Ice Sheet. The bursting of the ice dam that held in glacial Lake Missoula is truly one of the Earth's unique geologic events. Nothing like it is known. As we drive to Missoula, we will be crossing the site of the lake that was the source of floodwater for the colossal scabland scouring.

Drive 120 mi. from Wallace to Missoula, Montana.

STOP 5–1. GLACIAL LAKE MISSOULA SHORELINES

One of the more famous places to observe shorelines of glacial Lake Missoula is on the hillside just above the University of Montana in Missoula (Fig. 28). The lake that filled the drainage system of the Clark Fork of the Columbia River contained about 500 km^3 of water. At a number of places on the plateau, giant ripples (Fig. 29) occur on bars created by huge flood discharge. Channels cut by the flood show an anastomosing pattern on satellite imagery (Fig. 30).

Drive to Bozeman, Montana to spend the night.

QUATERNARY GEOLOGY OF THE GREATER YELLOWSTONE AREA

Kenneth L. Pierce, U.S. Geological Survey, Bozeman, MT 5971
Don Despain, U.S. Geological Survey, Bozeman, MT 5971
Cathy Whitlock, Univ. of Oregon, Corvallis, OR, 97403–1251
Kenneth P. Cannon, National Park Service, Lincoln, NE, 68508–3873

Figure 28. Shorelines of glacial Lake Missoula, Missoula, MT.

Figure 29. Giant ripples from Missoula flood. (Photo by D.A. Rahm, courtesy of EPIC)

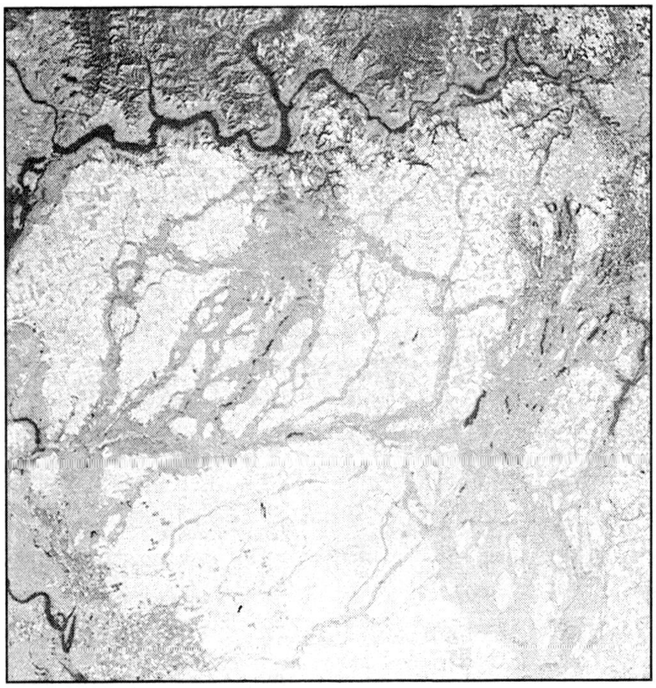

Figure 30. Channelled scabland (dark channels) across the Columbia Plateau. (NASA satellite image)

Grant Meyer, Univ. of New Mexico, Albuquerque, NM, 87131–1116
Lisa Morgan, U.S. Geological Survey, Denver, CO 80225

OVERVIEW OF YELLOWSTONE GLACIATIONS.

In the Rocky Mts., including the southern Yellowstone system, Bull Lake terminal moraines are typically farther downvalley than Pinedale moraines. However, in the northern and eastern Yellowstone system, Pinedale glaciers overrode Bull Lake terminal moraines.

Pleistocene Glaciation of Yellowstone was initiated in the surrounding mountains, built up to a thickness of 3,000 ft (~1 km) and flowed down the major valleys that drain Yellowstone. During deglaciation, the plateau icecap stagnated, and glaciers from the adjacent mountains flowed into terrain previously occupied by the plateau icecap (Pierce, 1979). This pattern resulted in flow differences up to 180°.

During full glacial conditions, the northern Yellowstone outlet glacier was fed from multiple sources in the Beartooth Range, the Absaroka Range, the Gallatin Range, the Washburn Range, and from the Yellowstone Plateau between the Washburn and Gallatin Ranges. This outlet glacier flowed 40 miles (65 km) downvalley from the northern park area to the Eightmile terminal moraines.

DAY 6. BOZEMAN VIA YELLOWSTONE VALLEY TO MAMMOTH; GLACIATION OF NORTHERN YELLOWSTONE (FIGS. 31, 32)

STOP 6–1. MALLARDS REST FISHING ACCESS

Northern Yellowstone outlet glacier; Eightmile moraines and outwash; local valley–glacier moraines; active faulting; Yellowstone River channel patterns.

An outwash fan of the northern Yellowstone outlet glacier emanates from the Eightmile moraines (Figs. 33, 34). The outwash–fan terrace is 200 ft (60 m) above the Yellowstone River at the Eightmile terminal moraines, 60 feet (18 m) above the river at this stop, and only 10–20 feet (3–6 m) above the river this side of the Livingston canyon. The Yellowstone River has eroded its post–glacial valley along the east edge of the outlet glacier fan. The scarp of an active fault with about 15 ft (5 m) of post–glacial offset occurs near the base of the range at Barney Creek.

Pull over before Emigrant and view flood deposits in the inner valley on the far side of the Yellowstone River. Just before Chico Hot Springs, take the road to the right and climb up onto the Chico outwash terrace and then onto the Chico moraines.

STOP 6–2. CHICO MORAINES

The Chico moraines (Fig. 34) of the northern Yellowstone outlet glacier are bounded on the north by the Chico meltwater channel outwash terraces. Cosmogenic ages from boulders on the Chico moraines averages 15.7 ± 0.5 ^{10}Be ka. Eightmile moraines

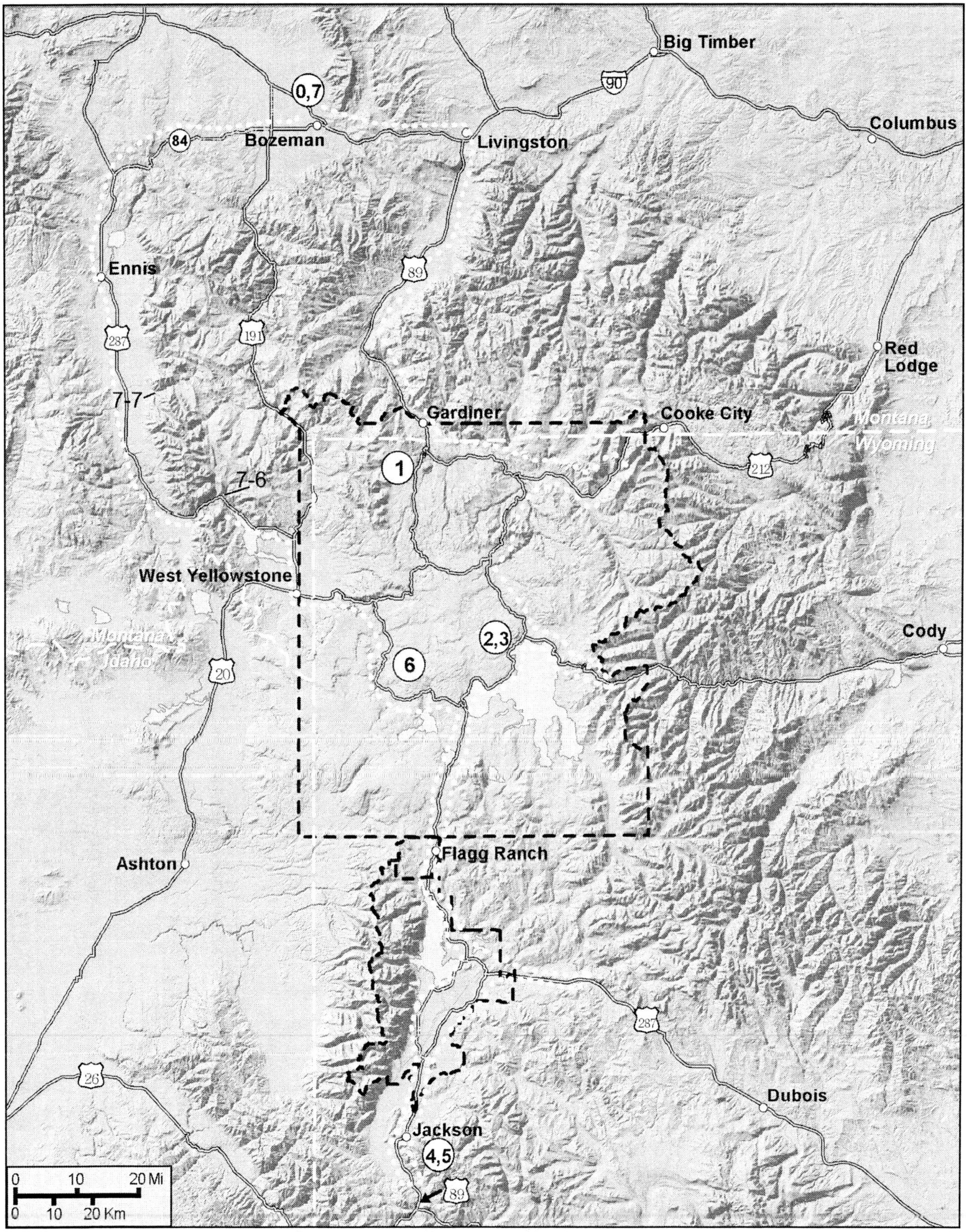

Figure 31. Shaded relief map of the Greater Yellowstone area showing the route of the field trip.

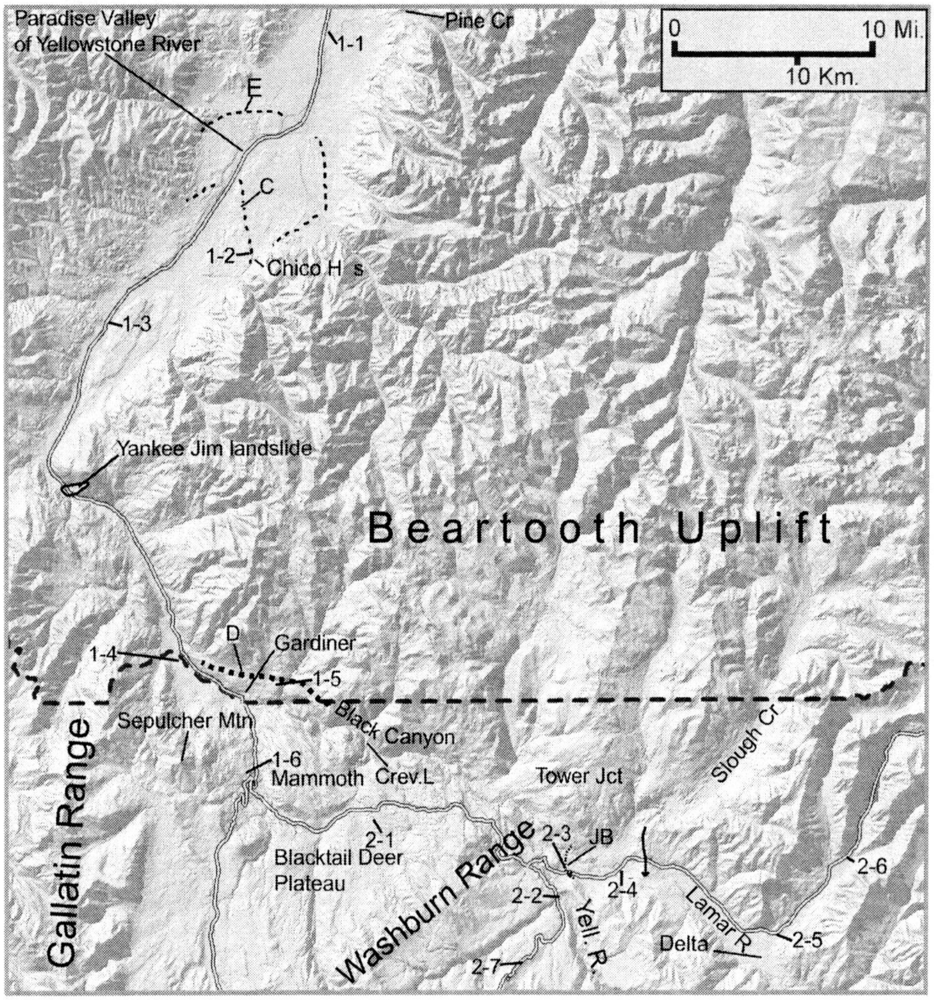

Figure 32. Northern Yellowstone area.

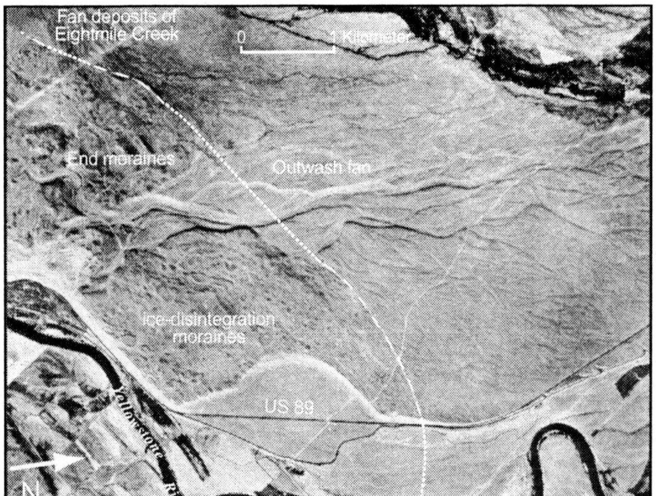

Figure 33. Eightmile Creek Pinedale moraines and outwash fan of the northern Yellowstone outlet glacier.

average 16.5 ± 0.4 ^{3}He ka and 16.2 ± 0.3 ^{10}Be ka (Fig. 35; Licciardi and others, 2001). These cosmogenic ages are younger than those based on radiocarbon dating (Licciardi and others, 2001; Pierce, 2003).

The Eightmile moraines (Figs. 33, 34) are well expressed just to the north on the other side of the Chico meltwater channel, and, from there, they arc around to their terminus 4.5 miles (7 km) beyond the Chico moraines. Across the valley, ice–marginal channels 150 ft (50 m) deep were eroded into bedrock at the Eightmile and Chico ice margins.

Continue upvalley and cross Yellowstone River at Corwin Springs.

STOP 6–3. MID–CHANNEL FLOOD BAR, STEVENS CREEK

This mid–channel flood bar has "giant ripples" spaced about 15 m apart and up to 2 m high, with boulders up to 2 m in

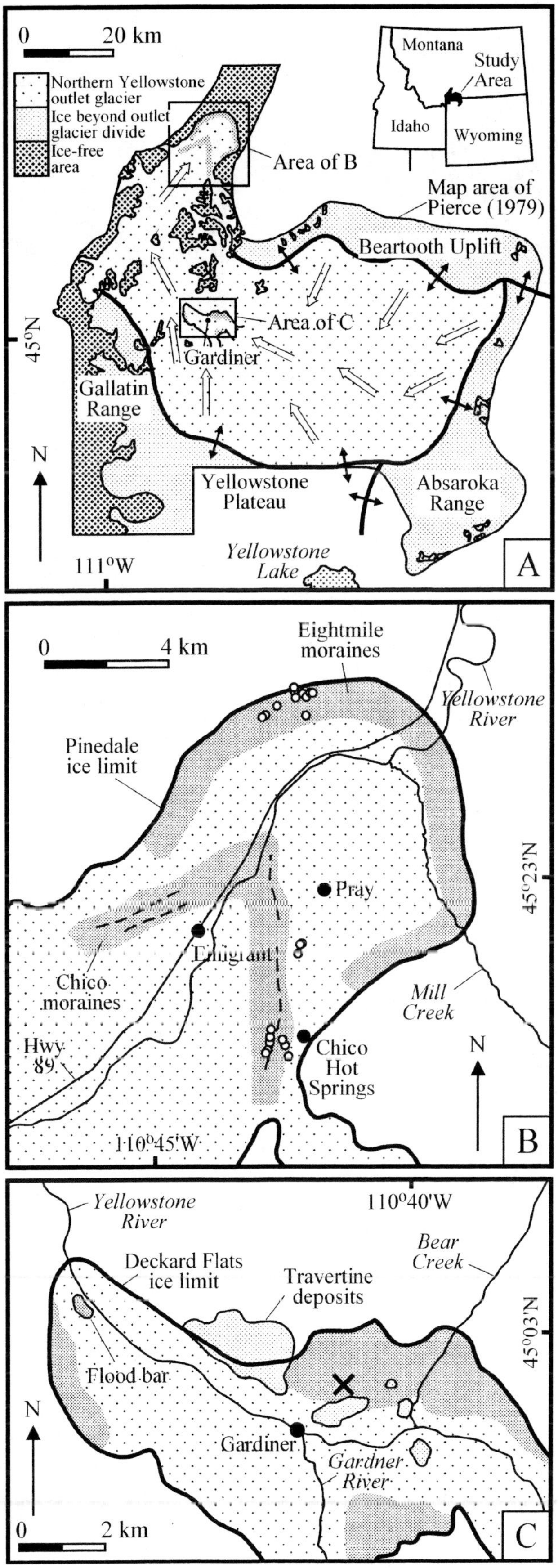

Figure 34. North Yellowstone moraines Glacial deposits in northern Yellowstone area (from Licciardi et al., 2001; adapted from Pierce, 1979). A: Northern Yellowstone ice cap. Open arrows indicate major ice flow directions. B. Eightmile and Chico moraines. C. Deckard Flats moraines.

diameter. Cosmogenic ages on the boulders average 13.7 ± 0.5 ^{10}Be ka (Fig. 35). The ridges are slightly asymmetric, with the boulders concentrated on the downflow side of the crests. The alluvial fan of Reese Creek 0.5 km to the northwest provides evidence of at least two floods 150–200 ft (45–60 m) deep. First an alluvial fan was deposited and then a flood down the Yellowstone valley eroded the fan front and left a "flood–ripped" fan front. Then a younger fan was built and, later, another flood undercut and eroded the fan front (Pierce, 1979). These floods were probably from release of glacially dammed lakes upstream, most likely from a lake dammed in the Lamar Valley by the Slough Creek glacier.

STOP 6–4. MAMMOTH OVERLOOK

Hike ~200 m up the gravel road behind Mammoth Hotel

This bench is on the Deckard Flats margin of glaciers originating to the southwest in the Gallatin Range. The bench is prob-

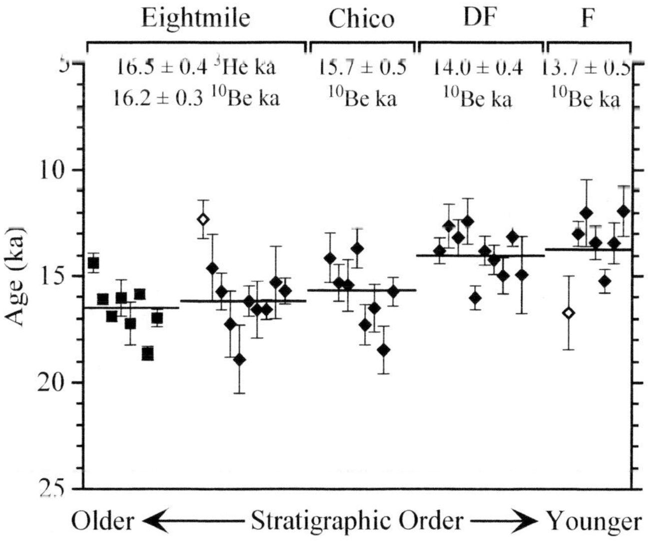

Figure 35. Ages of Yellowstone moraines, plotted in stratigraphic order from oldest (left side of graph) to youngest (right side) (from Licciardi et al., 2001). Solid squares are ^{3}He ages, and solid diamonds are ^{10}Be ages. Error bars on each age represent 1s analytical uncertainty only, and do not include errors due to production rate, scaling, and other uncertainties. Horizontal lines and quoted ages indicate weighted means of each landform. Open diamonds indicate two outliers not included in weighted means. DF – Deckard Flats moraines; F – late-glacial flood deposits.

ably mostly kame gravel and includes boulders of travertine. Capitol Hill is the conical mound to the south and is thought to be a thermal kame formed where hotsprings melted a cavity in the ice, which filled with kame gravel. Other conical hills in the area are thought to have a similar origin. A grassy, rounded hill of sub–glacial till to the north above the Deckard Flats bench also bears erratics of travertine.

The full–glacial, northern Yellowstone outlet glacier was about 3,500 ft (1 km) thick in the north Yellowstone valley near Gardiner. Under full–glacial conditions, ice from the Beartooth Range to the east, the Yellowstone Plateau to the south, and the Gallatin Range to the west converged in this area to form the northern Yellowstone outlet glacier.

Mt. Everts to the east exposes readily erodible Cretaceous shale and sandstone on its west face. The southern part of Mt. Everts is capped by 2.1 Ma Huckleberry Ridge tuff, with small, local valley fills of 0.64 Ma Lava Creek tuff. The upland on top of Mt. Everts is glacially scoured.

Mammoth area travertine terraces. Terrace Mt., the highest travertine near Mammoth Hot Springs, rests on 2.1 Ma Huckleberry Ridge tuff and has U–Th age of 406 ± 30 ka (Pierce et al., 1991). Pinedale glacial erratics rest on this travertine, which was scoured by northward–flowing ice that reached an altitude of almost 9,000 feet (2750 m) on Sepulcher Mt. The inactive and forested Pinyon terrace is post–glacial, with U–Th ages near 10 ka (Pierce et al., 1991). The lowest travertine terraces are post–glacial. A deep research hole was drilled in the 1960's in the main terrace and a U–Th age of 7.72 ± 0.88 ka was obtained on travertine from a depth of 239 ft (72.8 m).

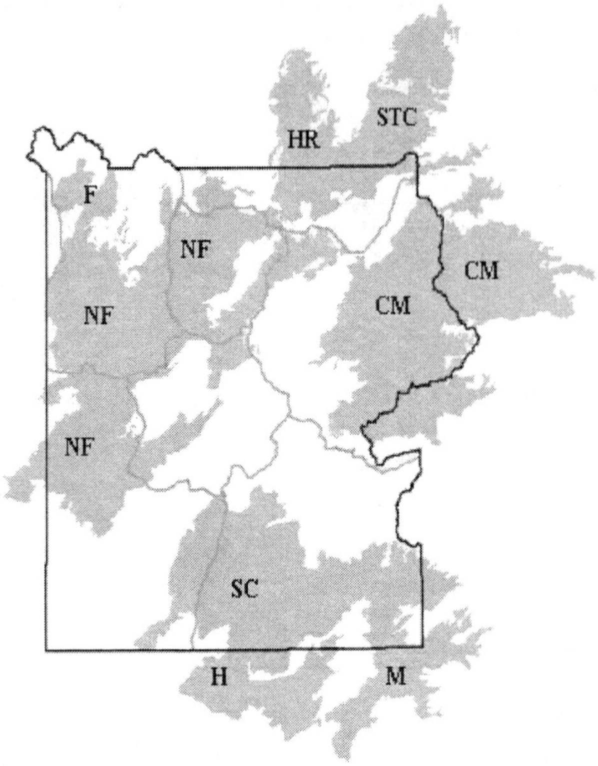

Figure 36. Perimeters of 1988 Yellowstone fires. **CM**= Clover Mist complex, **F**= Fan fire, **H**= Huck fire, **HR**= Hellroaring fire, **M**= Mink fire **NF**= North Fork fire, **SC**= Snake complex, **ST**= Storm Creek fire. The Clover Mist, Fan, Mink, Snake Complex and Storm Creek fires were all lightning caused fires that were allowed to burn early in the season.

From the Mammoth area eastward to the Lamar Valley, we will look at recessional deposits of the last glaciation.

Pull over to view High Bridge moraines deposited during the recession from the Deckard Flats position.
Pull over to view Lava Creek, which is the type area of the 0.64 Ma Lava Creek tuff.
Pull over to view recessional meltwater channel with Blacktail Ponds at the east end

STOP 6–5. BLACKTAIL DEER PLATEAU AT FROG ROCK

The Blacktail Deer Plateau is formed by moraines of Deckard Flats age that extend about 3 mi (5 km) south and west from this viewpoint. The successively lower morainal benches, locally including kame gravel, define about 10 recessional ice margins.

In 1988, extreme drought and high winds from mid–June to mid–September resulted in forest fires that eventually burned approximately 1.1 million acres (450,000 ha) in and around the park (Fig. 36). The total area encompassed by the fire was 1.7 million acres (690,000 ha). About 50% of the burned area was burned by crown fire, 40% by mixed–severity fire, and 10% by grassland/shrubland fires. Most of the sagebrush areas visible to the west of this stop were burned in the fires and burned forest is visible on the mountains forming the horizon to the south.

Pull over to view Phantom–Lake meltwater channel eroded more than 200 feet into bedrock by the ice–marginal "Yellowstone River".
Pull over to view Petrified–Tree–area fire ecology. This area was burned in 1988 by a severe crown fire. The convection column rose to 20,000–30,000 ft (6000–9000 m) with flames shooting more than 200 ft (60 m) above the crown. All the organic matter on the soil surface (litter and duff) was turned to ash and it truly was like the bottom of a barbecue pit after it burned. The pre–burn forest was a 200–300–year old Douglas–fir stand that had not burned for at least 150 years and probably not for 300 years.

STOP 6–6. JUNCTION BUTTE MORAINES

The Junction Butte moraines are the most prominent in the late Pinedale recession from the Deckard Flats readjustment. Large boulders of Precambrian crystalline rocks stud their surface and permanent and seasonal ponds occupy common depressions. A large glacier flowing down Slough Creek deposited these moraines and blocked the Lamar valley upstream from Slough Creek.

STOP 6–7. TOWER FALLS VIEWPOINT, 1.3 MA BUMPUS GLACIATION

The Narrows Canyon of the Yellowstone River was eroded during only about 12–14 ka of postglacial downcutting and, in the process, also exhumed several paleocanyons and their volcanic and sedimentary fill. A complex assemblage of Pleistocene basalt and glacial, fluvial, and lacustrine sediments known as the "sediments and basalts of The Narrows" are exposed in the steep, narrow canyon.

Pinedale till caps the section, locally resting on Lava Creek tuff. These rest on the eroded top of the upper basalt, which in turn is underlain by the till of Bumpus Butte, containing erratics of locally derived basalt and Precambrian rocks from sources to the north. Underlying the till is lacustrine sediment of an ice–dammed lake, containing ice rafted stones of Precambrian crystalline rock and a coarse, silicic, volcanic ash from a nearby source. The argon–argon age of the tephra, 1.3 ± 0.1 Ma, and chemistry suggest correlation with the Mesa Falls tuff (Obradovich and Izett, 1991). The Mesa Falls is the middle tuff of the Yellowstone sequence, erupted from a caldera (Island Park) mostly west of Yellowstone (Christiansen, 2001). The conformable relation of the ice–dammed lake sediments containing the ash and the till indicates the age of 1.3 ±0.1 Ma also closely dates the glaciation.

STOP 6–8. DUNRAVEN PASS, CALDERA AND PLATEAU OVERVIEW

Park in Dunraven Pass and hike a few hundred meters up the old road. This south–facing escarpment is formed in Absaroka volcanic rocks of an Eocene volcano whose southern part has dropped down onto the 640 ka caldera of the Lava Creek tuff. At the base of the escarpment is the northern margin of the caldera, and farther south is the Grand Canyon of the Yellowstone, inside the caldera. The caldera extends from here to the Red Mts. about 30 miles to the south. To the southwest, rhyolite flows form the nearly flat surface on the distant horizon. Most flows in and near the caldera are 70,000 to 160,000 years old.

Drive to Lake Village for the night

DAY 7. YELLOWSTONE LAKE, GLACIATION OF YELLOWSTONE

STOP 7–1. SHORELINES AND ARCHEOLOGY

Shoreline designations are after Meyer and Locke (1986), starting with the modern shoreline as *S1*. The discussion begins with the *S4* barrier beach on the east side of the outlet at Fishing Bridge (Fig. 37). Walk across Fishing Bridge fault with 1 m offset into the *S4* lagoon, then past Hamilton store on the *S4* shoreline and across highway down to *S2* shoreline near the Fishing Bridge museum.

The level of Yellowstone Lake has decreased and increased through time (Pierce et al., 2002). Lidar images (Fig. 38) show the shorelines and other geomorphology important to deciphering the lake level history, tilting, and faulting. The *S4* shoreline forms a barrier beach and lagoon about 10.7 cal ka B.P., based on buried charcoal. The *S2* shoreline here is a wave–cut shoreline about 8 cal ka B.P., as dated by buried–charcoal, both on the shoreline and on the Pelican Creek terrace that truncates the *S2* shoreline. When viewed from the 8 cal ka B.P. *S2* shoreline on the lake side of the Museum, all post–8 ka history of the lake is either at present lake level or is now underwater (Pierce et al., 2002). To the southwest in Bridge Bay, a 3.8 cal ka B.P. beach is submerged in

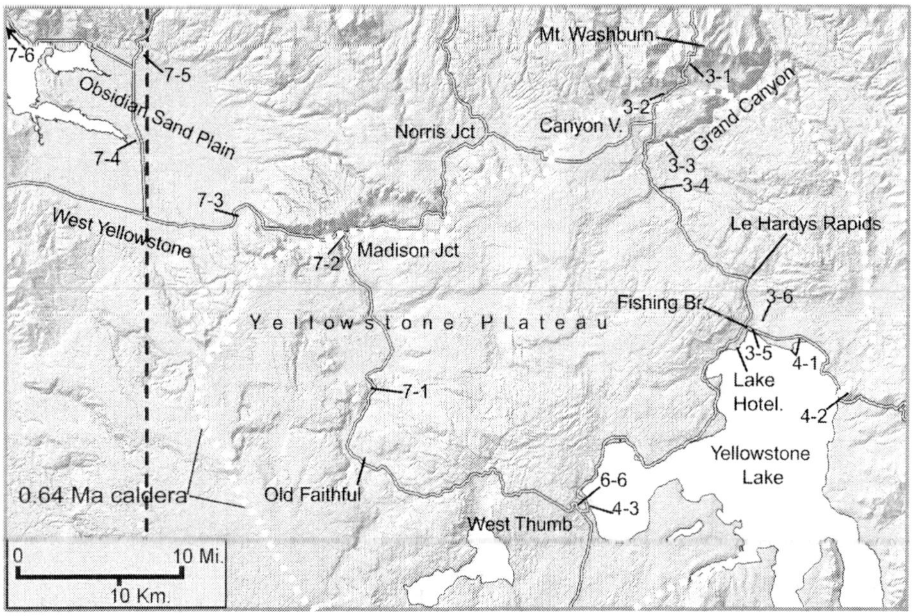

Figure 37. Central Yellowstone Park.

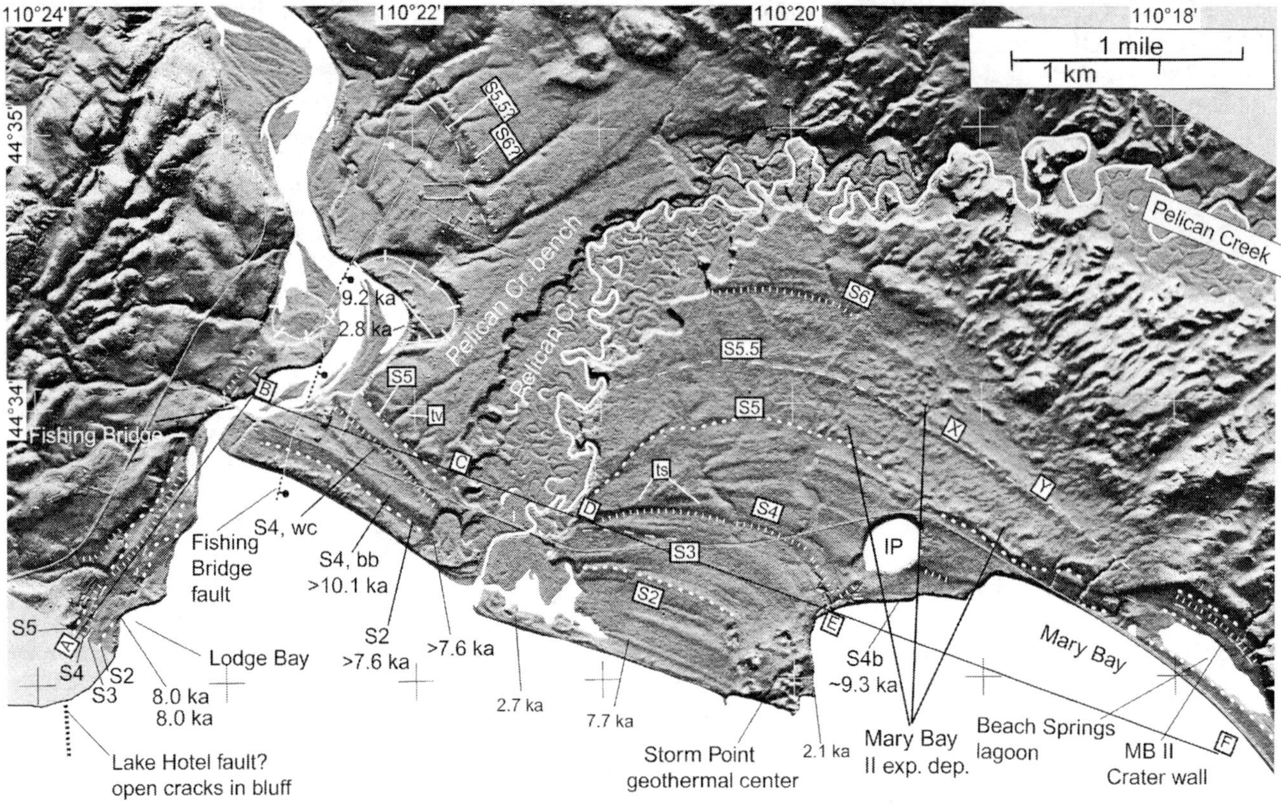

Figure 38. Lidar image of shorelines along the north shore of Yellowstone Lake (from Pierce et al., in press).

5 m of water. To the east, Pelican Creek lies in a drowned valley with gravel down to 5 m below the present lake.

STOP 7–2. DROWNED CHANNELS OF THE YELLOWSTONE RIVER

From S4 lagoon area, open the locked gate, go north, and park in the area of S–meander. Walk to the riverbank.

Cores in the abandoned channel here went through lake mud into gravel at a depth of ~5 m that is 2.9 cal ka B.P. The steep river bank at this site was undercut by the Yellowstone River at a time when the river was much more vigorous than it is now. The outlet reach of the Yellowstone River is now very low gradient, but about 3 ka it was a gravel–transporting river undercutting its banks, and at this site, depositing eolian sand at the top of the eroding river bank. Historic and older inflation of Le Hardys rapids diminished the river gradient such that the gravel at 5 m below the surface here is now well below the level of the bedrock threshold 4 km downstream at Le Hardys rapids (Fig. 37). Accounting for this and a 1m/km original gradient of the gravel–transporting river, Le Hardys rapids has come up 7 m relative to this site, and 8 m relative to the lake outlet 1 km farther upstream (Pierce et al., 2002).

Walk back across the road to the channel of the S–meander. Four backhoe pits were dug across the old channel of the Yellowstone River here. The outside of this meander is a high, steep bank formed at a time when the Yellowstone River was undercutting its banks and carrying gravel up to 3 cm. Then, as the river lost its competence, charcoal accumulated about 9 cal ka B.P. and was buried by shoreface gravel. The water level rose until this reach of the Yellowstone River became a northern arm of Yellowstone Lake and the *S3* and *S2* shorelines were cut.

Overview of post–glacial shorelines and caldera unrest (after Pierce et al., 2002).

Between 1923 and 1985, the center of the Yellowstone caldera rose nearly one meter along an axis between its two resurgent domes (Pelton and Smith, 1979; Dzurisin and Yamashita, 1987). From 1985 until 1995–6, it subsided at about two cm/yr (Dzurisin et al., 1990). More recent radar interferometry studies show renewed inflation (Wicks et al., 1998). Doming along the caldera axis reduces the gradient of the Yellowstone River from Le Hardys rapids to the Yellowstone Lake outlet and ultimately causes an increase in lake level. Meyer and Locke (1986) and Locke and Meyer (1994) identified the modern shoreline as *S1* (1.9 ± 0.3 m above the lake gage datum). Mapping of prominent paleoshoreline terraces *S2* to *S6* suggests that the shorelines were cut during intracaldera uplift episodes that produced rising water levels. Shoreline *S5* is about 46 ft (14 m) above the reference datum and is estimated to be ~12.6 cal ka because it post–dates a large hydrothermal explosion deposit from the Mary Bay area (MB–II) that occurred ~13 cal ka. *S4* formed about 26 ft (8 m) above the reference datum and is ~10.7 cal ka as dated by archeological material and radiocarbon, and was accompanied by offset on the Fishing Bridge fault. About 9.7 cal ka, the Yellowstone River eroded the "S–meander", followed by a ~ 16 ft (5 m) rise in lake level to *S3* and *S2*, the lowest generally recognizable subaerial shoreline. It is ~16 ft (5 m) above the reference datum, (10 ft, 3 m) above *S1*, and is ~8 cal ka, as dated on both sides of the outlet. Yellowstone Lake and the river near Fishing Bridge were 16–20 ft (5–6 m) below their present level about 3–4 cal ka, as indicated by ^{14}C ages from submerged beach deposits, drowned valleys, and submerged Yellowstone River gravel. Thus, the lake in the outlet region has been below or near its present level for about half the time since a 3,300 ft (1 km)–thick icecap melted from the Yellowstone Lake basin about 16 cal ka.

Based on reconstructions of the river gradient from the lake outlet to Le Hardys rapids and indicators of former lake and river levels, Le Hardys rapids was uplifted about 8 meters relative to the lake outlet at least twice, once between ~9.5 cal ka and ~8.5 cal ka and again after ~3 cal ka. Following these uplifts, it appears to have subsided a similar amount. Older possible rises in lake level are suggested by locations where the ~10.7–cal–ka *S4* truncates older shorelines and where valleys are truncated by the ~12.6–cal–ka *S5* shoreline. Using these controls, a plot of lake level through time shows 5–7 millennial–scale oscillations since 14.5 cal ka. Le Hardys rapids to Fishing Bridge spans only the central 25% of the historic caldera doming, so that using historic doming as a model, the total projected uplift would be fourfold greater, or ~32 m. This "heavy breathing" of the central part of the Yellowstone caldera may reflect a combination of several processes, (1) magmatic inflation, (2) tectonic stretching and deflation, and (3) inflation from hydrothermal fluid sealing, followed by cracking of the seal, pressure release, and deflation. Over the entire postglacial period, subsidence has balanced or slightly exceeded uplift, as shown by older shorelines that *descend* towards the caldera axis. We favor a hydrothermal mechanism for inflation and deflation because it provides for both inflation and deflation with little overall change. Other mechanisms such as inflation by magma intrusion and deflation by extensional stretching require two separate processes to alternate and yet result in no net elevation change.

In addition to inflation and deflation, new LIDAR data demonstrates previously unrecognized local deformation along the north shore of Yellowstone Lake. The newly recognized Fishing Bridge fault shows a progressive increase in offset from 1.5 ft (0.5 m) for the ~8–cal–ka *S2* to perhaps 15 ft (5 m) for the ~12.6–cal ka *S5*. Uplift of the Storm Point hydrothermal center has tilted shorelines westward as much as 20 ft (6 m) over one kilometer. A local anticline has as much as 10 ft (3 m) relief in 1600 ft (0.5 km). LIDAR data also shows that the Mary Bay II hydrothermal explosion debris has a surface relief of about 3 ft (1 m) over 300 ft (100 m), and that it overlies *S5.5* and *S6* shorelines, but not *S5*.

STOP 7–3. HAYDEN VALLEY

Hayden Valley was probably the site of glacially dammed lakes at least twice during its Pleistocene history: (1) before the last glacial maximum when an advance from the north reached the Canyon Village area, and (2) during recession from the last glacial maximum when the icecap on the Yellowstone Plateau had melted, but glaciers from the Beartooth Range blocked the Yellowstone River east of Mount Washburn (Pierce, 1979; Richmond, 1976, 1977).

Pull over at the Le Hardys rapids parking area. Le Hardys rapids serves as the outlet threshold of Yellowstone Lake. It is also on the crest of the historic axis of uplift and subsidence of the Yellowstone caldera.

Pull over to view wide, "pool reach" of the Yellowstone River between Le Hardys rapids and outlet at Fishing Bridge. In this reach, the Yellowstone River is quite wide, tranquil, and very low gradient. Recent uplift has diminished the river's drop to only 0.25 m in the 5 kilometers reach between Le Hardys rapids and Yellowstone Lake.

STOP 7–4. GRAND CANYON AT ARTIST POINT

Take the bridge across the Yellowstone River and park at the Artist Point parking area.

Upstream from Artist Point is Lower Falls (Fig. 39). The canyon near Artist Point is deeply altered and the alteration diminishes up canyon to leave more resistant rock at Upper and Lower Falls. The hydrothermally altered rhyolite in the deeper part of the canyon near Artist Point experienced pyrite mineralization before the canyon was cut, which was then oxidized to yield the yellow and red colors. Similar extensive alteration probably also exists elsewhere at depth within the caldera. Emplacement of rhyolite built up the Yellowstone Plateau and created steep gradients at the margin of this plateau. Stream erosion has incised the margins of this plateau creating canyons. The largest river, the Yellowstone, has eroded farthest into the plateau and created the deepest canyon. Across the canyon and extending about halfway down into the canyon is a paleovalley of the Yellowstone River filled with sediment. Glacial advances from the Beartooth Range early and late in a glacial cycle have blocked the Yellowstone River.

Pull over into the parking area for Canyon horseback stables. This is part of an extensive area of thick, glacial till that contains Precambrian rocks carried up the Yellowstone River from the Beartooth Range. This probably occurred early in the last glacial cycle before an icecap became established on the Yellowstone

Figure 39. Grand Canyon of the Yellowstone and lower Yellowstone falls. (Photo by K. Hamblin)

Plateau. This south–flowing glacier dammed the Yellowstone River and created a lake in the Hayden Valley–Yellowstone Lake area.

STOP 7–5. OLD FAITHFUL–WEST YELLOWSTONE BASIN–MADISON VALLEY

Madison Junction. Madison Junction is along the 0.64–Ma caldera margin. National Park Mt. to the southwest is the northeast margin of the large, West Yellowstone, rhyolite flow that was emplaced between the Bull Lake and Pinedale glaciations. Take the road to West Yellowstone through the Madison Canyon, which obliquely crosses the caldera margin.

Pinedale terminal moraine

Park in off–road parking area on north (right) side of the road.

Hike up onto the Pinedale moraine of a glacier that flowed down the Madison Canyon from the Yellowstone Plateau icecap. This moraine consists mostly of rhyolite from the Madison Canyon. The soils are weakly developed, except that the porous till has gray, clay lamellae translocated as deep as 7 ft. (2 m). The obsidian hydration thickness on glacial cracks is about 7 m. These moraines probably correlate with the Eightmile moraines north of Yellowstone and the Hedrick Pond moraines in Jackson Hole because the Yellowstone Plateau icecap was a key source in the glacial system that deposited these three end moraines. Across the Madison River from here, the end moraine complex is well expressed with more than five moraine positions. Outwash channels from the inner moraines trend westward through the outer moraines.

Bull Lake deposits with a loess mantle occur north of the Madison Canyon moraines. Farther north, Pinedale end moraines were built near the margin of the West Yellowstone Basin by ice flowing down the valleys of Cougar, Maple, and Gneiss Creeks.

The Madison Canyon moraines are at the head of the West Yellowstone obsidian sand–plain, a low–gradient, outwash fan rich in granule–sized obsidian that floors most of the West Yellowstone basin. The thickness of hydration on pebbles from the obsidian sand plain is about 7 m, the same as that in the terminal moraines. The hydration on recessional deposits is about 5 m. The thickness of weathering rinds on basalt is about 0.4 mm, whereas that on Bull Lake deposits is 0.8 mm (Colman and Pierce, 1981).

STOP 7–6. BULL LAKE MORAINES

Drive north on US 191 from Duck Creek and turn right into the old cemetery. Park and walk south up onto the broad moraine crest in the open meadow.

These Bull Lake moraines consist of glacially transported debris from the southern Gallatin Range and the Yellowstone Plateau. The moraines are broad, subdued, and capped with loess, commonly about 1.5–3 ft. (1/2–1 m) thick. A moderately developed, clayey–to–loamy soil has a high water–holding capacity. Although boulders protruding from the moraines are uncommon, boulders nearly flush with the surface are readily found. In soil pits, contortions of soil horizons and their local absence suggest congeliturbation or solifluction that also locally obliterated textural B–horizons and buried boulders into the moraines.

The West Yellowstone Basin was nearly filled by Bull Lake ice from the Yellowstone Plateau. Bull Lake moraines from west–flowing ice drape around Horse Butte, which forms a prominence in the western part of the basin. Two grassy moraines descend westward to the level of Hebgen Lake and connect with the moraines near this stop. The Bull Lake terminal moraines are 14 miles (22 km) beyond the Pinedale terminal moraines of STOP 7–5. This is a large separation but less than the 30 miles (50 km) in Jackson Hole. The Bull Lake moraines do not appear to be significantly tilted northward into the Hebgen–Red Canyon fault, suggesting that few offsets of 15 ft. (5 m) have occurred in the last 140 ka.

On the southern skyline is the West Yellowstone rhyolite flow that overlaps and post–dates the Bull Lake moraines. When the Bull Lake Glaciation occurred, the West Yellowstone flow and several more south of it were not present. Hydration rinds on obsidian pebbles from moraines in the West Yellowstone Basin averaged 14.7 m thick. Hydration was measured in thin–sections across pebble corners with pressure cracks produced by glacial abrasion. Based on 12.3 m hydration on the 123 ± 3 ka West Yellowstone flow and 16.2 m on the 183 ± 3 ka Obsidian Cliff flow, the 14.7 hydration yields an age of about 140 ka for glacial abrasion prior to deposition of obsidian clasts in the Bull Lake moraines.

STOP 7–7. MIDWAY GEYSER BASIN

Take the footbridge across the Firehole River to view the steamy Excelsior geyser crater and colorful Grand Prismatic spring, which are two of the largest geothermal pools in Yellowstone. Southwest along the skyline is the profile of the Mallard Lake resurgent dome. This is the southwestern of the two resurgent domes of the 0.64 Ma caldera. The skyline profile shows the pattern of keystone faults at the crest of the dome. These faults strike NW, are on the trend of the Midway Geyser Basin, and are likely to serve as conduits for thermal fluids.

Overnight at Old Faithful. *Stroll through upper Geyser Basin on your own in the evening, during which time you will have time to see an eruption of Old Faithful (Fig. 40).*

DAY 8. GLACIATION OF JACKSON HOLE

STOP 8–1. JACKSON LAKE OVERLOOK

Glacial scour excavated Jackson Lake. South of here, the lake is 400 ft (120 m) deep, with as much as 400 additional feet of sedimentary fill (Smith et al., 1993). A dam was built at the outlet of the lake about 100 years ago so the reservoir level could be raised about 35 ft (10m) above the original natural level. Before damming, the Snake River had built a delta southward past this stop. Post–glacial scarps on the Teton fault are locally apparent near the base of the Teton Range. Because the outlet of Jackson Lake is farther away (east) of the Teton fault than any other part of the lake, fault offsets would result in submergence of shorelines. Events here in the natural delta area are considered to represent the last two offsets of the Teton fault: (1) transgression of the lake up onto the riverine delta and construction of beaches on top of the scroll–work pattern of the riverine delta; this can be explained by ~1 m down–faulting of the delta relative to the lake outlet, dated by habitation of the beaches ~1.6 cal ka, and (2) formation of ridge–and–basin terrain apparently by shaking during fault offset, followed by Native American occupation of these ridges about 4,000 years ago. Across the reservoir from here, the bench about 1,500 ft (500 m) above the lake is mantled by hummocky moraines of the last glaciation, deposited by the Snake River lobe from Yellowstone. The terminus of this glacial lobe was 13 mi (20 km) to the south beyond the south shore of Jackson Lake.

Figure 40. Old Faithful geyser eruption.

Flagg Ranch Rest Stop

The ridge to the east is Huckleberry Mt., which is the type section of the 2.1 Ma Huckleberry Ridge tuff (Christiansen, 2001). The tuff forms a dip slope inclined about 9° to the west. On the western skyline, the same tuff caps ridges at the northern end of the Tetons and dips west. East–dipping normal faults are responsible for explain these offsets and tilts (Christiansen, 2001). Thirteen miles (20 km) northeast of here on the east side of Mt Sheridan, post–glacial scarps as high as 70 ft (20 m) represent the main postglacial faulting near the southern Yellowstone National Park boundary.

On Huckleberry Ridge, a well–defined Pinedale lateral moraine occurs at an altitude of 9,100 ft (2774 m) and defines the maximum of the Snake River lobe in Hedrick Pond time. Southward about 9 miles (14 km), the ice surface descends about 600 feet (180 m) to the high bench across the lake from STOP 4–6 where the upper limit of the Snake River lobe is at an altitude of 8,500 feet (2591 m).

STOP 8–2. SIGNAL MT.

A deep scour basin was excavated between here and Donoho Point, northwest across the lake, by the Pacific Creek lobe during the time of deposition of Burned Ridge (Fig. 41) to the south ('Burned Ridge time'). The north–south axis of Jackson Lake reflects another scour basin with water 400 feet (122 m) deep, and a sediment fill as much as 400 more feet (122 m) thick (Smith et al, 1993). The Snake River lobe excavated the north–south axial trough of Jackson Lake during the time of deposition of Hedrick Pond and Jackson Lake moraines. The excavated material was carried to the head of the outwash at the north end of the Spaulding Bay outwash fan, and deposited to build the large, thick, outwash fan that forms Baseline Flats between Spaulding Bay, Jenny Lake (Fig. 42) and Moose.

If offset on the Teton fault were to occur today, the western shoreline of Jackson Lake would move down more than the natural outlet at Jackson Lake dam, and when the water in the lake

Figure 41. Jackson Hole and Grand Tetons. The potholes in foreground; dark ridge in the center is Burned ridge; dark arc on the right side is Jenny Lake. (Photo by D.A. Rahm, courtesy of EPIC)

Figure 42. Jenny Lake moraine and Grand Tetons. (Photo by Austin Post)

rose to spill at the outlet, shorelines on the west side of the lake would be submerged. During rebuilding of the Jackson Lake dam in the late 1980's, the lake was at times lowered to its natural level. From a canoe in calm water, we could see deeply enough to distinguish two submerged shorelines at depths of 3.5–4 feet (1–1.2 m) and 8–9 feet (2.4–2.7 m). Combined side–scan sonar and fathometer records gathered in 1986 by Colman and Pierce show shorelines east of Bearpaw Bay about 2 km east of the Teton fault at depths below the natural shoreline of about 3, 8, 15, 20, 24, 27, 31, and 36 feet (1, 2.4, 4.6, 6.1, 7.3, 8.2, 9.4, and 11 m). These shorelines are inferred to be submergence events related to downfaulting of Jackson Hole, and if so, suggest at least 8 post–glacial faulting events.

STOP 8–3. POTHOLES CHANNELWAY VIEWPOINT

Park on the north side of the road in Mt. Moran scenic turnout (or pullout 0.4 miles to the east) and hike across outwash flats to the top of the bare slope.

After the Burned Ridge recession and prior to Hedrick Pond deposition, the Triangle X–1 lake occupied the glacial scour basin that extended 20 miles (32 km) from Deadman's bar up the Snake–Buffalo Fork drainage system. We are looking down the Potholes channelway, named for the abundant glacial kettles that represent ice blocks of Hedrick Pond age buried by outwash of Jackson Lake age. The kettles form alignments that indicate several ice–margin positions of the Snake River lobe of Hedrick Pond age. To the left of the Potholes channel, the highest outwash terrace is of Jackson Lake age and has linear trends of kettles that suggest the west margin of buried blocks of the Pacific Creek lobe of Hedrick Pond age.

Looking down The Potholes channelway and across the Snake River, the higher alluvial bench is Burned Ridge (BR–3) outwash and moraines, which climb to the left where the ice–marginal Spread Creek followed the southern margin of the Buffalo Fork lobe. This ice margin ascends toward Black Rock Meadows and Togowotee Pass.

The Hedrick Pond bench is 100–200 feet (30–60 m) below the Burned Ridge bench and climbs more gently to the left. This gentler, ice–margin gradient may result from the Hedrick Pond advance moving across Triangle X–1 lake sediments with basal sliding aided by shearing in the lake sediments. The biggest difference in the extent of ice between Hedrick Pond and Jackson Lake time is that the Triangle X lake basin was entirely filled in Hedrick Pond time, whereas in Jackson Lake time, the Pacific Creek lobe extended only to its northeast margin and the Snake River lobe to its northwestern margin.

The glacial geology of the "Burned Ridge" landform is puzzling. The ridge is far from a simple moraine of one age and orientation. The northern half of Burned Ridge has Hedrick Pond moraines of the Snake River lobe draped on its east side and part of its west side. An advance of Hedrick Pond ice through a breach in Burned Ridge occurred in a gap from 0.9 and 1.5 mi (1.4–2.3 km) north of the Snake River. The margins of this breach are marked by west–sloping small moraines of Hedrick Pond age plus a loop of aligned kettles that mark the advance through this breach.

The escarpment at this stop is one of a set of kame moraines of probable Hedrick Pond age that descend to the east. The distal slope of the moraine is a gently sloping (5–8°) bench, rather than a ridge.

Pull over into parking area on north side of South Landing channelway, which is the head of an outwash fan of Jackson Lake age that has remarkably well–preserved channel morphology with downstream branching channels. This is quite similar to the Potholes channelway, except it has many fewer kettles.

STOP 8–4. CATHEDRAL GROUP OVERLOOK

With low sun angles late in the day, this is a good place to view the Teton fault. Moraines of Pinedale slab–glaciers occur on the steep slopes above the fault. The turnout is on Jackson Lake outwash, rich in quartzite from the greater Yellowstone glaciation. The slope and channel pattern of outwash trends obliquely toward the Teton front, defining a Spaulding Bay source. Postglacial tilting of this surface is much less than the original depositional slope.

Forested moraines of Jackson Lake? age with large boulders from the Teton Range occur 200 yds (180 m) north of the turnout. Quartzite cobbles are common in the outwash in front of the moraine, but are uncommon in the moraine itself, indicating glacial flow from the Tetons was fronted by outwash from the Spaulding Bay head of the outwash.

STOP 8–5. SNAKE RIVER OVERLOOK

This is a famous photographic site, particularly of a photograph by Ansel Adams. We are still on west–sloping, Burned Ridge outwash. Across the Snake River, the forested moraine is the Burned Ridge moraine, mantled or modified by Hedrick Pond deposits. The outwash to the left of that is the Spaulding Bay outwash fan of Hedrick Pond age, which is younger than the outwash on which we are standing. It has distinctly fresher channel morphology on aerial photographs.

To the right, the forested moraine is outwash of Jackson Lake age, which is about 100 ft (30 m) lower than the Hedrick Pond outwash. The Jackson Lake outwash heads at the Potholes channelway, in which ice blocks of the northward–receding glacier of Hedrick Pond age were buried by outwash of Jackson Lake age. Jackson Lake outwash prograded into a lake basin excavated in Burned Ridge time and across which the Hedrick Pond ice briefly advanced before retreating to the Jackson Lake position. The outlet to this lake was just northwest of the overlook at the head of the inset terrace, which is the head of a "flood flume". Repeated flooding lowered the outlet from the ~6840 ft (2085 m) Hedrick Pond spill point to 6720 ft (2048 m), the inset terrace level. The floods may have come from damming a lake in the lower part of the Buffalo Fork drainage by the combined Pacific Creek–Snake River lobes. Looking north beyond the Potholes channelway is the position of Jackson Lake, and, beyond that, the source of the Snake River glacial lobe on the Yellowstone Plateau. The eroding bluffs above the Snake River expose, from bottom to top: gray gravel mantled by loess (?), yellow gravel (from the northeast?), grey gravel of Jackson Lake age.

STOP 8–6. TETON POINT TURNOUT

The highest peak is Grand Teton, altitude 13,771 ft (4197 m), south of which are Middle Teton, South Teton, and Buck Mt. North of Grand Teton are Mt. Owen and Teewinot. The Snake River floodplain is down in the cottonwoods and is characterized by branching and continuously changing channels.

For the following discussion, assume the clock position of 12:00 is the upvalley continuation of the bluff on which we are standing.

This stop is near the onlap edge of the Spaulding Bay outwash fan (11:00) of Hedrick Pond age onto the Antelope flats outwash fan (1:00) of Burned Ridge age. The Snake River follows the approximate boundary between these two, very large, outwash fans of different ages and discharge points. The inset terrace along the Snake River is of Jackson Lake age. We call it a flood flume, for it displays large–scale bedforms produced by flood waters 30–45 ft (10–15 m) deep and ~1 mile (1.8 km) wide. Across the valley, flood bars occur along the valley wall and up to the margin of the fill terrace. Looking downvalley along the length of the Snake River, this Jackson Lake terrace converges with the Hedrick Pond terrace that we are standing on and may become the highest Pinedale terrace below Moose and at the airport.

Burned Ridge moraines occur at the head of outwash between 12:00 and 1:00 and represent the youngest of three sets of Burned Ridge moraines (BR–3) recognized. The BR–2 moraines locally protrude through the outwash at 1:30, and where the grassy ridge at 2:00 comes down to the outwash level, kettles represent the BR–3 maximum advance. Along the Teton front at 8:00, moraines of Pinedale valley glaciers form large mounds that encompass Bradley and Taggart Lakes. The Teton fault scarp is along the front above these moraines.

Along the southern margin of the greater Yellowstone glacial system, three Pinedale glacial lobes poured into northern Jackson Hole. The dominance of these lobes shifted through time from southeast to northwest in the following order: Buffalo Fork lobe (1:00–2:00, from the Absaroka Range), Pacific Creek lobe (12:30 from the Two Ocean Plateau) and Snake River lobe (12:30, from the Yellowstone Plateau). This northwestward migration is towards the source of moisture moving up the Snake River Plain.

STOP 8–7. BULL LAKE GLACIAL TERRAIN NEAR CURTIS CANYON CAMPGROUND

Drive east through the Elk Refuge.

At this site is well–preserved glacial terrain of Bull Lake age and a view of much of the extent of Bull Lake glaciation that filled Jackson Hole. The strong, southwesterly grain of the local topography is the result of both the southwest flow of Bull Lake ice, whose upper limit is shown by till and erratics 1,200 ft (360 m) vertically higher than here, and ice–marginal, meltwater erosion forming multiple ice–marginal valleys that reach depths of 30–300 ft (10–100 m). The upper limit of Bull Lake deposits descends southward and near Jackson, it is at the top of the chair lift on Snow King Mt. Along the southern margin of the Greater Yellowstone glacial system, Bull Lake terminal moraines and erratics drape the middle and lower slopes of the mountains at the southern end of Jackson Hole, 30 miles (50 km) beyond the Pinedale terminal moraines.

The view from here includes: (1) Burned Ridge moraines and outwash, overlapped downvalley by successively younger Hedrick Pond outwash on antelope Flats, and Jackson Lake (?) outwash at the airport, (2) Blacktail Butte and the hills of the National Elk Refuge overridden by Bull Lake ice, with the loess–mantled, Bull Lake fan of the Gros Ventre River southwest of the Elk Refuge hills, (3) below that, the large, Pinedale, outwash fan of Flat Creek, (4) Pinedale moraines at the mouths of seven canyons along the front of the Teton Range, (5) the Teton fault, forming a scarp on the lower part of the range and offsetting the lateral moraines, (6) Gros Ventre Buttes streamlined by overriding Bull Lake glacier, and (7) Phillips Ridge, above these buttes on the far side of Jackson Hole, where the Bull Lake glacial limit is well defined at 8,100 ft (2,470 m). Within the area occupied by Bull Lake ice, all of the moraines of local valley glaciers have a fresh Pinedale appearance.

If the large glaciation filling all of Jackson Hole was pre–Bull Lake, the local valley glaciers would be expected to display Bull Lake and Pinedale moraines. No such older moraines have been recognized for the seven valley glaciers from the Teton Range or for several valley glaciers from the Gros Ventre Ranges, such as Sheep Creek just east of here. The Bull Lake glaciation near West Yellowstone is dated as about 140 ka by combined obsidian–hydration and K–Ar dating. Bull Lake moraines are well beyond Pinedale moraines on the west (West Yellowstone), southwest (Ashton), and south (Jackson Hole) sides of the greater Yellowstone glacial system.

Where slopes are gentle, Bull Lake deposits are characterized by several meters of loess with a well–developed, buried soil near its base. A 26–ft–(8 –m)–thick section of loess with buried soils was described and sampled at a site on a Bull Lake outwash terrace 400 ft (120 m) above the Snake River and about 2.5 mi (4 km) inside the Bull Lake terminus.

Return to Jackson to spend the night.

DAY 9. PINEDALE AND BULL LAKE GLACIATIONS, FREMONT LAKE, WY

Drive 77 mi. (128 km) from Jackson to Pinedale and continue on to Fremont Lake to see Pinedale and Bull Lake moraines.

COSMOGENIC NUCLIDE GLACIAL GEOCHRONOLOGY IN THE WIND RIVER RANGE, WYOMING

Gosse, J.C., Dalhousie University, Halifax, Nova Scotia, Canada
Evenson, E.B., Lehigh University, Bethlehem, PA
Klein, J., University of Pennsylvania, Philadelphia, PA
Sorenson, C., University of Kansas, Lawrence, KS

Overview

Terrestrial cosmogenic nuclide (TCN) exposure dating of Quaternary glacial history in the Wind River Range has provided useful insights into the timing of initial retreat following the LGM, piedmont lobe synchronicity, and timing of many Rocky Mt. glaciations. Moraines in the Wind River Range can be dated precisely because (1) the low fracture density in the Precambrian granodiorite has permitted many large boulders (up to 6 m high) to be plucked and deposited by glaciers, (2) the granodiorite is reasonably resistant to erosion (< 1mm kyr $^{-1}$) (Gosse, 1994; Zimmerman et al., 1994), (3) multiple lobes record each glaciation, and (4) in all instances, well–preserved recessional moraines are nested within terminal lobes, so stratigraphic relationships are straightforward. TCN measured in erratics and bedrock surfaces between recessional moraines provide additional constraints on ice marginal history and on the effectiveness of a glacier to have eroded enough bedrock (usually >3 m) to remove most of the previously exposed rock.

The chronology presented in this guidebook includes ages for the following glaciations: Sacagewea Ridge, Bull Lake, and Pinedale. Moraines for these glaciations were dated at the type localities and in other locations in the Wind River Range. Additional exposure ages on moraines provide estimates for the age of the Titcomb Basin moraines correlated with the Temple Lake advance, the Gannett Peak advance, and moraines in the Titcomb Basin correlated with the Audubon Advance of the Colorado Front Range. We include only the exposure histories on moraines, terraces, and bedrock provided by Gosse (1994), Gosse et al. (1995a, b), Phillips et al. (1997) and subsequent unpublished data. None of the ^{26}Al data presented here have been previously published. All ages have been recalculated according to the latest scaling approach (Stone, 2000) using a ^{26}Al/^{10}Be production ratio of 5.94 (Gosse et al., submitted). Other exposure chronologies on moraines (Chadwick *et al.*, 1997; Dahms, 2002; Phillips et al., 1997; Zreda, 1994), terraces (Anderson et al., 1996; Hancock et al., 1999; Phillips et al., 1997, Chadwick et al., 1997), and bedrock surfaces in the Wind River Range (Small et al., 1997) are not included in this guidebook.

Stratigraphic context and soils geomorphology

The exposure age glacial chronology presented in this guidebook is placed in context with numerous soils studies and glacial chronologies in the Wind River Range and adjacent mountains.

Previous glacial chronology

Blackwelder (1915) identified, named, and described evidence for the Pinedale and Bull Lake glaciations in the Wind River Range and designated the Bull Lake and Pinedale areas as their type localities for the Rocky Mts. He considered the deposits to correspond with early and late Wisconsinan deposits. The Bull Lake glaciation has since been correlated with the late Illinoian Glaciation (Richmond, 1986; Richmond and Fullerton, 1986), mostly on the basis of age assignments on deposits outside the Wind River Range (see below). The pre–penultimate Sacagawea Ridge Drift (Richmond, 1964, 1965), whose type locality is near Dinwoody Lakes, and scarce remains of older glaciations have been assigned relative ages based on geomorphologic evidence and soil character. The existence of early Wisconsin drift has never been proven in the Wind River Range, although several soil investigations have suggested the possibility (Hall, 1990; Hall and Horn, 1993; Hall and Shroba, 1993; Richmond, 1987).

Soils geomorphology

Interpretation of TCN data as exposure durations assumes an understanding of the stability of the moraine (Hallet and Putkonen, 1994) or terrace being dated. One means of evaluating landform stability is through soils geomorphology. The moraines in the Wind River Range had been mapped by Richmond in the 1970's, although Hall et al. have refined the Bull Lake quadrangle map to reflect details brought out in the soils and exposure chronology. Quaternary soils have been analysed by, among others, Sorenson and Dahms on the western side of the Wind River Range, and Hall on the eastern side.

Western side of the Wind River Range. The soils of the Fremont Lake South quadrangle were mapped in the early 1980s (Sorenson, 1987) following the surficial geology mapping of Richmond (1973). Soils on moraines in the vicinity of Fremont Lake formed from granitic and granodioritic lithologies, but include strong components of eolian material partly derived from Green River basin sediments (Dahms et al., 1997). Richmond's geologic map used soil characteristics as one, among several, means to identify individual map units, and on the 1981 INQUA field trip he stated that "soils could be used as a first approximation to separate Pinedale and Bull Lake moraines." A strong degree of correspondence exists between his geologic map units and soil development. Substages of Pinedale or Bull Lake glaciations could not be distinguished in the soil record at the level of detail required for mapping at the 1:24,000 scale (Sorenson, 1987). The Pinedale equivalent soils are non–calcareous A/C profiles, many of which display a Bw horizon. Bw horizons usually are found where glacial action incorporated Tertiary sediments in the till and provided a readily available clay source for subsequent translocation and/or where these soils have formed in relative stability during post–glacial time. In contrast, Bull Lake soils commonly include a Bt and/or a Bk horizon with $CaCO_3$ distributed below the solum into the uppermost parent material. The soils formed on these older Pleistocene deposits are located on surfaces of generally lower relief than their Pinedale counterparts and are more distant from cirques dominated by the igneous and metamorphic rock from which their glaciers emanated. They appear to contain more Tertiary sediments, which may account for their generally higher clay content and perhaps some of their carbonates. Soils formed on piedmont Pinedale and Bull Lake tills typically have mollic epipedons and have been enriched by the addition of loess to the profile. Soil profiles in Bull Lake moraines show clear indications of surface erosion. However the thickness of erosion is believed to be less than 1 m in most instances. In many places, the broad ridge crests of the Bull Lake moraines show indications of accumulation of eolian cover.

Geomorphic processes, acting subsequent to deposition of the glacial drift of various ages, have had a profound effect on the variability in soils found on surfaces of different ages as well as those located in different landscape positions on surfaces that are presumably of the same or similar age. For example, inter–morainal soils typically have pachic epipedons owing to erosion and accumulation of sediments from steep adjacent moraine slopes. In some instances, evidence of mass wastage exists and buried paleosols have been identified beneath adjacent debris accumulations. In one instance, a buried soil lying between two Pinedale lateral moraines is dated at 6070±100 years B.P., indicating early post glacial soil formation followed by active Holocene slope alteration and soil burial and subsequent soil development in the colluvial sediment. Elsewhere a basal date of 9305±75 years B.P. from peat in a former lake bed gives a minimum date for the development of histosols in the map area. Clearly these A/C profile soils lack strong soil development, so surface processes must result in post–depositional alteration or inhibition of soil development. Dahms (1991) offered compelling evidence that eolian deposition has had a profound affect on the piedmont and alpine glacial soils of the region wherein soil horizon development is retarded by the combination of pedoturbation and high amounts of eolian sedimentation. Add wind and water erosion to that and the setting is consistent with a low level of soil development in the study area during post–glacial time. The discovery of numerous buried soils, revealed in soil cores from low catenal positions in the landscape, provide ample evidence of the removal of sediment from moraine crests and slopes and its redeposition in kettles, and depressions throughout the region (Dahms, 1991; Sorenson, 1987). More gentle slopes and older, more clay–rich, sediments of Bull Lake tills result in soils that contrast with those on the younger Pinedale till.

Eastern side of the Wind River Range. The major differences between the Bull Lake (western) and Dinwoody Lakes (eastern) areas soil geomorphology for correlated moraines can be attributed to differences in local climate (the east side is slightly drier and windier) and lithology (in addition to the Precambrian granitic and granodioritic gneisses, Paleozoic sedimentary rocks comprise a substantial portion of the parent material). In the Bull Lake and Dinwoody Lakes region, more Pinedale and Bull Lake moraines occur. On the bases of soils in moraines and terraces correlated with moraines, Hall (Chadwick et al., 1997) subdivided Richmond's Bull Lake moraines and proposed that they represented two or more separate glaciations. On the eastern side of the Wind River Range at Dinwoody Lake, a single moraine ridge of Sacagewea Ridge glaciation is preserved. Erosion of more than 1 m from the crest of the ridge is evident from the soils record. As with the moraines and terraces on the western side, pre–Pinedale soils show significant eolian accumulation.

METHODS AND UNCERTAINTY OF COSMOGENIC DATING

Target preparation was completed at the University of Pennsylvania and the University of Kansas. Elemental chemical analyses was completed by DCP at Rutgers University, by ICP.AES at Los Alamos National Laboratory, and by ICP.OES at the University of Kansas. AMS analyses for ^{10}Be in all samples and ^{26}Al in many were performed at the University of Pennsylvania (WY–97–620 was measured at Lawrence Livermore National Laboratory). The Al carrier was from SPEX ICP standard. Process blank subtractions were <5% of the measured ratios. Production rate is based on scaling according to Lal (1991), modified according to Stone (2000), using a high latitude, sea level, production rate of 5.1 ^{10}Be atoms g^{-1} a^{-1} (Gosse and Stone, 2001; Stone, 1999), scaled with 2.2% and 2.6% of sea level production from muons for ^{10}Be and ^{26}Al respectively, and adjusted for a small deviation in atmospheric pressure from the standard atmosphere used by Lal (1991). The ^{26}Al high–latitude, sea level, production rate in quartz is 5.94 times faster than ^{10}Be (Gosse and Phillips, 2001; Gosse et al., submitted). Concentrations are integrated over the sample thickness assuming simple exponential decrease in nucleonic flux. The production rates are adjusted for thickness (adjustment was less than 5% in all samples except WY–91–003—11%, WY–91–024—8%, and WY–93–337, 335, 341—7%) and shielding (negligible, except for Titcomb Basin—3%). No geometric corrections were necessary because the centers of large, flat–topped surfaces (typically >2 m high except for Titcomb Lakes boulders, which were lower) were collected to minimize error due to edge effects and foreshortening. However, Masarik (unpublished data) has shown that geometry corrections for boulders may be greater than is calculated using the normal formulation (Gosse and Phillips, 2001). Shielding by snow is possible in all instances where boulders are less than 1 m and on bedrock ridges. However, we estimate that this would average < 3% decrease in age in worse case scenarios (high density snow, 4 months of continuous cover). Due to high boulder heights and sampling only on ridge crests, cover by till is unlikely in all cases except for the Sacagewea Rigde moraine samples where the thickness of eroded sediment may exceed boulder height (discussed below).

AMS measured ratios are adjusted for isobars, background, and chemical blank corrections. The AMS error is 1σ precision and includes uncertainty in background and boron corrections. Random error in the concentration is the precision of the concentration, considering random errors associated with the sample radiation (shielding, thickness, etc.), preparation (interchange, weighing, carrier, addition of Al, B, and Be), and AMS error. This is the 1σ total analytical precision of a single measurement.

Unless indicated, the exposure ages in the text and tables are calculated assuming a continuous exposure with no adjustment for erosion or burial. The ages are therefore minimum limited ages of total exposure duration. For samples exposed for less than 15 ka the effects of erosion are negligible, and, therefore, the ages represent the average exposure duration. Furthermore, exposure of the bedrock and autochthonous blocks may have been interrupted by one or more glacial events. The age uncertainty combines the analytical precision plus propagated systematic errors (including uncertainty in half–life and production rate scaling and temporal variations). It provides a $1\ \sigma$ estimate of accuracy of an individual age when comparing with other ages from non–TCN chronometers.

Reproducibility of the method on a single boulder can be evaluated by comparing the results of duplicate samples that have been taken through complete chemistry twice (duplicates of WY–91–003 yield ages of 20.5 and 20.7 ka). Reproducibility of an age of a moraine can be evaluated by comparing TCN concentrations from multiple boulders on a single landform (e.g. the upper Titcomb Lakes moraine).

EXPOSURE AGE GLACIAL CHRONOLOGY

Sacagawea Ridge Glaciation

The age of the Sacagawea Ridge till is very poorly constrained. Terraces near Dinwoody Lakes have been correlated with outwash interpreted to be of Sacagawea Ridge age and assigned ages of 620 and 660 ka B.P. based on the presence of Lava Creek B tephra in association with terrace gravel (Richmond and Murphy, 1965; Richmond, 1976a; Izett et al., 1992; Chadwick et al., 1997). However, the stratigraphic relationship between the gravel and the ash, and the outwash terraces and the moraines is still a subject of debate. Richmond (1986) bracketed ages for Sacagawea Ridge deposits between 350 and 610 ka B.P. Chadwick et al. (1997) argued that Sacagawea Ridge deposits near Bull Lake may pre–date the Lava Creek B ash.

The cosmogenic nuclide ages for the Sacagawea Ridge moraine are straightforward to interpret. The ages (Fig. 43) are interpreted to be minimum estimates of moraine deposition. The ages from the ^{10}Be and ^{26}Al measurements agree within 10% for all samples, indicating that interruption of the cosmic ray flux due to burial by loess or snow must have been insignificant. The ages of three similar–sized boulders are concordant, but the age of a boulder twice as high is older (Fig. 43), implying that the larger boulder was exposed before the smaller boulders. Simple continuous erosion of the moraine at a rate of between 0.3 and 0.4 cm ka^{-1} would make the ages of the smaller boulders equivalent to the age of the higher boulder. Correction for this amount of erosion would yield a ca. 650 ka age for the moraine. In 650 ka, an erosion rate of 0.35 cm kyr^{-1} would remove more than 2 m of till from the crest. This seems plausible and corresponds to the minimum erosion rate estimated from soils data on the Sacagawea Ridge moraine by Hall.

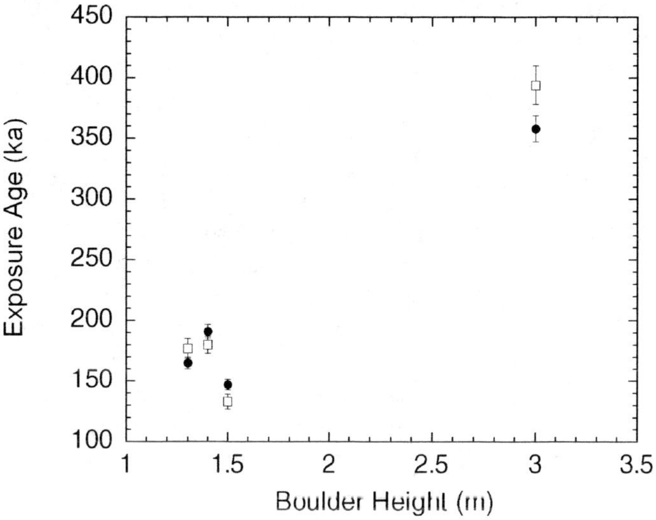

Figure 43. Ages of Sacagawea Ridge moraine. Solid boxes are ^{10}Be, open boxes are ^{26}Al. No correction was made for erosion.

Bull Lake Glaciation

Age constraints of the Bull Lake Glaciation are only slightly better than those for the Sacagawea Ridge Glaciation. Pierce et al. (1976) assigned the first numerical ages to Bull Lake moraines of 135 ka B.P. and 160 ka B.P. (mean 150 ka) based on obsidian hydration samples from West Yellowstone (Pierce, 1979). Obsidian hydration rates were calibrated using the West Yellowstone rhyolite flow, stratigraphically above the moraines, dated by K–Ar at 117.5 ± 8.0 ka B.P (corrected in Pierce, 1979) and the 179.0 ± 3.0 ka Obsidian Cliff rhyolite flow (Pierce et al., 1976), stratigraphically below the moraines. The only other dates that closely bracket the age of Bull Lake till are uranium–trend ages of 130.0 ± 40 ka B.P. in the Colorado Front Range, and 160.0 ± 50.0 ka B.P. in the Fremont Lake region, reported by Rosholt et al. (1985). Based on this evidence, the Bull Lake Glaciation has been correlated with marine oxygen isotope stage 6, although Colman and Pierce (1986) noted that the age of some Bull Lake Drift could be older and may correlate with marine oxygen isotope stage 8. In the Bull Lake type locality, soils and ^{36}Cl measurements on boulders have been interpreted by Hall and Phillips to indicate that the Bull Lake moraines represent at least two different glaciations, spanning 300 to 70 ka B.P.

Our cosmogenic nuclide measurements on Bull Lake age moraines were mostly from samples collected in the Fremont Lake region (N=13) with only two from the Bull Lake type locality. Boulders in the Fremont Lake region were larger and moraines were broader, which made the Fremont Lake lobe the preferable site to date the Bull Lake Glaciation.

No obvious correlation between boulder height and age exists on the west side of the Wind River Range (Fig. 44), so we consider that these boulders have always been exposed since their deposition. Soils work in the Fremont Lake region also indicates that more than 1 m of erosion from the moraines is unlikely after soil formation began. However, on the east side of the Wind River Range, erosion appears to be significant. In the case of the 67 ka age of the Bull Lake moraine in the Bull Lake type locality, the sample was a short (1.4 m) boulder on a very narrow, steeply–sloped, terminal moraine. In the same area, on a steep sided QBLt12 moraine (the 11th recessional moraine as mapped by Hall), a 4 m boulder yielded an age of 130 ka. As with the Sacagawea Ridge samples, this suggests that the Bull Lake moraines on the east side have lost more than 1.4 m of till from their crests.

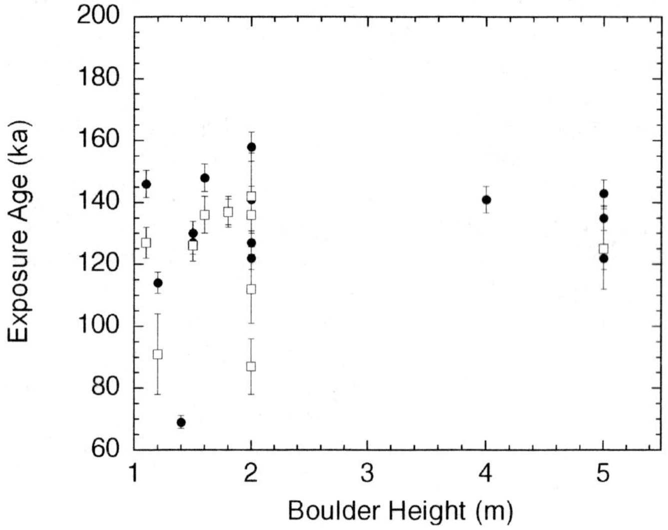

Figure 44. Ages of boulders on Bull Lake moraines.

The boulder ages also do not show a strong dependence on moraine position (Fig. 45). The range in ^{10}Be ages is 115 to 160 ka B.P. on the west side of the Wind River Range. ^{26}Al ages agree to within 10% of the ^{10}Be ages, except on three samples with high ^{26}Al AMS measurement error. The reason why the ^{26}Al ages are lower than ^{10}Be and more scattered is most likely due to the relative difficulty in precisely measuring ^{26}Al/^{27}Al with AMS. The weighted mean age of the ^{10}Be and ^{26}Al dates is 140 ka and 130 ka . These ages are minimum ages for the Bull Lake Glaciation. If a boulder was deposited 150 ka B.P. and eroded slowly at 0.8 mm kyr^{-1}, the exposure age would be 135 ka B.P.

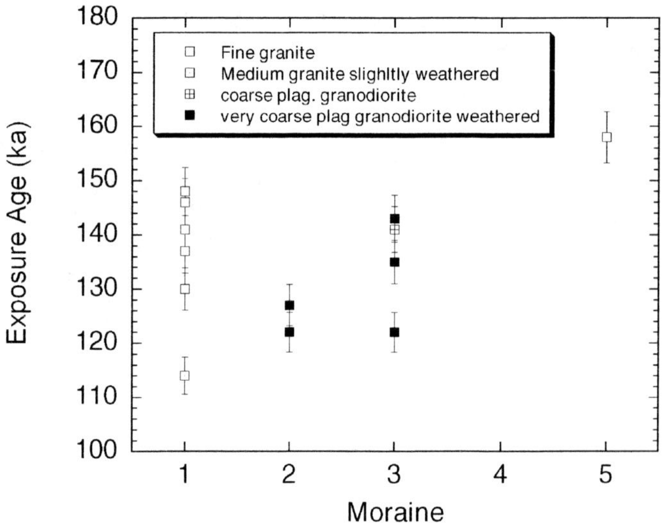

Figure 45. Ages of boulders on Bull Lake moraines, in order of moraine position. 1=terminal, 2=first recessional. Symbols represent the field classification of lithologies.

Gosse (1994) found that Bull Lake ages were correlated with boulder lithology. This may partly be due to the distribution of boulders on specific moraines (fine–to–medium–grained granite tends to occur on terminal moraines). The relatively finer grained granitic boulders yield higher ages than the coarse, "horse–tooth"–feldspar granodiorites that are more easily weathered and exhibit more and deeper gnammas and rillen. We altered our sampling strategy after discovering this trend to use finer–grained and harder–to–sample boulders. We interpret the scatter of the ages to reflect differential erosion rates among the boulders.

The ^{26}Al/^{10}Be ratio plot (Fig. 46) reveals that the samples have not likely experienced any interruption in their exposure due to burial or episodic deep erosion. The two low ^{26}Al/^{10}Be ratios were measured on samples with high uncertainty in the ^{26}Al/^{27}Al measurement, so we do not consider the low ratios to reflect significant burial.

Pinedale Glaciation

Despite the dating of numerous Pinedale deposits throughout the Rocky Mts., very few tightly bracket the timing of the last Pinedale glacial maximum (LPGM) or the duration of the glaciation. The age of Pinedale Glaciation at the type locality was constrained locally by ^{14}C dates of 6.1 ± 0.1 and 9.3 ± 0.08 ^{14}C ka B.P. (uncalibrated) on bog–bottom peat samples (Sorenson, 1987) that provide a minimum for the age of the late Pinedale glacial maximum. Approximately 21 miles NNW of Fremont Lake, the vegetational change associated with the Pinedale glaciation is bracketed between 20.8 and 10.2 ^{14}C ka B.P. (uncalibrated) (Ives et al., 1964)). Numerous relative–age determinations of local glacial deposits (Dahms, 1992; Dahms, 1994; Hall and Horn, 1993; Hall and Shroba, 1993; Mahaney, 1978) provide a basis for correlation with more tightly age–constrained deposits elsewhere in the Rocky Mts. Considering the Rocky Mts. as a whole, the timing and duration of the Pinedale Glaciation appears to have been variable, typically beginning prior to 40 ka B.P. (Nelson et al., 1979; Pierce, 1979; Pierce et al., 1976; Porter et al., 1983) and continuing throughout the time approximately equivalent to marine oxygen isotope stage 2, with evidence for several glacial advances in some locations. The majority of numerical chronological constraints reported on these Pinedale deposits are bracketing ^{14}C ages.

^{14}C dates reported here are not calibrated except where explicitly stated. Other dating methods, including calibrated obsidian hydration (Pierce, 1979; Pierce et al., 1976), thickness of weathering rinds (Colman and Pierce, 1981, 1986) and measurements of U–series disequilibrium in travertine (Sturchio et al., 1994), have provided additional dates for Pinedale advances. The last maximum expansion of Pinedale ice is bracketed by dates from a number of sites, including in the Colorado Front Range between 23.5 and 15 ^{14}C ka B.P. (Madole, 1986) and in Northern Yellowstone before 18.6 ^{14}C ka B.P. (Sturchio et al., 1994). From these dates, Porter et al. (1983) concluded that the last time the Rocky Mt. Pinedale glaciers were at their maximum extent was 20 to 15 ^{14}C ka B.P. By now, over a hundred dates from cosmogenic nuclide measurements in the northern Rocky Mts. (plus others in the Wind River Range) generally fall within this range (e.g. Licciardi et al., 2001).

Four significant aspects of the cosmogenic nuclide measurements of Pinedale moraines in the Wind River Range include: (1) a large scatter occurs among ages on terminal moraines, (2) three adjacent lobes appear to have synchronous histories, (3) the timing of retreat is tightly documented, (4) individual boulder ages on five recessional moraines fall within 4% of their mean.

The ages of the terminal moraine are scattered, ranging from 24.8 to 16.5 ka B.P. The standard deviation is 2.4 kyr (N=17), or 12% variation about their mean age. This scatter could be interpreted to indicate the true scatter of multiple measurements from a single moraine. However, we show that in two other locations with N>5 measurements on a landform (Titcomb Lakes moraine and Halls Lake recessional moraines) the variation about the mean age of the landform is much closer to 4%, the average 1σ random error in the $^{10}Be/^{9}Be$ AMS measurement. For samples that were completely reprocessed through chemistry, the reproducibility is high: WY–91–035 duplicate ^{10}Be ages are 22.8 and 22.8 ka, and WY–91–003 duplicate ^{10}Be ages are 20.5 and 20.7 ka.

So how do we interpret the scatter in the ^{10}Be ages on the QPt1–FL terminal moraine? Two alternative explanations are suggested: (1) The scatter could indicate that the large Pinedale terminal moraine took about 8 ka to build. This is supported by the observation (Gosse et al., 1995a) that the ages decrease from the front of the moraine to the back (Fig. 47). This explanation is also supported by the observation that the two dates (WY–91–010, WY–92–107) on the back side of the moraine are consistent with the ages on the Pinedale recessional moraines QPt2–5–HL (mean age = 18.2 ka). (The mean age is 18.2 ka for the two boulders nearest the back of the terminal crest. The anomalously young age of a boulder that is not on the crest but on the back slope, WY–91–030 (16.5 ka), may reflect boulder movement due to sediment mobilization above melting buried ice.

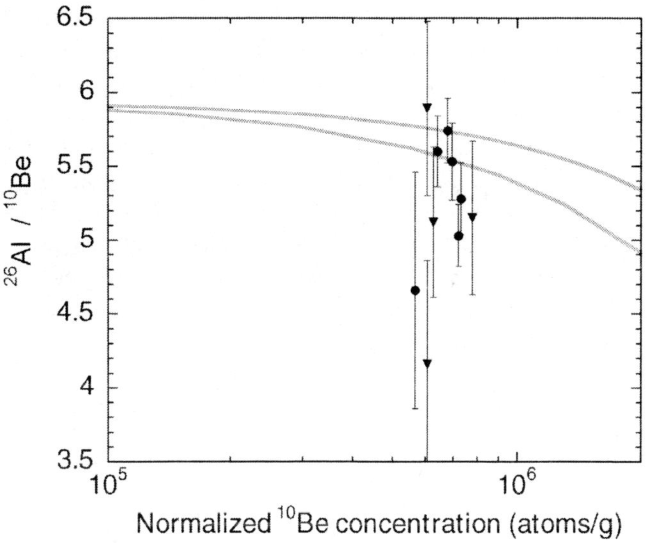

Figure 46. $^{26}Al/^{10}Be$ isotope ratio plot for all samples with both isotopes measured for the Bull Lake moraines on the west side of the Wind River Range. Circles are terminal moraine, triangles are recessional moraines.

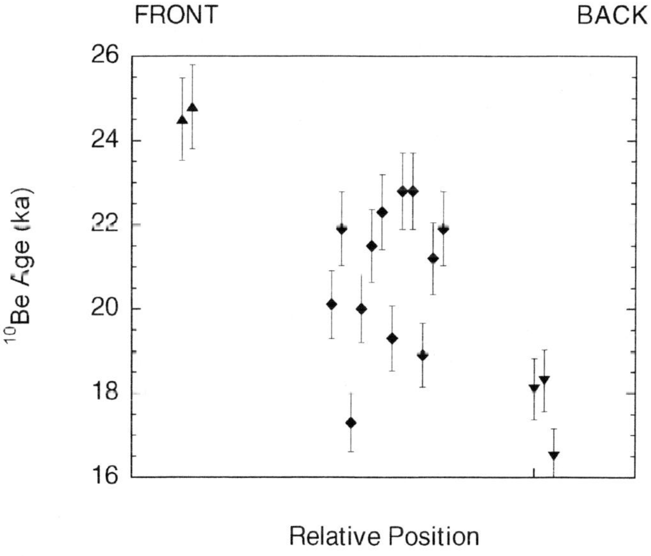

Figure 47. Ages of boulders on the Fremont Lake Pinedale terminal moraine, plotted according to approximate position on the moraine ('back' is toward the cirque).

The large scatter in ages on the QPt1 terminal moraine would also occur if the first boulders to arrive had a significant inherited component. In other words, the glacier first eroded regolith that had been previously exposed. With continued glacier erosion, only sediment that had no previous exposure would be delivered. This explanation is consistent with the trend in ages (oldest to youngest) from front to back of the moraine. It may also be apparent in the $^{26}Al/^{10}Be$ plot, which would indicate if there was an interruption in the rock exposure history. If a rock was exposed in a valley prior to glaciation, then buried by ice and till over a few glaciations, then eventually plucked and delivered to the terminal, the ratio would reflect this burial. This trend, although weak, is apparent in the ratio plot (Fig. 48). The samples with older ages (higher ^{10}Be concentrations) tend to have lower $^{26}Al/^{10}Be$ ratios. If the apparent trend is real, it could only be due to either ^{10}Be AMS measurements that were too high (unlikely, considering reproducibility) or to inheritance in the oldest (front) samples (i.e. shielding interrupted the exposure history of the sample.

Ages from boulders on terminal moraines of other lobes in the vicinity, Soda Lake and Halls Lake, are similar to the mid–position boulders on the larger Fremont Lake terminal. Soda Lake boulders yield ages of 24.9, 20.7, 20.5 and 21.3 ka (three ^{10}Be and one ^{26}Al). On the Halls Lake terminal moraine, the ^{10}Be ages are 22.3 and 21.9 ka, ^{26}Al ages are 19.7 and 21.4 ka. The $^{10}Be/^{26}Al$ indicates a simple exposure history, with the exception of WY–92–117 on Halls Lake moraine with ratio of 5.2 (we need to revisit the boulder to determine if it is situated toward the front of the terminal moraine). The similar elevation of the terminal positions and the similar exposure ages indicate that the three lobes occupied their maximum positions approximately synchronously. However, we do not have the resolution to determine precisely when either lobe initially arrived at their maximum positions.

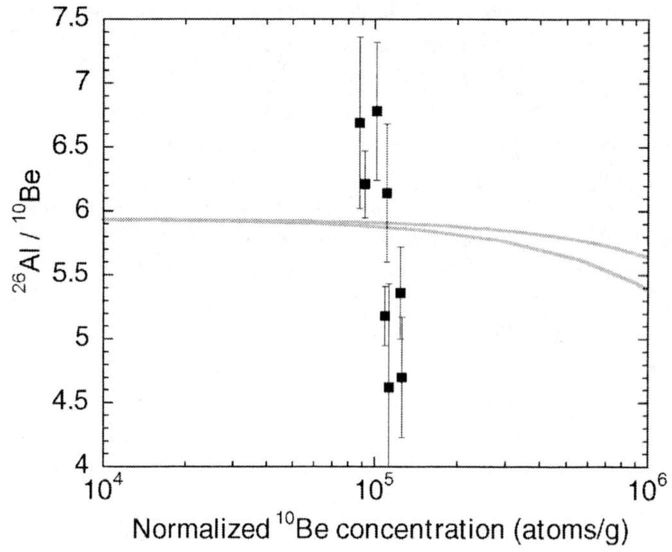

Figure 48. $^{26}Al/^{10}Be$ isotope ratio plot for boulders on the Fremont Lake lobe Pinedale terminal moraine.

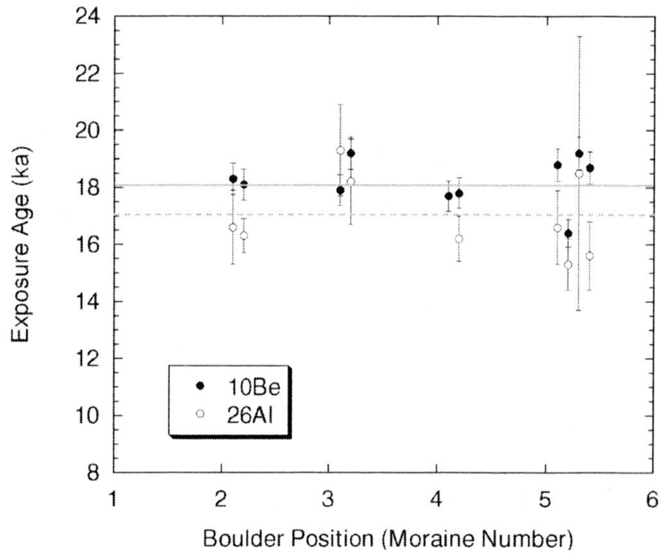

Figure 49. Exposure ages of boulders measured on Halls Lake (Mud Lake) recessional moraines. Solid and dashed lines are mean ages for ^{10}Be and ^{26}Al ages, respectively, on all Halls Lake recessional moraines.

Timing of retreat and bedrock erosion by glaciers

The timing of initial retreat can be determined precisely. As indicated earlier, boulders on the back side of the Fremont Lake terminal moraine (^{10}Be ages 18.3 and 18.1 ka) are the same age as the mean of the ages on the first recessional moraine (18.2 ka), and as the mean of all of the ages from the Halls Lake recessionals (Fig. 49), ca. 18.2 ka.

The $^{26}Al/^{10}Be$ of the recessional moraine boulders around the Fremont Lake region ranges from 4.9 to 5.7. This is intriguing because, taken at face value, it would imply that all of the boulders measured had a small amount of inherited isotopes carried over from exposures prior to one or more burial events before moraine deposition. However, more likely, we have simply met the limit of precision at which ^{26}Al can be measured in young rocks. Since these measurements were made (1994), we no longer measure ^{26}Al in such shortly exposed rocks.

Outcrops of granodiorite occur between Half Moon Lake and Halls Lake. Their stratigraphic position is above the fifth Halls Lake recessional moraine. The exposure ages should therefore be younger than the mean age of the recessional moraines (Fig. 49), ca. 17.5 ka for both isotopes. Two bedrock samples were dated, yielding 14.1 and 17.4 ka (^{10}Be) and 16.7 and 16.5 ka (^{26}Al). Two conclusions can be made from these bedrock exposure ages: (1) the bedrock ages are younger than or concordant with the recessional moraine chronology, indicating that ice marginal retreat from the Pinedale terminal was after 18 ka, and (2) because the bedrock ages are concordant with, or younger than, the moraine ages, an inherited isotope component in the bedrock is not likely. This indicates that at least 3 m of rock had been removed from these bedrock outcrops in order to remove any cosmogenic ^{10}Be or ^{26}Al accumulated since the Bull Lake glaciation. These piedmont surfaces are most definitely below the paleo–ELA.

Bedrock within Titcomb Basin was also sampled for ^{10}Be analysis. At the entrance to the basin, 300 m south of the southern end of the Lower Titcomb Lake, a bedrock sample had an exposure age of 14.0 ka. On the same surface, a perched erratic boulder age was 12.4 ka (^{10}Be). Closer to the Titcomb Lakes moraine, on the northwest corner of the Lower Titcomb Lake, a bedrock sample was dated at 14.5 ka. Two possible explanations of this age are suggested: (1) If the 12.4 ka boulder represents the time that the boulder was exhumed from a retreating Pinedale glacier, then the older bedrock ages reveal 2–ka worth of inherited ^{10}Be. However, equally plausible is that the Pinedale ice retreated back to the Titcomb Basin by 14 ka, and re–advanced by 12.4 ka. An age on a 30 cm high boulder on a bedrock ridge on the north side of Island Lake, 3 km south of the southern end of the Lower Titcomb Lake, has a minimum exposure age of 11.3 ka. Unfortunately, the boulder is short and the age probably reflects when it was exhumed from till.

Temple Lake Advance

The age of the Temple Lake glaciation at its type locality in the Wind River Range is tightly constrained by bulk ^{14}C ages between 11.8 ± 0.7 ^{14}C ka B.P. and 11.4 ± 0.6 ^{14}C ka B.P. (Zielinski and Davis, 1987). Because the Temple Lake ages are both minimum dates (lake bottom sediments and residence time effects), Zielinski and Davis (1987) suggest that the age of the Temple Lake moraines 12.0 ^{14}C ka B.P. (approximately 13.5 cal ka B.P.). The Titcomb Lake moraines have been correlated with the Temple Lake moraines (Hall, 1989). Mahaney (1988) reported minimum bog–bottom ages of ~8.0 ^{14}C ka B.P. for the Upper Titcomb Lake moraine in Titcomb Basin, but assigns it to the late Pinedale, based on morphologic, soil, and stratigraphic characteristics (Mahaney, 1978, 1988). In the late 1990's, Gosse, Davis, and colleagues cored the upper and lower Titcomb Lakes and demonstrated on the basis of the lake sediment records, pollen stratigraphy, and more than 30 plant–macrofossil, bulk, and humate ages that the glacier last occupied the Titcomb Lakes moraine site during the Younger Dryas chron (Davis et al., 1999; Gosse et al., 1999; Gosse et al., submitted). The Lower Titcomb Lake sediment cores have a two part stratigraphy, with a sandy silt bottom section and a finer organic–rich upper unit. The transition between the two is abrupt in both color, sedimentology, and loss–on–ignition (LOI) character. The age of the transition is about 11.4 ^{14}C ka B.P. (calibrated ^{14}C age, Gosse et al., submitted). The Titcomb Lakes paleo–vegetation record is similar to that recorded in the Rapid Lake–Temple Lake region, and the chronology is believed to be more robust due to the number of macro AMS dates (Davis, personal communication).

Ten boulders from the innermost Titcomb Lakes moraine have been dated with ^{10}Be and ^{26}Al (Fig. 50). Excluding one ^{10}Be and two ^{26}Al statistical outliers, the mean of the ages is 12.7 ka and 12.9 ka for ^{10}Be and ^{26}Al, with respective coefficients of variation of 4.8% and 8.8%. If corrected for a 4–month 2–m cover of dense snow on the moraine (probable maximum effect), the mean ages are 12.9 ka B.P. and 13.1 ka B.P. and still within the Younger Dryas chron. These ages are concordant with exposure ages from the Stough Basin on the southeastern flank of the Wind River Range (Dahms, 2002).

Post–Temple Lake glaciation

Younger cirque glaciations in the Wind River Range have been sporadically ^{14}C dated, soil characteristics in till, lichenometry, and geomorphic position. A three–fold Holocene glacial stratigraphy has been identified throughout the range (Burke and Birkeland, 1983).

The youngest deposits in the Titcomb Basin are correlated with the deposits of the most recent (and perhaps continuing) advance, the Gannett Peak Stade (Richmond, 1957) and, based on a lack of soil development and lichens, is considered to be only centuries old. Two ^{10}Be measurements from the Gannett Peak moraine on the Twins Glacier in Titcomb Basin are 0.66 ka B.P. and 0.98 ka B.P. These ages date to the Little Ice Age and indicate the lack of inheritance in the boulders.

Moraines immediately in front of the Gannett Peak moraines have been correlated with the Audubon Glaciation in Colorado (Miller and Birkeland, 1974), which has been assigned a 1.9 to 1.0 ^{14}C ka B.P. age (Benedict, 1973). However, without adequate numerical dates, such long–distance correlations must be considered questionable (Burke and Birkeland, 1983). We sampled two ridges on the west shoulder of the Titcomb Basin, approximately

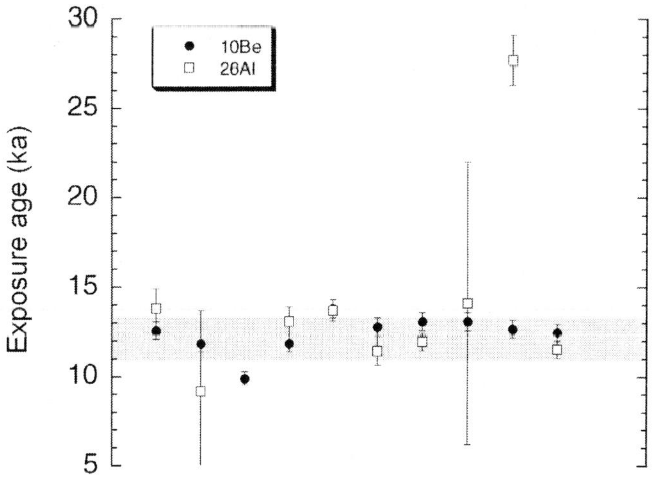

Figure 50. ^{10}Be and ^{26}Al ages for boulders on the innermost Titcomb Lakes moraine. Ages are corrected for topography but not for erosion or snow cover. Shaded region represents the duration of the Younger Dryas chron as measured in Greenland ice cores.

700 km above the north shore of the inner Titcomb Lake. The ages of the two boulders are 12.4 ka B.P. and 11.2 ka B.P. This is an unexpected result. Either the moraines have been correlated incorrectly and they are in fact recessional moraines from the Titcomb Lakes moraine advance, or the boulders have large (10 ka) inheritance, which has not occurred in some 80 measurements from the Wind River Range. We hope to re–evaluate the soils data and also the locations of our samples relative to the soils sample locations of Mahaney, Hall, and Blum. Based on the few cosmogenic nuclide data at hand, we tentatively interpret the moraines within 1 km above the Titcomb Lakes to be recessional moraines from the Younger Dryas advance, not a late Holocene readvance. Moraines between these and the Gannett Peak moraines have not been sampled for cosmogenic nuclide dating due to higher snow cover.

Return to Pinedale to spend the night.

DAY 10 GEOLOGY OF WESTERN UTAH

Ken Hamblin, Dept. of Geology, Brigham Young University, Provo, UT

Drive from Pinedale, WY to Salt Lake City, UT.

THE WASATCH MTS.

The Wasatch Mts. are a large, tilted fault block composed of an unusual assemblage of igneous, sedimentary, and metamorphic rocks. The internal structure of the mountains is likewise complex, consisting of folds and thrust faults formed during the collision of the North American and Pacific plates between 100 and 60 million years ago. Many of the rock formations exposed in the Wasatch Mts. were originally deposited in western Utah and thrust eastward about sixty miles. Foremost among these is the Oquirrh Fm., made up of thousands of beds of limestone, sandstone, and shale. The total thickness of this formation is truly amazing–as much as 28,000 feet of sediment were deposited during the Pennsylvanian and Permian periods to form this remarkable formation. Large segments of the Wasatch Range, such as Mt. Timpanogos, Cascade Mts., Spanish Peaks, Loafer Mountain, and Mt. Nebo. are made up almost entirely of the Oquirrh Fm. The sequence of Oquirrh rocks are especially well–exposed in Provo Canyon where thousands of feet of interlayered gray limestone and sandstone make up the entire canyon walls.

The general north–south trend of the Wasatch Mts. is determined by the Wasatch fault, one of the major structural features in the western United States. The fault extends as a continuous structure from Malad, Idaho southward past Mt. Nebo, a distance of more than 225 mi. (375 km). The total displacement along the Wasatch fault is up to seven miles. Movement began about 18 million years ago and continues today.

The Wasatch Mts. may at first appear to be like any other mountain range–but they are different. The ridges or spurs extending down the mountain front terminate abruptly in triangular–faceted spurs (Fig. 51). Indeed, the face of the Wasatch Mts. is a face with a thousand triangles. The triangular facets record repeated recurrent movement) and periods of stability.

The uplifted mountain range is only part of the story. The down–dropped block to the west is buried under thousands of feet of debris derived from erosion of the Wasatch Mts. Geophysical studies indicate that the sediment thickness in the basins or valleys ranges from 0.5 to 2.3 mi. (1–3.8 km). If this sediment could be removed, a series of great steep–walled terraces would be revealed and the Wasatch Mts. would appear like a huge wall nearly five miles high.

The front of the Wasatch Mts. has been sculptured exclusively by running water, but east of the crest line, ample evidence of glaciation may be seen. Numerous cirques, U–shaped valleys, and moraines occur on many of the higher peaks.

LAKE BONNEVILLE

The changes in global climate that produced the great ice age also increased precipitation in the Basin and Range, partly filling many of the basins to form more than 100 fresh water lakes. The largest was Lake Bonneville, which occupied the basins in much of western Utah and parts of adjacent Nevada and Idaho. Lake Bonneville grew until it was as large as Lake Michigan and at its highest level was 300 mi. (500 km) long and 145 mi. (242 km) wide and covered an area of approximately 20,000 mi^2. It was a deep lake, with a maximum depth of 1,050 ft. (320 m). Standing almost anywhere in the Salt Lake valley, you can see the ancient shorelines of Lake Bonneville etched along the face of the Wasatch Mts. a thousand feet above the valley floor. The lake was decorated with 30 islands, the mountain ranges that now rise above the ancient lake beds of the West Desert. Lake Bonneville developed a remarkable variety of coastal features including beautiful beaches, spits, bars, deltas, sea–stacks, caves, and wave–cut terraces (Fig. 52). It was a magnificent coast that rose and fell with changes in precipitation associated with the expanding and receding glaciers.

Figure 51. Triangular facets along mountain–front fault. (Photo by K. Hamblin)

The great variety of coastal features are now preserved on the dry desert floor and on the flanks of mountain ranges.

The highest Bonneville shoreline (5,090 feet) is easy to identify because it forms a well–defined terrace along the base of the mountains and separates two distinctive types of landforms. Above the Bonneville level, the mountain face is carved into faceted spurs, and the major stream and valleys are widely separated. Below the shoreline, the soft sediment of Lake Bonneville is commonly easily eroded, and numerous closely–spaced gullies dissect the terrace front.

Lake Bonneville ultimately overflowed near the present Red Rock pass between Preston and Downing, Idaho and then spilled into the Snake River. Discharge of water through the pass created huge catastrophic floods downstream, resulting in erosion of broad valleys (Marsh Valley near Downing) and fields of large boulders farther downstream.

The overflow occurred 14,500 ^{14}C yrs. B.P. and rapidly cut through the unconsolidated sediments in the pass down to a stable threshold on solid bedrock. This almost instantaneously lowered the lake level 350 feet. The new level of the lake is known as the Provo Level and is easily identified from large deltas and terraces built along the base of the Wasatch Mts. and prominent wave–cut and wave–built terraces along the base of the ranges to the west. Especially significant are the deltas built by the Spanish Fork River (Mapleton of Spanish Fork), the Provo River (Orem, BYU), the Weber River (Hill Air Field), and the Bear River (Logan).

With the end of the glacial period, the climate of the western United States became much drier and the lake began to shrink. In the end (approximately 8,000 ^{14}C yrs. B.P.), Lake Bonneville dried up, leaving Sevier Lake, Utah Lake, and the Great Salt Lake as its only remnants.

In detail, the history of Lake Bonneville is complex–rise and fall, rise and fall again, and finally gradual dessication. The major levels from highest to lowest were: Bonneville, elevation 5090 feet, about 16,000 to 14,500 ^{14}C yrs. B.P.; Provo, 4740 feet, about 14,500 to 13,500 ^{14}C yrs. B.P.; Stansbury, approximately 4,500 feet, about 23,000 to 20,000 ^{14}C yrs. B.P.; and Gilbert, approximately 4,250 feet, about 11,000 to 10,000 ^{14}C yrs. B.P. In addition, more than 17 minor shorelines can be counted on Fremont Island and the Hogup Mts. of northern Utah, each marked by narrow beaches, barrier bars, wave–cut terraces, and sea cliffs.

LITTLE COTTONWOOD CANYON – LONE PEAK

This section of the Wasatch Mts. is unique in that it exposes both the upper Precambrian Big Cottonwood Fm. and Tertiary granitic intrusions. The Big Cottonwood Fm. consists of a variety of quartzites, with interbedded shale and siltstone that have a cumulative thickness of approximately 16,000 ft. These Precambrian rocks are well exposed in Mt. Olympus just south of Little Cottonwood Canyon.

The Little Cottonwood stock is exposed on both walls of Little Cottonwood Canyon and is the great mass of rock that forms Lone Peak (elevation 11,926 feet). The age of the Little Cottonwood stock is approximately 26 Ma.

Glaciation has significantly modified this part of the Wasatch Mts., forming cirques and horns in the upper peaks and classic U–shaped glacial valleys along the mountain front. Little Cottonwood canyon is the longest glaciated valley in the Wasatch extending from Alta down to the mountain front, a distance of twelve miles. Bells canyon, immediately to the south, is another classic U–shaped glacial valley with a well developed terminal moraine which has been offset by recurrent movement along the Wasatch fault (Fig. 53).

Recurrent movement along the Wasatch fault, Little Cottonwood Canyon

The most recent expressions of recurrent movement along the Wasatch fault are low scarps that displace the ground surface

Figure 52. Lake Bonneville shoreline against the Wasatch Range south of Provo, Utah (Photo by D.A. Rahm, courtesy of EPIC)

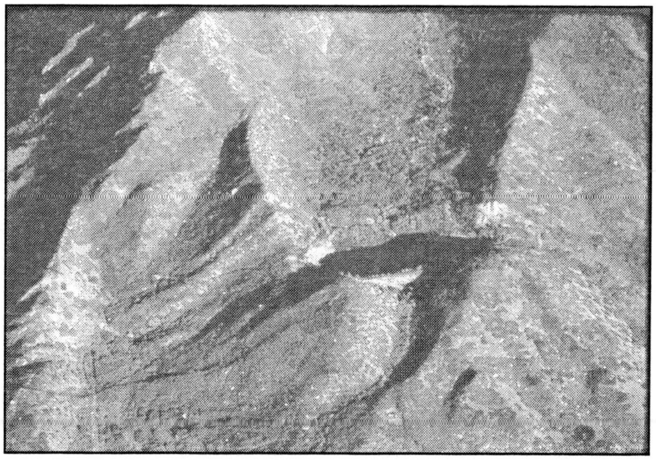

Figure 53. Faulted moraines at Bells Canyon , Utah.

of alluvial fans or Lake Bonneville sediments. These ruptures are not continuous, but extend along the Mountain front for several miles and then die out.

One of the most obvious scarps is near the mouth of Little Cottonwood Canyon. Along Wasatch Boulevard the scarp is 130 ft. (40 m) high and at the mouth of Bells Canyon the zone of deformation encompasses a series of parallel faults that displace the Lake Bonneville delta formed by Little Cottonwood Creek, in addition to the glacial moraines at Bells Canyon (Fig. 53). These sediments are 16,000 to 18,000 years old. Some of the scarps are over 130 ft. (40 m) high, indicating they were formed by repeated earthquakes, possibly as many as 10 events in the past 18,000 years. In other areas the ground ruptures extend out from the mountain front such as in Salt Lake Valley where Highland Drive parallels a fault scarp formed on the recent sediments on the valley floor.

Geologic studies indicate that at least 19 earthquakes which ruptured the ground surface have occurred in the past 6000 years. The recurrence interval of large earthquakes is once in 350 years. The last large earthquake happened 600 yrs. ago. The largest earthquake expected on the Wasatch fault has a magnitude of 7.5.

In several areas, recent excavation has removed sediment on the downthrown block to expose large areas of the fault surface, showing grooves gouged out of the rock as the fault blocks ground past each other. The excavation at Seven Peaks Resort in Provo is an excellent example.

North of Nephi near Mona, the fault cuts across ancient landslides 5,000–10,000 years old at the mouth of Pole Canyon but does not displace the youngest alluvial fan which is probably only a few hundred years old. Geologic studies show that three major earthquakes producing surface ruptures occurred in this segment of the Wasatch fault. One about 5,300 years ago, another about 3,800 years ago, and the last, approximately 1,000 years ago. This suggests that for the Nebo segment of the Wasatch fault, the average recurrence interval for major earthquakes is about 1,200 years. At the mouth of American Fork Canyon the fault displaces the Lake Bonneville delta 60 to 70 feet and at Provo the terraces of Lake Bonneville at the mouth of Rock Canyon are displaced about 60 feet.

LAKE BONNEVILLE SHORELINE–POINT OF THE MOUNTAIN

One of the most impressive shoreline features of Lake Bonneville is the terrace at the south end of Salt Lake valley. The terrace represents an ancient wave–cut platform carved on the Oquirrh Fm. and covered by a thick veneer of beach gravel, sand, and silt distributed by longshore currents, produced by waves moving from east to west. At the west end of the Traverse Range, the beach extends westward and forms a huge spit.

Striking differences in drainage patterns occur at the ancient shoreline. Above the shore, resistant bedrock has permitted only a few minor gullies to develop, but numerous deep gullies have formed on the loose, unconsolidated, Lake Bonneville sediments, all approximately the same size and spaced at strikingly regular intervals.

For 15,000 years, the shorelines of Lake Bonneville have been preserved as a ribbon–like string of beaches etched on the mountain side. But now, perhaps in a generation, one of the most visible records of the Pleistocene's 20,000–mi^2–inland sea could disappear. It is being covered by foothill subdivisions or dug up and hauled away for construction material.

UTAH VALLEY

Utah Valley shows many of the features characteristic of the Great Basin. The ranges to the west appear like warships floating on the sediments of the surrounding basins. Off to the south is West Mt., pointing northward into the southern end of Utah Lake. Lake Mountain rises above Utah Lake to the right. Both are composed mostly of the thick Oquirrh Fm. Utah Lake is the largest fresh–water lake west of the Mississippi River. Like other playa lakes it is extremely shallow, with an average depth of only 9.5 ft. Lake level is at an elevation of 4,489.3 feet, which is artificially maintained by controlling the outlet to the Jordan River that drains into Great Salt Lake.

FACETED SPURS, SPANISH PEAK

Faceted spurs are some of the most distinctive features of fault–block mountains and nowhere are they better expressed than along the Wasatch Mts. between Hobble Creek and Spanish Fork Canyon (Spanish Peak). It was here that the William Morris Davis, nearly 100 years ago (1905), first described and illustrated faceted spurs of the "Spanish Wasatch" as evidence of fault block mountains.

The spurs appear to have been cut off abruptly so that they terminate in triangular faces (facets) whose steepness ranges up to 40°. The small stream valleys between the spurs are characteristically V–shaped and become progressively wider from the base to the top of the mountain. As a result, they have been referred to as "wine–glass valleys."

Also clear in this area is that the large faces of the triangular facets between the major valleys are cut by smaller streams, which produce a series of smaller, or secondary, triangular facets. Furthermore, a flat or gently–inclined ridge extends back toward the mountain at the apex of each facet. These ridges are believed to be remnants of erosional surfaces generated during periods of tectonic stability when the mountain front receded a short distance back from the fault line. Patterns of recurrent movement along the Wasatch fault are indicated by the size and spacing of these small erosional remnants preserved at the apexes of successive generations of faceted spurs. A preliminary study suggests that movement occurred along the fault until 200 to 300 feet of vertical displacement was produced. This was followed by a period of tectonic stability. In the Spanish Peak area, eight periods of active movement were separated by significant periods of stability. Inasmuch as Spanish Peak rises 5,792 feet above the valley floor, each period of active movement produced approximately 700 feet of displacement before the periods of stability.

MT. NEBO

Mt. Nebo is the highest peak in the Wasatch Mts., reaching an elevation of 11,928 ft. (Mt. Timpanogas is 11,750 ft.). Two major structural features occur in Mt. Nebo. One is the gigantic 100–Ma–old Nebo thrust. The other is the 15–million–year–old Wasatch fault. Recent movement along the Wasatch fault has left a fresh scarp at the base of the mountains which appears as a thin gray line. Note that the scarp cuts across both alluvial fans and older debris flows which testifies to continued recurrent movement along the fault.

Three well–developed cirques are cut into the high summit of Mt. Nebo, a product of the last ice age. They are distinctive in that they are free from vegetation. These glaciers were almost "ice falls" which extended down the mountain front only a short distance to an elevation of about 9000 feet. Some moraines occur near the sides and front of the cirques.

Several large debris flows and landslides occur along the base of Mt. Nebo and form a hummocky topography which extends out into the basin. One of the oldest landslides is near the base of the mountain directly below the three cirques where shales of the Manning Canyon Fm. have slid down from the mountain front. This mass of debris is dissected into a distinctive rolling topography which is in marked contrast to the alluvial fans. Another distinctive feature of Mt. Nebo is that the base of the mountain is higher than the shoreline of Lake Bonneville so that there are no deltas, spits, and terraces along the mountain front. Instead, the basin margin is marked by numerous alluvial fans, some of which are quite large and well–developed.

Mt. Nebo–South Face

The basic elements of the structure of Mt. Nebo are well shown on the south side of Mt. Nebo. The Nebo thrust occurs about a third of the way up from the base of the mountain and marks the boundary between two contrasting rock types and landforms. Jurassic shale (Arapien) (light reddish color) form the rolling hills in the lower one–third of the mountain mass. The upper thrust sheet contains Paleozoic strata, largely Oquirrh Fm. Some of the rocks near the base of the overthrust sheet have been overturned. The Wasatch fault closely follows the base of the mountains.

LEVAN DEBRIS FLOWS

High rainfall, together with rapid melting of a thick snow pack, in the spring of 1983 combined to generate numerous landslides and debris flow along the front of the Wasatch Mts. The area is just east of the town of Levan is a good example. It shows more than fifty debris flows, all of which occurred during a period of two months. The abundant loose weathered rock and soil in the tributaries and gullies along the steep mountain front were saturated with water, and in many gullies the regolith became unstable and flowed rapidly down the hillside.

CANYON RANGE

The Canyon Range northwest of Scipio, Utah has the best exposures of a thrust fault we will see on this trip. The darker rocks on the skyline are Precambrian quartzite (Caddy Canyon quartzite) that have been thrust eastward over the lighter–colored Devonian carbonate rocks marked by two bold cliffs of light–gray dolomite. The Canyon Range thrust is part of a narrow belt of deformation (thrusts and folds) formed during the late Cretaceous (65–70 Ma) called the Sevier Orogeny. Deformation during the Sevier Orogeny trends diagonally across the state from Cache County on the north to Iron County on the south. It is part of the same thrust belt exposed in Mt. Nebo and between Ogden and Brigham City.

Continue to Nephi to spend the night.

DAY 11. EROSIONAL AND VOLCANIC FEATURES OF THE HIGH PLATEAUS

THE DOMES OF COVE FORT

Compared to the classic volcanic cinder cones in southwestern Utah, the eruptions which form the silicic domes of Cove Fort may seem uninteresting, even trivial, but this small field represents a distinctive type of silicic volcanic eruption similar to those in the Mono Crater area of California.

Black Rock Desert

Numerous basaltic lava flows and cinder cones indicate that volcanic eruptions occurred in Utah's West Desert before, during, and after Lake Bonneville. In the last million years, at least nine volcanic events have occurred during the same general time interval as those in the St. George area. But the basaltic flows in the West Desert are distinctive because of the terrain over which they flowed. In the St. George area, the lava flowed down preexisting drainage systems and ultimately evolved into elongate inverted valleys. Those in the West Desert flowed out onto the flat surface of the sedimentary basins and were free to spread out in all directions. They thus accumulated as large circular puddles of lava forming gently sloping pancake–like sheets, commonly surrounding the vents over which they originated. Their movement over the surface was somewhat like that of pouring syrup on a flat plate. Slight uplift and tilting of the basin, together with wind deflation has caused some topographic inversion, however. The older flows now form low black mesas (inverted basins), some of which are tilted to the east so that part of their original margins descend under the sediments of the basin floor. The youngest flows were extruded after Lake Bonneville and are essentially unmodified by weathering and erosion.

Six separate eruptive centers, each with a number of individual flows, are recognized in the Black Rock Desert. Two are pre–Lake Bonneville–970,000 to 250,000 years old.

THE PINK CLIFFS – CEDAR BREAKS

The upper–most step in the Grand Staircase is the spectacular Pink Cliffs, a remarkable landscape of pinnacles, and spires carved on the colorful pink and white limestone, shale, and sandstone of the Claron Fm. These rocks cap the High Plateaus of Utah and form an escarpment more than 800 ft. high that surrounds most of the southern margins of the High Plateaus of Utah. The most outstanding feature of these rocks is the incredible, colorful topography they form. Bryce Canyon National Park and Cedar Breaks National Monument are the best known and are world famous for the exotic pinnacles and spires of blazing colors of red, pink, and white.

Selective erosion produces a corrugated effect up and down on the columns giving each a destructive form. Thus, although the columns may appear to be similar, if not identical, each is unique; much like the "Terra–Cotta Warriors" of China. As a result, one seldom tires of viewing the scene. Many thousands of multicolored spires, varying in height up to several hundred feet, populate the steep slopes of the amphitheaters of the Pink Cliffs. Delicately touched with color that varies in intensity from place to place, from hour to hour, and from day to day, the Pink Cliffs are truly a masterpiece of landscape beauty.

The rate at which slope retreat occurs in the Pink Cliffs can be measured by tree–ring dating. Rapid removal of soil on the slopes expose the deep roots of many trees so that the trees appear to be standing on their toes. The process has been so rapid that many trees have had their roots undermined and have fallen. Tree–ring dating shows that the slopes of the amphitheaters are retreating at the rate of 9 to 48 inches per hundred years.

THE WEST TEMPLE AND TOWERS OF THE VIRGIN

When the first scientific reports and photographs of Zion Canyon were published, southern Utah was a little–known, wild, rugged country. Many people back east scoffed at the fantastic landsacpe and refused to believe that such a place existed. But the vertical cliffs, massive buttes and mesas, and deep slot canyon are real. The famous early American geologist, Clarence E. Dutton, wrote in 1880,

> *"There is an eloquence in their forms which stirs the imagination....Nothing can exceed the wondrous beauty of Zion."*

Many would agree. The view from behind the old Visitor's Center certainly underscores Dutton's impression. West Temple (right) and Towers of the Virgin (left) form a huge vertical wall. Here the Navajo sandstone attains its greatest thickness (2,000 ft.) and is capped by 275 ft. of Temple Cap Fm. The terrace to the left is old landslide debris. The red streaks down the face of West Temple consists of hemitite derived from red shale at the base of the Temple Cap Fm.

The Navajo sandstone was formed in a great desert and the internal structure of the rock consists of huge cross–beds formed as the sand dunes moved across the area. The sweeping giant cross–beds of the formation are now etched out in subtle relief marking the direction of the ancient prevailing wind. The cross–beds can be seen in every canyon wall and impart a unique style to the local details of the landscape in the canyon.

The geology of Zion Canyon was also affected by recent volcanic activity in which basaltic lava flowed down ancient stream valleys toward the Virgin River. The black volcanic rock and conical cinder cones stand out in sharp contrast to the white and red sheer cliffs of the Navajo and underlying strata. An excellent example is the cinder cone and lava flows just east of the town of Virgin. More extensive lava flows can be seen on the road to Kolob Reservoir where cones such as Firepit Knoll and lava flows such as those at Lava Point can be seen.

The Portal of Zion Mesa

The entrance to Zion Canyon from the west presents a spectacular scene of Mesozoic rocks and the colorful landscape eroded on them. Here the great panorama of Zion is spread out before you with all its wonder, but it takes time to absorb this visual feast.

Look to the northeast and you see the West Temple rising as a high white plateau capped by a small remnant of red sandstone of the Temple Cap Fm. If you look northward you see a different aspect of Zion. The high red cliffs are the Kolob section of the Park, an area where the Navajo sandstone has a deep red color. Eastward, the Navajo is its typical white, but black lava flows occupy several of the canyon interrupting the normally–continuous White Cliffs. This is the area of extensive, but little known, volcanic activity in the park. In the central foreground, relatively recent basaltic lava flowed down the Virgin River but has been transformed into an inverted valley with black basalt forming cliffs above the Virgin River. Weathered debris from the basalt forms the rubble that covers much of the slopes beneath the basalt cap rock.

What makes the Portals of Zion so impressive is that the steps of the Grand Staircase are telescoped together and the rocks in each step are much thicker than they are to the east. The Shinarump conglomerate–the top of the Chocolate Cliffs–is a good reference horizon in the discussion of the geology in this area. It forms a bold white cliff separating the red units of the Chinle and Moenave Fms. from the Moenkopi shale below. Directly in front of the West Temple, the horizontal red beds of the Moenave can be seen, below which is the purple shale of the Chinle, which forms a hummocky topography. Elsewhere, these rocks (Moenave and Chinle) are separated as a distinct step–the Vermillion Cliffs.

The Moenkopi Fm. is the antithesis of the Navajo. It erodes into badland topography with a fine network of closely spaced sharp ridges and gullies. The thinly bedded red and white Shnabkaib Member is especially impressive. It almost appears like a giant strip of bacon stretching across the countryside. The Shnabkaib and Upper Red Members of the Moenkopi extend across the entire central part of the area but are especially striking in the spur from the mesa in the right foreground.

Continue to Springerville to spend the night.

DAY 12. STRUCTURAL AND IGNEOUS FEATURES AT THE COLORADO PLATEAU, GREAT BASIN MARGIN

THE VOLCANIC FIELD AT HURRICANE

The little town of Hurricane, Utah is the site of some of the most interesting interplay of volcanic activity, faulting, and erosion in the state. The oldest eruptions occurred approximately 250,000 years ago. They filled the valley of the Virgin River and spread out into the lowland at the base of the Hurricane Cliffs. The towns of Hurricane and La Verkin are built on the flows. These flows dammed the Virgin River for a brief period of time, but soon a new gorge was cut through the basalt, forming the steep canyon (250 feet deep). Subsequent movement along the Hurricane fault has displaced this flow 400 feet and a narrow sliver of the basalt forms a narrow black terrace on the Hurricane Cliffs above the town of Hurricane.

At least three subsequent eruptions occurred in the area west of here, with the lava ultimately flowing into the Virgin River, each forming a temporary lava dam. The Virgin River has cut a new gorge through the lava dams. Several volcanic cones were formed during one of these more recent eruptions, as was Volcano Mt., the large cinder cone west of the town of Hurricane. To the north, salmon–pink Navajo sandstone forms high rugged ridges on either side of the Virgin anticline.

PINE VALLEY MTS.

The Pine Valley Mts. north of St. George rise 10,324 feet above sea level and are made up of "granitic" rock, believed by most geologists to be a laccolith. If so, it is one of the largest laccoliths in the world. The exposed igneous rock covers an area of more than 70 mi^2. The sedimentary cover of the intrusion has been completely removed by erosion, so that a mass of granite, more than 3,000 feet high, remains exposed on top of the mountain and along its southern face. The general form, color, and rugged massive Pine Valley Mts. stand out in stark contrast to the mesas, buttes, and cliffs of the plateau country of southern Utah.

The Pine Valley Mts. rest upon a broad, shallow, structural basin and are nearly surrounded by strata inclined toward the central part of the intrusion. Of the many layers of sedimentary rock still visible at its base, the youngest are the brightly colored layers of the Claron Fm., so vividly exposed in the Pink Cliffs. Below are gray Cretaceous sandstone and shale, which weather into slopes at the base of the mountain, but the most prominent sedimentary strata dipping northward below the Pine Valley Mts. are the Kayenta, Moenave, and Navajo Fms., which form the high, rugged, salmon–pink cliffs in the right half of this scene.

Recent studies indicate that the Pine Valley Mountain laccolith is truly remarkable, not only because of its huge size, but because of the rapid rate of intrusion and the effect it had on the landscape. The laccolith was like a huge blister arching up the overlying rock more than 3000 ft. (910 m). Field evidence and experimental studies indicate that the inflation of the laccolith occurred during a period of less than 2,500 years: and possibly less than 100 years. This is instantaneous in a geologic time frame and had a profound effect on the surrounding landscape. The rapid inflation of this huge dome resulted in a number of giant landslides that slide off the flanks of the dome and out into the flat area to the north. The larger landslides were more than 1800 ft. thick and covered an area of more than 60 mi^2 and extended more than 20 mi. from the flanks of the laccolith.

THE VIRGIN ANTICLINE

The Virgin anticline (Fig. 54) is the main structural feature in the St. George basin and many landforms in this area are related to it, either directly or indirectly. The structure extends from an area southwest of St. George northeastward to Cedar City, a distance of about 50 miles. In the St. George Basin, three structural highs along the hinge line form small domes in which Paleozoic limestones are exposed. From Toquerville northward however, the fold is displaced by the Hurricane fault so that only the east limb is exposed in the high Hurricane cliffs. The west limb on the downthrown block is covered by thick alluvium. Locally, near Kanarraville, the east limb is overturned.

The best surface expression of the Virgin anticline is between the towns of Bloomington and Hurricane, where the resistant Triassic Shinarump conglomerate and the nonresistant Moenkopi shale are exposed. The Shinarump forms a classic, sharp, resistant, hogback that rises 100 to 200 ft. above the surrounding area. The underlying, weak, Moenkopi shale is not very resistant and has been eroded into a colorful lowland.

Looking to the north, beyond the Shinarump hogback, you will note a definite northeast strike of several sandstone ridges, the highest and most rugged of which is the Navajo sandstone. These ridges are on the northwest flank of the anticline. The lower ridges in this area, beyond the large white building, are the sandstone of the famous Silver Reef mining area, a unique rich deposit of silver in a sandstone member of the Chinle Fm. This is the only known occurrence of commercial bodies of silver ore occurring in sandstone in the U.S. An estimated production of $7.5 million makes it one of Utah's great historic mining districts.

A recent construction in the Virgin anticline is Quail Creek reservoir, the large, blue, water body in the bowl–shaped basin at the nose of the north–east plunging fold where the Shinarump ridge makes a sharp 180° hair–pin turn. The construction of the

Figure 54. Virgin anticline near St. George, Utah. (Photo by K. Hamblin)

reservoir was made by sealing of this natural basin and piping in water from the Virgin River a few miles east of the town of Hurricane. Other small rectangular reservoirs can be seen to the right of Quail Creek Reservoir and are part of the regional water conservation project.

In the early 1900's, the major traps for oil exploration were anticlines, and the Virgin anticline attracted much attention from petroleum companies. A number of holes were drilled but all were dry. But oil was found in a small field north of La Verkin and was the earliest oil production in Utah. It continued to produce small quantities of oil until recently.

INVERTED VALLEYS NEAR ST. GEORGE

Spectacular changes in Utah's landscape have occurred during the last two million years as a result of volcanic eruptions. The most obvious are the numerous cinder cones which dot the landscape in the St. George basin and adjacent areas. Other expressions of recent volcanic activity are elongate lava–capped mesas characterized by their dark colored rims with their slopes below scattered with black basalt boulders. Some ridges stand more than 1,000 feet above the surrounding area. These elongate mesas are classic inverted valleys, among the finest in the world.

The long narrow lava ridges in the St. George basin are usually less than a mile wide and some are more than 10 miles long. Their sinuous pattern is almost identical to the present streams along their margins. Examination of the base of the lava flow invariably reveals remnants of river silt, sand, and gravel in ancient stream channels before the extrusion of lava.

In the St. George area, several of the inverted valleys are the same height above the surrounding area and are thus roughly the same age. Others are higher and older; some are lower and younger. Together, the system of inverted stream valleys record a series of volcanic eruptions punctuating the uplift and erosion. The most recent lava flow was extruded about 1000 years ago in the Diamond Valley area about ten miles north of St. George. South of St. George, in the Arizona Strip, a number of black lava–capped mesas form the margins of the St. George basin and constitute a much older and higher inverted topography than the black ridges of St. George.

The inverted valleys provide an important way to calculate strain rates by comparing the differences in elevation of the present stream gradient with the elevation of the paleo–stream gradient of the dateable basalt flow.

Continue to Pahrump, Nevada to spend the night.

QUATERNARY GEOLOGY OF DEATH VALLEY AND THE SIERRA NEVADA RANGE

Alan Gillespie, University of Washington, Seattle, WA
Bud Burke, Humboldt State University, Arcata, CA
Doug Clark, Western Washington University, Bellingham, WA

OVERVIEW OF DEATH VALLEY

Today, we cross the western margin of the Basin and Range geomorphic province, from Pahrump to the Sierra Nevada (Figs. 55, 56). We begin with an overview of Death Valley from the top of the Black Mts. on its eastern side and then stop at the lowest point of the basin to see evidence of Quaternary faulting and ancient lakes. Continuing westward, across Panamint Valley to Owens Valley, the westernmost basin of the geomorphic province, we will wind our way north through Owens Valley, looking at evidence of faulting and Pleistocene glaciation in the Sierra Nevada. Evening will find us in Mammoth Lakes, some 2600 m higher than the floor of Death Valley and distinctly cooler!

Historically, the origin of the name "Death Valley" has different versions, but all relate to the hardships faced by earlier settlers making the dangerous crossing (cf. Chalfant, 1933). In an era in which every arid, non–vegetated scene seemed to remind easterners (used to lush forests) of visions of Hell itself, the name seemed reasonable, and it stuck.

Perhaps the best overall reference to the geology and geologic history of Death Valley is given by Hunt and Mabey (1966).

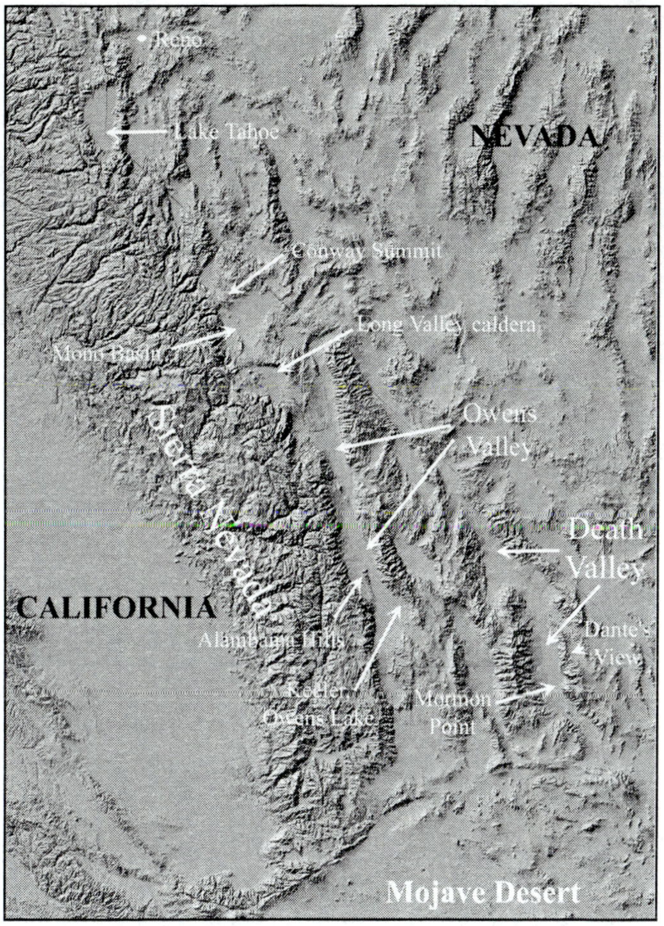

Figure 55. Shaded relief map of eastern California and western Nevada, including Death Valley, Owens Valley, and the Sierra Nevada (source: USGS digital elevation data, courtesy of Ralph Haugerud).

Blackwelder (1933) is a good early reference to the lacustrine history, amplified recently by Hooke (1999) and Knott et al. (2002). Hooke (1972) is a good reference to the detailed response of the alluvial fans to tectonic warping.

STOP 13–1. DANTE'S VIEW, DEATH VALLEY

Depart Pahrump, Nevada and drive southwest 42 km on SR–178 across the Resting Spring Range to the town of Shoshone, California. Turn right on SR 127 and drive 45 km to Death Valley Junction, turn west on SR 190, and cross the Greenwater Range to the headwaters of Furnace Creek (~30 km). Turn southeast ~20 km, climbing to the crest of the Black Mts. Stop at the end of the road, at Dante's View (1669 m).

Dante's View is a few km southeast of Badwater (STOP 13.2), which is the lowest point in North America. The view of Death Valley from here is both scenically and geologically spectacular (Fig. 57). The imposing ridges on the west side of Death Valley are the Panamint Mts. Death Valley is the lowest of more than 150 generally north–trending basins comprising the Great Basin.

The floor of Death Valley is a down–dropped fault block, tilted to the east. Death Valley near STOP 13–1 is bounded by high–angle normal faults at the edge of the Panamint Range to the west, and the Black Mts. on the east. The right–lateral strike–slip Death Valley fault zone extends up the basin on its eastern side. A second, right–lateral, strike–slip fault, the Furnace Creek Fault, enters the basin at the Furnace Creek fan and joins the fault zone there. This fault zone is over 220 km long.

Drive down the Furnace Creek fault to the floor of Death Valley.

The Death Valley fault zone is thought to be capable of generating M–7 earthquakes every 6500 years or so. The slip rate is ~3 mm/yr near its southern end, decreasing to ~0.7 mm/yr north of the Furnace Creek fan.

The rocks that comprise Death Valley and its neighboring ranges are mainly metamorphosed Precambrian–Paleozoic marine sediments, both silicate and carbonate. The ancient sea in which these rocks were deposited shallowed towards the end of the Paleozoic and withdrew during the Jurassic, at which time the Sierra Nevada to the west was an island arc of volcanoes. The metamorphic rocks have been intruded by Cretaceous and Tertiary granitic plutons.

Extension began in the Miocene, and a thick sequence of volcanic rocks developed by Pliocene time. Remnants of these rocks are now preserved on the sides of Death Valley. During the Pliocene, Death Valley began deepening and by the early Pleistocene, it had an essentially modern appearance, although it has continued to deepen throughout the Quaternary.

The subsurface topography of Death Valley is irregular. Death Valley itself is divided into three basins, and the contact between the southern two is at Mormon Point (STOP 13–3). Near Badwater, the depth to bedrock is <1 km, but south of Mormon Point, it increases to 1.5 km. The maximum depth is in the northern basin, where it may be as great as 5 km (Blakely et al., 1999).

During the Quaternary, the deepening basin filled with sediments from the intermittent lakes in its center and with bouldery alluvium shed from the bordering ranges deposited in large alluvial fans. Alluvial fans in the Great Basin have a strong debris–flow component, and some are dominated by debris–flow deposits. In Death Valley, the fan deposits show a preponderance of stratified and well–sorted fluvial sand and gravel, interspersed with lenses of cobbles and boulders.

The fans flanking Death Valley are significantly larger on the west side than on the east side of the east–dipping basin, and the locus of deposition there has shifted eastward, or down–fan, with time. This has led to fan heads consisting of ancient relict surfaces, and the distal ends of fans being covered by Holocene flood deposits. In contrast, the fans on the east side of the valley are small, and Holocene deposition occurs right at the fan heads, with few or no older surfaces. This asymmetry has been interpreted as a consequence of the eastward tilting.

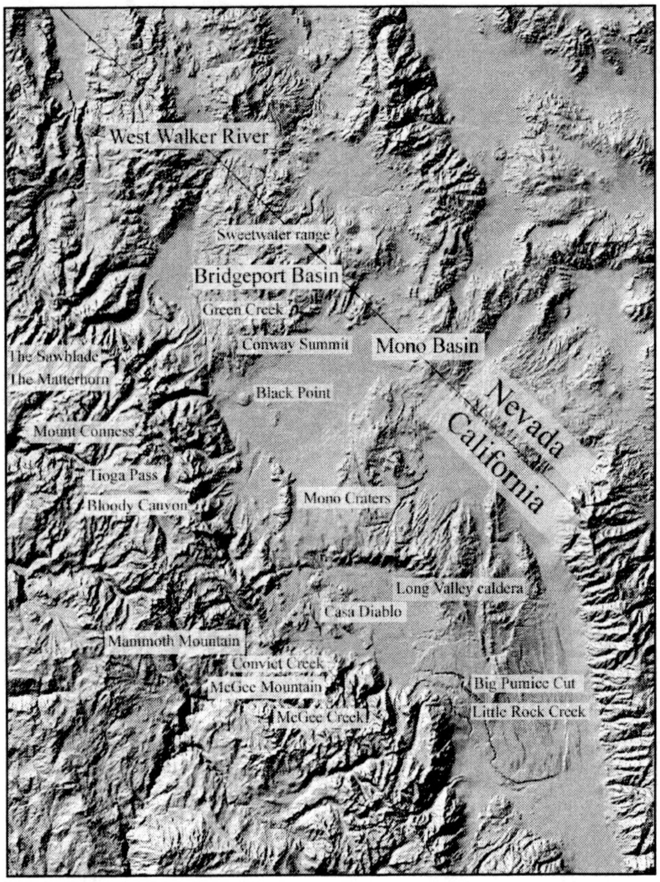

Figure 56. Map showing the locations of Stops 13–1, 2, & 3 (Dante's View, Badwater, and Mormon Point), the Death Valley Fault Zone (DVFZ), and the turtlebacks in the western escarpment of the Black Mts. (After Knott et al., 2002).

Figure 57. Perspective view of a shaded relief map calculated from 7.5' USGS digital elevation data (source: U.S. ASTER Science Team, Jet Propulsion Laboratory). Contrast–stretched natural–color Landsat TM image is "draped" over the topographic data, simulating the view north from a point over southern Death Valley and the Black Mts.

STOP 13–2. BADWATER

From Dante's View retrace the route to Furnace Creek, turn left and descend the arroyo or wash ~15 km to Death Valley National Park Headquarters on the Furnace Creek fan. Just across the range front into Death Valley, turn south and drive ~20 km along the east side of the basin to Badwater.

At –86 m, Badwater is the lowest point in North America and the 3rd lowest point in the world after the Dead Sea in Israel and Jordan and Turfan Basin in China. Here, the water table generally intersects the surface to create a shallow, perennial pond. Badwater was named for the mineralized water of this shallow lake.

Today, water from the Owens and Mojave Rivers never makes it as far as Death Valley. During Pleistocene pluvial periods, however, Owens River filled a series of lakes, the last of which (Lake Panamint) overflowed into Death Valley through a gap at the south end of the Panamint Mts. at Wingate Wash (Smith and Street–Perrott, 1983). The Mojave River reached Death Valley through Silver Lake, near the town of Baker, California. The Amargosa River carries little water.

The history of pluvial Lake Manly, which intermittently filled Death Valley, is debatable, but during the late Wisconsin, it appears to have risen at most only to *ca.* –30 m above present sea level. Sediments from this stand of Lake Manly have been studied by Anderson and Wells (1996). Earlier lakes (186,000–120,000 yr. B.P.) were deeper, with a relatively brief high stand at + 90 m (Knott et al., 2002). At this high stand, Lake Manly was 145 km long and 10–18 km across. Evidently, the effective precipitation was greater then than at any time since.

Early in the day, the temperature at the floor of Death Valley is likely near 40°C (104 F°), so we will not linger long. Ground temperatures as high as 99°C (210°F) have been reported in the valley floor sand, and rock temperatures commonly reach 70°C (158°F) in the heat of the day.

Rock surfaces here, as in other arid and semiarid areas at temperate and tropical latitudes, are blackened. From Badwater, the fans at the foot of the Panamint Mts. appear dark, even if they are composed of light–colored quartzite, dolomite or granite. The agent responsible for this darkening is "desert varnish," a thin (≤ 100 μm) coating consisting largely of illite clay and lesser amounts of manganese and iron oxide minerals. The source of this coating is apparently windblown dust, for it is compositionally similar, regardless of substrate, and develops even on quartzite, which contains no Mn and little Fe.

The processes by which this coating forms remain poorly understood. It may require microbes that live in the "varnish" to reduce and mobilize Mn and Fe from windblown dust, and somehow secrete oxidized Mn and Fe which is then incorporated into the thickening coating and cemented with an unidentified organic "glue."

STOP 13–3. MORMON POINT

Continue south on the east side of Death Valley another ~24 km to Mormon Point. Turn around and park on the side of the road. Because of the potential for high air temperatures, we will not hike far from the road and will make this a short stop.

At Mormon Point, the history of the high stand of Lake Manly and its decline is recorded by a series of distinct horizontal benches that resemble old shorelines. The highest of these is at +158 m (Trull et al., 1995); the most prominent (from the "Blackwelder stand") is at +90 m. A dozen or so recessional or other strandlines appear between the road on the valley floor and the Blackwelder strandline. Hooke (1999) relates the Blackwelder strandline to shoreline features in the Mojave Desert to the south, features that are now at much higher elevations. Hooke argues that tectonic warping and the lowering of Death Valley has caused this discrepancy. This interpretation, the timing, and the nature of the "strandlines" are still being debated.

The strandline features at Mormon Point are topographic benches and terraces developed in the Cenozoic conglomerate of the Mormon Point Fm. The conglomerate contains the ~510–ka Dibekulewe ash bed and must be younger. It underlies tufa–bearing lake sediments, and alluvial fan gravel of Hunt and Mabey's (1966) Qg2 unit that has never been submerged. McFadden et al. (1989) showed the Qg_2 fans to be late Pleistocene in age.

Uranium–series analysis of tufa at 57–73 m (Ku et al., 1998) yields ages of ~150–216 ka. Hooke and Dorn (1992) present U/Th tufa dates of ~186–128 ka for the Blackwelder shoreline. Cosmogenic exposure ages of quartzite cobbles at the Blackwelder strand (Trull et al., 1995) are 17–69 ka, and almost certainly are minima. Other minimum "dates" on the rock varnish that covers the cobbles on Mormon Point have been published. However, the techniques used to estimate ages from varnish have been discredited (Watchman, 2000; Beck et al., 1998).

Knott et al. (2002) and Hooke (1999) both conclude that the Blackwelder stand dated from MIS 6. Knott et al. further conclude that most of the recessional "strandlines" below the Blackwelder level are found where normal faults intersect the surface and are fault scarps. Therefore, the MIS 2 Lake Manley apparently was not very deep, and even its maximum shoreline was at or below the road at Mormon Point, no higher than –30 m. This is consistent with the observation that the late Pleistocene Qg2 fan was never under water. From sediment cores extracted from the lake beds at Badwater, Anderson and Wells (1996) suggest that Lake Manley had retreated to –84 m by 17,600 ^{14}C yrs B.P.

A curving, convex slope known as a "turtleback" (for its resemblance to the carapace of a tortoise) lies north of Mormon Point. Geologists view this as a low–angle normal fault plane dipping 20°N, eroded only lightly in the arid environment of Death Valley (Curry, 1938). Precambrian and Cenozoic rocks are juxtaposed by the fault, which merges with the steeper normal faults separating the Black Mts. and the Death Valley graben. These are the faults that cut through Mormon Point.

Between Badwater and Furnace Creek, fresh–appearing, late Holocene scarps on the range–front fault can be seen immediately east of the road. Also visible are textbook–quality examples of "wineglass" canyons, caused when rapid incision of streams following episodes of increased uplift along the range–front fault leads to a narrow notch cut into the floor of an ancestral V–shaped canyon.

The latest rupture along the Death Valley is interesting in that it apparently occurred during human occupation and offset encampments. Later Native Americans built campfires against the escarpment. Archaeological dates are in the range of 2000 yrs.

The road from Stovepipe Wells to Panamint Springs traverses the floor of Panamint valley. Water from Owens River had to fill the valley to a depth of 274 m (596 m) before overflow through Wingate Wash was possible. The area of Lake Panamint was 703 km^2.

STOP 13–4. OWENS VALLEY

Return to the vehicles and return to the Furnace Creek fan. Continue north 29 km, past the old Harmony Borax "works" (mines in the ancient lake beds) visible west of the road. At the junction with SR 90 across from the northern end of the Panamint Range, turn southwest and cross the floor of Death Valley 27 km to Stovepipe Wells on the way to Owens Valley. West of Panamint valley, SR 90 climbs another desert range and descends to Owens Valley at its deepest end. STOP 13–4 is at an overlook of the playa that fills the southern end of Owens valley, 82 km from Stovepipe Wells.

Like Death Valley, Owens Valley is an extensional basin bounded on both sides by normal faults, with an oblique–slip fault, the Owens valley fault zone, in the middle. Mt. Whitney, visible across Owens Valley to the northwest, is the highest point in the conterminous United States at 4418 m. The streams from the eastward–draining canyons of the Sierra Nevada south of Mono Basin combine to form Owens River, which flows down the axis of the Owens Valley. Owens River during interpluvial times terminates in Owens Lake. The elevation of the alluviated floor of Owens Valley near here is ~1067 m. During pluvial periods, Owens Lake filled to a depth of about 78 m and overflowed to the south in what is now an abandoned channel. This channel circled the mountains we have just crossed to fill Panamint Valley. Pleistocene Owens Lake had an area of 622 km^2.

Until 1913, Owens Lake was just that–a lake filled by flow from the Sierra Nevada. However, the city of Los Angeles diverted the Owens River through an aqueduct to water the run–away growth that southern California has experienced in the last century. The effect in Owens Valley was to create the playa that you now see, a source of dust that is especially hazardous because of the small (<2 μm) particle sizes. Now, millions of dollars are being spent in an effort to stabilize the playa floor. A recent ruling requires the Los Angeles Department of Water and Power to "divert" a fraction of the water from its aqueduct to Owens Lake, intended to reduce atmospheric haze. This practice may also partly restore the once–spectacular reflections of the distant Sierra Nevada in the waters of the lake as viewed from its eastern shores.

The view to the north shows the basic structure of Owens Valley: the giant east face of the Sierra Nevada, alluvium–covered bedrock benches between the Sierra Nevada escarpment and the valley axis (just east of US 395), and the flat, alluvium–filled graben between the western bajada and the Inyo Range.

As in Death Valley, the fans of the western bajada in Owens valley are much larger than those on the east. However, the cause in this instance is not tectonic, but climatic. The fans on the west are fed from the crest of the Sierra Nevada by abundant precipitation and heavy winter snows, whereas the fans on the east are fed from the Inyo Range, deep in the rain shadow of the Sierra Nevada.

The low bedrock knobs beyond Owens Lake are the Alabama Hills and are part of one of the bedrock blocks that emerges from the bajada. The normal range–front fault separating Owens Valley and the Sierra Nevada has about 2 km of vertical displacement plus another 2 km between the Alabama Hills and the floor of the graben to the east, for a total of 4 km. The floor of the graben is below sea level at its southern end, but rises to the north. Total horizontal slip across the right–lateral Owens Valley fault zone, which essentially bisects Owens Valley on its long (north–northwest) axis, is about 8–40 km, estimated by the offset of the Tinnemaha granodiorite and of dikes oblique to the fault. North of the next stop, we will see some of these dikes against the light marble bedrock of the Inyo Range, near the "town" of Swansea.

The Sierra Nevada is a major north–south mountain range that separates the internally drained basins and Desert Ranges to the east from the Central Valley to the west. It consists of a batholith of Mesozoic plutons, commonly granodiorite, intruded into marine sedimentary rocks and island–arc volcanic rocks, both now preserved as roof pendants. The range is a large crustal block that has been tilted to the west and broken by extensional basin–and–range faulting on the east. The timing of these events is debatable, but tilted Miocene lava flows west of the crest and faulted Pliocene basalt flows in the ranges east of the Sierra Nevada show that much of the early tectonic activity was Neogene, although the faulting, at least, continues today.

The basalt flows we have been driving across at the southern end of the Inyo range have been K/Ar dated at about 5–6 Ma, and they appear to predate major depression of the valley floor (Duffield and Bacon, 1981). Bacon et al. (1979) and Giovanetti (1979) suggest that the recent subsidence of Owens Valley began around that time, although a shallower north–south depression may already have developed by that time (Matthes, 1950). Long–term slip rates must be roughly 0.3–0.4 mm/yr (vertical) and at least 1.7 mm/yr (horizontal). Late–Quaternary rates for the range–front fault on the west side of Owens Valley are typically about 0.1–0.15 mm/yr; this value rises to around 1 mm/yr near Mammoth Lakes (Gillespie, 1991).

Although the Sierra Nevada extends from the Tehachapi Mts. on the northern edge of the Mojave Desert to Mt. Lassen (35–40.5° N), the crest with elevations of 4300 to 3400 m, lies between ~36 and ~38° N. The entire reach of the High Sierra was glaciated during the Pleistocene, and small glaciers occur today, sheltered in cirques high in the mountains, even though modern climatic snowline is at ~4500 m elevation at 37° N. The glaciers occurred at progressively lower elevations at northern latitudes. Even so, the average snowline, or equilibrium–line altitude (ELA), for modern glaciers rises above the topographic crest north of Bridegeport. During the Pleistocene, however, the ELA was far lower, and Pleistocene glaciers occurred on the ridges and peaks of the northern Sierra Nevada.

The glaciers of the Sierra Nevada expanded and shrank in response to changes in climate, especially changes in summer temperatures and winter precipitation. In modern times, the Sierra Nevada receives moisture mainly from winter low–pressure systems brought from the Pacific Ocean by westerlies guided by the jet stream, which descends in latitude from Canada to middle California as summer yields to winter. During the major Pleistocene glaciations, the jet stream was forced south so that precipitation was increased, and temperatures overall were reduced, although the two were not necessarily in phase.

In recent years many articles have been published that emphasize the strong synchroneity between the timing of mountain glaciations around the world and the advance and retreat of the high–latitude ice sheets to which the sawtoothed marine sediment–core $\delta^{18}O$ records are especially sensitive. Gillespie and Molnar (1995) suggested that the record left by the mountain glaciers themselves, however, tells a different story. In many ranges, including the Sierra Nevada, the glaciers appear to have advanced repeatedly to a maximum limit, regardless of the timing within the major 100–ka glacial–interglacial cycle. Whether this is due to dating inaccuracies or to some fundamental characteristic of the climate system, in the Sierra Nevada maximum depressions of ELAs during recognized glaciations seem to cluster around 1000, with MIS–4 ELAs about 100 m lower than MIS–2 ELAs (Gillespie, 1991). Gillespie and Molnar (1995) emphasized that the relative timing was not necessarily out of synch, but the ice volumes tied up the ice sheets and mountain glaciers were not strongly correlated.

By the 1970's, the modern framework of the Sierra Nevada glaciations was well understood (e.g., Wahrhaftig and Birman, 1965), but numerical age control, was weak or missing entirely. The advances of the past few decades have been focused on developing and applying rigorous dating techniques. These began with K/Ar dating of lava flows interbedded with till and ^{14}C dating of latest Pleistocene sediments in bogs that could be related stratigraphically to glacial deposits. Beginning in the 1980's, advances in accelerator mass spectrometry (AMS) led to the measurement of trace concentrations of exotic isotopes created by cosmic ray bombardment of rocks exposed at the Earth's surface, and the calculation of exposure ages of these rocks.

The glacial evidence of the eastern Sierra Nevada has been organized into allostratigraphic units assigned to glaciations via relative dating and correlation, augmented more recently by numerical dating. Much of this was due to the pioneering effort of Blackwelder, (1931) (cf. Gillespie et al., 1999) The glacial history for the Sierra Nevada is summarized in Table 1. Much of the history is anachronistic, and will no doubt be revised substantially in the future as better and more complete information becomes available.

The view west to the Sierra Nevada is especially interesting from here because the effect of repeated glaciation on the land-

TABLE 1. RECOGNIZED GLACIATIONS OF THE SIERRA NEVADA AND THEIR AGES

Glaciation	Mean age (10^3 yr)	Age references
Matthes (Little Ice Age)	0.6–0.1	Stine (1994)
Recess Peak	14.2–13.1*	
Tioga (retreat)	15–14	
Tioga (start)	21–20*	
Tioga ("Tioga 2-4")	25–16	Phillips et al. (1996)†
Tenaya (="Tioga 1"?)	31	Phillips et al. (1996)†; Benson et al. (1996)
	30*	Bursik and Gillespie (1993)
Tahoe II**	50–42	Phillips et al. (1996)
Tahoe I**	140–200	Phillips et al. (1990)
Casa Diablo	50–150	Bailey et al. (1976)
pre-Tahoe/post-Mono Basin**	190–220	Phillips et al. (1990), discussed in text
Mono Basin	80–60	Phillips et al. (1990)†
Walker Creek**	~ 550	A. Sarna-Wojcicki, (personal commun., 2002)
Sherwin	~820	Sharp (1968); Birkeland et al. (1980); Nishiizumi et al. (1989)
pre-Sherwin	~920	Birkeland et al. (1980)

* 10^3 cal yr B.P.
† Revised
** Nomenclature from this field guide.

scape is possible to assess from contrast in the appearance of the range front northwest and southwest of Owens Lake. West of STOP 13–4, a road zig–zagging up the steep escarpment can be seen. This is the road to Horseshoe Meadows, the southernmost glaciated drainage on the eastern escarpment. Cirque glaciers occurred on Olancha Peak to the south, but no large valley glaciers. Even at Horseshoe Meadows, little of the ice in Cottonwood basin spilled down Cottonwood Canyon, the steep valley below. Therefore, Cottonwood Canyon, to the left of the Horseshoe Meadows road, is essentially unmodified by glaciers, as is the face on which the road zig–zags. Likewise, the escarpment and peaks south of Cottonwood Canyon are unglaciated.

In contrast, the peaks and canyons cutting the escarpment from Mt. Whitney north were heavily glaciated, and moraines were left close to most of the fan heads (elevation ~1950 m). This close to the southern limit of glaciation, however, the sculpting is more pronounced on the southern, shaded sides of the canyons and even the peaks and ridges. Therefore, the effects of glaciation are even stronger viewed from a more northerly vantage point.

STOP 13–5. KEELER

Complete the descent into Owens Valle, and drive north along the eastern shore of the now–dry Owens Lake~10 km to the town of Keeler. East of the highway are the ruins of a mill that once processed ore from the Cerro Gordo silver mines.

This is the site of a mill that processed silver ore from the Cerro Gordo mines. Cerro Gordo is a peak on the crest of the Inyo Range just east of here. The mine was first discovered in 1865 and later worked by a few Mexican families. Soon, it was taken over by mining interests from San Francisco. Ore and supplies were moved via a toll road. Silver ingots produced at the mill originally were shipped across Owens Lake on a 26–m boat from Swansea, 5 km north of Keeler, to Cartago, on the far side of the lake. Miners are said to have lived in shacks made from silver bullion awaiting transport via the new rail system. For a time in the last third of the 19th century, this mining district was the richest source of silver in the U.S.

The pre–1872 historical shoreline of Owens Lake is very close to the modern highway. In 1872 an earthquake with a moment magnitude of 7.5–7.7 on the Owens Valley fault zone, shifted the eastern shoreline of Owens Lake hundreds of meters to the west. The dock for the steamer had to be extended, which was done with slag from the smelters.

The level of Owens lake has fluctuated with the climate, and it was probably several meters lower than in 1872 during droughts in the early part of the 2nd millennium (Stine, 1994). Remnants of Native American camps near the lowered shoreline are still visible west of the old Swansea dock. These must have been covered by several meters of water during the historic period, until drying of the lake in the early part of the 20th century exposed them.

Drive north along the eastern shore of the playa. Near Keeler, the Pleistocene lake shores have been eroded by flash floods, but near the town of Swansea, 5 km north of Keeler, the old beaches are well preserved and are visible just east of the highway. The beaches date to the latest pluvial period, probably during and just after the retreat of the Tioga glaciers at the end of the LGM.

STOP 13–6. INTERAGENCY VISITOR CENTER

From Keeler continue north and the west about 16 km, crossing the now–dry bed of Owens River to the intersection with US 395. The Visitor Center at this intersection is STOP 13–6. The Visitor Center has a bookstore with maps and books about the Death Valley, the Mojave Desert, and the Great Basin.

We have now crossed the northern end of paleo–Owens Lake. What follows is a summary of the climatic record gleaned from sediment cores, with an emphasis on the glacial history.

During times of decreased water flow, Owens Lake was closed, saline, alkaline, and biologically productive. During times of increased water flow, it was fresh and biologically unproductive. During glaciations, rock flour in the lacustrine sediment increased, total organic carbon decreased, and magnetic susceptibility peaked (Bischoff et al., 1997). Therefore, oscillations in the chemical, mineralogical, and magnetic attributes of sediment cores from Owens Lake can provide clues to the climate and glacial response in the Sierra Nevada headwaters.

The 759–ka–B.P–Bishop ash (Sarna–Wojcicki et al., 1997) was encountered at a depth of 304 m in core OL–92 (Smith and Bischoff, 1997). Several 100–ka glacial cycles are recorded in the core. However, detailed analysis has focused on the upper core and the last two glaciations.

Bischoff et al. (1997) analyzed oscillations of lake salinity and rock–flour content back to 155 ka B.P. with 1.5–ka resolution. They discovered evidence for two stades at 155–146 and 140–120 k B.P.

Dates for the MIS4–2 stades correspond well to the ^{36}Cl cosmogenic dates of Phillips et al. (1996) for boulders from Sierra Nevada moraines. According to Bischoff and Cummins (2001), rock–flour data from the core suggests that the glacial advances occurred at 49.0–45.1, 42.8–39.0, and 30.5–15.5 ka.

High–resolution magnetic susceptibility data of Benson et al., 1998a), however, indicate that the period 30.5–24.5 ka was characterized by four brief stades, and the LGM (Tioga glaciation in the Sierra Nevada) did not begin in earnest until 24.5 ka. The magnetic susceptibility data, together with total organic carbon data, showed an additional 12 stades in the interval 52.5–31.0 ka, at the base of the studied core section.

The dated moraine record in the Sierra Nevada shows much less detail and must be incomplete. Many of the stadial advances recorded in the core were likely restricted to the vicinity of the Sierra Nevada crest and left little evidence there that survived the larger Tioga glaciers.

The amount of rock flour in core OL–92 decreased abruptly at 118 ka B.P. Bischoff et al. (1997) interpreted this to show the onset of the last interglaciation. However, δ^{18}O data from the same core show a gradual transition following ~140 ka B.P., parallel to isotopic evidence from carbonate deposited by groundwater near Death Valley (Winograd et al., 1992), that suggest an earlier climatic amelioration. Neither date agrees with the 128–ka B.P. transition according to orbitally–tuned data from North Atlantic marine sediment cores (Martinson et al., 1987). Provided the age control for OL–92 is accurate, the glaciers in the Sierra appear to have remained active for thousands of years after the regional climate began changing, and even after strong changes in the global climate and shrinking of the high–latitude ice sheets.

High levels of $CaCO_3$ from OL–92 showed that the period 118–53 ka, which includes the "glacial" MIS 4 as well as the "interglacial" MIS 5, was largely dry, punctuated with brief periods of overflow. Bischoff and Cummins (2001) found evidence in the rock–flour data for four periods of glaciation between 80 and 53 ka B.P., at 78–77, 76–68, 63–61, and 58–56 ka.

The carbonate and rock–flour records are not perfectly correlated, and tell a different story. Bischoff and Cummins (2001) concluded that at least some glacial advances (for example, during MIS 5) began during dry, cold conditions, but the activity of the glaciers certainly increased during wet periods. For example, the MIS–2 LGM was a time of enhanced precipitation, as well as low temperatures.

STOP 13–7. THE 1872 FAULT SCARP

Continue north on US 395 through the town of Lone Pine. The foothills to the west are the Alabama Hills. Near their northern end about 7 km north of town, US 395 rises up a post–pluvial scarp of the Owens valley fault zone.

The Alabama Hills are a remnant of a tectonic block between the Owens Valley fault zone and the normal range–front faults at the base of the Sierra Nevada. The earthquake of 1872 ruptured strands of the Owens Valley fault zone for about 110 km, from Owens Lake north to the town of Big Pine and beyond. The damage was greatest in Lone Pine, where it destroyed most of the adobe buildings and killed 29 people. The 1872 earthquake was caused by about 1–2 m of vertical displacement, and 4–6 m of right–lateral displacement, on the Lone Pine fault, a strand of the Owens Valley fault zone near the town (Lubetkin and Clark, 1988).

Three earthquakes are recorded in fan gravel offset by the Lone Pine fault. Because the fan was deposited over tufa from Owens Lake, ^{14}C dated at ~21,000 ± 1300 ^{14}C yrs. B.P., Lubetkin and Clark (1988) were able to estimate average interseismic intervals of ~5000–10,500 years. This estimate was refined to 5800–8000 years by cosmogenic dating of boulders from the fan by Bierman et al. (1995).

STOP 13–7 is on the escarpment of the Owens Valley fault. Here, Bacon and Burke (2000) and Bacon (2003) excavated and inspected seven trenches and four exploration pits, finding direct stratigraphic evidence for only two earthquakes since ~15,000 cal yrs. B.P., including the 1872 earthquake, and indirect evidence for a third, earlier earthquake. The third event must predate ~15,000±600 cal yr B.P., the date of the oldest sediment in the trench exposures that is not deformed by faulting of that earthquake.

The escarpment at STOP 13–7 is about 10 m high, but only $1.4 + 0.4/- 0.3$ (2σ) m of that resulted from the 1872 earthquake, and only 0.9 ± 0.3 m resulted from the one before that. ^{14}C dates on charcoal and tufa, plus sedimentation rates, suggest that the penultimate event occurred at 9300 ± 300 cal yrd. B.P. The third event must have occurred between 15,000 cal yrs. B.P. and 21,000 ^{14}C yrs. B.P., and Bacon (2003) estimated that it was probably at *ca.* 18,000 cal yrs. B.P.

Using the average or maximum aspect ratios of horizontal to vertical slip of 6:1 or 10:1, Bacon (2003) calculated the 2σ ranges for the average oblique slip rate near STOP 13–7 as 0.4 –1.3 or 1.3–1.6 mm/yr, respectively (see also Lee et al., 2001).

Stratigraphic analyses of fluvio–deltaic and lacustrine sediments exposed in the trenches and along 1.5 km of Owens River bluffs reveal a relatively complete record of lake history between *ca.* 15,500 cal yrs. B.P., when overflow of Owens Lake ended, and 7000 cal yrs. B.P. The lake was low from *ca.* 15,500 to 11,000 cal yrs. B.P., and then rose from 11,000 to 8000 cal yrs. B.P. A tephra from the lake deposits suggests that the Holocene high stand occurred after ~8000 cal yrs. B.P., when the lake briefly attained an elevation of ~1135 m, ~10 m below the overflow sill. The lake–margin evidence augments the core records from the deeper part of the Pleistocene lake.

Prior to contact, Native Americans camped along the springs localized by this scarp. Clam shells from middens here have been radiocarbon dated at about 1100 ^{14}C yrs. B.P. (unpublished data, Gillespie). They must have been brought from the lake or from Owens River.

Drive north on US 395, up the axis of Owens Valley. Notice the uniform front of the Sierra Nevada and its bajada. The bajada is made from coalescing late–Pleistocene alluvial fans. Although the fans were deposited over a span of 400,000 years or more, their surface elevations vary by only a few meters. Evidently, the stability of the bajada results from a balance among aggradation, erosion of exposed clasts, reworking of fine sediments downfan, and deepening of the graben to accept the sediment.

The fans result largely from debris flows occurring with greatest frequency during the maximum advances of the glaciers, which reached down to the fan heads at elevations of about 1950 m, and during their retreat, when large volumes of water and loose, fine–grained sediment were available. Today, smaller debris flows occur occasionally during thunderstorms.

Repeated Pleistocene glaciations are responsible for the broad, sculpted, U–shaped canyons and the sharp crest of the Sierra Nevada. Nested moraines can be seen at the mouths of many canyons. Several ages have been recognized, but these are concentrated in the latest Pleistocene and represent only a fraction of the glaciations–19 from 50,000 to 23,500 yrs. B.P. have been detected in sediment cores from Owens Lake (Benson et al., 1996). Most of the moraines that are visible date from the "Tioga" glaciation of the last glacial maximum (about 21,000 yrs. B.P.). The ELA was depressed about 900 m then relative to today. The only glaciers found in the Sierra Nevada today are in protected cirques under the highest peaks. Compare the appearance of the glaciated canyons looking back to the south with the view of the exposed, sun–facing slopes to the north!

Northward through Owens Valley, basalt flows and cinder cones of the Big Pine volcanic field have been dated with K/Ar, Ar–Ar, and cosmogenic techniques, but many of the results remain unpublished. The earliest lavas remaining on the surface are about 1 Ma, and the youngest are latest Pleistocene. A single rhyolite dome was dated at about 1 Ma (Cox et al., 1963). The lavas were extruded along strands of the range front faults on both sides of Owens Valley.

STOP 13–8. FISH SPRINGS

Continue north on US 395 ~53 km, through the town of Independence and the Poverty Hills, to Fish Springs, on a strand of the Owens Valley fault zone. Between the Poverty Hills and Fish Springs, turn west onto a road that leads to Birch Creek. Stop 13–8 is on a low ridge of basalt just north of this road.

This stop is on a push–up between strands of the Owens Valley fault zone that cut through the eastern margin of the Poverty Hills. The western edge of the Poverty Hills is aligned with the Fish Springs fault (Martel, 1984), the escarpment of which can be seen west of STOP 13–8. The normal fault bisects the prominent basaltic Fish Springs cinder cone, ^{40}Ar/^{39}Ar dated at 314 ± 36 (2σ) ka B.P. Fish Springs are 3 km north of here, and arise from the continuation of the Owens Valley fault zone.

The middle Pleistocene age of the cinder cone was surprising to some, because of its generally well–preserved morphology, and some geologists have even argued that this cinder cone may have been of latest Pleistocene age. In contrast, cinder cones of comparable age in the Cima volcanic field of the Mojave Desert are more eroded, perhaps because the flux of windblown dust is less here than in the Mojave Desert, resulting is less reduction of permeability and therefore less erosion here in Owens Valley.

Careful inspection shows that the foot of the cinder cone, which should flare and gradually lose slope away from the vent, instead plunges directly into the fan gravel on either side. This suggests that it has been buried, which can be confirmed by inspection of the fault scarp, in which the distal cinders can be seen under tens of meters of fan deposits. Martel (1984) measured the vertical displacement of the cinder cone at 76 ± 8 m, which is consistent with its radiometric age.

Note that the cinder cone on the up–thrown block is black, whereas the cinder cone on the down–thrown block is red. Such color differences are common in cinder cones, but in this case the separation of the color zones by the fault plane has led some to suspect that there are two different cinder cones, and the 76–m offset is only apparent. The radiometric date was on the cinder cone east of the fault.

Martel (1984) measured offsets for different fan units distinguished by the relative weathering of exposed boulders, and Zehfuss et al. (2001) described the degree of soil development. The oldest fans were offset 31 ± 3 m, and younger fans were offset 3.3 ± 0.3 m. Martel (1984) found no evidence for horizontal slip along this strand of the right–lateral, oblique–slip, Owens Valley, fault zone. He attributed the partitioning of strain here to a difference in the azimuth of the N–S Fish Springs fault relative to the NW–trending fault zone.

Cosmogenic ^{10}Be and ^{26}Al dates for tens of boulders from the different fan units were measured by Zehfuss et al. (2001), who inferred minimum surface ages from averaged values calculated, assuming a boulder erosion rate of 2 ± 1 mm/ka. Using the revised production rate values of Larsen (1996), the fans ages calculated from the average boulder ages ranged from 171 to ~5 ka, with the older fans being offset more. The MIS–6 fan is essentially on the up–thrown block, and the late Pleistocene fans (here, 16 ± 1 to 35.7 ± 0.1 ka) are found east of the fault and north of the cinder cone. The fans south of the cinder cone are Holocene and modern.

Zehfuss et al. (2001) averaged boulder ages to estimate the landform age, but if the approach of Phillips et al. (1996) is used instead, fan aggradation periods at 185 ± 43, 36 ± 2, 17 ± 1,9 ± 2, and 3±1 ka result. Large uncertainties on the MIS–6 fan result from the 50% uncertainty in the erosion rate, but the assumed erosion rate changes the ages for the MIS–3 and younger fans little.

The Fish Springs fault broke during the historic 1872 earthquake, with vertical displacement of 1 m just north of the cinder cone. Martel (1984) and Zehfuss et al. (2001) agree that the late Quaternary slip rate has averaged 0.24 ± 0.04 mm yr^{-1}.

STOP 13–9. ANCIENT TILL AT LOWER ROCK CREEK

Continue north on US 395 ~32 km, through the towns of Big Pine and Bishop, where US 395 turns west. After another 15 km, turn left across the highway and onto a paved road at Pine Creek. Turn right Immediately onto Little Rock Creek Road and follow it past Paradise Inn and across Little Rock Creek. The stop is at a road cut just before the road climbs out of the canyon to rejoin US 395.

The drive north from of Bishop rises out of Owens Valley, up the flanks of an ancient volcano. The Bishop tuff erupted 600 km^3 759 ± 2 ka B.P.(^{40} Ar//^{39}Ar) (Bailey et al., 1976; Sarna–Wojcicki *et al.*, 1997). The tuff is normally polarized, and the eruption shortly postdated the Brunhes–Matuyama magnetic reversal. The age of the tuff is currently the best estimate of the age of the reversal.

Our stop is on the east bank of Little Rock Creek, which has eroded deeply through the tuff on the flank of the volcano to the underlying granitic rocks and sediments. Two especially interesting outcrops of till, described by Sharp (1968), are exposed in the road cut here. The uppermost is the Sherwin till, long regarded as the oldest of the commonly exposed glacial sediments in the Sierra Nevada. The matrix of pulverized granite is still gray, and granitic claslts have partially disintegrated or turned to grus.

The reason this stop is interesting is that a second, older till underlies the Sherwin till in this exposure. Sharp called this informally the "old red till" because its matrix has oxidized to a reddish hue. It is the oldest known till at the foot of the Sierra Nevada, but its age is unknown. Weathering characteristics have been used in an effort to correlate older tills here with those in the Bridgeport basin. These weathering characteristics suggest the "old red till" of Sharp is at least 100,000 years older than the Sherwin till (Birkeland et al., 1980).

Although we know today that the Sherwin Till underlies the Bishop tuff (Sharp, 1968), this was a controversial inference decades ago, and the controversy was finally resolved when the road cut we will see next was first exposed.

STOP 13–10. TYPE AREA FOR SHERWIN TILL AT THE BIG PUMICE CUT

Climb out of the canyon to the junction with US 395 and turn right. The Big Pumice cut (elevation ~2125) is at a big turnout on the right side.

125 m above STOP 13–9 is US–395. STOP 13–10 is on the west side of the highway across from the Big Pumice cut where the controversy about the Bishop tuff and the Sherwin till was finally resolved. In the road cut, the bouldery, weathered till is overlain by airfall pumice (Fig. 58) from the early stages of the eruption. The tuff caps the pumice.

The till was already heavily weathered when it was buried. Sharp (1968) and Birkeland et al. (1980) both estimated that the relative weathering was similar to that of till of the Tahoe glaciation, now thought to be between about 50 and 190 ka B.P. These age estimates are in partial agreement with cosmogenic dates that suggest the pre–eruption exposure time of the Sherwin till was about 53–67 ka B.P. (Nizhiizumi et al., 1989). Thus, the Sherwin Glaciation here was around 800–900 ka.

STOP 13–11. LONG VALLEY CALDERA

During a lull in traffic, cross US 395 to the northbound lanes and continue back across the caldera rim. A short distance farther is a turnoff to the right and STOP 13–11.

We have crossed into the southern end of Long Valley, the 32 x 19 x 3 km caldera of the volcano from which the Bishop tuff was ejected. The rhyolite tuff is the extrusive equivalent of batholithic rocks of the Sierra Nevada.

Figure 58. Sherwin till overlain by Bishop tuff at Sherwin Grade.

Crowley Lake is a reservoir, but the caldera was filled by a lake until it drained ≤100,000 years ago. Hills north of Crowley Lake are a resurgent dome, emplaced about 600 ka (Bailey et al., 1976). Since about 100 ka, most volcanic activity has been localized northwest of Long Valley. As recently as 1980, however, seismic evidence suggests that magma is intruding as dikes in the ancient caldera.

West of Crowley Lake, the moraines of McGee Creek were deposited by the Tahoe, Tenaya, and Tioga glaciations across the wall of the Long Valley caldera and faulted. Perhaps the most dramatic indication of this is not the scarps themselves (visible here where the right–lateral moraine crosses the range front), but the lowering and partial burial of successively older moraine crests.

The youngest moraines here, the Tioga moraines, form the crest of the composite moraine, and the Tenaya moraine buries the older, reddish Tahoe moraine that crops out near the base. The crest of the Tenaya moraine is ~45 m lower than the crest of the Tioga moraine, and the highest remnants of the Tahoe moraine are ~80 m lower still.

The trace of the Hilton Creek fault is visible near where the right–lateral moraines cross the range front. However, it is made less conspicuous by three breaches in the moraine crest caused by erosion from ice–marginal streams at the time of maximum advance of the glaciers. Fans have formed below the moraine where the suddenly reduced gradient caused deposition of debris carried by these streams.

Soil development and rock weathering are best assessed in locations such as these "lateral" fans, which are essentially unmodified by debris–flow deposition after the glaciers have retreated. In contrast, the axial fans have been subject to occasional debris flows down the modern streams throughout the Holocene.

The large, broad peak just north of the McGee Creek and its moraines is McGee Mt. Its shoulder of metamorphic rocks and the remnant of a 2.6 Ma basalt flow (Dalrymple, 1963, 1964) are mantled by large, erratic, granitic boulders derived from the headwaters of McGee Creek. Evidently, the time of transport was long ago (1.5 Ma) (Huber, 1981) because McGee Creek has cut down 1200 m since ice flow across the summit of McGee Mt. from the headwaters of McGee Creek was last possible (e.g., Putnam, 1962).

The origin of the boulders has long been debated. Are they glacial erratics or boulders carried by a mudflow? This controversy has not been renewed for years, but the consensus was that they were glacial, and the eroded deposit was once the McGee till.

STOP 13–12. MAMMOTH LAKES

From Crowley Lake, continue northwest on US 395 past the airport to a major intersection. Drive west on SR–203 to the town of Mammoth Lakes, where we will spend the night.

North of Mammoth Lakes large, well–preserved, sequences of moraines have been deposited on the west side of Long Valley along Hilton Creek, McGee Creek, and Convict Creek, mostly during the last 200,000 years. The moraines have been faulted, especially at McGee Creek.

Convict Creek hosts one of the most–complete, best–preserved moraine sequences in the Sierra Nevada (Fig. 59). This is partly because the glaciers changed course, flowing northward some time before the early part of the Tioga glaciation, perhaps around 30 ka B.P..

DAY 14. GLACIATION IN LONG VALLEY AND MONO BASIN

Today we will consider evidence of Pleistocene glaciations near Mammoth Lakes and in Mono Basin. We will discuss interpretations of cosmogenic 36Cl dates from moraines in Mono Basin and will conclude the field trip with a stop at a scenic overview of Mono Basin and the central Sierra Nevada from Conway Summit on US–395.

STOP 14–1. MINARET SUMMIT

From Mammoth Lakes, continue west a few km on SR 203. A short distance beyond a ski area at Mammoth Mt., the highway crosses the divide into the valley of the San Joaquin River, which drains to the Central Valley and the Pacific Ocean. STOP 14–1 is on this divide.

At 2797 m, Minaret Summit is a low point on the crest of the Sierra Nevada. The Minarets are a ridge of metavolcanic rock to the west across the glaciated valley of the San Joaquin River. This stop is mainly to get an idea of the spectacular scenery of the high Sierra. The long, latite–covered ridge we are on leads to Deadman Pass. East–draining Deadman Creek was named for an early

Figure 59. Convict Creek moraines. (Photo by K. Hamblin)

mining fraud that went awry. A miner lured investors from Bishop to this area with false claims of ledges of gold and then shot the investors and blamed the affair on the Native Americans. In this instance, the local sheriff exposed the false claims and justice was served.

Geological interest in Deadman Pass a few decades ago focused on a bouldery deposit thought to be an ancient till cropping out from under the latite. In the end, the granitic boulders were concluded to have been transported by a lahar. The peak to the south is Mammoth Mt., a popular ski resort. It is a volcano constructed on the rim of the ancient caldera after about 185 ka B.P. (Bailey et al., 1976).

STOP 14–2. CASA DIABLO

From Minaret Summit return to US 395. Just west of the intersection, stop at the side of the road and walk a north short distance to a dry channel cut through basalt.

Long after the resurgent dome was emplaced in Long Valley, basalt and andesite were erupted into the low "moat" around the dome. Casa Diablo, named for the hot springs and associated sulfurous fumes found there, was the site of some of these flows, which are interbedded with till or outwash from glaciers of the Mammoth embayment. Some of the granitic clasts in the till are coated with welded tuff composed of curved glass shards and angular, lithic fragments of pumice, rhyolite and glass (Ellis, 2001).

Curry (1968) originally named the Casa Diablo till and assigned an age of ~400 ka using limiting K/Ar dates on the basalt flows that stratigraphically bracket it. However, new K/Ar analysis provided improved dates of 126 ± 25 and 62 ± 13 ka (Bailey et al., 1976), which are now accepted as the age range for the till. On the basis of relative dating techniques, Burke and Birkeland (1979) suggested that the Casa Diablo till was correlative with the older Tahoe and/or Mono Basin tills to the north, a correlation not universally accepted. The Casa Diablo till, however, may be correlative with the younger Tahoe till or Mono Basin till at Bloody Canyon. Ages on the welded tuff will help constrain this age assignment and may be available by the time of the field trip.

STOP 14–3. PANNUM CRATER

Turn northwest on US 395, passing from Long Valley caldera to the flanks of the ancient volcano. Near Deadman Summit (2450 m) is a Highway rest stop. Beyond, US 395 descends past the June Lake turnoff and into Mono Basin, 32 km from Casa Diablo. Mono Craters, a series of rhyolite domes, begin near June Lake and terminate near the southern shore of Mono Lake. STOP 14–3 is near the northernmost volcano, Pannum Crater, just north of the Benton turnoff.

Mono Craters (Fig. 60) are the surface expression of silicic intrustions along faults linking Mono Basin and Long Valley caldera to the south. Pannum Crater, a rhyolite volcano with an obsidian dome in the crater, was last active around A.D. 1100. About 1.3 km^3 of ash was erupted in the final event about 605 ± 20 ^{14}C yrs. B.P. (Sieh and Bursik, 1986). This is a little more than the volume of the 1980 Mt. St. Helens eruption, and a tiny fraction of the volume of the Bishop ash.

Mono Basin is a north–dipping tectonic depression that apparently post–dates the Sherwin till. Mono Lake is alkaline, but supports tiny brine shrimp on which gulls feed. Tufa growing on telephone cables installed in the 1940's has an apparent radiocarbon age of several thousand years.

The lake today is ~225 m lower than at its last Pleistocene high stand, an artificially low level due to the ongoing diversion since 1941 of water into the Los Angeles aqueduct from the streams that feed Mono Lake. However, Stine (1994) showed that even this low level is similar to low stands that occurred during severe droughts between A.D. 900 and 1350.

Across the lake to the north is Black Point (2120 m), a sublacustrine basaltic volcano that was active during the late Pleistocene pluvial high stand of Mono Lake around 13,600 ^{14}C yrs. B.P. (Benson et al., 1998b). Only the very top of the volcano was above the lake level then. The high stand peaked at ~13,000 ^{14}C yrs. B.P. (Lajoie et al., 1982). During previous high stands Mono Lake may have overflowed to the east, becoming the source of Owens River at those times (Reheis et al., 2002).

STOP 14–4. WALKER CREEK MORAINES

The western side of Mono Basin has several moraine sequences, preserved together with the ancient shorelines of Mono Lake. The shorelines occur inside the moraines, suggesting to Lajoie (1968) that the pluvial high stands lagged glaciations by about 1000–2000 years, according to evidence from sediment cores (Benson et al., 1998b). Mono Lake was probably high during glacial advances at Lee Vining Creek (Fig. 61) where Tioga outwash formed a delta in the ancient lake.

The canyon of Walker Creek (Fig. 62) is known as Bloody Canyon, either because of the red metamorphic rocks of the roof pendant or because mules making the crossing of the Sierra Nevada at Mono Pass at the head of Bloody Canyon fell from the steep cliffs. During Pleistocene glaciations, ice flowed eastward from the Tuolumne ice sheet across Mono Pass. Bloody Canyon hosts one of the best–studied and most–dated alpine moraine sequences in the western U. S.

More than a dozen moraine crests in two geographic groups are preserved at Bloody Canyon. The older group clusters around Sawmill Canyon, a dry canyon south of Walker Creek that was beheaded by younger glaciers whose right–lateral moraines parallel the present course of Walker Creek.

The youngest, most obvious, moraines of Sawmill Canyon were defined as the type Mono Basin moraines by Sharp and Birman (1963) and were thought to predate the Tahoe glaciation. The moraine that truncates Sawmill Canyon (Fig. 63) was mapped as Tahoe by Sharp and Birman, but field inspection suggests that it is a composite "onionskin" moraine of at least two

Figure 60. Mono Craters. (Photo by D.A. Rahm, courtesy of EPIC)

Figure 62. Bloody Canyon moraines. (Photo by D.A. Rahm, courtesy of EPIC)

Figure 61. Lee Vining Creek moraines. (Photo by Austin Post)

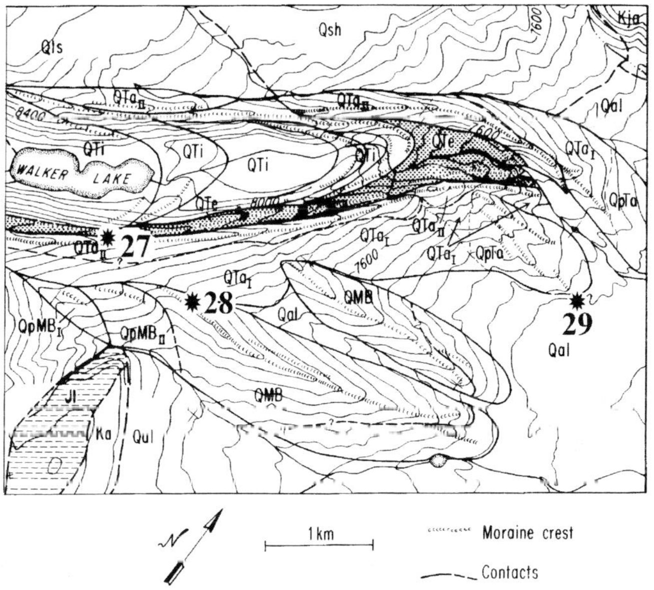

Figure 63. Map of the Bloody Canyon moraines (after Gillespie, 1982). Jl, Ka, and Kja are plutonic rocks. Qal is Quaternary alluvium and Qls shows colluvial deposits. Qsh is Sherwin till of Sharp and Birman (1963). $QpMB_I$ and $QpMB_{II}$ are pre–Mono Basin moraines; QMB is Mono Basin moraines. QpTa is the oldest set of moraines (pre–Tahoe) along Walker Creek; QTa_I and QTa_{II} are the older and younger Tahoe moraines of Gillespie (1982). QTe (shaded for clarity) is the Tenaya moraine, and Qti are the undifferentiated Tioga moraines. Topographic contour interval is 24 m (80 ft).

ages. The older moraine crops out about a third of the way down the southern flank and has a reddish soil with a few exposed boulders. The younger moraine supports a less–developed tan soil, and the crest is occupied with somewhat fresh–appearing granitic boulders. These moraines clearly cross–cut and, therefore, post–date the Mono Basin moraines.

The Tahoe moraine overlies another, older sequence of moraines exposed at the end of the right–lateral complex, at and a few tens of meters above a high shoreline of Mono Lake (2145 m). These moraines are also considered to post–date the Mono Basin moraines. Nested inside the big Tahoe moraine are younger, latest Pleistocene, Tenaya and Tioga moraines.

Key elements of this interpretive history, developed from field relations and relative dating, have been challenged by cosmogenic ^{36}Cl ages that appear to show that the old, largely buried moraines at the terminus of the right–lateral Tahoe composite are older than the boulders of the Mono Basin moraines. This interpretation requires that the glaciers changed course more than once.

Phillips et al. (1990, 1996) measured a large number of ^{36}Cl cosmogenic ages from boulders on the Bloody Canyon moraines. The data in Figure 64 and show a great deal of scatter, attributed by Phillips et al. (1996) to exposure of fresh boulders as soil is slowly stripped from the moraines. Most researchers agree that the dates for the younger Tahoe and Tioga moraines are fairly accurate, probably within a few thousand years at least. These edited ages suggest that the younger Tahoe moraine dates from just 49 ± 2 ka. Its relative youth may partly explain the small differences found by Burke and Birkeland (1979) between it and the Tioga moraines. However, Phillips et al. (1990) concluded that the Mono Basin moraines at 115 ka post–dated, not pre–dated, the pre–Tahoe moraines we are standing on (145 and 200 ka B.P.). The ages and even the relative age relations among the older moraines of Bloody Canyon remain unresolved today.

STOP 14–4. CONWAY SUMMIT

Continue north on US 395. Past the town of Lee Vining, the highway ascends highlands on the north side of Mono Basin. STOP 14–4 is at the top of this grade.

We have driven up the north side of Mono Basin to Conway Summit, which affords one of the better views of the Sierra Nevada. Two features are of interest at this stop. The road cut to the north and east of US 395 consists of Sherwin till and outwash. The Sherwin glaciers are thought to have been much larger than their successors due to the discovery of putative Sherwin till far outside the younger moraines. Ideas to explain this have included decreased rain–shadow effects–the Sierras may have been lower 800,000 years ago, but, in fact, no one knows. Some of the supposed Sherwin till may not be till, but rather, outwash or debris–flow deposits. This may be the case at Conway Summit.

Conway Summit (2480 m) is the dramatic northern edge of Mono Basin. The main points of interest here are (1) the outcrops of Sherwin till, and (2) the excellent view of the old upland surface, now gently dipping to the north, and the history of the subsidence of Mono Basin and the alkaline lake it contains.

Sherwin till. The road cut north of US 395 at Conway Summit reveals a thick deposit of diamicton blanketing deeply weathered granitic bedrock. Sharp (1972) considered these deposits to be Sherwin till and/or outwash. The till here, seemingly far from the late Pleistocene moraines, has provided some of the fuel for arguments that the Sherwin glaciers were of much greater extent than their successors. However, many of the outcrops identified as Sherwin till may, in reality, be outwash or till reworked by debris flows, in which case, the Sherwin ELAs need not be much lower than those of the late Pleistocene glaciers.

On the other hand, if the diamicton consists of primary till, then the Sherwin or other pre–Tahoe glaciers must have been more extensive than their successors, although here in Bridgeport Basin an additional ELA depression of as little as ~100 m might explain the difference in extent.

Upland surface. Conway Summit was once a on a large upland surface, now disrupted by the down–faulting of Mono Basin. Remnants of this old surface can be seen near Lundy Canyon, the northernmost canyon draining the Sierra Nevada into Mono Basin. Remnant benches of this old surface project onto the north–dipping piedmont at Virginia and Green Creeks, and dip north at about the same angle.

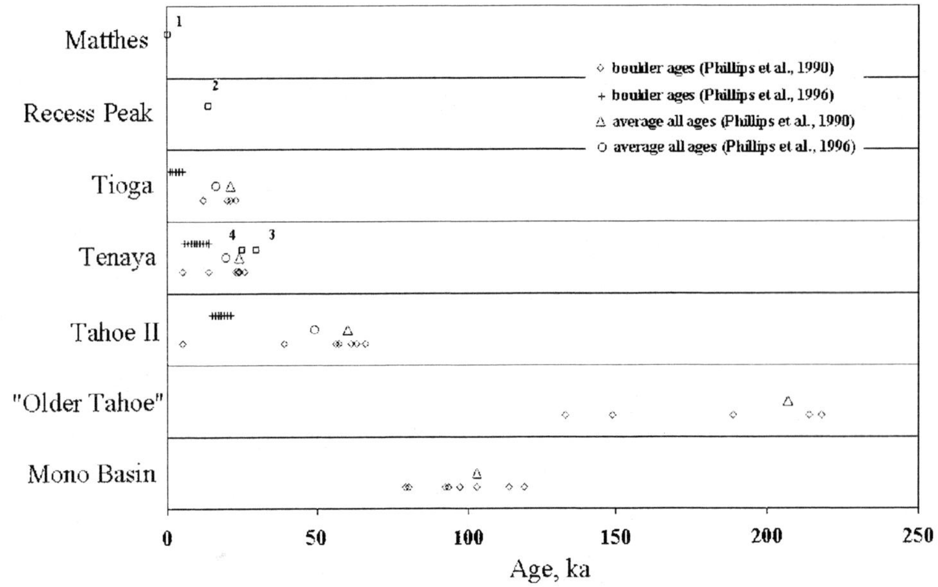

Figure 64. ^{36}Cl dates for Bloody Canyon moraines (Phillips et al., 1990, 1996) with the following independent age control:
(1) 100–600 cal yrs. B.P.
(2) 13,100–14,200 cal yr B.P. (11,200–12,200 ^{14}C yr).
(3) 30,000 cal yr B.P. (Bursik and Gillespie, 1993).
(4) >25,000 on ^{10}Be and ^{26}Al (Gillespie, unpublished data).

Subsidence of Mono Basin. Evidence that subsidence of Mono Basin had begun by Sherwin time is found north of Walker Creek on the northwestern slope of the basin where Sharp and Birman (1963) mapped eroded Sherwin till down to ~2300 m, 140 m below Conway Summit. Evidently, the ancient broad upland between Owens Valley and Carson Valley was disrupted from the south because Mono Basin appears to have been deepening since Sherwin time, especially relative to Bridgeport Basin.

If Mono Basin began subsiding before 850 ka, the long–term rate has been >0.6 mm/yr, based on the elevation difference between Conway Summit and the surface of Mono Lake (currently 1948 m). Normal faulting rates between the basin and the Sierra Nevada block to the west, based on ages of offset moraines, are 1–2 mm/yr (Bursik and Sieh, 1989), greater than faulting rates in Owens Valley, to the south.

Paoha Island in Mono Lake is composed of lake sediments that were uplifted by magma rising below the surface, and Negit Island is built from dark dacite (Lajoie, 1968). Stine (1987) reported that Paoha Island emerged from Mono Lake prior to A.D. 1690.

Drive north ~206 km on US 395 to Reno and the end of the field trip.

REFERENCES

Anderson, D. E., and Wells, S. G., 1996, Latest Quaternary lacustrine events of lake Manly: A record from ten shallow cors along a 75 km transect in southern Death Valley basin: Geological Society of America, Abstracts with Program, v. 28, p. 83.

Armstrong, J. E., Crandell, D. R., Easterbrook, D. J., and Noble, J. B., 1965, Late Pleistocene stratigraphy and chronology in southwestern British Columbia and northwestern Washington: Geological Society of America Bulletin, v. 76, p. 321–330.

Bacon, C. R., Giovanetti, D. M., Duffield, W. A., and Dalrymple, G. B., 1979, New constraints on the age of the Coso Formation, Inyo County, California: Geological Society of America, Abstracts with Program, v. 11, p. 67.

Bacon, S. N., 2003, Latest Quaternary Fluvio–Deltaic and Lacustrine Stratigraphy in Owens Valley and Paleoseismology on the Owens Valley Fault near Lone Pine, eastern California [Master's thesis]: Arcata, CA, Humboldt State University.

Bacon, S. N., and Burke, R. M., 2000, Preliminary paleoseismic investigations on the Owens Valley fault zone and latest Quaternary stratigraphy in Owens Valley near Lone Pine, eastern California: EOS Transactions, American Geophysical Union, v. 81, p. F856–F857.

Bailey, R. A., Dalrymple, G. B., and Lanphere, M. A., 1976, Volcanism, structure, and geochronology of Long Valley caldera, Mono County, California: Journal of Geophysical Research, v. 81, p. 725–744.

Beanland, S., and Clark, M. M., 1994, The Owens valley fault zone, eastern California, and surface faulting associated with the 1872 earthquake: U. S. Geological Survey Bulletin 1982, 29 p.

Beck, W., Donohue, D. J., Jull, A. J. T., Burr, G., Broecker, W. S., Bonani, G., Hajdas, I., and Malotki, E., 1998, Ambiguities in direct dating of rock surfaces using radiocarbon measurements: Science, v. 280, p. 2132–2135.

Benedict, J. B., 1973, Chronology of cirque glaciation, Colorado Front Range: Quaternary Research, v. 3, p. 584–599.

Benson, L. V., Burdett, J. W., Kashgarian, M., Lund, S. P., Phillips, F. M., and Rye, R. O., 1996, Climatic and hydrologic oscillations in Owens Lake basin and adjacent Sierra Nevada, California: Science, v. 274, p. 746–751.

Benson, L. V., Lund, S. P., Burdett, J. W., Kashgarian, M., Rose, T. P., Smoot, J., P., and Scwarz, M., 1998b, Correlation of Late–Pleistocene lake–level oscillations in Mono Lake, California, with North Atlantic climate events: Quaternary Research, v. 49, p. 1–10.

Benson, L. V., May, H. M., Antweiler, R. C., and Brinton, T. I., 1998a, Continuous lake–sediment records of glaciation in the Sierra Nevada between 52,600 and 12,500 ^{14}C yr B.P.: Quaternary Research, v. 50, p. 113–127.

Berger, G. W., and Easterbrook, D.J., 1993, Thermoluminescence dating tests for lacustrine, glaciomarine, and floodplain sediments from western Washington and British Columbia: Canadian Journal of Earth Science, v. 30, p. 1815–1828

Bierman, P. B., Gillespie, A. R., and Caffee, M., 1995, Cosmogenic age estimates for earthquake recurrence intervals and debris–flow fan deposition, Owens Valley, California: Science, v. 270, p. 447–450.

Birkeland, P. W., Burke, R. M., and Walker, A. L., 1980, Soils and sub–surface weathering features of Sherwin and pre–Sherwin glacial deposits, eastern Sierra Nevada, California: Geological Society of America Bulletin, v. 91, p. 238–244.

Bischoff, J. L., and Cummins, K., 2001, Wisconsin glaciation of the Sierra Nevada 79,000–15,000 yr B.P. as recorded by rock flour in sediments of Owens Lake, California: Quaternary Research, v. 55, p. 14–24.

Bischoff, J. L., Meisling, K. M., Fitts, J. P., and Fitzpatrick, J. A., 1997, Climatic oscillations 10,000–155,000 yr B.P. at Owens Lake, California reflected in glacial rock flour abundance and lake salinity in Core OL–92: Quaternary Research, v. 48, p. 313–325.

Blackwelder, E., 1915, Post–Cretaceous history of the mountains of central western Wyoming: Journal of Geology, v. 23, p. 97–117, p. 193–217, p. 307–340.

Blackwelder, E., 1931, Pleistocene glaciation in the Sierra Nevada and Basin Ranges: Geological Society of America Bulletin, v. 42, p. 865–922.

Blackwelder, E., 1933, Lake Manly, an extinct lake of Death Valley: Geographical Review, v. 23, p. 464–471.

Blakely, R. J., Morin, R. L., and Faunt, C., 1999, Looking beneath the surface, a three–dimensional geophysical view of the Death Valley region, California and Nevada: U.S. Geological Survey Open–File Report 99–153, p. 47–51.

Blunt, D., Easterbrook, D. J., and Rutter, N. A., 1987, Chronology of Pleistocene sediments in the Puget Lowland, Washington: Washington Division of Geology and Earth Resources Bulletin 77, p. 321–353.

Booth, D.B. 1986, The formation of ice–marginal embankments into iced–dammed lakes in the eastern Puget Lowland, Washington, U.S.A., during the late Pleistocene: Boreas, v.15, p.247–263.

Booth, D. B., 1987, Timing and processes of deglaciation along the southern margin of the Cordilleran ice sheet: *in* Ruddiman, W. F., and Wright, H. E., Jr., eds., North America and adjacent oceans during the last deglaciation, Geological Society of America, The Geology of North America, K–3, p. 71–90.

Bretz, J H., 1913, Glaciation of the Puget Sound region: Washington Division of Mines and Geology Bulletin, v. 8, p.244

Bretz, J H., 1969, The Lake Missoula floods and the Channeled Scabland: Journal of Geology, v. 77, p.505–543.

Burke, R. M., and Birkeland, P. W., 1979, Re–evaluation of multiparameter relative dating techniques and their application to the glacial sequence along the eastern escarpment of the Sierra Nevada, California: Quaternary Research, v. 11, p. 21–51.

Burke, R. M., and Birkeland, P. W., 1983, Holocene glaciation in the mountain ranges of the Western United States: *in* Wright, H. E., ed., Late–Quaternary environments of the United States, University of Minnesota, Minneapolis, MN, p. 3–11.

Bursik, M. I., and Gillespie, A. R., 1993, Late Pleistocene glaciation of Mono Basin, California: Quaternary Research, v. 39, p. 24–35.

Bursik, M.I., and Sieh, K., 1989, Range front faulting and volcanism in the Mono Basin, California: Journal of Geophysical Research, v. 94, p. 15,587–15,609.

Carlstad, C.A., 1992. Late Pleistocene deglaciation history at Point Partridge, central Whidbey Island, Washington: MS thesis, Western Washington University, Bellingham, WA, 140 p.

Cary, A. S., and Carlston, C. W., 1937, Notes on Vashon stage glaciation of the South Fork of the Skykomish River valley, Washington: Northwest Science, v. 11, p. 61–62.

Chadwick, O. A., Hall, R. D., and Phillips, F. M., 1997, Chronology of Pleisto cene glacial advances in the Central Rocky Mountains: Geological Society of America Bulletin, v. 109, p. 1443–1452.

Chalfant, W. A., 1933, The Story of Inyo: Chalfant Press, revised edition, Bishop, CA, 430 p.

Christiansen, R.L., 2001, The Quaternary and Pliocene Yellowstone Plateau Volcanic Field of Wyoming, Idaho, and Montana: U.S. Geological Survey Professional Paper 729–G, 145 p.

Clague, J.J., Saunders, I.R., and Roberts, M.C., 1988, Ice–free conditions in southwestern British Columbia at 16,000 years B.P.: Canadian Journal of Earth Sciences, v. 25, p. 938–941.

Colman, S.M., and Pierce, K.L., 1981, Weathering rinds on andesitic and basaltic stones as a Quaternary age indicator, Western United States: U.S. Geological Survey Professional Paper 1210, 56 p.

Colman, S. M., and Pierce, K. L., 1986, Glacial sequence near McCall, Idaho: Weathering rinds, soil development, and other relative–age criteria: Quaternary Research, v. 25, p. 25–42.

Cox, A. Doell, R. R., and Dalrymple, G. B., 1963, Geomagnetic polarity epochs and Pleistocene geochronometry: Nature, v. 198, p. 1049–1051.

Curry, H. D., 1938, "Turtleback" fault surfaces in Death Valley, California: Geological Society of America Bulletin, v. 49, p. 1875.

Curry, R. R., 1968, Quaternary climatic and glacial history of the central Sierra Nevada, California: PhD thesis, University of California, Berkeley, CA. 204 p.

Dahms, D. E., 1991, Eolian sedimentation and soil development on moraine catenas of the Wind River Mountains, West–Central Wyoming: PhD thesis, University of Kansas, Lawrence, KS.

Dahms, D. E., 1992, Buried soils on Late Quaternary moraines of the Wind River Mountains, Wyoming: Geological Society of America, Abstracts with Program, Annual Meeting, Cincinnati, p. A348.

Dahms, D. E., 1994, Mid–Holocene erosion of soil catenas on moraines near the type–Pinedale till, Wind River Range, Wyoming: Quaternary Research, v. 42, p. 41–48.

Dahms, D. E., 2002, Glacial stratigraphy of Stough Creek basin, Wind River Range, Wyoming: Geomorphology, v. 42, p. 59–83.

Dahms, D. E., Shroba, R. R., Gosse, J. C., Hall, R. D., Sorenson, C. J., and Reheis, M. C., 1997, Relation between soil age and silicate weathering rates determined from the chemical evolution of a glacial chronosequence: Comment and Reply, Geology, p. 381–382.

Dalrymple, G. B., 1963, Potassium–Argon ages of some Cenozoic volcanic rocks of the Sierra Nevada, California: Geological Society of America Bulletin, v. 74, p 379–390.

Dalrymple, G. B., 1964, Cenozoic chronology of the Sierra Nevada, California: University of California Publications in Geological Science, v. 47, 41 p.

Davis, P. T., Gosse, J. C., Reasoner, M., and Zielinski, G., 1999, Late Pleistocene and Holocene alpine glacier climate proxies from proglacial lacustrine sediments, Wyoming and Colorado, U.S.A. Evidence for a Younger Dryas event: Eos, Transactions, American Geophysical Union, v. 80, p. 321.

Deeter, J. D., 1979, Quaternary geology and stratigraphy of Kitsap County, Washington: MS thesis, Western Washington University, Bellingham, WA, 175 p.

Duffield, W. A., and Bacon, C. R., 1981, Geologic map of the Coso volcanic field and adjacent areas, Inyo County, California: U. S. Geological Survey I Map–1200.

Dzurisin, Daniel and Yamashita, K.M., 1987, Vertical surface displacements at Yellowstone Caldera, Wyoming, 1976–1986: Journal of Geophysical Research, v. 92, p. 13,753–13,766.

Dzurisin, D., Savage, J.C., and Fournier, R.O., 1990, Recent crustal subsidence at Yellowstone caldera, Wyoming: Bulletin of Volcanology, v. 52, p. 247–270.

Easterbrook, D.J., 1963, Late Pleistocene glacial events and relative sea–level changes in the northern Puget Lowland, Washington: Geological Society of America Bulletin, v. 74, p. 1465–1483.

Easterbrook, D. J., 1966a, Radiocarbon chronology of late Pleistocene deposits in northwest Washington: Science, v. 152, p. 764–767.

Easterbrook, D. J., 1966b, Glaciomarine environments and the Fraser Glaciation in northwest Washington: Guidebook for First Pacific Coast Friends of the Pleistocene Field Conference, 52 p.

Easterbrook, D. J., 1968, Pleistocene stratigraphy of Island County, Washington: Washington Division of Water Resources Bulletin, v. 25, p. 1–34.

Easterbrook, D.J., 1969, Pleistocene chronology of the Puget Lowland and San Juan Islands, Washington: Geological Society of America Bulletin, v. 80, p. 2273–2286.

Easterbrook, D. J., 1976. Quaternary geology of the Pacific Northwest: *in* Mahaney, W. C., ed., Quaternary stratigraphy of North America, Dowden, Hutchinson, and Ross, Stroudsburg, PA, p. 441–462.

Easterbrook, D. J., 1979, The last glaciation of northwest Washington: *in* Armentrout, J. M., Cole, M. R., and Terbest, H., eds., Cenozoic paleogeography of the western United States, Pacific Coast Paleogeography Symposium 3, Society of Economic Paleontologists and Mineralogists, Los Angeles, CA, p.177–189.

Easterbrook, D. J., Briggs, N.D., Westgate, J.A., and Gorton, M., 1981, Age of the Salmon Springs Glaciation in Washington: Geology, v. 9, p. 87-93.

Easterbrook, D. J., 1986, Stratigraphy and chronology of Quaternary deposits of the Puget Lowland and Olympic Mountains of Washington and the Cascade Mountains of Washington and Oregon: *in* Sibrava, V., Bowen, D. Q., and Richmond, G. M., Quaternary glaciations in the Northern Hemisphere, Quaternary Science Reviews, v. 5, p. 145–169.

Easterbrook, 1992, Advance and retreat of Cordilleran ice sheets in Washington, USA: Geographie Physique et Quaternaire, v. 46, p. 51–68.

Easterbrook, D.J., 1994, Stratigraphy and chronology of early to late Pleistocene glacial and interglacial sediments in the Puget Lowland, Washington: *in* Swanson, D.A. and Haugerud, R.A., eds., Geologic field trips in the Pacific Northwest, Geological Society of America, p. 1J23–38.

Easterbrook, D.J., 2003, Comment on the paper "Determinaton of 36Cl production rates from the well–dated deglaciation surfaces of Whidbey and Fidalgo Islands, Washingoton" by T.W. Swanson and M.C. Caffee: Quaternary Research, v. 59, p.132–134.

Easterbrook, D. J.; and Kovanen, D.J., 1998, 'Pre–Younger Dryas resurgence of the southwestern margin of the Cordilleran Ice Sheet, British Columbia, Canada: Comments, Boreas, v. 27, p. 225–230.

Easterbrook, D. J., and Rutter, N. A., 1981, Amino acid ages of Pleistocene glacial and interglacial sediments in western Washington: Geological Society of America, Abstracts with Program, v. 13, p. 444.

Easterbrook, D. J., and Rutter, N. A., 1982, Amino acid analysis of wood and shells in development of chronology and correlation of Pleistocene sediments in the Puget Lowland, Washington: Geological Society of America, Abstracts with Programs, v. 14, p. 480.

Easterbrook, D. J., Blunt, D. J., and Rutter, N. A., 1982, Pleistocene glacial and interglacial chronology in western Washington: Geological Society of America, Abstracts with Program, v. 14, p. 161.

Easterbrook, D. J.; Crandell, D. R.; Leopold, E. B., 1967, Pre–Olympia stratigraphy and chronology in the central Puget Lowland, Washington: Geological Society America Bulletin, v. 78, p. 13–20.

Easterbrook, D.J., Rahm, D.A., 1970, Landforms of Washington: Union Printing Co., 156 pp.

Ellis, C. C., 2001, Determining the age of the Casa Diablo till through correlation of ash flow tuffs: Senior thesis, Humboldt State University, Arcata, CA, 16 p.

Gillespie, A. R., 1982, Quarternary Glaciation and Tectonism in the Southeastern Sierra Nevada, Inyo County, California: Ph.D. thesis Pasadena, CA., Caltech, 695 p.

Gillespie, A. R., 1991, Testing a new climatic interpretation for the Tahoe glaciation: *in* Hall, C. A. Jr., Doyle–Jones, V., and Widawski, B., eds., Natural history of eastern California and high–altitude research, Proceedings of the White Mountain Research Station Symposium, v. 3, p. 383–398.

Gillespie, A. R., and Molnar, P., 1995, Asynchronism of maximum advances of mountain and continental glaciations: Reviews of Geophysics, v. 33, p. 311–364.

Gillespie, A. R., Burke, R. M., and Clark, M. M., 1999, Eliot Blackwelder and the Alpine Glaciations of the Sierra Nevada: *in* Moores, E. M., Sloan, D., Stout, D. L. eds., Classic California Cordilleran centennial concepts: A View from California, Geological Society of America, v. C5, p. 443–452.

Giovanetti, D. M., 1979, Volcanism and sedimentation associated with the formation of southern Owens Valley, California: Geological Society of America, Abstracts with Program, v. 11, p. 79.

Gosse, J. C., 1994, "Alpine glacial history reconstruction: 1. Application of the cosmogenic ^{10}Be exposure age method to determine the glacial chronology of the Wind River Mountains, Wyoming, USA; 2. Relative dating of Quaternary deposits in the Rio Atuel Valley, Mendoza, Argentina: PhD thesis, Lehigh University, Bethlehem, PA.

Gosse, J.C., Phillips, F.M., 2001, Terrestrial in situ cosmogenic nuclides; theory and application: Quaternary Science Reviews, v.20, p.1475–1560.

Gosse, J. C., and Stone, J. O., 2001, Terrestrial cosmogenic nuclide methods passing milestones toward palaeo–altimetry: EOS, American Geophysical Union, v. 82, p. 86–89.

Gosse, J.C., Klein, J., Evenson, E.B., Lawn, B., Middleton, R., 1995a, Beryllium–10 dating of the duration and retreat of the last Pinedale glacial sequence: Science, v.268, p.1329–1333.

Gosse, J.C., Evenson, E.B., Klein, J., Lawn, B., Middleton, R., 1995b, Precise cosmogenic super ^{10}Be measurements in western North America; support for a global Younger Dryas cooling event: Geology, v.23, p. 877–880.

Gosse, J. C., Davis, P. T., Burr, G., Jull, A. T. J., Bozarth, S., Sorenson, C., Klein, J., Lawn, B., and Dahms, D., 1999, Late Pleistocene/Holocene glacial and paleoclimate history of Titcomb Basin, Wind River Range, Wyoming based on lake sediments and cosmogenic nuclide dating: Geological Society of America, Abstracts with Program v. 31, p. A–56.

Gosse, J. C., Klein, J., Davis, P. T., Evenson, E. B., Jull, A. J. T., and Burr, G. submitted, Cosmogenic 10Be and 26Al production rates at mid–latitude high altitude sites for exposures of 10 to 15 kyr: Earth and Planetary Science Letters.

Hall, K.J., 1989, Dating late Quaternary deposits in the alpine of western Wyoming by soil stratigraphy: Geomorphology, v. 2, p. 405–410.

Hall, R. D., 1990, Evidence for glaciation during the early Wisconsin at Fremont Lake, Wind River Range, Wyoming: Geological Society of America, Abstracts with Program, Annual Meeting, Dallas, TX, p. A175.

Hall, R. D., and Horn, L. L., 1993, Rates of hornblende etching in soils in glacial deposits of the northern Rocky Mountains Wyoming–Montana, U.S.A.: Influence of climate and characteristics of the parent material: Chemical Geology, v. 105, p. 17–29.

Hall, R. D., and Shroba, R. R., 1993, Soils developed in the glacial deposits of the type areas of the Pinedale and Bull Lake glaciations, Wind River Range, Wyoming, U.S.A: Arctic and Alpine Research, v. 25, p. 368–373.

Hallet, B. and Putkonen, J., 1994, Surface dating of dynamic landforms: young boulders on aging moraines: Science, v. 265, p. 937–940.

Hancock, G. S., Anderson, R. S., Chadwick, O. A., and Finkel, R. C., 1999, Dating fluvial terraces with ^{10}Be and ^{26}Al profiles: application to the Wind River, Wyoming: Geomorphology, v. 27, p. 41–60.

Hansen, B. S., and Easterbrook, D. J., 1974, Stratigraphy and palynology of late Quaternary sediments in the Puget Lowland, Washington: Geological Society of America Bulletin, v. 85, p. 587–602.

Hansen, H. P., and Mackin, J. H., 1949, A pre–Wisconsin forest succession in the Puget Lowland, Washington: American Journal of Science, v. 47, p. 833–855.

Heusser, C. J.; Heusser, L. E., 1981, Palynology and paleotemperature analysis of the Whidbey Formation, Puget Lowland, Washington: Canadian Journal of Earth Science, v. 18, p. 136–149.

Hooke, R. LeB., 1972, Geomorphic evidence for late–Wisconsin and Holocene tectonic deformation, Death Valley, California: Geological Society of America Bulletin, v. 83, p. 2073–2098.

Hooke, R. LeB., 1999, Lake Manly ? shorelines in the eastern Mojave Desert, California: Quaternary Research, v. 52, p. 328–336.

Hooke, R. LeB., and Dorn, R. I., 1992, Segmentation of alluvial fans in Death Valley, California: New insights from surface exposure dating and laboratory modeling: Earth Surface Processes and Landforms, v. 17, p. 557–574.

Huber, N. K., 1981, Amount and timing of Cenozoic uplift and tilt of the central Sierra Nevada, California–Evidence from the upper San Joaquin River: U. S. Geological Survey Professional Paper 1197, 28 p.

Hunt, C. B., and Maybe, D. R., 1966, Stratigraphy and structure, Death Valley, California: U. S. Geological Survey Professonal Paper 494–A.

Ives, P. C., Levin, B., Robinson, R. D., and Rubin, M., 1964, U.S. Geological Survey Radiocarbon dates VII: Radiocarbon, v. 6, p. 59–60.

Izett, G.A., Pierce, K.L., Naeser, N.D., Jaworowski, C., 1992, Isotopic dating of Lava Creek B Tephra in terrace deposits along the Wind River, Wyoming; implications for post 0.6 ma uplift of the Yellowstone hotspot: Geological Society of America, Abstracts with Programs, v. 24, p. 102.

Johnson, S.Y., Dadisman, S.V., Mosher, D.C., Blakely, R.J., and Childs, J.R., 2002, Active tectonics of the Devils Mountain fault and related structures, northern Puget Lowland and eastern Strait of Juan de Fuca region, Pacific Northwest: U.S. Geological Survey Professional Paper 1643.

Johnson, S.Y., Potter, C.J., Armentrout, J.M., Miller, J.J., Finn, C., and Weaver, C.S., 1996, The southern Whidbey Island fault, an active structure in the Puget Lowland, Washington: Geological Society of America Bulletin, v. 108, p. 334–354.

Kienast, S. S. and McKay, J. L., 2001, Sea surface temperatures in the subarctic Northeast Pacific reflect millennial–scale climate oscillations during the last 16 kyrs: Journal of Geophysical Research Letters, v. 28, p. 1563–1566.

Knott, F. R., Tinsley, J. C. III, Wells, S. G., 2002, Are benches at Mormon Point, Death Valley, California, scarps or strandlines?: Quaternary Research, v. 58, p. 352–360.

Kovanen, D. J., 2002, Morphologic and stratigraphic evidence for Allerød and Younger Dryas age glacier fluctuations of the Cordilleran Ice Sheet, British Columbia, Canada and Northwest Washingont, U.S.A.: Boreas, v.31, p. 163–184.

Kovanen, D. J., and Easterbrook, D. J., 1996, Extensive readvance of Late Pleistocene Y.D.? Alpine Glaciers in the Nooksack River Valley, 10,000 to 12,000 years ago, following retreat of the Cordilleran Ice Sheet, North Cascades, Washington: Friends of the Pleistocene, Pacific Coast Cell Field Trip Guidebook, 74 p.

Kovanen, D. J. and Easterbrook, D. J. 2001, Late Pleistocene, post–Vashon, alpine glaciation of the Nooksack drainage, North Cascades, Washington: Geological Society of America Bulletin, v. 113, p. 274–288.

Kovanen, D. J., and Easterbrook, D. J., 2002a, Paleodeviations of radiocarbon marine reservoir values for the NE Pacific: Geology, v. 30, p. 243–246.

Kovanen, D. J., and Easterbrook, D. J., 2002b, Timing and extent of Allerod and Younger Dryas age ca. 12,500–10,000 ^{14}C yr B.P. oscillation of the Cordilleran Ice Sheet in the Fraser Lowland, western North America: Quaternary Research, v. 57, p. 208–224.

Ku, T., Luo, S., Lowenstein, T. K., Li, J., and Spencer, R. J., 1998, U–series chronology of lacustrine deposits in Death Valley, California: Quaternary Research, v. 50, p. 261–275.

Lajoie, K. R., 1968, Quaternary stratigraphy and geologic history of Mono Basin, eastern California: PhD thesis, University of California, Berkeley, CA, 217 p.

Lajoie, K. R., Robinson, S. W., Forester, R. M., and Bradbury, J. P., 1982, Rapid climatic cycles recorded in closed–basin lakes: 7th Biennial Conference, American Quaternary Association, Abstracts.

Lal, D., 1991, Cosmic ray labeling of erosion surfaces: in situ nuclide production rates and erosion models: Earth and Planetary Science Letters, v. 104, p. 424–439.

Larsen, P. L., 1996, In situ production rates of cosmogenic ^{10}Be and ^{26}Al determined form the terminal moraine of the Laurentide Ice Sheet, north–central New Jersey: MS thesis, University of Vermont, Burlington, VT, 129 p.

Lee, J., Spencer, J., and Owen, L., 2001, Holocene slip rates along the Owens valley Fault, California: Implications for the recent evolution of the eastern California shear zone: Geology, v. 29, p. 819–822.

Leopold, L. B., Nickmann, R., Hedges, J. L., and Ertel, J. R., 1982, Pollen and lignin records of late Quaternary vegetation, Lake Washington: Science, v. 218, p. 1305–1307.

Licciardi, J. M., Clark, P. U., Brook, E. J., Pierce, K. L., Kurz, M. D., Elmore, D., and Sharma, P., 2001, Cosmogenic ^{3}He and ^{10}Be chronologies of the late Pinedale northern Yellowstone ice cap, Montana, USA: Geology, v. 29, p. 1095–1098.

Locke, W.W., and Meyer, G.A., 1994, A 12,000–year record of vertical deformation across the Yellowstone Caldera margin: The shorelines of Yellowstone Lake: Journal of Geophysical Research, v. 99, p. 20,079–20,094.

Lubetkin, L. K. C., and Clark, M. M., 1988, Late Quaternary activity along the Lone Pine fault, eastern California: Geological Society of America Bulletin, v. 100, p. 755–766.

Mackin, J. H., 1941, Glacial geology of the Snoqualmie–Cedar area, Washington: Journal of Geology, v. 49, p. 449–481.

Mahaney, W. C., 1978, Late Quaternary stratigraphy in the Wind River Mountains, western Wyoming: *in* Mahaney, W. C., Quaternary Soils, Geo Abstracts, Norich, UK, p. 223–264.

Mahaney, W.C., 1988, Holocene glacier fluctuations: *in* Davis, P T. and Osborn, G., eds., Quaternary Science Reviews, v.7, p. 211–225.

Martel, S. J., 1984, Late Quaternary activity on the Fish Springs fault, Owens Valley fault zone, California: MS thesis, Stanford University, Stanford, CA, 112 p.

Martinson, D. G., Pisia, N. G., Hayes, J. D., Imbrie, J., Moore, T. C., Jr., and Shackleton, N. J., 1987, Age dating and the orbital theory of the ice ages; development of a high–resolution 0 to 300,000–year chronostratigraphy: Quaternary Research, v. 27, p. 1–29.

Matthes, F. E., Fryxell, F., ed., 1950, Sequoia National Park: Berkeley, University of California Press, 136 p.

McFadden, L. D., Ritter, J. B., and Wells, S. G., 1989, Use of multiparameter relative–age methods for age estimation and correlation of alluvial fan surfaces on a desert piedmont, eastern Mojave Desert, California: Quaternary Research, v. 32, p. 276–290.

Meyer, G.A., and Locke, W.W., 1986, Origin and deformation of Holocene shoreline terraces, Yellowstone Lake, Wyoming: Geology, v. 14, p. 699–702.

Miller, C. D., and Birkeland, P. W., 1974, Probably pre–Neoglacial age of the type Temple Lake moraine, Wyoming: Discussion and additional relative–age data: Arctic and Alpine Research, v. 6, p. 301–316.

Mullineaux, D. R., Waldron, H. H., and Rubin, M., 1965, Stratigraphy and chronology of late interglacial and early Vashon time in the Seattle area, Washington: U. S. Geological Survey Bulletin 1194–O, 10 p.

Nelson, A. R., Millington, A. C., Andrews, J. T., and Nichols, H. 1979, Radiocarbon–dated upper Pleistocene glacial sequence, Fraser Valley, Colorado Front Range: Geology, v. 7, p. 410–414.

Nishiizumi, K., Winterer, E. L., Kohl, C. P., Klein, J., Middleton, R., Lal, D., and Arnold, J.R., 1989, Cosmic ray production rates of ^{10}Be and ^{26}Al in quartz from glacially polished rocks: Journal of Geophysical Research, v. 94, p. 17907–17915.

Obradovich, J.D., and Izett, G.A., 1991, ^{40}Ar/^{39}Ar ages of upper Cenozoic Yellowstone Group tuffs: Geological Society of America, Abstracts with Program, v. 23, p. 84.

Page, B. M., 1939, Multiple glaciation in the Leavenworth area, Washington: Journal of Geology, v. 47, p. 785–815.

Pelton, J.R., and Smith, R.B., 1979, Recent crustal uplift in Yellowstone National Park: Science, v. 206, p. 1179–1182.

Phillips, F. M., Zreda, M. G., Benson, L. V., Plummer, M. A., Elmore, D., Sharma, P., 1996, Chronology for fluctuations in late Pleistocene Sierra Nevada glaciers: Science, v. 274, no. 5288, p. 749–751.

Phillips, F. M., Zreda, M. G., Smith, S. S., Elmore, D., Kubik, P. W., Sharma, P., 1990, Cosmogenic chlorine-36 chronology for glacial deposits at Bloody Canyon, eastern Sierra Nevada: Science, v. 248, p. 1529–1532.

Phillips, F.M., Zreda, M.G., Gosse, J.C., Klein, Jeffrey, Evenson, E.B., Hall, R.D., Chadwick, O.A., and Sharma, Pankaj, 1997, Cosmogenic 36Cl and 10Be ages of Quaternary glacial and fluvial deposits of the Wind River Range, Wyoming: Geological Society of America Bulletin, v. 109, p. 1453–1463.

Pierce, K. L. 1979. History and dynamics of glaciation in the Northern Yellowstone National Park area: U.S.G.S Professional Paper 729F, 91 p.

Pierce, K.L., 2003, Pleistocene Glaciations of the Rocky Mountains: *in* Gillespie, A., Porter, S.C., and Atwater, B. eds., Quaternary glacial ages in the United States–advances since 1965 in the Study of the Quaternary Period: Elsevier.

Pierce, K.L., Adams, K.D., and Sturchio, N.C., 1991, Geologic setting of the Corwin Springs Known Geothermal Resource area–Mammoth Hot Springs area in and adjacent to Yellowstone National Park: *in* Sorey, M.L. ed., Effects of potential geothermal development in the Corwin Springs Known Geothermal Resource Area, Montana, on the thermal features of Yellowstone National Park, U.S. Geological Survey Water Resources Investigations Report 91–4052, p. C1–C37.

Pierce, K.L., Obradovich, J.D., and Friedman, Irving, 1976, Obsidian hydration dating and correlation of Bull Lake and Pinedale glaciations near West Yellowstone, Montana: Geological Society of America Bulletin, v. 87, p. 703–710.

Pierce, K.L, Cannon, K.P., Meyer, G.A., Trebesch, M.J., and Watts, R., 2002, Post–glacial inflation–deflation cycles, tilting, and faulting in the Yellowstone caldera based on Yellowstone Lake shorelines, U.S. Geological Survey Open–File Report 02–0142, 30 p.

Porter, S.C., 1976, Pleistocene glaciation on the southern part of the North Cascade Range, Washington: Geological Society of America Bulletin, v.87, p.61–75.

Porter, S. C. and Swanson, T. W., 1998, Radiocarbon age constraints on rates of advance and retreat of the Puget lobe of the Cordilleran Ice Sheet, during the last glaciation: Quaternary Research, v. 50, p. 205–213.

Porter, S. C., Pierce, K. L., and Hamilton, T. D., 1983, Late Wisconsin mountain glaciation in the western United States: *in* Wright, H. E. ed., Late–Quaternary Environments of the United States, University of Minnesota Press, Minneapolis, MN, p. 71–111.

Putnam, W. C., 1962, Late Cenozoic geology of McGee Mountain, Mono County, California: University of California Publications in Geological Science, v. 40, p. 181–218.

Reheis, M. C., Stine, S., and Sarna–Wojcicki, A. M., 2002, Drainage reversals in Mono Basin during the late Pliocene and Pleistocene: Geological Society of America Bulletin, v. 114, p. 991–1006.

Richmond, G. M., 1957, Three pre–Wisconsin glacial stages in the Rocky Mountain region: Geological Society of America Bulletin, v. 68, p. 239–262.

Richmond, G. M., 1964, Three pre–Bull Lake tills in the Wind River Mountains, Wyoming–A reinterpretation: U.S. Geological Survey Professional Paper 501D, p. D104–D109.

Richmond, G.M., 1965, Glaciation of the Rocky Mountains: *in* Wright Jr HE, Frey DG eds., The Quaternary of the United States, Princeton University Press, Princeton, NJ p. 217–230.

Richmond, G.M., 1973, Geologic map of the Fremont Lake South Quadrangle, Sublette County, Wyoming: U. S. Geological Survey Geologic Quadrangle Map GQ–1138.

Richmond, G.M., 1976a, Pleistocene stratigraphy and chronology in the mountains of western Wyoming: *in* Mahaney, W. C., ed., Quaternary stratigraphy of North America, p. 353–379

Richmond, G.M., 1976b, Surficial geologic history of the Canyon Village Quadrangle, Yellowstone National Park, Wyoming, for use with map I–652 U. S. Geological Survey Bulletin, Report: B 1427, 35 p.

Richmond, G.M., 1977, Surficial geologic map of the Canyon Village quadrangle, Yellowstone National Park, Wyoming: U.S. Geological Survey Misc. Geol. Inv. Map I–652.

Richmond, G. M., 1986, Stratigraphy and correlation of glacial deposits of the Rocky Mountains, the Colorado Plateau, and the ranges of the Great Basin: *in* Sibrava, V. ed., Quaternary glaciations in the Northern Hemisphere, Quaternary Science Reviews, p. 99–127.

Richmond, G. M., 1987, Type Pinedale till in the Fremont Lake area, Wind River Range, Wyoming: Geological Society of America Centennial Field Guide—Rocky Mountain Section p. 201–204.

Richmond, G. M., and Fullerton, D. S., 1986, Summation of Quaternary glaciations in the United States of America: Quaternary Science Reviews, v. 5, p. 183–196.

Richmond, G.M., Murphy, J.F., 1965, Geologic map of the Bull Lake East Quadrangle, Fremont County, Wyoming: U. S. Geological Survey Geologic Quadrangle Map GQ–0431.

Rigg, G. B., and Gould, H. R., 1957, Age of Glacier Peak eruption and chronology of postglacial peat deposits in Washington and surrounding areas: American Journal of Science, v. 255, p. 341–363.

Rosholt, J.N., Bush, C.A., Shroba, R.R., Pierce, K L, Richmond, G.M., 1985, Uranium–trend dating and calibrations for Quaternary sediments: U. S. Geological Survey Open File Report OF 85–0299, 48 p.

Sarna–Wojcicki, A. M., Meyer, C. E., and Wan, E., 1997, Age and correlation of tephra layers, position of the Matuyama–Brunhes chron boundry, and effects of Bishop ash eruption on Owens Lake, as determined from drill hole OL, southeast California: *in* Smith, G. I., and Bischoff, J. L., eds., An 800,000–year paleoclimatic record from Core OL–92, Owens Lake, Southeast California: Geological Society of America Special Paper 317, p. 79–90.

Sharp, R. P., 1968, Sherwin Till–Bishop Tuff relationship, Sierra Nevada, California: Geological Society of America Bulletin, v. 79, p. 351–364.

Sharp, R. P., 1972, Pleistocene glaciation, Bridgeport Basin: Geological Society of America Bulletin, v. 83, p. 2233–2260.

Sharp, R. P., and Birman, J. H., 1963, Additions to the classical sequence of Pleistocene glaciations, Sierra Nevada, California: Geological Society of America Bulletin, v. 74, p. 1079–1086.

Siegfried, R. T., 1978, Stratigraphy and chronology of raised marine terraces, Bay View Ridge, Skagit County, Washington: MS thesis, Western Washington University, Bellingham, WA, 52 p.

Sieh, K. E., and Bursik, M. I., 1986, Most recent eruption of the Mono Craters, eastern central California: Journal of Geophysical Research, v. 91, p. 12,539–12,571.

Small, E. E., Anderson, R. S., Repka, J. L., and Finkel, R., 1997, Erosion rates of alpine bedrock summit surfaces deduced from in situ ^{10}Be and ^{26}Al: Earth and Planetary Science Letters, v. 150, p. 413–425.

Smith, G. I.,, and Bischoff, J. L., 1997, Core OL–92 from Owens Lake: Project rationale, geologic setting, drilling procedures, and summary: *in* Smith, G. I., and Bischoff, J., eds., An 800,000–year geologic and climatic record from Owens Lake, California: Core OL–92: Geological Society of America Special Paper 317, p. 1–8.

Smith, G. I., and Street–Perrott, F. A., 1983, Pluvial lakes in the western United States: *in* Wright, H. E., Jr., ed., Late Quaternary Environments of the United States, University of Minnesota Press, Minneapolis, MN, p. 190–211.

Smith, R.B., Pierce, K.L., and Wold, R.J., 1993, Seismic surveys and Quaternary history of Jackson Lake, Wyoming: *in* Snoke, A.W., Steidtmann, J.R., and Roberts, S.M., Geology of Wyoming, Wyoming Geological Survey Memoir No. 5, p. 668–693.

Sorenson, C. J., 1987, Soils Map of the Fremont Lake South Quadrangle, Sublette County, Wyoming: U. S. Geological Survey Miscellaneous Investigation Series I–1800.

Stine, S., 1987, Mono Lake: the past 4,000 years: PhD thesis, University of California, Berkeley, CA, 615 p.

Stine, S., 1994, Extreme and persistent drought in California and Patagonia during mediaeval time: Nature, v. 369, p. 546–549.

Stoffel, K. L., 1980, Stratigraphy of pre–Vashon Quaternary sediments applied to the evaluation of a proposed major tectonic structure in Island County, Washington: U. S. Geological Survey Open File Report, 81–292, 161 p.

Stone, J. O., 1999, A consistent Be–10 production rate in quartz – muons and altitude scaling: AMS–8 Proceedings, Abstract Volume, Vienna, Austria.

Stone, J. O., 2000, Air pressure and cosmogenic isotope production: Journal of Geophysical Research, B 105, p. 23,753–23,759.

Sturchio, N. C., Pierce, K. L., Murrell, M. T., and Sorey, M. L., 1994, Uranium–series ages of travertines and timing of the last glaciation in the northern Yellowstone area, Wyoming–Montana, U.S.A: Quaternary Research, v. 41, p. 265–277.

Trull, T. W., Brown, E. T., Marty, B., Raisbeck, G. M., and Yiou, F., 1995, Cosmogenic ^{10}Be and ^{3}He accumulation in Pleistocene beach terraces in Death Valley, California, U.S.A.: Implications for cosmic–ray exposure dating of young surfaces in hot climates: Chemical Geology, v. 119, p. 191–207.

Warhaftig, C., and Birman, J. H., 1965, The Quaternary of the Pacific mountain system in California: in Wright, H. E., and Frey, D. G., eds., The Quaternary of the United States, Princeton University Press, p. 299–340.

Watchman, A., 2000, A review of the history of dating rock varnishes: Earth Science Reviews, v. 49, p. 261–277.

Weber, S., 2001, Late Pleistocene littoral deposits in the Deming sand at Bellingham Bay, Washington, and their implications for relative sea level changes: MS thesis, Western Washington University, Bellingham, WA, 158 p.

Westgate, J. A.; Easterbrook, D. J.; Naeser, N. D.; Carson, R. J., 1987, Lake Tapps tephra—An early Pleistocene stratigraphic marker in the Puget Lowland, Washington: Quaternary Research, v. 28, p. 340-355.

Wicks, Charles Jr., Thatcher, Wayne, and Dzurisin, Daniel, 1998, Migration of fluids beneath Yellowstone Caldera inferred from satellite radar interferometry: Science, v. 282, p. 458–462.

Winograd, I. J., Coplen, T. B., Landwehr, J. M., Riggs, A. C., Ludwig, K. L., Szabo, B. J., Kolesar, P. T., and Revesz, K. M., 1992, Continuous 500,000–year climate record from vein calcite in Devils Hole, Nevada: Science, v. 258, p. 255–260.

Zehfuss, P. H., Bierman, P. R., Gillespie, A. R., Burke, R. M., and Caffee, M. W., 2001, Slip rates on the Fish Springs fault, Owens Valley, California, deduced from cosmogenic ^{10}Be and ^{26}Al and soil development on fan surfaces: Geological Society of America Bulletin, v. 113, p. 241–255.

Zielinski, G A, Davis, P.T , 1987, Probable revised late–Pleistocene and Holocene chronology of alpine glaciation in the Wind River Range, Wyoming, U.S.A: Geographie Physique et Quaternaire, v 41, p 397–401

Zimmerman, S. G., Evenson, E. B., Gosse, J. C., and Erskine, C. P., 1994, Extensive boulder erosion resulting from a range fire on the type–Pinedale moraines, Fremont Lake, Wyoming: Quaternary Research, v. 42, p. 255–265.

Quaternary stratigraphy, geomorphology, soils, and alpine archaeology in an alpine-to-plains transect, Colorado Front Range

Authors and Leaders: **David P. Dethier,** *Department of Geosciences, Williams College, Williamstown, MA 01267*
James B. Benedict, *Center for Mountain Archaeology, 8297 Overland Rd., Ward, CO 80481*
Peter W. Birkeland, *Department of Geosciences, University of Colorado, Boulder, CO 80309*
Nel Caine, *Department of Geography, University of Colorado, Boulder, CO 80309*
P. Thompson Davis, *Department of Geology, Bentley College, Waltham, MA 02452*
Richard F. Madole, *U.S. Geological Survey, MS 966, Box 25046, Denver Federal Center, Denver, CO 80225*
Penny E. Patterson, *Exxon Production Research Co., P.O. Box 2189, Houston, TX 77252*
Alan B. Price, *Natural Resources Conservation Service, 655 Parfet St., Rm. E200C, Lakewood, CO 80215*
Taylor F. Schildgen, *Department of Earth, Atmospheric and Planetary Sciences, Massachusetts Institute of Technology, Cambridge, MA 02139*
Ralph R. Shroba, *U.S. Geological Survey, MS980, Box 25046, Denver Federal Center, Denver, CO 80225*

Organizers: **Peter Birkeland,** *Department of Geosciences, University of Colorado, Boulder, CO 80309*
David P. Dethier, *Department of Geosciences, Williams College, Williamstown, MA 01267*
Ralph R. Shroba, *U.S. Geological Survey, MS980, Box 25046, Denver Federal Center, Denver, CO 80225*

INTRODUCTION

This five-day trip introduces the geomorphology and soils of the Front Range on a transect from the piedmont near Boulder, Colorado (Figs. 1, 2) to the summit of the Indian Peaks Wilderness along the Continental Divide some 40 km to the west. We will examine late Pleistocene glaciogenic and colluvial deposits downstream along deep canyons (DAYS 1 and 2), the classic, early-to-late-Pleistocene, alluvial sequence of the Colorado Piedmont (DAY 3), and glacial and periglacial deposits, geomorphology and archeology of the Niwot Ridge/Arapaho Pass area on day hikes (DAYS 4 and 5). Soil development and the most recent age estimates for deposits are discussed at many of the stops (leaders responsible for each stop are noted parenthetically). No attempt has been made to make soil descriptions by the different authors comparable to each other, nor to correct ^{14}C or cosmogenic radionuclide ages to a calibrated calendar time scale (cf. Stuiver and Reimer, 1993), except where noted. Readers are urged to consult the extensive list of references at the end of the chapter or to contact individual authors for additional information about the material presented below.

DAY 1. PINEDALE AND BULL LAKE GLACIATIONS, COSMOGENIC AGES AND SOILS ALONG THE PEAK TO PEAK HIGHWAY

Travel south from Nederland on Highway 119, and in about 0.5 mi (0.8 km) turn right onto County Road 130. The road cuts through Pinedale moraines. About 0.8 mi (1.3 km) west of the high school, turn left on the road that goes up to Eldora Mountain Resort, and about 1.3 mi (2.2 km) beyond the last turnoff pull off on the large shoulder on the right side of the road. Discussion of erosion surface is in Birkeland et al. (2003).

STOP 1-1. OVERVIEW OF GLACIATION IN THE REGION (Madole and Birkeland)

Most Front Range glaciers originated in cirques and flowed down deep, narrow valleys that were cut primarily by streams in late Miocene and Pliocene time. Valley deepening during the Pleistocene, by both glacial and fluvial processes, was probably not much more than 100 m (Madole et al., 1999).

Most Front Range valley glaciers were between 12 and 20 km long and 180 and 350 m thick. None of the glaciers reached as far east as the mountain front, and most terminated at elevations between 2,700 m and 2,450 m.

The glacial geology of Middle Boulder Creek is typical of the valleys on the east slope of the Front Range. The Middle Boulder Creek glacier was a little more than 18 km long during the most recent glaciation (Pinedale), 19 km long during the preceding glaciation (Bull Lake), and as much as 330 m thick during both glaciations. It terminated at an elevation of about 2530 m (8350 ft.) during the Pinedale Glaciation and 2490 m (8200 ft.) during Bull Lake time.

In the Front Range, glacial deposits are composed chiefly of material derived from Precambrian granite and gneiss, and, to a lesser extent, Tertiary intrusive and extrusive rocks. The typical Front Range till consists chiefly (75-85 percent) of sediment less

Dethier, D.P., Birkeland, P., and Shroba, R. R., 2003, Quaternary stratigraphy, geomorphology, soils, and alpine archaeology in an alpine-to-plains transect, Colorado Front Range: *in* Easterbrook, D.J., ed., Quaternary Geology of the United States, INQUA 2003 Field Guide Volume, Desert Research Institute, Reno, NV, p. 81–104.

Figure 1. View northwest toward the Front Range near Boulder, Colorado. Photo by Peter Almond.

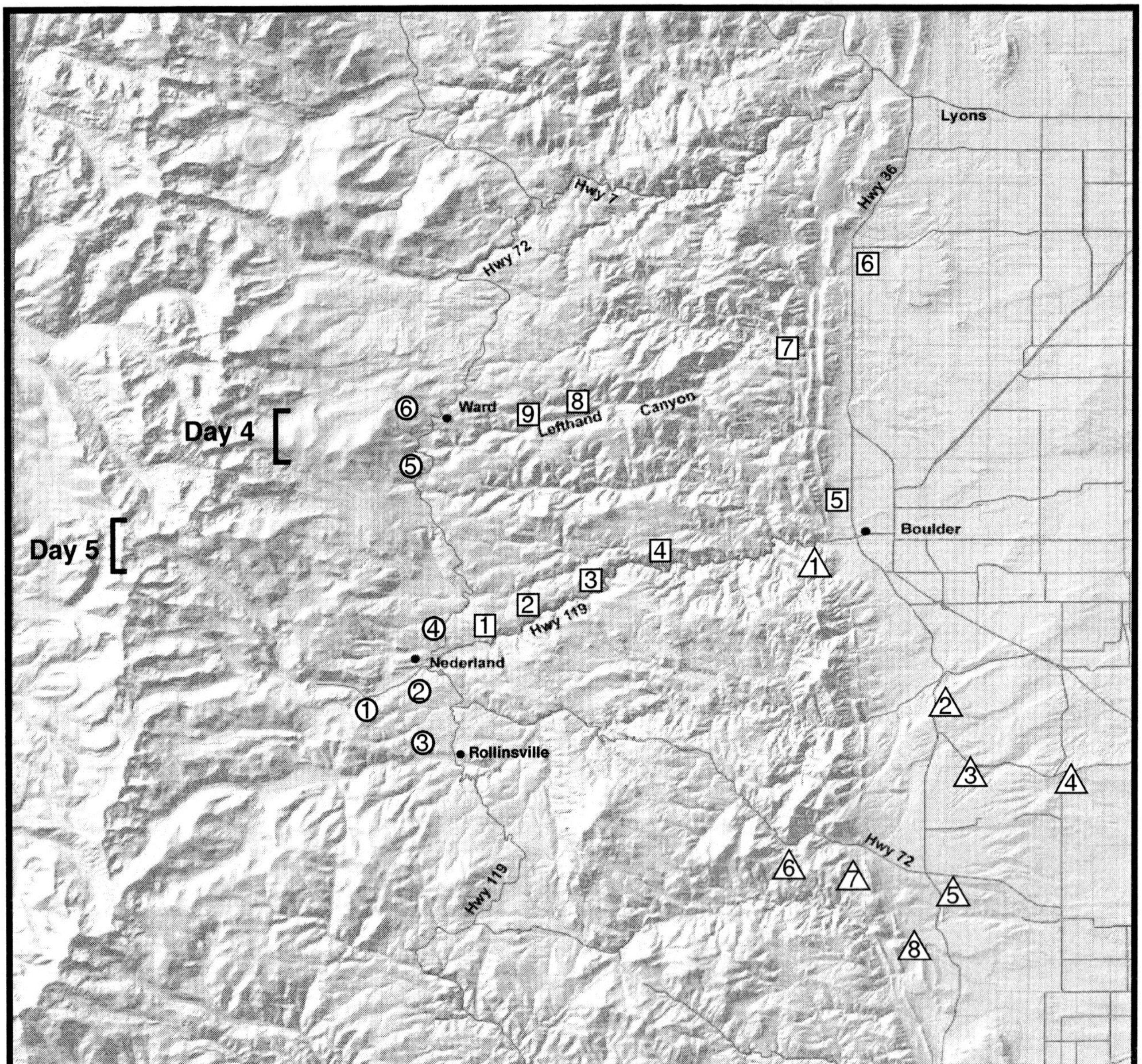

Figure 2. Digital elevation model (DEM) of the Front Range and adjacent piedmont showing the location of field stops. Figure courtesy of John Roesink. Circles–DAY 1; squares–DAY 2; triangles–DAY 3.

than 4 mm in size, most of which is sand. Subangular to subrounded pebbles, cobbles, and boulders make up 15-25 percent of the till.

In contrast to till, glaciofluvial deposits of the region are composed predominantly of pebbles and cobbles, and, in places, boulders. In some areas, glaciofluvial sand and gravel as much as 40 m thick underlie flat-floored, grassy meadows (such as the one below STOP 1-1) on the upvalley sides of Pinedale terminal moraines. Elsewhere, deposits of glaciofluvial sediment are generally thin (1-3 m).

Downvalley from terminal moraines, minor amounts of glaciofluvial sediment are preserved in small, narrow, highly discontinuous terrace deposits on the flatter reaches of canyon floors. During glaciation, the deep canyons between the glaciated areas and the Great Plains acted more as conduits than as depositories for glaciofluvial sediment. Most glaciofluvial sediment apparently was flushed through the canyons and deposited at the west edge of the Colorado Piedmont (DAYS 2 and 3).

Glacial till of at least three ages is recognized near the lower limit of glaciation (Madole, 1976a). From youngest to oldest, the deposits are referred to as Pinedale, Bull Lake, and pre-Bull Lake. Deposits of different ages are distinguished and correlated primarily on the basis of differences in (1) areal and stratigraphic position, (2) topographic expression, and (3) degree of weathering (Fig. 3).

About 75-90 percent of the drift-covered area is underlain by Pinedale till (Fig. 4), which is easily recognized by its (1) topographically distinct morphology, (2) weakly to moderately developed soils (STOP 1-2), and (3) thin weathering rinds on clasts. Pinedale till commonly has high (20-30 m), sharp-crested moraines, and knob-and-kettle topography with numerous unfilled and undrained kettles. The solum of soils developed in this till ranges from as little as 50 cm in places near the downvalley limit of glaciation to as much as 100 cm under subalpine forest at elevations above 2,900 m (9500 ft.) (Madole, 1969, 1976b; Shroba, 1977; Madole and Shroba, 1979). Weathering rinds on about 75 percent of the clasts in the soil are less than 3 mm thick, and decomposed clasts are rare (Madole, 1969, 1976a).

Pinedale glaciation on the east slope of the Front Range probably began after 30,000 ^{14}C yrs. B.P. and ended prior to 12,000 ^{14}C yrs. B.P. (Madole, 1986). A thin bed of lake sediment, deposited near a terminal moraine on the west slope of the Front Range about 30,000 ^{14}C yrs. B.P. (Nelson et al., 1979), is interpreted as providing a limiting date for the beginning of the Pinedale glaciation in this region (Madole, 1986). The pollen in the lake sediment records a climate similar to that of the present, indicating that Pinedale glaciers either had not yet formed by 30,000 ^{14}C yr B.P. or were just beginning to form.

The inception of glacial Lake Devlin near the lower limit of glaciation in the valley of North Boulder Creek (Fig. 4) provided a limiting date for the maximum expansion of Pinedale ice. Two ^{14}C dates, 21,180 +2360/-3360 (DIC-574) and 22,400 +1070/-1230 (DIC-870) ^{14}C yrs. B.P., indicate that this ice-marginal lake was dammed by the North Boulder Creek glacier sometime between 23,470 and 17,820 ^{14}C yrs. B.P. More recently, cosmogenic nuclide dates were obtained from boulders on Pinedale terminal moraines in the valleys of Middle Boulder Creek and North St. Vrain Creek (Fig. 4). Boulders in the valley of Middle Boulder Creek have ^{36}Cl ages of 19.5 and 18.1 ka, and a boulder on the terminal moraine in the valley of North St Vrain has a ^{36}Cl age of 18.5 ka and a ^{10}Be age of 19.6 ka (Benson et al., unpublished data).

Pinedale glaciers probably disappeared from the Front Range sometime between 14,000 and 12,000 ^{14}C yrs. B.P. The older limit is based on a single minimum ^{14}C age of 13,680 ± 110 ^{14}C yrs. B.P. (W-5028) for deglaciation of the Buffalo Pass area, which is in the Park Range about 115 km northwest of Lake Devlin (Madole, 1986). The younger limit is based on radiocarbon ages that date late-glacial advances in the Front Range (DAY 5).

Bull Lake till occurs at the surface in only a small part of the glaciated area (Fig. 4). This till has a more subdued, smoother topographic expression than Pinedale till (Fig. 3), and kettles, if present, rarely contain ponds. Soil development in Bull Lake till is markedly stronger than in Pinedale till, but

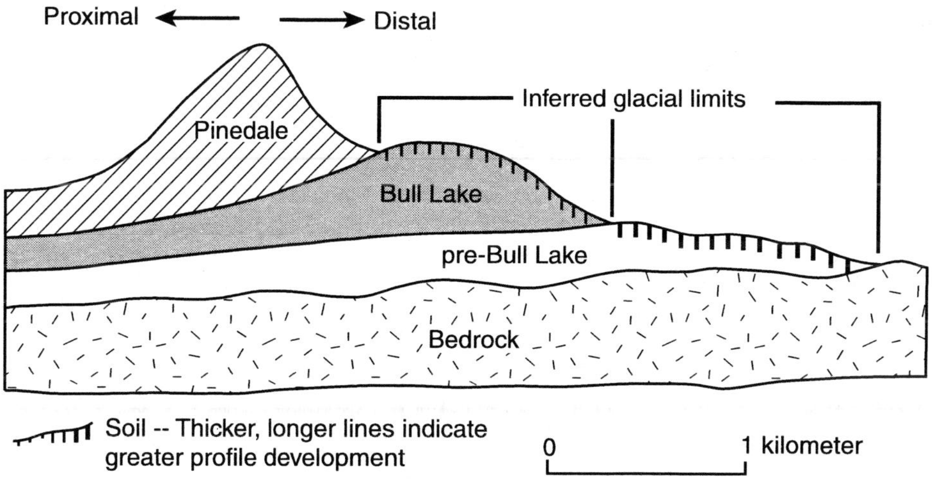

Figure 3. Differences in position, topographic expression, and soil development of Pinedale, Bull Lake and pre-Bull Lake tills (after Madole et al., 1999, Fig. 8).

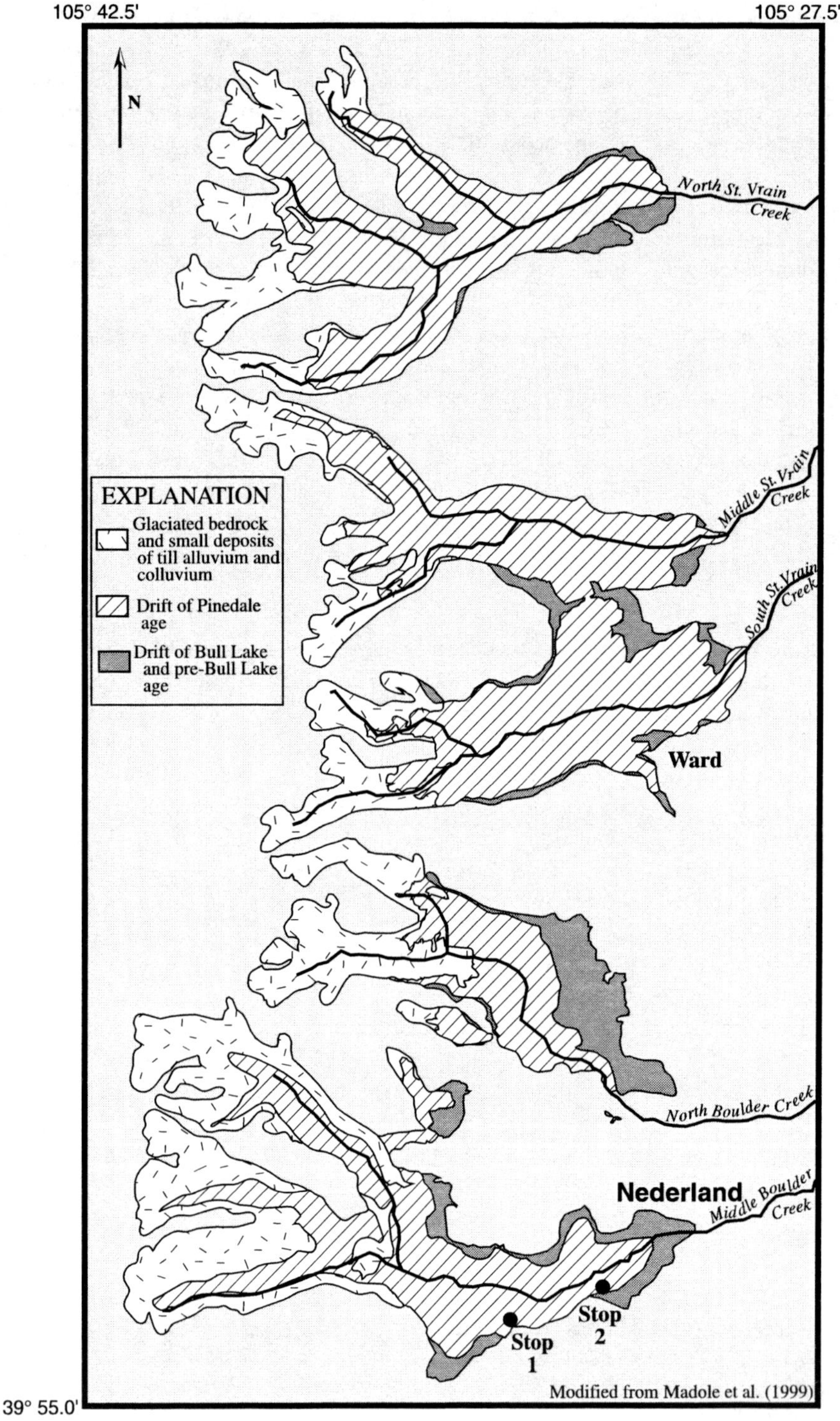

Figure 4. Spatial relations of glacial deposits in five glaciated valleys of the Front Range. Light grey areas were not glaciated; bedrock is not shown. (Modified from Madole et al., 1998, Fig. 9).

commonly does not appear to be much different than that in pre-Bull Lake till.

A uranium-trend age of 130 ± 40 ka (Shroba et al., 1983) indicates that till mapped as Bull Lake in the valley of North St. Vrain Creek (Madole, 1969; Madole and Shroba, 1979) is, at least in part, equivalent to the Illinoian glaciation of central North America. Recent cosmogenic dating of boulders in the North St. Vrain Creek area (Benson et al., unpublished data) supports this age assignment. An exceptionally large boulder (nearly 3 m tall) provided a ^{36}Cl date of 128 ka. However, a boulder in another nearby deposit mapped as Bull Lake has a ^{36}Cl age of only 88 ka.

Pre-Bull Lake till is preserved mainly as small remnants near the downvalley limit of glaciation. Diagnostic attributes of this till include (1) its sheet form and lack of end-moraine topography, (2) its position beyond the limits of younger till that has definite morainal form, and (3) the thick, reddish-brown, clayey soil developed in it. Soils in Bull Lake and pre-Bull Lake deposits may differ more in degree of development of B horizons (chiefly in clay content and color) than in profile thickness (STOP 2). Clasts that are rotted and weathered to grus are abundant in pre-Bull Lake till. Typically, about 40 percent of clasts in B horizons are highly decomposed. Pre-Bull Lake till may include deposits of more than one glaciation.

Climate and vegetation vary with altitude in the area over which this field trip will travel. John Marr, first head of INSTAAR, collected data that has since been updated (Fig. 5). The Colorado Piedmont is warm and dry with grassland vegetation. Westward into the mountains, the mean annual temperature decreases, mean annual precipitation increases, and the vegetation varies accordingly. Most of the range is in forest, except for tundra above about 3450 m (11,400 ft.).

Many soil properties follow the climate and vegetation variation with altitude (reviewed in Birkeland et al., 2003). A/Bw or Bt/Bk or K profiles are common in the Colorado Piedmont, O or A/E/Bw or Bt/C profiles in the forest, and A/Bw /C in the tundra. Soil pH decreases with elevation from 8 in the grassland to 5 and less in the tundra. A wide variety of clay minerals are present in the soils of the transect. Clays that are best associated with the overall bioclimate and geochemical conditions are smectite in the plains, vermiculite in the forest, and mixed layer, 1.0-1.8 μm clay above treeline.

Walk a short distance up the road from STOP 1-1. The soil is on the left side of the road near a gas pipeline marker.

STOP 1-1A. BURIED SOIL, POSSIBLY BETWEEN PINEDALE AND BULL LAKE TILLS (Birkeland)

Pinedale till (2.5Y5/3m; sandy loam) overlies a buried soil. The buried soil consists of a Bt horizon (40 cm; 7.5YR5/4m; sandy loam; common, distinct ped face clay films) over a Cox horizon (100+ cm; 10YR4/3m; sandy loam). The parent material of the buried soil is unknown. If it is till, the buried soil is similar enough to those on Bull Lake moraines to suggest a correlation. At any rate, the soil suggests a long period (100 ky?) of landscape stability prior to burial during the Pinedale Glaciation.

Return to Highway 119 and turn right (south) at the top of the first hill, then turn right onto West Magnolia Road and drive to the west end of it.

STOP 1-2. PINEDALE AND BULL LAKE MORAINES, SOIL DEVELOPMENT AND COSMOGENIC AGES, W. MAGNOLIA AREA (Birkeland and Dethier)

Note the weathered tors along the road. Alayne Street (1973) (now Street-Perrott) did a master's thesis on tors in relation to glaciated areas, mainly north of here, and found that they only exist in areas beyond those covered by Pinedale and Bull Lake glaciers. Soil profile development varies with the age of till (Birkeland et al., 2003). The characteristic soil on Pinedale till has an O and (or) A/Ej or E/Bw or Btj/Cox profile. The characteristic soil on Bull Lake till has an O and (or) A/E/Bt/Cox profile. Although some overlap occurs, horizon thickness, clay content, and color serve to differentiate the two characteristic soils (Table 1). Clast weathering in soil also helps in the differentiation of tills. Biotite-rich, granitic and gneiss clasts are mostly unweathered in Pinedale soils, and about 20% are weathered to grus in Bull Lake soils.

Netoff (1977; profile BC-1) studied the soil here as part of his soil transect of Boulder Creek Canyon. The profile consists of an O horizon (5 cm) over a discontinuous E horizon (3 cm; pH 5.5) over a Cox1 horizon (10 cm; 10YR3/4m; 5 % clay; pH 5.5) over a Cox2 horizon (38 cm; 10YR4/3m; 4 % clay; pH 6.0) over a Cu horizon (33 cm; 2.5Y4/3m; 2 % clay; pH 6.0). This relatively weakly developed soil is characteristic of many soils on Pinedale moraine crests in this area.

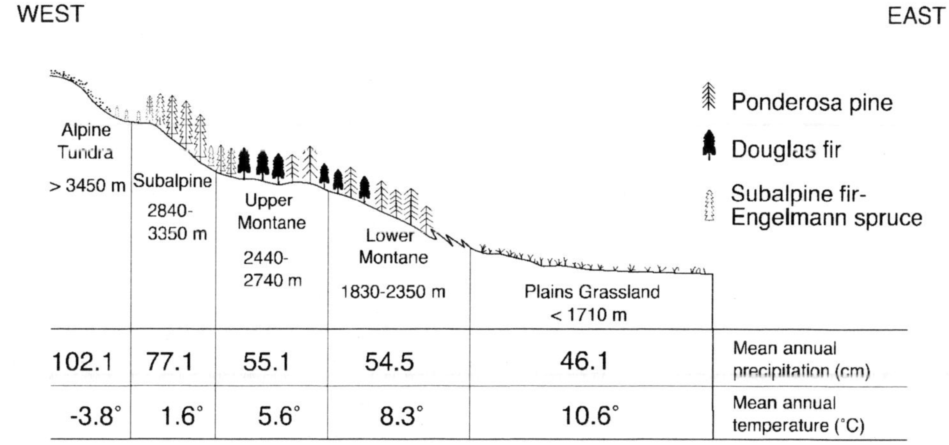

Figure 5. Variation of climate and vegetation with altitude in the Front Range area (from Birkeland et al., 2003, Fig. 2).

TABLE 1. CLAY AND COLOR DATA FOR SOILS FORMED IN TILLS OF PINEDALE, BULL LAKE, AND PRE-BULL LAKE AGE AT VARIOUS LOCALITIES IN MONTANE AND SAGEBRUSH-COVERED AREAS IN THE ROCKY MOUNTAINS*

Till parent materials	Soil horizons	Average B-horizon thickness (cm)	Average maximum percent clay (<2 μm)	g/cm^2 secondary clay (<2 μm) per horizon	Dry Munsell color
Pinedale	Bw	25	<6	0.1	10YR 6/4-5/6 7.5YR 5/3-7/3
	Bt	25	9	0.9	10YR 5/3-5/4 7.5YR 5/3-6/3
	Cox		5		2.5YR 6/3 10YR 7/3-6/4
Bull Lake	Bt	55	16	4.5	10YR 5/3 7.5YR 5/3-6/4 5YR 5/6
	Cox		7		2.5YR 7/3 10YR 7/3-6/4
Pre-Bull Lake	Bt	90	28	15	7.5YR 6/6-6/8 5YR 5/6
	Cox		11		7.5YR 7/4

* Modified from table 1 in Madole and Shroba (1979). Values for average B-horizon thickness are from Shroba and Birkeland (1983). All other data are from Madole and Shroba (1979). Values for maximum percent clay are weight percent of the < 2 mm-size fraction.

Schildgen and Dethier (2000) collected samples from large granitic boulders exposed at the crest of the Pinedale moraine about 100 m northeast of here and from the rolling surface of the Bull Lake moraine about 350 m to the southeast of where we parked. Cosmogenic exposure ages calculated from the ^{10}Be and ^{26}Al content in quartz from the moraines were 16.8± 3.4 ka (Pinedale) and 121.9± 25.8 ka (Bull Lake), assuming that the boulder surfaces had not eroded since they were deposited on the moraine. These were the first cosmogenic ages from Colorado moraines (Schildgen and Dethier, 2000). Examination of local boulders suggests that "no erosion" is a plausible assumption for Pinedale moraines and unlikely for Bull Lake boulders; cosmogenic ages thus are minima. As is true with all dating techniques, cosmogenic ages should be consistent with and supplement age criteria derived from geologic mapping, soil development, and relative-weathering.

Return to highway 119 and turn right (south). The first drainage crossed is Beaver Creek, which appears to be underfit. Discharge might have been much larger during the Pinedale glaciation as water from the right lateral portion of Pinedale ice spilled into a tributary of this creek.

At Rollinsville, turn right (west) toward East Portal, go about 0.75 mi (1.2 km) and stop at a south-facing roadcut just east of where a dirt road extends to the right and short of a railroad underpass.

STOP 1-3. OLD COLLUVIUM AND SOIL JUST EAST OF THE PINEDALE ICE LIMIT (Birkeland)

The slope here is about 20° south and mantled with colluvium derived from gneiss (Gable, 1969). The soil consists of about 5 cm of possible eolian material (historic?) at the surface over an A horizon (44 cm; 10YR3/2m; sandy loam) over an Ej horizon (20 cm; 10YR 5/4m; sandy loam) over a Bt1 horizon (45 cm; 7.5YR4/5m; sandy loam to sandy clay loam; common, distinct clay films as coats and bridges) over a Bt2 horizon (similar to Bt1, except it could be as red as 5YR4.5/6m). The pH is about 6 at the surface and 7 at depth (Thom Davis, class report). Colleagues and students of Birkeland have debated over the years whether or not the Bt color reddens with depth and what it means. Birkeland has suggested that the redder Bt might be a buried B horizon.

This soil is discussed here because we do not have a nearby representative, clay-rich soil on a Bull Lake moraine that is easily accessible. The soil profile has many properties common to those on Bull Lake moraines (Table 1) and suggests deposition of colluvium at least 100 ka ago and landscape stability since then. If the Bt2 horizon is part of a buried soil, then two long periods of landscape stability have occurred. This site is about 2 mi. downvalley from the Pinedale ice limit, yet does not seem to have experienced substantial erosion during the Pinedale glaciation. Finally, the U.S. Forest Service has mapped soils with Bt horizons across the erosion surface (Birkeland et al., 2003, Fig. 6), and if this soil is representative of those on interfluves, they could not sustain an erosion rate commensurate with Boulder Canyon incision (0.15 m/ky, Schildgen et al., 2002). This leads to the conclusion that canyon relief is increasing.

Return to Rollinsville, and turn left (north) onto Highway 119 toward Nederland (0.5 miles (0.8 km) north of Nederland roundabout on Highway 72)

STOP 1-4. WEATHERING STAGES OF BEDROCK (Birkeland)

The weathering of bedrock in the Front Range can be analyzed according to the 7 classes of Clayton et al. (1979):

Class 1—unweathered.

Class 2—oxidized rock with angular joints.

Classes 3 and 4—marked by joint junctions becoming subangular (3) and rounded (4).

Class 5—the rock can be broken into small fragments with hands.

Class 6—fragments disaggregate into sand-size particles (grus).

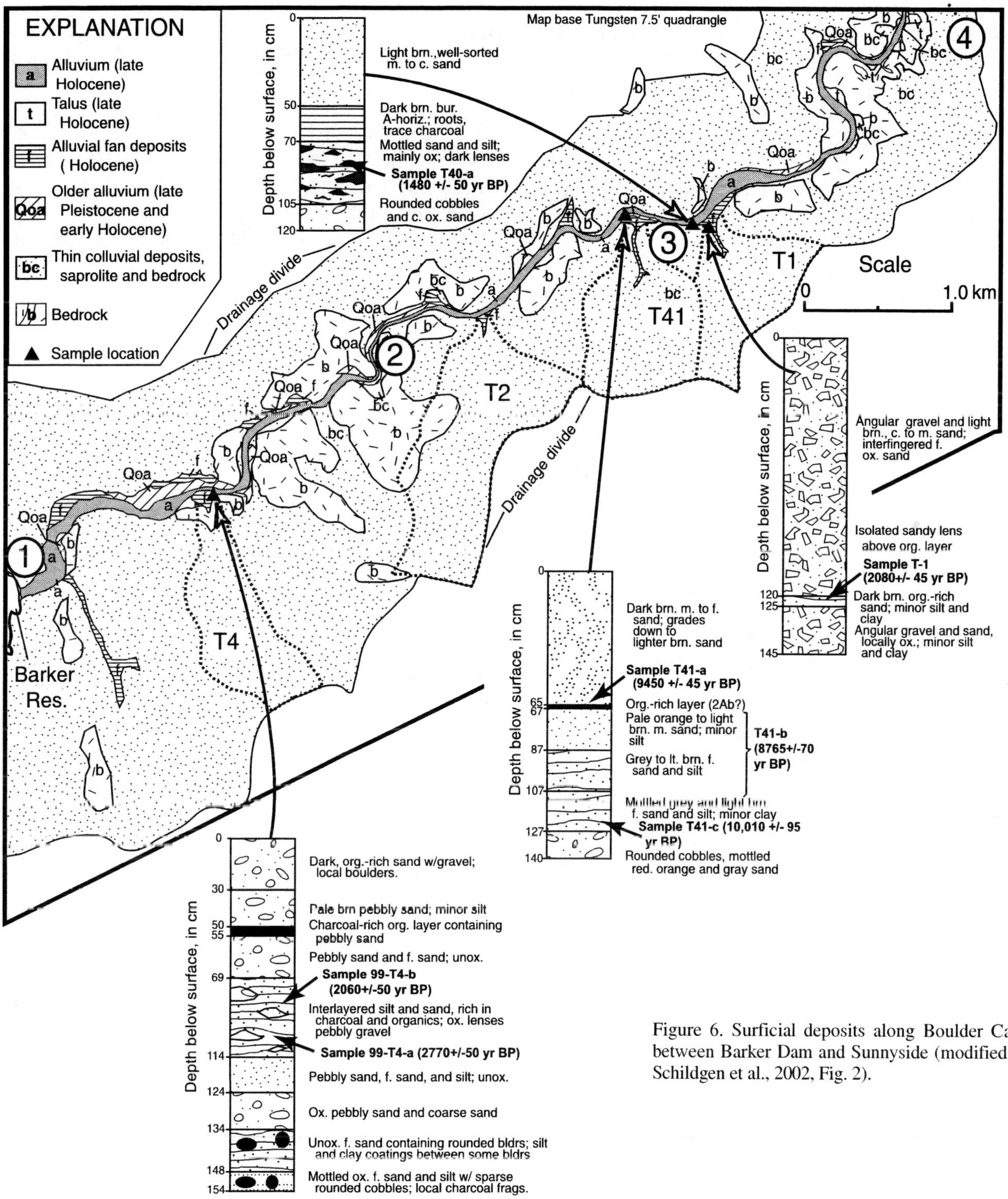

Figure 6. Surficial deposits along Boulder Canyon between Barker Dam and Sunnyside (modified from Schildgen et al., 2002, Fig. 2).

Class 7—has evidence of clay formation (material is plastic).

The rock here is biotite gneiss (Gable, 1969). Because the rock is oxidized and has angular joint junctions, it is considered class 2. Birkeland has biked along most of the major mountain roads in the area and suggests that bedrock weathering in most of the range beyond the areas covered by ice is class 2 (Birkeland et al., 2003). Rock weathered to classes 5 and 6 seems to be associated with the pre-Fountain paleosol (see STOP 3-1) or tributary drainages with low incision rates (for example, Sunshine Canyon, west of downtown Boulder).

About 8.6 mi (14.1 km) north of the Nederland roundabout and 0.4 mi (0.7 km) north of the junction of road to Gold Hill, pull off on a wide shoulder to the right as you begin a left turn.

STOP 1-5. OLD RED SOIL AND LONG- TERM LANDSCAPE STABILITY (Birkeland)

Here, several meters of gravelly colluvium rest on gneiss. The surface slopes 10° north and is located on the edge of the Tertiary erosion surface. Stone lines and relatively sandy mottles more or less parallel the surface. The soil here is one of the best developed that we have seen in the mountains of Colorado. The soil profile consists of an A horizon (11 cm; 10YR5/3d; sandy loam) over an E horizon (21 cm; 10YR8/3d; sandy loam) over a Bt1 horizon (105 cm; 2.5YR5.5/6m; clay; many distinct ped face clay films) over a Bt2 horizon (63+ cm; mottled with 2.5YR4.4/4m and 2.5-5YR4/6m; sandy loam to sandy clay loam; many, distinct ped face clay films). At greater depth is what might be Cox material (10YR5/6d; sandy loam) and it may contain some inherited clay. Some of the browner mottles are sandier than the rest of the same horizon and will be a subject of discussion.

Soils of this exceptional development indicate long-term landscape stability. This profile is so much redder and clay rich than the other soils on the erosion surface that it must have taken much more time to form. If the more common soils on the erosion surface with browner and less clay-rich Bt horizons took about 100 ka to form, this soil surely took over 200 ka, or perhaps as much as 500 ka, to form. It also suggests that this relatively small piece of landscape is not undergoing much denudation and supports the idea that canyon relief is increasing (see STOP 1- 3).

Just beyond the turnoff to the right to Ward, about 11.5 mi (18.9 km) north of the Nederland roundabout, turn left onto the Brainard Lake area road; park just before the summer fee station and walk about 300 m up the dirt road toward Left Hand Reservoir.

STOP 1-6. FORESTED SOIL ON PINEDALE MORAINE (Birkeland)

This stop is a 3-m-high prominent exposure of soil formed in Pinedale till (Gable and Madole, 1976) on the left-hand side of the road, about 100 m before the junction with the South Ski Trail. The surface slopes to the north.

The soil here is one of the better-expressed ones in Pinedale till and one for which we have laboratory data. The profile consists of an O horizon (17 cm; pH 5.2) over an E horizon (21 cm; 10YR6/1d; 1.7 % clay; pH 4.9) over a Bt horizon (25 cm; 10YR5/5d; 12.9 % clay; pH 5.2) over a Cox or Bt horizon (20 cm; 10YR6/3.5d; 11.1 % clay; pH 4.7) over a Cu horizon (24 cm; 2.5Y6/3d; 5.7 % clay; pH 4.7). The E horizon has 7-8 % more silt than the underlying horizons so could be slightly enriched in eolian materials. Dithionite- and oxalate-extractable Fe is low in the E horizon (0.39 %, 0.10 %, respectively), increases in the Bt horizon (1.31 %, 0.71 %), and then progressively decreases in the Cox or Bt horizon (0.85 %, 0.36 %) and the Cu horizon (0.62 %, 0.28 %).

If all of the clay increase from the Cu to the Bt horizon is pedogenic, this is one of the greater increases Ralph Shroba has noted in his regional study of Rocky Mountain soils (Table 1). The Fe profiles surely are pedogenic in origin.

DAY 2. COSMOGENIC AND 14C DATES FROM ALLUVIAL DEPOSITS OF BOULDER CANYON, LANDSLIDE IN BOULDER, RANGE FRONT STREAM CAPTURE, AND DEPOSITS AND SOILS IN LEFT HAND CANYON

From Nederland roundabout, go 1.7 mi (2.7 km) east on Highway 119 and pull off to the right at the dam.

STOP 2-1. OVERVIEW OF BOULDER CANYON FROM NEAR BARKER RESERVOIR (Dethier and Madole)

In 1999, Taylor Schildgen (then an undergraduate at Williams College) and Dethier mapped surficial deposits along Middle Boulder Creek from here to the canyon mouth and used cosmogenic and ^{14}C dating of these deposits to help constrain the late Quaternary evolution of Boulder Canyon. Mapping (summarized by Madole et al., 1999) shows that Bull Lake and Pinedale glaciers reached the west end of Barker Reservoir, building moraines (mapped by Madole) and sending large (?) volumes of outwash down Boulder Canyon. The canyon, which is straight, relatively wide, and of low relief here, narrows, becomes more sinuous, and locally steepens downstream, reaching a maximum depth of ~850 m between STOPS 3 and 4, locally referred to as the "Narrows". Outwash deposits, mainly Pinedale, are preserved locally near valley walls where they are capped by boulder-rich colluvial debris. Small, late Holocene fan deposits rich in charcoal are common for ~10 km downstream from here. As we travel down the canyon, we will visit several localities where exposures of soils have been examined in detail and where dating constrains the age of the deposit. We follow a convention of referring to terraces in terms of their height (in m) above Boulder Creek, which provides a good initial estimate of their age. Surficial deposits mapped along our route in the canyon in the Tungsten Quadrangle (Fig. 6) and a schematic cross-section of surficial deposits in the canyon and their approximate age (Fig. 7) emphasize thin, scattered deposits in a mainly bedrock landscape.

4.0 mi (6.6 km) below Nederland roundabout; turn off to the right on an old road and look for places to park near Middle Boulder Creek.

STOP 2-2. LATE PINEDALE SURFACE ON COARSE ALLUVIUM, OXIDIZED SOIL, AND POTHOLE LINES (Dethier)

Gravel-rich late Pinedale (?) alluvium here underlies a terrace ~4 m above Boulder Creek and is mixed with large, locally

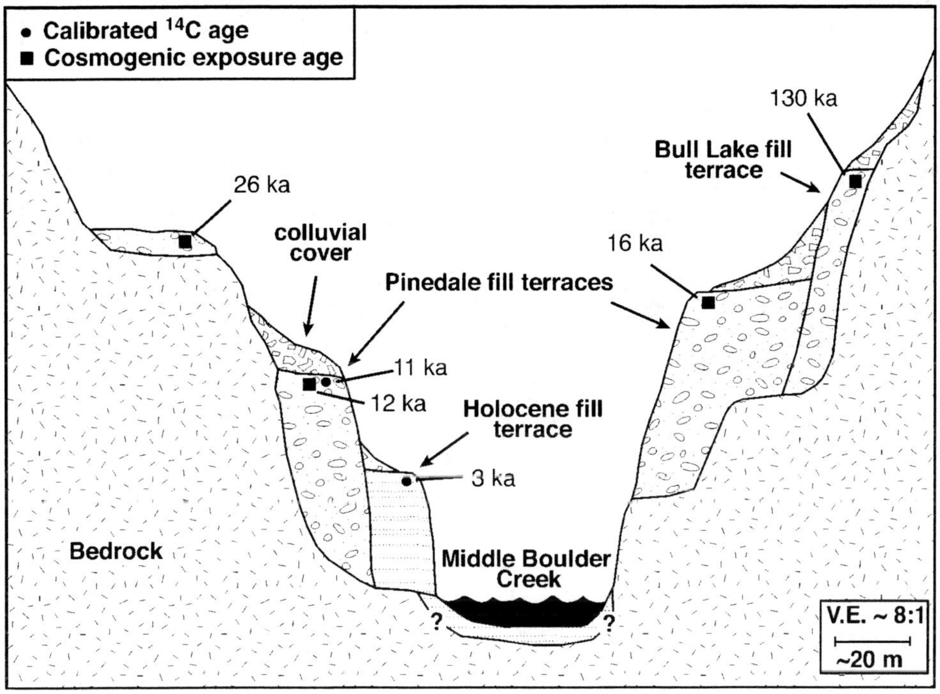

Figure 7. Surficial deposits along Boulder Canyon and their approximate ages (modified from Schildgen et al., 2002, Fig. 7).

derived, river-smoothed boulders. The relatively strong oxidation of the deposit may be a groundwater effect or reflect the large volumes of water that flow from the rock face above and pour through the deposit. Lines of potholes and river-polished bedrock extend at least 7 m above Boulder Creek, suggesting that stream incision here was through bedrock. Walk downstream between remnants of ~4 m terraces. Schildgen et al. (2002) collected a sample from the prominent saddle-shaped rock on top of the terrace, hoping it would provide a cosmogenic age for the terrace. However, both the measured age (35.6 ± 8.1 ka) and the $^{26}Al/^{10}Be$ ratios of ~5 are anomalous and demonstrate that the boulder had a complex burial and exposure history.

Continue downstream from STOP 2-2 on the dirt road, then turn right onto the highway and drive downhill 1.3 mi (2.1 km); turn right off road and park in a large pullout.

STOP 2-3. LATE HOLOCENE CHARCOAL-RICH ALLUVIUM AND HOLOCENE FAN/LATEST PLEISTOCENE ALLUVIUM (Dethier)

Wade across Boulder Creek and walk across the old ski jump parking lot to two sites to look at late Holocene, charcoal-rich alluvium and Holocene fan or latest Pleistocene alluvium. Low surfaces in this relatively wide part of the canyon consist of latest Pleistocene and Holocene, Boulder Creek alluvium, covered locally by alluvial fan deposits and boulders derived from small tributary catchments and adjacent bedrock slopes. Low surfaces here are quite extensive, but their original morphology has been modified by activities associated with the ski jump and with mining (?). The first site (Fig. 6), on a 1-m-high, riverbank terrace, exposes a buried, weak (A/C) soil developed in Boulder Creek alluvium rich in charcoal fragments (1480 ± 50 ^{14}C yrs. B.P.), many of them rounded. At a nearby site, charcoal in a buried A-horizon in alluvial fan debris in the old ski jump area gave a date of 2080 ± 45 ^{14}C yrs. B.P. That fan was graded to a low terrace level. Terrace heights and ^{14}C ages here demonstrate that Boulder Creek had cut down to its modern level before the late Holocene. From this site, walk upstream 0.5 km on low terraces and adjacent colluvium-mantled slopes, cross a small channel, and stop on a relatively extensive Holocene fan that overlies latest Pleistocene alluvium. Exposures where the fan has been eroded by Boulder Creek show 0.5 to 1.2 m of sandy alluvial-fan deposits containing one or more buried soils overlying oxidized, gravelly, Boulder Creek alluvium. Charcoal from fine-grained deposits 15 cm above the alluvium gave an age of 10,010 ± 95 ^{14}C yrs. B.P. Local stratigraphic and temporal relations thus suggest that Middle Boulder Creek cut back down through the latest Pleistocene fill sometime after about 9000 ^{14}C yrs. B.P.

Boulder Canyon here is relatively wide, bedrock is mantled in many places with colluvium, and the channel flows in alluvial deposits at a gradient of about 2.5 m/km. As we travel east from STOP 2-3, the valley narrows between bedrock cliffs, and the channel is developed in bedrock and thin alluvial deposits at a gradient of about 67 m/km (Fig. 8). This knickzone, which is typical of features found along most Front Range drainages north of Denver, extends down to Sunnyside (STOP 2-4), where the channel flattens to a gradient of about 26 m/km.

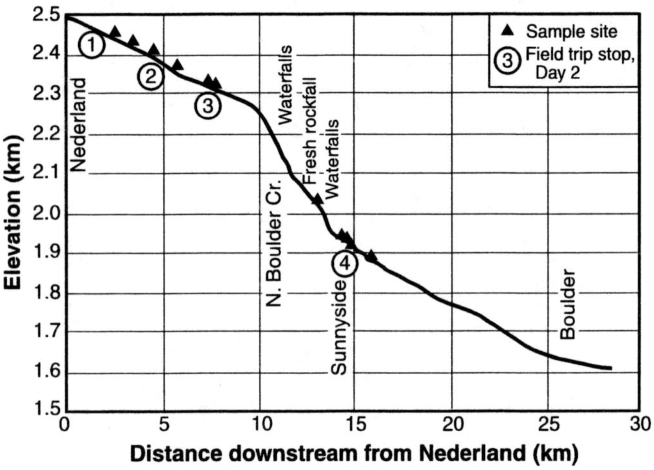

Figure 8. Longitudinal profile of Middle Boulder Creek between Barker Dam and Boulder, showing Stops 2-1 through 2-4 (modified from Schildgen et al., 2002 Fig. 2).

9.3 mi (15.2 km) from the Nederland roundabout and 4.6 miles (7.5 km) downstream from the ski jump area, pull off onto the large shoulder on the right and prepare to dash across the road.

STOP 2-4. BULL LAKE, EARLY (?) AND LATE PINEDALE ALLUVIUM, SOILS AND COSMOGENIC AGES, SUNNYSIDE AREA (Dethier and Birkeland)

The road and parking lot here are developed on a ~5-m-high terrace, and terrace remnants extend to a height of 21 m above the modern channel. If time permits, we'll look from afar at this tiny Bull Lake terrace remnant. Schildgen et al. (2002) mapped terrace deposits in this area and used cosmogenic dating to separate Bull Lake (above 16 m; 135.5 ± 27.8 ka), early Pinedale (30.4 ± 6.3 ka and 22.6 ± 4.5 ka at about 12 m), and late Pinedale alluvial deposits (16.0 ± 3.2 ka at about 12 m height, 2 km upstream). About 1.5 km downstream, a pothole 6.7 m above Middle Boulder Creek gave a cosmogenic age of 13.8 ± 2.8 ka. Using the cosmogenic and ^{14}C ages, constraints derived from soil development and regional stratigraphy suggest that middle Boulder Creek here is cutting down at a rate of about 0.15m/ka (Fig. 9).

We will examine soil development in this part of the canyon at the easternmost outcrop where Netoff (1977; profile BC-12) described a soil on colluvium/gravelly alluvium that forms a terrace about 12 m above Boulder Creek. The soil profile consists of an O horizon (5 cm) over an A horizon (31 cm; 10YR2-3/2-3m; 5 % clay; pH 6.0) over a 2Bt horizon (50 cm; 10YR4/4m; 11 % clay; pH 7.0) over a 2Cox horizon (117 cm; 10YR4/4; 4 % clay; pH 7.0). Biotite-rich, granitic clasts commonly are rotten. These soil properties suggested a Bull Lake age to Netoff for the gravel. The cosmogenic ages, however, indicate that the deposit stabilized at 30 to 23 ka. Soils on slightly lower surfaces at Sunnyside lack a Bt; their height and degree of soil development suggest that they are likely of late Pinedale age.

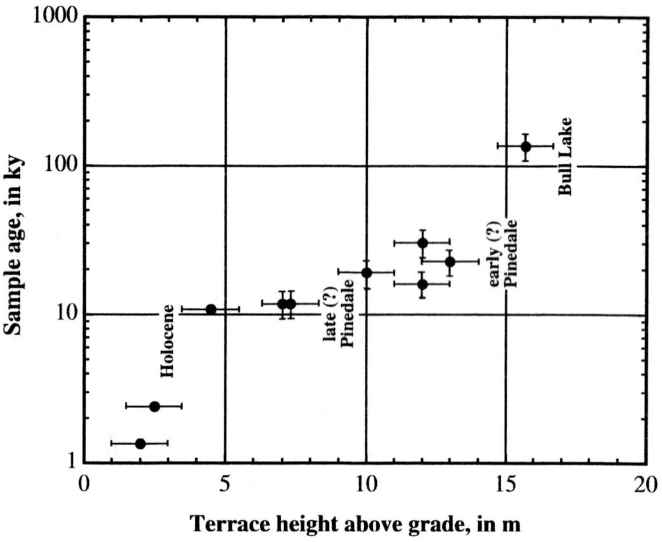

Figure 9. Downcutting along Middle Boulder Creek in Boulder Canyon during the past 130,000 years (modified from Schildgen et al., 2002, Fig. 5).

STOP 2-5. TRANSLATIONAL ROCK SLIDE IN DAKOTA GROUP ROCKS (Birkeland)

Small tributary canyon at 3rd St. and Hawthorn in NW part of Boulder; from trailhead parking walk due west on a trail on the north side of the drainage to a trail intersection on the first flat.

Here we will view a large translational rock slide with a buckle-like fold in Dakota Group rocks Braddock and Eicher (1962) and Braddock (1978) have described glide blocks (old term) in Dakota Group rocks farther north in the Front Range. There the rocks dip about 20°, and the up-dip sandstone slab broke loose from, or within, the underlying shale and rode along the landslide surface over the same *in situ* sandstone located down dip. With subsequent erosion, some of these features crop out as ridge top synclines. This locality differs in that we see what happens when the dip is steep, and the overriding slab buckles to form a fold. Bill Braddock, a retired structural geologist in the Univ. of Colorado geology department, helped with the interpretation below.

The Dakota Group in this area consists of the basal Lytle Fm. (sandstone and conglomerate) and the overlying South Platte Fm. From bottom to top, the South Platte consists of the Plainview member, the Skull Creek shale, and the Muddy sandstone. The Lytle Fm. makes up the top of the hogback here and dips 56-86° E. In contrast, Muddy sandstone crops out in the slopes on the east side of the hogback, and dips 15-81° W. Ripple marks in beds at a few localities indicate that the Muddy sandstone is overturned. We interpret this as a buckle in a large translational rock slide that extends 0.5 km north and over 1.0 km south of here. The lower part of the buckle is best exposed north of the drainage and the upper part south of the drainage. Figure 10 is a schematic diagram of the feature. Regional erosion has to be sufficient to allow space for the buckle to form as a surface feature. We can estimate the time in which regional lowering reached the site of the buckle by projecting the heights of terraces to the site (Wrucke and Wilson, 1967; Colton, 1978). This suggests formation near Verdos or Rocky Flats time (640 ka to 2 Ma; see DAY 3). The reconstructed buckle-fold axis is higher in elevation north of the drainage than it is south of the drainage, suggesting that the former might have moved independently before the latter.

Go east on Hawthorn to Broadway and turn left (north). Proceed out of town on this road, which becomes Highway 36.

STOP 2-6. STREAM CAPTURE AT THE FRONT RANGE-GREAT PLAINS BOUNDARY (Madole)

Highway 36 about 0.3 mi (0.5 km) north of the Left Hand Canyon road. Pull off to the right where a gate leads into a pasture.

The landscape seen to the east is the end result of a major episode of denudation that occurred in the late Cenozoic. An estimated thickness of 300-430 m of Cenozoic and Upper Cretaceous rocks were eroded from the western edge of the Great Plains in the Denver-Boulder area (Madole, unpublished report, 2000). Topographically-inverted, early and middle Pleistocene fan and floodplain deposits cap mesas and interfluves. Similar deposits are widely preserved in a narrow band, 3-5 km wide along the entire east flank of the Front Range (see DAY 3). Names proposed by Scott (1960) were widely adopted after the 7th INQUA Congress meeting in Boulder. Although Scott's nomenclature has been extended throughout the Colorado Piedmont (DAY 3), the ages he assigned to some units have been changed as better dating techniques have become available (Madole, 1991). Similarly, regional correlations of many deposits will change because many have been based chiefly on height above stream level without regard for: (1) important differences in geology and geomorphology between drainage basins; (2) influence of tectonism; or (3) isostatic response to denudation. Furthermore, some deposits may have no correlatives because they are products of non-synchronous channel avulsions and stream piracies.

Stream piracy is common in piedmont regions as a result of differences in flow regimes, load characteristics, and gradients between stream channels draining from uplands and those originating on lowlands. The highest surface east of here is a remnant of a broad alluvial fan (Fig. 11) that was deposited by an ancestral Left Hand Creek whose course passed south of the fan remnant. Later, Left Hand Creek was beheaded and flowed northeastward in a valley that was deeply incised into the northern part of the fan; then another episode of stream piracy redirected Left Hand Creek around the southern end of the fan remnant. The highest surface here was believed to be of middle Pleistocene age and correlative with others that contained Lava Creek B volcanic ash (Madole, 1963). Colton (1978) made the same correlation for this deposit, but Madole et al. (1998) assigned the fan remnant to the early Pleistocene, here inferred to be 1806-778 ka

Reverse course and drive up Left Hand Canyon. Mileages given are for distances west of Highway 36.

The incision rate of Left Hand Creek is 0.05-0.1 m/ka at the range front, given as a range because it depends on which terrace is used as a datum. Despite this rate, the soils in the canyon are intermediate in development between the young soils of Boulder Canyon and the old soils of Coal Creek Canyon, with range-front incision rates of 0.04 m/ka and 0.006 m/ka, respectively (Birkeland et al., 2003).

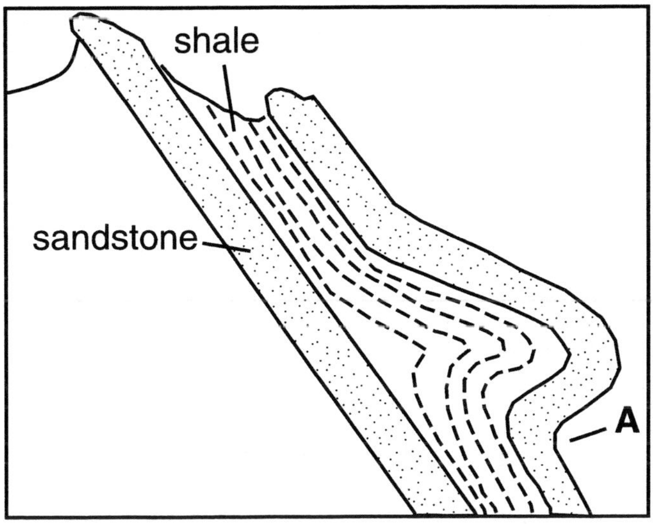

Figure 10. Schematic diagram of translational rockslide at Stop 2-5, looking north at Dakota Group rocks. 'A' is the level of erosion required to allow the buckle to form.

STOP 2-7. ZERO-ORDER DRAINAGE BASIN, AND USE OF SOILS IN DATING EVACUATION OF SEDIMENTS (Birkeland)

3.3 mi (5.4 km) up Left Hand Rd from junction with route 36. Park at a large picnic area parking lot on the east (left) side of the road.

Across the highway is a zero-order drainage basin in Boulder Creek granodiorite. Following the findings of Dietrich et al. (1986) and Tonkin and Basher (1990), such hollows probably go through cycles of sediment accumulation and stability, followed by evacuation. Welter (1995) has extensively studied similar basins and associated sediments and soils to the south in the Front Range.

The soil here consists of a Cu/2Ab/2Ejb/2Btb/3Coxb profile (Machette et al., 1976a, b). The Cu horizon appears to be predominantly recent eolian sand (5% clay). The buried soil profile consists of an A horizon (38 cm; 10YR3-4/2d; 8 % clay) over an Ej horizon (28 cm; 10YR5.5/3d; 7 % clay) over a Bt horizon (117 cm; 10YR5/4d; 13-15 % clay; few, faint ped face clay films

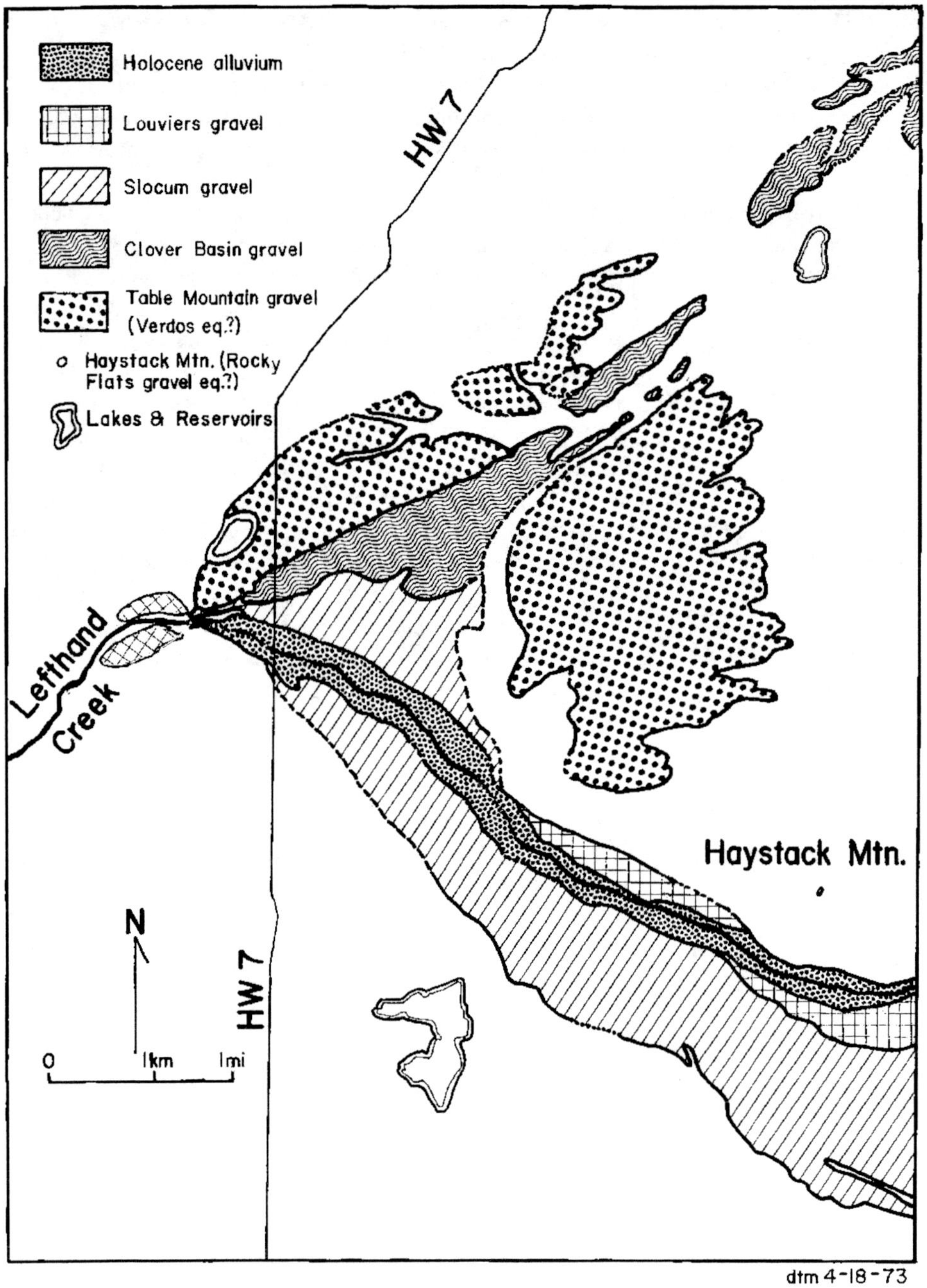

Figure 11. Drainage and Quaternary deposits along the Front Range near Left Hand Creek (R. Madole, unpublished data, 1973).

and some clay lamellae). The underlying 3Cox horizon has 15% clay. This clay is either inherited from the basin soils and deposits when this deposit was laid down or has formed pedogenically and is the lower part of the Bt horizon.

This soil helps limit the time this basin last filled with sediment and was followed by a period of stability and soil formation. This soil seems better developed than most soils on Pinedale moraines and is more akin to those on Bull Lake moraines. Thus, we suggest that deposition here last took place about 100 ka. Problems to be discussed here are identification of inherited clay, as well as the origin of the clay-rich gravelly sediments to the south in the roadcut.

Proceed up the canyon and take the left hand fork to Ward.

STOP 2-8. HOLOCENE ALLUVIAL FAN, SLOPE DEPOSITS, BURIED SOILS, AND INCISION RATE

(Birkeland)

11.1 mi (18.0 km) from junction with route 36, just west of the 11 mile road marker. Park in the large pullout on the right (north) side of the road.

The roadcut exposes colluvium and a small portion of a sandy alluvial-fan deposit at the base of a steep bedrock slope underlain with gneiss. The sediments here consist of relatively dark (10YR3.5-4/3m), organic-matter-rich, sandy loam sediments. The surface horizon is a relatively dark (10YR3/3m) A horizon, underlain by a darker buried A horizon (25-50 cm; 10YR3/2m). At 145-195 cm is another buried A horizon (10YR3/2m).

We interpret the sediments to be derived from the adjacent hillslopes and that sedimentation was halted or slowed at least twice for hundreds to several thousand years to account for the buried A horizons. Although erosion and deposition following burning is an attractive hypothesis, too much organic material seems to be present in the sediments, and a short search for charcoal was not successful.

Walk west up the road a short distance to an outcrop on the right where about 2.5 m of gravel rest on a cut surface on gneiss. The cut surface is 6.4 m above Left Hand Creek. The gravel is imbricated, the matrix is weathered, and biotite-rich granitic rocks are weathered to grus. Such weathering features are similar to deposits of Bull Lake age, and this in turn would suggest an incision rate here of 0.05 m/ka.

STOP 2-9. OLD COLLUVIAL DEPOSIT AND BURIED RED SOIL (Birkeland)

12.5 mi (20.5 km) from junction with route 36, just west of 12-mile road marker; huge cut on the right. The best parking is the wide shoulder on the left side of the road.

This deposit is thick colluvium composed of clasts of Silver Plume granite. Access is difficult as the slope is so steep. We will discuss soils to the left and right of the prominent pine tree.

The soil profile left of the pine tree consists of an A horizon (60 cm; 10YR4/3.5d; sandy loam) over a Bt1 horizon (60 cm; 7.5YR5/5d; sandy loam to sandy clay loam; common, distinct, ped face films) over a Bt2 horizon (90 cm; same color as Bt1, less clay; less developed films). To the right of the pine, the Bt horizon is more like the Bt2 horizon above.

Two to 3 m below the surface is a prominent buried red soil. It consists of a Bt1b horizon (100 cm; 5YR4/6m; sandy clay loam; many, distinct ped face clay films) over a Bt2b horizon (50 cm; 5YR3/4m; sandy loam +; films less developed than above) over a Bt3b horizon (60 cm; 10YR5/7m; sandy loam; films less developed).

The soil sequence points to long-term stability interspersed with sedimentation. The red, buried soil must have taken several 100 ka to form during a long period of landscape stability. Soil development was followed by renewed deposition of colluvium. This latter depositional event seems to have taken place at least 100 ka ago to account for a surface soil at least as developed as those on Bull Lake moraines. Perhaps the long-term stability can be accounted for by the low incision rate of Left Hand Creek (STOP 2-6), the relatively low discharge of Left Hand Creek, and the fact that Pleistocene glaciers barely spilled into the headwaters area of this drainage.

DAY 3—SOIL-GEOMORPHIC RELATIONSHIPS NEAR ROCKY FLATS, BOULDER, AND GOLDEN, AND THE PRE-FOUNTAIN PALEOSOL OF WAHLSTROM (1948)

This day will be spent in the Colorado Piedmont, mainly discussing alluvial units and related soils. A large exposure of buried soils helps us put age limits on the Rocky Flats alluvium. The low incision rate of Coal Creek Canyon is used to explain the existence of old, red soils on bedrock and colluvium near the canyon mouth. Finally, we use Wahlstrom's (1948) paleosol to discuss pitfalls in identifying and interpreting paleosols. All of this is in a reprint that we will hand out (Birkeland et al., 1999).

DAY 4. TIMBERLINE DYNAMICS, HYDROLOGY, SOILS, AND PERIGLACIAL LANDFORMS ON NIWOT RIDGE AND IN THE GREEN LAKES VALLEY

Drive north on Hwy 72 from Nederland. Just beyond the turnoff to the right to Ward, turn left onto the Brainard Lake area road and proceed to the Long Lake trailhead. Park at the trailhead and hike the short distance to Long Lake.

Niwot Ridge extends eastward from the summit of Navajo Peak for about 9 km before reaching the forest margin (Fig. 12). The ridge is a system of interconnected knolls and saddles, underlain by Precambrian gneiss and granite and by Laramide intrusive syenite and quartz monzonite. Its crest is unglaciated, although a till-like diamicton occurs near its western end (Madole, 1982). Elsewhere, its bedrock is largely obscured by periglacial colluvium. Ecological studies were begun on Niwot Ridge in 1951 when John W. Marr established the Institute of Arctic and Alpine Research. The D-1 weather station, at an altitude of 3744 m, began operation the following year and remains a significant source of climatic data. Greenland (1987) cites a mean annual air temperature of –3.7 °C, cold enough to maintain permafrost in favorable settings (Ives and Fahey, 1971). Annual precipitation averages 930 mm. Winter is characterized by strong westerly winds, redistribution of snow, and frequent poor visibility. Large areas of alpine tundra remain snow-free for much of the winter, freezing as deeply as 2 m. Where meltwater seepage from late-lying snow banks irrigates such areas, significant thicknesses of segregation ice can develop. Approximately 90–100 freeze-thaw cycles occur annually at 2-cm depth in bare, winter-snow-free soils (Benedict, 1992, Fig. 2). Because of these factors, miniature landforms characteristic of the equatorial high mountains occur in close proximity to landforms typical of the arctic coastal plain. No landscape in North America has a greater diversity of solifluction and patterned-ground features.

Our hike begins at a trailhead near Long Lake and takes us through old-growth forests dominated by Engelmann spruce (*Picea engelmannii*), subalpine fir (*Abies bifolia*), and limber pine (*Pinus flexilis*). As we approach the upper limit of Pinedale glaciation, the Navajo and Isabelle glaciers are visible to the west. Small glaciers survive in this relatively warm, dry environment

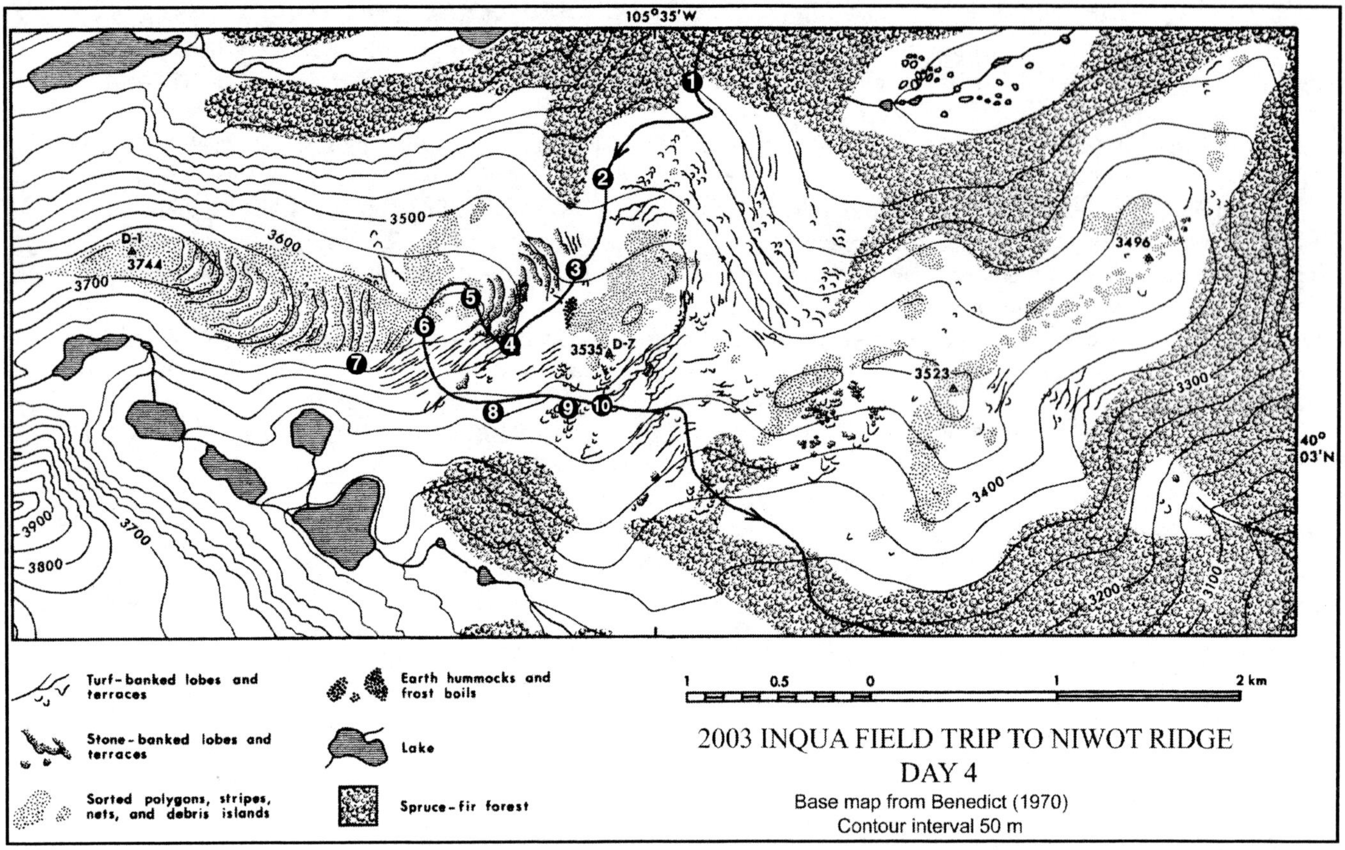

Figure 12. Location map showing Day 4 stops in the Niwot Ridge/Green Lakes area.

due to topographic shading and the four-to-eight-fold concentration of snow by wind in east-and northeast-facing cirques (Outcalt and MacPhail, 1965).

STOP 4-1. UPPER LIMIT OF PLEISTOCENE GLACIATION (Madole)

The altitude reached by the late Pleistocene valley glacier at this locality is about 3405 m. The area above us to the south was never glaciated. STOP 4-1 is essentially at the present-day timberline, but during the last full glacial, it is estimated to have been about 550 m higher. This estimate is based on the *Picea/Pinus* ratio in strata of glacial Lake Devlin (Legg and Baker, 1980) in the next drainage basin to the south. Little wonder that periglacial features, both relict and active, are widespread on the summit and slopes of Niwot Ridge.

The till at this stop was initially mapped as a Pinedale equivalent (Madole, 1960), then as Bull Lake (Madole, 1969), and then as Pinedale (Gable and Madole, 1976). The Bull Lake age assignment relied heavily on the greater weathering of boulder surfaces in the vicinity of STOP 4-1 relative to that in forested areas below. The fact that the soil here was more weakly developed than in most deposits of Bull Lake age was attributed to erosion and other periglacial processes. Subsequently, greater emphasis was given to the probable positions of late Pleistocene ice margins, based on estimates of Pinedale equilibrium-line altitudes. Research also showed that boulders in forested areas typically are less weathered and pitted than boulders of comparable age in areas above timberline. Weathered surfaces of boulders are removed by fire-induced spalling more frequently in forests than in tundra. Thirty ^{14}C ages of charcoal in a small glaciated drainage basin in Rocky Mountain National Park about 40 km to the north, show that large parts of the basin have been burned by stand-replacing forest fires at least 13 different times in the last 6250 ^{14}C yrs. B.P. (Madole, 1995).

STOP 4-2. TREE-ISLAND DYNAMICS IN THE FOREST-TUNDRA ECOTONE (Benedict)

The principal tree and shrub species of the subalpine forest continue into the forest-tundra ecotone (3350–3500 m) as krummholz. Tree islands in the ecotone become packed with snow in winter and develop snow banks to leeward. They provide sheltered microenvironments in which subalpine shrubs and herbs can grow where conditions would otherwise be prohibitively harsh. Increased effective precipitation beneath tree islands, combined with accumulation of windblown dust and a conifer needles, favor development of cambic B horizons (Holtmeier and

Broll, 1992). E horizons occur in the lower part of the forest-tundra ecotone (Benedict, 2000a).

Because of tree-island migration, podzolization at this altitude is a dynamic process. Krummholz spruce and fir trees reproduce vegetatively by layering—an adaptation that allows them to survive for millennia in an environment where reproduction by seed occurs infrequently (Holtmeier, 2000). As needles and twigs on the windward sides of tree islands are killed by freezing or desiccation, low branches on protected, leeward sides become buried by litter, develop roots, and send up new stems. Thus, spruce and fir tree islands migrate downwind across the ecotone, leaving trails of dead wood as long as 15 m to record their passage (Marr, 1977). Radiocarbon dating suggests that tree islands on Niwot Ridge move eastward at average rates of 1–2 cm yr^{-1} (Benedict, 1984). Conditions favorable for podzolization accompany the tree islands, influencing the landscape beneath them for many centuries.

STOP 4-3. ACTIVE SORTED STRIPES IN A LATE-LYING SNOWBED (Benedict)

The distribution of active patterned ground in the Front Range is controlled by soil-moisture availability during the autumn freeze. Areas of late-melting snow are particularly favorable for frost processes. During the period 1964-67, Benedict measured movement of marked stones in a line across two sorted stripes on the floor of a snowbed that does not melt completely every year (Benedict, 1970). Average rates of displacement on the 12–13° slope varied from 1.2 to 3.9 cm yr^{-1}. Stones embedded in fine-textured earth moved twice as rapidly as stones in the rocky stripes. Excavation showed that the stripes of coarse debris were only 20–25 cm deep. The long axes of stones lay more or less horizontally. Microfabric studies of elongate particles in oriented thin sections (Benedict, 1992, Fig. 8a) suggest that silt and fine sand are displaced downward and laterally beneath the stone stripes, rising upward toward the ground surface in the bands of fine material.

STOP 4-4. STONE-BANKED TERRACES (Benedict)

Some of the best-developed, stone-banked terraces in the world occur on Niwot Ridge. With fronts as high as 2.7 m, the terraces are draped like strands in a necklace across an east-to-southeast-facing, early-melting, snow-accumulation slope. The terraces move slowly today, with average annual surface velocities of 0.2–0.7 cm yr^{-1}. Fronts are virtually immobile. Slow rates of movement demonstrate that saturation during the spring snowmelt period does not in itself ensure rapid displacement (Benedict, 1970).

A trench across the front of a stone-banked terrace revealed a surface layer of openwork cobbles and boulders, 60-90 cm thick, overlying a relatively stone-free layer of fine-textured, dark brown to very dark grayish brown (10YR 3/3-3/2) sediment with strongly developed platy structure. These units constitute the moving portion of the terrace. Beneath them was nonsorted colluvium with a very dark brown (7.5YR 2/2) Ab horizon and an oxidized (7.5YR–10YR 3/4) Bw horizon. The tops of rocks that protruded through the Ab horizon displayed cavernous weathering, indicating lengthy exposure at the ground surface prior to burial (Benedict, 1970). This is consistent with presence of a well-developed Bw horizon.

The buried humus layer could be traced upslope for 8.35 m behind the terrace front. Radiocarbon dates for samples from this stratigraphic unit, corrected for the estimated mean residence time of carbon at time of burial, suggest that the terrace developed during a period of greatly increased snowcover that ended about 2470 ± 110 ^{14}C yrs. B.P. (I-1792). As the snow-free season lengthened, vegetation became reestablished on the slope, and the A horizon began to form. Frontal advance was especially rapid (2.4 cm yr^{-1}) between 1250 ± 120 (I-1697) and 1140 ± 90 ^{14}C yrs. B.P. (I-4044) (Benedict, 1976). This probably occurred in response to an interval of moderately increased snowfall when the ground was saturated, but snow-free, at the beginning of autumn freezing.

STOP 4-5. SORTED NETS ON A DRY TUNDRA KNOLL (Benedict)

At the top of the stone-banked-terrace slope, plugs of fine-textured earth have intruded a frost-sorted blockfield from below, forming well-defined sorted nets 1.5 to 4.3 m in maximum dimension. The centers of the nets are vegetated, and the stones in their borders are heavily lichen covered, suggesting that they are fossil forms. Inactive sorted nets are common in the Front Range on exposed knolls and ridge crests that receive much less effective precipitation than a shielded rain gauge would collect. Massive freezing, rather than ice segregation, occurs in the soils of these localities today. Existing theories of patterned ground formation are inadequate to explain the sorted nets. One possible explanation (Benedict, 1992) involves (1) formation of large volumes of clear ice in the soil due to freezing of meteoric water as the permafrost table rose during cold phases of the Pleistocene, (2) vertical frost sorting in the active layer, causing development of a metastable profile in which dense, rocky, surface material was underlain by less-dense silty sediment from which stones had been ejected (Czeppe, 1961), and (3) collapse of the system at the close of the Pleistocene when rapidly melting ground ice encouraged liquefaction of the fines. Small-scale laboratory experiments show that when such systems are disturbed, plugs of low-density sediment rise upward into the high-density surface layer, forming netlike patterns. Coarse particles move laterally from the up-domed centers, accumulating in the troughs between them (Dzulynski, 1963; Benedict, 1992).

STOP 4-6. MIOCENE (?) DIAMICTON ON THE WESTERN PART OF NIWOT RIDGE (Madole)

From this stop to just beyond the high point on Niwot Ridge, a distance of about 2 km, bouldery diamicton, 3-36 m thick, mantles the ridge. The diamicton is composed almost entirely of material derived from Precambrian granitic and gneissic rocks, but overlies monzonite and syenite of a stock that was emplaced in the early Cenozoic. The geometry and composition of the diamicton show clearly that it was transported to its present location from a source area to the northwest. For many years, the diamicton was believed to be till of a pre-canyon glaciation (Wahlstrom, 1947; Richmond, 1957; Madole, 1960, 1963). It, together with similar deposits on Tungsten Mountain and Winiger Ridge about 11-15 km to the southeast, was used to argue that the Front Range once had an ice cap that was considerably larger and extended much farther east than all of the valley glaciers that came later. Also, these deposits were used to argue that the Flat Top and Rocky Mountain erosion surfaces (also called summit and subsummit surfaces) were parts of a single erosion surface that had been offset by faulting (Wahlstrom, 1947). Madole (1982) investigated four possible origins for the diamicton—periglacial mass-wasting, landsliding (rockfall—rockslide avalanche), glaciation, and fluvial deposition—and concluded that the diamicton is most likely alluvium and colluvium that were deposited on a valley floor prior to regional uplift in the late Cenozoic. Similar diamictons occur in several mountain ranges in the Southern Rocky Mountains (Madole, unpublished data), many in locations that preclude a glacial origin.

The soils developed in the diamicton are considerably younger than the parent material. Most probably postdate the last glaciation (Pinedale). Osburn and Cline (1967) described soils at two localities on the western part of the diamicton. The sola at the two localities are 46 and 36 cm thick, and the profiles consist of O/A/Bs/BC/C horizon sequences. The clay content of the A1 horizon (25 and 20.9 percent) is 2.3 to 6.0 times greater than that of the Bs horizon. Mahaney and Madole (unpublished data, 1972) described soil profiles, similar to those reported by Osburn and Cline (1967), in four 2-m deep excavations in the diamicton at altitudes between 3640 m (12,000 ft.) and 3730 m (12,300 ft.). Field sketches and laboratory data from one of the profiles is shown in Figure 13.

STOP 4-7. HYDROLOGY AND RATES OF EROSION, GREEN LAKES OVERLOOK (Caine)

The hydrology of Green Lakes Valley above the old mining camp of Albion has been studied since 1981 as part of the Long Term Ecological Research Program. This is part of the City of Boulder watershed and supplies about 15% of the city's water supply. Flows at all sites within the basin are clearly dominated by snowmelt (Fig. 14a), although convective rainstorms in summer introduce transient responses that may be important for sediment transport in the basin. Specific discharges through the basin fit a regional pattern of flow depths that decline with increases in basin area and thus lower elevation (Fig. 14b). In the highest parts of the basin, annual discharges exceed precipitation totals at the D-1 station (1144 mm for the 1971-2000 period). This excess is explained by the concentration of snow from wind drifting in the higher part of the valley and, in the past few years, by a negative mass balance (4.5m water equivalent) on the Arikaree Glacier (Figure 15). Sediment budgets for Green Lakes Valley above Green Lake 4 (Caine, 1986) are revised here (Table 2).

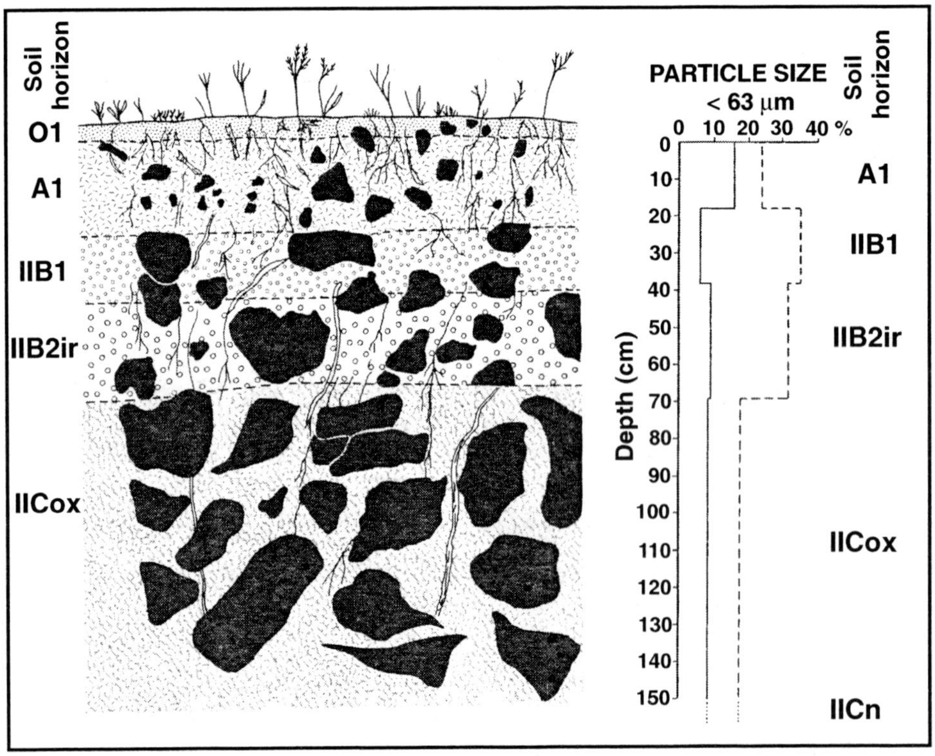

Figure 13. Graphic log and particle size from soil profile NR-3 in the diamicton along western Niwot Ridge.

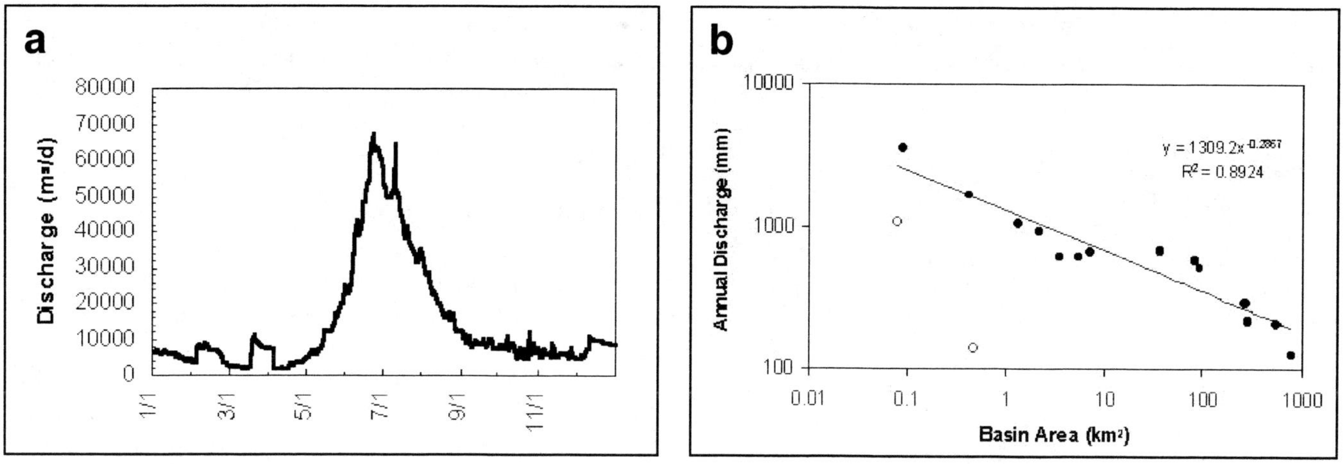

Figure 14. (a). Average daily water discharge (m^3) at Albion, based on the 1982-2002 period. February and March flows show the effect of intermittent reservoir management at Lake Albion. (b). Specific discharge (mm/yr) from drainage basins in Boulder County. Sites without significant flow abstraction or addition and with more than 10 years of record (not always the same years) are shown. The open symbols represent the Green Lake 1 and Martinelli drainages in Green Lakes Valley and have not been included in the regression.

The coarse debris system accounts for over 95% of all geomorphic work accomplished in the valley at the present. Within that system, rockfall dominates because of the relatively great (vertical) distance of transport that it entails, especially in the larger falls, which often traverse the entire talus slope. Geomorphic work from talus shift is less than this by a factor of three or four, despite the large mass of material involved. By comparison, other processes of coarse debris movement are relatively slight contributors to work done in the basin, although they may be important on individual slopes.

As in other coarse debris systems (e.g. Rapp, 1960), contemporary rates of talus accumulation and shift in Green Lakes Valley are too low to account for the volume of coarse debris accumulated on the valley walls since deglaciation (Caine, 1986). About 0.8 x 10^6 m^3 of talus and an additional 1.0 x 10^6 m^3 of other debris have accumulated in the upper valley over 12 ka, i.e. at an average rate of almost 150 m^3 yr^{-1}, an order of magnitude greater than the rate shown by contemporary measurement (Table 3). This disparity argues that coarse debris was produced at rates up to two orders of magnitude greater than the present during the late Pleistocene and in Neoglacial periods (Harbor, 1985). As Wahrhaftig (1987) pointed out with respect to rock glacier systems, this coarse material is likely to remain stored in alpine areas until the onset of another major glaciation.

The large mass of material involved in soil creep and solifluction accounts for almost half of the work done in the fine-sediment system. Fluvial transport of silt and clay to lake basins involves a smaller mass but a greater vertical transport, which allows it to contribute more than 50% to the work done in the system. The major uncertainty in this budget concerns eolian transport, for which the volume of sediment involved and distances of transport are still only poorly estimated.

In contrast to the systems involving clastic sediment and debris, the geochemical system is one in which the coupling of slope and stream systems is closest. For that reason, it is estimated with greater precision as an average for entire drainage basins. In terms of total mass and work estimates, it has relatively low magnitude, accounting for only 16% of the total geomorphic work done in the basin and a much smaller proportion of the mass involved in geomorphic exchanges (Table 3). Nevertheless, it dominates the estimate of basin denudation, accounting for more than 92% of the work done in sediment and solute export and in lake sedimentation.

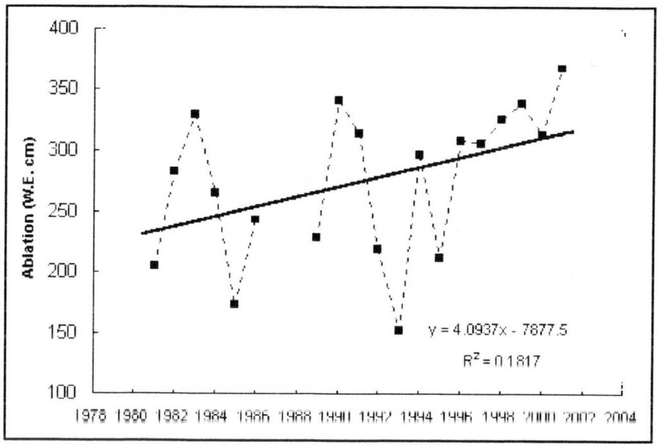

Figure 15. Time series of summer mass balance on Arikaree Glacier, 1981-2002.

STOP 4-8. MARTINELLI BASIN (Caine)

In good winters, the Martinelli Basin accumulates more than 15 m of snow depth in two distinct hollows arranged in series on the south side of Niwot Ridge. This has been the site of studies of snow hydrology and denudation in "nivation hollows" since the early work of Thorn (1976).

TABLE 2. SEDIMENT BUDGETS FOR THE UPPER GREEN LAKES VALLEY

System	Volume# (m^3 yr^{-1})	Work# ($*10^6$ J yr^{-1})
Coarse Debris		
Rockfall	10.0	44.6
Talus Accumulation	1.5	2.87
Talus Shift	206500	14.43
Debris Flow	9.4	15.32
Rock Glacier Flow	373000	3.59
Subtotal	579521	80.81
Fine Sediment		
Soil Creep and Solifluction	95440	0.90
Surficial Wasting	240	0.03
Lake Sedimentation	13.5	1.00
Fluvial Sediment Export	0.38	0.05
Eolian Transport	20	?
Subtotal	95714	2.25
Geochemical		
Solute Transport	1.9	13.36

Values are estimated for Green Lakes Valley above Green Lake 4. Solute transport, corrected for solutes in precipitation, is based upon the sum of major cations and Si concentrations and an assumed bedrock density of 2600 kg m^{-3}. From Caine (1986) with selected updates and additions.

TABLE 3. GEOCHEMICAL DENUDATION: GREEN LAKES VALLEY

	T.D.S.	Cations	Si	Cl	Water*	Sediment
Precipitation	5.63	0.58	0.01	0.23	1144	1.00†
Arikaree	13.5	2.25	1.27	0.27	3689	
Navajo	8.76	1.69	1.42	0.14	1510	0.98
Green L. 5	8.14	1.74	1.04	0.10	1070	
Green L. 4	9.22	1.94	1.25	0.12	927	1.01
Albion Inlet	7.60	1.81	0.73	0.10	594	
Spillway	8.24	1.98	0.71	0.08	574	
Albion	11.68	2.67	1.42	0.12	684	1.94
Green L. 1	5.04	1.20	0.50	0.02	147	
Martinelli	10.01	1.95	1.78	0.14	1070	1.78

* Except for water (mm), all values are g m^{-2} yr^{-1} and are based on 15 to 20 years of record (1981–1994).

† Dustfall estimate includes only the silt+clay fraction.

Water flows from the 0.08 km^2 basin reflect the seasonal pattern of snow accumulation and melt, rising rapidly to a peak that may exceed 70 liters/sec in late June and then receding to cease in August or September. Superimposed on the seasonal pattern for much of the summer is a marked diurnal cycle that shows a forward phase shift produced by decreasing snow depths, which determine the timing of water transfers to the basin outlet. Diurnal flow variations also are reflected in water quality and can be used to separate flows into surface and subsurface components, suggesting that more than half the flow is routed through surficial deposits even when the basin is snow-covered.

Rates of clastic sediment transport into the Snowpatch hollow and out of the basin show a marked disparity (Figure 16, upper), which suggests that the basin ought to be filling. This conclusion is supported by mid-Holocene paleosols at about 1 m depth in similar snow accumulation sites in the Front Range (Burns, 1979). Distributing the solute flux of major cations and Si (corrected for atmospheric deposition) from the basin according to differences in snow accumulation within the basin (Fig. 16) may be sufficient to slow such infilling and so maintain the hollow form over time.

STOP 4-9. SADDLE STREAM (Caine)

This stream drains the basin to the east of the Martinelli catchment and has an area of about 0.25 km^2, about three times that of Martinelli Basin. Hydrologic and related studies have been conducted since about 1997. Despite its larger area, annual discharge through the Saddle Stream is lower (284 mm) and more variable than that at Martinelli, in response to a shallower and more variable winter snow cover.

Despite lower flows and reduced flow duration, suspended sediment concentrations and sediment transport rates through Saddle Stream are about twice those at Martinelli. This contrast appears to be related to the contrast in snow depths between the two basins. A shallower snow cover means that the basin floor is exposed to summer rainstorms for longer periods and allows pocket gopher activity, as a sediment source (Thorn, 1982), close to the channels.

STOP 4-10. TURF-BANKED LOBE (Benedict)

The turf-banked lobe at this stop remains largely snowfree in winter, and receives groundwater recharge from melting snow banks to the northeast. Freezing is deep and effective, with seasonal frost heave as great as 36 cm—one of the largest values recorded in any environment (Benedict, 1976). Movement is greatest along the axis of the lobe about 15 m behind its front, where test-well data suggest that the water table lies within 20 cm of the ground surface at the beginning of the autumn freeze (Benedict, 1970). Regular monitoring of frost heaving and downslope movement from October 1965 to December 1966 showed that *potential frost creep* in autumn and winter varied from 1.0 to 2.2 cm yr^{-1}, but that *retrograde movement* due to non-vertical settling during thaw partially nullified this displacement. Downslope movement attributable to *solifluction* exceeded 2.2 cm yr^{-1} in spring and early summer. The data indicate that solifluction is an important process near the center of the lobe, but that lobe margins experience only frost creep. A column of cement rods installed in a vertical auger hole at the lobe axis in 1963 was excavated four years later while in a partially frozen state. Differential heaving was restricted to the upper 70 cm of the column, and measurable downslope movement to the upper 50 cm (Benedict, 1970).

This lobe, and others like it in the region, are thought to have formed at the close of the Pleistocene, when melting of icy permafrost triggered rotational landslips that later were modified by frost creep and solifluction (Benedict, 1970). A profile pit in a turf-banked lobe directly upslope from this site revealed an Ab horizon 100–110 cm below ground surface. A ^{14}C date of 10,400 ± 400 ^{14}C yrs. B.P. (Gak-3823, Mahaney and Fahey, 1980) provided a maxi-

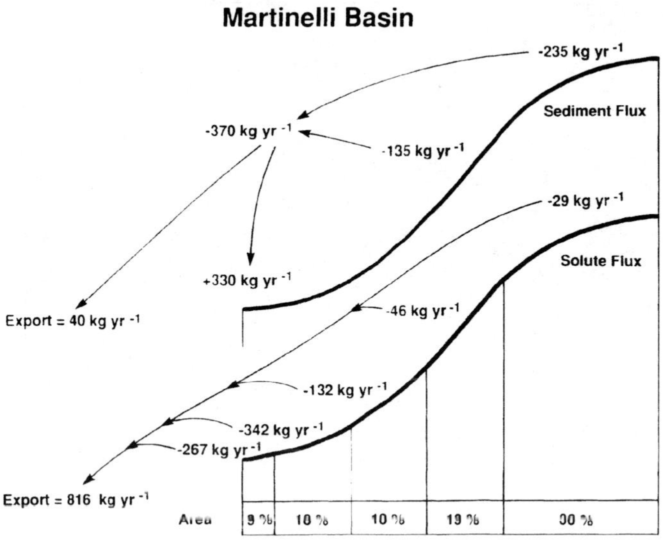

Figure 16. Materials budgets for the Martinelli Basin. Sediment flux based on estimates of mass wasting and clastic sediment transport rates. Solute flux estimates based on average yields, weighted for the varying snow depths in the basin.

mum age for burial of the sample by the advancing lobe and is consistent with formation of the lobe during the late Pleistocene.

The lobe surface looks strikingly different today than it did during the early 1960s. Repeat photography during the forty-year period documents (1) encroachment of vegetation into oriented ponds near the rear of the lobe, (2) revegetation of bare areas, (3) formation of new earth hummocks, (4) accumulation and subsidence of stones in a dilation crack near the lobe front, and (5) eastward expansion of wind-erosion scarps on the lobe's windward flank. This is a truly dynamic environment.

DAY 5. GLACIAL GEOLOGY, ROCK GLACIERS, SOILS, AND ARCHEOLOGY OF THE ARAPAHO PASS AREA

Travel south out of Nederland on Highway 119 and in about 0.5 mi (0.8 km) turn right onto County Road 130. Stay in the valley (don't go to the ski area). At the W edge of Eldora, the road turns to dirt. At a fork in the road, turn right to the Fourth of July Campground parking area, shown on Fig. 17.

STOP 5-1. FOURTH OF JULY MINE (Benedict, Birkeland)

The Fourth of July Lode claim was staked on July 4, 1872, and was worked intermittently until 1937, producing small quantities of gold, silver, copper, lead, and zinc. A ridge that slopes eastward from the mine resembles a lateral moraine, but is best interpreted as an erosional remnant of the till-covered valley floor, left as an interfluve by stream downcutting. This ridge has been used as a campsite since at least 9500–8000 yrs. B.P., judging from Scottsbluff and Eden projectile points found on the surface. Two stone-filled roasting or steaming pits and three shallow-basin hearths exposed in the bed of the wagon road that serves as the modern trail were excavated in 1992 and 1993. Based on the wagon-road transect, at least 300 cooking features are estimated to be present on the ridgecrest, making this one of the largest campsites in the region (Benedict, 2000b). A reliable water supply, large areas of level ground, and a sheltered location enroute to a major pass across the Continental Divide account for the site's popularity. Glacial deposits in the cirque north of the mine were described and interpreted by Mahaney (1973), Williams (1973), and Anderson (1978).

STOP 5-2. FACIES RELATIONS OF VALLEY-HEAD DEPOSITS (Madole)

A variety of morphostratigraphic units (body of rock identified by landform), such as talus, rock glaciers, and moraines, are widespread in the valley heads of the region. At first glance, these deposits may seem like chaotic masses of rock rubble and raise questions such as posed by Roots (1954), who wondered why adjacent talus cones and rock glaciers, although composed of the same material, assumed different forms. In much of the study area, adjacent talus and rock-glacier deposits are of different ages and may have formed under different conditions.

Madole (1972) proposed that various valley-head deposits could be treated as facies, each representing a different depositional environment. Frost riving and rockfall produce detritus over broad areas, but depending upon the environment in which the detritus comes to rest, it becomes part of talus, a lobate rock glacier, a tongue-shaped rock glacier, or a moraine. Differences in topoclimate, which are controlled chiefly by altitude and aspect, influence the distribution of facies. Oscillations of climate cause facies to shift up or down valley and also to contract or expand. These oscillations result in the vertical stacking of facies analogous to the facies sequences of marine transgressions and regressions. Thus, geologic history can be inferred from the vertical succession or horizontal distribution of valley-head deposits according to the stratigraphic axiom known as Walther's Law, which states that facies occur in the same order horizontally as vertically.

STOP 5-3. ARAPAHO PASS (Benedict)

Arapaho Pass, at an altitude of 3629 m, is one of a handful of easy travel routes between the east slope of the Front Range and Middle Park, a resource-rich basin west of the Continental Divide. As herds of elk (*Cervus elephus*) and bands of bighorn sheep (*Ovis canadensis*) crested the pass during their seasonal migrations, hunters waited in ambush (Benedict, 1985). Rock walls and cairn lines with an aggregate length of 1.2 km were used in conjunction with topographic barriers and twenty circular and semicircular blinds to drive and kill the animals. Similar systems were used for hunting caribou in the Canadian Arctic as recently as the 1920s. Size-frequency analysis of subgenus *Rhizocarpon* thalli

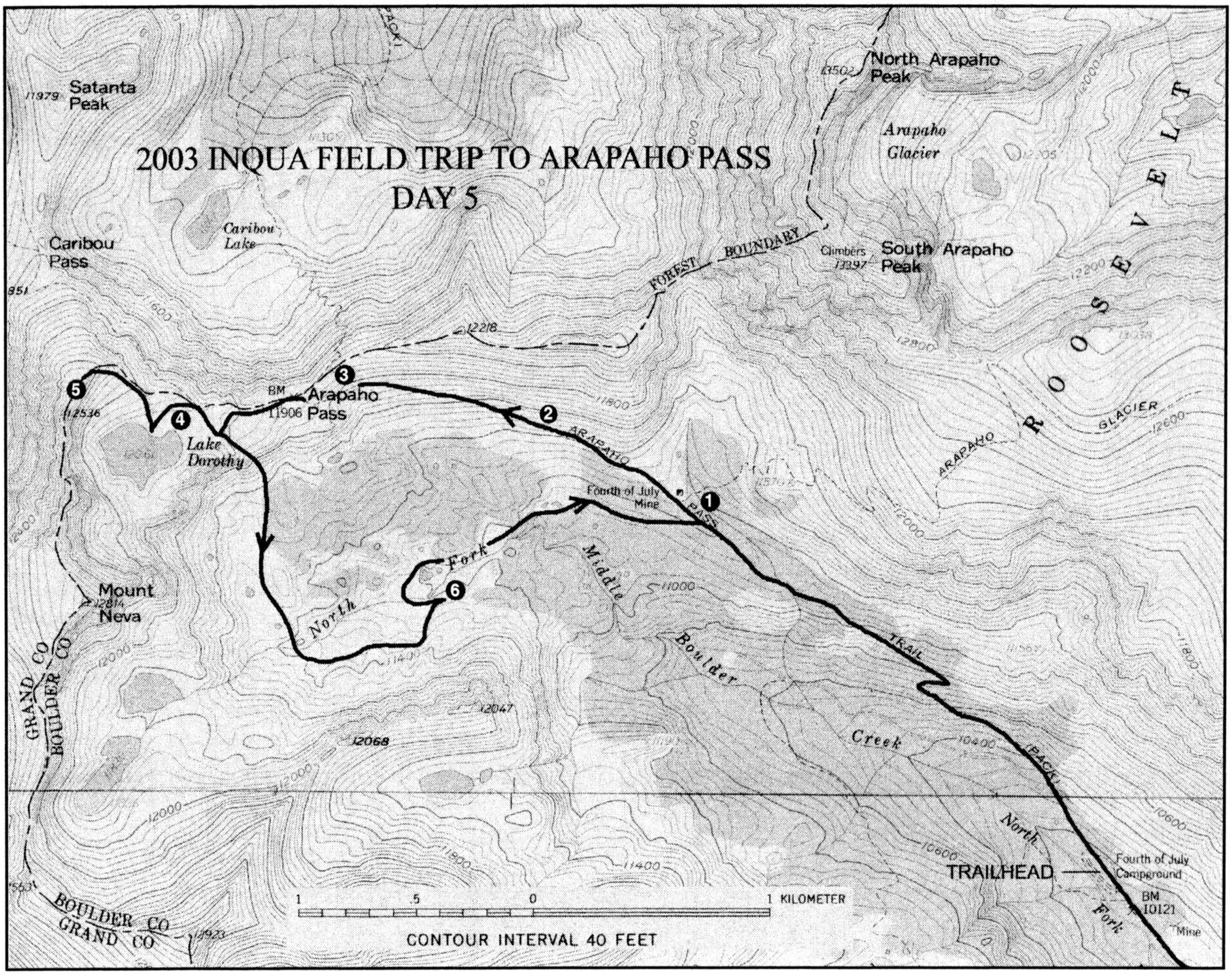

Figure 17. Day 5 stops in the Arapaho Pass area.

suggests two periods of wall construction. The oldest walls predate an episode of lichen snowkill that ended 1050–1040 ^{14}C yrs. B.P. The youngest walls were built between 850 and 780 ^{14}C yrs. B.P. (Benedict, 1985).

The pass provides an overview of the Satanta Peak (cf. Younger Dryas) type locality, and of an important paleoindian campsite (site 5GA22) on the Satanta Peak moraine that impounds Caribou Lake. Excavation of a 76-m^2 area at the site in 1970 and 1971 produced paleoindian artifacts in association with a basin hearth dated at 8460 ± 140 ^{14}C yrs. B.P. (I-5449, Benedict, 1985). This date was the first evidence that Front Range moraines previously thought to have formed about 2800 ^{14}C yrs. B.P. (Richmond, 1960) might be late Pleistocene in age (Benedict, 1974). Pitblado (2000) reopened the site in 1995 and 1996, recovering additional late paleoindian artifacts, and obtaining dates of 7940 ± 70 (AA-26255) to 9080 ± 75 ^{14}C yrs. B.P. (AA-18821). Midge-larva mandibles from proglacial-lake sediments beneath 2 m of peat in a fen separating the two outermost Satanta Peak moraines gave an AMS radiocarbon date of 11,700 ± 60 ^{14}C yrs. B.P. (Beta-101399, Benedict, unpublished). This date and others from the vicinity suggest an age range of 12,000–10,000 ^{14}C yrs. B.P. for Satanta Peak advances.

STOP 5-4. GLACIAL HISTORY OF THE LAKE DOROTHY AREA (Benedict, Birkeland, Davis,)

Lake Dorothy (3676 m; 12,100 ft.) occupies a bedrock-and moraine-dammed basin in a cirque on the north flank of Mount Neva. Lacking a surface outlet, the lake drains to both sides of the Continental Divide through subsurface channels. Water levels are lowest at the beginning of the summer melt season. Two former lake floors, 1 m and 3 m above the modern high-water line, are separated by an escarpment. Sorted nets on the upper lake floor formed at a time of increased meltwater production, when the surface was submerged in summer, but exposed to deep freezing in winter (Benedict, 1985).

Using lichenometry, Benedict (1985) dated the moraine that impounds Lake Dorothy at about 3945 yrs. B.P. and the lake floor about 3 m above the modern late-summer high-water at about 3785 yrs. B.P. Benedict (1985) also ^{14}C-dated organic material recovered from a bog in the channel of the former stream that drained Lake Dorothy as 3450 +/- 75 ^{14}C yrs. B.P. (I-12,841; weighted average of two ages on humic acid and humin fractions). Thus, Benedict (1985) suggested that the moraine confining Lake Dorothy is correlative with the Triple Lakes moraines in Arapaho cirque, which he believed are early Neoglacial in age (Benedict, 1973a). However, Davis (1987; 1988) dated organic material from a pond dammed by the type Triple Lakes moraines at about 10,000 ^{14}C yrs. B.P. (based on two ages of 10,410 +/- 520 ^{14}C yrs. B.P. [GX-11774] and 9915 +/- 380 ^{14}C yrs. B.P. [GX-11365]) and suggested that the type Triple Lakes moraines are Younger Dryas or older in age.

During the winter of 1991, Davis et al. (1992) recovered sediment cores 2.7 m long from Lake Dorothy (~35m water depth, ~3675m, 12,000 ft.) and 3.1 m long from Blue Lake (~37 m water depth, ~3445m, 11,400 ft.), 11 km to the north. A lightweight, percussion, coring system suspended from perlon ropes was used because of sediment thicknesses, water depths, and ski-backpacking requirements. Lake ice provided a stable coring platform. The sediment core from Lake Dorothy did not penetrate gray inorganic diamict and is entirely organic-rich. A bulk sample from 2.65-2.7m depth (Fig. 18) yielded a minimal conventional ^{14}C age of 10,910 +/-320 ^{14}C yrs. B.P. (GX-17863). The moraines impounding Lake Dorothy are thus late Pleistocene in age and 2 to 3 times older than suggested by Benedict (1985). The Blue Lake core bottomed in sandy, gray, inorganic sediment, presumably glacial diamict. A bulk sample from 2.8-2.9 m depth yielded a conventional ^{14}C age of 12,275+/-345 ^{14}C yrs. B.P. (GX-17878). Thus, ice retreated from the cirque occupied by Blue Lake more than 12,000 years ago. Davis believes that the sediment core from Lake Dorothy did not penetrate gray, inorganic sediment because a false signal in the coring operation suggested sediment refusal. Nevertheless, ^{14}C ages from sediments in Lake Dorothy, Blue Lake, and the pond in Arapaho cirque suggest that moraines impounding these lakes are all late Pleistocene in age.

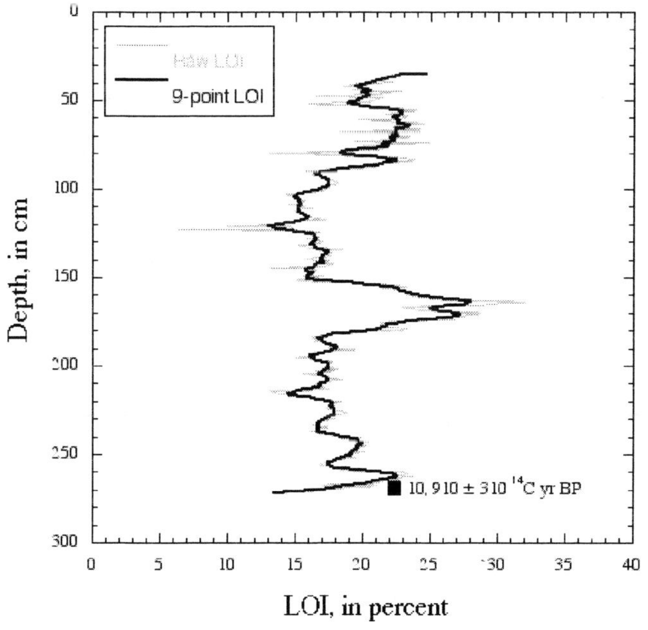

Figure 18. Loss on ignition plotted against depth for 474 contiguous sediment slices 0.5 cm thick from 2.7m sediment core from Lake Dorothy. Radiocarbon age indicated by solid box.

Nearly all 474 oven-dried (105° C) sediment samples ignited for 2 hrs at 450° C have loss-on-ignition (LOI) values between 14 and 28% (Fig. 17). Missing LOI values for the top 35 cm were the result of overdriving the core barrel, so the Little Ice Age part of the core is missing. The LOI trend is unusual in that low values typically representative of early Neoglaciation are lacking. However, a double spike in high LOI values centered at about 165 cm depth could represent a mid-Holocene thermal maximum (about 6700 ^{14}C yrs. B.P., if the sediment accumulation rate is uniform).

Soils in the Arapaho Glacier area were described and analyzed from a sequence of moraines mapped by Benedict (1973). Soil-profile morphology changes with time as follows: thin A/Cox (0.1-0.35 ky), A/AC/Cox (0.95-2.4 ky), A/Bw/Cox (3-5.2 ky), and A/Bw or Bt/Cox (10-12 ky). Soils 3 ka and older seem to have surface-horizon textures influenced by airborne dust, and some of the latter may account for some of the translocated clay in the Bt horizons of the oldest soils. Most soil laboratory properties change progressively with time. The most strongly developed 10-12-ka soil consists of an A horizon (15 cm; 7.5YR2/2d; 13.9 % clay; pH 6.0; 13.9 % org. C; 0.75 % Fed) over a 2Bt horizon (38 cm; 7.5YR4.5/4d; 9.1 % clay; 5.2 pH; 2.09 % org. C; 0.72 % Fed) over a 2Cox horizon (62+ cm; 10YR5/3.5d; 3 % clay; pH 5.3; 0.48 % org. C; 0.51 % Fed). Soils like this are not uncommon above treeline, suggesting that substantial soil formation has taken place in the harsh climate.

STOP 5-5. COLUMBINE ROCK GLACIER (Benedict)

Two morphologically different types of rock glaciers (*cirque-floor* and *valley-wall)* occur in the Front Range (Outcalt and Benedict, 1965). *Cirque-floor rock glaciers* are restricted to the heads of glaciated valleys. Tongue-shaped in plan view, they display arcuate transverse ridges and longitudinal furrows. Spoon-shaped depressions separate them from the glaciers or permanent snowfields at their heads. Active cirque-floor rock glaciers in the Front Range are cored with glacial ice. Surface layers of openwork rubble, resting on layers of ablation till, insulate the ice, allowing it to survive for millenia (Benedict, 1973b). Long-term surface-movement rates approach 20 cm yr^{-1}, but rock-glacier fronts advance more slowly (Benedict et al., 1986).

An optional climb to the Continental Divide west of Lake Dorothy (3820 m, 12,600 ft.) provides a spectacular aerial perspective of a cirque-floor rock glacier at the head of Columbine

Creek. Five, lithologically-distinct, depositional units on the rock-glacier surface suggest that boulders were supplied by landslides or massive rockfalls originating in different parts of the cirque headwall. A subgenus *Rhizocarpon* thallus 243 mm in diameter grows on the outermost lobe, indicating an age greater than 7000 years.

STOP 5-6. GEOARCHAEOLOGY OF THE FOURTH OF JULY VALLEY SITE (Benedict)

Leaving the trail, our route down the south side of the Fourth of July valley gives us a close look at a group of *valley-wall rock glaciers*, the other morphological type that occurs in the Front Range (Outcalt and Benedict, 1965). Such landforms develop at the bases of steep cliffs flanking glaciated valleys. Lobate in shape, they lack most of the surface features found on their tongue-shaped counterparts. Rock-glacial sorting and bare, oversteepened (42–44°) fronts suggest that some of the lobes are advancing today. The internal ice needed for their movement is likely to have multiple origins. Debris covers are the result of rockfalls and landslides-some caused by the collapse of sackungen (sagging fault blocks caused by expansion of valley walls. Although relative dating is complicated by late Holocene debris-flow and avalanche material, at least two generations of rock glaciers are present in this group (Benedict, 1981).

The Fourth of July Valley site (5BL120) is directly in front of a rock glacier attributed to the younger generation. Martin and Whalley (1987, Fig. 3) illustrated this landform as an example of the valley-wall, or lobate, type of rock glacier. First excavated in 1971, the site occupies an abandoned spillway at the crest of the valley's outer Satanta Peak terminal moraine. As ice retreated from the moraine, proglacial lake silt and clay accumulated against its inner slope. Advance of a second Satanta Peak glacier to within 100 m of the site caused the lake sediments to become buried by gravelly outwash. A shift to warmer conditions about 10,000 ^{14}C yrs. B.P. caused ice to melt from the moraine core, leading to localized slumping of lake sediments and outwash. Not long thereafter, prehistoric hunters camped in the abandoned spillway. Their chipped-stone implements and waste flakes suggest tasks such as replacement of broken projectile points, sharpening of use-dulled tools, processing of game and hides, and working of wood or bone (Benedict, 1981). Lanceolate and stemmed projectile points with ground, concave bases and oblique, parallel flaking share traits with James Allen (late paleoindian) and McKean (middle Archaic) points. ^{14}C dates from the site suggested an Early Archaic age of ca. 6000 yrs. B.P. This led to the hypothesis that the James Allen complex persisted along the well-watered crest of the Front Range for 1500 years after its disappearance from dry foothills and plains environments at the onset of the Altithermal Interval—evolving at high altitude into what later became known as the McKean complex (Benedict, 1979, 1981).

Laboratory study of a sediment column collected from the site in 2001 sheds doubt on this hypothesis. Large, charred sclerotia (the resting stages of mycorrhizal fungi) show that tree islands grew in the excavation area during the early and middle Holocene, raising the possibility that 6000-^{14}C-year-old charcoal at the site is the result of wildfire. Microflakes (flakes small enough to pass through fine sieves) were concentrated in loess-rich, reworked outwash *below* the 6000-yr-old charcoal layer, and were associated with charcoal particles dated at about 9000 ^{14}C yrs. B.P. Because microflakes are too small to participate in vertical frost sorting, they record the stratigraphic positions of occupation surfaces more accurately than large flakes and artifacts, which in periglacial environments are commonly upfrozen into overlying sediments. The waste flakes and chipped-stone tools recovered in 1971 may have been frost-sorted upward, accumulating on the ground surface that existed when wildfire consumed tree islands at the site and forming a pseudo-occupation layer.

Sometime following occupation, the valley-wall rock glacier advanced to the site's southern margin, bombarding the ground surface with angular and subangular boulders. The 14-m-high front of the rock glacier altered wind patterns, creating a bolster eddy in which snow accumulates to depths of 8–10 m. Tree islands were unable to survive on the moraine crest due to the short growing season. Tundra vegetation was almost totally excluded from the area of deepest snow, exposing artifact-bearing sediments to needle-ice activity and shallow frost creep. Much of site's cultural material was eroded from upper-slope positions and redeposited in natural sediment traps on the lower, inner slope of the moraine (Benedict, 1981).

Annual deposition of windblown sediment in the area affected by the bolster eddy led to development of a 20-cm thick layer of dark brown (10YR 3/3), humus-rich, *snowbed loess*. The basal 2 cm of the unit, which began to accumulate when the rock glacier arrived at the site, has a ^{14}C age of 3330 ± 40 ^{14}C yrs. B.P. (Beta-170170). Analysis of snow samples collected just prior to meltout show that loess currently accumulates on the site at a rate of about 2.8 g m^{-2} yr^{-1}. On a dry-weight basis, half of the newly deposited loess is organic (leaves, insects, conifer needles), and half is clastic (sand, silt, clay). A date of 3340 ± 65 ^{14}C yrs. B.P. (I-11091 avg.) for the onset of clastic deposition in a fen at the rock-glacier front east of the site (Benedict, 1981) confirms arrival of the rock glacier at this time. Nothing in the stratigraphic record suggests that its advance was gradual or uniformitarian: to the contrary, it appears to have been rapid, and perhaps catastrophic. An arcuate ridge of intricately folded and brecciated gravel, sand, and silt, 2.1 to 2.7 m high, formed at the west margin of the rock glacier when saturated lake and outwash sediments were squeezed from beneath the advancing lobe of ice and debris (Benedict, 1981).

To envision the Fourth of July Valley site when hunters camped on the moraine crest about 9000 ^{14}C years ago, the observer must mentally erase the rock glacier from the landscape. The occupation area was smoother and less bouldery than it is today, and the view to the south was unobstructed. A scarcity of winter snow and lack of topographic shading at seasons of low sun angle resulted in longer snow-free summers than the present one- to two-month period, allowing tree islands and herbaceous vegetation to grow where the ground today is bare or sparsely vegetated. Burrowing rodents, now absent, were active in the ancient spillway. No on-site refrigerator cooled the beverages of paleoindian field-trip participants.

ACKNOWLEDGMENTS

We acknowledge many helpful discussions over the years with Bill Bradley, John Pitlick, and Sid White and the patience and thoughtful editorial comments of Don Easterbrook.

REFERENCES

Anderson, L. W., 1978, Late Quaternary chronology of Fourth of July Cirque, Front Range, Colorado: Geological Society of America Abstracts with Programs, 31st Annual Meeting, p. 209.

Benedict, J. B., 1970, Downslope soil movement in a Colorado alpine region. Rates, processes, and climatic significance: Arctic and Alpine Research, v. 2, no. 3, p. 165-226.

________, 1973a, Chronology of cirque glaciation: Quaternary Research, v. 3, no. 4, p. 585-599.

________, 1973b, Origin of rock glaciers: Journal of Glaciology, v. 12, no. 66, p. 520-522.

________, 1974, Early occupation of the Caribou Lake site, Colorado Front Range: Plains Anthropologist, v. 16, no. 63, p. 1-4.

________, 1976, Frost creep and gelifluction features: a review: Quaternary Research, v. 6, no. 1, p. 55-76.

________, 1979, Getting away from it all: a study of man, mountains, and the two-drought Altithermal: Southwestern Lore, v. 45, no. 3, p. 1-12.

________, 1981, The Fourth of July Valley–glacial geology and archeology of the timberline ecotone: Center for Mountain Archeology Research Report 2, 139 p.

________, 1984, Rates of tree-island migration, Colorado Rocky Mountains, USA: Ecology, v. 65, no. 3, p. 820-823.

________, 1985, Arapaho Pass: Glacial geology and archeology at the crest of the Colorado Front Range: Center for Mountain Archeology, Research Report 3, 197 p.

________, 1992, Field and laboratory studies of patterned ground in a Colorado alpine region: University of Colorado, Institute of Arctic and Alpine Research Occasional Paper, v. 49, p. 1-38.

________, 2000a, Game drives of the Devil's Thumb Pass area: *in* Cassells, E. S., ed., This Land of Shining Mountains, Archeological Studies in Colorado's Indian Peaks Wilderness Area, Center for Mountain Archeology Research Report 8, p. 18-94.

________, 2000b, Excavations at the Fourth of July Mine site: *in* Cassells, E. S., ed., This Land of Shining Mountains, Archeological Studies in Colorado's Indian Peaks Wilderness Area, Center for Mountain Archeology Research Report 8, p. 159-188.

Benedict, J. B., Benedict, R. J., and Sanville, D., 1986, Arapaho rock glacier, Front Range, Colorado, U.S.A.: a 25-year resurvey: Arctic and Alpine Research, v. 18, no. 3, p. 349-352.

Birkeland, P. W., Miller, D. C., Patterson, P. E., Price, A. B., and Shroba, R. R., 1999, Soil-geomorphic relationships near Rocky Flats, Boulder and Golden, Colorado area, with a stop at the Pre-Fountain Formation paleosol of Wahlstrom (1948): Geological Society of America Field Trip #18, 13 p.

Birkeland, P. W., Shroba, R. R., Burns, S. F., Price, A. B., and Tonkin, P. J., 2003, Integrating soils and geomorphology in mountains—an example from the Front Range of Colorado: Geomorphology, in press.

Braddock, W. A., 1978, Dakota Group rockslides, northern Front Range, Colorado: *in* Voight, B., ed., Rockslides and avalanches, 1: Elsevier Science Publishing Company, N.Y., p. 439-479.

Braddock, W.A., and Eicher, D. L., 1962, Block-glide landslides in the Dakota Group of the Front Range foothills, Colorado: Geological Society of America Bulletin, v. 73, p. 317-324.

Burns, S. F., 1979, Buried soils beneath alpine snowbanks may date the end of the Altithermal: Geological Society of America, Abstracts with Programs, v. 11, p. 267-268.

Caine, N., 1986, Sediment movement and storage on alpine hillslopes in the Colorado Rocky Mountains: *in* Abrahams, A. D., ed., Hillslope Processes: Allen & Unwin, Winchester, Mass, p. 115-137.

Colton, R. B., 1978, Geologic map of the Boulder-Fort Collins-Greeley area, Colorado: U.S. Geological Survey Miscellaneous Investigations Map I-855-G, scale 1:100,000.

Clayton, J.L., Megahan, W.F., and Hampton, D., 1979, Soil and bedrock properties: weathering and alteration products and processes in the Idaho batholith: U.S. Dept. of Agriculture Forest Service Research Paper INT-237, 35 p.

Czeppe, Z., 1961, Roczny przebieg mrozowych ruchów gruntu w Hornsundzie (Spitsbergen) 1957-1958: Prace Geograficzne, Seria Nowa, v. 3, p. 1-74.

Davis, P. T., 1987, Late Pleistocene age for type Triple Lakes moraines, Arapaho Cirque, Colorado Front Range: Geological Society of America Abstracts with Programs, v. 19, no. 5, p. 270.

________, 1988, Holocene glacier fluctuations in the American Cordillera: Quaternary Science Reviews, v. 7, p. 129-157.

Davis, P. T., Birkeland, P.W., Caine, N., and Rodbell, D.T., 1992, New radiocarbon ages from cirques in Colorado Front Range: Geological Society of America, Abstracts with Programs, v. 24, no. 7, p. A347.

Dietrich, W. E., Wilson, C. J., and Reneau, S. L., 1986, Hollows, colluvium and landslides in soil-mantled landscapes: *in* Abrahams, A. D., ed., Hillslope Processes: Allen & Unwin, Winchester, Mass, p. 361-388.

Dzulynski, S., 1963, Polygonal structures in experiments and their bearing upon some periglacial phenomena: Bulletin de L'Académie Polonaise des Sciences, Série des sci. géol. et géogr., v. 11, no. 3, p. 145-150.

Gable, D. J., 1969, Geologic map of the Nederland quadrangle, Boulder and Gilpin Counties, Colorado: U.S. Geological Survey Geologic Quadrangle Map GQ-833. Scale 1:24,000

Gable, D. J., and Madole, R. F., 1976: Geologic map of the Ward Quadrangle, Boulder County, Colorado: U.S. Geological Survey Map GQ-1277, Scale 1:24,000.

Greenland, D., 1987, The climates of the Long-term Ecological Research sites: University of Colorado, Institute of Arctic and Alpine Research, Occasional Paper, v. 44, p. 1-81.

Harbor, J. M., 1985, Problems with the interpretation and comparison of Holocene terrestrial and lacustrine deposits: An example from the Colorado Front Range, U.S.A.: Zeitschrift fur Gletscherkunde und Glazialgeologie, v. 21, p. 17-24.

Holtmeier, F.-K., and Broll, G., 1992, The influence of tree islands and microtopography on pedoecological conditions in the forest-alpine tundra ecotone on Niwot Ridge, Colorado Front Range, U.S.A.: Arctic and Alpine Research, v. 24, no. 3, p. 216-228.

Holtmeier, F.-K., 2000, Die Höhengrenze der Gebirgswälder: Arbeiten aus dem Institut für Landschaftsökologie, Westfälische Wilhelms-Universität (Münster), v. 8, p. 1-337.

Ives, J.D., and Fahey, B., 1971, Permafrost occurrence in the Front Range, Colorado Rocky Mountains, U.S.A.: Journal of Glaciology, v. 10, no. 58, p. 105-111.

Legg, T. E., and Baker, R. G., 1980, Palynology of Pinedale sediments, Devlins Park, Boulder County, Colorado: Arctic and Alpine Research, v. 12, p. 319-333.

Machette, M. N., Birkeland, P.W., Markos, G., and Guccione, M.J., 1976a, Soil development in Quaternary deposits in the Golden-Boulder portion of the Colorado Piedmont: *in* Epis, R.C., and Weimer, R.J., eds., Studies in Colorado field geology: Professional Contributions of Colorado School of Mines, no. 8,p. 217-259.

Machette, M.N., Birkeland, P.W., Burke, R.M., Guccione, M.J., Kihl, R., and Markos, G., 1976b, Field descriptions and laboratory data for a Quaternary soil sequence in the Golden-Boulder portion of the Colorado Piedmont: U.S. Geological Survey Open-file Report 76-804, 21 p.

Madole, R. F., 1960, Glacial geology of upper South St. Vrain valley, Boulder County, Colorado: M.S. thesis, The Ohio State University, 109 p.

________, 1963, Quaternary geology of St. Vrain drainage basin, Boulder County, Colorado: Ph.D. thesis, The Ohio State University, 288 p.

________, 1969, Pinedale and Bull Lake glaciation in upper St. Vrain drainage basin, Boulder County, Colorado: Arctic and Alpine Research, v. 1, no. 4, p. 279-287.

________, 1972, Neoglacial facies in the Colorado Front Range: Arctic and Alpine Research, v. 4, no. 2, p.119-130.

________, 1976a, Glacial geology of the Front Range, Colorado: *in* Mahaney, W. C., ed., Quaternary stratigraphy of North America: Dowden, Hutchinson, and Ross, Stroudsburg, Pennsylvania, p. 297-318.

________, 1976b, Bog stratigraphy, radiocarbon dates, and Pinedale to Holocene glacial history in the Front Range, Colorado: U. S. Geological Survey Journal of Research, v. 4, no. 2, p. 163-169.

________, 1982, Possible origins of till-like deposits near the summit of the Front Range in north-central Colorado, U. S. Geological Survey Professional Paper 1243, 31 p.

________, 1986, Lake Devlin and Pinedale glacial history, Front Range, Colorado: Quaternary Research, v. 25, no. 1, p. 43-54.

________, 1991, Colorado Piedmont: *in* Wayne, W. J., ed., Quaternary geology of the Northern Great Plains, chap. 15: Quaternary nonglacial geology—conterminous United States, Geological Society of America, The Geology of North America, v. K-2, p. 444-445, p. 456-462.

________, 1995, Fire frequency during the Holocene in a small, forested drainage basin in the Southern Rocky Mountains: Geological Society of America Abstracts with Programs, v. 27, no. 6, p. 171.

Madole, R. F., and Shroba, R. R., 1979, Till sequence and soil development in the North St. Vrain drainage basin, east slope, Front Range, Colorado: *in* Ethridge, F. G., ed.: Fort Collins, Geological Society of America, Rocky Mountain Section Annual Meeting, 32d, p. 124-178.

Madole, R. F., Braddock, W. A., and Colton, R. B., 1998, Geologic map of the Hygiene quadrangle, Boulder County, Colorado: U.S. Geological Survey Geologic Quadrangle Map GQ-1772, scale 1:24,000.

Madole, R. F., VanSistine, D. P., and Michael, J. A., 1999, Pleistocene glaciation in the upper Platte River drainage basin, Colorado: U.S. Geological Survey Geologic Investigations Series I-2644.

Mahaney, W. C., 1973, Neoglacial chronology in the Fourth of July Cirque, central Colorado Front Range: Geological Society of America Bulletin, v. 84, p. 161-170.

Mahaney, W. C., and Fahey, B. D., 1980, Morphology, composition and age of a buried paleosol, Front Range, Colorado, U.S.A.: Geoderma, v. 23, p. 209-218.

Marr, J. W., 1977, The development and movement of tree islands near the upper limit of tree growth in the Southern Rocky Mountains: Ecology, v. 58, p. 1159-1164.

Martin, H. E., and Whalley, W. B., 1987, Rock glaciers, part 1: rock glacier morphology: classification and distribution: Progress in Physical Geography, v. 11, no. 2, p. 260-282.

Nelson, A. R., Millington, A. C., Andrews, J. T., and Nichols, H., 1979, Radiocarbon-dated upper Pleistocene glacial sequence, Fraser Valley, Colorado Front Range: Geology, v. 7: 410-414.

Netoff, D. I., 1977, Soil clay mineralogy of Quaternary deposits in two Front Range-Piedmont transects, Colorado: Ph.D thesis, University of Colorado, 169 p.

Osburn, W., and Cline, A.J., 1967, Penitente loam: *in* Soil survey laboratory data and descriptions for some soils of Colorado: Soil Survey Investigations Report No. 10, Soil Conservation Service, U.S. Department of Agriculture in cooperation with Colorado Agriculture Experiment Station, p. 158-161.

Outcalt, S. I., and Benedict, J. B., 1965, Photo-interpretation of two types of rock glacier in the Colorado Front Range, U.S.A.: Journal of Glaciology, v. 5, no. 42, p. 849-856.

Outcalt, S. I., and MacPhail, D. D., 1965, A survey of Neoglaciation in the Front Range of Colorado: University of Colorado Studies, Series in Earth Sciences, v. 4, p. 1-124.

Rapp, A., 1960, Talus slopes and mountain walls at Tempelfjorden, Spitsbergen: Norsk Polarinst. Skrifter 119, 96 p.

Richmond, G.M., 1957, Three pre-Wisconsin glacial stages in the Rocky Mountain region: Geological Society of America Bulletin, v. 68, p. 239-262.

Richmond, G. M., 1960, Glaciation of the east slope of Rocky Mountain National Park, Colorado: Geological Society of America Bulletin, v. 71, p. 1371-1382.

Roots, E.P., 1954, Geology and mineral deposits of the Aiken Lake map-area, British Columbia: Geological Survey of Canada Memoir 274, 246 p.

Schildgen, T. F., and Dethier, D. P., 2000, Fire and Ice: using isotopic dating techniques to interpret the geomorphic history of Middle Boulder Creek, Colorado: Geological Society of America Abstracts with Programs, v. 32, no. 7., p. A-18.

Schildgen, T., Dethier, D. P., Bierman, P., and Caffee, M., 2002, 26Al and 10Be dating of late Pleistocene and Holocene fill terraces: a record of fluvial deposition and incision, Colorado Front Range: Earth Surface Processes and Landforms, v. 27, p. 773-787.

Scott, G. R., 1960, Subdivision of the Quaternary alluvium east of the Front Range near Denver, Colorado: Geological Society of America Bulletin, v. 71, no. 1541-1543.

Shroba, R. R., 1977, Soil development in Quaternary tills, rock-glacier deposits, and taluses, Southern and Central Rocky Mountains: Ph.D. thesis, University of Colorado, 424 p.

Shroba, R. R., and Birkeland, P. W., 1983, Trends in late-Quaternary soil development in the Rocky Mountains and Sierra Nevada of the Western United States: *in* Porter, S. C., ed.: Late-Quaternary environments of the United States, v. 1: Minneapolis, University of Minnesota Press, p. 145-156.

Shroba, R. R., Rosholt, J. N., and Madole, R. F., 1983, Uranium-trend dating and soil B horizon properties of till of Bull Lake age, North St. Vrain drainage basin, Front Range, Colorado: Geological Society of America Abstracts with Programs, v. 15, p. 431.

Street, A, 1973, A study of tors in the Front Range of the Rocky Mountains in Colorado, with special reference to their value as an indicator of non-glaciation: MA thesis, University of Colorado, 241 p.

Stuiver, M., and Reimer, P. J., 1993, Extended ^{14}C database and revised CALIB radiocarbon calibration program: Radiocarbon, v. 35, p. 215-230.

Thorn, C. E., 1976, Quantitative evaluation of nivation in the Colorado Front Range: Geological Society of America Bulletin, v. 87, p. 1169-1178.

Thorn, C. E., 1982, Gopher disturbance: its variability by Braun-Blanquet vegetation units in the Niwot Ridge alpine tundra zone, Colorado Front Range, U.S.A.: Arctic & Alpine Research, v. 14, p. 45-51.

Tonkin, P.J., and Basher, L.R., 1990, Soil-stratigraphic techniques in the study of soil and landform evolution across the Southern Alps, New Zealand: Geomorphology v. 3, 547-575.

Wahlstrom, E. E., 1947, Cenozoic physiographic history of the Front Range, Colorado: Geological Society of America Bulletin, v. 51, p. 1789-1820.

Wahlstrom, E. E., 1948, Pre-Fountain and recent weathering on Flagstaff Mountain near Boulder, Colorado: Geological Society of America Bulletin, v. 59, p. 1173-1189.

Wahrhaftig, C., 1987, Foreword, in Giardino, J. R., Shroder, J. F., and Vitek, J. D., eds., Rock Glaciers: Allen & Unwin, Winchester, Mass, p. vii-xii.

Welter, S. P., 1995, Source-area/gradient threshold analysis for gully initiation in a Colorado montane forest: Association of American Geographers 91st Annual Meeting, abstracts, p. 322.

Williams, J., 1973, Neoglacial chronology of the Fourth of July Cirque, central Colorado Front Range: discussion: Geological Society of America Bulletin, v. 84, no. 11, p. 3761-3766.

Wrucke, C.T., and Wilson, R.F., 1967, Preliminary geologic map of the Boulder quadrangle: U.S. Geological Survey Open-file no. 67-281.

Soil geomorphological studies in the Mojave Desert: Impacts of Quaternary tectonics, climate, and rock type on soils, landscapes, and plant-community ecology

Authors and Leaders: **M. C. Eppes,** *University of North Carolina, Charlotte, 450 Mc Eniry, Charlotte, NC 28223*
E. V. McDonald, *Desert Research Institute, 2215 Raggio Parkway, Reno, NV, 89512*
L. D. McFadden, *University of New Mexico, Room 122 Northrop Hall, Albuquerque NM, 87131*

DAYS 1 AND 2. THE NORTH FLANK OF THE SAN BERNARDINO MOUNTAINS (Eppes and Mcfadden)

INTRODUCTION

Soils are an important component of a wide variety of Quaternary studies. They are typically employed to establish relative ages of geomorphic surfaces or to semi-quantitatively determine the age of deposits. Chronosequence studies, however, as well as the geologic studies to which they might apply, are limited to certain landscapes. These landscapes favor the preservation of geomorphic surfaces that, after their formation, are abandoned and become isolated from significant continued modification by geomorphic processes. Furthermore, in many settings, geomorphic and geologic variables contribute to soil formation, such that compelling time-trends in soil development are obscured. In some of these instances soil characteristics are controlled by individual factors other than time, and other univariant functions (climofunctions, topofunctions, lithofunctions, etc.) of soil development can be derived. Many workers have evaluated the individual effects of variables, such as parent material, climate, and relief on soil landscapes (e.g. McFadden and Tinsley, 1985; McDonald, 1994; Birkeland, 1999). Of course, in many cases, a landscape will be unavoidably subject to the simultaneous influences of several variables that can potentially affect soil development. Although researchers are beginning to address complicated issues of spatial variability and soil landscape complexity (e.g. Phillips, 1998; Eppes et al., 1999), the typical response of many researchers is to forgo soil-based studies to estimate landform ages or to study other aspects of landscape evolution. Rarely have researchers recognized or exploited the possibility that when soils may not be providing unambiguous evidence about the age of a given landform(s), their character reflects other important soil geomorphic processes. These processes are not only time dependent, but also contribute to produce the soil landscape that we see today.

The primary goals of this trip are to (1) evaluate a soil landscape that does not exhibit simple time-dependent or other univariant trends in soil development, and (2) demonstrate how combined soil and geomorphic studies can none the less greatly contribute to the understanding of that landscape. The northern flank of the San Bernardino Mountains (Fig. 1) is an ideal setting in which to evaluate the influence of multiple factors on soil landscape evolution. The north flank is an alluvial fan piedmont whose soil stratigraphy and morphology is governed by combined factors of parent material, climate, vegetation, relief, and time. Specifically, soil morphological, mineralogical, and chemical studies of the piedmont were integrated with studies of surficial processes, geology, drainage basin attributes, climate, and tectonics in order to document the interrelationships and feedbacks that characterize the landscape of that piedmont. By recognizing and describing multivariate trends in soil development and by relating these trends to geomorphology and landscape evolution, this study makes an important step forward in our use of soils to aid in the overall understanding of alluvial piedmonts.

During this field trip we will (1) explore general trends in soil development as they relate to deposit composition, age, and surface processes along the north flank piedmont, and (2) explore some of the feedbacks between soil development and the response of this landscape to tectonic and climatic forcing.

GENERAL GEOLOGIC BACKGROUND OF THE NORTH FLANK OF THE SAN BERNARDINO MOUNTAINS

The San Bernardino Mountains constitute the eastern portion of the Transverse Ranges of southern California (Fig. 1). The north flank piedmont (field area generally confined to U.S.G.S. quadrangles, Lucerene, Cougar Buttes, Big Bear City, and Fawn-

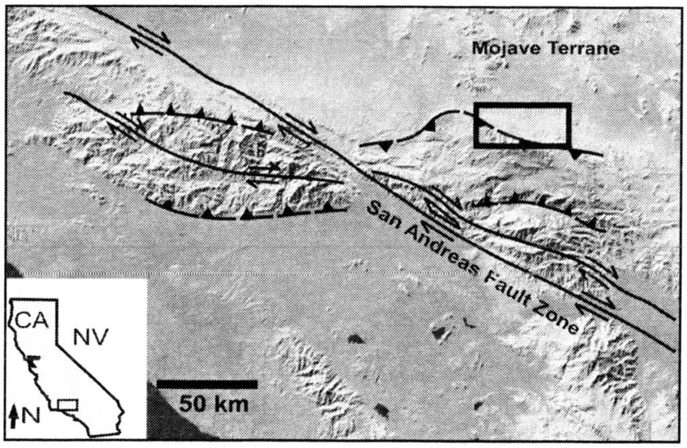

Figure 1. Location map of the field area in southern California.

Eppes, M.C., McDonald, E.V., and McFadden, L.D., 2003, Soil geomorphological studies in the Mojave Desert: Impacts of Quaternary tectonics, climate, and rock type on soils, landscapes, and plant_community ecology: *in* Easterbrook, D.J., ed., Quaternary Geology of the United States, INQUA 2003 Field Guide Volume, Desert Research Institute, Reno, NV, p. 105–122.

skin) forms the rim of a closed desert basin. Rock types exposed in drainage basins along the north flank range from primarily Mesozoic batholithic granite in the western half of the field area, to primarily Precambrian to Paleozoic metasedimentary roof pendants, particularly marble, in the eastern half of the area. Consequently, alluvial fan composition varies from more than 95% granite in the west to more than 85% carbonate rocks in the east, with mixed-composition fans (generally 25-50% limestone) in center portions of the area (Fig. 2). (In order to avoid confusion between the distinction of carbonate rocks and sediment and pedogenic carbonate that forms in those sediments, pedogenic carbonate will hereafter be referred to as carbonate, and carbonate rocks will be referred to as limestone).

STRATIGRAPHY

The stratigraphy of the north flank piedmont consists of weathered granite overlain by 100-500 m of Miocene and Pliocene basin sediments. The basal unit of these sediments is a reddish arkose with a northern provenance, known locally as the Old Woman Springs Fm. It represents continental deposition by a paleodrainage system flowing southward across what is now the Transverse Ranges (Meisling and Weldon, 1989). The Old Woman Springs Fm is unconformably overlain by gray and tan, fluvial and alluvial fan, limestone-rich sandstone grading upward into conglomeratic sandstone and conglomerate called the Cushenbery Springs Fm. (Pliocene-early Pleistocene). This unit is a synorogenic fan shed north off the uplifting San Bernardino block (Shreve, 1968). The contact between these units represents a regional reversal in major drainage direction. Unconformably overlying the Cushenbery Springs Fm. are middle-to-late Pleistocene and Holocene, poorly sorted, gravelly-to-bouldery, Quaternary alluvial-fan sand (e.g. Powell et al., 2000). Alluvial fan deposits of the San Bernardino piedmont have been assigned a stratigraphy that is based on the surficial geologic-map units proposed by SCAMP (2002) for southern California: Very old fans (Qvof, 400–700+ ka), Old fans (Qof, 12–400 ka), and Young fans (Qyf, 0–12 ka; Fig. 3). Each of these major units contains three or more sub units. In the San Bernardino piedmont, these units are distinguished primarily by landscape position, sediment size and composition, consolidation, overall morphology, and to some degree, soil development (SCAMP, 2002; Powell et al., 2000). We will be using this stratigraphy as a backdrop to the soil geomorphology discussed during the trip. Stratigraphy will not be a major focus of the field trip.

DRAINAGE BASIN GEOMORPHOLOGY

Drainage basin area does not vary by more than an order of magnitude in the basins that contribute to alluvial fans in the

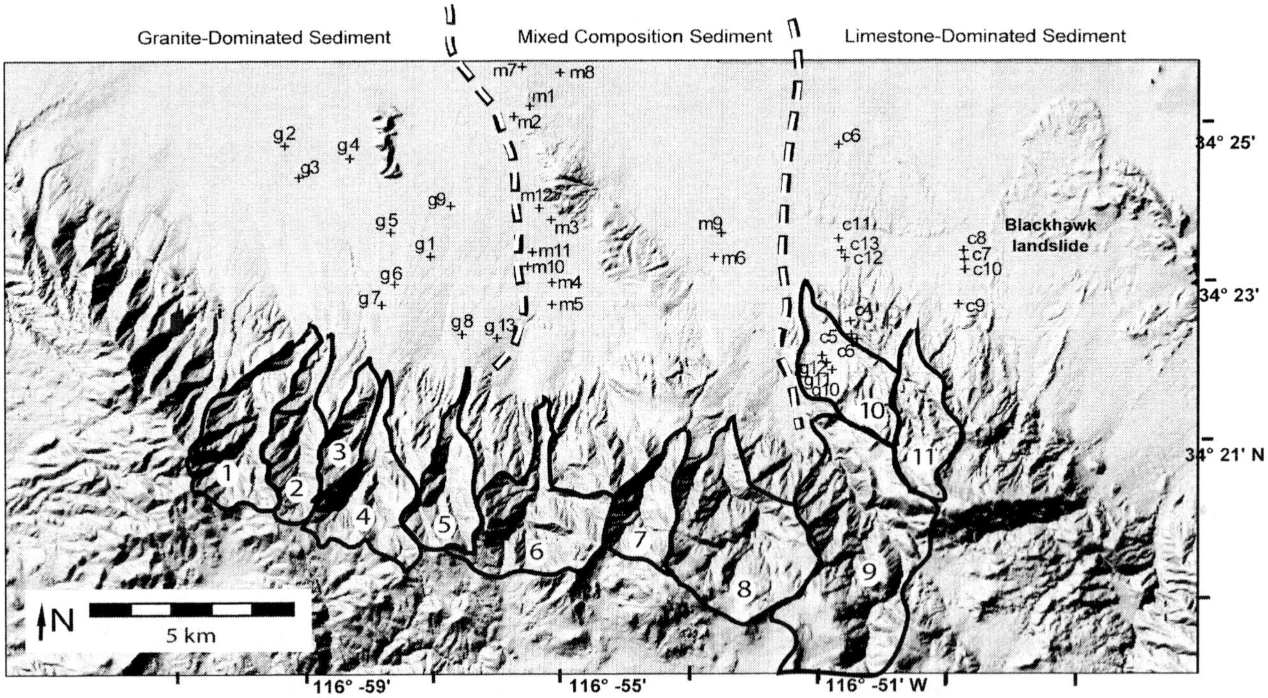

Figure 2. Shaded-relief image of a 30–m-spaced digital elevation model (DEM) from the north flank of San Bernardino Mountains, modified in areas to reduce DEM data-error effects. White and black dashed lines represent the transition in alluvial fan composition from limestone-dominated to mixed composition to granite-dominated sediments. Soil pit locations are noted by G-#, M-# and C-# designations. Drainage basins 1-11 that contribute to alluvial fans on the piedmont are outlined. This DEM is generally a composite from 4 U.S.G.S. quadrangles, Lucerne Valley (NW), Cougar Buttes (NE), Fawnskin (SW), and Big Bear City (SE). The 7.5 minute versions of these quads may be employed in conjunction with this DEM to locate specific areas of interest within the field area.

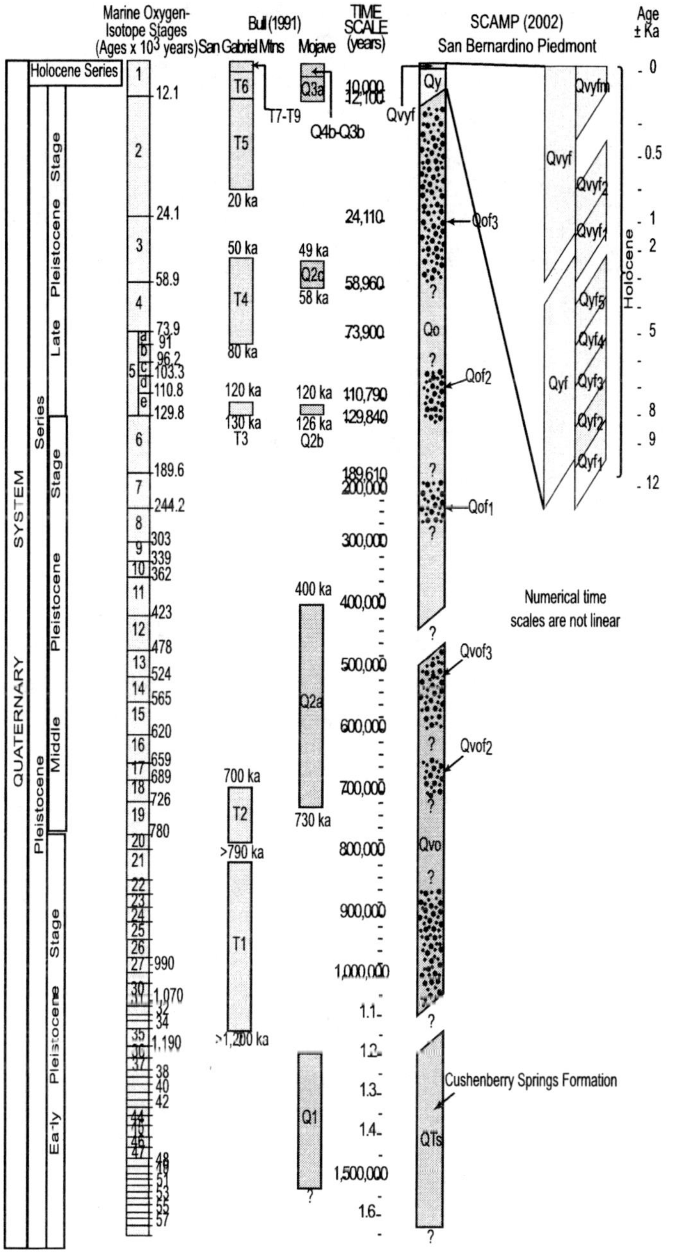

Figure 3. Stratigraphic position of Quaternary mapping units for the San Bernardino piedmont (Powell et al., 2000, SCAMP, unpublished data), the San Gabriel Mountains (Bull, 1991), and the Mojave Desert (Bull, 1991) relative to Quaternary time divisions and marine oxygen-isotope stages (modified from SCAMP, 2002).

Figure 4. (A) typical coarse colluvium associated with carbonate rock types in drainage basins of the field area. Field notebook is ~20 cm in length. (B) Saprolitized granite and grus colluvium that is typical of many hillslopes in drainage basins of the field area.

piedmont (Fig. 2). Granite is typically highly fractured and/or mantled by significant thickness (20 cm–3 m+) of saprolite or grus-dominated colluvium (Fig. 4). Carbonate bedrock is also highly fractured, but typically mantled by relatively thin (<1m), coarse, colluvial cover (Fig. 4). Limestone colluvium is coarse and characterized by development of petrocalcic soils. Granite colluvium is sand-and-grus-dominated and characterized by weak soil development. Drainage basins contributing to alluvial fans in the easternmost limestone portions of the piedmont are somewhat smaller than those to the west. These limestone-dominated basins head at lower elevations than those associated with the granite and mixed portions of the piedmont. Wind gaps, unusual stream geometries (Fig. 5 c and d), and paleocurrent directions derived from analysis of alluvial fan units (SCAMP, unpublished data) suggest that mountain front drainage systems have been reorganized.

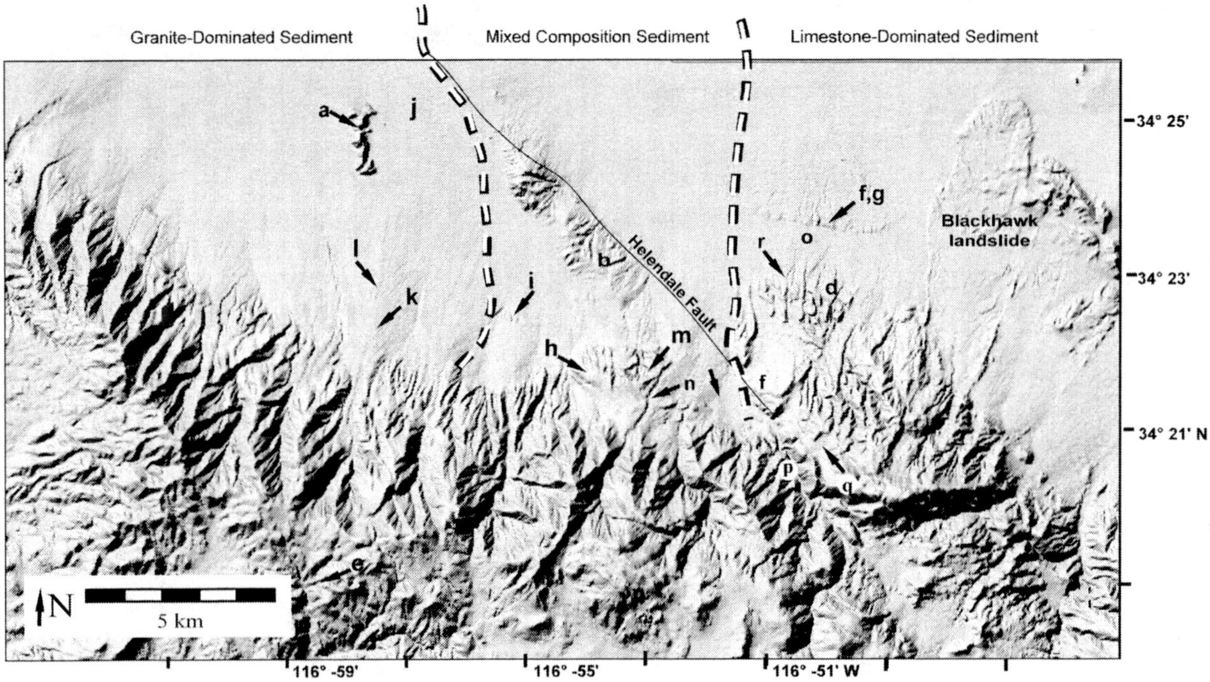

Figure 5. Points of interest in the piedmont located on a shaded-relief image of a 30-m-spaced digital elevation model (DEM) from north flank of San Bernardino Mountains, modified in areas to reduce DEM data-error effects. See text for explanation of lettered areas a-l. Other marked points of interest may be discussed during the field trip. This DEM is generally a composite from 4 U.S.G.S. quadrangles, Lucerne Valley (NW), Cougar Buttes (NE), Fawnskin (SW), and Big Bear City (SE). The 7.5 minute versions of these quads may be employed in conjunction with this DEM to locate specific areas of interest within the field area.

TECTONICS AND STRUCTURE

The northern flank of the San Bernardino Mountains has been uplifted to its present elevation by convergence along the North Frontal thrust system that began in the late Pliocene-early Pleistocene (Meisling and Weldon, 1989; Spotila et al., 1998). Deformation has continued into the late Pleistocene, with decreasing deformation during the Holocene. Inselberg-like landforms (Fig. 5, a and b), predominantly unrelated to uplift of the range and associated with the Mojave block provenance to the north, punctuate the piedmont. The North Frontal thrust system of the San Bernardino Mountains consists of complicated, overlapping, east-striking, reverse faults (e.g. Fig. 5, e) and fault cored folds (e.g. Fig. 5, f). These structures incorporate basement rocks and the entire Neogene stratigraphic section of the piedmont (Miller et al., 1998; Powell et al., 2000; Newland, 2001). Geomorphic indicators (e.g. degree of dissection) provide evidence that folds adjacent to the mountain front are older than those in the medial and distal piedmont (Powell et al., 2000). Recent studies have demonstrated a correlation between the topographic expression of these folds and soil development associated with alluvial fan composition across the piedmont (Eppes et al., 2002, see discussion below).

CLIMATE AND VEGETATION

The northern flank of the San Bernardino Mountains is a climatic transition zone between the sub-humid to semi-arid Mediterranean climate of the southern flanks of the Transverse ranges and the arid climate of the eastern and northern Mojave. Similar vegetation communities at similar elevations suggest that climate does not vary significantly along the piedmont within the area. However, an elevation-controlled climate gradient exists from the basin (~2300 ft. elevation, mean annual temperature, 64° F, mean annual precipitation 4 inches), to the mountain front (~4500 ft. [1360 m], mean annual temperature 57° F, mean annual precipitation 6-9 in. [2.4-3.5 cm]). This climate gradient is marked by a shift in vegetation from primarily Mojave desert scrub vegetation at lower elevations to a piñon juniper woodland at higher elevations. In addition to glacial/interglacial-scale climate changes throughout the Quaternary (summarized in Bull, 1991), the San Bernardino Mountains have also been subject to shorter periods (decadal–centennial scale) of increased precipitation during the Holocene (Enzel et al., 1989, 1992; Biondi et al., 2001).

SOIL GEOMORPHOLOGY

Approximately 40 soils have been examined in the San Bernardino piedmont. The Lucerne Valley and Cougar Buttes 7.5 minute quadrangles will be helpful in locating yourself in the field area. Soil pit sites are positioned primarily on Qyf and Qof deposits for granite and mixed-parent material fans, and Qyf and Qvof for limestone-parent material fans (Fig. 2). During the field trip, refer to this map to locate soil pits that we will be observing.

Field morphologic data (Soil Survey Staff, 1992) was collected for all soils (Eppes, 2002).

DAY 1

STOP 1-1. QY GRANITE SOIL (SBG2)

From Highway 18 west of the Ace Motel in Lucerne Valley, take Buona Vista Rd. south. The soil pit is 0.1 mi south of the intersection of Buona Vista Rd. and Sutter Rd. ~30 m west of the road.

Qyf deposits are extensive and thick across the granite portions of the field area (Fig. 6). Discontinuous, low terrace deposits of Qy sediments are sometimes present in larger incised drainages. However, Qyf deposits often bury Qof deposits and create broad alluvial fans (Fig. 6). Bar and swale topography is apparent on these fans, particularly at higher elevations where deposits are coarser. Qyf alluvial fans are characterized by active washes that head both on the fan surfaces and in adjacent drainage basins. Washes that head on alluvial fans are somewhat incised in upper reaches. Consequently, the argillic horizons of Qof deposits are frequently exposed along drainages as erosional fenêtres in Qyf deposits (a in Fig. 6). Vegetation composed almost entirely of black brush marks the areas where these horizons are exposed close to the surface. Washes that head in adjacent drainage basins are often incised into Quaternary alluvium in valley floors but grade into broad, braided, alluvial-fan washes as they exit the mountain front.

Qyf deposits are typically dominated by stacked packages of sandy grus sediment with occasional gravel and debris flow facies. Individual packages of sediment are punctuated by weak soil development. Surface soils in these deposits are consistently characterized by similar Av-Bw-Bwk-Ck profiles. Maximum development in soils associated with buried stratigraphy ranges from Bw to Bk horizonation. Weak structure, and minimal reddening and pedogenic carbonate characterize B-horizons in both surface and buried soils accumulation (< stage I–stage I development). Deposits with these relatively weak soils commonly bury older deposits with strong (Bt, Btk) soil development. Discontinuous, non-oriented coats of carbonate on larger clasts in Qyf soils provide evidence of the reworking of older soils.

STOP 1-2. QO GRANITE SOIL (EXPOSURE OFF WHITE KNOB ROAD, JUST SOUTH OF SBG 7)

From Highway 18 in Lucerne Valley, take Crystal Creek Rd. south. Cross the railway tracks and turn right on White Knob Rd. (NOTE: You MUST have permission from the mining company to access this road. Constant HEAVY EQUIPMENT traffic occurs.) Exposure is ~100 m north of road in the major drainage that crosses at mile marker #55.

The morphologic properties of soils (SBG 4,5 and 6 and the White Knob exposure) developing in Qof deposits are variable.

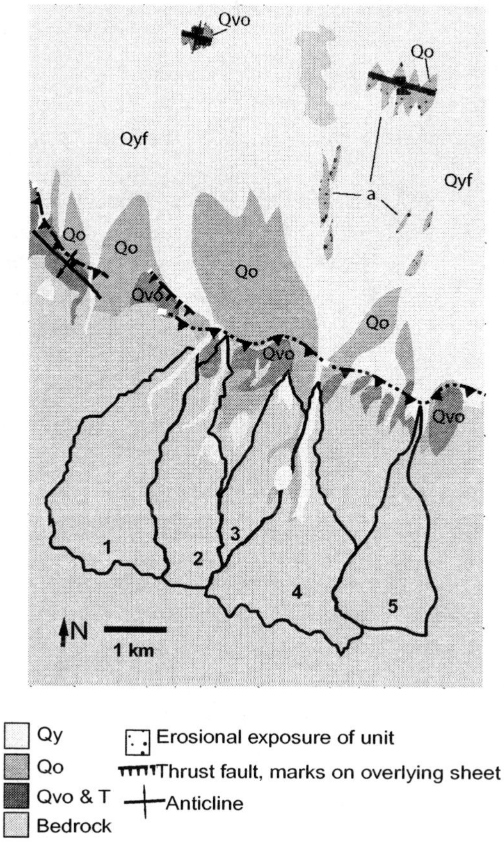

Figure 6. Generalized geologic map of granite-dominated piedmont. (Modified from Miller et al., 1998, 2000). See text for explanation of lettered areas.

The Bt horizons of the SBG 5 soil (down fan from the White Knob exposure) exhibited the most soil development of these three profiles, as shown by strong ped structure (3 c sbk), reddish brown colors (7.5 YR 5/4), clay films (2 p co/br), and chemistry. Thin (<1mm) black to iridescent manganese coatings are present on some ped faces and clast bottoms in the SBG 5 soil. These types of coatings in desert soils are attributed to higher past soil moisture conditions (Weitkamp et al., 1996). The argillic horizon of the SBG 6 soil is considerably less developed (2 m sbk structure, 10 YR 6/4, color, 1 f co clay films) than that of the SBG 5 soil, and the SBG 4 soil even less so (2 f/m sbk, 10 YR 6/4, no clay films). Minimal, if any, pavement formation occurs on alluvial fan surfaces in the granite-derived piedmont, regardless of age. Surface horizons of soils developing in deposits of all ages are weakly developed, dominated by grus, and typically bioturbated (Eppes et al., 2002). Grussification of granite clasts at depth is common in exposures of both Qof and Qvof granite deposits. Subaerialy exposed Qof granite-derived deposits contain minimal carbonate, accumulating only at great depths (>4 m). However, carbonate is often observed as a coating on clay films in buried Qof soils (e.g. SBG 10, 11, and 12).

SOIL – GEOMORPHIC RELATIONSHIPS

In the granite-dominated piedmont, strong evidence occurs for feedbacks between soil development, the weathering of granite into grus, and surficial processes acting on this portion of the San Bernardino Piedmont. Grus dominates sediment supply in the granite piedmont. Accordingly, channels and mountain front hillslopes are transport-limited environments that are very susceptible to processes such as creep, bioturbation, and raindrop splatter. Additionally, argillic horizons developing in granite alluvial fans are capable of producing significant runoff. Abundant active and inactive, backfilled, small drainages developing on argillic horizons provide evidence of this runoff that favors drainage migration over fan surfaces (Eppes et al., 2002). Reworking of sediments derived from older soils on the piedmont by these small channels and by hillslope processes accounts for the chemical and morphological characteristics of many soils observed in medial and distal portions of the piedmont. One of the most perplexing characteristics of soils along the north flank of the San Bernardino Mountains is the fact that the carbonate content of soils developing in granite parent materials decreases with increasing deposit age. The maximum stage of carbonate development observed for granite-parent-material soil (only stage I) is for a Qyf deposit (SBG 2). Although similar trends have been observed in other studies (e.g. McFadden, 1988; McDonald, 1994), this observation is in complete contrast to results of all conceptual (e.g. Gile et al., 1981) and numerical (McDonald, 1994; McFadden and Tinsley, 1985) models of carbonate accumulation in desert soils. The data collected in this study are not sufficient to unequivocally address this phenomenon. However, a number of observations from this study and others in southern California and elsewhere suggest that no one process linked to one soil forming factor can explain the observed lack of pedogenic carbonate accumulation in older granitic fan deposits. Instead, a combination of factors including dust trapping efficiency, interactions between elevation, vegetation and climate, surface erosion, and past climatic influences produce the unusual time-dependant pattern of carbonate accumulation observed along the north flank of the San Bernardino Mountains (Eppes, 2002).

STOP 1-3. QYF LIMESTONE SOIL (SBC 7)

From Highway 247 east of Lucerne Valley, turn south on Santa Fe Fire Rd. Go 2.8 miles (stay right at fork), and turn west on a small, unmarked road. Go 0.15 mi, and a soil pit is 5 m south of the road.

During the field trip we will see that limestone-derived alluvial fans are characterized by soils dominated by the development of calcic horizons (K horizons) that vary from stage I development in Qyf deposits, to stage IV in Qof deposits, to Stage V in Qvof deposits. Progressive stages of carbonate accumulation are thought to reflect increasing age (e.g. Gile et al., 1981; Machette, 1985). The rate of development of calcic horizons, however, is accelerated in soils developing in limestone parent materials. In general, limestone Qy deposits such as SBC 7 are characterized by color and structure Bw horizons and weak stage I carbonate development. Horizonation is generally similar to that of Qyf granite-dominated soils. Incipient desert pavements and biologic crusts are common.

STOP 1-4. QVOF LIMESTONE SOIL (SBC 11)

From Highway 18 south of Lucerne Valley, turn north on Camp Rock Rd. Go 1.4 miles, turn east on Bauer Rd., and continue 0.85 miles. Pit is ~55 m north of the road.

Several notable morphologic characteristics of soils are described in Qof and Qvof deposits in the limestone piedmont. K horizons exhibit stage III to stage V morphology. All of the soils associated with Qvof units and some of the Qof units described in the limestone piedmont are characterized by thin (15-30 cm) surface horizon soils with weak A/Bw development. Desert pavements are moderately well developed. Eppes et al. (2002) suggest that these surface horizons are largely genetically unrelated to underlying petrocalcic horizons and represent episodic reworking of sediment across the K horizon surface. Pisolitic horizons, as well as carbonate lined and/or filled vertical and horizontal fractures, were also commonly noted in Qvof soils. This morphology is similar to that observed by Sowers (1985) and Reheis et al. (1992) in soils of the Kyle Canyon fan. In the limestone soils of the San Bernardino piedmont, fractures result in a large macroscale permeability, and fracture flow is thought to contribute to discharges of incised channels (Eppes et al., 2002). Large areas (10 × 10 m) of K horizons associated with Qvof deposits are occasionally exposed at the surface in proximal portions of the limestone piedmont.

SOIL–GEOMORPHIC RELATIONSHIPS

Many aspects of limestone parent material, soil morphology, mineralogy, and chemistry can be explained by the interaction of soil formation and surficial processes. One of the most striking aspects of soil morphology in the limestone piedmont is the weak (A/Bw) development of surface soils associated with Qvof deposits. These surface soils can be understood, however, as a function of periodic stripping and reworking of sediment on the surfaces of these deposits. Typically, K horizons with a stage III+, IV, or V morphology are associated with relatively thick overlying Av/Btk horizons (e.g. Sowers, 1985; Reheis et al., 1992; McDonald, 1994). After Btk horizons develop in this environment, however, in a manner similar to argillic horizons of the granite piedmont, their infiltration capacity decreases. Given sufficient precipitation they will produce runoff and erode. Incision into these deposits is greatly retarded when incipient channels encounter the rock-like petrocalcic horizon. Instead of incising, streams rework available sediment across the alluvial fan surface. The episodic sealing of fractures in petrocalcic horizons may also contribute to the development of runoff and the subsequent reworking or removal of surface soils (Eppes, 2002). Sealed petrocalcic horizons could locally produce sufficient runoff to remove overlying sediments. K horizons currently exposed at the surface with little or no alluvial cover represent sites where this process is happening today. As a consequence of this reworking, the upper 'horizons' of Qvof soils are episodically rejuvenated by new sediment. The presence of surface horizons with minimal soil development in most of the soils examined for this study suggests that parent material rejuvenation of surface soils is common in the limestone piedmont.

On the field trip, we will see that another minor, but interesting, soil characteristic of the limestone piedmont is the character of desert pavements forming on fold flanks. Although only moderately well developed as compared to other desert pavements in southern California (e.g. Wells et al., 1985, 1995), some of the most strongly developed desert pavements found in the medial portions of the limestone dominated piedmont are found on the steep, south flanks of folds. On the flanks of folds, desert pavement development is fostered by a combination of factors, including limited catchment area, macroporosity of tectonically fractured petrocalcic horizons, aspect-related soil moisture, and vegetation conditions, and dust trapping efficiency (Eppes, 2002).

DAY 2

STOP 2-1. OVERLOOK SITE IN SAN BERNARDINO MT FRONT (TO BE DETERMINED BASED ON MINE ROAD ACCESS)

Influence of climate change on soil spatial and stratigraphic variability in the granite piedmont

The spatial and stratigraphic distribution of Qyf soils in the granite piedmont is largely controlled by the response of adjacent drainage basins to Holocene climate changes. Currently, no direct numerical age control of Qyf deposits in the San Bernardino piedmont with which to directly link these deposits to known Holocene climate changes. The presence of stacked buried soils in these deposits, however, indicates that deposition of Qyf deposits has been episodic. In general, the soil development in these stacked packages of sediment in Qy deposits suggests an episodicity of deposition and stabilization since the early Holocene on the order of a few thousand years per package within these deposits. Soil development in these packages is consistent with the scale of time between periods of increased precipitation that have been noted in southern California (e.g. 3000 ka as observed by Enzel et al., 1989, 1992). If increases in precipitation are the cause of deposition of the stacked sediments in Qy deposits, then the presence of more than one package of sediment within these deposits is consistent with episodic increases in precipitation in the Transverse Ranges at the millennial scale since the early Holocene. Again, no evidence exists for tectonic activity during this time, suggesting that deposition is not being triggered by uplift of the mountain front.

Geomorphologists have long recognized the strong link between slope sediment production and erodibility that is strongly influenced by rock type. Based on the soil stratigraphic record of the north flank piedmont, Eppes (2002) hypothesizes that weathering and erosion of ultra sensitive, weathered granite exposed in these basins can produce significant thicknesses of grus-dominated colluvium on hillslopes in relatively short periods of time. The large resultant grus and sand sediment supply of these basins, combined with the semi-arid climate, small size, and low elevations, make them highly sensitive to these episodic increases in precipitation. Eppes (2002) also suggested, as other workers (Bull, 1991; McDonald, 1994; Yair and Kossovsky, 2002) have for other regions, that the moderate vegetation density associated with semi-arid hillslopes, such as those in the drainage basins of the San Bernardino Mountains, also facilitates sediment mobilization during wetter and cooler periods, particularly when these periods do not persist long enough to significantly increase in density of vegetation communities.

STOP 2-2. COUGAR BUTTES ANTICLINE (NORTH OF SBC 11, G IN FIG. 5)

From Highway 18 south of Lucerne Valley, turn north on Camp Rock Rd. Go 1.4 miles, turn east on Bauer Rd. and go 0.85 miles. Exposure is approximately 0.3 miles north of road in arroyo.

Influences of soil development and tectonics on geomorphic processes and the development of macroscale landforms

Eppes et al. (2002) make the observation that the development of petrocalcic horizons in limestone alluvial fan deposits strongly modulates the geomorphic characteristics of those alluvial fans. They propose a model for the evolution of landforms associated with blind thrust faults in the limestone piedmont: After petrocalcic horizons form, they are extremely resistant to erosion, and only major channels are able to incise through the deposits in which they are developing. In addition, fracture-controlled macroporosity decreases runoff on these surfaces. Consequently as carbonate-cemented deposits are folded by blind thrust faults, they form prominent anticlinal ridges that punctuate the limestone piedmont (e.g., g, h and i in Fig. 5 and the Cougar Buttes anticline). As deformation continues, the petrocalcic-cemented, alluvial fan surface is abandoned and preserved as an elevated, undulating surface with high relief. Inset relationships of Qof and Qvof deposits in the proximal to medial piedmont provide evidence for at least two cycles of alluvial fan deposition, cementation, and folding. After abandonment, geomorphic processes acting on these older fans are limited to the reworking of locally-derived sediments across the petrocalcic surface.

STOP 2-3: PITZER BUTTES ANTICLINE (NORTH OF SBG 10, J IN FIG. 5)

From Highway 18, take Crystal Creek Road south to Agate Road (appx. 1.8 miles). Turn east on Agate and go 0.2 miles.

The exposure is a broad outcrop approximately 0.5 miles north of road.

Influences of soil development and tectonics on geomorphic processes and the development of macroscale landforms

In contrast to the limestone-dominated piedmont, deformation of alluvial fan surfaces in the granite piedmont produces only subdued landforms (e.g., j and k in Fig. 5, Eppes et al., 2002). Eppes et al. (2002) also attribute much of the character of thrust related landforms in the granite piedmont to soil-controlled hydrologic and erosion characteristics of alluvial fan deposits.

Eppes et al. (2002) show that as a fold develops in the granite piedmont, it is buried by the enormous volume of sediment available on the piedmont and/or beveled by fluvial processes. Growing folds in granite-derived deposits thus maintain a subdued convex morphology (j in Fig. 5, Newland, 2001) unless uplift rates significantly exceed erosion rates. Streams crossing these structures appear to be primarily antecedent. However numerous small drainages (l in Fig. 5) head on the flanks of these folds and show evidence of active tectonic influence (Newland, 2001). These studies show that although fold features in the granite piedmont will to some degree locally influence erosion and deposition across alluvial fan surfaces, they do not significantly contribute to the morphology of granite derived alluvial fans.

CONCLUSIONS

The geomorphic and soil variability observed along the north flank of the San Bernardino Mountains can be viewed as a function of both independent and dependent geomorphic variables acting at different scales along the piedmont. At the landscape scale, the primary independent variable affecting both soils and geomorphology across the piedmont is rock type. The bedrock composition of drainage basins along the north flank strongly modulates the response of these basins to climate change. Consequently, the aggradation of alluvial fans along the piedmont varies as a function of the rock types exposed in adjacent drainage basins. The sediment composition of these alluvial fans strongly influences soil spatial variability at the scale of the entire landscape. Soils in the limestone piedmont are dominated by petrocalcic horizons, whereas those of the granite piedmont are dominated by argillic horizons with no accumulation of carbonate. Only at very small spatial scales and relatively short temporal scales, such as a sequence of Holocene terraces exposed in an incised drainage, do we see definite time dependent trends in soil development along the north flank. Instead, the overall pattern of soils strongly reflects interactions and feedbacks at different temporal and spatial scales between climate, tectonics, geomorphology, and soil development. The accurate interpretation of the landscape evolution of this piedmont is hinged on the understanding of *both* soil and geomorphic relationships across the piedmont.

DAYS 3 AND 4. THE PROVIDENCE MOUNTAINS (Mcdonald and Mcfadden)

INTRODUCTION

Soil landscape relations are widely used to provide fundamental stratigraphic and age information for Quaternary geologic and climatic history. Many studies have shown that processes controlling soil development and landscape modification in desert environments are extremely susceptible to the (1) effects of climate change, (2) accumulation of aerosolic dust, and (3) lithology of source rocks. The influence of these variables on Quaternary soil landscape processes have rarely been integrated into one comprehensive study that addresses both the individual effects, as well as the combined effects, on the origin and development of soils and desert landforms. Several key attributes of the Providence Mts piedmont (PMP) provide an unprecedented opportunity to learn how the factors listed above control processes of soil development and landscape modification: (1) a well defined record of Holocene and late Pleistocene paleoclimatic and eolian activity in the Mojave Desert, (2) established Quaternary stratigraphy for alluvial fans along nearby mountain piedmonts, and (3) the juxtaposition of four lithologically different sequences of alluvial fan deposits. Days 3 and 4 of this field trip present an overview of an on–going study of Quaternary soil geomorphic processes common to the western piedmont of the Providence Mountains.

GEOMORPHIC SETTING

The Providence Mountains are a prominent feature along the eastern edge of the Mojave Desert (Fig. 7). Elevations range from about 600 m in distal fan areas to about 1000 m in proximal fan areas and exceed 2000 m at the highest mountain peaks. Fan deposits along the western margin grade westward to an extensive area of sand sheets associated with the Kelso Dunes. The modern climate is typical of the Mojave Desert with annual rainfall ranging from about 150 mm annually along the piedmont to about 250 mm annually at the mountain tops. Most of the annual precipitation is associated with winter Pacific frontal storms. Modern vegetation along the fans is predominantly Mojave Desert scrub dominated by *Larrea tridentata* (creosotebush) and *Ambrosia dumosa* (white bursage). The vegetation along the mountain hillslopes above 1000 m (3300 ft.) ranges from blackbrush (*Coleogyne ramosissima*) and Mojave yucca (*Yucca schidigera*) to juniper/pinyon woodland (predominantly on north-facing slopes above 1000 m).

DAY 3

STOP 3-1. OVERVIEW OF QUATERNARY STRATIGRAPHY

From the California/Nevada state line on Interstate 15, proceed south 9.9 miles to the Nipton Rd. exit, then east 3.5 miles to Ivanpah Rd., then south 3.2 miles to Morning Star Rd., then south

14.8 miles to Cima, then continue south 19.0 miles to Kelso. Proceed south 3.6 miles along the Kelbaker Rd. (toward Interstate 40) and then east 2.8 miles along the abandoned Vulcan Mine Rd. (no sign). Hike 50 m south to the top of a fan remnant (Qf2 VX).

Alluvial fan and eolian units are defined based on relative development of soils and desert pavements and stratigraphic relationships among depositional units (Table 1, Fig. 8). Four separate fan sequences are defined, based upon the dominant rock types (Miller et al., 1985) that make up the deposits in each sequence: (1) *PM*: leucocratic-to-mesocratic, mixed plutonic rocks (mostly syenite and syenogranite), (2) *QM*: quartz monzonite, (3) *LS*: limestone and marble, and (4) *VX*: volcanic-mixed (rhyolitic tuff and rhyodacite with lesser amounts of PM and LS). Fault scarps along the mountain front are not evident, indicating a lack of activity along range bounding normal faults during late to middle Quaternary.

Age control along the PMP is derived from (1) infrared stimulated luminescence (IRSL) ages of sand associated with alluvial and eolian deposits (McDonald, 1994; Clarke, 1994), (2) soil stratigraphic correlations to the Soda Mountains piedmont (SMP) and Cima volcanic field (CVF) alluvial units using a soil development index, (3) ^{14}C dating of pedogenic carbonate (Wang et al., 1996), and (4) identification of the 0.74 Ma Bishop Tuff near the base of the oldest fan unit (McDonald, 1994). Radiocarbon ages are reported as measured AMS ^{14}C dates and modeled ^{14}C ages. Modeled ages are measured ^{14}C modified to compensate for possible age underestimates due to the time required to form a carbonate coating of adequate thickness for dating (Wang et al., 1996). Neither modeled nor measured ^{14}C dates agree with age estimates based on IRSL dates, current knowledge of soil-forming rates in the Mojave desert, and soil stratigraphic correlations to nearby and relatively well–dated piedmont deposits. Inconsistencies of measured and modeled radiocarbon dates are considered to reflect the

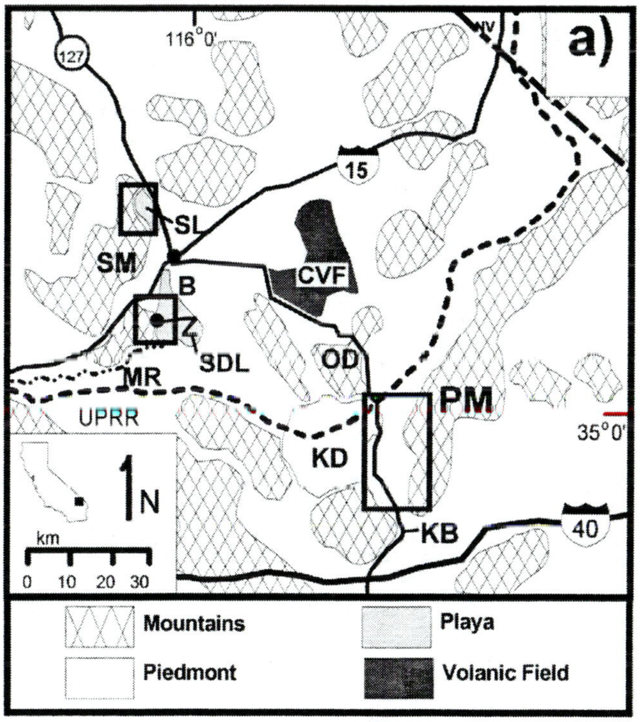

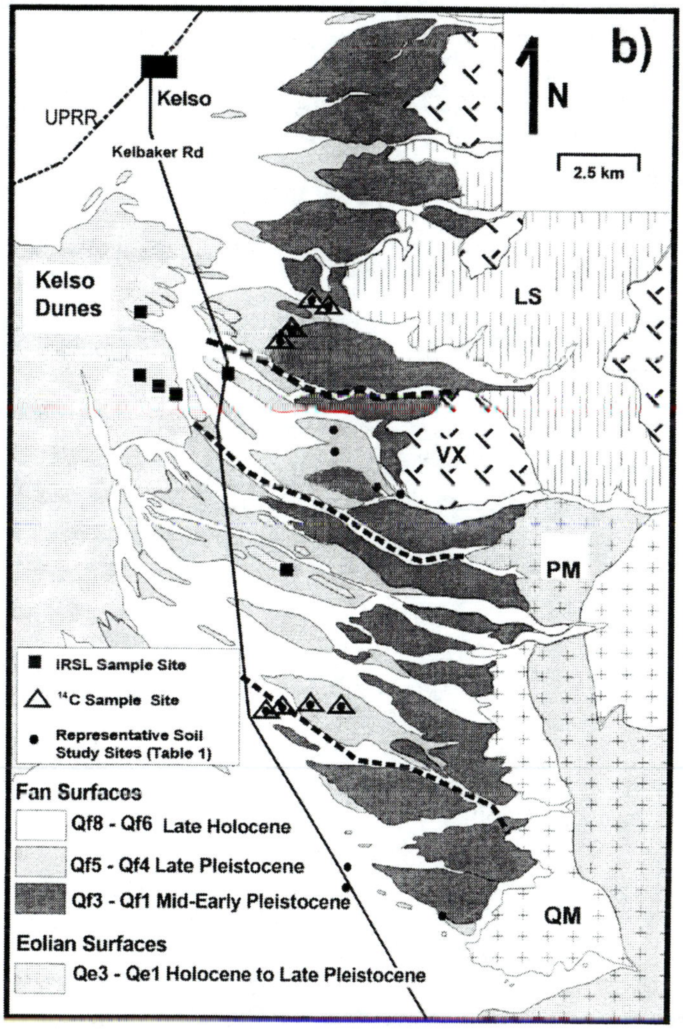

Figure 7. (a) Location of study area (open box) along the Providence Mountains (PM) piedmont in relation to study areas along the Soda Mountains (SM) adjacent to the Silver Lake Playa (SL) and Soda Lake playa (SDL). Other map abbreviations: Mojave River (MR, dot-dot-dash line), Union Pacific Railroad (UPRR, dash line), Nevada (NV)-California state line (short dash-long dash line), Baker (B), Cima volcanic field (CVF), Old Dad Mountain (OD), Kelso Dunes (KD), and the Kelbaker road (KB). (b) General distribution of Quaternary depositional units along the Providence Mountains piedmont, sample sites for ^{14}C and IRSL age control, and locations of representative soil study sites (Table 1). Piedmont deposits further subdivided (dotted line) into sequences based upon the dominant rock types that make up the alluvium: (1) PM – mixed-plutonic rocks, (2) QM – quartz monzonite, (3) LS – limestone and marble, and (4) VX – volcanic-mixed. Figure after McDonald et al. (2003).

TABLE 1. TYPICAL SOIL CHARACTERISTICS OF THE FOUR FAN SEQUENCES

Fan Unit*		Fan-Sequence			
		PM	QM	VX	LS
Qf8 Late Holocene	Horizon†:	no soil	no soil	no soil	no soil
	Texture#:	sand	sand	sand to loamy sand	sand to loamy sand
	Stage§:	0	0	0	0
	Pavement**:	none	none	none	none
Qf7 Late Holocene	Horizon:	Avk-AC-Ck-C	Av-A-AC-Ck-C	Avk-ACk-Ck-C	Avk-ACk-Ck-C
	Texture:	loamy sand	loamy sand	loamy sand	sandy loam
	Stage:	I-	I-	I	I
	Pavement:	none	none	very weak to weak	very weak to weak
Qf6 Late Holocene	Horizon:	Avk-Bwk-Ck-C	Av-A-Bw-Ck-C	Avk-Bwk-Bk-Ck-C	Avk-Bwk-Bk-Ck-C
	Texture:	sandy loam	loamy sand	loamy sand	sandy loam
	Stage:	I	I	I	I-II
	SB‡:	cambic	cambic	cambic	cambic
	Pavement:	weak	very weak to none	weak to moderate	weak to moderate
Qf5 Early Holocene to Latest Pleistocene	Horizon:	Avk-Bwk-Btk-Ck-C	fan unit not found	Avk-Bwk-Bk-Ck-C	Avk-Btvk-Bk-Ck-C
	Texture:	sandy loam		sandy loam	loam
	Stage:	I-II		I-II	II-III
	SB:	cambic or argillic		cambic/calcic	calcic
	Pavement:	moderate to weak		moderate to strong	moderate to strong
Qf4 Late Pleistocene	Horizon:	Avk-Btk-Bk-Ck-C	Av-ABv-Bt-Bk-Ck-C	Avk-BAv-Bwk-Bkm-Ck-C	Avk-Btvk-Bwk-Bkm-Bk-Ck-C
	Texture:	sandy clay loam	clay loam	sandy loam	loam
	Stage:	III	I-II	III-IV	III+
	SB:	argillic	argillic	petrocalcic	petrocalcic
	Pavement:	moderate to strong	weak to moderate	strong	strong
Qf3 Middle Pleistocene	Horizon:	Avk-BAvk-Bt-Bkm-Ck-C	Av-BAv-Bt-BCk-Ck-C	Avk-BAv-Btk-Bk-Bkm-Bk-Ck-C	Avk-Btvk-Bwk-Bkm-Bk-Ck-C
	Texture:	clay loam	loam	sandy clay loam	loam
	Stage:	III-IV	I-	IV	IV
	SB:	argillic/petrocalcic	argillic	argillic/petrocalcic	petrocalcic
	Pavement:	strong	moderate to weak	strong	strong to very strong
Qf2 Mid-Early Pleistocene	Horizon:	Avk-Btk/Bkm-Ck-C	Av-Bt-Btk-BCk-Ck-C	Avk-BAvk-Btk-Bk-Bkm-Ck-C	Avk-Bkm-Bk-Ck-C
	Texture:	sandy clay loam	sandy clay loam	sandy clay loam	sandy loam
	Stage:	IV	I-	IV	IV-V
	SB:	petrocalcic	argillic	argillic/petrocalcic	petrocalcic
	Pavement:	moderate to weak	weak to moderate	strong to moderate	moderate to strong

PM—mixed-plutonic; QM—quartz Monzonite; VX— mixed-volcanic; LS—limestone.
* Age estimations of fan units in Figure 1. Unit Qf1 not shown because unit is highly eroded.
† Typical soil horizon sequence.
# Finest texture of any B or AC horizon.
§ Soil carbonate stage.
** General quality of desert pavement where either best developed or where best preserved.
‡ Strongest diagnostic B horizon.

likely influence of older, detrital carbonate and noncontinuous deposition of pedogenic carbonate (McDonald et al., 1996, 2003).

STOP 3-2. SOILS AND GEOMORPHOLOGY OF THE VX FAN SEQUENCE

Examples of soil geomorphic processes for each of the four soil chronosequences are visible at STOP 1-2 (VX), STOP 1-4 (LS) on DAY 1 and STOP 2-1 (PM) and STOP 2-2 (QM) on DAY 2. From this stop, a series of Qf8 to Qf2 VX fan remnants can be accessed along the two active washes from just south of the first stop and continuing ca. 1.5 km west along these washes. The role of desert dust and lithologic control on Quaternary soil-geomorphic processes, summarized below, can be viewed for each of the four soil chronosequences.

Origin and evolution of desert pavements

Comparison among desert pavements and Av horizons across the piedmont demonstrates how geomorphic factors control pavement formation. A strong systematic trend exists between increases in the development of desert pavement and the thickness of underlying Av horizons (Fig. 9a, b). These trends indicate that the formation of pavements and Av horizons begins soon after deposition of the alluvial deposits and coevolve simultaneously, and pavements on alluvial fans form by the upward

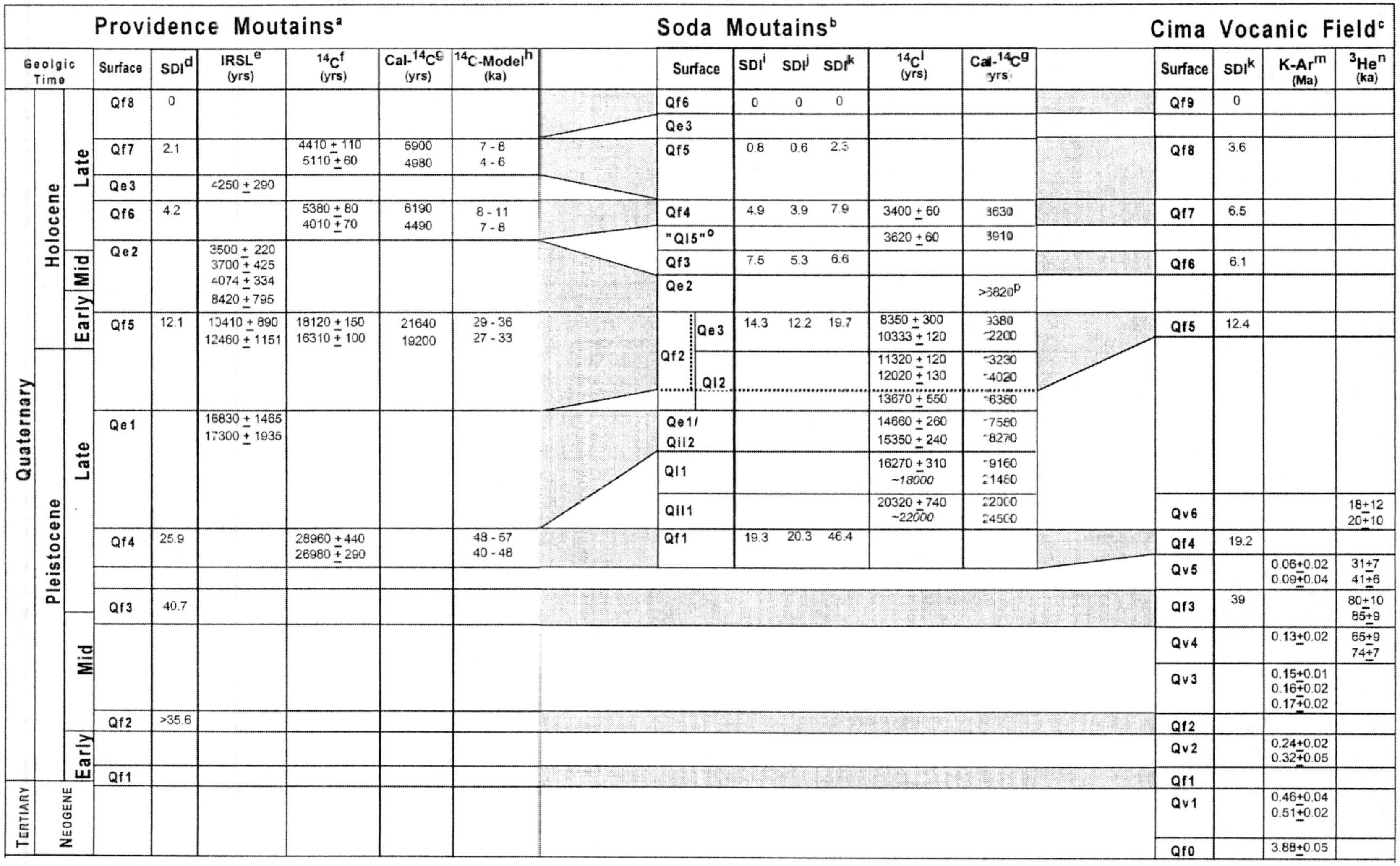

Figure 8. Regional correlation and age estimations among Quaternary fan, lacustrine, and eolian deposits in the eastern Mojave Desert. Figure modified from McDonald et al. (1994, 2003) and McDonald and McFadden (1995).

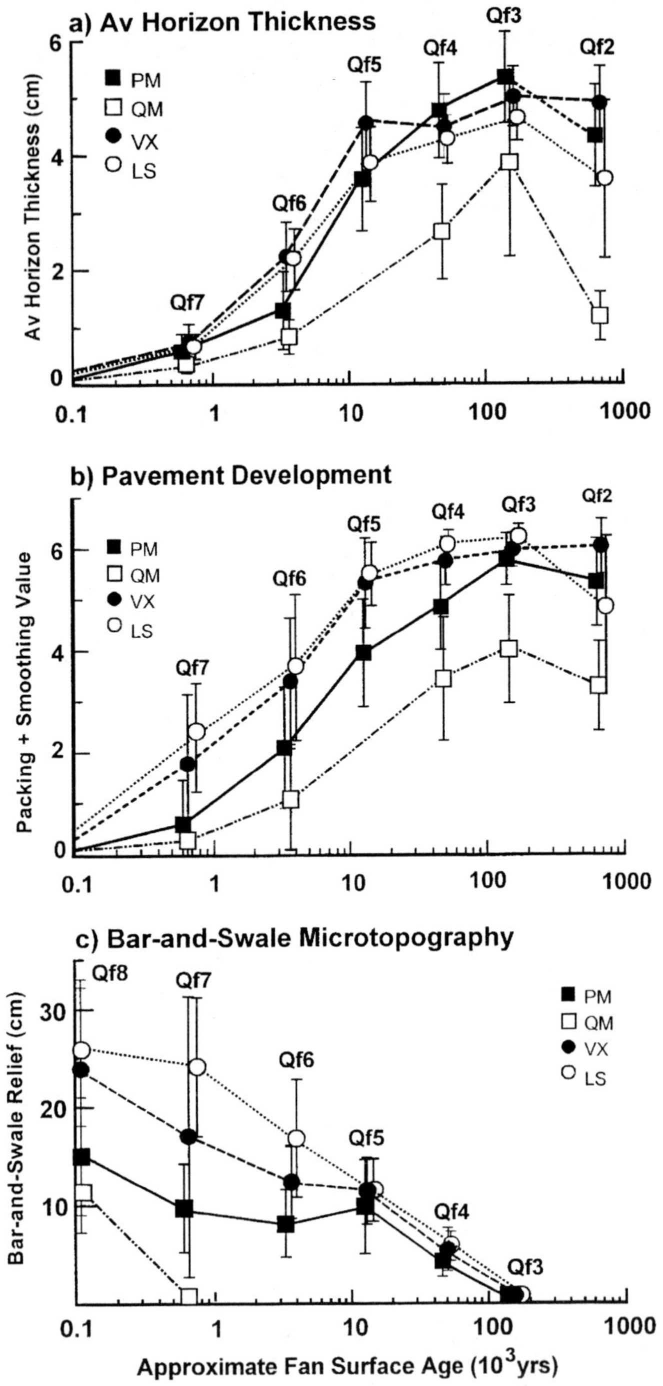

Figure 9. Temporal trends in mean (a) Av horizon thickness in cm, (b) pavement development based on mean alignment (packing) and smoothing (horizontal character) values, and (c) bar-and-swale microtopography. Each point represents between 20 and 50 measurements for each particular fan sequence. Bars represent ± one standard deviation. Displayed age for each set of fan units represents best age approximations. Ages of individual fan-units within each set of fan units are slightly offset to increase visibility of error bars. Figure after McDonald and McFadden (1995).

lifting of surface clasts from the underlying accumulation of eolian fines and not from deflation or upward migration of clasts, as has been generally hypothesized. These results indicate that the dust accretionary model developed by McFadden et al. (1987) for desert pavements on basalt flows also applies to pavement formation of alluvial fans. The Av horizons on alluvial fans may also rise vertically over time due to the continued accumulation of dust that is translocated through the Av horizon and into underlying soil horizons creating a stable platform for uplift of pavement clasts (McDonald, 1994).

Variation in the rate of pavement formation occurs among deposits derived from different rock types due to variations in dust-trapping efficiency and bar and swale microtopography (Fig. 9c). Av horizons and desert pavements develop fastest on fan deposits composed of resistant rock types that yield coarse texture deposits (cobble, gravel) with an initially open framework and a strong bar-and-swale microtopography—features that increase dust-trapping efficiency. By contrast, Av horizons and pavements form more slowly on fan surfaces composed of poorly resistant rock types, such as quartz monzonite, that produce finer grained deposits (grus) and have a poor bar and swale microtopography.

Soil formation

Soil formation in all parent materials show (1) the greatest concentration in clay and silt in near surface A and B horizons, and (2) systematic decreases in profile depth functions of silt, clay, major oxides, and cutans (McDonald, 1994). Depth-profile relations indicate that the accumulation of desert dust is a primary process controlling pedogenesis across all parent materials. The lithologic composition of the alluvium, however, strongly controls B horizon development and the rate of dust accumulation (Table 1).

Well-developed Bt horizons are common in Pleistocene soils formed on PM and QM fan deposits with the first occurrence of weakly developed Bt horizons on unit Qf5 PM. By contrast, the overall degree of development of Bt horizons in soils formed on the VX fan deposits is noticeably less than that in the PM and QM soils, with the first appearance of weakly developed Bt horizons on unit Qf3. Soils formed in the LS fan deposits lack Bt horizons except for a thin transitional Bvkt horizon in soils on Pleistocene fans.

Systematic variation in Bt horizon development among the four chronosequences is due in part to the presence of abundant limestone clasts in the parent material of the VX and LS soils. The strong influence of calcareous parent materials on limiting the development of fine textured B horizons was observed for soils in southern New Mexico (Gile et al., 1981). The connection between strongly calcareous parent material and limited Bt horizon development has been largely attributed to high ionic concentrations of Ca and Mg. High concentrations of these cations promote an increase in the flocculation of clay, which in turn, decreases the mobility and translocation of clay into subsurface horizons (Birkeland, 1999).

Variation in Bt horizon development is also due to the control of pavement formation on the accumulation of aerosolic silt and clay. Soils developed on the VX and LS fan sequences have faster rates of pavement formation and slower rates of dust accumulation relative to soils developed on the QM and PM fan sequences (Fig. 10). A reduction in dust–trapping efficiency would coincide with development of desert pavements. Variations in the rate of dust accumulation may also be due to high calcium carbonate content of pavement clasts and Av horizons in soils formed in the VX and QM deposits. High ionic strength of Ca in percolating soils water enhances the flocculation of clay, limiting the mobility and subsequent incorporation of clay from dust into developing soil profiles.

Development of Bk and Bkm horizons is strongest in the LS and VX soils relative to the PM and QM soils due to the abundance of limestone detritus in the VX and LS deposits. Weakly developed Bk horizons first occur in unit Qf6 in the LS and VX deposits. By contrast, Bk horizons first occur in unit Qf5 or Qf4 in the PM and QM deposits. Development of Bkm horizons in soils formed on the LS and VX surfaces is greater relative to the Bkm horizons on the PM fan surfaces. Bkm horizons are not present in Pleistocene soils on QM fan surfaces. Bkm horizons in VX soils develop more rapidly and are better developed than Bkm horizons in LS soils because VX soils have a higher soil water–holding capacity due to a greater accumulation of dust relative to LS soils. Much of the pedogenic carbonate in the VX and LS soils is derived from primary carbonate from limestone detritus. However, dust provides the primary source of calcium carbonate in the PM and QM soils.

Significant differences occur in rates of accumulation of pedogenic carbonate (Fig. 10b). Temporal increases in carbonate in the LS and VX soils are nearly 2 to 3 times the rate of carbonate accumulation in the PM soils reflecting the greater abundance of parent material carbonate in the LS and VX soils. Increases in carbonate in the PM soils generally reflect temporal increases in dust–derived carbonate. Limited increases in carbonate content in the QM soils probably reflects the greater leaching of carbonate through the QM soil profile due to greater effective soil moisture, due in part to a combination of higher elevation, a highly permeable sandy pebble texture (grus), and poor development of pavements and Av horizons, all of which favor greater infiltration.

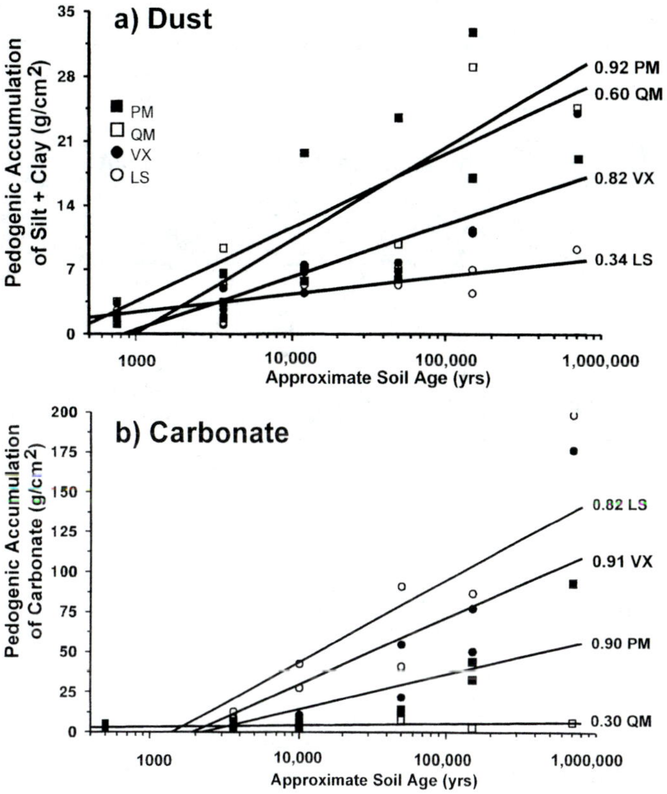

Figure 10. Linear regression estimates of rates of pedogenic accumulations of (a) silt + clay (g/cm²) and (b) carbonate (g/cm²) in the four soil-chronosequences. Best age estimates for soils are used in regression analysis. Figure modified from McDonald and McFadden (1995).

STOP 3-3. THE INFLUENCE OF QUATERNARY CLIMATE CHANGE AND LITHOLOGIC CONTROL ON PERIODS OF FAN DEPOSITION

Turn around and drive 0.6 miles west along the Vulcan Mine Rd. and park. Hike about 1.3 km north to the top of a large (>10 m high) wash cut (exposes Qf2 LS alluvium). Bishop Tuff (ca. 0.74 Ma) is exposed in the lowest east end of the cut.

Late Quaternary eolian and alluvial fan stratigraphy appears to be correlative across the east Mojave Desert, indicating that deposition is due to some aspect of regional climate change and related control of sediment supply and transport (McDonald, 1994; McDonald et al., 2003). The best evidence for linking climate change with regional alluvial fan deposition is demonstrated by geomorphic and age relations among fan units Qf5 (PMP) and Qf2 (SMP). Stratigraphic and age relations demonstrate that a regional period of major alluvial fan deposition occurred between about 9.4 and 14.0 ka (Fig. 8, 11). This age range indicates that deposition coincided with the Pleistocene to Holocene climatic transition and not simply a result of greater effective moisture and channel discharge during the last glacial maximum (Bull, 1991; Wells et al., 1987; Harvey et al., 1999). Pronounced fan aggradation is linked to increases in sediment yield with a concomitant increase in hillslope instability and monsoon storm activity. McDonald et al. (2003) argue that changes in vegetation cover during the Pleistocene to Holocene climatic transition may have had a limited impact on hillslope instability and sediment yield. An apparent lack of hillslope instability is due to the (1) inherently high infiltration capacity of coarse textured soil along hillslopes, which limits surface runoff, and (2) the modern vegetation (i.e. arid) cover that provided enough stability for the buildup of soils and colluvium during the Holocene. Together, these factors indicate that climate-driven changes in hillslope vegetation are not likely to have a strong impact on regional sedimentation. An

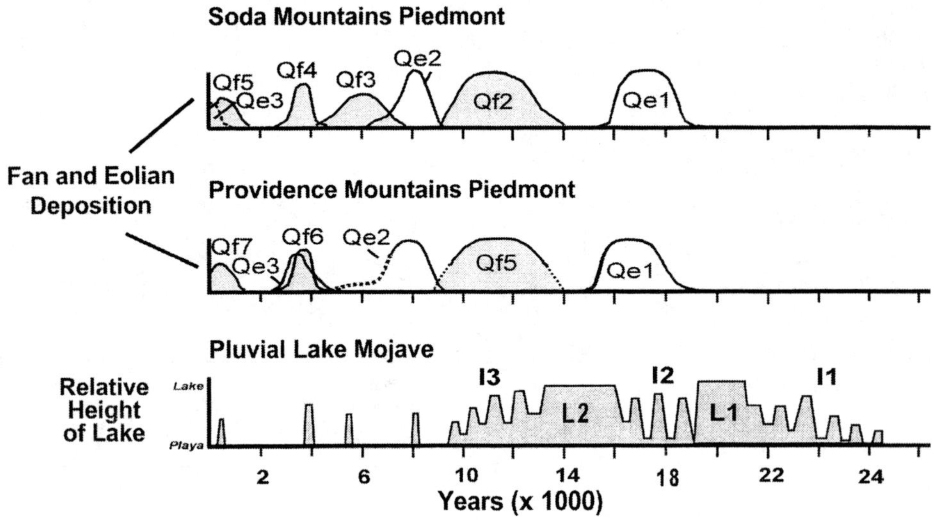

Figure 11. Summary of periods of fan and eolian deposition along the PMP and SMP and fluctuations of pluvial Lake Mojave during the last 23,000 years. Plots of geomorphic events imply general interval of time and levels of activity but do not imply relative volumes of sediment. Figure modified from McDonald et al. (2003).

increase in sediment yield may instead be primarily due to an increase in extreme storm events, perhaps an increase in tropical cyclones. These storms can yield several days of heavy rainfall across wide regions as they dissipate over land. Tropical cyclones have been associated with some of the largest flood events in southern Utah and Arizona (Ely, 1997). Extreme storms provide the high rainfall intensity and long duration required to mobilize permeable sediments from mountain catchments and into distal fan areas.

Other intervals of fan deposition also appear to be related to climate change. Regional fan deposition occurred between about 4.5 to 3.5 ka and appears to coincide with a brief interval of late Holocene pluvial lake activity at 3.6 ka (Figs. 8, 11). At least two intervals of fan aggradation occurred during the late-to-middle Pleistocene, with one occurring between about 22 and 90 ka and the other occurring between about 36 to 130 ka. These ages generally overlap with intervals of climate change associated with marine oxygen isotope stages 4-3 (~ 70 ka) and 6-5 (~130 ka). If these intervals of fan aggradation are related to glacial-interglacial transition or due to other aspects of regional climatic variation is not clear. The most extensive fan deposit (Qf2) is underlain by a tephra correlative with the 0.74 Ma Bishop Tuff (McDonald, 1994). We surmise that the extensive size of unit Qf2 and the general presence of well-bedded alluvium suggest that deposition occurred during a sustained interval of climate that was much wetter relative to subsequent pluvial periods. Periods of increased eolian deposition along the PMP appear to coincide with fluctuations in pluvial Lake Mojave and concomitant increases in the supply of eolian sediment during periods of lake desiccation (Fig. 11).

The influence of the drainage basin rock type on periods of fan deposition appears to have been secondary to the influence of climate change. Regional correlations indicate that fan aggradation occurred regardless of lithology, size, elevation, or topographic relief of drainage basins (McDonald, 1994; McDonald et al., 2003). First, similar stratigraphic sequences of fan deposits along the PMP have been derived from drainage basin source areas consisting of diverse rock types. Second, considerable variation exists between the geomorphology and ecology of the PMP and SMP mountain catchments. Drainage basins in the Providence Mts. are considerably higher and larger than basins in the Soda Mts. For example, one of the largest drainage basins in the Providence Mts. has an area of 8.8 km^2 and an elevation of 1200-2000 m (4000-6600 ft.). By comparison, one of the largest basins along the Soda Mts. has an area of 1.9 km^2 and an elevation from 100 to 300 m (339-1000 ft.). Similar deposition intervals across the region imply that the mechanisms driving deposition are extrinsic factors related to climatic variation and not intrinsic factors, such as complex response related to crossing inherent geomorphic thresholds or the instability of basin sediments due to tectonic activity.

STOP 3-4. SOILS AND GEOMORPHOLOGY OF THE LS FAN SEQUENCE

From Stop 3, continue 1.5 miles west along Vulcan Mine Rd. An abandoned dirt mine road (Rex Mine) extends ca. 2.5 km northeast away from the Vulcan Mine Rd. and traverses a series of Qf8 to Qf2 LS fan remnants.

DAY 4

STOP 4-1. SOILS AND GEOMORPHOLOGY OF THE PM FAN SEQUENCE

From Kelso, proceed 8.0 miles south along the Kelbaker Rd. Turn east and proceed 0.3 miles along a dirt road that follows the power lines. The dirt road from this point east for 4 km traverses a series of Qf8 to Qf2 PM fan remnants (mostly south of the dirt road). The PM soils provide the setting for extensive work, summarized below, on interrelations among soil properties, hydrology, and plant ecology.

Relation between soil formation and soil hydrology: Infiltration and surface evolution

The development of Av horizons and desert pavements has a profound impact on infiltration rates. Measured infiltration rates are reduced from about 35% to 1% of the infiltration rate for coarse textured deposits that do not have Av horizons (McDonald et al., 1996). Accumulation of aerosolic dust into soil B horizons that underlie the Av horizon also have a strong effect on soil water balance by increasing the soil water holding capacity. Increases in surface runoff that coincide with decreases in infiltration enhance surface instability with time. This, in turn, leads to the degradation of soils and desert pavements, eventually leading to the establishment of surface drainage and the ultimate widespread dissection of alluvial fan surfaces. The systematic development of soils and desert pavements, therefore, directly controls long term surface stability, promoting the eventual self destruction of alluvial fan surfaces. The soil surface and desert pavements show extensive degradation on the Qf3 and Qf2 surfaces that are underlain by argillic and/or petrocalcic horizons that impede infiltration.

Holocene climate change and the distribution of soil carbonate

Carbonate distribution in soils formed on Pleistocene surfaces is commonly bimodal, with carbonate accumulation occurring within the upper and lower parts of the soil profile. The bimodal distribution of carbonate is generally thought to correspond to a shift to a drier climate after the Pleistocene-to-Holocene climatic transition. Paleoclimate records derived from nearby pluvial lakes, however, indicate that the Holocene has been punctuated by at least four brief episodes (50-400 yrs; Fig. 9) of wetter climate (Enzel et al., 1989, 1992). Historic flooding of desert playas in the Mojave Desert appears to be associated with significant regional increases in rainfall from Pacific frontal storms as a result of a southern shift in the jet stream and frontal storm track during the winter and spring. McDonald et al. (1996) used a process-based, soil–water balance model (SHAW: Flerchinger and Pierson, 1991) to simulate soil water movement under varied conditions of climate, vegetation, and soil texture to determine if these episodic periods of extreme climate during the Holocene influenced soil water movement, and in turn, the profile distribution of soil carbonate. Simulations were run with actual daily weather data associated with relatively "wet" (historic playa flooding years with extreme regional increases in rainfall) and "dry" climate (historic average annual rainfall) to simulate the affects of wet and dry climate.

Simulations of soil water balance strongly suggest that the bimodal distribution of carbonate in soils on pre-Holocene surfaces can be linked to sizable changes in effective annual precipitation (Fig. 12). The depth of soil water flow only reached the lower zone of carbonate accumulation during a "wet" year when extreme increases in winter/spring storm activity resulted in a nearly 100% increase in annual precipitation. By comparison, the depth of soil water movement during a "dry" year corresponds to the upper zone of carbonate accumulation. Model simulations combined with measured infiltration rates also indicate that an increase in low intensity, winter and spring frontal storms would be required to account for changes in soil water balance rather than an increase in high intensity, summer convective storms. Increases in frontal storm precipitation can be linked to forced changes in atmospheric circulation patterns over the North Pacific Ocean and the formation of perennial lakes in the Mojave Desert.

Simulation results and calculations of pedogenic carbonate flux rates imply that Holocene climatic events, characterized by sizable increases in annual precipitation and the annual downward flux of soil water, resulted in the partial leaching of carbonate from the upper zone of accumulation in soils formed on pre-Holocene surfaces (McDonald et al., 1996). Episodic leaching of carbonates within the upper approximately 75 cm of soils, therefore, may not reflect accumulation that has occurred throughout the entire Holocene, but instead reflect a balance between episodic intervals of carbonate leaching and accumulation during the Holocene. This relation suggests that carbonate within the upper zone of soils may be susceptible to short term fluctuations in climate.

PLANT COMMUNITY ECOLOGY, SOILS, AND SOIL WATER BALANCE

Desert landscapes are complex soil mosaics with high spatial variation in soil properties due to the diversity of parent materials and ages of alluvial deposits. Knowledge of Quaternary landscape evolution and landscape relations is essential for investigating responses of key desert shrubs to contrasting soil conditions (McAuliffe, 1994; McAuliffe and McDonald 1995; Hamerlynck et al., 2002).

Two desert shrubs, *Larrea tridentata* (cresotebush) and *Ambrosia dumosa* (white bursage) account for most of total perennial plant canopy volume across the PMP (McAuliffe and McDonald, 1995). The growth responses and frequency of these two shrubs, however, varies considerably across the piedmont and clearly reflect edaphic controls associated with the degree of soil development and surface age. *Larrea* attains considerable variation is size among different soil geomorphic settings. *Larrea* reaches the greatest height, between 2-3 m, on the Qe3 late Holocene sand sheets and between 0.5 and 1.5 m on the late Holocene PM Qf6 and Qf7 surfaces (Fig. 13). By comparison, the height of *Larrea* on Pleistocene PM Qf4 and Qf3 surfaces is less than 1 m. Likewise, basal diameters of *Larrea* are greatest on the Qe3 and PM Qf7 and Qf6 (Fig. 13) with diameters up to nearly 3 m. Basal diameters are less than 1 m on the PM Qf4 and Qf3 surfaces. Plant density and canopy volumes display similar trends

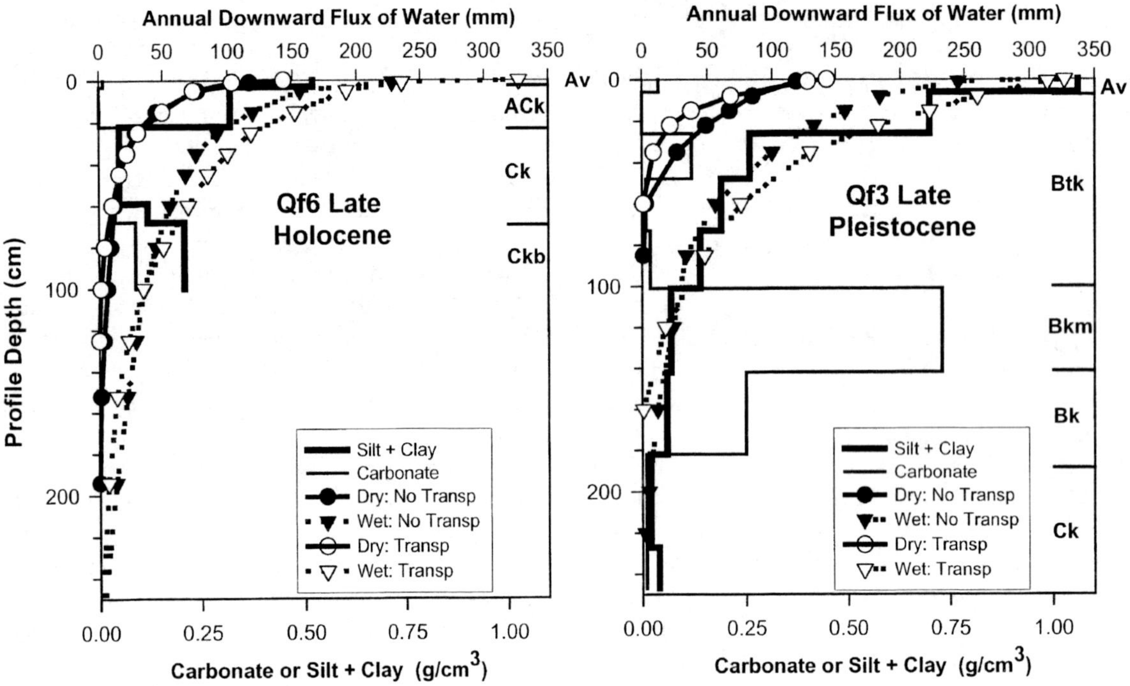

Figure 12. SHAW simulation results of the annual downward flux of soil-water for wet and dry years and for simulations run with and without transpirational-water loss. The downward flux of water represents the total amount of soil water that has infiltrated downward through each soil node for the entire year that was simulated. The downward flux of soil water is superimposed over profile mass (g/cm^3) distributions of total carbonate and silt + clay. Soils shown: Qf6 PM (a) and Qf3 PM (b). Depths of major horizon boundaries shown along the left margin. Figure modified from McDonald et al. (1996).

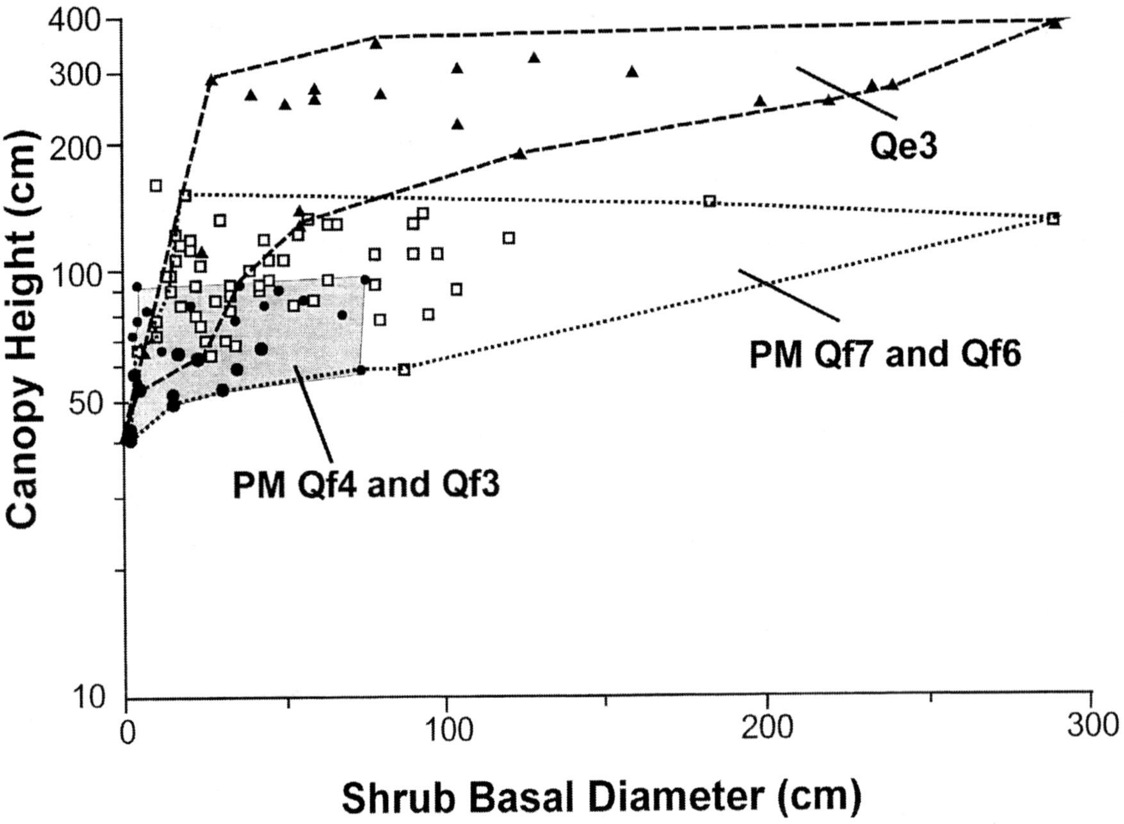

Figure 13. Relationship between *Larrea tridentata* canopy height (log scale) and basal diameter. Lines represent dimensional spaces achieved by *Larrea* for Holocene Qe3 and Qf7-Qf6 surfaces and Pleistocene Qf4-Qf3 surfaces. Figure modified from Hamerlynck et al. (2002).

between Holocene and Pleistocene surfaces with plant abundance and volume considerably greater on the Qe3, Qf7, and Qf6 relative to mean total volume on the Qf4 and Qf3 surfaces (Hamerlynck et al., 2002). *Ambrosia* does not have a similar response in plant form or abundance, but does increase in abundance relative to *Larrea* on Pleistocene surfaces.

The different responses of *Larrea* and *Ambrosia* among Pleistocene and Holocene soils is due to developmental and phonological adaptations of these two species and concomitant variation in soil hydrology and the degree of soil formation (McAuliffe, 1994; Hamerlynck et al., 2002). *Ambrosia* is a drought deciduous shrub that has the capacity for dormancy during periods of low soil moisture. Moreover, *Ambrosia* has a shallow root system (most roots above 50 cm depth). These characteristics allow *Ambrosia* to effectively use available soil moisture in near surface horizons and to better survive extended periods of low soil moisture. By contrast, *Larrea* is a perennial evergreen that lacks the ability for drought dormancy. *Larrea* has an extensive root system, with roots propagating laterally and deep (> 1 m), depending upon soil conditions. Unlike *Ambrosia*, *Larrea* remains photosynthetically active during extended periods of dry soil conditions and must rely on limited soil moisture from deep soil horizons. Extreme drought can result in extensive leaf loss and death of entire branches.

As discussed above, *Larrea* performs best on weakly developed soils (Qe3, Qf7, and Qf6 surfaces) that have high rates of infiltration, allow deep percolation of soil water, and have a low moisture-holding capacity (McDonald et al., 1996; Hamerlynck et al., 2002). Numerical simulations indicate that, at least during most years, one or two storms will provide a sufficient flux of moisture well below 50 cm (Fig. 12). Deep percolation limits evaporative loss and limits use of soil water from shallow rooting plants such as *Ambrosia*. Annual deep recharge provides *Larrea* with a limited, but more seasonally constant, source of available water that can sustain this shrub through dry summer and fall months. Therefore, *Larrea* responds favorably in weakly developed soils that allow the recharge of water at greater depths due to its relatively deep and extensive root system.

The greater development of fine-textured Av and Bt horizons on Pleistocene surfaces inhibits deep infiltration of water resulting in noticeable decrease in the performance of *Larrea*. Well developed soils have low rates of infiltration and high moisture holding capacity. Numerical simulations indicate that an annual recharge below 50 cm does not occur during most years (Fig. 12). Although near surface horizons have a greater flux of soil moisture, evapotranspirational loss, especially water use by competing and abundant shallow rooted plants, results in greater seasonal changes in plant–available water. Field observations indicate vertical root distributions of *Larrea* are constrained to shallow depths on soils with strong Bt horizons, but are much deeper in soils lacking such horizons (McAuliffe and McDonald, 1995). A lack of soil moisture below 50 cm inhibits shrub growth resulting in diminished size, frequency, and increased evidence of prolonged water stress reflected as stem mortality.

STOP 4-2. SOILS AND GEOMORPHOLOGY OF THE QM FAN SEQUENCE

Turn around and return to the Kelbaker Rd. Proceed south 3.7 miles and then east 0.45 miles along a small dirt road. A series of Qf8 to Qf2 PM fan remnants are exposed south and east for ca. 2 km to the large wash that lies ca. 100 m to the east.

ACKNOWLEDGMENTS

Funding for the research presented in Days 3 and 4 was provided through grants from NASA (NAG 5 1828), NSF (EAR 9118335), NSF (EAR 9973195), and scholarships to McDonald from the University of New Mexico, Sigma Xi, and the Geological Society of America.

REFERENCES

Biondi, F., Gershunov, A., and Cayan, D.R., 2001, North Pacific decadal climate variability since 1661: Journal of Climate, v. 14(1), p. 5-10.

Birkeland, P., 1999, Soils and geomorphology: Oxford University Press, NY, 430 p.

Bull, W.B., 1991, Geomorphic responses to climatic change: Oxford University Press, NY, 326 p.

Clarke, M.L., 1994, Infra-red stimulated luminescence ages from aeolian sand and alluvial fan deposits from the eastern Mojave Desert, California: Quaternary Geochronology, v. 13, p. 533-538.

Ely, L.L., 1997, Response of extreme floods in the southwestern United States to climatic variations in the late Holocene: Geomorphology, v. 19, p.175-202.

Enzel, Y., Cayan, D.R., Anderson, R.Y., and Wells, S.G., 1989, Atmospheric circulation during Holocene lake stands in the Mojave Desert: evidence of regional climate change: Nature v. 341, p. 44-48.

Enzel, Y., Brown, W.J., Anderson, R.Y., McFadden, L.D., and Wells, S.G., 1992, Short-duration Holocene lakes in the Mojave River drainage basin, southern California: Quaternary Research, v. 38, p. 60-73.

Eppes, M., Harrison, B., and McIntosh, W., 1999, Spatial variability of soils developing on basalt flows in the Potrillo volcanic field: prelude to a soil chronosequence study. Earth Surface Processes and Landforms, v. 24, p. 1-16.

Eppes, M.C., 2002, Soil Geomorphology of the north flank of the San Bernardino Mountains, California: Ph.D. Dissertation, University of New Mexico, 214 p.

Eppes, M.C., McFadden, L.D., Matti, J., and Powell, R., 2002, Influence of soil development on the geomorphic evolution of landscapes; an example from the Transverse Ranges of California: Geology, v. 30(3), p. 195-198.

Flerchinger, G.N. and Pierson. F.B., 1991, Modeling plant canopy effects on variability of soil temperature and water: Agricultural and Forest Meteorology, v. 56, p. 227-246.

Gile, L., Hawley, J.W., and Grossman, R.B., 1981, Soils and geomorphology in the Basin and Range area of southern New Mexico- Guidebook to the Desert Project: New Mexico Bureau of Mines and Mineral Resources Memoir 39, 222 p.

Hammerlynck, E.P, McAuliffe, J.R., McDonald, E.V, and Smith, S.D, 2002, Impacts of desert soil processes and drought on contrasting Mojave Desert shrubs: Ecology, v. 83, p. 768-779.

Harvey, A.M., Wigand, P.E., and Wells, S.G., 1999, Response of alluvial fan systems to late Pleistocene to Holocene climatic transition: contrasts between the margins of pluvial Lakes Lahontan and Mojave, Nevada and California, USA: Catena, v. 36, p.255-281. McGraw-Hill, New York, 281 p.

Machette, M., 1985, Calcic soils of the southwestern United States: *In* D. Welde (ed.), Soils and Quaternary Geomorphology of the Southwestern United States, Geological Society of America Special Paper 203, p. 1-21.

McAuliffe J.R., 1994, Landscape evolution, soil formation, and ecological patterns and processes in Sonoran Desert bajadas: Ecological Monographs 64:111-148.

McAuliffe, J.A., and McDonald, E.V, 1995, Soil-Vegetation Relations for Calcic Soils, East Mojave Desert: *in* R.E. Reynolds and J. Reynolds (eds.), Ancient surfaces of the east Mojave Desert, San Bernardino County Museum Association Quarterly, v. 42(3), p. 53-68.

McDonald, E.V., 1994, The relative influences of climatic change, desert dust, and lithologic control on soil-geomorphic processes and hydrology of calcic soils formed on Quaternary alluvial-fan deposits in the Mojave Desert, California: Ph.D. Dissertation, University of New Mexico, 383 p.

McDonald, E.V., and McFadden, L.D., 1995, Quaternary stratigraphy of the Providence Mountains Piedmont and preliminary age estimates and regional stratigraphic correlations of Quaternary deposits in the eastern Mojave Desert, California: *in* S.G. Wells, J.C. Tinsley, L.D. McFadden, and N. Lancaster, (ed.) Quaternary stratigraphy and dating methods: Understanding geologic processes and landscape evolution in southern California, Geological Investigations of an Active Margin, 1994 Geological Society of America Cordilleran Section Guidebook, p. 205-210.

McDonald, E.V., Pierson, F.B., Flerchinger, G.N. and McFadden, L.D., 1996, Application of a process-based soil-water balance model to evaluate the influence of Late Quaternary climate change on soil-water movement in calcic soils: Geoderma, v. 74, p. 167-192.

McDonald, E.V, McFadden, L.D., and Wells, S.G., (in press), Regional response of alluvial fans to the Pleistocene-Holocene transition and alluvial fan deposition in Providence Mountains: Geological Society of America Special Paper 368.

McFadden, L., and Tinsley, J., 1985, Rate and depth of pedogenic-carbonate accumulation in soils: Formulation and testing of a compartment model: Geological Society of America Special Paper 203, p. 23-41.

McFadden, L.D., Wells, S.G., and Jercinovic, M.J., 1987, Influence of eolian and pedogenic processes on the origin and evolution of desert pavements: Geology, v. 15, p. 504-508.

McFadden, L., 1988, Climatic influences on rates and processes of soil development in Quaternary deposits of southern California: *in* Reinhadt, J., and Sigleo, W., (eds.), Paleosols and weathering through geologic time: principles and applications, Geological Society of America Special Paper 206, p. 153-177.

Meisling, K. and Weldon, R., 1989, Late Cenozoic tectonics of the northwestern San Bernardino Mountains, southern California: Geological Society of America Bulletin, v. 101, p. 106-128.

Miller, D.M., Glick, L.L., Goldfarb, R., Simpson, R.W., Hoover, D.B., Detra, D.E., Dohrenwend, J.C., and Munts, S.R., 1985, Mineral resources and resource potential map of the south Providence Mountains Wilderness Study Area, San Bernardino County, California: United States Geological Survey Miscellaneous Field Map, MF-1780-A.

Miller, F.K.; Matti, J.C.; Brown, H.J.; and Powell, R.E., 1998, Digital geologic map of the Fawnskin 7.5' quadrangle, San Bernardino County, California: U. S. Geological Survey, Reston, VA, United States, Open-File Report OF 98-0579.

Miller, F.K., Matti, J.C., Brown, H.J., and Cossette, P.M. (digital compiler), 2000, Digital Geologic Map of the Butler Peak 7.5' Quadrangle, San Bernardino County, California: U.S. Geological Survey Open-File Report USGS OF 00-145, U.S. Geological Survey, Menlo Park, California.

Newland, S., 2001, The hydraulic geometry of ephemeral channels in a tectonically active landscape: north flank of the San Bernardino Mountains, California: M.S. thesis, Lehigh University, Bethlehem, PA.

Phillips, J.D., 1998, Enriched: On the relations between complex systems and the factorial model of soil formation: Geoderma, v. 86(1-2) , p.1-21.

Powell, R.E., and Matti, J.C., and Cossette, P.M. 2000, Geologic map and digital database of the Cougar Buttes 7.5' quadrangle, San Bernardino County, California: U.S. Geological Survey Open-File Report OF 00-0175, scale 1:24,000, 34 p.

Reheis, M., Sowers, J., Taylor, E., McFadden, L., and Harden, J., 1992, Morphology and genesis of carbonate soils on the Kyle Canyon fan, Nevada, U.S.A: Geoderma, v. 52, p. 303-342.

SCAMP (Southern California Areal Mapping Project), 2002, Surficial geologic materials in Southern California: General categories, and issues related to their geologic-map classification, version 1.0, U.S. Geological Survey Open File Report 01.

Shreve, R.L., 1968, The Blackhawk landslide: Geological Society of America Special Paper 108, 47 p.

Soil Survey Staff, 1992, Keys to Soil Taxonomy: Soil Conservation Service Technical Monograph No. 19, 5th edition, 541 p.

Sowers, J.M., 1985, Pedogenic calcretes of the Kyle Canyon Alluvial Fan, Southern Nevada-Morphology and Development: Ph.D. Thesis, Berkeley, University of California, p. 160.

Spotila, J., Farley, K., and Sieh, K, 1998, Uplift and erosion of the San Bernardino Mountains associated with transpression along the San Andreas fault, California, as constrained by radiogenic helium thermochronometry: Tectonics, v. 17, p. 360-378.

Wang, Y., McDonald, E.V., Amundson, R., McFadden, L.D., and Chadwick, O., 1996, Paleoenvironmental implications of carbon and oxygen isotopes in alluvial soils, Providence Mountains, California: Geological Society of America Bulletin, v. 108, p. 379-391.

Weitkamp, W., Graham, R., Anderson, M., and Amrhein, C., 1996, Pedogenesis of a vernal pool entisol-alfisol-vertisol catena in southern California: Soil Science Society of America Journal, v. 60, p. 316-323.

Wells, S., Dohrenwned, J., McFadden, L., Turrin, B., and Mahrer, K., 1985, Late Cenozoic landscape evolution on lava flow surfaces of the Cima Volcanic field, Mojave Desert, California: Geological Society of America Bulletin, v. 96, p. 1518-1529.

Wells, S.G., McFadden, L.D., and Dohrenwend, J.C., 1987, Influence of late Quaternary climatic changes on geomorphic and pedogenic processes on a desert piedmont, eastern Mojave Desert, California: Quaternary Research, v. 27, p. 130-146.

Wells, S.G., McFadden, L.D., Poths, J., and Olinger, C., 1995, Cosmogenic (super 3) He surface-exposure dating of stone pavements; implications for landscape evolution in deserts: Geology, v. 23(7), p.613 – 617.

Yair, A., and Kossovsky, A., 2002, Climate and Surface properties: Hydrological response of small arid and semi-arid watersheds: Geomorphology, v. 42, p. 43-57.

The Palouse loess and the Channeled Scabland: A paired Ice–Age geologic system

Authors and Leaders: **David R. Gaylord,** *Department of Geology, Washington State University, Pullman, WA 99164–2812*
Alan J. Busacca, *Department of Crop and Soil Sciences, Washington State University, Pullman, WA 99164–6420*
Mark R. Sweeney, *Department of Geology, Washington State University, Pullman, WA 99164–2812*

INTRODUCTION

The Palouse loess and Channeled Scabland of the Columbia Plateau are genetically–paired geomorphic wonders whose evolution is tied to global–scale, Quaternary climate shifts. Cataclysmic glacial–outburst floods that created the scabland supplied the bulk of loose sediment from which a *second* 'paired' system of eolian deposits–sand dunes and loess–arose. In order to appreciate these pairings, we have designed a three–day field trip that ties together the various parts of these sedimentary–geomorphic associations.

The starting point, Spokane, Washington (Fig. 1), lies on the southwestern edge of the Rathdrum Prairie, a gravel plain whose thick sedimentary cover accumulated during multiple, glacial, outburst floods. The rolling hills immediately south and east of Spokane are, by contrast, mantled with variably thick loess, much of it derived from eolian reworking of fine–grained Missoula flood sediment that bypassed the Spokane area. Over the next three days,we will trace a circuitous path that mimics the transport route of Missoula flood sediment that passed through the Spokane area, was transported to the south–central reaches of the Columbia Basin, and then carried back to Spokane by the prevailing wind.

FIELD TRIP ROUTE: THEMES AND TOPICS

The field trip route and stops highlight two themes: (1) The role that Quaternary events had on Columbia Plateau landscapes, and (2) the timing and development of the genetically–paired, dune–loess system that dominates the central and eastern plateau. Other topics include the: wind energy potential of the region, role of Pleistocene and Holocene landforms and soils in modern agriculture (including a world–class wine industry), and contemporary environmental problems associated with dust aerosols.

HISTORY OF THE CHANNELED SCABLAND

The Channeled Scabland in eastern Washington (Fig. 2) resulted primarily from cataclysmic outburst floods from glacially dammed Lake Missoula (Bretz, 1923; Bretz et al., 1956) both during and prior to the last glaciation. Outburst floods resulting from failures of ice dams on the Columbia and Spokane Rivers (Atwater, 1984, 1986, 1987) also may have helped sculpt the scabland topography. These outburst floods constituted some of the largest documented floods on earth, overwhelming the Columbia River drainage with up to 2500 km^3 (500 mi^3) of water per flood event (Baker, 1973, 2002; Baker and Nummedal, 1978; Waitt, 1985). Mean peak flood discharges estimated at 7 x 10^6 ft^3/sec (2 x 10^7 m^3/sec) and mean flow velocities as high as 68 miles/hr (110 km/hr) (Baker, 1973, 1978) attest to the enormity of these flood events.

The Channeled Scabland is characterized by an anastomosing complex of now–largely–abandoned, steep–walled channels or coulees that were incised into bedrock and loess. This scabland terrain includes km–long cataracts with >100 m cliffs (now recognized as dry falls), loess islands (streamlined erosional remnants of thick loess separated by former flood channels), ice–rafted boulders, and immense gravel bars. In south–central Washington, the many paths taken by the onrushing floods converged on Wallula Gap (Fig. 2), a narrow water gap along the Columbia River. Temporarily impounded behind this constriction, sediment–laden waters backflooded tributaries of the lower Columbia River

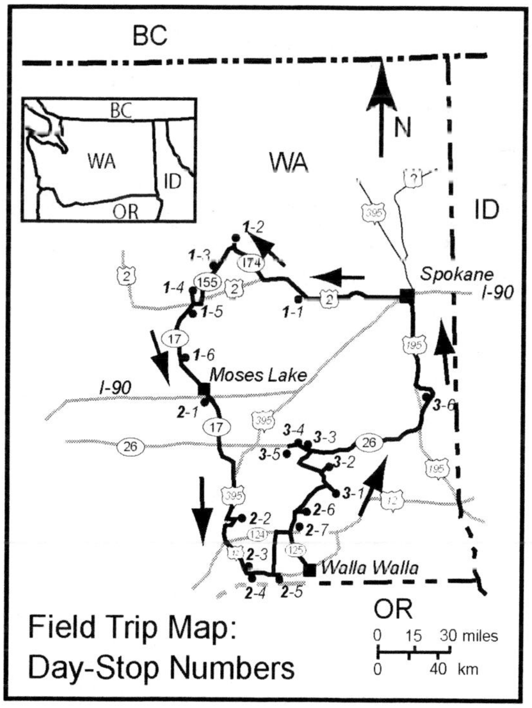

Figure 1. Field trip road guide and map. Stops are identified by day numbers (bold) and stop numbers.

Gaylord D. R., Busacca, A. J., and Sweeney, M. R., 2003, The Palouse loess and the Channeled Scabland: A paired Ice–Age geologic system: *in* Easterbrook, D.J., ed., Quaternary Geology of the United States, INQUA 2003 Field Guide Volume, Desert Research Institute, Reno, NV, p. 123–134.

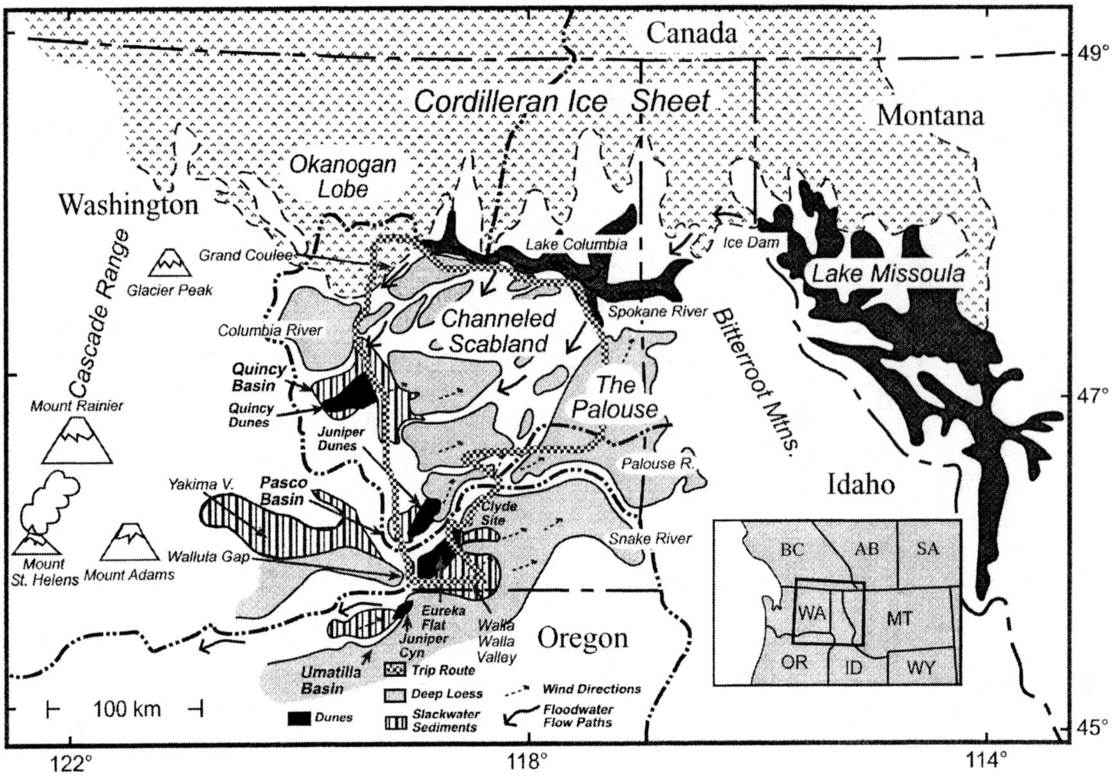

Figure 2. Schematic map of the Pacific Northwest at the height of the Last Glacial Maximum (LGM). Field trip route shown with broad, cross–hatched line.

including the Snake, Yakima, and Walla Walla Rivers. Backflooding resulted in accumulations of normally graded, fine sand, silt, and clay, slackwater rhythmites, 10's of m thick.

The interpretation of a giant–flood origin for the features of the Channeled Scabland made by J Harlen Bretz in journal articles beginning in 1923, (Bretz, 1923, 1925, 1928, 1932, 1969; Bretz *et al.*, 1956), precipitated one of the most celebrated scientific debates in American geology (Allen *et al.*, 1991). Bretz's ideas of giant floods were in direct conflict with the uniformitarian principles that had brought geology to its modern era. Even after additional evidence was found for a floodwater source (Pardee, 1942), not until 1956 (Bretz *et al.*, 1956) did his ideas become universally accepted. In recent decades, researchers have estimated the hydraulics of giant flood flows and explained the origin of numerous scabland features (Baker, 1973; Baker and Nummedal, 1978; Baker and Bunker, 1985), and identified multiple last–glacial floods (Atwater, 1984, 1986, 1987; Waitt, 1980, 1984, 1985), a number that may be as high as 89 individual events (Atwater, 1986, 1987). While the number and origins of outburst floods associated with late Wisconsin glaciation have been the focal point for continued debate (Shaw *et al.*, 1999, 2000; Atwater *et al.*, 2000), late Wisconsin outburst floods were not unique within the Pleistocene. In fact, a number of episodes of pre–late Wisconsin, glacial–outburst floods on the Columbia Plateau have been recognized (Bretz *et al.*, 1956; Patton and Baker, 1978; McDonald and Busacca, 1988; Bjornstad *et al.*, 2001).

DAY 1

The first day is largely dedicated to examining the glacial and outburst flood record of the northern Columbia Plateau.

Depart Spokane, and drive west on U.S. Highway 2 to Airway Heights. Set odometers to zero at Airway Heights and drive 39 miles (63 km) to the rest area on the south side of the road. The highway climbs in and out of a number of steep–sided, outburst flood coulees that incise the Columbia River Basalt Group (CRBG). Readers interested in learning more about the CRBG are directed to Hooper (1984; 2002). The smooth, cultivated hills and plateaus are generally loess–capped islands that escaped erosion by the floods. The Cordilleran Ice Sheet advanced within about 12 miles (19 km) (north) of U.S. 2.

STOP 1–1. CHANNELED SCABLAND

The first stop provides an excellent opportunity to view scabland topography and contemplate the broader history of the outburst flooding. This site is located within the Creston–Davenport scabland (Kiver and Stradling, 1989) where a relatively low–relief, butte–and–basin topography is developed on basalt. Note the typical scabland angular basalt gravel and scoured basalt hollows in which ponds now reside.

Because this setting is primarily erosional, the flood record here is sparse. However, ~25 miles (40 km) to the northwest, varves, outburst flood and glacial deposits exposed along Manila Creek, a tributary of the Sanpoil and Columbia Rivers, has preserved one of the longest and most detailed chronologies of flooding and glaciation anywhere on the Columbia Plateau. Glacial Lake Columbia, generated when the Okanogan lobe of the Cordilleran Ice Sheet blocked the Columbia River, influenced the sedimentation along Manila Creek from at least ca. 15.5 to 13 ^{14}C yrs. B.P (Atwater, 1984, 1986. 1987). This ~2,500–year interval essentially covers the late Wisconsin history of Missoula outburst flooding (Waitt, 1980, 1985; Baker *et al.*, 1991). Varve counts at Manila Creek (Atwater, 1986, 1987) indicate that floods triggered by ice–dam collapses were smaller and relatively more frequent when glacial Lakes Columbia and Missoula first began to form. Until the last glacial maximum (LGM), the volumes of these floods increased while their frequency decreased, with 50 to >65 yrs passing between flood events (Atwater, 1986). After the LGM, their frequency increased and volume decreased. The 89 individual floods that Atwater (1986) identified at Manila Creek more than double the approximately 40 outburst floods identified by Chambers (1971) in Montana and by Waitt (1980; 1985) in southern Washington. Atwater (1986, 1987) attributed the greater number of floods to the relative nearness of the Manila Creek site to local base level, a factor that enhanced the preservation of even small volume flood deposits. By contrast, sedimentary deposits from small–volume floods in the southern plateau tended not to be preserved (Baker and Bunker, 1985; Waitt, 1985; Atwater, 1986). Ice–rafted erratics at elevations of up to 750 m (Atwater, 1986), indicate that STOP 1 (695 m) was inundated frequently during outburst flooding.

Exit the rest stop and continue west on U.S. Highway 2 for 17.1 miles (27.5 km) to its junction with State Highway 174. Turn right onto Highway 174 and drive toward Grand Coulee. The road passes through channeled scabland and loess islands for approximately 30 miles (48 km) before descending into the steep, basalt–walled Columbia River valley. On the outskirts of the town of Grand Coulee, watch for the 'Business District sign'; turn right at this sign onto Spokane Way, driving 0.5 mile (0.9 km) to its junction with State Highway 155. Turn right onto Highway 155 and proceed 1.7 miles (2.7 km) to the Grand Coulee Dam Visitors Center parking area.

STOP 1–2 GRAND COULEE DAM VISITORS CENTER

Construction of Grand Coulee dam began in 1933 and was completed in 1942. It produces about 6,800 megawatts annually (Pitzer, 1994). The dam is operated by the U.S. Bureau of Reclamation, and backs up the Columbia River (Lake Franklin D. Roosevelt) nearly to the Canadian border, a distance of ~150 river miles (241 km). Rising ~550 feet (182 m) above its granite bedrock base, the ~4200–foot–(1380 m) –wide dam is a key element of the Columbia Basin Irrigation Project. A portion of the electricity produced by the dam is used to pump water from Lake Roosevelt ~95 m *up* to Banks Lake, a ~27–mile–(43.5 km) –long and 1–4.5–mi. (1.6–7.2 km)–wide reservoir in the upper Grand Coulee, ~2 mi. (3.2 km) to the southwest. Water from Banks Lake feeds a canal system that supplies more than 250,000 acres of irrigated cropland (Pitzer, 1994).

Turn left onto State Highway 155 and drive 13 miles (21 km) south toward Grand Coulee and Banks Lake. As the road climbs out of the Columbia River Valley it traverses numerous granite and gneiss exposures that are capped in the near distance by Miocene Columbia River basalt flows.

STOP 1–3. GRAND COULEE AND STEAMBOAT ROCK

Grand Coulee is the deepest and arguably the most spectacular of the scabland coulees, ~1.6–7.2 km wide, 300 m deep, 70 km long. It is subdivided into incised upper and lower Grand Coulee segments separated by an unconfined, kms–long scabland (Salisbury, 1901; Bretz, 1932, 1969).

The history of the Grand Coulee was first described in detail by Bretz (1932). After the Okanogan lobe dammed the Columbia River in the late Wisconsin, creating Lake Columbia, Grand Coulee provided the principal drainage outlet for the lake (Bretz, 1932; Richmond *et al.*, 1965; Atwater, 1986; Waitt and Atwater, 1989). Glacial ice blocked the coulee at the present Grand Coulee dam when the Okanogan lobe was at its maximum (Bretz, 1932; Atwater, 1986, 1987; Waitt and Atwater, 1989). Glacial drift that caps Steamboat Rock probably was deposited during a pre–late Wisconsin glaciation (Bretz, 1932; Atwater, 1986, 1987; Waitt and Atwater, 1989). Lake Columbia continued to occupy Grand Coulee for several centuries following the last Missoula flood (Atwater, 1986). Varved, silt and clay lake sediments, some containing dropstones, are intercalated with thin flood deposits (Atwater, 1987). Following the last flood, several decades of ponding led to continued silt–clay deposition here (Atwater, 1987; Waitt and Atwater, 1989).

Turn right (south) at Steamboat Rock State Park and drive 15.6 miles (25 km) toward Coulee City State Highway 155. At the junction of State Highway 155 and U.S. Highway 2 follow U.S. 2 west 5.9 miles (9.5 km) to its junction with State Highway 17. Turn right onto State Highway 17. Two miles (3.2 km) north of the junction, note the hummocky, rubble–strewn ridge to the northwest (on your left). This ridge marks the southern terminus of the Withrow end moraine (Easterbrook, 1976, 1979). Stop 5.5 miles (8.8 km) from the junction of Highways 17 and U.S. 2, where a basalt 'haystack' rock rises from the prairie on the right (east) side of the road. This enormous glacial erratic was transported by the Okanogan lobe. (Optional) Drive 8.5 miles (13.7 km) north from this stop to Sims Corner (Junction of Highways 17 and 172). Driving west on Highway 172 takes one past an array of recessional glacial topography, including kames, eskers, and kettles (Easterbrook, 1979).

STOP 1–4. WITHROW END MORAINE AND THE WATERVILLE PLATEAU

This stop provides an opportunity to examine glacial drift left by the Okanogan lobe after it crossed the Columbia River

~30 miles (48 km) to the north. As the Okanogan lobe emerged from the Okanogan valley, it radiated outward across the Waterville Plateau. The terminus of this glacial advance is marked by a distinctive end moraine, several kilometers wide and 30–70 m high. The hummocky moraine extends approximately 35 miles (56 km) from near Lake Chelan on the west to upper Grand Coulee (Bretz, 1932; Flint, 1935). Termed the Withrow moraine (Easterbrook, 1976, 1979) for the settlement where it achieves its maximum height (Flint, 1935), this prominent feature records a prolonged ice front presence (Easterbrook, 1979; Waitt and Thorson, 1983).

The glacial drift of the Withrow moraine overlies glacially polished, striated, and scoured basalt (Bretz, 1932; Flint, 1935, 1936; Easterbrook, 1976, 1979). Behind the terminal moraine, the low, rolling topography bears the imprint of glacial recession, with numerous stagnation–related kettles, kames, and eskers (Bretz, 1932; Flint, 1935, 1936; Easterbrook, 1976, 1979). The timing of Waterville Plateau glaciation has been obscured by a lack of in–place, datable materials (Easterbrook, 1976, 1979). Atwater's (1986, 1987) reconstructions of outburst flood activity at Manila Creek, suggest that the Okanogan lobe had retreated north of the Columbia River by a few centuries after the final Missoula outburst flood, ca. 12.5 ka (Waitt, 1985).

Drive 5.5 miles (8.8 km) south on Highway 17 to the junction with U.S. Highway 2. Turn left (east) on Highway 2 and drive 1.7 miles to the junction with Highway 17; turn right and drive 1.9 miles (3 km) to the Dry Falls overlook.

STOP 1–5. DRY FALLS OVERLOOK

This spectacular, dry waterfall marks the northern margin of the lower Grand Coulee. From this overlook, only the western edge of a continuous 5.5–km–wide, 120–m–high cataract, created by the progressive recession of the cliff face, is visible (Bretz, 1932; Baker, 1973, 1989). Bretz (1969) attributed the Dry Falls cataract to highly focused flood flows caused by downstream incision of weaker basalt in the Coulee monocline. Deep, plunge–pool lakes, like Dry Falls Lake, attest to the erosive power of the flood flows. Falls–related, plunge–pool scouring and subaqueous plucking of the coulee bottom generated many of the cataracts in the lower Grand Coulee (Baker, 1989). Subaqueous plucking was facilitated by vortices and near–bed fluid turbulence that excavated the less–resistant, columnar–jointed portions of the basalt flows (Baker, 1987).

Turn left (south) on State Highway 17 and drive 23.2 miles (37.3 km) south past Soap Lake. Turn left on Hatchery Road; at 0.7 miles (1.2 km) note the giant basalt megaclast amidst the boulder field on the left. Continue another 1.9 miles (3 km) to the stratified and cross–stratified gravel exposure along the road leading out of the Rocky Ford Creek valley.

STOP 1–6. EPHRATA FAN

The concentrated outflow of sediment–laden water from lower Grand Coulee near Soap Lake into the unconfined Quincy Basin generated a broad, 200 mi^2 (520 km^2), fan–shaped accumulation of coarse–to–fine gravel. Reaching a maximum thickness of 56 m (Bretz, 1932), Ephrata fan deposits generally fine upward from thick, high–angle–cross–bedded, cobble–boulder gravel to horizontally stratified, coarse sand and granule–pebble gravel. The giant basalt megaclast along the road has a maximum dimension of 18 m (Baker, 1987) and is part of an armored surface generated by waning–stage, outburst flood incision (Baker, 1973). The northwest–oriented, crescent–shaped depression surrounding the megaclast records turbulent scour (Baker, 1973, 1987). Rocky Ford Creek has incised the fan surface to an average depth of 15 m revealing the coarse–grained, fines–depleted nature of the fan interior. Sand, silt, and clay transported here by outburst floods largely bypassed this region, ultimately accumulating in slackwater deposits or leaving the plateau via the Columbia River.

Drive 2.6 miles (4.2 km) to the junction with State Highway 17. Turn left (south) and drive 3.8 miles (6 km) to the junction with State Highway 282; bear left and continue south on Highway 17 for 9.4 miles (15 km) to Moses Lake. ***End of Day 1.***

DAY 2

Much of the second day is dedicated to examining the paired, eolian, sand–dune–and–loess systems and their relation to paleoclimatic fluctuations on the Columbia Plateau. Special tours of a wind energy facility and slackwater flood stratigraphic succession have been arranged.

SAND DUNES ON THE COLUMBIA PLATEAU

Numerous stabilized–to–active dune fields, sand sheets, deflation flats, and areas of deflationary gravel lags mantle the plateau. Dune fields cover from 20 km^2 at the Sand Hills Coulee (Schatz, 1996; Gaylord *et al.*, 1999) to >1000 km^2 in the Quincy dunes (Petrone, 1970). Relatively large dune fields exist in the southern portion of the scabland on the Hanford site (Gaylord *et al.*, 1991; Smith, 1992; Gaylord and Stetler, 1994; Stetler and Gaylord, 1996), near Smith Canyon (Schatz, 1996) and within the Juniper dunes Wilderness area (Gaylord *et al.*, 1997; Gaylord *et al.*, 1998; Gaylord *et al.*, 2001). Extensive sand sheets mantle many areas of the plateau including the important loess generating region of Eureka Flat (Sweeney et al., 2001). All of these dune fields have SW–NE orientations and are located upwind from loess. Sediments from the dunes commonly are linked compositionally to outburst–flood deposits and Palouse loess (Busacca and McDonald, 1994; Gaylord *et al.*, 1991; Gaylord *et al.*, 2001); but at Sand Hills Coulee, dune sand compositions reflect a Miocene–Pliocene, Ringold Formation (paleo–Columbia River ancestry) provenance (Schatz, 1996; Gaylord *et al.*, 1999).

Tephra chronology and limited geoarcheology from the Hanford and Smith Canyon dunes suggests that sand dune activity on the plateau peaked during the early and middle Holocene (Gaylord and Stetler, 1994; Gaylord *et al.*, 2001). The prevalence of vegetation since the start of the Holocene has favored the development of parabolic and blowout dune types. The 20% of dunes that are actively migrating today also include barchan and barchanoid ridge types. These active dunes tend to form in elevated locations subject to higher wind speeds, in zones of low subsurface moisture, or within areas subject to persistent anthropogenic disturbance.

LOESS ON THE COLUMBIA PLATEAU

The loess that blankets the Columbia Plateau is one of the more prominent and widespread geologic features produced during the Quaternary in the Pacific Northwest. The core area of the deepest and most continuous loess in eastern Washington is termed the 'Palouse' (Fig. 2), a region that covers >50,000 km^2 with deposits that are from a few meters to 75 m thick. Palouse loess accumulated at elevations ranging from about 200 m along the eastern margin of the Pasco Basin to approximately 800 m near Moscow, Idaho. The loess on the plateau largely accumulated on the Miocene Columbia River Basalt. Incised perennial streams within the Palouse reveal basalt–floored valleys. First–and–second–order intermittent streams tend to bottom in loess rather than basalt (Fryxell, 1968). The drainage pattern of major streams on parts of the Palouse not affected by floods probably was established before outburst flooding began (Ringe, 1970).

Forty six natural and artificial loess exposures on the Columbia Plateau have been described, mapped, and correlated during the past 20 years (McDonald and Busacca, 1992; Busacca *et al.*, 1992). Region–wide correlations using paleosol and tephra markers have revealed two loess units that span approximately the last 70,000 yrs. These are informally named L1 and L2 (McDonald and Busacca, 1992) (Fig. 3). Two review articles (Busacca, 1989; 1991) summarize much of what had been written about the Palouse between the 1880's and 1980's. Three articles since 1991, McDonald and Busacca, 1992; Busacca *et al.*, 1992; Busacca and McDonald, 1994, add considerable detail concerning late Quaternary loess and paleosols.

The Palouse loess has attributes that have made it attractive for paleopedologic and paleoclimatic research, including: (1) a depositional record that spans much of the Quaternary , (2) its relatively great thickness per unit time, especially in areas proximal to outburst flood (source) sediments, (3) numerous distal tephra layers that provide chronostratigraphic control, and (4) paleoecologic features preserved in surface and buried soils. Research on the significance of cicada–burrow fabrics (Tate, 1998; King, 2000; O'Geen and Busacca, 2001) and on plant opal phytoliths (Blinnikov *et al.*, 2001, 2002) indicates that an *Artemisia*–dominated, periglacial steppe that supported huge populations of sagebrush–dependent, soil–dwelling, cicada nymphs and the above–ground, adult forms was widespread on the Columbia Plateau during development of the Washtucna and other cold–phase soils. The striking visual and morphologic characteristics of the majority of buried soils in loess across the plateau largely reflect the abundance of cicada burrows that have been cemented by carbonates. The widespread periglacial, shrub–steppe was replaced by a pure, perennial, bunchgrass–steppe at the opening of the Holocene (except that sage persisted in the driest core area of the plateau). Lacking a shrub component, the bunchgrass soils that developed during the Holocene did not support cicada nymphs. Consequently, the L1 loess is all but devoid of large burrows that typify the Washtucna soil in L2 (O'Geen and Busacca, 2001). In summary, significant soil development on the Palouse seems to have occurred during cold–climate phases, while warm, interglacial phases were characterized by rapid loess deposition and relatively little soil development, except in distal areas along the loess field margin that have received higher rainfall.

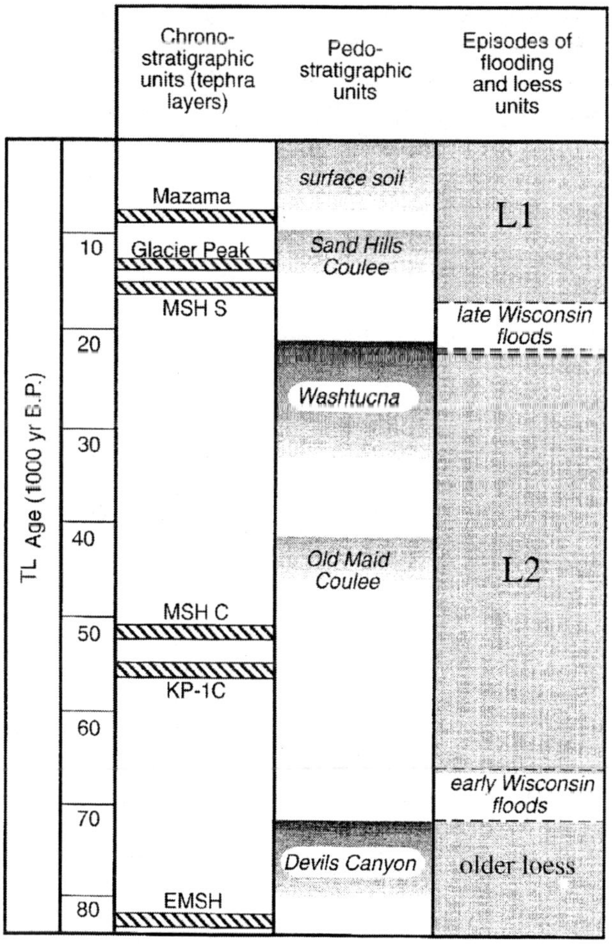

Figure 3. Composite schematic stratigraphic diagram of late Quaternary chronostratigraphic, pedostratigraphic, and loess units in Palouse loess (revised from McDonald and Busacca, 1992) and episodes of cataclysmic flooding. Correlated tephras from Busacca *et al.*, (1992); TL age scale after Berger and Busacca (1995).

EOLIAN TRANSPORT AND DEPOSITION: CONTROLS ON THE THICKNESS AND DISTRIBUTION OF LOESS ON THE PLATEAU

Field and wind tunnel evidence demonstrate that saltating sand is critical to entrainment of silt–and–clay–sized particles (Bagnold, 1941; Gillette *et al.*, 1974; Shao et al., 1993; Saxton et al., 2000). Saltating sand grains impact the surface and eject the smaller grains that otherwise would be difficult to entrain because of their cohesive nature and low profiles to the wind. Once ejected, particles of silt and clay may remain in suspension for prolonged periods of time, riding turbulent eddies well into the atmosphere. Ejected particles of silt and clay remain in suspension until: (1) the wind velocity falls below the settling velocity of the particles, (2) electrostatic bonding of aggregates produces particles with sufficient settling velocities to fall, or (3) individual dust particles or aggregates become incorporated in rain or snowfall (Pye, 1995). Typically, loess deposits are thickest and coarsest near their sources and become progressively thinner and finer–grained downwind.

Preliminary data collected by the authors from four test sites on the plateau indicate that the thickness and distribution of loess have been controlled by three primary factors, topography, bioclimate (vegetation and moisture), and source sediment supply (Fig. 4; Sweeney et al., 2000, 2001; Gaylord et al., 2002).

Topographic Controls

Distribution and thickness of loess is a function of topography, such as incised stream valleys or other topographic impediments that can trap saltating particles (Mason et al., 1999).

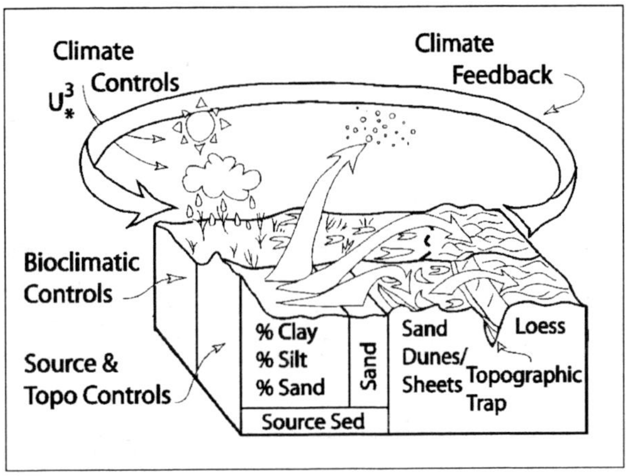

Figure 4. Conceptual diagram of pathways and fates of near–source eolian sediment and atmospheric dust (adapted from C. Zender, personal communication). Source sediment and topographic trap controls on dust and loess production (via saltation–induced ejection of silt and clay) are depicted in the front half of diagram. Bioclimatic (moisture and vegetation) controls on loess production (in the absence of a topographic trap) are depicted in the rear half of the diagram.

Saltating sand activity on a mixed–texture, source sediment inhibits the accumulation of thick loess upwind from incised valleys by persistently re–entraining particles of silt and clay or by temporarily promoting the storage of sand in dunes or sand sheets. However, thick loess can accumulate immediately downwind from incised valley 'topographic traps' because the saltating sand grains are removed from the system as they migrate into traps from which they do not escape, a situation well described for late Pleistocene loess deposits from the U.S. upper Midwest by Mason et al. (1999).

Bioclimatic Controls

Distribution and thickness of loess is a function of precipitation and vegetation density, or bioclimate. If vegetative cover or soil moisture increase, then the potential exists to maximize loess deposition. At the same time, as the loess area expands, saltating activity shifts upwind; if the bioclimatic factors expand to the point where saltation is not a major eolian process then loess generation and deposition ceases. Conversely, sand dunes and sheets will expand during times of increased aridity, effectively pushing loess deposition farther downwind.

Source Sediment Controls

Distribution and thickness of loess is also a function of the source sediment textural character. Bioclimatic and eolian conditions permitting, source sediments cannot produce thick loess deposits downwind if the source is so depleted in silt and clay (e.g., the Ephrata Fan; Stop 1–6) that even if saltating sand particles are present insufficient fine–grained particles are ejected to produce thick loess. Sand–and–gravel–rich source sediment instead tends to be reworked into dunes or sand sheets.

EOLIAN SOILS AND CONTEMPORARY WIND EROSION

The Columbia Plateau of eastern Washington, northern Oregon, and the Idaho Panhandle have experienced numerous exceedences of the mean 24–hr PM_{10} health standard (PM_{10}= airborne particles < 10–mm in diameter) since measurements were started in 1985. Evidence from recent research, largely concerning occupational health risks, has strongly implicated airborne PM_{10} as a cause of respiratory problems such as pneumoconiosis. A broad–scale research project, called the "Columbia Plateau Wind Erosion/Air Quality Project" ("CP3" or PM_{10} project for short) has provided a significantly enhanced understanding of the wind erosion and potential dust emission control strategies for the region. For more information on outcomes from this research the reader is directed to Saxton *et al.* (2000).

Starting at the I–90 underpass, drive 3.6 miles (5.8 km) south on State Highway 17 to Road 2 SE. Turn right on Road 2SE and drive 1.2 miles (1.9 km) west to a ridge crest with a view of the Quincy dunes on the horizon.

STOP 2–1. QUINCY DUNES

The Quincy dune field encompasses the largest concentration of dune sand (>1000 km^2) on the Columbia Plateau (Petrone, 1970). Although formerly more active, these dunes now are relatively dormant because extensive irrigation in the area has promoted high surface moisture and widespread growth of stabilizing vegetation. The nearby Potholes Reservoir (completed in 1952) (Pitzer, 1994) drowns a formerly active dune field. Present–day islands in this reservoir are vegetated and many retain original barchan and barchanoid dune shapes.

Drive 1.2 miles (1.9 km) to the junction with State Highway 17; turn right and drive 47.6 miles (76.7 km) south to the junction with U.S. Highway 395, passing through scabland, loess, and cultivated fine–grained outburst flood deposits. Near the town of Mesa, the road briefly descends into the steep–sided Esquatzel Coulee. At the junction with U.S. Highway 395, turn right (south) and drive 15.5 miles (25 km), to Phend Road. Turn left (east) on Phend Rd until it ends (4.2 miles; 6.9 km). Turn right onto Frontier Road and drive 1 mile (1.6 km) south to its junction with Eltopia Canal and Falls Roads. Turn left onto the south–side Eltopia irrigation canal road. Drive 1.6 miles east.

STOP 2–2. ELTOPIA CANAL

This site reveals an unusually well–exposed stratigraphic section of late Pleistocene, paleosol–capped, slackwater deposits, overlain by Holocene fluvial and eolian strata. The ~14–m–thick, Eltopia stratigraphic section (Fig. 5) preserves the most complete record of late Pleistocene and Holocene sand–dune and fluvial sedimentation yet described from the Columbia Plateau (Gaylord *et al.*, 2001). Notable features in the Eltopia Canal exposure include Mazama tephra clasts that accumulated on the slip face of an ancient sand dune (during the mid–Holocene), cicada–burrowed, dune sediments, and mixed eolian–fluvial deposits. The Mazama tephra clasts (ca. 6845 ^{14}C yrs. B.P.) (Bacon, 1983) provide the most definitive temporal control. The sedimentary deposits exposed at the eastern end of Eltopia Canal are subdivided into five informal units that span deposition from the time of outburst flooding to the Holocene (Gaylord *et al.*, 2001).

Drive Back to the junction of the Eltopia Canal Road and Falls Road and proceed along Falls Road for 5.1 (8.2 km) to the junction with U.S. Highway 395. Turn left (south) on U.S. 395 and drive 6.5 miles (10.5 km) south to junction with U.S. Highway 12. Turn southeast on U.S. 12 toward Walla Walla, driving along the Columbia River. Continue on U.S. 12 for 14.7 miles (23.6 km) to the town of Wallula. Turn left into town to the parking area across from the post office.

STOP 2–3. WALLULA GAP

Wallula Gap is a narrow canyon in the Horse Heaven Hills through which the Columbia River now passes. Outburst flood waters ponded behind this flow constriction caused backflooding of tributary valleys and widespread deposition of fine–grained, slackwater sediment. High–divide channels eroded into basalt above Wallula Gap indicate that the maximum flood stage was at least 350 m (1150 ft), or ~ 240 m above present day Wallula. Loess scarps at the entrance to Wallula Gap suggest that the maximum flood stage could have reached as high as 365 m (1200 ft) (O'Connor and Baker, 1992), or close to the highest elevations at which ice–rafted erratics are found around the Pasco Basin. Calculations of maximum flood discharges based on the high–water evidence suggest that about 10 million m^3s^{-1} passed through Wallula Gap (O'Connor and Baker, 1992), an amount that is 300 times the maximum flows of the 1993 Mississippi River flood (O'Connor and Waitt, 1994)!

Drive 1.8 miles (2.9 km) south on U.S. Highway 12 to Wallula Junction, and turn east toward Walla Walla, driving < 0.1 miles to the stop sign. Cross the road to reach the parking area. A special arrangement has been made for the INQUA group to receive a guided tour of the FPL Stateline wind farm. ***Note: No permission has been given to others who might be following this field***

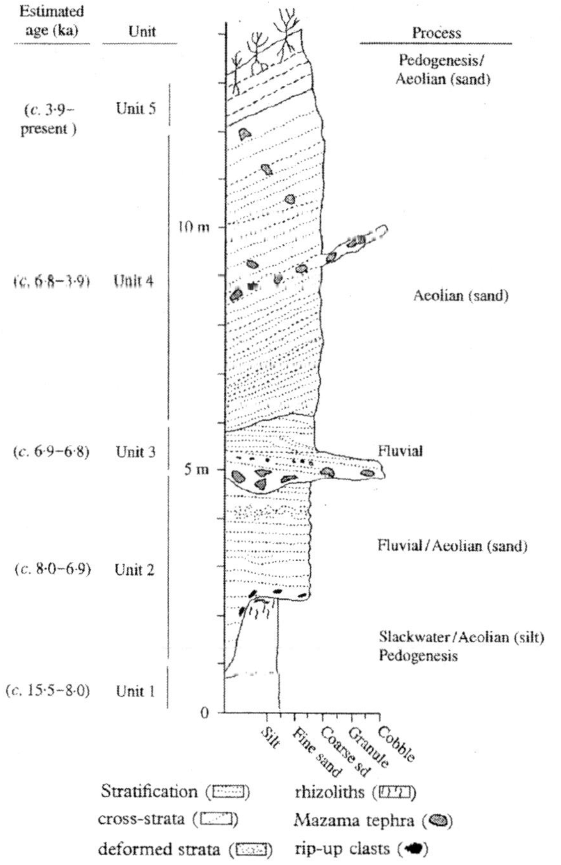

Figure 5. Eltopia Canal composite stratigraphic section (after Gaylord *et al.*, 2001). Ages loosely constrained by slackwater deposits (minimum age ca. 12.5 ka) and Mazama tephra (ca. 6.8 ka).

guide after the time of the initial trip; others who wish to visit the Stateline property must first contact FPL.

STOP 2–4. HORSE HEAVEN HILLS WIND FARM

Developed, owned, and operated by FPL Energy, the 263–megawatt, Stateline Energy Center provides clean, renewable energy to PacificCorp Power Marketing (PPM) for its Pacific Northwest customers. With more than 200 wind turbines, this site became fully operational in December, 2001.

Drive east on U.S. Highway 12 to the east side of the town of Touchet (13.2 miles; 21.3 km). A representative from the Gardena Farms Irrigation District will escort the INQUA group onto their property. ***Note: The Burlingame Canyon exposure is on private property of the Gardena Farms Irrigation District and persons are allowed to visit during this field trip only by special permission. Trespassing will result in access being denied to all.***

STOP 2–5. BURLINGAME CANYON: THE "LITTLE GRAND CANYON": AN EOLIAN SEDIMENT SOURCE

The thick exposures of stacked slackwater rhythmites at sites such as Burlingame Canyon formed the basis for Richard Waitt's hypothesis for multiple, Missoula outburst floods during the last glaciation (Waitt, 1980, 1984, 1985). Others have argued that these stacked rhythmites may instead reflect multiple flood surges during a few large floods (Baker, 1973; Bjornstad, 1982; Bunker, 1982; Baker and Bunker, 1985). This interesting debate is central to discussions of outburst flood dynamics but less important to the paired eolian system that we are examining because regardless of their origin, these fine–grained sedimentary deposits were an ideal (and abundant) source for the eolian deposits downwind.

Turn north from U.S. Highway 12 onto Touchet Road at the east end of Touchet and drive north 21 miles (33.8 km) to the junction with State Highway 124. Turn right (east) onto Highway 124 and continue 5.3 miles (8.5 km) to the junction with Lyons Ferry Road. Turn left (north) on Lyons Ferry Road and drive 9.5 miles (15.3 km) to Smith Springs Road. Turn right (east), drive 2 miles (3.2 km) and pull off the road.

STOP 2–6. BABCOCK ROAD SECTION

The exposures on Babcock Road are located in the north–central part of Eureka Flat, a deflationary plain that has been a persistent source of loess–producing, wind–blown silt throughout the Pleistocene. These roadcuts reveal an overlying blanket of L1 loess (~1m) capping eolian–sand–sheet, outburst–flood Pleistocene sediments and older paleosols. Eureka Flat was inundated during the late Wisconsin by backflooding from the lower, southern end of the flat and possibly by flood surges that breached the Snake River divide to the northeast. Some of the subdued hills nestled within Eureka Flat, including this exposure, appear to be deeply eroded, loess–island remnants mantled by outburst flood and sand sheet sediment and loess. Flood sediments in this part of Eureka Flat are characterized by poorly sorted silt and sand, with occasional weakly imbricated, pebble–to–cobble soil carbonate clasts. These deposits record a transition from semi–arid and arid conditions during sand sheet deposition to progressively less arid conditions during loess accumulation.

Return to the junction with Lyons Ferry Road and turn left (south), driving 7.7 miles (12.4 km) to the Clyde Site. The Skyrocket Hills, a repository for some of the thickest Palouse loess, are visible on the horizon to the left. The substantial accumulation of loess on these hills attests to the persistent loess–generating "engine" that Eureka Flat represents. ***Note: This exposure is on a blind curve of a road traveled by large grain trucks. Caution is warranted.***

STOP 2–7. CLYDE SITE–1 (CLY–1)

This roadcut exposes the thickest known Pleistocene loess in the Palouse. The exposure includes loess units L1 and L2 as well as the Devils Canyon buried soil complex in the core of the exposure and several meters of loess beneath that. Tephras identified here include the Mt. St. Helens sets S, C, and a pre–set–C Mt. St. Helens tephra. The loess at road level has a probable age of >100 ka (Busacca and McDonald, 1994) based on stratigraphic position. The site lies immediately downwind from Eureka Flat where accumulation of loess since the LGM is almost 6 m. CLY–1 is representative of late Quaternary loess stratigraphy for sites that are proximal to major sources of eolian sediment. This section has a much thicker sequence of the same buried soils and loess layers than at sites farther downwind from their source areas, such as site WA–5 (DAY 3, STOP 3). The area that includes the CLY–1 site has the thickest known occurrences of L1 and L2 loess on the plateau (Fig. 3). Regional thickness trends for L1 loess strongly imply that Eureka Flat and the Walla Walla valley were the source of voluminous latest Pleistocene and Holocene loess.

Continue south on Lyons Ferry Road to the junction with State Highway 124. Drive 7 miles (12.3 km) east on Highway 124 to the junction with State Highway 125. Turn right (south) on Highway 125 and drive 16 miles (25.7 km) to Walla Walla. ***End of Day 2.***

THE WASHINGTON WINE INDUSTRY

Washington State is an agricultural region with world class wines whose character is a function of the complex interplay between climate, soil, and geology, a concept known as *terroir*. In the Walla Walla area the terroir is strongly influenced by the soils that have been derived from weathering of the basalt and from slackwater and eolian deposits that have been central to this field trip. To learn more about the terroir of Washington state wines, readers are directed to Meinert and Busacca (2002).

DAY 3

The third day further explores the record of outburst flood and eolian activity as the route continues towards Spokane.

From the junction of U.S. Highway 12 and State Highway 125, north side of Walla Walla, drive 16 miles (25.7 km) north on Highway 125 to the junction with Highway 124. Turn left on Highway 124 and drive 7 miles (12.3 km) west to the junction with the Lyons Ferry Rd. Turn right (north) onto Lyons Ferry Road and drive 16.2 miles (26 km) to the junction with State Highway 261. Turn right on Highway 261 and drive 4.8 miles (7.7 km) to a turnout on the left side of the road, just before the Little Goose Dam turnoff. ***Note: Be extremely cautious at this road cut because of blind corners and heavy traffic.***

STOP 3–1. STARBUCK SITE: PROTOLOESS AND PROTODUNES

This stop provides an opportunity to examine sedimentary structures and texturally diverse deposits generated during multiple, late Wisconsin, outburst floods whose waters flowed upstream into the Tucannon River valley. Although flood deposits along the Tucannon River have been known for many years, only Smith (1993) has conducted a detailed sedimentologic examination of them. His strategy was to describe and interpret sedimentary facies in measured stratigraphic sections. The deposits here are dominated by granule–to–cobble, flood–surge sediments intercalated with very–fine–sand–to–clay slackwater deposits. Six distinct flooding events are preserved at this site (Smith, 1993). The abundance of sand, silt, and clay in these flood deposits again attests to their importance as sand dune, sand sheet, and loess sources.

Drive back down State Highway 261 for 13.4 miles (21.4 km) past Lyons Ferry and the confluence of the Palouse and Snake Rivers to the turnoff for Palouse Falls State Park. On the right side of the road, 2.5 miles (4.0 km) from the last stop numerous, elongated, gravel ridges are oriented perpendicular to the road. These ridges are giant ripples or 'megaripples' (Bretz *et al.*, 1956, Bretz, 1969; Baker, 1973) that record relatively late–stage, outburst–flood, traction–transport, rounded pebble–cobble–and– boulder clasts, most of which are basalt. Closer inspection will reveal a few rounded quartzite and granitic cobbles derived from plutonic and metamorphic sources in Canada, Montana, and Idaho.

STOP 3–2. PALOUSE FALLS

The road into Palouse Falls provides classic views of butte–and–basin scabland topography. At the falls, the Palouse River flows over an ~60 m high cataract cut into multiple basalt flows during the last episode of outburst flooding. The floods that sculpted the surrounding terrain flowed down the Cheney–Palouse scabland tract, overtopped the Snake–Palouse–River divide south of Washtucna Coulee and excavated a new route for the Palouse River to the Snake. Now simply an underfit stream dwarfed by its valley, the perennial Palouse River is nevertheless a major resource for this area.

Turn right (northeast) at the Park Road entrance onto State Highway 261 and drive 8.2 miles (13.1 km) to the junction with Highway 260. Turn right and drive northeast 6.5 miles (10.5 km) to Washtucna and the junction with State Highway 26. Turn left (west) on Highway 26, drive 4 miles (6.4 km) and pull off the highway adjacent to the staired–stepped exposure.

STOP 3–3. WASHTUCNA SITE–5 (WA–5)

This roadcut contains a distinctive sequence of loess layers, buried calcic soils, sand–sheet deposits, and marker tephras that define the late Quaternary loess stratigraphy of the Channeled Scabland. It is the same sequence, although much thinner, seen at CLY–1 (STOP 2–7). Buried soils and tephras at this site were first examined by Lucy Foley and Henry Smith, who established correlations to eruptions of Mount St. Helens for several of the tephras (Foley, 1982). This site was central in developing the regional framework of late Quaternary soil–stratigraphic units (McDonald and Busacca, 1992; Busacca *et al.*, 1992). These workers defined four soil–stratigraphic units within the upper 7 m of the WA–5 section that are recognized at most of the more than 100 sites they described across the Channeled Scabland and Palouse: the Sand Hills Coulee soil, Washtucna soil, Old Maid Coulee soil, and the Devils Canyon soil.

At this stop the L1 loess overlies the Mt. St. Helens set S tephra and contains the Sand Hills Coulee soil. The L2 loess layer as noted earlier consists of eolian sediment that was deposited following the penultimate episode of flooding. The L2 layer here contains the Mt. St. Helens set C tephra and the Old Maid Coulee and Washtucna soils.

Continue west on State Highway 26 for 1.5 miles (2.4 km) and pull off the road.

STOP 3–4. WASHTUCNA SITE– 9 (WA–9)

This site exposes evidence for multiple episodes of outburst flooding prior to the late Wisconsin (McDonald and Busacca, 1988). Evidence gathered at sites such as this one indicate that loess hills adjacent to scabland coulees can contain one–to– several, large, angular unconformities (Fig. 6) that interrupt the layer–cake stratigraphy typical of sites away from scabland channels. These unconformities, along with the sedimentology of material overlying the unconformities, led McDonald and Busacca (1988) to conclude that at least six episodes of Missoula floods occurred in the Channeled Scabland during the Quaternary. All of the loess units at this site are normally magnetized (McDonald and Busacca, 1988), suggesting up to five episodes of flooding during the Brunhes Normal Magnetic Polarity Chron, or more recently than 780,000 yrs. B.P. A flood–cut unconformity at the nearby Benge site lies stratigraphically within reversely magnetized strata, suggesting at least one episode of outburst flooding prior to 780,000 yrs. B.P. (McDonald and Busacca, 1988).

Drive 6.5 miles (10.4 km) west on State Highway 26 to the east edge of Sand Hills Coulee dunes. ***Note: These dunes are on private property; permission must be granted to gain access.***

STOP 3–5. SAND HILLS COULEE DUNES

The Sand Hills Coulee dune field is a relatively small concentration of semi–active–to–stabilized, Holocene, sand dunes that partially fill a former outburst flood channel incised into Palouse loess (Schatz, 1996; Gaylord *et al.*, 1999). Blowout exposures reveal interstratified sand dune and interdune deposits, paleosols, and Mt. Mazama (ca. 6845 ^{14}C yrs. B.P.) and Mt. St. Helens set D (1980) tephras. These dunes cover ~20 km^2, have SW–NE alignments, and consist of iron–stained, quartzo–feldspathic sediment that is compositionally distinct from virtually all other Quaternary dune sands on the plateau. Unlike the Quincy, Hanford, Smith Canyon, and Juniper dunes that were derived from Missoula flood sediment, the Sand Hills Coulee dunes originated from reworking of the Miocene(?)/ Pliocene Ringold Formation that underlies flood deposits <30 km to the west (Schatz, 1996; Gaylord *et al.*, 1999). Except for occasional sand–sheet deposits intercalated with loess, such as at WA–9 (Stop 3–4), eolian sand deposits on the plateau largely post–date the LGM. Dune histories at the Hanford site (Gaylord and Stetler, 1994; Stetler and Gaylord, 1996), Juniper dunes (Gaylord *et al.*, 2001), and here (Gaylord *et al.*, 1999) all indicate that the main phase of sand dune mobility on the plateau occurred during the mid–Holocene. The absence of older, Pleistocene, sand dune deposits attests to their relative impermanence and fragility as landforms.

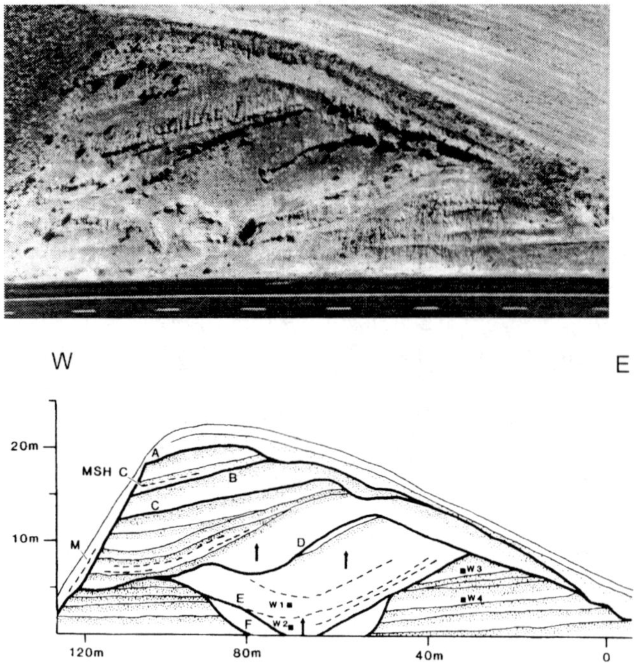

Figure 6. Low–angle–oblique photo and interpretive map of the Washtucna–9 (WA–9) roadcut. View is to the north. Heavy lines on the diagram highlight interpreted outburst–flood cut unconformities (lettered A through F); line and stipple patterns represent calcic paleosols; M is ca. 6,845 ^{14}C yrs. B.P., Mazama tephra (Bacon, 1983); C is the 36,000 ^{14}C yrs B.P., Mt. St. Helens set C tephra. Square boxes indicate samples for magnetostratigraphy; all samples were normally magnetized. Arrows represent fining–upward from sand to silt. See also Foley (1982) and McDonald and Busacca (1988).

Drive 61.1 miles east on State Highway 26 to Colfax. Turn left (north) on U.S. Highway 195 and drive 6.6 miles (10.6 km) to the McCroskey Road turn off (on the right). This turn–off is 0.4 miles past Dry Creek Road. Follow McCroskey and Hume Roads for 4 miles (6.4 km) to the turn–off for Steptoe Butte State Park. The top is 2.5 miles (4 km) from the park entrance.

STOP 3–6. STEPTOE BUTTE

The final stop at the top of Steptoe Butte offers panoramic views of eastern Washington and northern Idaho. Rising ~300 m above the surrounding terrain, Steptoe Butte is cored by late Proterozoic metaquartzite from the Belt Supergroup. During the Miocene, Columbia River Basalt fissure flows engulfed the lower flanks of the butte, leaving it isolated from the more contiguous metamorphic and crystalline terrains to the east. During the Pleistocene, loess, locally >30 m thick, accumulated on the basalt but did not bury Steptoe Butte, likely because the steep slopes prompted relatively high rates of loess erosion.

The origin of the distinctive, steeply–rolling, complex, hill patterns of the Palouse (Kaiser *et al.*, 1951) has been the subject of considerable debate among geomorphologsits and pedologists. The drainage pattern is dendritic in the eastern Palouse but distinctly trellis–like with strict alignment of linear loess ridges in the southwestern Palouse (Lewis, 1960). The Palouse landscape here lies in a transition zone between the linear loess hill landscapes of the southwestern Palouse and the rounded loess hills of the eastern Palouse. Many visitors to this area misinterpret the asymmetric hill shapes of the rounded loess hills immediately below the butte as those formed by migrating sand dunes. The fine–grain sizes of the sediment making up the hills (Busacca and McDonald, 1994) and the absence of sand–dune–specific sedimentary structures precludes this possibility. We regard the asymmetric loess hill morphologies to have resulted from water erosion and seasonally controlled mass wasting. Accumulation of snow on north–facing slopes, followed by slow thawing and water saturation contributes to over–steepening of these slopes via rotational slumps and liquefied flows during the late spring. The contact between the L1 loess and underlying paleo–argillic horizons often is the plane of failure for these slumps and flows.

At the entrance to the State Park, turn left on Hume Rd and drive 1 mile (1.6 km). Turn left again onto Ragon Road and fol-

low it 6.5 miles (10.5 km) to the junction with U.S. Highway 195. Drive north 48 miles (76.8 km) to reach Spokane.

ACKNOWLEDGMENTS

Many people have helped make this field tour possible, and an even larger group have contributed to research projects on the geologic, pedologic, and paleoecologic history of the Palouse during the last two decades. We particularly acknowledge the free exchange of ideas with Eric McDonald, Toby O'Geen, and, Charlie Zender. We also thank Mike Pope and Kurt Wilkie for their constructive reviews and assistance during the compilation of this manuscript. Major funding for Palouse research has come from the: National Science Foundation, Columbia Plateau Wind Erosion and Air Quality Project, Washington State Dept. of Ecology, U.S. Environmental Protection Agency, U.S. Department of Agriculture Cooperative Extension Service, and Washington State University.

REFERENCES

Allen, J.L., Burn, M., and Sargent, S., 1991, Cataclysms on the Columbia: Timber Press, Portland, OR, 248 p.

Atwater, B.F., 1984, Periodic floods from glacial Lake Missoula into the Sanpoil arm of glacial Lake Columbia, northeastern Washington: Geology, v. 12, p. 464–467.

Atwater, B. F., 1986, Pleistocene glacial–lake deposits of the Sanpoil River valley, northeastern Washington: U.S. Geological Survey Bulletin 1661, 39 p.

Atwater, B.F., 1987, Status of glacial Lake Columbia during the last floods from glacial Lake Missoula: Quaternary Research, v. 27, p. 182–201.

Atwater, B.F., Smith, G.A., Waitt, R.B., 2000, The Channeled Scabland: Back to Bretz ?: Comment: Geology, v. 28, p. 574–576.

Bacon, C. R., 1983, Eruptive history of Mount Mazama and Crater Lake caldera, Cascade Range, U.S.A.: Journal of Volcanology and Geothermal Research, v. 18, p. 57–115.

Bagnold, R.A., 1941, The physics of blown sand and desert dunes: Methuen and Co., London, 265 p.

Baker, V. R., 1973, Paleohydrology and sedimentology of Lake Missoula flooding in eastern Washington: Geological Society of America Special Paper 144, 79 p.

Baker, V.R., 1978, Paleohydraulics and hydrodynamics of scabland floods, *in* Baker, V.R., and Nummedal, D., eds., The Channeled Scabland: National Aeronautics and Space Administration, p. 59–79.

Baker, V.R., 1987, The Spokane Flood debeate and its legacy: *in* Graf, W.L., ed., Geomorphic systems of North America: Geological Society of America, Centennial Special Volume, p. 416–423.

Baker, V.R., 1989, The Grand Coulee and Dry Falls: *in* Breckenridge, R., ed., Glacial Lake Missoula and the Channeled Scablands: Guidebook for Field Trip T310, 28th International Geologic Congress, American Geophysical Union, Washington D.C., p. 51–56.

Baker, V.R., 2002, High–energy megafloods: planetary settings and sedimentary dynamics: *in* Martini, I.P., Baker, V.R., and Garzon, G., eds., Flood and megaflood processes and deposits: Recent and ancient examples, International Association of Sedimentologists Special Publication 32, p. 3–16.

Baker, V. R., and Nummedal, D., 1978, The Channeled Scabland. A guide to the geomorphology of the Columbia Basin, Washington: NASA, Washington, D. C.

Baker, V. R., and Bunker, R. C., 1985, Cataclysmic late Pleistocene flooding from glacial Lake Missoula–a review: Quaternary Science Reviews, v. 4, p. 1–41.

Baker, V.R., Bjornstad, B.N., Busacca, A.J., Fecht, K.R., Kiver, E.P., Moody, J.G., Rigby, J.G., Stradling, D.F., and Tallman, A.M., 1991, Quaternary geology of the Columbia Plateau: *in* Morrison, R.B., ed., Quaternary nonglacial geology: Conterminous U.S., Geological Society of America, v. K–2, p. 216–250.

Berger, G. W., and Busacca, A. J., 1995, Thermoluminescence dating of late Pleistocene loess and tephra from eastern Washington and southern Oregon, and implications for the eruptive history of Mount St. Helens: Journal of Geophysical Research, v. 100, p. 22,361–22,374.

Bjornstad, B.N., 1982, Catastrophic flood surging represented in the Touchet Beds, Walla Walla Valley, Washington: American Quaternary Association, Program and Abstracts of the 7th Biennial Conference, Seattle, WA, p. 72.

Bjornstad, B. N., Fecht, K. R., and Pluhar, C. J., 2001, Long history of pre–Wisconsin, Ice Age cataclysmic floods: Evidence from southeastern Washington State: Journal of Geology, v. 109, p. 695–713.

Blinnikov, M., Busacca, A., Whitlock, C., 2001, A new 100,000–yr phytolith record from the Columbia Basin, Washington, USA: *in* Meunier, J.D., and Colin,F., eds., Phytoliths–Applications in Earth Science and Human History, A.A. Balkema, Dordrecht. P. 27–55.

Blinnikov, M., Busacca, A., Whitlock, C., 2002, Reconstruction of the late–Pleistocene grassland of the Columbia Basin, Washington, USA based on phytolith records in loess: Paleogeography, Palaeoclimatology, Paleoecology, v. 177, p. 77–101.

Bretz, J H., 1923, The channeled scablands of the Columbia Plateau: Journal of Geology, v 31, p. 617–649.

Bretz, J H., 1925, The Spokane flood beyond the channeled scablands Pts I and II: Journal of Geology, v. 33, p. 97–115, and p. 236–259.

Bretz, J H., 1928, Alternative hypothesis for channeled scabland, Pts I and II: Journal of Geology, v. 36, p. 193–223, and p. 312–341.

Bretz, J H., 1932, The Grand Coulee: American Geographical Society Special Publication 15, 89 p.

Bretz, J H., 1969, The Lake Missoula floods and the channeled scabland: Journal of Geology, v. 77, p. 505–543.

Bretz, J H., Smith, H.T.U., and Neff, G.E., 1956, Channeled Scabland of Washington: new data and interpretations: Geological Society of America Bulletin, v. 67, p. 957–1049.

Bunker, R.C., 1982, Evidence of late Wisconsin floods from glacial Lake Missoula in Badger Coulee, Washington: Quaternary Research, v. 18, p. 17–31.

Busacca, A. J., 1989, Long Quaternary record in eastern Washington, U. S. A. interpreted from multiple buried paleosols in loess: Geoderma Special Issue on Climatic and lithostratigraphic significance of paleosols, v. 45, p. 105–122.

Busacca, A. J. 1991, Loess deposits and soils of the Palouse and vicinity. pp. 216–228, *in* Baker, V. R., *et al.* The Columbia Plateau, Ch. 8, *in* Morrison, R. B. ed., Quaternary Non–Glacial Geology of the United States, Geological Society of America Geology of North America, Vol. K–2, p. 216–227.

Busacca, A. J., Nelstead, K. T., McDonald, E. V., and Purser, M. D., 1992, Correlation of distal tephra layers in loess in the Channeled Scabland and Palouse of Washington state: Quaternary Research, v. 37, p. 281–303.

Busacca, A. J., and McDonald, E. V., 1994, Regional sedimentation of late Quaternary loess on the Columbia Plateau: Sediment Source Areas and Loess Distribution Patterns: Regional Geology of Washington State, Washington Division of Geology and Earth Resources Bulletin, v. 80, p. 181–190.

Chambers, R.L., 1971, Sedimentation in glacial Lake Missoula: M.S. thesis, University of Montana, Missoula, MT, 100 p.

Easterbrook, D.G., 1976, Quaternary geology of the Pacific Northwest: *in* Mahaney, W.C., ed., Quaternary stratigraphy of North America, Dowden, Hutchinson, and Ross, Stroudsburg, PA, p. 441–462.

Easterbrook, D.J., 1979, The last glaciation of northwest Washington: *in* Armentrout, M.R., Cole, M.R., and TerBest, H. Jr., eds., Cenozoic paleogeography of the western United States, Society of Economic Paleontologists and Mineralogists, Pacific Section, Los Angeles, p. 177–189.

Flint, R.F., 1935, Glacial features of the southern Okanogan region: Geological Society of America Bulletin, v. 46, p. 169–194.

Flint, R.F., 1936 Stratified drift and deglaciation in eastern Washington: Geological Society of America Bulletin, v. 47, p. 1849–1884.

Foley, L. L., 1982, Quaternary chronology of the Palouse loess near Washtucna, Washington, M.S. Thesis, Western Washington University, Bellingham, WA, 137 p.

Fryxell, R., 1968, Origin and age of Palouse Hills topography, eastern Columbia Plateau: Geological Society of America Special Paper 101, 74 p.

Gaylord, D.R., Stetler, L.D., Smith, G.D., and Chatters, J.C., 1991, Holocene and recent eolian activity at the Hanford Site, Washington: PNW American Geophysical Union Meeting, Richland, WA, September, 18–20, Abstracts with Programs.

Gaylord, D.R., Schatz, J.K., Coleman, A.J., and Foit, F.F. Jr., 1997, Holocene to recent sedimentary, stratigraphic, and geomorphic development of the Juniper dune field, Washington: Geological Society of America Annual Meeting, Abstracts with Programs, v. 29, p. 218.

Gaylord, D.R., Foit, F.F. Jr., Schatz, J.K., and Coleman, A.J., 1998, Sand dune activity within sensitive agricultural lands, Columbia Plateau, USA: *in* Busacca, A.J., ed., Dust aerosols, loess soils and global change, Washington State University College of Agriculture and Home Economics, Miscellaneous Publication no. MISC0190, Pullman, WA. p. 145–148.

Gaylord, D.R., Foit, F.F. Jr., Schatz, J.K. & Huckleberry, G. 1999, The Sand Hills Coulee dune field, southeastern Washington: sedimentary and stratigraphic evidence for Late Pleistocene and Holocene eolian–climatic activity: Geological Society of America Annual Meeting, Abstracts with Programs., v. 31, p. 54.

Gaylord, D.R., Foit, F.F. Jr., Schatz, J.K., and Coleman, A.J., 2001, Smith Canyon dune field, Washington, U.S.A: relation to glacial outburst floods, the Mazama eruption, and Holocene paleoclimate: Journal of Arid Environments, v. 47 p. 403–424.

Gaylord, D.R., Sweeney, MR., and Busacca, A.J., 2002, Geomorphic development of a late Quaternary paired eolian sequence, Columbia Plateau, Washington, Geological Society of America Annual Meeting, Abstracts with Programs, v. 34, p. 245.

Gaylord, D.R., and Stetler, L.D., 1994, Eolian–climatic thresholds and sand dunes at the Hanford Site, south–central Washington, USA: Journal of Arid Environments, v. 28, p. 95–116.

Gillette, D. A., Blifford, J. H. Jr., and Fryrear, D. W., 1974, The influence of wind velocity on the size distributions of aerosols generated by the wind erosion of soils: Journal of Geophysical Research, v. 79, p. 4068–4075.

Hooper, P.R., 1984, Physical and chemical constraints on the evolution of the Columbia River basalts: Geology, v. 12, p. 495–499.

Hooper, P.R., 2002, Chemical discrimination of Columbia River basalt flows: Geochemistry, Geophysics, Geosystems, v. 1, Paper Number 2000GC000040.

Kaiser, V. G., Starr, W. A., and Johnson, S. B., 1951, Types of topography as related to land use, Whitman County, Washington: Northwest Science, v. 25, p. 59–97.

King, M., 2000, Late Quaternary loess–paleosol sequences in the Palouse, Northwest USA: Pedosedimentary and paleoclimatic significance: Ph.D thesis, Royal Holloway, University of London, 231 p.

Kiver, E.P., and Stradling, D.F., 1989, The Spokane Valley and northern Columbia Plateau: *in* Breckenridge, R., ed., Glacial Lake Missoula and the Channeled Scablands: Guidebook for Field Trip T310, 28th International Geologic Congress, American Geophysical Union, Washington D.C., p. 23–36.

Lewis, P. F., 1960, Linear topography in the southwestern Palouse, Washington–Oregon: Annals of the Association of American Geographers, v. 50, p. 98–111.

Mason, J. A., Nater, E. A., Zanner, C. W., and Bell, J. C., 1999, A new model of topographic effects on the distribution of loess: Geomorphology, v. 28, p. 223–236.

McDonald, E. V., and Busacca, A. J., 1988, Record of pre–late Wisconsin giant floods in the Channeled Scabland interpreted from loess deposits: Geology, v. 16, p. 728–731.

McDonald, E. V., and Busacca, A. J., 1992, Late Quaternary stratigraphy of loess in the Channeled Scabland and Palouse regions of Washington state: Quaternary Research, v. 38, p. 141–156.

Meinert, L.D., and Busacca, A.J., 2002, Geology and wine: Terroir of the Red Mountain appelation, central Washington State, USA: Geoscience Canada, v. 29, p. 149–168.

O'Connor, J. E., and Baker, V. R., 1992, Magnitudes and implications of peak discharges from glacial Lake Missoula: Geological Society of America Bulletin, v. 104, p. 267–29.

O'Connor, J. E., and Waitt, R. B., 1994, Beyond the Channeled Scabland: Guidebook for the Friends of the Pleistocene, Pacific Northwest Cell Meeting, May 13–15, 1994.

O'Geen, A.T., and Busacca, A. J., 2001, Faunal burrows as indicators of paleo–vegetation in eastern Washington, U.S.A.: Palaeogeography, Palaeoclimatology, and Palaeoecology, v. 169, p. 23–37.

Pardee, J. T., 1942, Unusual currents in glacial Lake Missoula, Montana: Geological Society of America Bulletin, v. 53, p. 1569–1600.

Patton, P. C., and Baker, V. R., 1978, New evidence for pre–Wisconsin flooding in the Channeled Scabland of eastern Washington: Geology, v. 6, p. 567–571.

Petrone, A., 1970, The Moses Lake sand dunes: M.S. thesis, Washington State University, Pullman, WA, 89 p.

Pitzer, P.C., 1994, Grand Coulee: Washington State University Press, Pullman, WA, 504 p.

Pye, K., 1995, The nature, origin and accumulation of loess: Quaternary Science Reviews, v. 14, p. 653–667.

Richmond, G. M., Fryxell, R., Neff, G. E., and Weiss, P. L., 1965, The Cordilleran Ice Sheet of the northern Rocky Mountains and related Quaternary history of the Columbia Plateau: *in* Wright, H. E., Jr., and Frey, D. G., eds., The Quaternary of the United States, Princeton University Press, Princeton, NJ, p. 231–242.

Ringe, D., 1970, Sub–loess basalt topography in the Palouse Hills, southeastern Washington: Geological Society of America Bulletin, v. 81, p. 3049–3060.

Salisbury, R.D., 1901, Glacial work in the western mountain in 1901: Journal of Geology, v. 9, p. 718–31.

Saxton, K, Chandler, D., Stetler, L., Lamb, B., Claiborn, C., and Lee, B. H., 2000, Wind erosion and fugitive dust fluxes on agricultural lands of the Pacific Northwest: Trans. Am. Soc. Agric. Eng., v. 43, p. 623–630.

Schatz, J.K., 1996, Provenance and depositional histories of the Smith Canyon and Sand Hills Coulee dunes, south–central, Washington, M.S. thesis, Washington State University, Pullman, WA, 87 p.

Shao, Y., Raupach, M. R., and Findlater, P. A., 1993, Effect of saltation bombardment on the entrainment of dust by wind: Journal of Geophysical Research, v. 98 (D7), p. 12719–12726.

Shaw, J., Munro–Stasiuk, M., Sawyer, B., Beaney, C., Lesemann, J.–E., Musacchio, A., Rains, B., and Young, R. R., 1999, The Channeled Scabland: Back to Bretz?: Geology, v. 27, p. 605–608.

Shaw, J., Munro–Stasiuk, M., Sawyer, B., Beaney, C., Lesemann, J.–E., Musacchio, A., Rains, B., and Young, R. R., 2000, The Channeled Scabland: Back to Bretz?, Reply: Geology, v. 28, p. 576.

Smith, G.A. 1993. Missoula flood dynamics and magnitudes inferred from sedimentology of slack–water deposits on the Columbia Plateau, Washington. Geological Society of America Bulletin v. 105, p. 77–100.

Smith, G.D., 1992, Sedimentology, stratigraphy, and geoarcheology of the Tsulim site, on the Hanford Site, Washington: M.S. thesis, Washington State University, Pullman, 169 p.

Stetler, L.D. and Gaylord, D.R., 1996, Evaluating eolian–climatic interactions using a regional climate model from Hanford, Washington (USA): Geomorphology, v. 17, p. 99–113.

Sweeney, M.R., Busacca, A. J., and Gaylord, D. R., 2000, Eolian facies of the Palouse region, Pacific Northwest: clues to controls on thick loess accumulation: Geological Society of America Abstracts with Programs, v. 32, p. 22.

Sweeney, M.R., Gaylord, D. R., and Busacca, A. J., 2001, Geomorphic, sedimentologic and bioclimatic models for loess accumulation on the Columbia Plateau, Pacific Northwest: Geological Society of America Abstracts with Programs, v. 33, p. 317.

Tate, T.A., 1998, Micromorphology of loessial soils and paleosols on aggrading landscapes on the Columbia Plateau M.S. Thesis, Washington State University, Pullman, WA, 192 p.

Waitt, R.B., 1980, About forty last–glacial lake Missoula jökulhlaups through southern Washington: Journal of Geology, v. 88, p. 653–679.

Waitt, R.B., 1984, Periodic jökulhlaups from Pleistocene glacial Lake Missoula—new evidence from varved sediment in northern Idaho and Washington: Quaternary Research, v. 22, p. 46–58.

Waitt, R. B., 1985, Case for periodic, colossal jökulhlaups from Pleistocene glacial Lake Missoula: Geological Society of America Bulletin, v. 96, p. 1271–1286.

Waitt, R.B., and Thorson, R.M., 1983, The Cordilleran ice sheet in Washington, Idaho, and Montana: *in* Porter, S.C., ed., v. 1, *of* Wright, H.E., Jr., ed., The Late Pleistocene, Late–Quaternary Environments of the United States, University of Minnesota Press, p. 53–70.

Waitt, R.B. and Atwater, B.F., 1989, Stratigraphic and geomorphic evidence for dozens of last–glacial floods: *in* Breckenridge, R., ed., Glacial Lake Missoula and the Channeled Scablands: Guidebook for Field Trip T310, 28th International Geologic Congress, American Geophysical Union, Washington D.C., p. 37–50.

Contrasting glacial landscapes created by ice lobes of the southern Laurentide Ice Sheet

Authors and Leaders: **Carrie J. Patterson,** *Minn. Geol. Surv., 2642 Univ. Ave., St. Paul, MN 55114*
A. K. Hansel, *Illinois State Geological Survey, 615 East Peabody Drive, Champaign, IL. USA 61820*
D. M. Mickelson, *Dept. of Geol. and Geophys., Univ. of Wisc.,1215 West Dayton St. Madison, WI 53706*
D. J. Quade, *Iowa Dept. of Natl. Res., Iowa Geol. Survey, 109 Trowbridge Hall, Iowa City, IA 52240*
E. A. Bettis III, *Univ. of Iowa; and T.J. Kemmis, Ill. Geol. Survey;*
P. M. Colgan, *Dept. of Geology, Grand Valley State University, Allendale, MI 49401*
E. D. McKay, *Illinois State Geological Survey, 615 East Peabody Drive, Champaign, IL. USA 61820*
A. J. Stumpf, *Illinois State Geological Survey, 615 East Peabody Drive, Champaign, IL. USA 61820*

Organizers: **Carrie J. Patterson,** *Minn. Geol. Surv., 2642 Univ. Ave., St. Paul, MN 55114*
A. K. Hansel, *Illinois State Geological Survey, 615 East Peabody Drive, Champaign, IL. USA 61820*

During the Wisconsin glaciation, the southern edge of the Laurentide Ice Sheet formed several lobes that flowed down regional topographic lows into the north-central United States. The late–glacial history in the mid–continent is mainly a story of ice lobes advancing, stagnating, retreating, and blocking rather than a story of uniform ice sheet advance and retreat. Colgan (1999) noted the variety, complexity, and distribution of landforms associated with the Laurentide Ice Sheet, and suggested that glacial dynamics and bed conditions varied through space and time.

The trip route provides the opportunity to view geomorphic differences among the southern lobes of the Laurentide Ice Sheet. We will traverse deposits of the Lake Michigan lobe in Illinois, the Des Moines lobe in Iowa and Minnesota, the Grantsburg sublobe in Minnesota, the Superior lobe in Minnesota and Wisconsin, and the Chippewa and Green Bay lobes in Wisconsin, returning through the Lake Michigan lobe to Illinois.

DAY 1. THE LAKE MICHIGAN LOBE IN ILLINOIS

(Hansel and Stumpf)

Overview

During DAY 1 of the field trip, we will traverse late Wisconsin end moraines, till plains, and lake plains of the Lake Michigan lobe in northern Illinois. We will visit two classic sections (Wedron silica quarry and Clear Creek) that expose multiple glacigenic sequences. There, we will discuss the glacial processes and ice sheet dynamics associated with the Lake Michigan lobe between about 25,000 and 14,000 ^{14}C years ago.

Whether made up of end moraine, till plain, or lake plain, the landscape of the Lake Michigan lobe in Illinois is relatively flat, especially when compared to other glacial landscapes we will visit. Although only subtle variations are observable, the landforms are significant features (Fig. 1) and have important implications for understanding glacial processes and ice sheet dynamics. In till plain areas, the Wisconsin drift in Illinois averages about 15–20 m thick, whereas in large end moraines, it averages 30–60 m thick.

Most of the end moraines on our route consist of ridges of till deposited at an ice margin during the last interval of till deposition in the area. The broader, multiple–crested moraine systems (Lake Border, Valparaiso, Marseilles, Bloomington, Fig. 2a) are made up of tills of multiple readvances or stillstands during which the ice built a new moraine on the proximal slope of an older moraine or, in some cases, overrode the older end moraine.

In transverse cross section, the end moraines are asymmetrical; distal slopes are steeper and more prominent than long, proximal slopes, which grade into a till plain or lake plain in the up–ice direction (Fig. 1). Both distal and proximal slopes are low (generally less than 2% and 1%, respectively). Moraine height ranges from about 10 to 60 m, and moraine width ranges from about 2 to 20 km. Local relief in the end moraines is generally less than 5 m and is primarily the result of dissection. Advance or readvance moraines (e.g., Marengo, Shelbyville/Bloomington, Dover, Arlington, Valparaiso) are more broadly arcuate and are more likely to have an apron of outwash associated with their distal slopes, whereas recessional moraines (Mt. Palatine, Farm Ridge, Marseilles) are generally more lobate and lack outwash aprons.

The late Wisconsin glacial succession in Illinois consists of off–lapping drift sheets (Johnson and Hansel, 1999; Hansel et al., 1999). The drift sheets (mostly diamictons) pinch out beneath younger drift sheets in the direction of Lake Michigan. Locally, tongues of proglacial fluvial and lacustrine sediment separate till units of different drift sheets and provide evidence for a fluctuating ice margin (Hansel and Johnson, 1992, 1996). The base of the Wisconsin drift is marked by a fairly abrupt contact with paleosols and older drift that filled bedrock valleys prior to the Wisconsin Glaciation. The last interglacial soil (Sangamon Geosol) and in some cases the cold-climate, Wisconsin, interstadial soil (Farmdale Geosol) are preserved beneath proglacial outwash and lake sediment and/or till in much of the outer 50 to 80 km of the Wisconsin drift (Kempton and Gross, 1971; Hansel et al., 1999).

Patterson, C.J., Hansel, A.K., Mickelson, D.M., Quade, D.J., Bettis E. A.III, Colgan, P.M., McKay, E.D., and Stumpf, A.J., 2003, Contrasting glacial landscapes created by ice lobes of the southern Laurentide Ice Sheet: *in* Easterbrook, D.J., ed., Quaternary Geology of the United States, INQUA 2003 Field Guide Volume, Desert Research Institute, Reno, NV, p. 135–153.

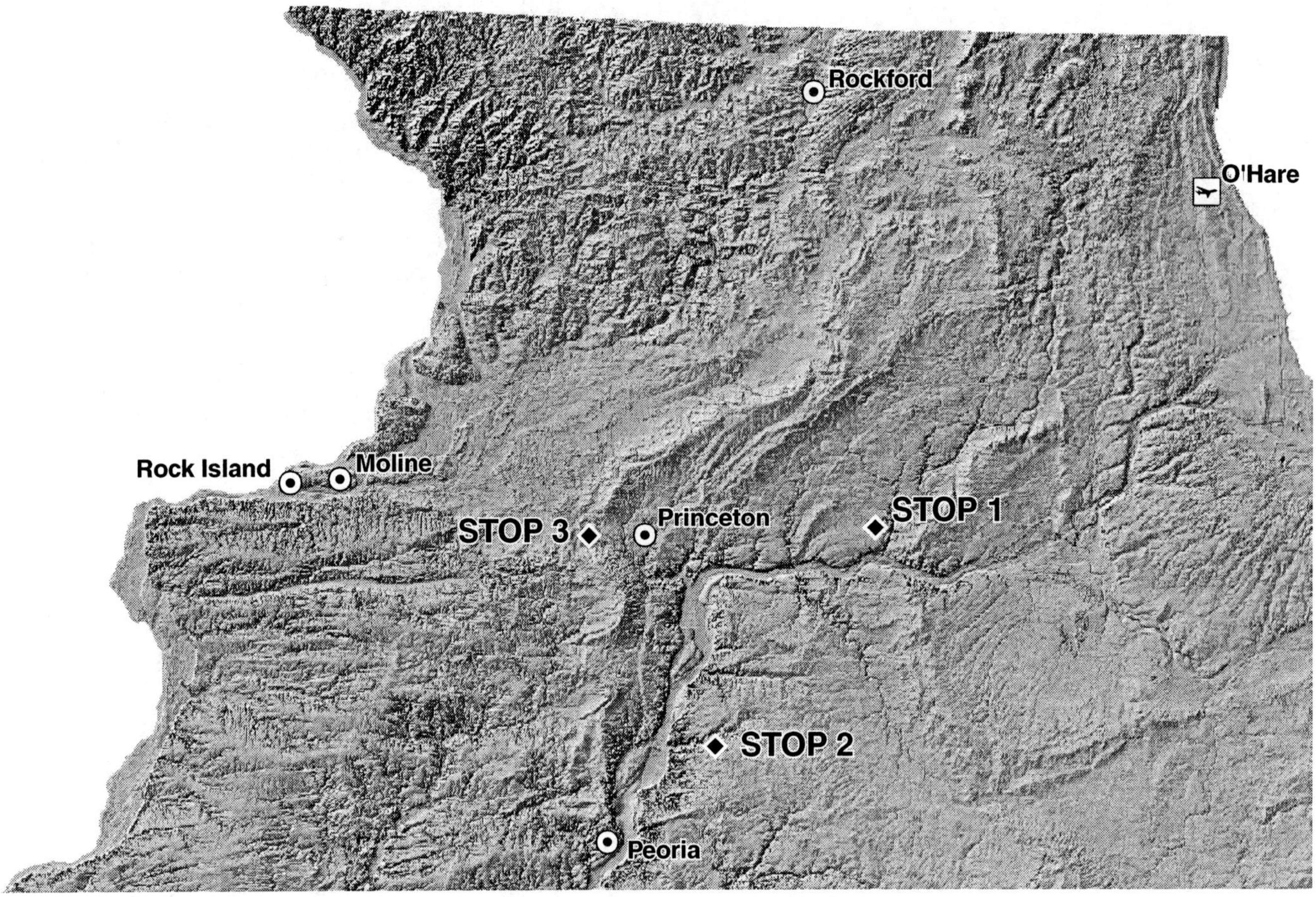

Figure 1. Surface topography of northern Illinois showing stop locations for DAY 1. From Luman et al., 2003.

Figure 2b is a time-distance diagram from Peoria to Chicago for the Lake Michigan lobe in Illinois. It is based on study of exposures and cores across the lobe and radiocarbon ages. The Lake Michigan lobe advanced into Illinois about 26,000 ^{14}C years ago (Hansel and Johnson, 1992, 1996). Remnants of the Farmdale Geosol indicate that a boreal spruce forest existed over the landscape of northeastern Illinois as the lobe advanced to its last glacial maximum (LGM) position near Peoria to the west and Shelbyville to the south (about 19,500 ^{14}C yrs. B.P.). On the basis of ice–wedge casts and relict patterned ground on the Illinoian till plain and late Wisconsin surface, Johnson (1990) concluded that a relatively narrow, but continuous zone of permafrost and tundra vegetation may have been present along the ice margin as the glacier retreated from the LGM until about 16,000 ^{14}C yrs. ago. The permafrost zone, which probably lasted less than 1000 years at any one place, had a southern limit of about 38° 30' N. latitude and was likely discontinuous to the south. The lack of radiocarbon ages for wood in Illinois between about 18,000 and 15,000 ^{14}C yrs. B.P. is consistent with a very unstable landscape with little woody vegetation during that time. Recently acquired AMS ages on twigs and leaves in lake sediment have helped to further constrain some events between about 17,500 and 14,000 ^{14}C yrs. B.P. in Illinois (Fig. 2b).

Discussion

The end moraines that formed in Illinois during late Wisconsin deglaciation requires (1) actively flowing ice to deliver debris to the glacier terminus and (2) conditions whereby the ice margin was stationary for tens to hundreds of years to build the 30–60 m of sediment in the end moraines. The large number of true end moraines (15–20) that formed between the LGM position and Lake Michigan between about 19,000 and 14,000 ^{14}C yrs. B.P. indicates that the ice margin stabilized numerous times during the overall retreat. Whereas, most of the end moraines represent recessional stillstands or slight readvance positions of the ice margin, others represent readvances of the order of tens of kilometers (Hansel and Johnson, 1992). Figure 2b illustrates that the ice readvanced at least six times during the overall retreat.

The thick, widespread, uniform till sheets and numerous, large, broad, end moraines of Illinois have generally been attributed to a wet–based, fast-moving, Lake Michigan lobe that had a low ice-surface profile and remained active during retreat from the LGM position (e.g., Willman and Frye, 1970; Mickelson et al., 1981, 1983; Clayton et al., 1985; Begét, 1986; Clark, 1992, 1994, 1997; Johnson and Hansel, 1990, 1999; Hansel and Johnson, 1992; Hansel et al., 1999; Alley, 1991; Boulton, 1996a,

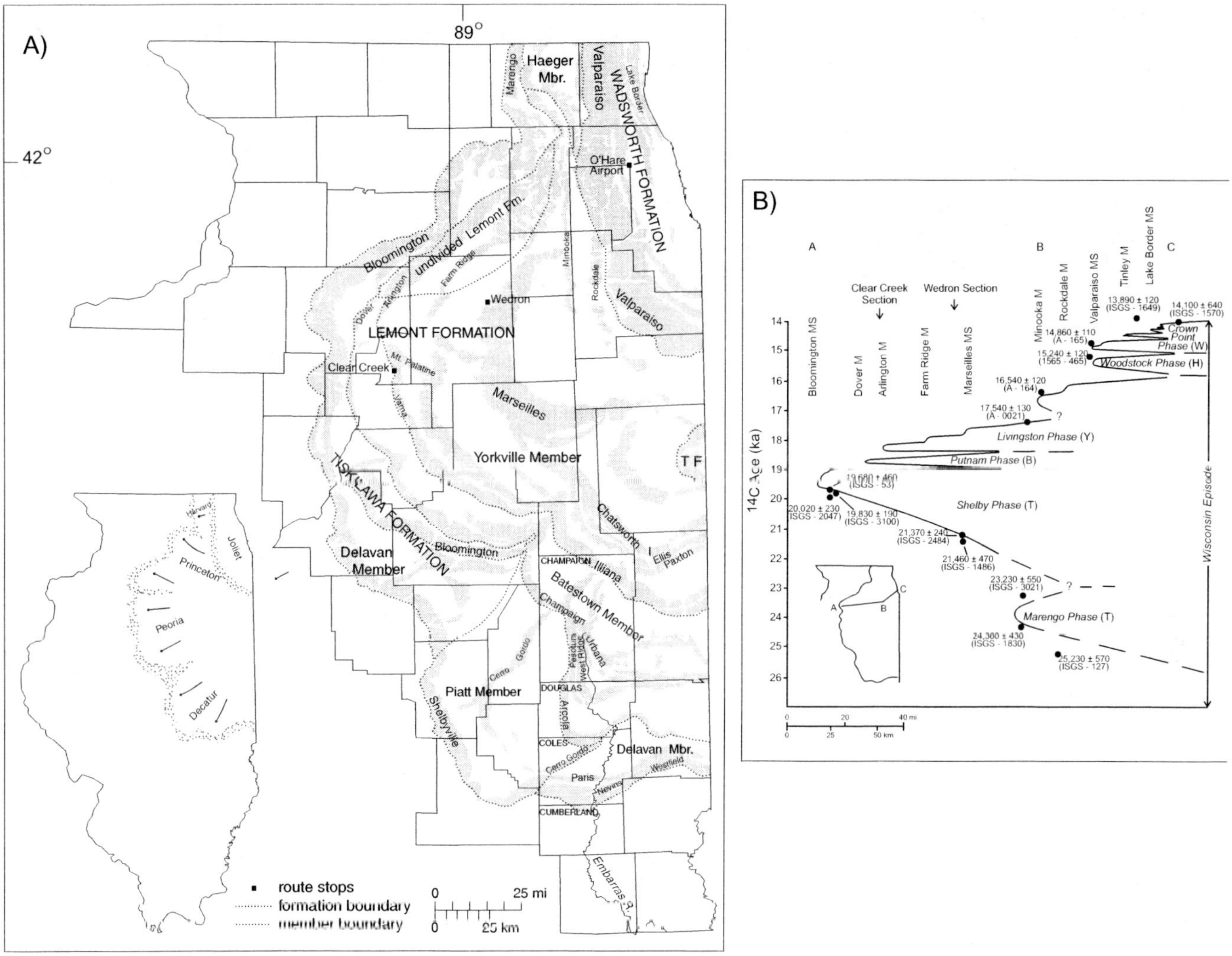

Figure 2. Spatial and temporal relationships of end moraines, lithostratigraphic units, and ice margin fluctuations of the Wisconsin Lake Michigan lobe in Illinois: (a) location of lithostratigraphic units, end moraines (gray), and sublobes (inset map) and (b) time-distance diagram from Peoria to Chicago showing relationship of end moraines and the Tiskilwa (T), Batestown (B), Yorkville (Y), Haeger (H), and Wadsworth (W) tills to glacial phases. Key ^{14}C control is also indicated.

1996b). Johnson and Hansel (1999) concluded that both the low-relief landscape and sediment succession are consistent with subglacial deposition from basal ice and/or a deforming bed.

Uniform till beds with erosional basal contacts overlying relatively undeformed substrate, such as are common in Illinois, are consistent with what Boulton (1996a, b) and Alley (1991) have predicted for till deposited from a deforming bed. The general lack of supraglacial sediment and ice–contact landforms (kames, kettles, eskers, ice–walled lake plains) is more consistent with deposition from a deforming bed beneath relatively clean ice than from meltout from debris–rich ice (Mickelson, 1987; Johnson and Hansel, 1999). Nevertheless, other origins for the till and the landforms cannot be ruled out because the deforming bed model of glacial erosion, transport, and deposition remains largely theoretical and the formation of uniform till is not limited to a unique process.

STOP 1-1. WEDRON SILICA QUARRY

Quarrying of the Ordovician St. Peter sandstone began at Wedron in 1898, and study of the Quaternary succession there began early in the 20th century (Sauer, 1916). The Wedron section has proved to be an important stratigraphic and type section for deposits of the last glaciation (Willman and Payne, 1942; Leighton and Willman, 1953; Wilman and Frye, 1970; Johnson et al., 1985; Hansel and Johnson, 1996). In 1990, Johnson and Hansel identified three late Wisconsin glacigenic sequences in pits at Wedron quarry. Figure 3, a composite section for Wedron, is modified only slightly to update the stratigraphic classification. The glacial deposits of Illinois have been classified into lithostratigraphic units primarily on the basis of till and loess lithology (color, texture, composition) and stratigraphic position, particularly with respect to intertonguing, proglacial, sorted sediments

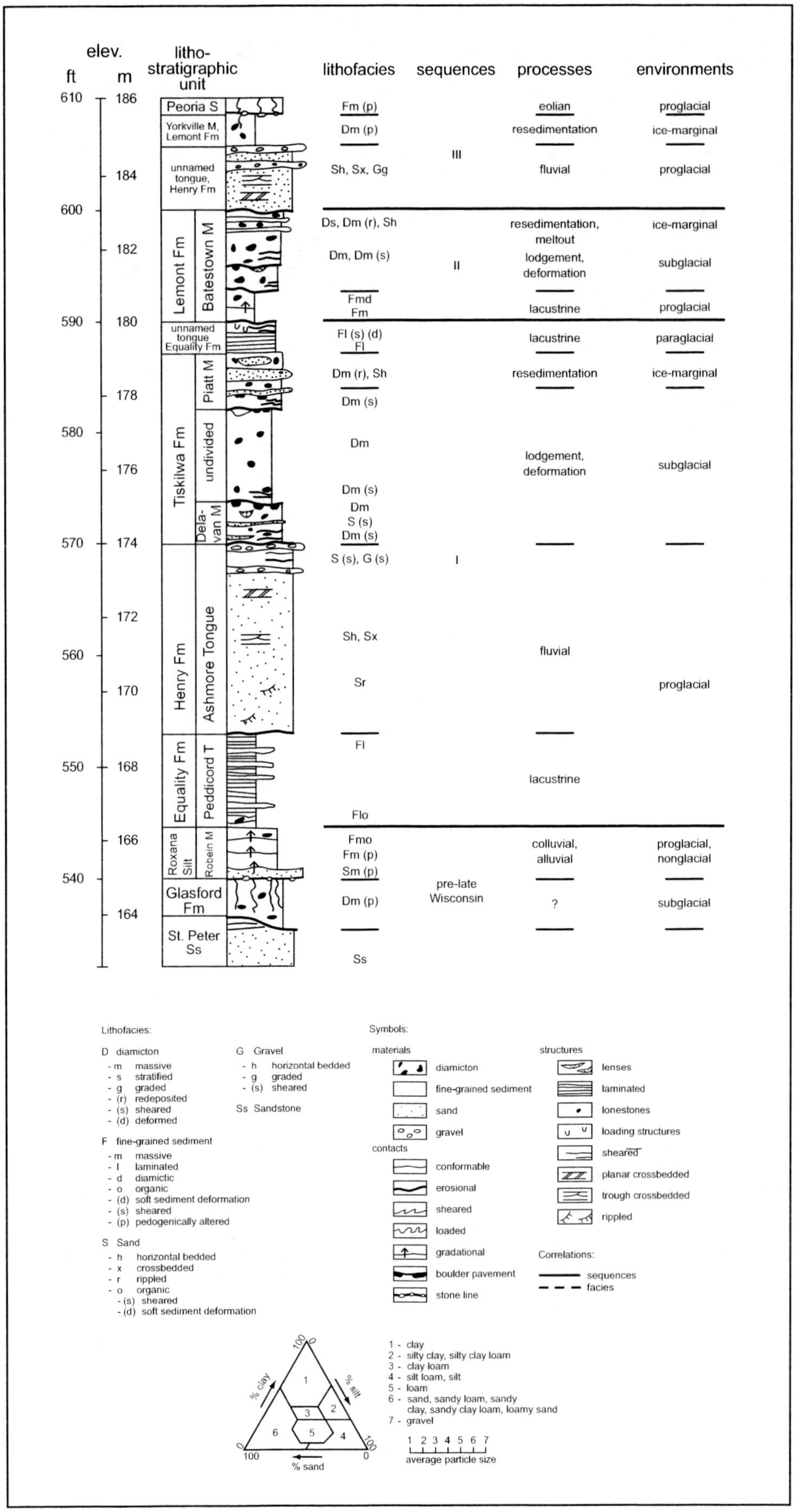

Figure 3. Composite lithofacies and interpreted sequences, processes, and glacial environments for the Wedron section. Modified from Johnson and Hansel, 1990.

and buried soils. The lithostratigraphic classification used here is from Hansel and Johnson (1996).

The composite section shows lithofacies associated with the three glacigenic sequences and inferred processes and glacial environments. Each glacigenic sequence represents an advance/retreat event of the Lake Michigan lobe at Wedron. The sequences are made up of lithofacies that reflect proglacial, subglacial, and ice–marginal environments associated with the events. For example, the lowermost Tiskilwa sequence consists of proglacial lacustrine and fluvial sediment overlain by three, lithologically distinct, subglacial tills, separated by erosion surfaces. The uppermost till of the Tiskilwa sequence is overlain by ice–marginal, reworked sediment and proglacial/paraglacial lacustrine sediment. The main (middle) till of the Tiskilwa Fm. has a distinct red hue and loam to clay loam texture that can be mapped over many square km, whereas the underlying and overlying tills of the sequence are grayer and less clayey. Striated clast pavements and truncated subglacial channel–fill deposits locally are present along the erosional contacts between the till beds.

Topics for discussion at the Wedron section include the origin of erosion surfaces and truncated channel–fill deposits between lithologically distinct till beds, processes of till deposition, provenance of the three till beds of the Tiskilwa sequence, and implications of the succession for ice sheet dynamics.

STOP 1-2. CLEAR CREEK SECTION

The Clear Creek section was first studied in the 1950s by Horberg (1953), who interpreted the bulk of the succession to be pre-Wisconsin in age. Johnson, Hansel, Kerosotis, Voorhees, and others began sedimentological and stratigraphic studies of the section in the late 1980s and early 1990s. On the basis of the similarities of the glacigenic sequences at Clear Creek with late Wisconsin sequences at Wedron and other nearby localities, Hansel et al. (1993) concluded that the sequences correlated. In 2002, Hansel, Stumpf, and McKay studied the section. Figure 4, a composite sketch for the Clear Creek section, is based on their study and earlier observations. The sand at the base of the section (Ashmore Tongue of the Henry Fm.) is interpreted to be a distal, valley–train deposit of the ancient Mississippi River, which flowed from Rock Island east to Princeton and then south to St Louis before it was diverted about 20,000 ^{14}C years ago to its present course along the western border of Illinois (Glass et al., 1964; McKay, 1979; Grimley et al., 1998; Curry, 1998). This diversion resulted from blockage of the Princeton Valley stretch of the ancient Mississippi Valley and the creation of glacial Lake Milan when the Lake Michigan lobe advanced to its western margin to form the Bloomington morainal system. Diversion of the river occurred when the glacial lake overflowed across a former drainage divide southwest of Rock Island.

The organic-rich, Oakland facies of the Tiskilwa Fm. overlies the sand at Clear Creek. This facies consists of stratified and poorly homogenized diamicton containing sand and organic silt beds, many of which show evidence of deformation. Locally along its lower erosional contact, the Oakland facies is folded with the Ashmore sand. A piece of wood collected by McKay from the organic–rich Oakland facies at the base of the Tiskilwa sequence yielded a finite radiocarbon age of 41,010 ± 970 ^{14}C yrs. B P., which is consistent with a Wisconsin age for the glacigenic sequences at Clear Creek. Previously, samples of the abundant wood in the lower part of the Tiskilwa Fm. have yielded ages beyond the limit of radiocarbon dating, leaving the age of the deposits in question. Some of the Oakland facies appears to contain, or be intertongued with, an organic-rich silt derived from the Farmdale Geosol (Robein Member, Roxana silt), which ranges from >40,000 to 20,000 ^{14}C yrs. B P.

The tills of the Tiskilwa sequence at Clear Creek are separated by erosional contacts, sometimes marked by truncated channel-fill deposits and locally by stone concentrations. The tills appear to be more homogeneous upward, and lenses of underlying units are common within the basal till of each unit. Topics for discussion at the Clear Creek Section include its similarities to, and differences from, the Wedron Section, processes of till formation and deposition, and the influence of buried valleys on subglacial drainage and sedimentation.

STOP 1-3. PROVIDENCE MORAINE OVERLOOK

This stop is at the crest of the Providence moraine, one of the moraines of the Bloomington morainal system. The stop offers a view from the LGM terminus across the Illinoian drift plain to the west-southwest and the Wisconsin till plain and younger moraines to the east.

DAYS 2 AND 3. THE DES MOINES LOBE IN IOWA (Quade and Bettis)

The Des Moines lobe (DML) extended from central Canada through the Dakotas and Minnesota into Iowa, terminating at what is now the city of Des Moines in central Iowa (Fig. 5). The age of the DML is well established by ^{14}C dates. It entered Iowa shortly before 15,000 ^{14}C yrs. B.P. and reached the terminal position at Des Moines about 13,800 ^{14}C yrs. B.P., after which the glacier stagnated. The lobe readvanced to the position of the Altamont ice margin just north of Ames and Boone about 13,500 ^{14}C yrs. B.P., then stagnated. At that time, proglacial meltwater was carried from the central portion of the glacier front by Beaver Creek and the Skunk River (the Des Moines River had not yet originated on the Lobe) (Bettis and Hoyer, 1986; Bettis et al., 1988). Between 13,500 and 12,600 ^{14}C yrs. B.P., four minor readvances were marked by the Clare, Unique (recently mapped), Renwick, and West Bend moraines (Fig. 6). The morainal topography associated with these advances is discontinuous, and only the terminal margins are recognized. The final advance into Iowa to the Algona ice margin occurred about 12,300 ^{14}C yrs. B.P. This advance, too, was rapidly followed by stagnation and wastage of the glacier. During the Algona advance, the upper Des Moines River, the major axial

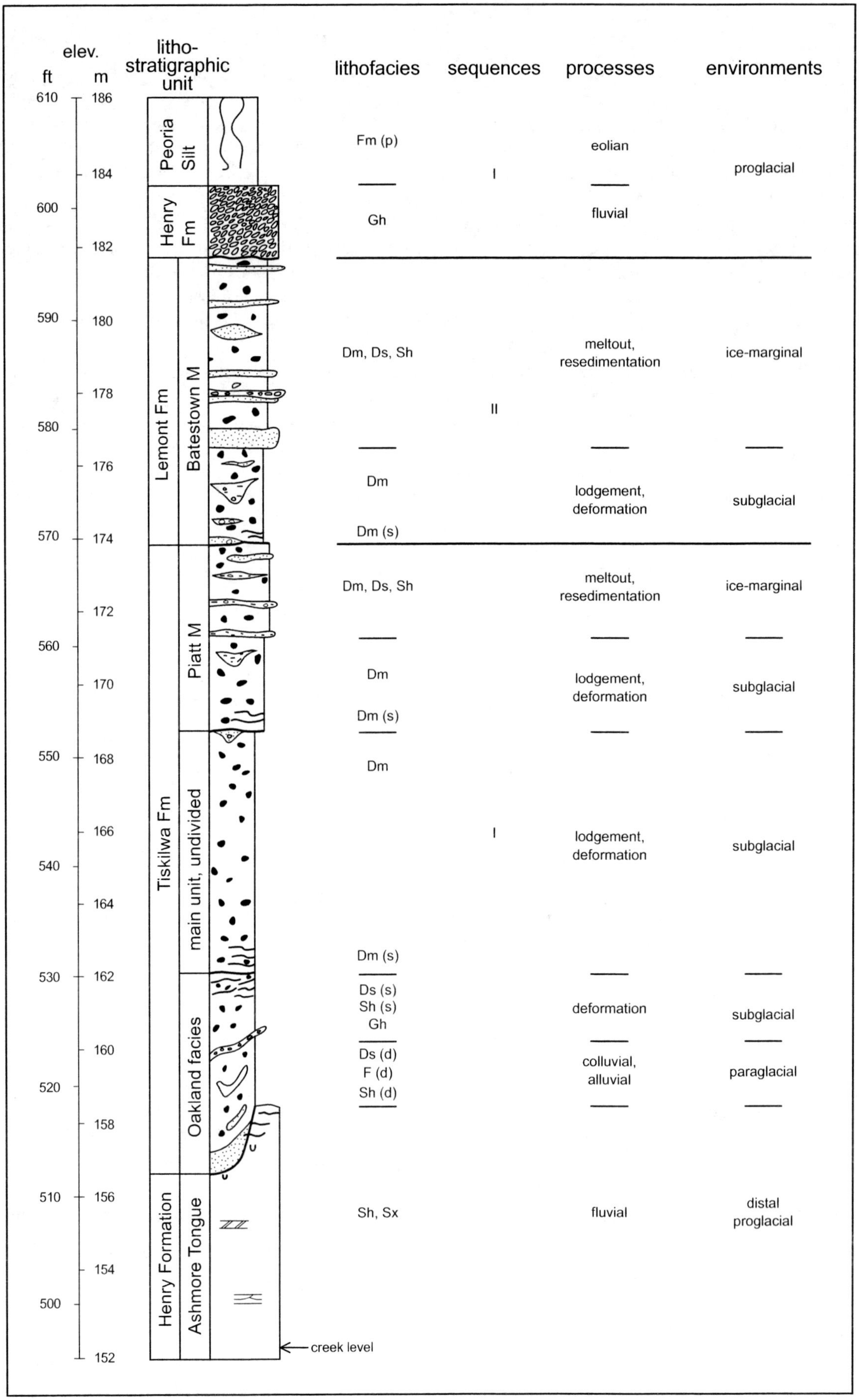

Figure 4. Composite lithofacies and interpreted sequences, processes, and glacial environments for the Clear Creek section.

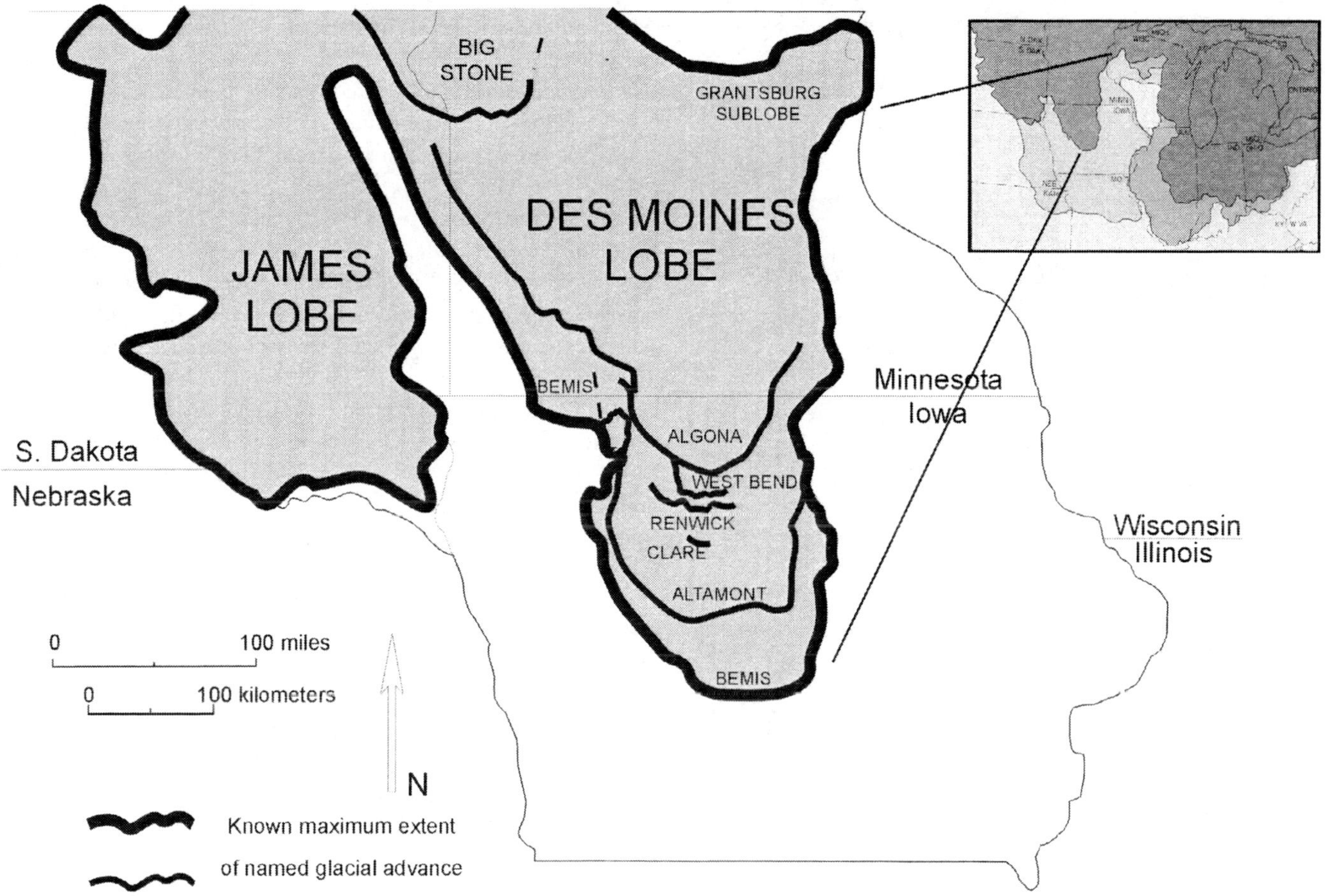

Figure 5. Location of the Des Moines lobe in the northern Midwest. The lobe entered Iowa shortly before 15,000 ^{14}C yrs B.P. and reached the terminal position at Des Moines about 13,800 ^{14}C yrs B.P. The inset shows the limits of glacial advances in the midcontinent by age. Dark gray represents Wisconsin and successfully lighter shades represent Illinois and pre-Illinois limits.

drainage of the Lobe, originated (Bettis and Hoyer, 1986; Bettis et al., 1988).

The DML was active in Iowa between about 15,000 and 12,000 ^{14}C yrs. B.P., about 5,000 to 8,000 radiocarbon years later than glacial lobes to the east made their southernmost maximum advance (Johnson, 1986; Fullerton, 1986). The advance occurred well into a period of regional warming and was thus climatically out of equilibrium (Kemmis et al., 1994). Shortly after 12,000 ^{14}C yrs. B.P., the glacier was gone from Iowa. Flow conditions were rapid and unstable, thus inducing regional stagnation, thin glacial margins and a suite of complex landform sediment assemblages dominated by englacial and supraglacial deposits.

The DML advanced over a previously glaciated landscape. At least seven episodes of pre-Illinoian glaciation occurred in this region from approximately 2.2 Ma to 500,000 yrs. B.P. Regionally extensive upland units were not deposited in the area between about 500,000 and 50,000 yrs. B.P. During this period, several episodes of landscape development resulted in the formation of an integrated drainage network, slope evolution, and soil development on stable land surfaces. The Sheldon Creek Fm. overlies this pre-Wisconsin surface. ^{14}C ages from the Sheldon Creek Fm. indicate at least two glacial advances into central Iowa during the period ca. 40,000 to 26,000 ^{14}C yrs. B.P. (Bettis, 1997). Recent drilling data indicates that Sheldon Creek deposits do not extend south of the DML's Altamont I moraine.

Two Wisconsin loess units were deposited across Iowa, the older Pisgah Fm. and the younger Peoria loess. Pisgah Fm. loess was deposited from about 45,000 to 24,000 ^{14}C yrs. B.P. (Bettis, 1990; Bettis et al., in press). The Pisgah Fm. is buried by Peoria Fm. loess. In central Iowa the Pisgah Fm. is thin and includes loess and related slope sediments that have been altered by hillslope, periglacial and pedogenic processes. The Farmdale Geosol is developed through the unit, and is usually welded to a Sangamon Geosol formed in older slope sediments or pre-Illinoian glacial diamicton.

Peoria Fm. loess accumulated on stable land surfaces in eastern Iowa from about 21,000 to 12,000^{14}C yrs. B.P. Repeated observations of the lower Dows contact show little evidence of significant erosion, which suggests till deposition by a "warm bedded" glacier rather than a frozen bed glacier.

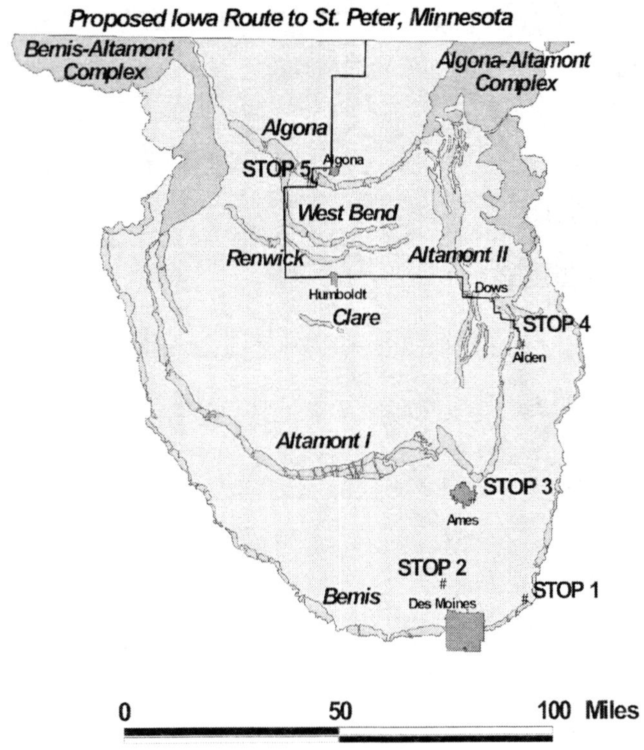

Figure 6. (FKA Figure 2.) The Des Moines lobe landform region, former ice margins associated with this glacial advance into north-central Iowa, and the trip route and stop locations.

MORPHOLOGIC FRAMEWORK FOR THE DES MOINES LOBE

The stratigraphic framework of Wisconsin and Holocene deposits of the DML includes four formations: the Dows, Noah Creek, Peoria, and DeForest. The Dows Fm. consists of upland glacial deposits. The Noah Creek Fm. is composed predominantly of coarse-grained, glaciofluvial, and fluvial deposits in stream valleys and on outwash plains. The Peoria Fm. consists primarily of loess beneath the southern part of the Des Moines Lobe and eolian sand locally blanketing the surface of the Lobe. The DeForest Fm. includes post-glacial sediments that are primarily fine-grained alluvial, colluvial, and palludal deposits. The Dows and DeForest Fms. are further subdivided into lithologically different members. Discussion is here limited to Dows Fm.

Dows Formation

The Dows Fm. includes all upland glacigenic deposits on the DML in Iowa. It is subdivided into four members. The Alden Member consists predominantly of massive, dense, compositionally uniform diamicton. The Morgan Member consists of diamictons interbedded with generally thin, discontinuous beds of sorted sand, silt, silty clay, and gravel. The Lake Mills Member is predominantly massive to laminated silt and silty clay, frequently with a thin basal zone of sand and gravel. The Pilot Knob Member consists predominantly of upland sand and gravel, occasionally interbedded with thin, discontinuous diamictons.

The Dows Fm. is continuous across uplands on the DML in Iowa. The formation is typically 15 to 30 m thick across most of the Lobe. Members of the formation are distinguished by their characteristic lithologic properties. Although these properties are not defined by the origin of the deposits, the members are usually associated with distinctive glacial depositional environments. The massive diamicton of the Alden Member is usually till that has been deposited in a subglacial environment. The interbedded diamicton and sorted deposits of the Morgan Member were probably deposited in ice-marginal and supraglacial environments. The fine-grained, generally pebble-free deposits of the Lake Mills Member were deposited in glacial lakes, and the coarse-grained, sand and gravel deposits of the Pilot Knob Member are found in the core of kame and esker landforms deposited in association with glacial meltwater.

The Dows Fm. was deposited by advances of the DML dating from approximately 15,000 to 12,000 ^{14}C yrs. B.P. (Kemmis et al., 1981; Ruhe, 1969) and is correlative with the New Ulm till of Minnesota (Hallberg and Kemmis, 1986).

Categories of end moraine and ground moraine do not adequately describe the complex landforms of the DML, and the stratigraphic sequences do not indicate active-ice sedimentation during the final phases of glaciation. Along former ice- marginal ("end moraine") areas, landforms change and include several types of hummocky and non-hummocky ridges and escarpments. Behind ice-marginal zones (in "ground moraine" areas), the distribution of upland landforms is not the systematic arrangement expected from an actively receding glacier. Most of the deglaciated surface consists of aligned ridges and hummocks, circular hummocks, disintegration ridges, former ice-walled lake plains, drainage channels, valley-train outwash, and networks of linked depressions. The linked-depression systems have been interpreted to result from the development and collapse of stagnant ice. Such systems do not develop in actively moving ice because compressive stresses inhibit pervasive tunnel development (Fig. 7). Although the size, depth, and density of linked-depression systems were greatest in former ice-marginal areas, the systems extend across the entire area of each advance. At almost all studied localities, the upper depositional sequence is composed of deposits from stagnant-ice environments (re-sedimented diamictons and interbedded sorted sediment) that are commonly 1-5 m thick and may be up to 15 m thick near former ice margins. Both drainage/depressional and upland landforms, then, clearly indicate regional stagnation following each advance.

STOP 2-1. BEMIS MORAINE

The Bemis moraine marks the southeastern margin of the DML and the maximum reach of the late Wisconsin ice into Iowa. The hummocky landscape north and west of here is characteristic of terrain associated with an ice-marginal setting that

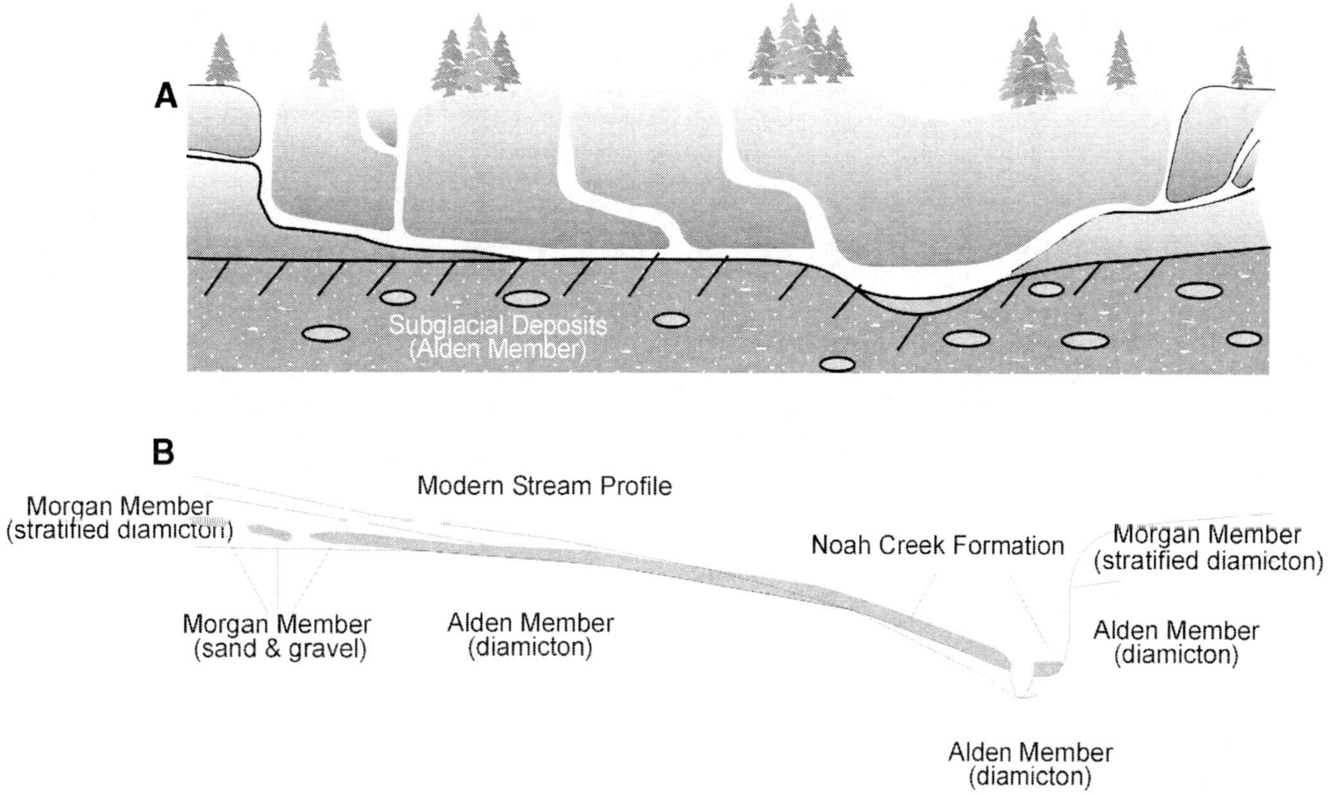

Figure 7. (FKA Figure 3.) (a) Cross section through the wasting Des Moines lobe glacier showing numerous tunnels developed in the glacier karst system and the linkage to a central drainage tunnel. (b) Idealized profile along a linked depression system illustrating the distribution of sand and gravel within the former glacial karst system.

developed approximately 14,000 ^{14}C yrs. B P. Recent drilling investigations along and behind the Bemis moraine in Dallas and Polk counties reveals a thinner package of DML sediments, commonly 9-15 m on the margin, and in the low-relief area extending 20 km north behind the moraine, generally 9-12 m thick. Supraglacial deposits are variable in thickness but generally are 1-5 m thick. The basal Alden Member is a less uniform, variably-textured diamicton, with increasing silt content at depth, and commonly with inclusions of the underlying Peoria loess near the base of the Dows Fm. At ice-marginal positions, Bettis et al. (1985) attributed the thinning of DML sediments and evidence of shearing and deformation to compressive flow and thrusting as the warm-based DML glacier advanced onto rising bedrock and pre-Illinoian glacial sediments.

STOP 2-2. ACORN VALLEY SECTION—SAYLORVILLE LAKE

Des Moines Lobe glacial deposits associated with the Bemis advance form the uppermost Quaternary stratigraphic unit across the uplands of the area. Dows Fm. deposits partially fill a bedrock valley on the Pennsylvanian surface in this area. The Dows Fm. sediments here are primarily diamictons with thin interbedded sand and fine gravel, as well as inclusions of Pennsylvanian bedrock and Wisconsin loess. The lower portion of the DML diamicton is considerably siltier and appears to have fewer and smaller pebbles, suggesting incorporation of the underlying loess. Overall, the matrix textures of the Alden Member diamicton is higher in silt content and lower in sand content than basal tills found north of the Altamont moraine in the central portion of the DML. The diamictons at Acorn Valley are interpreted as basal tills based on evidence of shearing, and consistent pebble fabrics in the diamicton. The DML deposits overlie Peoria Fm. loess, and Pisgah Fm. pediment deposits which originated by slopewash and solifluction processes (Ruhe, 1969). The Farmdale Geosol, developed in the Pisgah Fm., is welded to the Sangamon Geosol developed in hillslope deposits and the underlying pre-Illinoian till.

STOP 2-3. WHATOFF'S BARROW PIT

This site is located at the edge of a tract of aligned ridges associated with Bemis advance. We have made numerous visits to this section and with each visit the section description changes to some extent. Kemmis et al. (1981) described this section as entirely basal till (Alden Member). More recent visits to this section by the authors, as well as cores taken in the vicinity, suggest that a thin mantle of the Morgan Member overlies the Alden Member in this area. Lake Mills Member glacial lake sediments have also been observed here. Kemmis et al. (1981) concluded that aligned hummocks in Iowa were subglacial in origin. However, he did note that the bedded nature of the deposits is worthy of discussion. This is an excellent site to discuss the origin and sediment assemblages associated with aligned ridges and elongated hummocks, i.e., "minor moraines".

STOP 3-1. ALDEN QUARRY

The Alden section is within the Altamont advance and is located in a tract of moderate-relief, curvilinear, aligned ridges ("minor moraines"). This section is the type section for the Alden Member (basal melt-out till). The section illustrates the uniformity and density of the matrix-dominated, subglacial member of the Dows Fm.. Near the base of the Dows Fm., the Alden Member is siltier and the Peoria loess appears to be thoroughly incorporated and dispersed thoughout the till matrix. Kemmis (1981) theorized that repeated regelation at the base of the ice caused a progressive dilution upward through the basal till package. The DML deposits are underlain by Peoria loess. The loess overlies unweathered glacial diamicton of the Sheldon Creek Fm.. The Sheldon Creek Fm. buries a deformed Sangamon Geosol developed in pre-Illinoian Wolf Creek Fm. till that overlies Mississippian limestone.

STOP 3-2. ALGONA MORAINE

We are here on the toe of the Algona moraine, the final advance of the DML into Iowa that occurred about 12,300 ^{14}C yrs. B P. The Algona front at this location is a steep, non-hummocky front over 30 m high, with a well-defined crest. The frontal margin of the Algona moraine is dominated by basal melt-out till. In contrast, Kemmis et al. (1981) noted that the lateral margins were characterized by high-relief, hummocky topography and a thick package of supraglacial sediments and therefore grouped the lateral margins into the Altamont-Algona moraine complex. Kemmis suggested that ice moving down a broad trough would be more active along the central axis and that ice along the lateral margins was most likely moving slower and advancing upslope, therefore promoting compressive flow in the glacier and moving subglacial debris to englacial and supraglacial positions. Stretching to the south is a low-relief plain composed of Noah Creek Fm. outwash (adjacent to the moraine front); followed by a broad plain composed Lake Mills Member proglacial lake sediments (Glacial Lake Jones) extending south to the West Bend moraine.

DAY 3 AND 4. GLACIAL LANDFORMS IN MINNESOTA (Patterson)

Late Wisconsin ice lobes

During the late Wisconsin Glaciation, at least five main ice lobes (the Superior, Rainy, Itasca/Wadena, Des Moines, and Red River lobes) and their sublobes (the Grantsburg and St. Louis sublobes) affected Minnesota (Fig. 8). We will see landforms and sediment from the Superior and Des Moines lobes and the Grantsburg sublobe. The lobes had multiple advances of varying extent. Although discrimination of similar tills of a single lobe by subtle unconformities in an outcrop is possible, correlating them to other outcrops is difficult. However, individual lobes carried distinctive sediment. These systematic variations in till composition among the lobes have allowed recognition of specific provenance regions (Riding Mountain, Winnipeg, Rainy, and Superior) (Meyer, 1997).

The lobes are probably the result of ice-streaming within the ice sheet. Their asynchronous movement would therefore reflect ice dynamics and substrate conditions in the part of the ice sheet that fed them (Patterson, 1997). The concept that ice lobes were the outlets of ice streams having very narrow source regions has not been entirely incorporated into the interpretations of provenance. For example, present-day West Antarctic ice streams have discrete, but shifting, ice sheds. Ice streams erode headward and capture drainage areas of nearby ice streams, thereby potentially changing the nature of the sediment they carry without any discernible change in ice-lobe behavior or ice margins.

Ice Margins

The ice margins of the Superior lobe (Wright, 1972; Johnson and Mooers, 1998), Des Moines lobe (Patterson, 1997), and Grantsburg sublobe (Johnson, 2000; Patterson and Knaeble,

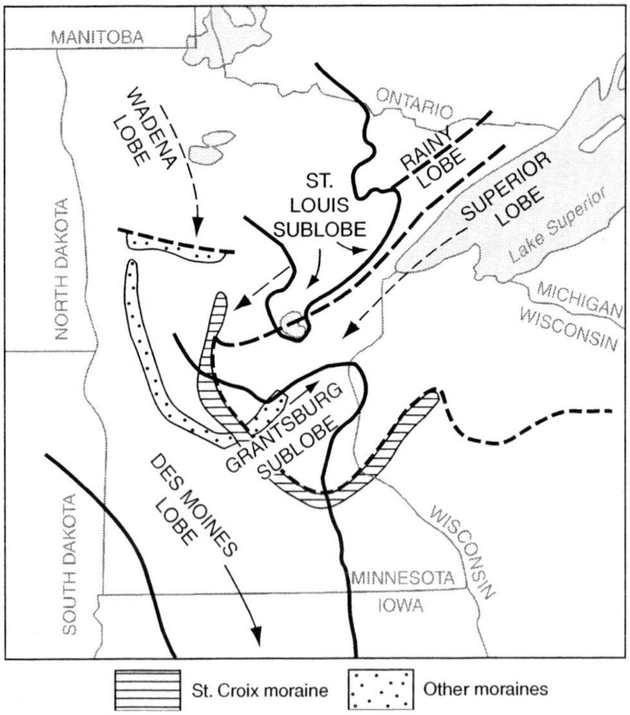

Figure 8. (FKA Figure 1.) The maximum extents of the late Wisconsin icelobes and sublobes and associated moraines. Solid lines represent the known extent of the ice lobe, and dashed lines indicate uncertainty. Solid arrows indicate the known direction of ice advance; dashed arrows indicate some uncertainty in inferred ice-movement direction.

2001) are well preserved and contrasting in style. The Grantsburg sublobe is delimited on the north and east margins by a series of single-crested, moraine ridges. More commonly, ice margins are a broad complex of hummocky topography (such as the Des Moines lobe) comprised of till and sediment sorted by flowing and still water that developed as ice stagnated (Wright, 1972; Meyer, 1985; Patterson, 1997). Margins may also be marked by areas of glacially thrust terrain (such as the Superior lobe) (Mooers, 1990; Knaeble, 1998), remnants of ice-marginal stream deposits and channels (Superior and Des Moines lobes and Grantsburg sublobe), terminations of subglacial valleys (Superior and Des Moines lobes; Wright, 1973), subglacial stream deposits (Superior and Des Moines lobes) or valley-mouth fans (Superior and Des Moines lobes) (Patterson, 1994; Johnson and Mooers, 1998), or by edges of till sheets (Superior and Des Moines lobes and Grantsburg sublobe).

Glacial Lakes

We will see direct evidence (lake plains and sediment) for four proglacial lakes (glacial Lakes Lind, Grantsburg, Anoka, and Minnesota) and evidence for the draining of two other lakes (glacial Lakes Agassiz and Duluth). Glacial Lake Lind was related to the retreat of the Superior lobe and existed for over 1,000 years (Johnson et al., 1999). Glacial Lake Grantsburg occupied parts of the same basin as glacial Lake Lind, but fronted the Grantsburg sublobe, which advanced into the basin from the southwest. It existed for probably only 100 years (Johnson and Hemstad, 1998; Meyer, 1998). Following the retreat and stagnation of the Grantsburg sublobe, several levels of glacial Lake Anoka formed. In southern Minnesota, the retreating Des Moines lobe was fronted by lakes at least twice. Of these, we will see evidence from glacial Lake Minnesota, which recorded less than 20 years of sedimentation. The largest and most significant lake in Minnesota was glacial Lake Agassiz, which developed in front of the retreating Red River lobe as ice left northwestern Minnesota (see Teller and Clayton, 1983). Glacial Lake Agassiz existed for over 3,500 years in Minnesota, and its outlet, River Warren, was responsible for carving the valley now occupied by the Minnesota and Mississippi Rivers. Finally, glacial Lake Duluth formed as the Superior lobe retreated into the Lake Superior basin. Glacial Lake Duluth drained to the south through the Kettle and the St. Croix River valleys, forming prominent spillways.

OVERVIEW

Des Moines Lobe

Entering Minnesota from Iowa (Fig. 9), we will continue to see a flat, subglacial landscape with broad drainageways interpreted to have been subglacial tunnels. Low-relief, stagnation landforms mark the multiple ice margins of the Des Moines lobe. We will descend into the trench created by River Warren as it drained glacial Lake Agassiz. Tributary streams continue to adjust their gradients to this downcutting event, exposing Quaternary sections. We will see this either by canoe or on foot. In some places, an intertill boulder concentration ("boulder pavement") may separate the till sheets (Matsch, 1972). We will then travel eastward along Minnesota Highway 19 to cross the lateral moraines of the Des Moines lobe. The similar breadth of the ice lobe during several advances caused stacking of debris, resulting in high-relief, hummocky topography along the margins.

Superior Lobe

Our tour of the Superior lobe in central Dakota County (Fig. 9) begins with the low relief "extra-morainic" features that represent the first late Wisconsin activity of the Superior lobe, the Emerald phase. It was an advance 10-15 km south of the widely-recognized, St. Croix phase (Johnson and Savina, 1987; Johnson, 2000). We will cross the St. Croix moraine, formed at the time of the maximum of marine oxygen isotope stage 2, approximately 20,000 to 18,500 ^{14}C yrs. B P. (Fig. 8; Wright, 1972; Clayton and Moran, 1982; Attig et al., 1985; Johnson, 2000). Beyond the moraine, we enter a basin that held glacial meltwater at least three times. Glacial Lake Lind was a proglacial lake of the Superior lobe that may have formed as early as 18,500 ^{14}C yrs. B P. (Johnson, 1992; Meyer, 1998; Johnson et al., 1999). It is no longer expressed at the surface because glacial lakes Grantsburg and Anoka occupied nearly the same basin and their sediment overlies glacial Lake Lind deposits.

Grantsburg Sublobe

Near Pine City, we will cross the simple moraine ridges created by the Grantsburg sublobe of the Des Moines lobe. Two closely spaced, well-defined moraines, the Pine City and West Rock moraines, are paralleled by a series of smaller retreatal moraines. The till of the Grantsburg sublobe is similar to that of the Des Moines lobe, but in some places, the till includes material incorporated from the underlying red till of the Superior lobe. The incorporated till is visible as millimeter-to centimeter-thick bands, decimeter to meter-scale beds, or is in some place homogenized. Some outcrops are visibly banded at many scales. Wright (1972) estimated that the sublobe advanced prior to 12,000 ^{14}C years B.P., possibly as early as 16,000 ^{14}C yrs. B.P. The lobe had to advance after glacial Lake Lind had been filled with sand. That stratigraphic relationship is well established in outcrop. The Grantsburg sublobe did not create the tunnel valleys and eskers within its limits. It covered two Superior lobe tunnel valleys that extend beyond the margin of the sublobe (Meyer and Lusardi, 2001). The Grantsburg sublobe was fronted by glacial Lake Grantsburg, which occupied the same basin as the northern portion of glacial Lake Lind (Johnson and Hemstad, 1998; Meyer, 1998). It existed for approximately 80 to 100 years (Johnson and Hemstad, 1998; Johnson, 2000). The lake deposits are at the surface in the southern part of Pine County immediately north of the Granstburg sublobe margin.

Northern Superior Lobe

Once beyond the deposits of glacial Lake Grantsburg north of Hinckley, we will re-enter the landscape of the Superior lobe, which was buried farther south. The sandy, Automba-phase deposits are estimated to be 17,000 to 15,000 ^{14}C years old (Wright and Watts, 1969). Drumlins, tunnel valleys, and eskers formed during this phase. The tunnel valleys and eskers are best developed in the area underlain by the Fond du Lac sandstone probably because this is a topographic low, and the sediment availability and bedrock erodibility are high. Smaller eskers occur in the area, underlain by basalt. The area underlain by Hinckley sandstone seems to have had minor eskers, but subglacial water also apparently flowed through bedrock joints.

The Beroun phase, the next youngest phase, was initially recognized by Leverett (1932) and later by Johnson and Mooers (1998). This was a readvance of the lobe that resulted in a well-defined, thrust moraine. Till of this phase is sandy, like the till of the Automba phase. Discharges of water associated with this advance left a bouldery lag deposit immediately down-ice from the thrust moraine. The moraine associated with the still-younger Askov phase, and underlain by the Hinckley sandstone, is a narrow and striking feature. In outcrop, it appears to be composed primarily of large blocks of sandstone. Depressions are aligned with this margin. The ice may have advanced rapidly, owing to high subglacial water pressure. The subglacial water discharged through the fractures in the weathered bedrock in a path consistent with the hydraulic gradient and widened existing joints by excavation and/or dissolution of weathered bedrock, leading to development of depressions immediately beyond the ice margin.

Kettle and St. Croix Rivers

Meltwater from the Superior lobe drained through the Moose and Kettle Rivers (Winchell, 1899; Leverett, 1932). The Kettle River has varied reaches because it is a patchwork of valleys of different age. We will see the whitewater stretch of the Kettle River from Banning to Sandstone that was cut into the rock by meltwater from the Superior lobe. Glacial Lake Duluth was the lower-level proglacial lake that drained out the Brule (originally referred to as the Bois Brulé; Winchell, 1899) and St. Croix Rivers. We will see the potholes in the basalt that developed during the occupation of the glacial St. Croix River valley by the outlet stream.

STOP 3-3. DES MOINES LOBE TILL OVERLYING OLDER GLACIAL DEPOSITS

Depending on water levels and access, we will visit an outcrop on either the Le Sueur or Watonwan Rivers, tributaries to the Minnesota River, south of Mankato. The general stratigraphy exposed is: till of the Des Moines lobe over older glacial sediment including the Granite Falls till (Matsch, 1972), outwash and lacustrine sediment, and thick, much older, gray, clay-loam till (Gramstad et al., 1997). Some exposures also have a red, pre-late Wisconsin northeastern-source till. Is more than one till of the Des Moines lobe present here? What is the nature of the contacts and what do they suggest about the method of deposition? The Des Moines lobe has been interpreted as a lobe with a deforming substrate. What evidence in the outcrop do you see to support or refute this?

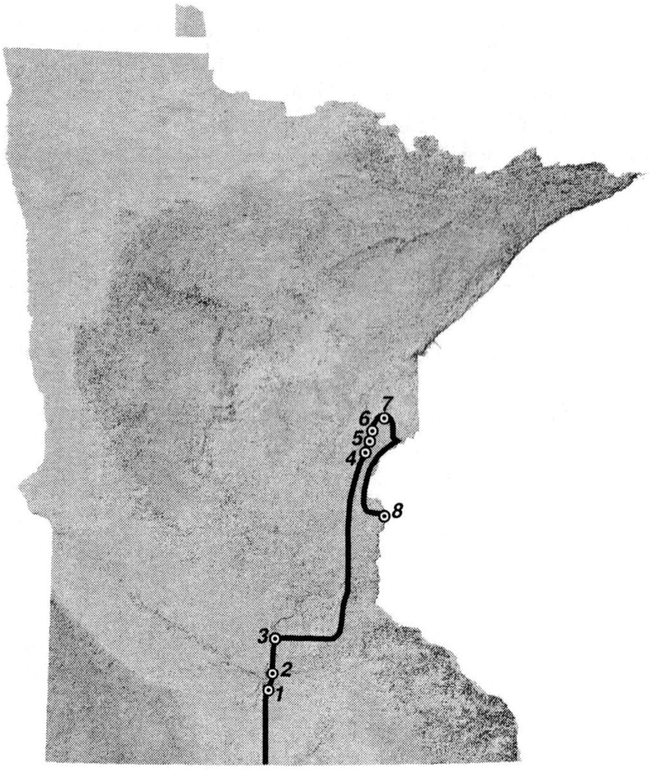

Figure 9. (FKA Figure 2.) Field stops within Minnesota on a shaded relief backdrop. Note the prominent trough of the Des Moines lobe.

STOP 3-4. BOULDER PAVEMENT

A boulder pavement is a common, but not continuous, feature associated with the Des Moines lobe. It had been thought to exist exclusively at the lower contact of the Des Moines lobe till, but has recently been found at other stratigraphic positions within the Des Moines lobe tills and in older units. It is typically one boulder thick, with evidence of some ploughing and flow of material around boulder obstacles. The boulder tops are planed off and when exposed at the surface, the pavement resembles a stone road. What does this feature tell us about subglacial processes occurring in the Des Moines lobe?

STOP 4-1. DES MOINES LOBE "LATERAL" MORAINE, SALISBURY HILL

As many as five glacial units are exposed in this road cut. In the upper 10 m of the section, at least two Des Moines lobe tills occur. Subtle unconformities and horizons with concentrations of sorted sediment occur in pod-like bodies. Units underlying the Des Moines lobe tills are most likely pre-late Wisconsin (Lusardi, 1998). This section is located within the zone interpreted as having stacked stagnation sequences. The sand bodies have been interpreted as subglacial, englacial, and proglacial channels. What type of evidence would be useful to make a distinction?

STOP 4-2. GRANTSBURG TILL AND MORAINES

In this outcrop, two layers of till are separated by deformed lake sediment. We interpret this as representing a readvance (possibly seasonal) of the lobe that overrode a small proglacial pond. The till overlies red sand that filled in glacial Lake Lind. In places, the sand is incorporated by shear into the lower parts of the till. This sedimentary evidence is consistent with that observed at the Russell glacier, an outlet of the Greenland Ice Sheet, where cold-based conditions predominate. Alternatively, a wet-based, deforming substrate has been proposed. How could we tell the difference?

STOP 4-3. CROSS LAKE TUNNEL VALLEY AND ESKER

This lake occupies a tunnel valley created by the Superior lobe. Lake bathymetry maps show an esker as a nearly continuous feature along the lake bottom. The tunnel valley was then covered by the Grantsburg sublobe (Meyer, 1993), while Superior-lobe ice was still within the valley, as indicated by the collapsed Grantsburg sublobe till and the well preserved Superior lobe esker. Why do some ice margins develop tunnel valleys whereas others do not?

STOP 4-4. BEROUN THRUST MORAINE

A gravel pit in this feature has exposed recumbent folds, in places interrupted by low-angle thrusts, and penetrated by the injection of fine-grained dikes. The sediment involved is till, lake sediment, and outwash. The structural relationships indicate an advance of ice from the north. Can such failures be used to assess the speed of glacial advance and/or the rheology of proglacial sediment?

STOP 4-5. SINKHOLES RELATED TO THE ASKOV-PHASE ADVANCE OF THE SUPERIOR LOBE

First we will see vertical joints in the Hinckley sandstone in Robinson Park in the town of Sandstone. Then we will visit a series of sinkholes that have been active recently. They formed beneath a beaver-dammed pond in an area with thin sediment over bedrock. Is the evidence for subglacial drainage through bedrock plausible and compelling?

STOP 4-6. POTHOLES AT TAYLOR'S FALLS

On our way to or from the riverboat cruise, we will see potholes created by the glacial St. Croix River in the basalt of the mid-continent rift system. They are unusually deep and well-formed. How is turbulence transmitted so deeply in such narrow, deep bedrock holes? Are other explanations possible for the origin of these features?

DAY 5 AND 6. CONTRASTING GLACIAL LANDSCAPES IN WISCONSIN (Mickelson and Colgan)

During the next 2 days, we will see landform assemblages left by the Superior, Chippewa, and Green Bay lobes of the southern Laurentide Ice Sheet. The wealth of glacial features in Wisconsin has led to the development of Wisconsin's Ice Age Scientific Reserve, a series of nine parks jointly administered by the Wisconsin Department of Natural Resources and the National Park Service. The purpose of these parks is to preserve and illustrate features produced by glaciation, and it is the first of its kind in the United States. The Ice Age Trail, which follows the outermost moraine of the late Wisconsin glacier throughout most of its length, connects the parks. We will visit 3 or 4 units of the Ice Age Reserve on the field trip. Although not the focus of the trip, we will cross pre-late Wisconsin deposits and the border of the Driftless Area.

No record exists of when the earliest advance of ice took place in Wisconsin, but it was probably more than 780,000 years ago. How much earlier is not known, but evidence of glaciation by 2.2 million years ago occurs in the western plains as far south as Nebraska and Iowa. The conclusion that early glaciations entered Wisconsin more than 780,000 years ago is based on lake and glacial sediment west of Eau Claire that has reversed magnetic polarity (Hersey member and parts of the Kinnikinnic member of the Pierce Fm.) (Baker, et al., 1994).

Before the turn of the century, many glacial geologists (including T.C. Chamberlin and R.T. Salisbury) recognized multiple glaciations in the mid-continent region. They suggested (probably based mostly on what was at that time becoming accepted in the European Alps) that four glaciations had occurred, each separated by interglacial periods. The four glaciations were named the Nebraskan, Kansan, Illinoian, and Wisconsin, after states where deposits of each glaciation were exposed. We now know that continental ice sheets may have entered this area between 12 and 15 times during the last two million years, and much work remains to be done to interpret the very incomplete stratigraphy.

The Wisconsin Glaciation, was the final, and in some places in southern Wisconsin, the most extensive ice cover of all. Although evidence exists of early Wisconsin ice advances farther east in Ontario, little evidence of the extent of Isotope Stage 3 or 4 ice occurs in the western Great Lakes area. The lack of evidence probably does not mean that glaciers failed to enter the region. Rather, almost everywhere, earlier deposits probably were mostly eroded and the remnants covered by younger materials laid down by the late Wisconsin glaciers. Ice almost certainly entered the Mississippi River basin because loess, associated with glaciation blew off the Mississippi valley and into southwestern Wisconsin during this time (Leigh and Knox, 1994). It was evidently a very cold period, with permafrost and barren, windswept tundra plains in northern and central Wisconsin.

Another ice sheet began to advance as early as 28,000 ^{14}C yr B.P., and probably entered Wisconsin soon after (Attig et al., 1985). Because ice flow was controlled by the underlying topography, the southern margin of the ice sheet was lobate (Fig. 10, inset). The Lake Michigan lobe extended far down the Lake Michigan basin into central Illinois, but westward only a few tens of kilometers into Wisconsin, where it joined the Green Bay lobe. The Green Bay lobe was funneled down the Green Bay and Lake Winnebago lowland to just south and west of Madison, and ice of this lobe flowed toward the northwest in north-central Wisconsin, where it met the southwest-flowing Langlade lobe. The interlobate zone between the Lake Michigan and Green Bay lobes (over what is now the Kettle moraine) was lower than the ice in the axis of the lobes on either side and formed a trough that sloped southwestward. Our last stop of the trip will be close to the southern Kettle moraine.

Because of the east-west trend of the >300 m deep Lake Superior trough and the upland south of Lake Superior, ice flowing from the north was diverted toward the southwest to beyond Minneapolis and toward the southeast into the Langlade and Green Bay lobes (Figure 10, insert). Model results suggest that the ice was very slow to cross Lake Superior into Wisconsin because of calving losses in the lake (Cutler, et al., 2002). In fact, this inability of the ice to rapidly advance over Lake Superior is probably why the Driftless Area of southwest Wisconsin was never glaciated. The east-west trending Lake Superior basin has always diverted glaciers to the east and west, and allowed only limited ice to cross the highland of the Upper Peninsula and northwest Wisconsin into the Wisconsin Valley and Chippewa lobes (Fig. 10, inset).

Landform assemblages

Mickelson et al.,(1983) considered the area we will cross in the next two days to be two landform assemblages, but we have combined them here because changes in relief of moraines appear gradational from north to south, and streamlined hills dominate the landscape behind the marginal zone.

Ice marginal zone

This landscape has been mapped as an end moraine. The zone is 2-20 km wide and consists mostly of moderate-to- high-relief hummocks in end moraines. In the end moraine, this topography stands 5-40 m above the surrounding landscape. Both distal and proximal sides of the end moraine have slopes greater than 2% (20 m/km). The internal relief of moraine hummocks is much higher than the broad, smooth moraines seen in Illinois and Iowa earlier in the trip. In most places, the moraines are narrower (less than 2 km), but about the same height as moraines in Illinois and Iowa. Erratic boulders are scattered on the moraine surface. In many moraine exposures, beds of gravel and till dip steeply in the up-ice direction, indicating bulldozing, ice push, and localized thrusting at the margin. Large flat plateaus of sorted sediments and glaciolacustrine sediment that have been interpreted as ice-walled lake plains (Ham and Attig, 1996; Johnson, 1986; Johnson, et al. 1995 occur within the wide, hummocky, end moraine areas. These ice-walled lake plains will be seen at stops 1 and 2. Farther south in the Green Bay lobe of Wisconsin, the end moraine is narrower and in many places layers of sand and gravel, supraglacial sediment, and basal till are stacked up in this zone (Johnson et al., 1995). Some of the gravel in the moraine appears to have been deposited in subglacial tunnels close to the ice margin (Cutler, et al., 2001).

Large volumes of outwash were deposited as fans from the mouths of tunnel channels. The tunnel channels cut through the outermost moraine every 5-10 km in Wisconsin and Minnesota. Much of the sediment in the outwash plains in front of the ice appears to have been delivered by these tunnel channels (Attig et al. 1989; Clayton et al. 1999; Cutler et al., 2001). The tunnel channels are about 0.5 to 1 km across where they intersect the moraine. Sometimes a steep-sided channel is present, but often it is filled with hummocky sand and gravel, which collapsed as ice melted. Mapping of the channels in the Green Bay lobe indicates that water was driven up 80 m in elevation as it approached the ice margin (Clayton et al., 1999). Some of the sediment within the fans is extremely coarse. Intermediate axes of boulders of up to 2 m have been described in the proximal part of the fans, grading into coarse to fine sand the distal part of the fan (Cutler et al., 2001).

Subglacial Zone

A narrow zone (a few km) occurs between the moraine discussed above and the distal side of the drumlin field. In this area, a thin diamicton, probably much of it basal till, commonly covers palimpsest features. In places where bedrock is close to the surface, bedrock hills with only a thin cover of till dominate (Clayton and Attig, 1990; Colgan, 1999). A few eskers are present, as well as hummocky sand and gravel.

The landforms in most of the subglacial zone of Wisconsin and adjacent Minnesota are strikingly different than in Illinois and Iowa (Johnson and Hansel, 1999). In Wisconsin most of the surface has been streamlined by ice flow. Drumlins and megaflutes up to 50 m high and as long as 9 km occur throughout this zone. Thousands of drumlins occur in Minnesota and Wisconsin. The composition of most of the drumlins is not known, although some detailed studies have been done in Wisconsin. Some of the drumlins are composed of sand and gravel (Upham, 1894; Alden, 1905; Whittecar and Mickelson, 1979; Stanford and Mickelson, 1985). Others are composed of diamicton, some have bedrock cores, and some contain lake sediment (Alden, 1905; Colgan and Mickelson, 1997). The drumlins appear to be composed to a great extent of pre-existing material or sediments deposited in front of or at the advancing ice margin. This consists of bedrock, sand and gravel, older diamictons, and lake sediment.

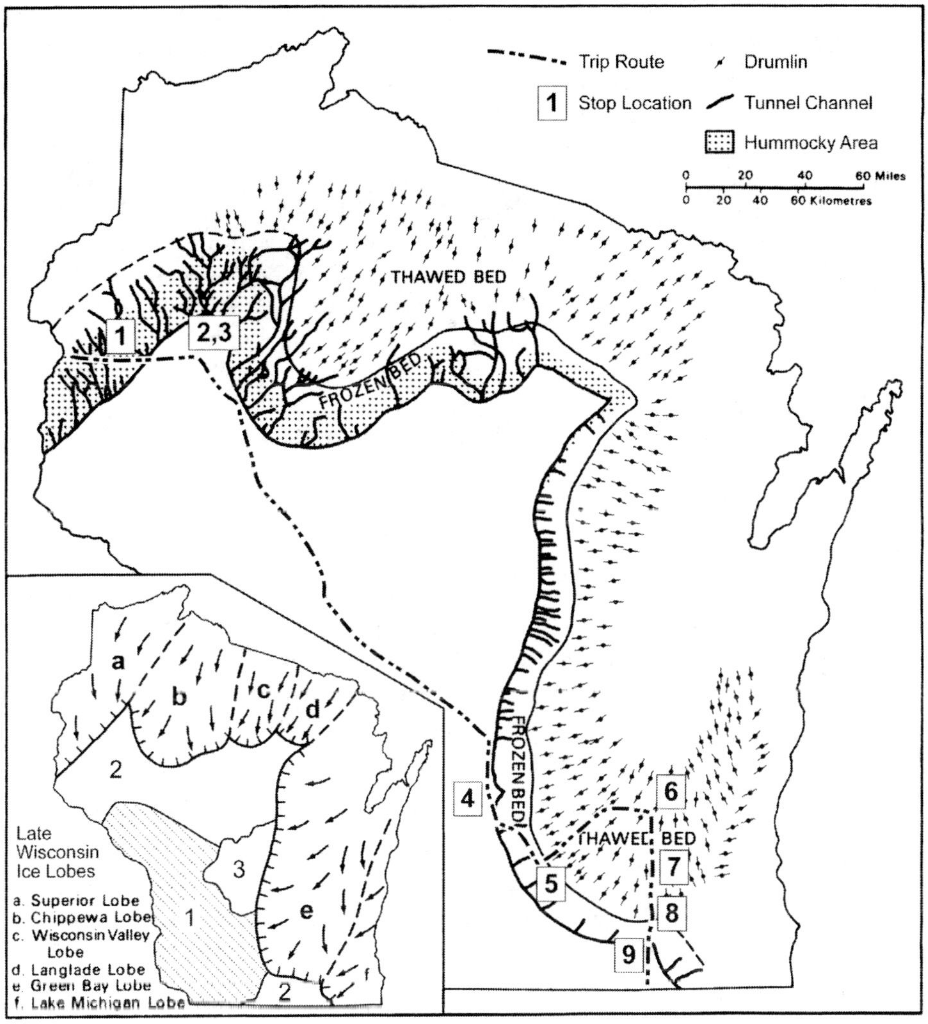

Figure 10. (FKA Figure 1.) Trip route, ice lobe margins and landform assemblages in Wisconsin.

The drumlins appear to be partly erosional features, although certainly part of their height and length is attributable to the growth beneath the ice by movement of sediment towards the axis of the drumlin and the deposition of diamicton and down-ice accretion of eroded material (Boulton, 1987).

Other Observations

Throughout most of Wisconsin, diamictons are fairly thin and, in many places, diamictons of the last glacial maximum lie directly on bedrock. Abundant local bedrock seems to have been incorporated into the till. This is distinctly different than the nature of till in Illinois and Iowa and indicates much more local erosion in the marginal zone and just behind it. Striated rock surfaces are present even in the marginal areas, but striated boulder pavements are rare. Many bedrock exposures show indications of plucking and removal of large clasts as well as abrasion. Although signs of erosion of the bedrock in the marginal zone are present, the amount of erosion was probably very limited, based on cosmogenic dates of striated surfaces (Colgan et al., 2002).

In summary, major differences exist between the glacial landforms in northern Wisconsin and to the south in Illinois. The low-relief till plain, so common in Illinois, Indiana and Ohio, is completely absent, and in its place is a landscape with striking streamlining in the direction of ice flow.

Interpretation of the high-relief moraines and drumlins

We believe that the following observations are consistent with the hypothesis that the glacial assemblages in Wisconsin developed because the glacier had a frozen bed in a zone near the margin when ice was at its maximum position (Fig. 10). Much of the volume of moraines in this assemblage was developed by the stacking of basal till slabs on top of layers of sand and gravel and supraglacial sediment (Lundqvist et al., 1993). This indicates that for some time when ice was at its maximum extent, outwash was being generated along the ice margin as well as from tunnel channels. We interpret the tunnel channels as indicators of large discharges of subglacial water, perhaps stored as an extensive subglacial layer and then released during at least one large discharge event (Cutler et al., 2001). Tunnel channels were clearly active in the Green Bay lobe when the ice was as its outermost margin because fans head at the distal side of moraines where tunnel channels intersect (Attig et al., 1989). Where they have been studied farther west in Minnesota, they are interpreted to have formed during ice retreat as well.

The abundance of local material and the prevalence of streamlined forms suggest that erosion was the major process occurring behind the ice margin. Another indication of the prevalence of subglacial erosion here compared to Illinois is the rarity of paleosols in the north. Wood is not present in the LGM tills in Wisconsin as it is in Illinois. We believe the absence of wood is one indication that continuous permafrost and tundra conditions existed when ice advanced to its maximum position. Wood was simply not sufficient on the landscape to be preserved. Both outside and inside the outermost margin of the ice in Wisconsin indications of former permafrost occur (Clayton et al., 2001). Patterned ground can be observed on aerial photographs, especially on outwash surfaces created during the LGM (Black, 1976; Johnson et al., 1995; Clayton et al., 1997). In Wisconsin and Minnesota, this permafrost apparently lasted until about 13,000 ^{14}C yrs. B. P. in the south and then slowly disappeared northward about 3,000 years later (Clayton, et al., 2001). Not only was permafrost overridden, it must have reformed as ice retreated between 17,000 and 13,000 B.P. At least discontinuous permafrost may have also been present in some areas of Illinois (Johnson, 1990), but we believe this permafrost was not extensive enough to have the same impact as it did in Wisconsin, and that it likely developed along the ice margin after the ice reached its maximum extent.

STOP 5-1. TURTLE LAKE ICE-WALLED LAKE PLAIN

This stop is in the outer part of the St. Croix moraine, where high hummocks and ice-walled lake plains are common (Johnson, 2000). The low section contains foreset beds and offshore sediment common in the ice-walled lake plains of the St. Croix moraine where a stream off the ice margin presumable flowed into the standing water. Surrounding hummocks are a mix of hummocks composed of supraglacial sediment and meltout till (Johnson et al., 1995) and ice walled lake plains. How did the hummocks originate? Are they mostly composed of meltout till? Is thick supraglacial sediment necessary for the formation of ice-walled lake plains? Can flowing water be involved in eroding the landscape, as proposed in other areas?

STOP 5-2. ICE AGE VISITOR CENTER IN CHIPPEWA MORAINE

This area is classic, high-relief, hummocky, topography similar to that in the St. Croix moraine. The Chippewa Moraine Visitor's Center was dedicated in the fall of 1992, and exhibits designed by the National Park Service were installed during the summer of 1997. The views from here include ice-walled lake plains, hummocks composed of diamicton, kettle lakes, and an outwash plain. The well for the Chippewa Moraine Visitor's Center penetrated 86 m of sediment, much of it fine-grained and not good for producing water, and did not encounter bedrock (Syverson, 1998). The Visitor's Center ice-walled lake plain has two crests. This suggests that two ice-walled lakes coalesced as the ice walls confining the lakes slowly melted back, and then more lacustrine sediment was deposited at lower levels within the new ice-walled lake (Syverson, personal communication., 2002).

STOP 5-3. BOB LAKE ICE-WALLED LAKE PLAIN

The pit displays well-developed deltaic foreset beds and silt drapes, as well as a vertical exposure of reddish-brown, sandy-loam, Copper Falls Fm. diamicton (supraglacial sediment) in the north face of the pit. The diamicton is quite uniform, and Johnson et al. (1995) proposed that this is supraglacial meltout till formed as the sediment melted out slowly at the glacier surface beneath a thick insulating layer of supraglacial sediment (Syverson, personal communication, 2002). Is this meltout till? Was thick meltout till required for the development of ice–walled lake plains?

STOP 5-4. JOHNSTOWN MORAINE AT DEVILS LAKE

We will walk the moraine along its crest. Note the lower internal relief (about 10 m) here than at other stops. The glacial history of this area has been interpreted by John Attig and Lee Clayton (Attig at al., 1990; Clayton and Attig, 1990). If exposure is good, we will see meltout till that is dipping steeply up-ice, interbedded with sand and gravel that was presumably deposited just in front of the ice margin and overridden. Note the steep dip of the gravel beds and the large amount of variability in the exposed section. The moraine was built by stacking of debris-rich ice, supraglacial sediment, and stream sediment during several very minor readvances (Lundqvist et al., 1993). Such variability is typical in this part of the Green Bay lobe and contrasts with the uniformity in clay till moraines close to Lake Michigan and with the moraines in Illinois that are made up mostly of basal till. Supraglacial sediment is characterized by an abundance of gabbro and basalt that was derived more than 300 km up ice. In places, very coarse tunnel deposits are present, surrounded by the meltout till. These are especially rich in gabbro, probably indicating a supraglacial source for the water (Lundqvist et al, 1993). Was the supraglacial sediment thinner here than at stops 1, 2, and 3? Is this meltout till?

STOP 6-1. OBSERVATORY AND BASCOM HILLS

These drumlins lie within a few km of the outermost edge of the Green Bay lobe drumlin field and about 15 km behind the Johnstown moraine. Upham (1894) described the drumlins when he visited Madison in 1892 and saw exposures of thin till over sand and gravel. He named sand-and-gravel-cored drumlins "Madison type drumlins" based on his observations here. Length-to-width ratios of these hills and the drumlin on which the State Capitol stands are about 1:1, contrasting greatly with the drumlins elsewhere. The low, flat surface west of Observatory Hill is the former lake bed of Glacial Lake Yahara.

STOP 6-2. VIEW STOP ON HIGHWAY M DRUMLIN

This drumlin is in the central part of the Green Bay lobe drumlin field. Length-to-width ratios of the drumlins here are well over 10:1. These may be gravel or till cored. The length of the drumlins could be related to ice flow velocity and that the ice was slowing down from here toward the ice margin. Is velocity the reason for the differences in the extent of streamlining? If not, is it length of time the area was ice covered?

STOP 6-3. WATERLOO DRUMLIN

The exposed part of this well-developed drumlin is composed entirely of Holy Hill Fm. diamicton. Note the uniformity of the till and its local source (mostly dolomite), suggesting that it is till deposited at the base of the ice and not let down from a supraglacial position. Was this diamicton deposited in the drumlin form by subglacial deformation or lodgement, or was it deposited during ice advance and then overridden and shaped into the drumlin form? Was it deposited directly beneath the ice as a streamlined feature? Compare with the drumlins at STOPS 5 and 8 that have gravel cores. The drumlin is oriented nearly north-south and is in the middle of the Green Bay Lobe.

STOP 6-4. SULLIVAN DRUMLINS

If well exposed, these low drumlins are among the best to illustrate advance and retreat tills (Whittecar and Mickelson, 1979). Note that gravel is folded (overturned) and truncated by an erosion surface and till in the northernmost Sullivan pit. Above the basal till (mostly in the soil) is an abundance of Precambrian sandstone that was probably transported englacially. The Stephan pit is in a poorly streamlined drumlin, as is typical of those close to the outer edge of the drumlin field.

STOP 6-5. WOODWARD PIT

This deposit is ice-contact, stratified sand and gravel in a hill that is not streamlined. Note that the bedding indicates deposition of sand and gravel as foreset beds into standing water. This formed as ice withdrew from the Kettle moraine and traces of diamicton near the surface may be mudflow deposits off the ice surface. Alternately, it may be a proto-drumlin, consisting of sediment that pre-dated the drumlin-forming advance that was overridden, but not streamlined significantly. In this case, the thin diamicton near the top of the section may be basal till.

REFERENCES

Alden, W.C., 1905. Drumlins of Southeastern Wisconsin: U.S. Geological Survey Bulletin, 273, 111 pp.

Alley, R. B., 1991, Deforming bed origin for southern Laurentide till sheets?: Journal of Glaciology, v. 37, no. 125, p. 67-76.

Attig, J.W., Clayton, L., and Mickelson, D.M., 1985, Correlation of Late Wisconsin glacial phases in the western Great Lakes area: Geological Society of America Bulletin, v. 96, no. 12, p. 1585-1593.

Attig, J.W., Clayton, Lee, Lange, K.I. and Maher, L.J., 1990, Ice age geology of Devils Lake State Park: Wisconsin Geological and Natural History Survey, Educational Series 35, 28 p.

Attig, J.W., Mickelson, D.M., and Clayton, Lee, 1989, Late Wisconsin landform distribution and glacier-bed conditions in Wisconsin: Sedimentary Geology, v. 62, p. 399-405.

Baker, R.W., 1994, Pre-Wisconsin glacial stratigraphy, chronology, and paleomagnetics of west-central Wisconsin: Geological Society of America Bulletin, v. 94, p. 1442-1449.

Begét, J. E., 1986, Modeling the influence of till rheology on the flow and profile of the Lake Michigan lobe, southern Laurentide ice sheet, U.S.A.: Journal of Glaciology, v. 32, no. 111, p. 234-241.

Bettis, E.A., III, 1990, Holocene alluvial stratigraphy of western Iowa: *in* Bettis, E.A., III, ed., Holocene alluvial stratigraphy and selected aspects of the Quaternary history of western Iowa: Midwest Friends of the Pleistocene Field Trip Guidebook, 72 p.

Bettis, E.A. III, 1997, Late-middle and early Wisconsin glaciation in north-central Iowa: Geological Society of America Abstracts with Program, v. 29, p. 5.

Bettis, E.A., III, and Hoyer, B.E., 1986, Late Wisconsinan and Holocene landscape evolution and alluvial stratigraphy in the Saylorvile Lake area, central Des Moines River Valley, Iowa: Iowa Geological Survey Open File Report 86-1, 71 p.

Bettis, E.A., III, Pearson, J., Edwards, M., Gradwohl, D., Osborn, N., Kemmis, T., and Quade, D., 1988, Natural history of Ledges State Park and the Des Moines Valley in Boone County: Geological Society of Iowa Guidebook 48, 71 p.

Bettis, E.A., III, Kemmis T.J., and Witzke, B.J., 1985, After the Great Flood: Exposures in the Emergency Spillway, Saylorville Lake: Geological Society of Iowa Guidebook 43, 81 p.

Bettis, E.A. III, Muhs, D.R., Roberts, H.M., and Wintle, A.G., in press, Last glacial loess in the conterminous U.S.A.: Quaternary Science Reviews.

Black, R.F., 1976, Quaternary geology of Wisconsin and contiguous Upper Michigan: in Mahaney, W.H., ed., Quaternary Stratigraphy of North America: Dowden, Hutchinson, and Ross, Stroudesburg, PA. ,p. 93-117.

Boulton, G. S., 1987, A theory of drumlin formation by subglacial deformation: *in* Menzies, J, and J. Rose (eds.), Drumlin Symposium, Balkema, Rotterdam, p. 25-80.

Boulton, G. S., 1996a, The origin of till sequences by subglacial sediment deformation beneath mid-latitude ice sheets: Annals of Glaciology, v. 22, p. 75-84.

Boulton, G. S., 1996b, Theory of glacial erosion, transport, and deposition as a consequence of subglacial sediment deposition: Journal of Glaciology, v. 42, no. 140, p. 43-62.

Clark, P. U., 1992, Surface form of the southern Laurentide Ice Sheet and its implications to ice sheet dynamics: Geological Society of America Bulletin 104, p. 595-605.

Clark, P. U., 1994, Unstable behavior of the Laurentide Ice Sheet over deforming sediment and its implications for climate change: Quaternary Research, v. 41, p. 19-25.

Clark, P. U., 1997, Chapter 6, Sediment deformation beneath the Laurentide Ice Sheet: *in* Martini, I. P., ed., Late Glacial and Postglacial Environmental Changes: Quaternary, Carboniferous-Permian, and Proterozoic: Oxford University Press, NY.

Clayton, L. and Attig, J.W., 1990, Geology of Sauk County, Wisconsin: Wisconsin Geological and Natural History Survey, Information Circular 67, 68 pp.

Clayton, L., and Moran, S., 1982, Chronology of Late Wisconsinan glaciation in middle North America: Quaternary Science Reviews, v. 1, no. 1, p. 55-82

Clayton, L., Attig, J.W., Mickelson, D.M., 1997, Conditions around the margin of the Green Bay lobe during the height of the Wisconsin glaciation: *in*: Mudrey, M.G., Jr., ed., Guide to Field Trips in Wisconsin and adjacent areas of Minnesota, Annual meeting of the North-central section, Geological Society of America: Wisconsin Geological and Natural History Survey, p. 23-30.

Clayton, L, Attig, J.W., and Mickelson, D.M., 1999, Tunnel channels formed in Wisconsin during the last glaciation: *in* Mickelson, D.M. and Attig, J.A., eds., Glacial Processes Past and Present: Geological Society of America Special Paper 337, p. 69-82.

Clayton, L., Attig, J.W., and Mickelson, D.M., 2001, Effects of late Pleistocene permafrost on the landscape of Wisconsin, USA: Boreas, v. 30, p. 173-188.

Clayton, L., Teller, J. T., and Attig, J. W., 1985, Surging of the southwestern part of the Laurentide Ice Sheet: Boreas, v. 14, no. 3, p. 235-241.

Colgan, P.M., 1999, Reconstruction of the Green Bay Lobe, Wisconsin, United States, from 26,000 to 13,000 radiocarbon years B.P.: *in* Mickelson, D.M. and Attig, J.A., eds., Glacial Processes Past and Present: Geological Society of America Special Paper 337, p. 137-150.

Colgan, P.M. and Mickelson, D.M., 1997, Genesis of streamlined landforms and flow history of the Green Bay lobe, Wisconsin, USA: Sedimentary Geology, v. 111, p. 7-25.

Colgan, P.M., Bierman, P.R., Mickelson, D.M., and Caffee, Marc, 2002, Variation in glacial erosion near the southern margin of the Laurentide Ice Sheet, south-central Wisconsin, USA: Implications for cosmogenic dating of glacial terrains: Geological Society of America Bulletin, v. 114, p. 1581-1591.

Curry, B. B., 1998, Evidence at Lomax, Illinois, for Mid-Wisconsin (~40,000 yrs. B.P.) position of the Des Moines Lobe and for diversion of the Mississippi River by the Lake Michigan Lobe (20,350 yr B.P.): Quaternary Research, v. 50, no. 2, p. 128-138.

Cutler, P.M., Mickelson, D.M., Colgan, P.M., MacAyeal, D.R., Parizek, B.R., 2001, Influence of the Great Lakes on the Dynamics of the southern Laurentide Ice Sheet: Numerical Experiments: Geologv, v.29, no. 11, p. 1039-1042.

Cutler, P.M., Colgan, P.M., and Mickelson, D.M., 2002, Sedimentologic evidence for outburst floods from the Laurentide Ice Sheet margin in Wisconsin, USA: implications for tunnel-channel formation: Quaternary International, v. 90, p. 23-40.

Fullerton, D.S., 1986, Stratigraphy and correlation of glacial deposits from Indiana to New York and New Jersey: *in* Sibrava, V., Bowen, D.Q., and Richmond, G.M., eds., Quaternary Glaciations in the Northern Hemisphere: Quaternary Science Reviews, v. 5, p 23-37.

Glass, H. D., Frye, J. C., and Willman, H. B., 1964, Record of Mississippi River diversion in the Morton Loess of Illinois: Transactions, Illinois Academy of Sciences, v. 57, p. 24-27.

Gramstad, S.D., Mann, J.E., Moe, A.P., and Palmsten, M.L., 1997, Pleistocene till stratigraphy of the Mankato bend area of the Minnesota River valley: Geological Society of America Abstracts with Programs, v. 29, p. 17.

Grimley, D. A., Follmer, L. R., and McKay, E. D., III. 1998, Magnetic susceptibility and mineral zonations controlled by provenance in loess along the Illinois and central Mississippi River valleys: Quaternary Research, v. 49, p. 24-36.

Hallberg, G.R. and Kemmis, T.J., 1986, Stratigraphy and correlation of the glacial deposits of the Des Moines and James Lobes and adjacent areas in North Dakota, South Dakota, Minnesota, and Iowa: *in* V. Sibrava, D.Q. Bowen, and G.M. Richmond, eds., Quaternary Glaciations of the Northern Hemisphere: Quaternary Science Reviews, v. 5, p. 11-16.

Ham, N.R., and Attig, J.W., 1996, Ice wastage and landscape evolution along the southern margin of the Laurentide Ice Sheet, north-central Wisconsin: Boreas, v. 25, p. 171-186.

Hansel, A. K., and Johnson, W. H., 1992, Origin and significance of a layered diamicton facies at Wedron, Illinois: Geological Society of America Programs with Abstracts, v. 24, no. 4, p. 19.

Hansel, A. K., and Johnson, W. H., 1996, Wedron and Mason Groups: Lithostratigraphic reclassification of the deposits of the Wisconsin Episode, Lake Michigan lobe area: Illinois State Geological Survey Bulletin 104, 116 p.

Hansel, A. K., Johnson, W. H., and Voorhees, D. H., 1993, Subglacial till of deformation origin from the last glacial episode in central Illinois: Geological Society of America Abstracts with Programs, v. 25, no. 6, p. A-393.

Hansel, A. K., Berg, R. C., Phillips, A. C., and Gutowski, V. G., 1999, Glacial sediments, landforms, paleosols, and a 20,000-year-old forest bed in east-central Illinois: Illinois State Geological Survey Guidebook 26, 31 p.

Horberg, C. L., 1953, Pleistocene deposits below the Wisconsin drift in northeastern Illinois: Illinois State Geological Survey Report of Investigations 165, 61 p.

Johnson, M.D., 1986, Pleistocene geology of Barron County, Wisconsin: Wisconsin Geological and Natural History Survey Information Circular 55.

——— 1992, Glacial Lake Lind: A long-lived precursor to glacial Lake Grantsburg in western Wisconsin and eastern Minnesota: Geological Society of America Abstracts with Programs, v. 24, no. 6, p. 23.

———2000, Pleistocene geology of Polk County, Wisconsin: Madison, Wis., Wisconsin Geological and Natural History Survey Bulletin 92, 70 p.

Johnson, M.D., Addis, K.L., Ferber, L.R., Hemstad, C.B., Meyer, G.N., and Komai, L.T., 1999, Glacial Lake Lind, Wisconsin and Minnesota: Geological Society of America Bulletin, v. 111, no. 9, p. 1371-1386.

Johnson, M.D., and Hemstad, C.B., 1998, Glacial Lake Grantsburg: A short-lived lake recording the advance and retreat of the Grantsburg sublobe: *in* Patterson, C.J., and Wright, H.E., Jr., eds., Contributions to Quaternary studies in Minnesota: Minnesota Geological Survey Report of Investigations 49, p. 49-60.

Johnson, M.D., Mickelson, D.M., Clayton, Lee, and Attig, J.W., 1995, Composition and genesis of glacial hummocks, western Wisconsin: Boreas, v. 24, p. 97-116.

Johnson, M.D., and Mooers, H.D., 1998, Ice-margin positions of the Superior lobe during Late Wisconsinan deglaciation: *in* Patterson, C.J., and Wright, H.E., Jr., eds., Contributions to Quaternary studies in Minnesota: Minnesota Geological Survey Report of Investigations 49, p. 7-14.

Johnson, M.D., and Savina, M., 1987, The southern margin of the Superior lobe during the latter part of the Wisconsin glaciation exceeded the St. Croix moraine by 10 to 15 kilometers: Geological Society of America Abstracts with Program, v. 19, p. 206.

Johnson, W.H., 1986, Stratigraphy and correlation of the glacial deposits of the Lake Michigan Lobe prior to 14 ka BP: *in* Sibrava, V., Bowen, D.Q., and Richmond, G.M., eds., Quaternary Glaciations of the Northern Hemisphere: Quaternary Science Reviews, v. 5, p.17-22.

Johnson, W. H., 1990, Ice-wedge casts and relict patterned ground in central Illinois and their environmental significance: Quaternary Research, v. 33, no. 1, p. 51-72.

Johnson, W. H., and Hansel, A. K., 1990, Multiple Wisconsinan glacigenic sequences at Wedron, Illinois: Journal of Sedimentary Petrology, v. 60, p. 26-41.

Johnson, W. H., and Hansel, A. K., 1999, Wisconsin Episode glacial landscape of central Illinois: A product of subglacial deformation processes?: *in* Mickelson, D. M., and Attig, J. W., eds., Glacial Processes Past and Present: Geological Society of America Special Paper 337, p. 122-135.

Johnson, W. H., Hansel, A. K., Socha, B. J., Follmer, L. R., and Masters, J. M., 1985, Depositional environments and correlation problems of the Wedron Formation (Wisconsinan) in northeastern Illinois: Illinois State Geological Survey Guidebook 16, 91 p.

Kemmis, T.J, Bettis, E.A. III, and Quade, D.J., 1994, The Des Moines Lobe in Iowa: a surging Wisconsinan glacier: American Quaternary Association, Program and Abstracts, 13th Biennial Meeting, Minneapolis, p. 112.

Kemmis, T.J. Hallberg, G.R. and Lutengger, A.J., 1981, Depositional environments of glacial sediments and landforms on the Des Moines Lobe, Iowa: Iowa Geological Survey Guidebook No. 6, Iowa City, 132 p.

Kempton, J. P., and Gross, D. L., 1971, Rate of advance of the Woodfordian (Late Wisconsinan) glacial margin in Illinois—Stratigraphic and radiocarbon evidence: Geological Society of America Bulletin, v. 82, no. 11, p. 3245-3250.

Knaeble, A.R., 1998, Superior-lobe glacial thrusting of sediment and bedrock along the St. Croix moraine, Stearns County, Minnesota: *in* Patterson, C.J., and Wright, H.E., Jr., eds., Contributions to Quaternary studies in Minnesota, Minnesota Geological Survey Report of Investigations 49, p. 15-26.

Leigh, D.S. and Knox, J.C., 1994, Loess of the upper Mississippi Valley Driftless Area: Quaternary Research, v. 42, p.30-40.

Leighton, M. M., and Willman, H. B., 1953, Basis of subdivisions of Wisconsin glacial stage in northeastern Illinois: Guidebook 4th Biennial State Geologists Field Conference, part 1, Illinois State Geological Survey and Indiana Geological Survey, 98 p.

Leverett, F., 1932, Quaternary geology of Minnesota and parts of adjacent states: U.S.Geological Survey Professional Paper 161, 149 p.

Luman, D. L., Smith, L. R., and Goldsmith, C. C., 2003, Illinois Surface Topography: Illinois State Geological Survey Map 11, 1:500,000 scale.

Lundqvist, J., Clayton, L., and Mickelson, D. M, 1993, Deposition of the late Wisconsin Johnstown moraine, southcentral Wisconsin: Quaternary International, v. 18, p. 53-59.

Lusardi, B.A., 1998, Salisbury Hill Road cut: A deformed multiple-till section near Henderson, Scott County, Minnesota: *in* Patterson, C.J., and Wright, H.E., Jr., eds., Contributions to Quaternary studies in Minnesota, Minnesota Geological Survey Report of Investigations 49, p. 141-153.

McKay, E. D., III, 1979, Wisconsinan loess stratigraphy of Illinois: Illinois State Geological Survey Guidebook 13, p. 75-108.

Matsch, C.L., 1972, Quaternary geology of southwestern Minnesota: *in* Sims, P.K. and Morey, G.B., eds., Geology of Minnesota, A Centennial Volume: Minnesota Geological Survey, St. Paul, p. 548-560.

Meyer, G.N., 1985, Quaternary geologic map of the Minneapolis–St. Paul urban area: Minnesota: Minnesota Geological Survey Miscellaneous Map M-54, scale 1:48,000.

———1993, Quaternary geologic map of Chisago County, Minnesota: Minnesota Geological Survey Miscellaneous Map M-78, scale 1:100,000.

———1997, Pre-Late Wisconsinan till stratigraphy of north-central Minnesota: Minnesota Geological Survey Report of Investigations 48, 67 p.

———1998, Glacial lakes of the Stacy Basin, east-central Minnesota and northwest Wisconsin: *in* Patterson, C.J., and Wright, H.E., Jr., eds., Contributions to Quaternary studies in Minnesota: Minnesota Geological Survey Report of Investigations 49, p. 35-48.

Meyer, G.N., and Lusardi, B.A., comps., 2001, Surficial geology, pl. 1 of Meyer, G.N., and Lusardi, B.A., comps., Aggregate resource potential of Chisago County, Minnesota: Minnesota Geological Survey Miscellaneous Map M-116, scale 1:100,000.

Mickelson, D. M., 1987, Central Lowlands: *in* Graf, W, L., ed., Geomorphic systems of North America: Geological Society of America Centennial Special Volume 2, p. 111-118.

Mickelson, D. M., Acomb, L. J., and Bentley, C. R., 1981, Possible mechanism of for rapid advance and retreat of the Lake Michigan lobe between 13,000 and 11,000 years B.P.: Annals of Glaciology, v. 2, p. 185-186.

Mickelson, D. M., Clayton, L., Fullerton, D. S., and Borns, H. W., Jr., 1983, The Late Wisconsin glacial record of the Laurentide ice sheet: *in* Wright, H. E., Jr., ed., Late-Quaternary Environments of the United States, Volume 1, The Late Pleistocene, The Univ. of Minnesota Press, Minneapolis, Minn. p. 3-37.

Mooers, H.D., 1990, Ice-marginal thrusting of drift and bedrock: Thermal regime, subglacial aquifers, and glacial surges: Canadian Journal of Earth Sciences, v. 27, no. 6, p. 849-862.

Patterson, C.J., 1994, Tunnel valley fans of the St. Croix moraine, east-central Minnesota, U.S.A: *in* Warren, W.P., and Croot, D.G., eds., Formation and deformation of glacial deposits: , A.A. Balkema, Rotterdam, p. 69-87.

———1997, Surficial geology of southwestern Minnesota: in Patterson, C.J., ed., Contributions to the Quaternary geology of southwestern Minnesota: Minnesota Geological Survey Report of Investigations 47, p. 1-45.

Patterson, C.J., and Knaeble, A.R., 2001, Surficial geology, pl. 4 of Boerboom T.J., project manager, Geologic atlas of Pine County, Minnesota: Minnesota Geological Survey County Atlas C-13, scale 1:100,000.

Ruhe, R.V., 1969, Quaternary Landscapes in Iowa: Iowa State University Press, Ames, IA, 255 p.

Sauer, C. O., 1916, Geography of the upper Illinois Valley and history of development: Illinois State Geological Survey Bulletin 27, 208 p.

Stanford, S.D. and Mickelson, D.M., 1985, Till fabric and deformational structures in drumlins near Waukesha, Wisconsin, U.S.A.: Journal of Glaciology, v. 31, no. 109, p. 220-228.

Syverson, K.M., 1998, Glacial geology of the Chippewa Moraine Ice Age National Scientific Reserve Unit, Chippewa County, Wisconsin: *in* Syverson, K.M., and Havholm, K.G., eds., Geology of Western Wisconsin: Guidebook for 61st Annual Tri-State Geological Field Conference and University of Wisconsin System Geological Field Conference, p. 57-65.

Syverson, K.M., in preparation, Pleistocene geology of Chippewa County, Wisconsin: Wisconsin Geological and Natural History Survey Bulletin and 1:100,000-scale map.

Teller, J.T., and Clayton, L., 1983, Glacial Lake Agassiz: Geological Association of Canada Special Paper 26, 451 p.

Upham, W., 1894, The Madison type of drumlins: American Geologist, v. 14, p. 69-83.

Whittecar, G.R. and Mickelson, D.M., 1977, Sequence of till deposition and erosion in drumlins: Boreas, v. 6, p. 213-218.

Whittecar, G.R. and Mickelson,D.M., 1979, Composition, internal structures and an hypothesis of formation for drumlins, Waukesha County, Wisconsin, U.S.A.: Journal of Glaciology, v. 22, p. 357-372.

Willman, H. B., and Frye, J. C., 1970, Pleistocene stratigraphy of Illinois: Illinois State Geological Survey Bulletin 104, 204 p.

Willman, H. B., and Payne, J. N., 1942, Geology and mineral resources of the Marseilles, Ottawa, and Streater Quadrangles: Illinois State Geological Survey Bulletin 66, 388 p.

Winchell, N.H., 1899, The geology of Carlton County: *in* Winchell, N.H., The geology of Minnesota: Geological and Natural History Survey of Minnesota, Final Report, v. 4, p. 1-24.

Wright, H.E., Jr., 1972, Quaternary history of Minnesota: *in* Sims, P.K., and Morey, G.B., eds., Geology of Minnesota A centennial volume: Minnesota Geological Survey, p. 515-547.

———1973, Tunnel valleys, glacial surges and subglacial hydrology of the Superior lobe, Minnesota: *in* Black, R.F., Goldthwait, R.P., and Willman, G.P., eds., The Wisconsinan stage: Geological Society of America Memoir 136, p. 251-276.

———1998, A sketch of the history of glacial investigations in Minnesota: *in* Patterson, C.J., and Wright, H.E., Jr., eds., Contributions to Quaternary studies in Minnesota: Minnesota Geological Survey Report of Investigations 49, p. 1-6.

Wright, H.E., Jr., and Watts, W.A., 1969, Glacial and vegetational history of northeastern Minnesota: Minnesota Geological Survey Special Publication 11, 59 p.

Pliocene to Holocene lakes in the western Great Basin: New perspectives on paleoclimate, landscape dynamics, tectonics, and paleodistribution of aquatic species

Authors: **Marith Reheis,** *U.S. Geological Survey, MS-980, Federal Center, Box 25046, Denver CO 80225*
Joanna Redwine, *Department of Geology, Humboldt State University, Arcata, CA 95521*
Kenneth Adams, *Desert Research Institute, 7010 Dandini Boulevard, Reno NV 89512*
Scott Stine, *Department of Geography, California State University-Hayward, 25800 Carlos Bee Boulevard, Hayward, CA 94542*
Kathy Parker, *Department of Chemistry, Geology, and Physics, California State University-Bakersfield, 9001 Stockdale Highway, Bakersfield, CA 93311*
Rob Negrini, *Department of Chemistry, Geology, and Physics, California State University-Bakersfield, 9001 Stockdale Highway, Bakersfield, CA 93311*
Raymond Burke, *Department of Geology, Humboldt State University, Arcata, CA 95521*
Gabrielle Kurth, *Department of Earth and Environmental Sciences, New Mexico Tech, Socorro, NM 87801*
John McGeehin, *U.S. Geological Survey, National Center MS-926A, 12201 Sunrise Valley Drive, Reston, VA 20192*
James Paces, *U.S. Geological Survey, MS-421, Federal Center, Box 25046, Denver CO 80225*
Fred Phillips, *Department of Earth and Environmental Sciences, New Mexico Tech, Socorro, NM 87801*
Andrei Sarna-Wojcicki, *U.S. Geological Survey, MS-975, 345 Middlefield Road, Menlo Park CA 94025*
Joseph Smoot, *U.S. Geological Survey, National Center MS-926A, 12201 Sunrise Valley Drive, Reston, VA 20192*

Organizers: **Marith Reheis,** *U.S. Geological Survey, MS-980, Federal Center, Box 25046, Denver CO 80225*
Kenneth Adams, *Desert Research Institute, 7010 Dandini Boulevard, Reno NV 89512*
Scott Stine, *Department of Geography, California State University-Hayward, 25800 Carlos Bee Boulevard, Hayward, CA 94542*

Co-Leaders: **Joanna Redwine,** *Department of Geology, Humboldt State University, Arcata, CA 95521*
Rob Negrini, *Department of Chemistry, Geology, and Physics, California State University-Bakersfield, 9001 Stockdale Highway, Bakersfield, CA 93311*

INTRODUCTION

The pluvial lakes of the Great Basin have been fertile ground for Quaternary studies since the 1870's (well over 100 published papers and books), and are also a topic of fascination for the interested public. Imagine huge lakes (10^2–10^4 km^2), now dry playas, fed by perennial rivers that are now only intermittent streams. Despite the attention they have received, these lakes continue to provide new evidence and inspiration for new generations of scientists—not only through more sophisticated analyses of isotopes and micro–stratigraphy, but also by revealing new field relationships that enhance and revise our understanding of climate history and geomorphic responses to deformation and climate change.

This field trip will focus on long–standing and recently developed information on large lakes that existed in the western Great Basin from the late Pliocene to the Holocene. Significant advances have been made in our understanding of the chronology of lake–level high–stands, timing of drainage integration, distribution of aquatic organisms, effects of river diversion on lake basin behavior, controls on, and effects of, isostatic rebound, and timing of minor fluctuations in lake level during the last lake cycle. We will emphasize key outcrops and geomorphic features that contain stratigraphic and chronologic evidence for lake fluctuations and drainage changes, with particular emphasis on tephrochronology, paleomagnetic data, paleosols, and tectonics. Among the topics covered in this field trip (Figs. 1–3) are:

(1). Evidence for rapid climate changes during the Holocene that produced, then desiccated relatively large lakes in Carson Sink, terminus of the Carson River (STOPS 1, 17).

(2). Synchronous changes in runoff documented by near–desiccation of the West Walker River and lowstands of Mono Lake during the Holocene (STOPS 3, 5, 6).

(3). Changes in river drainages induced by variable amounts of isostatic rebound after desiccation of the Sehoo (late Pleistocene) high–stand of Lake Lahontan, among which may have been diversion of the Walker River between Walker Lake and Carson Sink (STOP 1).

Evidence for large lakes in the early and middle Pleistocene that stood at higher levels than the well–documented, late Pleisto-

Reheis, M., Redwine, J., Adams, K., Stine, S., Parker, K., Negrini, R., Burke, R., Kurth, G., McGeehin, J., Paces, J., Phillips, F., , Sarna–Wojcicki, A. and Smoot, J., 2003, Pliocene to Holocene lakes in the western Great Basin: New perspectives on paleoclimate, landscape dynamics, tectonics, and paleodistribution of aquatic species: *in* Easterbrook, D.J., ed., Quaternary Geology of the United States, INQUA 2003 Field Guide Volume, Desert Research Institute, Reno, NV, p. 155–194.

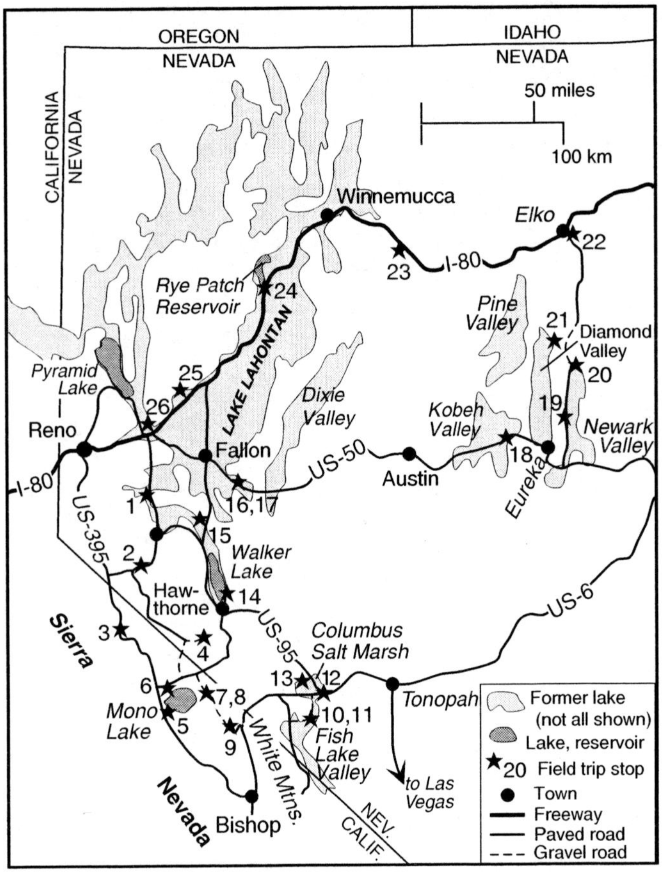

Figure 1. Index map of area of field trip in Nevada and eastern California.

cene, pluvial lakes in several basins, including the Lahontan basin, Mono Basin, Columbus–Fish Lake basin, and Newark basin (STOPS 7, 8, 12, 13, 15, 20, 21).

Pliocene and younger lake basins that became integrated by large–lake overflows and drainage captures to form present–day drainage basins, such as the Humboldt River (STOPS 11, 18, 22). Reconstructing these drainage–basin changes has important implications for paleoclimate, local tectonic effects, and migration of fish and other aquatic species in the Great Basin.

As an aid to traveling geologists and field courses, the guidebook emphasizes the road log and features that are visible from vehicles. Because many details of the histories of Lakes Lahontan and Russell (Mono Basin) have been published repeatedly, we do not reiterate this information but cite the key published literature. We will distribute copies of recent publications on specific field trip stops, especially those with information on little–known lake basins, pre–late Pleistocene stratigraphic records of most basins, and re–interpretations of certain sites. In sections following the road log, the guidebook includes summaries of studies and interpretations that are in progress or not yet published. Information not available before the manuscript deadline will also be made available on the trip.

PREVIOUS WORK AND GENERAL DESCRIPTION

During its well–studied, late Pleistocene (Sehoo) high–stand, Lake Lahontan (Figs. 1–3) reached maximum altitudes of about 1330 m and extended more than 350 km, from the Nevada–Oregon border to beyond the southern end of Walker Lake (e.g., Russell, 1885; Morrison, 1991; Adams and Wesnousky, 1999). Many smaller lakes occupied other basins in the western Great Basin (e.g., Mifflin and Wheat, 1979), although their late Pleistocene (Sehoo–equivalent) high–stands are generally not well dated. Until recently, little information was available on older pluvial lakes and their shorelines (Reheis, 1999; Reheis et al., 2002a, 2002b; Redwine, this guidebook). New evidence that previously mapped "Eetza" deposits (currently thought to be ~220–130 ka, but in part as old as 300 ka; Morrison, 1991) may in fact be as young as ~70 ka in some locations will be discussed at STOPS 2 (Lake Wellington) and 14 (Thorne bar). This interpretation reverts to previous views that the "Eetza" is at least in part early Wisconsin (e.g., Morrison, 1964).

Previous workers observed that altitudes of unfaulted Sehoo shorelines range from about 1318 m at the northeastern margin of Lake Lahontan to as much as 1341 m in the Carson Desert (Adams et al., 1999) due to isostatic rebound and tilting. This crustal response, coupled with rapid fall in lake level at the end of the Pleistocene, may have diverted some of the major rivers from one terminal subbasin to another. For example, the Walker River has switched from the Walker Lake subbasin to the Carson subbasin (STOP 1, Adrian Valley), and the Humboldt River has sometimes drained north to the Black Rock–Smoke Creek Deserts (STOP 24, Rye Patch Dam).

Much of Lake Lahontan, as well as Lakes Russell, Columbus–Rennie, and Wellington, are within the Walker Lane tectonic belt. These lake basins were locally faulted and warped throughout the Quaternary. Quaternary faults have deepened basins and displaced shorelines in other basins on the field trip (e.g., STOPS 20–22, Lakes Newark and Diamond). The changes in topography due to tectonism make fully outlining the highest strandlines of early Pleistocene deep–lake episodes impossible. The highest vertical faulting rates are along the Wassuk fault (Demsey, 1987), which bounds the west side of the Walker Lake sub–basin (Day 4). Extensional faulting has changed the outlet of Lake Russell in Mono Basin, redirecting lake discharge (STOPS 4, 7, 8; Reheis et al., 2002b) and providing a pathway for fish to migrate along the eastern Sierra Nevada.

Morrison (1991) summarized the previously known ages and altitudes of early–middle and early Pleistocene near–shore and deltaic deposits of Lake Lahontan (Fig. 4). We will visit two of the four sites where incision has exposed outcrops of older deposits buried by Sehoo and Eetza sediment (STOP 15, Walker River canyon; STOP 24, Humboldt River canyon). The pre–Eetza lake cycles identified by Morrison (1991) are the Rye Patch Alloformation, about 700–600 ka, and unnamed deposits containing and underlying the ~1,000–ka Glass Mt. tephra. He believed that the 760–ka Bishop ash bed was deposited during a long period of

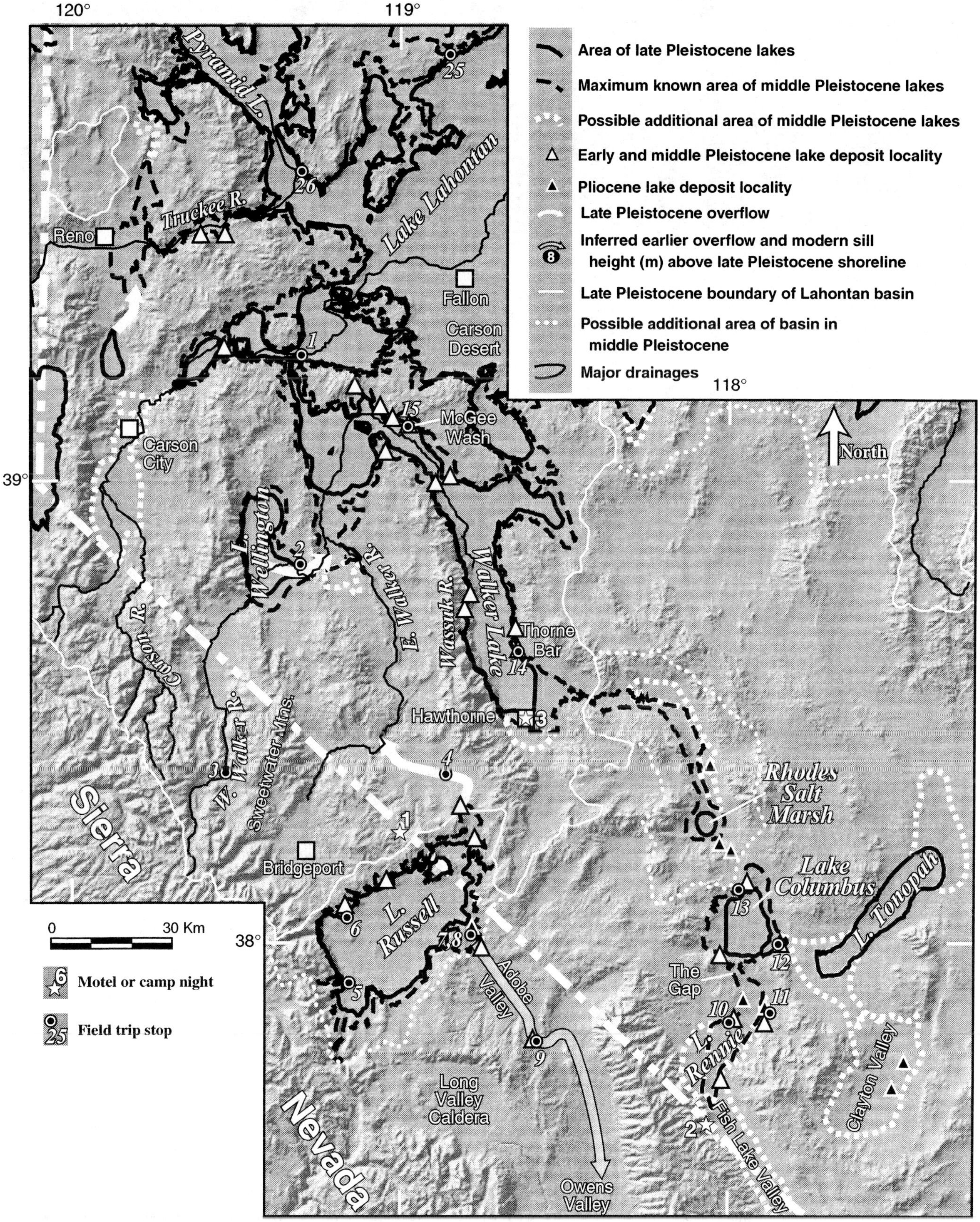

Figure 2. Shaded-relief map of western part of field trip (Stops 1-15, 25 and 26), showing shorelines of pluvial lakes and overflow pathways.

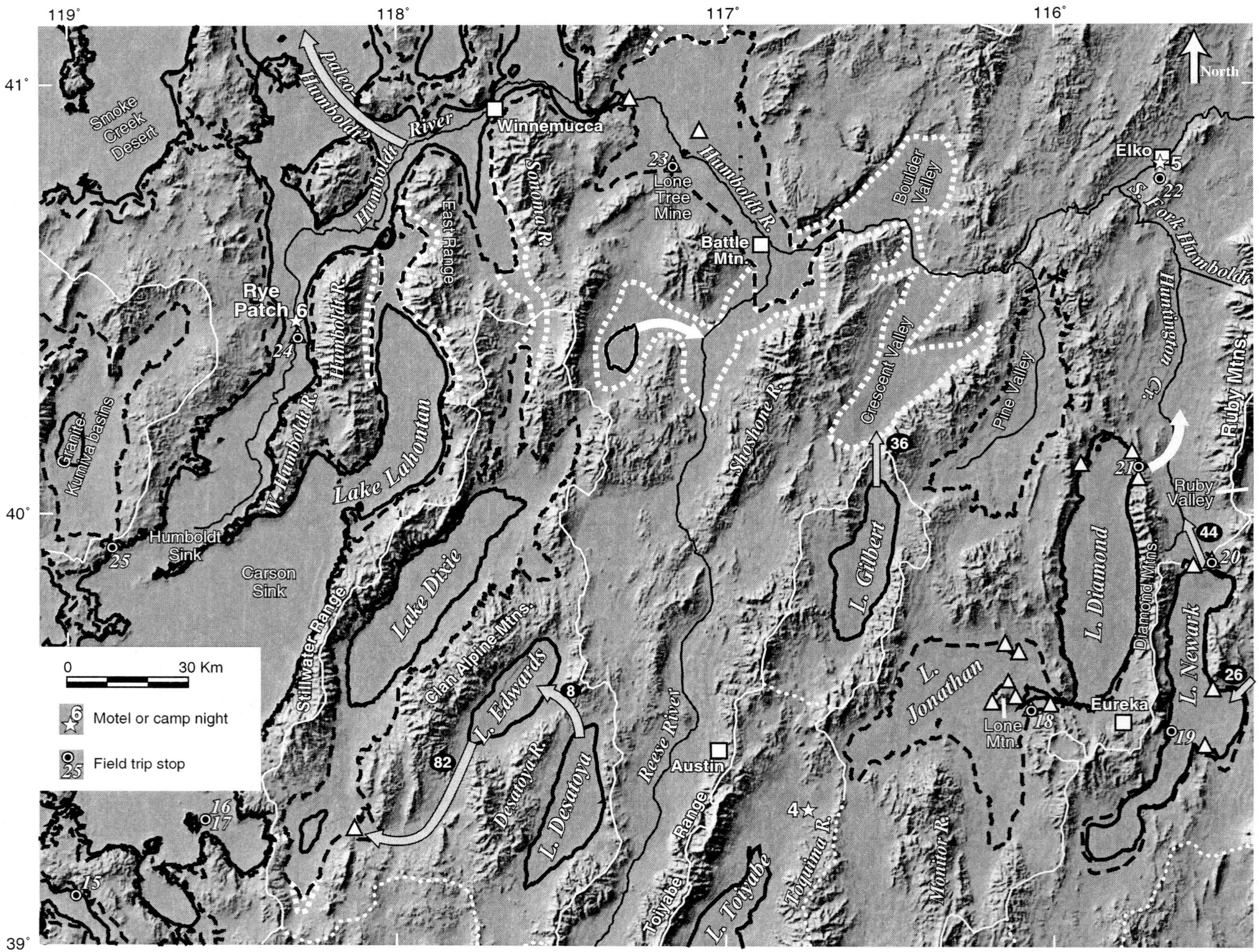

119°
118°
117°
116°
41°
40°
39°
North
Smoke Creek Desert
paleo-Humboldt?
Humboldt River
Winnemucca
Sonoma R.
East Range
Humboldt R.
Rye Patch 6
24
W. Humboldt R.
Granite-Kumiva basins
Lake Lahontan
Humboldt Sink
25
Carson Sink
Stillwater Range
Lake Dixie
Clan Alpine Mtns.
L. Edwards
82
8
Desatoya R.
L. Desatoya
Reese River
Austin
Toiyabe Range
L. Toiyabe
4
Toquima R.
Monitor R.
23
Lone Tree Mine
Battle Mtn.
Shoshone R.
Boulder Valley
Crescent Valley
Pine Valley
36
L. Gilbert
L. Jonathan
Lone Mtn.
18
Eureka
L. Diamond
Diamond Mtns.
21
Huntington Cr.
Ruby Valley
44
20
Ruby Mtns.
L. Newark
26
19
Elko
5
22
S. Fork Humboldt
16
17
15
0
30 Km
6
Motel or camp night
25
Field trip stop

alluvial and eolian sedimentation (the Lovelock Alloformation) with no deep lakes in the Lahontan basin. However, the Bishop ash bed is contained within lacustrine deposits in Fish Lake Valley (STOP 10) and other basins to the south and was also identified in beach gravel exposed in a fault scarp below the Sehoo shoreline north of Walker Lake (Reheis et al., 2002a).

Many localities record beach gravel at altitudes higher than the Sehoo and Eetza shorelines of Lake Lahontan and their equivalents in other lake basins. We will visit several of these sites (STOP 14, Thorne bar; STOP 15, McGee Wash; STOP 23, Lone Tree Mine; STOP 20, Lake Newark; STOP 21, Lake Diamond), which are discussed in detail in Reheis et al. (2002a), to see the evidence for higher lake stands and their ages. The highest known shorelines in the western Great Basin are thought to be about 700–600 ka, equivalent in age to marine oxygen–isotope stage (OIS) 16 (Reheis et al., 2002a). Lakes in several basins, including Kobeh, Diamond, Pine, Newark, and Clover Valleys, probably overflowed their sills during this intense, pluvial–glacial episode and became either temporarily or permanently part of the Lahontan drainage basin.

Reheis et al. (1993) documented several Pliocene to middle Pleistocene cycles of pluvial Lake Rennie in Fish Lake Valley (Fig. 1, STOPS 10–11) and proposed that at about 760 ka, Lake Rennie merged with Lake Columbus to the north. In the Columbus basin, several sites preserve a record of at least three shorelines that are older and higher than that of the late Pleistocene high–stand (STOPS 12–13). Ongoing cosmogenic ^{36}Cl (Kurth et al., 2002) and TIMS U–series dating is providing ages on associated tufa deposits and beach gravel here and in Newark and Walker Lake valleys.

Other pluvial lakes in western Nevada have been little studied until recently. Mifflin and Wheat (1979) identified late Pleistocene shorelines for Nevada pluvial lakes and briefly noted higher, older, shoreline features of some lakes. Tackman (1993) dated the late Pleistocene shorelines of Lake Diamond, located pre–late Pleistocene shorelines, and described control of lake levels by overflow from Kobeh Valley and incision of sills (STOP 21, Diamond Valley sill). Redwine has now documented and obtained age estimates for several shorelines at altitudes higher than the late Pleistocene high–stand of Lake Newark (STOPS 19–20, Newark Valley). We have also developed a rough understanding of the early–to–middle Pleistocene history of Lake Jonathan (STOP 18, Kobeh Valley).

Finally, many studies have shown that the terminal–basin lakes of the western Great Basin respond sensitively to changes in Holocene climate, but this Holocene record has only been well documented for Mono Lake (STOP 5; Stine, 1990). Ongoing work in Carson Sink (STOPS 16–17; Rhode et al., 2000; Adams, 2002) is providing new details on the size and timing of Holocene lakes in the Carson sub–basin, including correlation to the eastern Sierra drought record of Stine (1994; STOP 3, West Walker River).

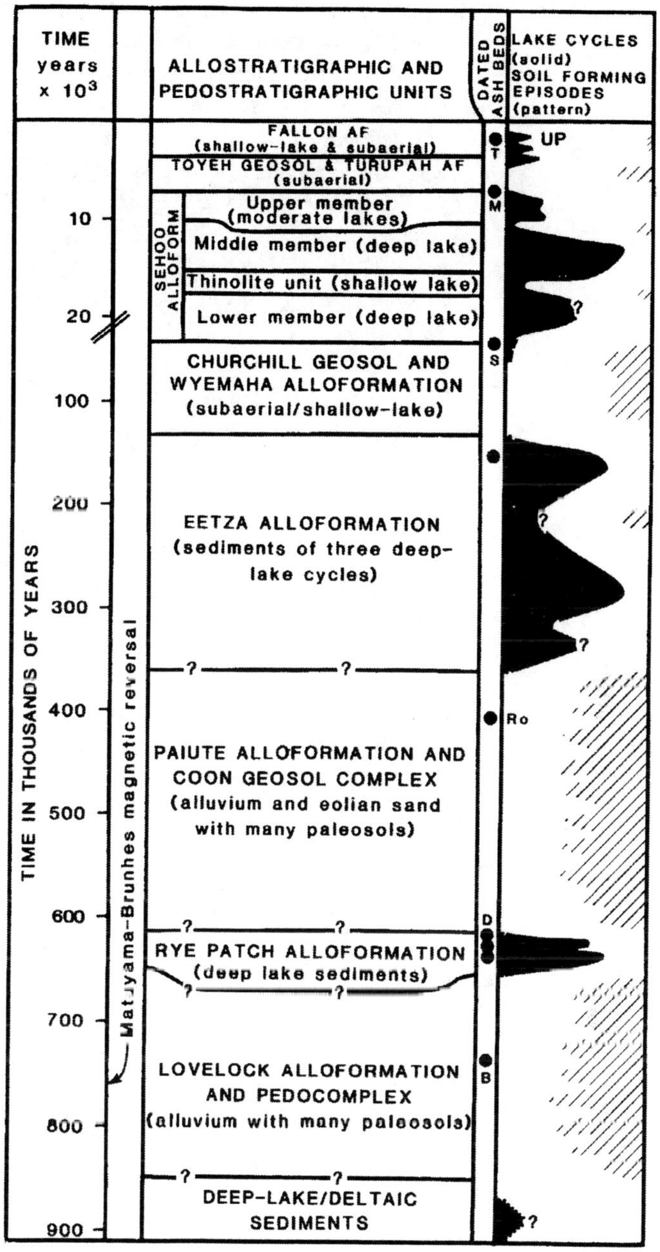

Figure 4. Stratigraphic units of Lake Lahontan (from Morrison, 1991).

DAY 1. RENO TO MONO LAKE

Miles **Km**

0.0 mi *0.0 km* *Drive east on I–80, starting at Exit 21 (Vista Blvd., beneath overpass) on the east side of Reno–Sparks metro area.*

Figure 3. Shaded-relief map of eastern part of field trip (Stops 16-24), showing shorelines of pluvial lakes and overflow pathways. For geomorphic features, see explanation on Figure 2. Overlaps Figure 2 in area of geomorphic legend (note Stops 15 and 25 on both maps).

4.3	6.9	*At Mustang, jökulhlaup deposits from glacial outburst floods from Lake Tahoe interfinger with Lake Lahontan sediments (Birkeland, 1965; Bell and Bonham, 1987).*
11.3	18.2	*On the left (north) side of highway are interbedded fluvial deltaic deposits and fan gravels derived from canyon walls. From here to Fernley are many outcrops of similar sediments. Although Birkeland (1965) mapped the fine–grained sediment as Eetza deposits, they are undated. Many of these outcrops may be Sehoo deposits because this canyon was backfilled during Sehoo transgression and subsequently dissected by the Truckee River.*
23.4	37.7	*To the left (north) are fluvial–deltaic deposits that display large deltaic foreset bedding dipping east (downstream).*
25.8	41.5	*Cross the Truckee River and ascend a large, dissected, deltaic complex. Wadsworth Amphitheater (STOP 26), a few km downstream, has excellent exposures of this depositional feature.*
27.1	43.6	*Exit at West Fernley–US 95A (Exit 46); turn right at the end of the ramp and pass beneath railroad tracks.*
28.6	46.0	*Turn right (south) at the traffic light on US 95A.*
29.8	48.0	*Crossing Truckee River Canal, which diverts much of the river flow to the Fallon area. Built in 1904, this was one of the first projects of the U.S. Bureau of Reclamation.*
42.4	68.3	*Passing the high shoreline of Lake Lahontan, which here is a small, dissected, gravel barrier bar.*
43.1	69.4	*Intersection with US 50 at Silver Springs. Continue south on US Alt. 95.*
49.4	79.5	*Here the highway cuts through a high–stand, gravelly, spit complex studied by Blair (1999).*
52.1	83.9	*Crossing the Carson River.*
54.1	87.1	*Crossing the Sehoo high–stand (~1335 m), marked by borrow pits in beach gravel. The Road now ascends the divide between Carson Valley and Walker Lake subbasin.*
54.6	87.9	*Turn right onto a graded dirt road. Pass through the gate and continue uphill on the main road.*
57.0	91.8	*Pass the crest of hill and descend into the relatively flat–floored Adrian Valley, through which Walker River has periodically flowed.*
57.4	92.4	*The road crosses railroad tracks. Turn right (north) about 100 ft (30 m) past the tracks.*
58.5	94.2	*Turn hard right (about 160 degrees). Leave plenty of room for vehicle in front to turn. Park beneath the cottonwoods just ahead. Walk a few hundred meters to the north–northwest on cattle trails, passing through a barbed–wire fence, to banks of Carson River.*

STOP 1. ADRIAN VALLEY–CARSON RIVER CONFLUENCE

Walker River now flows into Walker Lake, but several times during the late Pleistocene and Holocene, the river was diverted into Carson Sink via Adrian Valley (Fig. 5). During diversions, Walker Lake lost its primary source of water and Carson Sink gained a significant new source of water. At the confluence of Adrian Valley and Carson River, cutbanks expose several paleochannels of the Walker River. The base of section has been dated at 965 ± 58 ^{14}C yrs. B.P. and two paleochannels have been dated at 195 ± 46 and 223 ± 47 ^{14}C yrs. B.P. Thus, this record only reveals the late Holocene history of a dynamic system. Complementary studies of Holocene lake–level fluctuations in Walker Lake and Carson Sink and of paleochannels in Adrian and Mason valleys are currently underway with the goal of understanding timing, causes, and effects of diversions on hydrologic histories of Walker Lake and Carson Sink.

58.8 mi	94.7 km	*Turn around and return to US 95A. Turn sharp left, again leaving space for the vehicle in front.*
59.8	96.3	*Bear left (east) at the road junction, cross the railroad tracks, and continue east to the highway.*
62.6	100.8	*Intersection with US 95A. Turn right (south).*
64.3	103.5	*View of Adrian Valley to the right (west).*
66.9	107.7	*The low point on the highway is at the 1400–m level; roadcuts ahead expose Tertiary gravel and volcanic rock.*
69.7	112.2	*Descending past Sehoo high–stand of Lake Lahontan in Walker Lake sub–basin. Ahead is the intersection of US 95A with the road to Wabuska Hot Springs (east) and Adrian Valley (west). Walker River currently flows east of Mason Butte, a few miles to the south–southeast. The highway follows the former river channel, only slightly higher than the present channel of Walker River. The former channel drains through Adrian Valley into the Carson River (King, 1993; STOP 1).*
75.1	120.9	*To the east is Mason Butte, which would have been submerged by 1400–m high–stand of Lake Lahontan in the early middle Pleistocene. Farther northeast, two other buttes would have projected above this high–stand.*
81.8	131.7	*US 95A turns left (east) to Yerington. Continue straight on NV 339, ascending past Sehoo high–stand of Lake Lahontan. The road lies on the floodplain of the combined*

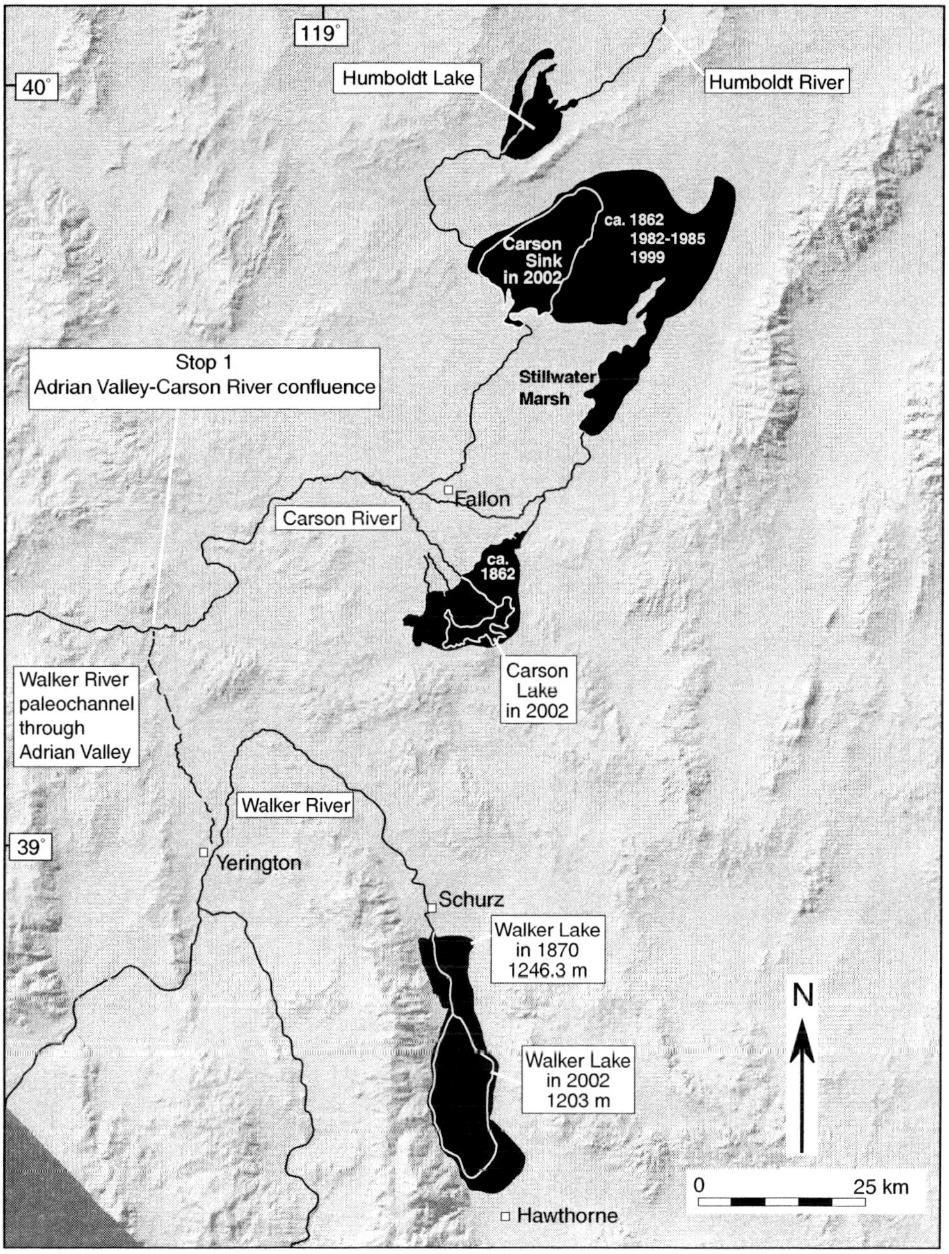

Figure 5. Map of Carson Sink and Walker Lake areas (see fig. 2 for location) with historic hydrographic features and localities in text. In average years, Carson Sink is a playa and Carson Lake is shallow and marshy. Walker River presently flows to Walker Lake but in the past has been diverted to Carson River, thus providing a significant new source of water to Carson Sink.

East and West Walker Rivers, here very wide and flat because the rivers built deltas during lacustrine episodes. Huge, open–pit of the Yerington mine lies ahead to the west.

86.7 139.6 The gravel pit to the west is excavated in alluvial–fan deposits shed from Singatse Range, bounded on the east by a Quaternary fault.

93.6 150.7 NV 340 ends, merging with NV 208. Continue south and west on NV 208, approaching Wilson Canyon on West Walker River. The canyon mouth is presently at ~1400 m, so the highest known level of Lake Lahontan would have flooded to at least this point in Mason Valley.

95.4 153.6 Pull off into the wide gravel space on the left between the highway and the river for brief STOP 2A. Stay at vehicles.

STOP 2A. FLOOD DEPOSITS FROM LAKE WELLINGTON

High on the north canyon rim, 8–10 m of well–rounded, fluvial gravel contains boulders as much as 2 m in diameter. The deposit may represent breaching of Lake Wellington in Smith Valley to the west after or during the 1487–m high–stand; the deposit is at about 1485 m and is capped by locally–derived, fan gravel. Slopes are unstable–use caution if you climb up.

96.2 mi 154.9 km Continue west, crossing West Walker River. From here west to a rest area, west–tilted deposits of probable Pliocene age (based on vertebrate fauna) occur on the north side of the river. The tilt is likely due to uplift on a fault bounding the west side of Mason Valley east of the canyon.

97.3 156.7 Turn right toward the river on dirt pulloff. Park along the fence line for a brief view at STOP 2B.

STOP 2B. LAKE WELLINGTON DEPOSITS

Across the river, ~10–m–thick, pale gray deposits cap the bluff and overlie dipping Tertiary rocks. Beach gravel is exposed at 1487 m, marking the shoreline of the last high–stand of Lake Wellington in Smith Valley. If this beach gravel is the same age as those farther northwest in Smith Valley, then present incision of West Walker River postdates this time.

98.6 158.7 Continue west, climbing out of the valley of West Walker River. The roadcut at the top of the bluffs exposes lake sediment capping older Pliocene deposits. The shoreline was about 10 m higher than the road.

102.4 164.9 At the town of Smith where the road bends sharply south, stay on NV 208. Sediment of Lake Wellington is visible to the north along the West Walker River. Heidi Stauffer (oral communication, 2002) reports that the probable age of this high–stand is ~70–80 ka as dated by a tephra layer analyzed by Andrei Sarna–Wojcicki (written communication, 2002).

106.7 171.8 Where NV 208 bends sharply right (northwest); enter Wellington (watch your speed!).

107.5 173.1 Cross West Walker River, approaching the active fault bounding west side of Smith Valley. To the north, a roadcut exposes shallow lake or marsh deposits interbedded with fault–scarp colluvium, indicating two different fault events. The road ascends the divide between Smith Valley and Antelope Valley.

111.0 178.7 Antelope Valley extends west and south. The canyon of West Walker River is visible to the southeast. Lacustrine and fluvial sediments, well exposed in the canyon, indicate that the valley was a closed basin during the Pliocene.

117.7 189.5 Holbrook Junction; intersection of NV 208 and US 395. Turn left (south) on US 395.

123.8 199.3 Intersection of US 395 and CA 89 (Monitor Pass). Denuded slopes along this stretch burned in the summer of 2002. The range–front fault along the west side of Antelope Valley forms the western margin of the Basin and Range province.

134.7 216.9 Pass through Coleville and Walker; enter West Walker River Canyon (in 1844, John C. Fremont abandoned his howitzer along this reach). The stream divides the Sierra Range on the west from the Sweetwater Mtns. on the east. Deposits from debris flows and rock falls are conspicuous along this narrow, steep–walled canyon. At several sites, lacustrine sediments testify to damming episodes caused by mass wasting. Fresh flood debris in and beyond the canyon was deposited during this stream's largest flood–of–record, January 1–3, 1997.

143.6 231.2 Pull off and park at Chris Flat.

STOP 3A. LUNCH, CHRIS FLAT, WEST WALKER RIVER

Along this reach and elsewhere in the canyon, rooted stumps and trunks of trees protrude from the stream bed. Death dates on these specimens fall into two distinct groupings of ~1130 AD and ~1350 AD. They, like other evidence that we will see at Mono Lake (and discussed in Adrian Valley), point to severe and persistent droughts in the Sierra region during Medieval time (Stine, 1994).

145.0 233.5 Continue south. In about 1.4 miles, pull over for brief STOP 3B.

STOP 3B. WEST WALKER RIVER CROSSING

The river is here constricted by Tioga glacial deposits (Clark, 1967), the left lateral moraine of which is nicely exposed in a road cut. Despite the narrowness of the stream canyon at this constriction, Medieval stumps of both generations are rooted at stream level on the right margin.

146.7 236.2 Intersection of US 395 and CA 108 (Sonora Pass).

163.7 263.6 Intersection of US 395 and CA 182 on the east side of Bridgeport. Turn left on CA 182. Bridgeport Reservoir has been a natural lake in the past, possibly dammed by movement on the east–side fault. G. I. Smith (oral communication, 1999) reports beach gravel around the north end of lake.

168.4 271.1 Descend East Walker River.

176.0 283.4 California–Nevada state line; continue on NV 338.

178.6 287.5 Intersection of NV 338 and the gravel road to Aurora and Hawthorne. Turn right (east) on a gravel road. The road first follows, then

		crosses, East Walker River (stay on the main road), then climbs away from the river.
183.3	295.1	*Roadcuts are in Pliocene(?), tuffaceous sedimentary rocks. Crossing a saddle, the view west is of the Sweetwater Range separating West Walker and East Walker Rivers. Descend to river level and then climb again.*
185.4	298.5	*Cross over a low saddle. The confluence of East Walker River and Rough Creek is due north but not visible from the road. Rough Creek carried discharge from Lake Russell during the early Pleistocene (STOP 4; Reheis et al., 2002b).*
189.9	305.7	*Floodplain of Rough Creek. Broad surfaces away from valley bottoms are mapped as pediments underlain by late Miocene and Pliocene sediments; stream terraces locally occur along watercourses.*
191.2	307.8	*Confluence of Bodie Creek and Rough Creek. The road fords Bodie Creek at a house.*
193.4	311.4	*4–way intersection of gravel roads. Continue across and east up Mud Spring Canyon.*
194.1	312.5	*The right fork leads to an old ranch. The 1.3–Ma paleovalley ahead to right merges with the present valley of Mud Spring Canyon.*
196.2	315.9	*Park at the top of the bluff above Mud Spring Canyon.*

STOP 4A. MUD SPRING PALEOVALLEYS

The road lies atop the north wall of a paleo–valley (Mud Spring 1) cut into Pliocene(?) fan gravel and filled by a 1.3–Ma basalt flow to the south. Note how the road is higher than the flow top to the west. The base of the flow is exposed just east of the stop. Note that Mud Spring Canyon is greatly overfit relative to its drainage area (Reheis et al., 2002b).

198.6	319.7	*Continue east on the graded gravel road, crossing Mud Spring Canyon and ascending the pediment surface to the north. Note the roadcut exposure of latest Miocene deposits capped by gravel and intercalated paleosols. Park on the right.*

STOP 4B. MUD SPRING PALEO–VALLEYS

The view to the south shows the feather–edge of a young basalt flow (<0.04 Ma) that filled against, and lapped over, the north wall of a paleo–valley (Mud Spring 2). Pale sediments below the basalt flow are 2–2.5 Ma, dated by correlation of tephra layers (Reheis et al., 2002b). Mt. Hicks, the highest cone to the south, is the source of one flow (1.6 Ma) that filled and flowed down Mt. Hicks spillway. Mt. Hicks was a major source of obsidian for Native Americans.

200.3	322.5	*Return to Mud Spring Canyon. Turn left (east) at the base of the slope on a small, unimproved, dirt road. Take the left fork just after turning on this side road. Turn around where the road deteriorates at the west end of a young basalt flow and park.*

STOP 4C. (OPTIONAL) MUD SPRING PALEOVALLEYS

Examine the base of a young basalt flow where it fills a paleo–valley (MS2) incised on the north side of a 1.3–Ma basalt flow, which previously had filled an older paleo–valley (MS1). The young flow thickens from north to south toward the axis of MS2.

202.7	326.3	*Backtrack to the graded gravel road and turn left.*
206.2	332.0	*Ascend the south side of Mud Spring Canyon and return to 4–way intersection. Turn left (south) toward Bodie. The road ahead is rough and rocky, so be careful with the tires!*
208.1	335.0	*The left fork leads to the ghost town of Aurora and mine on the northwest side of Aurora Crater, dated at ~350 ka. Bear right and continue toward Bodie. Aurora was founded in 1861. In 1863, it had 5000 people and was the county seat for both Esmeralda County, Nev., and Mono County, Calif. (due to confusion over the location of the boundary line), and was virtually abandoned by 1865.*
220.9	355.6	*Entering the ghost town of Bodie; the visitor parking lot lies to the left of the road. Optional stop for 1–2 vans for a stroll through abandoned (but restored) buildings. To go to DAY 1 camp, turn right (west) up a graded gravel road.*
223.1	359.2	*DAY 1 camp on the ridge on the southwest side of the road, with view north to Sweetwater Range and a spectacular view of the Sierra Nevada to the southwest.*

DAY 2, MONO LAKE TO FISH LAKE VALLEY

2.2	3.5	*Return to the Bodie visitor parking lot and turn right (south). Just south of Bodie, fork left (south) on graded gravel road toward Mono Lake and climb to the divide between Bodie valley and Mono Basin.*
9.7	15.6	*Descend Cottonwood Canyon. Holocene terraces locally contain white tephra layers from eruptions of Mono Craters that lie south of Mono Lake. At the mouth of the canyon, the broad slope to the east is a fan graded to a late Pleistocene high–stand of Lake Russell at*

		2155 m, whereas a higher trimline with more abrupt change in slope is the older, last–overflow shoreline at about 2188 m.
12.6	*20.3*	*At a bend in the road is a view west of Lake Russell shorelines. A faint break in slope where trees thin is the last–overflow shoreline at ~2188 m. Abundant beach gravel also forms a preserved berm (tombolo) at 2225 m at the top of the hill directly above this break in slope, and beach gravel extends as lag on bedrock upward to 2274 m (Reheis et al., 2002b).*
13.5	*21.7*	*Intersection with CA 167 and road up Cottonwood Canyon. Turn west on CA 167.*
23.9	*38.5*	*Proceed to the junction with US 395 and park briefly. To the west, CA 167 points toward Lundy Canyon. At the canyon mouth, defined by the Sierra Nevada frontal fault, moraines of Tioga age (20 ka) are offset down to the east ~25 m. The shoreline (measured at 2155 m west of the fault) from Lake Russell's last high–stand (~13 ka) has been dropped ~20 m.*
31.1	*50.1*	*Turn left (south) on US 395, passing through Lee Vining, to the junction with CA 120 (Tioga Pass road through Yosemite National Park), and park. Mono Lake dominates the view to the east and the Sierra Nevada to the west. Mono Lake is the hydrographically closed remnant of Lake Russell. Diversion of its feeder streams by Los Angeles since 1940 has lowered the lake surface by about 15 m to an altitude of 1945 m. Because it has not overflowed in tens of thousands of years and because of volcanic eruptions emanating from its floor (STOP 5C), the lake is highly saline (presently about 88 g/l total dissolved solids). With a pH of 9.7, and supersaturated with respect to calcium, it readily precipitates $CaCO_3$ at spring sites, forming tufa towers visible in several groves along the western shore.*

STOP 5A. SHORELINES AND DELTAS OF LAKE RUSSELL

To the west, high shorelines of Lake Russell are associated with a deltaic complex of pro–glacial and post–glacial Lee Vining Creek. The general surface on which we stand (altitude 2073 m) is part of an extensive delta plain deposited over a long interval of the last glacial and lake cycle. Plains of smaller, higher (to altitude 2155 m), and younger "onset deltas" (Stine, 1987) are visible to the southwest at the mouth of Lee Vining Canyon. These were deposited when Lake Russell rose to its high–stand in latest "Wilson Creek time" (~13 ka–see below).

35.1	*56.5*	*Continue south 0.75 mile (1.2 km) on US 395. Turn left (east) on an unmarked road that crosses Wilson Creek Fm. (see STOP 6) to the western shore of Mono Lake. Drive 1.0 mile (1.6 km) to the road fork, then bear left (north) 2.3 miles (3.7 km) to where the road disappears beneath Lee Vining Creek and park.*

STOP 5B. LEE VINING CREEK

Walk eastward ~0.3 mile (0.5 km) to view two generations of Medieval–age stumps, coeval with those in West Walker River canyon. Trees grew during two multi–decade intervals just prior to 1130 AD and just prior to 1350 AD when Mono Lake dropped to its lowest Holocene level. Immediately to the north is the "delta trench" of Lee Vining Creek. Exposed in this cut are lake–transgressive and lake–regressive sedimentary sequences, with intervening volcanic ashes and buried soils, which record lake fluctuations over the past several thousand years (Stine, 1990).

39.4	*63.4*	*Return to vehicles, then backtrack 2.3 miles (3.7 km) on the county road to the fork. Bear left, heading southeast, for 2.0 miles (3.2 km) to just beyond Rush Creek road crossing; park at the toe of Panum block avalanche.*

STOP 5C. PANUM BLOCK AVALANCHE

This feature was formed by eruption of Panum Dome, the last event in the most recent eruptive sequence (~605 ^{14}C yrs. B.P.) of the Mono Craters and Domes (Sieh and Bursick, 1986). We have seen tephra from this and an older eruption (~1190 yr BP) along our route. Inscribed onto the toe of a block avalanche is Clover Ranch shoreline (~AD 1600; Stine, 1990). This shoreline and the youngest Mono Craters tephra constitute geomorphic and stratigraphic time–lines that constrain ages of the Mono islands. These timelines and other types of evidence indicate that at least eight different lake–bottom eruptions have occurred during the past 1800 years. The most recent resulted in emergence of Paoha Island just before AD 1690 (Stine, 1987).

46.5	*74.9*	*Proceed east on county road 2.4 miles (3.8 km) (bearing right at all road forks) to the intersection with CA 120. Turn right (west) on CA 120, 4.7 miles (7.5 km) to the junction with US 395. Over the last several miles of this reach, the road points just south of Bloody Canyon. Arguably, more glaciations are represented in the morainal record of this canyon than at any other valley on the continent, though ages obtained by different dating methods conflict. Bloody Canyon is*

		the type locality for the Mono Basin Glaciation of Sharp and Birman (1963).
57.4	92.4	*Turn right (north) on US 395 and proceed 9.3 mi. (14.9 km) to the junction with unmarked "Cemetery Road." Turn right (east) on Cemetery Road and proceed 1.6 mi. (2.6 km) to just beyond Mill Creek crossing and park.*

STOP 6 WILSON CREEK FM. EAST OF MILL CREEK DELTA TRENCH

Roadcuts east of Mill Creek expose lacustrine and deltaic sediments and tephra beds that make up the Wilson Creek Fm. of Lajoie (1968). Sediments accumulated during the last high–stand cycle. Grossly convoluted tephra near top of the section is Black Point ash, used by Lajoie (1968) to date the last Lake Russell high–stand. Numerous workers interested in glacial and lake chronology and in paleomagnetism have studied the type section (0.5 mile east) in a cut of the artificial Wilson Creek. The base of section has yielded conflicting dates (from 23,000 to 36,000 ^{14}C yrs. B.P.) and the top has been dated at ~12,000 ^{14}C yrs. B.P. (e.g., Lajoie, 1968; Benson et al., 1990).

61.5	99.0	*Return to US 395. Turn right (north) and proceed 2.5 miles (4 km) to the junction with CA 167.*
77.5	124.8	*Turn right (east) on CA 167. In about 16 miles (25 km), Kirkwood Spring may be seen at 9:00. On the slope above the spring is beach gravel that extend upward to about 2252 m, much higher than the last overflow shoreline (2188 m). Prepare to turn right on a large, unmarked, gravel road.*
78.0	125.6	*Turn right (south) on the county road on the east side of Mono Lake. For the next several miles, the road has loose sand, so drive cautiously (but not too slowly).*
89.0	143.3	*Crossing ~2155–m, is the late–Pleistocene high–stand of Lake Russell in the southeast part of the basin. The roadcut ahead is incised in tufa–cemented beach–rock and snail coquina of the previous high–stand, the last time that Lake Russell rose high enough to overflow its sill and discharge through Adobe Hills spillway into Adobe Valley (Putnam, 1949).*
89.2	143.6	*At top of cut, park on left in a small borrow pit.*

STOP 7. LUNCH. ENTRANCE TO ADOBE HILLS SPILLWAY

Examine a roadcut and discuss the age and nature of the spillway. North–striking normal faults and NE–striking, left–lateral faults displace 3–4–Ma. basalt flows, forming a pull–apart that over time lowered the rim of Mono Basin below the altitude of Mt. Hicks spillway and eventually permitted overflow of Lake Russell southward into Adobe Valley (Reheis et al., 2002b).

90.7	143.6	*Continuing south along the road, pass first through a broad saddle that is the spillway lip into the northern subbasin described by Reheis et al. (2002b), then across another low sill into the southern sub–basin. Park where the road emerges from a small canyon between sub–basins.*

STOP 8. OLD LAKE RUSSELL DEPOSITS

Hike due east about 1.5 km to a powerline. Outcrops there are well above the last overflow level. The lowest deposits exposed consist of tuffaceous lacustrine mud with an age of ~2–2.5 Ma, overlain by lacustrine gravel and sand. Higher up slope are beach rock and snail coquina exposed in roadbed and power pole footings. Beach pebbles are lagged on basalt slopes above the powerline to altitudes as high as 2243 m (Reheis et al., 2002b).

91.7	147.6	*Continue south along Adobe Hills spillway. At an intersection, a rough road to the east leads to the powerline outcrops. Beach gravel of Lake Russell that predates the opening of Adobe Hills spillway lies atop ridge to west.*
93.3	150.2	*Beach gravel in an embayment on the right (west) shows that before the channel ahead was incised into basalt, Lake Russell stood at this level (2200 m). The road ahead descends through a meandering channel into Adobe Valley. White volcanic ash from Holo cene eruptions of Mono and Inyo Craters drapes over basalt and obscures fault scarps.*
99.6	160.4	*Entering Adobe Valley. At a bend in the road, the view eastward to the crest of White Mt. shows a Tertiary erosion surface sloping north. Stay on the main road through cattle loading pens.*
101.2	162.9	*Intersection with CA 120;. turn left (south).*
102.0	164.2	*Forest Service road to Taylor Canyon, where the 2–Ma tuffs of Taylor Canyon were named. View pink, welded Bishop Tuff (0.76 Ma) along this stretch. On the far west skyline is the crest of the Sierra Nevada.*
107.2	172.6	*At a curve in road, on the western skyline is Glass Mt. on the northeast flank of Long Valley caldera, source of the Bishop Tuff. The road lies on alluvial fans derived from erosion of these and older tuff sheets.*
111.7	179.8	*Black Lake (remnant of Pleistocene Lake Adobe) lies to the east. Fan slopes above the lake are underlain by Bishop ash and (or)*

		1–Ma–old, Glass Mt. ash. A pale outcrop crowned by two trees just above the lake marks the base of beach gravel of Lake Adobe, probably deposited at the time of the last outburst flood. The altitude of this gravel matches a knickpoint at the upper end of the canyon draining eastward into Blind Spring Valley (Reheis et al., 2002b).
112.4	*181.0*	*At 9:00 is a stunning view down–canyon from Adobe Valley into Blind Spring Valley, through which discharge was carried from Lake Adobe into Owens River drainage.*
114.1	*183.7*	*Round a curve and descend into Blind Spring Valley. The canyon we saw from above, which carried discharge east out of Adobe Valley, lies to the north but is not visible. After crossing under a powerline, a pile of loose dirt at a pumice mine on the north side of the road yields a glimpse of white Bishop ash, which underlies the entire slope.*
116.0	*186.8*	*Park on right (east) side of road on wide shoulder.*

STOP 9. FLOOD DEPOSITS AND CANYON CUT BY FLOODS OUT OF ADOBE VALLEY

Cross the road and walk west up pumice–covered slopes. Observe airfall and welded deposits of Bishop ash overlying fluvial conglomerate in a small valley draining north toward the canyon from Adobe Valley. The ridge on the south side of this canyon is capped by giant granite boulders deposited by a catastrophic flood (Reheis et al., 2002b).

117.7	*189.5*	*Continue northwest on CA 120 to Benton Hot Springs, built as a stagecoach stop in 1852. Hot spring bathing is available in privately owned facilities. Drive along the marshy floor of Blind Spring Valley. View the upper end of a slot canyon cut in a fault–block ridge between Blind Spring Valley and Queen Valley at 2:00. The slot was deepened, if not cut entirely, by flood(s) out of Adobe Valley. A fan–shaped deposit of flood gravel lies at the mouth of canyon.*
118.2	*190.3*	*Intersection of CA 120 with Foothills Road. The left fork ascends a high surface, underlain by Bishop ash, that predates floods out of Adobe Valley and eventually leads to views of the upper canyon and younger flood deposits.*
120.2	*193.5*	*Intersection of CA 120 and US 6 in Benton. Turn left (north). To the east is Mt. Montgomery and an excellent view of the sheer front of the White Mtns. bounded by the White Mtns. fault zone.*
129.0	*207.7*	*Passing Janey's, the first little brothel in Nevada, but alas no longer open.*
130.5	*210.1*	*Vista point for Boundary Peak, the highest peak in Nevada. The peak to the right is Mt. Montgomery in California. Prominent scarps of the NE–striking, Queen Valley fault cut alluvial fans at the mouth of Queen Canyon. Ahead, the road climbs through late Tertiary volcanic rocks to Montgomery Pass.*
141.6	*228.0*	*Mineral County–Esmeralda County line. Abundant diatomite beds are exposed from here eastward.*
147.1	*236.8*	*Intersection of US 6 and NV 264. Turn right (south) on NV 264 toward Fish Lake Valley. Grefco mine exploits diatomite interbedded with 3–4–Ma basalt flows and tephra layers. Diatoms from these sediments indicate fresh-water marsh and stream environments. Kennedy Flat at 3:00 is a northeast–tilted, grassy surface underlain by 3–4–Ma basalt flows, defining a former erosion surface on the north end of the White Mtns..*
151.1	*243.3*	*Intersection of NV 264 and NV 663. Continue straight. 3.0–3.4–Ma–old basalt caps a flat–topped surface at 3:00 along the White Mtns., and glaciated cirques are seen in the distance. At 9:00, view sedimentary rocks and volcanic tuffs east of NV 663. This is a well–known, late–Miocene, vertebrate fossil locality.*
165.1	*265.8*	*On the left (east) is Fish Lake Valley Community School. Prepare to turn right.*
174.0	*280.1*	*Turn right (west) ona gravel road at a sign for Fish Lake Valley Community Park for DAY 2 camp. Flush toilets and hot water available! After dinner, vans can shuttle interested participants to: (1) The Boonies, a 'very Nevada' small–town bar 1.5 miles south in the town of Dyer, or (2) Fish Lake hot spring for soaking and swimming (about a 20–minute drive north).*

DAY 3. FISH LAKE VALLEY TO HAWTHORNE

0.0	*0.0*	*Drive south to Dyer for an optional brief stop to view 100–m–high, late–Pleistocene, fault scarp on Perry Aiken fans (vertical slip rate ~2–5 mm/yr). Return to Community Park turnoff and reset mileage to 0.*
1.0	*1.6*	*Intersection with a road to a cemetery. This large alluvial fan, deposited by glaciated Leidy Creek, extends to the east side of the valley and interfingers with marsh deposits. A relict population of Lahontan tui chub*

		(Hubbs and Miller, 1948), long thought extinct but recently rediscovered (D. Sada, oral communication, 2002), lives in springs near the marsh. Coring on this fan showed that alluviation occurred during relatively dry periods (Reheis et al., 1993). E. L. Cord, inventor of the Cord automobile, built the ranch house to the east.
3.4	*5.5*	*At 10:00 are prominent, east–facing scarps forming the west side of a complex graben on the Indian Creek fan. This graben at the north end of Fish Lake Valley fault zone is part of the horsetail splay that transfers slip eastward across the valley to the Emigrant Peak fault zone (Reheis and Sawyer, 1997).*
8.9	*14.3*	*Past a grove of trees by Chiatovich Creek, turn right (east) on the gravel road toward Mineral Ridge mine.*
12.9	*20.8*	*Ahead is the Emigrant Peak fault zone. Prominent scarps from 12:00 to 1:00 cut dissected, middle–to–early Quaternary, fan gravel with intercalated beach sand. From 1:00 south, scarps cut Holocene to late Pleistocene fan deposits. The Holocene vertical slip rate is estimated at ~3–4 mm/yr (Reheis and Sawyer, 1997). This N–striking, normal fault zone is the right–stepping complement of the right–oblique Fish Lake Valley fault zone, forming a pull–apart.*
14.5	*23.3*	*The road bends left at rock outcrops. Park on the right and walk uphill through siliceous hot–spring deposits.*

STOP 10. SINTER MOUND, BISHOP ASH, AND 1460–M SHORELINE OF PLUVIAL LAKE RENNIE

Outcrops consists of reworked Bishop ash and opaline silica beds that overlie altered green mudstone (Fig. 5 in Reheis et al., 1993). Silica–rich beds are sinter formed by primary hot–spring discharge at the lake edge.

16.7	*26.9*	*Continue east past the turnoff to the hot spring. At an intersection, turn right (southeast) and cross the valley south of the playa. To the south, the Emigrant Peak fault zone passes south out of fan deposits to bound a dark brown bedrock escarpment. Here, the fault merges with a detachment fault forming the gently curved surface of the hill.*
17.0	*27.4*	*Intersection; turn left (northeast).*
18.2	*29.3*	*Sediment mounds to left are waste piles mined from Fish Lake playa for ulexite. The burned area on the right is the remains of a Chinese workers' camp. Due east, the presently active trace of the Emigrant Peak fault zone bounds fan gravel and lake beds of the 2–Ma Emigrant Pass section (STOP 11).*
20.5	*33.0*	*Turn right (south) at the triangular intersection and drive along the scarp of the Emigrant Peak fault zone. Weathered badland sediments are older Pliocene lakebeds containing Putah and Nomlaki tephra layers, ~3.4 Ma. These beds are separated from the late Pliocene Emigrant Pass lakebeds by a small fault.*
20.9	*33.6*	*Park where a wash crosses road. Hike east to see a section of near–shore lacustrine and alluvial deposits.*

STOP 11. EMIGRANT PASS SECTION

Lacustrine mudstone and sandstone overlie Tertiary gravel and grade upward into fan gravel (Fig. 7 in Reheis et al., 1993). Of the 13 tephra layers present, most are correlated with ~2.0–Ma–old compound tuff of Taylor Canyon erupted from the Long Valley area. The uppermost ash bed in lacustrine deposits is the ~2.0–Ma Huckleberry Ridge ash from Yellowstone. Note the long erosional slope of White Mtns. capped by 11–Ma basalt flows.

21.3	*34.3*	*Return to the triangle intersection and continue straight (north). The road skirts the east side of the playa, a temporary stopping place for water flowing north out of Fish Lake Valley through The Gap into Columbus valley.*
24.3	*39.1*	*Slight rise. For the next 2 mi., the road passes outcrops of Pliocene lacustrine sediments and fan gravel containing tephra layers similar in character and age to those at STOP 11.*
26.9	*43.3*	*Turn right (north) at the T–intersection, following drainage from Fish Lake playa into Columbus Salt Marsh. The Gap transects a colorful complex of Miocene volcanic rocks and is boggy throughout.*
28.5	*45.9*	*A bench at 3:00 (east) is underlain by blocky rhyolite (altitude 1457 m). The bench may be a landslide, although intact bedrock seems to underlie it, or it may be a wave–cut platform of pluvial Lake Columbus–Rennie.*
29.3	*45.9*	*At the mouth of The Gap to the left (west) are prospects for hydrothermal minerals in a massive spring mound.*
30.4	*48.9*	*S–curves through sandy, pebbly sediments that are late Pleistocene beach ridges (1400–m level) of Lake Columbus–Rennie. At this time, the lake did not extend into Fish Lake Valley. Just past the curves at 3:00 (east) are low, active, spring mounds.*

Travertine beds extending north from these mounds interfinger with late Pleistocene shoreline sediment.

31.6 50.9 Turn right (north) at the intersection with NV 773. Before turning, note the low, basalt hill due west, which has a prominent bench that is a shoreline marked by well–rounded basaltic clasts set in a pavement. The altitude of the bench is 1430 m, coincident with the berm at STOP 11. The lower brown hill farther west preserves evidence of at least four lacustrine units higher and older than the late Pleistocene shoreline of Lake Columbus–Rennie (Figs. 21 and 22 in Reheis et al., 2002a).

31.9 51.4 Intersection of NV 773 and US 6. Turn right (east) on US 6 toward Tonopah.

38.0 61.2 Intersection of US 6 and US 95. Continue east, but prepare for immediate right into Coaldale Junction (abandoned). Park on the east side of Coaldale Junction parking lot.

STOP 12. BEACH GRAVEL AT COALDALE JUNCTION

East and south of the buildings is beach gravel of a shoreline at about 1430 m, 30 m higher than the late Pleistocene shoreline of Lake Columbus–Rennie. This lake would have extended at least to the south end of Fish Lake playa. A landfill behind the buildings is cut into a beach remnant. East of the parking area are exposures of tufa–cemented beach rock. Trench CM01–3 was excavated in beach gravel 100 m to the east for cosmogenic profile dating and soil description.

38.9 62.6 Backtrack to the junction of US 6 and US 95. Follow US 95 northward, skirting the east side of Columbus Salt Marsh. The playas here at Rhodes Salt Marsh and at Soda Springs to the north were mined for ulexite, a borate salt. Based on these concentrated salts, Mifflin and Wheat (1979) speculated that they once had been part of a much larger body of water.

47.5 76.0 Intersection of US 95 and gravel road to Columbus ghost town (historic marker) on the north side of Columbus Salt Marsh. Turn left (west). View to the southwest is of Volcanic Hills and the crest of the White Mtns. Stop at the large open pit just past the fenceline and cattle guard.

STOP 13A. LUNCH. THE BIG PIT

The large pit intersects the groundwater table and provided water to nearby mines. The exposure includes three alluvial–fan units interbedded with two lacustrine units (Fig. 21 in Reheis et al., 2002a). The lower lake unit consists of a beach ridge and marl containing Bishop ash on the lakeward side. The upper lake unit extends upward to an altitude of 1400 m and includes tufa dated at 14,860 ^{14}C yrs. B.P. The youngest fan deposits contain Holocene tephra from Mono Craters.

47.9 77.1 Continue west on Columbus road. 100 m upslope (north) of the metal shack near road is brown gravel plastered on dark gray bedrock. The gravel marks a pre–late Pleistocene shoreline of Lake Columbus–Rennie at about 1417 m.

48.1 77.4 A small conical hill at 1:00 has a long backslope to the north that was wave–eroded during the 1417–m high–stand. The late Pleistocene shoreline at 1400 m on the south flank of the hill and extending south of the road was sampled for cosmogenic profile dating.

49.4 79.5 Park at small jeep track leading north. Walk to top of gray beach barrier 0.2 mile to north.

STOP 13B. TRENCH SITE CM01–2 IN BEACH GRAVEL AT THE 1417–M HIGH–STAND

The trench was sampled for cosmogenic profile dating and soil description (Fig. 8). Tufa–cemented beach rock at the same altitude to the north yielded a TIMS U–series age of about 150 ± 2.5 ka.

51.3 82.6 Return east along Columbus Salt Marsh. At the intersection with US95, turn left (north) toward Hawthorne.

56.3 90.6 Turnoff to Candelaria, a large, open–pit, gold mine lies to the west. We just crossed Redlich Summit, the divide between Columbus Salt Marsh and Rhodes Salt Marsh. The saddle to the west is the present threshold at an altitude of 1495 m. It is 35 m above the highest known shoreline of Lake Columbus–Rennie.

57.9 93.2 Lacustrine deposits crop out across a wide area north of Redlich Summit (Figs. 19 and 20 in Reheis et al., 2002a). These deposits represent a late Tertiary lake basin, possibly extending from Columbus Salt Marsh north through Rhodes Salt Marsh to Walker Lake. This basin was disrupted in late Pliocene by faulting and (or) compression.

60.4 97.2 A roadcut exposes S–dipping lacustrine sediment, including two sequences of sand, silt, and tephra separated by gravel (Fig. 19 in Reheis et al., 2002a). These 2–Ma lake deposits may represent the upper portion of the Tertiary lacustrine sequence to the south.

		Lake Lahontan at 1400 m would have flooded south to this point.
63.3	*101.9*	*Intersection with NV 360 and US 95. Skirting the west side of Rhodes Salt Marsh, which contained a small lake (about 20 m deep) during late Pleistocene.*
68.1	*101.9*	*Desert lobster available at this natural spring complex. View at 7:00 is of Boundary Peak and Mt. Montgomery at the north end of White Mt.*
70.8	*114.0*	*Town of Mina. To the northeast are light–colored late Pleistocene spring deposits and Miocene lake beds, cut by the NW–striking, right–oblique, Benton Spring fault zone, one of the Walker Lane faults.*
79.7	*128.3*	*Town of Luning. To the east is Luning playa, a remnant of a very small late Pleistocene lake.*
94.1	*151.5*	*Divide between Soda Springs Valley and Walker Lake valley. This pass and valley to the right are below 1400 m.*
97.6	*157.1*	*Crossing the high–stand of Sehoo (late Pleistocene) shoreline of Lake Lahontan in the Walker Lake basin.*
104.9	*168.9*	*Town of Hawthorne and El Capitan Hotel–Casino. We will return here after STOP 14. Continue north on US 95.*
108.1	*174.0*	*Intersection of Thorne Road and entrance to US Army Ammunition Depot at a flashing light. Turn right (east) and proceed through Ammunition Depot. Gillis Range lies on the east side of Walker Lake.*
113.5	*182.7*	*Thorne siding. Cross railroad tracks and bear left (north) at "Y"; pavement ends; continue north.*
118.6	*190.9*	*Road bears right and climbs Thorne bar.*
119.4	*192.2*	*Cross Sehoo high–stand shoreline. In 0.3 mile (0.5 km), park on the crest of the bar.*

STOP 14. THORNE BAR

A large, V–shaped, barrier bar complex (Fig. 10 of Adams and Wesnousky, 1999). Morphology, soils, cosmogenic dating, and U–series ages on tufa indicate the presence of four lacustrine units above the historic limit of Walker Lake. The barrier bar at the road has a moderately preserved surface at ~1350 m altitude. A younger barrier bar (OIS 4–but OIS 2 according to Adams and Wesnousky, 1999) encloses a small playa below and to the north. Sehoo bar (≤1330 m) extends far to the west. East of the cars is a higher, more eroded bar extending as high as 1365 m. The trimline on basalt hills 1–1.5 km northwest of the vehicles marks the 1402–m shoreline.

134.5	*216.5*	*Retrace route into Hawthorne to El Capitan Hotel and Casino. End of DAY 3.*

DAY 4. HAWTHORNE TO SPENCER HOT SPRINGS

0.0	*0.0*	*Leave El Capitan on US 95 north along the west side of Walker Lake. For the next 35 miles (55 km), the road follows the base of Wassuk Range, the most active range front fault in the Lahontan basin. The summit of Mt. Grant at 3426 m is >2100 m above the surface of Walker Lake. Demsey (1987) identified two Holocene ruptures totaling 6–7 m of vertical separation. Based on her altitude surveys of shoreline scarps cut into alluvium on Wassuk Range and across basin on presumed stable Gillis Range, Wassuk Range has risen about 1.5 m in the Holocene, or about 20 to 25% of total fault displacement.*
7.5	*12.1*	*Large gravel pit on Rose Creek fan. Prospect pits north of the pit expose a thin plastering of probable Sehoo–lake gravel atop a paleosol formed on older lake gravel, likely equivalent to an older high–stand preserved at Thorne bar.*
13.5	*21.7*	*Town of Walker Lake. Parts of the town lie on the historic shorelines formed before Walker River was diverted for irrigation. Tufa and beach–rock overlying wave–beveled, fault–sheared bedrock are abundant from here north along the lake.*
17.0	*27.4*	*For the next few miles, the highway lies on historic shorelines of Walker Lake. These shorelines are less vegetated and have virtually no soil development when compared to Sehoo and older lake features.*
20.5	*33.0*	*Ascending a fan and approaching 20–Mile Beach. The pyramidal hill on the left (west) exposes alternating layers of gray beach gravel and reddish fan gravel (Fig. 6B in Reheis et al., 2002a). The top of the hill at 1475 m is capped with beach gravel, which may be similar in age to lacustrine deposits north of Thorne bar but have been uplifted along the Wassuk fault.*
21.5	*34.6*	*The ridge at 10:00 to the road is capped by orange–weathering, beach rock at 1452 m (Fig. 6A in Reheis et al., 2002a). Beach gravel and tufa are better preserved north of this ridge in a bench between 1378 and 1396 m. U–series data suggest minimum ages of ~90–150 ka on both remnants, indicating significant uplift of the footwall block of the Wassuk fault. On the right (east) is the Walker River. Walker Lake has dropped 45 m since 1885 due to diversion and damming of the river for agriculture. Increasing salinity*

of the lake is threatening existence of native Lahontan cutthroat trout. Rapid headward incision of the river as it adjusts to its new base level has caused serious problems for the Walker River Paiute Tribe.

31.0 49.9 The fault scarp at 3:00 (west) exposes ledges of cemented lacustrine gravel. Light–colored layer near the top of the exposure contains 760–ka Bishop ash (Davis, 1978; Reheis et al., 2002a) as matrix around clasts.

34.3 55.2 Intersection of US 95 with US 95A. Bear left (north–northwest) on US 95A toward Yerington.

41.3 66.5 Intersection. Turn right (east) on a gravel road to Weber Dam and Reservoir.

41.7 67.1 Merge with gravel road. Bear left. The road crosses Reservation Hill, a horst of old fan gravel, then descends a bench underlain by Sehoo beach deposits. Badlands to north expose early Pleistocene lacustrine deposits (Morrison, 1991). ***Pause before descending to dam for overview.***

43.9 70.7 Bear left at the fork and cross Weber Dam. A photo of the bluffs ahead is on the cover of DNAG Volume K–2, Quaternary Non–Glacial Geology: Conterminous U.S. (Morrison, 1991). Continue up the hill.

45.1 72.6 Turn right (southeast) on a graded side road. Bear left at a fork in the road and continue south.

46.0 74.1 The road crosses McGee Wash. If conditions are favorable, turn left (east) and drive up the wash as far as practical (usually about 0.3 mile).

STOP 15. MCGEE WASH

McGee Wash, named for W H McGee of the Russell expedition, exposes deposits of four, pre–late Pleistocene lakes above the Sehoo level (1330 m, about where the cars are parked). Walk up the wash about 2 km to see these deposits (Figs. 7 and 8 in Reheis et al., 2002a). To the west, the wash is shown as a stratigraphic cross section in Figures 2–7 of Morrison and Davis (1984).

46.9 75.5 Retrace route to Weber Dam road. Turn right (east) ascending away from dam and reservoir.

49.3 79.4 Outcrops on the right (south) expose Miocene volcanic rocks overlain by Miocene and Pliocene alluvium and gypsiferous, Pliocene lacustrine deposits. These are in turn overlain by Pliocene to Pleistocene alluvial and eolian deposits of unit QTt in McGee Wash.

50.6 81.5 Intersection of Weber Dam road with US 95. Just 0.5 mile south are extensive outcrops of Pliocene lacustrine and alluvial deposits (Fig. 7 in Reheis et al., 2002a). Turn left (north) on US 95. Ascend to the divide between Walker Lake and Rawhide sub–basins. This divide is too high to have been overtopped by Lake Lahontan (even at 1400 m), but Walker Lake and the northern sub–basins could have been connected by a pair of sills to the east that lie between 1360 and 1390 m.

56.0 90.2 Sehoo high–stand of Lake Lahontan in Rawhide subbasin. Thin sheets of lacustrine deposits lying on wave–eroded bedrock surfaces are locally preserved.

58.7 94.5 View east into Rawhide playa, now a gunnery range for Fallon Naval Air Station. Note tremendous amount of young eolian sand blown up against the flanking hills, named Blow Sand Mt. by Harriet Morrison.

63.9 102.9 Divide between Rawhide subbasin and Carson subbasin. West of the road is a classic site called Russell Spit, studied first by Israel Russell, then by Roger Morrison and several others, most recently by Ken Adams. This is one of many controversial sites where Sehoo and Eetza high–stands appear very close in altitude.

66.5 107.1 Wildcat scarp, a 25–km–long, semi–circular feature around the southern and eastern edges of Carson Desert. The scarp here is 1–2 m high; farther east it is as much as 10 m high. Morrison (1964) interpreted Wildcat scarp as a wave–modified fault scarp with two faulting events in post–Sehoo time.

68.0 109.5 At a bend in road, Carson Lake, one of the Carson River sinks, lies to the northeast. To the west and southwest are prominent strandlines and beach gravel banked against the Desert Mt. This region of Lake Lahontan was studied and mapped by Roger Morrison (1964), the first Ph.D. graduate of University of Nevada in any field.

80.7 129.9 Intersection of US 50 and US 95 in downtown Fallon. Turn right (east) on US 50 toward Austin.

85.6 137.8 Turnoff to Stillwater Wildlife Refuge, the current low point of Carson Sink along Stillwater Range to north.

89.6 144.3 Approaching Grimes Point, note Sehoo Mt. at 10:00 and Eetza Mt. at 11:00. Morrison (1964) named and identified these hills as the type localities of alloformations repre-

senting two major lake cycles of Lake Lahontan.

90.9 — 146.3 — Turn left (north) on the paved road leading to Grimes Point. In 0.4 mile (0.7 km), park in visitor lot.

STOP 16. LUNCH GRIMES POINT ARCHAEOLOGICAL SITE

Take a walking path through numerous petroglyphs on basalt boulders. Most glyphs date between 5000 BC and 1500 AD. Nearby are wave–eroded caves with numerous excavated remains including burials. Currently, the most important to studies of the earliest immigrants to the Western Hemisphere is Spirit Cave man, recently dated at ~10,000 ^{14}C yrs. B.P. After lunch, return to US 50 and turn left (continue east).

92.0 — 148.1 — Macari Road intersects on the right (south). In 0.2 mile (0.3 km), turn right (southeast) on a dirt road.

92.6 — 149.1 — At an intersection with another dirt road near a beach barrier, make a hard left and park.

STOP 17. SALT WELLS BEACH BARRIER AND HOLOCENE LAKES IN CARSON SINK

This beach ridge (1204 m) was mapped as a Fallon 1 shoreline by Morrison (1964), who estimated an age of 2–4 ka. Davis (1978) placed a maximum age on this feature of about 2 ka, based on the presence of Turupah Flat tephra. Although they did not specifically mention this site, Currey (1988) and Benson et al. (1992) reinterpreted the Fallon 1 lake cycle to date from 11.3 to 10.3 ka. Based on the presence of the Turupah Flat Bed within this deposit and a ^{14}C age of 812 ± 72 ^{14}C yrs. B.P. on charcoal found directly beneath the tephra, this beach ridge is late Holocene in age (Fig. 6). A lake with a surface altitude of 1204 m would have a surface area of about 3300 km^2 and inundate most of Carson Sink. The rise of this lake was not an isolated event, as a lake rose to at least 1198 m (~2800 km^2) at about 1510 ± 40 ^{14}C yrs. B.P. We will discuss the paleoclimatic implications of large, late Holocene lakes in the Carson Sink.

93.0 — 149.8 — Return to US 50, turn right and continue east.

93.3 — 150.2 — Salt Wells "Villa," originally a watering stop for the Fallon–Fairview stage line, is now a renowned brothel.

94.4 — 152.0 — The windmill at 9:00 is near the southern end of the Rainbow Mt. fault zone. Two earthquakes, M=7.0 and 6.3, occurred on this fault in 1954 (Tocher, 1956). Fault trenches to the north exposed ^{14}C–dated subaerial and lacustrine deposits that record Sehoo lake–level fluctuations as well as evidence for two pre–1954 earthquake events (Ramelli et al., 2002).

104.8 — 168.8 — Turnoff to "Loneliest Phone" at Sand Mt., an active dune 180 m high (and ORV playground). Views to the south and east of late Pleistocene shorelines cut on basalt as the road climbs to Sand Springs Pass. The bench on the nearest outcrop north of the highway may be an older shoreline about 15 m higher than the Sehoo high–stand. In 1963, the Atomic Energy Commission detonated a 12–kt underground nuclear device near the pass to compare seismic waves from recent earthquakes to those of nuclear tests.

106.3 — 171.2 — Passing Sehoo high–stand of Lake Lahontan at 1335 m. The site of a Pony Express and stage station is ahead at 9:00.

108.4 — 174.6 — Sand Springs Pass; entering Fairview Valley. This is the lowest pass over which discharge from Lake Lahontan into Fairview and Dixie Valleys might have occurred. It is only ~8 m higher than the 1400–m high–stand of Lake Lahontan.

110.6 — 178.1 — Intersection with NV 839. To the north, light–colored beds at the base of the Stillwater Range are diatomite and mudstone capped by Pliocene basalt flows.

113.1 — 182.1 — Entering Frenchman Flat, the playa of Lake Labou, which overflowed northward into Lake Dixie. To the southeast are targets for "Top Gun" school at Fallon Naval Air Station.

117.0 — 188.4 — A "radar" station is atop a hill to north. Sediments there and in a similar hill to the northeast may be fluvio–deltaic deposits. If so, they must have been deposited when Lahontan discharged over Sand Springs Pass into Dixie–Fairview Valley.

118.4 — 190.7 — Intersection with the road north into Dixie Valley. Continue east, rising up to Drumm Summit.

120.2 — 193.6 — Sign points to "Earthquake Scarps;" turn right (south) for a brief optional stop at 1954 Fairview Peak fault scarp. Then continue east on US 50. Miocene lacustrine deposits lie north of Drumm Summit (Reheis et al., 2002a).

126.0 — 202.9 — Junction of US 50 with NV 361 south to Gabbs. Old Middlegate station is just off the highway. Ahead are dissected Miocene alluvial and lacustrine deposits.

138.9 — 223.7 — Pass between Eastgate–Fairview–Dixie Valley to the south and Edwards Creek Valley to the north. This divide is profoundly higher than the late Pleistocene shoreline of Lake

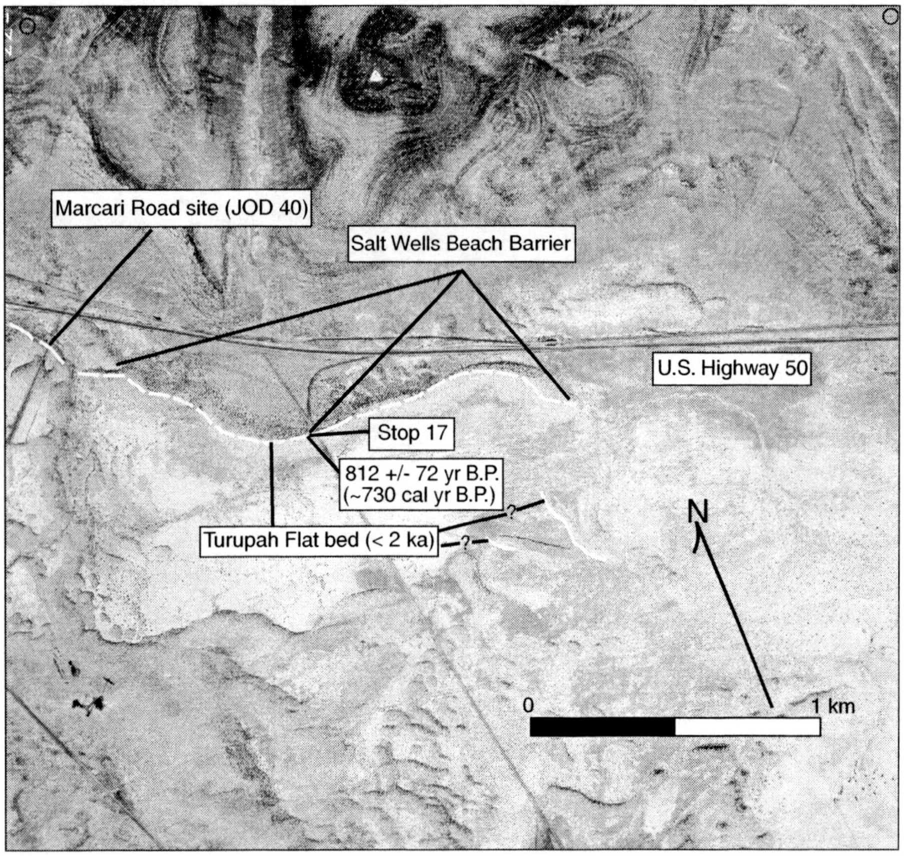

Figure 6. Vertical aerial photograph of Salt Wells beach barrier and vicinity (see fig. 3 for location). The Turupah Flat ash bed outcrops in a very consistent manner, suggesting that overwash processes on the backsides of beach barriers deposited it when water filled the Carson Sink to 1204 m. Radiocarbon age of detrital charcoal from within Salt Wells beach barrier indicates that it was deposited about 730 cal yrs B.P.

Edwards. Fault scarps bound the Desatoya Range east of road.

145.5 234.3 Clan Alpine Mt. on the west, with an active range–front fault shown by faceted spurs and fault scarps on alluvial fans; Desatoya Range on the east.

156.9 252.6 Ascend over indistinct late Pleistocene shoreline of Edwards Creek Valley. No one has looked for a higher–than–late–Pleistocene lake in this valley. However, if Smith Creek Valley (next valley to east) overflowed westward, such a shoreline must exist, because combined drainage areas would produce a much bigger lake.

160.8 258.9 Climb east into New Pass Gorge. View to the west (behind) of dry playa in Edwards Creek Valley.

163.4 263.1 View at 2:00 is up drainage that could have carried westward overflow from Lake Desatoya in Smith Creek Valley. The sill is slightly southwest of highway and 5–7 m higher than the late Pleistocene shoreline. Detailed work that might prove such discharge occurred has not been done.

166.1 267.5 Due south of the road, the pile of dirt is a borrow pit excavated in late Pleistocene shoreline deposits of Lake Desatoya. Ascend to divide between Smith Creek Valley and Reese River Valley.

179.7 289.3 Cross Pony Express Trail (sign). Roadcuts expose lacustrine and alluvial deposits assumed to be Tertiary in age. Reese River valley is a long drainage tributary to the Humboldt River. It may have been internally drained and later integrated with Humboldt River in the late Pliocene or early Pleistocene. Lacustrine deposits have been poorly dated if at all, and no geomorphic work has been done.

187.5 301.9 Intersection of US 50 and NV 305. Enter Austin, a historic mining town and Pony Express station, renowned for beautiful old churches and newly popular for mountain biking.

Optional stop for interested participants

Espresso and cappuccino (yes, really!) available at T–Ricks, a mountain biking center on the east edge of Austin. Just west of

T–Ricks is the Austin Historical Museum in the old Gridley Store, and Pony Express House.

188.4	*303.4*	*T–Ricks; ascend to Austin Summit.*
193.1	*310.9*	*Intersection with the county road leading north into Grass Valley.*
199.1	*320.6*	*Entering Big Smoky Valley, bounded by the Toquima Range on the east and Toiyabe Range on the west. The valley held Lake Toiyabe in late Pleistocene. No detailed work has been done here. Springs in the northern part of valley contain two endemic fish related to Lahontan tui chub and speckled dace that may have arrived as long ago as the Pliocene, and have since been isolated. However, Hubbs and Miller (1948) thought that fish had migrated south from Lake Gilbert in Grass Valley into Big Smoky Valley over a low alluvial–fan divide by drainage switching.*
199.4	*321.1*	*Intersection of US 50 and NV 376. Turn right (south) on NV 376 and in 0.3 mile (0.5 km), turn left (southeast) on the gravel road to Spencer Hot Springs.*
204.1	*328.6*	*Triangular intersection; north fork goes to the main hot springs. The south fork is the exit in the morning.* ***End of DAY 4; camp for the night. Enjoy the hot springs!***

DAY 5. SPENCER HOT SPRINGS TO ELKO

0.0	*0.0*	*Begin mileage at triangular intersection on south side of Spencer Hot Springs. Drive south on the graded road.*
0.6	*1.0*	*Turn left (east) up the fan on a dirt road for a shortcut back to US 50. In a mile, the road curves northward. Continue northward past the intersections of two dirt roads from the left, curving around a big pit.*
2.1	*3.4*	*1948 Plymouth beside the road. Continue north on the main road winding through mining areas and pits. Pass the remains of an ore–crushing mill. Descend through prospect pits and continue north through the crossroads.*
6.2	*10.0*	*Junction of the gravel road leading to Spencer Hot Springs with US 50. Turn right (east).*
13.2	*21.3*	*Cross over Hickison Summit. Just over the pass is the access road to Hickison petroglyph site.*
14.6	*23.5*	*The view opens out into Kobeh Valley. At 1:00 is Lone Mt., an island in Lake Jonathan, completely encircled by lacustrine deposits. Lake Jonathan, named for J. O. Davis who first identified deposits of this lake, was internally drained until it overtopped its sill and incised the outlet. It is now tributary to Diamond Valley to the east and in turn to Humboldt River and Lake Lahontan in pluvial periods.*
17.7	*28.5*	*Approximate high–stand of Lake Jonathan. View south is up Monitor Valley, the largest tributary to the lake.*
21.8	*35.1*	*Lander–Eureka county line.*
28.6	*46.0*	*Crossing a low hill. Kobeh Valley has a very large drainage basin. To the south, the Monitor Range divides Antelope and Monitor valleys. Antelope Valley is the eastern of these two tributaries. To the north, Simpson Park Range separates Kobeh Valley from Grass Valley (Lake Gilbert).*
38.4	*61.8*	*To the northeast, a prominent light–colored outcrop against Lone Mt. is the site of one measured section in Figure 24 of Reheis et al. (2002a).*
43.5	*70.0*	*Dissected pale–colored alluvium, fan, spring, and lacustrine deposits crop out along the south side of Lone Mt. On the east side of Lone Mt., just before the intersection of a county road leading north, turn right (south) into a graveled pulloff and park.*

STOP 18. LAKE JONATHAN DEPOSITS

Several tephra layers are exposed in deposits east of Lone Mt. The youngest, and closest to the top of lacustrine sediments, is Rye Patch Dam ash bed, dated at ~670 ka. In a pale colored knoll to north–northwest are spring discharge deposits containing Huckleberry Ridge ash, ~2 Ma. The long, flat bench extending east from Lone Mt. is beach gravel capping older lacustrine deposits. Walk uphill to the southeast to see some of this stratigraphy and the Bishop(?) and Rye Patch Dam ashes (Figs. 23 and 24 in Reheis et al., 2002a).

49.3	*79.4*	*Continue east on US 50. The hills to the south are capped by lake gravel and locally underlain by fan gravel. Enter Devil's Gate, the spillway from Kobeh Valley into Diamond Valley. Slightly after 670 ka, Lake Jonathan apparently overtopped its eastern sill and began cutting Devil's Gate.*
51.2	*82.4*	*Exit Devil's Gate into Diamond Valley. The road descends a delta composed of sediment derived from the erosion of Kobeh Valley.*
54.6	*87.9*	*Crossing the late Pleistocene shoreline of Lake Diamond. To the north is a small hill with a faint shoreline scarp. Diamond Valley preserves shorelines of higher and older lakes at the north end of the valley (STOP 21).*

58.0 93.4 *Entering Eureka. After passing through the town, ascend to Pinto Summit.*

68.4 110.2 *Newark Valley to the east. Lake Newark extended 13 km south of US 50 at the late Pleistocene high–stand.*

68.9 110.9 *The Pancake Range at 1:00. 6 backhoe pits were excavated into 2–3 ages of shorelines at the northern end of the range (Redwine, in preparation). In 1867, a blacksmith ran off an Indian eating his last beans. The Indian, Paiute Jim, returned with a chunk of silver as a peace offering. The blacksmith forced Paiute Jim to take him to the source of the silver, Treasure Hill, just east of Mt. Hamilton (distant skyline 12:00)—a 305–m long, 1–3–m wide, 7–m deep vein of almost pure silver. Hamilton quickly became a town of 25,000 people and produced over 300 tons of silver from 1867 to 1901. By 1885, it was a ghost town (Smith, 1976).*

70.0 112.7 *At 2:00, a late Pleistocene shoreline is notched into Black Point.*

72.1 116.2 *A late Pleistocene cuspate barrier bar to the left (north) is partially buried.*

73.0 117.5 *Turn left (north) on NV 892 (a green sign points to Strawberry).*

73.9 118.9 *A gravel pit lies to the left; the road is cut through a late Pleistocene recessional cuspate(?) bar/spit complex. The road is parallel to the Diamond Mt., whose crest ranges from 2590 to 3235 m. Lone Mt. lies ahead to right.*

78.2 125.9 *A barrier bar ~1.2 km to east at 1869 m has tufa dated 13,690 ± 40 ^{14}C yrs. B.P. The dated barrier lies east of the playette to the right.*

79.3 127.6 *Turn right (east) on a gravel road south of Lone Mt. for STOP 19. Go 0.1 mile and park in a gravel pit.*

STOP 19. LUNCH. LONE MT. GRAVEL PIT

The pit is excavated into a barrier bar at 1845 m that, based on soil morphology and altitude, is correlative with the late Pleistocene high–stand of Lake Newark. This deposit contains an older gravel unit below an unconformity delineated by fine sand and silt containing fresh water snails. AAR ratios on snails suggest a middle Pleistocene age (Kaufman, 2001, written communication).

79.5 127.9 *Turn right (N) on NV 892.*

80.5 129.7 *Harper Ranch. From this point on, many shorelines and faults are visible in alluvial fans on the west side of the highway. Some locations have been mapped in detail to identify faults that offset alluvial fans of varying ages and total displacements. Springs line the valley east of NV 892, suggesting that the range–front fault system is stepping basinward. Shorelines older than late Pleistocene are not preserved on the west side of valley, likely due to active faulting and fan deposition.*

82.8 133.4 *Wine glass valleys to the left, Mining and Water Canyons, with short stems. Cuspate barrier bars to the left are hard to identify on the west side of the valley due to burial by fan sediments along faulted valley margin. Barrier bars are mostly cemented, which complicates distinction of shorelines and faults because barrier bars hold a steep scarp angle and are commonly covered with fan sediment.*

83.3 134.1 *Sand dunes at 3:00. Holocene sand dunes fringe playa perimeter. Dune morphology indicates wind from south. In contrast, cuspate barriers, bars, and spits are primarily directed to south, showing longshore current and dominant wind direction from north during the late Pleistocene.*

83.8 135.0 *To left (west), the range–front fault offsets older fans and is covered by younger fans; patterned ground covers portions of youngest alluvial fans. Road cut through large cuspate barrier at 1:00.*

86.0 138.4 *Roadcut in a big cuspate barrier bar. The cemented layer the near top of the cut is at 1804 m and has an age of 13,025 ± 40 ^{14}C yrs. B.P. Due west is Diamond Peak, the highest peak in the Diamond Mts. at 3235 m.*

86.3 138.9 *Goicoechea Ranch raises sheep, cattle, and wild horses (Circle Ranch on maps). Just ahead, the large scarp to the left is suspected to be a fault scarp modified by shoreline erosional processes. The base of scarp is at altitude of the late Pleistocene high–stand.*

87.1 140.2 *Shoreline and fault scarps on the left. Just ahead, late Pleistocene strandlines are visible on east side of the valley, against the southern end of the Ruby Mt. The Elko–Hamilton Stage Line ran along these shorelines. Buck Station, directly across the valley, was the halfway point and hotel for those transporting silver from Hamilton to Elko in the late 1860's. The lowest part of the basin at 1778 m continues from here north to the first Cold Creek Ranch entrance.*

90.6 145.8 *Road cut in bar. The bedrock lithology is distinct enough near these large bars that unique lithologies can be traced in exposures. Limited distributions suggest that longshore transport distance is short.*

91.6 147.4 *Late Pleistocene shorelines cut on bedrock. Light–colored patch behind Strawberry Ranch, just ahead, is from a fire in 2000. That summer, fires in Nevada covered ~3237 km² (800,000 acres). Many ranchers in Diamond and Huntington valleys went bankrupt when 100% of their grazing land burned.*

93.2 150.0 *Late Pleistocene recessional cuspate barrier bar.*

95.7 154.1 *Shoreline and faults to the left have been mapped at a reconnaissance level. The big fault scarp has been modified by the late Pleistocene high–stand.*

98.1 158.0 *We suspect that the bedrock scarp to the left is a fault scarp subsequently re–occupied by a late Pleistocene shoreline. Home Ranch (Goicoechea Ranch on maps) is part of Silver State Ranch, which includes 6475 km² (1.6 million acres) in Newark, Ruby, Long, and Duckwater Valleys. They supported much of this project with donated backhoe work.*

99.0 159.4 *A late Pleistocene shoreline occurs to the left. The low spots on both sides of the low hill in the valley center to the north are potential overflow locations for Lake Newark. We will visit shorelines on the east side of this hill.*

100.7 162.2 *First Cold Creek entrance to the right and the northern end of the lowest part of basin. The maximum relief in the valley is 1457 m.*

101.0 162.6 *Second Cold Creek entrance is to the right. On the left, the reservoir is at the range front fault. Directly across the valley is Warm Springs Ranch and a nearby gravel pit, where Lymnaea snails in late Pleistocene high–stand shoreline yielded a ^{14}C age of 13,780 ± 50 ^{14}C yrs. B.P. On Dec. 21, 1932, a M=7.2 earthquake centered in Gabbs, Nevada, ~250 km to the southwest, stopped discharge of Warm Springs and caused dry Barrel Springs, located farther south on the east side of the valley, to discharge at such a high volume that flow reached US 50. After several weeks, the springs both went back to normal, much to the joy of folks living and ranching at Warm Springs. This earthquake also changed groundwater patterns closer to Elko.*

102.8 165.6 *Turn right on a well–maintained gravel road at signs for Bald Mt. Mine, Warm Springs, and Placer Dome US Inc. Be alert for very large, fast trucks and heavy equipment.*

103.3 166.4 *The light green–and–white bluff near the hills to left (north) is the source of distinctive volcanic lithologies found in the beach gravel east of the low hill near STOP 20. These clasts are only found as far as the wide channel to the east, suggesting limited longshore transport. The headwaters of the channel just west of the bluff are the location of the eastern potential overflow at 1887 m. Farther east, subtle late Pleistocene recessional shorelines occur on the left and right sides of the road. A very subtle, NE–striking fault along the south end of the low hill cuts these shorelines at an oblique angle. The level of activity and sense of motion on this fault is unknown. Backhoe pits in pre–late Pleistocene shorelines cut by the fault exposed fractures with small offsets.*

104.3 168.0 *Crossing cattle guard, shadowed fault scarp may be visible to the left (north).*

105.4 169.7 *Turn left (north) on major gravel road at signs for Big Bald Mt. Mine and Placer Dome US Inc. This is the old Elko–Hamilton stage route.*

106.4 171.2 *Prominent late Pleistocene recessional shoreline. In 0.4 mile (0.6 km), turn left onto a faint, but drivable, seismic line road to STOP 20. Continue to a lone juniper tree and park.*

STOP 20. SHORELINES ON THE NORTH END OF LAKE NEWARK

Walk ~0.5 km southwest to look at the backhoe pit excavated into a late Pleistocene shoreline, then work back to vehicles looking at soil development on shorelines of two ages and discussing age implications of soil properties, PDI, Cs, and clay index values. Note the preservation differences in berm crests and in concentrations of rounded stones at the surface.

107.4 172.8 *Return to the main gravel road and turn left (north).*

107.7 173.4 *The road merges from the right–watch for traffic. The low spot to the left (11:00) is the eastern potential (and most likely) overflow location at 1890 m.*

108.7 175.0 *Elko–Hamilton Stage Line turns left. Bear right on the main road toward Big Bald Mt. Mine. The fan complex is composed of fans of many ages. The oldest fan is mapped as older than the pre–late Pleistocene shorelines but younger than deposits in the potential overflow channel.*

111.3 179.2 *Turn left on a major gravel road. Use caution for the next 35 miles–heavy equipment and roadwork are possible at any time. After a small hill, the view is into Huntington valley from a sweeping right turn. The lowest point on the distant skyline (11:00) in Diamond Mts. is Railroad Pass (STOP 21). The*

		Pony Express trail crossed Diamond Mts. through Overland Pass (9:00). Little or no work has been done on the geomorphology and lake sediments in Huntington valley.
118.5	*190.7*	*Pony Express trail crosses the road.*
120.0	*193.1*	*Intersection with a smaller dirt road on left (west); stay straight (north). In 5.4 miles (8.6 km), turn left (northwest) onto another main gravel road. Proceed 6.4 miles (10.2 km).*
131.8	*212.2*	*T–intersection. Turn left on the gravel road and cross Huntington Creek. The creek may have carried Lake Newark overflow and does carry Lake Diamond overflow into Humboldt River and Lahontan Basin. At the intersection just across the creek, turn right.*
132.6	*213.4*	*Intersection. Turn right. Entering the spillway of Lake Diamond through Railroad Pass. Non–catastrophic overflow and downcutting occurred during OIS 6(?) and 2 (Mifflin and Wheat, 1979) and possibly earlier. Overflow during OIS 2 began ~23,400 ^{14}C yrs. B.P. and continued until ~13,500 ^{14}C yrs. B.P. (Tackman, 1993). Bedrock is Tertiary basalt over tuff, unconformable on Pennsylvanian Ely limestone.*
136.1	*219.1*	*Entering Diamond Valley. Just before the gate and cattleguard, turn left (south) and continue along the fence line. Ascend across younger, OIS 2 shorelines toward the older OIS 6(?) shorelines (STOP 21).*
136.7	*220.1*	*Green fence post and road coming in from the right (south). Turn around and backtrack far enough to park.*

STOP 21. LAKE DIAMOND OVERFLOW

Shorelines of late Pleistocene and older ages curve into Railroad Pass spillway, which conducted discharge from Lake Diamond into Lahontan Basin during the middle and late Pleistocene pluvial episodes. Tackman's (1993) thesis on Lake Diamond focused on late Pleistocene history. From slope morphometry, he tentatively correlated the higher shorelines to oxygen–isotope stage (OIS) 6. Observations of soil development on these shorelines suggests an age older than OIS 6 (Mifflin and Wheat, 1979; Reheis, oral communication, 1999).

137.3	*221.1*	*Return to the main road, turn right (east) and head down the canyon.*
140.6	*226.4*	*Bear left at the Y. In 0.7 mile (1.1 km), veer left (N) at the intersection with the road from right (south). In 0.3 mile (0.5 km), cross Huntington Creek and bear left (north) at the intersection with the road from right (S).*
145.6	*234.4*	*Railroad Pass discharge channel enters Huntington Creek at the intersection with the main road. Turn left (north) (sign for Big Bald Mt. Mine points right). In 1.5 mile (2.4 km), turn right (northeast) at the intersection.*
148.8	*239.6*	*Harrison Pass (1:00) into Ruby Valley marks the south end of the high, glaciated Ruby Mt.. The OIS 2 high–stand of Lake Franklin in Ruby Valley is dated at about 16 ka (Lillquist, 1994).*
158.5	*255.2*	*The gravel road ends at pavement. Turn left (north) on NV 228 toward Jiggs and Elko. Wilden and Kistler (1969) mapped large terraces (glacial outwash?) to the north as "ancient gravels". Jiggs store has refreshments, bathrooms, and ice.*
167.0	*268.9*	*The large channel draining Ruby Mtns. is occupied by an underfit stream.*
168.3	*271.0*	*The glacial trough of Rattlesnake Canyon was initially mapped by Sharpe. Work here by Bud Burke (Humboldt State University) is ongoing. Ahead are more large terraces (glacial outwash?).*
172.3	*277.4*	*South Fork Humboldt River. Floodplain to the left. The outcrop along the river exposes old lake sediments.*
172.7	*278.0*	*Intersection with the road to Lee on the right (east). Stay straight (north). Tilted and faulted tephra occur in indurated Pliocene(?)–Miocene(?) lake sediments. The deposits are unstudied, as are most Tertiary and Quaternary deposits between Newark Valley and Elko.*
177.5	*285.8*	*The tribal land of the South Fork Indian Reservation extends 3 miles north from here.*
183.7	*295.8*	*Turn left (north) at Spring Creek, staying on NV 228 (Lamoille highway).*
185.7	*299.0*	*View into Elko basin from Lamoille Summit.*
187.8	*302.3*	*Intersection of NV 228 and Pinyon Rd. Tomorrow's first stop (22) is at Elko landfill on the right. Past a right curve and through Wilson Avenue light, NV 228 becomes South 5th St. Cross Humboldt River.*
189.4	*304.9*	*Turn left at first stop light on Silver St., then immediately turn right on 3rd St.*
189.6	*305.2*	*Stockmans Hotel is 1/2 block to the right. Turn right on Commercial St.* ***We will have a traditional Basque family–style dinner at the nearby Star Hotel.***

DAY 6. ELKO TO RYE PATCH DAM

0.0	*0.0*	*Return to 5th Street and turn right (south) toward Spring Creek, Lamoille, and Jiggs.*

Cross the railroad tracks and the Humboldt River.

1.1 *1.8* *Road bends left along the flank of poorly exposed late Tertiary deposits. Continue; road bends right (south).*

2.1 *3.4* *Turn east at the sign for Buggy Bath and City of Elko landfill sign on Pinyon Avenue. In 0.3 mile (0.5 km), enter Elko landfill and proceed to the southeast corner.*

STOP 22. ELKO LANDFILL AND PLIOCENE LAKEBEDS

This cut and others in the landfill expose fine–grained alluvial and lacustrine deposits (Fig. 7) that suggest Elko Basin was internally drained until at least the late Pliocene. Oligocene oil shale was mined during the early history of Elko, and Miocene tuffaceous alluvial and lacustrine deposits are widespread in the region. Previous maps identified possible younger deposits south of the river, but these were poorly exposed until excavation began here. Correlated ages of two tephra layers place these deposits between 2.5 and 2.2 Ma (A. Sarna–Wojcicki, 1999, written communication). Ostracode species (*Elkocythereis bramlettei* and *Limnocythere ceriotuberosa*, sample S1–EK, Fig. 7) are consistent with a late Pliocene age. Their abundances indicate alkaline, slightly–to–moderately–saline lake water, and their environmental tolerances suggest seasonal to annual dominance of subtropical air with occasional incursions of westerly, polar air (R. Forester, 2002, written communication).

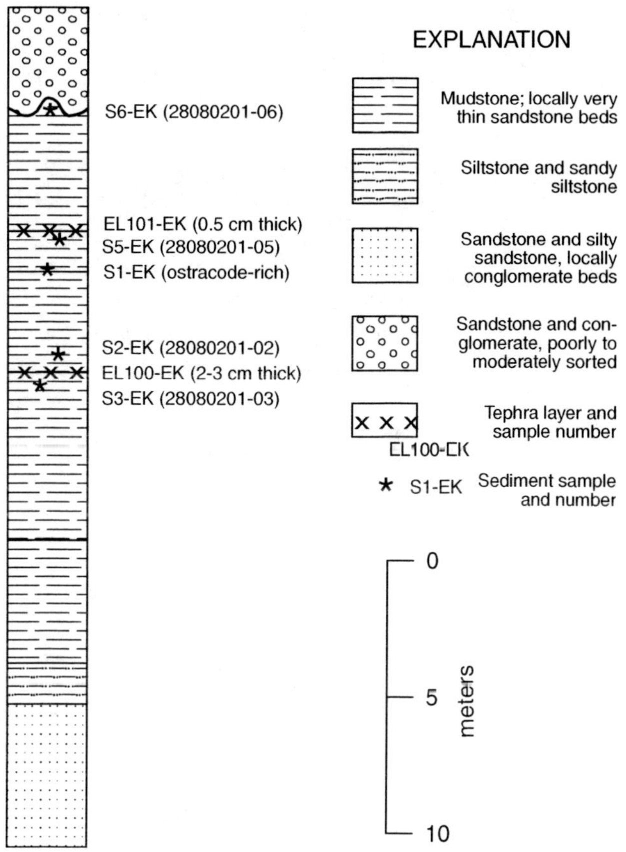

Figure 7. Stratigraphic sketch of section exposed in southeast corner of Elko Landfill.

4.8 *7.7* *Retrace route to downtown Elko. Turn left (west), following NV 228.*

5.5 *8.9* *Turn right (north) toward I–80. In 0.6 mile (0.9 km), cross overpass and turn left (west) onto I–80.*

14.3 *23.0* *Hunter exit. Due south is the canyon of the South Fork Humboldt River; beyond is the glaciated crest of the Ruby Range. The South Fork carries about half the drainage from Ruby Range into Humboldt River, as well as overflow from Lakes Jonathan, Diamond, and Newark(?), thus providing much of Humboldt River flow in peak pluvial periods. Pediments south of the river obscure outcrops of pale sediment, probably similar to Tertiary deposits at the Elko landfill.*

21.0 *33.8* *Cross Humboldt River and enter tunnel that bypasses Carlin Canyon, a choke point for the river between Elko and Carlin basins.*

27.6 *44.4* *West exit to Carlin on NV 278. The brown slopes to the south are covered by cheatgrass, an invasive annual from Central Asia, in post–fire takeover. To the west in a gorge (not visible), Pine Valley Creek flows into the Humboldt River. Pine Valley has been mapped (Mulhern, 1982; Gordon and Heller, 1993) because it contains one of two known oil fields in Nevada. However, little work has been done on the geomorphology or lake history. The basin was breached by overflow or captured by the Humboldt River after deposition of the Lava Creek ash bed (~650 ka) in Pine Valley lake.*

34.1 *54.9* *Beautiful 10–m–thick, silvery tephra exposed in roadcuts on the east side of Emigrant Pass was dated by Glen Izett (1999, written communication) at ~10 Ma*

36.1 *58.1* *Summit of Emigrant Pass, used by wagon trains forced out of the impassable Humboldt River gorge to the south. Note the absence of sagebrush due to range fires in the 1990's fueled by cheatgrass (dry brown grass everywhere).*

45.4 *73.1* *Beowawe exit. The highway descends westward to the confluence of the Humboldt River valley with Crescent Valley to the south. This area was likely the upper end of Humboldt River delta at the 1400–m high–stand.*

49.3 79.4 Cross the Humboldt River at the confluence with Boulder Valley to the north. The 1400–m lake may have covered most of the flat valley floors in this area, which likely have aggraded by fluvial deposition in the past ~600 kyr.

53.9 86.8 Eureka–Lander County line. The 1400–m contour crosses floor of Humboldt Valley.

68.9 110.9 Confluence of Reese River with Humboldt River (we crossed Reese River three days ago near Austin).

73.4 118.2 Town of Battle Mt. Nearly all buildings are younger than ~1960; the older buildings were destroyed by a flood. The town was submerged for days until entrepreneurs blew up the railroad tracks that were impounding the water.

84.4 135.9 North flank of Battle Mt. The strong color change on fans reflects fire scarring, not a geologic contact. The area is extensively mined for gold.

88.6 142.6 Valmy exit and roadside rest.

91.4 147.2 At 1:00–2:00 is Treaty Hill, with a power plant to the northwest. Treaty Hill has two summits. The ridge at 1402 m that connects them consists of beach gravel. A pebble lag extends to about 1414 m. Treaty Hill is bounded by the same two faults as Lone Tree Hill (next stop) and may have been uplifted relative to the former 1400–m lake level.

93.1 149.9 Exit I–80 at Stonehouse, Exit 212, and turn left (south), passing under freeway. Immediately turn right (west) on the access road to Lone Tree Mine. The hill ahead was a waypoint on the Overland Trail. The spring has long since dried up due to mine dewatering. In dry years, water entering the river from this dewatering makes up as much as 90% of flow of Humboldt River to west.

94.0 151.3 Turn left into parking lot of the Lone Tree mine and ***WATCH FOR LEFT–HAND TRAFFIC!!***

STOP 23. LONE TREE MINE

Here, we will tour an open–pit gold mine that exposes Pliocene-to-middle–Pleistocene lake and fan deposits in the upper 5–50 m of the mine walls (Figs. 15 and 16 in Reheis et al., 2002a). A cut northeast of the water tanks against the hill exposes well–rounded, wave–sorted gravel. Pre–mining topography shows a small projection from the hill at about 1350 m, the same as the altitude of the best–preserved pre–Sehoo berm at Thorne Bar.

94.9 152.8 Exit Lone Tree Mine and return to interstate. Go under the highway and enter I–80 west (or backtrack east to roadside rest for lunch).

102.2 164.5 Ascend through the 1400–m level. The exit at Pumpernickel Valley leads north to Emigrant Trail and tracks where wagon trains climbed over the flank of basalt flows forming Iron Point. Rounded pebbles scattered on basalt extend as high as ~1375 m and represent either former positions of the river as it slipped northward down the basalt slope, or pre–Sehoo Lahontan high–stands. The hills in which this canyon is cut may have been the divide between the upper Humboldt drainage and the middle Humboldt drainage (Winnemucca to Rye Patch area). Pre–0.76 Ma lake deposits at Lone Tree Mine may represent a lake confined to east of this former barrier.

107.0 172.3 Golconda Summit. Descend west toward the Humboldt River. At the base of the hill, enter a broad deltaic complex of the Humboldt River in late Pleistocene, just at and beneath the Sehoo high–stand level (1330 m).

112.7 181.4 Golconda–Midas exit. North of the freeway, the 1400–m high–stand would have lain where alluvial fans abut bedrock on the south flank of Osgood Mt..

117.4 189.0 Old gravel pit on the left (south) lies at the Sehoo shoreline.

126.2 203.2 East Winnemucca exit, on the Holocene floodplain of the Humboldt River, so close to the apex of the Sehoo river delta that shorelines are subtle. The view north is up Paradise Valley, a beautiful ranching area. The long, flat tableland projecting out to a small basalt butte would have been under the 1400–m high–stand, and the town submerged under ~80 m of water.

129.2 208.0 North of the freeway is a borrow pit in late Pleistocene beach gravel on the south flank of Winnemucca Mt.. The 1400–m high–stand would have lain at the bedrock–alluvium contact above the pit. From here westward, the highway skims along the Sehoo shoreline.

156.9 252.6 South of the road, just past the exit for Mill City and Unionville, is strongly dipping travertine of unknown age against a low bedrock hill. They defend a gap that opens into Buena Vista Valley. The valley was flooded by Lake Lahontan, but this and an adjacent gap were not under water during the Sehoo lacustral.

161.9 260.7 The area south of the road contains borrow pits excavated in a large point bar at the north end of the Humboldt Range. The valley to the north is the direction the Humboldt River took when it flowed into Smoke Creek Desert.

166.9	268.7	*Passing spoil piles of Florida Canyon Mine on the left (east). Large bare flats northwest of the highway are spillover reservoirs for water storage during wet periods.*
176.3	283.8	*The highway now follows the west side of the Humboldt Range. Northwest of the river are several small ranges that define the eastern side of Granite Springs and Kumiva basins, which occupied a large island in the middle of Lake Lahontan and were not flooded during Sehoo time. Alluvial fans coalescing between the Humboldt Range and these low ranges may form a barrier that at times turned the Humboldt River westward from Winnemucca to empty into Smoke Creek and Black Rock Deserts, thus tributary to Pyramid Lake basin rather than to Humboldt Sink. Exit right (west) at Rye Patch Dam.*
177.1	285.1	*The road drops westward down slopes incised into (sequentially), Sehoo Alloformation (AF), Churchill soil and Wyemaha AF, gray mud of Eetza AF, alluvial and eolian deposits of Paiute AF, and finally tufa–cemented, deltaic foreset beds of Rye Patch Dam AF (Fig. 4). On the east side of the bridge was an outcrop of Lava Creek tephra (Morrison and Davis, 1984). Downstream are sections exposing, in sequence downward, Rockland ash in Paiute AF and Lava Creek and Rye Patch Dam ash beds in Rye Patch Dam AF.*
177.5	285.8	*Entering Rye Patch Dam State Park at the dam. Across the dam, turn right (north) to the Group Camp. Optional activities: swimming, snoozing, hiking, and excursion south to optional stop 24.*

STOP 24. PRE–EETZA LAKE DEPOSITS, TEPHRA AND PALEOMAGNETIC STUDIES

DAY 7. RYE PATCH DAM TO RENO

0.0	0.0	*Reset mileage at Rye Patch Dam; return to I–80, turn right, and head west.*
15.1	24.3	*To the south is the West Humboldt Range. To the northwest, outcrops mark the south end of Humboldt canyon.*
24.6	39.6	*Exit to Lovelock and railroad bridges. Wave–cut terraces are visible on Lone Mt. to the northwest, and also on the west flank of the West Humboldt Range to the south.*
32.6	52.5	*Wave–cut surface is north of highway. Granite outcrops above this surface are coated with thick tufa.*
42.6	68.6	*Humboldt Sink, the modern terminus of Humboldt River, lies to the south.*
44.5	71.6	*Crest of Humboldt Bar. This bar has a drainage cut (not visible here) through it due to overflow between Carson Sink and Humboldt Sink. In 2 miles (3.3 km), pass Fallon exit.*
50.6	81.5	*Exit at Jessup. Turn right at the stop sign at the end of the ramp and proceed west on the dirt road. North of Jessup is an embayment in Trinity Range. Beyond that opening, Granite and Kumiva basins were not flooded by Lake Lahontan in Sehoo time. However, the 1400–m high–stand would have flooded these basins via Jessup embayment.*
51.5	82.9	*Pass through an active gravel mining operation and 1235–m barrier bar complex. Proceed forward.*
52.4	84.4	*Pass straight through the road intersection. In about 0.6 mile (1 km), fork right (north) onto the road in the wash. In 0.9 mile (1.4 km), park along the road.*

STOP 25. JESSUP EMBAYMENT

The embayment has a well–preserved, three–dimensional record of lake sediments deposited by Sehoo and older lakes. The Sehoo high–stand was dated here at about 13,000 ^{14}C yrs. B.P. using camel bones preserved in a lagoon enclosed by a small high–stand barrier bar (Adams and Wesnousky, 1998). Other radiocarbon ages on gastropods from deposits near the high–stand altitude suggest a rapid rise to the high–stand. We will discuss shoreline processes, the resultant stratigraphic record, and isostatic rebound of the Lahontan Basin (Adams et al., 1999).

56.0	90.2	*Return to I–80 and turn right (west) onto freeway.*
60.0	96.6	*Carson Sink is in view to the southeast. The Stillwater Range is visible in distance.*
65.8	105.9	*Passing through area of diatomite quarries in Tertiary rocks, thinly blanketed by Lahontan sediments.*
69.6	112.1	*Nightingale Hot Springs is now co–opted by a geothermal plant.*
76.0	122.4	*Shorelines incised in basalt. Note a lower bank of shorelines, thickly blanketed with tufa, which is the so–called "thinolite terrace," in contrast to the upper slopes with little tufa. The terrace is interpreted to reflect a lengthy high–stand of Lake Lahontan slightly below the Sehoo maximum level.*
80.5	129.6	*To the northwest are low–lying outcrops of Sehoo sediment and shorelines cut on basalt flows forming the south flank of the Truckee*

		Range. In contrast, the Hot Springs Mt. southeast of the highway are thickly blanketed with dune sand, obscuring the shorelines.
83.9	*135.1*	*On the north side of road, basalt flows are thickly covered with sheets of tufa.*
88.0	*141.7*	*West exit to Fernley and Wadsworth. Exit north on NV 454 toward Wadsworth.*
89.0	*143.3*	*Just after the big blue mill, turn right (north) and follow the east bank of the Truckee River. Note the barrow pits excavated in Sehoo deltaic gravel. The road drops onto a Holocene river terrace.*
93.6	*150.7*	*Wadsworth Amphitheater. Two pre–fab houses lie in the southern part of the amphitheater in front of the Paiute AF outcrop. The curve just northwest of the houses has a wide shoulder for parking.*

STOP 26. WADSWORTH AMPHITHEATER

A classic site, important because deposits contain the so–called "Wadsworth tephra," one of very few age controls on Eetza AF. This tephra is currently thought to be about 200 ± 50 ka (Berger, 1991; A. Sarna–Wojcicki, 1997, written communication). Pre–Sehoo lake sediments here cannot be tied directly to the shoreline altitudes and ages of Reheis (1999). Morrison (1965, 1991) established a broad stratigraphic interpretation (Fig. 4), recently reinterpreted by Smoot (1993) using detailed sedimentology (Fig. 8). John Bell and Kyle House (Nevada Bureau of Mines and Geology) are currently mapping the Wadsworth quadrangle and have new observations bearing on these spectacular exposures.

99	*160*	*Turn around and retrace the route to NV 454. Turn right (W) toward town of Wadsworth.*
100	*161*	*At the main intersection in Wadsworth, turn right (north) on NV 447 toward Pyramid Lake. The remainder of the trip will be a scenic drive down Truckee River to Pyramid Lake, returning to Reno over Mullen Pass. We will stop at one of several vantage points for an overview of Pyramid Lake, a jewel of the Great Basin, and a swim if we have enough time. From Wadsworth on, the mileages are approximate.*
116	*187*	*Leaving Wadsworth, the road ascends a broad terrace (Dodge Flat) underlain by deltaic deposits of the Truckee River and deep–lake deposits. Passing through Pyramid Lake Paiute Reservation, is one of the few tribes of North America never displaced from their home territory. The tribe controls access to the lake and surroundings. In about 8 miles (13 km), the highway descends onto Holocene terraces along the Truckee downstream from its deep trench. Just south of Nixon, turn left (west) on NV 446.*
129	*208*	*In about 5 miles (8 km), the highway closely follows the southwestern shore of Pyramid Lake. We will stop somewhere along here for views and swimming. The large island to the north is Anajo Island, a bird sanctuary, and farther north just visible between Anajo Island and the eastern shore is The Pyramid. Both islands are covered with tufa mounds and pinnacles. About 11 miles (18 km) from the last intersection, another cluster of pinnacles, including Indian Head Rock, is east of the road. Two miles (3 km) past the pinnacles, turn left (south) on NV 445 toward Reno.*
157	*253*	*The road ascends Mullen Pass into Warm Springs Valley, mostly submerged by the Sehoo high–stand. Across the valley, the road again ascends past the Sehoo shoreline and up to the low saddle leading into the Spanish Springs valley. This saddle is only a few meters higher than the highest–known Lahontan shoreline at 1400 m. Most of this valley, as well as the Reno metro area, lie below 1400 m and was likely flooded by the highest lake. Return to Reno through rapidly growing suburbs. At McCarran Boulevard intersection, either continue south to I–80 or turn right (west) to reach US 395.*

COLUMBUS SALT MARSH AND LAKE COLUMBUS–RENNIE (Reheis, Kurth, Phillips, Paces, and McGeehin)

Reheis et al. (1993) showed that a large lake, informally named Lake Rennie, existed in Fish Lake Valley at intervals during the Pliocene to middle Pleistocene. They speculated that Lake Rennie and Lake Columbus at times were contiguous through The Gap, the current drainage between Fish Lake Valley and Columbus Salt Marsh (Fig. 2). Deposits in Columbus Salt Marsh indicate at least four lacustrine units higher and older than the late Pleistocene shoreline (altitude 1400 m) of Lake Columbus–Rennie. Reheis et al. (2002a, Figs. 21 and 22) showed that the two oldest lakes rose to at least 1420 m and one of these to 1457 m, an intermediate–age lake to about 1435 m, and a younger lake to 1417 m, followed by the late Pleistocene lake to 1400 m. One of the oldest two lakes probably occurred about 760 ka, based on altitudes of the 1457–m shoreline and the 1460–m sinter mound (STOP 10) and other evidence in Fish Lake Valley. The ages of the other lake units are unknown. New studies have provided ages using cosmogenic ^{36}Cl profile modeling, ^{14}C, and TIMS U–series dating on lacustrine tufa.

The strategy for obtaining cosmogenic profile ages follows the methods of Ayarbe et al. (1998) by measurement of detailed depth profiles in trenches excavated to 2.5–3 m. The shape of

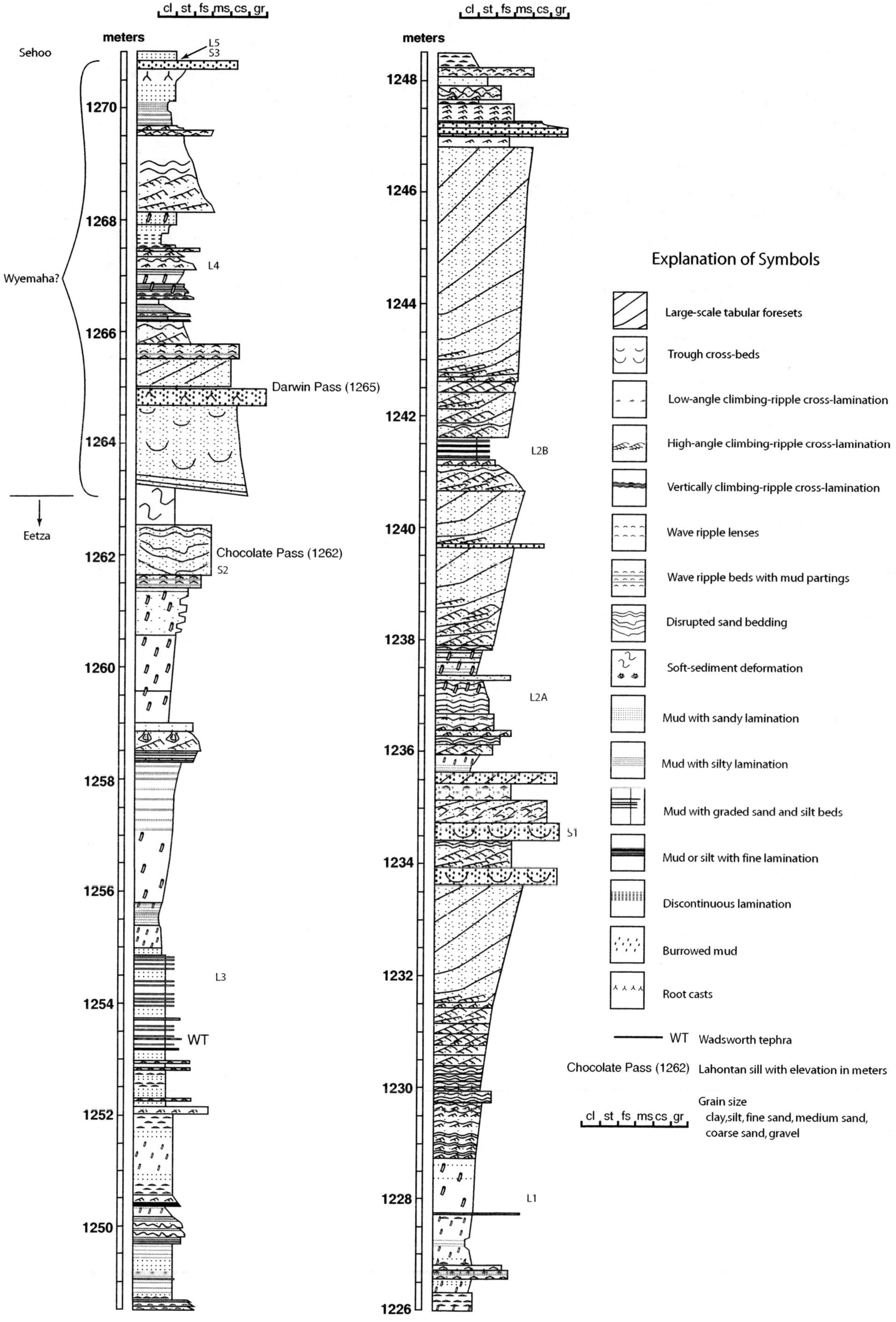

Figure 8. Stratigraphic section exposed at Wadsworth Amphitheater (revised from Smoot, 1993). WT indicates position of Wadsworth tephra. Paiute AF not shown; present only on south side of outcrop. L, lake unit; S, subaerial unit.

such profiles is strongly sensitive to erosion rate and, because the cosmogenic production rate is low below about 2 m depth, the background inherited from exposure prior to deposition can be separated from the production since deposition. Shielding of the old beach surfaces by younger deposition is possible, but is restricted to deposits of desert dust (loess) or transitory coppice dunes. The maximum amount of inflation of surfaces by incorporation of dust seems to be about 50 cm, estimated by comparison of older and younger soil profiles. Thus, shielding of the shoreline surfaces is a minor effect that can be accounted for in the data interpretation model.

The sampling areas in Columbus Salt Marsh are in gently sloping valley margins. The slopes at trenches CM01–1 (late Pleistocene; numbers increase with shoreline age and altitude) and CM01–3 are less than one degree and erosion has been very slight (in fact, some of the berms have nearly been buried by alluvial fans). Erosion has likely been greater on the steeper slopes at trench CM01–2. About 10 samples were collected from each pit for ^{36}Cl analysis, and soil profiles were also described and sampled. The ^{36}Cl samples were screened to separate the ~2 to ~0.5 cm fraction, then >150 individual clasts of various lithologies were combined and ground together to average the differences in clast exposure histories and compositions. Certain samples were also selected for ^{10}Be analysis because (1) it can give more accurate ages for relatively old (>5 x 10^5 yr) samples, and (2) used in combination with ^{36}Cl, it can greatly improve the ability to distinguish the effects of exposure age from those of erosion (Liu et al., 1994; Dep et al., 1994).

^{36}Cl model ages in Columbus Salt Marsh are in preparation and will be discussed on the field trip. Selected soil–profile data are shown in Figure 9. Values of the profile–development index (Harden and Taylor, 1983) increase from 14.4 to 26.1 to 30.1 with shoreline altitude. Although soil properties such as solum thickness and content of pedogenic silt and clay are not significantly better developed on the highest and presumably oldest shoreline (CM01–3) compared to the next younger profile (CM01–2), the older deposit is only 2.1 m thick atop Tertiary mudstone and is probably much more eroded than the others. In addition, the CM01–3 profile is greatly disrupted by salt crystals compared to the younger soils (fig. 9).

The lake unit corresponding to CM01–1, the youngest and lowest shoreline (1400 m), has an uncorrected ^{14}C age of 14,860 ± 50 ^{14}C yrs. B.P. from a tufa head in the Big Pit (STOP 13A). In trench CM01–2 (STOP 13B), a thin, dirty, tufa layer at a depth of 1.4 m has a ^{14}C age of 38,580 ± 410 ^{14}C yrs. B.P.—in our opinion, a minimum age. Tufa–cemented beach rock from a prospect pit on a shoreline at the same altitude (~1417 m) to the northwest yielded TIMS $^{230}Th/U$ ages of 144 ± 2 ka and 156 ± 3 ka for detritus–free, banded carbonate rinds from a single hand specimen. Tufa–cemented beach rock was also sampled from the cut below the power pole at Coaldale Junction, associated with a shoreline whose deposits extend as high as about 1432 m. Detritus–free carbonate clast rinds yielded U–series isotope ratios with large excesses of ^{230}Th ($^{230}Th/^{238}U$ activity ratios of 5 to 10) indicating significant post–depositional U losses. Measured $^{234}U/^{238}U$ activity ratios of about 1.5 in these samples preclude calculation of $^{234}U/^{238}U$ model ages because the initial $^{234}U/^{238}U$ activity ratios obtained from the two analyses with ~150 ka ages, considered fairly reliable, were much lower—about 1.26.

THORNE BAR AND THE WALKER LAKE SUB–BASIN OF LAKE LAHONTAN (Reheis, Kurth, Phillips, and Paces)

The Thorne bar (STOP 14, Fig. 2; Fig. 10 in Adams and Wesnousky, 1999) is a large V–shaped barrier bar complex of shore gravel built at the mouth of a large canyon. This area is described in detail in Reheis et al. (2002a, Figs. 2–4). The bar consists of four morphologic units. The lowest and youngest is the group of historic shorelines below 1250 m in altitude, formed during the last 100 years. The next higher group of shorelines consists of sharp, well–preserved barrier bars marked at the top by the Sehoo high–stand shoreline at about 1330 m. Between 1330 m and about 1370 m, the bar complex consists of two nested, eroded, V–shaped barrier bars. The lower (unit $Ql6_T$) is better preserved than the upper (unit $Ql5_T$). Between 1370 m and 1402 m, the complex has no preserved morphology, but deep arroyos expose cemented shore gravel of unit $Ql4_T$. An outcrop north of Thorne bar exposes four sequences of lacustrine gravel and sand, separated by paleosols and alluvium, well above the Sehoo shoreline. At this site, lake unit $Ql3_T$ contains a tephra that is chemically correlative with the Bishop–Glass Mt. family of tephra erupted from Long Valley, California. $Ql4_T$, correlative with the highest shoreline at Thorne bar, overlies unit $Ql3_T$.

Since Reheis et al. (2002a), new studies have provided ages using cosmogenic ^{36}Cl profile modeling and TIMS U–series dating on lacustrine tufa. These new ages produced one big surprise that potentially has significant implications, both for the interpretation of "Sehoo" vs. "Eetza" in the Lahontan basin and for the rapidity and magnitude of lake response to climate change. Selected soil–profile data (Fig. 10, table 1) from the trenches excavated for cosmogenic sampling indicate that the soils accumulate silt with time due to incorporation of eolian dust and that the overall soil development increases with age as reflected in the horizon index values. Solum thickness does not significantly change with time, however, probably due to erosion on the steep–sided, beach barrier bars.

A small loop bar at 1333 m on the north side of the main barrier bar had previously been interpreted to have formed at the Sehoo high–stand, based on its altitude and soil morphology in a small hand–dug pit (Adams and Wesnousky, 1999). A deeper backhoe pit (TB00–1) revealed a minor unconformity at about 1.1 m with reddish clay films (not clearly pedogenic in origin) on clasts below the unconformity. However, the cosmogenic age profile portrays a smooth curve that is best modeled with an age of ~72 ka (Table 1; Kurth et al., 2002). One interpretation of these data is that the small bar was formed during OIS stage 4 and was minimally reworked, if at all, during stage 2 (Sehoo), "resetting" the soil properties but not the cosmogenic age profile. Pit

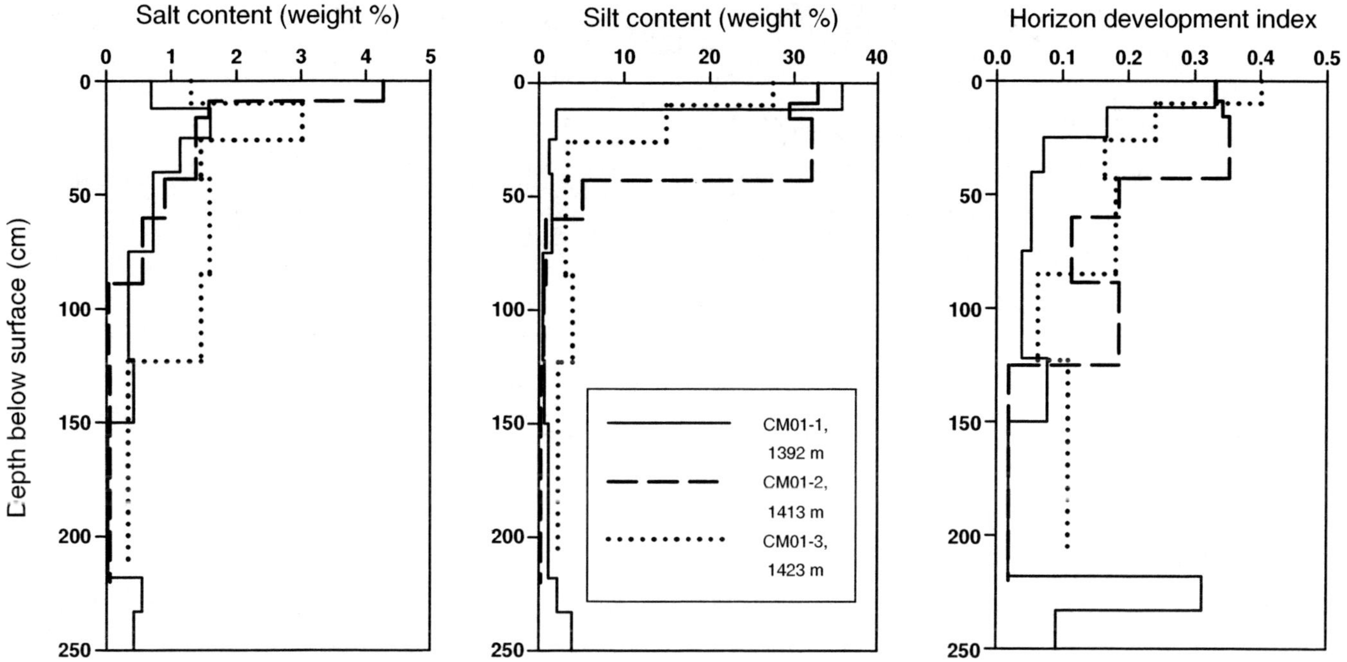

Figure 9. Depth profiles of trenches on shorelines of different age in Columbus Salt Marsh. Horizon development index (Harden and Taylor, 1983) calculated for 6 soil properties—paling, lightening, texture, structure, dry consistence, and clay films—using template of Taylor (1988).

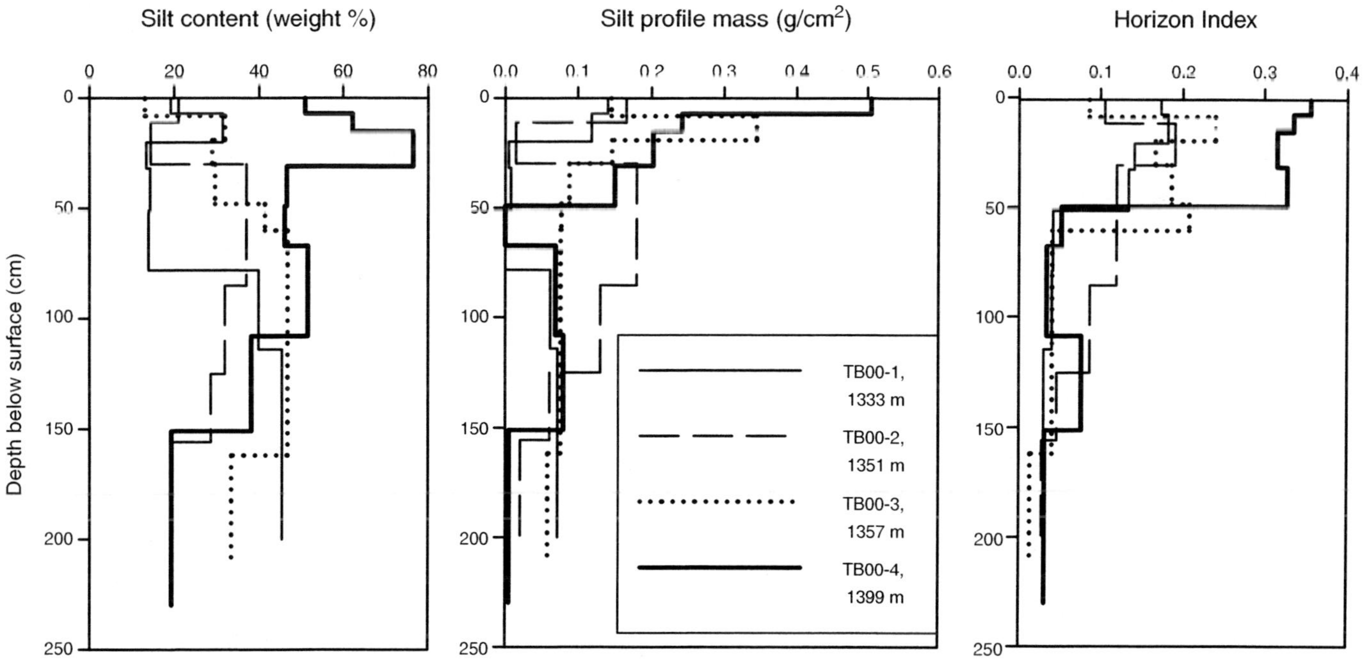

Figure 10. Depth profiles of trenches on shorelines of different age on Thorne Bar. Horizon development index (Harden and Taylor, 1983) calculated for 6 soil properties—paling, lightening, texture, structure, dry consistence, and clay films—using template of Taylor (1988).

TABLE 1. MAP UNITS, NUMERICAL AGES, AND PROFILE DEVELOPMENT INDEX VALUES OF SAMPLING TRENCHES ON THORNE BAR

Map unit	Shoreline alt (m)	Trench number	Trench altitude (m)	^{36}Cl model age (ka)*	Other age (ka)†	Index value
$Ql7_T$	~1250–1330	TB02-00	1325	~17	N.D.	N.D.
$Ql7_T$	~1250–1330	TB00-01	1333	~72 (65–80)	N.D.	13.1
$Ql6_T$	~1335–1352	TB00-02	1351	~165 (130–200)	128 ± 6	17.6
$Ql5_T$	~1352–1370	TB00-03	1357	~285 (230–450)	167 ± 17 210 ± 37	15.8
$Ql4_T$	~1370–1402	TB00-04	1399	400–900	<760§	23.3

N.D.—no data.
* Data of Kurth et al. (2002); age range estimates based on assumed differences in erosion rate.
† U-series data from J. Paces, U.S. Geological Survey, unless annotated.
§ Contains reworked disseminated shards of Bishop ash, identified by glass chemistry (A. M. Sarna-Wojcicki, written commun., 2001) and normal polarity.

TB02–00, dug by hand at 1325 m on the western extension of the main barrier bar, showed a less–developed soil profile (Av/Ck compared to Av/Bw/Bwk/Ck of TB00–1) and yielded a model age of ~17 ka (table 1), clearly a Sehoo age.

Trenches on the higher, older barrier bars at and east of the road yielded ^{36}Cl model ages that suggest correlations of shorelines to OIS stages 6, 8 or 10, and possibly 16 (Table 1). Model ages of older shorelines are not well constrained due to cumulative errors in estimation of erosion rates, but they are comparable to U–series ages on tufa samples collected nearby at similar altitudes (Table 1). We hope to refine the ages of the older two barrier bars using ^{10}Be data.

PALEOCLIMATE IMPLICATIONS OF THE MIDDLE TO LATE QUATERNARY PLUVIAL HISTORY OF NEWARK VALLEY, EAST–CENTRAL NEVADA (Redwine)

Introduction

Newark Valley, in east–central Nevada (Figs. 1 and 3), has at least two, possibly three, suites of shoreline deposits that decrease in altitude with decreasing age (Hubbs et al., 1974; Mifflin and Wheat, 1979; Reheis, 1999; Redwine, submitted). This lake–level record is similar to that of Lake Lahontan to the west (Reheis, 1999), and different from that of Lake Bonneville to the east (Scott et al., 1983; Jarrett and Malde, 1987; Benson et al., 1990; Kaufman et al., 2001). Lake Newark shorelines are assigned to age groups based upon geomorphologic preservation of shoreline features, topographic profiles of beach barriers, soil profile development, laboratory analyses of soils, and index calculations of 26 soil profiles. Investigations were concentrated at six sites within Newark Valley (Fig. 11). Age criteria, in particular soil development, will be viewed and discussed at STOP 20 (site N).

Four soil indices, including the Profile Development Index (PDI; Harden and Taylor, 1983), the secondary $CaCO_3$ index (Cs) (Machette, 1985), the clay accumulation index (Birkeland, 1999), and the mass accumulation of clay index (Birkeland, 1999), were calculated for each soil profile (Table 2, Fig. 12). PDI calculations included 8 soil properties: paling, lightening, texture, clay films (except in Av horizons), dry and moist consistence, carbonate morphology, and structure (cementation was substituted for structure in Bkm horizons (Redwine, in preparation). The four indices helped (1) determine how many suites of shoreline deposits are discernible from one another, (2) correlate soil development with similar soil development on deposits with numerical age control, and (3) establish relative ages of lacustrine features throughout Newark Valley. Changes in parent material with depth and multiple buried soils complicate the use of these indices, but index values do appear to be consistent among shorelines of similar altitude and geomorphic preservation around the basin (Table 2).

Relative age data are supported by AMS ^{14}C ages of snail shells and tufa, and by amino acid racemization (AAR) age estimates on snails from the lacustrine deposits using methods of Kaufman and Manley (1998). The ages of these suites of shorelines are correlated to the marine oxygen isotope (OIS) curve of Bassinot et al. (1994). The correlations are based upon numerical age control where available, and elsewhere upon relative soil index values. Synchroneity is not implied with these correlations to the OIS terminology, and others have shown that pluvial and glacial maxima are somewhat out of phase (Benson et al., 1990; Adams and Wesnousky, 1998). These proposed correlations to the OIS chronology are utilized to provide an age–assignment framework and to prevent unnecessary introduction of provisional nomenclature. Stratigraphic interpretations with numerical ages of lacustrine samples permit construction of a lake level curve for part of the late Pleistocene history of Newark Valley. Stratigraphic context of some of these dated samples and the implications of ages for lake level history will be discussed at STOP 19 (figs. 3 and 11, site LMt).

Hydrologic index values (Mifflin and Wheat, 1979) were calculated to obtain estimates of the magnitude of climate change

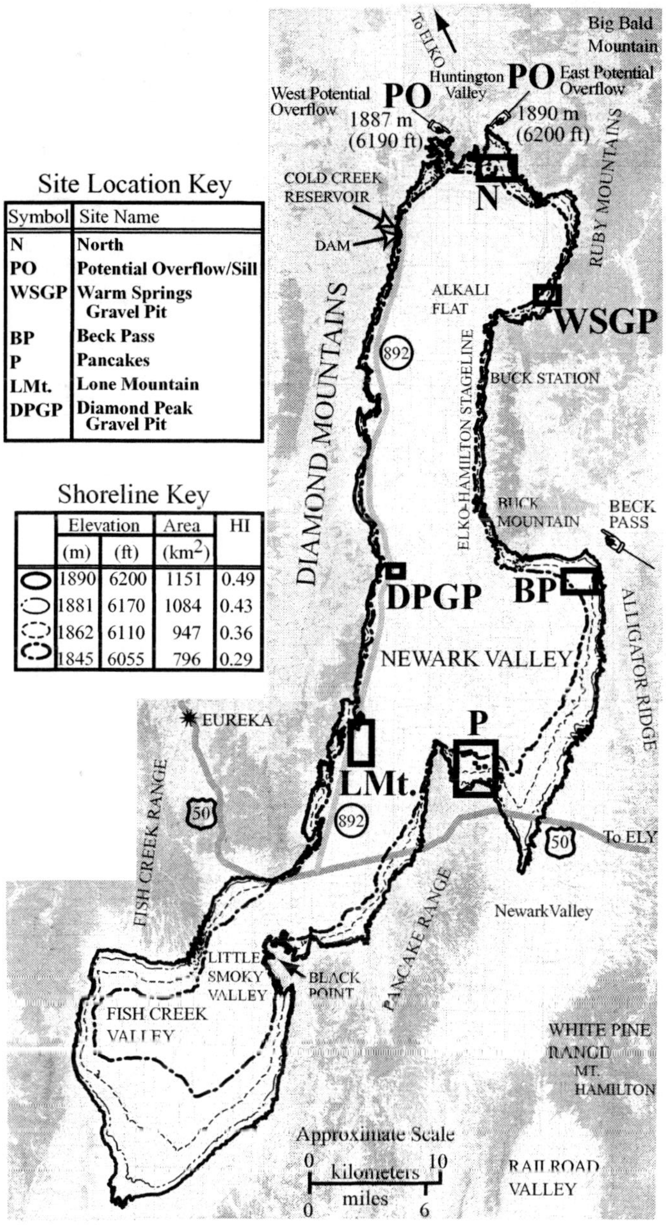

Figure 11. Geographic setting of Newark Valley, showing study sites and shorelines with associated areas and elevations.

from the present using a comparison of two dimensionless ratios. One ratio, HI, is calculated using direct measurements of observed features (area of the pluvial lake measured from shorelines of known altitudes (Fig. 11) ÷ area of tributary basin) (Mifflin and Wheat, 1979). The second ratio, Z, defines a relation among climate parameters that have been estimated using historical records, in particular mean annual temperature, mean annual precipitation, evaporation, and runoff amounts (Z = HI = runoff from tributary basin ÷ (lake evaporation – lake precipitation)). Mifflin and Wheat (1979) and Redwine (in preparation) discuss the assumptions, climate variable relations, and potential shortfalls of this methodology. This paper focuses on estimates of the magnitude of change in temperature and precipitation associated with observed lake level fluctuations in Newark Valley.

Characterization and interpretation of shoreline deposits

The oldest and highest probable shoreline is at or near the altitude of the sill threshold level at the north end of Newark Valley (Fig. 11, site PO; 1890 m, 6200 ft) and is identified primarily by the presence of rounded and flattened gravels of exotic lithologies relative to the local bedrock. Although such gravel clasts are not definitive indicators of a shoreline, similar gravel lags in the Lahontan Basin have been demonstrated to be of lacustrine origin (Reheis, 1999; 2002a). Features resembling wavecut benches also lie at this altitude around the valley, and one site at the south end of the valley preserves a remnant of lacustrine deposits at 1881 m. These features are provisionally assumed to represent the highest Lake Newark shoreline.

The sill threshold is inset within alluvial fans that have been mapped, based on geomorphic relations, as coeval to or older than the older suite of shorelines (Redwine, in preparation). A soil profile developed on a fine–grained alluvial deposit in the sill threshold has soil index values (Fig. 12) similar to those of the older suite of shorelines, described below, which are correlated with a marine oxygen–isotope stage (OIS) 6 age. This fine–grained deposit most likely postdates the postulated overflow. Thus, the time of overflow likely predates OIS 6. Similar remnant shoreline deposits in the Lahontan Basin and in Kobeh Valley (Lake Jonathan, Figs. 2 and 3) have been correlated with OIS 16, based on stratigraphic relations with dated tephra layers (Reheis, 1999; 2002a) and an assumption that the relatively large climatic shifts demonstrated by extreme sea level changes in the marine oxygen isotope curve (Bassinot et al., 1994) may be responsible for the largest of the observed pluvial lakes. The age of the likely overflow and highest shoreline of pluvial Lake Newark is clearly very poorly constrained, but is very tentatively correlated with a possible OIS 16 age. The altitude of this potential overflow yields a lake area of 1151 km^2 and an HI value of 0.47, compared with an HI of 0.29 for the late Pleistocene level.

The next–lower suite of shorelines occupies altitudes from 1849 m (6065 ft) to 1862 m (6110 ft) (Fig. 11). These shorelines are distinguishable from lower, younger shorelines by discontinuous preservation, topographic profiles showing comparatively broader barrier crests with gentler slopes (Fig. 13), and fewer round and flat beach clasts on the surface. The shorelines are characterized by Av/Bwk1/Bwk2/2Bkm or 2Km/2Coxk soil profiles, the uppermost 40–60 cm of which are developed in fine eolian sand, silt, and clay that overlie beach gravel with a maximum $CaCO_3$ morphology from stage III+ to IV (Machette, 1985). Soil indices (Table 2, Fig. 12) and most soil properties, such as depth to the C horizon and maximum sand, silt, clay, and $CaCO_3$ values, were larger for soil profiles on shorelines at these altitudes compared to those at lower altitudes. This suite of shorelines is considered to represent an OIS 6 pluvial lake by comparison of geomorphic

TABLE 2. RANGE OF SOIL INDEX VALUES FOR SHORELINES OF DIFFERENT ALTITUDES AND AGES

OIS*	No. of profiles	Altitude range (m)	Altitude range (ft)	Profile Devel. Index typical	Profile Devel. Index range	Cs (g)[†] Summed	Cs (g)[†] Normalized	Clay (g) a[§]	Clay (g) b[§]
2	11	1820–1847	5970–6060	0.07–0.16	0.01–0.16	4–19	6–15	3–16	167–618
6	15	1848–1866	6065–6123	0.20–0.34	0.26–0.34	14–43	20–33	7–48	288–2368
16	1	1881	6170	0.23	ND	16	33	34	1923

N.D.—no data.
* Correlation to marine oxygen isotope stage (OIS) using numerical ages and soil data of this study.
[†] Secondary $CaCO_3$ index of Machette (1985).
[§] Indices of Birkeland (1999): a—clay accumulation index, b—mass accumulation of clay index.

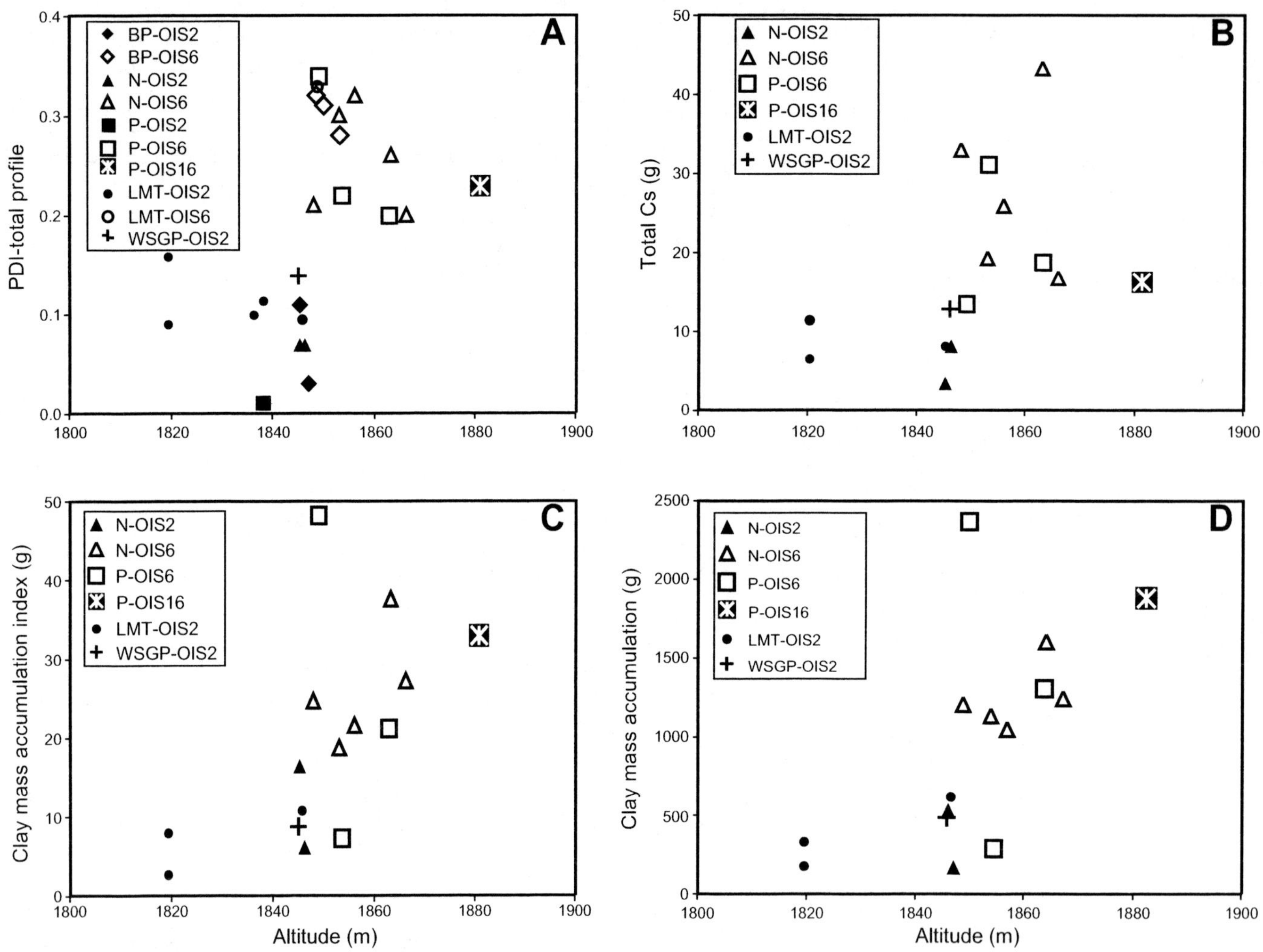

Figure 12. Soil index values for soil profiles developed in shorelines of different altitudes in Newark Valley. In legends, first letters indicate site locations in figure 11; OIS# indicates assignment to marine oxygen-isotope stage based on all index data and shoreline morphology. A, Profile Development Index (Harden and Taylor, 1983); B, Secondary CaCO3 index (Machette, 1985); C, Clay mass accumulation; D, Clay mass accumulation index (Birkeland, 1999).

preservation and soil properties to those of dated deposits elsewhere in the Great Basin (Harden and Taylor, 1983; Reheis and Sawyer, 1997). The maximum lake areas associated with these shorelines range from 947 km^2 (for 1862 m) to 1084 km^2 (for 1881 m) (Fig. 11), yielding HI values of 0.36 to 0.43 respectively.

The most prominent suite of shorelines reach a maximum altitude of 1845 m (6055 ft) and are distinguishable from higher shorelines by their nearly continuous geomorphic preservation around the basin and by comparatively narrow barrier bar crests with steeper slopes (Fig. 13B). These deposits are characterized by Av/Bwk1/Bwk2/2Coxk soil profiles. The uppermost 30–50 cm is developed in eolian fine sand, silt, and clay that overlie beach gravel with a maximum $CaCO_3$ morphology of stage I to I+. Based on soil indices and differences in soil properties from those of shorelines at higher altitudes (Table 2, Fig. 12) compared to those on dated deposits (Harden and Taylor, 1983; Reheis and Sawyer, 1997), these shorelines are considered to represent an OIS 2, Sehoo–equivalent, pluvial lake. The maximum altitude of this suite of shorelines (Fig. 11) yields a lake area of 796 km^2 and a HI value of 0.29.

Numerical Ages of Lacustrine Deposits

The correlation of the youngest shorelines with OIS 2 is supported by AMS ^{14}C ages and AAR analyses of snail shells from a high–stand barrier bar at an altitude of ~1845 m (~6055 ft) in the northeastern part of the basin (Fig. 11, site WSGP; Fig. 14A). Two genera of snails, *Lymnaea* and *Gyraulus*, were found in lacustrine sediment exposed in a gravel pit near Warm Springs. AMS ^{14}C analysis of *Lymnaea* yields an age of 13,780 ± 50 ^{14}C yrs. B.P. (16,540 ± 350 cal yrs. B.P., table 3). The D/L ratios of aspartic and glutamic acid from *Lymnaea* and *Gyraulus* (Table 4) suggest an age of 25 cal ka when correlated with AAR analyses on samples from Bear Lake Valley, Utah (Kaufman, personal communication, 2001). The analyzed shells have an intershell variability of >5 %, suggesting the AAR age estimate has an error of ±10 ka (Kaufman, personal communication, 2001). This is a large uncertainty, but the two independent techniques yield similar results, suggesting the samples are not likely contaminated or likely to be in gross error. The AMS ^{14}C age of the snail fossils is a maximum for the high–stand, because the snails were in a beach gravel deposit that predates the last high–stand by an unknown amount of time, represented by ~1 m of overlying beach and lagoonal deposits. Because these deposits are thin and unconformities or buried soils are not apparent, not much time is inferred to have passed between deposition of the snail–bearing sediments and the last beach deposit left by the lake at this location. Therefore, this numerical age most likely closely dates the high–stand of pluvial Lake Newark in latest Pleistocene time.

Sediments exposed in three gravel pits and one backhoe trench near Lone Mt. on the west side of Newark Valley (Fig. 11, site LMt; Fig. 14B) also provide numerical age control for lacustrine deposits. We will visit gravel pit LMt–GP–A (STOP 19), excavated into a barrier bar at 1845 m (6055 ft). Based on both soil development and shoreline altitude, this barrier bar is correlative with the latest Pleistocene high–stand of pluvial Lake Newark. Within this exposure are three coarse–gravel units. The uppermost two units are separated by a sub–horizontal unconformity underlain by fine sand and silt (Fig. 14C). Due to the similarity in freshness of clasts in these two gravel units, they were initially assumed to both be late Pleistocene. However, ^{14}C and AAR analyses of snails within the fine–grained unit show that this unconformity represents a significant time hiatus.

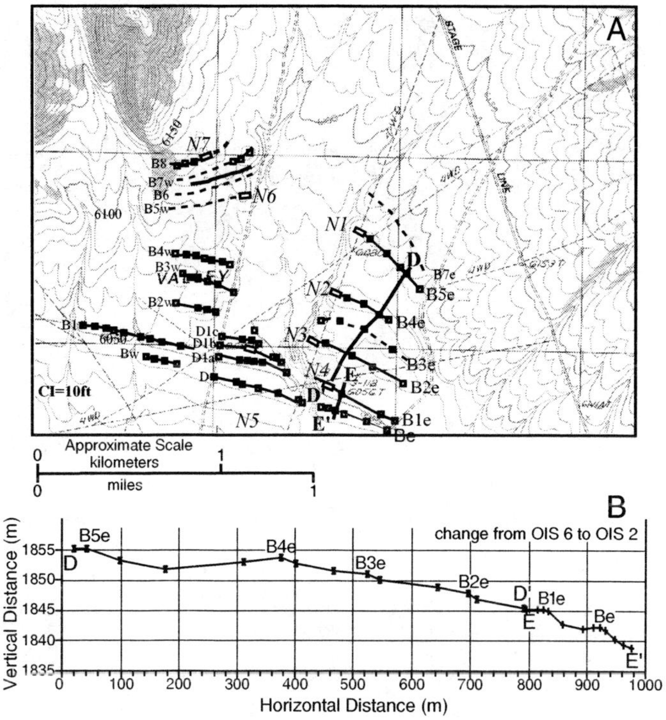

Figure 13. A, Location map of topographic profiles at Stop 20. B, Topographic profiles of older and younger shorelines.

Molluscs from the fine–grained deposit below the unconformity (Fig. 14C) were analyzed by AAR to derive approximate ages (Table 4). Four samples, each composed of five individual shells of three different genera, *Lymnaea* [Stagnicola], *Pisidium*, and *Gyraulus*, from two locations, were analyzed for aspartic and glutamic acid racemization (Table 4). The *Lymnaea* shells showed the most consistent results and have been used previously in the Great Basin for age estimates (e.g., McCoy, 1987; Laabs and Kaufman, 2003). The results indicate these shells are early middle Pleistocene (190–130 ka) in age, and roughly of OIS 6 age, because the D/L ratios are similar to those from the type locality of the Little Valley Alloformation of Lake Bonneville in Utah (Oviatt et al., 1999; Scott et al., 1983). Older or younger ages are possible depending on the temperature history of Newark Valley (D. Kaufman, 2001, written communication). *Lymnaea* shells from the same sample sites, dated by Jack McGeehin (U.S. Geol. Survey), yielded ages of about 38,000–39,000 ^{14}C yrs. B.P. (Table 3). These ages are near the limit of radiocarbon analysis capability and conflict with the older age estimated by AAR techniques. The AMS ^{14}C ages may be anomalously young; however, the possibility of accurate ^{14}C ages, and thus an OIS 3 age correlation, cannot be ruled out. Based on stratigraphic relations, Bowers (2001) interpreted this fine–grained deposit to grade from lagoonal to deep–water lacustrine sediment along the length of the exposure. This interpreta-

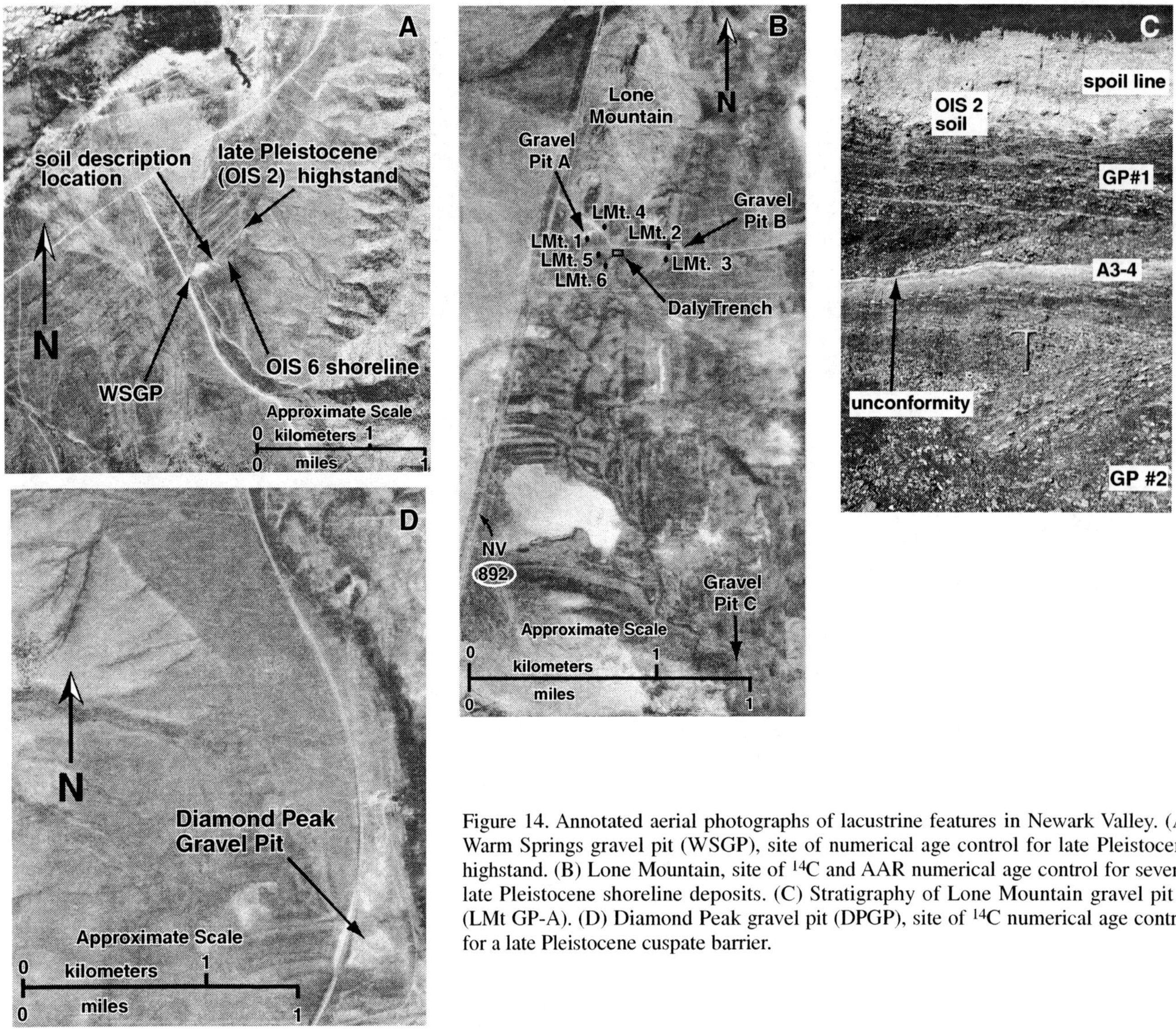

Figure 14. Annotated aerial photographs of lacustrine features in Newark Valley. (A) Warm Springs gravel pit (WSGP), site of numerical age control for late Pleistocene highstand. (B) Lone Mountain, site of ^{14}C and AAR numerical age control for several late Pleistocene shoreline deposits. (C) Stratigraphy of Lone Mountain gravel pit A (LMt GP-A). (D) Diamond Peak gravel pit (DPGP), site of ^{14}C numerical age control for a late Pleistocene cuspate barrier.

tion, combined with the AAR data, suggests that the fine–grained deposit is related to the higher, older lake level rather than to that of OIS 2, here represented by directly overlying, late Pleistocene high–stand gravels (Fig. 14C, GP#1).

Approximately 200 m southeast of LMt–GP–A, a 2–m deep backhoe trench was excavated across a lacustrine erosional escarpment (Fig. 14B). The Daly trench, at an altitude of 1833 m (6014 ft), exposes fine–grained sediments containing three genera of freshwater molluscs—*Pisidium*, *Lymnaea*, and *Gyraulus*—and one ostracode genus. AMS ^{14}C analysis of seven *Gyraulus* shells from a 20–cm–thick interval yielded an age of 14,450 ± 60 ^{14}C yrs. B.P. (17,310 ± 380 cal yrs. B.P.) (Table 3). Thus, these relatively deep–water sediments were deposited at about the same time as the nearshore deposits of barrier bars ~5–12 m higher at site LMt–GP–A..

About 500 m east of LMt–GP–A, a second gravel pit, LMt–GP–B (Fig. 14B), is excavated into a barrier bar at 1820 m (5970 ft). Porous tufa within lake sediments yielded an age of 14,480 ± 40 ^{14}C yrs. B.P. (17,340 ± 500 cal yrs. B.P.). This tufa lies near the base of a two–fold lake–recessional sequence of deposits (Bowers, 2001).

About 1.7 km to the south, a third gravel pit, LMt–GP–C (Fig. 14B), exposes beach gravel and a tufa layer in a barrier spit at 1809 m (5935 ft). The basal platy tufa (tufa–a) is 1–2 cm thick and is primarily composed of ostracodes (as seen in thin section); it yielded an age of 13,690 ± 40 ^{14}C yrs. B.P. (16,430 ± 460 cal yrs. B.P.). The overlying porous tufa (tufa–b) is 5–10 cm thick and is composed of a $CaCO_3$ micritic matrix with silt–sized particles and few ostracodes; it yielded an age of 12,375 ± 40 ^{14}C yrs. B.P. (14,330 ± 680 cal yrs. B.P.) (Table 3). These two ages

TABLE 3. SUMMARY OF AMS ^{14}C NUMERICAL AGES

Site	Material dated	Alt. (m)	Alt. (ft)	^{14}C yr B.P.	Cal yr B.P.	Laboratory Sample ID
WSGP	*Lymnaea*	1845	6055	13,780 ± 50	16,550 ± 350	Beta Analytic-148046
LMt-TD	*Gyraulus*	1833	6014	14,450 ± 60	17,310 ± 380	Beta Analytic-160009
LMt-GP-A -(1-4)	*Lymnaea*	1843	6047	39,150 ± 560	NA	USGS-WW 2157
LMt-GP-A-(3-4.1)	*Lymnaea*	1842	6044	37,740 ± 460	NA	USGS-WW 2758
LMt-GP-C-(a)	tufa-porous	1809	5935	12,375 ± 40	14,330 ± 680	USGS-WW 3875
LMt-GP-C-(b)	tufa-platy	1809	5935	13,690 ± 40	16,430±460	USGS-WW 3876
LMt-GP-B-(2-R1-C)	tufa-porous	1820	5970	14,480 ± 40	17,340± 500	USGS-WW 3877
DPGP	tufa-porous	1804	5920	13,025 ± 40	15,560± 750	USGS-WW 3878

NA—not available.

TABLE 4. SUMMARY OF AMINO ACID RACEMIZATION AGE ESTIMATES

Site	Genera	Elevation		AAR Ratios				Age Estimation (cal. yr B.P.)	Sample ID#
				Aspartic Acid		Glutamic Acid			
		(m)	(ft)	D/L[1]	st. dev.	D/L[1]	st. dev.		
WSGP	*Lymnaea*	1845	6055	0.40	0.05	0.15	0.01	25,000 ± 10,000	UAL-3411
WSGP	*Gyraulus*	1845	6055	0.38	0.05	0.15	0.02	25,000 ± 10,000	UAL-3412
LMt-GP-A-(1-4)	*Lymnaea*	1843	6047	0.55	0.01	0.37	0.01	middle Pleistocene-OIS 6?-younger	UAL-3041
LMt-GP-A-(3-4)	*Lymnaea*	1842	6044	0.57	0.02	0.38	0.00	middle Pleistocene-OIS 6?-older	UAL-3042
LMt-GP-A-(3-4-2)	*Pisidium*	1842	6044	0.54	0.03	0.38	0.04	middle Pleistocene-OIS 6?-older	UAL-3043
LMt-GP-A-(3-4-3)	*Gyraulus*	1842	6044	0.52	0.03	0.33	0.03	middle Pleistocene-OIS 6?-older	UAL-3044

AAR data from D. Kaufman, Northern Arizona University.
[1] D, D-alloisoleucine; L, L-isoleucine

suggest that either (1) this deposit has within it an unconformity representing ~1300 ^{14}C yrs., (2) one or both of these ages are in error, or (3) tufa dates may vary somewhat due to variable effects of reservoir carbon in lake and ground water.

About 11 km north of Lone Mt., the Diamond Peak gravel pit (Fig. 11, site DPGP; Fig. 14D) is excavated into a cuspate barrier bar at 1804 m (5920 ft) and exposes relatively coarse beach gravel and tufa. The tufa yielded an age of 13,025 ± 40 ^{14}C yrs. B.P. (15,560 ± 750 cal yrs. B.P.) (Table 3). Thin sections of this tufa show that it is composed primarily of ostracodes within a $CaCO_3$ micritic matrix. Recrystallization within some of the ostracodes shows some alteration of this tufa.

Lake level history of Newark Valley

The numerical ages and stratigraphic correlations were used to construct a lake level curve vs. ^{14}C age (Fig. 15; data in Table 3). Pluvial Lake Newark stood relatively high from at least 14,480 ^{14}C yrs. B.P. to at least 13,025 ^{14}C yrs. B.P. During this time, a brief, poorly constrained recession after 14,480 ^{14}C yrs. B.P. is recorded by stratigraphy at site LMt–GP–B, followed by a relatively quick transgression to near high–stand altitudes at 14,450 ^{14}C yrs. B.P., after which lake level gradually rose until the maximum, Sehoo equivalent, OIS 2 high–stand was attained at 13,780 ^{14}C yrs. B.P. Following the high–stand a quick, major recession at 13,690 ^{14}C yrs. B.P. took place. A steady lake level then persisted until 13,025 ^{14}C yrs. B.P., after which the desiccation record is unknown.

A larger and older lake occupied Newark Valley, provisionally during OIS 6 time. This age assignment is supported by soil morphology, soil index calculations, AAR age estimations, and stratigraphic correlations from LMt–GP–A. In addition, comparison of the degree of preservation of shoreline deposits above and

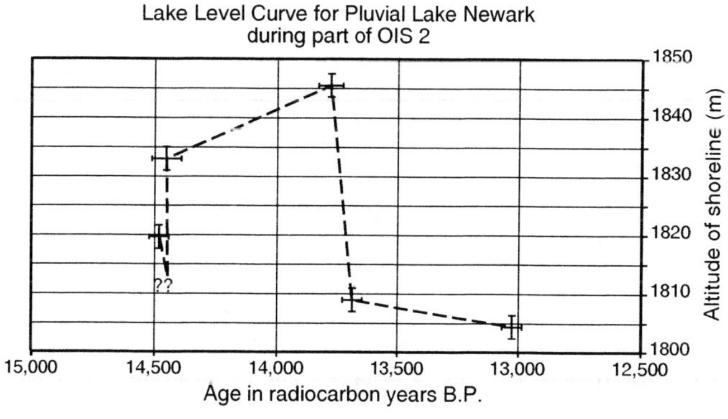

Figure 15. Lake level curve for Newark Valley.

below 1862 m (6110 ft) suggest that a lake reached a maximum altitude of about 1881 m, possibly 1890 m, (6200 ft) some time prior to OIS 6. Thus, the maximum altitude attained by the OIS 6(?) lake may have been 1862 m (6110 ft). No conclusive shoreline evidence for a pluvial lake of OIS 4 age was found in the basin, suggesting that the altitude of any lake during OIS 4 was at or below the altitude of the OIS 2 high–stand.

Comparison of Lake Newark with other pluvial lakes

In the Lahontan basin, higher and older shorelines than those representing OIS 6 have been recognized. Old shorelines above the OIS 6 suite in Newark Valley lack shoreline morphology, like the very old shorelines in other basins (Reheis et al., 2002a). Soil data for those with associated deposits (Table 2, Fig. 12) do not uniquely differentiate these deposits from the OIS 6 suite (similar to soil data for pre–late Pleistocene shorelines of several ages in the Lahontan and Columbus basins; see Figs. 9 and 10). Older lakes with shorelines higher than those of the Bonneville strandline (OIS 2 age) in the Bonneville Basin have not been reported, implying that either Lake Bonneville does not have the same hydrologic history as Lake Lahontan, or such deposits have not yet been recognized. The Bonneville lake history is complicated by the diversion of the Bear River into the Bonneville Basin (Jarrett and Malde, 1987; Bouchard et al., 1998). Thus, the change in volume of water due to paleoclimatic differences cannot be quantified. In the Lahontan Basin and in Newark Valley, the OIS 6 lake was higher than the OIS 4 and OIS 2 lakes. In the Bonneville Basin, the lake level during OIS 4 time appears to have been lower than lakes during OIS 6 time, and both were lower than, or equal to, lakes that existed during OIS 2 time. Lake Lahontan had a late Pleistocene high–stand at ~13 ka (Benson and Thompson, 1987) or 13,100 ^{14}C yrs. B.P. (Adams and Wesnousky, 1998) and Lake Newark had a high–stand at 13,780 ± 50 ^{14}C yrs. B.P. Both basins lag behind the youngest high–stand of Lake Bonneville at ~14.5 ka and the high–stand of Lake Franklin (Ruby Valley) at ~16 ka (Lillquist, 1994), northeast of Newark Valley (Fig. 3) and east of Lake Bonneville. Diamond Valley, directly west of Newark Valley (STOP 21, Fig. 3), overflowed from about 22,000 to 13,500 ^{14}C yrs. B.P. Although the timing of the actual high–stand of Lake Diamond is unknown, it is within the range of nearby pluvial lakes (Tackman, 1993). These comparisons suggest that because Lake Newark has a lake level history similar to Lake Lahontan, yet lies very close to Lake Bonneville, the differences between the shoreline records of the two largest pluvial lakes in the Great Basin is not due to regional paleoclimate differences.

Implications of paleoclimate parameters calculated from hydrologic index (HI)

The HI method of examining paleoclimate parameters (Mifflin and Wheat, 1979) permits: (1) a quantitative estimate of paleoclimate parameters during pluvial periods; and (2) a relative measurement of response of a basin and/or a region to climate change. Changes in temperature or precipitation can be difficult to compare among studies because small differences in the initial assumptions of modern mean annual temperature (MAT) and mean annual precipitation (MAP) can yield significant differences in the estimated climate change (Table 5). This study used data from the Desert Research Institute in Reno, Nevada (http://www.wrcc.dri.edu/summary/), whereas Mifflin and Wheat (1979) extrapolated from regional climate averages and Reheis (1999) used National Climatic Data Center information for Eureka.

The results from the present study suggest that small paleoclimate differences between lake levels of different ages, here correlated with OIS 2, OIS 6, and pre–OIS 6 can potentially cause significant differences in basin response (Table 5). These estimates are similar to estimates by Mifflin and Wheat (1979) for climate change related to OIS 2 lake levels in Nevada and to estimates for an OIS 16 lake by Reheis (1999). The apparent differences among studies in the estimations of change in climate based on Hydrologic Index calculations are partially due to differences in estimates of modern mean annual temperature and precipitation. Mifflin and Wheat (1979) noted that the changes they suggested during pluvial times are within or very near the range of historic variability. Due to uncertainties associated with the HI methodology, to which some investigators have taken exception (Benson and Thompson, 1987), concentrating on the relative magnitude of change suggested by the HI analysis rather than estimates of quantitative change may be a better approach.

Isostasy and Quaternary faulting in Newark Valley

Mifflin and Wheat (1979) recognized pre-late–Pleistocene lacustrine deposits in Newark Valley and elsewhere in Nevada and acknowledged that previous investigations had alluded to older and more extensive lakes in the Great Basin. However, Mifflin and Wheat (1979) thought the older lacustrine deposits were only preserved on the southern end of these basins and inter-

TABLE 5. COMPARISON OF CLIMATE-CHANGE ESTIMATES BASED ON HYDROLOGIC INDEX VALUES

Modern MAT (°C)	Modern MAP (cm)	MAT decrease (°C)	MAP increase (%)	Age (OIS)[1]	HI[2]	Source
28*	7.4*	2	45-55	2	0.29	This study
18*	7.8*	2.2-3.3	~70	2	0.28	Mifflin and Wheat, 1979
28	7.4	~2.5	55-75	6	0.36-0.43	This study
28	7.4	2.8	~75	16	0.47	This study
33	7.6	5	70	16	0.38	Reheis, 1999
33	7.6	>5	<70	16	0.38	Reheis, 1999

MAT—mean annual temperature; MAP—mean annual precipitation.

[1] Correlation to marine oxygen isotope stage using numerical ages and soil data of this study.

[2] HI—hydrologic index = lake area ÷ drainage basin area (Mifflin and Wheat, 1979)

* Data from http://www.wrcc.dri.edu/summary/

preted this as evidence for a northward regional tilt, not remnants of higher and older lacustrine deposits. Since then, investigations (Reheis, 1999; Reheis et al., 2002a) have produced evidence that these higher and older lacustrine deposits exist in a variety of locations around many basins (e.g., Lakes Newark, Hubbs, Gale, Diamond, Lahontan). Thus, the preservation of these deposits cannot be caused by a northward regional tilt. Results from this study show consistent altitudes of shoreline suites around Newark Valley, suggesting no significant tilt in this region, nor any deformation of shorelines around the basin that may be attributed to isostatic rebound of Newark Valley or to far–field effects of Lake Bonneville. This is contrary to data collected from Diamond Valley to the west (Tackman, 1993) and Ruby Valley to the northeast (Lillquist, 1994) (Fig. 3), which suggest a far–field hydro–isostatic signal in this region, likely related to rebound of the Bonneville basin, but o rebound effects from the individual, local lakes.

Quaternary faults are evident in Newark Valley, particularly on the west side, but their influence on shoreline altitudes is not clear. Data and interpretations of slip rates and age of faulting from ongoing studies within the valley may be available by the time of the field trip.

MAGNETOSTRATIGRAPHY OF THE HUMBOLDT RIVER CANYON, PERSHING COUNTY, NEVADA

(Parker and Negrini)

Introduction

Pliocene and Pleistocene sediments are exposed along hundreds of kilometers of the Humboldt River Canyon and its tributaries between the towns of Lovelock and Winnemucca, Nevada (Fig. 16). The exposures are particularly impressive south of Rye Patch Dam where more than 50 meters of layer–cake stratigraphy reflect the cycling between pluvial (light–colored sediments) and interpluvial (darker–colored sediments) climatic stages. The formalized stratigraphy of these sediments has been summarized by Morrison (1991; see Fig. 4). The age of the Rye Patch through Eetza Alloformations is fairly well constrained due to the presence of several tephra layers and a few U–Th dates on older tufas (see Negrini, 2002, for review). The younger Wyemaha and Sehoo Alloformations are, in places, exceptionally well dated (e.g., Benson et al., 1990; 1995). In contrast, the ages of the lowermost alloformations are poorly constrained (Morrison, 1991;

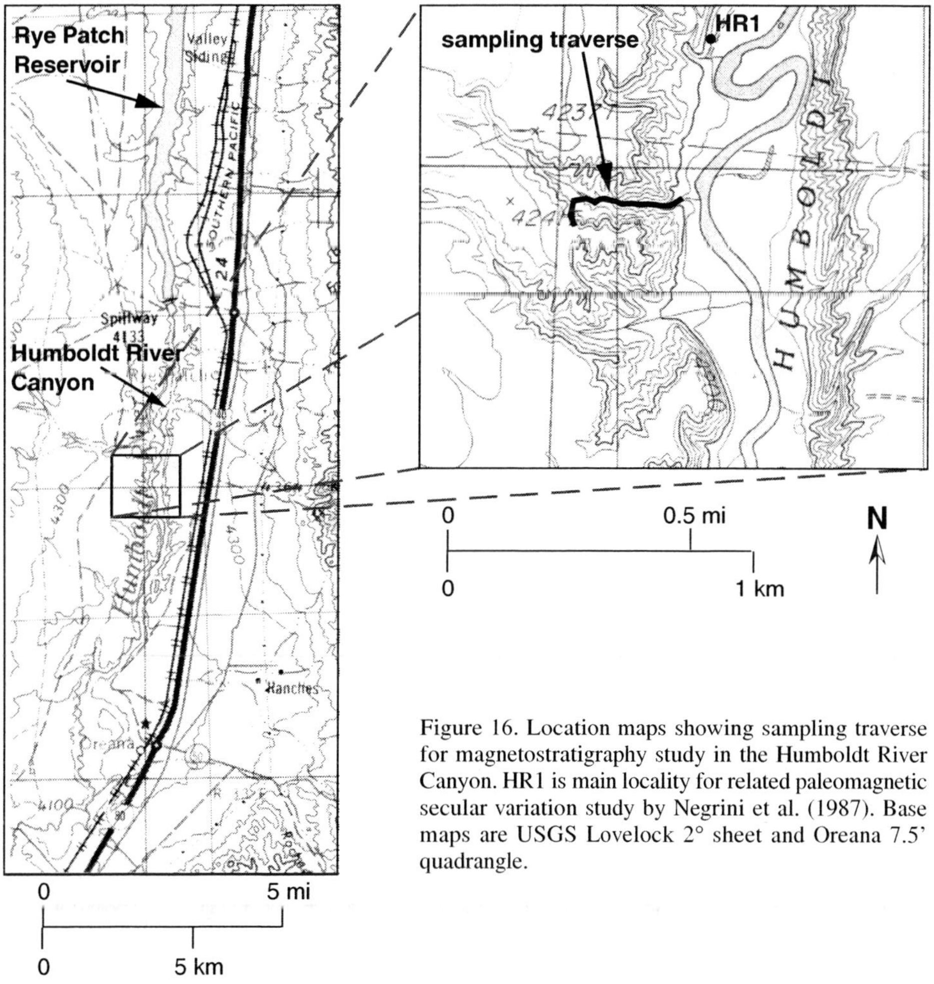

Figure 16. Location maps showing sampling traverse for magnetostratigraphy study in the Humboldt River Canyon. HR1 is main locality for related paleomagnetic secular variation study by Negrini et al. (1987). Base maps are USGS Lovelock 2° sheet and Oreana 7.5' quadrangle.

Negrini, 2002). Here we report on a magnetostratigraphic study which yields an upper limit of ~2.0–2.2 Ma for the base of the Lovelock Alloformation.

Methods

Samples were collected in a tributary canyon on the west side of the main Humboldt River Canyon approximately 5 km south of Rye Patch Dam (see "sampling traverse" in Fig. 16). This tributary canyon was chosen because it contained three prominent ash layers and because it exposed about 25 m of the Lovelock Allormation. Fifty–one horizons were sampled at l–m intervals (Fig. 17). At least three samples were collected from each horizon using the technique described in Negrini et al. (1987). Depths were measured relative to six control stakes using a Jacob's staff and Brunton compass. The control stakes were surveyed using a laser total station instrument and were spaced evenly along the traverse. The uppermost absolute depth (0) corresponds to the top of the tributary canyon.

At least one sample per horizon was subjected to a suite of magnetic measurements designed to determine the stability of the magnetic remanence (Parker, 1993). These included (1) the Lowrie–Fuller test for multi–domain vs. single–domain magnetizations, (2) a ten–step, alternating–field demagnetization (AFD) from 0 to 60 mT, and (3) acquisition of saturated isothermal remanent magnetization. Magnetizations were measured in a Molspin spinner magnetometer.

Results

Seven samples with multiple magnetic domains failed the Lowrie–Fuller test (Parker, 1993). Results of all samples from these seven horizons were discarded because their multi–domain nature makes them more likely to have significant viscous remanence overprints that would alter the original magnetization. The results of the AFD analysis demonstrate that the remaining samples possess a strong magnetization usually in the range of 0.05 to 0.10 A/m. Furthermore, after removal of a viscous remanence by AFD levels of 20–30 mT, the magnetization directions were stable for the rest of the AFD (Parker, 1993). Because isothermal remanence saturated at <400 mT, the magnetic carrier likely belongs in the magnetite class of minerals (Parker, 1993).

The paleomagnetic vector components of all samples possessing stable remanences plotted versus depth are distinctly separated into four zones of alternating magnetic polarity (Fig. 17). We interpret the two reversed zones to be the Matuyama Polarity Chron, the bottom normal zone to be the Olduvai Subchron, and the top normal zone to be the Brunhes Chron.

We do not entirely preclude the possibility that the observed polarity zones belong to older chrons and subchrons. However, our interpretation is consistent with estimated ages of chron and subchron boundaries obtained by linear extrapolation using the ages of the tephra layers. For example, such an extrapolation would date the sediments corresponding to the bottom and top of the Olduvai Subchron at ~1.8 and 1.45 Ma, respectively, which is

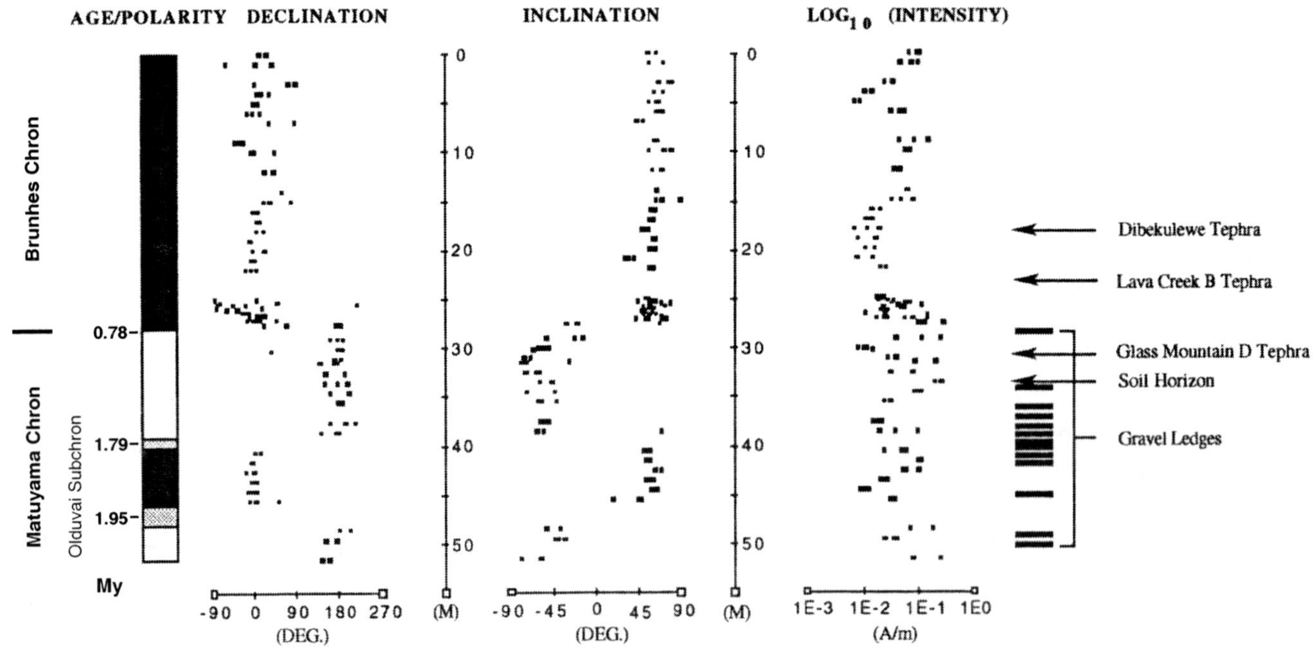

Figure 17. Paleomagnetic directions of Humboldt River Canyon sediments along traverse shown in Figure 16. Dibekulewe tephra layer (~0.51 Myr) is in base of Paiute AF; Glass Mountain D tephra layer (0.9-1.0 Myr) is near top of Lovelock AF (fig. 4). Tephra ages and associated references are reviewed in Negrini (2002). Polarity boundary ages are from Hilgen (1991) and Tauxe et al. (1996). Contact between Rye Patch AF and Lovelock AF is located above Brunhes-Matuyama polarity transition (after Parker, 1993).

considerably younger than the accepted boundary ages of this subchron (Fig. 17). Because the Olduvai Subchron is the youngest extended interval of normal magnetic polarity in that age frame, any other interpretation would require even greater discrepancies than that of the age model in Figure 17.

DISCUSSION AND IMPLICATIONS FOR LAHONTAN–AREA PLIOCENE AND PLEISTOCENE STRATIGRAPHY

Based on the reversal stratigraphy model (Fig. 17), the oldest sediments exposed in the Humboldt River Canyon in the vicinity of the sampling traverse are between 2.0 and 2.2 Ma in age. This represents a minimum age for the base of the Lovelock Alloformation.

Morrison (1991) describes "pre–Lovelock lacustrine deposits" underlying the Lovelock Alloformation in the Humboldt River Canyon near Rye Patch Dam. These older lacustrine deposits are currently undated. If further study shows that they are of regional significance and are younger than the base of the Lovelock Alloformation as identified in the results of this paper (2.0–2.2 Ma), then the base of the sediments of this study will belong to a new alloformation lying below a bounding unconformity separating them from another new alloformation containing the "pre–Lovelock lacustrine deposits."

ACKNOWLEDGEMENTS

The National Science Foundation (NSF) supported material presented in the road log (STOPS 1, 17) by Adams under Grant No. 0087840. The NSF funded cosmogenic dating by Kurth and Phillips (STOPs 12–14, 20) under Grant No. EAR–0001063. The U.S. Geological Survey Earth Surface Dynamics Program funded contributions of Reheis, McGeehin, Paces, Sarna–Wojcicki, and Smoot.

The late Jonathan O. Davis of the Desert Research Institute, Reno, was instrumental in the early stages of this research. His contributions include locating the ash layers and surveying. Andrei Sarna–Wojcicki of the U.S. Geological Survey at Menlo Park is responsible for the ash correlations along with Davis. Jonathan Parker assisted with sample collection.

The J. O. Davis Scholarship fund, Geological Society of America Student Research Grants, and EDMAP funds from the U.S. Geological Survey supported this study.

REFERENCES

Adams, K. D., 2002, Late Holocene lakes in the Carson Sink: Age and paleoclimatic significance: *in* Caskey, J., ed., Historical faulting, chronostratigraphy, and paleoseismicity of the central Nevada seismic belt: Field Guide, 2002 Pacific Cell Friends of the Pleistocene, privately published, p. 52–62.

Adams, K. D., and Wesnousky, S. G., 1998, Shoreline processes and the age of the Lake Lahontan high stand in the Jessup embayment, Nevada: Geological Society of America Bulletin, v. 110, p. 1318–1332.

Adams, K. D., and Wesnousky, S. G., 1999, The Lake Lahontan high stand: age, surficial characteristics, soil development, and regional shoreline correlation: Geomorphology, v. 30, p. 357–392.

Adams, K. D., Wesnousky, S. G., and Bills, B. G., 1999, Isostatic rebound, active faulting, and potential geomorphic effects in the Lake Lahontan basin, Nevada and California: Geological Society of America Bulletin, v. 111, p. 1739–1756.

Ayarbe, J. P., Phillips, F. M., Harrison, J. B. J., Elmore, D., and Sharma, P., 1998, Application of cosmogenic nuclides to fault–scarp chronology: Preliminary results from the Socorro Canyon Fault: *in* Harrison, J. B. J., ed., Soil, Water, and Earthquakes around Socorro, New Mexico: Friends of the Pleistocene, Rocky Mt. Cell, 1998 Fall Field Trip, privately published, 11 p.

Bassinot, F. C., Labeyrie, L. D., Vincent, E., Quidelleur, X., Shackleton, N. J., and Lancelot, Y., 1994, The astronomical theory of climate and the age of the Brunhes–Matuyama magnetic reversal: Earth and Planetary Science Letters, v. 126, p. 91–108.

Bell, J. W., and Bonham, H. F., 1987, Geologic map of the Vista Quadrangle: Nevada Bureau of Mines and Geology Map 4Hg, scale 1:24,000.

Benson, L. V., Currey, D. R., Dorn, R. I., Lajoie, K. R., Oviatt, C. G., Robinson, S. W., Smith G. I., and Stine, S., 1990, Chronology of expansion and contraction of four Great Basin lake systems during the past 35,000 years: Palaeogeography, Palaeoclimatology, Palaeolimnology, v. 6, p. 205–255.

Benson, L. V., Currey, D. R., Lao, Y., and Hostetler, S. W., 1992, ^{14}C yr B.P. Lake–size variations in the Lahontan and Bonneville basins between 13,000 and 9,000 ^{14}C yr B.P: Palaeogeography, Palaeoclimatology, Palaeoecology, v. 95, p. 19–32.

Benson, L.V., Kashgarian, M., and Rubin, M., 1995, Carbonate deposition, Pyramid Lake subbasin, Nevada: 2. Lake levels and polar jet stream positions reconstructed from radiocarbon ages and elevations of carbonates (tufas) deposited in the Lahontan Basin: Palaeogeography, Palaeoclimatology, Palaeoecology, 117, pp. 1–30.

Benson, L. V., and Thompson, R. S., 1987, The physical record of lakes in the Great Basin: *in* Ruddiman, W. F., and Wright, H. E., Jr., eds., North America and Adjacent Oceans During the Last Deglaciation, The Geology of North America, Geological Society of America, v. K–3, p. 241–260.

Berger, G. W., 1991, The use of glass for dating volcanic ash by thermoluminescence: Journal of Geophysical Research, B, Solid Earth and Planets, v. 96, no.12, p.19705–19720.

Birkeland, P. W., 1965, The Truckee River Canyon between Wadsworth and Vista, Nevada: *in* Wahrhaftig, C., Morrison, R. B., and Birkeland, P. W., eds., Guidebook for Field Conference I, Northern Great Basin and California, International Association of Quaternary Research, 7th Congress, U.S.A., 1965, p. 35–38.

Birkeland, P. W., 1999, Soil and Geomorphology: Oxford University Press, NY, 448 p.

Blair, T. C., 1999, Sedimentology of gravelly Lake Lahontan high stand shoreline deposits, Churchill Butte, Nevada, USA: Sedimentary Geology, v. 123, p. 199–218.

Bouchard, D. P., Kaufman, D. S., Hochberg, A., Quade, J., 1998, Quaternary history of the Thatcher Basin, Idaho, reconstructed from the ^{87}Sr/^{86}Sr and amino acid composition of lacustrine fossils—implications for the diversion of the Bear River into the Bonneville basin: Palaeogeography, Palaeoclimatology, Palaeoecology, v. 141, p. 95–114.

Bowers, R. J., 2001, Stratigraphic and geomorphic investigation of late Pleistocene lacustrine deposits near Lone Mt. in Newark Valley, Nevada: Senior thesis, Humboldt State University, Arcata, CA, 52 p.

Clark, M. M., 1967, Pleistocene glaciation of the drainage of the West Walker River, Sierra Nevada, California: Ph.D. thesis, Stanford University, 170 p.

Currey, D. R., 1988, Isochronism of final Pleistocene shallow lakes in the Great Salt Lake and Carson Desert regions of the Great Basin: Program and Abstracts of the Tenth Biennial Meeting, American Quaternary Association, p. 117.

Davis, J. O., 1978, Quaternary tephrochronology of the Lake Lahontan area, Nevada and California: University of Nevada, Nevada Archeological Survey Research Paper 7, 137 p.

Demsey, K., 1987, Holocene faulting and tectonic geomorphology along the Wassuk Range, west–central Nevada: MS thesis, University of Arizona, Tucson, 64 p.

Dep, L., Elmore, D., Fabryka–Martin, J., Masarik, J., and Reedy, R. C., 1994, Production rate systematics of in–situ produced cosmogenic nuclides in terrestrial rocks: Monte Carlo approach of investigating ^{35}Cl(n,γ)^{36}Cl: Nuclear Instruments and Methods in Physics Research, v. B 92, p. 321–325.

Gordon, Ian, and Heller, P. L., 1993, Evaluating major controls on basinal stratigraphy, Pine Valley, Nevada: Implications for syntectonic deposition: Geological Society of America Bulletin, v. 105, p. 47–55.

Harden, J. W., and Taylor, E. M., 1983, A quantitative comparison of soil development in four climatic regimes: Quaternary Research, v. 20, p. 342–359.

Hilgen, F. J., 1991, Astronomical calibration of Gauss to Matuyama sapropels in the Mediterranean and implication for the Geomagnetic Polarity Time Scale: Earth and Planetary Science Letters, v. 104, p. 226–244.

Hubbs, C. L., and Miller, R. R., 1948, The Great Basin. II. The zoological evidence: University of Utah Bulletin, v. 38, p. 17–166.

Hubbs, C. L., Miller, R. R., and Hubbs, L. C., 1974, Hydrographic history and relict fishes of the north–central Great Basin: California Academy of Sciences, Memoirs, v. VII, 259 p.

Jarrett, R. D., and Malde, H. E., 1987, Paleodischarge of the late Pleistocene Bonneville Flood, Snake River, Idaho, computed from new evidence: Geological Society of America Bulletin, v. 99, p. 127–134.

Kaufman, D. S., Forman, S. L., and Bright, J., 2001, Age of the Cutler Dam Alloformation (late Pleistocene), Bonneville Basin, Utah: Quaternary Research, v. 56, p. 322–334.

Kaufman, D. S., and Manley, W. F., 1998, A new procedure for determining enantiometric (D/L) ratios in fossils using reverse phase liquid chromatography: Quaternary Science Reviews (Quaternary Geochronology), v. 17, p. 987–1000.

King, G. Q., 1993, Late Quaternary history of the lower Walker River and its implications for the Lahontan paleolake system: Physical Geography, v. 14, p. 81–96.

Kurth, G. E., Phillips, F. M., and Reheis, M. C., 2002, The relationship between lake–filling cycles in the western Great Basin and Pleistocene Epoch climate changes: Eos, Transactions of American Geophysical Union, v. 83(47), Fall Meeting Supplement, Abstract PP71B–0399.

Laabs, B. J. C. and Kaufman, D. S., 2003, Quaternary high–stands in Bear Lake Valley, Utah and Idaho: Geological Society of America Bulletin, v. 115, p. 463–478.

Lajoie, K. R., 1968, Late Quaternary stratigraphy and geologic history of Mono basin, eastern California: PhD. thesis, University of California, Berkeley, 271 p.

Lillquist, K. D., 1994, Late Quaternary Lake Franklin: lacustrine chronology, coastal geomorphology, and hydro–isostatic deflection in Ruby Valley and northern Butte Valley, Nevada: PhD thesis, University of Utah, Salt Lake City, UT, 268 p.

Liu, B., Phillips, F. M., Fabryka–Martin, J. T., Fowler, M. M., and Stone, W. D., 1994, Cosmogenic ^{36}Cl accumulation in unstable landforms, 1. Effects of the thermal neutron distribution: Water Resources Research, v. 30, p. 3115–3125.

Machette, M. N., 1985, Calcic soils of the southwestern United States: *in* Weide, D. L., ed., Soils and Quaternary geology of the southwestern United States, Geological Society of America Special Paper 203, p. 1–21.

McCoy, W. D., 1987, Quaternary aminostratigraphy of the Bonneville basin, western United States: Geological Society of America Bulletin, v. 98, p. 99–112.

Mifflin, M. D., and Wheat, M. M., 1979, Pluvial Lakes and Estimated Pluvial Climates of Nevada: Nevada Bureau of Mines and Geology, Bulletin 94, Mackay School of Mines, University of Nevada, Reno, NV, 55 p.

Morrison, R. B., 1964, Lake Lahontan: Geology of the southern Carson Desert: U.S. Geological Survey Professional Paper 401, 156 p

Morrison, R. B., 1965, Quaternary Geology of the Great Basin: *in* Wright, H. E., Jr., and Frey, D. G., eds., The Quaternary of the United States: Princeton, Princeton University Press, New Jersey,p. 265–285.

Morrison, R. B., 1991, Quaternary stratigraphic, hydrologic, and climatic history of the Great Basin, with emphasis on Lakes Lahontan, Bonneville, and Tecopa: *in* Morrison, Roger, B. ed., Quaternary Nonglacial Geology, Conterminous U.S., The Geology of North America, Geological Society of America, v. K–2, p. 283–320.

Morrison, R. B., and Davis, J. O., 1984, Supplementary guidebook for GSA field trip 13, Quaternary stratigraphy and archeology of the Lake Lahontan area, a re–assessment: Reno, Desert Research Institute, Social Science Center Technical Report 41, 50 p.

Mulhern, M.E., 1982, Lacustrine, fluvial, and fan sedimentation: A record of Quaternary climate change and tectonism, Pine Valley, Nevada: Zeitschrift fur Geomorphologie, Suppl. Bd. 42, p. 117–133.

Negrini, R. M., 2002, Pluvial lake sizes in the northwestern Great Basin throughout the Quaternary Period: *in* Currey, D. Madsen, D., and Hershler, R., eds., Great Basin Aquatics Systems History: Smithsonian Contributions Series, Smithsonian Institute Press, p. 11–52.

Negrini, R. M., Verosub, K. L., and Davis, J. O., 1987, Long–term nongeocentric axial dipole directions and a geomagnetic excursion from the middle Pleistocene sediments of the Humboldt River Canyon, Pershing County, Nevada: Journal of Geophysical Research, v. 92, p. 10617–10627.

Oviatt, C. G., Thompson, R. S., Kaufman, D. S., Bright, J., and Forester, R. M., 1999, Reinterpretation of the Burmester core, Bonneville Basin, Utah: Quaternary Research, v. 52, p. 180–184.

Parker, K.E., 1993, Magnetostratigraphy of Plio–Pleistocene Sediments from the Humboldt River Canyon, Pershing County, Nevada: M.S. thesis, California State University, Bakersfield, CA, 91 p.

Putnam, W. C., 1949, Quaternary geology of the June Lake district, California: Geological Society of America Bulletin, v. 60, p. 1281–1302.

Ramelli, A. R., Caskey, John, and Bell, J. W., 2002, Paleoseismicity and Lake Lahontan shoreline fluctuations along the Rainbow Mt. fault: *in* Caskey, J., ed., Historical faulting, chronostratigraphy, and paleoseismicity of the central Nevada seismic belt, Field Guide, 2002 Pacific Cell Friends of the Pleistocene, privately published, p. 34–42.

Redwine, J. L., (submitted), Quaternary pluvial history and paleoclimatic implications of Newark Valley, east–central Nevada: M.S. thesis, Arcata, Humboldt State University.

Reheis, M. C., 1999, Highest pluvial–lake shorelines and Pleistocene climate of the western Great Basin: Quaternary Research, v. 52, p. 196–205.

Reheis, M. C., and Sawyer, T. L., 1997, Late Cenozoic history and slip rates of the Fish Lake Valley, Emigrant Peak, and Deep Springs fault zones, Nevada and California: Geological Society of America Bulletin, v. 109, p. 280–299.

Reheis, M. C., Slate, J. L., Sarna–Wojcicki, A. M, and Meyer, C. E., 1993, A late Pliocene to middle Pleistocene pluvial lake in Fish Lake Valley, Nevada and California: Geological Society of America Bulletin, v. 105, p. 953–967.

Reheis, M. C., Sarna–Wojcicki, A. M., Reynolds, R. L., Repenning, C. A., and Mifflin, M. D., 2002a, Pliocene to middle Pleistocene lakes in the western Great Basin—Ages and connections: *in* Hershler, R., Currey, D., and Madsen, D., eds., Great Basin Aquatic Systems History: Smithsonian Contributions Series, Smithsonian Institute Press, p. 53–108.

Reheis, M. C., Stine, Scott, and Sarna–Wojcicki, A. M., 2002b, Drainage reversals in Mono Basin during the late Pliocene and Pleistocene: Geological Society of America Bulletin, v. 114, p. 991–1006.

Rhode, D., Adams, K. D., and Elston, R. G., 2000, Geoarchaeology and Holocene landscape history of the Carson Desert, western Nevada: *in* Lageson, D. R., Peters, S. G., and Lahren, M. M., eds., Great Basin and Sierra Nevada, Boulder, Geological Society of America Field Guide 2, p. 45–74.

Russell, I. C., 1885, Geological history of Lake Lahontan, a Quaternary lake of northwest Nevada: U.S. Geological Survey Monograph 11, 288 p.

Scott, W. E., McCoy, W. D., Shroba, R. R., and Rubin, M., 1983, Reinterpretation of the exposed record of the last two cycles of Lake Bonneville, Western United States: Quaternary Research, v. 20, p. 261–285.

Sharp, R.P., and Birman, J.H., 1963, Additions to the classical sequence of Pleistocene glaciations, Sierra Nevada, California: Geological Society of America Bulletin, v. 74, p. 1079–1086.

Sieh, K.E., and M.I. Bursick, 1986, Most recent eruption of the Mono Craters, eastern central California: Journal of Geophysical Research, v. 91, p. 12539–12571.

Smith, R. M., 1976, Mineral resources: *in* Geology and mineral resources of White Pine County, Nevada: Nevada Bureau of Mines and Geology Bulletin 85, p. 36–105.

Smoot, J. P., 1993, Field trip guide: Quaternary–Holocene lacustrine sediments of Lake Lahontan, Truckee River Canyon north of Wadsworth, Nevada: U.S. Geological Survey Open–File Report 93–689, 35 p.

Stine, Scott, 1987, Mono Lake: The Past 4,000 Years: PhD thesis, University of California, Berkeley, CA, 615 p.

Stine, Scott, 1990, Late Holocene fluctuations of Mono Lake, eastern California: Palaeogeography, Palaeoclimatology, Palaeocology, v. 78, p. 333–381.

Stine, Scott, 1994, Extreme and persistent drought in California and Patagonia during mediaeval time: Nature, v. 369, p. 546–549.

Tackman, Gary, 1993, Pleistocene hydrologic history of Diamond Valley, Nevada, with climatic and isostatic implications: M.S. thesis, Univ. of Utah Limneotectonics Laboratory, 182 p.

Tauxe, L., Herbert, T., Shackleton, N. J., and Kok, Y. S., 1996, Astronomical calibration of the Matuyama/Brunhes boundary: Consequences for magnetic remanence acquisition in marine carbonates and the Asian loess sequences: Earth and Planetary Science Letters, v. 140, p. 133–146.

Taylor, E. M., 1988, Instructions for the soil development index template—LOTUS 1–2–3 (and program disk): U.S. Geological Survey Open–File Report 88–233, 23 p.

Tocher, D., 1956, Movement on the Rainbow Mountain fault: Bulletin of the Seismological Society of America, v. 46, p. 10–14.

Willden, Ronald and Kistler, R. W., 1969, Geologic Map of the Jiggs Quadrangle, Elko County, Nevada: U.S. Geological Survey, Map GQ–859, scale 1:62,500.

Cenozoic eolian sedimentary systems of the USA midcontinent

Authors and Leaders: **E. Arthur Bettis III,** *Department of Geoscience, University of Iowa, Iowa City, Iowa 52242.*
Joseph P. Mason, *Conservation and Survey Division, Institute of Agriculture and Natural Resources, University of Nebraska, Lincoln, Nebraska 68588–0517.*
James B. Swinehart, *Conservation and Survey Division, Institute of Agriculture and Natural Resources, University of Nebraska, Lincoln, Nebraska 68588–0517.*
Xiaodong Miao, *Department of Geosciences, University of Nebraska– Lincoln, Nebraska 68588–0517.*
Paul R. Hanson, *Department of Geosciences, University of Nebraska, Lincoln, Nebraska 68588–0517.*
Ronald J. Goble, *Department of Geosciences, University of Nebraska, Lincoln, Nebraska 68588–0517.*
David B. Loope, *Department of Geosciences, University of Nebraska, Lincoln, Nebraska 68588–0517.*
Peter M. Jacobs, *Department of Geography and Geology, University of Wisconsin, Whitewater, Wisconsin 53190–1790.*
Helen M. Roberts, *Luminescence Dating Laboratory, Institute of Geography and Earth Sciences, University of Wales, Aberystwyth, Wales ST23 3DB, UK*

INTRODUCTION

Eolian sediments cover vast areas of the North American mid–continent beyond the Wisconsin (marine oxygen isotope stages 3 and 2) glacial border (Fig. 1) and contain a record of cratonic sedimentation spanning the Cenozoic Era. Increasing interest in eolian sediments and associated soils as climate proxies has spurred several recent studies of Quaternary loess and eolian sand in the region (Muhs et al., 1999a; Arbogast and Muhs, 2000; Mason and Kuzila, 2000; Muhs and Bettis, 2000; Mason, 2001; Forman and Pierson, 2002; Loope and Swinehart, 2000; Mason et al., in press). Pre–Quaternary records of climate and environmental change are preserved in Tertiary eolian sediments intercalated with alluvial and lacustrine sequences extending from the Rocky Mountains eastward across the Great Plains (Diffendahl, 1982; Retallack, 1983). Loess, eolian sand, and volcanic ash provide direct records of atmospheric circulation that can be used to test atmospheric general circulation models (Muhs et al., 1999a; Muhs and Bettis, 2000). Buried soils within eolian sequences provide detailed records of environmental change spanning time intervals of centuries to glacial–interglacial cycles. A proliferation of studies over the last two decades have led us to recognize the dynamic nature of eolian landscapes and the complex linkages among geologic, climatic, and biotic systems in the history of the central Great Plains' Cenozoic eolian systems.

On this trip, we will visit localities in western Iowa and throughout Nebraska that contain key stratigraphic sequences of Cenozoic eolian deposits (Fig. 2). Many of these localities have a long history of investigation dating back to the late 1800's and early 1900's. Recent studies at several localities have yielded new insights on several aspects of the region's late Quaternary loess and eolian sand record, most notably the chronology and sources of the sediments. Many of the data in this guidebook are derived from previous studies, but some are presented here for the first time.

STRATIGRAPHIC FRAMEWORK

A detailed treatment of central Great Plains Cenozoic stratigraphy is beyond the scope of this guidebook. Presented here are salient elements of the region's eolian stratigraphy that relate to localities visited on this trip. Extensive deposits of wind–deposited, tuffaceous silt and sand with interbedded tuff occur in the wedge of Tertiary sediment extending eastward from the Rocky Mountain Front Range across the western Great Plains. These sediments represent a range of eolian depositional environments that formed as sediment was shed from the rising Rocky Mountains via volcanic and fluvial processes during the Oligocene, Miocene, and Pliocene. Extensive sand sheets and thin loesses with associated paleosols occur in the Pliocene Fullerton Fm. of Nebraska (Reed and Dreeszen, 1965; May et al., 1995) and its correlative, the upper part of the Blanco Fm. in Kansas (Frye and Leonard, 1952; Holliday, 1988). Pearlette–family volcanic ashes from Yellowstone caldera provide three time lines for Great Plains stratigraphy: the 2.059–Ma Huckleberry Ridge ash, 1.285–Ma Mesa Falls ash, and 0.639–Ma Lava Creek B ash (Boellstorff, 1978; Izette and Wilcox, 1982; Lanphere et al., 2002). As-yet-poorly understood, loess–paleosol sequences occur below the Lava Creek B ash in the Pleistocene Walnut Creek and Sappa Fms. of Nebraska and Kansas and above the Lava Creek B ash, but below Illinoian (marine oxygen isotope stage 6) Loveland loess in the region (Reed and Dreeszen, 1965; May et al., 1995).

Four middle-to-late Pleistocene loess units occur on the central Great Plains. From oldest to youngest, these are Loveland loess, Gilman Canyon Fm. (equivalent to Iowa's Pisgah Fm.), Peoria loess, and Bignell loess. Loveland loess is usually only a few meters thick and sometimes has an eolian sand facies. The last interglacial Sangamon geosol is developed in the upper part of the Loveland. This paleosol is relatively thick (1.5–2.5 m), exhibits 7.5YR hue colors, and commonly contains clay coatings in the upper B-horizon and carbonate coatings or nodules in the

Bettis E.A. III, Mason J.P., Swinehart, J.B., Miao, X., Hanson, P. R., Goble, R.J., Loope, D.B., Jacobs, P.M., and Roberts, H.M., 2003, Cenozoic Eolian Sedimentary Systems of the USA Midcontinent: *in* Easterbrook, D.J., ed., Quaternary Geology of the United States, INQUA 2003 Field Guide Volume, Desert Research Institute, Reno, NV, p. 195–218.

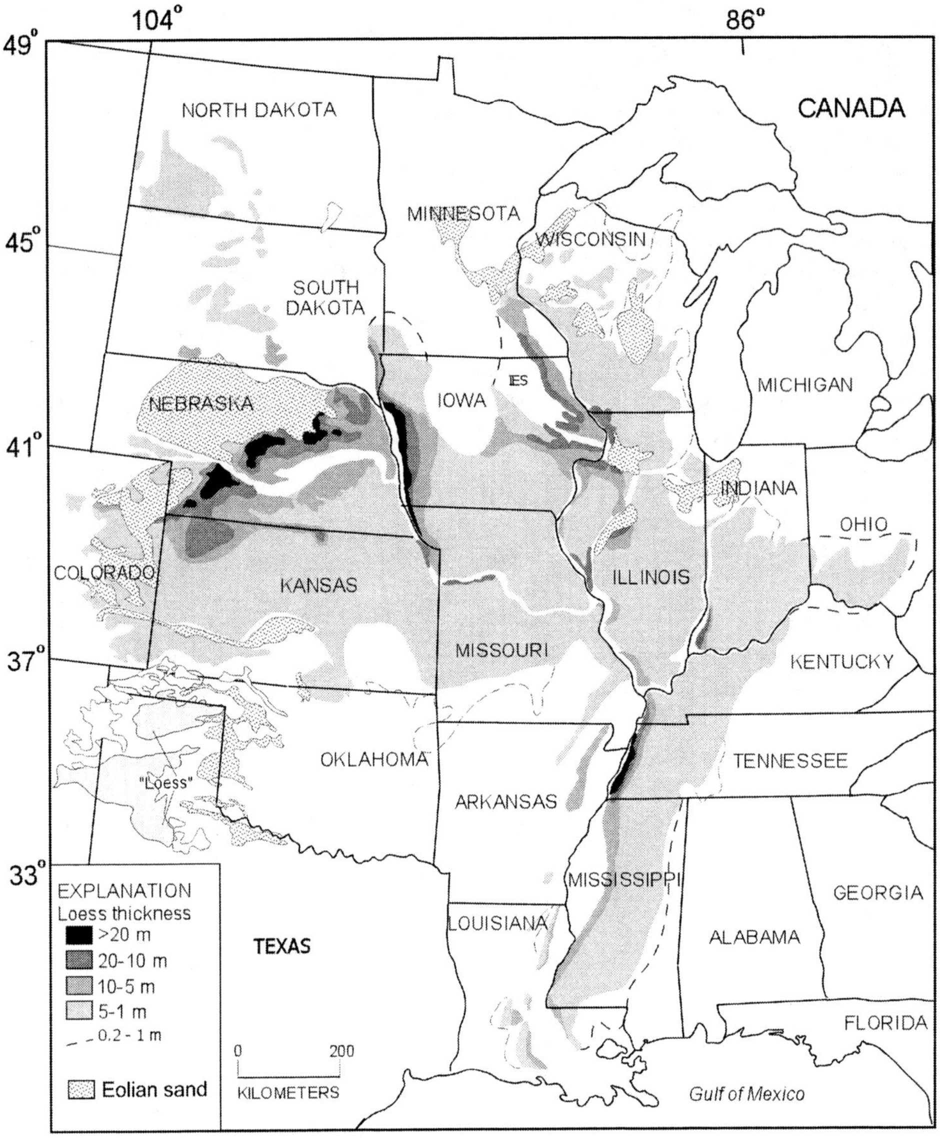

Figure 1. Midcontinental United states showing thickness of Peoria loess and generalized distribution of eolian sand. Note thick loess accumulations along the eastern side of major valleys and southeast of the Nebraska Sand Hills and the Iowan erosion surface (IES).

lower part of its profile. The Gilman Canyon Fm. is also thin (usually <2 m) and is usually modified throughout by organic accumulation and biological structures (burrows, etc.). The soil developed in the Gilman Canyon Fm. (equivalent to the Farmdale geosol east of the Missouri River valley) is usually welded to the underlying Sangamon geosol developed in the Loveland loess. In rare sections where the Gilman Canyon Fm. is thick, two or more soils are developed in the unit. Peoria loess buries the Gilman Canyon Fm. The Peoria is the thickest (up to 50m) and one of the most aerially extensive, last glacial loesses in the world (Muhs and Bettis, in press). At some near–source localities, eolian sand is interstratified with the loess and bedding structures are present. Throughout most of the Great Plains, the Bignell loess buries Peoria loess and the dark–colored Brady soil is developed in the upper part of the Peoria. The Bignell is usually less than 2 m thick, although thicker sections, such as one visited on this trip, do exist. The Bignell is also patchy in its distribution and appears not to occur as a once–continuous sheet as most of the other loess units in the region do. The Bignell is not present (or discernable) east of the Missouri River valley where the modern soil is developed in Peoria loess.

Recent applications of thermoluminescence (TL) dating to loess in the central Great Plains are refining the region's loess chronology, and sections like Loveland, Bignell Hill, and Wauneta Roadcut (all visited on this trip) have revealed details about the accumulation history of individual loess units. TL ages from Eustis in south central Nebraska and TL and infrared–stimulated luminescence (IRSL) ages from the Loveland paratype locality in western Iowa (STOP 8 on this trip) suggest that the Loveland loess accumulated during the penultimate glacial period, equivalent to marine oxygen isotope stage 6 (Forman et al., 1992; Maat and Johnson, 1996; Forman and Pierson, 2002).

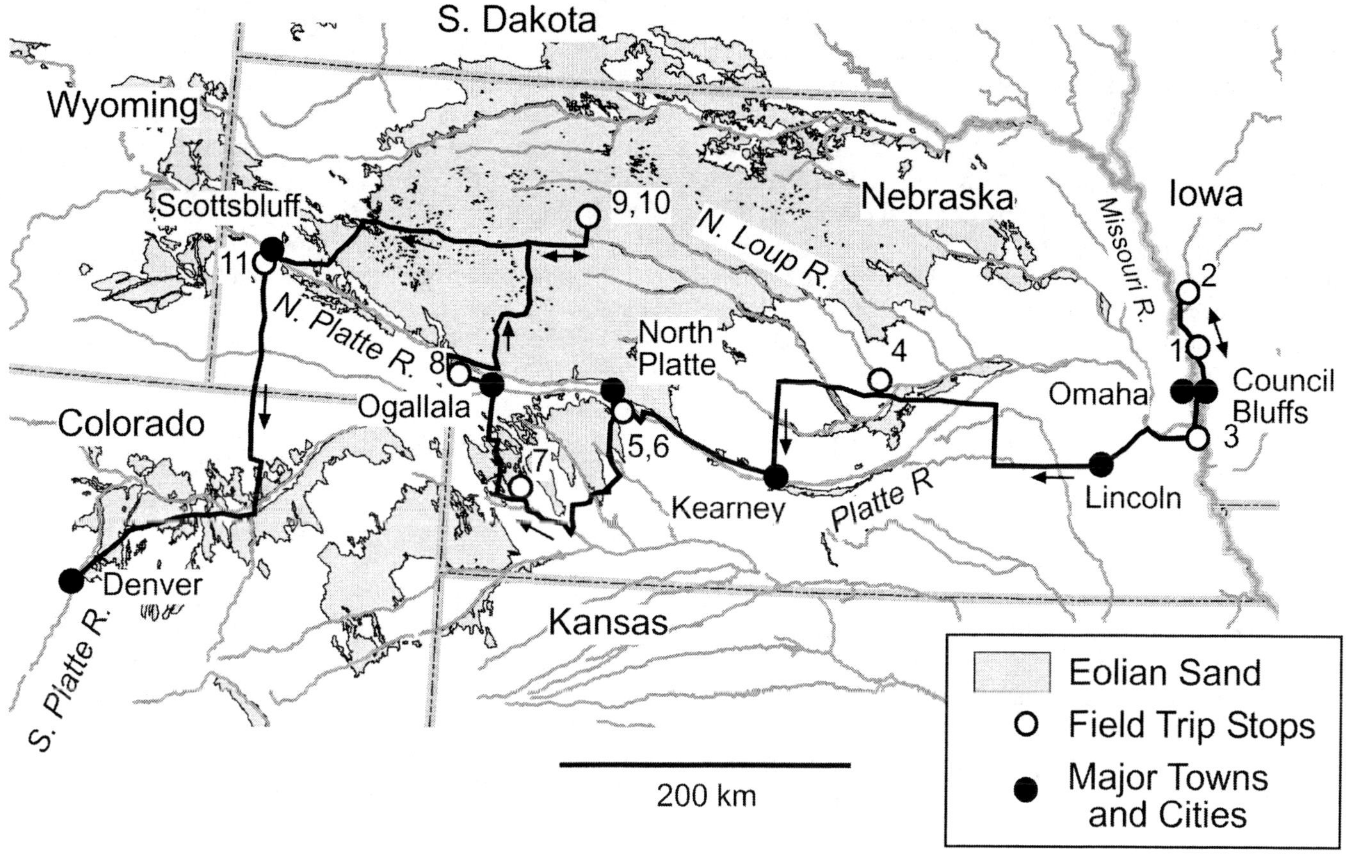

Figure 2. Field trip route map showing stops (open circles), major towns and cities (closed circles), and the distribution of eolian sand in the map area.

TL and radiocarbon ages on soil organic matter and charcoal bracket accumulation of the overlying Gilman Canyon Fm. and its Iowa equivalent, the Pisgah Fm. to between ~45,000 and 25,000 cal yrs. B.P. (Forman et al.,, 1992; Martin, 1993; May and Holen, 1993; Pye et al., 1995; Maat and Johnson, 1996; Muhs et al., 1999a, b; Forman and Pierson, 2002; Bettis et al., in press). Thus, in the central Great Plains, the Sangamon paleo–geomorphic surface, marked by the Sangamon geosol, may have developed over a time span that corresponds to all of oxygen isotope stages 5 and 4. A large body of ^{14}C and TL ages indicates that Peoria loess began to accumulate near its source areas around 23,000 cal yrs. B.P. and continued to accumulate across the region until about 12 cal yrs. B.P. (Muhs et al., 1999a; Forman and Pierson, 2002; Bettis et al., in press). These ages and others from east of the Great Plains indicate that Peoria loess accumulated during the late Wisconsin Glaciation (oxygen isotope stage 2). Suites of TL, IRSL, and optically–stimulated luminescence (OSL) ages through the Peoria at several Great Plains sections demonstrate that the rate of accumulation varied through time in a systematic fashion across the region (see Stops 1 and 3 on this trip and Roberts et al., in press).

Maximum limiting ages of the Bigness loess are based on ^{14}C ages of soil organic matter from the Brady soil developed in the upper part of the Peoria loess at several Great Plains localities. These ages, in combination with other ^{14}C, TL, and OSL ages within the Bignell loess, indicate that the unit accumulated episodically, although not necessarily across the entire region, through the entire Holocene (Martin, 1993; Pye et al., 1995; Maat and Johnson, 1996; Muhs et al., 1999 a and b; Mason and Kuzila, 2000; Mason et al., 2002).

The most extensive eolian sands in North America occur in the central and southern Great Plains. Eolian sand occurs throughout the Cenozoic section as extensive sand sheets and dune fields, as well as more localized environments. The combined results of a number of investigations during the last two decades have significantly improved our understanding of the chronology of sand movement and the relationships among dune fields and loess in the late Pleistocene. Radiocarbon ages indicate that eolian sand sheets were deposited over large areas of the western part of the central Great Plains during the last glacial period while large barcanoid dunes were forming in the Nebraska Sand Hills (Swinehart and Diffendal, 1990; Loope et al., 1995; Madole, 1995; Muhs et al., 1996; Muhs et al., 1999a; Loope and Swinehart, 2000).

The middle Holocene (ca 8000-5000 ^{14}C yrs. B.P.) was the driest and warmest part of the post-glacial period, and abundant evidence for drought and vegetation density reduction during this period exists in the Great Plains (Webb et al., 1993; Laird et al.,

1996; Grimm, 2001). Although this seems to be an optimum time for eolian sand movement, chronologic evidence for extensive sand movement has been rather elusive, with only a few localities indicating some probable mid–Holocene eolian activity (Loope et al., 1995; Stokes and Swinehart, 1997). Mason et al., (2002) suggest that Bignell loess sections adjacent to dune fields, such as the Wauneta Road Cuts (STOP 7), may provide a clearer and more complete picture of mid–Holocene eolian sand movement than is preserved in the dune fields. Abundant chronologic and soil morphology evidence exists for several episodes of late Holocene eolian sand movement in the region (Albrandt et al., 1983; Arbogast, 1996; Muhs et al., 1996; 1997; Stokes and Swinehart, 1997; Muhs et al., 1999a; Mason et al., in press). Extensive sand movement under essentially modern climatic conditions underscores the climatically sensitive nature of Great Plains eolian systems.

Studies of provenance tracers (zircon ages, Pb isotopes in K–feldspar, and loess geochemistry) as well as thickness trends have demonstrated that the region's last glacial eolian sediments are derived from several sources, some glaciogenic in origin and others nonglaciogenic (Aleinikoff et al., 1999; Muhs and Bettis, 2000; Mason, 2001; Bettis et al., in press). Volcani-clastic siltstone of the White River Group (and possibly the Arikaree Group) were a significant nonglaciogenic source of last-glacial loess, while the Platte and Missouri River valleys were the major sources of glaciogenic loess in the region. Although large dune fields, such as the Nebraska Sand Hills, were the areas of last entrainment for much of the loess in Nebraska, Pb isotopic composition and the ages of zircons in the loess indicate that most of the silt was ultimately derived from Tertiary bedrock sources. Nevertheless, eolian sand was an essential component of the regional loess sedimentary system as it served as the tool in ballistic impacts that brought silt and clay particles into suspension in the atmosphere.

Thickness patterns and grain–size and geochemical trends indicate loess–and–sand–transporting winds were dominantly from the northwest or west across the region (Smith, 1942; Hutton, 1947; Snowden and Priddy, 1968; Mason et al., 1994; Muhs et al., 1999a; Muhs and Bettis, 2000; Mason, 2001). Thick deposits of loess are located immediately southeast of large dune fields in the central Great Plains and the Iowan erosion surface (IES) in southeastern Minnesota and northeastern Iowa, and east of large valleys that drained the Laurentide Ice Sheet (Fig. 1).

OMAHA/COUNCIL BLUFFS TO STOP 1

The field trip headquarters is located on a Historic Missouri river channel belt (Hallberg et al., 1989). The alluvial fill of the Missouri River valley ranges from 25 to about 40 m in thickness in this area (Miller, 1964; Iowa Geological Survey file data) and consists of a thin veneer of fine–grained alluvium (2–5 m) grading downward to medium-to-coarse sand with some meter-thick basal gravel. Near the bluff line, the stratigraphy becomes much more complex as wedges of tributary alluvium and slope deposits interfinger with the main–valley sediments. The valley floor east of the present river is Holocene in age, but on the west side of the river, Wisconsin interstadial floodplain deposits (ca. 22,500-20,000 ^{14}C yrs. B.P., are preserved beneath the Peoria-loess-mantled Fort Calhoun terrace. In the Omaha/ Council Bluffs area, the bedrock surface (Pennsylvanian) is deepest on the east side of the Missouri River valley and rises in a series of steps westward. The bedrock surface beneath the Fort Calhoun terrace is higher than the base of the modern Missouri River channel. Several large quarries in the area extract Pennsylvanian limestone, primarily for use as aggregate.

This area was last glaciated during the pre–Illinoian, about 500,000 yrs. B.P. Multiple glaciations, beginning about 2 Ma, affected the central North American midcontinent and deposited a series of matrix–supported, glacial diamictons (Boellstorff, 1978). During the last one or two of these glaciations the Missouri River Valley probably evolved into its present position and major tributary configuration (Wayne, 1985).

During the last glacial period, the Missouri River valley drained the western margin of the Laurentide Ice Sheet and also was fed by rivers draining unglaciated parts of the central and northern Great Plains. The Platte River, which carried outwash from Rocky Mountain Front Range glaciers and also drained a large portion of the unglaciated central Great Plains, joins the Missouri just south of Omaha.

Bluffs along both sides of the valley are mantled with loess. Pre–Illinoian glacial diamicton generally crops out low along the bluff and is overlain by up to 50 m of late-Pleistocene loess.

STOP 1. LOESS STRATIGRAPHY AT LOVELAND, IOWA

The three most widespread loesses in the North American mid–continent (Peoria, Gilman Canyon/Pisgah, and Loveland) are present at the Loveland section in western Iowa (Fig. 3). Glacial till deposited during one of the last pre–Illinoian glaciations ("A" till of Boellstorff, 1978) underlies the loess sequence here.

This is the reference locality (paratype) for the Loveland loess, which is thought to date from the penultimate ("Illinoian") glaciation. TL and infrared–stimulated luminescence (IRSL) ages from the Loveland here range from 165,000 to 125,000 cal yrs. B.P. and are in good agreement with TL ages of Loveland loess from Nebraska (Maat and Johnson, 1996). The Sangamon geosol is developed in the upper part of the Loveland loess and represents weathering, soil development, and landscape evolution that occurred through the last interglacial and until the overlying Pisgah Fm. loess began to accumulate. Charcoal 3 cm above the base of the Pisgah Fm. dates 42,150 $\pm$ 2350 ^{14}C yrs. B.P. (CAMS 6359) and 44,790 $\pm$ 3250 ^{14}C yrs. B.P. (CAMS 6638) at this section. Luminescence ages (TL, IRSL, and OSL) are in general agreement and bracket accumulation of the Pisgah Fm. between about 46,000 and 23,000 yrs. B.P. Carbonate nodules collected from the loess 5.5-6.5 m below the top of the Sangamon geosol give U–series ages ranging from 57,000 $\pm$ 6000 to 68,000 $\pm$ 8000 yrs. (Table 1), suggesting that they formed during the latter part

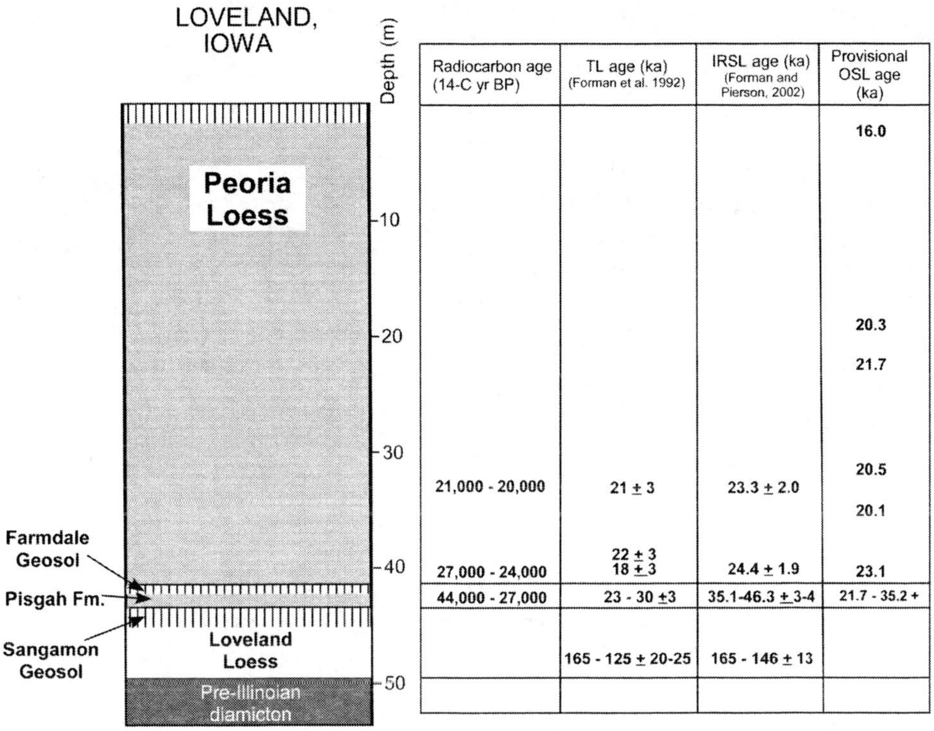

Figure 3. Stratigraphy and radiocarbon and luminescence chronology of the Loveland paratype locality in western Iowa.

of marine oxygen isotope stage 4 and at the stage 4/3 boundary, rather than during stage 5 (the last interglacial). Carbon isotopic data from speleothems in Crevice Cave, near St. Louis, Missouri, indicate that the interval during which these carbonate nodules formed was one of general warming in the mid–continent that culminated between 55,000 and 59,000 cal yrs. B.P. (Dorale et al., 1998). At Loveland, the Sangamon paleo–geomorphic surface, marked by the Sangamon geosol marks a ca.-80,000-year-long depositional hiatus spanning the last interglacial period and most of the early and middle Wisconsin (oxygen isotope stages 5, 4 and the first half of stage 3).

The morphology of the Sangamon geosol at this locality (solum thickness < 2 m, weak to moderate grade soil structure, and a lack of clay coatings) appears at first to be inconsistent with an extended period of weathering covering a full interglacial period. One needs to keep in mind that this is a "bluff line" locality and that local relief on the Sangamon paleogeomorphic surface probably inhibited soil horizonation through runoff, surficial erosion, and reduced infiltration. A short distance east of the bluff line where upland divides are wider, the Sangamon geosol has thick, well-expressed, soil horizons with strong structure and abundant clay coatings. Similar patterns in soil development are seen in transects of modern soils extending from the bluff line eastward.

At Loveland, Pisgah Fm. loess accumulated between about 46,000 and 23,000 yrs. B.P. (Fig. 3) as the first Wisconsin advance of the Laurentide Ice Sheet into the mid–continent occurred (Bettis et al., 1996; Curry, 1998; Dorale et al., 1998). The Pisgah (stratigraphically equivalent to the Gilman Canyon Fm. in Nebraska) here is exceptionally thick, 4 m, and is one of the few examples of relatively unaltered, isotope stage 3 loess in the region. Small pieces of *Picea* charcoal and land snails within the unit, as well as rare pollen, indicate that this loess accumulated on a landscape occupied by vegetation communities that probably shifted from parkland to boreal forest. The Farmdale geosol is developed in the upper 50–60 cm of the Pisgah Fm. This soil marks a significant interglacial period of very slow or no loess accumulation. Abundant pollen and macrofossil evidence indicate that the western Iowa landscape was occupied by boreal forest as this soil formed. The Farmdale geosol is commonly par-

TABLE 1. URANIUM–SERIES ISOCHRON–PLOT AGES OF CARBONATE NODULES FROM THE LOVELAND LOESS AT THE LOVELAND PARATYPE AND MURRAY HILL LOCALITIES, WESTERN IOWA

Sample	Age (ka)
Loveland paratype	
LN–2	63 ± 6
LN–5	68 ± 8
LN–6	57 ± 6
Murray Hill	
LN–3	82 ± 2

Note: U–series ages were calculated from half-lives of 230^{Th} and ^{234}U of 75,200 and 244,000 years, respectively; errors of ages are reported at 1 standard deviation. Ages were determined by B. Szabo, US Geological Survey.

tially eroded and/or deformed, and at Loveland, the Farmdale is a 50-60-cm-thick AC soil profile that is discontinuously present across the section. At STOP 1–3, we will discuss the reasons for widespread erosion of the Farmdale geosol.

During the last decade, this section has been the focus of several investigations addressing the chronology, sedimentology, and geochemistry of Missouri–Valley–source, Peoria loess. About 41 m of Peoria loess occurs above the Farmdale geosol at Loveland. The Peoria is a massive, pale brown-to- light-yellowish-brown, calcareous, silt loam that extends to the modern land surface. A large number of ^{14}C ages from across the region suggest that Peoria loess began to accumulate between 23,000 and 18,000 ^{14}C yrs. B.P. (Bettis et al., in press). When the Peoria loess stopped accumulating is unknown, but estimates based on a few ^{14}C ages of gastropod shells and soil organic matter range from 11,000 to 12,000 ^{14}C yrs. B.P. (Wang et al., 2000; Bettis et al., in press).

Grain–size and geochemical analyses demonstrate that Peoria loess is not uniform in composition through the section (Muhs and Bettis, 2000). Three grain–size zones are present; an upper fine-grained zone, a coarse-grained middle zone, and a lower zone where grain size is more variable (Fig. 4). The upper and middle zones also differ geochemically. Muhs and Bettis (2000) showed that Peoria loess at Loveland has geochemical properties intermediate between nonglaciogenic loess of Nebraska and glaciogenic loess of western Illinois. Eastern Nebraska loess has higher K_2O content (from K–feldspar and micas), and generally higher Na_2O content (from plagioclase) than loess in western Illinois, differences that reflect different silicate mineralogy in the source rocks of these loesses (Fig. 5). Overall, the Peoria at Loveland is intermediate in composition between Nebraska and Illinois loess, indicating that it could contain a mixture of both Laurentide–derived sediments (glaciogenic) and nonglaciogenic sediments. The composition of the upper grain–size zone of the Peoria loess at Loveland is more like that of nonglaciogenic (eastern Nebraska) loess, while the middle zone's composition is more like the glaciogenic loess of western Illinois. This suggests that a larger component of nonglaciogenic, Nebraska-type–source loess may be present in upper Peoria loess than in middle Peoria loess at Loveland.

OSL ages determined using a modified single aliquot (SAR) procedure on 30–50 μm quartz reveals the age structure of the Peoria loess at Loveland (Fig. 3). The lower 7m of the Peoria accumulated between about 23,000 and 20,000 yrs. B.P., slow enough that carbonate was partially leached from this zone as the loess accumulated (Fig. 4; Muhs and Bettis, 2000). Ages from the middle increment of the Peoria (~33–19 m) are statistically indistinguishable, agreeing to within 1–sigma error. The change from the middle to upper zone of the Peoria loess (at 19m depth), occurred abruptly about 20,300 yrs. B.P., and the upper Peoria accumulated over a 4,300 year period until about 16,000 yrs. B.P. (Fig. 3). Mass accumulation rates (MAR) calculated based on the central values for the OSL ages and assuming a typical loess bulk

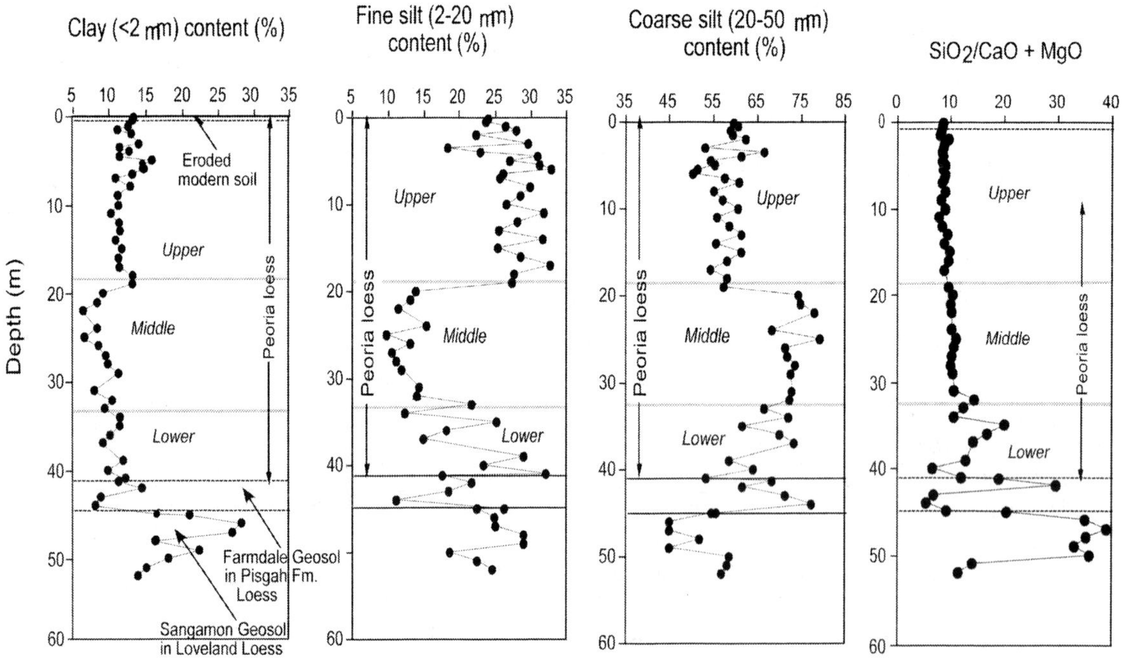

Figure 4. Grain–size and silica/carbonate depth plots for the loess sequence at the Loveland site. Note the presence of three distinct grain size zones in the Peoria loess, and upper fine–grained zone, a coarser middle zone and a lower zone of variable grain size. The silica/carbonate plot indicates that the lower zone is depleted in carbonate relative to the upper zones. Grain–size analysis was done using the pipette method; chemistry was done by wavelength–dispersive X–ray fluorescence. Data from Muhs and Bettis (2000).

density of 1.45 g cm^{-3} show that, based on the uppermost and lowest OSL ages alone, the average mass accumulation rate over the last glacial period is extremely high ~8169 g m^{-2} yr^{-1}. This value is one to two orders of magnitude higher than those calculated for last-glacial, eolian silt elsewhere in the world (Roberts et al., in press). At Loveland MAR varies from 3383 g m^{-2} yr^{-1} in the lower Peoria to 6070 g m^{-2} yr^{-1} in the upper Peoria. In the middle zone of the Peoria MAR were more rapid than OSL dating allows us to resolve (>10,000 g m^{-2} yr^{-1}).

BETWEEN STOPS 1 AND 2

We will drive along the base of the eastern bluff of the Missouri River valley for the next ~30 km (18.5 mi). The stratigraphic sequence beneath this bluff line is similar to that we observed at 1–1. We cross the Boyer River about 4 km north of STOP 1–1. This river carried outwash from the Des Moines lobe of the Laurentide Ice Sheet from about 15,000–13,500 ^{14}C yrs. B.P. The late-glacial floodplain is 10–15 m below the level of the modern floodplain, but rises above the modern floodplain level about 12 km up the valley. Last-glacial gradients typical of valleys that drained the Laurentide Ice Sheet are steeper than modern gradients. The Missouri River valley widens northward and the depth to bedrock increases. This is a region of classic "loess topography" produced by a combination of a thick loess mantle and deeply entrenched stream systems. Bohumil Shimek proposed an eolian origin for loess on the basis of the occurrence of terrestrial gastropod fauna in the loess and the distribution of the loess across the highest parts of the landscape in this region (Shimek, 1896; 1898).

About 30 km north of STOP 1–1, we proceed up the Soldier River Valley, which did not carry outwash during the last glacial period. As we enter the Soldier River valley, the road rises about 25 m (83 ft) up onto a loess–mantled terrace. Loess is more than 30m thick on this terrace and the alluvial surface beneath the terrace is below the level of the modern floodplain (Quade, Iowa Geological Survey, personal communication, 2002). We cross the Soldier River in the town of Pisgah and proceed west toward the Missouri River valley.

Murray Hill is located on the Missouri River valley bluff-line and affords a spectacular view of the valley, which is about 20 km wide here. Exposures along Murray Hill have attracted Pleistocene geologists for a long time. Several ^{14}C ages from soil organic material associated with the Farmdale geosol developed in the Pisgah Fm. have been obtained from exposures along the road descending the bluff. These average about 23,000 ^{14}C yrs. B.P. and are in agreement with those from the same stratigraphic position at Loveland (Bettis, 1990). The Pisgah Fm. overlies the Sangamon geosol developed in Loveland loess at Murray Hill. A large septarian carbonate nodule from the lower part of the Sangamon geosol here yielded a U–series age of 82,000 ± 2,000 yrs. B.P. (Table 1). A complex sequence of pre–Illinoian glacial diamictons and interbedded sand and gravel is beneath the Loveland loess at Murray Hill.

STOP 2. MIDDLE PLEISTOCENE TEPHRA AT THE COUNTY LINE ASH SITE, IOWA

Lenses and beds of Quaternary tephra (volcanic ash) occur in the in the pre–Illinoian glacial, fluvial, and eolian sequences of the Great Plains and adjacent prairies. The tephras occur more commonly and are thicker in the central and western parts of the Great Plains and become rare and thin eastward. The most widely distributed of these tephras are the Pearlette family ashes, which consist of three, geochemically distinct tephra whose source is the Yellowstone caldera complex (Izett and Wilcox, 1982). K–Ar ages of ash–flow tuffs (proximal correlatives of the distal ash beds) and fission–track dating of the tephras provide a firm chronology (Izette and Wilcox, 1982; Lanphere et al., 2002):

0.639 Ma	Lava Creek ash
1.285 Ma	Mesa Falls ash
2.059 Ma	Huckelberry Ridge ash.

The most widely distributed of these tephra, the Lava Creek B, extends from southern California northeastward into southern Alberta, eastward to central Iowa and Missouri, and southward to the Gulf of Mexico (Izette and Wilcox,1982).

The tephra exposed at the County Line site is the Lava Creek B ash (Izett and Wilcox, 1982). This locality has a long history of investigation and illustrates some of the problems inherent in interpretation of pre–Illinoian stratigraphic sequences in the mid–continent. The tephra (about 1200 km downwind from the source) is about 40 cm thick and has an abrupt lower boundary with laminae of silty clay and fine sand. The upper contact of the tephra is gradational, and thin lenses of ashy, silt loam are stratified with silt-loam alluvium. This grades upward to an approximately 14 m thick sequence of planar-bedded, silt loam with thin

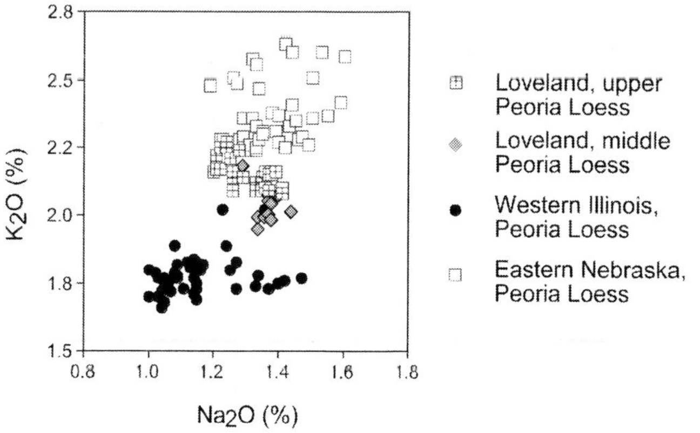

Figure 5. Concentrations of K_2O and Na_2O in bulk samples of upper and middle Peoria loess at Loveland, Peoria loess in eastern Nebraska (Plattsmouth, Elba, and Lincoln sections), and Peoria loess in western Illinois (Morrison and Rapids City B sections). The plot shows that upper Peoria loess has geochemical properties similar to those of nonglaciogenic eastern Nebraska loess, while the geochemistry of the middle Peoria Loes at Loveland is more like glaciogenic (Laurentide-source) loess of Illinois. Adapted from Muhs and Bettis, 2000.

lenses and beds of sand and pebbles. This unit contains several normally graded intervals, and has been interpreted as a pro–glacial sequence unrelated to the Missouri River valley (Bettis, 1990). Massive silt that has been called the Loveland loess by most workers caps the section. A core obtained by the Nebraska Survey and Conservation Division (Core 9–A–76) indicates that the tephra is underlain by about 10 m of sand and gravel that overlie pre–Illinoian glacial diamicton ("A3" till of Boellstorff, 1978). The till overlies 16.7 m of sand and gravel that rests on Cretaceous bedrock.

The Pearlett ashes provide the only currently available means of assigning numerical ages to pre–Illinoian (pre–isotope stage 6) deposits in the Great Plains and adjacent prairies. In addition, because of their wide distribution across the central part of the continent, these tephra and paleomagnetic studies (Easterbrook and Boellstorff, 1981) can provide time lines for linking events and processes in the unglaciated central Plains with glaciated areas to the east.

STOP 3. MAAS DRIVE SECTION, BELLEVUE, NEBRASKA

This locality is about 7 km north of the Platte River valley and about 5.5 km east of the Missouri River valley (Fig. 2). Last–glacial loess, the Sangamon geosol developed in Loveland loess, and older loess and paleosols are present here (Fig. 6). Peoria loess is thicker at localities, like Bellevue, along the east side of the Missouri River valley, than to the immediate west (Mason, 2001; Bettis et al., in press). Mason (2001) estimated that the Missouri and Platte valleys may have contributed from 25 to no more than 50% of the loess deposited in eastern Nebraska. Missouri River valley–source loess east of the valley in Nebraska indicates that although loess–transporting winds were primarily westerlies, easterly or northeasterly winds at times carried significant quantities of dust (Mason, 2001).

The lowest 1–1.5 m of the Peoria is distinctly different from the loess above it. This basal zone has more mottles, contains streaks of deoxidized (light gray) silt and secondary accumulations of manganese oxide, exhibits coarse horizontal parting, and has prominent wavy "streaks" of organic–rich silt (including charcoal) that are sometimes overturned into recumbent folds. AMS ^{14}C ages on humic acids extracted from organic matter in three of the organic–rich zones yielded ages of 22,040±160 (CAMS–10181; 32 cm above base of Peoria), 20,810±160 (CAMS–10187; 36 cm above base of Peoria), and 22,210±170 (CAMS–10188; 38 cm above base of Peoria) ^{14}C yrs. B.P. These ages are out of stratigraphic sequence and older than would be expected from the base of the Peoria loess. This can be explained by considering the genesis of the basal part of the Peoria. Several investigators have noted features attributable to periglacial activity in the upper mid–continent (Péwé, 1983; Muhs and Bettis, in press), and paleo–ecological studies point to the period from about 21,000 to 16,500 ^{14}C yrs. B.P. as one that brought arctic fauna and flora into Iowa, Illinois, and eastern Nebraska (Baker et al., 1986; Schwert et al., 1997). The wavy topography and discontinuous nature of the organic–rich beds and the presence of overturned folds and horizontal partings in the basal Peoria sediments suggest that solifluction played a role in the accumulation of this zone. If the basal zone contains sediment deposited by solifluction during the initial phase of Peoria loess accumulation, then the organic–rich zones were probably derived from the A horizons of the surface soil (Farmdale geosol) upslope. The ^{14}C ages, therefore, may not date the time of deposition, but rather the age (mean residence time) of carbon in the soil being eroded upslope.

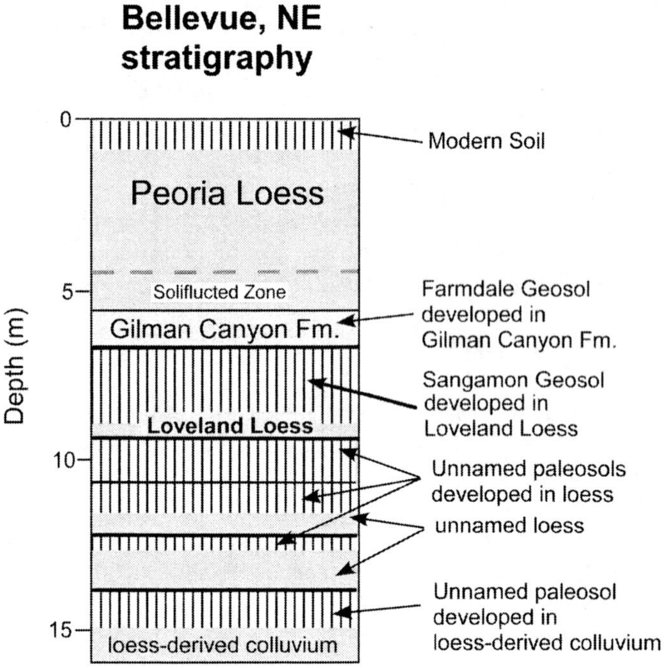

Figure 6. Loess stratigraphy at the Bellevue site south of Omaha, Nebraska.

The Peoria abruptly overlies the Farmdale geosol developed in the Gilman Canyon Fm. (Fig. 6). Lithologic characteristics of the Gilman Canyon Fm. here are very similar to the Pisgah Fm. across the river in Iowa. The unit is modified by the Farmdale geosol which lacks the strong biofabric common to the Gilman Canyon pedocomplex we will see at later stops. Charcoal collected 10 cm below the top of the Gilman Canyon Fm. yielded an AMS radiocarbon age of 25,340±260 ^{14}C yrs. B.P. (CAMS–10190).

The Gilman Canyon Fm. has a gradational boundary with the underlying Sangamon geosol developed in Loveland loess. Here the Sangamon is a morphologically well-expressed soil with a 2 m thick A–AB–Bt profile, thick continuous clay films, strong structure, and strong rubifaction. The base of the Sangamon geosol is not exposed, but a hand boring below the exposed section indicates that older silt (loess) and paleosols are present beneath the Loveland (Fig. 6). The ages of these units are presently unknown.

STOP 4. ELBA CUT, FULLERTON CANAL, NEBRASKA

Quaternary and Pliocene eolian and fluvial sediments, including a Pearlette family ash, are exposed at the Elba cut, a 20–m high exposure through a divide between the North Loope River and Cedar Creek (Figs. 2 and 7; May et al., 1995). The package of late Pleistocene loesses we have seen at previous stops (Peoria, Gilman Canyon, and Loveland) buries older loesses, sand sheets(?), fluvial sand and paleosols. May et al. (1995) suggested that deposits below the Loveland represent three major aggradation episodes capped by paleosols, and apply the provisional names of Reed and Dreeszen (1965) to the sequence. A 5 cm–thick tephra lens near the base of the Red Cloud Fm. (contained in a fluvial channel cut into the Fullerton Fm. and buried by Walnut Creek Fm. west of the area shown in Fig. 7) is either the Mesa Falls (1.285 Ma) or Lava Creek (0.639 Ma). The Walnut Creek Fm. in this exposure appears to consist of at least two loess units with paleosols developed in the upper part of each. These soils are welded into a thick Ab–Bt profile. The lower part of the Walnut Creek is either sandy alluvium or eolian sand that has been modified by pedogenesis. The Red Cloud Fm. underlies the Walnut Creek Fm. and fills channels cut into the Fullerton Fm. A soil (truncated Bt horizion) is formed on the top of the Red Cloud Fm. The Fullerton Fm. consists of at least four loess units with paleosols formed in the upper part of each. The soils formed in the lowest two exposed loess units have Bk horizons.

The Huckelberry Ridge ash (2.059 Ma) is probably present in the upper part of the Fullerton Fm. at the Acadia Canal Site (Boellstorff, 1978; May et al., 1995). As originally conceived, the Fullerton Fm. represents the periglacial equivalent of the earliest glacial advance into the midcontinent, and the upper part of the formation, as exposed at the Elba Cut, may indeed fit that concept.

STOP 5. BIGNELL HILL

Bignell Hill is the best–known loess section in Nebraska (Schultz and Stout, 1945; Dreeszen, 1970; Feng et al., 1994a; Feng et al., 1994b; Maat and Johnson, 1996; Muhs et al., 1999a). It contains the thickest Peoria loess discovered to date (approx. 50 m), which may represent the most rapid, sustained rate of loess deposition recorded anywhere in the world (3–5 m/1000 yr. average). Bignell Hill is also the type section of Bignell loess

Figure 7. A portion of the Elba cut showing the succession of loess and formation-bounding paleosols present at the site. Note person (D. Muhs) in center of photo just above the Loveland/Walnut Creek contact.

(Schultz and Stout, 1945). The section occurs in a series of roadcuts along a dirt road that climbs to the summit of Bignell Table, an isolated remnant of a formerly more extensive loess tableland now destroyed by headward extension of an intricate network of loess canyons and their tributary gullies (Fig. 8). A large, closed depression occupies the center of Bignell Table, but both the roadcut sections and the borehole drilled in 2000 (see below) are located near the outer rim of this depression and do not represent anomalous sections overthickened by depression fill.

Despite the significance of the roadcut section, key questions regarding the stratigraphy at this site remained unanswered until recently. First, because the section extends horizontally over almost 0.5 km, Peoria loess may possibly be draped on a pre–existing slope descending to the Platte River valley and is not as thick as suggested by measuring down the roadcuts. Second, a zone of fine sand apparently crops out within the Peoria loess in the roadcuts, but whether or not a similar sand zone was interbedded with loess directly beneath the table summit is uncertain.

To resolve these questions, Goeke (Conservation and Survey Division, University of Nebraska) arranged for a borehole to be drilled from the table summit in 2000. Although continuous core samples were not recovered, the drill cuttings, together with geophysical logs, allowed accurate identification of the Brady soil, a sand zone within Peoria loess, and the contact between Peoria loess and underlying Gilman Canyon Fm. loess and sand (Fig. 9). Comparison of the borehole log and a measured roadcut section confirms that the latter accurately represents the vertical section of loess beneath Bignell Table. Sand interbedded with Peoria loess is also present in the test hole, but it is thinner, suggesting that the sand may form a wedge that thins southward under the table. Peoria loess, including interbedded sand, is only slightly thinner in the borehole (ca. 49 m) than in the roadcut section (52 m). Previously reported ^{14}C and TL ages from the Bignell Hill section are summarized in Fig. 9. New TL dating by two research groups is in progress and results may be available at the time of the field trip.

In seeking explanations for the enormous thickness of loess at Bignell Hill, the first instinct of many visitors to the site is to look no farther than the Platte River valley immediately to the north. Identifying the Platte as the major source of Peoria loess on the central Plains is consistent with the key role played by other glacially influenced rivers as loess sources in the Midwestern U.S. Although the Platte did not drain the Laurentide Ice Sheet, it drained numerous alpine glaciers during the last glaciation. However, two major problems exist with this explanation. First, loess provenance tracers indicate that Peoria loess across western and central Nebraska has a large component of silt from the White River Group and other Tertiary rocks (Aleinikoff et al., 1998; Aleinikoff et al., 1999). These silt–rich bedrock units crop out northwest of the main Peoria loess deposits and on the unglaciated western Great Plains, but not in the glaciated headwaters of the Platte. Second, a regional view of Peoria loess thickness trends reveals that the association of thick loess with the Platte Valley at Bignell Hill is in large part a coincidence. Peoria loess is thickest in a southwest–northeast trending band from northwestern Kansas to northeastern Nebraska (Fig. 10; Mason, 2001). This band of thick loess is immediately southeast of the Nebraska Sand Hills and other dune fields, and crosses the

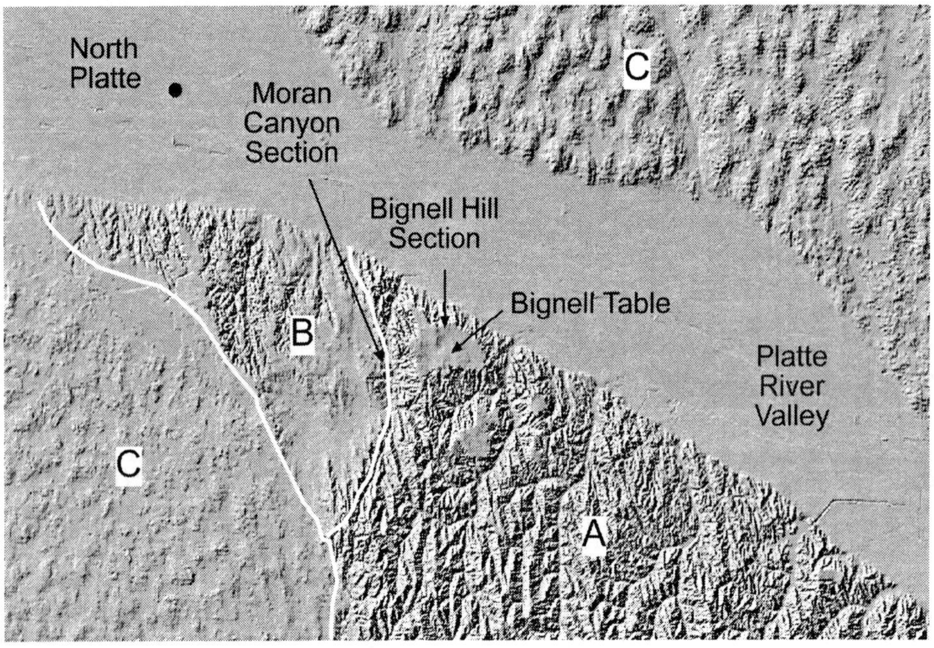

Figure 8. Shaded relief images of Bignell Table and surrounding area, from USGS 30–m digital elevation models. Letters indicate surficial geology: (A) very thick Peoria loess (>30 m), overlain by Bignell loess on stable summits; (B) moderately thick Peoria loess, locally overlain by Bignell loess, with some areas of eolian sand sheets and dunes; (C) dunes and sand sheets with patches of thin loess.

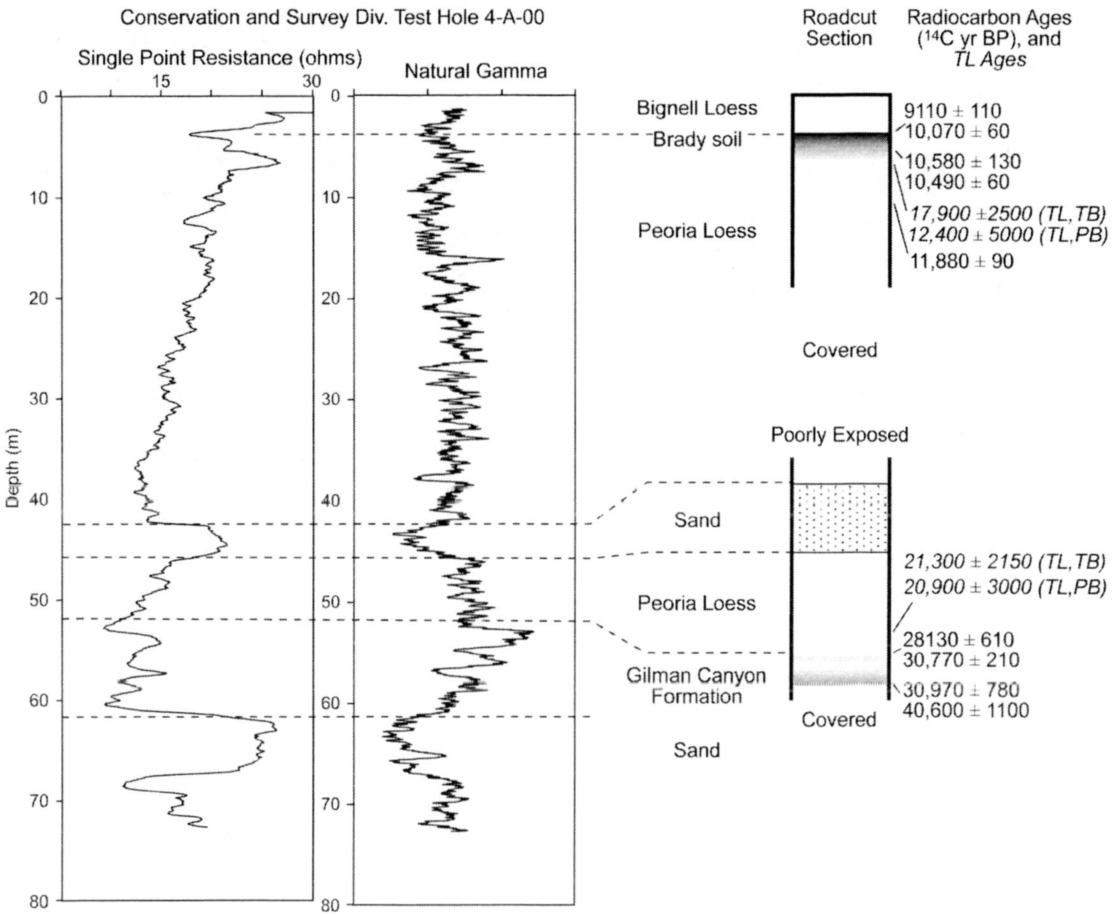

Figure 9. Stratigraphy at Bignell Hill, based on measured roadcut section and borehole 4–A–00. The borehole was about 400 m east of top of roadcuts, at sharp bend in dirt road. Stratigraphy in borehole interpreted from drill cuttings and geophysical logs. Single point resistance is higher when either grain size is coarser or porosity is greater. Brady soil and Gilman Canyon Fm. are finer grained and denser than Peoria loess, and have lower resistance. Natural gamma logs record K, U, and Th content in sediment. Gamma log is higher in the Gilman Canyon Fm. than in Peoria loess, probably because of higher clay content. Resistance is high and the gamma log is low in sand. The roadcut section was measured by J. Mason and X. Miao, using a total station. The most recently reported radiocarbon ages from the roadcut section are at right (Feng et al., 1994b; Maat and Johnson, 1996; Muhs et al., 1999a). TL ages are from Maat and Johnson (1996); see that publication for additional ages from Bignell loess and Gilman Canyon Fm..

Platte valley near Bignell Hill. Although most dune forms in these fields are younger than Peoria loess, evidence exists for eolian sand sheets or dunes in the same general areas during the last glaciation (Mason et al., 1997; Stokes et al., 1999). Peoria loess at Bignell Hill and other sites bordering the dune fields has a modal diameter between 60 and 90 μm, based on laser diffraction particle size analysis, too coarse to have been carried in suspension directly from White River Group outcrops >100 km away (Pye, 1987). Mason (2001) proposed that the dune fields acted as a surface of transport for coarse dust that was ultimately derived from Tertiary bedrock sources. This dust was carried by northwesterly winds across the dune fields, by repeated re-entrainment. This hypothesis is based on wind–tunnel and field studies indicating that impacts from saltating sand are highly effective in ejecting fine–grained material from the surface, resulting in large–scale entrainment of suspended dust. In contrast, a surface covered by uniform silt or clay, but no "saltators," may produce little dust even at high wind speeds (Shao, et al., 1993; Nickling, 1994). Thus, where saltating sand is present, little long–term loess accumulation will occur, whereas thick loess can accumulate where saltating sand is absent (Mason et al., 1999). In this interpretation, Peoria Loess at Bignell Hill is so thick because it is immediately downwind from the dune field where the loess was last entrained. This is analogous to the thick Peoria loess at sites proximal to river valleys in the Midwestern U. S., but in this case the proximal source is not the nearby river valley but the dune field just beyond.

Local Peoria loess thickness anomalies do occur along parts of the Platte Valley in Nebraska, which suggest that the floodplain of that river may have been a secondary source. Based on

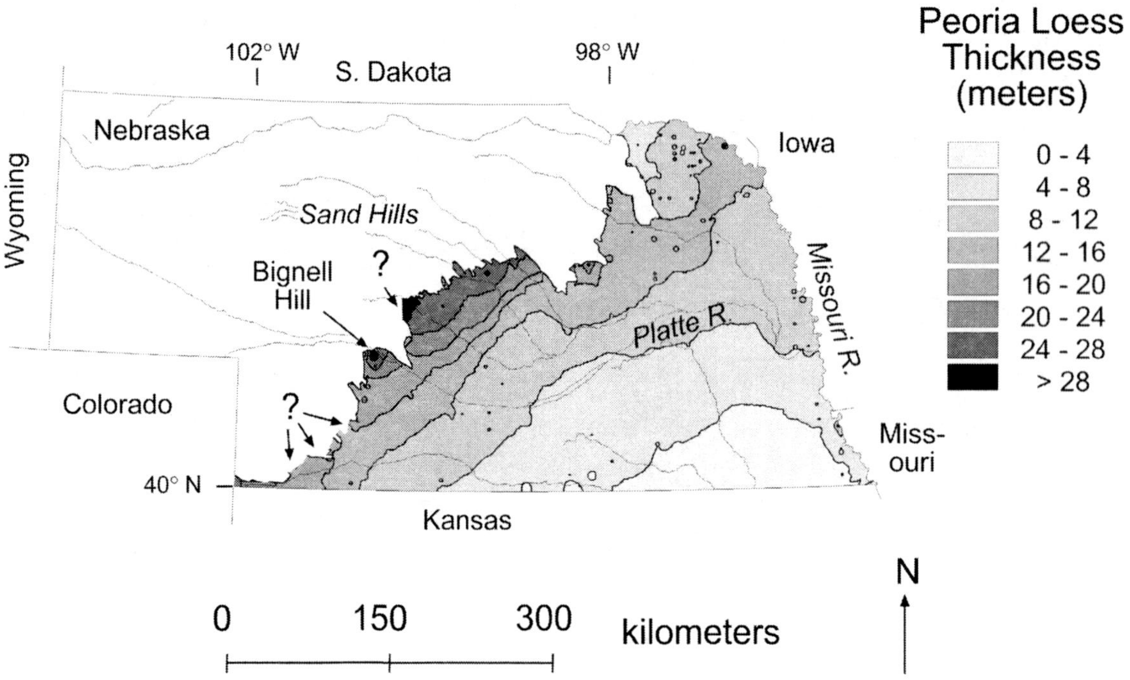

Figure 10. Setting of Bignell Hill relative to regional Peoria loess thickness trends, adapted from Mason (2001). The map is intended to represent an idealized surface of "optimal" loess thickness on the most stable upland summits. See Mason (2001) for details on the data sources and analysis. Question marks indicate areas where the map as originally published may not be valid because of the complex shape of the eolian sand–thick loess boundary.

the regional pattern, however, at least 30 m of Peoria loess would be expected at a dune field–proximal site like Bignell Hill, even far from secondary river valley sources (Fig. 10). In northeastern Colorado, provenance data indicate a more significant contribution from the South Platte valley at times during the last glaciation (Aleinikoff et al., 1999).

"Typical loess" is often described as massive silt with a modal diameter somewhere between 20 and 30 µm (Pye, 1987). In contrast, Peoria loess at Bignell Hill is laminated and has a modal diameter >60 µm, based on laser diffraction particle size analysis. Lamination and a modal diameter in the very fine sand range are typical of thick, proximal Peoria loess across western Nebraska and northwestern Kansas. Laminated Peoria loess at Bignell Hill and similar sites is a primary eolian deposit; that is, lamination does not result from reworking of loess into alluvial or lacustrine deposits. Several localities occur where lamination is ubiquitous in the numerous Peoria loess outcrops distributed along loess tableland escarpments for distances of 10 km or more. Laminated loess is also consistently encountered in cores collected beneath upland summits in the same areas. Thus, laminated loess is not confined to former lake basins, closed depressions, or valley fills.

Although the lamination in Peoria loess appears to be of eolian origin, it differs in two important ways from lamination produced by migrating wind ripples in eolian sand, where transport is predominantly by saltation (Hunter, 1977). Wind ripple laminae commonly coarsen upward (Hunter, 1977), whereas the laminae in Peoria loess fine upward, with a distinctly fine–grained layer capping each lamina. In addition, wind–ripple laminae can be quite evenly spaced at a local scale But when they are traced laterally, low angle truncations are commonly observed. In contrast, laminae in Peoria loess fine upward and can sometimes be traced for 10 m or more across an outcrop, with no evidence of truncation. These contrasts with sedimentary structures typically produced by saltation favors an alternative explanation. The laminae may represent increments of suspended dust that formed thin sheets across a nearly–level land surface. Feng et al., (1994a) proposed that the lamination in Peoria loess is annual, given that the typical thickness of laminae at Bignell Hill is 3–7 mm, and the average sediment rate of Peoria loess in this section is probably within the range of 3–7 mm yr^{-1}. This scenario requires that the laminae were not destroyed by bioturbation, perhaps because of the high deposition rate. Observations in the Nebraska Sand Hills suggest that grass root growth alone does not destroy sedimentary structures. Animal burrowing is required to accomplish this, and perhaps burrowing was not extensive enough to destroy lamination in rapidly accumulating loess. Thin sections of laminated loess indicate that finer–scale grain size variation occurs within the macroscopic laminae, possibly reflecting pulses of dust deposition during individual storms.

Typical wind–ripple lamination does occur in fine sand interbedded with lower Peoria loess in one part of the Bignell Hill section. Sedimentary structures in this sand are generally consistent with an eolian sand-sheet origin. Abundant calcite rhizoliths suggest intermittent vegetation cover. Soft sediment deformation features are also present within the sand, at least superficially similar to those reported from niveo–eolian sand deposits and periglacial coversands (e.g. Ruegg, 1983), where deformation occurs by cryoturbation or through melting of trapped snow or ground ice. Possible sources of sand interbedded with or underlying Peoria loess are discussed in the following section.

STOP 6. MORAN CANYON ROADCUT

This roadcut exposes Bignell loess, the Brady soil, laminated Peoria loess, and eolian sand directly underlying the Peoria loess. The site is immediately across a canyon from Bignell Table (Fig. 8). Eolian sand sheets and dunes also occur within a few km west of the site. The original land surface at the top of the roadcut climbs eastward out of a closed depression; the rim of the depression drops directly into the canyon to the east. Optical luminescence ages from the roadcut and an adjacent core are in progress and results will be reported on the field trip.

About 2.5 m of Bignell loess overlies the prominent Brady soil in the roadcut. Note that the Bignell loess does not thicken appreciably toward the lowest part of the closed depression (i.e. toward the west end of the roadcut). Coring in the pasture north of the roadcut confirmed that both the Brady soil and overlying Bignell loess are present even on the highest part of the depression rim where it drops off into the canyon, although the Bignell loess is thinner there than in the roadcut (ca. 1 m). These observations lead to the following scenario, which also applies to other closed depressions that have been investigated in areas proximal to the Sand Hills and other dune fields: Bignell loess was deposited across the upland landscape, as demonstrated by preservation of at least some Holocene loess even on narrow convex depression rims. Some Bignell loess was reworked by slopewash, thickening sediment above the Brady soil in the depressions and on gentler slopes surrounding them (e.g. the part of the landscape exposed by the Moran Canyon roadcut), and thinning the loess on the least stable depression rims.

Bignell loess stratigraphy in a core directly above the Moran Canyon roadcut is illustrated in Fig. 11. The stratigraphy here is broadly similar to that observed at other sites. This similarity suggests that any overthickening of the Bignell section here by slopewash from the depression rim has not significantly altered the stratigraphic record of Holocene dust deposition. The three zones that can consistently be identified in thick Bignell loess sections are better expressed at the Wauneta roadcuts (STOP 3–2A), where they will be discussed in detail.

Peoria loess here is laminated but apparently much thinner than beneath Bignell Table immediately to the east (although more Peoria loess may be present beneath the eolian sand exposed in the lower part of the roadcut). The sand forms a clearly recognizable dune, with large scale cross–strata deposited on a slip face as the dune migrated eastward. Preliminary results of OSL dating indicate that the dune was active sometime in the interval 20–25 ka. Correlation of this dune sand with the eolian sand interbedded with Peoria loess at Bignell Hill is a distinct possibility, which is currently being tested with additional OSL dating. If eolian sand at both sites are about the same age, they may record a widespread, but temporary, advance of dunes and sand sheets eastward and/or southward into the region where loess was accumulating. eolian sand activity may have been shorter lived at Bignell Hill than at the Moran Canyon site. If dunes remained active for much longer on the west side of Moran Canyon, this could explain the much smaller thickness of Peoria loess that is exposed there.

Likely sources of the sand include sand bars exposed at low flow in the Platte River system (including both North and South Platte branches, which join just west of Moran Canyon), and the Sand Hills dune field north of the Platte system. Muhs et al., (2000) investigated the provenance of sand in dune fields of southwestern Nebraska, south of the Platte River system Contents of K and Rb in this dune sand are consistent with mixed provenance, from both Platte River sediment and the Sand Hills dune field north of the Platte. Muhs et al. (2000) proposed that dune sand migrated across the Platte valley from the Sand Hills at a time when the Platte River channel belt was dry, and suggested that this occurred when the river was dammed by dune sand. The sand below or within Peoria loess at Moran Canyon and Bignell Hill may record such an episode of eolian sand migration across the valley. The modern Platte River frequently experiences low flows in late summer, and long reaches of the river have occasionally gone dry, including during the dry summer of 2002. However, discharge of both North and South Platte rivers is now reduced by irrigation water diversions upstream, so complete cessation of flow may have required more extreme dry conditions in the upstream basin than observed in historical times.

A final question for discussion at the Moran Canyon roadcut: Is the thick A horizon on the lower part of the slope dropping to Moran Canyon a remnant of the "Brady paleolandscape," that is, a remnant of a stable landscape present when the Brady soil formed, which has been partially removed by erosion of steeper slopes?

BETWEEN STOPS 6 AND 7

We will drive along the irregular, southwest-northeast- trending border between eolian sand and thick loess, from the Bignell Hill area toward the southwest corner of Nebraska. The exact route will be based on weather and road conditions. The eolian sand of southwestern Nebraska has been the subject of a few previous studies (e.g. Madole, 1995; Muhs et al., 1997; Muhs et al., 2000), but no comprehensive effort to map dunes, sand sheets and loess in detail, comparable to the work done by Smith (1965) and Swinehart (1990) in the Nebraska Sand Hills has been undertaken. The following brief description and Fig. 12 are based on soil surveys, air photos, satellite imagery, and field reconnaissance in 2001–2002.

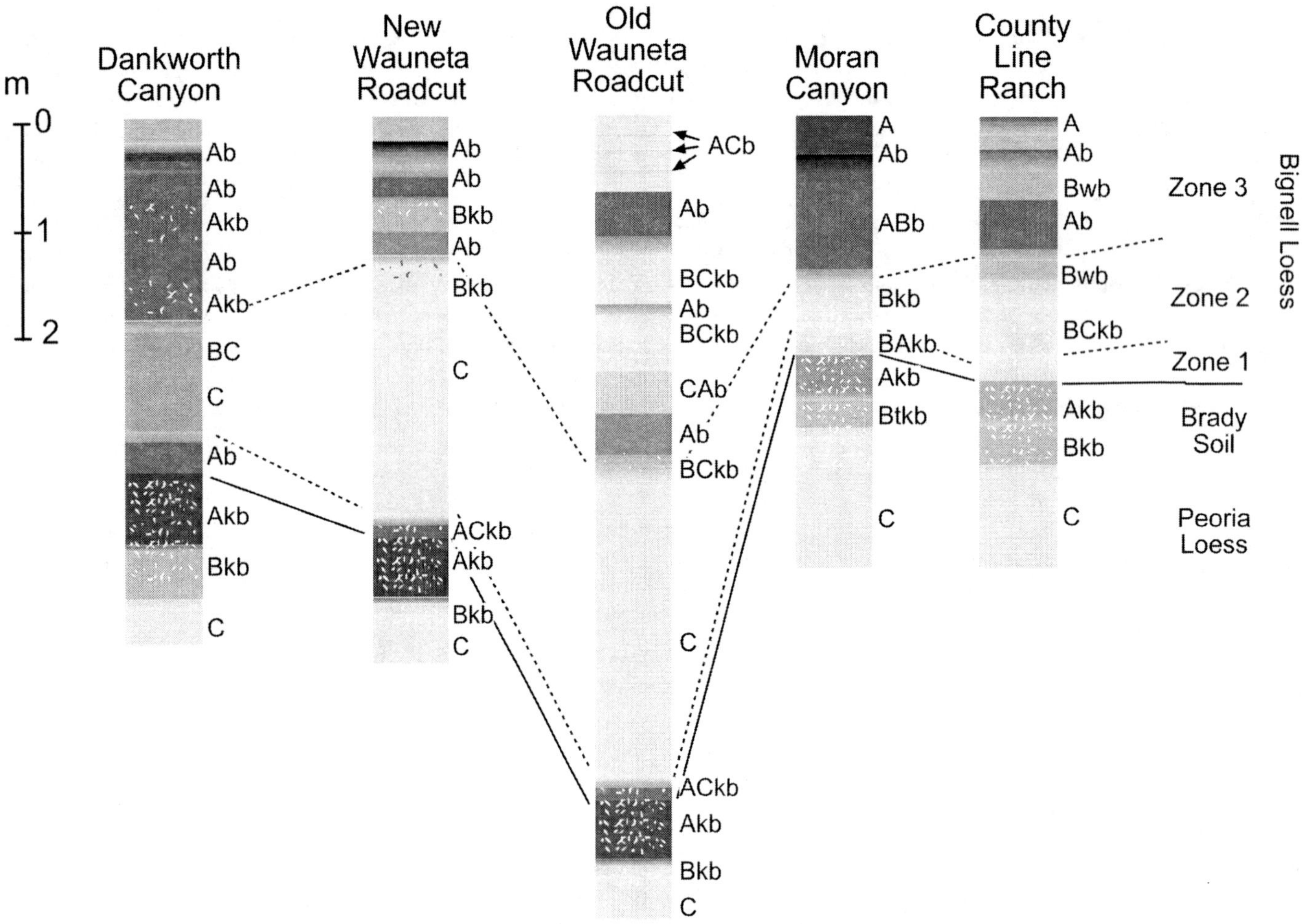

Figure 11. Stratigraphy and paleopedology of thick Bignell loess at representative sites. County Line Ranch is located north of the Platte River valley at the edge of the Sand Hills and will not be visited on this field trip. Zones are discussed in the text.

Thick loess is mainly Peoria, with a thin, but extensive, veneer of Bignell loess on stable uplands. Older loess is present in some areas, but in other localities Peoria loess rests directly on bedrock of the Ogallala Group (Miocene) or Pierre shale (Cretaceous). Peoria loess reaches at least 30 m thickness in areas near the northwestern border of thick loess. Bignell loess is generally <1 m thick, except near the eolian sand/loess border where it can be 4–6 m thick. The thick loess landscape has a distinctive character, with either gently sloping ridgetops or broader "loess tables" bounded by highly dissected slopes descending to narrow canyons (e.g. Fig. 8). The surface of the loess tables is often marked by shallow closed depressions.

Thin loess occurs mainly north of our route from STOP 3–1 to STOP 3–2, in areas immediately south of the South Platte River, and separated from thick loess to the southeast by dune fields and sand sheets (Fig. 12). Loess appears to be generally <2 m (commonly <1 m) thick, and often interspersed in a complex pattern with eolian sand sheets and dunes. The thin loess generally overlies Ogallala Group bedrock or gravel of the Broadwater Fm. (Pliocene). Much of it may be of Holocene age because it overlies a buried soil with morphology similar to the Brady soil.

Sand sheets and low–relief dunes cover large areas northwest of the thick loess. These areas have only modest relief on the ground, but display a prominent northwest–southeast lineation, and some long, narrow, parabolic dunes on air photos and satellite images. The sand sheets typically contain at least a little more silt and clay than higher-relief dunes. In some cases, the sand sheet deposits are relatively rich in silt and approach the texture of the coarsest, most proximal, Peoria or Bignell loess. Soils in the sand sheets often have Bw or even Bt horizons, although some sand sheet soils display only A–C profiles. Little age control is available for sand sheets in the field trip area, but similar sand sheets in which soils with Bt horizons have formed are of late Pleistocene age in northeastern Colorado (Muhs et al., 1996). The sand sheet deposits appear to be only a few meters thick in most exposures.

High–relief dunes have a variety of forms, including simple and compound parabolic and linear dunes. The parabolic dunes

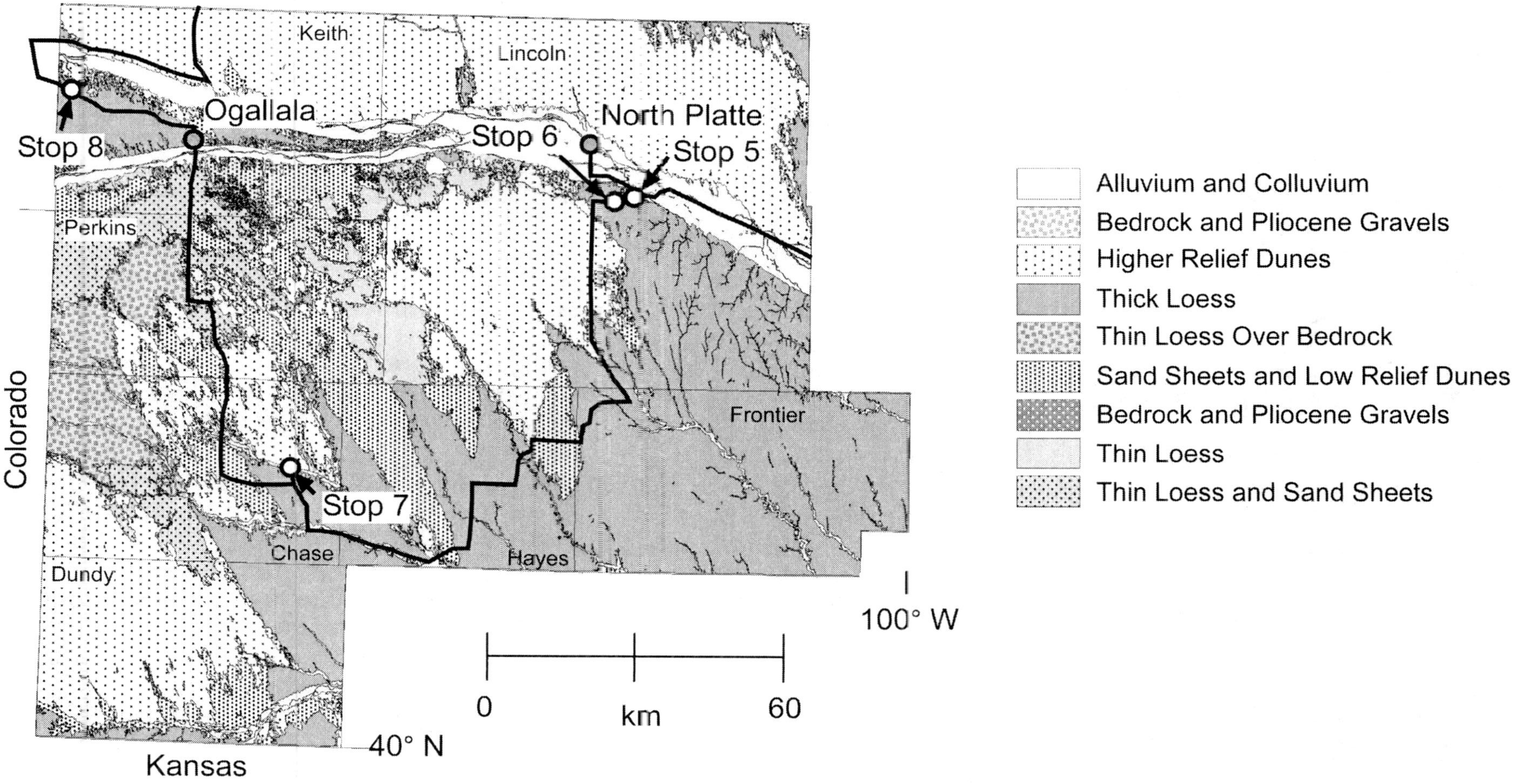

Figure 12. Surficial deposits (mainly eolian) in southwestern Nebraska, based mainly on soil surveys. County lines and streams are shown for reference.

were formed by northwesterly winds. Compound parabolic dunes consist of large parabolic forms hundreds of meters long, covered with much smaller parabolic forms. Compound parabolic dunes are much better developed to the west in northeastern Colorado. Linear dunes are oriented west to east or west–northwest to east–southeast and are similar to linear dunes representing the final phase of widespread sand movement in the central and eastern Sand Hills (see discussion of STOP 4–2). Sand in high-relief dunes generally has very low silt and clay content. These dunes typically have A–C surface soil profiles and may contain similar weakly developed buried soils at a shallow depth. In some of the dune fields, discerning a surficial A horizon at all is difficult, suggesting recent activity. The thickness of eolian sand in areas of high-relief dunes is probably not much greater than the height of present dune forms because bedrock crops out along the margins of shallow valleys within the dune fields, and forms the floor of some interdune depressions.

In southwestern Nebraska, the boundary between eolian sand sheets or dune fields and thick Peoria loess has a distinctive sawtooth–like form, typically extending toward the west or northwest along stream valleys, then veering abruptly toward the southeast until reaching another stream valley (Fig. 12). The sand to thick loess transition is remarkably abrupt in some areas, with an escarpment of thick loess rising high above adjacent sand sheets or dunes (e.g. Fig. 13). Along other segments of the boundary, however, eolian sand either interfingers with, or overlies Peoria loess. The landscape near the boundary is often characterized by pronounced northwest–southeast "flutes" formed in loess (erosional features, similar to yardangs?).

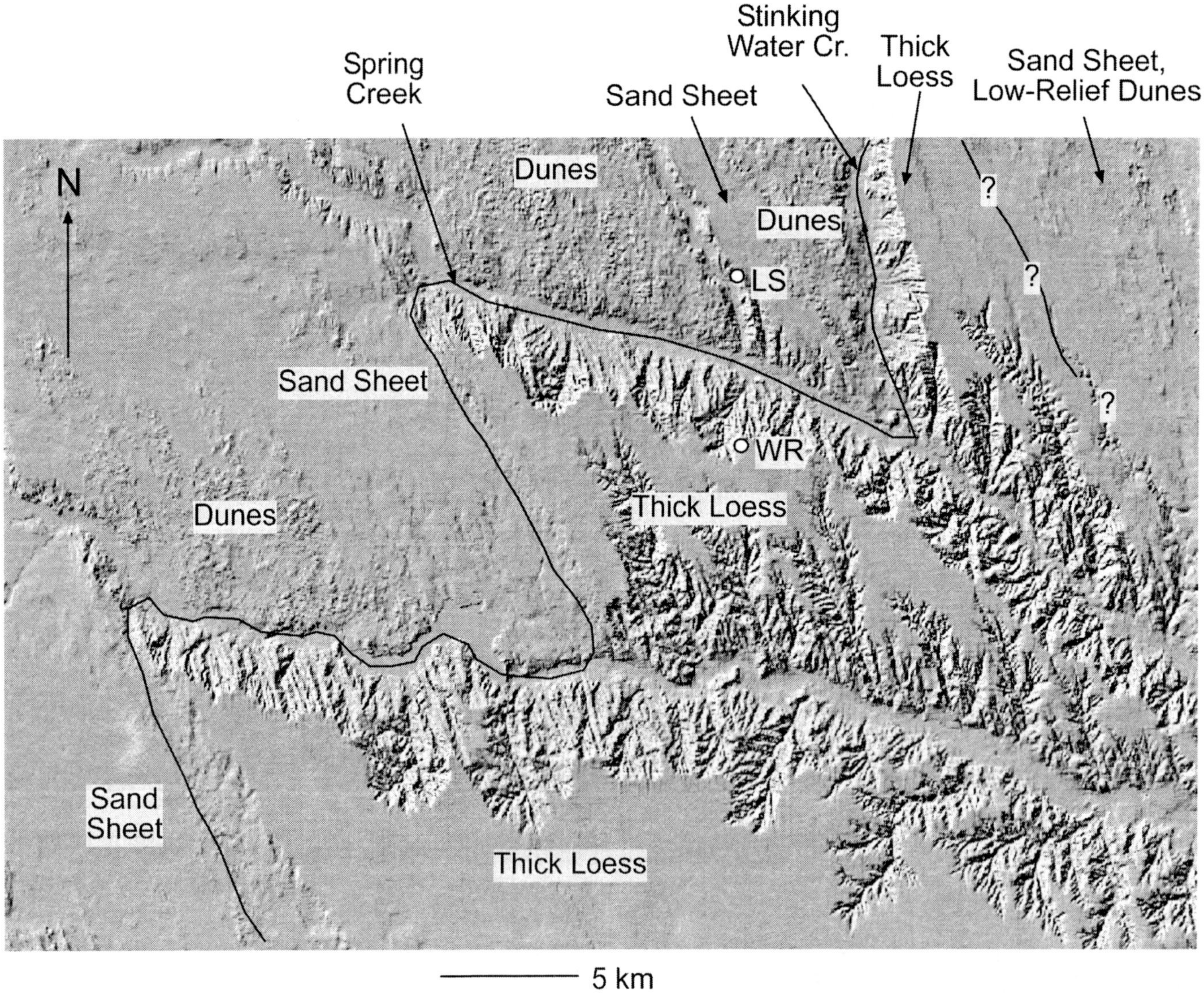

Figure 13. Shaded relief image of the region surrounding the Wauneta roadcuts, showing surficial geology in relation to topography and stream valley locations. From USGS 30-m digital elevation models. WR = Wauneta roadcuts; LS = Lewis site.

Mason et al., (1999) described a very similar sawtooth–like windward limit of thick loess in southeaster Minnesota and northeastern Iowa. They explained the location of the abrupt proximal edge of thick loess as the result of narrow, steep–sided valleys that were topographic barriers to the downwind migration of saltating sand. Noting that saltating sand can be very effective in entraining silt (as discussed above in the section on Bignell Hill), Mason et al. (1999) proposed that long–term loess accumulation was prevented upwind from these barrier valleys because eolian sand was moving across uplands there. Downwind from the barrier valleys, thick loess accumulated. Moving upstream along each stream valley, a point occurred beyond which it was no longer incised deeply enough to act as a topographic barrier to sand movement. At that point, the sand–loess boundary veered to the southeast toward the next deeply incised stream.

In southwestern Nebraska, the relationship between the eolian sand to thick loess boundary and stream valleys is very similar (Fig. 12, 13). However, the valleys in Nebraska appear too shallow, with sides that are too gently sloping, to have acted as topographic barriers to eolian sand movement. Instead, the barrier may have been provided by dense vegetation and perennially flowing streams on the valley floors, both fed by discharge from the High Plains aquifer (= "Ogallala Aquifer"). The points where the sand-loess border veers to the southeast are probably the upstream limits of significant groundwater discharge during accumulation of Peoria loess, that is, during formation of the abrupt boundary between sand and thick loess. This concept is illustrated in Fig. 14. Along most streams in the region today, a point occurs at which springs appear on the valley floor and stream flow becomes perennial. This explanation depends on the concept that the wind speed threshold for re–entrainment of loess is very high in the absence of a saltating sand supply. The applicability of this concept to extremely coarse, western Nebraska loess could be questioned; as usual, more work is needed...

STOP 7A. WAUNETA ROADCUTS

At this stop, we will examine a pair of adjacent roadcuts, New Wauneta roadcut on the present county road and Old Wauneta roadcut on the former road just to the east. Old Wauneta roadcut contains the thickest Bignell loess identified to date (6 m). Both cuts expose the stratigraphy beneath the north edge of a broad loess table, which faces dunes and sand sheets across the valley of Spring Creek. This stop illustrates (1) typical Brady soil morphology, (2) the sources, typical geomorphic setting, and stratigraphy of thick Bignell loess, and (3) the nature of the eolian sand–thick loess boundary where it is associated with a stream valley.

The Brady soil is well-exposed in the Wauneta roadcuts, and its morphology here is typical of many other exposures. The Brady soil has an A–Bk profile almost everywhere it is exposed in western Nebraska. The A horizon is thick and dark, and the Bk has relatively strong fine blocky structure that is distinctly different from any B horizon morphology observed in overlying Bignell loess. The most distinctive Brady soil characteristic, however, is the ubiquitous occurrence of cylindrical burrows, 4–10 mm in diameter, throughout the lower A and Bk horizons. In some cases, most of the total soil volume in the A to Bk transition is a mass of crosscutting burrows. We believe the burrows are the work of insects, probably cicadas (Homoptera: Cicadidae). Their paleo–environmental significance is the subject of ongoing work.

Most previous work on Bignell loess suggested that it is thickest near large river valleys and was derived from floodplain sources (e.g. Dreeszen, 1970; Johnson and Willey, 2000; Mason and Kuzila, 2000). At first glance, the type section and several other localities where thick Bignell was reported fit this conceptual model. More extensive investigation of Bignell loess distribution has revealed a very different story (Mason et al., 2002). Most Bignell loess >1 m thick in Nebraska and northwestern Kansas occurs in a similar setting to the Wauneta roadcuts-the top of a table–edge escarpment facing north or west toward dune fields and sand sheets. Bignell loess 3–6 m thick is present along at least 15 km of the table edge where the Wauneta roadcuts are located, and on other table edges visible to the northeast and southwest. The small streams separating the loess tables from the dune fields (Spring Creek and Stinking Water Creek) are not plausible loess sources (we will cross Spring Creek, allowing trip participants to make this judgment for themselves).

This leaves the dune fields and sand sheets as the only viable source for the Holocene loess. The type section at Bignell Hill and other thick Bignell loess sections, such as Dankworth Canyon (Johnson and Willey, 2000), are also on table edges facing north toward dune fields, but the presence of the Platte valley between the loess and the dune fields has made the source–sink relationship more difficult to recognize. However, thick Bignell loess is not present at bluff top sites along the Platte Valley where no dune fields occur nearby (Mason et al., 2002). While a floodplain source for some Holocene loess cannot be ruled out, the weight of the evidence clearly favors dune fields and sand sheets as the predominant source.

Bignell loess thickness at table edge sites, like the Wauneta roadcuts, is enhanced by local topographic effects. As northerly winds sweep up over the table edge, flow separation probably occurs for a short distance downwind from the brink. If the winds were driving a dust plume southward from the dune fields, much of the coarsest dust would drop out in this zone of flow separation, creating a thick loess deposit that thinned rapidly downwind in the area where the wind flow reattached. Consistent with this scenario, Bignell loess thins rapidly with distance south or east from the table edges. In a core transect a few km east of the Wauneta roadcuts, Bignell loess thinned from 5.5 m at the table edge to 0.9 m within 1.5 km distance south from the edge (J. A. Mason and H. Loope, unpublished data). The rate of thinning then becomes much less as shown by Bignell loess 0.5–1 m thick on stable uplands over large areas downwind from the dune fields. Rapid thinning southward from the northernmost points along the table edge may explain the contrast in Bignell loess thickness between Old and New Wauneta roadcuts.

Although Bignell loess at table–edge sites is anomalously thick, we believe that its internal stratigraphy is representative of variations in dust deposition rate and pedogenesis across the entire region downwind from the dune field sources. The three basic zones observed in the thickest sections can be identified even in Bignell loess <1 m thick at more distal sites. These zones are illustrated in Fig. 11 and discussed in more detail by Jacobs and Mason (in press). Zone 1 has *A* horizon characteristics that become weaker with distance above the Brady soil. Zone 2 is light–colored, with relatively little pedogenic alteration. In the thickest Bignell loess sections, Zone 2 looks much like unweathered Peoria loess. In thick sections, Zone 3 is a sequence of buried A–C, A–Bw, or A–Bk profiles, each interpreted as representing an episode of more rapid loess deposition, followed by an interval of slower deposition and more effective pedogenesis. Where Bignell loess is thinner, the multiple buried soils of Zone 3 merge into a single, over-thickened, A horizon. The uppermost part of Zone 3 is light–colored, suggesting that we are still in the latest interval of more rapid deposition. In the Old Wauneta roadcut, three very weakly developed buried A horizons can be distinguished within this upper light–colored part of Zone 3.

Preliminary results of OSL dating at Wauneta roadcuts and several other sites indicate that Zone 1 is of early Holocene age (9–10 ka) and Zone 2 accumulated between 9 ka and 6 ka. Most of Zone 3 postdates 4 ka, and the upper light–colored part of Zone 3 was deposited in the last 1000 years with a cluster of ages around 0.7 ka. The ages obtained from Zone 2 provide strong evidence for an early–mid Holocene dry period on the central Great Plains, a point that has been difficult to establish using radiocarbon ages from dune fields. Zone 3 records episodic aridity and dust deposition in the late Holocene as well. More definite ages and paleoclimatic interpretations will be presented on the field trip.

Stops 7A and B provide excellent examples of the evidence for control of the eolian sand-thick loess boundary by belts of dense vegetation along wet valley floors (Fig. 13, 14). Gully exposures and the few well logs available suggest that Peoria loess may be 40–50 m thick at the north edge of the table. Gully exposures also suggest that Peoria loess in this area rests almost directly on bedrock of the Ogallala Group. To the north, across Spring Creek, eolian sand a few meters thick also rest directly on bedrock. Higher–relief dunes are used for pasture, while fields of row crops are located on sand sheets. The high ground visible to the northeast is also thick loess, separated from the eolian sand by the valley of Stinking Water Creek.

In the reaches of Spring and Stinking Water creeks visible from this site, both have wet valley floors, with many springs feeding perennial stream flow. The dunes on the north side of Spring Creek descend to the valley floor, but no evidence exists that eolian sand has migrated across the valley to the thick-loess side. About 8 km west of the Wauneta roadcuts, the sand-thick–loess boundary swings abruptly southeastward, away from Spring Creek (Fig. 13). A short distance upstream from that point, groundwater discharge ceases, and both stream channel and valley floor are dry. Eolian sand occurs on both sides of this dry upstream reach, and dunes have apparently blocked several tributary valleys a few km farther west.

The present relationship between the eolian sand–thick loess boundary and wet valley floors supports the hypothesis outlined above. However, we don't know where the upstream limit of groundwater discharge in these valleys was located at the time Peoria loess was accumulating. In fact, even the present upstream limit may be an artifact of recent water table drawdown by pumping for irrigation, which has been as much as 15 m in this area.

STOP 7B. LEWIS SITE

The Lewis site is a roadcut exposing a dark colored, organic-matter-rich-swale fill in eolian very fine sand and silt, which is in turn buried by younger sand that contains less silt. Bedrock (Ogallala Group.) crops out just down the hill from the roadcut, indicating that all eolian deposits here are much thinner than the loess across the valley to the south. Preliminary OSL ages indicate that the swale fill formed between 15 and 10 ka, approximately the same time that the Brady soil was developing in the Wauneta roadcuts just to the south. The very fine sand underlying the swale fill is of late Pleistocene age, based on preliminary OSL chronology. It is fine to very fine and has a high silt content. In fact, it is not much coarser than Peoria loess in the Wauneta roadcuts. This sediment could be interpreted as a thin outlying remnant of coarse Peoria Loess, or as a fine–grained sand sheet deposit, laterally continuous with sand sheets east of the roadcut. Surface soils in those sand sheets have either Bw or Bt horizons, consistent with a late Pleistocene age. The swale fill contains even more silt than the underlying sand. We interpret the swale fill as the product of slow sedimentation, allowing syndepositional organic matter accumulation, pedogenic structure development, and possibly bioturbation, which destroyed any sedimentary structures. The source of the sediment in the fill could have been dust generated by limited eolian activity in the surrounding landscape, slopewash, or a combination of the two. More extensive eolian activity would have filled the swale with light-colored sand; thus, the period of Brady Soil formation was a time of limited eolian activity at this site within the dunefields as well as in the loess–mantled landscape downwind. The lower part of the fill is darker than the upper part, suggesting slower sedimentation or higher plant productivity early in the period of time represented by the fill.

The silty sand underlying the fill is of late Pleistocene age, based on preliminary OSL chronology and may be laterally continuous with sand-sheet deposits east of the roadcut. Surface soils in these deposits have either Bw or Bt horizons, consistent with a late Pleistocene age. The sand overlying the swale fill began accumulating in the early Holocene, but the age of the upper part of this sand is not known at present. West of the road, high-relief dunes with minimally developed A–C surface soils are probably late Holocene. The younger dunes west of the road may be the product of reactivation and reworking of the late Pleistocene silty sand sheets still preserved east of the road. This process could

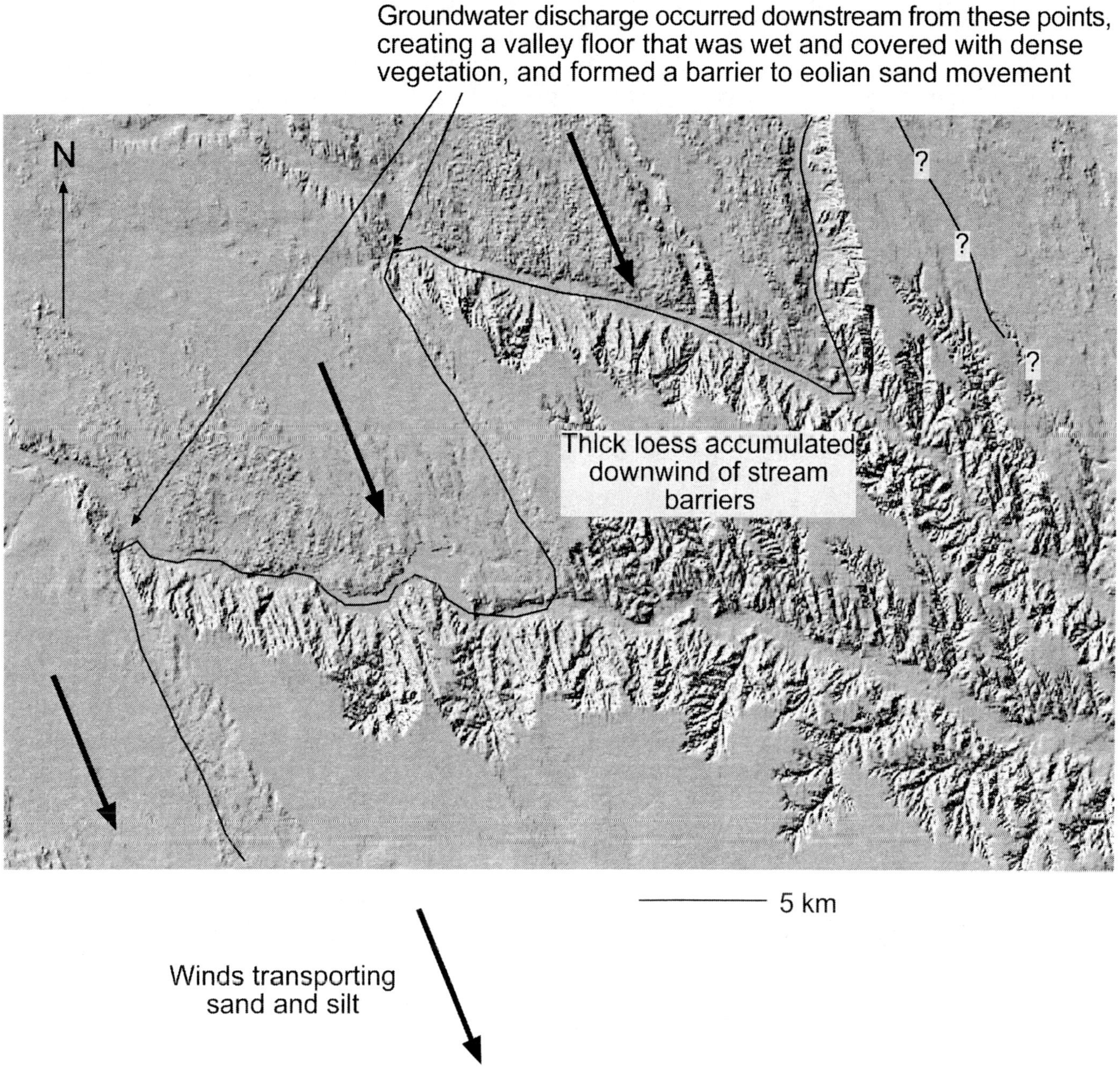

Figure 14. Same as Fig. 7, but labeled to illustrate conceptual model of how groundwater discharge to stream valleys influences location of eolian sand–thick loess boundary.

have produced abundant coarse dust that was carried south to become Bignell loess.

BETWEEN STOP 7B AND OGALLALA, NEBRASKA

We will drive west from the Wauneta roadcuts and descend the remarkably abrupt and straight edge of thick loess where it trails southeast from Spring Creek. The point where this edge leaves Spring Creek was presumably the upstream limit of late Pleistocene groundwater discharge and dense valley floor vegetation, if the hypothesis outlined above is valid. Note the fluted appearance of the landscape at the west edge of the table. After we drop off the table, bedrock is just below the land surface, with a patchy cover of eolian sand sheets and thin loess. After passing through the town of Imperial, we will drive north to the South Platte River valley through a mixture of high–relief dunes, sand sheets, and thin loess (the latter two will be hard to distinguish from the road) (Fig. 12). Note the frequent Ogallala Group outcrops. This is essentially a wind–scoured bedrock surface on which thin, patchy, eolian sediment is preserved.

STOP 8. OVERLOOK AT DANKWORTH CANYON

This stop will be made if time permits, and is mainly an "arm–waving" overview. At the overlook, we will be in an outlying area of thick Peoria loess, forming a wedge between the North and South Platte River valleys. To the north, across the North Platte valley (now filled with the Lake McConaughey reservoir), lie the western Sand Hills, through which we will travel for most of the day. Peoria loess here is about 33 m thick, measured both in outcrop in the canyon and in a borehole about 0.5 km to the west. The entire outlier of thick loess in this area is cut by northwest–southeast oriented flutes, similar in form to those at the west edge of the Wauneta table, but much larger. Bignell loess 3.5 m thick overlies the Brady soil near the canyon rim. This is a similar geomorphic setting to the Wauneta roadcuts and Bignell Hill. The internal stratigraphy of Bignell loess here is similar to other sites (Fig. 11).

Down the canyon and out of sight are outcrops of lake sediments, first discovered by R. F. Diffendal, Jr., and later attributed by Loope and Swinehart (2000) and Muhs et al., (2000) to dune-damming of the North Platte valley. The lake sediments have yielded late Pleistocene radiocarbon ages, suggesting that damming occurred at that time.

We will pass near the Kingsley Dam, one of the largest earth–fill dams in the world, while traveling toward the next stop. The core of the dam is built of compacted "loess soil" (probably Peoria loess from south of the dam site); its location was chosen to allow the dam to rest on the low–permeability Brule Fm. of the White River Group, which is itself volcaniclastic loess of Oligocene age.

BETWEEN STOPS 8 AND 9: THE NEBRASKA SAND HILLS

The Nebraska Sand Hills form the largest dune field in North America, but are almost completely grass–covered and stable at present. For many years, this dune field was interpreted as a relict of cold, dry conditions in the late Pleistocene, although the possibility of minor Holocene activity was acknowledged (Smith, 1965). Ahlbrandt et al., (1983) challenged this view with evidence of large–scale Holocene dune activity. More-recent studies have documented extensive Holocene sand transport and dune migration (Muhs, 1985; Swinehart, 1990; Muhs and Holliday, 1995; Muhs et al., 1997; Stokes and Swinehart, 1997). However, evidence for late Pleistocene eolian activity also exists, and some of the larger dunes may have initially formed at that time (Mason et al., 1997; Stokes et al., 1999). Many recent developments in the geologic history of the Sand Hills are reviewed by Loope and Swinehart (2000), along with some speculative ideas on interactions between climate, vegetation, groundwater, and dune activity that should be explored in future work. Here, we only briefly review points of particular interest for the field trip stops.

Dune forms range from mega–barchans and barchanoid ridges up to 130 m high in the central Sand Hills (Fig. 15A) to smaller linear dunes that predominate in the southeastern Sand Hills (Fig. 15B). One key aspect of dune morphology not emphasized in earlier work is the presence of relatively small linear dunes superimposed on larger dunes across much of the central and eastern Sand Hills (Fig 15C). The linear dunes are often oriented west-northwest to east-southeast. Where they are present, they clearly represent the final major phase of eolian activity, and they may signal a major change in wind regime. Current work on upland dunes of the Sand Hills is focused on applying OSL dating to the linear dunes at multiple sites to determine their age and paleoclimatic significance.

In addition to their value for interpretation and dating of dunes, the Sand Hills interdune valleys hold valuable paleoenvironmental records, recognized by Swinehart and Loope in the early 1990's. The High Plains aquifer underlies the entire dune field. Where high-relief dunes are present, local shallow groundwater flow systems are superimposed on the regional eastward flow of the aquifer, conveying water from recharge areas beneath the dunes to discharge in adjacent interdune sites (Loope and Swinehart, 2000; Mason et al., in press). Where groundwater discharge is particularly strong, peat has accumulated through much of the Holocene. Sand layers occur within the peats. These are interpreted as eolian sand sheets that extended into the interdune areas at times when groundwater recharge was temporarily reduced, the local flow system was shut down or greatly reduced in volume, and the peat surface was desiccated (Mason et al., in press). The sand layers within the peat can be ^{14}C dated, using bracketing plant macrofossils, to provide a high–resolution record of dry periods during which eolian activity occurred. The uppermost sand layer in several fens formed between 950 and 650 yrs. B.P., which is also the time of most recent significant sand movement at several upland sites distributed across the central Sand Hills (Mason et al., in press). It also appears to be a time of Bignell loess deposition at several sites (see discussion of Wauneta roadcuts, Stop 3–2A). One or more episodes of dry climate in this time interval are recorded at sites elsewhere in the Great Plains.

STOP 9. MILEPOST 81 ROADCUT

This stop is to examine a recent roadcut through one of the linear dunes representing the most recent major episode of eolian activity in this part of the Sand Hills (Fig. 16). Cross–strata exposed here indicate a bimodal wind regime, with both southeasterly and northerly components, in contrast to the prevalence of northerly to northwesterly paleowinds inferred from older dunes. Preliminary OSL ages suggest that this linear dune formed during the dry episode recorded by the uppermost sand layer in interdune peatlands.

STOP 10. BIG CREEK FEN

This is a brief stop to allow a view of a typical interdune peatland. A vibracore collected in a similar peatland will be examined at this stop or later in the day.

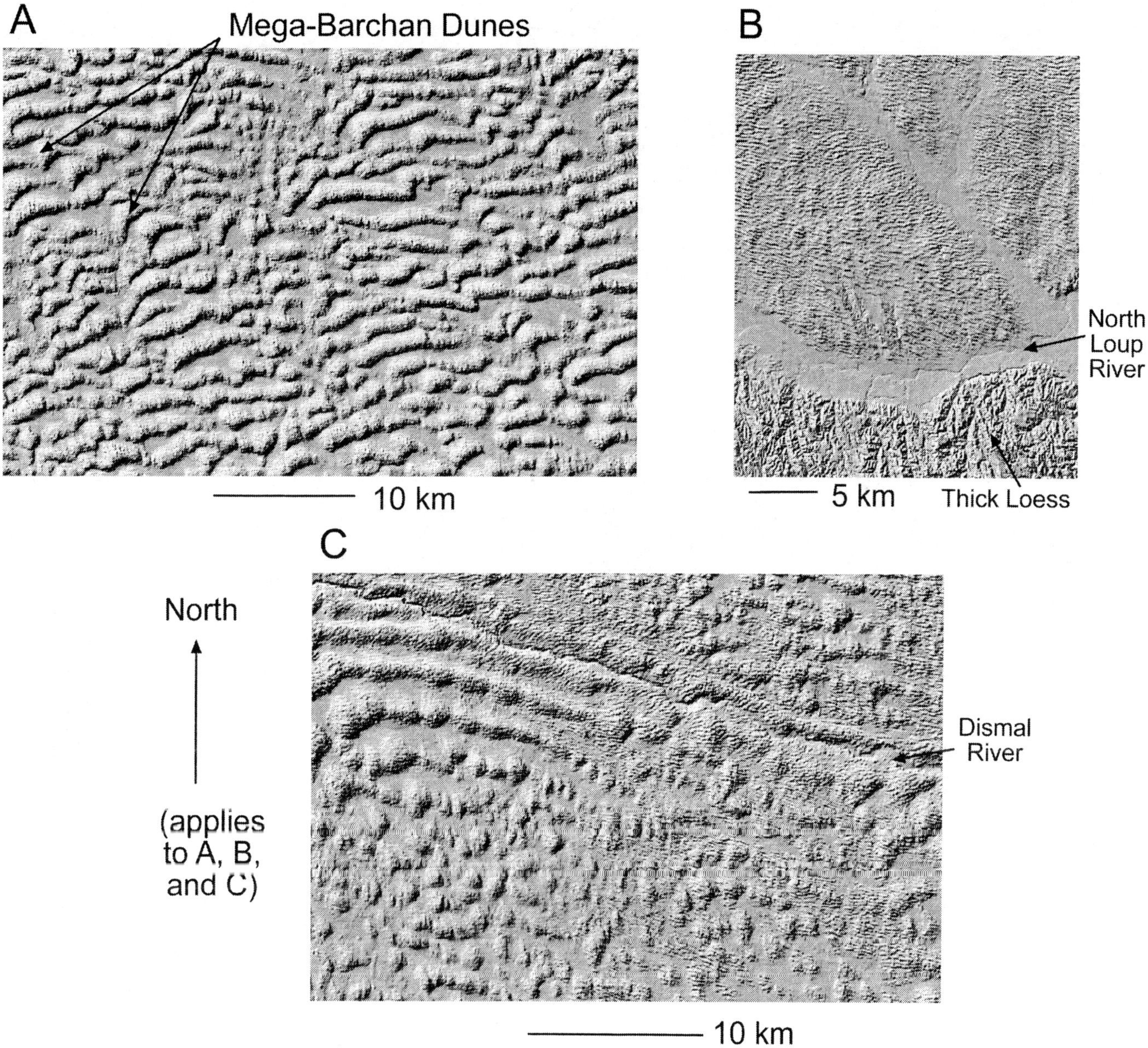

Figure 15. Examples of dune forms in the Nebraska Sand Hills, using shaded relief images based on USGS 30-m digital elevation models. (A) Area in central Sand Hills dominated by barchanoid ridge dunes, but with some mega–barchan dunes in upper left. Field trip route will pass through the east or west side of the area shown, depending on time available. (B) Area dominated by linear dunes in eastern Sand Hills. (C) Linear dunes superimposed on older barchanoid ridge and dome–like dunes (Swinehart, 1990), southeast of STOPS 4–2 and 4–3.

BETWEEN STOP 10 AND ALLIANCE, NEBRASKA

We will follow US Highway 2 for much of the way between STOP 4–3 and Alliance. Because the highway follows east-west-oriented interdunes, it does not provide the best view of the large–scale transverse ridge dunes that we are passing through. The route does, however, offer excellent examples of alluvial fans at the base of the steep leeward (south) faces of the dunes. These fans are built by sediment eroded from gullies on the dune slopes during extreme rainstorms, transported mainly in hyperconcentrated flows (Sweeney and Loope, 2001). They provide clear evidence that surface runoff can occur on the Sand Hills dunes, and, in fact, can significantly degrade dune forms during humid periods when eolian activity is minimal.

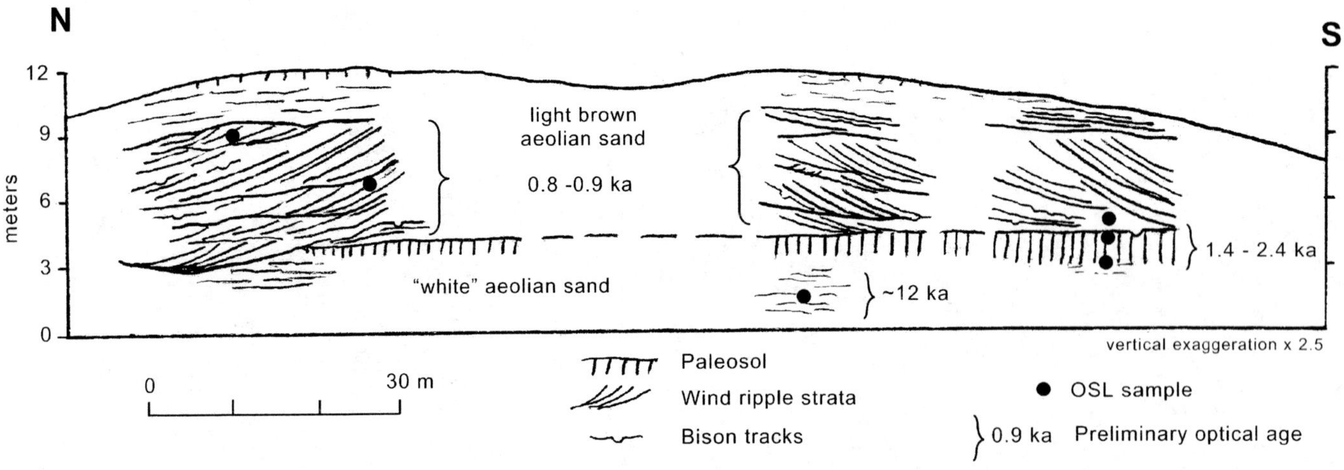

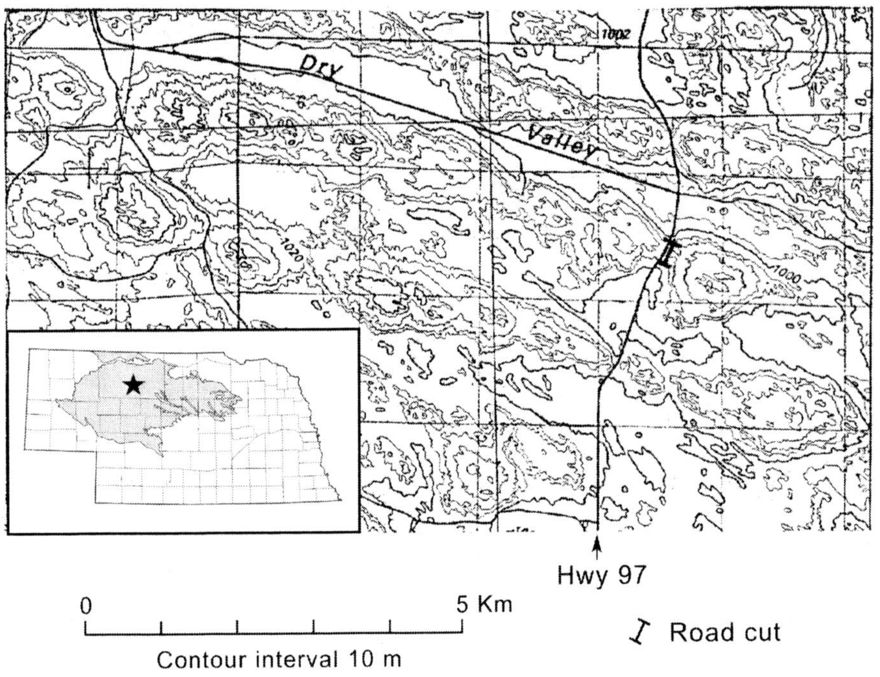

Figure 16. Milepost 81 (Highway 97) roadcut through a linear dune, Cherry County, Nebraska (STOP 4–2). (A) Sketch map of a roadcut with preliminary optical ages on aeolian sand. Wind ripple strata on south side of exposure have dip azimuths between 90° and 150° (E–SE), while those on the north side have dip azimuths from 290° and 25° (NW–NNE). Note that the basal paleosol underlying the linear dune is truncated at the north end of the cut. (B) Topographic map of the Milepost 81 road cut in a saddle between two 50 m high mega–barchan dunes that are part of a 10 km long barchanoid ridge dune. ESE-trending linear dunes (with relief of 6–9 m) are superimposed on the big dunes but are not apparent on the topographic map. Base map from USGS Mullen 1:100,000 quadrangle.

STOP 11. SCOTTS BLUFF NATIONAL MONUMENT

A reprint of "Swinehart, J.B. and Loope, D.B., 1987, Late Cenozoic geology along the summit to museum hiking trail, Scotts Bluff National Monument, western Nebraska: *in*, Biggs, D.L., ed., Geological Society of America Centennial Field Guide –North Central Section, p. 13–18." will be provided to field trip participants.

ACKNOWLEDGMENTS

We thank the landowners of the sites for permission to work on their property and to allow our group access for this trip. We greatly appreciate the considerable contributions of Dan Muhs, U.S. Geological Survey, to the sampling and interpretation of several stops on this trip. Josh Been, USGS, and Geoff, Duller and Ann Wintle of the Luminescence Dating Laboratory, Institute of Geography and Earth Sciences, University of Wales, Aberystwyth assisted with field sampling and laboratory analyses. Tom Stafford processed the AMS ^{14}C samples from the Loveland and Bellevue sites. This work was partially supported by NSF grants EAR–00–87572 to Bettis, BCS–0079252 and EAR–0087916 to Mason, BCS–0079320 to Jacobs, EAR97–09942 to Loope and Swinehart, DEB–0084075 to D. Wedin and others at UNL, and the Conservation and Survey Division and Department of Geosciences, UNL.

REFERENCES

Arbogast, A.F., 1996, Stratigraphic evidence for late–Holocene eolian sand mobilization and soil formation in south-central Kansas, U.S.A.: Journal of Arid Environments, v. 34, p. 403–414.

Arbogast, A.F., and Muhs, D.R., 2000, Geochemical and mineralogical evidence from eolian sediments for northwesterly mid–Holocene paleowinds, central Kansas, USA: Quaternary International, v. 67, p.107–118.

Albrandt, T.S., Swinehart, J.B., and Maroney, D.G., 1983, The dynamic Holocene dune fields of the Great Plains and Rocky Mountain basins, U.S.A.: *in* M.E. Brookfield and T.S. Albrandt, eds., Eolian sediments and processes: Elsevier, NY, p. 379–406.

Aleinikoff, J. N., Muhs, D. R., and Fanning, C. M., 1998, Isotopic evidence for the sources of Late Wisconsin (Peoria) loess, Colorado and Nebraska: implications for paleoclimate: *in* Busacca, A. J., ed., Dust Aerosols, loess soils, and Global Change: College of Agriculture and Home Economics Miscellaneous Publication MISC0190: Washington State University, Pullman, Washington, p. 124–127.

Alienikoff, J.N., Muhs, D.R., Sauer, R.R., and Fanning, C.M., 1999, Late Quaternary loess in northeastern Colorado II: Pb isotopic evidence for the variability of loess sources: Geological Society of America Bulletin, v. 111, p. 1876–1883.

Baker, R.G., Rhodes, R.S., II, Schwert, D.P., Ashworth, A.C., Frest, T.J., Hallberg, G.R., and Janssens, J.A., 1986, A full–glacial biota from southeastern Iowa, USA: Journal of Quaternary Science, v. 1, p. 91–107.

Bettis, E.A. III, 1990, Holocene alluvial stratigraphy and selected aspects of the Quaternary history of western Iowa: Midwest Friends of the Pleistocene 37th Field Conference Guidebook, 197p.

Bettis, E.A., Muhs, D.R., Roberts, H.M., and Wintle, A.G., in press, Last glacial loess in the conterminous U.S.A.: Quaternary International.

Bettis, E.A., Quade, D.J., and Kemis, T.J., 1996, Hogs, bogs, and logs: Quaternary deposits and environmental geology of the Des Moines Lobe: Iowa Department of Natural Resources, Geological Survey Bureau Guidebook Series No. 18, Iowa City.

Boellstorff, J.D., 1978, A need for redefinition of North American Pleistocene stages: Transactions Gulf Coast Association of Geological Societies, v. 28, p.65–74.

Curry, B.B., 1998, Evidence at Lomax, Illinois for Mid–Wisconsin (~40,000 yr B.P.) position of the Des Moines Lobe and for diversion of the Mississippi River by the Lake Michigan Lobe (20,350 yr B.P.): Quaternary Research, v. 50, p. 128–138.

Dorale, J.A., Edwards, L.A., Ito, E., and González, L.A., 1998, Climate and vegetation history of the midcontinent from 75 to 25 ka: A speleothem record from Crevice Cave, Missouri, USA: Science, v. 282, p. 1871–1874.

Diffendahl, R.F., Jr., 1982, Regional implications of the geology of the Ogallala Group (Upper Tertiary) of southwestern Morril County, Nebraska, and adjacent areas: Geological Society of America Bulletin. v.93, p. 964–976.

Easterbrook, D.J. and Boellstorff, J., 1981, Age and correlation of early Pleistocene glaciations based on paleomagnetic and fission–track dating in North America: International Geological Correlation program Project 24 Report no. 6, p. 189–207.

Feng, Z. D., Johnson, W. C., and Diffendal, R. F., Jr., 1994a, Environments of aeolian deposition in south–central Nebraska during the last glacial maximum: Physical Geography, v. 15, p. 249–261.

Feng, Z. D., Johnson, W. C., Lu, Y., and Ward, P. A., III, 1994b, Climatic signals from loess–soil sequences in the central Great Plains, USA: Palaeogeography, Palaeoclimatology, Palaeoecology, v. 110, p. 345–358.

Forman, S.L., Bettis, E.A. III., Kemmis, T., and Miller, B.B., 1992, Chronologic evidence for multiple periods of loess deposition during the late Pleistocene in the Missouri and Mississippi River Valleys, U.S.: Implications for the activity of the Laurentide Ice Sheet: Palaeogeography, Palaeoclimatology, Palaeoecology. v. 93, p. 71–83.

Forman, S.L. and Pierson, J., 2002, Late Pleistocene luminescence chronology of loess deposition in the Missouri and Mississippi river valleys, United States: Palaeogeography, Palaeoclimatology, Palaeoecology. v. 186, p. 25–46.

Frye, J.C. and Leonard, A.B., 1952, Pleistocene geology of Kansas: Kansas Geological Survey Bulletin 99.

Grimm, E.C., 2001, Trends and paleoecological problems in the vegetation and climate history of the northern Great Plains, U.S.A.: Proceeding of the Royal Irish Academy. v. 101B, p. 47–64.

Holliday, V.T., 1988, Mt. Blanco revisited: Soil–geomorphic implications for the ages of the upper Cenozoic Blanco and Blackwater Draw Formations: Geology, v. 16, p. 505–508.

Hunter, R. E., 1977, Basic types of stratification in small eolian dunes: Sedimentology, v. 24, p. 361–387.

Hutton, C.E., 1947, Studies of loess–derived soils in southwestern Iowa: Soil Science Society of America Proceedings, v. 12, p. 424–431.

Izette, G.A., and Wilcox, R.E., 1982, Map showing localities and inferred distributions of the Huckelberry Ridge, Meas Falls, and Lava Creek ash beds (Pearlette Family ash beds) of Pliocene and Pleistocene age in the western United States and southern Canada: U.S. Geological Survey, Miscellaneous Investigations Series, Map I–1325.

Jacobs, P. M., and Mason, J. A., In Press, Paleopedology of soils in thick Holocene loess, Nebraska, USA: Revista Mexicana de Ciencias Geologicas.

Johnson, W. C., and Willey, K. L., 2000, Isotopic and rock magnetic expression of environmental change at the Pleistocene–Holocene transition in the central Great Plains: Quaternary International, v. 67, p. 89–106.

Laird, K.R., Fritz, S.C., Grimm, E.C. and Mueller, P.C., 1996, Century–scale paleoclimate reconstruction form Moon Lake, a closed–basin lake in the northern Great Plains: Limnology and Oceanography, v. 41, p. 890–902.

Lanphere, M.A., Champion, D.E., Christiansen, R.L., Izett, G.A., and Obradovich, J.D., 2002, Revised ages for tuffs of the Yellowstone Plateau volcanic field: Assignment of the Huckelberry Ridge Tuff to a new geomagnetic polarity event: Geological Society of America Bulletin, v. 114, p. 559–568.

Loope, D.B., Swinehart, J.B., and Mason, J.P., 1995, Dune–dammed paleovalleys of the Nebraska Sand Hills: Intrinsic versus climate controls on the accumulation of lake and marsh sediments: Geological Society of America Bulletin, v. 107, p. 396–406.

Loope, D. B., and Swinehart, J. B., 2000, Thinking like a dune field: Geologic history in the Nebraska Sand Hills: Great Plains Research, v. 10, p. 5–35.

Maat, P.B. and Johnson, W.C., 1996, Thermoluminescence and new ^{14}C age estimates for late Quaternary loesses in southwestern Nebraska: Geomorphology, v. 17, p. 115–128.

Madole, R.F., 1995, Spatial and temporal patterns of late Quaternary eolian deposition, eastern Colorado: Quaternary Science Reviews, v. 14, p. 155–177.

Martin, C.W., 1993, Radiocarbon ages on late Pleistocene loess stratigraphy of Nebraska and Kansas, central Great Plains: Quaternary Science Reviews, v. 12, p. 179–188.

Mason, J. A., 2001, Transport direction of Peoria loess in Nebraska and implications for loess sources on the central Great Plains: Quaternary Research, v. 56, p. 79–86.

Mason, J.A., Jacobs, P.M., Hanson, P.R., Miao, X.–D., and Goble, R.J., 2002, Bignell loess in the central Great Plains provides a continuous record of Holocene dune field activity and climate change: American Quaternary Association Program and Abstracts of the 17th Annual Meeting. p. 90.

Mason, J.A., and Kuzila, M.S., 2000, Episodic Holocene loess deposition in central Nebraska: Quaternary International, v. 67, p. 119–131.

Mason, J.A., Nater, E.A., and Hobbs, H.C., 1994, Transport direction of Wisconsinan loess in southeastern Minnesota: Quaternary Research, v. 41, p. 44–51.

Mason, J. A., Nater, E. A., Zanner, C. W., and Bell, J. C., 1999, A new model of topographic effects on the distribution of loess: Geomorphology, v. 28, p. 223–236.

Mason, J. A., J. B. Swinehart, R. J. Goble, and D. B. Loope, In press, Late Holocene dune activity linked to hydrologic drought, Nebraska Sand Hills, USA: The Holocene.

Mason, J. P., Swinehart, J. B., and Loope, D. B., 1997, Holocene history of lacustrine and marsh sediments in a dune–blocked drainage, southwestern Nebraska Sand Hills, USA: Journal of Paleolimnology, v. 17, p. 67–83.

May, D.W. and Holen, S.R., 1993, Radiocarbon ages of soils and charcoal in late Wisconsinan loess, south–central Nebraska: Quaternary Research, v. 39, p. 55–58.

May, D., Swinehart, J.B., Loope, D. and Souders, V., 1995, Late Quaternary fluvial and eolian sediments: Loop River Basin and the Nebraska Sand Hills: *in* Flowerday, C.A., ed., Geologic Field trips in Nebraska and Adjacent Parts of Kansas and South Dakota, Conservation and Survey Division, Institute of Agriculture and Natural Resources, University of Nebraska–Lincoln, Guidebook no. 10, p. 13–31.

Miller, R.D., 1964, Geology of the Omaha–Council Bluffs area, Nebraska–Iowa: U.S. Geological Survey Professional Paper 472.

Muhs, D.R., 1985, Age and paleoclimatic significance of Holocene sand dunes in northeastern Colorado: Annals of the Association of American Geographers, v. 75, p. 566–582.

Muhs, D.R., Aleinikoff, J.N., Stafford, T.W., Jr., Kihl, R., Been, J., Mahan, S.A., and Coward, S.D., 1999b, Late Quaternary loess in northcentral Colorado, I: Age and paleoclimatic significance: Geological Society of America Bulletin, v. 111, p. 1861–1875.

Muhs, D.R. and Bettis, E.A. III, 2000, Geochemical variations in Peoria loess of western Iowa indicate paleowinds of midcontinental North America during last glaciation: Quaternary Research, v. 53, p. 49–61.

Muhs, D.R. and Bettis, E.A. III, in press, Quaternary loess–paleosol sequences as examples of climate–driven sedimentary extremes: *in* Chan, M., and Archer, A.W., (eds.), Sedimentary Extremes: Geological Society of America Special Paper.

Muhs, D. R., and Holliday, V. T., 1995, Evidence of active dune sand on the Great Plains in the 19th century from accounts of early explorers: Quaternary Research, v. 43, p. 198–208.

Muhs, D.R., Stafford, T.W., Jr., Cowherd, S.D., Mahan, S.A., Kihl, R., Maat, P.B., Bush, C.A., and Nehring, J., 1996, Origin of late Quaternary dune fields of northeastern Colorado: Geomorphology, v. 17, p. 129–149.

Muhs, D.R., Stafford, T.W., Jr., Swinehart, J.B., Cowherd, S.D., Mahan, S.A., Bush, C.A., Madole, R.F., and Maat, P.B., 1997, Late Holocene eolian activity in the mineralogically mature Nebraska Sand Hills: Quaternary Research, v. 48, p. 162–176.

Muhs, D.R., Swinehart, J.B., Loope, D.B., Aleinikoff, J.N., and Been, J., 1999a, 200,000 years of climate change recorded in eolian sediments of the High Plains of eastern Colorado and western Nebraska: *in* Lageson, D.R., Lester, A.P., and Trudgill, B.D., eds., Colorado and Adjacent Areas: Geological Society of America Field Guide 1, Boulder, Colorado, p. 71–91.

Muhs, D. R., Swinehart, J. B., Loope, D. B., Been, J., Mahan, S. A., and Bush, C. A., 2000, Geochemical evidence for an eolian sand dam across the North and South Platte Rivers in Nebraska: Quaternary Research, v. 53, p. 214–222.

Nickling, W. G., 1994, Aeolian sediment transport and deposition: *in* Pye, K., ed., Sediment transport and depositional processes: Oxford, Blackwell. p. 293–350.

Pewe, T.L., 1983, The periglacial environment in North America during Wisconsin time: *in* Wright, H.E. and Porter, S.C., eds., Late Quaternary Environments of the U.S., University of Minnesota Press, Minneapolis, MN, p. 157–189.

Pye, K., 1987, Aeolian dust and dust deposits: Academic Press, London, UK, 334 p.

Pye, K., Winspear, N.R., and Zhou, L.P., 1995. Thermoluminescence ages of loess and associated sediments in central Nebraska, U.S.A.: Palaeogeography, Palaeoecology, Palaeoclimatology. v. 118, p. 73–87.

Reed, E.C. and Dreeszen, V.H., 1965, Revision of the classification of the Pleistocene deposits of Nebraska: Nebraska Geological Survey Bulletin 23.

Retallack, G.J., 1983, Late Eocene and Oligocene paleosols from Badlands National Park, South Dakota: Geological Society of America Special Paper 193.

Roberts, H.M., Muhs, D.R., Wintle, A.G., Duller, G.A.T., and Bettis, E.A. III, in press, Unprecedented last glacial mass accumulation rates recorded in loess from midcontinental North America: Quaternary Research.

Ruegg, G. H. J., 1983, Periglacial evenly laminated sandy deposits in the late Pleistocene of NW Europe, a facies unrecorded in modern sedimentological handbooks: *in* Brookfield, M. E., and Ahlbrandt, T. S., eds., Eolian sediments and processes/ 11th International Association of Sedimentologists Congress: Amsterdam, Elsevier, p. 455–482.

Schultz, C.B., and Stout, T.M., 1945, Pleistocene loess deposits of Nebraska: American Journal of Science, v. 243, p. 231–244

Schwert, D.P., Torpen–Kreft, H.J., and Hajic, E.R., 1997, Characterization of the late–Wisconsinan tundra/forest transition in midcontinental North America using assemblages of fossil beetles: Quaternary Proceedings, v. 5, p. 237–243.

Shao, Y., Raupach, M., and Findlater, P. A. 1993, Effect of saltation bombardment on the entrainment of dust by wind: Journal of Geophysical Research, v. 98, p. 12719–12726.

Shimek, B., 1896, A theory of the loess: Iowa Academy of Science Proceedings, v. 3, p. 82–89.

Shimek, B., 1898, Is the loess of aqueous origin?: Iowa Academy of Science proceedings, v. 5., p. 32–45.

Smith, G.D., 1942, Illinois loess: Variations in its properties and distribution, a pedologic interpretation: University of Illinois Agricultural Experiment Station Bulletin 490, p. 139–184.

Smith, H. T. U., 1965, Dune morphology and chronology in central and western Nebraska: Journal of Geology, v. 73, p. 557–578.

Snowden, J.O., and Priddy, R.R., 1968, Geology of Mississippi loess: Mississippi Geological Survey Bulletin, 111, p. 13–203.

Stokes, S., Rich, J., Swinehart, J. B., and Loope, D. B., 1999, Holocene timing of megabarchan dune construction in the Nebraska Sand Hills: Geological Society of America Abstracts with Programs, v. 31, p. A231.

Stokes, S and Swinehart, J.B., 1997, Middle–and–late–Holocene dune reactivation in the Nebraska Sand Hills, USA: The Holocene. v. 7, p. 263–272.

Sweeney, M. R., and Loope, D. B., 2001, Holocene dune–sourced alluvial fans in the Nebraska Sand Hills: Geomorphology, v. 38, p. 31–46.

Swinehart, J. B., 1990, Wind–blown deposits: *in* Bleed, A., and Flowerday, C., eds., An Atlas of the Sand Hills: Resource Atlas No. 5, University of Nebraska–Lincoln, p. 43–56.

Swinehart, J.B. and Diffendal, R.F, Jr., 1990, Geology of the pre–dune strata: *in* AS. Bleed and C. Flowerday, eds., An atlas of the Sand Hills: Resource Atlas No. 5, University of Nebraska–Lincoln, p. 29–42.

Wang, H., Follmer, L.R., and Liu, J.C., 2000, Isotopic evidence of paleo–El Niño–Southern Oscillation cycles in loess–paleosol record in the central United States: Geology, v. 28, p. 771–774.

Wayne, 1985, Drainage patterns and glaciations in eastern Nebraska: Nebraska Academy of Sciences Institute for Tertiary–Quaternary Research Symposium Series, v. 1, p. 111–117.

Webb, T., III, Bartlein, P.J., Harrison, S.P., and Anderson, K.H., 1993, Vegetation, lake levels and climate in eastern North America for the past 18,000 years: *in* Wright, H.E., Jr., Kutzbach, J.E., Webb, T., III, Ruddiman, W.F., Street–Perrott, F.A. and Bartlein, P.J., eds., Global climates since the last glacial maximum: Minneapolis, University of Minnesota Press, p. 415–467,

Margin of Laurentide Ice to the Atlantic Coastal Plain: Miocene–Pleistocene landscape evolution in the Central Appalachians

Authors and Leaders: **Duane D. Braun,** *Dept. of Geography and Geosciences, Bloomsburg University, Bloomsburg, PA, 17815*
Frank J. Pazzaglia, *Dept. of Earth & Environmental Sciences, Lehigh University, Bethlehem, PA, 18015*
Noel Potter, Jr., *Dept. of Geology, Dickinson College, Carlisle, PA, 17013*

DAY 1. NEWARK AIRPORT TO BLOOMSBURG, PA

Exit Newark airport to I–95 (NJ Turnpike toll road) to exit 15W. Proceed on I–280 west to I–80. Continue west on I–80.

Field trip A–9 begins with a journey across northern New Jersey and northeastern Pennsylvania from the Newark International Airport to Bloomsburg, Pennsylvania. The trip starts just within the late Wisconsin (20 ka) glacial limit and will go along the edge of that limit on the way to STOP 1 (Fig. 1). This journey crosses several physiographic provinces of the Appalachian Mts.. It starts on the Fall Zone, named for the falls and rapids on streams flowing from the consolidated rocks of the Appalachians onto the unconsolidated sediments of the Coastal Plain. The eastern U.S. urban centers are aligned along the Fall Zone, the upstream limit of river navigation. Typically, the rocks west of the Fall Zone are part of the Piedmont province, which exposes the metamorphic core of the Appalachian Mts. exhumed by tectonic uplift and erosion. At least four major phases of deformation are preserved in Piedmont rocks, three Paleozoic convergent events that closed Iapetus, followed by Mesozoic extension that opened the Atlantic Ocean. A record of Cretaceous to Quaternary exhumation of the Appalachians is preserved as Coastal Plain sediments. Late Triassic and Jurassic erosion is preserved in fault basins, such as the Newark basin, or is buried beneath Coastal Plain sediments (Fig. 1). The trip proceeds northwest from the airport, across the Fall Zone, and into the Newark basin. Several road cuts expose late Triassic and Jurassic, fluvial, red sandstone,

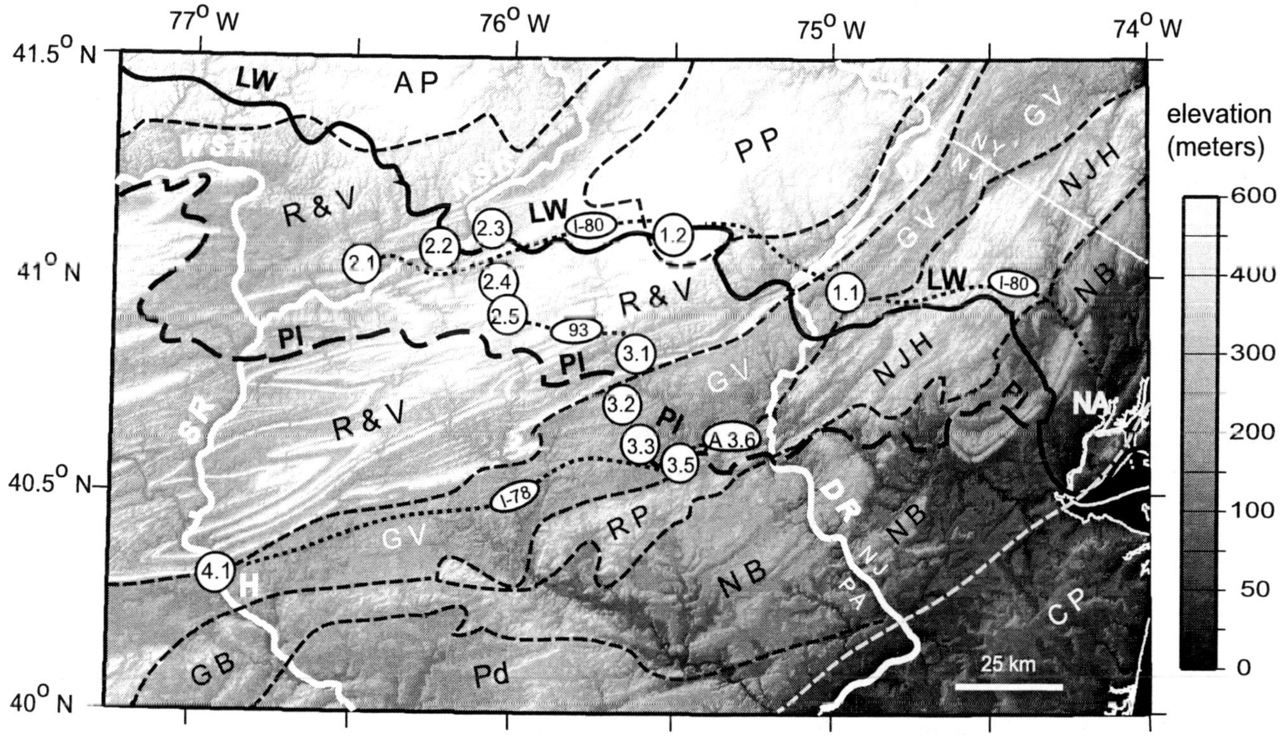

Figure 1. Digital relief map showing the trip route and stops for the first three days. Circles denote stops. First number is day and second number is stop. Trip route shown as dotted line. Physiographic Provinces (boundaries shown as dashed lines) – A P = Appalachian Plateau, C P = Coastal Plain, G B = Gettysburg Basin, G V = Great Valley, N B = Newark Basin, N J H = New Jersey Highlands, Pd = Piedmont, P P = Pocono Plateau, R P = Reading Prong, R & V = Ridge & Valley; Glacial Limits – LW = Late Wisconsinan, PI = Pre–Illinoian; Rivers – D R = Delaware, N S R = North Branch Susquehanna, S R = Susquehanna W S R = West Branch Susquehanna; CITIES – H = Harrisburg, NA = Newark Airport.

Braun, D.D., Pazzaglia, F.J., and Potter, N.,Jr., 2003, Margin of Laurentide Ice to the Atlantic Coastal Plain: Miocene-Pleistocene landscape evolution *in* the Central Appalachians: in Easterbrook, D.J., ed., Quaternary Geology of the United States, INQUA 2003 Field Guide Volume, Desert Research Institute, Reno, NV, p. 219–244.

lacustrine gray shale, and basalt. The structural basin is also a topographic basin that held a large proglacial lake during ice recession. The Newark basin is separated from the New Jersey Highlands by the Ramapo fault, a fault with contemporary seismicity. The New Jersey Highlands are the northern extension of Blue Ridge and represent a great thrust sheet that was emplaced from the southeast during the Alleghenian orogeny (Permian). The New Jersey Highlands summits are broad and accordant. Davis (1889) projected that accordance westward to the summits of the Ridge and Valley to define his highest and oldest peneplain–the Schooley peneplain. The field trip route takes us just north of Schooley Mt.. West of the Highlands is the Great Valley section of the Ridge and Valley Province (Fig. 1), here called the Minisink valley. This section is underlain by lower Paleozoic carbonate, shale, and slate folded and faulted during the lower Paleozoic Taconic orogeny. The prominent ridge west of the Minisink valley is Kittatinny ridge, the first ridge of the Ridge and Valley Province. The folded and faulted sedimentary rocks of the Appalachian foreland basin, deformed during the Alleghenian orogeny. Drainage during most of the Paleozoic was to the northwest, bringing detritus into the Appalachian foreland basin. With the opening of the Atlantic Ocean in the Triassic, the drainage started to reverse with southeast–flowing streams establishing courses transverse to the strike of resistant rocks, like the Silurian Shawangunk Conglomerate holding up Kittatinny Ridge.

Exit I 80 at the rest stop overlook near milepost 6.5.

STOP 1–1. DELAWARE WATER GAP OVERVIEW

The purpose of this stop is to introduce the general geologic and geomorphologic transect for DAY 1 and the landscape that spawned a major paradigm of long–term landscape evolution, the Geographic Cycle (Davis, 1889). From this vantage point, the Delaware Water Gap, carved by the Delaware River flowing across Kittatinny ridge, can be viewed. The topics of drainage reversal, both its initiation and progression, the carving of these water gaps, and the meaning of the accordant summits of ridges like Kittatinny ridge are central topics in our field trip. The original interpretation of the gaps is that they formed as streams were superimposed across the ridges from courses established on peneplains at or above the current ridge tops (Davis, 1889; Johnson, 1931). The alternate explanation is that the gaps represent zones of weakness in the resistant ridges that headward stream erosion took advantage of during drainage reversal following the opening of the Atlantic Ocean (Thompson, 1949).

At the Delaware Water Gap, the Shawangunk Fm. has a transverse flexure, forming a zone of weakness exploited by the river (Epstein, 1966).

When the entire assemblage of wind and water gaps in the region (Fig. 2) are examined and tied to the Coastal Plain sedimentary record, the Delaware River appears to be a relative latecomer, occupying the gap in Miocene to even early Pleistocene time (Campbell and Bascom, 1933; Stanford, 1997; Witte, 1997, 2001). Southeast of the Delaware Water Gap, a series of wind–gaps in the New Jersey Highlands line up with the Raritan River (Fig. 2), suggesting that the gap was originally cut by the Raritan River. More significantly, the upper Cenozoic stratigraphy and sedimentology of the lower Delaware valley and Coastal Plain indicates that the Hudson River drained through that region and the Delaware was a very minor tributary of that system until the Plio–Pleistocene (Owens and Minard, 1979; Newell et.al., 1989).

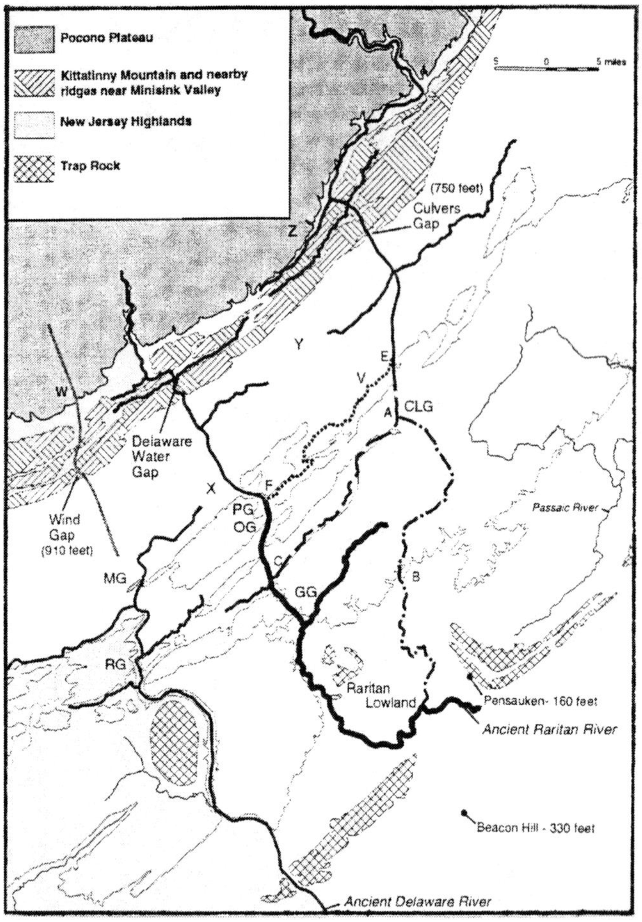

Figure 2. The Delaware Water Gap in mid–Miocene times, occupied by a branch of the Raritan River. Successive captures by the Delaware cut off the branches of the Raritan and the Wind Gap headwaters to produce the present Delaware. Other lettered sites are present day windgaps in the New Jersey Highlands. (From Witte, 2001)

West of the Delaware Water Gap, I–80 climbs the Glaciated Pocono Plateau section of the Appalachian Plateaus Province, underlain by nearly horizontal Devonian sandstone. I–80 crosses the plateau on or next to the 20 ka glacial terminus. Reconstructing the record of repeated advances of the Laurentide ice sheet during the Pleistocene is the primary goal of first two days of the field trip.

Exit I–80 onto PA Rt 115 south. Turn right onto Rt 903 . Turn right onto Rt 534 west. Follow the signs to Hickory Run State Park and to the boulder field.

STOP 1–2. THE HICKORY RUN BOULDER FIELD AND OVER–VIEW OF PLEISTOCENE GLACIAL AND PERIGLACIAL EVENTS

The purpose of this stop is to outline the sequence of Pleistocene glacial episodes and to observe the dramatic effects of periglacial climates on the Appalachian landscape. During the Quaternary, eastern Pennsylvania was affected by at least ten cold, glacial-periglacial episodes, separated by short intervals of warm, humid-temperate climate (Braun, 1989a, 1994a, 1999). Glacial limits of three ages have been mapped across the region (Fig. 1) (Braun, 1994a; Sevon and Braun, 1997) with a fourth limit now being separated out (Braun, 1999 and ongoing work). The farthest southwest and oldest glacial limit, with sediments containing reversed paleomagnetism is considered to be of pre–Illinoian–G age (850 ka, Stage 22 or older) (Braun, 1994a; Stanford, 1997). The next younger glacial limit has deposits with a degree of weathering and erosion that is consistant with either late Illinoian (150 ka, Stage 6) or pre–Illinoian–B (450 ka, Stage 12) age and is only about 10-15 km beyond the most recent, late Wisconsin (20 ka, Stage 2) glacial limit. Most other cold episodes brought glaciers close to Pennsylvania and caused severe periglacial activity in Pennsylvania (Braun, 1989a, 1994b, 1999).

Most of our field trip will be within the periglacial fringe. A boulder colluvium mantle, with strips of unvegetated boulder fields on the floor of zero–order drainageways, is a ubiquitous part of the landscape. Intense freeze-thaw, coupled with down-slope gelifluction is the preferred model for boulder colluvium and boulder field genesis (Peltier, 1949, Smith, 1953; Potter and Moss, 1968; Sevon, 1987; Braun, 1989a, 1994). Hickory Run is a particularly large (500 x 125 m) and gently sloped (1°) example of a boulder field surrounded by boulder colluvium (Fig. 3, and 4).

Figure 3. Picture of the Hickory Run Boulder Field, view eastward up the axis of the field.

Other origins for the boulder field have been suggested that involve fluvial, glaciofluvial, or glacial agents. Figure 5 shows that the last glaciation and its meltwater did not quite reach the site and the field is in a first–order hollow with discharge insufficient to move the boulders. Gelifluction is the only process capable of moving boulders on such a gentle slope in such a topographic setting. Other features that support a periglacial frost heave and gelifluction origin are that the largest boulders are at the surface, tabular clasts are on edge, "fitted surface" boulder contacts, patterned ground micro-topography with sorted stone rings, and stone stripes that can be traced downslope from source ledges.

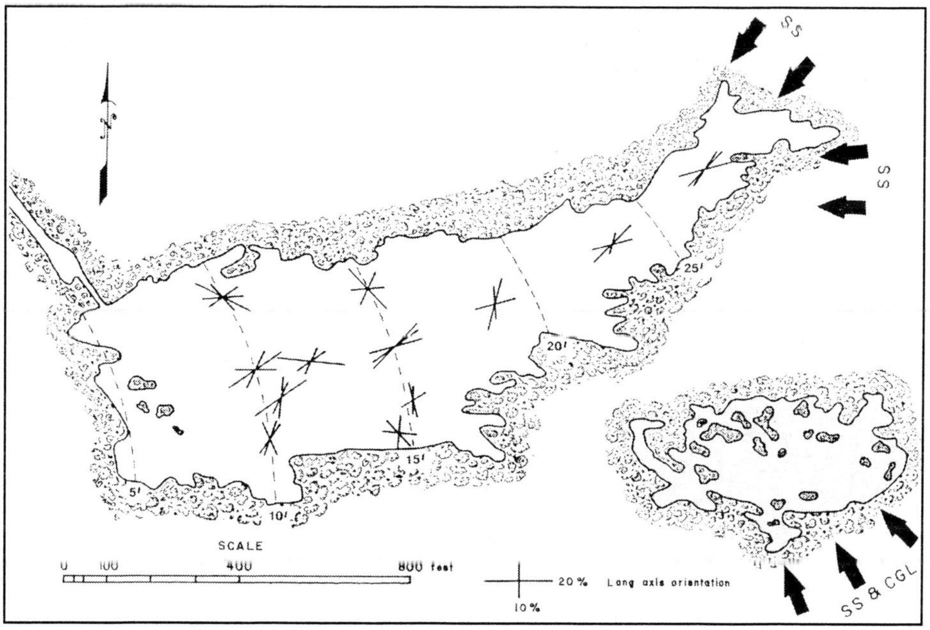

Figure 4. Hickory Run boulder field showing the slope, boulder orientation, and source ledges for the boulders in the field (From Smith , 1953).

The boulder field is in the middle of a much more extensive boulder colluvium deposit that covers the toe–slopes in the area (Fig. 5). The boulder colluvium is mostly matrix-supported diamict with clast-supported stripes or rings. This matrix-rich material would have readily crept and flowed down these gentle slopes during the summer thaw season.

The unvegetated boulder fields lie on the floors of first-or zero-order valleys where Holocene groundwater sapping can remove the fines (Smith, 1953; Braun, 1994b). Such removal is indicated by "boiling sand" springs down-valley from the boulder fields. Also the clasts under the surface are polished on all sides, requiring fines to do the polishing.

To re–emphasize an important point, colluvium of variable texture mantles much the landscape from the glacier margin to Washington D.C. That colluvium is best explained, as noted above, as a product of repeated, intense periglacial episodes. Periglacial erosion has dominated over interglacial fluvial erosion during the Pleistocene in this part of the Appalachians (Braun, 1989a).

Exit boulder field to Rt 534 W, back to I–80. Go west on I–80.

The trip continues west, crossing the Lehigh River and entering the Anthracite Upland section of the Ridge and Valley Province. I–80 follows a strike valley on the Mauch Chunk Fm. redbeds. Pennsylvanian Pottsville Fm. conglomerate holds up the strike ridge to the south, while the Mississipian Pocono Fm. sandstone holds up the ridge to the north. The late Wisconsin terminal moraine is crossed where I–81 crosses I–80. Beyond, I–80 goes through a water gap in the Pocono ridge to continue west on rolling hills underlain by the Devonian Catskill Fm. redbeds. I–80 then crosses the abandoned course of the Susquehanna River just before it crosses the outwash terraces of the present Susquehanna (Fig. 6). The Susquehanna valley was dammed by a head-of-outwash of the late Illinoian or older terminus. A head–of–outwash or frontal kame is a 30–70–m thick deposit of sand and gravel at the front edge of the glacier. This thick mass of sand and gravel was higher in the Susquehanna Valley to the left than in the strike–valley cut in shale and limestone that lies straight ahead (Braun and others, 1984). As the ice receded, drainage went down the strike valley instead of the original, now blocked, Susquehanna Valley. At the Bloomsburg exit, the road crosses the late Illinoian terminal

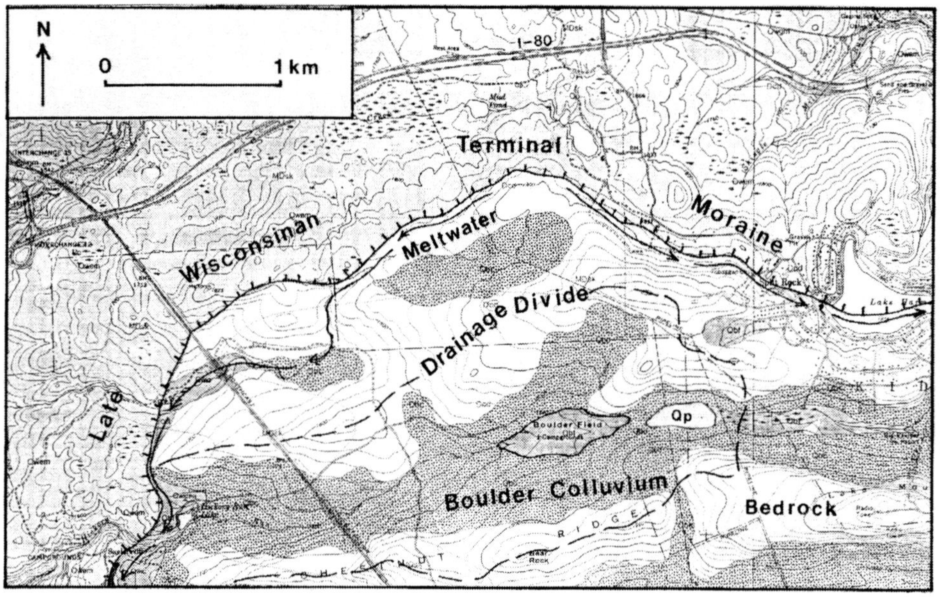

Figure 5. Topographic and surficial deposit map of the area around the Hickory Run boulder field (From Sevon, 1975).

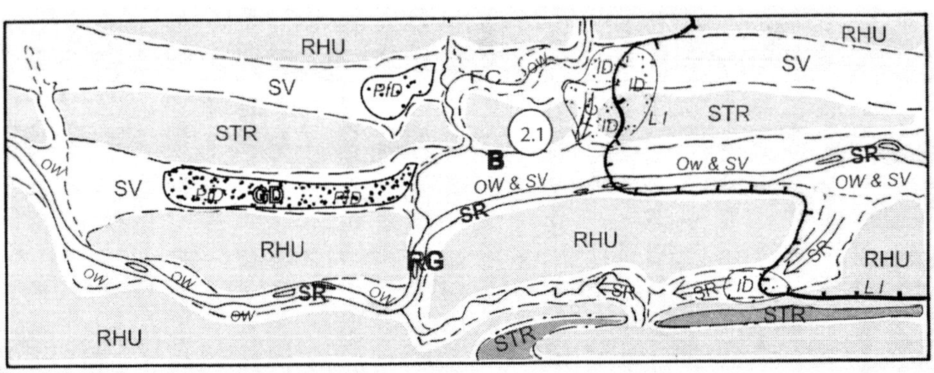

Figure 6. The Bloomsburg area showing the present Susquehanna River (SR), its blocked and abandoned course (SR arrows), buried course of Fishing Creek (FC arrow) and the late Illinoian or older glacial limit (LI). Other features are: B = Bloomsburg, CC = Catawissa Creek, FC = Fishing Creek, GD = Grovania Divide, ID = Illinoian glacial deposits, OW = outwash terraces and alluvium, PID = Pre–Illinoian glacial deposits, RG = Rupert water gap, RHU = Rolling hill uplands, SV = Strike valley, STR = Strike ridge.

head-of-outwash that blocked the Fishing Creek valley. An exposure to the right of I–80 at Buckhorn consists of intensely oxidized, red, pre-Illinioian outwash.

DAY 2. LATE WISCONSINAN TO PRE–ILLINOIAN GLACIAL DEPOSITS DRAPED OVER A THREE–TIER LANDSCAPE

Leave the hotel, turn left on Rt.42, cross I–80, take exit for US 11 east, go through Bloomsburg to Bloomsburg University, and take 2nd St. to Hartline Science Center. roof-top vantage point.

STOP 2–1. OVERVIEW OF THE SUSQUEHANNA VALLEY AND THE QUESTION OF CATASTROPHIC, SUB-GLACIAL FLOODS

The purpose of this stop is to discuss the evidence for catastrophic, channeled-scabland-scale floods coming down the Susquehanna valley. Shaw (1989) proposed that the Susquehanna valley be examined for evidence of catastrophic discharges on the order of 4.8 x 10^6 m^3/s. Such discharges supposedly come from 14-16 ka, sub-glacial floods that formed the drumlins in New York State.

Directly below our vantage point are the North Branch Susquehanna River floodplain and outwash terraces covering a shale and limestone strike valley (Fig. 6). The river undercuts the Trimmers Rock Fm. sandstone ridge on the south side of the valley. On the south skyline is Catawissa Mtn., underlain by Pocono Fm. sandstone. The Trimmers Rock ridge to the east is cut by the 1.5-km-wide, abandoned, North Branch Susquehanna valley, blocked by late Illinoian or older Bloomsburg ice margin deposits (Fig. 6). Between there and here was the location of the pre-glacial, low divide in the strike valley that separated the Susquehanna River from Fishing Creek. Pre-glacial Fishing Creek once passed through the ridge behind us in a water gap 2 km east of here that was also buried by Bloomsburg ice-marginal deposits.

To the southwest, the river turns 90° and enters the 0.5-km-wide, Rupert water gap (Fig. 6). The narrowness of the gap reflects its origin as the former gap of the 30-m-wide Fishing Creek. It has been occupied by the river only since about 150 ka when the late Illinoian glacier (Braun, and others, 1984, Braun, 1994a) diverted the river down the strike valley below us and into the Rupert gap.

To the right of the Rupert water gap is the continuation of the strike valley. To the west at Grovania is a low divide that is only 50 m above the floor of Rupert gap. The strike valley is a "straight ahead" by-pass for catastrophic floods ponding at the constriction presented by Rupert gap (Fig. 6). The Rupert water gap is also exceptionally shallow, 140 m, because the river cuts the Trimmers Rock ridge, a lower ridge compared to the higher, strike-ridges, like nearby Catawissa Mt.. Due to this combination of a low strike ridge and a uniquely youthful age, the Rupert water gap has less than one-half the cross–sectional area (0.74 x10^5 m^2) of any water gap upriver. This "undersized" Rupert water gap is the critical "choke–point" for hypothesized catastrophic floods. Braun (1990) estimated only 25% of the proposed catastrophic flood would fit through Rupert gap and the rest would flow down the adjacent strike valley at a depth of 50 m or more. Those sites should show dramatic evidence of scabland-like erosion features, but no such features are observed. Instead, deeply weathered pre-Illinoian glacial deposits (Fig. 6) mantle the valley floor, often capped by Wisconsin colluvium or loess (Leverett, 1934; Peltier, 1949; Braun, 1990). No evidence for catastrophic flooding exists.

Leave B.U. on Rt.487 N, take I–80 east and immediately cross the buried valley of Fishing Cr., cross the Susquehanna River, exit I–80 onto Rt.339 to the left through Mifflinville. Travel on Susquehanna River outwash and alluvial terraces to Nescopeck. Turn left, then right onto Rt.93, leave Nescopeck and bear right on Wapwhallopen Rd. Ascend the riser onto the outwash terrace and stop at the gravel pit on right.

STOP 2–2. LATE WISCONSIN TERMINAL HEAD-OF-OUTWASH IN THE SUSQUEHANNA VALLEY

The purpose of this stop is to examine the landforms and deposits that characterize the terminus of the Laurentide ice in the Ridge and Valley Province of eastern Pennsylvania. The site is a head-of-outwash or frontal kame in the Susquehanna River valley. The sequence of deposits coarsens upward and becomes enriched in locally derived clasts as ice gradually approached the site (Fig. 7). The ground surface exhibits knob and kettle topography with a thin loess mantle. The kame was gradually incised 35 m by increasing meltwater flow as ice retreated northward and increased the ice-front drainage area. Meltwater drainage ceased at about 14 ka, when ice retreated north of the Susquehanna basin. Since then, the river has incised 6 m more and is now incising into bedrock. This site shows that the river rapidly goes from aggradation during glacial advance to degradation during and after glacial retreat. The river is now cutting a new channel in bedrock, leaving terraces capped by outwash to either side. We will discuss the formation of these terraces more on DAY 4 where we will examine a flight of terraces that represent the entire Pleistocene record.

Continue east along Susquehanna. River to Rt. 239 and Wapwallopen. Turn right onto Hobbie Rd., ascend the slope and follow signs to Council Cup park, making several turns.

Note: If the weather does not permit seeing the view at the next stop, the trip will proceed directly to STOP 2.4. The discussion about long-term evolution of the Appalachian landscape will be done on day 3 at STOP A 3.7.

STOP 2–3. COUNCIL CUP OVERVIEW OF THE THREE–TIER LANDSCAPE: DAVIS, HACK, AND MORE–RECENT HYPOTHESES ON IT'S ORIGIN

The purpose of this stop is to view the three-tier landscape in the Ridge and Valley (Fig. 8) and review past and current explana-

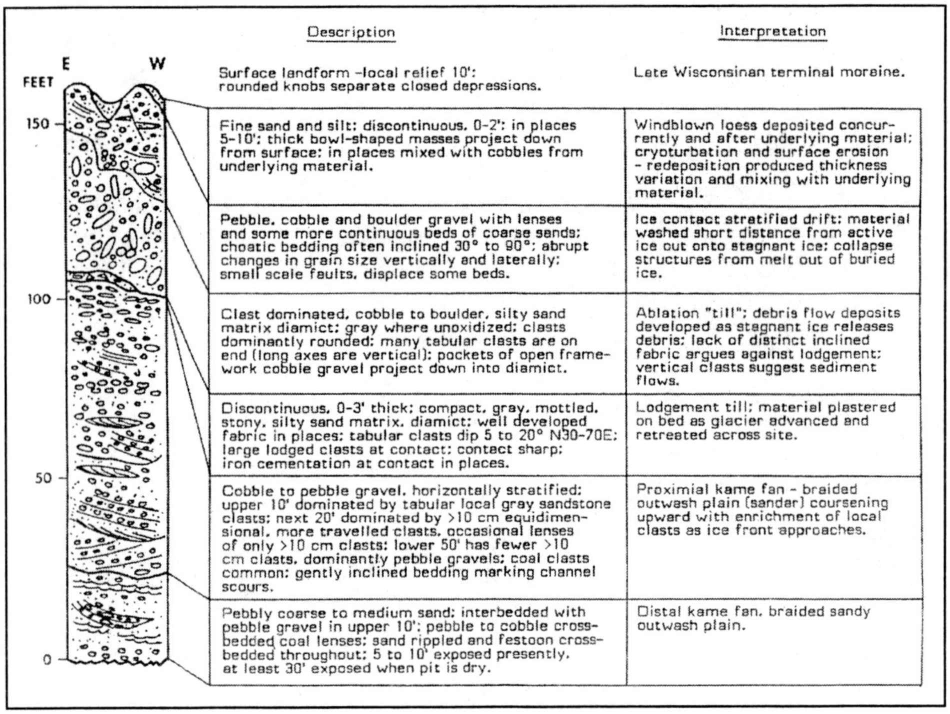

Figure 7. Stratigraphic section of the late Wisconsinan terminus in the Susquehanna valley. (From Braun and Inners, 1988)

tions of how that landscape developed. This landscape has inspired two diametrically opposed hypotheses: the geographic cycle (Davis, 1889, 1899) and dynamic equilibrium (Hack, 1960, 1965). Davis explained the three tier landscape as being remnants of three peneplains (low relief, fluvial erosion surfaces) of differing age (Fig. 8). Davis argued that the highest surface, visible here on the skyline, developed during the Cretaceous and named it the Schooley peneplain (Davis and Wood, 1890). That peneplain was uplifted in the early Tertiary. Then a partial peneplain was cut during the middle Tertiary (named the Harrisburg peneplain by Campbell, 1903), visible here as a prominent mid–elevation bench. A second uplift occurred in the late Tertiary forming a second partial peneplain developed locally on the weakest rock (named the Somerville by Davis and Wood, 1890). This explanation requires post–Cretaceous erosion to carve out the present relief of about 500 m, an average incision (not erosion) rate of about 6 m/Ma. Later workers largely introduced variations to the Davisian theme, such as adding to the number of peneplains (Bascom, 1921), extended the Coastal Plain cover (Johnson, 1931) atop the Schooley surface, or made the peneplains younger (Johnson, 1931; Ashley, 1935; Stanford, 1997, 2001).

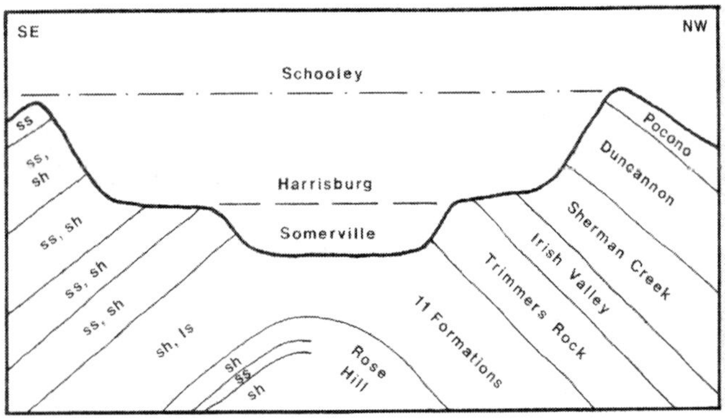

Figure 8. Schematic cross–section of the breached Berwick anticline showing the relationship of bedrock units to the three–tier landscape and postulated peneplains.

In contrast, Hack explained the three tier landscape as being a result of continuous erosion working on rocks of three differing resistances that produces the three different elevations. The hardest sandstone forms the highest ridges, interbedded sandstone and shale form an intermediate elevation rolling hill landscape, and the shale and limestone units form the gently rolling lowest elevations (Fig. 7). A concise comparison of various elements of the Davis and Hack theories is given in Table 1.

A crucial test of Davis' geographic cycle is to examine the amount of sediment eroded from the Appalachians and deposited in the Coastal Plain and offshore and see if that sediment can be matched to the evacuation of the valleys since the Schooley peneplain. Long-term rates of post-Triassic erosion have been reconstructed from the offshore sediment load in the Baltimore Canyon trough (Poag and Sevon, 1989) (Fig. 9). Exhumation was relatively fast in the late Jurassic and late Cretaceous (>30 m/Ma assuming constant drainage basins size) and very slow in the early–middle Cretaceous and most of the Tertiary (<30 m/Ma assuming constant drainage basin size)(Fig. 9). The Davisian concept of impulsive uplift, fol-

TABLE 1. DAVIS vs. HACK–APPALACHIAN LANDSCAPE EVOLUTION

Davis–Geographic Cycle	Hack–Dynamic Equilibrium
1. Time dependent landscape	1. Time independent landscape
2. Cyclic uplift and stillstand	2. Continuous uplift – varying rate
3. Episodic or cyclic erosion	3. Continuous erosion–varying rate
4. Sequential landscape–youth, maturity–old age, relief reduction	4. Non–sequential landscape change–perpetual maturity
5. Overall structure truncated by topography	5. Structure and topography closely adjusted to each other
6. Rocks of differing resistance preserve peneplain remnants of differing ages	6. Rocks of differing resistance develop differences in elevation and slope angle
7. Summit accordance–evidence for past peneplain	7. Summit accordance from sub–equal stream spacing on rocks of similar resistance

lowed by reduction of mean elevation and relief is supported by the two episodes of exponential decay in sediment yield to the Baltimore Canyon Trough, each on the order of 50 Ma long. Two major unconformities at the edge of the Coastal Plain, one underneath the late–Cretaceous sediments and the other under the middle Miocene sediments are generally coincident with the end of each major sediment pulse to the Baltimore Canyon Trough. We will see and discuss these unconformities on day five of this trip. By the beginning of the Miocene, some sort of low relief erosion surface (peneplain, pediplain, or etchplain ?) in the Appalachians was likely.

But then, very fast exhumation resumed in the Miocene, declining somewhat in the Plio–Pleistocene to near peak late Cretaceous rates (Fig. 9). This slug of Miocene-Recent sediment in the Baltimore Canyon Trough requires removal of at least 1–2 km of rock in the past 20 Ma (Braun, 1989b; Pazzaglia and Brandon, 1996). This amount of exhumation is greater than the present relief of the Appalachians and would effectively destroy the pre–Miocene erosion surface. The most likely place where a pre–Miocene erosion surface might be preserved on the present landscape is near the Fall Zone–Coastal Plain hinge–line between net erosion and net deposition.

The destruction of a pre–Miocene erosion surface by the Miocene–Recent erosion leaves the possibility that the supposed Schooley peneplain is of Miocene or even Pliocene age and that is what was suggested Ashley (1933, 1935) and Stanford (1997, 2001). The theoretical concept of the peneplain is that the landscape, through an exponential decline in relief, reaches a graded condition that produces little clastic sediment (Davis,1889; Schumm and Lichty, 1965; Ahnert, 1970; Pitman and Golovchenko, 1991). As noted above, that is what the Baltimore Canyon Trough sediment record shows during the middle Cretaceous and middle Tertiary. But the Miocene–Recent sediment pulse, while declining, remains far above the sediment yield rate of the mid–Cretaceous and mid–Tertiary (Fig. 9). This indicates that since the mid–Miocene, exhumation rates have not yet declined enough to permit development of a peneplain on the present landscape, let alone the development of a complete peneplain (the Schooley), and two partial peneplains (the Harrisburg and Somerville).

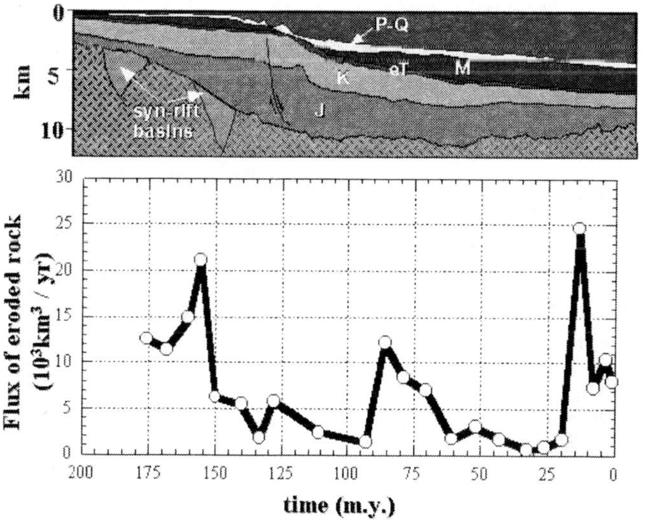

Figure 9. Cross–section of the Baltimore Canyon trough (BCT, above) and the flux of eroded rock to the BCT over the past 180 Ma (below). Note the pulses of sediment, followed by exponential decay as well as the large pulse of sediment in the past 20 Ma (From Pazzaglia and Brandon, 1996).

The high Miocene–Recent sediment yield rates are matched by measured erosion rates for the same time–frame. Historic river suspended sediment yields suggest erosion rates between ~20-40 m/Ma (Sevon, 1989a; Milliman and Syvitsky, 1992; Conrad and Saunderson, 1999), but these values are enhanced by human-induced soil erosion. Historic solute loads range between ~5-10 m/Ma (Cleaves and others, 1970). Quaternary periglacial erosion of ridge tops may be as high as ~80-100 m/Ma (Braun, 1989a). Carbonate dissolution rates typically range between ~8 and 30 m/Ma (White, 1984, 2000). River incision rates range from 5-40 m/Ma (Pazzaglia and Gardner, 1994; Granger et al., 1997, 2001; Mills, 2000). The erosion rate from cosmogenic–dated surfaces averaged across entire watersheds in the Smoky Mts. of the southern Appalachians is 27 m/Ma (Matmon et al., 2003). Overall the Appalachians are currently unroofing at rate of about 20-30 m/Ma, considerable above the rate suggested by the mid–Cretaceous and mid–Tertiary sediment yield lows. This erosion has removed the pre–Miocene erosion surface (peneplain) and has not yet slowed down enough to permit a new one to form.

While the Baltimore Canyon Trough sediment record and erosion rate measurements do not support Davis' hypothesis that the present landscape is a set of peneplain remnants, does it support Hack's dynamic equilibrium hypothesis? On one hand, the Hack (and Gilbert) concepts of form reflecting process, and an equality of action between driving and resisting forces is rarely so well expressed as it is in every ridge, valley, and stream course in the Appalachian landscape (Hack, 1960, 1965, 1982; Braun, 1983). The geology of Pennsylvania's Ridge and Valley is written into the topography with the hard rocks high standing and the soft rocks underlying the valleys. This fundamental topographic adjustment to rock type and structure is what lead Hack (1960; 1975) to argue for continuity of erosion and a landscape in dynamic equilibrium. On the other hand, uncertainty in measuring basin-wide erosion rates and river-specific incision rates makes uncertain whether the Appalachian landscape is maintaining relief in a truly dynamic equilibrium state (Hack, 1960) or relief is locally increasing or decreasing. Little separation between the straths of successive glaciofluvial terraces on the Susquehanna suggests that the river is incising slower than periglacial activity is eroding the hilltops, essentially reducing relief in the landscape. However, close to the Fall Zone, the river profile is clearly convex, indicating base level fall and active incision. Also, process rates are slow and have significant lag times in this landscape (Pazzaglia et al., 1998) making determination of whether the Appalachian landscape is in dynamic equilibrium or in some disequilibrium state even more difficult. The Baltimore Canyon Trough Miocene–Recent sediment pulse argues for rapid, continuous erosion at a varying rate that implies a landscape presently in disequilibrium. So the Baltimore Canyon Trough sediment record support Hack's concept of continuous erosion forming the present landscape but not his concept of equilibrium as the landscape formed.

Any new hypothesis of the evolution of the present Appalachian landscape, particularly the Susquehanna Basin, must first try to explain the Baltimore Canyon Trough Miocene–Recent sediment pulse. No known tectonic event helps explain the sediment out–pouring (Sevon, 1989b), so one must look elsewhere. Braun and Pazzaglia have each developed an opposing hypothesis for the rapid exhumation rate and sediment pulse for discussion during the trip. Pazzaglia suggests that the Susquehanna drainage divide remained near the Piedmont-Ridge and Valley boundary until the early Miocene and then rapidly shifted westward to its present position on the Appalachian Plateau in the middle Miocene. That rapid headward erosion would produce the sediment pulse off–shore. Braun argues that the Susquehanna divide migrated to near its present location in the Cretaceous or early Tertiary and then deep weathering occurred during the middle Tertiary sediment yield minimum. Then Miocene climate change caused rapid erosion of the weathered material and the Miocene–Recent sediment pulse.

Pazzaglia's suggestion that the Susquehanna drainage divide remained near the Piedmont-Ridge and Valley boundary until the early Miocene and then rapidly shifted westward to its present position is supported by new U–Th/He and existing apatite fission track (AFT) thermochronology which are difficult to reconcile otherwise. In summary, the fission track cooling ages are the same as the lower temperature U–Th/He cooling ages in the Ridge and Valley. Both thermochronometers indicate slow erosion of about 15 m/Ma, an apparent paradox because this is where most of the relief in the Appalachian landscape currently exists, it is where the fastest rates of erosion are currently measured (>> 15 m/Ma), and it is where Davisian theory would predict the greatest amount of incision below the Schooley peneplain. In contrast, apatite fission–track cooling ages in the Pennsylvania Piedmont are significantly older than U–Th/He data from the same region. These data also present an apparent paradox in that they indicate greater rates of long–term erosion (~40 m/Ma) for a part of the Appalachian landscape that has very low relief, low current rates of erosion, and are proximal to Davis's down–dip projection of the Schooley peneplain presumably much less post–Schooley incision has occurred. A recent jump in the drainage divide to its present location would result in a pulse of sediment to the BCT as the Ridge and Valley was exhumed. This recent exhumation would be expressed as a youthful–looking landscape with relatively high modern rates of erosion, but would not be of a great enough magnitude to be recorded by even the low temperature U–Th/He thermochronometer.

The cause of a westward jump in the drainage divide is not clear, but other geologic evidence exists that it did occur (Fig. 10 and 11) and is still occurring farther south in the Appalachians. Westward shift in the divide may actually be rolling southward along orogen strike as suggested by: (1) the location of the highest rates of Appalachian stream incision, (2) considerable separation of the crest of the range and the crest of the long-wavelength topography (defined as the topography low–pass filtered at wavelength of Airy isostatic compensation of the crust), and (3) a southwest sweep of Miocene to Quaternary depositional centers on the Atlantic Coastal Plain (Fig. 11). These lines of evidence suggest that even in "dead" orogens like the Appalachians, the continental divide is localized geodynamically and the integration of orogen-scale drainage transverse to the orogen may be long delayed. Atlantic-slope streams, like the Susquehanna River, that owe their origin to drainage reversal during the opening of the Atlantic ocean, may have remained short, steep drainages from the late Triassic until the Miocene. Then their headwaters were rapidly expanded westward by geodynamic reconfiguration of the flexural crest of the orogen.

Braun argues that the Susquehanna basin eroded headward to near its present divide by the beginning of the Tertiary or even the beginning of the Cretaceous as suggested by Judson (1975). The Susquehanna River is located where Triassic rifting most deeply cut into the Appalachians and faulted out the Blue Ridge-Reading Prong. During most of the Triassic, a southeast-draining river entered the north side of the rift basin and formed a major depositional center 50 km east of the present course of the river (Smoot, 1999). That river is the likely ancestor to the Susque-

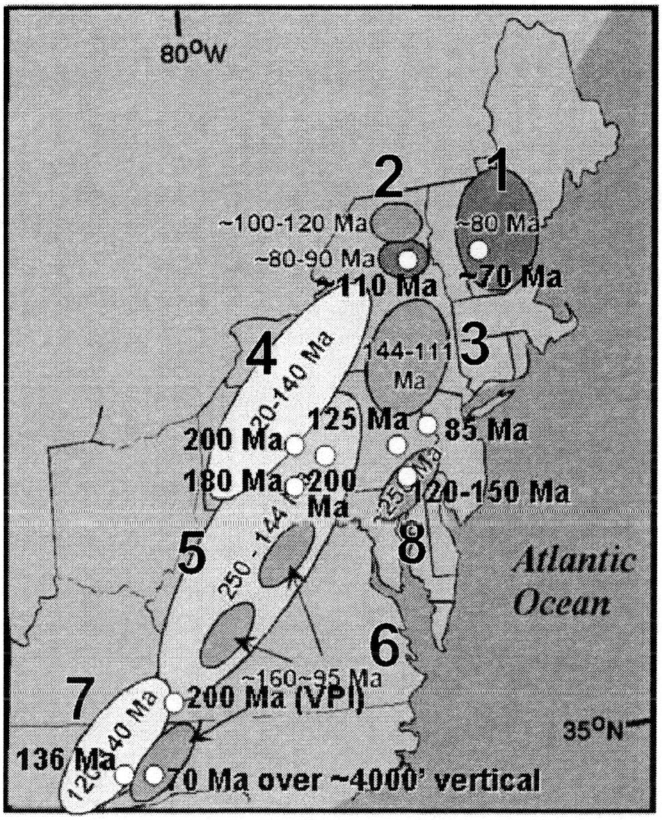

Figure 10. New U–Th/He thermochronology ages (white circles and large, dark text) superimposed on the apatite fission track (AFT) thermochronology for the Appalachian Mts. The average AFT unroofing rate is ~ 17 m/Ma The U–Th/He ages are younger in the Piedmont and Blue Ridge, but very similar in age to the AFT in the Ridge and Valley (Naeser et al., 1999, 2001). 1. Zimmerman et al. (1975); 2. Schofield el al. (1999); 3. Miller and Duddy (1989); 4. Roden and Miller (1989); Blackmer et al. (1994); 5. Roden (1989); Blackmer et al. (1994); 6. Naeser et al. (1999); 7. Boettcher and Millikan (1994); 8. Zimmerman (1979).

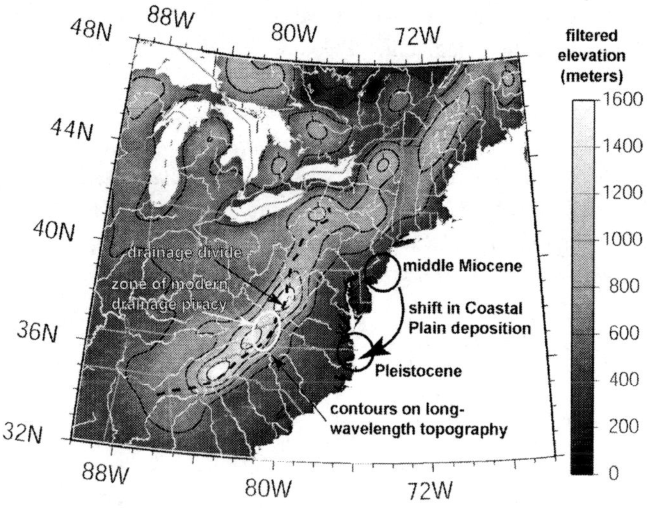

Figure 11. A possible explanation of the Mid–Miocene sediment pulse to the BCT. The drainage divide began migrating rapidly west and south in the Miocene, delivering the large slug of sediment by simply increasing the size of the contributing drainage basin. Where the drainage divide is coincident with the axis of highest standing topography (gray contours, determined by running the topography through a low pass filter) it is relatively stationary. Where the drainage divide diverges from the axis of highest standing topography, it is actively migrating westward. The overall southwest movement of deposition centers on the Coastal Plain might be linked to the divide migration.

hanna and marks the early initiation of the reversal of drainage from the northwest to the southeast as envisioned by Davis (1889). The volume of sediment in the rift basin implies significant headward erosion of the ancestral Susquehanna in the Triassic. Two other episodes of erosion in the Jurassic and Cretaceous then produced the sediment pulses shown on Figure 9. The Susquehanna was a primary source of sediment for the Jurassic pulse, a secondary source for the Cretaceous pulse, and a primary source for most of the Tertiary (Poag and Sevon, 1989). All of this sediment delivery to the Baltimore Canyon Trough is most easily explained as being from a basin similar in size to the present watershed of the Susquehanna River.

Judson, (1975) noted that while the entire length of the present Appalachian drainage divide does not coincide with the regional geologic structure or physiographic provinces, it does roughly parallel the edge of the Coastal Plain and the present shoreline. This parallelism, along with the juxtaposition of the drainage divide with the axis of greatest negative gravity anomaly implies that the overall length of the river systems from the Susquehanna southward has been similar since the Cretaceous. In other words, the divide rapidly became established near its current location atop the deepest gravity low because the crust is thickest there and isostasy keeps the elevation high. The Susquehanna's headward erosion was not slowed by the hard Blue Ridge rock types or thick, buoyant crust in contrast to rivers in the southern Appalachians who must erode headward across exceptionally thick, buoyant crust (greatest gravity negative) that also coincides with the widest area of hard Blue Ridge rock types.

Braun contends that by the mid–Miocene, global climate, which had been cooling throughout the middle Tertiary, passed a threshold that initiated relatively rapid erosion of the entire Susquehanna basin. The basin was underlain by a very deep weathering profile developed during the previous 40 Ma of low sediment yield (Fig. 9). Geologic evidence supporting a landscape with a former thick, compositionally mature residuum is present as saprolite containing laterite and bauxite weathering products in central Pennsylvania, and as the nearly pure quartz composition of middle Miocene, ancestral Susquehanna River fluvial deposits on the Fall Zone (Bryn Mawr Fm.; Pazzaglia, 1993). The amount of relief in the Miocene landscape at the time when the stripping initiated is not known. Garner's (1974) concept of a 'humid climax topography' would suggest that the landscape had a moderate relief equal to or greater than that of today. Sevon (1989b, 1999) argued for a landscape with lower than present relief based on a model that he developed for landscape evolution on the Appalachian Piedmont.

Return to Wapwhallopen, turn left onto Rt.239 then bear left, upslope on Rt.239. Cross a kettled kame terrace and then an "indistinct" terminal moraine. Turn right, continuing on Rt.239. Turn left onto Rt.93 and ascend Nescopeck Mtn. At the top, bear right onto Old Berwick Rd. and descend the mountain. Near the base of the mountain, turn left into the Phone Co. parking lot.

STOP 2–4. LATE WISCONSIN COLLUVIUM OVER ILLINOIAN COLLUVIUM

The purpose of this stop is to examine two different ages of colluvium. The site is on the toe of the dip-slope of a Pocono sandstone ridge. Late Illinoian or older ice covered the site (Sevon and Braun, 1997). The land surface is covered by a one-stone-thick cobble and boulder mantle that is sorted into crude stripes of stone-rich and stone-poor soil.

The upper half meter or so of material has a brown color and degree of soil development that is typical of late Wisconsin material in this region. The thinness of the brown material suggests that this location was a transport zone supplying colluvium to a deposition zone farther downslope. The upper colluvium is underlain by more intensely weathered reddish-brown colluvium derived from the underlying Illinioian glacial drift. The intensity of weathering and depth of soil formation in the lower colluvium is equivalent to that of nearby Illinoian till. The well-developed downslope clast fabric indicates the material is colluvium. Ice flow was parallel to the slope and would have left a fabric perpendicular to what is observed.

Turn left upon leaving. At Rt.93 turn right, then right again onto I–80 east. Cross I–81 and the late Wisconsin glacial terminus. Exit onto Rt.309 north and ascend the ridge, crossing the glacial terminus again. Turn left at the traffic light onto East Butler Road and ascend Green Mtn. Near the top, turn left onto Prospect Road, cross anthracite coal strip mines, go through upper Lehigh, cross other coal strips and park on the right shoulder beside the pond in the reclaimed strip mine. Walk east to the lake in the strip mine. Exposures are on the north side of the lake.

STOP 2–5. LATE ILLINOIAN TERMINUS DEPOSITS

The purpose of this stop is to examine older glacial deposits capped by late Wisconsin colluvium and discuss the amount of erosion since the glacier retreated. Figure 12 gives a detailed stratigraphic column of the deposits.

Continue ahead on Burback St., turn right onto Front St., and then left onto Centre St. and continue on Rt.940 south through Freeland. Just after Oakdale, turn left onto SR 3019, cross the railroad tracks and active strip mine area, go over an anticlinal ridge, cross another coal syncline, and ascend another ridge. Turn right onto Rt.93, then left onto 4–lane Rt.424. At a crossroads sign, park on the right shoulder. Cross the 4–lane. road and walk up a gated access road toward a water tank. Just before a fence, turn right into the woods to the tors.

STOP 2–6. TORS AND BOULDER COLLUVIUM DEVELOPED SINCE PRE–ILLINOIAN GLACIATION

The purpose of this stop is to examine the amount of weathering and erosion of a quartz conglomerate since it was glaciated in pre–Illinoian times. Strip mines in adjacent valleys expose striated bedrock pavement under pre–Illinioian till (Fig. 13). This indicates that the pre–Illinoian ice was erosive and should have removed any pre–glacial tors. The top surfaces of the tors show a variety of weathering pits, pans, and runnels. Downslope, the tors become more disrupted and form the head of the boulder colluvium that extends to the bottom of the valley. The volume of colluvium indicates 10 to 30 m of rock have been eroded from the ridge since ice covered the site. In the strike valley to the south of the site, reversed magnetic polarity glacial lake sediments are exposed while just two miles east of the site normal-polarity lake sediments are exposed (Sasowsky, 1994). This indicates that early Pleistocene ice covered the site and middle Pleistocene ice reached almost to the site.

Continue ahead, cross Rt.309 and turn right onto I–81 north. Exit right for Rt.93, bear right onto Rt.93. At the second traffic light, turn right into the Comfort Inn.

DAY 3. PRE–ILLIONIAN G OR OLDER GLACIAL DEPOSITS AND POST–GLACIAL LANDSCAPE LOWERING

Turn right on Rt.93 and go through Hazleton and Beaver Meadows. Cross Spring Mtn. and then Broad Mtn. Turn left onto US 209 and follow Lehigh R. through Jim Thorpe and Lehighton. After US 209 crosses the river, turn right onto Rt.248 and continue down the river to the Palmerton exit. Take the exit across the river and turn right onto a dead-end road. Park on the shoulder beside a redbed outcrop capped by bouldery sand.

STOP 3–1. BOULDERY OUTWASH AT THE LEHIGH WATER GAP

The purpose of this stop is to examine a peculiar bouldery sand deposit on a strath terrace and obtain an over–view of the Lehigh water gap. 1-2m-diameter boulders are surrounded by sand. How both sediment sizes were transported together is difficult to determine. The age of the deposit is also problematic but is thought to be Illinoian or older due to the degree of weathering of the material. This water gap is the last one the Lehigh River flows through as it enters the Great Valley section of the Ridge and Valley Province.

M	COL	DESCRIPTION OF MATERIAL	INTERPRETATION
6-9		Reddish brown; diamict with clasts of rock, tree, and anthropogenic debris; lower contact is a truncated ground surface with the original colluvial mantle removed in places.	Strip mine waste pile
1-3	11	Conglomerate boulder line underlaine by yellow brown diamict with abundant quartz pebbles; near base reddish brown diamict material encorporated from underlying material.	Late Wisconsinan colluvium
3	10	Reddish brown; bouldery, clayey sand to clayey silt matrix diamict; compact with tabular clasts showing slight N-NE dip; clasts are dominantly Mauch Chunk redbed material, angular to subrounded, striated, and many show some weathering, boulder concentration in lower 5 ft. ; lower contact is sharp.	Lodgement till, final readvance of late-Illinoian ice to terminus
3.5	9	Reddish yellow to reddish brown; pebbly to cobbly silty to clayey sand diamict grading downward into a pebbly to cobbly compact sand ; matrix in upper part displays a well developed near horizontal fissility or play structure, becomes blocky below; no boulders; no striated clasts; Mauch Chunk clasts common but not dominant; lower contact transitional over 2 to 3 cm.	Sub-lacustrine debris flows from advancing late Illinoian ice
1.5	8	Light brown in upper 3 ft., light brownish gray below; gravelly sand; friable; sparse redbed clasts; forms top surface of a bench; forms light brown stripe across the outcrop face; lower contact is sharp.	Sub-lacustrine outwash, initial ice readvance
1	7	Reddish brown; clayey silt; massive to laminated; at lower contact the material is draped over underlying boulders.	Proglacial lacustrine sediment
1.8	6	Reddish yellow to reddish gray with black laminae; bouldery sand to silty sand; silty material more grayish; boulder concentration near base; forms a lighter brown stripe across the outcrop face; lower contact gradational over 2 - 4 cm.	Sub-lacustrine outwash
2.7	5	Reddish brown to reddish gray; interbedded bouldery sand, sand, and silty sand; silty material more grayish with more redbed grains; cross-bedding dips SW; gray sandstone clast surfaces strongly oxidized, quartz conglomerate and redbed clast surfaces less oxidized; a few striated clasts; bedding curves under some clasts; lower contact gradational over 5 - 10 cm.	Sub-lacustrine outwash
1.5	4	Reddish brown; pebbly to cobbly, clayey sand to clayey silt matrix diamict; matrix becomes sandier upward; redbed clasts dominant near base; a few striated clasts; thickens to 18 f.t eastward 400 ft.; lower contact transitional over 2 - 5 cm.	Till from first late Illinoian advance to terminus
0.5	3	Reddish gray; clayey sand to clayey silt matrix diamict; forms a grayish band across the outcrop face; clasts are dominantly pebble sized; lower contact transitional over 5 -10 cm.	Debris flows from advancing ice
2.4	2	Reddish gray to brown; pebble gravel with sand stringers; rounded to platy clasts; horizontal strata with Clast imbrication indicating flow from N-NE; redbed pebbles dominant at base, subequal to quartz and sandstone pebble upward; Black bands and patches of Fe & Mn cemenation; lower contact transitional over 2 cm.	Colluvium deposited primarily by slope wash, lower part from removal of pre-Illinoian drift
0.6	1	Light gray to pinkish gray; pebbly to cobbly, clayey silt matrix diamict; moderate red to pale red purple, angular tabular clasts from underlying bedrock; light to medium gray quartz and sandstone clasts from upslope; lower contact sharp.	Mixed colluvium and residuum of pre-Illinoian age
		Micaceous sandstone, claystone, & coal; dips steeply to SE, upper surface weathered with widened joints.	Llewellyn Frm. bedrock

Figure 12. Stratigraphic column at Stop 2.4 showing late Illinoian or older glacial deposits overlying resdiuum, interbedded with lake sediments, and capped by late Wisconsinan colluvium. M column is thickness in meters.

Return to Rt. 248 and go through the water gap. Turn right onto Rt.873, cross the river and then bear right onto SR 4024. Go under I–476 and turn at the second left onto SR 4017. Turn right into the slate quarry.

STOP 3–2. PRE-ILLINOIAN TILL OVERLYING SLATE

The purpose of the stop is to examine pre–Illinoian–G–or–older till on top of non-soluble, slate bedrock and the amount of erosion since glaciation. In the rolling, slate-hills portion of the Great Valley, erosion since glaciation has removed all glacial deposits and some bedrock from all but the valleys. Colluvium from the till and bedrock always overlies the till (Fig. 14) and has a late Wisconsin weathering profile. Older, more weathered, colluvium in places underlies the surface colluvium before till is reached. The till contains many erratics from the north side of Blue Mt., the ridge through which the Lehigh water gap .crosses.

Retrace route to Rt.873, turn right through Slatington and take Rt. 873 to Rt. 309. Bear left onto Rt.309. Just before going under I–476, turn right onto Snowdrift Rd. and then left onto Blue Barn Rd. Cross railroad tracks, 4–lane Main Street, go under I–78, and turn left onto Oak View Drive. Take the 3rd right onto Worley Drive. Park on the right shoulder where fields with wetlands begin.

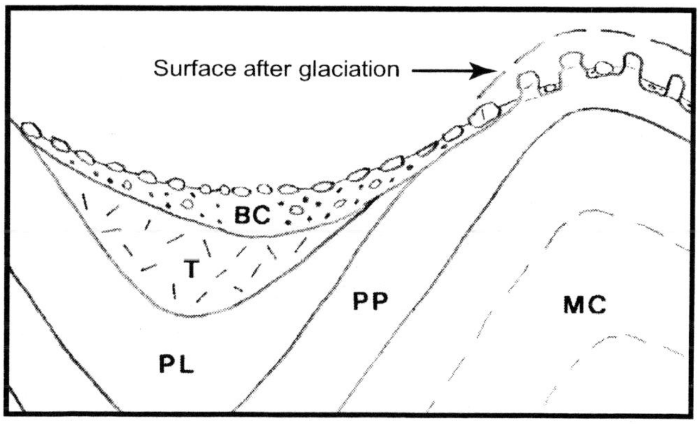

Figure 13. Schematic cross–section of the ridge–top tors and the adjacent boulder colluvium (BC) and pre–Illinoian till (T) mantled valley. PL = Pennsylvanian aged Llewellyn Fm.–coal bearing, PP = Pennsylvanian aged Pottsville cong;lomerate, MC = Mississipian aged Mauch Chunk redbeds.

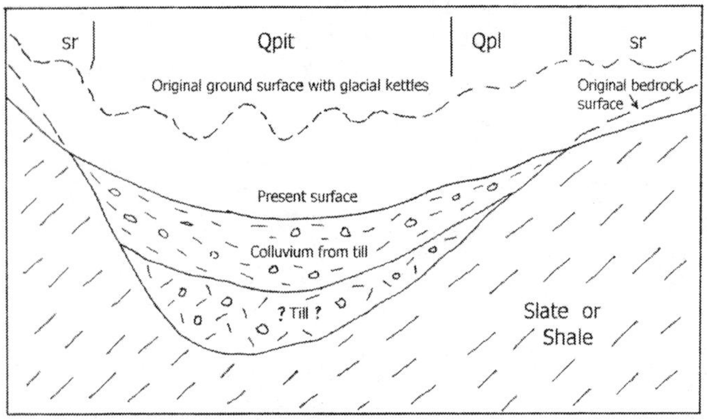

Figure 14. Schematic cross–section showing pre–Illinoian till overlain by colluvium in a valley in the slate hills portion of the Great Valley. Qpit = Pre–Illinoian till, Qpl = Pre–Illinoian lag.

STOP 3–3. KNOB AND KETTLE LANDSCAPE ON PRE-ILLINOIAN G OR OLDER, TILL-MANTLED LIMESTONE

The purpose of this stop is to view the depressions and knobs of a "pseudo–moraine" and discuss its origin. This morainic-looking landscape is not a moraine because the entire landscape has been lowered about 30 m since the area was glaciated more then 800,000 years ago (Fig. 15) (Braun, 1999). Immediately within the late Wisconsinan limit, glacial till lies directly on unweathered, striated limestone. Remnants of limestone residuum exist only deep down in fractures and steep–sided, collapse sinks. Within the pre–Illinoian limit here, about 1 m of limestone residuum has formed under the till (typically several meters to 10's of meters thick) since glaciation . Limestone dissolution rate studies in Pennsylvania (Ciolkosz et al., 1995) indicate that much residuum would come from the dissolution 30+ m of limestone since glaciation. So a moraine was here once, but as this carbonate landscape was lowered by repeated sinkhole collapse, the original moraine landform was totally destroyed. But the depressions here, typically a few 10's of meters across and less than a meter deep, are not themselves sinkholes. Nearby borehole records for the Interstate Highways show that the sinkhole depressions in the bedrock surface are much larger than these depressions at the ground surface.

Continue ahead. Turn left then left again, going under I–476 on Cetronia Road. and immediately turn right onto Devonshire Rd. Go under I–78 and cross US 222. Take the 3rd. right onto Minesite Road. and then left into Rodale Park parking lot.

STOP 3–4. EXPOSURE OF PRE-ILLINOIAN G OR OLDER TILL OVERLYING LIMESTONE RESIDUUM

The purpose of this stop is to examine the texture, and clast lithology of the "let–down" pre-Illinoian G or older till. The

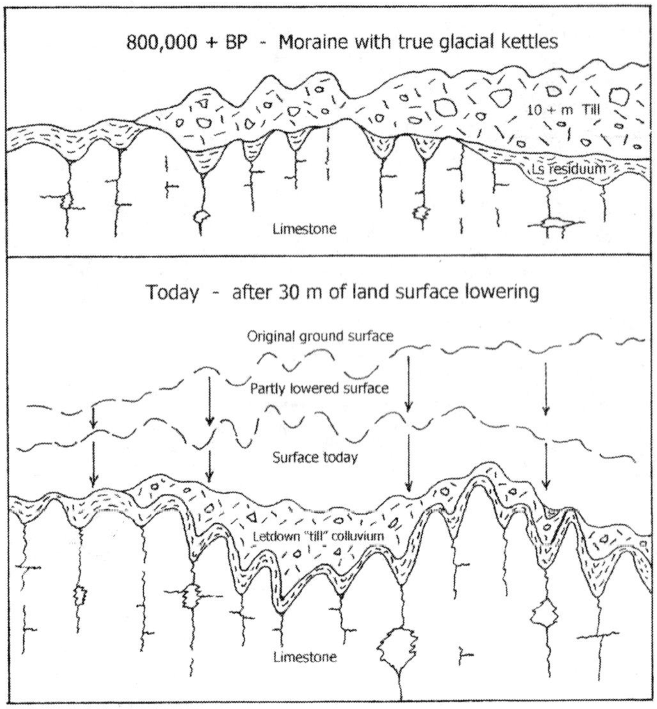

Figure 15. Schematic cross–section of the limestone lowland part of the Great Valley showing the landscape and deposits when glaciers had just retreated from the area, the amount of landscape lowering since glaciation, and the present landscape and deposits.

reworked till is a reddish-brown to reddish-yellow, clayey-silt-matrix diamict with erratic clasts from the north of the site.

What is unusual at this site is that the reddish till is overlain by a yellowish-brown, clayey silt with no clasts. The color of the material and the lack of sandstone clasts suggest the material was derived from carbonate residuum and washed onto the site from a nearby, once-higher, carbonate pinnacle. Alternately, the overlying material may be waste from an adjacent iron ore pit.

Return to Devonshire Rd., go right, cross Cedar Crest Blvd., and cross the incised Little Lehigh Cr. valley. Cross Country Club Road. and take 6th left onto Harrison St. Cross 4–lane State St. and railroad tracks to then turn right onto Lee St. Turn left onto Evergreen St. and then right onto Alpine St. Turn left into the parking lot at the end of street. Walk on the trail through the woods to the abandoned gravel pit.

STOP 3–5. EXPOSURE IN A DEEPLY WEATHERED, PRE-ILLINOIAN-G OR OLDER KAME AT THE BASE OF SOUTH MT.

The purpose of this stop is to examine the deep weathering of >788,000-year-old gravel and sand. On the walk to the pit, we will cross a colluvial mantle that overlies eroded glacial deposits (Fig.16). In places, just a few erratic clasts remain in limestone residuum that has developed since glaciation (Qpl on

Fig. 16). The upper 2 m of the material exposed in the pit is a round-stone diamict with a bright-reddish-yellow, clayey matrix. Below that, are 2–4 m of gravel with some fine matrix material and relict bedding. Yellow-brown sand is exposed in the floor of the pit. The entire 10 m high exposure shows evidence of weathering.

The ice lobe that occupied the Great Valley lowland blocked the east-draining Little Lehigh Creek and impounded a shallow proglacial lake, named Glacial Lake Packer by Williams (1893). The lowest available outlet col westward to the Schuylkill River system is at an elevation of about 500 ft (152 m) at Topton. The large mass of outwash deposited between the ice lobe and South Mt. at Emmaus Junction has a top elevation of 500 ft (149-152 m). This suggests that the outwash deposit represents a large kame delta built into Glacial Lake Packer and graded to the outlet at Topton. (These elevations relate to the present elevation of an area underlain by carbonate rock. Such elevations were probably 30 m or more higher at the time of glaciation.)

Proceed south on Alpine Drive, turn right onto Emmaus Ave (E. Main St.) and proceed northeast for approximately 5 miles.

For ALTERNATE STOP A 3.6, turn right onto Honeysuckle Road and proceed up to the top of the hill. At the top, turn left into the parking area adjacent to the green gate. Walk along the road past the gate to the abandoned sand pit.

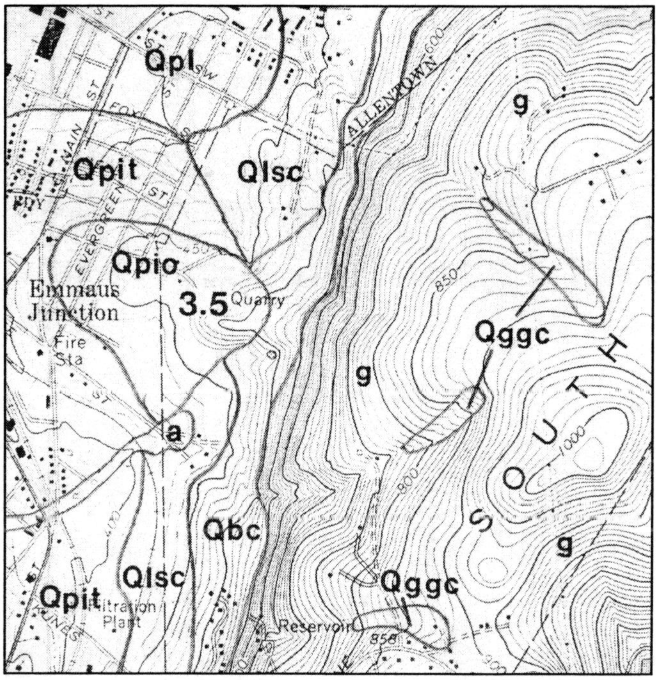

Figure 16. Surficial deposit map of the area around the pre–Illinoian–G or older kame at the base of South Mt. The top elevation of the deposit matches that of the outlet for Glacial Lake Packer, suggesting the deposit reprsents a kame delta. Qpio = Pre–Illinoian gravel, Qpit = Pre–Illinoian diamict (till or weathered gravel), Qpl = Pre–Illinoian lag, Qbc = Boulder colluvium, Qlsc = limestone and sandstone colluvium, Qggc = granitic gniess colluvium, g = granitic gniess.

ALTERNATE STOP 3–6A. SAPROLITE ON SOUTH MOUNTAIN

The purpose of this stop is to observe the irregular weathering profile developed on the granite and gneiss of South Mt. and discuss its implications for erosion and landscape evolution. Saprolite, 0-10+ m deep, forms an irregular weathered mantle in the sand pit. On ridgetops in the Ridge and Valley Province, the Tuscarora Fm quartz-arenite has weathered to a clean, white-quartz-sand saprolite. Locally, as on the Gatesburg ridge near State College, PA, a sandy dolomite has been saprolitized to depths of 30 + m and contains extensive laterite and bauxite deposits. Saprolite formation on the Piedmont is thought to be favored by an incised landscape where ground water flow is vigorous (Cleaves, 1993). The maximum rate of saprolite production has been estimated by cosmogenic methods by to be about 20 m/Ma (Pavich et al, 1989). Saprolite residence time on the landscape is estimated to be 1–5 Ma

Back out of parking area, continue up the hill, and turn right onto East Rock Road. Proceed to the intersection with Rt 309/I–78. Exit onto Rt 309/I-78 west towards Allentown. Continue on Rt 78 west for about 50 miles to I–81 south. Continue on I–81 south to Harrisburg. Take the last exit before the Susquehanna River (Exit 66) onto Front Street and proceed north toward Rockville and the Super 8 motel.

For ALTERNATE STOP 3–6B, turn right onto Honeysuckle Road, then in less than a mile, left onto Black River Road. Proceed on Black River Road to Rt 378. Turn left on Rt 378 and proceed up the hill. At the top of the hill, turn right onto Mountain Drive West. At the next intersection, bear right and proceed to the Mountaintop campus of Lehigh University. Park in the lot for Iacocca Hall.

ALTERNATE STOP 3–7A. LEHIGH UNIVERSITY OBSERVATION TOWER AT IACOCCA HALL

The purpose of this stop is to discuss the long-term evolution of Appalachian landscape if the Council Cup stop on day two was a wash-out and to review the topics of Days 2 and 3 at a spectacular 360–degree overview of the Lehigh and Saucon valleys. The topics of discussion are: (1) a review of the sequence of Pleistocene glaciations in eastern Pennsylvania, (2) a review of the drainage evolution in this part of the Appalachians, including discussion of the water and wind gaps, (3) discussion of the presence or absence of peneplains in the Appalachian landscape, (4) discuss possible explanations of the mid–Miocene sediment pulse, and (5) set the stage for the final two days of the trip.

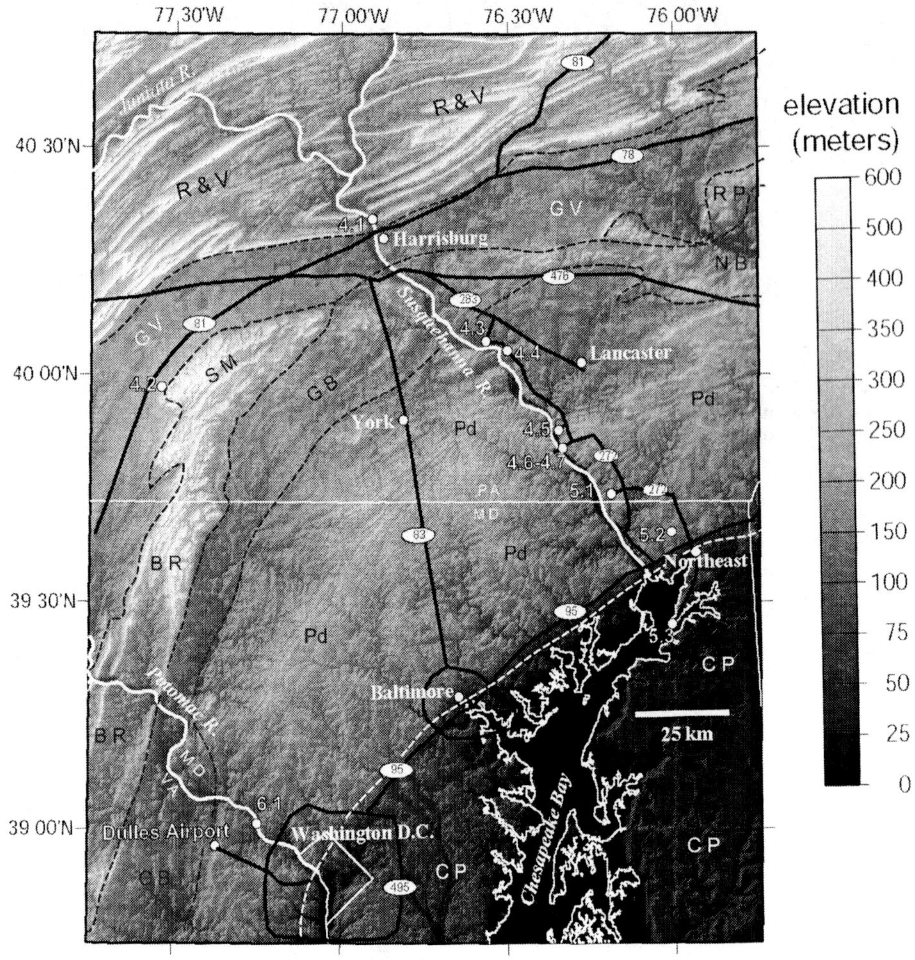

Figure 17. Digital relief map showing the trip route and stops for the last three days. White numbers denote stops. First number is day and second number is stop. Physiographic Provinces (boundaries shown as dashed lines)–B R = Blue Ridge, C B = Culpepper Basin, C P = Coastal Plain, G B = Gettysburg Basin, G V = Great Valley, N B = Newark Basin, Pd = Piedmont, R P = Reading Prong, R & V = Ridge & Valley; S M = South.

Depart Iacocca Hall and retrace path to Mtn. Drive South. Bear left onto Mountain Drive South and proceed to the bottom of the Hill. After the stop sign, turn left onto Friedensville Road and follow signs to Exit 67, I–78. Enter I–78 west to Allentown and go 67 miles to I–81. Continue with instructions after ALTERNATE STOP A3.6 to get to the hotel.

Start, Fort Hunter County Park, approximately 1 mile north on Front Street from hotel.

DAY 4. SUSQUEHANNA WATER GAPS, THE GREAT VALLEY, AND LOWER SUSQUEHANNA RIVER

STOP 4–1. SUSQUEHANNA WATER GAPS AT FORT HUNTER COUNTY PARK

The purpose of this stop is to view the classic Susquehanna River water gaps north of Harrisburg (Fig. 18) and to review the suggested origins for such water gaps. The park buildings occupy a high point on the floodplain that just projected above the largest historic flood in June 1972 from Tropical Storm Agnes (Page and Shaw, 1973). The high area is a remnant of a late Wisconsin, recessional, outwash-terrace level about 8 m (25 ft) above present low-water level, Peltier's (1949) Valley Heads terrace. Much of the city of Harrisburg lies on the 15-m-(50 ft)-high, outwash terrace developed during the late Wisconsin maximum (STOP 2.2). The entire Pleistocene terrace sequence will be examined at Stop 4.3. A post-glacial, abandoned channel separates this high point from the eastern bedrock wall of the water gap. A low berm or strath, 2 to 3 m high, runs along the edge of the present Susquehanna channel. The channel here is on bedrock, as it is almost entirely from the late Wisconsin terminus to Chesapeake Bay.

The geomorphic puzzle here is how the river became entrenched across the ridges of resistant rock and why the water gap developed where it did. The various hypotheses on the origin of this water gap can be place in several groups (modified from Sevon, 1989b): (1) Superposition from an estuary that projected up the valley on the Schooley peneplain from the edge of the sea at the Fall Zone (Davis, 1889), (2) from Cretaceous marine transgression (Johnson, 1931), (3) headward erosion along structural weakesses through the strike ridges (Ashley, 1935; Meyerhoff and Olmstead, 1936; Thiesen, 1983; Hoskins, 1987), (4) resistant sandstone strata were thinner here (Thompson, 1949), and (5) position determined by the front of a large overthrust sheet in the Anthracite region to the northeast (Sevon, 1986). Hypotheses 1 and 2 are not supported by what is now known about the Coastal Plain stratigraphy; to be discussed on DAY 5 (Pazzaglia, 1993). Proof or disproof of hypotheses 3 through 5 is difficult because on the order of 17 km of rock has been removed since the Alleghanian deformation that produced the folds here. Structures at the present depth of erosion do not necessarily reflect those at the level where erosion began (Sevon, 1989b).

From the water gap, we will travel down the Great Valley to the northern end of the Blue Ridge, here called South Mt. We will first cross folded Ordovician shale and then folded Cambrian-Ordovician carbonate as we approach South Mt., "held up" by resistant early Cambrian sandstone.

Continue north on Front Street, turn right onto PA Rt 443, then Rt onto US Rt 22/322 east. Proceed on US Rt 22/322 to the interchange with I–81 and take I–81 south towards Carlisle. Continue 38 miles to Exit 29 (King St). Turn left (south) to Cleversburg. Turn right (west) and go to Mainsville. Turn left (south) onto Linsay Lot Road and go 0.8 miles to the sign on right to Valley Quarries. Enter and turn immediately left on the paved road to the gravel driveway into the pit

STOP 4–2. MAINSVILLE QUARRIES AND GEOLOGY OF THE GREAT VALLEY

The purpose of this stop is to examine the intensely weathered gravel at the base of South Mt., the "trapping" of such gravel by carbonate dissolution, and the influence of that dissolution on drainage basin evolution. The Valley Quarries Mainsville pit is in a complex of alluvial-colluvial sand and gravel derived from South Mt. These deposits occur on top of carbonate rocks as fan–like aprons 1-5 km wide (Clark, 1991) that gently slope away from the mountain front. Similar deposits lie along the west edge of the Blue Ridge from here to central Virginia. The topography on the deposits is complex because karst processes have modified the aprons. This pit and two others nearby are a few of the places where these gravels are exposed. Much of the gravelly material is a diamicton developed both by emplacement by debris flows and by weathering of both debris-flow and fluvial material after deposition. The details of the geomorphology here and the relation of these deposits to the carbonates are summarized by Sevon (2001, DAY 2, Stops 2 and 3).

The overall stratigraphy of the aprons (Fig. 19) is almost entirely interpreted from water-well records. The thickness of the material on top of the carbonates is commonly 30 m (100 ft) and in places is as much as 120 m (400 ft) (Sevon, 1991). Thickness generally decreases away from the mountain front. In a few places, thick deposits of clay residuum from weathering of the carbonates are noted beneath the gravel. Beneath the gravel at Pond Bank, about 8 miles southwest of here, Pierce (1965) described 43 m (140 ft) of clayey residuum and a lignite deposit that contained late Cretaceous spores. Using an insoluble residue content of 10% for the carbonates, he inferred that 430 m (1400 ft) of carbonates have weathered beneath the gravel to produce the residuum. The processes that preserve the gravel and residuum are straightforward, but complex in detail (Fig 18). Acidic runoff from South Mt. percolates down through the gravel to dissolve the carbonate rocks. Voids of 2-3 m (5-10 ft) in the carbonates are not uncommon in well records. As the carbonates dissolve, the landscape above lowers, residuum is added to the bottom of the pile of surficial deposits, and karst depressions form on the gravel-apron surface. The gravel mantle protects the residuum from subaerial removal. Some karst depressions remain for long periods of time, for cores from ponds in two depressions have basal dates of 14,000 ^{14}C yrs. B.P. (Watts, 1979) and 16,000 ^{14}C yrs. B.P. (Delano, et al., 2002). The age of the gravel is uncertain, but weathering of some of the gravel is more intense (even quartzite cobbles disintegrate to the touch) then that observed in the oldest of the Pleistocene deposits on DAY 3. This suggests a late Tertiary age for the oldest gravel (Sevon, 2001).

Figure 18. The Susquehanna River water gap at Harrisburg, view to the north.

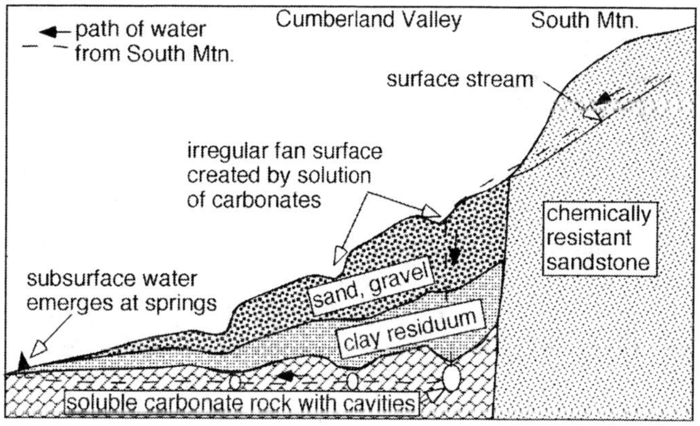

Figure 19. Schematic cross–section of the flank of South Mt., irregular fan like surface mantled by sand and gravel, and underlying limestone.

Carbonate dissolution controls a lot of the local landscape evolution in the Cumberland Valley portion of the Great Valley. For example, the level of karst development controls local accommodation space for the fan material, and determines if Yellow Breeches Creek will be the master drainage of the Great Valley. The carbonate under-drain appears to respond to base level fall faster and more efficiently than the part of the valley underlain by shale (Conodoquinet Creek). So when the Susquehanna River incises, Yellow Breeches Creek efficiently flushes detritus through it, the fans are short and steep, and Yellow Breeches has a competitive advantage for drainage in the Great Valley over Conodoquinet Creek. But when the Susquehanna River has reached its base level of erosion, so does Yellow Breeches. The carbonate under-drain backs up with detritus, the fans are broad and gentle, and the Conodoquinet has the competitive advantage. This is the case in the Cumberland Valley today. These roles are reversed in Virginia where shale drainages, like the Shenandoah headwaters, are loosing to the carbonate drainages, like the Maury, a James River headwater. That is because the base level fall there is happening NOW, and the base level fall on the Susquehanna happened in the past.

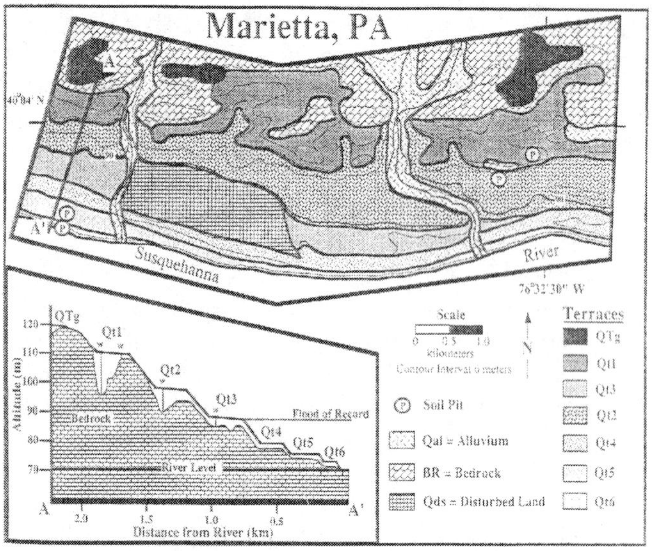

Figure 20. Pleistocene terraces at Marietta (from Engle et al., 1996). Note Qt3 and Qt2 labels are reversed in legend. w = water well. Stop 4.3 is located along cross–section A–A′.

STOP 4–3. PLEISTOCENE TERRACES AT MARIETTA

The purpose of this stop is to observe a flight of Pleistocene terraces and discuss their genesis. Pleistocene terraces are well preserved along this reach of the Susquehanna River. Pleistocene terrace stratigraphy along the Susquehanna River was first investigated by Peltier (1949) who demonstrated several alluvial fills underlying mappable treads extending all the way from the head of Chesapeake Bay to heads-of-outwash at the glacier margin. The precise stratigraphic and genetic relationship between these terraces was more recently investigated by Engle et al. (1996). At least six Pleistocene terraces and one possibly pre-Pleistocene terrace remnant preserved are here at Marietta (Fig. 20). Water-well data and former gravel pits confirm that treads Qt1 through Qt6 are underlain by several meters of stratified sand and gravel. QTg is not a stratified deposit, but rather a scattered lag of rounded clasts of Appalachian provenance, with an occasional diabase boulder. QTg can be traced downstream all the way to the Coastal Plain where it projects both in elevation and in composition to the upper part of the Pliocene Pensauken Fm. If this correlation is correct and QTg is late Pliocene (~2.5 Ma), the incision rate for the river at this point is 20 m/Ma Terraces Qt1 through Qt6 are distinguished compositionally by containing granite and gneiss, rock-types not found in the upper Susquehanna basin and that could only have been introduced by glaciation. For this reason, Qt1 through Qt6 are generally held to be genetically related to glacial outwash. Soil chronosequences establish a correlation between the terraces here and those at the heads-of-outwash 150 km upstream. Qt4 through Qt6 have soils consistent with late Pleistocene (Wisconsin) age. Qt3 has a significant loess cap and soil development consistent with a pre–Wisconsin, presumably Illinoian, age. Qt2 and Qt1 are thought to be pre-Illinoian age. A geophysical study is currently underway to more carefully investigate the nature of the straths underlying the Pleistocene terraces at this site. The strath geometry may help resolve two competing ideas regarding the preservation of Pleistocene terraces along the river. One possibility is that the Susquehanna River has been incising throughout the Pleistocene, producing the accommodation space to preserve the terraces. Each pulse of glacial outwash provides the tools to carve the strath and then aggrade the terrace alluvium. The alternative explanation is that the Susquehanna River incised to its current elevation in the Pliocene or early Pleistocene and the terrace represent the uneroded, "wings" of thick alluvial fills related to glacial outwash. The latter idea hinges on the fact that more recent glaciations were either less severe or transported less sediment than earlier glaciations.

Retrace route out to PA Rt. 441 (River Road) and turn right. Stay on River road for about 8 miles as it climbs Chickies Ridge and then begins to descend into Columbia. Turn right into Chickies Rock overlook.

STOP 4–4. LUNCH AT CHICKIES ROCK OVERLOOK

The lunch stop commands a view of the Susquehanna River as it turns south to cut across the resistant quartzite of Chickies Ridge. Note the ribs of bedrock exposed in the channel bottom.

Return to PA Rt 441 and continue south into Columbia. Follow signs to remain on Rt 441. South of Columbia, Rt 441 becomes the River Road and the trip follows it as it winds through Washington Boro, where another well-preserved flight of Pleistocene terraces is preserved, to Turkey Hill, the beginning of the High Piedmont, to Safe Harbor, where the road crosses Conestoga Creek, and finally to Martic Forge. At Martic Forge, turn left, then an immediate right to stay on River Road. Proceed across Tucquan Creek, which is a steep, incised drainage typical of Piedmont tributaries to the Susquehanna River. After climbing out of the Tucquan valley, turn right and follow signs to Pinnacle overlook.

STOP 4–5. PINNACLE OVERLOOK

The purpose of this stop is to observe the deepest part of the Piedmont gorge, discuss its origin, and introduce the terrace stratigraphy for the Piedmont reach. This stop overlooks the deepest portion of the Piedmont gorge (~ 200 m) at Holtwood. The river here is part of the Lake Aldred pool behind the Holtwood dam. Deep gorges like this one, and steep, convex, river long-profiles are typical of all of the major streams that traverse the Piedmont and Fall Zone of the middle Atlantic States (Fig. 21). These features probably represent river incision in response to some combination of eustatic fall and/or epeirogenic uplift. The elevation and incision of the Piedmont here has been used to argue for broad, recent arching of the Piedmont (Campbell, 1929; Fig. 22), an explanation foreshadowing a more recent flexural isostatic model of margin deformation (Pazzaglia and Gardner, 1994). The flexural isostatic model will be more fully developed at the end of DAY 5. Flat uplands flanking the river through the Piedmont reach are locally mantled by a rounded gravel lag representing the remnants of strath terraces. The trip will visit one of these localities at the beginning of DAY 5, but at this time, illustrating the general terrace stratigraphy and petrography is useful. Three upland gravel units, Tg1, Tg2, and Tg3 (Fig. 23) occur. They have similar texture, with the possible exception of Tg1, which tends to be coarser grained and perhaps a bit more angular. Upland gravel dominated by resistant rock types, such as vein quartz, quartzite, and quartz sandstone. However, the amount of quartz sandstone increases with decreasing stratigraphic age. Tg1, Tg2, and Tg3 are petrographically distinct enough to use clast counts as a means of correlation along the river. They are also much less heterolithic than QTg, which is only locally preserved through the Piedmont reach, typically at the confluence of major tributaries.

Retrace steps out to River Road and turn right. Cross Kellys Run, and proceed straight to PA Rt. 372. Turn right onto Rt 372 and proceed for about 2 miles as the road descends into the Holtwood gorge. Just before the bridge, slow and pull off into the wide parking area on the left hand side of the road.

STOP 4–6. NORMAN WOOD BRIDGE

The purpose of this stop is to observe the bedrock channel of the Susquehanna River and discuss the processes of strath genesis. Walk out on the bridge to observe the Susquehanna channel. The view from the bridge affords an outstanding perspective on the Holtwood Islands and the modern Susquehanna channel. At low water, this is an impressive active strath and it

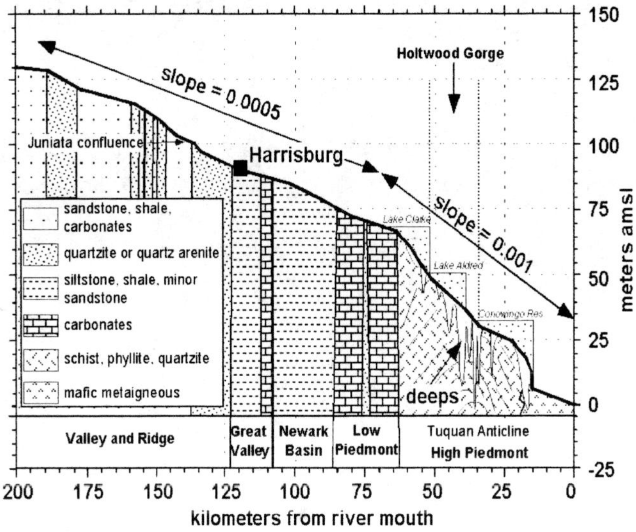

Figure 21. Long profile of the lower Susquehanna River through part of the Ridge and Valley, the great Valley, the Triassic basin, and Piedmont. Note the distinct convex profile. The deeps in the Piedmont are shown to scale, but are not located precisely (Pazzaglia and Gardner, 1993).

indicates this river's ability to carve out a wide valley bottom while still vertically incising. The extreme eastern channel bank is not a flat strath, but rather has been revealed as one of six "deeps" (Mathews, 1917) along the Piedmont reach of this river. Deeps are a series of unconnected, spoon-shaped excavations that tend to hug the eastern side of the modern channel. They are intensely pot-holed and sculpted. The deep here at Holtwood is over 40 m, which places its base approximately 6 m below sea level (Fig. 21 and 24

The many islands that dot the Holtwood area attest to relatively rapid incision of a paleo-strath through this reach (Thompson, 1990). Note that even though the island tops appear flat, a correlation based on the tops alone suggests an upstream dip (Fig. 24). The passage of a knickpoint at a rate similar to the vertical down-wearing of the channel would result in a time-transgressive, upstream-dipping strath, here represented by the island tops, that is old down stream, but become younger upstream where it merges with the modern strath just in front of the dam.

Strath genesis is probably linked to times when the river has more abrasive tools, such as during glacial outwash. Wide straths are favored by both the availability of tools as well as protracted periods of base level stability. In contrast, the strath is destroyed by vertical incision, concentrated in numerous rivulets, during times of low bedload flux and/or during times of base level fall. Upstream migration from knickpoints, born at the Fall Zone by impulsive, glacio-eustatic fall, are an additional mechanism for channel bed lowering. The strath exposed here, presumably carved during the late Pleistocene, is currently being destroyed by vertical incision of numerous small channels and rivulets. If the island are a paleo–strath, it was abandoned by both vertical incision, as well as the passage of a knickpoint.

fig 4.12

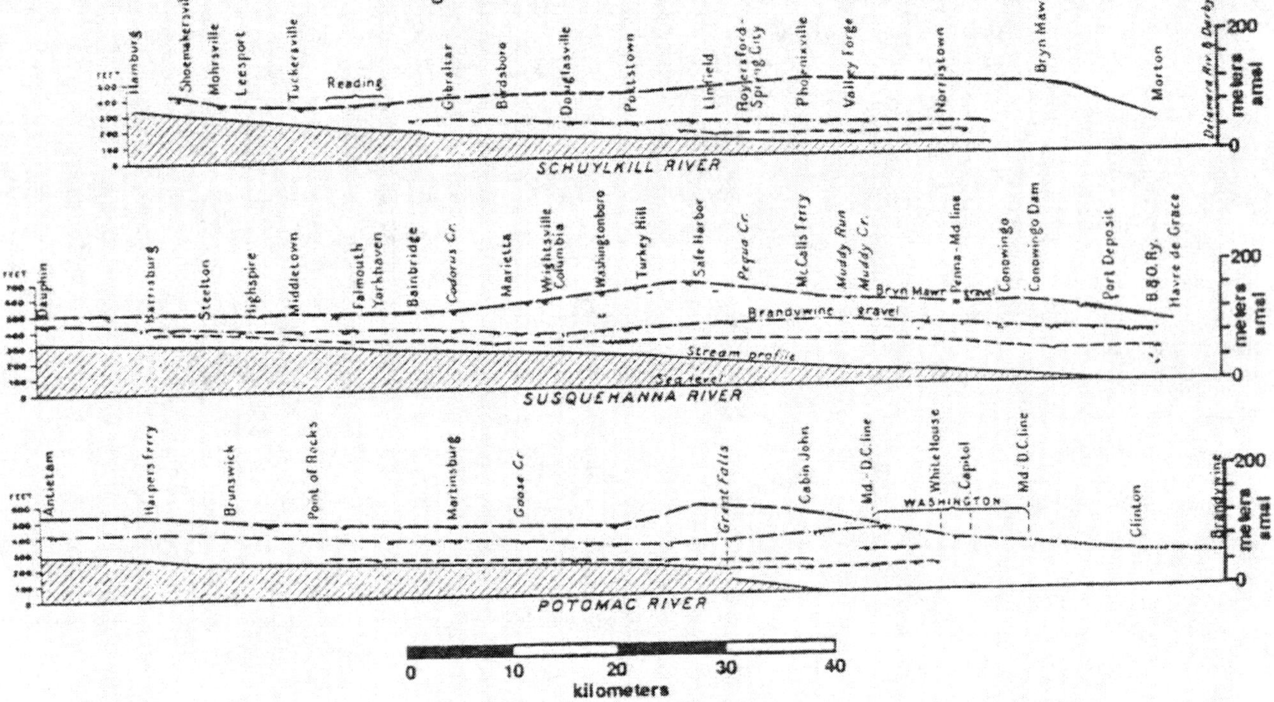

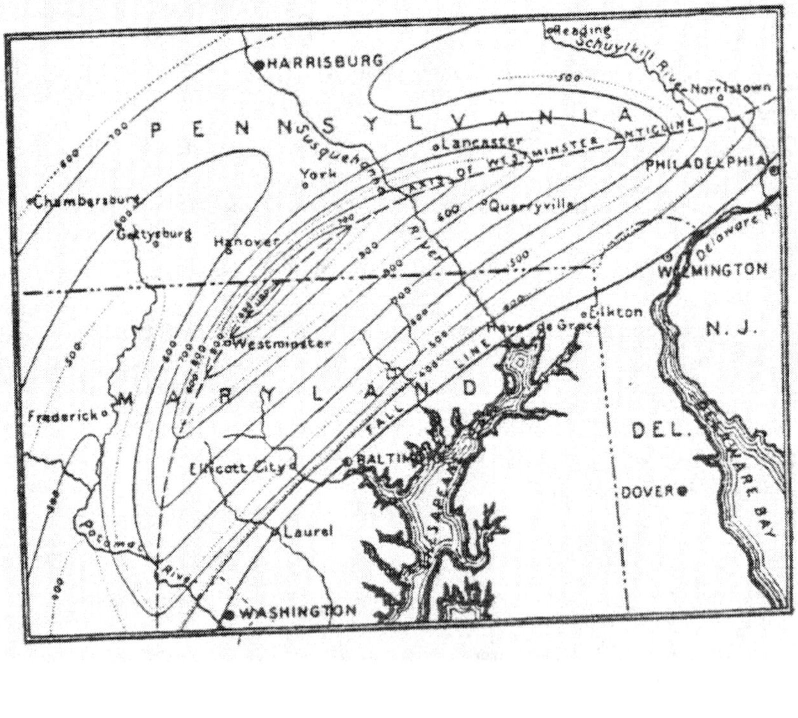

Figure 22. Arching of the Piedmont called the Westminster anticline proposed by Campbell (1929). The arching was based on correlation of peneplains, which locally were mantled with river gravel.

Continue west on part 372, cross theNorman wood bridge. Turn right immediately after the bridge into the lock 12 parking area.

STOP 4–7. HOLTWOOD ISLANDS

The purpose of this stop is to inspect the islands at Holtwood and discuss the mechanisms of bedrock channel erosion, including the formation of potholes and the deeps. The islands along the western channel bank are easily reached by following the blue blazes of the Mason–Dixon trail. Potholes are an important feature on the island tops and along their sides. The potholes, along with the deeps, have been used to argue for large, perhaps catastrophic, floods in the lower Susquehanna River during the Pleistocene (Thompson, 1985; 1990), in part because these features are similar to much larger features associated with the scabland floods of the Columbia Plateau. Two potential sources of catastrophic floods have been suggested. The first has an origin along the North Branch of the river as catastrophic, sub-glacial floods (Shaw, 1989). We reviewed the evidence on Day 2 for the unlikelihood of such floods coming down the North Branch of the river. The other possible origin of catastrophic floods is the failure of an ice dam along the West Branch of .Susquehanna River and the rapid draining of glacial Lake Leslie. The presence of Lake Leslie has been recently verified (Gardner et al, 1994; Ramage et al., 1998) as an early-Pleistocene, ice-dammed lake in the Bald Eagle valley of central Pennsylvania. Alternatively, the potholes and the deeps are not relict, but rather modern features actively being carved by the river. The presence of similar features on rivers like the Potomac or tributaries to the Susquehanna River that never experienced glaciation in their headwaters argue against catastrophic floods. Furthermore, the Susquehanna River has generated large historic discharges such as those resulting from hurricane Agnes (>10^6 cfs), and spring, rain-on-snow events have generated historic ice-chocked discharges in excess of several hundred thousand cfs.

The age of the island tops is currently being investigated by cosmogenic methods at the time of the writing of this guidebook. Should results be available by the time of the field trip, we will provide an addendum with the data.

Return to PA Rt 372 and travel east across the bridge. Continue for approximately 8 miles to PA Rt 272 at Buck. Turn right and follow Rt 272 into Wakefield. Stay on Rt 272 across Octoraro Creek, U.S. Rt 1, and enter the little crossroads of Chrome. Turn right, remaining on Rt 272, cross into Maryland, and proceed south for approximately 12 miles to the interchange with I–95. Turn left onto Lums Road and proceed to the Crystal Inn.

DAY 5. THE FALL ZONE, COASTAL PLAIN, AND FLEXURAL DEFORMATION OF THE CONTINENETAL MARGIN

Depart hotel to Rt 272 north. Continue on Rt 272 back into Pennsylvania, through Chrome, and over Octoraro Creek. One half mile after the creek, at the cross-roads of Kirk's Bridge, bear left off of Rt 272, proceed to Kirk's Mill, Eastland, and Wrightsdale. At Wrightsdale, stay right at the triangle and proceed toward New Texas. Stay straight at New Texas on Black Baron Road to Pleasant Grove and the intersection of Pilottown Road. Turn right onto Riverview Road and proceed one-half a km to the top of the hill.

STOP 5–1. KIRK FARM GRAVEL

The purpose of this stop is to visit an upland gravel locality to study the texture and composition and suggest a correlation through the Piedmont into the Fall Zone. This is a "famous" location on the Francis Kirk farm where large quantities of indisputably fluvial-origin, "potato stones" occur. Locally, the fluvial gravel is still in place, or nearly in place, as it is cemented by

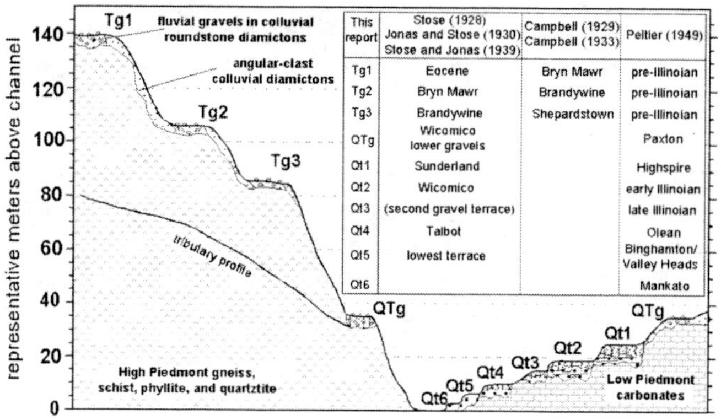

Figure 23. Representative cross section of the lower Susquehanna River valley showing the relationship between pre–Pleistocene upland gravels and Pleistocene terraces (from Pazzaglia and Gardner, 1993).

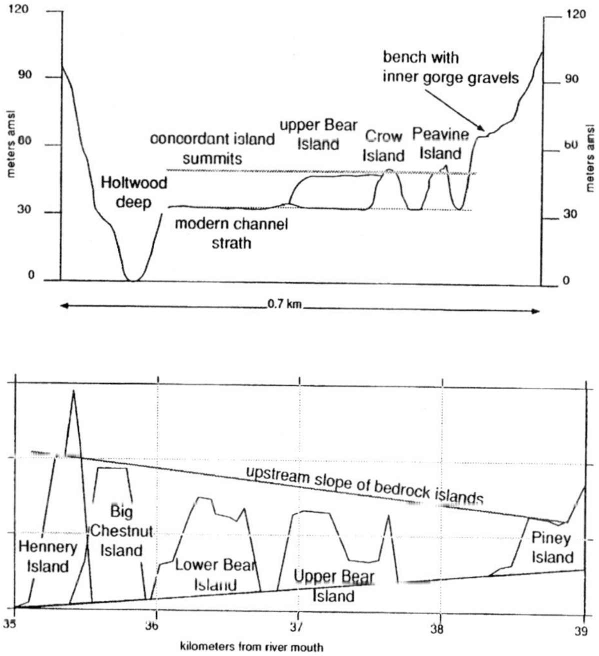

Figure 24. Cross–section (top) and long profile of the islands at Holtwood.

hematite. The site is at an elevation of 500 feet (150 m) on the Pennsylvania Piedmont. Petrography reveals that the lithology is virtually all vein quartz and very likely proximal fluvial equivalents to the Bryn Mawr Fm. A pit dug at this location showed that the gravel is part of a matrix-supported, colluvial deposit that is ubiquitous on the Pennsylvania Piedmont (Pazzaglia and Gardner, 1993). The gravel is found throughout at least 2-3 m of colluvium that lies atop saprolite and bedrock. Presumably, the gravel was once part of a stratified, more extensive deposit, but has been eroded by the intense periglacial activity that has repeatedly affected this region during the middle and late Pleistocene. Because this gravel deposit is highest in the landscape, it is mapped as Tgl, or the most ancient, recognizable terrace of the lower Susquehanna River. The change in gravel composition with respect to age is very well expressed in the surrounding area as the gravel exposed at the Brinton Farm only two miles from here, but at an elevation of 340 feet (104 m), contains much sandstone, siltstone, and chert not found at Kirk Farm. Petrography forms the basis for correlation of terraces throughout the Piedmont (Fig. 25; Pazzaglia and Gardner, 1993). Virtually every flat field in the Piedmont of York and Lancaster Counties has been walked and inspected. Fluvial gravel lags, like the one exposed here, are only found when the field is (1) within 2 km of the river, and (2) at a specific elevation corresponding to the three main terraces (here at 500, 400, and 340 feet).

Return to Pleasant Grove. Turn right onto Pilottown road and proceed to Pilot. Turn left toward Oakwood, and continue straight to U.S. Rt. 222. Turn right and proceed south on Rt 222 to its junction with U.S. Rt 1. Do not cross the dam! Continue south on Rt. 222 along the Susquehanna River to Port Deposit, than Craigtown, go under I–95, then proceed to Perryville. Turn left onto U.S. Rt. 40 and proceed for about 5 miles to Belvedere Road. Turn left onto Belvedere Road and proceed straight ~4 miles to the York Building products Belvedere plant weighing station on the left.

STOP 5–2. BRYN MAWR FM. AT THE BELVEDERE PIT

The purpose of this stop is to observe the down-dip, Fall Zone equivalent of the Piedmont upland gravel and reconstruct the environment, eustatic conditions, and age of its deposition. The large gravel pits here are operated by York Building Products and they quarry a quartzose sand and gravel called the Bryn Mawr Fm. (Pazzaglia, 1993). It is just one of several upland gravel deposits mapped at the head of Chesapeake Bay (Fig. 26).

The Bryn Mawr Fm. unconformably overlies the Cretaceous Potomac Group and a well developed saprolite in the underlying bedrock. The sand and gravel is both texturally and compositionally mature. The light fraction is virtually all quartz, and the heavy minerals are zircon, tourmaline, rutile, ilmenite, and leucoxene. Several lithofacies are present in the high walls at any time during the quarrying operations. The dominant facies is a large-channel, fluvial facies, characterized by large upper delta plain lateral accretion surfaces draped by laminated clay. Subordinate facies include channel thalweg gravel, oxbow lake deposits (pollen bearing), middle delta-plain shoreline, and lower delta-plain distributary-mouth bars. In 1994, the quarry operators broke into and collected a deposit of black clay very rich in plant remains and pollen. An analysis of the pollen reveals that it is a late-middle Miocene (~12 Ma) assemblage (Pazzaglia et al., 1997). Bryn-Mawr-Fm–like gravel occurs all along the mid-Atlantic Fall Zone where it is called the Bon Air Fm. (Virginia), Brandywine Fm. (Maryland), and Beacon Hill Fm. (New Jersey). Such widespread gravel deposition very likely occurred during middle and late Miocene eustatic sea-level highstands at a time when the Appalachians were shedding a deeply-weathered regolith. Higher on the Fall Zone, the Bryn Mawr gravel is represented by at least three river terraces. Out onto the Coastal Plain, they are generally equivalent to the Choptank Fm., and to a lesser degree, the Calvert and St. Mary's Fms. (Fig. 27).

Return to Rt 40 via Belvedere Road. Turn left on Rt. 40 and proceed approximately 6 miles to Northeast. Turn right on Rt 272 and continue for ~ 15 miles to the end of the Elk Neck Peninsula to Turkey Point and the parking area for Elk Neck State Park.

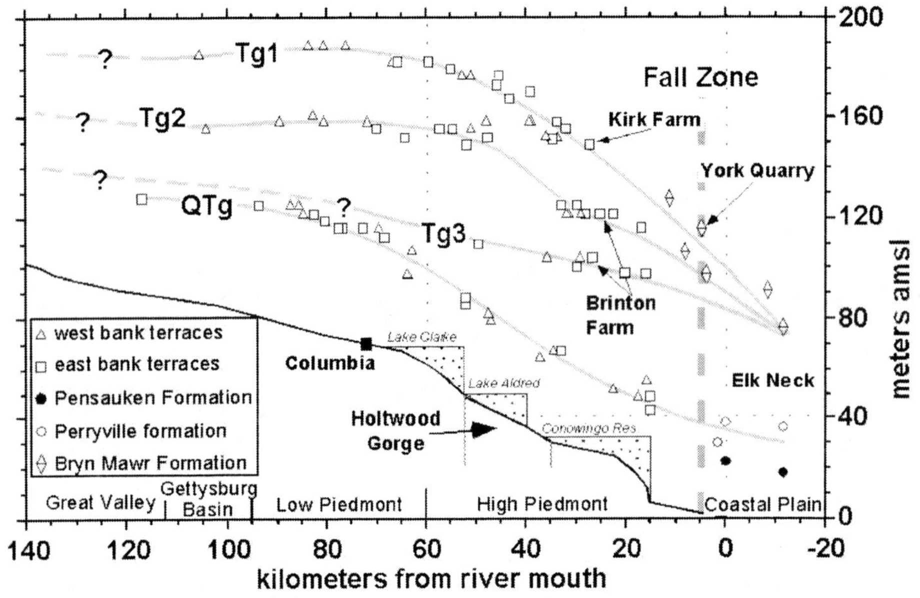

Figure 25. Proposed correlation of upland gravels with respect to the Susquehanna River long profile through the Pennsylvania Piedmont (From Pazzaglia and Gardner, 1993).

STOP 5–3. STRATIGRAPHY OF THE PENSAUKEN FM. AND TURKEY POINT BEDS AT TURKEY POINT

The purpose of this stop is to observe another Fall Zone/upper Coastal Plain site that contrasts with the Miocene Bryn Mawr Fm. in texture, facies, composition, and age. The sea cliffs at Turkey Point are underlain mostly by the red, arkosic, Pensauken Fm. This fluvial to fluvial-estuarine deposit unconformably overlies Cretaceous Potomac Group, and is unconformably overlain by a newly defined, but unofficial, unit called the Turkey Point beds (Fig. 28). Despite its location at the mouth of the Susquehanna River, the Pensauken Fm. is an ancestral Delaware–Hudson River deposit. Like all other central Appalachian streams, the Hudson River formerly made a sharp bend to the southwest at the Fall Zone. Beginning in the late Miocene, the paleo-Hudson swung across southern New Jersey, depositing a subarkosic sandstone called the Bridgeton Fm. with down-dip equivalents beneath the Delmarva Peninsula called the Bethany and Manokin Formations (Pazzaglia, 1993). Inset into this late Miocene, fluvial deposit is the Pliocene (age palynologically confirmed) Pensauken Fm. It covers the upper two thirds of Delmarva (where it is called the Columbia Fm.) and is equivalent down-dip to the Beaverdam and Yorktown Formations (Pazzaglia, 1993; Fig. 29). Here at Turkey Point, the Pensauken Fm. is predominantly fluvial in origin, with minor silty clay beds, perhaps indicative of local, limited, estuarine conditions. Reactivation surfaces between waxing/waning packages of 12–14 foresets indicates local tidal conditions (Duke, personal communication, 1991). The Pensauken Fm. is unconformably overlain by white-to-buff, fluvial gravel fining upward to estuarine mud called the Turkey Point beds. These deposits have a petrography consistent

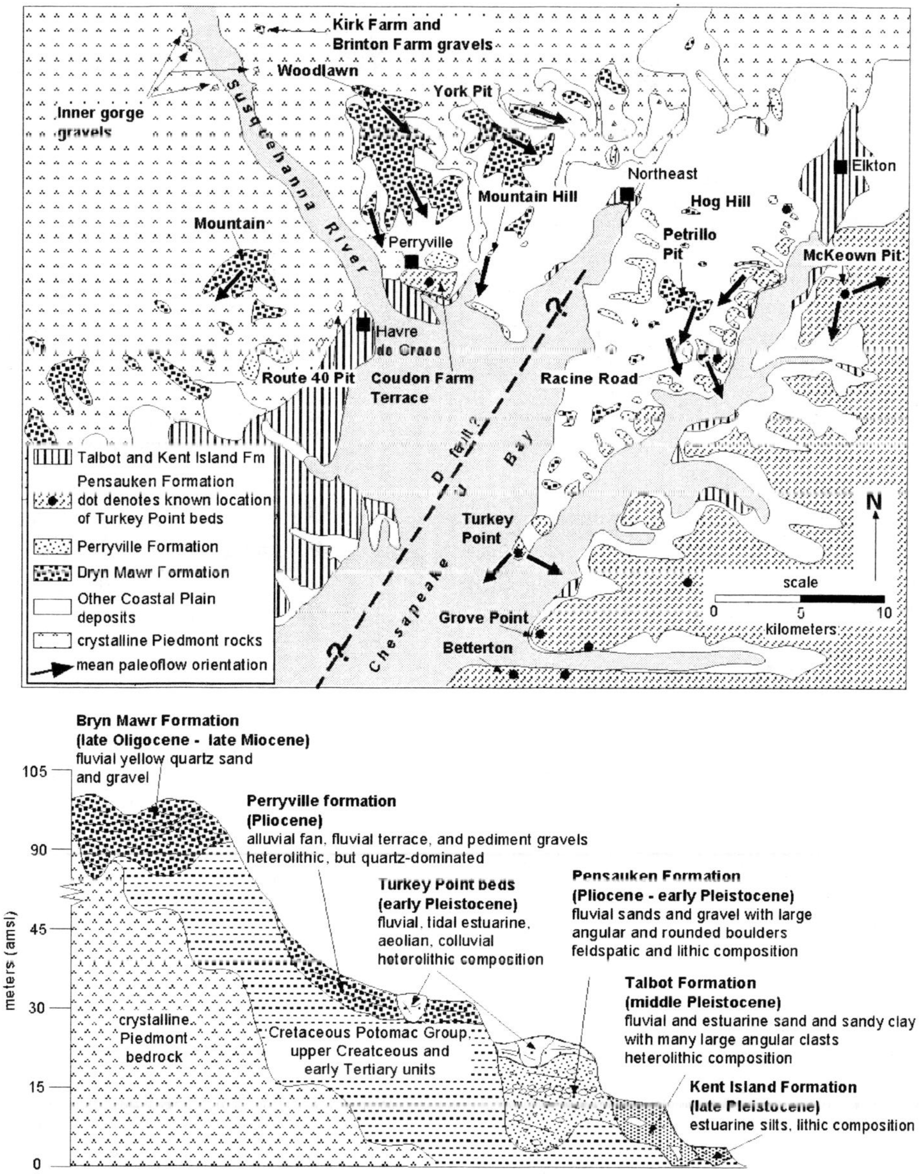

Figure 26. Map (above) and conceptual cross section (below) of the fluvial deposits on the Fall Zone and upper Coastal Plain at the head of Chesapeake Bay (From Pazzaglia, 1993)

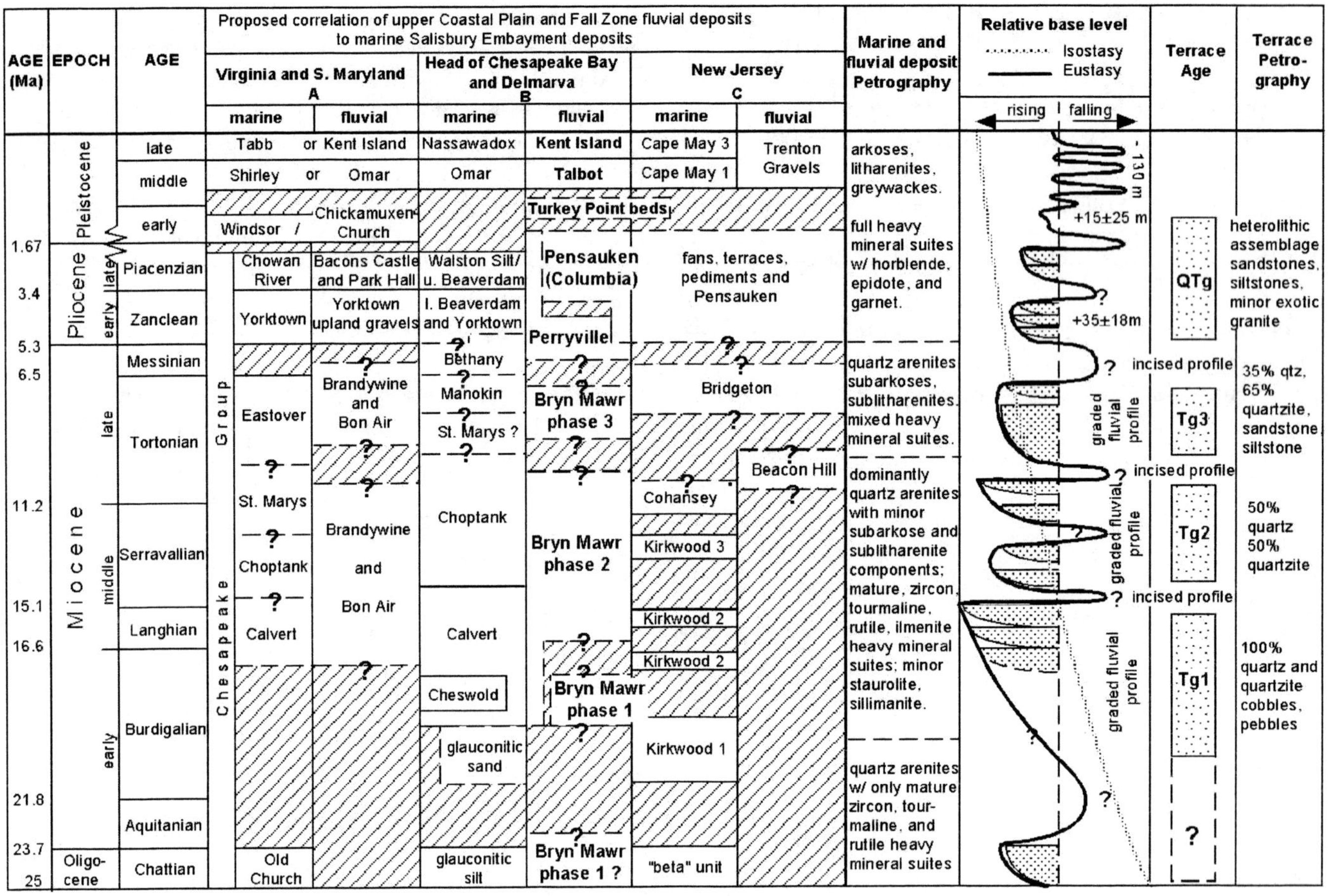

Figure 27. Correlation of Piedmont terraces, Fall Zone deposits, and Coastal Plain deposits at the head of Chesapeake Bay (From Pazzaglia, 1993).

with a Susquehanna River source. Turkey Point is one of the few places where the coarse-grained, bouldery facies is present. Elsewhere, the mud–rich facies, which is locally laminated, prevails. The muddy facies is interpreted as an ancestral, maybe early Pleistocene Chesapeake Bay. Clearly, the top of the Turkey Point beds shows at least three buried weathering profiles, the first two of which are developed in loess (Woodfordian and Illinoian ?). Magnetostratigraphy through the soils and into the estuarine deposits remains normal until a subtle unconformity is crossed, whereupon reversed polarity is found.

Return to Rt 272 and proceed north for approximately 3 miles through a private community to the main entrance to Elk Neck State Park. Turn left to proceed up to Mauldin Mt.. Park at the pavilion.

STOP 5–4. LUNCH AND SYNTHESIS OF THE LOWER SUSQUEHANNA RIVER LANDSCAPE EVOLUTION

The geomorphology and Coastal Plain stratigraphy of the lower Susquehanna River and upper Chesapeake Bay region provide the field constraints (Fig. 29) for a geodynamic model of late Cenozoic deformation of the middle Atlantic passive margin. The details of these models are presented in Pazzaglia and Gardner (1994, 2000) and the interested reader is directed to these papers for full treatment of the methodology and results. A synthesis of all of the observations of the field trip over the past several days can be summarized as follows:

The early Tertiary Appalachian landscape was more deeply-weathered and of lower local relief than the present topography. Climate change, epeirogenic uplift, or rapid increase in the size of the Atlantic-slope drainage basin, or some combination of all three factors, initiated the stripping of a mature regolith in the middle Miocene and its delivery to the Fall Zone. Deposits accumulated during eustatic highs, as the Bryn Mawr Fm., with the upstream equivalents being preserved as upland gravel terraces.

Increased sediment flux to the Baltimore Canyon trough, coupled with erosional unloading, caused flexure of the margin with the Fall Zone located at the flexural hinge. A one–dimensional, line–load model suggests an effective elastic thickness of 30–40 km and a long-term average erosion rate of about 10.m/Ma (Fig. 30; Pazzaglia and Gardner, 1994). A more sophis-

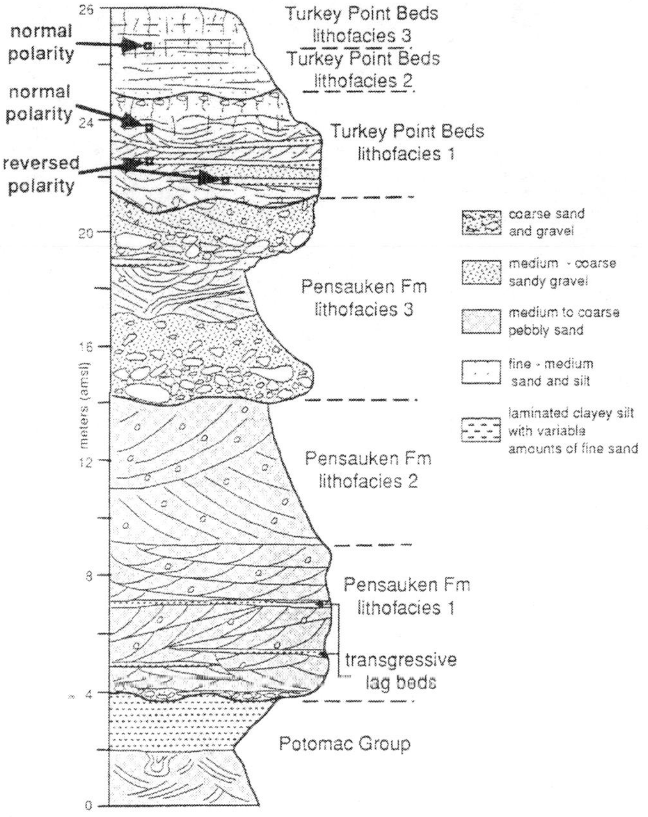

Figure 28. Stratigraphic section at Turkey Point.

ticated, 2–D, distributed-load model, using the same elastic thickness and erosion rates, confirms the Fall Zone as the flexural hinge and offers an explanation for a mechanism locating the drainage divide at the crest of the flexural fore-bulge (Fig. 31; Pazzaglia and Gardner, 2000).

Continued flexural warping in the Piedmont has arched the Miocene terraces and contributes, along with long–term eustatic decline, to the convexity and incision of the Susquehanna River channel.

The incised Appalachian landscape now delivers a more immature, heterolithic load to the Coastal Plain and Baltimore Canyon. This immature composition is a reflection of the incision, along with the effects of repeated Quaternary glaciations in the basin.

Return to Rt 272 and proceed north, through Northeast, and to the interchange with I–95. Enter I–95 south and continue for about 73 miles, through Baltimore and to the capitol beltway. At the beltway, enter I–495/95 south to Exit 25. Exit to Rt. 1 and proceed south for about 1 mile to Greenbelt Road. The hotel is in the southwest corner of the Greenbelt Road-Rt 1 intersection.

The late afternoon and evening of DAY 5 is left open for participants to relax and visit Washington, D.C. A Metro station is within walking distance of the hotel.

DAY 6. GREAT FALLS OF THE POTOMAC AND TRIP END

Follow Rt 1 north to the capitol beltway and proceed west on I–95/495 to I–495 to Exit 39. Exit on Rt 190 west towards Potomac. Proceed for about 4 miles to Rt 189, turn left on Rt 189 and proceed south towards the Potomac River for approximately 3 miles to the C&O Canal NHP and Great Falls of the Potomac.

STOP 6–1. GREAT FALLS OF THE POTOMAC

The purpose of this stop is to view a knickpoint in the channel of the Potomac River and compare/contrast it with the Holtwood Gorge area of the Susquehanna River. Great Falls is a knickpoint initiated probably by one or more Pleistocene eustatic drawdowns. The upstream lip of the knickpoint is analogous to the island tops at Holtwood. Eighteen samples for cosmogenic ^{10}Be and ^{26}Al analysis have been collected from exposed, fluvially eroded outcrops of quartz-bearing schist along strath terraces in Mather Gorge below Great Falls (Bierman et al., 2002). The most prominent strath terrace, a severa-km–long, bedrock feature 20 to 25 m above the current low water level, was apparently abandoned rapidly as the Potomac River incised about 30 ka. Nine samples, collected from

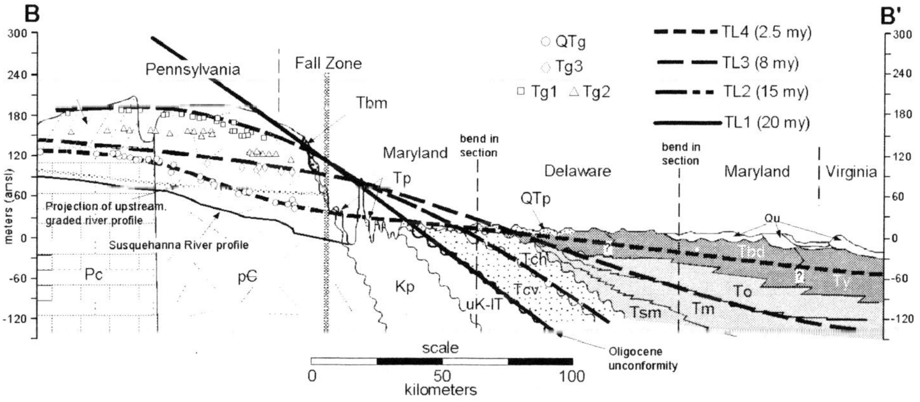

Figure 29. Stratigraphic timelines constraining geodynamic models of Atlantic margin flexural deformation (From Pazzaglia and Gardner, 1994).

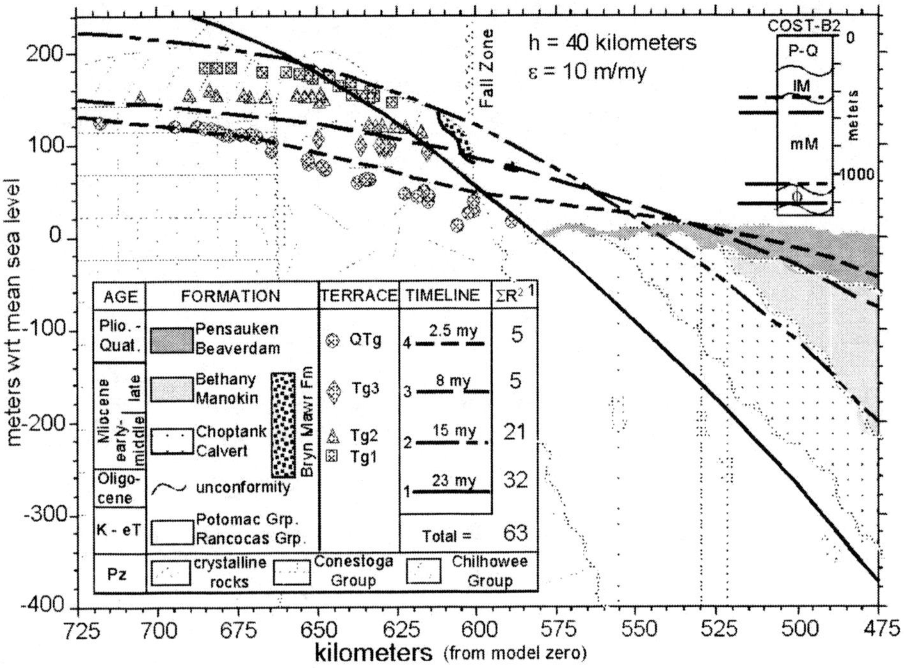

Figure 30. Results of a 1–D line load model (From Pazzaglia and Gardner, 1994).

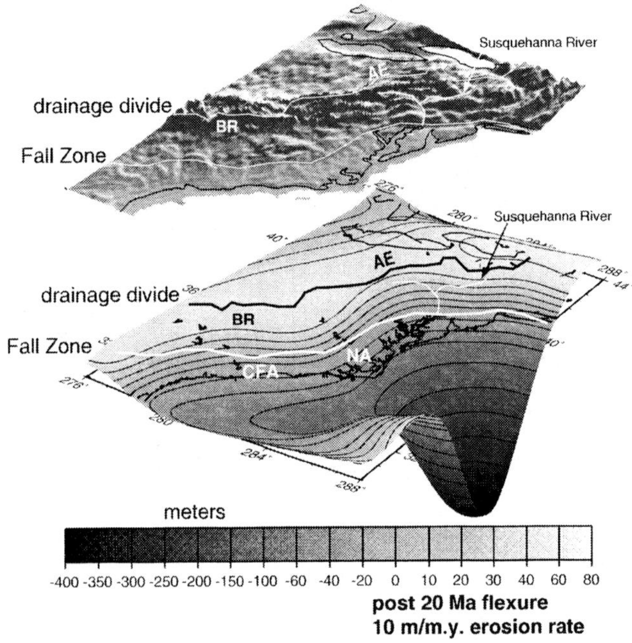

Figure 31. Results of a 2–D distributed load model (From Pazzaglia and Gardner, 2000).

water-polished rock surfaces down a cross-section from this terrace to just above the river, have decreasing nuclide activities consistent with a fluvial, bedrock incision rate of about 70 cm/ka (700 m/Ma) and an effective 6 ka exposure age at water level (corrected for cosmic–ray dosing at and just below the water's surface). The implication of these results is that strath abandonment by knickpoint retreat is rapid and local incision rates can be much greater than the long–term, integrated rates. The adjacent uplands on the Virginia side of the river are mantled with a late Miocene or Pliocene upland gravel.

REFERENCES

Ahnert, F., 1970, Functional relationship between denudation, relief, and uplift in large mid–latitude drainage basins: American Journal of Science, v. 268, p. 243–263.

Ashley, G. H., 1933, The scenery of Pennsylvania, its origin and development, based on recent studies of physiographic and glacial geology: Pennsylvania Geologic Survey, Bulletin G 6, 91 p.

Ashley, G.H., 1935, Studies in Appalachian mountain structure: Geologic Society of America Bulletin, v. 46, p. 1395–1436.

Bascom, F.,.1921, Cycles of erosion in the Piedmont province of Pennsylvania: Journal of Geology, v.29, p. 540–559.

Bierman, P., Reusser, L., Pavich, M., Zen, E–an, Finkel, R., Larsen, J., and Butler, E. M., 2002, Major climate–correlative incision of the Potomac River gorge at Great Falls about 30,000 years ago: Geological Society of America Abstracts with Programs, v. 34, p. 127.

Blackmer, G. C., Omar, G. I., and Gold, D. P., 1994, Post–Alleghenian unroofing history of the Appalachian basin, Pennsylvania, from fission track analysis and thermal models: Tectonics, v. 13, p. 1259–1276.

Boettcher, S. S. and Milliken, K. L., 1994, Mesozoic–Cenozoic unroofing of the southern Appalachian basin: apatite fission track evidence from Middle Pennsylvanian sandstones: Journal of Geology, v. 102, p. 655–663.

Braun, D.D., 1999, Pleistocene geology of the Allentown area:*in* Sevon, W. D. and Fleeger, G. M., eds., Economic and enrivonmental geology and topography in the Allentown–Bethelehem area: Harrisburg, 64th Annual Field Conference of Pennsylvania Geologists, Guidebook, p. 31–40.

Braun, D.D., 1994a, Late Wisconsinan to pre–Illionoian (G?) glacial events in eastern Pennsylvania: *in* Braun, D.D., ed., Late Wisconsinan to pre–Illinoian(G?) Glacial and periglacial events in eastern Pennsylvania: Bloomsburg, 57th Friends of the Pleistocene Field Conference, Guidebook, U.S.Geologic Survey Open File Report 94–434, p. 1–21.

Braun, D.D., 1994b, The ubiquitous boulder colluvium of eastern Pennsylvania, a relict of periglacial mobilization of the entire land surface: *in* Braun, D.D., ed., Late Wisconsinan to pre–Illinoian(G?) glacial and periglacial events in eastern Pennsylvania: Bloomsburg, Guidebook, 57th Friends of the Pleistocene Field Conference, U.S.Geologic Survey Open File Report 94–434, p. 34–36.

Braun,D.D., 1990, Negative evidence of late Wisconsinan catastrophic flooding down the Susquehanna River: Geological Society of America Abstracts with Programs, v. 22, p. 6.

Braun, D. D., 1989a, Glacial and periglacial erosion of the Appalachians: Geomorphology, v. 2, p. 233–256.

Braun, D.D., 1989b, The depth of post middle Miocene erosion and the age of the present landscape: *in* Sevon, W.D. ed., The rivers and valleys of Pennsylvania, then and now: Harrisburg, Harrisburg Area Geological Society and the 20th Annual Geomorphology Symposium, Guidebook, p. 16–20.

Bruan, D.D. and Inners, J. D. 1988, Stop 2. Honey Hole gravel pit: Glacial and economic geology of a late Wisconsinan frontal kame: *in* Inners, J. D. , ed., Bedrock and glacial geology of the North Branch Susquehanna lowland and the eastern middle Antharcite field: Harrisburg, 53 rd. Annual field conference of Pennyslvania geologists, Guidebook, p. 114–118.

Braun, D.D, Miller, D.S., Miller,J.C., and Inners, J.D., 1984, Abandoned valley of the North Branch Susquehanna River at Mifflinville, Pennsylvania: Evidence for a pre–late Wisconsinan ice margin: Geological Society of America Abstracts with Programs, v. 15, p. 5.

Braun, Duane D.,1983, Lithologic control of bedrock meander dimensions in the Appalachian Valley and Ridge Province: Earth Surface Processes and Landforms, v. 8, p. 223–237.

Campbell, M. R., 1903, Geographic development of northern Pennsylvania and southern New York: Geologic Society of America Bulletin, v. 14,. p. 277–296.

Campbell, M. R., 1929, Late geologic deformation of the Appalachian Piedmont as determined by river gravels: Proceedings of the National Academy of Sciences of the United States of America 15, p. 156–161.

Campbell, M. R. and Bascom, F., 1933, Origin and structure of the Pensauken gravel: American Journal of Science, v. 26, no. 153, p. 300–318.

Ciolkosz, E.J., Cronce, R.C., Sevon, W.D., and Waltman, W.J., 1995, Genesis of Pennsylvania's limestone soils, Pennsylvania State University Agronomy Series, no.135, p. 1–28.

Clark, G. M., 1991, South Mountain geomorphology: *in* Sevon, W. D. and Potter, N., Jr., eds., Geology of the South Mountain area, Pennsylvania: Harrisburg, 56th Annual Field Conference of Pennsylvania Geologists, Guidebook, p. 55–94.

Cleaves, E. T., 1993, Climatic impact on isovolumetric weathering of a coarse–grained schist in the northern Piedmont Province of the central Atlantic states: Geomorphology, v. 8, p. 191–198.

Cleaves, E. T., Godfrey, A. E., and Bricker, O. P., 1970, Geochemical balance of a small watershed and its geomorphic implications: Geological Society of America Bulletin, v. 81, p. 3015–3032.

Conrad, C. T. and Saunderson, H. C., 1999, Temporal and spatial variation in suspended sediment yields from eastern North America: *in* Smith, B. J., Whalley, W. B., and Warke, P. A., eds., Uplift, Erosion, and Stability: Perspectives on Long–term Landscape Development: Geological Society Special Publication 162, London, UK, p. 219–228.

Davis, W. M., 1889, The rivers and valleys of Pennsylvania: National Geographic Magazine, v. 1, no. 3, p. 184–253.

Davis, W. M., 1899, The geographical cycle: Geography Journal, v. 14, p. 481–504.

Davis, W. M. and Wood, J. W., Jr., 1890, The geographic development of northern New Jersey: Boston Society of Natural History, Proceedings, v. 24, p. 365–423.

Delano, H. L., Miller, N. G., and Potter, N., Jr., 2002, Plant fossil evidence for Late Pleistocene tundra conditions in South–central Pennsylvania: Geological Society of America Abstracts with programs, v. 34, no. 1, p. A–27.

Engel, S. A., Gardner, T. W., and Ciolkosz, E. J., 1996, Quaternary soil chronosequences on terraces of the Susquehanna River, Pennsylvania: Geomorphology, v. 17, p. 273–294.

Epstein, J. B., 1966, Structural control of wind gaps and water gaps and of stream capture in the Stroudsburg area, Pennsylvania and New Jersey: U. S. Geological Survey Professional Paper 550–B, p. B80–B86.

Gardner, T. W., Sasowski, I. D., and Schmidt, V. D., 1994, Reversed–polarity glacial sediments and revised glacial chronology, West Branch Susquehanna River valley, central Pennsylvania: Quaternary Research, v. 42, p. 131–135.

Granger, D. E., Fabel, D., Palmer, A. N., 2001, Pliocene–Pleistocene incision of the Green River, Kentucky, determined from radioactive decay of cosmogenic ^{26}Al and ^{10}Be in Mammoth Cave sediments: Geological Society of America Bulletin, v. 113, p. 825–836.

Granger, D. E., Kirchner, J. W., and Finkel, R. C., 1997, Quaternary downcutting rate of the New River, Virginia, measured from differential decay of cosmogenic ^{26}Al and ^{10}Be in cave–deposited alluvium: Geology, v. 25, p. 107–110.

Hack, J. T., 1982,Physiographic divisions and differential uplift in the Piedmont and Blue Ridge: U. S. Geological Survey Professional Paper 1265, 49 p.

Hack, J.T., 1975, Dynamic equilibrium and landscape evolution: *in* Melhorn, W.N. and Flemal, R.C., eds., Theories of landform development: Proceedings, 6th Annual Geomorphology Symposia, Binghamton, New York, p.87–102.

Hack, J.T., 1965, Geomorphology of the Shenandoah Valley, Virginia and West Virginia, and origin of the residual ore deposits: U.S. Geological Survey Professional Paper 484, 84 p.

Hack, J. T., 1960, Interpretation of erosional topography in humid temperate regions: American Journal of Science, v. 258–A, p. 80–97.

Hoskins, D. M., 1987, The Susquehanna River water gaps near Harrisburg, Pennsylvania: *in* Roy, D. C., ed., Northeastern Section of the Geological Society of America Centennial Field Guide, Volume 5: Geological Society of America, p. 47–50.

Johnson, D. W., 1931, Stream sculpture on the Atlantic Slope, a study in the evolution of Appalachian rivers: Columbia University Press, NY142 p.

Judson, S., 1975, Evolution of Appalachian topography: *in* Melhorn, W.N. and Flemal, R.C., eds., Theories of landform development: Proceedings, 6th Annual Geomorphology Symposia, Binghamton, NY, p.1–28.

Leverett, F., 1934, Glacial deposits outside the Wisconsin Terminal Moraine in Pennsylvania: Pennsylvania Geologic Survey, Bulletin G 7, 123p.

Mathews, E. B., 1917, Submerged "deeps" in the Susquehanna River: Geological Society of America Bulletin, p. 335–346.

Matmon, A., Bierman, P. R., Larsen, J., Southworth, S., Pavich, M. and Caffee, M., 2003, Temporally and spatially uniform rates of erosion in the southern Appalachian Great Smoky Mountains: Geology, v. 31, p. 155–158.

Meyerhoff, H. A. and Olmstead, E. W., 1936, Postorogenic development of the Appalachians: Geological Society of America Bulletin, v. 83, p. 1709–1728.

Miller, D. S. and Duddy, I. R., 1989, Early Cretaceous uplift and erosion of the northern Appalachian basin New York, based on apatite fission track analyses: Earth and Planetary Science Letters, v. 93, p. 35–49.

Milliman, J. D. and Syvitski, J. P. M., 1992, Geomorphic/tectonic control of sediment discharge to the ocean: The importance of small mountainous rivers: Journal of Geology, v. 100, p. 525–544.

Mills, H. H., 2000, Apparent increasing rates of stream incision in the Eastern United States during the late Cenozoic: Geology, v. 28, p. 955–957.

Naeser, N. D., Naeser, C. W., Morgan, B. A., III, Schultz, A. P., Southworth, C. S., 1999, Cooling history of the Blue Ridge Province, Virginia, North Carolina, and Tennessee, from apatite and zircon fission–track analysis: Geological Society of America, Abstracts with Programs, v. 31, p. 32.

Naeser, C. W., Naeser, N. D., Kunk, M. J., Morgan, B. A., Schultz, A. P., Southworth, C. S., Weems, R. E., 2001, Paleozoic through Cenozoic uplift, erosion, stream capture, and deposition history in the Valley and Ridge, Blue Ridge, Piedmont, and coastal plain provinces of Tennessee, North Carolina, Virginia, Maryland, and District of Columbia: Geological Society of America, Abstracts with Programs, v. 33, p. 312.

Newell, W. L., Wyckoff, J.S., Owens, J.P., and Farnsworth, J,. 1989, Cenozoic geology and geomorphology of southern New Jersey Coastal Plain: Guide book, 2nd annual Southeast Friends of the Pleistocene field conference, U.S. Geological Survey Open File Report 89–159, 51 p.

Owens, J.P. and Minard, J.P., 1979, Upper Cenozoic sediments of the lower Delaware Valley and northern Delmarva Peninsula, New Jersey, Pennsylvania, Delaware, and Maryland: U.S. Geological Survey Professional Paper 1067D, 47 p.

Page, L. V. and Shaw, L. C., 1973, Floods of June 1972 in the Harrisburg area, Pennsylvania: U.S. Geological Survey Hydrologic Investigations Atlas HA–530.

Pavich, M. J., Leo, G. W., Obermeier, S. F., and Estabrook, J. R., 1989, Investigations of the characteristics, origin, and residence time of the upland residual mantle of the Piedmont of Fairfax County, VA: U. S. Gelogical Survey Professional Paper 1352, 114 p.

Pazzaglia, F. J. and Brandon, M. T., 1996, Macrogeomorphic evolution of the post–Triassic Appalachian mountains determined by deconvolution of the offshore basin sedimentary record: Basin Research, v. 8, p. 255–278.

Pazzaglia, F. J. and Gardner, T. W., 1994, Late Cenozoic flexural deformation of the middle U.S. Atlantic passive margin: Journal of Geophysical Research, v. 99, n. B6, p. 12,143–12,157.

Pazzaglia, F. J. and Gardner, T. W., 2000, Late Cenozoic large–scale landscape evolution of the U.S. Atlantic passive margin: *in* Summerfield, M. ed., Geomorphology and Global Tectonics: John Wiley, New York, p.283–302.

Pazzaglia, F. J., 1993, Stratigraphy, petrography, and correlation of late Cenozoic middle Atlantic Coastal Plain deposits: Implications for late–stage passive margin geologic evolution: Geological Society of America Bulletin, v.105, p.1617–1634.

Pazzaglia, F. J., Gardner, T. W., and Merritts, D., 1998, Bedrock fluvial incision and longitudinal profile development over geologic time scales determined by fluvial terraces: *in* Wohl, E. and Tinkler, K., eds., Bedrock Channels: American Geophysical Union, Geophysical Monograph Series, v. 107, p. 207–235.

Pazzaglia, F. J., Robinson, R., and Traverse, A., 1997, Palynology of the Bryn Mawr Fm. (Miocene): Insights on the age and genesis of middle Atlantic margin fluvial deposits: Sedimentary Geology, v. 108, p. 19–44.

Peltier, L. C., 1949, Pleistocene terraces of the Susquehanna River, Pennsylvania: Pennsylvania Geologic Survey, Bulletin G 23, 158 p.

Pierce, K. L., 1965, Geomorphic significance of a Cretaceous deposit in the Great Valley of southern Pennsylvania: U.S. Geological Survey Professional Paper 525C, p. C152–C156.

Pitman, W. C. and Golovchenko, X., 1991, The effect of sea level changes on the morphology of mountain belts: Journal of Geophysical Research, v. 96, p. 6879–6891.

Poag, C. W. and Sevon, W. D., 1989, A record of Appalachian denudation in postrift Mesozoic and Cenozoic sedimentary deposits of the U.S. Middle Atlantic continental margin: Geomorphology, v. 2, p. 119–157.

Potter, N. and Moss, J. H., 1968, Origin of the Blue Rocks block field and adjacent deposits, Berks County, Pennsylvnia: Geological Society of America Bulletin, v. 79, p. 255–262.

Ramage, J. M., Gardner, T. W., and Sasowski, I. D., 1998, Early Pleistocene glacial Lake Lesley, West Branch Susquehanna River valley, central Pennsylvania: Geomorphology, v. 22, p. 19–37.

Roden, M. K. and Miller, D. S., 1989, Apatite fission-track thermochronology of the Pennsylvania Appalachian basin: Geomorphology, v. 2, p. 39–51.

Roden, M. K., 1989, Apatite fission-track thermochronology of the southern Appalachian basin: Maryland, West Virginia, and Virginia: Journal of Geology, v. 99, p. 41–53.

Sasowsky, I.D., 1994, Paleomagnetism of glacial sediments from three locations in eastern Pennsylvania: *in* Braun, D.D., ed., Late Wisconsinan to pre–Illinoian(G?) Glacial and periglacial events in eastern Pennsylvania, 57th Friends of the Pleistocene Field Conference, U.S.Geologic Survey Open File Report 94–434, p. 21–23.

Schofield, I. S., Michaud, D. P., Sents, R. C., Roden–Tice, M. K., Tice, S. J., 1999, Potential post–early Cretaceous unroofing in the southeastern Adirondack Mts., New York State using apatite fission–track analysis: Geological Society of America, Abstracts with Programs, v. 31, p. 65.

Schumm, S. A. and Lichty, R.C., 1965, Time, space, and causality in geomorphology: American Journal of Science, v. 263, p.110–119.

Sevon, W. D., 1986, Susquehanna River water gaps: many years of speculation: Pennsylvania Geology, v. 17, no. 3, p. 4–7.

Sevon, W. D., 1987, The Hickory Run boulder field, a periglacial relict, Carbon County, Pennsylvania: *in* Roy, D. C., ed., Northeastern Section of the Geological Society of America Centennial Field Guide, v. 5, Geological Society of America, p. 75–76.

Sevon, W. D., 1989a, Erosion in the Juniata River drainage basin, Pennsylvania: Geomorphology, v. 2, p. 303–318.

Sevon, W.D., 1989b, The rivers and valleys of Pennsylvania, then and now: *in* Sevon, W.D. ed., The rivers and valleys of Pennsylvania, then and now, Harrisburg, Harrisburg Area Geological Society and the 20th Annual Geomorphology Symposium, Guidebook, p. 1–16.

Sevon, W.D., 1999, Cenozoic History: *in* Shultz, C.H., ed., The geology of Pennsylvania: Pennsylvania Geologic Survey Special Publication 1, p. 450–455.

Sevon, W. D., 2001, Landscape evolution in the Cumberland Valley, southeastern Pennsylvania: in Potter, N., Jr., editor, The geomorphic evolution of the Great Valley near Carlisle, Pennsylvania: Carlisle, Southeast Friends of the Pleistocene 2001 Annual Meeting, Guidebook, p. 1–29.

Sevon, W.D. and Braun, D.D., 1997, Glacial deposits of Pennsylvania: Pennsylvania Geologic Survey Map 9.

Shaw, J., 1989, Drumlins, subglacial meltwater floods, and ocean responses: Geology , v. 17, p. 853–856.

Smith, H. T. U., 1953, The Hickory Run boulder field, Carbon County, Pennsylvania: American Journal of Science, v. 251, p. 625–642.

Smoot, J.P., 1999, Early Mesozoic sedimentary rocks: *in* Shultz, C.H., ed., The geology of Pennsylvania: Pennsylvania Geologic Survey Special Publication 1, p. 180–201.

Stanford, S. D., 1997, Pliocene–Quaternary geology of northern New Jersey: An overview: *in* Stanford, S. D. and Witte, R. W., eds., Pliocene–Quaternary geology of northern New Jersey: New Brunswick, 60th Annual Reunion of the Northeast Friends of the Pleistocene, Guidebook, p. 1.1–1.26.

Stanford, S. D., Ashley, G. M., and Brenner, G. J., 2001, Late Cenozoic fluvial stratigraphy of the New Jersey Piedmont; a record of glacioeustasy, planation, and incision on a low–relief passive margin: Journal of Geology, v. 109, p. 265–276.

Thiesen, J. P., 1983, Is there a fault in our gap?: Pennsylvania Geology, v. 14, no. 3, p. 5–11.

Thompson, H. D., 1949, Drainage evolution in the Appalachians of Pennsylvania: Annals of the New York Academy of Science, v. 52, p. 31–63.

Thompson, G. H., 1985, The preglacial great falls of the Susquehanna River: Geological Society of America Abstracts with Programs, v. 19, n. 7, p. 734.

Thompson, G. H., 1990, Geomorphology of the lower Susquehanna River: *in* Scharnberger, C. K., ed., Carbonates, Schists, and Geomorphology in the vivinity of the lower reeaches of the Susquehanna River: Harrisburg, 55th Field Conference of Pennsylvania Geologists, Guidebook, p. 86–106.

Watts, W. A., 1979, Late Quaternary vegetation of central Appalachia and the New Jersey coastal plain: Ecological Monographs, v. 49, p. 427–469.

White, W. B., 1984, Rate processes: chemical kinetics and karst landform development: *in* LaFleur, R. G., ed., Groundwater as a geomorphic agent: Allen and Unwin, Boston, MA, p. 227–248.

White, W. B., 2000, Dissolution of limestone from field observation: *in* Kimchouk, A. B., Ford, D. C., Palmer, A. N., and Dreybrodt, W., eds., Speleogenesis, evolution of karst aquifers: National Speleological Society January 2000 edition, p. 149–155.

Williams, E.H., Jr., 1893, Extramorainic Drift between the Delaware and the Schuylkill: Geologic Society of America Bulletin, v.5, p. 281–296.

Witte, R. W., 1997, Late history of the Culvers Gap River: a study of stream capture in the Valley and Ridge, Great Valley , and Highlands physiographic provinces, northern New Jersey: *in* Stanford, S. D. and Witte, R. W., eds., Pliocene–Quaternary geology of northern New Jersey: New Brunswick, 60th Annual Reunion of the Northeast Friends of the Pleistocene, Guidebook, p. 3.1–3.16.

Witte, R.W., 2001, Culvers Gap: *in* Inners, J. D. and Fleeger, G. M,.eds., 2001: A Delware River Odyssey: Harrisburg, 66st Annual field conference of Penyslvania geologists, Guidebook, p. 114–118.

Zimmerman, R. A., 1979, Apatite fission track age evidence of post–Triassic uplift in the central and southern Appalachians: Geological Society of America Abstracts with Programs, v. 11, p. 219.

Zimmerman, R. A., Reimer, G. M., Foland, K. A., and Faul, H., 1975, Cretaceous fission track ages of apatites from northern New England: Earth and Planetary Science Letters, v. 28, p. 181–188.

Glacial geology of upper Cook Inlet, Matanuska Glacier and Denali Highway, Alaska

Authors and Leaders: **Grahame J. Larson,** *Department of Geological Sciences, Michigan State University, East Lansing, Michigan 48823*
Edward B. Evenson, *Department of Earth and Environmental Sciences, Lehigh University, Bethlehem, Pennsylvania 18015*
Daniel E. Lawson, *U.S. Army Cold Regions Research and Engineering Laboratory, Anchorage, Alaska 99505*
Staci L. Ensminger, *Department of Geology and Geography, Northwest Missouri State University, Maryville, Missouri 64468*
Greg Baker, *Department of Geology, State University of New York at Buffalo, Buffalo, New York, 14260*
Richard B. Alley, *Environmental Institute and Department of Geosciences, The Pennsylvania State University, University Park, Pennsylvania 16802*

INTRODUCTION

Between Anchorage and the Matanuska Glacier, Alaska, evidence for multiple glaciation is abundant. Much of it is visible from the Glenn Highway that winds through some of the most spectacular and breathtaking scenery of south–central Alaska. The first day of the field trip will begin in Anchorage and end at the Matanuska Glacier, a distance of about 160 km. The second day we will examine recent fluctuations of the Matanuska Glacier and processes occurring along the glacier margin. The third and last day will be an excursion to the Alaska Range to examine glacial and periglacial landforms in an alpine terrain.

COOK INLET

The Cook Inlet basin (Fig. 1, 2) is located along the coast of south–central Alaska and occupies a large structural trough of the trench–arc type (Dickinson, 1971) that extends NNE–SSW from the Pacific Ocean. The trough is about 325 km long and 95 km wide and is bounded by the Castle Mts. and Bruin Bay faults on the north and west and the Kenai Mts. fault zone on the east. The basin occupies about 31,000 km^2, 70% of which is flooded by Cook Inlet and its two Arms, the Knik and Turnagain (Fig. 3). The Chugach and Kenai Mts. (2,000–2,500 m) rim the basin on the southeast, and on the northwest side of the basin lie the Neacola and Tordrillo Mts. (>3,350 m; 11,000 ft.) (Fig. 2). Toward the northwest and about 50 km beyond the Neacola and Tordrillo Mts. lie the Alaska Range, which includes Mt. Denali (6,178 m; 20,000 ft.).

Most of the lowland in the basin lies below 150 m (500 ft.) elevation and can be divided into the Kenai Lowland that flanks the western side of the Kenai Mts. and the Susitna Lowland that flanks the eastern side of Neacola and Tordrillo Mts. (Fig. 2). The Susitna Lowland, however, includes isolated uplands such as Mt. Susitna (1,332 m; 4400 ft.) and Beluga Mountain (1,127m; 3700 ft.). Major rivers flowing into Cook Inlet include the Matanuska and Knik, which drain westward into Knik Arm and the Susitna which drains southward along the axis of the structural trough and into upper Cook Inlet. Separating the Matanuska valley from the Susitna valley are the Talkeetna Mts. (<2,213 m).

Bedrock of the basin consists of basement igneous and metamorphic rocks primarily of Mesozoic age and overlying weakly consolidated clastic sedimentary rocks of Tertiary age that exceed 4000 m in thickness in the deeper parts of the structural trough. Locally, the sedimentary rocks contain petroleum and coal. Unconsolidated sediments up to 1,300 m also occur in the basin and include marine and glaciomarine deposits, as well as glacial and glaciofluvial deposits.

Glacial history

Evidence for glaciation of the Cook Inlet basin abounds in the form of Quaternary deposits and glacially eroded landscapes. Moraines and ice–contact complexes are common in upper Cook Inlet and, throughout much of the basin lowlands, glacial and glaciomarine deposits underlie late–glacial deposits. Abundant ice–rounded summits, ice–scoured benches, and truncated spurs show varying degrees of preservation. Despite the plethora of evidence, however, only a fragmentary record of the glacial history has been preserved and much debate as to the number and timing of glaciations remains.

Early interpretations of the Quaternary history of the Cook Inlet basin include work by Karlstrom (1953, 1955, 1957, 1964, 1965), Miller and Dobrovolny (1959), Trainer (1960) and Trainer and Waller (1965). The work by Karlstrom (1964) is perhaps the most significant because it establishes a framework for the Quaternary chronology of the basin that includes five glaciations: Mt. Susitna, Caribou Hills, Eklutna, Knik and Naptowne (Table 1). Subsequent work includes that of Boss et al. (1976), Schmoll and Yehle (1983), Schmoll et al. (1984), Reger and Updike (1983), Reger et al. (1995) and Schmoll et al. (1999). Of note is the work by Boss et al. (1976) who interpret a diamicton in drill holes within the Kenai Lowland as tillites of Miocene to Pliocene age, and the work by Schmoll et al. (1984) who assign a pre–Pleisto-

Larson, G.J., Evenson, E.B., Lawson, D.E., Ensminger, S.L., Baker, G., and Alley, R.B., 2003, Glacial geology of upper Cook Inlet, Matanuska glacier and Denali highway, Alaska: *in* Easterbrook, D.J., ed., Quaternary Geology of the United States, INQUA 2003 Field Guide Volume, Desert Research Institute, Reno, NV, p. 245–264.

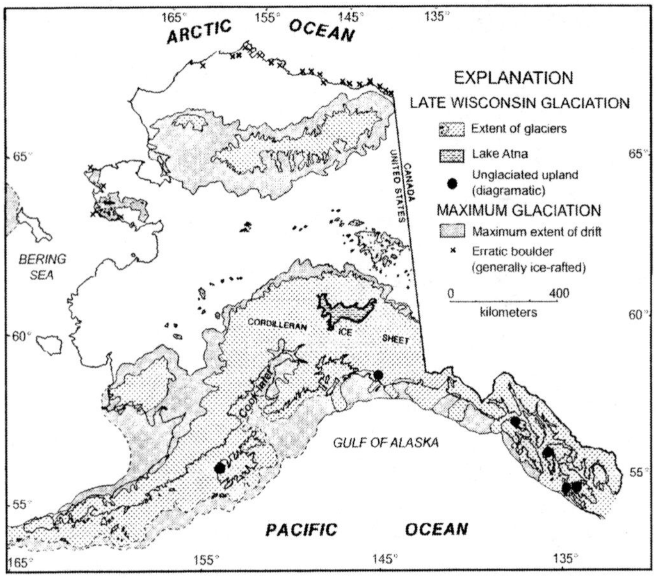

Figure 1. Maximum known advances of glaciers of late Wisconsin age and limits of older drift in Alaska (shown by dashed lines where inferred). (Modified slightly from Hamiltion 1994).

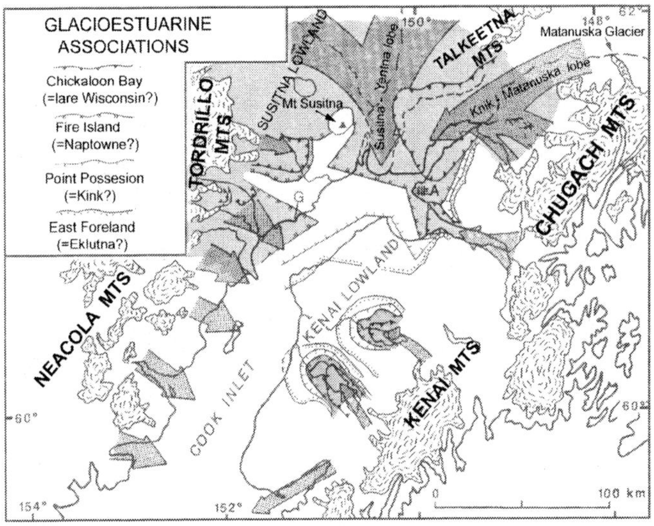

Figure 2. Late Pleistocene and present–day glaciation of Cook Inlet, Alaska. Modern glaciers shown by concentric dashed patters. Arrows show routes of principal glacier tongues; shaded areas are inferred extent of late Wisconsin glaciers in upper inlet. Principal glaciolacustrine associations are from Schmoll and Yehle (1986). (Modified slightly from Schmoll and Yehle, 1986 and Hamilton, 1994).

cene age to an exposed, slightly folded diamicton (Kaloa deposits) on the west side of Cook Inlet.

Kaloa Deposits

Diamictons, interpreted as tillites, occur in drill holes that penetrate through the Beluga and Sterling Fms. adjacent to the Kenai Mts. on the Kenai Peninsula (Fig. 4; Boss et al., 1976). They are as much as 120 m thick and intertongue with what has been interpreted as outwash sediments (Boss et al., 1976). Toward Cook Inlet they grade into floodplain shale and sandstone with minor interbedded coal seams. The age of the diamictons appear to be late Miocene to early Pliocene and, based on their distribution, Boss et al. (1976) propose that the Kenai Mts. were high in the late Miocene and that the diamictons were deposited by piedmont glaciers that flowed out into the Kenai lowland.

Miocene diamictons informally termed "Kaloa deposits" also occur along the shore at Granite Point on the west side of Cook Inlet (Schmoll et al., 1984). They have a total stratigraphic thickness of about 110 m and include interbeds of sandstone, siltstone and claystone, as well as a coal bed near the base. Beds within the diamicton dip 1–6 degrees and are disconformably overlain by Pleistocene tills. The origin of the diamictons is not fully understood and they may be the result of direct deposition by a glacier or derived from subaqueous and/or subaerial debris flows (Schmoll et al., 1984).

Mt. Susitna Glaciation

The Mt. Susitna Glaciation was named by Karlstrom (1953) for the ice–rounded summit of Mt. Susitna (Fig. 3). In addition to ice–scoured surfaces above the upper limit of the next younger (Caribou Hills) glaciation, scattered erratics occur within 30 m of the mountain summit. Other ice–scoured surfaces and weathered erratics believed to have been derived from Mt. Susitna glaciation also occur above 1,299 m (4300 ft.) elevation in the southwestern Talkeetna Mts., and ice–scoured surfaces in the western Kenai Mts. are possibly equivalent (Reger and Updike, 1983).

Karlstrom (1964, 1965) believed that during the Mt. Susitna Glaciation ice caps were present on the Kenai, Chugach, and Talkeetna Mts., as well as on the Alaska Range and that ice drained through coalescing tributary valleys to fill the Cook Inlet basin to the present elevation of 1,212 m (4000 ft.). How far the ice extended is not known, but it probably flowed south out of the basin to form an extensive ice shelf in the Gulf of Alaska (Karlstrom, 1964). The distribution of ice–sculptured surfaces and erratics indicate that glacier flow patterns during the Mt. Susitna Glaciation were unrelated to modern valley systems (Reger and Updike, 1983). The age of the Mt. Susitna Glaciation is unknown, but correlations with central Alaska Range glaciations suggest that it is late Oligocene or early Pleistocene (Péwé, 1975).

Caribou Hills Glaciation

The Caribou Hills Glaciation was also named by Karlstrom (1953) for weathered drift in the southern Kenai lowland that includes considerably modified lateral moraines and kame terraces, as well as glacial drainage channels. In the upper Cook Inlet region, the Caribou Hills Glaciation is represented by ice–scoured benches and truncated spurs between 818 and 970 m (2700–3200 ft.) along the western front of the Chugach Mts., between 818 and 970 m on

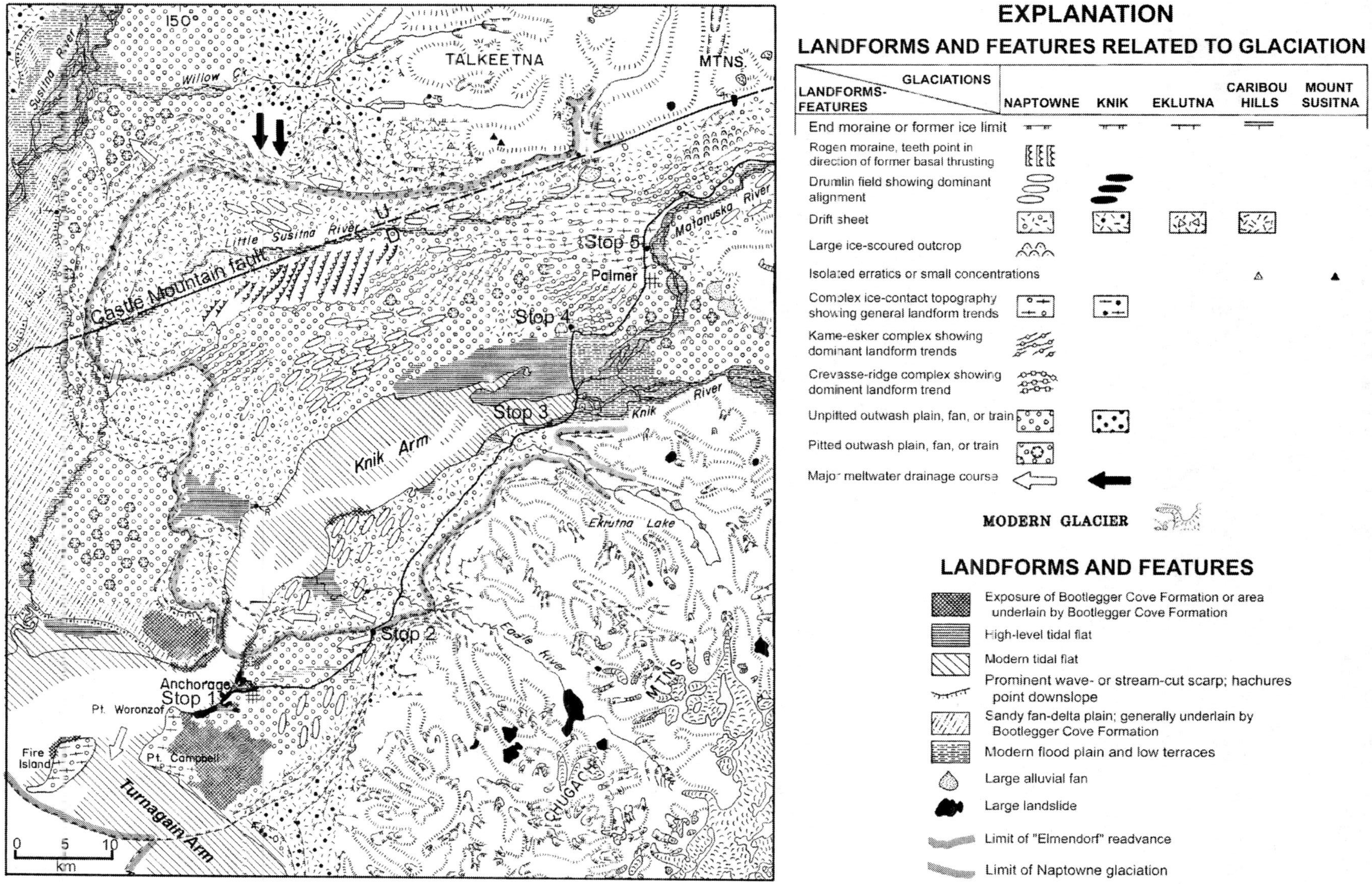

Figure 3. Physiographic map and field trip stops of upper Cook Inlet area, Alaska. (Modified from Reger and Updike, 1983).

Table 1: Chronology of Upper Cook Inlet Glaciations (Modified from Hamilton 1994)

Age	Event
Late Wisconsin	Naptowne glaciation
Penultimate (Illinoian) Glaciation	Knik glaciation
Middle Pleistocene	Eklutna glaciation
Early Pleistocene	Caribou Hills glaciation
Late Pliocene	Mount Susitna glaciation
Older Tertiary	Kaloa deposits (late Miocene and younger)

the upper slopes of Mt. Susitna, and up to 1,136 m (3750 ft.) on the southwestern flank of the Talkeetna Mts. (Reger and Updike, 1983). Bouldery ground moraine at 975 m (3200 ft.) near Mt. Magnificent about 27 km northeast of Anchorage may possibly also be related to the Caribou Hills Glaciation (Hamilton, 1994).

Karlstrom (1964) postulated that during the Caribou Hills Glaciation, ice caps formed on the highlands of south–central Alaska and again drained down major valleys to fill the upper Cook Inlet basin with ice to an elevation of about 909 m (3000 ft.). More–recent field work, however, suggests that the basin may actually have filled to an elevation between 1,061 and 1,136 m (3500–3750 ft.) (Reger and Updike, 1983). Karlstrom (1964) also believed that during the Caribou Hills Glaciation, ice flowed south out of lower Cook Inlet into the north Pacific and formed another extensive ice shelf.

The age of the Caribou Hills Glaciation is not certain, but Reger and Updike (1983) and Hamilton (1994) suggest it may be early Pleistocene. What is certain is that evidence for Caribou Hills Glaciation is better preserved then that for the Mt. Susitna Glaciation and that a considerable hiatus separated the two (Karlstrom, 1964).

Eklutna Glaciation

The Eklutna Glaciation was named by Karlstrom (1964) for weathered recessional outwash exposed in the canyon of Eklutna River about 40 km northeast of Anchorage. Subsequently, however, the outwash has been reinterpreted as fan–delta deposits formed during a later glaciation and later stained by iron–bearing groundwater (Reger and Updike, 1983). Nevertheless, the presence of obvious, notched, bedrock ridges and subdued lateral moraines between 712 and 818 m (2350–2700 ft.) along the western flank of the Chugach Mts. east of Anchorage provide strong evidence that a major glaciation occurred between the Caribou Hills and Knik Glaciations (Karlstrom, 1964; Reger and Updike, 1983). Additional evidence also occurs in the southwestern Talkeetna Mts. and on the flanks of Mt. Susitna where lateral moraines believed to be Eklutna in age have also been mapped (Karlstrom, 1964; Reger and Updike, 1983). Moraines of Eklutna age have also been mapped along the western flank of the Kenai Mts. (Karlstrom, 1964).

Karlstrom (1964) believed that during the Eklutna Glaciation, ice once again spread from ice caps developed on the highlands of south–central Alaska. He thought that ice was channeled down the Susitna, Matanuska and Turnagain Arm valleys to fill the upper two–thirds of Cook Inlet, where it coalesced with ice spilling off the western slopes of the Kenai Mts..

The age of the Eklutna Glaciation is not known for certain. However, Reger and Updike (1983) believe that Eklutna glacier advances must predate the last interglacial and suggest that it is probably Illinoian (oxygen isotope stage 6) in age.

Knik Glaciation

The term "Knik Glaciation" was introduced by Karlstrom (1964) for "till" exposed along the northwestern shore of Knik Arm. According to Karlstrom, the till is separated from a younger till (Naptowne) by outwash, glaciomarine diamicton, and two

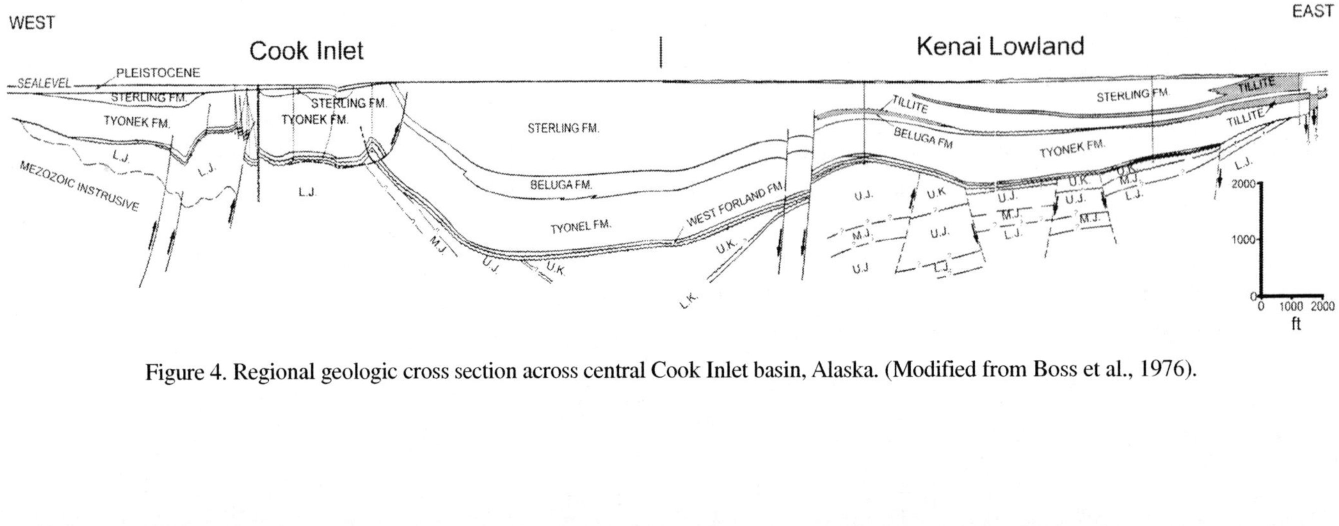

Figure 4. Regional geologic cross section across central Cook Inlet basin, Alaska. (Modified from Boss et al., 1976).

organic zones containing highly compressed peat (Goose Bay peat) that is older than the range of ^{14}C dating (Reger and Updike, 1983). According to Reger and Updike (1983), a diamicton of Knik age also extends continuously below the city of Anchorage and underlies drift of early Naptowne age. Slightly modified lateral moraines attributed to the Knik Glaciation occur between 636 and 455 m (2100–1500 ft.) on the western flank of the Chugach Mts. east of Anchorage and moraines tentatively attributed to the Knik Glaciation have also been identified on the western flank of the Talkeetna Mts. (Reger and Updike, 1983). Moraines of Knik age have also been identified on the flanks of Mt. Susitna and beyond glacially incised valleys along the western flank of the Kenai Mts. (Karlstrom, 1964).

During the Knik Glaciation ice caps, which discharged ice into trunk glaciers within the Susitna, Matanuska and Turnagain Arm valleys, are believed to have formed on the uplands of south–central Alaska. These glaciers coalesced over the Anchorage area (possibly forming an ice shelf) but did not extend far into Cook Inlet (Reger and Updike, 1983). At the same time, piedmont glaciers developed along the western flank of the Kenai Mts. and extended into lower Cook Inlet, possibly coalescing with ice from the Aleutian Range and producing an interior lake in the inlet (Karlstrom, 1964). In the Chugach and Talkeetna Mts., moraines document at least one and possibly two advances or standstills of Knik age and in the western Kenai Mts. and northern Kenai lowlands moraines document at least three or more advances or standstills (Reger and Updike, 1983).

The age of the Knik Glaciation is not certain. Reger and Updike (1983) proposed that it probably began about 75,000 yrs. B.P. However, an age older than oxygen isotope stage 5e is possible because diamicton of probable Knik age in the Anchorage area underlies peat with probable interglacial flora (Ager and Brubaker, 1985). Also, an $^{40}Ar/^{39}Ar$ age of 378 ± 0.64 ka for volcanic ash (Reger et al., 1995, 1996) within the Goose Bay peat along the northwestern shore of Knik Arm suggests that the Knik glaciation is middle Pleistocene in age.

Naptowne Glaciation

The term "Naptowne Glaciation" was introduced by Karlstrom (1953, 1964) for a series of moraines in the vicinity of Sterling, then called Naptowne, in the Kenai lowland. At least four lateral moraines of Naptowne age are also present along the Chugach Mt. front, and at the mouth of the Matanuska/Knik valley lies a prominent, bi–lobate end moraine complex of late Naptowne age (Schmoll et al., 1999). Drift of Naptowne age covers most of the Anchorage area and the Susitna lowland and extends up into the Matanuska Valley. It is typically little modified (Reger and Updike, 1983; Schmoll et al., 1999) and kettle, channel, kame, esker, crevasse–fill–ridge, outwash terrace and drumlin complexes are common (Reger and Updike, 1983; Schmoll et al., 1999). Glacioestuarine deposits (Bootlegger Cove Fm.) interfinger with Naptowne drift in upper Cook Inlet and about 0.3 m of loess mantles the Naptowne surface. In some places the loess approaches 3.6 m thick (Schmoll and Yehle, 1986; Reger and Updike, 1983; Schmoll et al., 1999).

During the early Naptowne Glaciation, ice flowing southwest out of the Matanuska/Knik valley contacted and perhaps joined ice flowing west out of Turnagain Arm and south out of the Susitna valley (Reger and Updike, 1983; Schmoll et al., 1999). Although no moraines of early Naptowne age have been recognized in the upper Cook Inlet region, ice likely flowed some distance southwest down Cook Inlet and terminated in standing water (Schmoll and Yehle, 1986; Reger and Updike, 1983; Schmoll et al., 1999). During the late Naptowne Glaciation, ice flowing southwest out of the Matanuska/Knik valley and west out of the Turnagain Arm valley again entered upper Cook Inlet and terminated in deep marine water, probably just beyond the mouth of Knik Arm (Reger and Updike, 1983; Reger et al., 1995) or possibly farther into Cook Inlet (Schmoll and Yehle, 1986). The latest late Naptowne event in the Anchorage area was a readvance of ice mainly out of the Knik valley that formed the Elmendorf moraine (Fig. 3) just north of Anchorage (Schmoll and Yehle, 1986; Reger and Updike, 1983; Schmoll et. al, 1999) and out of the Turnagain Arm valley to form the smaller Bird Point moraine about 33 km southeast of Anchorage (Schmoll and Yehle, 1986; Schmoll et al., 1999). The distance that ice retreated up Cook Inlet between early and late Naptowne Glaciation is uncertain (Reger and Updike, 1983). However, prior to the formation of the Elmendorf and Bird Point moraines, an estuary appears to have extended at least as far inland as the present head of the two arms of Cook Inlet and at least 70 km north into the Susitna lowland (Schmoll et al., 1999).

Radiocarbon dates on early Naptowne drift are few and contradictory but imply that the drift predates the limit of ^{14}C dating (Reger and Updike, 1983). However, possibly some, or all of early Naptowne drift is late Wisconsin in age (Schmoll and Yehle, 1986; Schmoll et al., 1999). In contrast, late Naptowne events are better constrained by a succession of finite ^{14}C dates, especially from shells within the Bootlegger Cove Fm. (Schmoll et al., 1972; Reger et al., 1995). These dates indicate that late Naptowne ice occupied much of the Anchorage area just prior to 14,900 ^{14}C yrs. B.P. The advance to the Elmendorf moraine culminated around 13,500 ^{14}C yrs. B.P. and the Bird Point moraine around 14,000 ^{14}C yrs. B.P. (Reger et al., 1995). Ice appears to have withdrawn from Knik Arm by about 11,400 ^{14}C yrs. B.P. and from Turnagain Arm by about 10,730 ^{14}C yrs. B.P. (Reger et al., 1995).

STOP 1–1. RESOLUTION PARK OVERLOOK

Resolution Park is located at the intersection of 3rd Avenue and L Street in Anchorage and overlooks Cook Inlet (Fig. 3). On a clear day, the park provides an unobstructed view of the Tordrillo Mts. and the Aleutian Range. Prominent peaks include Mt. Spurr (a volcano) and Mt. Susitna. The Alaska Range lies to the north.

Mt. Susitna (Fig. 5) rises to an elevation of 1,332 m (4400 ft.) and is composed mainly of quartz diorite. Within 30 m of the glacially rounded summit dissected by younger cirque development are glacial erratics of greywacke–argillite, basalt, sandstone, quartzite, and greenstone (Karlstrom, 1964). Ice–scoured benches and truncated spurs attributed to the Caribou Hills Glaciation occur on the upper slopes of the mountain between 818 and 970 m (2700–3200 ft.). Between 879 and 818 m (2900–3200 ft.), lateral moraines and ice–marginal features are attributed to the Eklutna Glaciation (Reger and Updike, 1983). A distinctive morainal belt that also surrounds the lower slops of the mountain between 545 and 485 m (1800–1600 ft.) has been attributed to the Knik Glaciation, although they were possibly formed during the early Naptowne Glaciation (Reger and Updike, 1983). The age of the Mt. Susitna Glaciation is uncertain, but it is believed to be late Pliocene to early Pleistocene (Hamilton, 1994).

The Bootlegger Cove Fm. (Fig. 6) reaches a thickness of 35 m (Updike et al., 1988) and consists of silty clay and clayey silt with minor interbedded silt, fine–to–medium sand, and thin beds of diamicton (Updike et al., 1988). Scattered pebbles and

Figure 5. Mt. Susitna looking northwest. (Photo by Staci Ensminger)

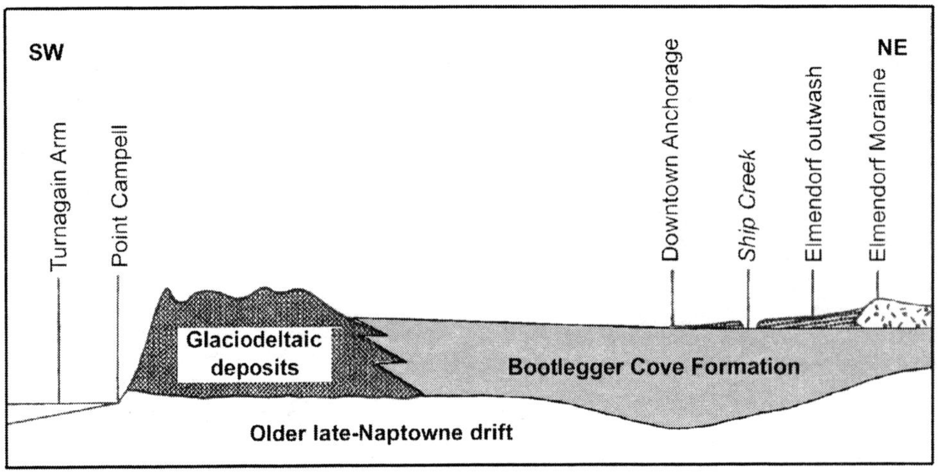

Figure 6. Sketch of late Naptowne stratigraphy in the Anchorage area. (Slightly modified from Reger et al., 1995).

cobbles are also present in widely varying concentrations and, throughout much of the formation, brackish–marine macro and microfossils are present (Schmidt, 1963, 1963; Hopkins, 1967). The formation is well exposed in the southern part of the Anchorage area (Hunter et al. 2000) and is conspicuously exposed in bluffs along the Knik Arm (Reger et al., 1995). Southwest of Anchorage (Fire Island–Pt. Campbell), it overlies and interfingers with relatively coarse–grained glacio–deltaic deposits, and northeast of Anchorage it underlies surficial deposits associated with the Elmendorf moraine (Miller and Dobrovolny, 1959; Cederstrom et al., 1964; Reger et al., 1995).

Karlstrom (1964) interpreted the occurrence of erratics on the ice–rounded summit and upper slopes of Mt. Susitna as evidence for the earliest recognized glaciation of the region, which he named after the mountain. He postulated that during this glaciation, ice drained from uplands in the region and eventually filled Cook Inlet to the present elevation of over 1,212m (4000 ft.). He also believed the ice flowed south out of the Cook Inlet trough into the Pacific Ocean and formed an extensive ice shelf. Originally, the Bootlegger Cove Fm. was thought to be of glaciolacustrine origin (Miller and Dobrovolny, 1959). However, with the discovery of marine mollusks, foraminifera, and ostracodes in place, a marine origin became clear (Schmidt, 1963; Hopkins, 1967) and ice–rafted debris in the base of the formation indicates that glaciers calved icebergs into the sea (Schmidt, 1963). The formation has been associated with the late Knik Glaciation (Miller and Dobrovolny, 1959), the late Knik and early Naptowne Glaciations (Karlstrom, 1964, Trainer and Waller, 1965) and a mid–Naptowne interstadial transgression (Hopkins, 1967), but numerous ^{14}C dates of marine shells now provide solid evidence that the formation is associated with the late Naptowne Glaciation (Schmoll et al., 1972, Reger et al., 1995) and was deposited between about 15,000 to 13,000 ^{14}C yrs. B.P. (Reger et al., 1995).

STOP 1–2. ELMENDORF MORAINE

The Elmendorf moraine (Figs. 3, 6) stop, just off the "Landfill–Eagle River" exit along the Glenn Highway, is an arcing end moraine that forms a prominent morphologic feature north of Anchorage and marks the terminal position of an ice lobe from the Matanuska/Knik valley (Miller and Dob rovolony, 1959; Schmoll et al., 1999). It can be traced westward from the mouth of Eagle River across the north edge of Elmendorf Air Force Base to the Susitna Lowland where it becomes bi–lobate and turns northward, then eastward toward the southern edge of the Talkeetna Mts.. In the vicinity of Eagle River, the moraine merges with lateral moraines on the lower slops of the Chugach Mountain front. In places within the Air Force Base, the crest of the moraine lies as much as 30 m above the head of an outwash plain (Mt. View fan) related to the moraine. From the proximity of the Air Force Base to its distal edge near downtown Anchorage, the outwash overlies the Bootlegger Cove Fm. (Fig. 6; Reger et al., 1995; Schmoll et al., 1999). Good exposures of sediments that underlie the Elmendorf moraine surface are few but generally show a polygenetic mixture of diamicton, sand, and gravel that probably interfinger with the strata of the outwash (Hunter et al., 2000).

The Elmendorf moraine appears to have formed as a result of a significant advance of ice from the Matanuska/Knik valley following retreat of ice from the southeast Anchorage area (Reger et al., 1995) to probably beyond the head of the Knik Arm (Schmoll et al., 1999). According to Hunter et al. (2000), the advance probably occurred behind a marine shoal that protected most of the glacier's terminus from tidewater conditions, and sand layers and general coarsening upwards in the upper most units of the Bootlegger Cove Fm. record the increasing proximity of the ice or shoaling. On the basis of ^{14}C dates from shells and barnacle plates in the Bootlegger Cove Fm. just beyond the limit of the Elmendorf moraine, Reger et al. (1995) placed the culmination of the Elmendorf advance at around 13,500 ^{14}C yrs. B.P.

STOP 1–3. TWIN PEAKS

The Twin Peaks stop is just off the "Old Glenn Highway" exit on the Glenn Highway (Fig. 3). On the northern slope of East Twin Peak (1,780 m; 5875 ft.) and West Twin Peak (1,637 m; 5400 ft.), Reger and Updike (1983) identified several lateral moraines deposited by ice advancing west through Knik Arm during the Naptowne and Knik Glaciations (Fig. 7). Farther up the slope, they identified a ridge capped by numerous erratics and scattered patches of till that they believe were scoured by ice during a pre–Knik Glaciation, perhaps during the Eklutna or Caribou Hills Glaciation. Holocene moraines and rock–glacier deposits occur in cirques near the summit of Twin Peaks and in some instances they are offset by faults (Reger and Updike, 1983).

Clearly, from the position of lateral moraines and ice–scoured surfaces on the slopes of Twin Peaks, at least three glaciations affected Knik Arm and the older glaciations involved thicker and more extensive ice. Other intervening glacial events as extensive or less extensive then those recorded on

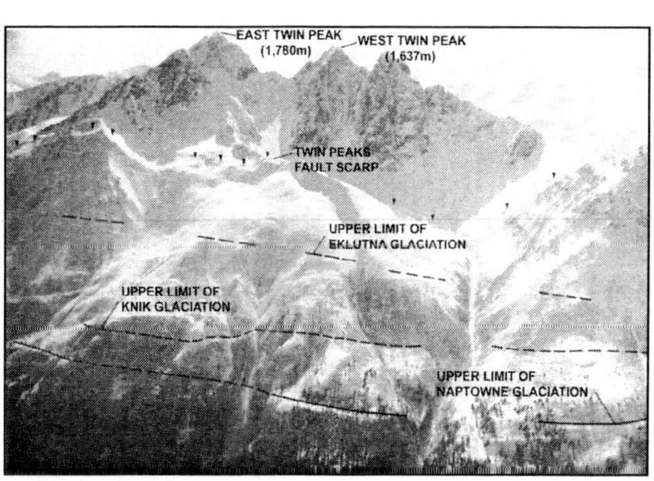

Figure 7. Aerial view (to the south) of Twin Peaks area, western Chugach Mountain front, Alaska. (Photo from Reger and Updike, 1983).

the slopes probably also occurred, but little or no evidence remains. This is undoubtedly also true for most of Cook Inlet (Schmoll et al., 1999).

Along the northwest shore of Knik Arm near Goose Cove Bay is a discontinuous peat up to 2m thick (Karlstrom, 1964; Reger and Updike, 1983). According to Karlstrom (1964), the peat occurs below till of Naptowne age and above till that he defines as the type deposit for the Knik Glaciation. The peat is generally very dense and fibrous and contains pollen suggestive of a spruce–birch forest (Ager and Brubaker, 1985). Several samples of the peat that have been dated lie beyond the limit of the ^{14}C method (Reger and Updike, 1983), and more recently, Reger et al. (1995, 1996) report an $^{40}Ar/^{39}Ar$ age of 378 ± 0.67 ka for volcanic ash within the peat, which presents a major problem for the glacial chronology of the region. Initially, the peat was believed to represent the last interglacial and the till beneath the peat to represent the penultimate or Knik glaciation (Karlstrom, 1964; Reger and Updike, 1983). The $^{40}Ar/^{39}Ar$ age, however, indicates that both the peat and the "Knik till" are much older, mid Pleistocene in age, and brings into question the age of some of the lateral moraines on the slopes of Twin Peaks.

Following the advance of late Naptowne ice to the Elmendorf moraine, the glacier system in the Knik Arm valley retreated eastward, exposing a landscape of drumlinized ground moraine (Schmoll et al., 1999). According to Reger and Updike (1983) and Reger et al. (1995), retreat was also accompanied by land emergence, ice stagnation, widespread meltwater activity, and rapid downcutting by streams. Several ^{14}C dates of organic sediments from boreholes in paludal and estuarine deposits that overlie the drumlinized ground moraine provide minimum ages for deglaciation of Knik Arm (Fig. 8). These dates range from 12,350 ± 350 to 9,255 ± 420 ^{14}C yrs. B.P. and have lead Reger et al. (1995) to conclude that the ice margin may have retreated at least 47 km from the Elmendorf moraine by 11,400 ^{14}C yrs. B.P.

STOP 1–4. CREVASSE MORAINE

The Crevasse moraine stop (T&G Gravel Products) is on the Parks Highway about 0.6 km west of the Glenn Highway and Parks Highway intersection (Fig. 3). An alternate stop is on the Glenn Highway about 6 km northeast of the Glenn Highway and Parks Highway intersection (Central Paving Products).

The ridge complex west and southwest of Palmer is locally known as the "Crevasse moraine" and covers an area of about 25 km^2. It consists of a series of arcuate, subparallel ridges as much as 45.5 m high that grade northeastward into a kame–esker complex and westward into a pitted outwash (Reger and Updike, 1983). Exposures within the ridges (Fig. 9) show them to be composed mainly of cross–bedded and planner–bedded, sandy gravel with concave–upward bedding. In some instances, beds of fine sand and silt up to 1 m thick are also present. Normal faulting and dip reversals are common along the ridge flanks and the tops of the sections are commonly capped by silt up to 0.3 m thick. In the southern part of the complex, oval–to–arcuate depressions occur between the ridges and sometimes contain lakes, but in the northern part the ridges are less conspicuous and are separated by a plain generally underlain by horizontally stratified sandy gravel to pebbly sand.

Reger and Updike (1983) attribute the ridges to infilling of open channels between vertical walls of stagnant glacier ice, probably from the Knik lobe about 10,000 ^{14}C yrs. B.P. They also attribute depressions between the ridges within the southern part of the ridge complex to sites formerly occupied by dead–ice masses that separated the channels. The plain between the ridges in the northern part of the complex is part of the "Palmer terrace," a fan–shaped body of pitted outwash with its apex near the mouth of the Matanuska Valley just north of Palmer (Reger and Updike, 1983). The silt that often caps the ridges is of eolian origin and is part of an extensive loess blanket that covers much of the Palmer area (Reger and Updike, 1983).

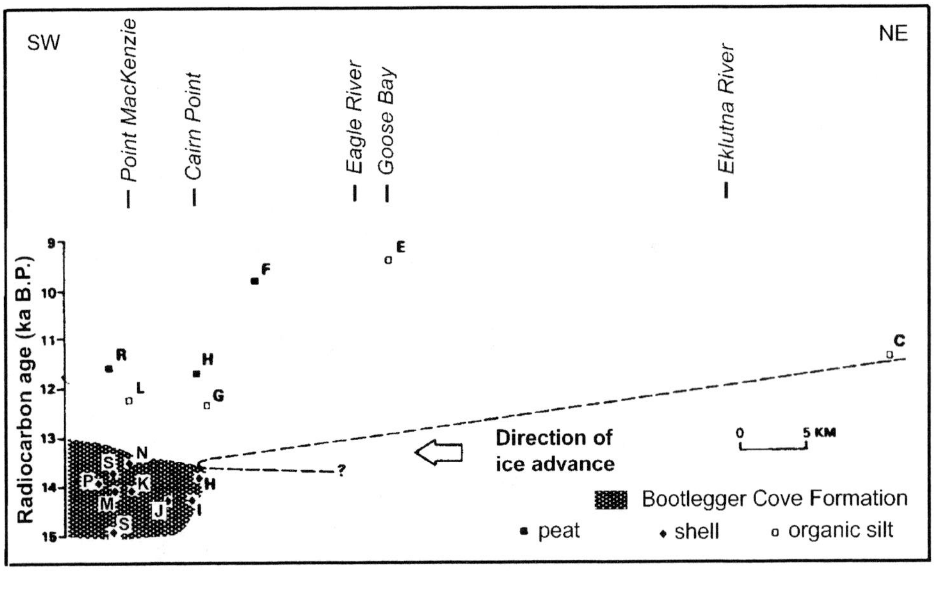

Figure 8. Tentative time–distance curve for the Elmendorf advance in Knik Arm. (Modified from Reger et al., 1995).

STOP 1–5. MATANUSKA RIVER OVERLOOK

The overlook stop is along the Glenn Highway approximately 3 km north of Palmer (Fig. 3). Exposed along bluffs on the west side of the Matanuska River is a 45–to–50 m–thick section of stratified sand and gravel capped by sand (Fig. 10). The sand and gravel is horizontally bedded, includes interbeds of pebbly sand, and underlies an extensive, high–level, pitted terrace surface 82 to 121 m above the modern floodplain of the Matanuska River. The sand cap overlying the sand and gravel is 0 to 5 m thick and is associated with cliff–head dunes and a thick eolian sand blanket in the Palmer area. It is composed mostly of medium to coarse sand and is characterized by planner bedding. Occasional small pebbles also occur within the sand.

The sand and gravel exposed in the buff is outwash deposited as the late Naptowne ice margin retreated into the Matanuska Valley. According to Reger and Updike (1983), the outwash surface initially formed as a broad, thick, sandy outwash fan deposited by the Matanuska River about 10,000 ^{14}C yrs. B.P. that was later deformed into a gently irregular surface when underlying stagnant ice melted.

The sand at the top of the section and in the cliff–head dunes is clearly of eolian origin and was deposited by winds blowing down the Matanuska valley from the northeast (Reger and Updike, 1983). According to Reger and Updike (1983), cliff–head dunes are still developing along the bluff and are built by winds capable of picking grains of sand and small pebbles from the underlying outwash and eolian sand blanket.

STOP 1–6. LATE–NAPTOWNE DIAMICTON

Late–Naptowne diamicton is exposed in a road cut along the north side of the Glenn Highway about 4.4 km east of Chickaloon River. The exposure is approximately 50m long and reveals about 2.5 m of horizontally stratified, poorly sorted, medium–to–course sand, overlain by as much as 3 m of matrix–supported diamicton. The stratified sand is characterized by beds 30 to 60 cm thick and includes pebbles and cobbles and occasional subrounded boulders <1m in diameter. The upper 1.5 m of the sand is deformed by abundant shearing and low–angle thrusts. The diamicton matrix consists mostly of silty coarse sand that supports angular to subrounded clasts ranging from <1 cm to 0.7 m in diameter. The contact between the stratified sand and diamicton is generally sharp. Locally, downglacier–dipping (west), wedge–shaped dikes or till wedges (Dreimanis, 1969) extend as much as 1.5 m into the top of the stratified sand (Fig. 11). About 1–1.5 m below the top of exposure is a band of caliche 5 to 20 cm thick, and at the western end of the section, the diamicton is overlain by <0.5 m of silt.

The stratified sand most likely represents proximal outwash or a subglacial channel–fill deposit, an interpretation supported by the presence of large boulders within the sand. The diamicton, on the other hand, probably represents a basal till deposited under temperate, flowing ice. Evidence for this interpretation includes

Figure 9. View (to southwest) of cross section through two crevasse fillings. (Photo by Ed Evenson)

Figure 10. View (to northeast) of Matanuska Valley. Note prominent outwash terrace and eolian cap. (Photo by Ed Evenson)

Figure 11. View (to the north) of matrix supported diamict along the Glenn Highway. Note till wedges penetrating into underlying stratified sand. (Photo by Grahame Larson)

the fact that the diamicton is injected into the underlying stratified sand and the injections are inclined in the direction of ice flow (Humlum, 1978; Dreimanis, 1992). Abundant shearing and low–angle thrusts within the sand just beneath the diamicton also support this interpretation. The silt capping the western end of the section is probably loess.

STOP 1–7. MATANUSKA GLACIER OVERLOOK

The Matanuska Glacier overlook stop is located on the Glenn Highway approximately 8 km east of Hicks Creek (Fig. 12). The Matanuska Glacier (Fig. 13) is one of the largest valley glaciers in south–central Alaska and occurs at the head of the Matanuska Valley, about 160 km northeast of Anchorage. It flows generally north for about 45 km from ice fields in the central Chugach Mts. and drains approximately 647 km^2 of the highest part of the range (Lawson, 1979). Flow rates, measured by surveying stakes set in the ice surface, range from 0.55 to 0.64 m per day about 10 km upglacier from the terminus (Lawson, 1983), and from 12–25 cm per day near the terminus (Ensminger et al., 1999b).

Near the terminus, ablation has produced a cover of supraglacial debris over about two–thirds of the lobe and the northernmost portion of the glacier is partly covered by a forest (Williams and Ferrians, 1961). The ice surface is generally free of debris only along the western margin. Debris is distributed unevenly within the glacier and is concentrated mainly in a thin basal zone that varies in thickness from about 3–15 m, but locally up to 35 m if stacked by overriding at the margin (Lawson and Kulla, 1978; Lawson, 1979; Lawson, 1983; Lawson et al., 1998; Evenson et al., 1999). Along the western area of the terminus (Fig. 14) subglacial water discharges from numerous vents during the melt season and is supercooled (Lawson et al., 1998). Also, ground–penetrating–radar data indicate that the basal–zone ice exists in areas peripheral to an ice–marginal overdeepening (Arcone et al., 1995).

In 1961, Williams and Ferrians mapped a series of discontinuous end moraines about 4–8 km beyond what was the terminus of the glacier at that time, and traced some of the moraines into

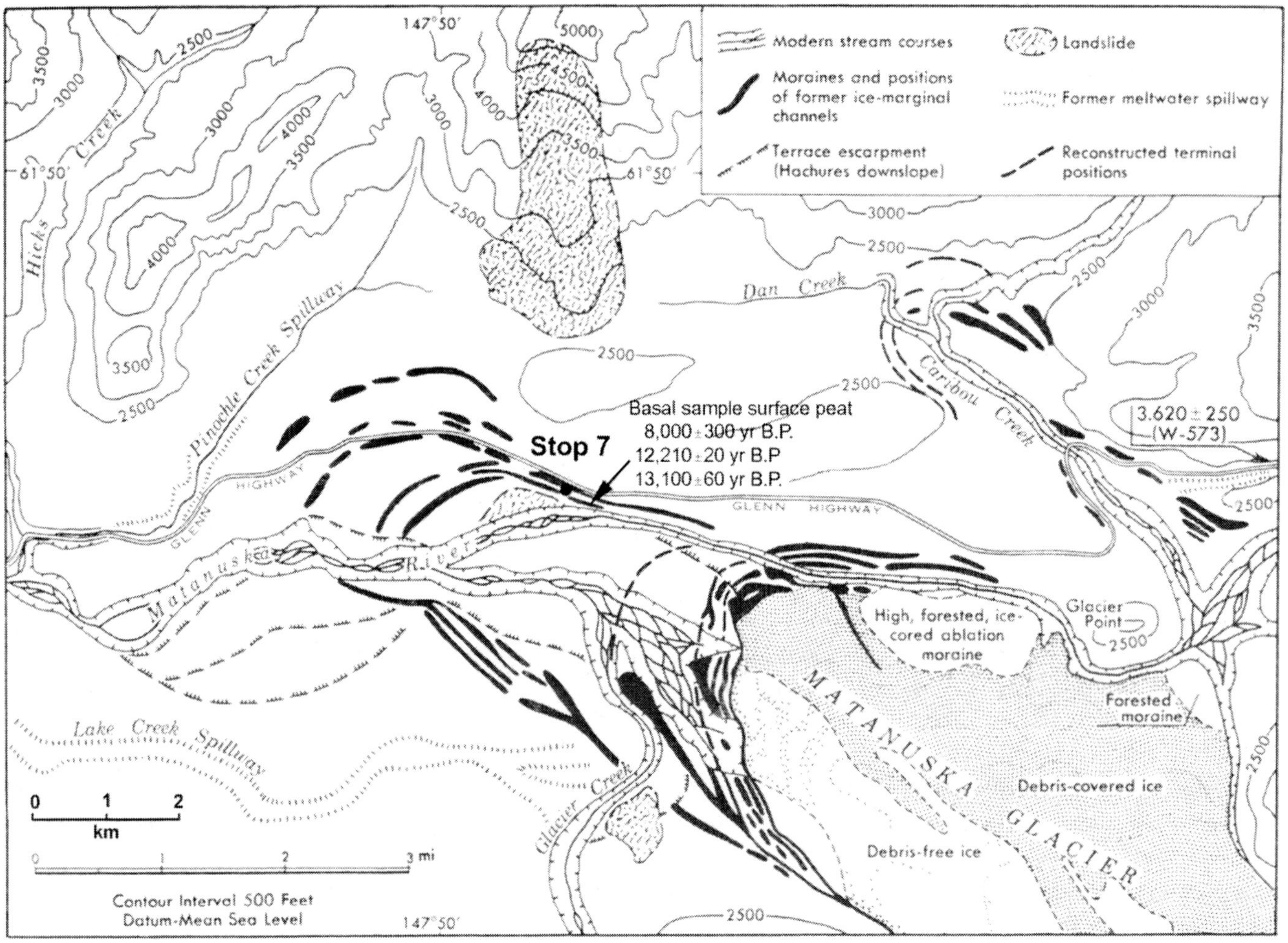

Figure 12. Sketch map of terminus of Matanuska Glacier in 1961 and vicinity showing former terminal positions. Note stops 8 and 9 are beyond the present glacier terminus. (Modified from Williams and Ferrians, 1961).

Figure 13. View (to south) of Matanuska Glacier (Photo by Ed Evenson).

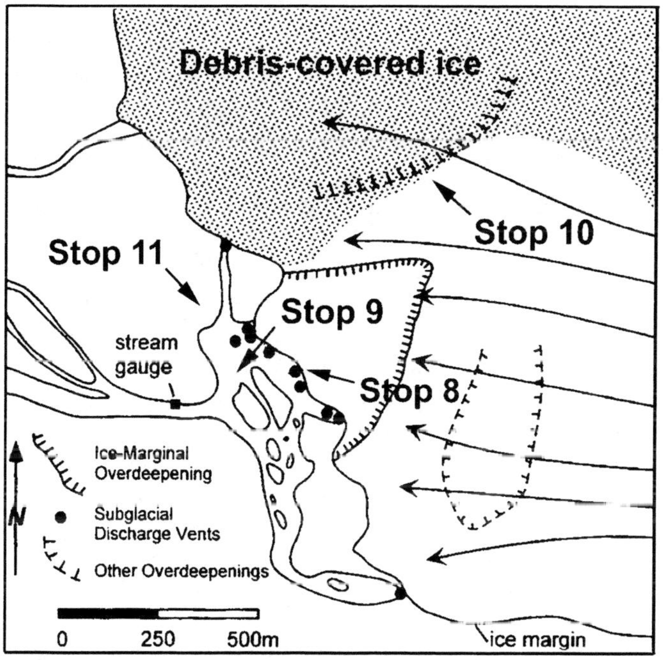

Figure 14. Map of the western margin of the Matanuska Glacier showing location of stops 8–11, overdeepenings, and subglacial–discharge vents. (Modified from Lawson et al., 1998).

well–developed lateral moraines on both sides of the Matanuska Valley (Fig. 12) that are clearly visible from the Glenn Highway. Peat and pond deposits as much as 6.6 m thick occur in a channel cut into the innermost of these moraines, just visible below this overlook stop. Samples from the base of the peat and pond deposits have yielded ^{14}C dates of 8,000 ± 300 (Williams and Ferrians, 1961), 12,210 ± 120, and 13,100 ± 60 ^{14}C yrs. B.P. (Williams, 1986). Williams and Ferrians (1961) also recognized a younger series of end moraines within 1.6 km of the 1961 terminus position and, on the basis of photographs taken by Mendenhall in 1898, determined that the innermost of these moraines was formed in part prior to 1898. Since 1961, the western terminus has retreated from the innermost moraine, leaving behind an ice–cored area 100–500m wide that parallels the lobe margin (Lawson, 1983).

Williams and Ferrians (1961) interpreted the ^{14}C date of 8,000 ± 300 ^{14}C yrs. B.P. from the peat and pond deposits as evidence that the late–Naptowne ice margin had retreated east up the Matanuska Valley and then readvanced to 4–8 km beyond the 1961 terminus prior to 8,000 ^{14}C yrs. B.P. However, the two more–recent dates of 12,210 ± 120 and 13,100 ± 60 ^{14}C yrs. B.P. caused Williams (1986) to abandon the interpretation offered by Williams and Ferrians (1961) in favor of one that advocates a readvance to 4–8 km beyond the 1961 terminus prior to 13,100 ^{14}C yrs. B.P. This new interpretation conflicts with the conclusion offered by Reger et al. (1995) that the culmination of ice advance in Knik Arm to the Elmendorf moraine occurred around 13,500 ^{14}C yrs. B.P. and that the Bootlegger Cove Fm. was deposited during the interval between 15,000 and 13,000 ^{14}C yrs. B.P. Clearly, additional data are needed from the middle and lower Matanuska Valley in order to resolve this conflict and clearly understand the late Naptowne glaciation of the Matanuska Valley.

STOP 2–1. Basal–ice zone, debris bands, and supercooling

The margin of the Matanuska Glacier (Fig. 12, 14) can be reached by an access road off the Glenn Highway leading to Glacier Park Resort and the western margin of the glacier. Three debris–bearing zones occur at the Matanuska Glacier, the supraglacial, englacial and basal (Fig. 15; Lawson, 1979). The basal zone contains a stratified and dispersed facies described in detail by Lawson (1979), Lawson et al. (1998) and Evenson et al. (1999). They are easily distinguished from one another by their debris content. The englacial zone contains a diffused and banded facies described in detail by Lawson (1979) and Lawson et al. (1998). They too are easily distinguished from one another by their debris content. The superglacial zone is principally an ice–surface cover 0 to 0.2 m thick which consists chiefly of very angular–to–angular gravel to sandy gravel (Lawson, 1979). According to Lawson (1979) it is mainly material weathered from valley walls and nunataks and mass transported onto the glacier surface.

The stratified facies is 3–15 m thick and has an average debris content of about 25% by volume, consisting of silt to sand to gravel. Stratification results from the alternation of layers or zones of ice containing abundant debris with those containing much less to no debris. Rounded to subrounded pebbles dominate the stratified facies and ice is very fine–grained, generally <2 mm in diameter. The dispersed facies is 0.2 to 8 m thick and has average debris content of 3.8% by volume, consisting almost entirely of silt and sand suspended as aggregates within the clean ice. Pebbles within the dispersed facies are mainly angular to subangular and ice grains tend to be 10–40 mm in diameter.

The boundary that separates the stratified and dispersed facies is irregular but well defined and often is marked by a con-

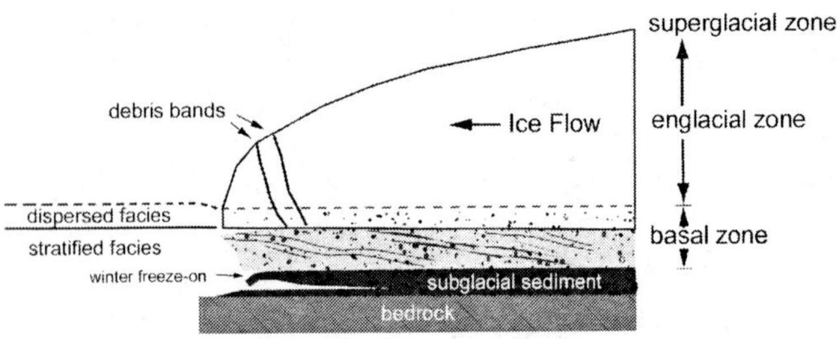

Figure 15. Schematic diagram of the englacial and basal ice zones at the Matanuska Glacier. The englacial ice zone is cross cut by numerous debris bands that have been interpreted as basal–crevasse–fills. The contact between the englacial and the dispersed facies is diffuse and the contact between the dispersed and stratified facies is sharp and somewhat planar. The nature of the facies contacts with one another reflects the processes that operate at the bed of the glacier.

TABLE 2. AVERAGE ISOTOPIC COMPOSITION OF ICE AND WATER, MATANUSKA GLACIER, ALASKA

Ice or water type	*T.U. (n)*	*δO18 (n)*	*δD (n)*
Englacial zone			
Banded facies of englacial zone			
Dirty laminated debris-band ice	7.0 (4)[1]	-22.4 (23)[1]	-170.8 (23)[1]
Clear laminated debris-band ice		-22.8 (6)[1]	-176.8 (6)[1]
Diffused facies	0.0 (10)[2]	-24.0 (19)[1]	-177.2 (19)[1]
Clear ice adjacent to laminae		-23.7 (17)[1]	-176.8 (17)[1]
Basal zone			
Dispersed facies	0.4 (5)[2]	-24.7 (6)[4]	-188.6 (6)[4]
Stratified facies	13.2 (13)[2]	-21.0 (158)[1]	-163.8 (158)[1]
Other			
Subglacial discharge	4.6 (33)[3]	-22.8 (15)[3]	-179.8 (16)[1]
Frazil and anchor ice	16.5 (1)[2]	-21.4 (10)[4]	-165.3 (10)[1]
Precipitation	7.8 (2)[2]	-20.9 (24)[4]	-165.5 (24)[1]

[1] Ensminger et al. (2001), [2] Strasser et al. (1996), [3] Titus et al. (1999), [4] Lawson et al (1998)

centration of pebble–to–cobble size clasts. A distinct, relatively planar contact, resulting from a rapid change in debris content, generally separates the dispersed facies from the englacial zone above (Lawson, 1979). In winter, along the margin of the glacier, subglacial sediment freezes onto the base of the glacier and is often thrust forward with basal–and englacial–zone ice to form overhanging ramps with grooves and striations extending along their base.

Stratified facies ice is characterized by elevated concentrations of hydrogen–bomb–produced ^3H compared to that of dispersed–and–diffused–facies ice (Table 2; Lawson and Kulla, 1978; Lawson, 1979; Strasser et al., 1996; Lawson et al., 1998; Titus et al., 1999; Ensminger et al., 2001). The average $\delta^{18}O$ value of stratified–facies ice also is about 3‰ greater than that of dispersed (and diffused) facies ice, and the average δD value is at least 13‰ greater. Recent isotopic analyses of stratified–and–dispersed–facies ice (Fig. 16) also show gradational isotopic compositions between the two facies, which suggests some isotopic diffusion is occurring across the contact (Ensminger et al., 2001).

The diffused facies of the englacial zone consists primarily of white bubbly to blue clear ice with a debris content consistently around 0.002% by volume. Generally, fine–grained debris is distributed uniformly in the facies. Occasionally, however, zones containing sand to granule size aggregates of clay–sized particles occur. Grains of white bubbly ice are about 2–20 mm in diameter and grains of blue clear ice are about 20–100 mm in diameter. Near the surface of the glacier, supraglacial debris incorporated into the diffused facies also results in locally abundant debris concentrations of angular pebbles, cobbles and boulders.

The banded facies is distributed randomly throughout the englacial zone in the form of debris bands 0.02 to 0.5 m thick, with debris concentrations generally up to 25% by volume and occasionally higher (Lawson 1979). Debris bands that lie conformably with ice layering typically consist of very angular–to–angular silt to gravel and represent material that was mass transported onto the snow in the accumulation area (Lawson 1979). Debris bands that are not compatible with ice layering generally consist of well–sorted, coarse silt or very fine sand that is often visibly concentrated in a centrally located seam or in several subparallel, millimeter–thick laminations separated by bands of clear ice (Ensminger et al., 1999a; Ensminger et al., 2001). Little difference exists between the diameter of ice grains of these debris bands and adjacent ice, and debris released by ablation often forms sediment flows on the ice surface (e.g. Lawson, 1981, 1988; Lawson, 1995), causing the bands to appear much wider than they really are (Ensminger et al., 2001). Also, the debris–band ice is commonly characterized by elevated concentrations of nuclear bomb tritium compared to ice from the diffused faces (table 2) and the average $\delta^{18}O$ value of ice from debris seams is about 2.4‰ greater than that of diffused–facies ice, while the average δD value is about 6.4‰ greater (Ensminger et al., 2001).

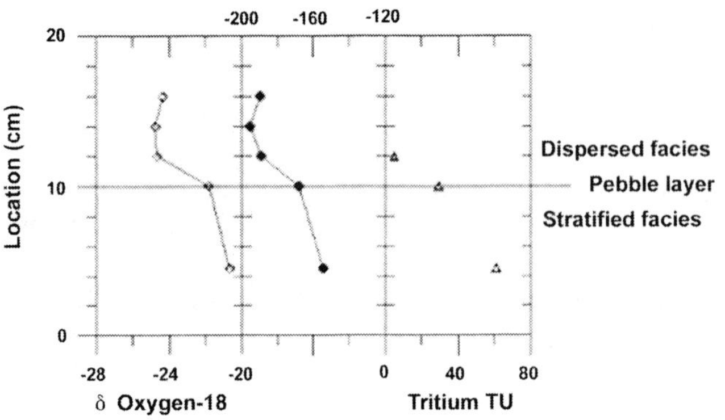

Figure 16. High–resolution isotopic data from ice samples that spanned across the stratified–dispersed facies contact. $\delta^{18}O$ and δD profiles mimic the ^3H, indicating isotopic concentrations in the dispersed–facies ice are being modified by water from the ^3H–rich stratified ice.

Lawson et al. (1998), building on a process model (Alley et al., 1998), conclude that stratified–facies ice is derived principally from glacio–hydraulic supercooling of water flowing in a distributed drainage system up the adverse slope of the overdeepening near the glacier terminus and accretion of debris–rich ice to the glacier sole. The criterion for supercooling is a large meltwater flux into a temperate glacier and an adverse bed slope approximately >1.2–1.7 times the magnitude of the surface slope (Alley at al., 1998). These conditions allow water ascending a sufficiently steep slope to supercool as the pressure–melting point rises, generating frazil ice which then accretes to the glacier bed (Alley at al., 1998).

The elevated concentrations of bomb–produced ^{3}H in the stratified–facies ice demonstrate that the ice is relatively young, consistent with a glacio–hydraulic supercooling origin (Strasser et al., 1996; Lawson et al. 1998; Titus et al., 1999). The δ^{18}O and δD value of stratified–facies ice compared to that from the rest of the glacier suggests fractionation during partial freezing of subglacial water in an open system, a process also consistent with glacio–hydraulic supercooling (Lawson et al. 1998; Titus et al., 1999). Lawson et al. (1998) argue that most of the debris in the stratified facies is suspended sediment that has been trapped by the growing ice. For a photographic atlas and brief explanation of ice growth and sediment entrapment observed at the Matanuska Glacier due to glaciohydraulic supercooling, the reader is referred to Evenson et al., (1999).

Ensminger et al. (1999a; 2001) attribute the origin of most of the debris bands that cross cut ice layering to basal crevassing and some combination of in situ freezing following injection of debris–laden basal water from below during the melt season, and glacio–hydraulic supercooling during injection and throughflow from the bed to englacial conduits or the surface. They cite the fact that grain–size of debris in these debris bands is distinct from that of supraglacial sediment, till and bedload of proglacial streams, but is similar to debris in the stratified facies and in frazil ice developed around discharge vents during the melt season as support for a basal–water source. The elevated concentration of bomb–produced ^{3}H in the debris–band ice also shows that at least some of the debris bands are of recent origin and must have been derived from refreezing of meltwater (Ensminger et al., 2001). The average δ^{18}O and δD value of banded–facies ice compared to that of diffused–facies ice also supports refreezing of meltwater (Ensminger et al., 2001). Recently, glacio–hydraulic supercooling under both ablation–dominated and jokulhlaup hydrologic regimes also has been invoked to explain the origin of intraglacial debris bands at two temperate Icelandic glaciers (Roberts et al., 2002).

The debris–laden stratified facies of the Matanuska Glacier is the source of nearly all sediment deposited in the ice–cored morainal zone along the western terminus (Lawson, 1979) and the ice–marginal processes that formed these moraines, especially sediment gravity flows, are documented and described in detail by Lawson (1979, 1982).

Sediment release from the basal zone and re–sedimentation processes may also have been important in ancient glacial systems where basal freeze–on processes operated (Lawson, 1979, 1995). In particular, glacio–hydraulic supercooling and basal freeze–on provide mechanisms by which warm–based glaciers may entrain and transport large volumes of sediment, and glaciers meeting the criteria for glacio–hydraulic supercooling can provide an abundant sediment supply, which may enable the generation of several types of landforms (Alley et al., 1998, 1999).

STOP 2–2. BURIED–ICE OVERLOOK

This overlook stop is above the "lake flats" about 200 m south from the parking lot of the Glacier Park Resort access road (Fig. 14). The lake flats extend to the glacier terminus and are exposed in winter (Fig. 17) but are covered by shallow, running water in summer. Exposures along stream channels cut 0.5–1 m into the drained flats during late Autumn reveal mainly horizontally bedded silts and discontinuous channel gravels.

The lake flats area consistently shows evidence of thermokarst (e.g., Fig. 18). Thermokarst sinkholes of this type are located up to 200 m in front of the active ice margin. In addition, mid–stream discharge fountains are located tens of meters in front of the active margin (e.g., a particular one we name "mega–vent"). Both of these phenomenon require the presence of ice buried beneath the lake flats.

A 2000–2001 geophysical study employed seismic reflection and refraction, ground–penetrating radar, and electrical resistivity techniques (see Fig. 19) to image the extent of the buried ice beneath the lake flats. Data were acquired in October and December, 2000, on the flats near the most active section of the glacier terminus. The thermokarst sinkholes and mid–stream discharge fountains indicate substantial buried stagnant ice with a persistent conduit system acting as part of the larger subglacial system. Geophysical data reveal relatively continu

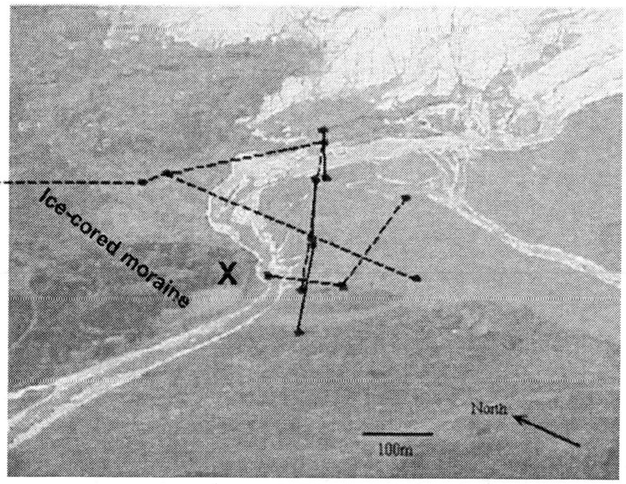

Figure 17. The "Lake Flats" in winter near the terminus of the Matanuska Glacier (marked with an "X") and overlay of the locations of coincident GPR and electrical resistivity profiles (dashed lines), and seismic reflection profiles (thick solid lines).

Figure 18. Exposed thermokarst feature in October 2000 on the "lake flats" just below the overlook. This feature was no longer visible in October 2002 (filled in by sediment). (Photo by Ed Evenson)

ous sheets of buried, stagnant ice up to 15 m thick within lake sediment and outwash that are organized as a series of upglacier–dipping, imbricated, "thrust sheets." In particular, seismic reflection data (Fig. 19) indicate three potentially different thrust "sheets" at increasing depths beneath the lake flat surface away from the terminus. The "mega–vent" discharge exists directly on strike with the edge of the "shallowest" sheet of buried ice, and subglacial water is thought to flow through an old meltwater conduit within the buried ice.

The thermokarst features in the lake flats are believed to form in a way more analogous to limestone karst (i.e., rapid melting along zones of fracturing and/or relict meltwater channels within the buried ice) rather than melting of discreet buried blocks of ice. The remaining ice appears to melt out gradually, allowing for semi–continuous deposition of sediment as accommodation space is generated. Also, the laterally extensive sheets of buried ice allow for semi–permanent, subglacial discharge pathways to exist past the current margin of the active ice.

STOP 2–3. SEISMICS ON ICE

The "seismics on ice" stop (Figs. 14, 20) is about 1 km southeast from the overlook parking lot of the Glacier Park resort access road, following the trail out onto the active ice. Once on the ice, the site is in the region to the southeast characterized by a broad, relatively flat expanse of ice with thin debris cover. Geophysical and hot–water drilling data have been collected here in order to determine the lateral extent of the debris–laden, stratified–facies basal ice of the Matanuska Glacier.

One critical piece of supporting evidence for the importance of supercooling and basal freeze–on is the distribution and extent of the debris–rich, stratified facies relative to the subglacial topography that to date is only well constrained near the glacier terminus. Logistical issues have significantly limited wide–coverage ground–truth by drilling. Ground–penetrating radar (GPR) has been successfully used to image the top of the debris–rich, basal zone. However, GPR could not adequately resolve the thickness or distribution of the zone because of limited radar penetration through the basal zone and the fact that variations in the distribution of liquid water and frozen zones within the glacier partially distort depth estimates and images. The alternative is to use seismic techniques because the basal zone does not negatively affect elastic wave propagation.

Several common–midpoint, seismic reflection profiles collected on the Matanuska Glacier (Fig. 21) clearly demonstrate the feasibility of collecting high–quality, high–resolution, near–surface reflection data on a temperate glacier. The results indicate that this technique can be used to accurately determine the thickness and horizontal distribution of debris–rich ice at the base of the glacier. The facies thickens about 30% over a 300–m distance as the glacier flows out of an overdeepening (Fig. 22). The reflection events ranged from 80–140 m depths along the longitudinal axis of the glacier. Two orthogonal profiles show variations perpendicular to ice flow as well (Fig. 23). The dominant reflection is from the contact between relatively clean englacial–zone ice and the underlying debris–rich, stratified–facies ice, but a strong characteristic reflection is also observed from the sole of the glacier.

The data yielded very accurate (± 1.0 m) estimates of depths to interfaces that were subsequently confirmed by hot–water

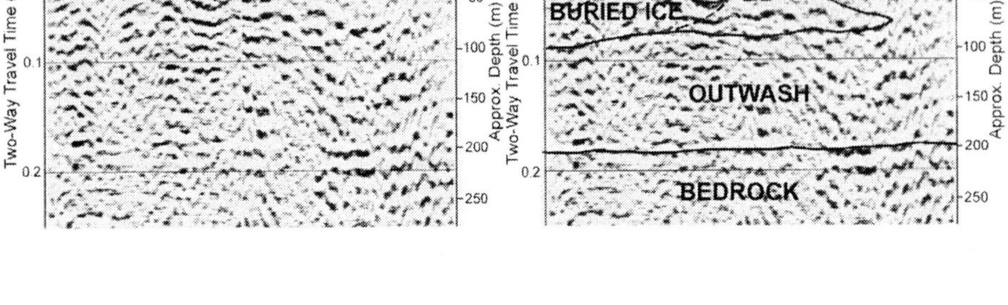

Figure 19. Uninterpreted (left) and interpreted (right) seismic reflection data from the "lake flats" region. The interpreted top of buried ice (as determine from the anomalously high p–wave velocities) steps down and outward from about 10 to 30m depth. Bedrock at this location is interpreted at 0.18 s, or approximately 100m depth. Note series of upglacier–dipping, imbricated, "thrust sheets."

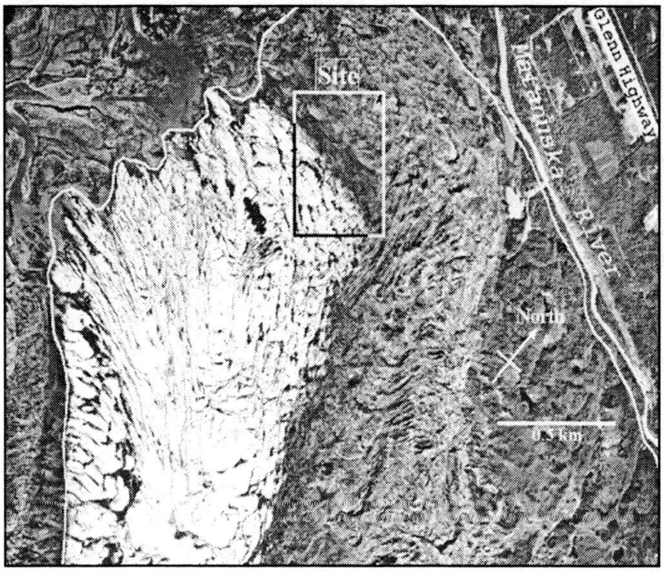

Figure 20. Terminus region of the Matanuska Glacier in 1999 (BLM photo). The terminus is outlined (white line). Seismic reflection data were collected in the relatively flat debris covered region outlined by the box .

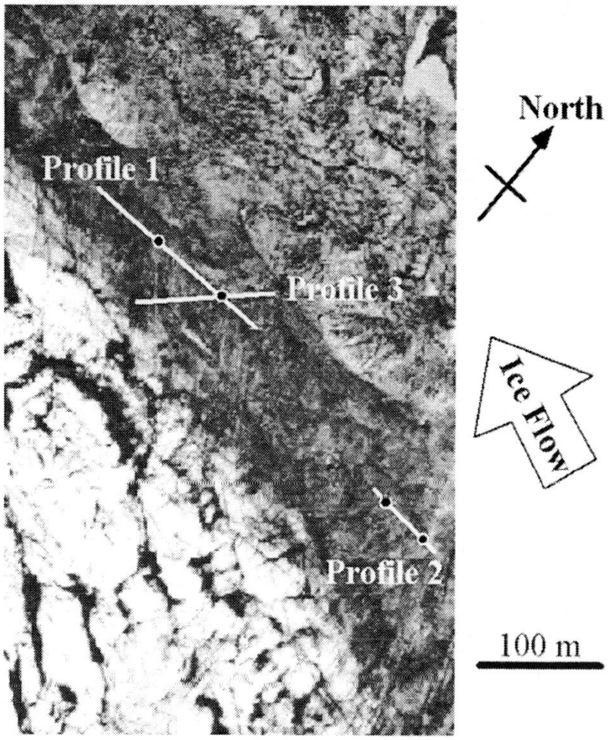

Figure 21. Close–up view of Summer 2000 seismic reflection profile locations. The black dots represent hot–water drilling stations.

drilling. The interpreted basal–ice thickening supports the hypothesized freeze–on mechanism, dictating that such thickening should occur when a glacier experiences glaciohydraulic supercooling as it flows out of an overdeepening.

STOP 2–4. ICE–CORED MORAINE

The "ice–cored moraine" stop is located at the western terminus of the Matanuska Glacier, about 100 m north from the overlook parking lot of the Glacier Park lodge access area, between the active ice and an end moraine escarpment (Fig. 24). The origin of the moraine is described by Lawson (1979).

Mechanisms for glaciotectonic sediment deformation are not universally accepted and often contradictory (i.e., field evidence versus known problems with the ability of saturated sediments to transmit stresses). Ground–penetrating radar data collected at the ice–cored moraine site (Fig. 25) demonstrate that during a recent re–advance (2001–2002) to its farthest position in about 15 years, the active ice impacted and deformed stagnant ice buried in the moraine and the overlying sediments subsequently passively deformed. This method of "passive" glaciotectonic deformation will be discussed at the site, and additional field evidence examined.

The ice–cored moraine site includes many facies, such as fine–grained, layered, debris–flow deposits of various types and primary diamicton (Lawson 1979, 1981, 1982). The GPR data allowed for quantitative determination and three–dimensional visualization of the geometry of the debris flows, the geometry of the top of the buried ice, the sediment thickness over the buried ice, and the thickness of the melt–out till. From 1.5 m to 3m of these sediments overly the stagnant ice, as determined from three–dimensional GPR data. Ground–truthing via hand–dug pits and auguring have confirmed the presence of the buried ice beneath the sediments at the depths predicted by the GPR data.

A detailed GPR study of a 100 x 50–m area at the site made in 2001 and 2002 resulted in moderate attenuation and generated high quality data (100 MHz antennas penetrate to >3 m). Prominent reflections were generated from interfaces between decimeter–scale debris flows, at the interface between the debris–flow package underlying actively developing melt–out till, and at the top of ice buried by these sediments. A comparison of data collected in August 2001 (prior to the readvance) and August 2002 (after the readvance) highlighted the sediment deformation resulting from the readvance. An examination of two typical profiles collected parallel to the ice flow direction also showed that the top surface of the buried ice was buckled and thrust, and subsequently the sediments overlying the buried ice were deformed.

Results of the study also indicated a method of glaciotectonic deformation in sediments, in which deformation of stagnant buried ice generated during impact of readvancing active ice causes the overlying sediments to "passively" deform. This mechanism may also be in play elsewhere and resolve some of the conflict related to mechanisms of glactiotectonic deformation.

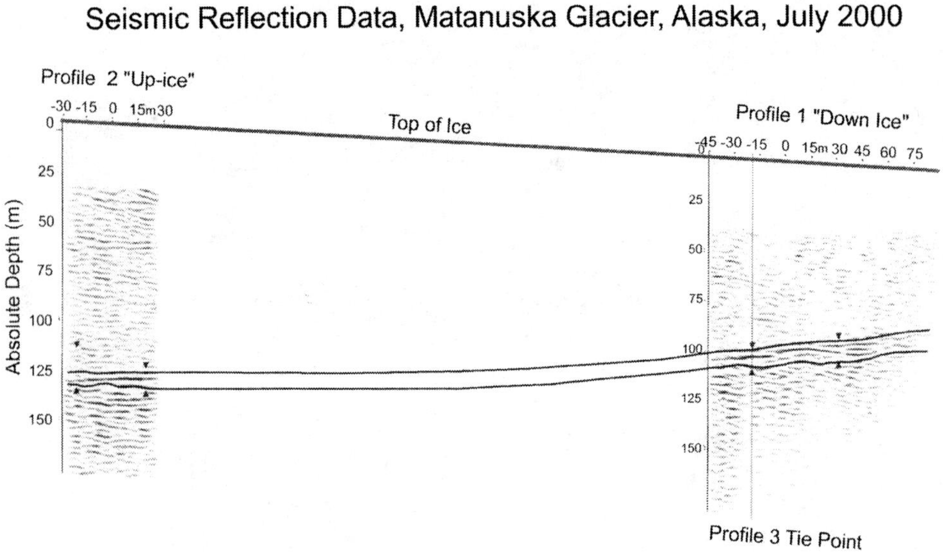

Figure 22. Seismic reflection profiles parallel to flow showing a 30% increase in basal ice thickness as the glacier flows out of an overdeepening. The triangles represent locations ground–truthed via hot water drilling.

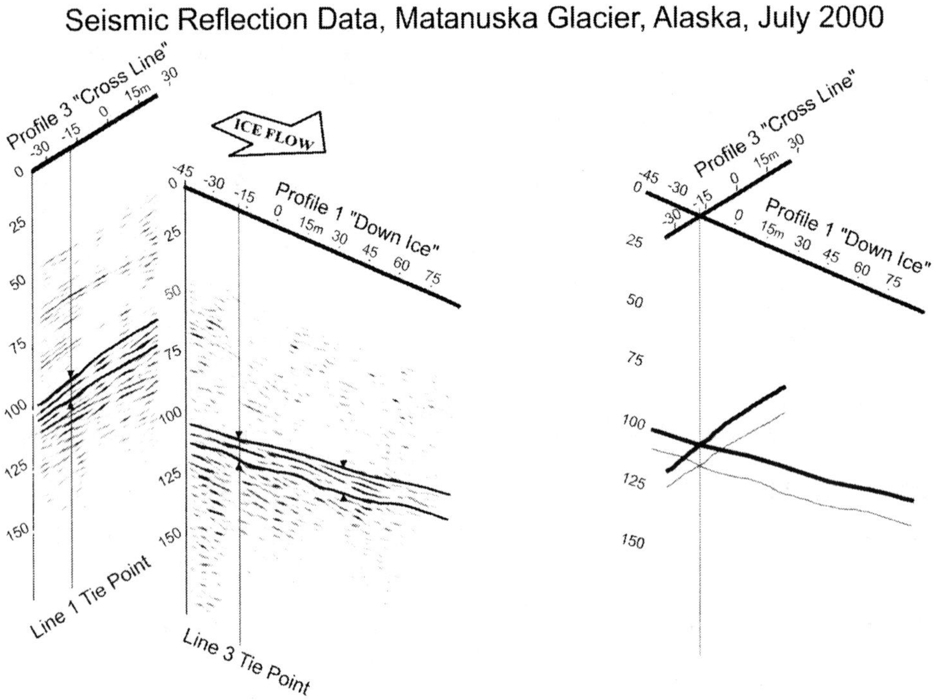

Figure 23. Interpreted seismic reflection data of crossed profiles showing the basal ice thickness variations parallel and perpendicular to ice flow.

Figure 24. An August 2001 photo of the ice–cored moraine site. The region surveyed using GPR is outlined by the white box.

STOPS 3–1 to 3–8. DENALI HIGHWAY

The excursion along the Denali Highway is designed to expose geologists to the Quaternary geology and spectacular scenery of the south flank of the Alaska Range. The trip begins at the junction of the Richardson and Denali Highways at Paxson, Alaska. Stop and viewpoint locations are Denali Highway mile markers that begin at "0" at Paxson and increase to the west toward the Maclaren River and the town of Cantwell. Much of the geology presented below is excerpted from Péwé et al. (1965).

0.0	*Paxson Lodge. Meals, gas, lodging available.*
2.5	*Ice–contact deposits of the Summit Lake advance, which hold in Summit Lake.*
6.0	*View south. Meltwater–escape gorge of Wisconsin age on Paxson Mountain.*
7.2	***STOP 3–1.** Alaska Range panorama (Fig. 26).*
20.4	***STOP 3–2.** Ice–contact deposits and Tangle Lakes. The drainage here goes north through the Alaska Range into the Delta–Tanana–Yukon–River system. Péwé reports a recent rock glacier on the west side of the river 13 km in the distance.*
29.0	***STOP 3–3.** High Valley altiplanation (cryoplanation) terraces. High Valley extends from milepost 28 to 36.5. At an elevation of 1,480 m (4,855 feet), the altiplanation terraces stand out above the upper limit of the Wisconsin ice.*
30.0	*Directly ahead is the terminal moraine of the glacial lobe that flowed through Glacier Gap and terminated in High Valley.*
34.5	*Alternate view of altiplanation terraces to north (Fig. 27).*
35.6	*Early Wisconsin right lateral moraine deposited when Maclaren Glacier pushed over the edge of Maclaren Valley into High Valley.*
37.0	***STOP 3–4.** Maclaren Summit Trail. View of Maclaren Valley and Maclaren Glacier. A short walk down Maclaren summit trail reveals many, now–dry, kettles in till with well–developed, patterned ground (sorted stone circles) on the kettle floors (Figs. 28–29).*
40.8	***STOP 3–5.** Pingos in the bog on both sides of the road. The ice core is exposed in the bog on the left. Lowest*

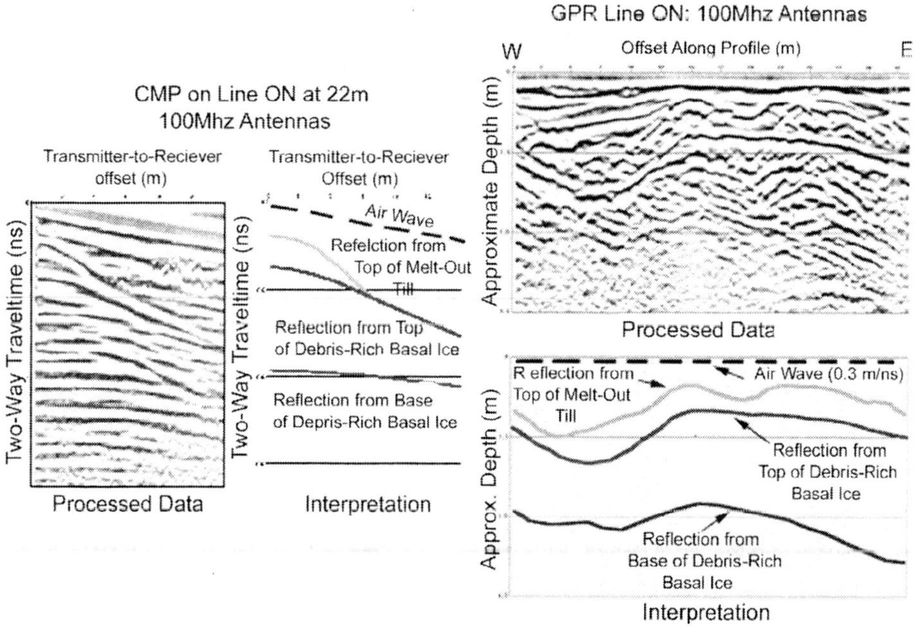

Figure 25. Typical GPR data from the ice–cored moraine site. (a) An uninterpreted (left) and interpretation (right) of a common–midpoint gather showing the major GPR phases. (b) Uninterpreted (top) and interpretation (bottom) GPR profile across the ice–cored moraine.

Figure 26. View of the south flank of the Alaska Range. (Photo by Ed Evenson)

Figure 28. Well developed sorted stone circles on floor of dry kettle (August 1994). (Photo by Ed Evenson)

Figure 27. Altiplanation Terraces. (Photo by Ed Evenson)

Figure 29. Detail of sorted stone circles. Note coarser clasts being "followed" by smaller clasts. (Photo by Ed Evenson)

peat in contact with ice dates 10,565 ± 225 ^{14}C yrs. B.P. A Pingo developed in post–glacial time provides a minimum age for late Denali advance.

41.1 ***STOP 3–6.*** *Large, kettle approximately 15 m (50 feet) deep developed in ice contact gravels (Fig. 30).*

42.0 ***STOP 3–7.*** *Maclaren River and Maclaren River lodge. Meals, gas, lodging, restrooms available.*

45.0 *Entrance to Crazy Notch cut into bedrock.*

46.0 *Exit of Crazy Notch. According to Péwé, a sub–aerial stream cut Crazy Notch, which was sub–ice both to the east and west. A large esker begins (or ends) at the exit of the notch. Two systems of eskers enter Crazy Notch from the west and drainage was to the east into the Maclaren Valley.*

49.0 *Pingos on the right.*

50.0 ***STOP 3–8.*** *Road on top of large esker (Fig. 31)*

53.0 *End of excursion*

Figure 30. 15 m deep kettle hole in ice contact gravel. (Photo by Ed Evenson)

Figure 31. Road along crest of Crazy Notch esker. (Photo by Ed Evenson)

REFERENCES

Ager, T. A. and Brubaker, L., 1985, Quaternary palynology and vegetational history of Alaska: *in* Bryant,V., Jr., and Holloway, R., eds., Pollen records of Late Quaternary North American sediments: American Association of Stratigraphic Palynologists Foundation, Dallas, Texas, p. 353–384.

Alley, R. B., Lawson, D.E., Evenson, E.B., Strasser, J.C., and Larson, G.J., 1998, Glaciohydraulic supercooling: a freeze–on mechanism to create stratified, debris–rich basal ice. 2. Theory: Journal of Glaciology, v. 44, no. 148, p. 563–569.

Alley, R. B., Strasser, J.C., Lawson, D.E., Evenson, E.B., and Larson, G.J., 1999, Some glaciological and geological implications of basal–ice accretion in overdeepenings: *in* Mickleson, D. M. and Attig, J. W., eds., Glacial Processes Past and Present, Geological Society of America, Special Paper 337, p. 1–9.

Arcone, S.A., Lawson, D.E., and Delaney, A.J., 1995, Short–pulse radar wavelet recovery and resolution of dielectic constrasts within englacial and basal ice of Matanuska Glacier, Alaska, U.S.A.: Journal of Glaciology, v. 41, p.68–86.

Boss, R.F., Lennon, R.B., and Wilson, B.W..1976, Middle Ground Shoal oil field, Alaska: *in* Braunstein, J., ed., North American oil and gas fields: American Association of Petroleum Geologists Memoir 24, p. 1–22.

Cederstrom , D.J., Trainer, F.W., and Waller, R.M., 1964, Geology and ground–water resources of the Anchorage area, Alaska: U.S. Geological Survey, Water–Supply Paper 1773, 108p.

Dickinson, W.R., 1971, Clastic sedimentary sequences deposited in shelf, slop, and trough settings between magmatic arcs and associated trenches: Pacific Geology, v. 3, p. 15–30.

Dreimanis, A., 1969, Till wedges as indicators of direction of glacial movements: Geological Society of America Annual Meeting Abstracts and Program, p. 52–53.

Dreimanis. A., 1992, Downward injected till wedges and upward injected till dikes: Sveriges Geologiska Undersökning, Ser. Ca 81, p. 91–96.

Ensminger, S. L., Evenson, E.B., Lawson, D.E., Larson, G.J., and Alley, R.B., 1999a, Preliminary study of laminated, silt–rich debris–bands: Matanuska Glacier, Alaska, USA: Annals of Glaciology, v. 28, p. 261–266.

Ensminger, S. L., Evenson, E.B., Alley, R.B., Larson, G.J., Lawson, D.E., and Strasser, J.C., 1999b, Example of the dependence of ice motion on subglacial drainage system evolution: Matanuska Glacier, Alaska, United States: *in* Mickleson, D. M. and Attig, J. W., ed., Glacial Processes Past and Present: Geological Society of America Special Paper 337, p. 11–22.

Ensminger, S. L., Alley, R.B., Evenson, E.B., Lawson, D.E., and Larson, G.J., 2001, Basal–crevasse–fill origin of laminated debris bands at Matanuska Glacier, Alaska, USA: Journal of Glaciology, v. 47, no. 158, p. 412–422.

Evenson, E. B., Lawson, D.E., Strasser, J.C., Larson, G.J., Alley, R.B., and Ensminger, S.L., 1999, Field evidence for the recognition of glaciohydrologic supercooling and basal ice accretion: *in* Mickleson, D. M. Attig, J. W., Glacial Processes Past and Present: Geological Society of America Special Paper 337, p. 23–36.

Hamilton, T.D., 1994, Late Cenozoic glaciation of Alaska, The Geology of North America: V. G–1, The Geology of Alaska, Geological Society of America, p. 813–842.

Hopkins, D.M., 1967, Quaternary marine transgressions in Alaska: *in* Hopkins, D.M., ed., The Bering Land Bridge, Standford University Press, Stanford, CA, p. 47–90.

Humlum, O., 1978, A large till wedge in Denmark:implications for the subglacial thermal regime: Geological Society Denmark Bulletin, v. 27, p. 63–71.

Hunter, L.E., Lawson, D.E., Bigl, S.R., and Schagel, J.D., 2000, Glacial geology and stratigraphy of Fort Richardson: A review of available data on the hydrology: U.S. Army Engineering Research and Development Center, Cold Regions Research and Engineering Laboratory, TR–00–03, 68p.

Karlstrom, T.N.V., 1953, Upper Cook Inlet region, Alaska: *in* Péwé, T.L. and others, eds., Multiple glaciation in Alaska: A progress report: U.S. Geological Circular 289, p. 3–5.

________, 1955, Late Pleistocene and Recent glacial chronology of south–central Alaska: Geological Society of America Abstract, v. 66, no. 3237, p. 73–74.

________, 1957, Tentative correlation of Alaskan sequences, 1956: Science, v. 66, no. 12, p. 1581–1582.

________, 1964, Quaternary geology of the Kenai Lowland and glacial history of the Cook Inlet region, Alaska: U.S. Geological Survey Professional Paper 443, 69 p.

________, 1965, Upper Cook Inlet area and the Matanuska River Valley: *in* Péwé, T.L., 1965: Ferrians, O.J., Jr., Nichols, D.R., and Karlstrom, T.N.V., Guidebook for field conference F, central and south–central Alaska, International Association for Quaternary Research, 7th Congess, Fairbanks: Lincoln, Nebraska Academy of Science, p. 114–141.

Lawson, D.E., 1979, Sedimentological analysis of the western terminus region of the Matanuska Glacier, Alaska: U. S. Army Cold Regions Research and Engineering Laboratory, 122p.

________, 1981, Distinguishing characteristics of diamictons at the margin of the Matanuska Glacier, Alaska: Annals of Glaciology, v. 2, p. 78–84.

________, 1982, Mobilization, movement and deposition of active subaerial sediment flows, Matanuska Glacier, Alaska: Journal of Geology, vol,. 90, p. 279–300.

________, 1988, Glaciogenic resedimentation: Classification concepts and application to mass–movement processes and deposits: *in* Goldthwaite, R.P. and Matsch, C.L., eds., Genetic Classification of Glaciogenic Deposits, Balkema, Rotterdam, :p. 147–172.

________, 1983, Overview of the Matanuska Glacier, *in* Glacial Geology and Geomorphology of North America: Richardson and Glenn Highways, Alaska: Guidebook to permafrost and Quaternary geology, Alaska Division of Geological and Geophysical Surveys Guidebook 1, p. 177–183.

________, 1995, Sedimentary and hydrologic processes within modern terrestrial valley glaciers: *in* Menzies, J., ed., Modern Glacial Environments: Processes, Dynamics, and Sediments, Butterworth–Heinemann, Ltd., Boston, MA, p. 337–363.

Lawson, D. E. and Kulla, J. B.,1978, An oxygen isotope investigation of the origin of the basal zone of the Matanuska Glacier, Alaska: Journal of Geology, v. 86, no. 6, p. 673–685.

Lawson, D. E., Strasser, J. C. Evenson, E.B., Alley, R.B., Larson, G.J., and Arcone, S.A., 1998, Glaciohydraulic supercooling: A mechanism to create stratified, debris–rich basal ice. 1. Field evidence and conceptual model: Journal of Glaciology, v. 44, no. 148, p. 547–562.

Miller, R.D., and Dobrovolny, E., 1959, Surfical geology of Anchorage and vicinity, Alaska: U.S. Geological Survey Bulletin 1093, 128 p.

Péwé, T.L., 1975, Quaternary Geology of Alaska: U.S. Geological Survey Professional Paper 835, 145 p.

Reger, R.D., and Updike, R.G., 1983, Upper Cook Inlet region and the Matanuska Valley: *in* Péwé, T.L. and Reger, R.D., eds., Richardson and Glenn Highways, Alaska, Guidebook to permafrost and Quaternary geology: Alaska Division of Geological and Geophysical Surveys Guidebook 1, p. 185–283.

Reger, R.D., Pinney, D.S., Burke, R.M., Wiltse, M.A., 1996, Catalog and intitial analyses of geologic data related to middle to later Quaternary deposits, Cook Inlet region, Alaska: Alaska Division of Geological and Geophysical Surveys Report of Investigations 95–6, 188p.

Reger, R.D., Combellick, R.A. and Brigham–Grette, J., 1995, Late–Wisconsinan events in the upper Cook Inlet region, southcentral Alaska: *in* Short Notes on Alaska Geology 1995, Combellick, R.A. and Tannian, F., eds., Alaskan Division of Geological and Geophysical Surveys Professional Report 117, p. 33–45.

Roberts, M.J., Tweed, F.S., Russell, A.J., Knudsen, _., Lawson, D.E., Larson, G.J., Evenson, E.B. and Bjornsson, H., 2002, Glaciohydraulic supercooling in Iceland: *Geology*, v. 30, p. 439–442.

Schmoll, H.R., Yehle, L.A. and Updike, R.G.,1999, Summary of Quaternary geology of the Municipality of Anchorage, Alaska: Quaternary International, v. 60, p. 3–36.

Schmoll, H.R., and Yehle, L.A., 1983, Glaciation in the upper Cook Inlet basin; A preliminary reexamination based on geologic mapping: *in* Thorson, R.M., and Hamilton, T.D., eds., Glaciation in Alaska, Extended abstracts from a workshop, Fairbanks, University of Alaska Museum Occasional Paper 2, p. 75–81.

Schmoll, H.R., Yehle, L.A., 1986, Pleistocence glaciation in upper Cook Inlet basin – A preliminary re–examination based on geologic mapping in progress: in Hamilton, T.D., Reed, K.M., Thorson, R.M. , eds., Glaciation in Alaska–geologic record: Alaska Geological Society, Anchorage, pp. 193–218.

Schmoll, H.R., Szabo, B.J., Rubin, Meyer, and Dobrovolny, Ernest, 1972, Radiometric dating of marine shells from the bootlegger Cove Clay, Anchorage, Alaska: Geological Society of America Bulletin, v. 83, no. 4, p. 1107–1114.

Schmoll, H.R., Yehle, L.A., Gardener, C.A., and Odum, J.K., 1984, Guide to surficial geology and glacial stratigraphy in the upper Cook Inlet basin: Anchorage, Alaska Geological Society Guidebook, 89 p.

Schmidt, R.A.M., 1963, Pleistocene marine microfauna in the Bootlegger Cove Clay, Anchorage Alaska: Science, v. 141, no. 3578, p. 350–351.

Strasser, J. C., Lawson, D.E., Larson, G.J., Evenson, E.B. and Alley, R.B.,1996, Preliminary results of tritium analyses in basal ice, Matanuska Glacier, Alaska, USA: Annals of Glaciology, v. 22, p. 126–133.

Titus, D. D., Larson, G.J., Strasser, J.C., Lawson, D.E., Evenson, E.B.,and Alley, R.B.,1999, Isotopic composition of vent discharge from the Matanuska Glacier, Alaska: *in* Mickleson, D. M. a. J. W. A., ed., Glacial Processes Past and Present: Boulder, CO, Geological Society of America, Special Paper 337, p. 37–44.

Trainer, F.W., 1960, Geology and ground–water resources of the Matanuska Valley agricultural area, Alaska: U.S. Geological Survey Water Supply Paper 1494, 116p.

Trainer, F.W., and Waller, R.M., 1965, Subsurface stratigraphy of glacial drift at Anchorage, Alaska: U.S. Geological Survey Professional Paper 525–D, p. D167–D174.

Updike, R.G., Olsen, H.W., Schmoll, H.R., Kharaka, Y.K., and Stokoe, K.H., 1988, Geologic and geotechnical conditions adjacent to the Turnagain Heights landslide, Anchorage, Alaska: U.S. Geological Survey Bulletin 1817.

Williams, J.R. and Ferrians, O.J., Jr., 1961, Late Wisconsin and Recent history of the Matanuska Glacier, Alaska: Arctic, v. 14, p. 82–90.

Williams, J.R.,1986, New radiocarbon dates from the Matanuska Glacier bog section: *in* Bartsch–Winkler, S., and Reed, K.R., eds., Geological studies in Alaska by the U.S. Geological Survey during 1985, U.S. Geological Survey Circular 978, p. 85–88.

Cordilleran Ice Sheet glaciation of the Puget Lowland and Columbia Plateau and alpine glaciation of the North Cascade Range, Washington

Don J. Easterbrook, *Dept. of Geology, Western Washington University, Bellingham, WA 98225*

THE CORDILLERAN ICE SHEET

The Puget Lowland (Fig. 1) contains widespread Pleistocene sediments, ranging in age from nearly 2 Ma to 10 ka, well exposed in hundreds of km of sea cliffs and roadcuts. During this four–day field excursion, we will examine all of the late Pleistocene stratigraphic units of the past 300 ka.

Figure 1. Field trip route map.

DAY 1. DOUBLE BLUFF, POSSESSION, AND VASHON GLACIATIONS; WHIDBEY AND OLYMPIA INTERGLACIATIONS; DISINTEGRATION OF THE CORDILLERAN ICE SHEET

STOP 1-1. DOUBLE BLUFF DRIFT AND WHIDBEY FM. TYPE SECTIONS AT DOUBLE BLUFF

Enroute to STOP 1-1, travel north from Seattle on I–5, take the ferry from Mukilteo to Clinton on southern Whidbey Island, then drive northwest to Freeland. Turn south from Freeland to Double Bluff. Walk the beach from Useless Bay to Double Bluff (about 1.5 km; 1 mi.).

Double Bluff Glaciation. Sea cliff exposures of glacial outwash, till, glaciomarine drift, and glaciolacustrine sediments at Double Bluff (Fig. 2) were designated as the type locality for Double Bluff Drift by Easterbrook et al. (1967). Double Bluff Drift is rarely exposed, but crops out in sea cliffs at or near sea level where it lies stratigraphically beneath the Whidbey Fm. of the last interglaciation (Easterbrook, 1968, 1969, 1986; Easterbrook et al., 1967).

Thermoluminesence (TL) dating of clay beneath Double Bluff Drift has yielded ages of 289 ± 74 ka and 291 ± 86 ka. A TL date of 177 ± 38 ka was obtained from Double Bluff glaciomarine drift, and a date of 320 ± 100 ka was obtained from clay associated with Double Bluff till east of here (Berger and Easterbrook, 1993). Amino–acid ages from mollusk shells in Double Bluff glaciomarine drift range from 111 to 178 ka at the type locality and from 150 to 250 ka elsewhere in the central Puget Lowland (Easterbrook et al., 1982; Blunt et al., 1987).

Whidbey Interglaciation. The type section of the Whidbey Fm. consists of silt and sand, interbedded with, clay, peat, and widely scattered gravel exposed in the ~2–km–long sea cliffs between Double Bluff and Useless Bay (Fig. 2) (Easterbrook et al., 1967). The Whidbey Fm. is underlain by Double Bluff Drift at the west end of the bluffs. Along the sea cliffs in the eastern part of

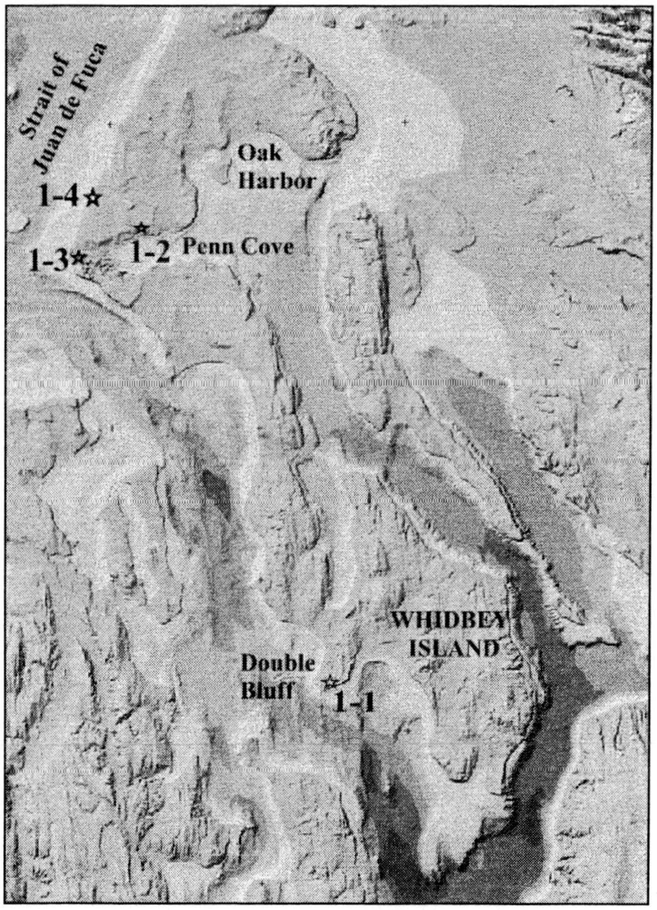

Figure 2. Whidbey Island showing field trip stops. (Modified from shaded digital topographic model by PRISM, University of Washingon.

Easterbrook, D.J., 2003, Cordilleran Ice Sheet glaciation of the Puget Lowland and Columbia Plateau and alpine glaciation of the North Cascade Range, Washington: *in* Easterbrook, D.J., ed., Quaternary Geology of the United States, INQUA 2003 Field Guide Volume, Desert Research Institute, Reno, NV, p. 265–286.

the section, Whidbey sediments are unconformably overlain by Olympia lacustrine silt, ^{14}C–dated at 23,600 ± 275 and 22,210 ± 530 ^{14}C yrs. B.P. (Stoffel, 1980), Vashon advance outwash and till, and Everson glaciomarine drift (Easterbrook, 1968).

Hansen and Mackin (1949) suggested that the Whidbey sediments originated as floodplain deposits formed from "very slow aggradation by meandering streams flanked by floodplain lakes and swamps" in which silty clay and peat were deposited. Pollen studies suggest that the Whidbey Fm. was an interglacial period with a climate somewhat warmer than the present (Hansen and Mackin, 1949; Hansen and Easterbrook, 1974; Heusser and Heusser, 1981).

More than 20 ^{14}C ages beyond the range of ^{14}C dating have been obtained from peat in the Whidbey Fm. Amino-acid ages of 97 ± 35, 96 ± 35 ka and 107 ± 9 ka have been obtained from shells in Whidbey marine sediments to the north (Easterbrook, 1968; Blunt et al., 1987). TL ages measured on clay from Whidbey fluvial sediments are shown below. (Berger and Easterbrook, 1993; Easterbrook, 1994).

TL Age (ka)	Locality
102 ± 38	Lagoon Point, Whidbey Island
106 ± 17	Blowers Bluff, Whidbey Island
142 ± 10	Northern West Beach, Whidbey Island
151 ± 43	Point Wilson, NW of Port Townsend

On route to STOP 1–2, return to Freeland and drive north on Highway 525 (which becomes Highway 20 northward) to the north side of Penn Cove. Turn right off Highway 20 at San de Fuca onto Penn Cove Road and continue to Penn Cove Park. Walk along the beach from here to Blowers Bluff.

STOP 1–2. BLOWERS BLUFF; POSSESSION DRIFT OVER WHIDBEY SILT/SAND; EVERSON GMD

Possession Glaciation. Massive, cliff-making, Possession till and glaciomarine drift overlie peat-bearing sand and silt of the Whidbey Fm. at mid-cliff and are overlain by Vashon advance outwash, till, and fossiliferous Everson gmd. (Easterbrook, 1968).

Amino-acid analyses of shells in massive, poorly sorted, till–like, Possession glaciomarine drift suggest a mean age of 80 ± 22 ka (Blunt and others, 1987; Easterbrook and Rutter, 1981, 1982).

Olympia Interglaciation. Nonglacial floodplain and lacustrine silt, clay, and peat, deposited in the lowland between the Possession and Fraser glaciations, provide the basis for definition of the Olympia Interglaciation (Armstrong et al., 1965) (also referred to as the Olympia nonglacial interval). Radiocarbon dates from Olympia sediments just east of here range from 22-28 ka ^{14}C yrs. B.P. (Hansen and Easterbrook, 1974; Easterbrook, 1976a, 1986; Johnson et al., 1996, Johnson et al., 2002).

Fraser Glaciation. The late Wisconsin Fraser Glaciation consists of several stades; (1) the Evans Creek Stade, an early alpine phase; (2) the Vashon Stade, the maximum advance of the Cordilleran Ice Sheet (CIS) (3) the Everson Interstade, an interval of deposition of glaciomarine drift (gmd) during deglaclation of the lowland; and (4) the Sumas Stade, consisting of several short readvances of the ice terminus before complete deglaciation (Easterbrook, 1963, 1969, 1992, Armstrong et al., 1965; Kovanen and Easterbrook, 2002a; Kovanen, 2002a).

Vashon Stade. The last major climatic episode during which drift was deposited by the CIS is defined as the Vashon Stade (Armstrong et al., 1965). It began with advance of the CIS into the lowland and ended with the beginning of marine and glaciomarine conditions of the Everson Interstade. Ahead of the advancing glacier, meltwater streams deposited an apron of outwash sand and gravel that was later overridden by the ice sheet and covered with till. Timing of the advance is limited by ^{14}C dates from sediments beneath Vashon till. These dates indicate that Vashon ice advanced southward across the Canadian border sometime after 17-18 ka ^{14}C yrs. B.P. and reached the Seattle area soon after 14.5 ka ^{14}C yrs. B.P. (Mullineaux et al., 1965; Deeter, 1979; Easterbrook, 1992; Porter and Swanson, 1998).

Grooves and striations carved on bedrock and elongate drumloidal hills and flutes indicate directions of movement of Vashon ice (Easterbrook, 1979). At its maximum, the Vashon ice extended to approximately 25 km (15 mi.) south of Olympia (Fig. 3). Ice was more than 1,830 m. (6,000 ft) thick near Bellingham and about 915 m. (3,000 ft) thick near Seattle.

Everson Interstade. The CIS underwent sudden, large–scale recession and downwasting about 14.5 ka ^{14}C yrs. B.P. The Puget lobe retreated northward from its southern terminus by backwasting past the Seattle area by about 14 ka ^{14}C yrs. B.P. (Fig. 4) (Rigg and Gould, 1957; Leopold et al., 1982). Rapid thinning of Vashon ice after the terminus had receded north of Seattle allowed marine water from the Strait of Juan de Fuca to flood the lowland (Fig. 4), floating the remaining ice and quickly disintegrating the remaining CIS all the way north to Canada. During this interval, Everson glaciomarine drift (gmd) was deposited over a large area in the central and northern Puget Lowland (Easterbrook, 1963, 1966a,b, 1968, 1969, 1971, 1976a,b, 1979, 1986, 1992, 2003; Dethier et al., 1995; Kovanen and Easterbrook, 2002a; Kovanen, 2002) and British Columbia (Armstrong and Brown, 1954; Armstrong, 1981, 1984; Armstrong and Hicock, 1980a, b).

The gmd consist largely of poorly sorted diamictons deposited when melting of floating ice released rock debris that accumulated on the sea floor. Unbroken, articulated, marine shells, some in growth positions, indicate that the gmd represents *in situ* deposition (Easterbrook, 1963, 1992). The origin of the gmd from floating ice in sea water was first recognized in Washington by Easterbrook (1963, 1966a, 1966b, 1969) and in British Columbia by Armstrong (1957, 1960, 1981, 1984). Since then, more than 150 ^{14}C dates from Washington and British Columbia fix the age of the Everson glaciomarine drift at 11,500 to 13,500 ^{14}C yrs. B.P., (Fig. 5) (Easterbrook, 1992, 2003), making it a valuable stratigraphic marker over the central and northern Puget Lowland.

Return to Highway 20 and drive west. Turn right on the Libby Road to and continue to West Beach.

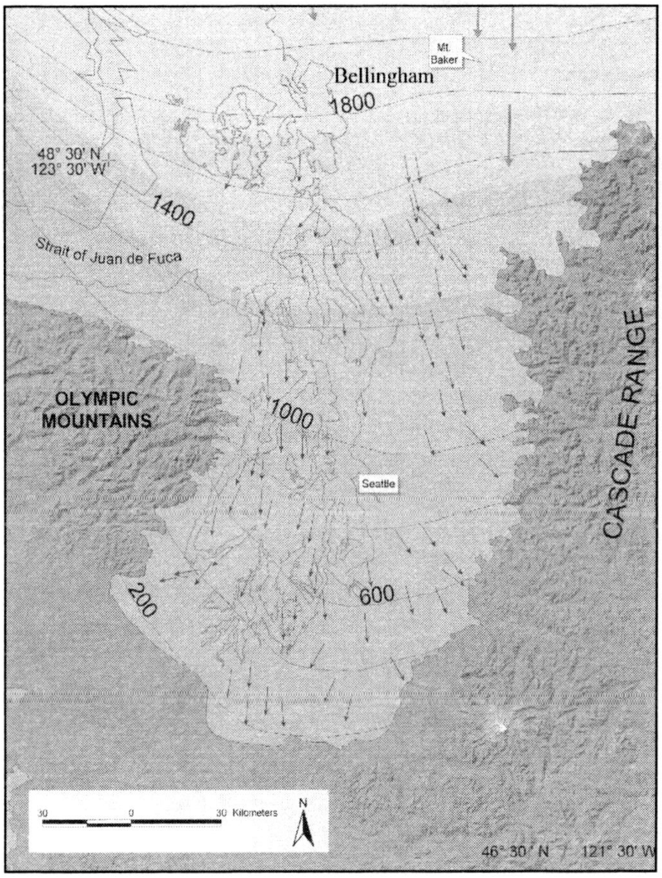

Figure 3. Shaded digital topographic model reconstruction of the Cordilleran Ice Sheet at its late Wisconsin maximum. (D.J. Kovanen and M. Price)

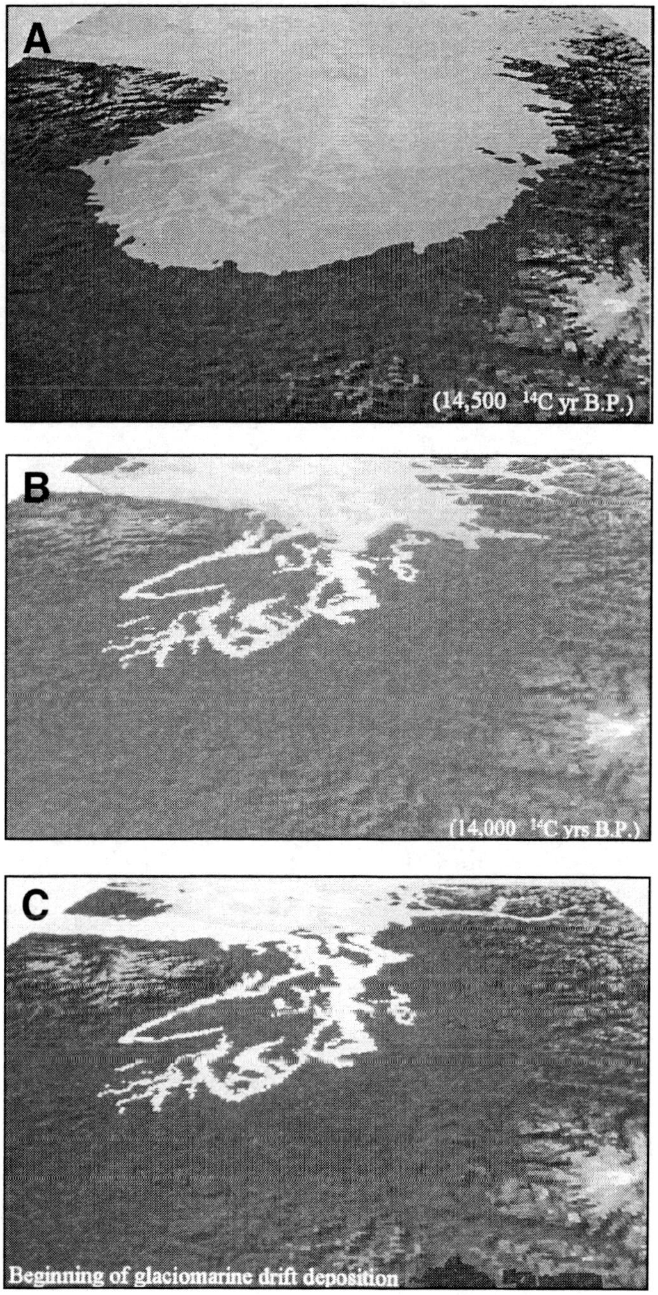

Figure 4. Retreat of the Cordilleran Ice Sheet. (Shaded digital topographic models by D. J. Kovanen and M. Price)

STOP 1–3. DEGLACIATION, EVERSON GLACIOMARINE DRIFT CHRONOLOGY, WEST BEACH

Sea cliffs just north of the park at West Beach consist of fossiliferous Everson gmd, dated at 12,535 ± 300 ^{14}C yrs. B.P. (uncorrected) (Easterbrook, 1968). Applying the new marine reservoir correction for this region (Kovanen and Easterbrook, 2002b) gives a corrected age of 11,400 ^{14}C yrs. B.P.

The first recognition and dating of glaciomarine drift on Whidbey and nearby islands in the 1960s were supported by 11 ^{14}C dates (Easterbrook, 1966a,b, 1968, 1969), and in the intervening years, 140 additional ^{14}C dates were obtained (Fig. 5). The inescapable conclusion from Figure 5, is that the distribution of dates indicates that glaciomarine conditions were essentially contemporaneous over the whole region and that the age of final emergence was about 11,500 ^{14}C yrs. B.P. for the entire area. Thus, the progressive, backwasting, calving terminus concept of Domack (1983) cannot be correct.

Notwithstanding all of these ^{14}C dates, Swanson and Caffee (2001) contend that deposition of the Everson ended 13,500 ^{14}C yrs. B.P., fully 2,000 years earlier. They ignore the 150 dates shown on Figure 5 and based their conclusion on only four ^{14}C dates. The significant implication of this is that they use this time as the basis for calculating a production rate of ^{36}Cl, which they then use for measuring the ages of erratic boulders in the region. Because the base of their production rate calculations is off by 2000 years, their cosmogenic dates are off by 20% (Easterbrook, 2003).

The surface topography in the Partridge Point area is deeply indented with kettles (Fig. 6), many >30 m deep. Just south of West Beach Park, wave erosion has breached a 30–m–deep ket-

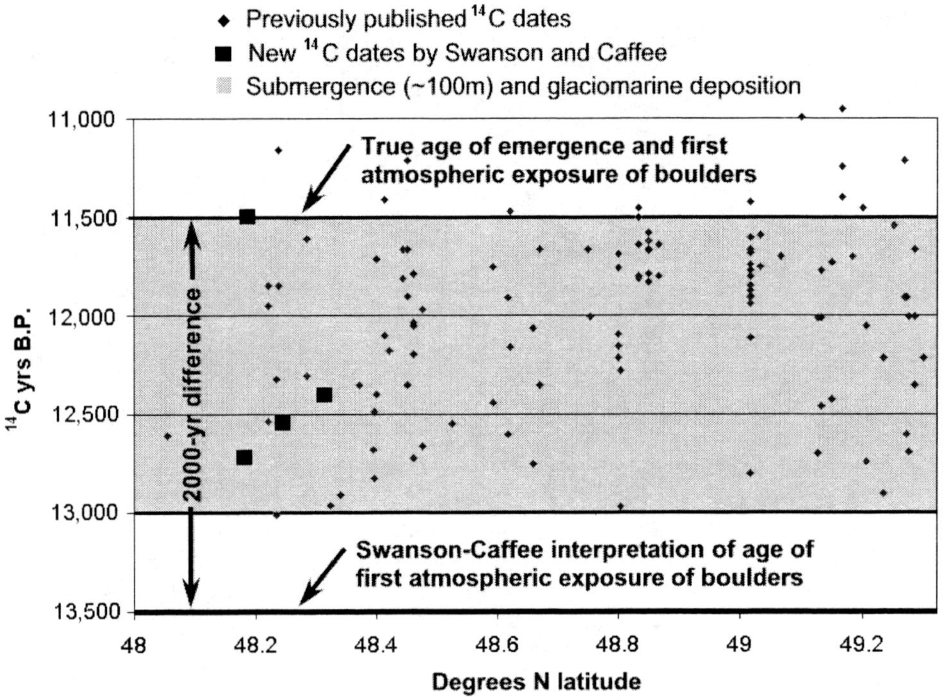

Figure 5. Chronology of Everson glaciomarine drift deposition. (Easterbrook 2003)

tle, exposing the Partridge Point gravel that makes up the kettled topography. Collapse structures are common around the margins of the kettles where sediments slumped into the cavities made by melting of the buried blocks of ice. The gravel typically has strong foreset bedding, with dips up to 20–25°, and contains pumice clasts and many small shell fragments (Easterbrook, 1966a,b, 1968; Carlstad, 1992).

Figure 6. 30m–deep kettles in west–dipping deltaic gravel at Partridge Pt. (Photo by D.A. Rahm, courtesy of EPIC)

The top of the Partridge Point gravel, where not indented by kettles, consists of a flat surface about 55–60 m above present sea level. Relict meltwater channels on this surface are shown on the remarkably detailed LIDAR imagery rendered by Kovanen (Fig. 8). Everson gmd caps the Partridge Point gravel in the bluffs just south of the park (Easterbrook, 1968).

SEA–LEVEL CHANGES, ISOSTATIC MOVEMENTS, AND ICE–FLOW INDICATORS ON WHIDBEY AND CAMANO ISLANDS, PUGET LOWLAND: GEOLOGIC DATA FROM LIDAR–AIDED OBSERVATIONS

Dori J. Kovanen, The University of British Columbia, Department of Geography, Vancouver, British Columbia V6T 1Z2, Canada, e–mail: dkovanen@geog.ubc.ca

Glacial and marine features

High–resolution, LIDAR (Light Detection and Ranging), digital data of Whidbey and Camano Islands reveal preliminary geologic evidence of former sea–levels and ice–flow directions (Fig. 7A). Raised shorelines in this area are found up to c. 88 m above present sea level and are believed to record glacio–isostatic movements related, in part, to glacial rebound (as well as eustatic and tectonic processes which operate on different spatial and temporal scales) of the Cordilleran Ice Sheet, which inundated the Puget Lowland during the last glaciation.

The timing of ice retreat on central Whidbey Island is constrained between 13,650 ± 350 to 11,850 ± 240 ^{14}C yr B.P. by the

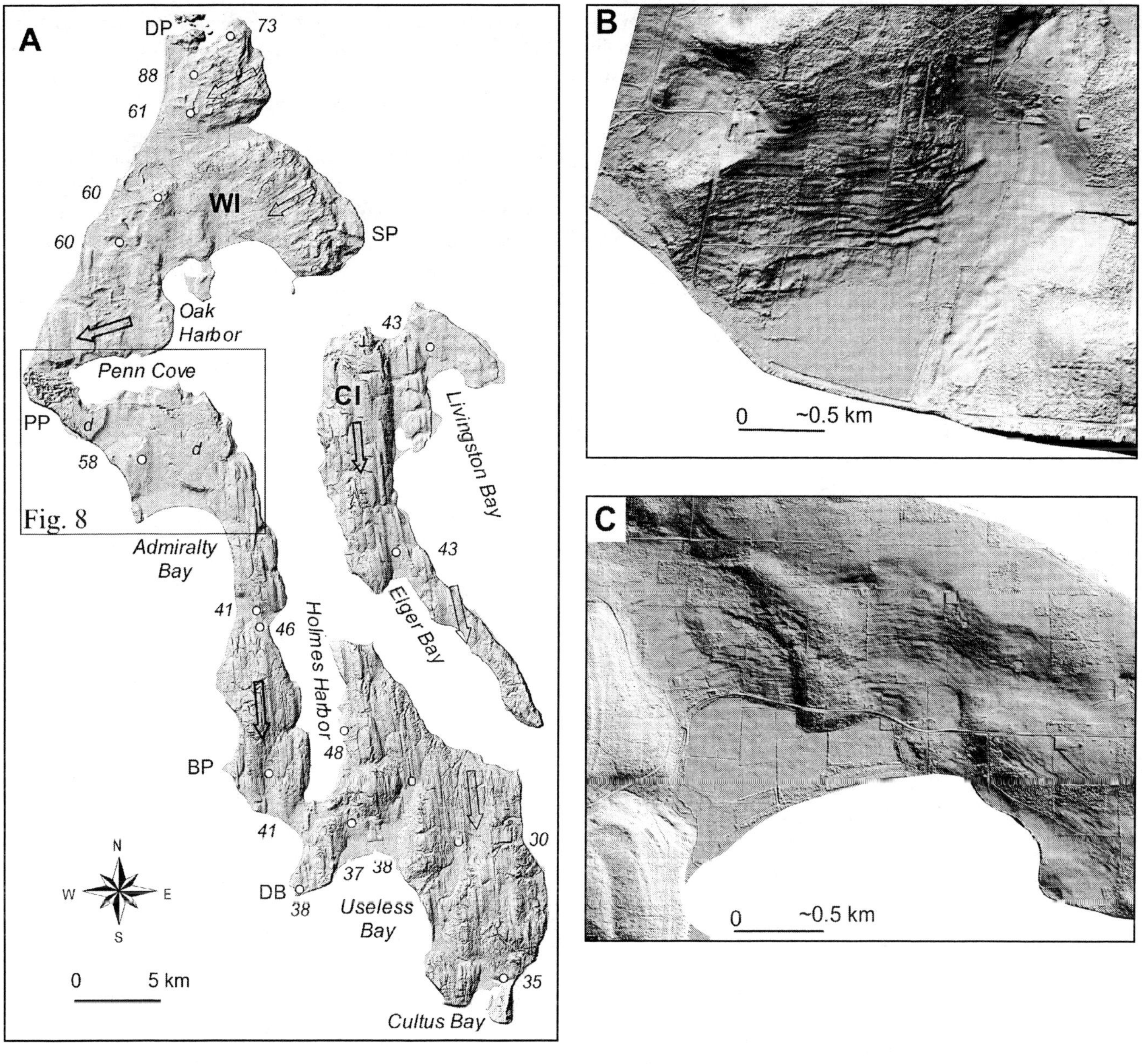

Figure 7. (A) Shaded digital topographic model of Whidbey and Camano Islands from LIDAR data (2 m resolution). White–centered dots indicate the location of raised shoreline sequences and the corresponding number is the approximate upper elevation (m). Arrows indicate inferred ice flow directions and "d"indicates the location of raised deltas. WI–Whidbey Island; CI–Camano Island; DP–Deception Pass; SP–Strawberry Point; PP–Partridge Point; BP–Bush Point; DB–Double Bluff. (B) Raised shoreline sequence northwest of Oak Harbor. (C) Raised shoreline sequence at Livingston Bay.

ages of marine bivalves in glaciomarine sediment (Easterbrook, 1969; Domack, 1983; Dethier et al., 1995; Swanson and Caffee, 2001). Correcting these shell dates for a radiocarbon marine reservoir value of –800 years (Robinson and Thompson, 1981; Southon et al., 1990), yields an age range of ~12,850–11,050 ^{14}C yr B.P. for deglaciation, which may be conservative (see Kovanen and Easterbrook, 2002b). Changes in relative sea level occurred rapidly following deglaciation. The shoreline sequences shown in Figure 7B and 7C have not yet been directly dated and are the subject of future work. Close examination of many of the gullies and raised deltas reveal that they are commonly graded to, and truncated at, an elevation that is equal to the highest, local, raised shoreline elevation.

Superimposed bedforms (i.e., drumlins, flutes, till lineations) record two flow lines with different orientation, suggesting more than one ice–flow event (Fig. 7A): (1) An "older" event with a north–south, ice–flow direction (Vashon advance?) and (2) a relatively younger event on northern Whidbey with a northeast–

southwest ice–flow direction (during deglaciation). These overlap the north–south ice–flow indicators and therefore did not form contemporaneously. A well–defined, marine, ice–margin exists along the south and west side of Penn Cove (Fig. 8) (see also Hewitt and Mosher, 2001 for paleoenvironments of the eastern Juan de Fuca Strait during the last deglaciation).

Significance and conclusions

The use of high–resolution LIDAR data allow the first recognition of raised shoreline sequences on Whidbey and Camano Islands. The preliminary marine–limit data are generally consistent with the maximum elevation of shell–bearing stony muds interpreted as glaciomarine drift (Easterbrook, 1968, 1992) and postglacial isostatic rebound (i.e., Thorson, 1989) that occurred late in the regional deglacial history. Based on the many closely–spaced shoreline sequences, the argument is made that the land response to transient ice loading was dominantly elastic and rebound is accomplished immediately upon unloading. In this active tectonic regime (Johnson et al., 1996, 2002), the relations between isostatic depression, postglacial rebound, and postglacial eustatic sea level rise are complex. These sites (Fig. 7A) reflect important aspects of uplift and the geologic controls, geomorphic processes, landscape evolution, and environmental change during the late Pleistocene and early Holocene. Future work includes establishing a detailed geochronology of these environmental changes so that they can be compared with local and regional variations in sea level (i.e., Easterbrook, 1963, 1992; Dethier et al., 1995) and with curves of glaciated North America (Dyke and Peltier, 2000) to better understand the geographic patterns of sea level changes during deglaciation.

ACKNOWLEDGEMENTS

The Puget Sound LIDAR Consortium (especially Jerry Harless and Diana Martinez) is thanked for providing the LIDAR data, as well as timely and kind assistance.

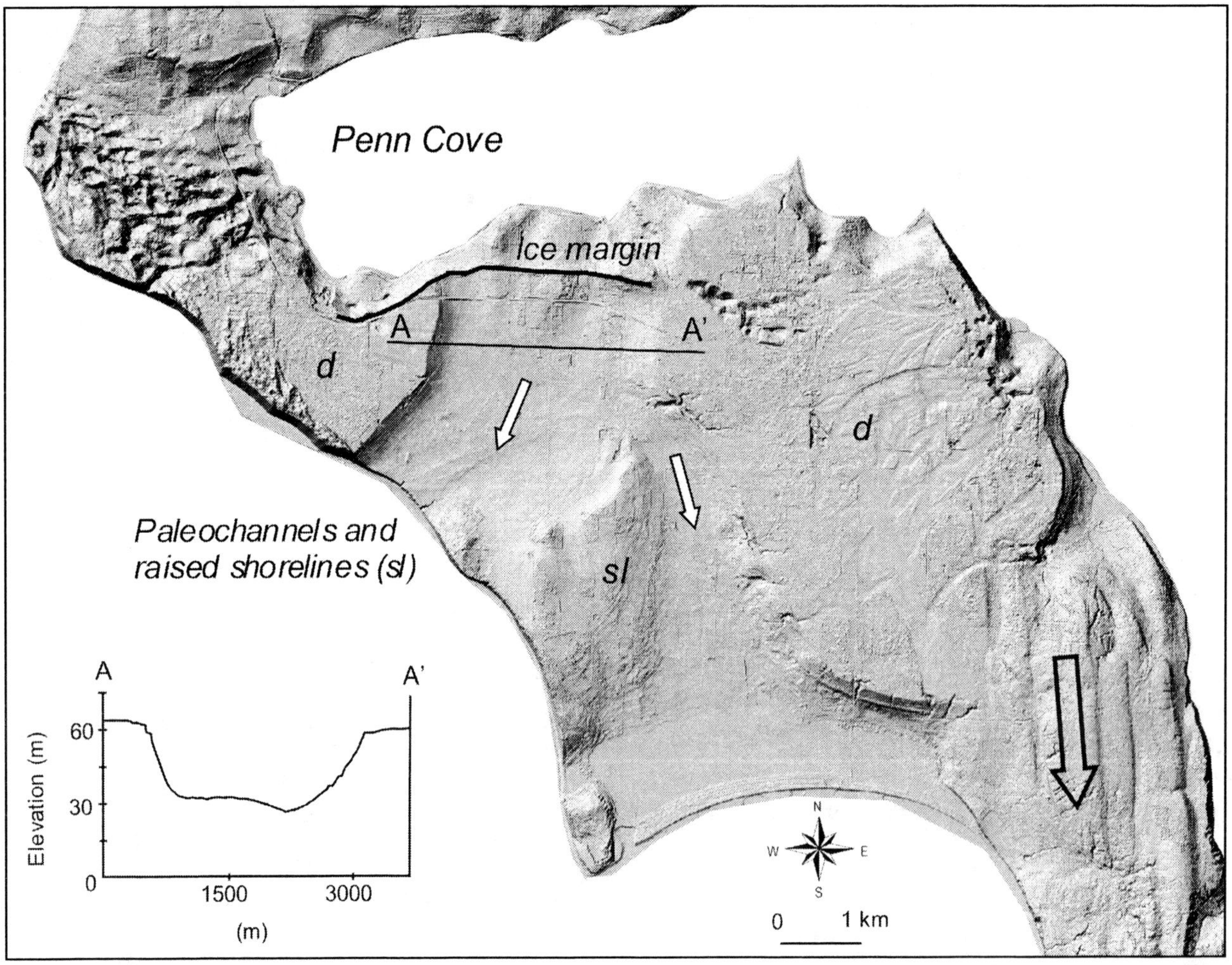

Figure 8. Shaded digital topographic model of the Penn Cove area showing the major geomorphic features; paleochannels (white arrows), raised deltas (d), late glacial ice margin, and raised shoreline sequences (SL). The inferred ice–flow direction in the early stages of Vashon glaciation is north–south.

STOP 1–4. LATE PLEISTOCENE RAISED SHORELINES

The remarkable detail shown on the LIDAR imagery derived by Kovanen (previous section) has opened a new chapter in the understanding of late Pleistocene glacial events and relative sea level changes. Except for the delta surfaces at Penn Cove, the raised shorelines (Fig. 7B,C) that she has discovered on the LIDAR imagery have never been observed before. Also striking is the dramatic shift in ice-flow direction northeast of Penn Cove (Fig. 7A).

That Penn Cove was occupied by an ice-mass that shed sediment into a marine delta and onto stagnant buried ice at Partridge Point has long been recognized (Easterbrook 1966, 1968; Carlstad, 1992), but the details of the ice margin and raised delta surfaces (Fig. 8) are now nicely shown on the LIDAR imagery. The change in ice-flow direction from N-S to NE-SW, shown so vividly on the LIDAR images, means that as the Vashon glacier thinned rapidly during deglaciation and began to disintegrate in the Strait of Juan de Fuca and northward, a remnant of ice was grounded on the eastern part of the lowland long enough to build the features seen near Penn Cove. Because it was no longer directly fed by ice flow from the north, the grounded remnant ice assumed a southwesterly flow direction transverse to the Vashon direction (Fig. 7A), terminating in the sea at the Penn Cove deltas.

Close inspection of the raised shorelines on the LIDAR images shows a fairly regular step-like progression from the highest down to present sea level. Thus, isostatic rebound in this region seems not to have been constant, but rather as a series of short upward jogs, leaving a shoreline at each one.

Drive to the north end the island. Turn right and drive east to I–5. Turn left and drive north on I–5 toward Bellingham.

STOP 1–5. SUMAS OUTWASH CHANNEL INCISED INTO EVERSON GMD ACROSS SHORELINES

Sumas Stade. Emergence of the northern lowland from the sea and readvance of the CIS over its own outwash defines the beginning of the Sumas Stade (Armstrong et al., 1965). Four Sumas phases, spanning the Younger Dryas and part of the Allerød, have now been recognized (Easterbrook and Kovanen, 1998; Kovanen and Easterbrook, 2002a; Kovanen, 2002). We will see the field evidence and chronology of these on DAY 2.

At the Samish River, gmd extends to elevations of 65 m, about the same as the highest shorelines in the area, and is deeply incised by the Samish River, which is badly underfit (Fig. 9). It flows with present–day meander radii of about 0.15 km (0.1 mi.) on a valley floor having meander radii >1.7 km (1 mi.). A much larger stream than the present Samish River obviously made the Samish valley. The meanders are incised into Everson gmd and cut across beach ridges at the marine limit of 65 m (215 ft.) before terminating at a marine delta about 30 m (100 ft.) above

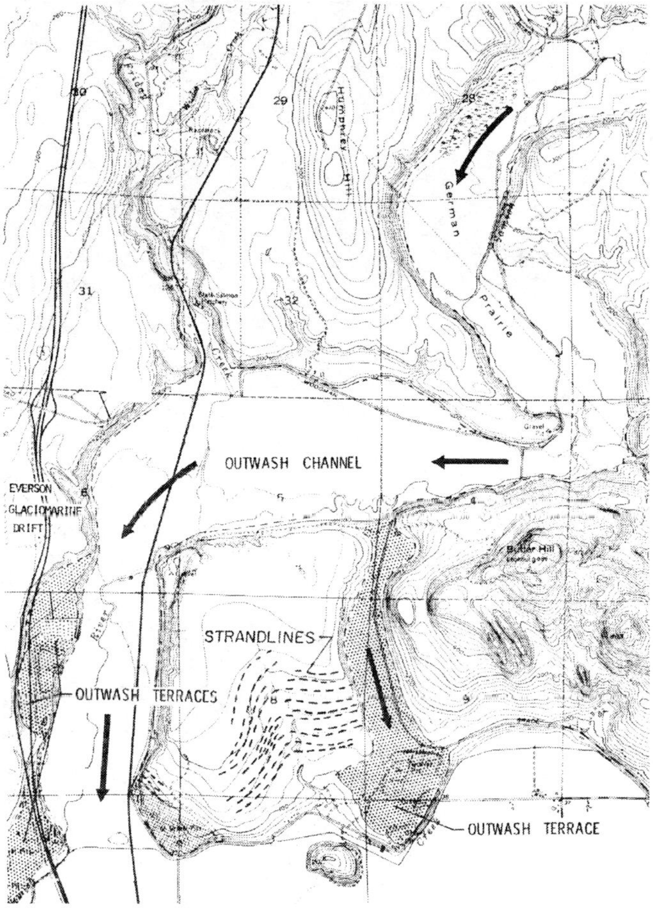

Figure 9. Sumas outwash channel cut into Everson glaciomarine drift and across late Everson shorelines.

present sea level (Easterbrook, 1979, 1994). The outwash was graded to a sea level ~35 m lower than that of the gmd. The outwash channel and terraces belong to the Sumas Stade, C^{14}–dated elsewhere at about 10–11,400 ^{14}C yrs. B.P. A limiting age for similar strandlines nearby is given by a date of 11,700 ± 110 ^{14}C yrs. B.P. from basal peat at 47 m (Siegfried, 1978).

Arrive in Bellingham. End of DAY 1.

DAY 2 EVERSON RELATIVE SEA LEVEL CHANGES AND SUMAS READVANCES

EVERSON INTERSTADE

Everson glaciomarine drift (gmd) was deposited from floating ice in the Fraser Lowland (Fig. 10) to elevations up to 180 m. (600 ft) above present sea level and extended more than 10 km (6 mi.) up the Nooksack valley into the North Cascades (Fig. 11) (Easterbrook, 1963, 1992, 1994; Kovanen and Easterbrook, 2001).

The Everson in the Fraser Lowland consists of two fossiliferous glaciomarine drifts, the Kulshan (lower) and Bellingham

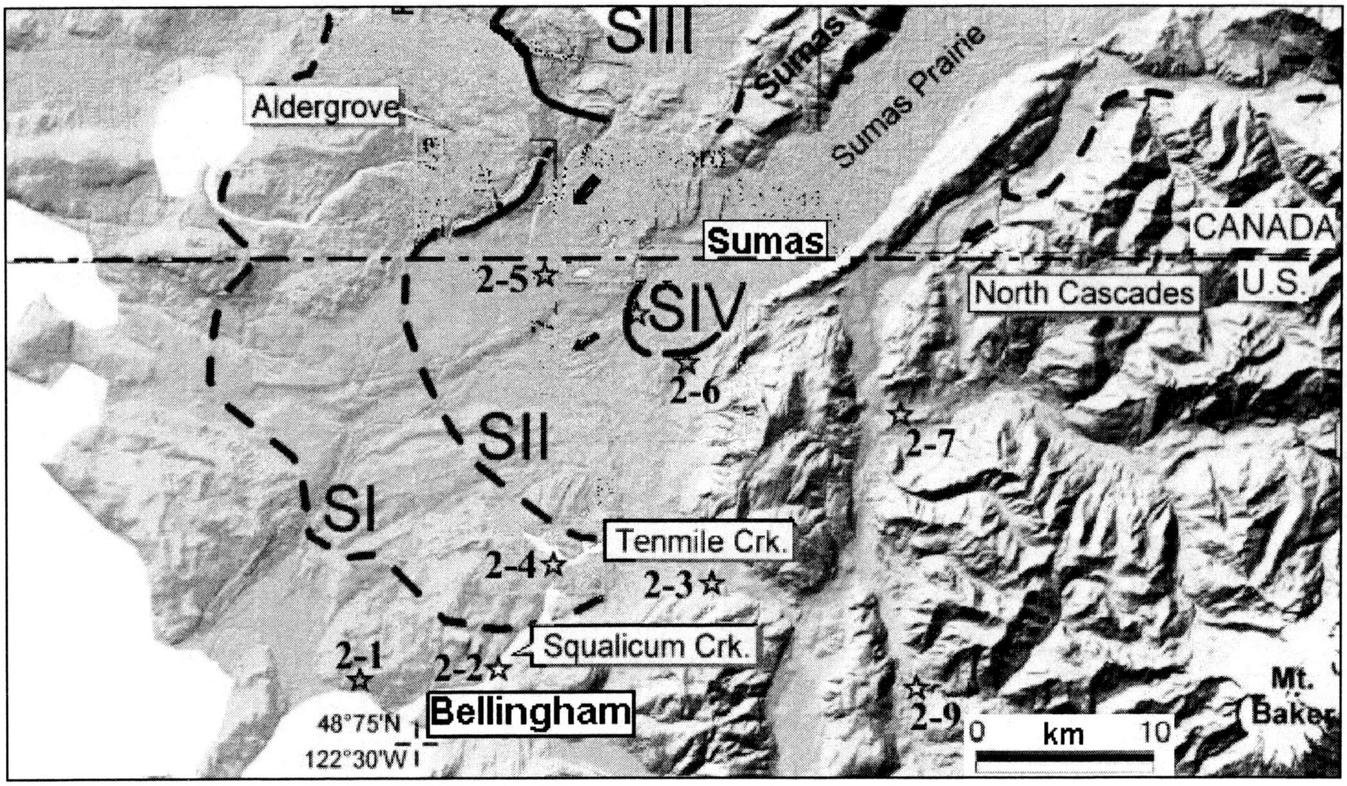

Figure 10. Location of stops in the Fraser Lowland and position of ice margins of the four Sumas phases. (Modified from Kovanen and Easterbrook, 2002a)

(upper), separated by subaerial, fluvial, Deming sand with paleochannels, crossbedding, *in situ* tree stumps (Fig. 12), and basal peat lacking pollen from any salt-tolerant plants. The sandwiching of the non–marine Deming sand between the two glaciomarine drifts demonstrates extraordinary fluctuations of relative sea level in this area over short periods of time. Relative sea level stood at ~90 m (300 ft) above present level during deposition of the Kulshan gmd, dropped to within 8-12 m (25-40 ft.) of present sea level, then rose again to more than 180 m (600 ft.), and finally dropped to present sea level by 11,400 ^{14}C yrs. B.P. during the Sumas Stade (Fig. 11) (Easterbrook, 1963, 1992; Easterbrook and Kovanen, 1996; Kovanen and Easterbrook, 2002a). Sumas outwash channels that cut deeply into Bellingham gmd were graded to a sea level close to that of the present, indicating that the 180 m (600 ft.) of post-Bellingham-gmd emergence took place quickly.

Clearly, a complicated combination of isostatic, eustatic, and tectonic events is needed to explain the ups and downs of relative sea level spanning about 2000 years in the Everson Interstade (Easterbrook 1963, 1992). They cannot be explained by any combination of only isostatic and eustatic changes because of submergences that occurred during a times of isostatic uplift. A strong local tectonic influence is required to adequately explain these late Pleistocene relative sea level changes. Interestingly, analysis of recent seismic records in this area show a remarkable pattern (more than 400 earthquakes since 1969).

SUMAS STADE

Following deposition of the Everson drift, the Cordilleran Ice Sheet readvanced several times during the Sumas Stade, depositing moraines, till, ice-contact sediments, and outwash (Easterbrook, 1963, 1969, 1971, 1976a, 1986, 1992). This readvance reversed the wholesale deterioration of the ice sheet that began shortly before ~14,500 ^{14}C–yrs B.P. The Sumas Stade began with emergence of the area from the Everson sea and readvance of the Cordilleran Ice Sheet (Armstrong et al., 1965). The multiple advances of Sumas ice have been documented by Easterbrook (1994), Easterbrook and Kovanen, (1998), Kovanen and Easterbrook, 2002a, Kovanen (2002), and Mark and Ojamma (1980).

Multiple moraines are associated with oscillations of the Cordilleran Ice Sheet in the Fraser Lowland. The paleogeography of the ice margin and timing of ice retreat during the Sumas Stade is reconstructed and bracketed by 70 radiocarbon dates, which are secured by morphologic and stratigraphic evidence. Four topographically distinct phases of the Sumas readvances have been documented (Fig. 13) (Easterbrook and Kovenan, 1998; Kovanen and Easterbrook, 2002a; Kovanen, 2002).

Leave WWU campus, drive northeast on Garden St., and turn left (northwest) on Holly St. Holly St. becomes Marine Drive to the north. Turn left on an unnamed secondary road north of

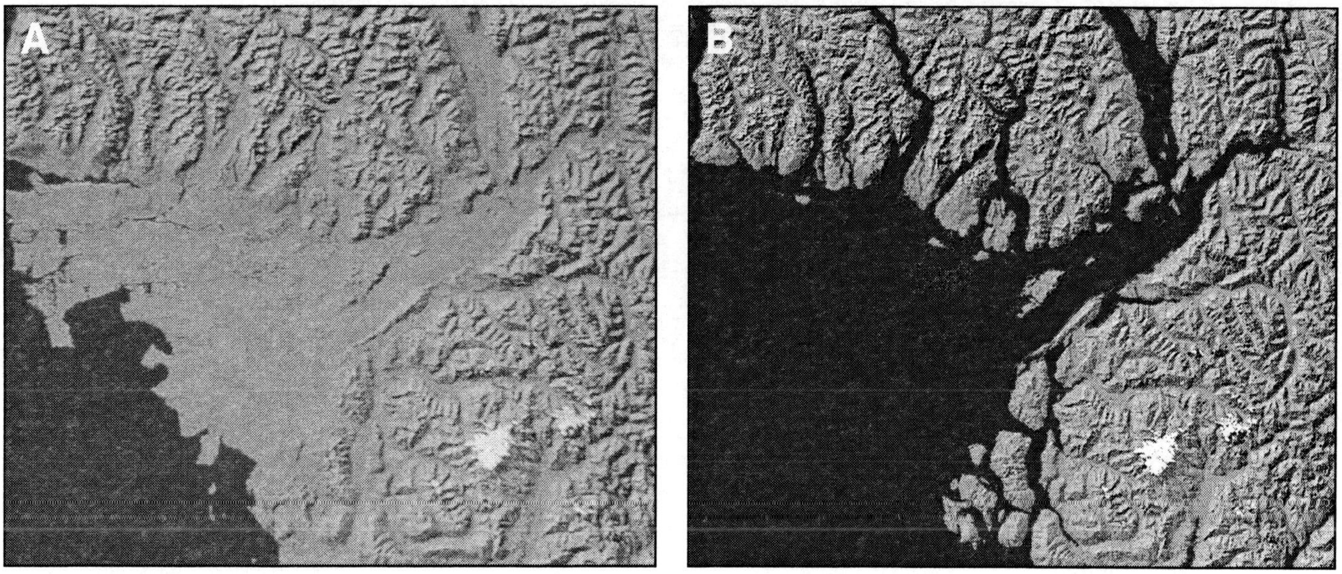

Figure 11. A. Present day shoreline. B. Everson shoreline at 180 m. (Shaded digital topographic model by D.J. Kovanen)

the old cement plant. Walk down the trail to the beach and proceed northward up the beach to sea cliff exposures.

STOP 2-1. BELLINGHAM GLACIOMARINE DRIFT OVERLYING BEACH DEPOSITS OF DEMING SAND, BELLINGHAM BAY

The purpose of this stop is to examine the Kulshan-Deming-Bellingham sequence and establish sea level during deposition of the Deming beach deposits. The sea cliff exposes Bellingham glaciomarine drift (gmd) lying on beach deposits of the Deming sand (Fig. 14) (Easterbrook, 1963). Kulshan gmd is exposed beneath Deming sand near beach level to the south.

The Bellingham gmd consists of stony silt/clay containing marine fossils and makes up the surface topography in the upland to the east where it extends to elevations up to 180 m (600 ft.) along the Cascade foothills. Six ^{14}C dates from wood and marine shells in the gmd range from 11,580 to 11,830 ^{14}C yrs. B.P. and marine fossils in beach deposits of the Deming sand here were dated at 11,685 ± 85 and 11,760 ± 85 (Kovanen and Easterbrook, 2002a). The fossiliferous Deming beach deposits range in elevation from 7.5 m (25 ft.) to 20 m (65 ft.) and establish sea level limits for the Deming sand. Evidence for the beach origin of these deposits is documented in Easterbrook (1963) and Weber (2001). Fossiliferous Kulshan gmd, exposed in bluffs just above the modern beach and in the wave platform seaward from the bluffs, underlies the Deming sand. Shells from the Kulshan have been dated at 12,210 ± 82 ^{14}C yrs. B.P.

The significance of the Kulshan-Deming-Bellingham sequence is that it replicates the same sequence exposed at the Everson type section (STOP 2–2) and establishes the position of sea level for the Deming sand between the two gmds.

Figure 12. Rooted stump in Deming sand between Kulshan and Bellingham gmd at the Everson type locality.

Drive east on Sunset Drive (Mt. Baker Highway). Turn left and drive north on the Hannegan Road. Across the Squalicum outwash channel (Sumas I)

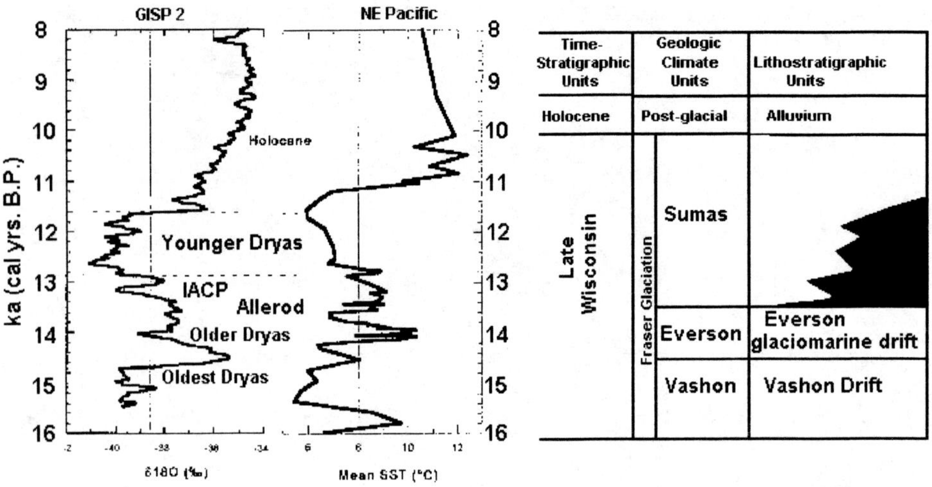

Figure 13. Correlation of the chronology of Sumas phases with the GISP2 ice core and sea surface paleo–temperatures in the northeast Pacific. (From Kovanen and Easterbrook, 2002a)

STOP 2-2

This outwash channel (Fig. 10) originated just to the east at a brief stand of the earliest phase of the Sumas Stade (Sumas I). The channel incises Bellingham gmd and is thus post–11,600 ^{14}C yrs. B.P. and is older than Sumas II (11,400 ^{14}C yrs. B.P.). No well–defined moraine occurs at the head of the channel, it just pinches out at the position of the glacier terminus.

Turn right on Van Wyck Road, cross Squalicum outwash channel, and continue east on the Mt. Baker Highway.

STOP 2-3. EVERSON INTERSTADE TYPE LOCALITY

The type locality of the Everson Interstade (Armstrong and others, 1965) includes fluvial Deming sand sandwiched between Kulshan and Bellingham gmd (Easterbrook, 1963, 1992) (Fig. 15), demonstrating at least two submergences and two emergences of the region in only 1500–2000 years (the "*yo–yo*" effect). The base of the section consists of 5 m (15 ft.) of stratified, crossbedded, pebbly sand interbedded with thin, silt beds (Fig. 15), probably fluvial overbank deposits. A twig and a thin, *in situ*, organic mat a few mm thick interbedded with silty-clay were dated at 12,070 ± 80 and 12, 185 ± 80 ^{14}C yrs. B.P., providing a lower limiting age for the Kulshan gmd and indicating that this area was above sea level before it was submerged to deposit Kulshan gmd. The significance of this is that it demonstrates that neither the Kulshan nor the Bellingham gmds could have resulted from a retreating, calving Vashon ice sheet.

Kulshan gmd consists of 9–12 m (30-40 ft.) of poorly sorted pebbly silt and clay containing widely scattered marine shells that overlies the basal sand. No ^{14}C dates have been obtained form the Kulshan here, but shells in the same unit 6 km (3.7 mi.) to the east were dated at 11,870 ± 380 years. The top of the Kulshan here is about 75 m (250 ft.) above sea level. Adding another 30 m for probable water depth gives a sea level about 100 m (350 ft.) higher than present sea level for the Kulshan.

Figure 14. Bellingham glaciomarine drift overlying Deming sand at Bellingham Bay.

About 10 m (33 ft.) of Deming sand overlies the Kulshan gmd (Figs. 15, 16). It consists mainly of crossbedded sand with some interbedded silt Prominent channeling may be seen in bluff exposures. A peat bed about 20-30 cm (8-12 in.) thick at the contact between the Kulshan and Deming units contains *in situ* rooted stumps (Fig. 12). Additional evidence of the nonmarine origin of the Deming sand has been documented in Easterbrook (1963, 1992).

Beach deposits in Deming sand at Bellingham Bay (STOP 2–1) indicate that sea level at this time was about 25–65 ft. above present sea level. ^{14}C ages from the Deming sand at

Figure 15. Type section of the Everson glaciomarine drift.

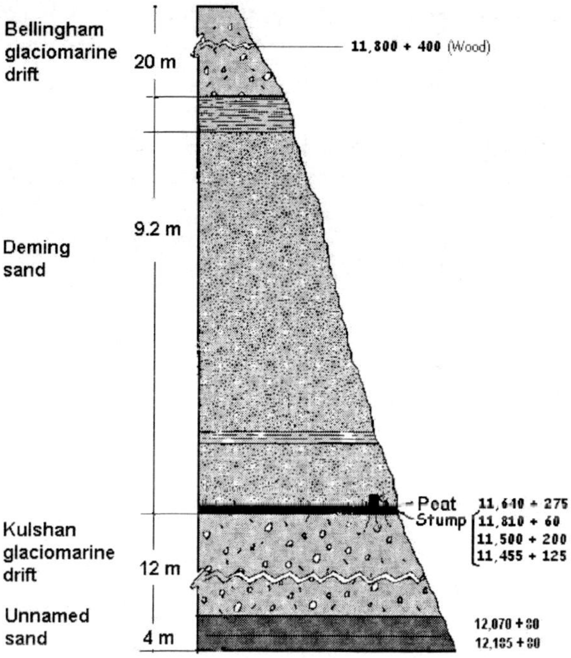

Figure 16. Stratigraphic section at the type locality of the Everson glaciomarine drift.

Bellingham Bay are 11,685 ± 85 and 11, 760 ± 85 ^{14}C yrs. B.P., and the rooted stump at the Everson type locality is 11,810 ±60 ^{14}C yrs. B.P. This means that the Deming sand at the Everson type locality must have been deposited ~60 m (200 ft,) above sea level at that time.

Deming sand is overlain by about 20 m of massive, poorly sorted, Bellingham gmd. Marine fossils, mostly *Nuculana* and *Clinocardium*, are more abundant than in the Kulshan gmd, but are widely scattered. An allocthanous piece of wood in the middle portion of the Bellingham gmd was dated at 11,800 ± 400 years (Easterbrook, 1963). Outcrops of Bellingham gmd extend to elevations of 180–210 m (600–690 ft.) above present sea level, and adding another 30 m for probable water depth gives a relative sea level of 210-240 m (700-800 ft.) above present sea level, a relative submergence of >200 m (>660 ft.) from Dem ing to Bellingham time. A similar 'sandwich' of fluvial sediment between two gmds occurs in British Columbia (Fig. 17), demonstrating that this is a regional feature. Ten ^{14}C dates from the Bradner pit in B.C. range from 11,600 to 11,800 ^{14}C yrs. B.P., almost identical to the ages from the Everson type locality and Bellingham Bay.

Post-Bellingham gmd relative sea level dropped abruptly about 11,500 ^{14}C yrs. B.P. Sumas outwash, graded to a sea level close to that of the present, was then deposited on Bellingham gmd and outwash channels were incised into the gmd, indicating yet another abrupt relative sea level change of about 200 m. Remarkable as they may seem, these ups and downs of relative sea level appear to be irrefutable–the physical evidence for them can be seen in the single riverbank exposure at the Everson type locality. No correlation of strata or extrapolation of data is required.

Return to Mt. Baker Highway and drive west to the Smith Road intersection. Turn right and continue to the Starry Road. Turn right and drive 1.7 km (1 mi.) to the Kulshan Golf Club on the right. Pull into the parking lot.

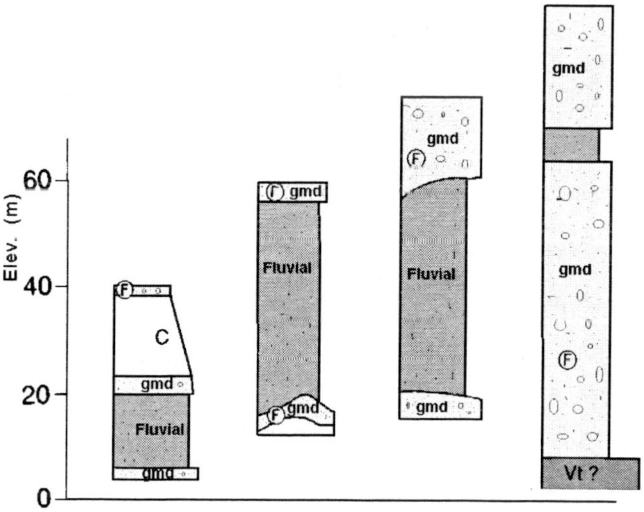

Figure 17. Fluvial sand between two glaciomarine drifts in southwestern B.C. The fluvial sand in B.C. contains rooted stumps that have the same ^{14}C age as the Deming at the type locality and at Bellingham Bay. (Modified from Armstrong, 1984)

STOP 2-4. OVERLOOK OF TENMILE CREEK SUMAS II OUTWASH CHANNEL.

The purpose of this stop is to verify that deep, Sumas, meltwater channels incise the Bellingham gmd, indicating that emergence closely followed deposition of the Bellingham gmd and providing a lower limiting age for the Sumas II phase. The Sumas outwash channel is incised 33m (100 ft.), exposing Bellingham gmd overlying Deming sand in the valley walls. Basal peat from a 6-9 m-(20-30 ft.)-thick bog that floors the channel was dated at 11,080 ± 100 and 11,113 ± 88 ^{14}C yrs. B.P. (Fig. 18). These dates correlate well with the age of a prominent Sumas II moraine in southwestern British Columbia. A basal bog date of 10,400 ± 85 ^{14}C yrs. B.P. was obtained from Fazon Lake, a large kettle just to the north.

Drive west on Axton Road, turn right on Hannegan Road. Cross peat-filled Sumas IV outwash channels, continue through Lynden to near the Canadian border.

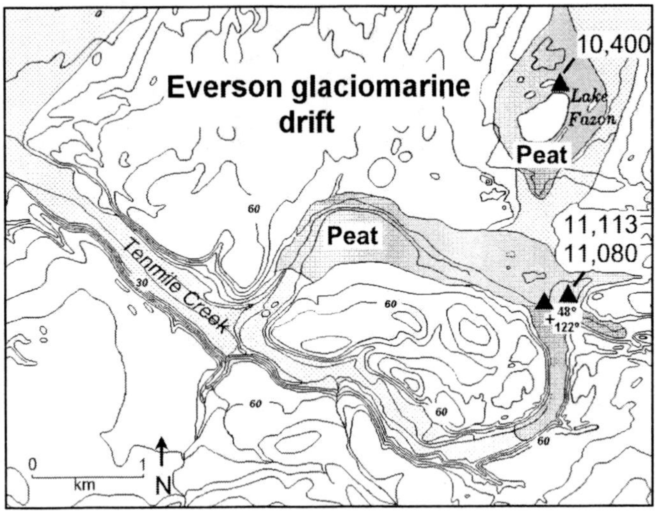

Figure 18. Sumas II outwash channel incised ~30 m in Bellingham gmd and filled with peat. (Kovanen and Easterbrook, 2002a)

Figure 19. Sumas II moraine at the Canadian border (road on the left side). (Photo by D.A. Rahm courtesy of EPIC)

STOP 2-5. SUMAS II MORAINE, SUMAS III OUTWASH PLAIN, ^{14}C LIMITING DATES

Just across the border is an elongate, tapering, Sumas II moraine (Figs. 19, 20) (Armstrong, 1981; Kovanen and Easterbrook, 2002a, Kovanen, 2002). The moraine rests on Everson gmd from which many ^{14}C dates on wood and shells have been obtained, clearly limiting the Sumas II advance at younger than 11,600 . ^{14}C yrs. B.P.

A ^{14}C date of 11,413 ± 75 ^{14}C yrs. B.P. from basal peat in an outwash channel from the moraine (Fig. 20) indicates that the glacier had retreated back from the end moraine by 11.4 ka, cutting off meltwater flow in this channel. A ^{14}C date of 10,980 ± 250 ^{14}C yrs. B.P. from basal peat in a channel *behind* the moraine indicates that the glacier had receded to the Sumas III moraines to the north by this time.

We are presently on the broad Abbotsford outwash plain that heads at Sumas III moraines farther north (Fig. 21). The significance of the position of the outwash plain is that lies behind the Sumas II moraine (Fig. 21b) and must therefore have retreated 20 km to the Sumas III moraines. ^{14}C dates of basal bogs on the outwash plain indicate that the Sumas III phase must be Younger Dryas.

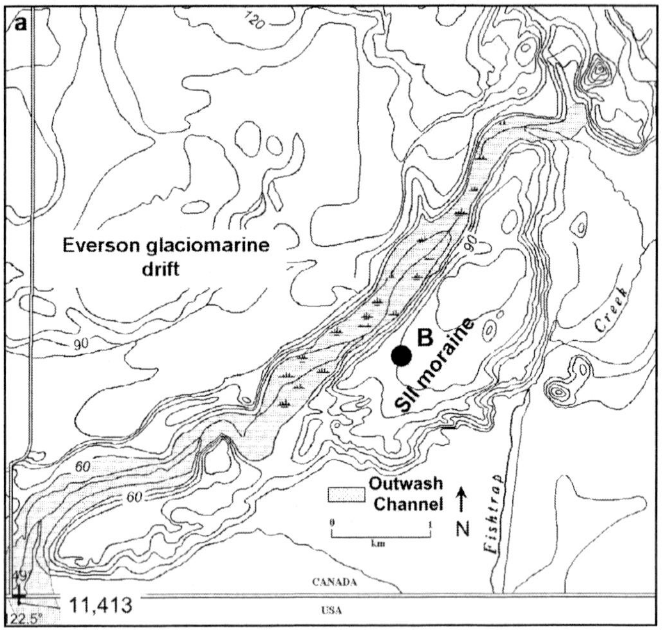

Figure 20. Sumas II moraine at the Canadian border. (Kovanen and Easterbrook, 2002a)

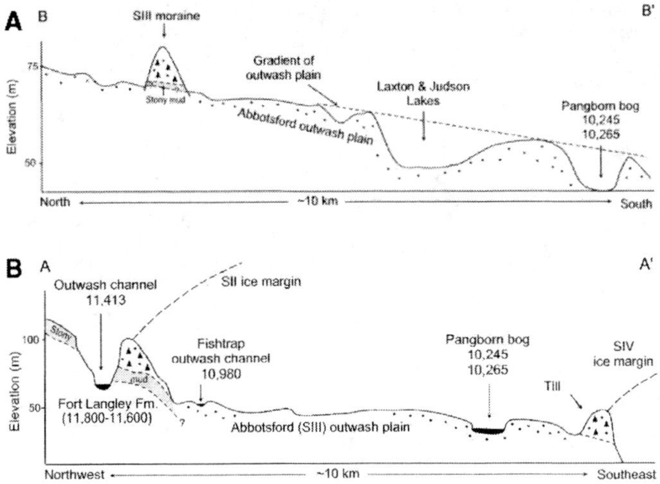

Figure 21. Profiles of the Sumas moraines. (Kovanen and Easterbrook, 2002a) A. N–S profile from Sumas SIII moraines to the Abbotsford outwash plain. B. E–W profile from the Sumas SII moraine to the Sumas SIV moraine.

Drive east on Halverstick Road past Pangborn bog to Clearbrook (Sumas IV moraine across Sumas III outwash channel).
Turn right off Halverstick Road onto road to Clearbrook, then to Van Buren Road. Turn left on Frick Road and take road across the front of the Sumas IV moraine.

STOP 2–6. SUMAS IV MORAINE AND OUTWASH TERRACE

The Sumas IV moraine consists of a hummocky, linear ridge that extends in an east-west direction away from the mountains for about 3.5 km (2.2 mi.) at the head of an outwash plain that slopes southward away from the moraine (Fig. 22).. Erosion has removed it from the central part of the Sumas Valley, but a remnant occurs on the west side of the valley, filling the distal end of a Sumas III meltwater channel (Fig. 22), showing that it represents a readvance of the glacier. This is the final phase of the Sumas advances and represents a late Younger Dryas readvance of the CIS.

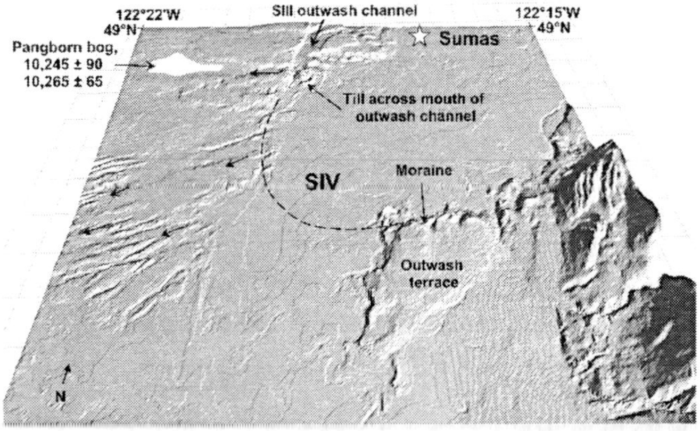

Figure 22. Sumas IV moraine and outwash terrace. (Kovanen and Easterbrook, 2002a).

Continue east on Telegraph Road-Easterbrook Road to Sumas. Turn right on road to Columbia Valley, drive to Kendall.

STOP 2–7. KENDALL MORAINE AND OUTWASH TERRACE

After the CIS had disappeared from the North Cascades, long valley glaciers, heading on Mt. Baker and fed by tributary glaciers, occupied the Nooksack drainage system (Figs. 23, 24) (Kovanen and Easterbrook, 2001). The purpose of this stop is to examine evidence for extension of a post-CIS, Cascade, alpine glacier ~45 km downvalley from its sources on Mt. Shuksan and Mt. Baker. The Kendall moraine forms a distinct ridge several hundred meters long whose axis is transverse to the valley (Fig. 25). It is composed of ~12 m (40 ft.) of till containing abundant, glacially-faceted, striated boulders and cobbles derived from the upper North Fork valley. The upper North Fork has a distinct U–shaped cross-profile at right angles to the direction of flow of the CIS.

The number of well-faceted and striated cobbles and boulders in the till is striking, probably because the valley glacier was so long. The lithology of cobble and boulders in the Kendall moraine is dominated by Chuckanut sandstone (54%), Mt. Baker andesite (12%), and graywacke/greenstone (14%)

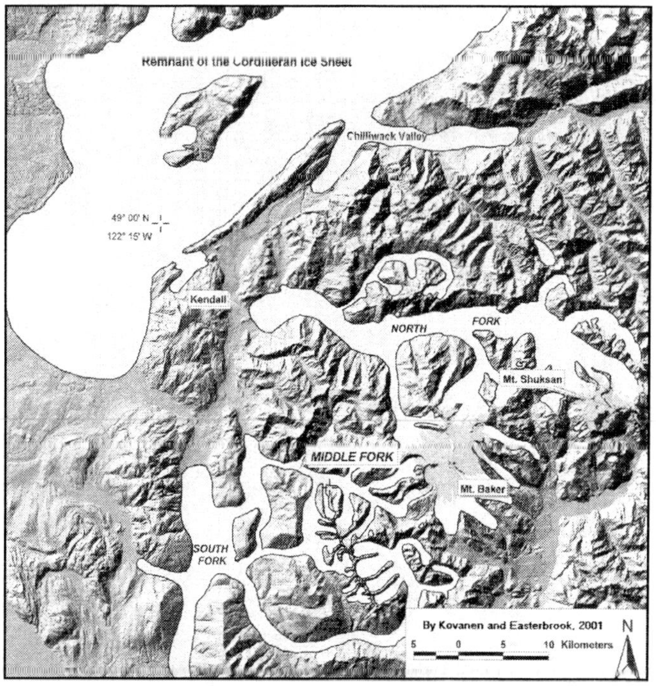

Figure 23. Reconstruction of Nooksack Alpine Glacier System. (Kovanen and Easterbrook, 2001)

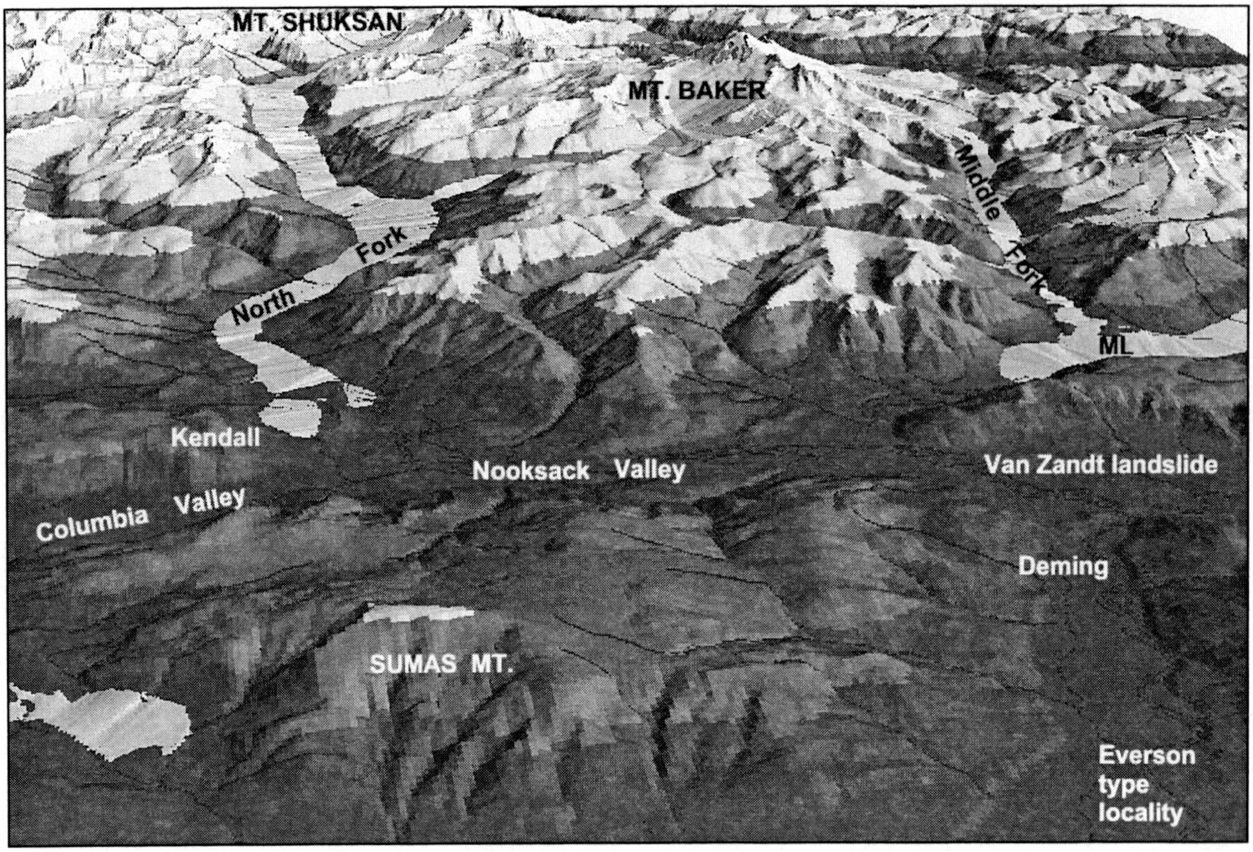

Figure 24. . Reconstruction of post–CIS alpine glaciers in the North Cascades. (DEM image by D.J. Kovanen).

Figure 25. The Kendall moraine and outwash terrace. (From Kovanen and Easterbrook, 2001).

(presumably from the Jurassic Nooksack Group) (Kovanen and Easterbrook, 2001). These lithologies demonstrate that the source of ice was from the North Fork valley to the east, rather than from the CIS to the north.

The Mt. Baker andesite was probably contributed from Mt. Baker via Bagley Creek, Glacier Creek and Wells Creek. Chuckanut sandstone was derived from the valley sides immediately upvalley, and the graywacke/ greenstone came from the valley sides near the town of Glacier.

Not far to the west, outwash from the Kendall moraine and the Maple Falls moraine just upvalley merge with the outwash terrace in the Columbia Valley. A few km downvalley, the outwash terrace rests on gmd dated at 11,910 ± 80 ^{14}C yrs. B.P. Distinct charcoal layers interbedded with sandy outwash terrace gravel gave dates of 10,788 ± 77 and 10,603 ± 69 ^{14}C yrs. B.P. (Fig. 26), indicating a Younger Dryas age of the outwash (Kovanen and Easterbrook, 2001).

Time and weather permitting, continue east on Mt. Baker Highway to Heather Meadows.

STOP 2–8 (OPTIONAL) ARTISTS POINT AND HEATHER MEADOWS, SCENIC VIEWS OF MT. BAKER, MT. SHUKSAN, AND NORTH CASCADES; CORDILLERAN ICE SHEET GROOVES; HOLOCENE CIRQUE GLACIATION; MAZAMA ASH

Table Mt., just west of the Artists Point parking lot, is an andesite flow that has become topographically inverted by erosion. Erratics of Paleozoic greenstone from Mt. Herman, the next peak to the north, indicates that glacial ice rode over Mt. Herman

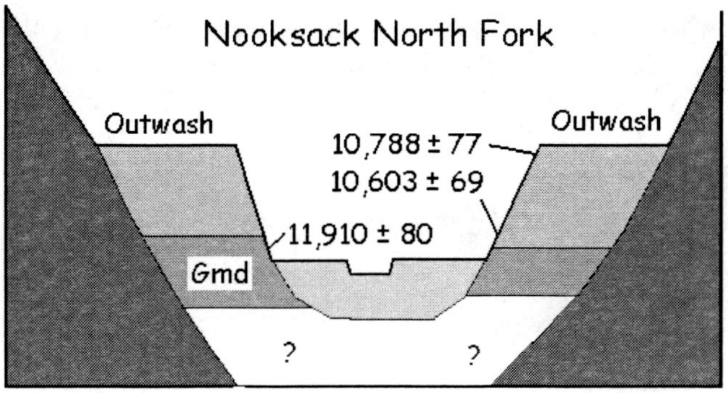

Figure 26. Cross section through the North Fork Sumas outwash

and deposited greenstone erratics on Table Mt. at an elevation of 1725 m (5700 ft.). Adding a few hundred feet for ice thickness means that the surface of the CIS here must have been more than 1800 m (6000 ft.). From the parking lot, walk eastward along the trail to Artists Point and note the beautifully developed grooves and striations carved into the andesite by the CIS. Significantly, all of the grooves are oriented north-south, indicating the direction of ice flow. Thus, the CIS did not flow westward from here down the Nooksack drainage toward the lowland, it flowed almost due south down the axis of the range.
Return to Picture Lake at Heather Meadows.

Picture Lake is contained by a moraine deposited by a cirque glacier. Mazama ash is exposed in a road cut beside the lake.

Return west on Mt. Baker Highway to Welcome. Turn left onto Mosquito Lake Road, cross Nooksack River, stop at overview of Deep Kettle bog.

Figure 27. Ice–contact drift, Deep Kettle bog, and lateral moraine, Nooksack Middle Fork. (Kovanen and Easterbrook, 2001)

STOP 2-9. DEEP KETTLE BOG, MOSQUITO LAKE KETTLE-KAME COMPLEX AND LATERAL MORAINE

The purpose of this stop is to examine evidence for the alpine ice origin of a post–CIS lateral moraine and kettle-kame complex. The kettle–kame complex is composed of poorly-to-moderately-sorted sand and gravel containing many glacially faceted and striated boulders and cobles. The lithology of the boulders and cobbles is dominated by Mt. Baker andesite (17.4%), dunite from the Twin Sisters Range (16.6%), and Chuckanut sandstone (45.2%) from upvalley, demonstrating that the glacier that transported them must have originated in the headwaters of the Nooksack Middle Fork (Kovanen and Easterbrook, 1996a, 2001).

The age of the kettle-kame complex is known from a 20 m core from a deep bog in a kettle (Fig. 27). Rootlets in gravel at the base of the bog were ^{14}C–dated at 12,356 ± 115 ^{14}C yrs. B.P. and had to be growing in rock debris on stagnant ice before the ice melted out to make the kettle. A date on wood lying on the basal gravel was 12,165 ± 115 ^{14}C yrs. B.P. (Kovanen and Easterbrook, 2001).

A lateral moraine drapes across the kettle–kame complex (Fig. 27), indicating that ice in the Middle Fork Valley readvanced following the stagnation that produced the kettle–kame complex. The lithology of the moraine is dominated by glacially faceted and striated Mt. Baker andesite and Twin Sisters dunite, proving that it was deposited by an alpine glacier from the Middle Fork. The age of the moraine must be younger than the 12,000 year–old kettle–kame complex because it overlies it, both morphologically and stratigraphically.

The strong dominance of glacially faceted boulders and cobbles from the Middle Fork headwaters 23 km (14.3 mi.) upvalley demonstrates that the kettle–kame complex and the lateral moraines were deposited by true alpine glaciers, not by remnants of the CIS and must therefore postdate the CIS. These alpine glaciers could not have been fed by Cordilleran ice because of the lithology of the drift and, as shown by gmd in the Nooksack drainage, the area to the north was free of Cordilleran ice at that time.

A shallow landslide exposes till of the lateral moraine overlying the ice-contact sediments of the kettle-kame complex, giving stratigraphic, as well as morphologic, evidence of the relationship of the lateral moraine to the kettle-kame complex. The lithology of the till is dominated by glacially faceted and striated Mt. Baker andesite and Twin Sisters dunite. The lateral moraine demonstrates that the Middle Fork alpine glacier readvanced over its previously–deposited kettle–kame complex.

Upvalley, logs in a lateral moraine have been dated at 10,680 ± 70 and 10,500 ± 70 ^{14}C yrs. B.P. (Kovanen and Easterbrook, 2001).

Continue west to I-5, turn right, return to Bellingham.

DAY 3. GLACIATION OF THE NORTH CASCADES

Depart Bellingham, drive 58 mi. (97 km) south on I–5 to Everett. Exit I–5 onto Highway 2, drive to Monroe, then turn right (south) to Snoqualmie falls.

ICE–MARGINAL FEATURES OF THE VASHON GLACIER.

The eastern margin of the Cordilleran Ice Sheet was banked against the Cascade Range as it flowed southward down the Puget Lowland. The glacier extended up major Cascade valleys, blocking drainage and impounding ice–marginal lakes into which large deltas were built (Cary and Carlston, 1937; Mackin, 1941; Booth, 1987). An ice–marginal embankment was constructed in the Snoqualmie valley near North Bend. It consists of a high, flat–topped ridge that extends across the Snoqualmie River valley. Deltaic sediments, consisting mostly of coarse sand and gravel, grading upvalley to sand, then to silt and clay, were deposited in an ice–dammed lake upvalley from the Snoqualmie embankment. Wood from rhythmically–bedded (varved?) lake sediments upvalley was ^{14}C dated at 13,570 ± 130 (UW–35) years. As the ice retreated to the north and west, the newly–opened, ice–free valleys held ice–marginal lakes that became successively lower as new outlets were uncovered (Mackin, 1941).

STOP 3–1. TOKUL CREEK DELTA AND SNOQUALMIE FALLS.

As the Puget lobe retreated northward from north–sloping valleys, an ice–marginal lake formed in the Snoqualmie valley south of the ice margin. Tokul Creek, a tributary of the Snoqualmie River brought large quantities of sand and gravel into the lake, building a delta that extended nearly all the way across the valley. As the ice front retreated northward, new, lower outlets were uncovered and the lake eventually drained. When the Snoqualmie River re–established its course, the delta sediments forced the river to the west side of the valley at the distal edge of delta. As the Snoqualmie cut down through these sediments, it encountered the bedrock of the west valley wall, and, once notched into the rock, couldn't get out. Downstream, the river quickly eroded the valley–filling sediments, but where the river was notched into the bedrock, it could not erode its bed as rapidly, and Snoqualmie falls were born. As the falls retreat upvalley, they will eventually erode through the bedrock spur and into the valley fill upstream. When this happens, the river will quickly cut down through the sediments and bring an end to the falls.

Continue through North Bend, turn left on I–90 and drive 22 mi. (37 km) to Snoqualmie Pass. Exit I–90 to the ski area.

STOP 3–2. LATEST PLEISTOCENE ALPINE MORAINES

Multiple, as-yet-undated, post-LGM moraines occur at Snoqualmie Pass in the North Cascades (Porter, 1978). Half a dozen, closely-spaced moraines at Snoqualmie Pass, well upvalley from the LGM, record several periods of glacial retreat and stillstand. No numerical ages have been published for any of these moraines, although a limiting date of 11,050 ± 50 ^{14}C yrs. B.P. was obtained from wood in gravel on till between two of the moraines (Porter, 1978).

Rejoin I–90 and drive east to Exit 70 to the Easton Campground for Lunch.
Drive east 16 mi. (27 km) on I–90 to Exit 85 to Blewett Pass. Contine 9 mi. (15 km) to Swauk Prairie.

STOP 3–3. SWAUK PRAIRIE MORAINE

Long valley glaciers extended down several tributaries of the Yakima River. At this stop we will examine the oldest moraines of this system.

Drive 37 mi. (62 km) to the junction with Highway 2. Turn left and continue 6 mi. (10 km) to Leavenworth. At the west end of town, turn left up Icicle Creek road and continue 6 mi. (10 km) to Rat Creek. Pull into the campground on the left.

STOP 3–4. YOUNGER DRYAS (?) MORAINE AT RAT CREEK

A double moraine in Icicle Creek (Page, 1939) from an alpine glacier in Rat Creek near Leavenworth has been correlated with the Sumas Stade of the CIS, which has been shown to include the YD (Kovanen and Easterbrook, 2002a). However, no numerical dates have yet been published for these moraines and this correlation is based on the position of the moraines upvalley from LGM moraines.

Return to Leavenworth, turn right on Highway 2 and drive 26 mi. (43 km) to Wenatchee. Exit to the right.

DAY–4. GLACIAL LANDFORMS OF THE OKANOGAN LOBE OF THE CORDILLERAN ICE SHEET; MISSOULA FLOOD FEATURES

The Cordilleran Ice Sheet extended down both sides of the North Cascades, overwhelming the crest of the range and connecting the Okanogan lobe with the Puget Lowland ice across the north Cascades. At the late Wisconsin maximum, the Okanogan lobe reached about 50 km south of the Columbia River onto the Columbia Plateau where it built a massive end moraine and left a virtual museum of glacial landforms beautifully preserved in the arid climate. Drumlins and flutes show the radial movement of ice, and eskers, kames, kame terraces, and recessional moraines were left during ice recession. The Okanogan lobe played an important role in directing giant floods from the bursting of ice dams and draining of glacial Lake Missoula. Glacial landforms are intimately associated with Missoula flood features.

GLACIAL GEOMORPHOLOGY AND ICE–FLOW INDICATORS OF THE OKANOGAN LOBE OF THE CORDILLERAN ICE SHEET: AN ARCHIVE OF GLACIAL FEATURES

Dori J. Kovanen and **Olav Slaymaker**, The University of British Columbia, Department of Geography, Vancouver, British Columbia V6T 1Z2, Canada

The Okanogan lobe flowed southward through the Okanogan trough from the Interior Plateau of British Columbia and invaded the Waterville Plateau during the late Wisconsin (Fraser Glaciation). This had a profound influence on the landscape and meltwater routing and flooding events in adjacent areas (e.g., Pardee, 1922; Bretz, 1959; Hanson, 1970; Easterbrook, 1979; Baker and Bunker, 1985; Waitt, 1985; Atwater, 1986). To help improve our understanding of the spatial distribution, morphology, processes, and dynamics of the Okanogan lobe, and because vertical stratigraphic sections showing the glacial sediments of the primary landforms are relatively sparse, we extracted glacial bedform information (i.e., drumlin, macro–flutes, till lineations, and eskers) from aerial photographs and 10–m digital elevation models (DEMs) obtained from the Washington State Geospatial Data Archive (University of Washington). More than sixty U.S.G.S. 7.5 minute quadrangles were merged together in the Geographic Information Systems (GIS) environment. The spatial distribution of glacial features and ice–flow pattern is shown on Figures 28 and 29. An advance phase is marked by the streamlining of flutes and drumlins and ended with the deposition of a broad morainal

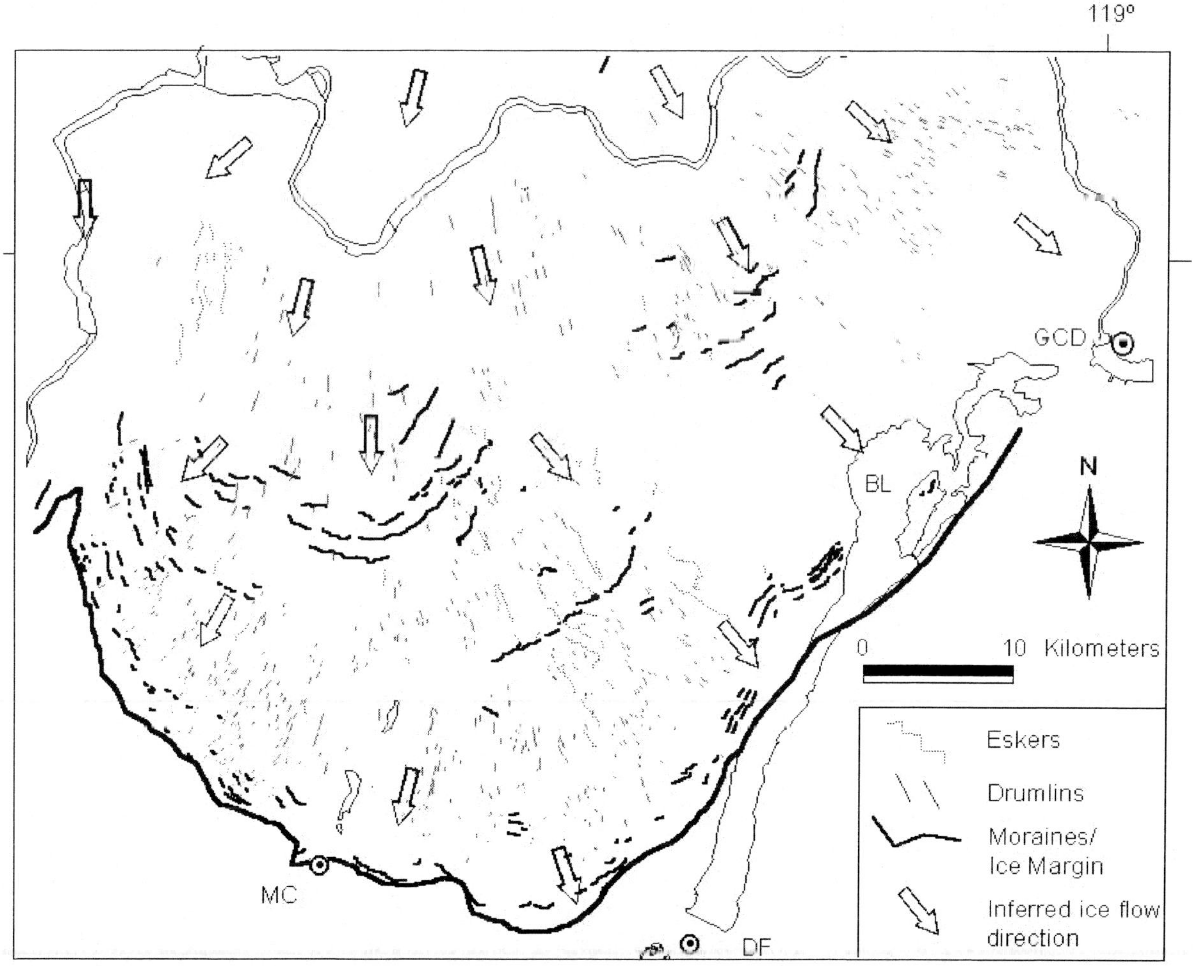

Figure 28. Distribution of glacial landforms and inferred ice–flow directions. Recessional moraines commonly overprint drumlins. DF–Dry Falls; BL–Banks Lake; MC–Moses Coulee; GCD–Grand Coulee dam.

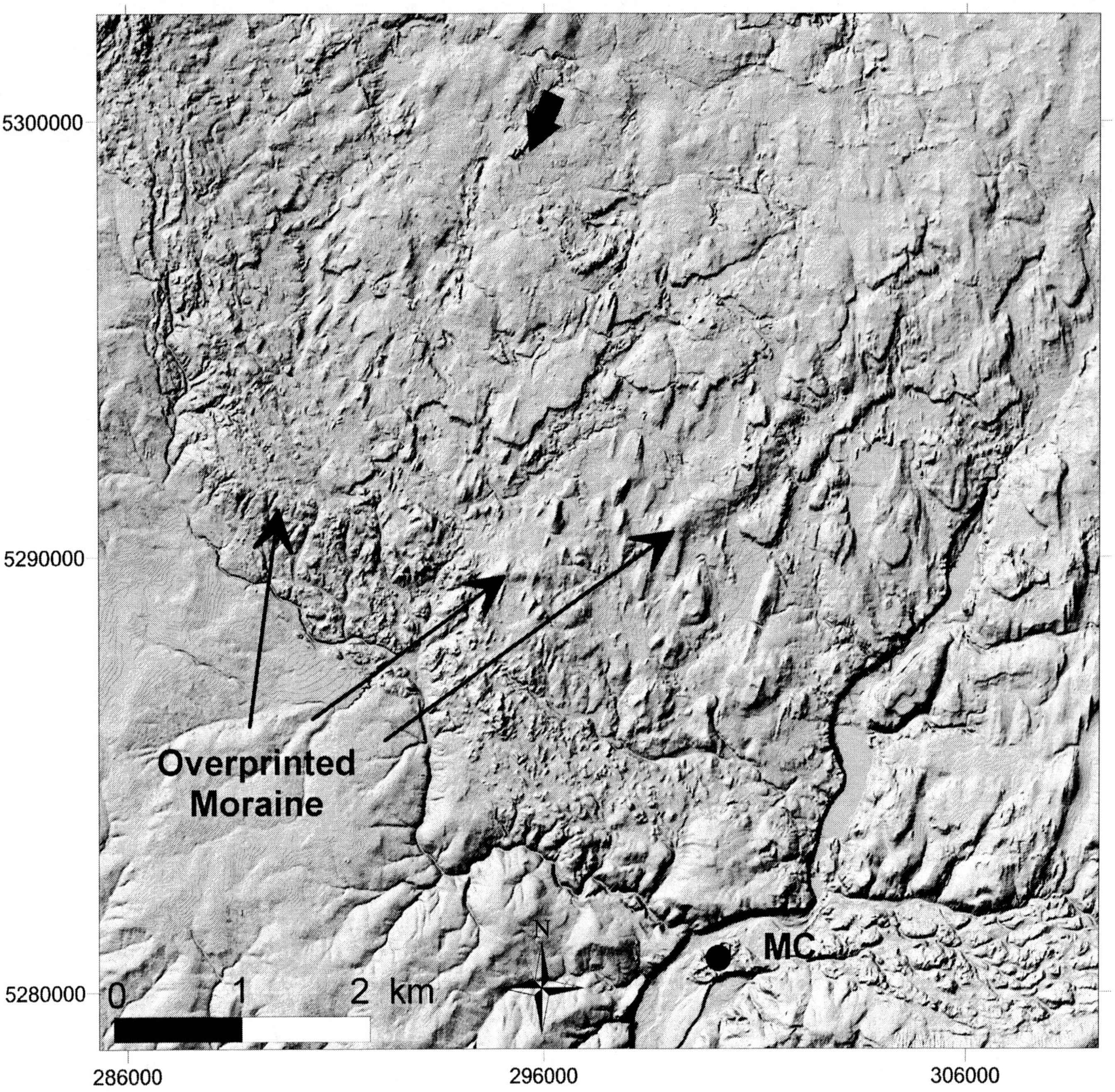

Figure 29. Shaded digital topographic model along the southwestern margin of the terminal area of the Withrow moraine. This image shows a lobate form that has been overprinted by till lineations. Arrows indicate inferred ice–flow direction.

complex which consisted of isolated kames composed of till and occasional, segmented, sharp–crested moraine ridges (the Withrow moraine). The retreat phase is marked by many moraines or morainal segments that are sub–parallel, arcuate in form, and transverse to ice–flow directions (see also Hanson, 1970; Easterbrook, 1979). Some recessional moraines can be traced almost continuously across the plateau and may be up to c. 60 m wide and rise 15–20 m above the local topography. Eskers are distinct, extensive, rise up to 15 m above the local relief, are up to 5 km long (60 m wide), and are typically found in association with back–wasting ice positions. Generally, two types of eskers were noted: (1) those associated with channels incised into bedrock and spatially fixed (i.e., "Nye Channels" or "N channels"; Nye, 1976), and (2) those that were formed when channels were melted upward into the glacier ice (i.e., "Röthlisberger channels" or "R channels"; Röthlisberger, 1972). Some glacial landforms overlap older forms, indicating more than one ice–flow event and possibly a switch in flow and/or thermal regimes.

Timing of ice–flow events and conclusions

The timing of the advance and retreat of the Okanogan lobe is not well constrained and this imposes caveats on the interpretation of the timing of meltwater routing and flooding events which were controlled by topographic features, ice margin configuration, and the opening of drainage routes. The main chronologic controls come from: (1) the presence or absence of volcanic ash in deposits which provide limiting estimates (e.g., Mullineaux *et al.*, 1978; Porter, 1978; Waitt and Thorson, 1983; Carrara *et al.*, 1996), (2) correlation with other major ice lobes, some of which may (Puget lobe) or may not (Columbia River and Purcell Trench lobes) be well–dated (i.e., Richmond, 1965; Easterbrook, 1979); and (3) counting–backwards techniques of dated varve sequences (e.g., Atwater, 1986; Waitt, 1985). The archive of glacial features in this area presents a dynamic picture of the Okanogan lobe. Reconstructions of the ice–surface morphology and estimated driving stresses (17–26 kPa) implied from ice thickness and surface slope in the terminal area suggest fast basal flow.

From Wenatchee, drive north on Highway 2 for 50 miles to Jameson Lake junction. Turn left and drive 2 ½ mi. to Moses Coulee moraine

STOP 4–1. WITHROW MORAINE AND OUTWASH TERRACE DRAPED ACROSS A GIANT MISSOULA FLOOD BAR, MOSES COULEE

Moses Coulee was scoured out by a giant flood from Lake Missoula prior to the last glaciation, leaving a large, dry gorge over 1.7 km (1 mi.) wide and 180 km (600 ft.) deep. During the late Wisconsin glaciation, the Okanogan lobe built the Withrow moraine, which drapes into Moses Coulee and across an earlier giant flood bar (Fig. 30).

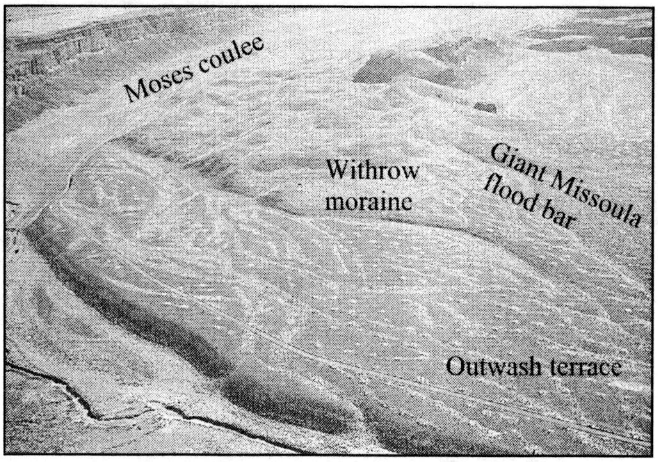

Figure 30. Withrow moraine and outwash terrace draped over a giant Missoula flood bar in Moses Coulee. (Photo by D.A. Rahm, courtesy of EPIC)

Return to Highway 2, turn right and return to Farmer. Turn right and drive 9 mi. to the Withrow moraine.

STOP 4–2. THE WITHROW MORAINE

The Withrow moraine (Fig. 31), is a hummocky moraine, 1–2 km wide and 30–70 m high and extends across the Waterville Plateau. Large basalt erratics commonly dot the surface. The Columbia Plateau here slopes gently northward so meltwater streams could not flow directly away from the ice terminus. They flowed parallel to the ice margin, then down Moses to the Columbia River (Easterbrook, 1979). Dutch Henry Draw, which we crossed just before this stop, is one such channel.

Drive 12 ½ mi. to Mansfield. Cross recessional moraines. Continue for 3 ½ mi. to huge erratic, then 1 mi. to large erratic. Continue for 4.3 mi. to Barnes Butte kame complex. Drive 4.8 mi. past several eskers to Sims Corner.

Figure 31. Drumlins behind the Withrow moraine (background). (Photo by D.A. Rahm, courtesy of EPIC)

STOP 4–3. ESKERS AND DISTAL FANS

The Columbia Plateau north of the Withrow moraine contains a profusion of eskers, kames, kettles, kame terraces, and related ice–contact deposits indicative of widespread glacier stagnation during deglaciation of the Okanogan lobe. Eskers which terminate at ice–marginal fans and low morainal ridges mark temporary halts in the recession of the glacier terminus. Three eskers end abruptly at ice–marginal fans which still retain the traces of abandoned braided channels on their surface (Fig. 32). The proximal side of the fans mark the former ice margin.

STOP 4–4. GRAND COULEE, DRY FALLS, MISSOULA FLOOD

Grand Coulee was described by Bretz (1969) as the "deepest, longest, and steepest in gradient. Its system possesses the

Figure 32. Esker terminating in a distal fan near Sims Corner. (Photo by D.A. Rahm, courtesy of EPIC)

Figure 33. Dry falls. (Photo by D.A. Rahm, courtesy of EPIC).

largest number of abandoned cataracts, the widest one, a record of the highest, the greatest cascade, and the largest number of distributary canyons. It is unique in that, at its initiation, it found no preglacial valley head but crossed 20 miles or so of the northern divide of the plateau before encountering such a valley, and it trenched that divide nearly 1000 feet deep".

Dry Falls (Fig. 33) extends about three miles across and is one of several falls excavated by spectacular quarrying of jointed basalt during catastrophic floods spilling down Grand Coulee when the Okanogan lobe blocked flow down the Columbia River near the site of Grand Coulee Dam and forced it to spill over into Grand Coulee. Much of the large size and depth of Grand Coulee is due to waterfall recession.

Drive 14 mi. to junction with Highway 2 and Dry Falls. Drive down Grand Coulee to Elphrata and return to Seattle.

REFERENCES

Armstrong, J.E., 1957, Surficial geology of New Westminster map–area, British Columbia: Geological Survey of Canada, Map 57–5, 25 p.

Armstrong, J.E., 1960, Surficial geology of the Sumas map–area, British Columbia: Geological Survey of Canada, Map 92G/1, 27 p

Armstrong, J.E., 1981, Post–Vashon Wisconsin glaciation, Fraser Lowland, British Columbia: Geological Survey of Canada, Bulletin 322, 34 p

Armstrong, J.E. 1984, Environmental and engineering application of the surficial geology of the Fraser Lowland, British Columbia: Geological Survey of Canada, Paper 83–23, 54 p.

Armstrong, J.E. and Brown, W.L., 1954, Late Wisconsin marine drift and associated sediments of the lower Fraser Valley, British Columbia, Canada: Geological Society of America Bulletin, v. 65, p. 349–364.

Armstrong, J.E., and Hicock, S.R., 1980a, Surficial geology of New Westminster map–area, British Columbia: Geological Survey of Canada, Map 1484A.

Armstrong, J.E., and Hicock, S.R., 1980b, Surficial geology of Vancouver map–area, British Columbia: Geological Survey of Canada, Map 1486A.

Armstrong, J. E., Crandell, D. R., Easterbrook, D. J., and Noble, J. B., 1965, Late Pleistocene stratigraphy and chronology in southwestern British Columbia and northwestern Washington: Geological Society of America Bulletin, v. 76, no. 3, p. 321–330.

Atwater, B.F., 1986, Pleistocene glacial–lake deposits of the Sanpoil River Valley, northeastern Washington: U.S. Geological Survey Bulletin, v. 1661, 39 p.

Baker, V.R., and Bunker, R.C., 1985, Cataclysmic late Pleistocene flooding from glacial Lake Missoula. A review: Quaternary Science Reviews, v. 4, p. 1–41.

Berger, G. W., and Easterbrook, D.J., 1993, Thermoluminescence dating tests for lacustrine, glaciomarine, and floodplain sediments from western Washington and British Columbia: Canadian Journal of Earth Science, v. 30, p. 1815–1828

Blunt, D., Easterbrook, D. J., and Rutter, N. A., 1987, Chronology of Pleistocene sediments in the Puget Lowland, Washington: Washington Division of Geology and Earth Resources Bulletin 77, p. 321–353.

Booth, D. B., 1987. Timing and processes of deglaciation along the southern margin of the Cordilleran ice sheet: *in* Ruddiman, W. F., and Wright, H. E., Jr., eds., North America and adjacent oceans during the last deglaciation, Geological Society of America, The Geology of North America, K–3, p. 71–90.

Bretz, J H., 1959, Washington's Channeled Scabland: Washington Department of Conservation, Division of Mines and Geology Bulletin, v. 45, 57 p.

Bretz, J H., 1969, The Lake Missoula floods and the Channeled Scabland: Journal of Geology, v. 77, p. 505–543.

Carlstad, C.A., 1992, Late Pleistocene deglaciation history at Point Partridge, central Whidbey Island, Washington: MS thesis, Western Washington University, Bellingham, Washington, 140 p.

Carrara, P.E., Kiver, E.P., and Stradling, D.F. 1996, The southern limit of Cordilleran ice in the Colville and Pend Oreille valleys of northeastern Washington during the Late Wisconsin glaciation: Canadian Journal of Earth Sciences, v. 33, p. 769–778.

Cary, A. S., and Carlston, C. W., 1937, Notes on Vashon stage glaciation of the South Fork of the Skykomish River valley, Washington: Northwest Science, v. 11, p. 61–62.

Deeter, J. D., 1979. Quaternary geology and stratigraphy of Kitsap County, Washington: M.S thesis, Western Washington University, Bellingham, WA, 175

Dethier, D.P., Pessl, F., Keuler, R.F., Balzarina, M.A., 1995, Late Wisconsinan glaciomarine deposition and isostatic rebound, northern Puget Lowland, Washington: Geological Society of America Bulletin, v. 107, p. 1288–1303.

Domack, E.W., 1983, Facies of late Pleistocene glacial–marine sediments on Whidbey Island, Washington: An isostatic glacial–marine sequence: *in* Molnia, B.F., ed., Glacial–Marine Sedimentation, Plenum, NY, p. 535–570.

Dyke, A.S., and Peltier, W.R., 2000, Forms, response times and variability of relative sea–level curves, glaciated North America: Geomorphology, v. 32, p. 315–333.

Easterbrook, D.J., 1963, Late Pleistocene glacial events and relative sea–level changes in the northern Puget Lowland, Washington: Geological Society of America Bulletin, v. 74, p. 1465–1483.

Easterbrook, D. J., 1966a, Radiocarbon chronology of late Pleistocene deposits in northwest Washington: Science, 152, 764–767.

Easterbrook, D. J., 1966b, Glaciomarine environments and the Fraser Glaciation in northwest Washington. Guidebook for First Pacific Coast Friends of the Pleistocene Field Conference, 52 p.

Easterbrook, D. J.; Crandell, D. R.; Leopold, E. B., 1967, Pre–Olympia stratigraphy and chronology in the central Puget Lowland, Washington: Geological Society America Bulletin, v. 78, p. 13–20.

Easterbrook, D. J., 1968, Pleistocene stratigraphy of Island County, Washington: Washington Division of Water Resources Bulletin, v. 25, p. 1–34.

Easterbrook, D.J., 1969, Pleistocene chronology of the Puget Lowland and San Juan Islands, Washington: Geological Society of America Bulletin, v. 80, p. 2273–2286.

Easterbrook, D. J., 1971, Geology and geomorphology of western Whacom County, Washington: Western Washington University Press, 68 p.

Easterbrook, D. J., 1976a, Quaternary geology of the Pacific Northwest: *in* Mahaney, W. C., ed., Quaternary Stratigraphy of North America. Dowden, Hutchinson, and Ross, Stroudsburg, PA, 441–462.

Easterbrook, D. J., 1976b, Geologic map of western Whatcom County, Washington: U. S. Geological Survey Miscellaneous Investigations. Map I–854B.

Easterbrook, D. J., 1979, The last glaciation of northwest Washington: *in* Armentrout, J. M., Cole, M. R., and Terbest, H., eds., Cenozoic Paleogeography of the Western United States. Pacific Coast Paleogeography Symposium 3, Los Angeles, Pacific Coast Section, Society of Economic Paleontologists and Mineralogists, p. 177–189.

Easterbrook, D. J., 1986, Stratigraphy and chronology of Quaternary deposits of the Puget Lowland and Olympic Mountains of Washington and the Cascade Mountains of Washington and Oregon: *in* Sibrava, V., Bowen, D. Q., and Richmond, G. M., Quaternary Glaciations in the Northern Hemisphere, Quaternary Science Reviews, v. 5, p. 145–169.

Easterbrook, D.J., 1992, Advance and retreat of Cordilleran ice sheets in Washington, USA: Geographie Physique et Quaternaire, v. 46, p. 51–68.

Easterbrook, D.J., 1994, Stratigraphy and chronology of early to late Pleistocene glacial and interglacial sediments in the Puget Lowland, Washington: *in* Swanson, D.A. and Haugerud, R.A., eds., Geologic Field Trips in the Pacific Northwest, Geological Society of America, p. 1J23–38.

Easterbrook, D.J., 2003, Comment on the paper "Determinaton of ^{36}Cl production rates from the well–dated deglaciation surfaces of Whidbey and Fidalgo Islands, Washingoton" by T.W. Swanson and M.C. Caffee: Quaternary Research, v. 59, p.132–134.

Easterbrook, D. J.; and Kovanen, D.J., 1998 'Pre–Younger Dryas resurgence of the southwestern margin of the Cordilleran Ice Sheet, British Columbia, Canada: Comments, Boreas, v. 27, p. 225–230.

Easterbrook, D. J., and Rutter, N. A., 1981, Amino acid ages of Pleistocene glacial and interglacial sediments in western Washington: Geological Society of America Abstracts with Programs, v.13, p. 444.

Easterbrook, D. J., and Rutter, N. A., 1982, Amino acid analysis of wood and shells in development of chronology and correlation of Pleistocene sediments in the Puget Lowland, Washington: Geological Society of America Abstracts with Programs, v. 14, p. 480.

Easterbrook, D. J., Blunt, D. J., and Rutter, N. A., 1982, Pleistocene glacial and interglacial chronology in western Washington: Geological Society of America Abstracts with Program, v. 14, p. 161.

Hansen, B. S., and Easterbrook, D. J., 1974, Stratigraphy and palynology of late Quaternary sediments in the Puget Lowland, Washington: Geological Society of America Bulletin, v. 85, p. 587–602.

Hansen, H. P., and Mackin, J. H., 1949. A pre–Wisconsin forest succession in the Puget Lowland, Washington. American Journal of Science, v. 47, p. 833–855.

Hanson, L.G., 1970, The origin and development of Moses Coulee and other scabland features on the Waterville Plateau, Washington: PhD thesis, University of Washington, Seattle, WA, 137 p.

Heusser, C. J.; Heusser, L. E., 1981, Palynology and paleotemperature analysis of the Whidbey Formation, Puget Lowland, Washington: Canadian Journal of Earth Science, v. 18, no. 1, p. 136–149.

Hewitt, A.T., and Mosher, D.C., 2001, Late Quaternary stratigraphy and seafloor geology of eastern Juan de Fuca Strait, British Columbia and Washington: Marine Geology, v. 177, p. 295–316.

Johnson, S.Y., Dadisman, S.V., Mosher, D.C., Blakely, R.J., and Childs, J.R., 2002, Active tectonics of the Devils Mountain fault and related structures, northern Puget Lowland and eastern Strait of Juan de Fuca region, Pacific Northwest: U.S. Geological Survey Professional Paper 1643.

Johnson, S.Y., Potter, C.J., Armentrout, J.M., Miller, J.J., Finn, Carol, and Weaver, C.S., 1996, The southern Whidbey Island fault, an active structure in the Puget Lowland, Washington: Geological Society of America Bulletin, v. 108, p. 334–354.

Kovanen, D. J., 2002, Morphologic and stratigraphic evidence for Allerød and Younger Dryas age glacier fluctuations of the Cordilleran Ice Sheet, British Columbia, Canada and Northwest Washingont, U.S.A.: Boreas, v.31, p. 163–184.

Kovanen, D. J., and Easterbrook, D. J., 1996a, Extensive readvance of late Pleistocene (Y.D.?) alpine glaciers in the Nooksack River Valley, 10,000 to 12,000 years ago, following retreat of the Cordilleran Ice Sheet, North Cascades, Washington: Friends of the Pleistocene, Pacific Coast Cell Field Trip Guidebook, 74 p.

Kovanen, D. J. and Easterbrook, D. J. 2001, Late Pleistocene, post–Vashon, alpine glaciation of the Nooksack drainage, North Cascades, Washington: Geological Society of America Bulletin v. 113, p. 274–288.

Kovanen, D. J., and Easterbrook, D. J., 2002a, Timing and extent of Allerod and Younger Dryas age (ca. 12,500–10,000 ^{14}C yr B.P.) oscillation of the Cordilleran Ice Sheet in the Fraser Lowland, western North America: Quaternary Research, v. 57, p. 208–224.

Kovanen, D. J., and Easterbrook, D. J., 2002b, Paleodeviations of radiocarbon marine reservoir values for the NE Pacific: Geology, v. 30, p. 243–246.

Leopold, L. B., Nickmann, R., Hedges, J. L., and Ertel, J. R., 1982, Pollen and lignin records of late Quaternary vegetation, Lake Washington: Science, v. 218, p. 1305–1307.

Mackin, J. H., 1941, Glacial geology of the Snoqualmie–Cedar area, Washington: Journal of Geology, v. 49, p. 449–481.

Mark, D. M. and Ojamaa, P. M., 1980, Late Pleistocene meltwater channels and Sumas deglaciation, Fraser Lowland, British Columbia: Proceedings from Western Division of Canadian Association of Geographers Meeting, p. 81–91.

Mullineaux, D. R., Waldron, H. H., and Rubin, M., 1965, Stratigraphy and chronology of late interglacial and early Vashon time in the Seattle area, Washington: U. S. Geological Survey Bulletin, 1194–O, 10 p.

Mullineaux, D.R.,Wilcox, R.E., Ebaugh, W.F., Fryxell, R., and Rubin, M., 1978, Age of the last major scabland flood of the Columbia Plateau in eastern Washington: Quaternary Research, v. 10, p. 171–180.

Nye, J.F., 1976, Water flow in glacier: jökulhlaups, tunnels and veirns: Journal of Glaciology, v. 17, p. 181–207.

Page, B. M., 1939, Multiple glaciation in the Leavenworth area, Washington: Journal of Geology, v. 47, p. 785–815.

Pardee, J.T., 1922, Glaciation in the Cordilleran region: Science, v. 56, p. 686–687

Porter, S.C., 1978, Glacier Peak tephra in the North Cascade Range, Washington: Stratigraphy, distribution, and relationship to late–glacial events; Quaternary Research, v. 10, p. 30–41.

Porter, S. C. and Swanson, T. W., 1998, Radiocarbon age constraints on rates of advance and retreat of the Puget Lobe of the Cordilleran Ice Sheet, during the last glaciation: Quaternary Research, v. 50, p. 205–213.

Richmond, G.M., 1965, Glaciation of the Rocky Mountains: *in* The Quaternary of the United States, Wright, H.E. Jr,, Frey, D.G., eds., Princeton University Press, Princeton, NJ, p. 217–230.

Rigg, G. B., and Gould, H. R., 1957, Age of Glacier Peak eruption and chronology of postglacial peat deposits in Washington and surrounding areas: American Journal of Science, 255, 341–363.

Robinson, S.W., and Thompson, G., 1981, Radiocarbon corrections for marine shells dates with application to southern Pacific Northwest Coast pre–history: Syesis, v. 14, p. 45–57.

Röthlisberger, H., 1972, Water pressure in intra–and subglacial channels: Journal of Glaciology, v. 11, p. 177–203.

Siegfried, R. T., 1978. Stratigraphy and chronology of raised marine terraces, Bay View Ridge, Skagit County, Washington: MS thesis, Western Washington University, Bellingham, WA, 52 p.

Southon, J.R., Nelson, D.E., and Vogel, J.S.A., 1990, Record of past ocean–atmosphere radiocarbon difference from the Northeast Pacific: Paleoceanography, v. 5, p. 197–206.

Stoffel, K. L., 1980, Stratigraphy of pre–Vashon Quaternary sediments applied to the evaluation of a proposed major tectonic structure in Island County, Washington: U. S. Geological Survey Open File Report 81–292, 161 p.

Swanson, T.W., and Caffee, M.L., 2001, Determination of ^{36}Cl production rates derived from the well–dated deglaciation surfaces of Whidbey Island, Washington: Quaternary Research, v. 56, p.366–382.

Thorson, R.M., 1989, Glacioistatic response of the Puget Sound area, Washington: Geological Society of America Bulletin, v. 101, p. 1163–1174.

Waitt, Jr, R.B., 1985, Case for periodic, colossal Jökulhlaups from Pleistocene glacial Lake Missoula: Geological Society of America Bulletin, v. 96, p. 1271–1286.

Waitt, Jr, R.B., and Thorson, R.M., 1983, The Cordilleran Ice Sheet in Washington, Idaho, and Montana: *in* Porter, S.C., ed., Late Quaternary environments of the United States, The late Pleistocene, University of Minnesota Press: Minneapolis, MN, v 1, p. 53–70.

Weber, S., 2001, Late Pleistocene littoral deposits in the Deming sand at Bellingham Bay, Washington, and their implications for relative sea level changes: M.S. thesis, Western Washington University, 158 p.

Mountain glaciations of the Sierra Nevada

Authors and Leaders: **Doug Clark,** *Dept. of Geology, Western Washington University, Bellingham, Washington 98225*
Alan R. Gillespie, *Dept. of University of Washington, Seattle, WA*
Malcolm Clark, *U.S. Geological Survey (retired)*
Bud Burke, *Dept. of Geology, Humboldt State University, Arcata, CA*

INTRODUCTION

Field Trip B–3 traverses an important segment of the glaciated Sierra Nevada from Lake Tahoe in the northern section, south to the town of Bishop, under the high ridges of the central Sierra Nevada (Fig. 1). The field trip is designed to visit key areas at which the stratigraphy and chronology of the Pleistocene glacial deposits were first developed and subsequently refined. West of the crest, the glaciers were substantially larger and more complex, and key evidence there is commonly obscured by heavy vegetation or poorly preserved in the steep canyons. Our field trip concentrates on the Sierra Nevada east of the crest because that is where the visibility, access, and scale of the deposits are best suited for reconnaissance introduction to the region. It is also where most of the research has been concentrated.

The Sierra Nevada is a major north–south mountain range that separates the internally drained basins and desert ranges to the east from the Central Valley to the west. Although it runs from the Tehachapi Mts. on the northern edge of the Mojave Desert to near Mt. Lassen, the highest part of the Sierra Nevada (crest elevations of 3400–4300 m; 11,000–14,000 ft.) lies between ~36°N and ~38°N, the approximate latitudes of the towns of Olancha (south) and Bridgeport (north). This entire reach was heavily glaciated during the Pleistocene, and small glaciers still occupy sheltered cirques high in the mountains. Enhanced accumulation and shading from adjacent peaks allow these modern glaciers to exist well below the regional climatic snowline (~4500 m; 14,800 ft; at 37°N) (Flint, 1957). Although the glaciers occur at progressively lower altitudes to the north, the average snowline, or equilibrium line altitude (ELA), for the modern–glacier ELA rises above the topographic crest north of Bridgeport. During the Pleistocene, however, the ELA was far lower (~900 m), and large Pleistocene glaciers emanated from the ridges and peaks of the northern Sierra Nevada (Fig. 2).

The Sierra Nevada consists of a batholith of Mesozoic plutons, commonly granodiorite, intruded into metamorphosed marine sedimentary rocks and island–arc volcanic rocks, both now preserved as roof pendants. The range is a large crustal block, tilted to the west and broken by extensional, basin–and–range faulting on the east. The timing of these events is controversial, but tilted Miocene lava flows west of the crest, and faulted Pliocene basalt flows in the ranges east of the Sierra Nevada indicate that much of the early tectonic activity initiated before the Quaternary Period, although the faulting, at least, continues today.

California in the 19th century was a long way from the European centers where geological theory was being developed. Nevertheless, the Sierra Nevada received a surprising amount of attention, and the understanding of mountain glaciation was developed in no small part there. For example, J. D. Whitney, the California State Geologist, and John Muir, prominent naturalist, debated famously the relative roles of fluvial, glacial, and tectonic processes in sculpting the spires, ridges, and deep valleys of the Sierra Nevada, with special attention given to Yosemite Valley. Joseph LeConte (1873) modified the positions taken earlier and derived an essentially modern view that glaciers had excavated the highlands and widened the upper reaches of river valleys, but that the basic landscape was nevertheless formed largely by fluvial activity. François Matthes (1930, 1965) undertook decades of study of the glaciated troughs of the western slopes of the Sierra Nevada, and Eliot Blackwelder (1931) similarly analyzed the glacial evidence east of the crest. Blackwelder, in particular, pioneered relative dating in an effort to correlate moraines from valley to valley, and to the glacial stratigraphy developed along the southern margin of the Laurentide Ice Sheet. Matthes and Blackwelder made an early attempt to reconcile their glacial sequences, but, in the absence of numerical age control, correlation was difficult, as successive generations of glacial geologists have confirmed while refining the glacial stratigraphy.

By the 1970's, the modern framework of the Sierra Nevada glaciations was well developed (e.g., Wahrhaftig and Birman, 1965), but numerical age control was weak or missing entirely. Advances in the past few decades have focused on developing and applying rigorous dating techniques to the glacial sequence. Early refinements included K/Ar dating of lava flows interbedded with till and ^{14}C dating of latest Pleistocene sediments in bogs that could be related stratigraphically to glacial deposits. Beginning in the 1980's, advances in accelerator mass spectrometry (AMS) led to the measurement of trace concentrations of exotic isotopes, created by cosmic ray bombardment of rocks exposed at the Earth's surface, and to the calculation of exposure ages of these rocks. From these dates, landform ages could be inferred, and glacial chronologies became logical and early targets of this new technology.

Ideas about glacial chronology in the Sierra Nevada were first advanced, and then hindered, by correlation to the Laurentide Ice Sheet, with its supposed, four, Pleistocene glaciations. Evidence from the Sierra Nevada glaciers, in common with evidence from mountain glaciers elsewhere, is frustratingly incomplete and difficult to piece together due to the overriding of earlier

Clark, D., Gillespie, A.R., Clark, M., Burke, R., 2003, Mountain glaciations of the Sierra Nevada: *in* Easterbrook, D.J., ed., Quaternary Geology of the United States, INQUA 2003 Field Guide Volume, Desert Research Institute, Reno, NV, p. 287–311.

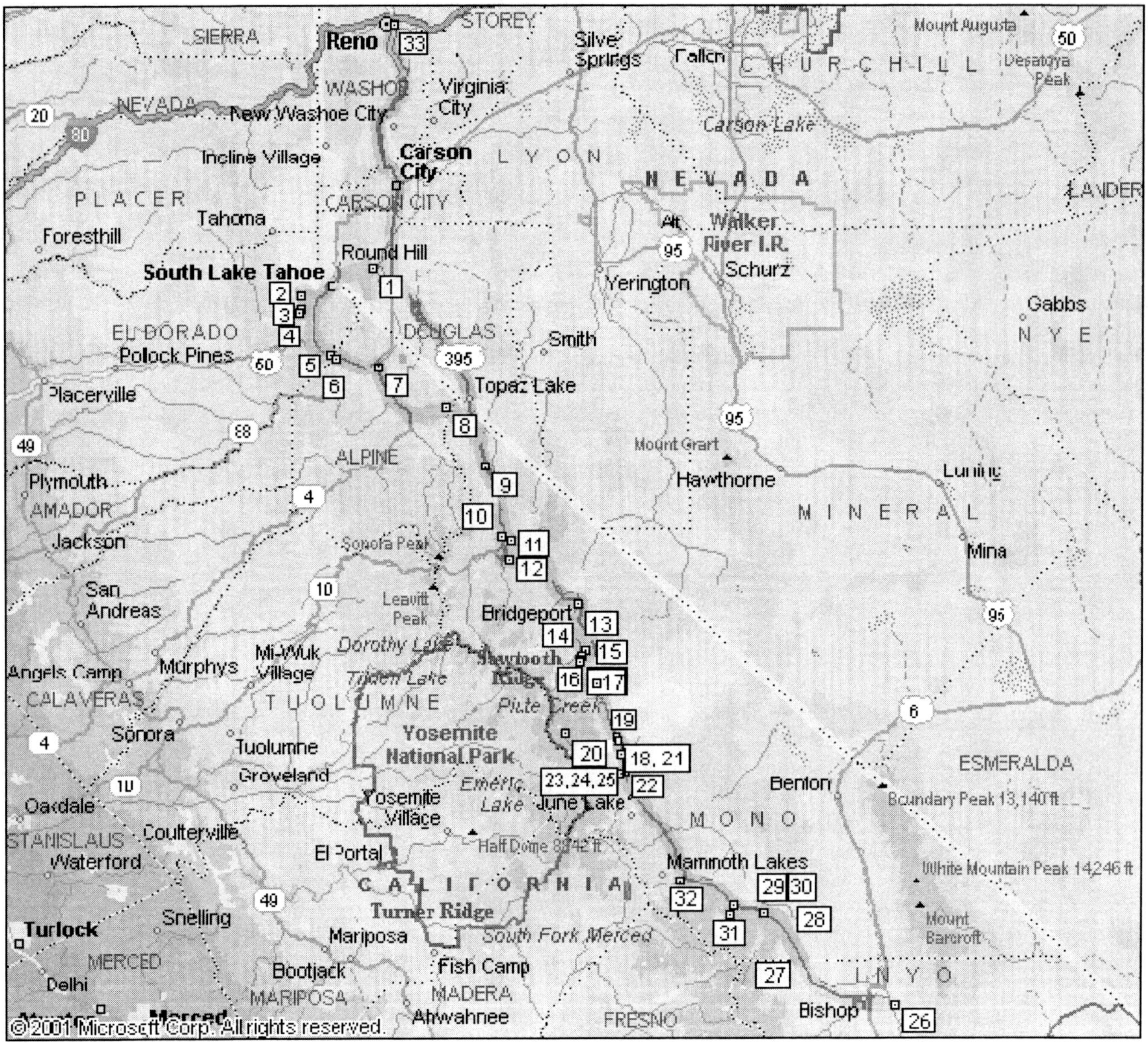

Figure 1. Road map of the Sierra Nevada, with route and numbered stops of the field trip.

deposits by later glaciers and the high potential for subsequent erosion in the steep canyons and slopes. The temptation to use distant evidence to illuminate local events is an understandable response to these frustrations. Even today, attempts are frequently made to link the timing of mountain glaciations by correlation to oxygen isotope data from marine sediment cores or to ice cores from Greenland or Antarctica. Hence, direct numerical ages, which are often imprecise or of questionable significance, are commonly augmented by assignments to marine oxygen isotope stages (MIS) or to discrete events recorded in the isotope records (e.g., Dansgaard–Oeshger events). This is apparently done with the assumption that such correlations, guided by the available direct dating, gives a broad–brush view of the timing that is an improvement or that such correlation facilitates the relation of Sierra Nevada glaciations to those elsewhere. It is with this latter goal in mind that we use MIS assignments in this field trip guide.

Sediment cores from lakes fed by streams from the glaciated drainages have also been used to infer details of the glacial chronology. In the Sierra Nevada, cores have been analyzed from several such lakes and the history inferred from these cores, although less direct, is indeed more complete than from studies so far from of the glacial deposits themselves.

The glaciers of the Sierra Nevada expanded and shrank in response to changes in climate, especially changes in summer temperatures and winter precipitation. In modern times, the Sierra Nevada receives moisture mainly from winter low–pres-

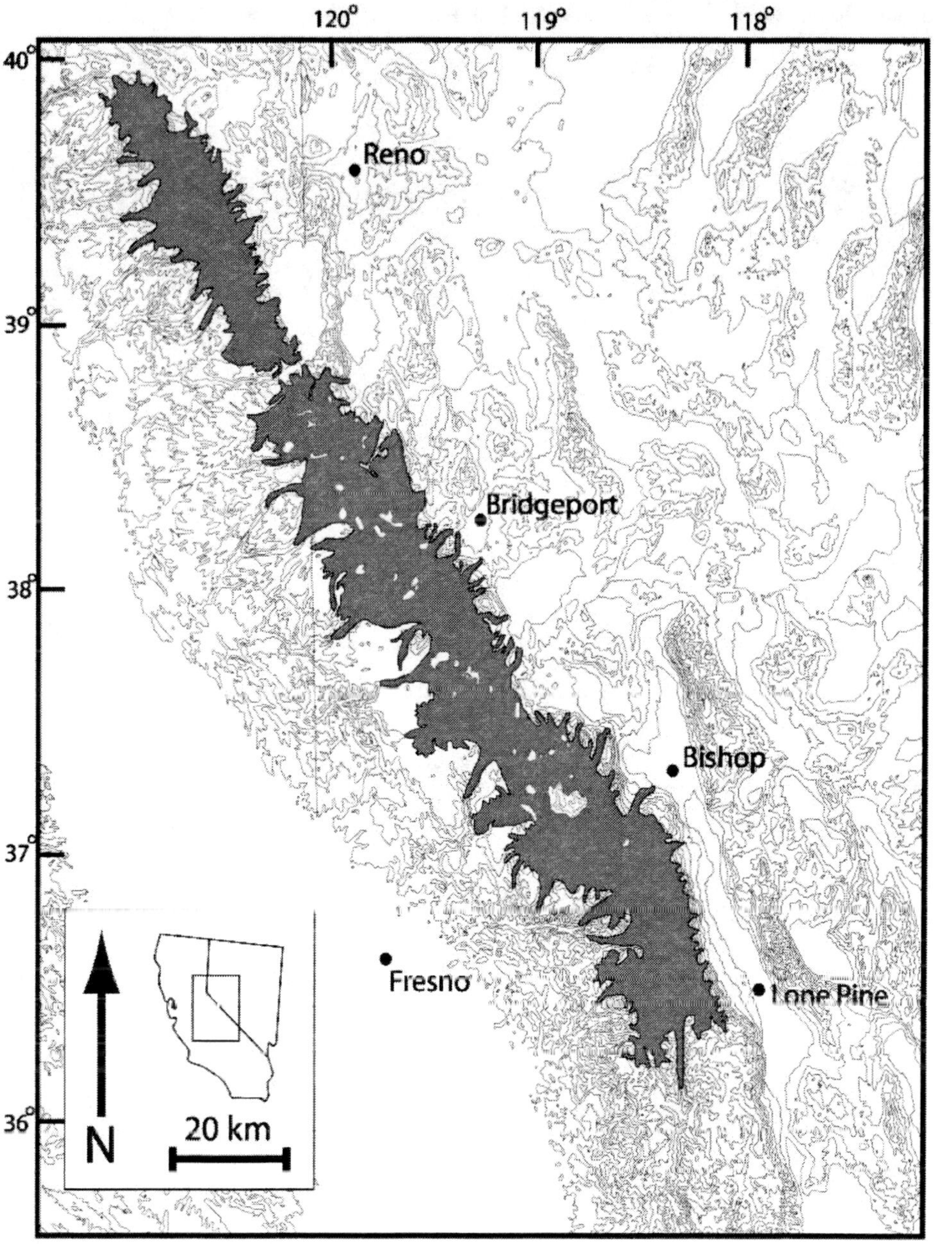

Figure 2. Extent of late Pleistocene glaciers in the Sierra Nevada (courtesy, Paul Zehfuss; modified from Wahrhaftig and Birman, 1965).

sure systems brought from the Pacific Ocean by westerlies guided by the jet stream, which descends in latitude from Canada to central California as summer yields to winter. Precipitation in the Sierra Nevada ranges from <50 to >200 cm/yr., with much of the range controlled by a strong topographic rainshadow. During the major Pleistocene glaciations, the jet stream was forced farther south so that precipitation was increased, and temperatures overall were reduced, although the two were not necessarily in phase.

The glacial evidence of the Sierra Nevada has been organized into allostratigraphic units that have been assigned to glaciations based on relative dating and correlation, augmented more recently by numerical dating. The glacial history for the Sierra Nevada is summarized in Table 1. Much of the history is anachronistic, and will no doubt be revised substantially in the future as better and more complete information becomes available.

Deeply weathered erratic boulders on mountaintops and diamictons on highland surfaces indicate the earliest glaciations. These deposits were obviously left before the terrain became as deeply sculpted as it is today, but the exact timing is unknown. The two oldest named tills are the till of McGee Mt. and the "old red till" on Lower Rock Creek. The story is clearly complex, and will take a lot of work before a more complete story can be gleaned from the scattered deposits. Details of the glaciations that left these deposits, including their ELAs and extents and even the elevation of the Sierra Nevada are unknown.

TABLE 1. RECOGNIZED GLACIATIONS OF THE SIERRA NEVADA, THEIR AGES, AND EAST-SIDE ELA DEPRESSIONS INTERPOLATED TO 37°N

Glaciation	Mean age (10^3 yr)	ΔELA (m)	Age references
Matthes (Little Ice Age)	0.6–0.1	480 ± 25	Konrad and Clark (1998); Stine (1994)
Recess Peak	14.2–13.1[§]	335 ± 20	Clark (1997)
Tioga (retreat)	15–14	125 ± 30	James et al. (2002); Clark and Gillespie (1997)
Tioga (start)	21–20[§]	0	Clark et al. (1996)
Tioga ("Tioga 2-4")	25–16	0	Phillips et al. (1996)[†]
Tenaya (="Tioga 1"?)	31	-45 ± 15	Phillips et al. (1996)[†]; Benson et al. (1996)
	30[§]	-45 ± 15	Bursik and Gillespie (1993)
Tahoe II**	50–42	-95 ± 10	Phillips et al. (1996)
Tahoe I**	140–200		Phillips et al. (1990)
Casa Diablo	60–125		Bailey et al. (1976)
pre-Tahoe/post-Mono Basin**	190–220		Phillips et al. (1990), discussed in text
Mono Basin	80–60	195 ± 50	Phillips et al. (1990)[†]
Walker Creek**	~ 550		Clark (1968); A. Sarna-Wojcicki, (personal commun., 2002)
Sherwin	~820	~200	Sharp (1968); Birkeland et al. (1980); Nishiizumi et al. (1989)
pre-Sherwin	~920		Birkeland et al. (1980)

Note: from Gillespie, 1991.
* Elevation differences are relative to the lowest Tioga ELA (~3040 ± 150 m) and interpolated to 37°N (Gillespie, 1991). Uncertainties are ± 1σ and are random and do not include the systematic uncertainties of ~150 m. Accuracy depends critically on assignment of moraines to the correct glaciation.
[§] 10^3 cal yr B.P.
[†] Revised (see discussions in James et al., 2002, and Phillips et al., 2001).
** Nomenclature from this field guide.

Both till remnants predate the Sherwin till, which is widely mapped in the Sierra Nevada and occurs in association with the better–understood, late–Pleistocene deposits. The Sherwin till in its type area is the earliest for which we have a reasonable date, ~820 ka. However, that all tills mapped as Sherwin really are primary glacial deposits, rather than proximal outwash, for example, is not clear. Until such "details" are resolved, even the Sherwin Glaciation will remain a mystery.

The gap between ~820 and ~190 ka (the beginning of MIS 6) is sparsely populated with glacial evidence. Two tills (in the Mohawk Valley, north of this field trip, and near Sonora Junction, which we will visit) are dated to this interval by tephras. Any systematic understanding of glaciations in this interval, however, lies in the future.

The detailed record of Sierra Nevada glaciations begins with MIS 6, and even so it is incomplete compared to the generalized history deduced from lake sediment cores. Named tills that probably date from MIS 6 are (1) the Mono Basin, pre–Tahoe/post–Mono Basin and older Tahoe (Tahoe I) tills of Bloody Canyon, (2) possibly the Casa Diablo till of Mammoth Lakes, and (3) the pre–Hogsback till of Sawmill canyon in Inyo County, south of Bloody Canyon.

The Tahoe glaciation, one of the four major Sierra Nevada glaciations recognized early in the 20th century, has commonly been assigned to MIS 6. With the advent of rigorous relative dating and numerical dating methods, however, even this assignment is suspect. Some Tahoe tills are now thought to date from MIS 3 or 4, even though Tahoe moraines are generally more massive than the later, MIS–2, Tioga moraines that date from the global, last–glacial maximum (LGM).

The controversy over the Tahoe glaciation is illuminating because it derives from incorrect stratigraphic correlations and from inaccurate dating. Thus, deposits from two or more different glaciations may both be called "Tahoe," assumed to be of the same age, and numerical age constraints on one Tahoe moraine are commonly taken to apply to all.

Only with the latest Pleistocene Tioga Glaciation does the picture snap into sharper focus. Blackwelder (1931) typically recognized two Tioga moraines in Sierra Nevada valleys, and Sharp and Birman (1963) and Birman (1964) recognized a third, "Tenaya" moraine in some valleys between Tioga and Tahoe in relative age. Debate since has considered the question of the age of the Tenaya glaciation (MIS 3?) on the one hand, and its very existence on the other. Do Tenaya moraines represent simply a Tioga (or Tahoe) advance for which evidence is preserved only locally? Fortunately, lake–core evidence (e.g., Benson, 1999; Benson et al., 1996) now suggests that the debate was fueled by reasonable but different interpretations of incomplete evidence, and that a better interpretation of late Pleistocene glacial history has a dozen or more short–lived advances beginning around 50 ka and ending around 14 ka. Moraines commonly represent the larger of these, and of course, the same partial record is not necessarily preserved in each canyon. Thus, both the Tahoe II and the Tenaya advances seem to have occurred during earlier and later parts of MIS 3, respectively, with the Tioga advances culminating around 19–25 ka and ending by 15 ka with retreat to the Sierra Nevada crest or even complete disappearance for a brief period. The Hilgard moraines, once thought to derive from a separate, post–Tioga glaciation, are now thought to represent recessional phases of the Tioga glaciation, or even in some cases Tioga medial moraines.

Following the retreat of the MIS–2 Tioga glaciers, another brief re–advance (Recess Peak) occurred, ending before the start of the Younger Dryas event in the North Atlantic. This glaciation has been well dated by ^{14}C on macrofossils as well as gyttja from lake cores, and has been confirmed by ^{36}Cl dating (Plummer, 2002). This event is important because it demonstrates the fact that not all glaciations world–wide are synchronous, and regional variations in timing or extent may occur because climate itself is a regional phenomenon (e.g., Gillespie and Molnar, 1995).

Moraines correlated with the Younger Dryas have been recognized in the Rocky Mts. and in the north Cascade Range, but appear to be absent in the central Cascade Range and the Sierra Nevada. This regional variability may have been correctly assessed and may hold the key to greater understanding of paleoclimate or, of course, future research may once again alter the evolving history of glacial advances and retreats.

Regardless of the timing and distribution of the latest Pleistocene glaciers, Neoglacial and modern advances seem to have a distinctive worldwide signature. In the Sierra Nevada, the climate, as assessed by δ^{18}O data from Bristlecone pine tree rings (Feng and Epstein, 1994), seems to have become gradually more "glacial" after 6 ka, and evidence from rock glaciers and lake sediment cores (Konrad and Clark, 1998) shows that glaciers reappeared in the Sierra Nevada cirques by ~3.4 ka, after a hiatus of several thousand years. However, the Matthes glaciers, thought to correlate with the Little Ice Age elsewhere, seem to have reached their maximum extents sometime late in the last millennium and have been in retreat for the past century.

During the field trip, we will inspect some of the evidence that led to the development of the above story of glacial advance and retreat. Practical constraints preclude visiting all locations, but care has been taken to select some of the most important within those constraints. The glacial history summarized above is likely to change as existing technologies are refined and as new technologies and evidence are developed. Thus, this summary and this field trip are best viewed as a guide to work in progress. All of the trip leaders have enjoyed the rare privilege of a lifetime of study in the Sierra Nevada (among us, over a century!), and we hope to share some of our findings and enthusiasm during the brief five days of this field trip.

DAY 1. GLACIATION NEAR LAKE TAHOE

On DAY 1 of the field trip, we will see some excellent examples of moraines in and near the Lake Tahoe Basin. These moraines were some of the first noted by early geologists who mapped evidence of glaciation in the Sierra Nevada and were considered informally by Blackwelder (1931) to be "type areas" for Tioga and Tahoe advances (LGM and penultimate glacial maximum, respectively). At Luther Pass, we will visit a unique bog, which contains the longest continuous record of sedimentation yet identified in the Sierra Nevada, including interglacial, stadial, and interstadial sediments. We will also pass along and across the eastern Sierra Nevada escarpment in several areas where we will view evidence for late–Pleistocene and Holocene tectonic activity. Figure 3 shows the sites and topography for DAY 1 of the field trip.

STOP 1–1. GENOA FAULT.

Depart Reno Hilton and drive south on US 395 through Carson City. Turn right on Genoa Lane and proceed to Jacks Valley Road (Nevada 206, at km 67 (mile 42; #1, Fig. 1).

Between Reno and Carson City, several range–front faults define the eastern escarpment of the Carson Range, tectonically part of the Sierra Nevada South of Carson City, the escarpment is steep, straight, and displays abundant geomorphic evidence of Holocene faulting, including faceted ridge spurs and faulted alluvial fans. Evidence from trenching of the fault zone at STOP 1–1 south of the village of Genoa demonstrates that this portion of the Genoa fault has a Holocene slip rate of 2.0 ± 1.3 mm/yr. and may have generated two *M 7* earthquakes in the past 1000 years (Ramelli et al., 1999).

The Carson Range reaches altitudes of just under 3350 m (11,000 ft) southwest of STOP 1–1, whereas the floor of Carson Valley is at ~1430 m (4700 ft), yielding a relief of ~1890 m (6200 ft). Despite the altitude of the Carson Range, it apparently only supported a few cirque or small valley glaciers on its east side during the most recent (Tioga) glaciation. Somewhat larger glaciers extended down the northwestern slope of the range into the Tahoe Basin. However, no glacial deposits here have been mapped in detail.

From STOP 1–1, we will drive along the range front for a few km before turning west and climbing the Kingsbury Grade across the Carson Range into the Lake Tahoe basin. As we descend, note the spectacular views of Lake Tahoe, and beyond it, the Crystal Range, which defines the western margin of the basin and the crest of the Sierra Nevada. In contrast to the small glaciers that occupied the Carson Range during the late Pleistocene, the Crystal Range supported a substantial ice field with large outlet glaciers, some of which calved into Lake Tahoe at their maxima. Near lake level, highway US 50 crosses a broad alluvial plain on which the city of South Lake Tahoe is built.

STOP 1–2. FALLEN LEAF LAKE TERMINAL MORAINES

At the west end of South Lake Tahoe, turn right (north) onto State Route 89 (SR 89) along the west side of Lake Tahoe. Several km north, turn left onto Fallen Leaf Lake Rd. and continue ~1 km up this road to STOP 2–2 at km 116 (mile 72; #2, Fig. 1). Fallen Leaf Lake Tioga end moraines are the first glacial stop.

Evidence for glaciation is abundant along the western side of Lake Tahoe. A large ice field occupied the Crystal Range during major Pleistocene glaciations, with outlet glaciers that flowed onto a broad upland to the west of the crest and toward Lake Tahoe to the east. One of the best–preserved examples of a termi-

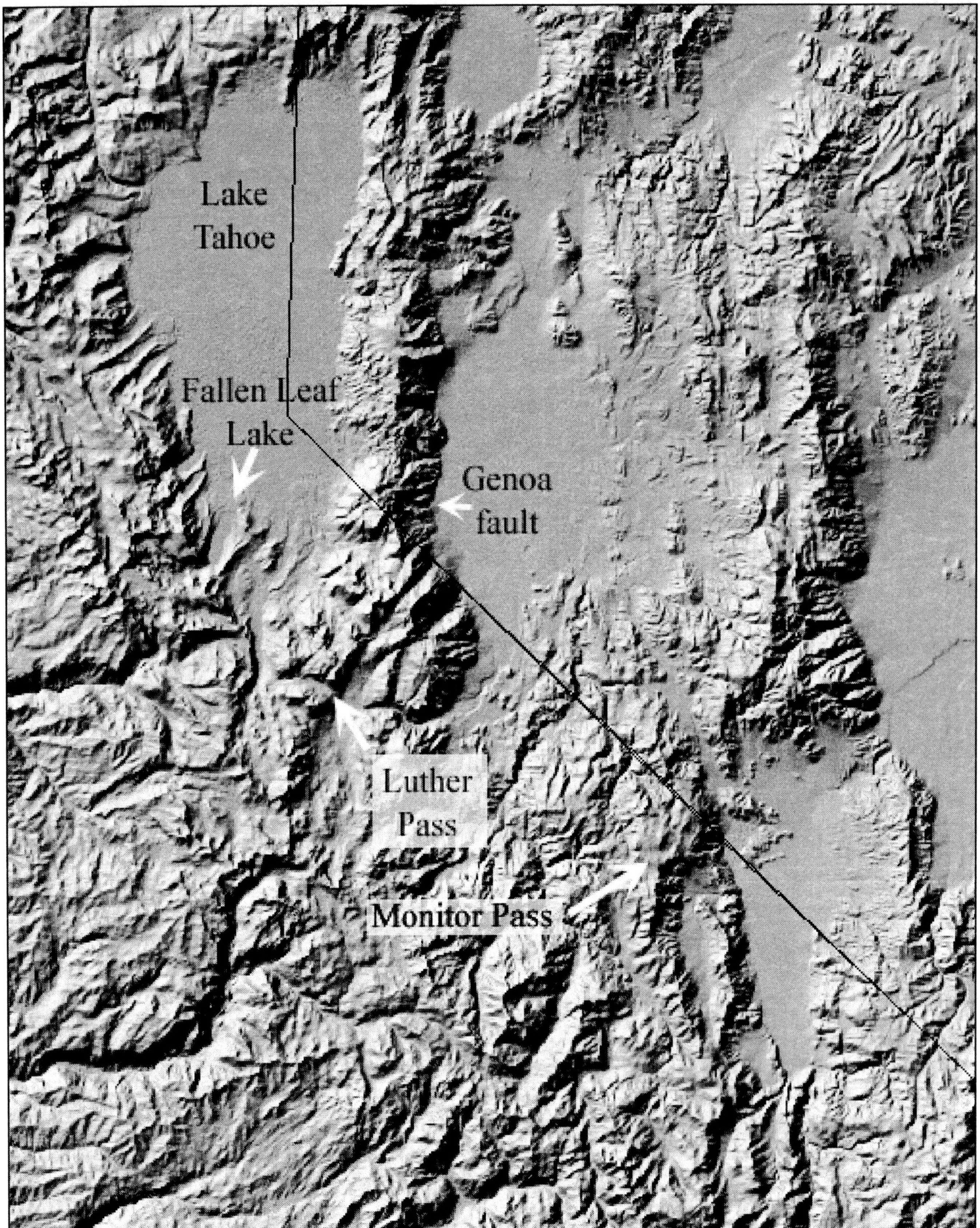

Figure 3. Shaded relief map of the Sierra Nevada from Lake Tahoe to the West Walker River (source: USGS digital elevation data, courtesy Ralph Haugerud, University of Washington).

nal moraine complex of the late–Wisconsin, Tioga glaciation occurs at the north end of Fallen Leaf Lake. The Fallen Leaf Lake moraines, and the moraines of several other drainages that extended from the eastern side of the Crystal Range, have a long history of study, dating back to Joseph LeConte (1873). These moraines were informally identified as type–areas by Blackwelder (1931) when he established the Tahoe (older) and Tioga (younger) glacial stratigraphy of the Sierra Nevada.

During the Tahoe and Tioga glaciations, glaciers flowed from altitudes as high as 3040 m (9980 ft.) and terminated in the lake at ~1900 m (6230 ft.), forming large compound lateral moraines (Fig. 4). At Fallen Leaf Lake, the Tioga glacier apparently terminated on a flat outwash plain that preserves numerous small end moraines. In contrast, moraines suggest the glacier may have had a calving margin during that event, an inference supported by evidence that the surface level of Lake Tahoe was up to 60 m higher than it is now (Birkeland, 1964),

The Tioga moraines at this stop are all well formed, but small in volume. Note the level of preservation of the moraines, the abundance and weathering of the boulders, vegetation cover, and the presence of small, local, closed depressions. All are typical of Tioga moraines in the Tahoe Basin. The geomorphic expression and soil development of the moraines are particularly useful in distinguishing Tioga from Tahoe tills, as will be seen on later stops.

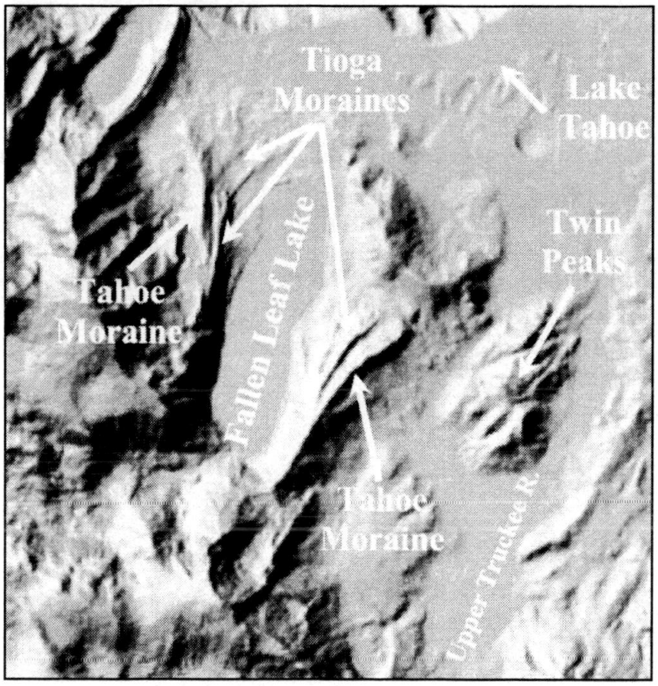

Figure 4. Shaded relief DEM of Fallen Leaf Lake, showing Tioga and Tahoe moraines of Fallen Leaf Lake glacier

STOP 1–3. FALLEN LEAF LAKE RIGHT–LATERAL MORAINES, TIOGA VS. TAHOE

From STOP 1–2, continue up Fallen Leaf Lake Road to Tahoe Mountain Road, turn left and continue ~2 km. At Angora Ridge Road, turn right and continue to small pullout in drainage between the Tioga and Tahoe moraines, at km 120 (mile 75; #3, Fig. 1).

The Tahoe Mt.–Angora Ridge road intersection lies immediately outside the Tioga maximum, right–lateral moraine. The small meadow to the left is a bog dammed between the Tioga moraine on the right and the Tahoe right–lateral moraine that lies outboard of the bog. At the far end of the bog, the road climbs steeply up the drainage between the Tioga and Tahoe moraines.

At this stop, we will compare adjacent Tioga and Tahoe lateral moraines. Be sure to note the differences in morphology, weathering, surface soil, and vegetation between the two moraines. This stop shows the differences that were first noted by Blackwelder (1931) and other early workers in the range.

We will hike up the short distance to the crest of the Tioga right–lateral moraine, which displays the characteristic steep slope and less–oxidized surface soils of Tioga moraines in the region. Also note the abundant manzanita and mountain mahogany, also common on Tioga moraines.

After inspecting the Tioga moraine, descend back past the vans and climb up the opposite ridge (to the east). This moraine is the Tahoe right–lateral moraine of the Fallen Leaf Lake glacier. Note the differences in slope, surface soil, weathering, boulder abundance, and vegetation. The differences are substantial and reflect significant differences in age. However, as is discussed later in this guidebook, the age of the Tahoe glaciation remains controversial, and moraines mapped as Tahoe throughout the range may relate to two or more different events. The Fallen Leaf Lake Tahoe moraines have not been dated.

STOP 1–4. ANGORA RIDGE LOOKOUT

Return to the vehicles and continue 0.5 km up Angora Ridge Road to the lookout at km 121 (Mile 75; #4, Fig. 1).

Lunch stop. The old fire lookout here has not been used since the 1970's, but is maintained by local hiking groups. The lookout is built on the crest of the Fallen Leaf Lake Tioga right–lateral moraine, just up–valley from where it merges with and buries the Tahoe right–lateral moraine that we visited at STOP 1–3. The site affords a spectacular view of Fallen Leaf Lake, and the headwaters in the Crystal Range to the west and south, in addition to the Upper Truckee River drainage and Lake Valley to the east. Note the sharp transition from erosional trim–lines to depositional moraines above Fallen Leaf Lake.

In contrast to the massive and distinct moraines formed by the Tioga and Tahoe Fallen Leaf Lake glaciers, the Upper Truckee glaciers left only small, indistinct moraines that are not visible from our lunch spot. In fact, the extent of the Tioga and Tahoe glaciers here remains debatable. Gardner et al. (1999) suggest that the Tahoe glacier there reached into present Lake Tahoe, making it bigger than the coeval Fallen Leaf Lake glacier. Conversely, mapping and modeling by Clark (1995) and Molitor and Clark (2002) indicates that the Tahoe glacier did not reach beyond Twin Peaks, the granitic hill in the center of Lake Valley that lies to the east of us.

After lunch, we will drive back down Angora Ridge Rd., turn right at Tahoe Mtn. Rd., left at Lake Tahoe Blvd., right at Sawmill Rd., and then west onto US–50 again, through Lake Valley. Along Sawmill Rd., note Twin Peaks to the left (north). The bedrock here is heavily weathered, with abundant grus and tafoni, suggesting that it was not overridden by either the Tioga or Tahoe advances of the Upper Truckee glacier.

STOP 1–5. GRASS LAKE BOG

After passing through the town of Meyers on US 50, turn left (south) onto SR 89, and drive up the axis of the Upper Truckee valley towards Luther Pass. The outermost mapped moraine in the valley is just south of the US 50/SR 89 intersection. Note how small it is, as are the subsequent recessional moraines that the highway crosses. The stop is a pullout on SR 89, adjacent to Grass Lake at km 145 (mile 90; #5, Fig. 1).

Grass Lake bog is an unusual basin in that it is perched at 2350 m (7720 ft) in a pass that comprises a doubly beheaded cross–drainage connecting the Upper Truckee Valley on the west with the West Fork Carson River Valley on the east. What makes the bog truly exceptional is that it lies outside the maximum Tioga ice limits of both drainages, and thus remained ice–free during the most recent (Tioga) glaciation. The bog is in fact dammed by moraines at either end: by a Tahoe or pre–Tahoe moraine at the western end (over which we just drove), and by a Tioga–maximum, left–lateral moraine from the West Fork Carson glacier at the eastern end (Fig. 5). This unusual situation has allowed the bog to preserve the oldest record of glacial sedimentation yet identified within the Sierra Nevada (Clark et al., in press).

Simple projections of the steep bedrock sidewalls north and south of the bog indicate that the basin is at least 50 m, and possibly as deep as 100 m. Resistivity measurements, ground–penetrating radar, and sediment coring support this projection (Clark et al., in press). The sediment cores show a distinct stratigraphy (Fig. 6). The upper 7.5 m are peat and organic lake mud, with organic mud becoming more prevalent with depth. At about 7.5 m depth, the sediments become coarser and less organic–rich, and at 8.7 m depth, a sharp transition occurs to laminated, inorganic sediments. This transition, dated at 17,200 ± 70 ^{14}C yrs. B.P. (21,130–19,850 cal. yrs. B.P.), apparently records the initial retreat of the Tioga glacier. West Fork Carson glacier from its maximum position; as soon as the glacier pulled back from the pass, direct outwash sedimentation in the bog would have ceased.

The laminated inorganic sediments continue downward for ~4 meters below the 8.7–m transition, and are in turn underlain by at least 5 m of peat and high–to–low organic silt and sand. The maximum depth cored (so far) in the bog is 17.7 m. These basal organic sediments reach ages greater than 52,100 ± 2800 ^{14}C yrs. B.P., making them the first pre–Tioga lake or bog sediments yet recorded in the Sierra Nevada. These dated sediments also confirm the contention (Clark et al., in press) that the bog lies outside Tioga ice limits and existed throughout much or all of the last glacial cycle. Further investigation of these sediments promises to give us a first look at pre–Tioga interstadial and interglacial paleo–environmental conditions in the heart of the Sierra Nevada. They may also help constrain the timing of the Tahoe Glaciation.

Figure 5. Shaded relief DEM of Luther Pass (USGS 10–m data, showing pixelation from the limits of the data). Note Tioga moraines damming right (east) side of bog, and rounded older till damming the west side.

STOP 1–6. LUTHER PASS TIOGA MORAINES (OPTIONAL)

Continue to a turnout at Luther Pass, at km 148 (mile 92; #6, Fig. 1).

This brief, optional stop is to inspect the Tioga–maximum, left–lateral moraines of the West Fork Carson glacier, both of which dam the eastern end of the Grass Lake bog. At this site, the road crosses two main sets of Tioga lateral moraines. The western set has low relief and defines the eastern end of the Grass Lake Bog. The eastern set of moraines is more prominent and forms Luther Pass itself, where a roadcut exposes the interior of the moraine. A smaller bog, informally named the Luther Pass bog, separates the two sets of moraines. These moraines are clear evidence that at its maximum, the West Fork Carson glacier flowed several hundred meters into, but not across Grass Lake. A subtle divide that crosses Grass Lake bog west of the westernmost Tioga moraine may indicate a buried Tahoe moraine, suggesting that even those glaciers may not have crossed the bog.

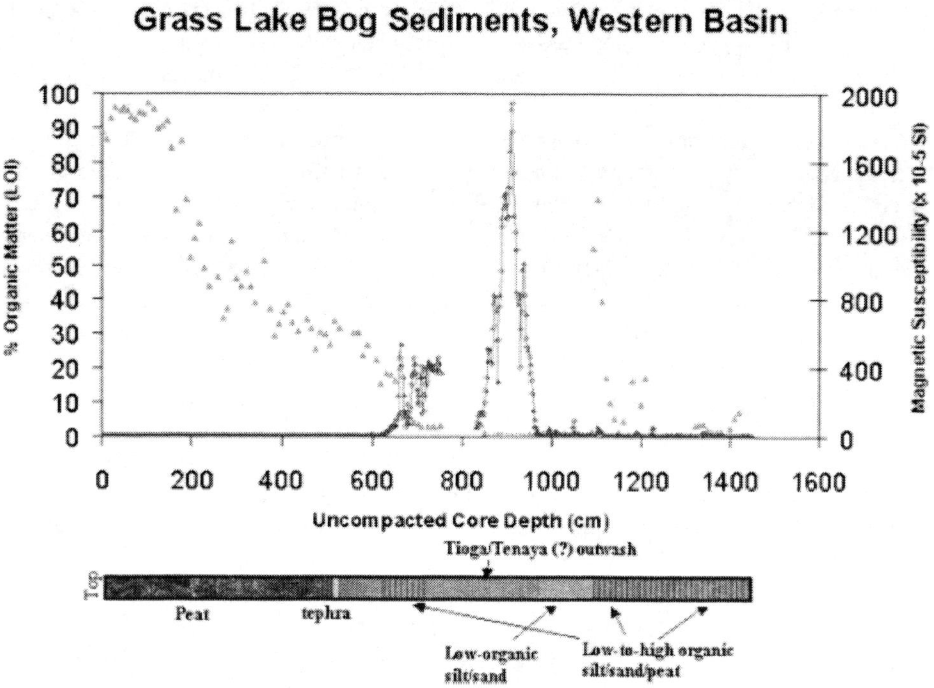

Figure 6. Sediments of Grass Lake bog. Generalized core log (bottom) and chart showing organic content and magnetic susceptibility. Note interval of high–susceptibility (coincident with Tioga glaciation) and rapid fluctuations in susceptibility and organic content (not entirely coincident) of pre–Tioga interstadial sediments. The significance of these fluctuations is not yet clear.

STOP 1–7. WEST FORK CARSON RIVER TAHOE MORAINE, WOODFORDS (OPTIONAL)

Continue south on SR 89, descend from Luther Pass into Hope Valley, where you get a sense of the broad scale of the Pleistocene West Fork Carson glacier that filled the valley. This glacier was a major outlet of the Mokelumne Ice Field, one of the largest Pleistocene ice fields in the Sierra Nevada (Clark, 1995). Note the numerous Tioga recessional moraines that cross the highway as we descend into the valley. At the intersection with SR 88, turn left (east) and head down the Carson River Canyon. The Tioga and Tahoe glaciers fed tongues of ice down the canyon; the Tioga glacier terminated near the mouth of the canyon, whereas the Tahoe glacier reached beyond the mouth. At the mouth of the canyon at Woodfords, turn right (south) onto SR–89. Note that we are immediately upstream from a Holocene scarp along the continuation of the Genoa Fault. The road crosses the scarp as it climbs out of the river valley. Pull off of the road after climbing out of the Tioga terrace of the West Fork Carson River at km 163 (mile 101; #7, Fig. 1).

At this optional stop, we visit the Tahoe right–lateral moraine exiting the West Fork Carson River Canyon. The moraine experienced a forest fire about 10 years ago, so that the morphology is still readily apparent. Note both the extent of weathering of the boulders, and the effects of the recent forest fire, causing spalling of the boulders. Such spalling adds to uncertainty with surface exposure dates on morainal boulders.

STOP 1–8. MONITOR PASS LOOKOUT

Continue south on SR 89, past the turnoff for SR 4 (Ebbetts Pass), climb to Monitor Pass, and turn left up the dirt road to the Monitor Pass lookout, km 195 (mile 121; #8, Fig. 1).

The fire lookout at Monitor Pass affords one of the better vistas of the Sierra Nevada crest, from northern Yosemite in the south to the Carson Range in the north. During the Pleistocene glaciations, much of the crest in this view was covered by ice from the Stanislaus and Mokelumne ice fields. In part, the present topography reflects the exhumation of a nonconformity between underlying Mesozoic granitic rocks and overlying Tertiary volcaniclastic deposits. The glaciers have stripped much of the poorly resistant volcanic sediments, exposing large expanses of low–relief, deeply weathered granite. The dark, angular peaks, many of which were nunataks within the ice fields, consist of the remaining volcanic material.

The effects of the Sierra Nevada rain shadow are apparent to the east where many of the ranges are unglaciated, despite altitudes that are similar to this part of the Sierra Nevada. One notable exception is the Sweetwater Range, the first large mountains to the southwest (Figure 3). This range, composed primarily of Tertiary extrusive rocks, apparently supported several small valley glaciers, extending primarily off its north and northeastern slopes.

The Sierra Nevada range front is complex through this region, with no single, defining fault as we saw in the Carson Valley. However, Meadowcliff Lodge, where we will stay tonight, lies at the foot of one locally–well–expressed branch of this fault system. The uplifted granitic rocks are spectacularly jointed where exposed adjacent to the fault behind the lodge.

Continue south off Monitor Pass on SR 89 until it intersects US 395. Turn right (south) and pull off at Meadowcliff Lodge at km 222 (mile 138; #9, Fig. 1).

Meadowcliff Lodge is our residence for the night. We will eat dinner tonight, and breakfast tomorrow morning at the restaurant here. Be sure to take some time to inspect the range front that lies directly behind the lodge. The tectonic activity on this fault has not been assessed.

DAY 2. GLACIATION FROM SONORA PASS TO CONWAY SUMMIT

On DAY 2, we will explore Pleistocene glacial deposits in the northern part of the High Sierra from the West Walker River to Bridgeport Basin. Here, the piedmont east of the Sierra Nevada Crest is much higher than to the north or south. This affords better preservation of moraines left by glaciers that here crossed the range front to terminate on the upland surface. It also provides a glimpse of how much of the eastern Sierra Nevada may have looked during the mid–Pleistocene before range–front faulting created the deep basins that presently occur along most of the rest of the eastern foot of the range. The main reference for glaciation of the West Walker River is Clark (1967) and for Bridgeport Basin is Sharp (1972). Figure 7 shows the sites and topography for Days 2–5 of the field trip.

STOP 2–1. RECENT FLOOD EVIDENCE ON THE WEST WALKER RIVER

From Meadowcliff, drive south on US 395 and soon enter the gorge of the West Walker River. The stop is at km 243 (mile 151; #10 on Fig. 1).

From the fault–bounded southern end of Antelope Valley where we stayed last night, U.S. 395 climbs to the south for 18 km up the canyon of the West Walker River to Sonora Junction, the location of the terminal moraines of the LGM (Tioga) glacier of the West Walker River. As we drive up Walker River Canyon, note the evidence of erosion and engineering of the river channel along this entire length. STOP 2–1 is one of the many exposures showing some of the severe damage wrought by a large flood in the spring of 1997. This site lies at an elevation of ~2000 m, only 100 m below the Tioga end moraines near Sonora Junction.

The 1997 flood damaged the highway in multiple locations, and severely eroded many of the alluvial fans that feed into the canyon from side drainages. The highway was entirely rebuilt and the stream course was rerouted in an attempt to keep the main channel away from the highway.

Numerous, in–place tree stumps emerge from the center of the river channel at this site and at several other locations along this portion of the canyon. Dated outer rings of these stumps indicate that they died in two distinct events, ~A.D. 1112 and ~A.D. 1350 (Stine, 1994). Tree ring counts on the stumps indicate that they grew for >200 years and >140 years, respectively. These periods of growth coincide with low–stands at Mono Lake, Lake Tahoe, and Tenaya Lake in Yosemite, supporting the concept that the region experience two prolonged droughts that were much more severe than any historical drought in the region (Stine, 1994). We will discuss the Mono Lake evidence for drought at STOP 3–1.

STOP 2–2. MORAINES OF THE WEST WALKER RIVER

Continue up the West Walker River until the valley widens and the end moraines are visible. Turn left from US 395 onto a dirt road and double back to an overlook of the end moraines at km 258 (mile 160; #11, Fig. 1).

The largest Pleistocene glaciers of the eastern slope of the Sierra Nevada invaded the West Walker River drainage repeatedly. The Tioga glacier reached 32 km from the Sierra Nevada crest to Sonora Junction, whereas the largest Pleistocene advance extended another 18 km north to Antelope Valley. Glacial deposits of six different ages have been identified in the West Walker drainage (Blackwelder, 1931; Clark, 1967, 1968).

From the viewpoint at this stop, ~1 km east of Sonora Junction and ~150 m above it, deposits from at least three different ages of Pleistocene West Walker glaciers are visible. We stand on Tahoe till (MIS 4 or 6), and below at Sonora Junction lie Tioga (MIS 2) lateral, recessional, and terminal moraines. Scattered erratic boulders of the largest and oldest glaciation yet identified can be identified with binoculars here. These granitic boulders lie as much as 470 m above us to the west, north, and east. Their upper limit describes the surface of a 130 km^2 "ice pool" of the West Walker glacier that occupied the basin around Sonora Junction. The main ice outflow from this impoundment was north down the canyon of the West Walker River to Antelope Valley, as shown by erratics high above the river all the way to Antelope Valley. Some ice also escaped eastward into the East Walker drainage through Devils Gate (we will pass through here later today on the drive to STOP 2–4) and two other gaps south of it. This oldest glaciation has not yet been correlated with old glaciations elsewhere in the Sierra Nevada, but on the basis of extent and amount of post–glacial erosion, it probably is correlative with the Sherwin glaciation (>758 ka) south of Mono Basin (Black-

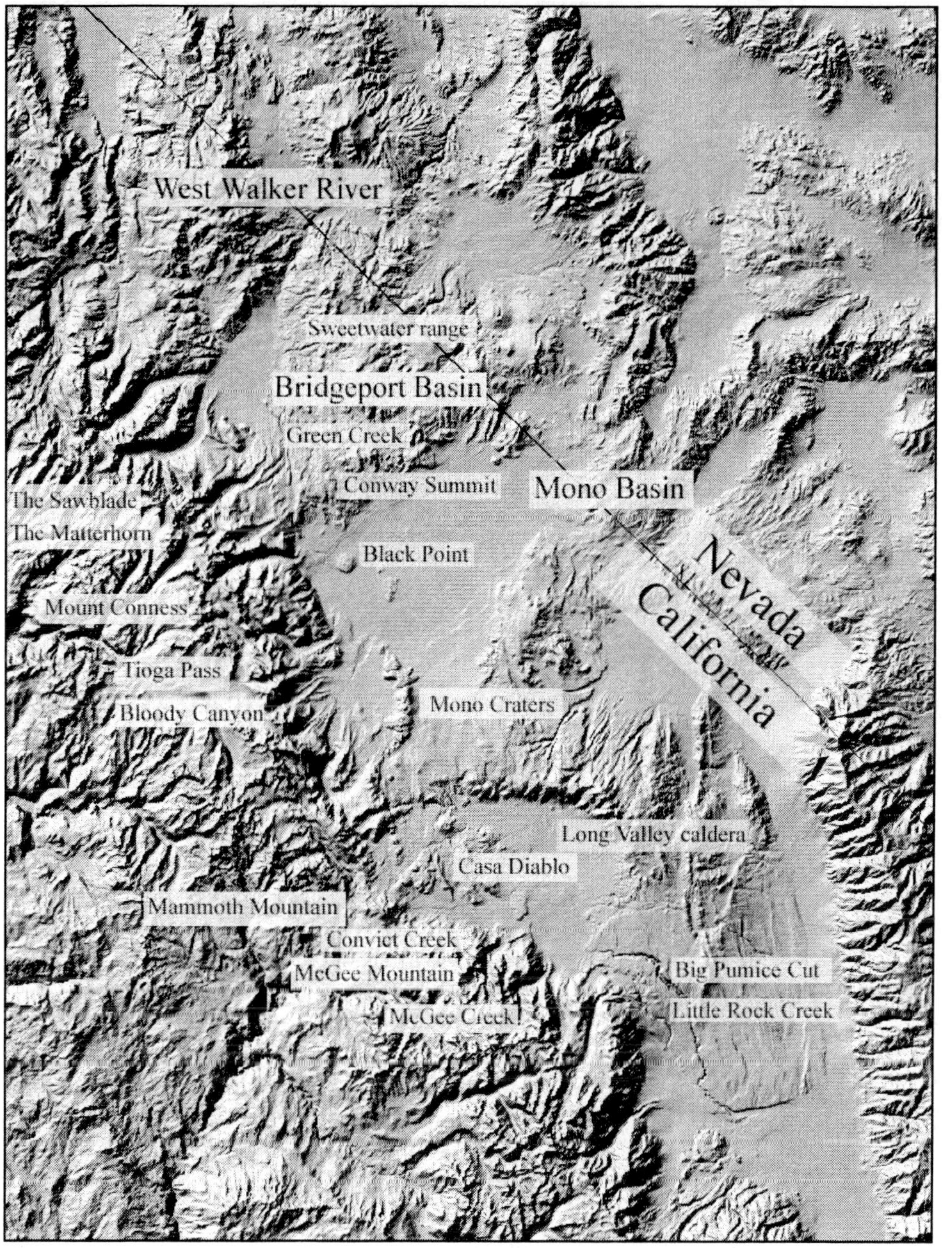

Figure 7. Shaded relief map of the Sierra Nevada from the West Walker River to northern Owens Valley (source: USGS digital elevation data, courtesy Ralph Haugerud, University of Washington).

welder, 1931, Sharp, 1968). It also probably correlates with the Sherwin (?) glaciation of Sharp (1968) in Bridgeport Basin.

The Tahoe till at this stop extends only a few kilometers to the east and north, and does not entirely cover the older fluvial or glaciofluvial gravel that underlies Burcham and Wheeler Flats. This older, pre–Tahoe gravel is at least 200 m thick and was deposited after two larger and older glaciers withdrew from here. The main gravel sources were apparently to the west in the upper canyon of the West Walker River. The gravel accumulated in a basin at least as deep as the present one, but before the Tahoe glaciation (Clark, 1967). A tephra within the gravel is the subject of STOP 2–3.

STOP 2–3. PRE–TAHOE TILL OF THE WEST WALKER RIVER

Return to US 395. The stop is a cut bank just northwest of the bridge across the creek, at km 263 (mile 163; #12, Fig. 1).

This roadcut exposes fluvial gravel of Wheeler Flats that contains a tephra identified as Rockland Ash–Tuff (Sarna–Wojcicki et al., 1985). The Rockland Ash–Tuff was erupted about 550 ka (A. M. Sarna–Wojcicki, personal communication, 2002) from near Mt. Lassen, a volcano 320 km to the north. The tephra dates the gravel deposit as post–Sherwin, but pre–Tahoe. If the gravel is glaciofluvial, as seems likely, the tephra may date an unnamed Sierra Nevada glaciation that was less extensive than Tahoe, but perhaps of greater duration.

The gravel deposit has a nearly flat upper surface that is only partly overlain by Tahoe till. The upper surface reaches nearly to the level of the sinuous bedrock channel of Devils Gate to the east, suggesting that overflow through the water gap limited the maximum thickness of the gravels.

STOP 2–4. OVERVIEW OF BRIDGEPORT BASIN

Continue south on US 395, which soon drops into Bridgeport Basin and the town of Bridgeport (founded in 1863). Turn right from US 395 entering town and drive about 1 km towards Twin Lakes. The stop is at km 289 (mile 180; #13, Fig. 1).

The Bridgeport Basin is a tectonic depression formed by a north–tilting block that is bounded on its northern margin by a normal fault. Subsidence here began more recently than on the margin of the Great Basin to the north. The alluviated floor of the basin is about 1970 m (6500 ft.), 600 m higher than Reno. At this elevation, the Pleistocene glaciers flowed far into the basin, and the low slope and relative tectonic stability have led to their exceptional degree of preservation.

The Recess Peak and Matthes glaciers, nevertheless, were all restricted to the cirques under the high crest of the Sierra Nevada, as can be seen in the Sawtooth range to the southwest. North of the Bridgeport Basin, the Sierra Nevada crest is lower, and the Matthes glaciers here are the northernmost active glaciers of the Sierra Nevada. Only a tiny glacier remnant remains on Mt Lassen. However, the giant stratovolcanos of the Cascades Range from Mt Shasta in California to Mt Garibaldi in British Columbia host many large, active glaciers.

The prominent peak south of the Sawblade, the long, thin arête in the Sawtooth Range, is known as "The Matterhorn," although at 3738 m (12,300 ft.) it is 740 m lower than its Swiss namesake. The peak is an easy climb from the southwest. The Matterhorn was Jack Kerouac's metaphorical goal in his 1950's novel, "The Dharma Bums." Kerouac's novels were important to a generation or two of Americans, and comparison of reality to his imaginative descriptions is interesting.

To the south, the moraine complex of Green Creek can be seen above the meadows of the basin's floor (Fig. 8). They are our next destination. As we drive south, keep in mind that Bridgeport Basin is probably a close modern analog for the appearance of much of the eastern Sierra Nevada in the early Pleistocene, with low relief between the desert uplands and the mountain crest.

STOP 2–5. TIOGA MORAINES AT DYNAMO POND, GREEN CREEK

Return to US 395 and drive south, through Bridgeport. Turn right at Green Creek Road, just before US 395 climbs to Conway Summit. Drive up this dirt road and pass through the end moraines, stopping at Dynamo Pond, at km 304 (mile 189; #14, Fig. 1).

Figure 8. Green Creek moraines. Stops are numbered. (Photo by C.C. Plummer)

Dynamo pond is a small, historic reservoir constructed in 1892 behind Tioga end moraines. It was built to provide electricity to Bodie, then a mining boomtown ~25 km to the east in the desert ranges. This utility was the first in California to transmit electricity over such a distance.

Comparison of the morphology, soil development, and boulder weathering of these Tioga moraines to the ones from Lake Tahoe (DAY 1) and also to the older members of the moraine sequence here at Green Creek (Fig. 8) is informative.

The Tioga glaciers advanced several times between about 30 and 14 ka (^{36}Cl and calibrated ^{14}C). Commonly, two large moraines are preserved. Within 4 km of Dynamo Pond are not only Tioga moraines, but also older moraines that have been assigned ages correlated to the rest of the Sierra Nevada sequence: Tenaya, Tahoe, Mono Basin, and Sherwin (Sharp, 1972). Relative dating is especially useful here because of the restricted range of elevations and the virtually identical climatic conditions from moraine to moraine.

STOP 2–6. TAHOE MORAINE CREST, GREEN CREEK

Return to the north ~4 km on the Green Creek road, then double back to the right on another dirt road that proceeds up the outside of the right–lateral moraines to an obvious point near the crest of the main moraine at km 306 (mile 190; #15, Fig. 1).

The elevation of the crest of this right–lateral moraine here is 2438 m (8050 ft.), 467 m above the town of Bridgeport. The flat, broad–crested moraine with few, but heavily eroded, boulders is typical for moraines of the Tahoe glaciation and is in stark contrast to the sharp–crested Tioga moraines with their numerous, fresh–appearing boulders. However, this moraine crest is deceptively simple, and careful observation shows that it is a composite landform consisting of Tahoe and younger moraines. A cross section would reveal a normal, nested sequence, but here on the south side of Bridgeport Basin, with so little tectonic subsidence, the moraine crests are within a few meters of vertical distance of each other. A traverse down the crest of the moraine passes from Tahoe to Tioga and back again, all in a few dozen meters, as the younger moraines overlap the Tahoe moraine. Robert Sharp considered many of the famously massive Tahoe moraines of Blackwelder (1931) to be composite, especially where basin down–faulting made burial of older moraines more probable. He referred casually to these as "onionskin" moraines.

The proximity to each other of the Green Creek moraines of different ages not only reflects the relative tectonic stability, but also the curious and unexplained fact that, for a period of ca. 100,000 years or more, the maximum lowering of regional snowline during different stades was about the same, ~900–1000 m. Furthermore, the equilibrium–line altitudes (ELAs), roughly the average snowlines, for these ancient glaciers tracked one another closely for 450 km from south to north along the crest of the Sierra Nevada. These observations (Gillespie, 1991; Gillespie and Molnar, 1995) suggest a remarkable similarity of climatic conditions and change, and also show that the history of ELA depressions in the Sierra Nevada differed significantly from the familiar sawtooth pattern of the high–latitude ice–sheet volumes.

STOP2–7. SHERWIN MORAINE, GREEN CREEK

Continue southwest 3 km to the junction of an old, rounded moraine remnant and the main moraine. Take a short spur road ~1 km to a high point on this remnant, at km 307 (mile 191; #16, Fig. 1).

The moraine at this stop was mapped by Sharp (1972) as Sherwin. Note that the only boulders remaining on the surface are what Sharp called "resistates," including fine–grained aplites that comprise only a few percent of the original population of tillstones. The rest have weathered away, and even those that might currently be exposed by mass wasting of the moraine crest are deeply weathered and will have a relatively short residence time on the surface. This is an indication of great age.

The Sherwin glaciation at its type locality south of Long Valley predates the eruption of the 758 ± 1 ka (Sarna–Wojcicki et al., 1997) Bishop tuff from Long Valley, ~50 km to the south. According to Blackwelder (1931), Sherwin till is so badly eroded that the original shape of the moraines has been lost, and most exposed boulders have disintegrated completely. Therefore, the "Sherwin" moraine at this stop is anomalous because something of its original shape appears to be preserved. Whether it is a true Sherwin moraine remnant, however, or a younger, unnamed one, it is clearly older than the Tahoe glaciation–that is, it is likely older than MIS 6, or older than 190 ka.

This ancient moraine remnant is recognizably part of the same nested sequence culminating in the LGM Tioga moraines. In other words, the "Sherwin" glacier at Green Creek was not much larger than its successors, in contradiction to common claims that Sherwin till is found far beyond the limits of the late Pleistocene moraines.

This is an important issue because, as noted above, for the Sierra Nevada, maximum glaciations tend to have similar ELA depressions, regardless of their timing with respect to the major ~100,000–yr. global glacial cycle. The Sherwin Glaciation stands out as an unexplained exception, with possibly important climatic implications if the extent of its glaciers is really significantly larger than the extents of younger glaciers. The Green Creek Sherwin moraine is evidence against this view, if it truly dates from the Sherwin glaciation.

STOP 2–8. OVERVIEW OF MONO BASIN FROM CONWAY SUMMIT

Continue southwest and then south to Virginia Creek. Turn left on the paved road to return to US 395, and drive south to a large parking area just where the highway begins to steeply descend into Mono Basin, at km 327 (mile 203; #17, Fig. 1).

Conway Summit (2480 m; 8200 ft.) is the dramatic northern edge of Mono Basin, a more developed version of Bridgeport Basin that is today hydrologically closed. Mono Basin appears once to have been the northernmost basin drained by the Owens River system (Reheis et al., 2002), which terminated far to the south in Death Valley. Today, the headwaters of Owens River are near Deadman Summit, and its the interpluvial terminus is Owens Lake, at the southern end of Owens Valley. The three main points of interest here are: (1) the voluminous outcrops of "Sherwin till"; (2) the excellent view of the old upland surface, now gently dipping to the north; and (3) the history of the subsidence of Mono Basin and the alkaline lake it contains.

Sherwin till. The road cut north of US 395 at Conway Summit reveals a thick diamicton blanketing the deeply weathered granitic bedrock. A similar diamicton blankets the slopes of the basin floor to the west between Virginia Creek and the moraines of Green Creek. Bedrock prominences locally protrude through this eroded deposit. Sharp (1972) considered these deposits to be Sherwin till and/or outwash. The till here, seemingly far from the late Pleistocene moraines, has provided some of the fuel for arguments that the Sherwin glaciers were of much greater extent than their successors.

What evidence can be seen here at Conway Summit that allows distinction of till and outwash? Especially near glaciers, fabric can be non–distinctive, and locally fluvial deposits are often found within moraines.

Some of the outcrops identified as "Sherwin till" may in reality be outwash and/or till reworked in debris flows, transported fluvially far beyond the end moraines (e.g., Birkeland et al., 1980). In this case, to infer that the Sherwin ELAs were much lower than those of the late Pleistocene glaciers is not necessary.

On the other hand, if this diamicton consists in part of primary till, then the Sherwin or other pre–Tahoe glaciers must have been more extensive than their successors, although here in Bridgeport Basin an additional ELA depression of as little as ~100 m might explain the difference in extent.

Upland surface. Conway Summit was once a non–distinctive location on a large upland surface, now disrupted by the down–faulting of Mono Basin. Remnants of this old surface can be seen near Lundy Canyon, the northernmost canyon draining the Sierra Nevada into Mono Basin. Remnant benches of this old surface project onto the north–dipping piedmont at Virginia and Green Creeks, and dip north at about the same angle.

Southern Bridgeport Basin is probably the best analog for the early Pleistocene upland surface east of the Sierra Nevada. The presence of deep layers of putative Sherwin gravel at Conway Summit, regardless of their origin, shows that this surface persisted at least partly intact until ~850 ka, the age estimated for the Sherwin till at its type area. Recall, however, that these diamictons have not been numerically dated at this locality.

If Mono Basin began subsiding at ~850 ka, the long–term rate has been >0.6 mm/yr., based on the elevation difference between Conway Summit and the surface of Mono Lake (currently 1948 m). Normal faulting rates between the basin and the Sierra Nevada block to the west, based on ages of offset moraines, are 1–2 mm/yr. (Bursik and Gillespie, 1993). This rate is greater than faulting rates in the Bridgeport Basin or in Owens Valley, to the south. Evidently, the broad ancient upland between Owens Valley and Carson Valley is being disrupted from the south.

Subsidence of Mono Basin. Evidence that subsidence of Mono Basin had begun by Sherwin time is found north of Walker Creek (STOP 22), on the northwestern slope of the basin, where Sharp and Birman (1963) mapped eroded Sherwin till down to ~2300 m (7600 ft.), 140 m below Conway Summit. (In the American West are many geographic features named after Joseph Walker, a widely traveled 19th century explorer and guide. Walker Creek and Walker River are otherwise unrelated.) Evidently, the ancient broad upland between Owens Valley and Carson Valley is being disrupted from the south because the Mono Basin appears to have been deepening since Sherwin time, especially relative to Bridgeport Basin.

If Mono Basin began subsiding before 850 ka, the long–term rate has been >0.6 mm/yr, based on the elevation difference between Conway Summit and the surface of Mono Lake (currently 1948 m). Normal faulting rates between the basin and the Sierra Nevada block to the west, based on ages of offset moraines, are 1–2 mm/yr. (Bursik and Sieh, 1989). This rate is greater than faulting rates in Owens Valley, to the south, and cannot have been maintained for long.

Mono Lake is the most obvious feature of Mono Basin. During and following the LGM, it was much deeper than today, and its ancient shorelines are widely preserved. Black Point (2120 m; 7000 ft.), a 15 ka sublacustrine, basaltic volcano on the northwest shore of the lake, rose above lake level only a few tens of meters then.

The lake today is artificially low, due to the ongoing diversion of water since 1941 into the Los Angeles aqueduct from the streams that feed Mono Lake. However, Stine (1994) showed that even this low level is higher than low stands that occurred during severe droughts between A.D. 900 and 1350.

South of Mono Lake is Mono Craters, a striking chain of volcanic domes. They are the surface expression of silicic dikes intruding faults linking Mono Basin and Long Valley caldera to the south. The northernmost vent, Pannum Crater, ejected >1 km^3 of tephra about 605 ±20 ^{14}C yrs. B.P. (Sieh and Bursik, 1986), and volcanic activity probably will be renewed here someday.

The islands in Mono Lake are remarkable: the larger light–colored one (Paoha Island) is composed of lake sediments that were uplifted by magma rising below the surface, and the other (Negit Island) is built from black, basaltic, cinder cones. Stine (1987) reported that Paoha Island emerged from Mono Lake prior to A.D. 1690.

Follow US 395 down the steep grade to Mono Lake and continue south along its western shore to the town of Lee Vining. The stop for the evening is at the Best Western Motel at km 346 (mile 215; #18, Fig. 1).

DAY 3. POST–LGM GLACIATION NEAR THE SIERRA NEVADA CREST

Day 3 is our most strenuous day. We start with an overview of Mono Lake, including discussions of the lake level history, and its relation to glaciations in the Sierra Nevada. We will then drive up Lee Vining Canyon, one of the most impressive glacial canyons on the east side of the range, and spend most of the day inspecting moraine and lake records of the latest Pleistocene and Holocene glaciations. These records place important constraints on climatic variations in the Sierra Nevada during the last ~15,000 years and provide an interesting contrast with similar records in Europe, Asia, and the Rocky Mts.. These detailed records of dated sediments are also important for testing ideas on production rates for cosmogenic nuclides that are gaining prominence as a tool for dating geomorphic and glacial surfaces and landforms.

STOP 3–1. MONO LAKE

A short distance north of Lee Vining is the Mono Lake Overlook, at km 348 (mile 216; #19, Fig. 1).

The Mono Lake Overlook is a good vantage point from which to view the Tioga delta of Lee Vining Creek, on which we spent the night, and the range–front. At Mono Lake County Park, a maintained walkway leads from the parking lot down through marked historic shorelines to the present–day lakeshore. This walk gives a sense for the dramatic lowering of the lake within a single lifetime. Although lake level is now ~13 m below the historic high in 1920, prehistoric droughts nearly a millennium ago resulted in even greater desiccation (Stine, 1994). Two droughts, lasting >200 and >140 years occurred at about A.D. 900–1112 and 1210–1350, just before the onset of the Matthes Glaciation in the Sierra Nevada (see STOP 3–2) and the Little Ice Age in Europe. These droughts are not only recorded in the geologic record, but also resulted in major discontinuities in the archaeological records, possibly including the end of the Anasazi or Pueblan Ancestors in present–day southern Colorado, New Mexico, and Arizona. Here in the eastern Sierra Nevada, one hypothesis envisions a wave of immigration following the droughts, around A.D. 1400. Recently, however, this inference has been questioned, and an alternative proposed: the Native peoples may have changed their living patterns and settlements during the drought instead of leaving and returning.

The droughts and neoglaciations of the past millennium are a reminder that our recent ancestors faced sudden and dramatic climatic oscillations on a scale we do not appreciate well. To this natural climatic baseline, we are potentially adding anthropogenic changes deriving from our unbridled activities, including deforestation and industrial development.

Mono Lake is home to brine shrimp that can tolerate the alkaline waters and to numerous birds, especially gulls, which nest on its islands. Environmentalists are concerned that any further lowering of Mono Lake will cause Negit Island to join the mainland via a narrow isthmus that will allow predation of gull chicks and eggs by coyotes. The lake level was stabilized at 1944 m by a court order in 1989.

Mono Lake is also home to algae, and the floor, at least near beaches, is locally covered with thick, black ooze. Tufa, a fibrous calcareous mat, is also deposited in the lake, and tufa towers built around underwater springs and seeps near the modern southern and western shorelines are prominent tourist attractions. Tufa is a possible candidate for ^{14}C dating (e.g., Lajoie, 1968). However, at Mono Lake, calcareous deposits on telephone cables gave a reservoir effect of thousands of years.

STOP 3–2. SADDLEBAG LAKE

From the Mono Lake stops return to Lee Vining and turn onto the Tioga Pass road, just south of town. The road climbs steeply up the scree slopes on the north side of Lee Vining Canyon. Before reaching Tioga Pass, turn north on a paved road to Saddlebag Lake, at km 366 (mile 227; #20, Fig. 1). We will spend the balance of DAY 3 here, hiking to lakes and ponds associated with the latest Pleistocene Recess Peak glaciers, and, optionally, we may continue to the Conness Glacier.

At Saddlebag Lake, we will board a water taxi to cross the lake (actually a reservoir) to the inlet creek. From the drop–off point, Greenstone Lake, the first stop is a short 0.5–km hike.

GREENSTONE LAKE

Greenstone Lake is a shallow tarn fed by meltwater from Conness Glacier, nestled in the cirque on the northeast face of Mt. Conness (Fig. 7). Several other small tarns above Greenstone Lake capture the majority of the glacial flour produced by the modern glacier. Below the modern glacier are two sets of moraines, the Matthes, or Little Ice Age moraines directly adjacent to the glacier, and the older, Recess Peak moraines, which are visible as low, tree–covered ridges ~0.4 km upstream from the inlet of Greenstone Lake (Fig. 9).

From Greenstone Lake, we will hike up the inlet creek to the Recess Peak end moraines (Fig. 9). The Recess Peak moraines above Greenstone Lake are typical of those that occur throughout the higher parts of the Sierra Nevada, within several kilometers of higher cirques and typically many kilometers above most Tioga recessional deposits. As first described by Birman (1964), Recess Peak moraines typically look "fresh," with abundant unweathered (but lichen–covered) boulders, poorly developed soils, and scarce vegetation. As with the ones we visit, the moraines are generally small in volume with stable slopes, but with only minor post–depositional modification. Because of their fresh character, most early workers concluded that the Recess Peak moraines related to a Neoglacial advance that occurred within the past 2000–3000 years (e.g., Birman, 1964; Curry, 1969; Scuderi, 1987).

Although soil work by Yount et al. (1982) suggested that at least some Recess Peak deposits were pre–Neoglacial, at that time most were thought to be Neoglacial, and firm numerical constraints on the moraines from sediment coring of nearby lakes were required to settle the controversy. Greenstone Lake provides an ideal coring site to test the age: (1) it is shallow and accessible, and thus easily cored, and (2) it lies directly downstream from well–formed Recess Peak moraines and therefore captures the entire package of outwash sediment related to the advance and retreat of the glacier during this event. Cores from Greenstone Lake preserve a distinct pulse of low–organic outwash related to emplacement of the Recess Peak moraines just upstream (Fig. 10) (Clark, 1997). This pulse of outwash occurs near the bottom of the lake sediments, indicating that the event occurred shortly after Tioga deglaciation of the basin. Radiocarbon dates of organic material from directly above and below the outwash package (both at Greenstone Lake and in other similar lakes) demonstrate that the Recess Peak advance began by ~12,200 ± 60 ^{14}C yrs. B.P. (~14,200 cal yrs. B.P.) and ended before ~11,190 ± 70 ^{14}C yrs. B.P. (~13,100 cal yrs. B.P.). The Recess Peak advance thus shortly precedes, and is distinct from, the Younger Dryas event.

From Greenstone Lake we hike west (and up) toward the Conness Glacier.

Matthes moraines and the Conness Glacier

No organized moraines occur upstream from the Recess Peak moraines until the small but voluminous moraines correlative to the Matthes advance (Fig. 9) (Birman, 1964). Moraines associated with the Matthes advance occur below high cirques and active glaciers throughout the higher parts of the Sierra

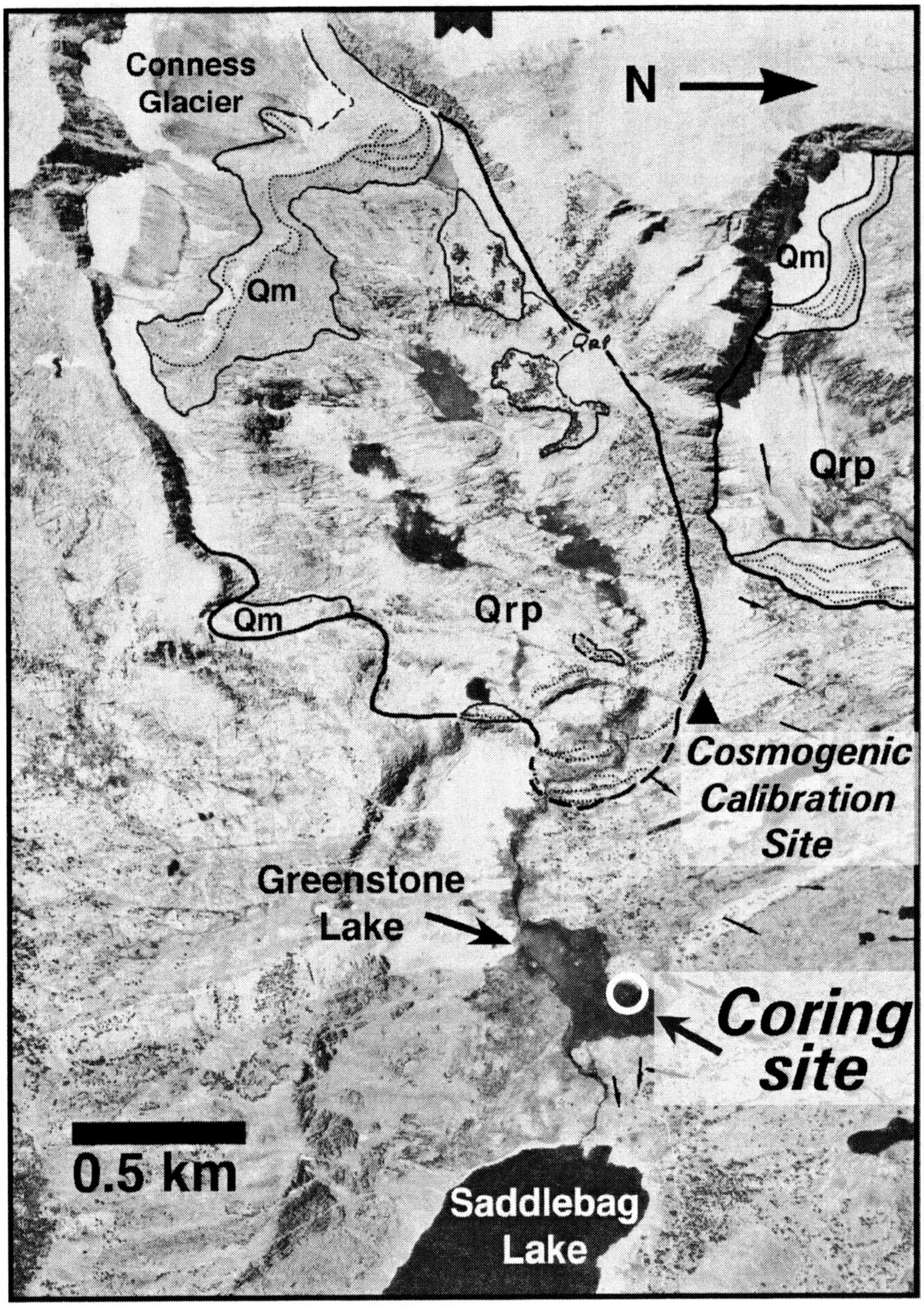

Figure 9. Aerial photograph of Conness cirque, showing coring site at Greenstone Lake, and Recess Peak (Qrp) and Matthes (Qm) moraines below Conness Glacier. Location of calibration sample site of Nishiizumi et al., 1989) is also indicated.

Nevada from Lake Tahoe to south of Mt. Whitney. The moraines below the Conness Glacier are characteristic of Matthes deposits: fresh, steep, unstable, angular clasts nearly free of vegetation, with boulders that support only small lichens. The absence of a ~1190 ^{14}C yrs. B.P. (~1100 cal yrs. B.P.) Mono Craters tephra blanketing Matthes moraines farther to the south, as well as lichenometry on the moraines, clearly indicate that Matthes glaciers reached their maximum positions in the past millennium, probably the past 200–300 years. Together, these constraints indicate that the Matthes advance correlates with the global Little Ice Age. Coring of the Conness Lakes indicates that Neoglaciation in the area began by ~3200 ^{14}C yrs. B.P. (3400 cal yrs. B.P.) (Konrad and Clark, 1998).

The absence of any moraines between the Recess Peak moraines and the Matthes moraines, as well as the absence of any periods of outwash between 13,100–3400 cal yrs. B.P., indicates that no significant glacial advances in the Sierra Nevada occurred during that time, including during the Younger Dryas interval (Clark and Gillespie, 1997). This finding, combined with evidence favoring Younger Dryas advances in the Rocky Mts. (e.g.

Reasoner et al., 1994) and the North Cascades (e.g., Kovanen and Easterbrook, 2001), suggests a complex regional climate during the late–glacial period. It also indicates that the climate during the Little Ice Age was the coldest and/or wettest (i.e., most glacial) in the Sierra Nevada of the past 13,000 years.

As we hike back down towards Greenstone Lake, we will cross bedrock immediately north of the Recess Peak terminal moraine. The location was one of the sample sites for the original production–rate calibration of *in situ* cosmogenic ^{10}Be and ^{26}Al (Fig. 9)(Nishiizumi et al., 1989). In the paper, Nishiizumi et al. assume a deglaciation age of 10,000 ^{14}C yrs. B.P. (~11,000 cal yrs. B.P.). The basal age from the Greenstone Lake cores instead demonstrates that the area was deglaciated by ~13,100 ^{14}C yrs. B.P. (~15,500 cal yrs. B.P.) (Clark, 1997). As a result of this and other similar findings, production rates for these isotopes have since been reduced by ~15–20%: current sea–level, high–latitude production rates in quartz are 5 and 31 atoms g^{-1} yr^{-1} for ^{10}Be and ^{26}Al, respectively, as compared to the original estimates by Nishiizumi et al. (1989) of 6.03 and 36.8 atoms g^{-1} yr^{-1} for ^{10}Be and ^{26}Al, respectively.

Return to the Best Western Motel in Lee Vining for a second night, at km 388 (mile 241; #21, Fig. 1).

DAY 4. DATED PLEISTOCENE MORAINE SEQUENCE AT BLOODY CANYON

On DAY 4, we will inspect field and cosmogenic dating evidence for the age of moraines at Bloody Canyon. This is one of the most complete and well–studied sequences in the Sierra Nevada, and is the type area for Mono Basin till, thought by Sharp and Birman (1963) to be stratigraphically between Sherwin and Tahoe on the basis of field relations and relative dating.

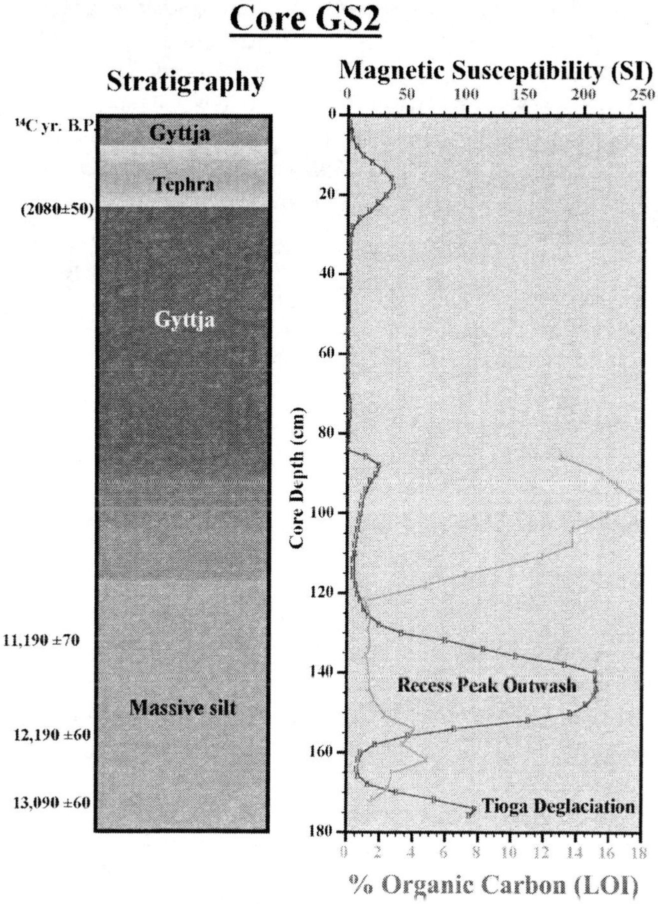

Figure 10. Simplified stratigraphy, organic content, and magnetic susceptibility of cores from Greenstone Lake. Note interval of low–organic, high–susceptibility sediments near bottom of core, related to formation of Recess Peak moraine upstream (Fig. 9).

STOP 4–1. BLOODY CANYON

Proceed south on US 395, and just at the point where the highway leaves the range front south of Lee Vining Canyon, exit onto old US 395. At km 393 (mile 244; #22, Fig. 1) is an excellent view of the moraines of Walker Creek.

From this position near the western edge of Mono Basin, view the moraine sequence of Walker Creek. The canyon of Walker Creek is known as "Bloody Canyon," either because of the red metamorphic rocks of the roof pendant or because mules making the crossing of the Sierra Nevada at Mono Pass, the low divide at the head of Bloody Canyon, fell from the steep cliffs. During Pleistocene glaciations, Mono Pass was occupied by ice flowing eastward from the Tuolumne ice sheet, and trim lines are clearly evident in the view up the canyon from our stop.

Bloody Canyon hosts one of the best–studied and most–dated moraine sequences in the western United States. Although studied by many workers, the main reference to the geology is Sharp and Birman (1963), and the main references to the dating are Phillips et al. (1990, 1996).

More than a dozen moraine crests in two geographic groups are preserved at Bloody Canyon. The older group clusters around Sawmill Canyon, a dry canyon south of Walker Creek that was beheaded by younger glaciers whose right–lateral moraines parallel the present course of Walker Creek.

The youngest, most obvious, moraines of Sawmill Canyon were defined as the type "Mono Basin" moraines by Sharp and Birman (1963) and were thought to predate the Tahoe glaciation. The left–lateral moraine is broad–crested and has few large boulders exposed. In contrast, the right–lateral moraine has a much narrower, somewhat bouldery crest, possibly due to post–depositional mass wasting. Birkeland and Burke (1988) estimated that 10 m or more of soil has been removed from the crests of these moraines.

The moraine that truncates Sawmill Canyon was mapped as "Tahoe" by Sharp and Birman (1963), but field inspection suggests that it is a composite "onionskin" moraine of at least two ages. The older crops out about a third of the way down the southern flank (indicated by a consistent inflection in the slope, Fig. 11), has a reddish soil and few exposed boulders. The younger moraine supports a less developed tan soil, and the crest is occupied with somewhat–fresh–appearing granitic boulders. These moraines clearly cross–cut and therefore post–date the Mono Basin moraines.

The Tahoe moraine overlies another, older sequence of moraines exposed at the end of the right–lateral complex, extending a few tens of meters above a high shoreline of Mono Lake at 2145 m (7080 ft.). Field geologists consider these to also post–date the Mono Basin moraines.

Nested inside the big Tahoe moraine are younger, latest Pleistocene moraines of the Tenaya and Tioga stades, thought to date from the LGM. The heavily eroded sheets of Sherwin till referred to at STOP 2–7 are mapped on the hillside north of Walker Creek's left–lateral moraines. The Sherwin (?) glaciers apparently deposited their till and outwash when Mono Basin was not as deep as it is today. In contrast, the post–Sherwin moraine crests of Walker Creek all record glaciers that must have appeared similar to each other, and advanced when Mono Basin looked quite "modern." However, the older moraine crests at Sawmill Canyon are topographically lower than the young ones at Walker Creek (Fig 11). We will discuss several explanations that have been offered to explain this elevation difference, including the possibility of down–faulting along the basin's western margin (Clark, 1972).

This interpretive history, developed from field relations and relative dating, has been challenged on the basis of cosmogenic ^{36}Cl exposure ages, which appear to show that the old, largely buried moraines at the terminus of the right–lateral Tahoe composite are older than the boulders of the Mono Basin moraines. This interpretation requires that the Bloody Canyon glacier changed course more than once. At the next few stops we will see the field evidence and discuss the dating evidence that underlie this debate.

STOP 4–2. LATEST PLEISTOCENE MORAINES ABOVE WALKER LAKE

Proceed on a dirt road toward the mouth of Walker Creek. A short distance south, a dirt road extends westward between two moraines. This dry canyon is Sawmill Canyon, and the dirt road continues up a larger and younger right–lateral moraine from Walker Creek that beheaded Sawmill Canyon. A short distance past the moraine crest is a parking area in the fosse between Tioga and Tahoe moraine crests, at km 405 (mile 252 #23, Fig. 1).

STOP 4–2 is in the Tioga/Tahoe fosse above Walker Lake. The view of Walker Lake shows a classical, glaciated, Sierra Nevadan valley. However, the point of this and the next two stops is to compare the appearance of the moraines of different ages and to get a feel for the amount of mass wasting that has occurred,

Figure 11. View of the Bloody Canyon moraines from Tahoe I end moraines. Tahoe II moraines comprise high ridge to right (north) and Mono Basin left–lateral moraine emerging from it in left middle–ground.

along with the degree of soil development and boulder weathering. The moraine north of the parking area is the Tioga right–lateral moraine, and the moraine just south of the parking area is the Tahoe II (younger Tahoe) moraine. Burke and Birkeland (1979) studied the soil development here and were surprised to observe that the Tahoe and Tioga soils were less distinct than expected. They also noticed that the boulders on the Tahoe crest appeared similar to the Tioga boulders, which they attributed in part to "freshening" of the boulders due to spalling during forest fires.

Spalling may diminish apparent weathering of a rock by removing the outer, weathered material, but heating in fires also creates microfractures throughout the outer part of a boulder, even where it does not spall. This microfracturing should change the elastic properties of the boulder, including the speed of sound through it. Therefore, while freshening the appearance of boulders, fires should also increase the degree of weathering as measured by the speed of sound. Crook and Gillespie (1986) showed that this was the case here and that the Tahoe and Tioga moraines were clearly separable on this basis.

A short distance down the outer flank of the Tahoe moraine is the color discontinuity that reveals the subtle change in the weathering of the till described at STOP 4–1. This difference in soil development is also detectable in the data of Birkeland and Burke (1988), who dug and described soils in five pits from the crest of the moraine to its outer base near here in a study of catena development.

Because the moraine crest is shedding material to form a catena, weathered material would be expected accumulate at the base of the moraine. In this case, however, the catena alone clearly cannot explain the observations: occasional tillstones exposed below the contact are decidedly more weathered than tillstones above it, and where the contact between the reddish and tan units is exposed in gullies, the tan till overlies the reddish till–the opposite of the relation expected for a catena. The reddish till was mapped by Gillespie (1982) as the "older Tahoe." The contact is best exposed a short distance east of the junction of the right–lateral Tahoe and left–lateral Mono Basin moraines.

STOP 4–3. MONO BASIN MORAINES TYPE AREA, SAWMILL CANYON

Return to the junction between Sawmill Canyon and the right–lateral moraine of Walker Creek and park on the south (right) side of the road, in a short turn–out at km 407 (mile 253; #24, Fig. 1).

STOP 4–3 is in Sawmill Canyon between the right–and left–lateral Mono Basin moraines. The left–lateral moraine clearly underlies the reddish older Tahoe till, discussed at STOP 4–2, which can be seen best from its color on the side of the composite, right–lateral Tahoe moraine. The crest is the tan, bouldery, younger Tahoe.

The crest of the left–lateral Mono Basin moraine is broad and smooth, with few boulders greater than 0.5 m in diameter. In contrast, the right–lateral Mono Basin moraine is relatively sharp–crested and somewhat bouldery. The obvious discrepancy between the appearance of the right– and left–lateral moraines is almost certainly not due to a difference in landform age, but is more likely related to undercutting of the right–lateral moraine by fluvial action at its outer base, and subsequent mass wasting. Relative dating distinguishes between the younger Tahoe and Mono Basin moraines, but soil development suggests that these moraines may both date from MIS 6.

STOP 4–4. PRE–TAHOE TERMINAL MORAINES OF WALKER CREEK

Return to the mouth of Walker Creek and park where the road skirts the ancient right–lateral end moraines at km 411 (mile 255; #25, Fig. 1).

STOP 4–4 is at the terminus of the right–lateral Tahoe moraine complex on an old shoreline of Mono Lake. Immediately west of the road are four sub–parallel moraines that crop out from beneath the older Tahoe moraine and therefore predate it. The outer three appear to be significantly lower topographically than the Tahoe moraine crests (Fig. 11, 12). This is consistent with a significant age difference in a setting such as this, i.e., with active basin subsidence. The topographic relation with the Mono Basin moraines is less clear, because the glaciers followed different paths. However, the inflection between the younger Tahoe and the underlying older till (Fig. 11) projects to near the pre–Tahoe moraine crests at this stop, suggesting a possible genetic connection.

The reason the glaciers followed different paths may be important to the debate over the ages of the moraines. The Mono Basin glaciers followed a more southerly path than the Tahoe and Tioga glaciers. Many ideas have been forward to explain this and similar "avulsions," including shadowing of the southern side of the east–flowing glaciers and tectonic tilting of the basin to the north. All such "systematic" explanations share in common the implication that, although the glacier path may shift, it does not shift back and forth repeatedly.

The pre–Tahoe and the Mono Basin moraines are not exposed in contact, although both are exposed in contact with and buried by the Tahoe moraines. Therefore, the stratigraphic relation between the pre–Tahoe and the Mono Basin moraines is strictly unknown. From Occam's Razor, the "avulsion" argument discussed above and the projection of the Tahoe moraine inflection, however, the pre–Tahoe moraines appear to post–date the Mono Basin moraines of STOP 4–3. Burke and Birkeland (1979) showed that soil development here was most likely equivalent to MIS 6. Relative–dating results, including measurements of sound velocity in tillstones, do not distinguish between boulders of the Mono Basin and pre–Tahoe moraines.

Phillips et al. (1990, 1996) have measured a large number of ^{36}Cl cosmogenic exposure ages from the Bloody Canyon moraines, and their results for the older moraines are unclear. The data (Fig. 13) show a great deal of scatter that is attributed by Phillips et al. (1996) to exposure of "fresh" (previously buried) boulders as soil is slowly stripped from the moraines. To infer the age of a moraine from the exposure ages of these tillstones, Phillips et al. used only the overlapping ages at the older end of the histogram for each moraine. Most researchers agree that the dates for the younger Tahoe and Tioga moraines are fairly accurate, probably within a few thousand years at least. These edited ages suggest that the younger Tahoe moraine dates from just 49 ± 2 ka. Its relative youth may partly explain the small differences found by Burke and Birkeland (1979) between it and the Tioga moraines, which have been dated to three or four episodes between about 32 and 15 ka, consistent with their usual assignment to the MIS–2 LGM. The debate arises with the dates of the older moraines. Phillips et al. (1990) concluded that the Mono Basin moraines at 115 ka post–dated, not pre–dated, the pre–Tahoe moraines we are standing on (145 and 200 ka).

Hallet and Putkonen (1994) showed by diffusion modeling of slope degradation that cosmogenic exposure ages for old moraines probably severely underestimate landform ages, a prediction that was supported by measurements of U/Th ages on calcite rinds on tillstones from Wyoming that were much older than the ^{36}Cl ages of the boulders there (Sharp et al., in press). Hallet and Putkonen further concluded that even apparent age reversals, such as observed for the Bloody Canyon moraine sequence by Phillips et al. (1990), were predicted from their models if boulder erosion was considered. Nevertheless, the ages and even the relative age relations among the older moraines of Bloody Canyon are unresolved today.

Retrace route to US 395, and turn south. We will shortly rise out of Mono Basin, cross Deadman Summit (Rest Stop) and drop into the Long Valley caldera. US 395 crosses the caldera floor, Crowley Lake, and then the caldera rim before dropping roughly 1000 m to the floor of Owens Valley and the town of Bishop. From the center of town, turn left (east) onto East Line

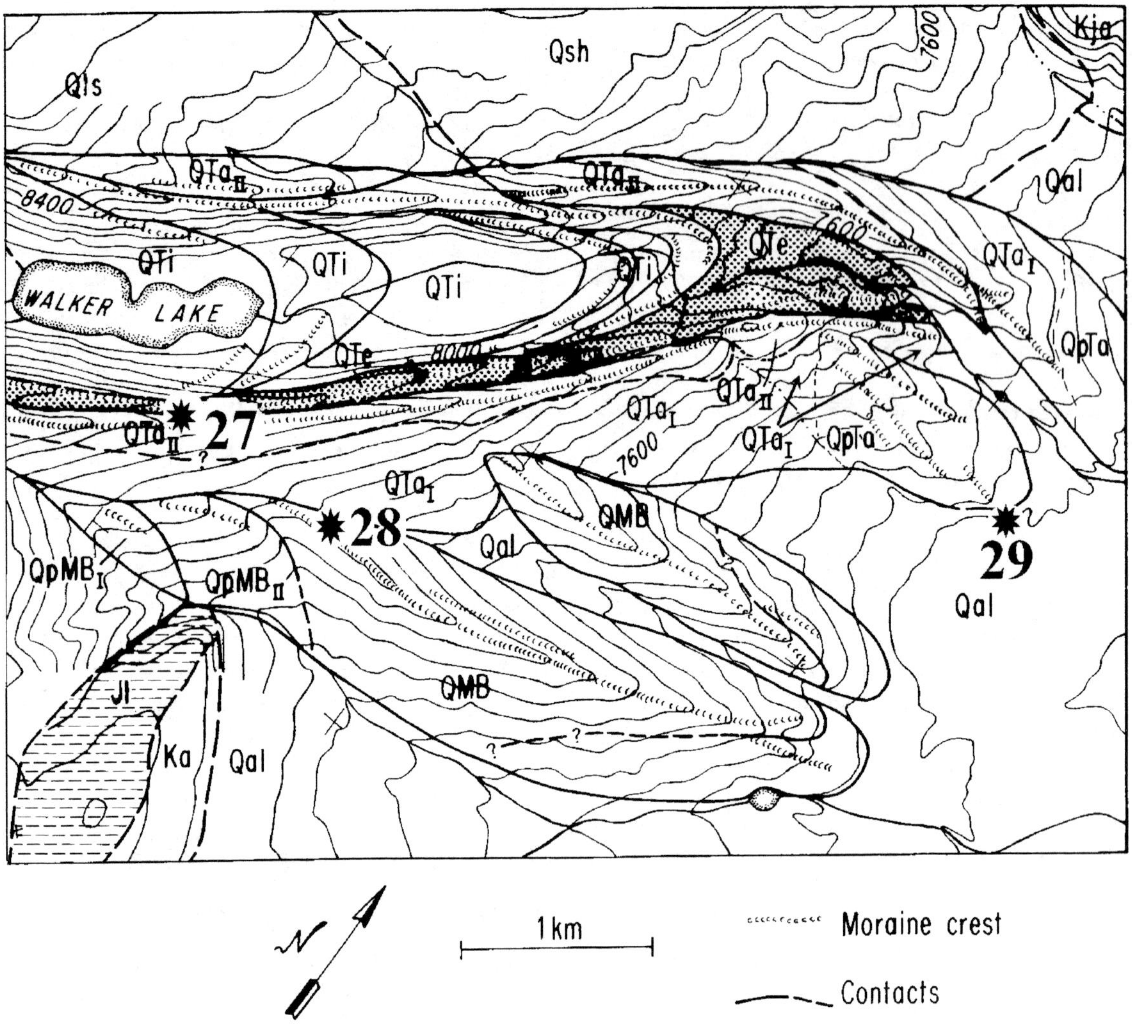

Figure 12. Map of the Bloody Canyon moraines (after Gillespie, 1982). Stops are shown by numbered star symbols. Jl, Ka, and Kja are plutonic rocks. Qal is Quaternary alluvium and Qls shows colluvial deposits. Qsh is Sherwin till of Sharp and Birman (1963). $QpMB_I$ and $QpMB_{II}$ are pre–Mono Basin moraines; QMB is Mono Basin moraines. QpTa is the oldest set of moraines (pre–Tahoe) along Walker Creek; QTa_I and QTa_{II} are the older and younger Tahoe moraines of Gillespie (1982). QTe (shaded for clarity) is the Tenaya moraine, and Qti are the undifferentiated Tioga moraines. Topographic contour interval is 24 m (80 ft).

Street and drive toward the White Mts. across Owens River. We will spend the night at the White Mountain Research Station, which is on the south side of the road, just before it turns south at the foothills of the White Mts.. The station is at km 517 (mile 321; #26, Fig. 1).

DAY 5. GLACIATION NEAR LONG VALLEY CALDERA

Today we will study outcrops of two ancient tills, ~850 ka and older, and see from afar the mountaintop location of a third. We will also see the Long Valley caldera, from which ~600 km^3 of Bishop tuff erupted 758 ka. We will look at faulting relations in late Pleistocene till on the western wall of the caldera at McGee Creek, and examine the type locality of the Casa Diablo till, originally thought to be nearly 400 ka, but now believed to be a MIS–4 or –6 till sandwiched between basalt flows dated by K–Ar at 126 ± 25 and 62±13 ka (Bailey et al., 1976). Casa Diablo is the final scheduled stop of the field trip.

STOP 5–1. ANCIENT TILL AT LITTLE ROCK CREEK

Take US 395 back to the base of the Sherwin Grade, the long sloping surface of Bishop tuff that emerges from the alluvium of Owens Valley north of Bishop, and turn left (west) across the highway and onto a paved road at Pine Creek. Immediately, turn right onto Little Rock Creek Road and follow it past Paradise Inn and across Little Rock Creek. The stop is at a road cut just before the road climbs out of the canyon to join with US 395 just south of the caldera rim. The road cut is at km 561 (mile 349; #27, Fig. 1).

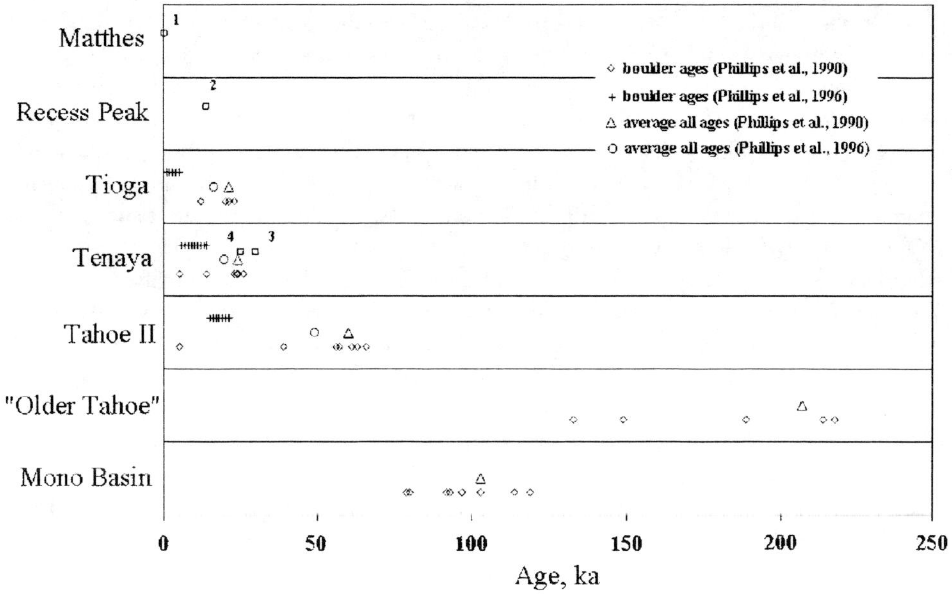

Figure 13. ^{36}Cl dates for Bloody Canyon moraines (Phillips et al., 1990, 1996), with the following independent age control: 1) 100–600 cal yrs. B.P. (Yount et al., 1982; Konrad and Clark, 1998; Clark and Gillespie, 1997). 2) 13,100–14,200 cal yrs. B.P. (11,200–12,200 ^{14}C yrs. B.P.) (Clark, 1997). 3) 30,000 cal yrs. B.P. (Bursik and Gillespie, 1993). 4) >25,000 yrs: ^{10}Be and ^{26}Al (Gillespie, unpublished data).

Driving north from the town of Bishop, we begin to rise out of Owens Valley, up the flanks of an ancient volcano. The rock here is the 758–ka Bishop tuff (Bailey et al., 1976). The tuff is normally polarized, and the eruption shortly postdated the Brunhes–Matuyama magnetic reversal. The age of the tuff is currently the best estimate of the age of the reversal.

Our stop is on the east bank of Little Rock Creek, which has eroded deeply through the tuff on the flank of the volcano to the underlying granitic rocks and sediments. Exposed in the road cut here are two especially interesting outcrops of till, described by Sharp (1968). The uppermost is the Sherwin till, long regarded as the oldest of the commonly exposed glacial sediments in the Sierra Nevada. The matrix of pulverized granite is still gray, and granitic tillstones have partially disintegrated or turned to grus.

The reason this stop is interesting is that a second, older till underlies the Sherwin till in this exposure. Sharp called this informally the "old red till" because its matrix has oxidized to a reddish hue. It is the oldest till described at the foot of the Sierra Nevada, but its age is unknown. Subsurface weathering characteristics have been used in an effort to correlate older tills here with those in the Bridgeport Basin. In addition, these weathering characteristics suggest the "old red till" of Sharp is at least 100,000 years older than the Sherwin till (Birkeland et al., 1980).

Although we know today that the Sherwin Till underlies the Bishop tuff (Sharp, 1968), this was a controversial inference a few decades ago. The controversy was finally resolved when the road cut we will see next was first exposed.

STOP 5–2. "TYPE AREA" FOR SHERWIN TILL AT THE BIG PUMICE CUT

Climb out of the canyon to the junction with US 395 and turn right. Immediately, a big turnout on the right side is present across from a large road cut at km 563 (mile 350; #28, Fig. 1). This is the Big Pumice Cut.

US 395 is 125 m above the Little Rock Creek stop. Our stop there is on the west side of the highway, across from the Big Pumice Cut where the controversy about the Bishop tuff and the Sherwin Till was finally resolved. In the road cut can be seen the bouldery, weathered till overlain by airfall pumice from the early stages of the eruption. The tuff itself caps the pumice. The till was already heavily weathered when it was buried by the pumice. Sharp (1968) and Birkeland et al. (1980) both estimated that it was weathered similarly to the till of the "Tahoe" glaciation, now thought to be from MIS 4 and/or MIS 6 (i.e., between about 50 and 190 ka). These age estimates are in partial agreement with a cosmogenic study that suggests the exposure time of the pre–Bishop Sherwin till was about 53–67 ka (Nishiizumi et al., 1989). Thus, the Sherwin glaciation here was around 800–900 ka.

STOP 5–3. OVERVIEW OF CROWLEY LAKE AND THE LONG VALLEY CALDERA

During a lull in traffic, cross US 395 to the northbound lanes and continue back across the caldera rim. A short distance farther is a turnoff to the right and our stop at km 567 (mile 352; #29, Fig. 1).

We have crossed into the southern end of Long Valley, the 32 x 19 x 3 km caldera of the volcano from which the Bishop tuff was ejected. The rhyolite tuff is the extrusive equivalent of batholithic rocks of the Sierra Nevada.

Crowley Lake is a reservoir, but the caldera was filled by a lake until it drained ≤100,000 years ago. Hills north of Crowley

Lake are a resurgent dome, emplaced about 600 ka (Bailey et al., 1976). Since about 100 ka, most volcanic activity has been localized north of Long Valley, where rhyolite plugs and domes have squeezed up along north–south faults. As recently as 1980, however, seismic evidence suggests magma intruding as dikes in the ancient caldera.

As we drive north by the town of Mammoth Lakes, we will pass well–preserved sequences of large moraines on the west side of Long Valley, deposited along Hilton Creek, McGee Creek, and Convict Creek, mostly during the last 200,000 years. The moraines have been faulted, especially well shown at McGee Creek. We will see the faulting evidence up close at the next two stops.

Convict Creek may host one of the most complete, best–preserved, moraine sequences in the Sierra Nevada. This is partly because the glaciers changed course, flowing northward some time before the early part of the Tioga glaciation, perhaps around 30 ka. Convict Creek received its name in 1871 when six convicts escaped from the jail in Carson City, Nevada and hid from searching posses here.

STOP 5–4. OVERVIEW OF MORAINES AT MCGEE CREEK

Continue north on US 395 to McGee Creek, and exit the highway to the left. This is the old Highway 395. We will later take the road up McGee Creek, but for now we double back south a kilometer to an excellent view of the moraines at km 577 (mile 359; #30, Fig. 1).

West of this stop are the moraines of McGee Creek (Fig. 14) that were left by the Tahoe, Tenaya, and Tioga glaciations, probably all since MIS 6. They have been deposited across the wall of the Long Valley caldera and are faulted. Perhaps the most dramatic indication of this is not the scarps themselves (visible here where the right–lateral moraine crosses the range front), but the lowering and partial burial of successively older moraine crests.

The youngest moraines here, the Tioga moraines, form the crest of the composite moraine, and the Tenaya moraine buries the older, reddish Tahoe moraine that crops out near the base. The crest of the Tenaya moraine is ~45 m lower than the crest of the Tioga moraine, and the highest remnants of the Tahoe moraine are ~80 m lower still.

The trace of the Hilton Creek fault is visible near where the right–lateral moraines cross the range front. However, it is made less conspicuous by three breaches in the moraine crest caused by erosion from ice–marginal streams at the time of maximum advance of the glaciers. Fans have formed below the moraine where the suddenly reduced gradient caused deposition of debris carried by these streams.

Soil development and rock weathering are best assessed in locations such as these "lateral" fans, which are essentially unmodified by debris–flow deposition after the glaciers retreated. In contrast, the axial fans have been subject to occasional debris flows down the modern streams throughout the Holocene.

The large, broad peak just north of the McGee Creek and its moraines is McGee Mountain. Its shoulder of metamorphic rocks

Figure 14. Right–lateral moraines of McGee Creek.

and the remnant of a 2.7 Ma basalt flow (Dalrymple, 1963, 1964) is covered by large, exotic, granitic boulders derived from the headwaters of McGee Creek. Evidently, the time of transport was long ago (1.5 Ma) (Huber, 1981), since McGee Creek has cut down 800 m or more since flow across the summit of McGee Mountain from the headwaters of McGee Creek was last possible (e.g., Putnam, 1962).

The origin of the boulders has been debated. Are they glacial erratics or boulders carried in a mudflow? This debate has not been renewed for years, but the consensus was that they were glacial, and the eroded deposit was till. It is known as the McGee till.

STOP 5–5. MCGEE CREEK

Return north to the junction with the McGee Creek road and turn left, following the road as it climbs to the range front northwest of the moraine complex, and then crosses the moraines at a low point to a campground along McGee Creek at km 581 (mile 361; #31, Fig. 1).

The stop is inside the end moraines of McGee Creek, just east of a prominent fault scarp that offsets the ground till and outwash from the retreating Tioga glacier that comprises the valley floor. We have driven through a number of latest Pleistocene end moraines, but the first one we encountered on McGee Creek deserves special comment. The road makes an "S" turn as it crosses the low crest, exactly where it emerges from the higher and younger, less faulted Tioga moraines. This moraine may be an early Tioga, or Tenaya, moraine. Its crest here is ~20 m lower than the Tioga crests, and this difference is probably a good example of down–faulting, but this close to the end moraines these differences are difficult to interpret as faulting rates.

The normal fault scarp that cuts across McGee Creek and the old bed of the glacier west of the parking lot is offset 16 m. The scarp can be traced across both the left– and right–lateral Tioga moraines. It crosses the right–lateral moraines just west of the windgaps cut by ice marginal streams.

The offset of the moraines is greater than the offset of the valley floor, probably because the ice persisted some time at this location after it achieved its maximum depth and extent. The vertical offset is easiest to measure on the left–lateral moraine, where it is 25 m.

If we accept 21 ka as the time of maximum extent of the Tioga glaciers and the formation of this moraine, the implied rate is about 1.2 mm/yr. At this rate, the valley floor would have been abandoned by ice about 14 ka, comparable to, although slightly younger than, what we know from other localities. These results do not require that the Tioga glaciers remained near their maximum positions for 7 ka, because the observations could reasonably be explained by a late re–advance that did not overtop the 21 ka moraines.

Figure 15. Oblique aerial photograph of the end moraines of Convict Creek. (Photo by D.A. Rahm).

If the faulting rate derived here is applied to the vertical offset of the Tenaya and Tahoe moraines, respective age estimates of 36 and 105 ka are calculated. The 36–ka estimate for the Tenaya moraine is greater than the 30–ka value reported by Bursik and Gillespie (1993) from ostracode dates at Mono Lake.

STOP 5–6. CASA DIABLO

We return to US–395, turn left and continue north, along the southwestern wall of the caldera. The first large drainage we pass is Convict Creek (Fig. 15). We will drive below the faulted terminus of its Tahoe moraines, and on outwash from its Tioga glaciers. At the major intersection past the airport, we leave the highway and drive a short distance towards the town of Mammoth Lakes, stopping at the side of the road at km 599 (mile 372; #32, Fig. 1).

Long after the resurgent dome was emplaced in Long Valley, basalt and andesite were erupted into the low "moat" around the dome. Casa Diablo, named for the hot springs and associated sulfurous fumes found there, was the site of some of these flows.

We have stopped here because of the chance to see the lavas interbedded with till or outwash from the glaciers of the Mammoth embayment. Some of the granitic clasts in the till are coated with a welded tuff composed of curved glass shards, and angular lithic fragments of pumice, rhyolite and glass (Ellis, 2001).

Curry (1968) originally named the Casa Diablo till and assigned an age of ~400 ka using limiting K/Ar dates on the basalts that stratigraphically bracket it. However, new K/Ar analysis provided improved dates of 126 ± 25 and 62 ± 13 ka (Bailey et al., 1976), which are now accepted as the age range for the till. On the basis of relative dating techniques, Burke and Birkeland (1979) suggested that the Casa Diablo till was correlative with the older Tahoe and/or Mono Basin tills to the north–a correlation not universally accepted. The Casa Diablo till, however, must date from MIS 4–6, and hence may be correlative with the younger Tahoe till or Mono Basin till at Bloody Canyon. Ages on the welded tuff will help constrain this age assignment and may be available by the time of the field trip.

Return to the cars and drive north on US 395 to Reno and the end of the field trip, at km 862 (mile 536).

REFERENCES

Bailey, R. A., Dalrymple, G. B., and Lanphere, M. A., 1976, Volcanism, structure, and geochronology of Long Valley caldera, Mono County, California: Journal of Geophysical Research, v. 81, p. 725–744.

Benson, L., 1999, Records of millennial–scale climate change from the Great Basin of the Western United States: *in* Clark, P. U., Webb, R. S., and Keigwin, L. D., eds., Mechanisms of Global Climate Change at Millennial Time Scales, Geophysical Monograph 112, American Geophysical Union, Washington, D.C., p. 203–226.

Benson, L. V., Burdett, J. W., Kashgarian, M., Lund, S. P., Phillips, F. M., and Rye, R. O., 1996, Climatic and hydrologic oscillations in Owens Lake basin and adjacent Sierra Nevada, California: Science, v. 274, p. 746–751.

Birkeland, P. W., 1964, Pleistocene glaciation of the northern Sierra Nevada, north of Lake Tahoe, California: Journal of Geology, v. 72, p. 810–825.

Birkeland, P. W., Burke, R. M., and Walker, A. L., 1980, Soils and sub–surface weathering features of Sherwin and pre–Sherwin glacial deposits, eastern Sierra Nevada, California: Geological Society of America Bulletin, v. 91, p. 238–244.

Birkeland, P. W., and Burke, R. M., 1988, Soil catena chronosequences on eastern Sierra Nevada moraines, California, U.S.A.: Arctic and Alpine Research, v. 20, p. 473–484.

Birman, J. H., 1964, Glacial geology across the crest of the Sierra Nevada, California: Geologic Society of America Special Paper, v. 75, 80 p.

Blackwelder, E., 1931, Pleistocene glaciation in the Sierra Nevada and Basin Ranges: Geological Society of America Bulletin, v. 42, p. 865–922.

Burke, R. M., and Birkeland, P. W., 1979, Re–evaluation of multiparameter relative dating techniques and their application to the glacial sequence along the eastern escarpment of the Sierra Nevada, California: Quaternary Research, v. 11, p. 21–51.

Bursik, M. I., and Gillespie, A. R., 1993, Late Pleistocene glaciation of Mono Basin, California: Quaternary Research, v. 39, p. 24–35.

Bursik, M. I., and Sieh, K., 1989, Range front faulting and volcanism in the Moo Basin, eastern California: Journal of Geophysical Research, v. 94, p. 15,587–15,609.

Clark, D. H., 1995, Extent, timing, and climatic significance of latest Pleistocene and Holocene glaciation in the Sierra Nevada: PhD thesis, University of Washington, Seattle, WA 193 p.

Clark, D. H., 1997, A new alpine lacustrine sedimentary record from the Sierra Nevada: implications for Late–Pleistocene paleoclimate reconstructions and cosmogenic isotope production rates [abstract]: EOS, Transactions American Geophysical Union, v. 78, p. F249.

Clark, D. H., and Gillespie, A. R., 1997, Timing and significance of late–glacial and Holocene cirque glaciation in the Sierra Nevada, California: Quaternary International, v. 38/39, p. 21–38.

Clark, D. H., Housen, B. A., and Filippelli, G. M., in press, A deep, pre–Late Wisconsinan bog in the central Sierra Nevada, California: Arctic, Antarctic, and Alpine Research.

Clark, M. M., 1967, Pleistocene glaciation of the drainage of the West Walker River, Sierra Nevada, California: PhD thesis, Stanford University, Stanford , CA, 170 p.

Clark, M. M., 1968, Pleistocene glaciation of the upper West Walker drainage, Sierra Nevada, California [abstract]: Geological Society of America Special Paper 115, p. 317.

Clark, M. M., 1972, Range–front faulting: cause of anomalous relationships among moraines of the eastern slope of the Sierra Nevada, California: Geological Society of America Abstracts with Program, v. 4, p. 137.

Crook, R. Jr., and Gillespie, A. R., 1986, Weathering rates in granitic boulders measured by P–wave speeds: *in* S. M. Colman and D. Dethier, eds., Rates of Chemical Weathering of Rocks and Minerals, Academic Press, p. 395–417.

Curry, R. R., 1968, Quaternary climatic and glacial history of the Sierra Nevada, California: PhD thesis, University of California, Berkeley, CA, 107 p.

Curry, R. R., 1969, Holocene climatic and glacial history of the central Sierra Nevada, California: *in* Shumm, S.A., and Bradley, W.C. (eds.), United States Contributions to Quaternary Research, Geological Society of America Special Paper 123, p. 1–47.

Dalrymple, G. B., 1963, Potassium–Argon ages of some Cenozoic volcanic rocks of the Sierra Nevada, California: Geological Society of America Bulletin, v. 74, p 379–390.

Dalrymple, G. B., 1964, Cenozoic chronology of the Sierra Nevada, California: University of California Publications in Geological Science, v. 47, 41 pp.

Ellis, Colleen C., 2001, Determining the age of the Casa Diablo till through correlation of ash flow tuffs: Senior thesis, Humboldt State University, Arcata, CA, 16 p.

Feng, X., and Epstein, S., 1994, Climatic implications of an 8000–year hydrogen isotope time series from Bristlecone pine trees: Science, v. 265, p. 1079–1081.

Flint, R. F., 1957, Glacial and Pleistocene Geology: Wiley, NY, 553 p.

Gardner, J. V., Mayer, L. A., and Hughs Clarke, J. E., 1999, Morphology and processes in Lake Tahoe, California and Nevada: Geological Society of America Bulletin, v. 112, p. 736–746.

Gillespie, A. R., 1982, Quaternary Glaciation and Tectonism in the Southeastern Sierra Nevada, Inyo County, California: PhD thesis, Caltech, Pasadena, CA., 695 p.

Gillespie, A. R., and Molnar, P., 1995, Asynchronism of maximum advances of mountain and continental glaciations: Reviews of Geophysics, v. 33, p. 311–364.

Gillespie, A. R., 1991, Testing a new climatic interpretation for the Tahoe glaciation: *in* Hall, Jr., Doyle–Jones, V., and Widawski, B., eds., Natural History of Eastern California and High–Altitude Research, Proceedings of the White Mountain Research Station Symposium, v. 3, p. 383–398.

Hallet, B. and Putkonen, J., 1994, Surface dating of dynamic landforms: young boulders on aging moraines: Science, v. 265, p. 937–940.

Huber, N. K., 1981, Amount and timing of Cenozoic uplift and tilt of the central Sierra Nevada, California–Evidence from the upper San Joaquin River: U. S. Geological Survey. Professional Paper 1197, 28 p.

Konrad, S. K., and Clark, D. H., 1998, Evidence for an early Neoglacial glacier advance from rock glaciers and lake sediments in the Sierra Nevada, California, U.S.A.: Arctic and Alpine Research, v. 30, no. 3, p. 272–284.

Kovanen, D. J, and Easterbrook, D. J., 2001, Late Pleistocene, post–Vashon, alpine glaciation of the Nooksack drainage, North Cascades, Washington: Geological Society of America Bulletin, v. 113, no. 2, p.274–288.

Lajoie, K. R., 1968, Quaternary Stratigraphy and Geologic History of Mono Basin, Eastern California: PhD thesis, University of California, Berkeley, CA, 217 p.

LeConte, J., 1873, On some of the ancient glaciers of the Sierra: American Journal of Science, 3rd series, v. 5, p. 325–339.

Matthes, F., 1930, Geologic history of the Yosemite Valley: U.S. Geological Survey Professional Paper 160, 137 pp.

Matthes, F., 1965, Glacial reconnaissance of Sequoia National Park, California: U.S. Geological Survey Professional Paper 504–A, p. A1–A58.

Molitor, B. R., and Clark, D. H., 2002, Extent and climatic implications of the Upper Truckee glacier during the Tahoe glaciation–assessing conflicting scenarios: Geological Society of America Abstracts with Program, v. 34(5), p. A79.

Nishiizumi, K., Winterer, E. L., Kohl, C. P., Klein, J., Middleton, R., Lal, D., and Arnold, J. R., 1989, Cosmic ray production rates of ^{10}Be and ^{26}Al in quartz from glacially polished rocks: Journal of Geophysical Research, v. 94, 17907–17915.

Phillips, F. M., Zreda, M. G., Smith, S. S., Elmore, D., Kubik, P. W., and Sharma, P., 1990, Cosmogenic chlorine–36 chronology for glacial deposits at Bloody Canyon, eastern Sierra Nevada: Science, v. 248, p. 1529–1532.

Phillips, F. M., Zreda, M. G., Benson, L. V., Plummer, M. A., Elmore, D., Sharma, P., 1996, Chronology for fluctuations in late Pleistocene Sierra Nevada glaciers: Science, v. 274, no. 5288, p. 749–751.

Plummer, M., 2002, Paleoclimatic conditions during the last deglaciation inferred from combined analysis of pluvial and glacial records: PhD thesis, New Mexico Institute of Mining and Technology, Socorro, NM.

Putnam, W. C., 1962, Late Cenozoic geology of McGee Mountain, Mono County, California: University of California Publications in Geological Science, v. 40, p. 181–218.

Ramelli, A. R., Bell, J. W., dePolo, C. M., and Yount, J. C., 1999, Large–magnitude, late Holocene earthquakes on the Genoa fault, west–central Nevada and eastern California: Seismological Society of America Bulletin, v. 89, no. 6, p. 1458–1472.

Reasoner, M. A., Osborn, G., and Rutter, N. W., 1994, Age of the Crowfoot advance in the Canadian Rocky Mountains: a glacial event coeval with the Younger Dryas oscillation: Geology, v. 22, p. 439–442.

Reheis, M. C., Stine, S., and Sarna–Wojcicki, A. M., 2002, Drainage reversals in Mono Basin during the late Pliocene and Pleistocene: Geological Society of America Bulletin, v. 114, p. 991–1006.

Sarna–Wojcicki, A. M., Meyer, C. E., Bowman, H. R., Hall, T., Russell, P. C., Woodward, M. J., Slate, J. L., 1985, Correlation of the Rockland ash bed, a 400,000–year–old stratigraphic marker in northern California and western Nevada, and implications for middle Pleistocene paleogeography of central California: Quaternary Research, v. 23, p. 236–257.

Sarna–Wojcicki, A. M., Meyer, C. E., and Wan, E., 1997, Age and correlation of tephra layers, position of the Matuyama–Brunhes chron boundry, and effects of Bishop ash eruption on Owens Lake, as determined from drill hole OL, southeast California: *in* Smith, G. I., and Bischoff, J. L., eds., An 800,000–year paleoclimatic Record from Core OL – 92, Owens Lake, Southeast California, Geological Society of America Special Paper 317, p. 79–90.

Scuderi, L. A., 1987, Glacier variations in the Sierra Nevada, California, as related to a 1200–year tree–ring chronology: Quaternary Research, v. 27, p. 220–231.

Sieh, K. E., and Bursik, M. I., 1986, Most recent eruption of the Mono Craters, eastern central California: Journal of Geophysical Research, v. 91, p. 12539–12571.

Sharp, R. P., 1968, Sherwin Till–Bishop Tuff relationship, Sierra Nevada, California: Geological Society of America Bulletin, v. 79, p. 351–364.

Sharp, R. P., 1972. Pleistocene glaciation, Bridgeport Basin: Geological Society of America Bulletin, v. 83, p. 2233–2260.

Sharp, R. P., and Birman, J. H., 1963, Additions to the classical sequence of Pleistocene glaciations, Sierra Nevada, California: Geological Society of America Bulletin, v. 74, p. 1079–1086.

Sharp, W. D., Ludwig, K. R., Chadwick, O. A., Amundson, R., Glaser, L. L., in press, Dating fluvial terraces by ^{230}Th/U on pedogenic carbonate, Wind River Basin, Wyoming: Quaternary Research.

Stine, S., 1987, Mono Lake: The past 4,000 years: PhD thesis, University of California, Berkeley, CA, 615 p.

Stine, S., 1994, Extreme and persistent drought in California and Patagonia during mediaeval time: Nature, v. 369, p. 546–549.

Warhaftig, C., and Birman, J. H., 1965, The Quaternary of the Pacific mountain system in California: *in* Wright, H. E., and Frey, D. G., eds., The Quaternary of the United States, Princeton University Press, p. 299–340.

Yount, J. C., Birkeland, P. W., and Burke, R. M., 1982, Holocene glaciation, Mono Creek, central Sierra Nevada, California: Geological Society of America Abstracts with Program, v. 14, p. 246.

Quaternary geology and ecology of the greater Yellowstone area

Authors and Leaders: **Kenneth L. Pierce,** *U.S. Geological Survey, Bozeman, MT 5971*
Don G. Despain, *U.S. Geological Survey, Bozeman, MT 5971*
Cathy Whitlock, *Univ. of Oregon, Corvallis, OR, 97403–1251*
Kenneth P. Cannon, *National Park Service, Lincoln, NE, 68508–3873*
Grant Meyer, *Univ. of New Mexico, Albuquerque, NM, 87131–1116*
Lisa Morgan, *U.S. Geological Survey, Denver, CO 80225*
Joe M. Licciardi, *University of New Hampshire, Durham, New Hampshire, 03824*

INTRODUCTION

This field guide focuses on the glacial geology, ecology, paleoecology, caldera unrest, and archeology in Yellowstone and Grand Teton national parks and vicinity (Fig. 1). Some previous field guides of Yellowstone are Locke et al. (1995) for the Yellowstone valley, Fournier et al. (1994) for hydrothermal and volcanic geology of Yellowstone, and Pierce and Good (1992) for the Quaternary of Jackson Hole. Non–technical overviews of Yellowstone and Grand Teton National Parks are Good and Pierce (1996) and Smith and Siegel (2000). Geologic maps are: Grand Teton (Love et al., 1992), and Yellowstone (bedrock and surficial geology (USGS, 1972a; 1972b). Christiansen (2001) extensively describes Yellowstone's volcanic geology, and Pierce (1979) describes the glacial geology of the northern Yellowstone region. We suggest that you obtain detailed maps.

GLACIAL AND ARCHEOLOGY OVERVIEW

In the Rocky Mts., Bull Lake terminal moraines are typically farther downvalley than Pinedale moraines, which are generally correlated with marine oxygen isotope stages (OIS) 6 and 2 respectively. However, for the Greater Yellowstone system, Bull Lake terminal moraines to the west, southwest, and south of Yellowstone are much farther downvalley than Pinedale moraines, whereas to the north in the Yellowstone valley and to the east, Pinedale glaciers have overridden Bull Lake terminal moraines.

Pleistocene glacial flow patterns for Yellowstone were complex. Glaciation was initiated in the mountains that surround the Yellowstone Plateau. Glaciers then built up on the plateau to about 3,000 ft (~1 km) thickness and flowed outward, down the major valleys that drain Yellowstone. During deglaciation, the plateau icecap stagnated, and glaciers from the adjacent mountains flowed into terrain previously occupied by the plateau icecap (Pierce, 1979). This pattern of buildup, full glacial conditions, and recessional changes resulted in shifts by up to 180º in direction of glacial flow and the transport of glacial erratics. Changing flow patterns sometimes dammed glacial lakes, including lakes on the Yellowstone Plateau. Outlet glaciers, augmented by ice from the Yellowstone Plateau, probably culminated later than typical mountain–valley glaciers, apparently because of both the interval needed to build up the plateau icecap and its self–amplifying nature once established.

Under full glacial conditions, the northern Yellowstone outlet glacier was formed by convergence from multiple sources in the northern Yellowstone area (Fig. 2). Sources, from east to west, were (1) glaciers from the Beartooth uplift, and the Absaroka Range, (2) ice flowing over the Washburn Range from the eastern Yellowstone Plateau, (3) ice flowing from the Yellowstone Plateau between the Washburn and Gallatin Ranges, and (4) ice from the Gallatin Range (Fig. 4A). From the northern park boundary, this outlet glacier flowed 40 miles (65 km) downvalley to the Eightmile terminal moraines.

Interest in the archeological record of Paradise Valley extends back to the 19th century with the description of rock alignments, bison traps, and quarry sites, which probably represent the earliest archeological work in Montana (Norris, 1880; Brackett, 1893). Barnum Brown (1932) conducted the first investigations at the Emigrant bison drives in 1932. At Rigler Bluffs (24PA401; Haines 1966) fired–rock hearths eroding from a forty–foot clay bluff dated 5.6 cal ka B.P. (4900 ± 300 ^{14}C yrs. B.P. and 5.8 cal ka B.P. (5040 ± 150 ^{14}C yrs. B.P.). Good (1964) interpreted the overlying fine silts as having been deposited in a lake dammed by the Yankee Jim landslide.

George Arthur (1966a) recorded 47 sites and classified them as occupations, bison drives (jumps and pounds), wickiups, stone alignments, stone circles, pictographs, burials, and stone piles. The most impressive of these sites is the late Holocene complex of drive lines and bison kills associated with the glaciated basalt benches near Emigrant. The Carbella site (24PA302) is located at the mouth of Yankee Jim Canyon. Arthur (1966a) noted that the artifacts were recovered from underneath and within large glacial boulders that may be the result of periodic or seasonal flooding or related to the rupture of the Yankee Jim dam. A middle Holocene age is suggested by the projectile point styles. The Eagle Creek site (24PA301) is located above the town of Gardiner within alluvial deposits. A ^{14}C date of 1200 cal yrs. B.P. (1230 ± 160 ^{14}C yrs. B.P.) was obtained from a fired rock hearth in tan, sandy loam (Arthur 1966b). Work by Arthur and others indicates a rather extensive archeological record in Paradise Valley that probably includes Paleo–indian occupation in latest Pleistocene time

Pierce, K.L., Despain, D.,Whitlock, C., Cannon, K.P., Meyer, G., and Morgan, L., 2003, Quaternary geology and ecology of the greater Yellowstone area: in Easterbrook, D.J., ed., Quaternary Geology of the United States, INQUA 2003 Field Guide Volume, Desert Research Institute, Reno, NV, p. 313–344.

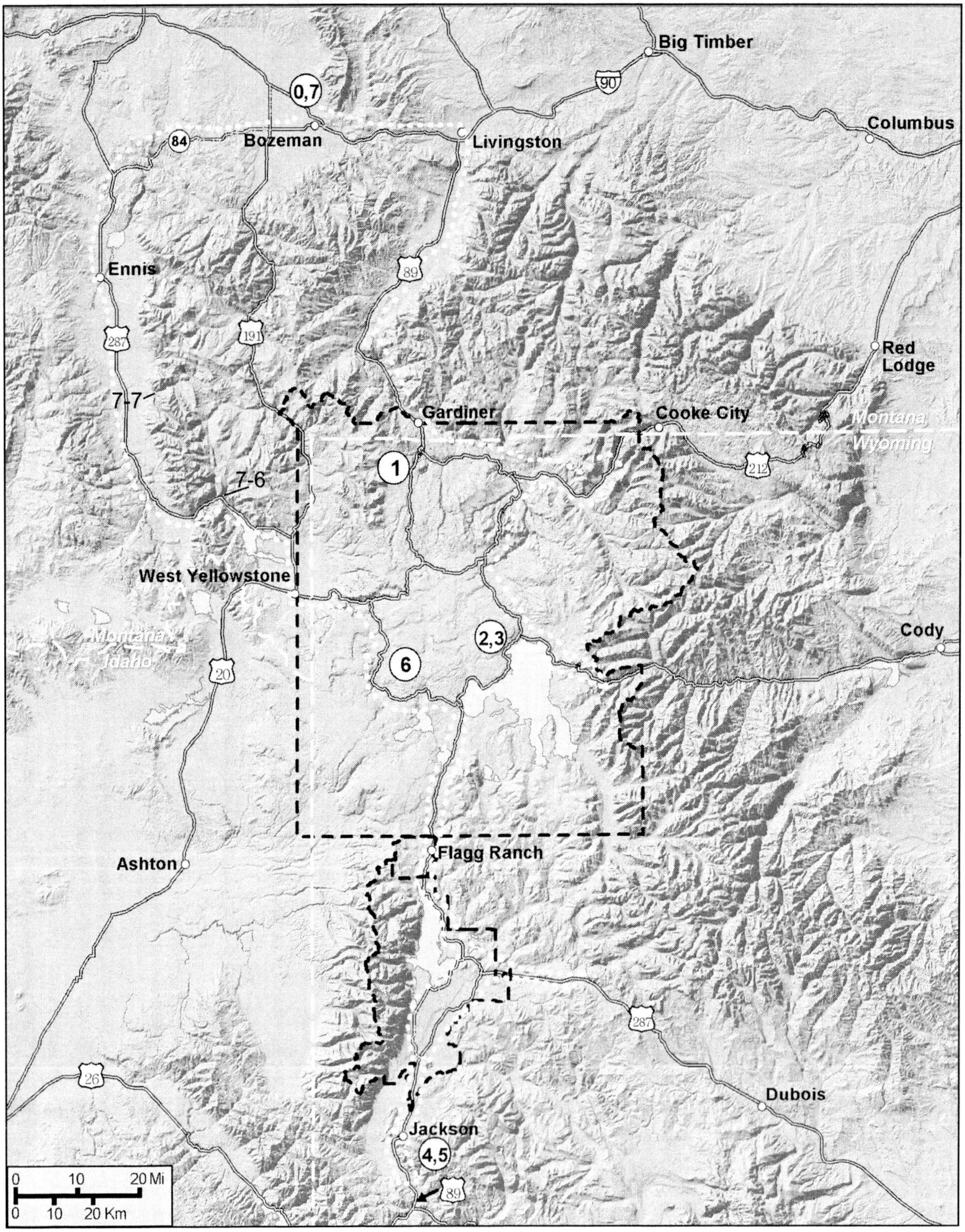

Figure 1. Shaded relief map of the Greater Yellowstone area showing the route of the Greater Yellowstone area field trip (white dots) and the place of at end of each day (number in white circle).

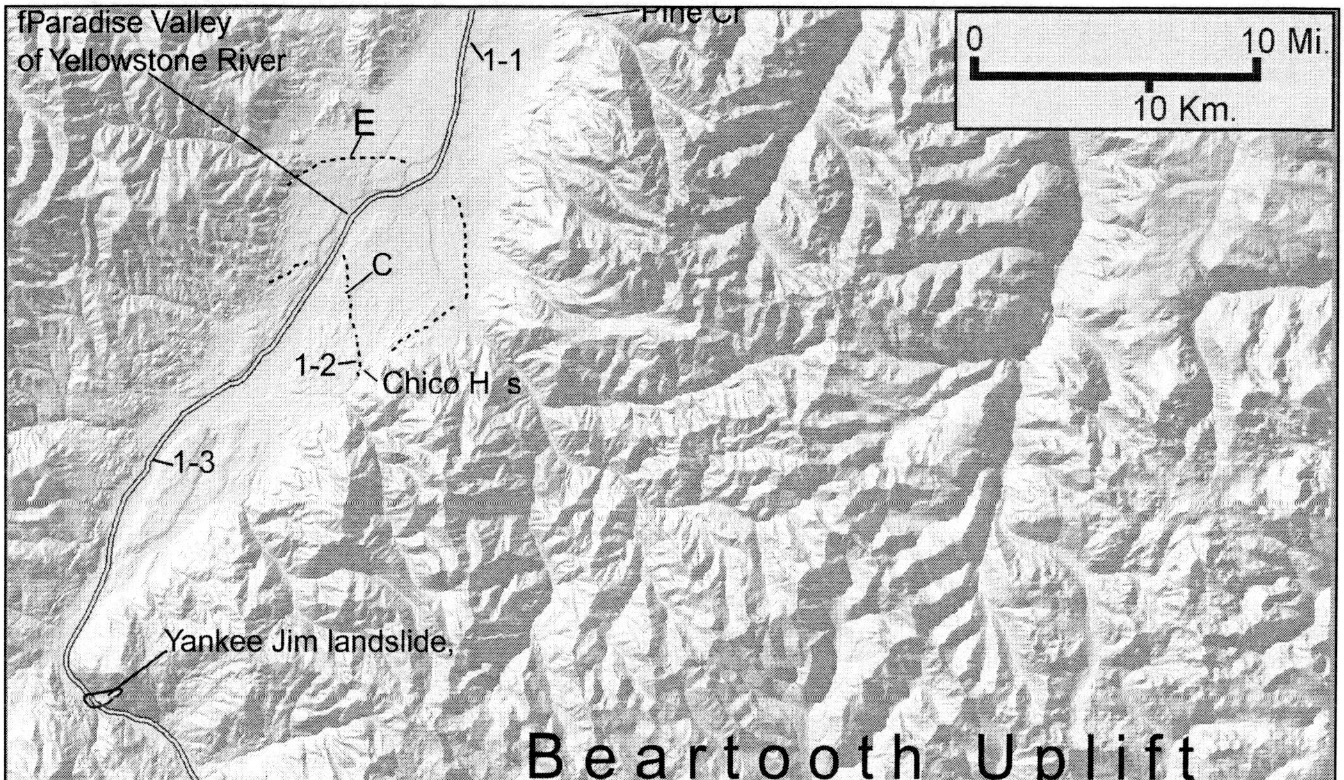

Figure 2. Map of the northern Yellowstone area showing location of stops on Day 1 and 2. C– Chico moraines, E–Eightmile moraines, D–Deckard Flats moraines, JB–Junction Butte moraines.

DAY 1. BOZEMAN TO MAMMOTH

STOP 1–1. MALLARDS REST FISHING ACCESS

West of the Yellowstone River, the outwash fan of the Northern Yellowstone outlet glacier emanates from the Eightmile moraines (Figs. 3, 4). This outwash fan is 200 ft (60 m) above the Yellowstone River at the Eightmile terminal moraines, 60 feet (18 m) above the river at this stop, and only 10–20 feet (3–6 m) above the river this side of the Livingston canyon. On the east side of the valley is the Beartooth uplift that supported local valley glaciers (Montagne and Locke, 1989), including Pine Creek with

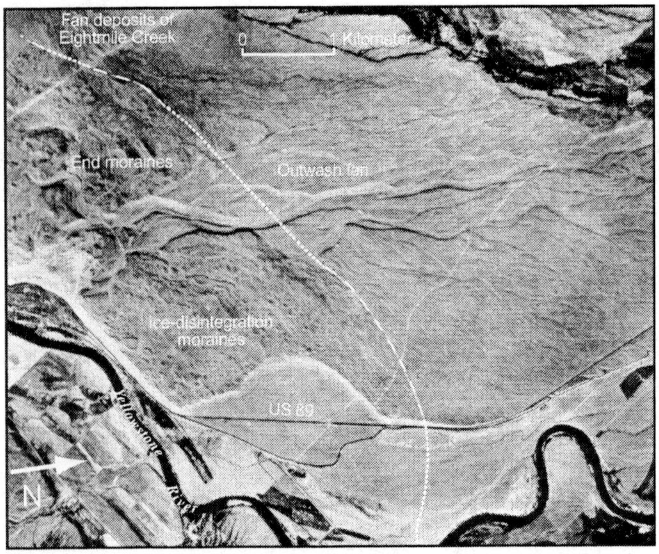

Figure 3. Pinedale end moraines and outwash fan of the northern Yellowstone outlet glacier, showing preservation of detailed braided channel pattern (from Pierce, 1979, Fig.16). The head of the fan is 200 ft above the Yellowstone River whereas it converges with the river level about 15 miles to the north (right).

Pinedale (~20 ka) and Bull Lake (~140 ka) moraines. The scarp of an active fault with about 15 ft (5 m) of post–glacial offset occurs near the base of the range at sites such as Barney Creek.

The modern Yellowstone River is eroding its banks downstream in the Livingston area, but upstream from here, it is remarkably stable. The history of glaciation and glacial floods exert important geomorphic controls on channel pattern and type. These, in turn, largely control the distribution of the three main alluvial types from here upstream to Gardiner, MT: (1) moderately to very stable, (2) incised to entrenched, and (3) single pool–riffle. Plane–bed channels form most of the channel. Several disturbance/sedimentation zones occur with multiple channels that range from anastomosing to anabranching. Downstream from Mallards Rest, multiple channels predominate and the Yellowstone River is a classic high–gradient, wandering, gravel–bed

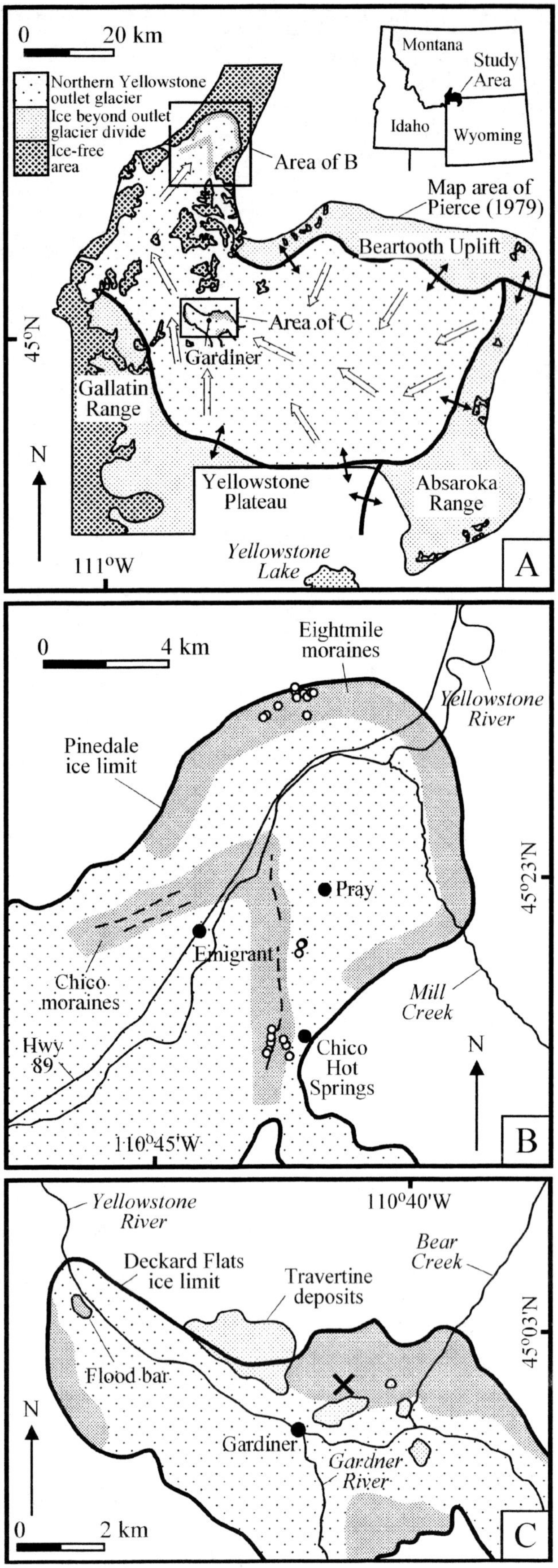

Figure 4. Pinedale glaciation of the in northern Yellowstone area (from Licciardi et al., 2001; adapted from Pierce, 1979). **A**: Northern Yellowstone ice cap. Thick black lines with double–pointed arrows indicate main ice divides of various ice masses that fed northern Yellowstone outlet glacier. Open arrows indicate flow directions of major ice drainageways. **B**: Sample sites (open circles) on Eightmile and Chico moraines. Dashed lines show prominent crests in Chico moraine complex. Three sampled boulders located 2 km south–southwest of the town of Pray are considered to belong to proximal part of Eightmile moraines. **C**: Sample sites on Deckard Flats moraines and late–glacial flood deposit. Individual sample sites on Deckard Flats moraines (dark stipple) are clustered in location marked ×, and sites on flood bar are clustered within outline of deposit. Travertine deposits are those used by Sturchio et al. (1994) to construct their U–series chronology.

river. In response to the floods of 1996 and 1997, the channel changed downstream from Mallards Rest, but changed little upstream. These contrasts have important management implications (River activity from Chuck Dalby, written communication 2002). Newly created gravel bars provide fresh substrates for primary plant succession, particularly by willow (*Salix exigua*) and cottonwood (*Populus angustifolia* and *P. trichocarpa*).

STOP 1–2. CHICO MORAINES

Just before Chico Hot Springs, turn right and climb up onto Chico outwash terrace and then Chico moraines. The Chico moraines (Fig. 4) of the Northern Yellowstone outlet glacier are bounded on the north by the Chico meltwater channel. Note the succession of outwash terraces that line the meltwater channel now occupied by the Chico road. Cosmogenic ages on boulders on the Chico moraines average 15.7 ± 0.5 ^{10}Be ka. Eightmile moraines average 16.5 ± 0.4 ^{3}He ka and 16.2 ± 0.3 ^{10}Be ka (Fig. 5) (Licciardi et al., 2001). These cosmogenic ages are younger than those based on ^{14}C dating (Licciardi et al., 2001: Pierce, 2003).

The Eightmile moraines (Figs. 3, 4) are well expressed just to the north on the other side of the Chico meltwater channel, where they arc around to their terminus 4.5 miles (7 km) beyond the Chico moraines. Across the valley, ice–marginal channels 150 ft (50 m) deep were eroded into bedrock at the Eightmile and Chico ice margins. To fit the Bull Lake–Pinedale paradigm, Horberg (1940) thought irregular stony deposits in the inner valley of the Yellowstone demonstrated a Pinedale age for the Chico moraines in contrast to the Eightmile moraines on the terrace seen at STOP 1–1, which he considered Bull Lake. But this inner valley deposit is more likely flood gravel rather than a morainal deposit because of the pattern of highs and lows and the gravelly, but not boulder studded, excessively drained highs.

STOP 1–3. HIGHWAY REST AREA

The low "terraces" of the Yellowstone River are actually flood deposits. In contrast to normal, flat–topped, terraces, the surface of these flood deposits has spoon–shaped forms: the lows are concave–up spoon–forms and highs are convex up spoon–forms.

Across the Yellowstone River is the basalt of Hepburns Mesa, whose age of 2.2 Ma (Smith et al., 1995) is consistent with the start of the Yellowstone volcanic field. The surface of this basalt is striated and polished. Locally, it forms cliffs that were used by Native Americans as buffalo jumps to procure bison.

STOP 1–4. FLOOD BAR, STEVENS CREEK

Cross Yellowstone River at Corwin Springs and continue just into Yellowstone Park. This mid–channel flood bar (Fig. 4C) has "giant ripples" spaced about 15 m apart and up to 2 m high, with boulders up to 2 m in diameter. Cosmogenic ages on the boulders average 13.7 ± 0.5 ^{10}Be ka (Fig. 5). The alluvial fan of Reese Creek 0.5 km to the northwest provides evidence of at least two

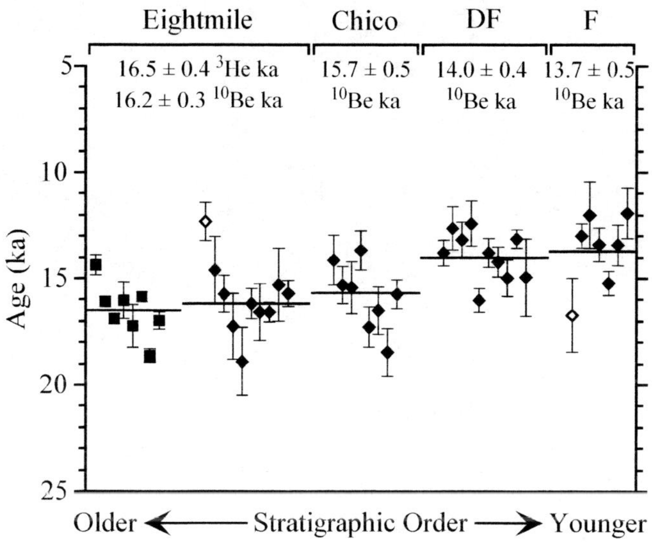

Figure 5. Cosmogenic ages of Yellowstone moraines and flood deposits, plotted in stratigraphic order from oldest (left side of graph) to youngest (right side) (from Licciardi et al., 2001). Solid squares are ^{3}He ages, and solid diamonds are ^{10}Be ages. Error bars on each age represent 1σ analytical uncertainty only, and do not include errors due to production rate, scaling, and other uncertainties. Horizontal lines and quoted ages indicate weighted means of each landform. Open diamonds indicate two outliers not included in weighted means. DF – Deckard Flats moraines; F – late–glacial flood deposits.

floods 150–200 ft (45–60 m) deep. First, an alluvial fan was deposited and then a flood down the Yellowstone valley eroded the fan front and left a "flood–ripped" fan front. The younger fan was then built and, later, another flood undercut and eroded the fan front (Pierce, 1979, Fig. 45). These floods were probably from release of glacially dammed lakes upstream, most likely from a lake dammed in the Lamar Valley by the Slough Creek glacier (STOP 2–3).

STOP 1–5. DECKARD FLATS MORAINES; TRAVERTINE BENCH; COSMOGENIC AGES

The 0.6 Ma Undine Falls basalt underlies the travertine bench and Deckard Flats bench (Christiansen, 2001) and this and other volcanic units date the incision of the Yellowstone valley (Fig. 6) The type area of the Deckard Flats moraines is south across Bear Creek. Large boulders in the moraines of the Deckard Flats 'readjustment' date 14.0 ± 0.4 ^{10}Be ka (Fig. 5). The Deckard Flats position is termed a 'readjustment' rather than an advance because it represents the time when the icecap on the Yellowstone Plateau ceased contributing to the northern Yellowstone outlet glacier, and the glaciers from adjacent mountains (Beartooth uplift, Absaroka Range, and Gallatin Range) readjusted to this loss and established a stable ice margin that can be traced over much of the northern Yellowstone area. In full–glacial time, the outlet glacier was about 3,500 ft (1 km) thick at Gardiner and terminated about 40 miles (65 km) downvalley,

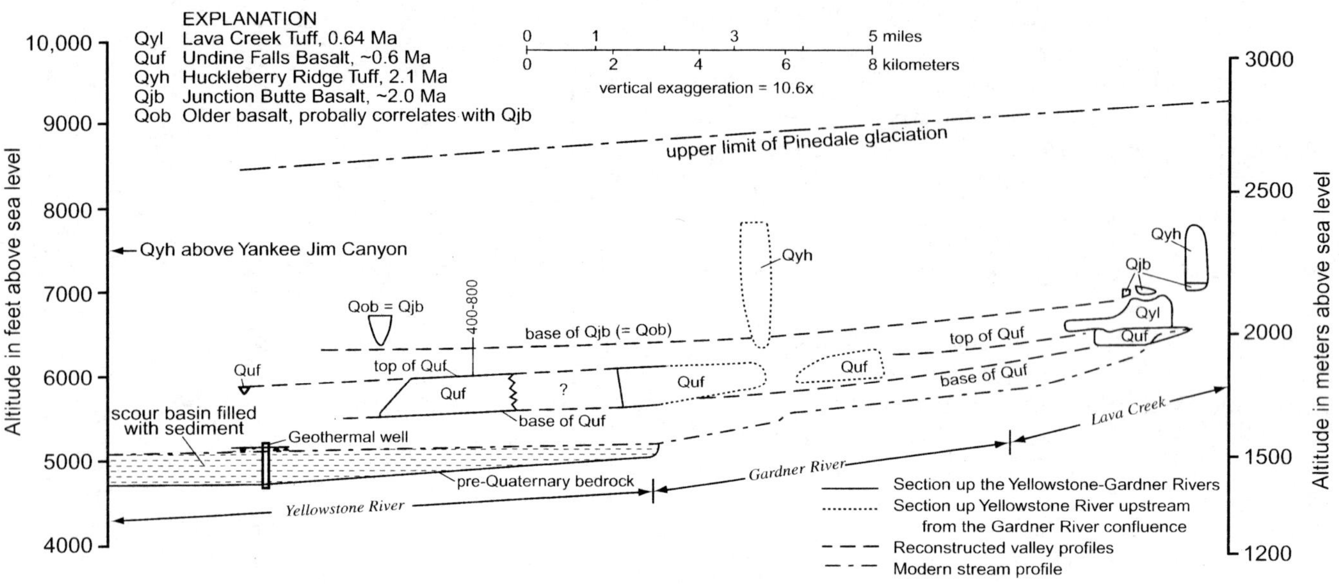

Figure 6. Valley deepening in Gardiner–Mammoth area based on incision of dated volcanic units (from Pierce et al., 1991). K–Ar ages from Obradovich (1992).

whereas during the Deckard Flats, it was about 900 ft (275 m) thick and terminated 4 miles (7 km) downvalley.

STOP 1–6. MAMMOTH OVERLOOK

The bench north of the Mammoth Hotel marks the Deckard Flats margin of glaciers originating to the southwest in the Gallatin Range. The bench is probably kame gravel and includes boulders of travertine. The conical hills to the south and east are thought to be thermal kames formed where hot springs melted cavities in the ice that filled with kame gravel. To the north above the Deckard Flats bench, a grassy, rounded hill of sub–glacial till also bears erratics of travertine. North in the Yellowstone valley near Gardiner, the full–glacial northern Yellowstone outlet glacier was about 3,500 ft (1 km) thick and created glacial streaming from northward flow. Under full–glacial conditions, ice from three points of the compass converged in this area to form the northern Yellowstone outlet glacier. Mt. Everts to the east exposes readily erodible, Cretaceous shale and sandstone on its west face. The southern part of Mt Everts is capped by 2.1–Ma Huckleberry Ridge tuff, with local, small, valley fills of 0.64–Ma Lava Creek tuff. The upland on top of Mt. Everts is glacially scoured with ridges and lakes forming a diverse wildlife habitat of lush meadows bordered by bands of forest.

Hydrothermal activity at the Mammoth terraces changes frequently. The following list shows the sequence of travertine deposits in the Mammoth area: (1) Terrace Mountain, the highest travertine near Mammoth Hot Springs, rests on 2.1–Ma Huckleberry Ridge tuff and has a U–Th age of 406 ± 30 ka (Pierce et al., 1991). Pinedale glacial erratics rest on this travertine, which was scoured by northward–flowing ice that reached an altitude of almost 9,000 feet (2750 m) on Sepulcher Mountain, (2) Above the active terraces, the inactive and forested Pinyon terrace is post–glacial with U–Th ages near 10 ka (Pierce et al., 1991). (3) The lowest travertine terraces are post–glacial. A deep research hole was drilled in the 1960's in the main terrace and a U–Th age of 7.72 ± 0.88 ka was obtained on travertine from a depth of 239 ft (72.8 m). We encourage you to tour the Mammoth Terraces in evening.

DAY 2. MAMMOTH TO LAKE VILLAGE

Vegetation and 1988 Fires

From the Mammoth area eastward to the Lamar Valley, we will look at recessional deposits of the last glaciation. The major mountain vegetation zones in this part of the Rocky Mts. are diagrammed in Fig. 7. Lower timberline is at about 6400 ft (1950 m) and upper timberline very close to 10,000 ft (3050 m). Below the forested zone, vegetation is dominated by semiarid sagebrush shrublands with basin big sage (*Artemisia tridentata* spp *tridentata)* and bluebunch wheatgrass (*Agropyron spicatum*). Alpine tundra above upper timberline has a large number of plant communities populated by species from the arctic tundra and species that have evolved from subalpine floras. Between these extremes are two forested zones: (1) the lower, montane zone, ca. 6400–7600 ft (1950–2300 m), dominated by Douglas–fir (*Pseudotsuga menziesii*) with occasional stands of limber pine (*Pinus flexilis*) and aspen (*Populus tremuloides*), and (2) the upper, subalpine zone, 7600–10,000 ft (2300–3050 m). The climax forest is dominated by Engelmann spruce (*Picea engelmannii*) and subalpine fir (*Abies lasiocarpa*) with lodgepole pine

(*Pinus contorta*) as the pioneer species in forest successions. Above 8600 ft. (2600 m), whitebark pine (*Pinus albicaulis*) is the dominant pioneer species but also forms climax stands on sites that are too cold and dry for spruce and fir. Areas underlain by fine soils are generally covered with shrublands or grasslands that provide most of the forage to support populations of large mammals. Those in the montane zone provide winter range and those in the subalpine and alpine zones provide summer range.

In 1988, extreme drought and high winds from mid–June to mid–September resulted in forest fires that eventually burned approximately 1.1 million acres (450,000 ha) in and around the park (Fig. 8). The total area encompassed by the fire was 1.7 million acres (690,000 ha) (Fig. 8). About 50% of the burned area was burned by crown fire, 40% by mixed–severity fire, and 10% by grassland/shrubland fires. Most of the sagebrush areas visible to the west of this stop were burned in the fires and burned forest is visible on the mountains forming the horizon to the south. Areas that were previously covered by lodgepole pine forests are now covered by lodgepole pine seedlings 4–8 ft (1.5–2.5 m) tall. Where Douglas–fir or spruce/fir dominated the forest stands the seedlings are sparser. Geological influences are also apparent in the vegetation response. The good soils developed from andesitic and sedimentary substrates have produced herbaceous growth and sparse tree seedlings while the poor soils derived from rhyolitic substrates are covered by dense lodgepole pine seedlings

The burned areas seen here are part of the North Fork fire, which started July 22, 1988 just west of the park boundary, due west of Old Faithful. It was immediately reported and crews were dispatched to it. High winds made it impossible to contain. It burned 111 acres (45 ha) the first day, 922 ac (373 ha) the second day, and 1317 ac (533 ha) the third day. The overhead team in charge of the fire requested bulldozers and park officials told them that they could use them if they would be effective. The overhead team decided that the long distance spotting (fire brands carried by winds) and high spread rate at the fire head would not be contained by bulldozers, and that also made having men on the fire lines at night too dangerous. For the next 57 days, for only five days it did not increase in size, for only 15 days it gained less than 1000 ac (400 ha), for 28 days between 1000 and 5000 acres (400–2000 ha) burned, and for 9 days more than 5000 acres (2000 ha) burned. On 10 September, fires burned 60,000 acres (24,300 ha). The fires overran the developments at Old Faithful, Madison Junction, and Norris Junction, and threatened Mammoth Hot Springs, West Yellowstone, and Canyon Village. Fires eventually burned over 215,000 ac (87,000 ha).

Pull over to view: (1) High Bridge moraines deposited during the recession from the Deckard Flats position, (2) Lava Creek, which is the type area of the 0.64 Ma Lava Creek Tuff, and (3) recessional meltwater channel with Blacktail Ponds at the east end.

STOP 2–1. BLACKTAIL DEER PLATEAU

The Blacktail Deer Plateau is formed by moraines of Deckard Flats age that extend about 3 mi (5 km) south and west

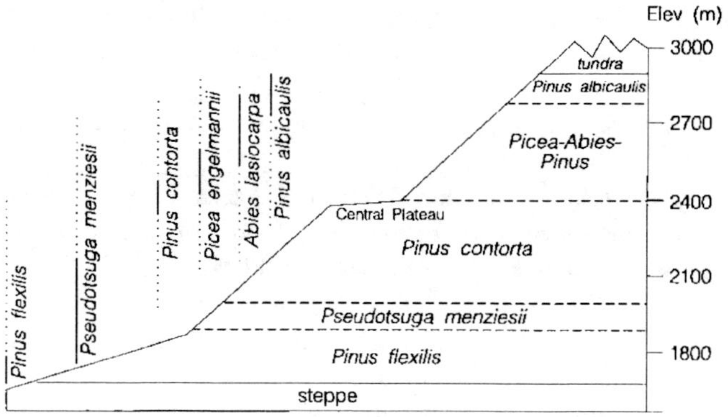

Figure 7. Vegetation zones of Yellowstone and Grand Teton national parks. Solid vertical lines show the altitudinal range of important species. Dotted vertical lines indicate where species is present (modified from Whitlock, 1993).

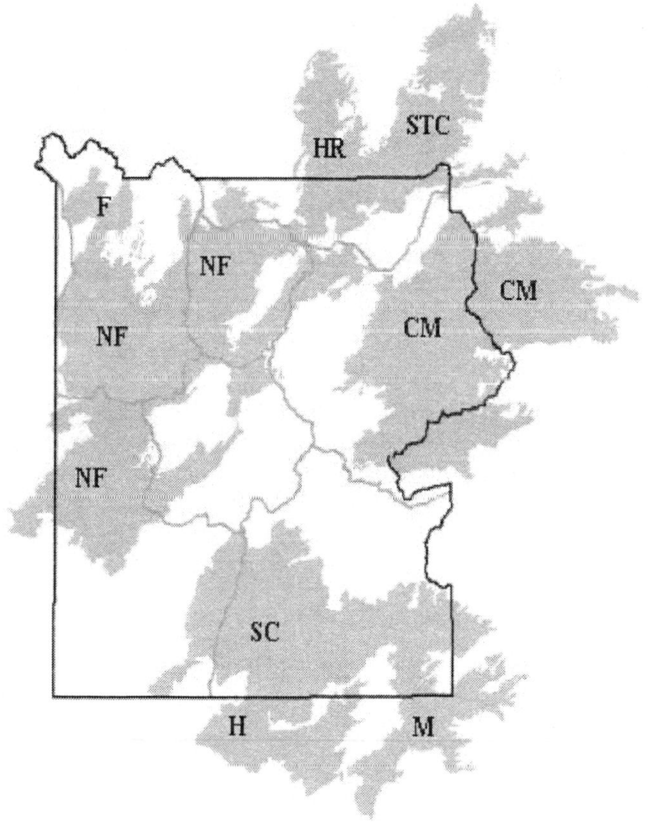

Figure 8. Areas (shaded) burned during 1988 Yellowstone fires. **CM**= Clover Mist complex, **F**= Fan fire, **H**= Huck fire, **HR**= Hellroaring fire, **M**= Mink fire, **NF**= North Fork fire, **SC**= Snake complex, **STC**= Storm Creek fire. The Clover Mist, Fan, Mink, Snake Complex and Storm Creek fires were all lightning caused fires that were allowed to burn early in the season. The Huck, Hellroaring, and North Fork fires were human caused and suppression action applied from first detection and throughout the season.

from this viewpoint. The successively lower morainal benches, locally including kame gravel, define about 10 recessional ice margins. Wind–drift of soil and snow makes for a loamier and moister soil on the northeast sides of ridges, whereas the south sides are drier due to greater solar insolation, wind exposure, and erosion of fine soil.

The Deckard Flats moraines of the Blacktail Deer Plateau are an important part of Yellowstone's northern winter range. Grasslands at the site are dominated by Idaho fescue (*Festuca Idahoensis*), western wheatgrass (*Agropyron smithii*), Junegrass (*Koeleria cristata*), bluebunch wheatgrass, horsebush (*Tetradymia canescens*), green rabbit brush (*Chrysothamnus viscidiflorus*), and mountain big sage (*Artemisia tridentata spp menziesii*). Local vegetation patterns are strongly controlled by the surficial geology (Despain, 1990). The sagebrush–grasslands are underlain by glacial deposits and contrast with the adjacent bedrock areas. Limestone supports limber pine, Douglas–fir, and aspen, depending on moisture availability. Andesite of the Washburn Range and northern Grand Teton region supports spruce/fir forest at the higher elevations and Douglas–fir and aspen at lower elevations, interspersed with meadows, while rhyolite plateaus beyond the Deckard Flats ice margin to the southwest are dominated by lodgepole pine. One way to appreciate the influence of geologic processes on the vegetation is to note that the Deckard Flats moraines form extensive grasslands. If glaciation had not occurred, rhyolite would have been at the surface and the region would have probably been covered with lodgepole pine forest. If rhyolite had not been emplaced, then andesite and sedimentary rocks would have been at the surface and the northern range would probably have been covered by Douglas–fir forest. These substrate differences have exerted long–term influences on the vegetation that are evident in the paleo–ecological record as well (Whitlock, 1993).

Blacktail Creek descends into the Black Canyon of the Yellowstone River and joins the river near Crevice Lake, a deep lake with a varved–sediment record. In February of 2001, scientists from the USGS and Universities of Oregon and Nebraska recovered an 8–m–long core that spans the last 11,000 ^{14}C years. The exceptional water depth (30 m), compared with its relatively small surface area prevents yearly mixing of the water column. The availability of a record with this high temporal resolution creates opportunities to examine the environmental history of Yellowstone with a precision not possible in most paleo–ecological studies. In addition, tree–ring studies in and around Crevice Lake offer an opportunity to compare hydrological and ecological variables with climate–related changes in tree growth. Limber pine (*Pinus flexilis*), Douglas–fir, and other species from climatically sensitive sites in the Greater Yellowstone ecosystem (GYE) have been used to develop a tree–ring–based hydrologic record that spans several centuries (Graumlich and Littell, 2001; Graumlich et al., in press). Analysis of fire–scars on Douglas–fir around Crevice Lake discloses past fire events in the watershed (Littell, 2002), which can be compared with events found in previous studies (Houston, 1973) and charcoal–based studies (Millspaugh and Whitlock, 2002).

Pull over to view Phantom–Lake meltwater channel eroded more than 200 feet into bedrock by the ice–marginal "Yellowstone River." A lake dammed by a small alluvial fan was commonly present here in the early part of the summer.

STOP 2–2. TOWER FALLS VIEWPOINT

The Narrows Canyon of the Yellowstone River was eroded in postglacial time. A complex assemblage of Pleistocene basalt and glacial, fluvial, and lacustrine sediment known as the "sediments and basalts of The Narrows" are exposed in the steep, narrow canyon. Pinedale till caps the section, locally resting on Lava Creek tuff. These units rest on the eroded top of the upper basalt, which in turn is underlain by the till of Bumpus Butte, containing erratics of Precambrian rocks from sources to the north. Conformably beneath the till is lacustrine sediment of an ice–dammed lake, containing ice–rafted stones of Precambrian crystalline rock and coarse, silicic volcanic ash from a nearby source. The argon–argon age of the tephra is 1.3 ± 0.1 Ma and the age and chemistry suggest correlation with the Mesa Falls Tuff (Obradovich and Izett, 1991). The close tie between the ice–dammed lake sediments containing the ash and the overlying till also date the glaciation as ~1.3 ± 0.1 Ma, an age two orders of magnitude older than the Pinedale.

STOP 2–3. JUNCTION BUTTE MORAINES

The Junction Butte moraines of the post–Deckard Flats recession are the most prominent and occupy a position typically referred to as late Pinedale in the Rocky Mts. Large boulders of Precambrian crystalline rocks stud their surface and permanent and seasonal ponds occupy common depressions. A large glacier flowing down Slough Creek deposited these moraines and blocked the Lamar valley upstream from Slough Creek. The till has enough fines for good water–holding capacity and adequate nutrients. The mineral soil is well covered by the vegetation. The vegetation at this site is mountain big sage, Idaho fescue, Junegrass, bluebunch wheatgrass, silvery lupine (*Lupinus argenteus*), Indian paintbrush (*Castilleja* sp.), and green rabbit brush. The morainal topography with ridges, slopes and swales produces a mosaic of vegetation assemblages. Near the pond, note willow (well browsed), sedge (*Carex* spp.), tufted hairgrass (*Deschampsia caespitosa*), spikerush (*Eleocharis* sp), and bulrush (*Scirpus* sp.). The water level in the ponds has diminished since the 1970's as a result of prolonged winter drought.

STOP 2–4. SLOUGH CREEK PITTED OUTWASH, ECOLOGY AND PALEOECOLOGY

Walk up onto kame–gravel bench south of road before the Lamar River bridge. Here, near the confluence of Slough Creek and the Lamar River, six, nested, pitted, outwash terraces formed during recession of the Slough Creek glacier. In contrast to the boulder–studded Junction Butte moraines, the surfaces of the kame gravel terraces are boulder–free. Lakes dammed by the Slough Creek glacier existed in the Lamar valley at the time of the nested kame terraces as well as when the Slough Creek glacier was even larger. Draining of such lakes is the prime candidate for the flood deposits near Gardiner. The kame gravel with its well–washed sand matrix forms well–drained soils with low moisture–holding capacity and sparse vegetation.

Walk southwest to northeast from a dry SW–facing slope to a moist, NE–facing slope, noting plant changes in this dry–to–moist transect. The dry, SW–facing slope has high solar insolation, wind deflation of snow, and a sandy soil at the surface, with a high proportion of mineral soil not covered by vegetation. The dry–slope vegetation is composed of needle–and–thread grass, bluebunch wheatgrass, Junegrass, Hood's phlox (*Phlox hoodii*), fringed sage (*Artemisia frigida*) but no Idaho fescue. The NW–facing slope accumulates snowdrifts and soil blown from the dry site and has lower solar insolation and a silty soil with a complete vegetation cover of the soil. The vegetation is silky lupine (*Lupinus sericeus*), Idaho fescue, bluebunch wheatgrass, (tall) prairie smoke (*Geum triflorum*), a non–native smooth brome (*Bromus inermis*), sticky geranium (*Geranium viscosissimum*), and wild buckwheat (*Eriogonum umbellatum*).

In this pitted outwash complex, Slough Creek Lake provides a vegetation and fire history for the last 16,000 cal yrs, B.P. (Fig. 9). Following deglaciation, the area supported tundra vegetation and conditions were cooler and drier than at present. A period of pine–juniper (*Pinus–Juniperus*) forest and low fire frequency was present between ca. 11,000 and 7000 cal. yrs. B.P. (10,000 and 6000 ^{14}C yrs. B.P.), when the climate was warmer and wetter than at present. Fire frequency increased from four fires/1000 yrs. in the early Holocene to >10 fires/1000 yrs. in the last 7000 cal. years. The establishment of Douglas–fir parkland occurred as a result of cooler drier conditions and increased fire activity in the late Holocene.

Slough Creek Lake and other northern sites record a different climate history than sites in the southern region (Fig. 10). Today, the northern part of the Park experiences higher summer and annual precipitation as a result of convectional storms associated with summer monsoonal circulation, while the rest of the Park is relatively dry and under the influence of the northeastern Pacific subtropical high–pressure system. In the early Holocene, greater–than–present summer insolation amplified the contrast between wet–summer and dry–summer regions (Fig. 11). Central and southern Yellowstone (the dry–summer area) became drier in the early Holocene as a result of a stronger–than–present northeast Pacific subtropical high–pressure system, while wet–

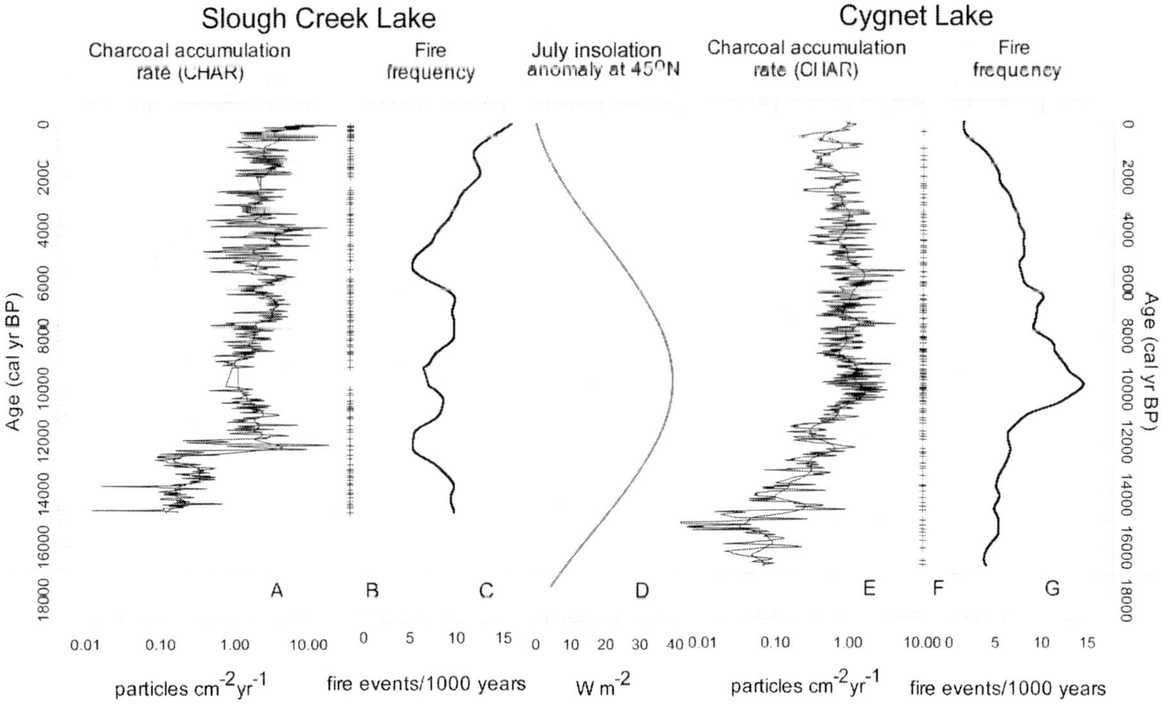

Figure 9. Comparison of charcoal data from Slough Creek Pond in the Yellowstone–Lamar region (summer–wet) and Cygnet Lake in the Central Plateau (summer–dry). The reconstructions are derived from accumulation rates of macroscopic charcoal particles (CHAR) in contiguous 1–cm–thick core samples. Results are plotted as log–transformed CHAR values (A, E). Charcoal peaks (+) above background levels define fire events (B, F), which are summarized as number of fires/1000 years (C, G). Long–term trend in fire frequency compares well with changes in precipitation regime associated with the July insolation anomaly (D, Berger, 1978).

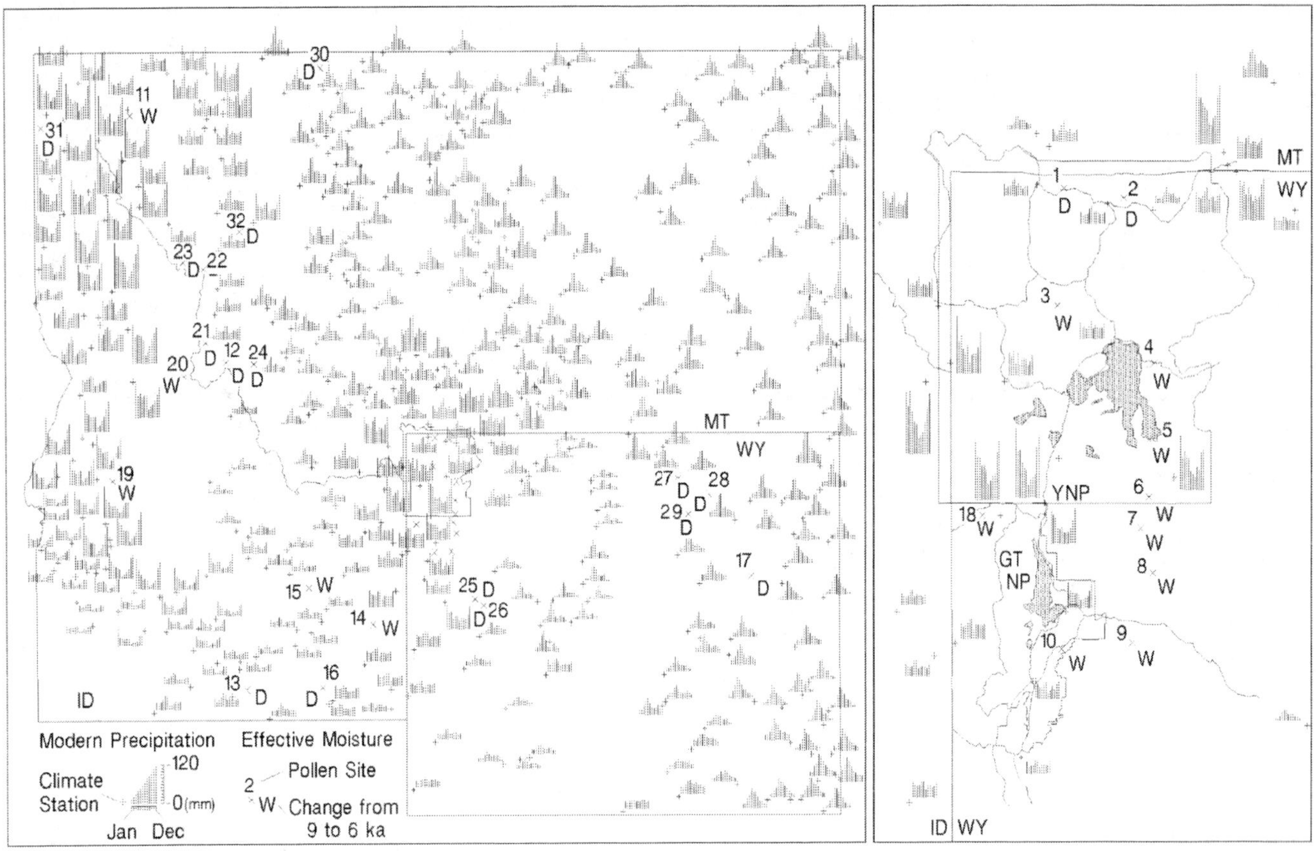

1. Gennett & Baker (1986) Palynology 10:61-71.
2. Whitlock & Bartlein (1993) Quaternary Research 39:231-238; Millspaugh & Whitlock (in press) In: After the Fires: The Ecology of Change in Yellowstone National Park, L. Wallace (Ed.).
3. Millspaugh *et al.* (2000) Geology 28:211-214.
4. Waddington & Wright (1974) Quaternary Research 4:175-184.
5. Baker (1976) USGS Professional Paper 729-E:E1-E48.
6, 7, 8, 9, 10. Whitlock (1993) Ecol. Monogr. 63:173-198
11. Mack *et al.* (1983) Quaternary Research 20:177-193.
12. Mehringer *et al.* (1977) Arctic and Alpine Research 9:345-368.
13. Davis *et al.* (1986) Quaternary Research 26:321-339.
14. Beiswenger (1991) Ecol. Monogr. 61:165-182.
15. Davis *et al.* (1986) Quaternary Research 26:321-339.
16. Bright (1966) Tebiwa 9:1-47.
17. Markgraf & Lennon (1986) Plains Anthropologist 31:1-12.
18. Whitlock *et al.* (1995) Quaternary Research 43:433-436.
19. Doerner & Carrara (1999) Arctic, Antarctic, and Alpine Research 31:303-311.
20, 21, 22. Brunelle-Daines (2002) Dissertation, Univ. of Oregon.
23. Karsian (1995) Master's Thesis, Univ. of Montana.
24. Brunelle-Daines (2002) Dissertation, Univ. of Oregon.
25. Fall *et al.* (1995) Quaternary Research 43:393-404.
26. Lynch (1998) Ecology 79:1320-1338.
27, 28, 29. Burkart (1976) Dissertation, Univ. of Iowa.
30. Barnosky (1989) Quaternary Research 31:57-73.
31. Mack *et al.* (1995) Quaternary Research 10:241-255.
32. Mehringer (1985) In: Pollen Records of Late-Quaternary North American Sediments, V. M. Bryant and R. G. Holloway (Eds.).

Figure 10. Annual distribution of precipitation at climate stations in Idaho, Montana, and Wyoming (World WeatherDisc Associates, n.d.), and inferred effective moisture changes from 9000 to 6000 cal yrs. B.P. at paleoecological sites. "W" indicates an inferred effective moisture change to wetter conditions; "D" indicates an inferred effective moisture change to drier conditions; and "–" indicates no inferred direction of change. (Right) The same information for the region of Yellowstone National Park (YNP) and Grand Teton National Park (GTNP) (adapted after Whitlock and Bartlein, 1993; Shafer et al., in press).

summer northern Yellowstone became wetter in response to enhanced monsoonal circulation (Fig. 11). The boundary between wet–summer and dry–summer regimes is controlled by topography and has not changed significantly during the Holocene (Whitlock and Bartlein, 1993).

STOP 2–5. FIRE–RELATED EROSION AND SEDIMENTATION

Glacial ice completely filled the Lamar and Soda Butte valleys. During deglaciation, glaciers in the upper Lamar drainage

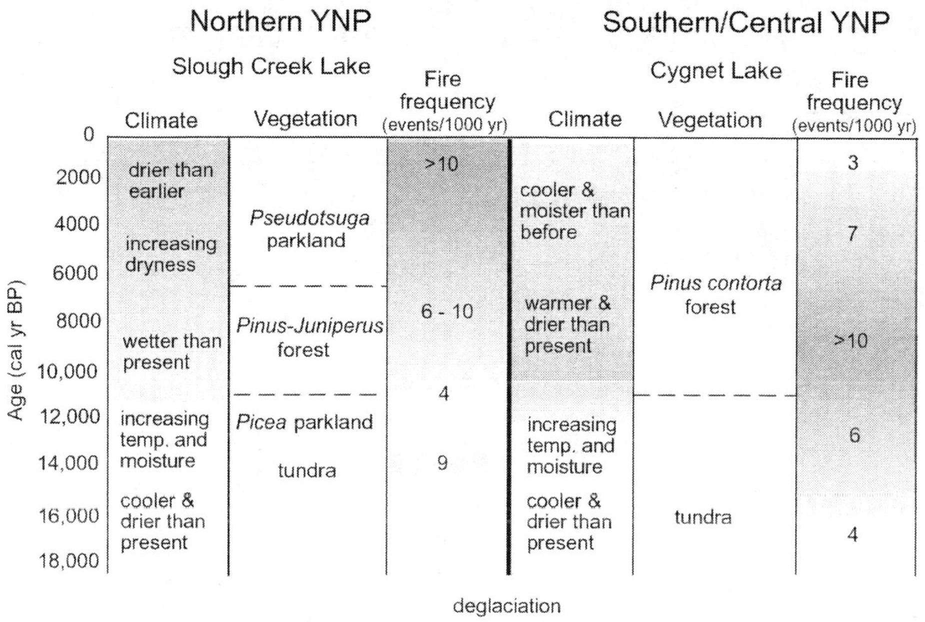

Figure 11. Summary of late–glacial and Holocene changes in climate, vegetation, and fire frequency for the Yellowstone–Lamar and Central Plateau provinces, based on pollen and charcoal data. Shading indicates intensity of aridity (darker shading=greater aridity) (from Millspaugh and Whitlock, in press).

downwasted and extensive kame gravel and lake sediments accumulated. Later, during deglaciation, the upper Lamar River built a sandy delta into glacial–lake Lamar that was dammed in the Lamar Canyon area by the Slough Creek glacier.

The dense, lodgepole–pine and mixed–conifer forests of northeastern Yellowstone tend to burn infrequently in large, stand–replacing fires, as in 1988. These severe burns promote debris flows and floods because of greatly enhanced runoff, especially in summer thunderstorms. Alluvial fans along the sides of glacial trough valleys, like those of Soda Butte Creek, were built in significant part by post–fire sedimentation. Approximately 30% of the stratigraphic thickness of these fans consists of fire–related deposits. Radiocarbon dating shows that fan aggradation was greatly accelerated by post–fire sedimentation during warm, drought–prone periods (e.g., Medieval warm period ca. 900–1200 AD). In contrast, fire–induced sedimentation is minimal at times that coincide with the ~1400–year Holocene cycle of cold episodes in the North Atlantic (Bond et al., 1997) (Fig. 12). During these cool, wet intervals, active lateral migration of lower Soda Butte Creek created broad floodplains cut on underlying glacial sediments, as shown by ^{14}C ages on overbank deposits. These surfaces were later preserved as fill–cut terrace treads. Millennial–scale climate change has had a major influence on both fires and geomorphic processes during the Holocene (Fig. 12).

STOP 2–6. ROUND PRAIRIE HISTORIC FLOODS

Infrequent, major floods are part of a suite of catastrophic geomorphic processes that exert a strong influence on valley–floor ecosystems in northeastern Yellowstone. Along Soda Butte Creek, these floods carried bedload high enough to deposit extensive overbank gravel that allowed conifers to replace floodplain meadows, as in upper Round Prairie. Tree–ring methods date these floods to 1918, ~1873, and possibly ~1790. Short–term gage records indicate that snowmelt runoff approached the 100–year flood in 1996 and 1997, but indirect estimates show that peak discharge was 2–3 times greater in 1918 than in 1996. Gravelly overbank deposits create well–drained, more xeric surfaces and also reduces bank stability.

Anthropogenic impacts to Soda Butte Creek include a mine–tailings dam failure at Cooke City in 1950. The resulting flood produced extreme peak discharges, but was too brief to cause much erosion. It did emplace a large volume of acidic, metal–contaminated sediment over high floodplain surfaces. Because these tailings deposits lie above the reach of most natural floods, they may persist over long periods, with documented impacts to vegetation and aquatic organisms.

Pull over at Lamar Ranger Station to view Lamar Valley. This area was once the Buffalo Ranch where a few bison were sheparded until they formed large herds. These in turn became the stock for much of the bison now surviving in the U.S. Downvalley across the Lamar River, large erratics of the Slough Creek glacier provide evidence for damming of a deep lake in the Lamar Valley. In the roadcut going into the Lamar Ranger Station, foreset bedding suggests a delta of Rose Creek built into glacial lake Lamar. The arcuate perimeter also suggests a delta. Absaroka volcanic rocks of Eocene age form both sides of the valley, and those across the valley are noted for multiple beds containing petrified trees. Continue back to Tower Junction and head south.

STOP 2–7. WASHBURN RANGE VIEWPOINT

During the last glaciation, the Washburn Range (Absaroka andesite) was overridden, scoured, and striated by the north–flowing icecap from the Yellowstone Plateau (Pierce, 1979). The crest of Specimen Ridge was also striated by glacial

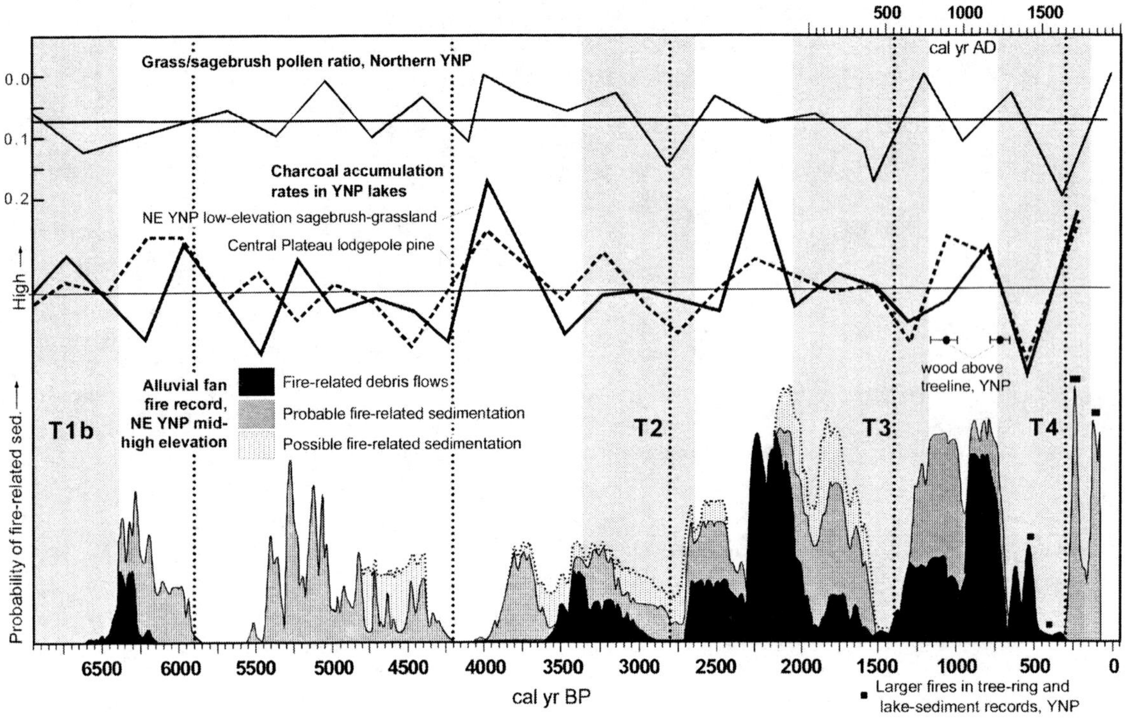

Figure 12. Comparison of records of middle to late Holocene fire–related alluvial–fan sedimentation (Meyer et al., 1995) with other fire and climate proxy records. Scale in cal yrs. B.P. indicates ^{14}C–calibrated calendar year ages before present, where present is defined by convention as 1950 (Stuiver and Reimer, 1993). The spectrum of probability of fire–related sedimentation is constructed by summation of calibrated probability distributions for radiocarbon dates on individual events. Note correspondence of fire–related sedimentation events with large fires identified in high–resolution lake sediment study of the last ~750 yr (black squares; Millspaugh and Whitlock, 1995). 500–yr average charcoal accumulation rates in Yellowstone lakes (Millspaugh, 1997; Millspaugh and Whitlock, this volume) show correspondence with major fire–related sedimentation peaks in the last 3000 years. Wood above present treeline (K.L. Pierce, written comm. 1993) indicates warmer temperatures ca. AD 800–1200 corresponding to increased fire sedimentation in the "Medieval Warm Period". Periods of overbank sedimentation on mainstem Soda Butte Creek terraces (gray shading, T1b–T4) correspond with generally wetter conditions indicated by high grass/sagebrush pollen ratios (note inverted scale; from Gennett and Baker, 1986). Vertical dotted lines indicate maximum North Atlantic cooling events identified by Bond et al. (1997), all of which correspond to a distinct minima in fire–related sedimentation, strongly suggesting control of large fires by hemispheric– and millennial–scale climatic variations.

ice flowing toward Gardiner. During glacial recession, as the plateau icecap stagnated, glaciers flowed south from the Beartooth uplift, left southerly striations on Specimen Ridge, and reached as high as 8800 ft (2680 m) on the north side of Mt Washburn. This Deckard Flats "readjustment" carried boulders of granite and limestone and locally left substantial deposits of till and kame gravel (Pierce, 1979).

This upland areas on Absaroka volcanic bedrock has soils of variable thickness. Most glaciated ridges have very thin soils with bedrock just below the surface. Such ridges form the dry meadows on the north side of Mt Washburn. Where the soil is thicker, particularly in concavities, are forests of spruce and subalpine fir. At this stop, till several meters thick supports lusher meadows, with mixed grass and forbs including Idaho fescue, bearded wheatgrass (*Agropyron caninum*), mountain brome (*Bromus ciliatus*), bluegrasses (*Poa* spp.), sticky geranium, cinquefoils (*Potentilla* spp.), little sunflower (*Helianthella uniflora*), balsamroot (*Balsamorhiza sagittata*), and mountain big sagebrush. Across Yellowstone Canyon, benches of Lava Creek Tuff support lodgepole pine. Above this bench on Specimen Ridge the Absaroka volcanic terrain is a mosaic of spruce/fir forest interspersed with meadows. Above the modern treeline on Mt Washburn, old stumps date from the Medieval warm period. Continue driving south to Lake Village.

DAY 3. CANYON–LAKE VILLAGE AREA

STOP 3–1. DUNRAVEN PASS, CALDERA, AND PLATEAU OVERVIEW

Park in Dunraven Pass and hike a few hundred meters up the old road. This south–facing escarpment is formed in Absaroka volcanic rocks of an Eocene volcano whose southern half was down–faulted into the 640 ka caldera. The caldera extends from here to the Red Mts. about 30 mi. to the south. To the southwest, rhyolite flows form the nearly flat surface on the distant horizon. Most flows in and near the caldera are 70,000 to 160,000 years old. The vegetation on rhyolite is lodgepole pine whereas that on andesite is spruce/fir with extensive meadows. Well above this site, the top of Mt Washburn was striated by north–flowing ice from the Yellowstone Plateau. The Washburn Range was rounded and sculpted by overriding ice during the last and earlier glaciations.

STOP 3–2 (OPTIONAL). DUNRAVEN PASS PICNIC AREA. GEO–ECOLOGY AND FIRE HISTORY

Park in the Dunraven Pass picnic area. We are near the contact between Absaroka volcanic rocks that form the Washburn Range to the north and rhyolite lava flows that form the Yellowstone Plateau to the south. A small drainage occupied by the paved road separates the two. This north–to–south traverse illustrates the strong geologic control of forest types. We will walk from Absaroka volcanic terrain with a spruce/fir forest southward up onto rhyolite terrain (Dunraven Road flow) with a lodgepole forest.

STOP 3–3. GRAND CANYON AT ARTIST POINT AND REST STOP

Take the bridge across the Yellowstone River and park at the Artist Point parking area. The hydrothermally altered rhyolite in the deeper part of the canyon near Artist Point experienced pyrite mineralization before the canyon was cut, which was then oxidized to yield the yellow and red colors. Similar extensive alteration probably also exists elsewhere at depth within the caldera. Alteration decreases upstream from Artist Point to firmer rock at Lower Falls. Emplacement of rhyolite built up the Yellowstone Plateau and created steep gradients at the margin of this plateau. Stream erosion has incised the margins of this plateau creating canyons. The largest river, the Yellowstone, has eroded farthest into the plateau and created the deepest canyon. Across the canyon and extending about halfway down into the canyon is a paleo–valley of the Yellowstone River filled with sediment. Glacial advances from the Beartooth uplift both early and late in a glacial cycle have dammed the Yellowstone River.

Pull over into the parking area for Canyon horseback stables. The smooth grassy slopes are part of an extensive area of thick, glacial till containing Precambrian rocks carried southward from the Beartooth uplift, probably early in the last glacial cycle before an icecap became established on the Yellowstone Plateau. This south–flowing glacier also dammed the Yellowstone River creating a lake in the Hayden Valley–Yellowstone Lake area.

STOP 3–4. HAYDEN VALLEY

Hayden valley is a productive grassland on glacial lake sediments surrounded by lodgepole pine on rhyolite. These meadows can exist here because of the accumulation of lake and other sediments that have both better water holding capacity and more nutrients than the surrounding rhyolite terrain. Hayden Valley was the site of glacially dammed lakes: (1) before the last glacial maximum when an advance from the north reached the Canyon Village area, and (2) during recession from the last glacial maximum when the icecap on the Yellowstone Plateau had melted but glaciers from the Beartooth uplift blocked the Yellowstone River east of Mount Washburn (Pierce, 1979; Richmond, 1976, 1977).

At this stop, a thin veneer of alluvium mantles lake sediments resulting in a high water table that favors silver sage (*Artemisia cana*) and Richardson's needlegrass (*Stipa richardsonii*). Varved lake sediments deposited during Pinedale recession are exposed in the Yellowstone River bank. Where the overlying sediments are thick enough to allow deeper water drainage, mountain big sage replaces the silver sage and Idaho fescue that is the dominant understory species. Other flora includes sticky geranium, cinquefoils, bluegrasses, and bearded wheatgrass. Lodgepole pine forests surrounding Hayden valley grow on glaciated rhyolite that has sandy soils with low nutrients and low water holding capacity.

A pollen study from Cygnet Lake (Fig. 9) in the Central plateau indicates that high percentages of sagebrush and grass (Poaceae) were present between 17,000 and 12,800 cal yrs. B.P. (14,800–10,900 ^{14}C yrs. B.P.) when the region was colonized by tundra. An increase in pine (*Pinus*) between 12,800 and 11,300 cal yrs. B.P. (10,900 and 10,100 ^{14}C yrs. B.P.) marks the establishment of lodgepole pine forest at the site. Unlike sites in non–rhyolitic regions, the pine forest at Cygnet Lake has persisted with little change for the last 11,300 cal yrs., despite changes in temperature and effective moisture (Whitlock, 1993). In contrast, fire frequency variations inferred from the charcoal record are well correlated with the July insolation anomaly and reach highest levels ca. 9900 cal yrs. B.P., when summer drought was most intense (Millspaugh et al. 2000).

Pull over at the Le Hardys Rapids parking area. Le Hardys Rapids serves as the bedrock threshold of Yellowstone Lake. It is also on the crest of the historic axis of uplift and subsidence of the Yellowstone caldera. Upstream from here, note that the Yellowstone River is quite wide, tranquil, and quite low gradient.

OVERVIEW OF CALDERA UNREST

Between 1923 and 1985, the center of the Yellowstone caldera rose ~1 m (Pelton and Smith, 1979, Dzurisin and Yamashita, 1987). From 1985 until 1995–6, it subsided at about two cm/yr (Dzurisin et al., 1994). More recent radar interferometry studies show renewed inflation (Wicks et al., 1998). Doming along the caldera axis reduces the gradient of the Yellowstone River from Le Hardys Rapids to the Yellowstone Lake outlet and ultimately causes an increase in lake level. Meyer and Locke (1986) and Locke and Meyer (1994) suggested that the shorelines were cut during intra– caldera uplift episodes that produced rising water levels. Figure 14 shows the history of lake level changes and other events (Pierce et al., 2002). The lake in the outlet region has been below or near its present level for about half the time

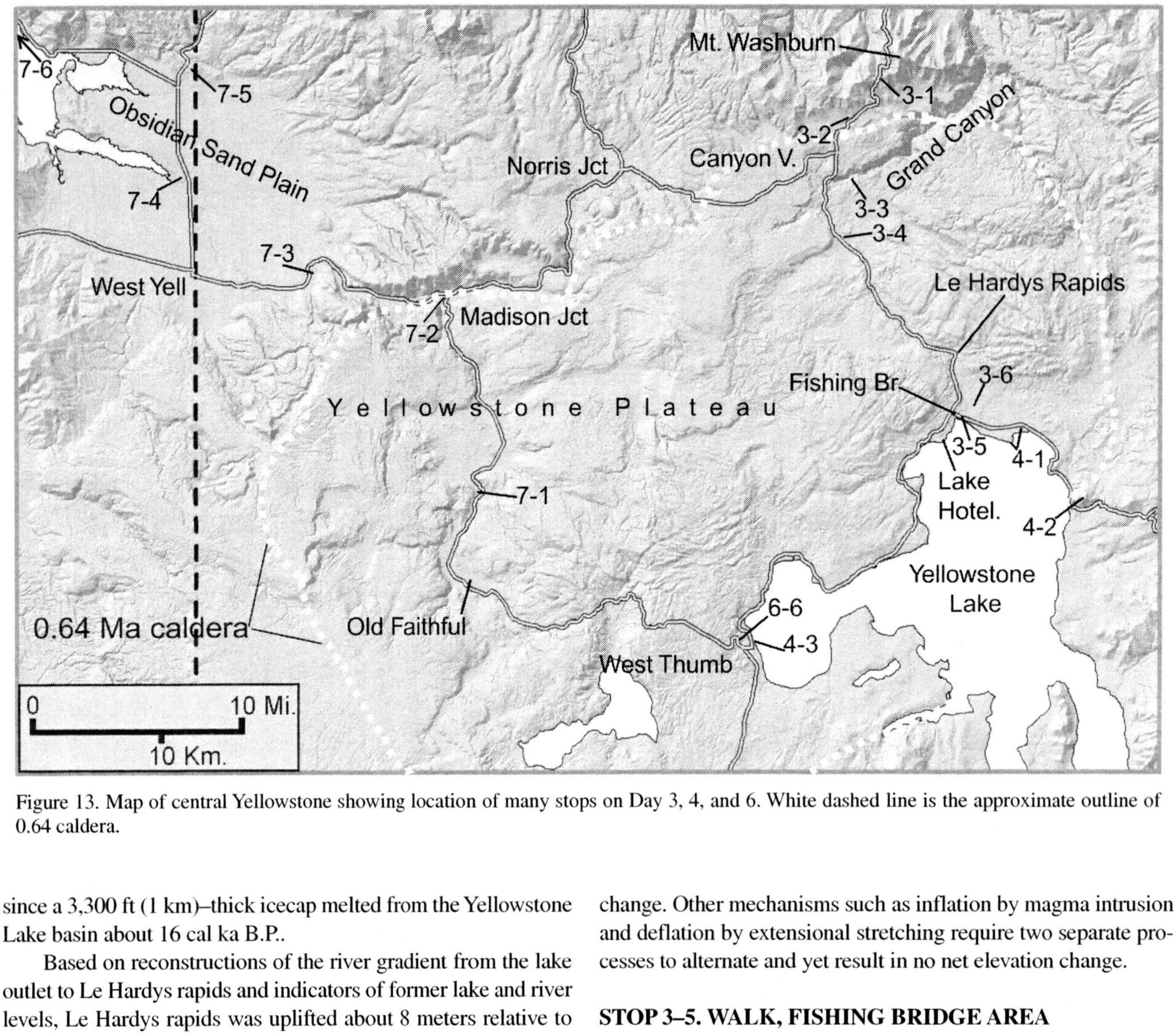

Figure 13. Map of central Yellowstone showing location of many stops on Day 3, 4, and 6. White dashed line is the approximate outline of 0.64 caldera.

since a 3,300 ft (1 km)–thick icecap melted from the Yellowstone Lake basin about 16 cal ka B.P..

Based on reconstructions of the river gradient from the lake outlet to Le Hardys rapids and indicators of former lake and river levels, Le Hardys rapids was uplifted about 8 meters relative to the lake outlet at least twice, once ~9.5–8.5 cal ka B.P. and again after ~3 cal ka B.P. (Pierce et al., 2002). Following these uplifts, it appears to have subsided a similar amount. Older possible rises in lake level are suggested by *S4* truncating older shorelines and *S5* truncating valleys. A plot of lake level through time shows 5–7 millennial–scale oscillations after 14.5 cal ka B.P. (Fig. 14). Le Hardys Rapids to Fishing bridge spans only the central 25% of the historic caldera doming, so if historic doming is used as a model, the total projected uplift is four–fold greater. This "heavy breathing" of the central part of the Yellowstone caldera may reflect a combination of several processes, (1) magmatic inflation, (2) tectonic stretching and deflation, and (3) inflation from hydrothermal fluid sealing, followed by cracking of the seal, pressure release, and deflation. Over the entire postglacial period, subsidence has balanced or slightly exceeded uplift, as shown by older shorelines that *descend* towards the caldera axis. We favor a hydrothermal mechanism for inflation and deflation because it provides for both inflation and deflation with little overall change. Other mechanisms such as inflation by magma intrusion and deflation by extensional stretching require two separate processes to alternate and yet result in no net elevation change.

STOP 3–5. WALK, FISHING BRIDGE AREA

Depart busses on the east side of Fishing bridge and drive over to Fishing Bridge Museum and walk to S4 barrier beach on the east side of the outlet. Shoreline designations are after Meyer and Locke (1986), starting with the modern shoreline as *S1* that is 1.9 m above datum (the Bridge Bay gauge). Walk east across Fishing Bridge fault with 1 m offset into the S4 lagoon and discuss archeological findings. Then hike south past Hamilton Store on S4 shoreline and across highway down to S2 shoreline near the Fishing Bridge Museum. Restrooms and busses will be near the museum.

The level of Yellowstone Lake has decreased and increased through time (Fig. 14; Pierce et al., 2002). Lidar images (Fig. 15) show the shorelines and other geomorphology important to deciphering this lake level history as well as tilting and faulting. The *S4* shoreline forms a barrier beach and lagoon and is about 10.7 cal ka B.P., based on buried charcoal. Cody Complex archeological material of late Paleoindian age also supports an age for *S4* of

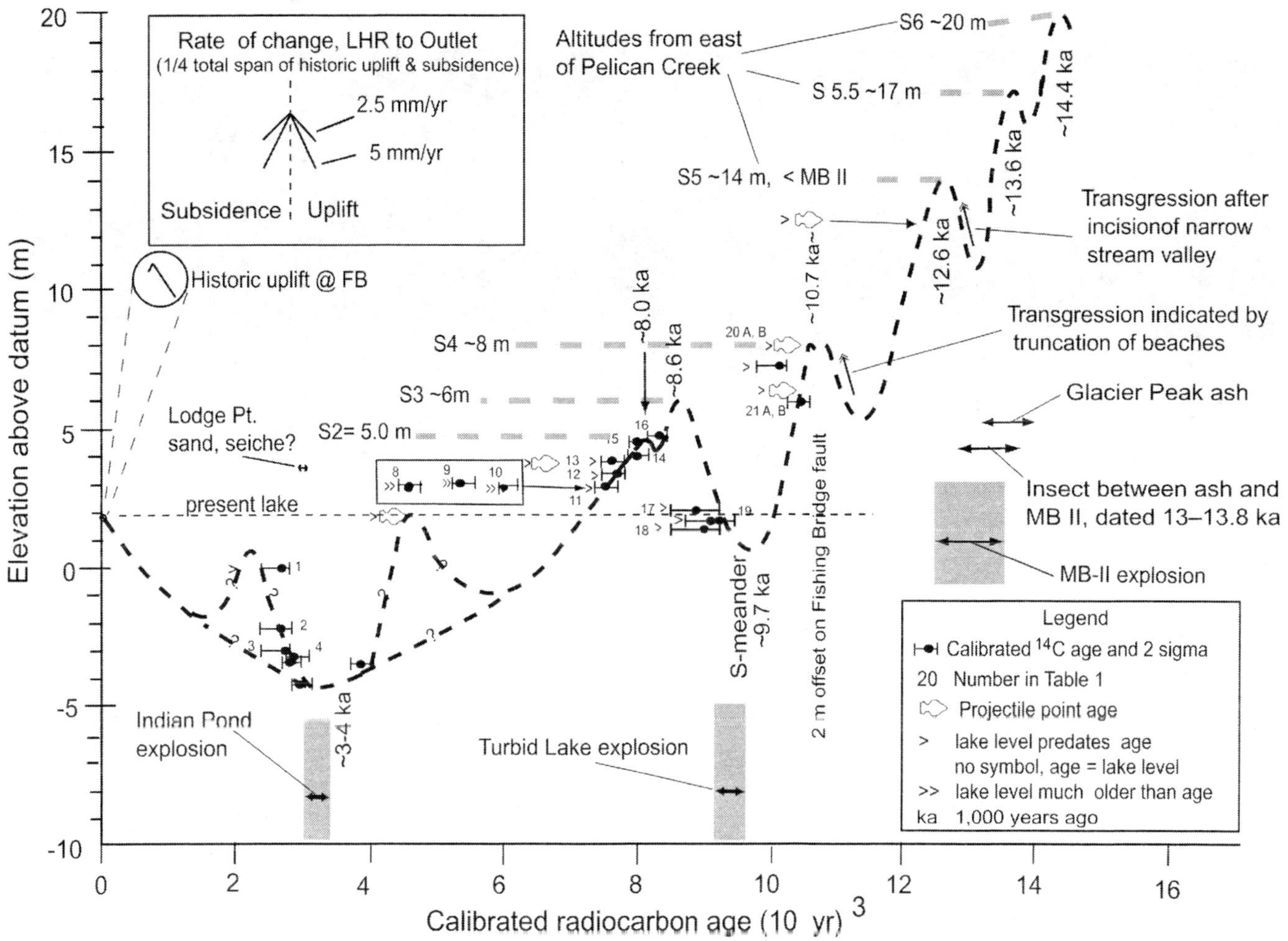

Figure 14. Holocene history of changes in Yellowstone Lake level suggesting cycles of inflation and deflation of the Yellowstone caldera (from Pierce et al., 2002). S2–S6, numbered shorelines following Meyer and Locke (1986).

10–11 ka (Cannon et al., 1994). The *S2* shoreline here is a wave–cut shoreline and is about 8 cal ka B.P. as dated by archeological material, and by charcoal ages on the *S2* shoreline and on the Pelican Creek terrace that truncates the *S2* shoreline (Cannon et al., 1994). When viewed from the *S2* shoreline, all post–8 ka history of the lake is either at present lake level or is now underwater (Pierce et al., 2002). To the southwest in Bridge Bay, a 3.8 cal ka B.P. beach is submerged in 5 m of water. To the east, Pelican Creek lies in a drowned valley with gravel down to 5 m below the present lake.

STOP 3–6. DROWNED RIVER CHANNELS

Drive north from the Hamilton store across the *S4* barrier beach and lagoon and 0.4 mi (0.7 km) beyond the locked gate. Park in the S–meander and walk over to the riverbank. The steep river bank here was undercut by the Yellowstone River at a time when the river was much more vigorous than it is now (Pierce et al. 2002). Cores in the abandoned channel here went through water and lake mud into gravel dated at 2.9 cal ka B.P. at a depth of ~5 m. The outlet reach of the Yellowstone River is now very low gradient, but about 3 ka B.P., it was a gravel–transporting river undercutting its banks, and at this site depositing eolian sand at the top of the eroding river bank. Inflation of Le Hardys rapids diminished the river gradient such that the gravel at 5 m below the surface here is now well below the level of the bedrock threshold 4 km downstream at Le Hardys Rapids. Accounting for this and a 1m/km original gradient of the gravel–transporting river, Le Hardys rapids has come up 7 m relative to this site (Pierce et al., 2002).

Across the road in the S–meander, we dug four backhoe pits across the old channel of the Yellowstone River (Pierce et al., 2002). The outside of this meander is a high, steep bank formed at a time when the Yellowstone River was undercutting its banks and carrying gravel up to 3 cm. Then, as the river lost its competence, charcoal accumulated at about 9 cal ka B.P., and was buried by shoreface gravel. Water level rose until this reach of the Yellowstone River became a northern arm of Yellowstone Lake and the *S3* and *S2* shorelines were cut. Return to Lake Village for second night and possible evening talks.

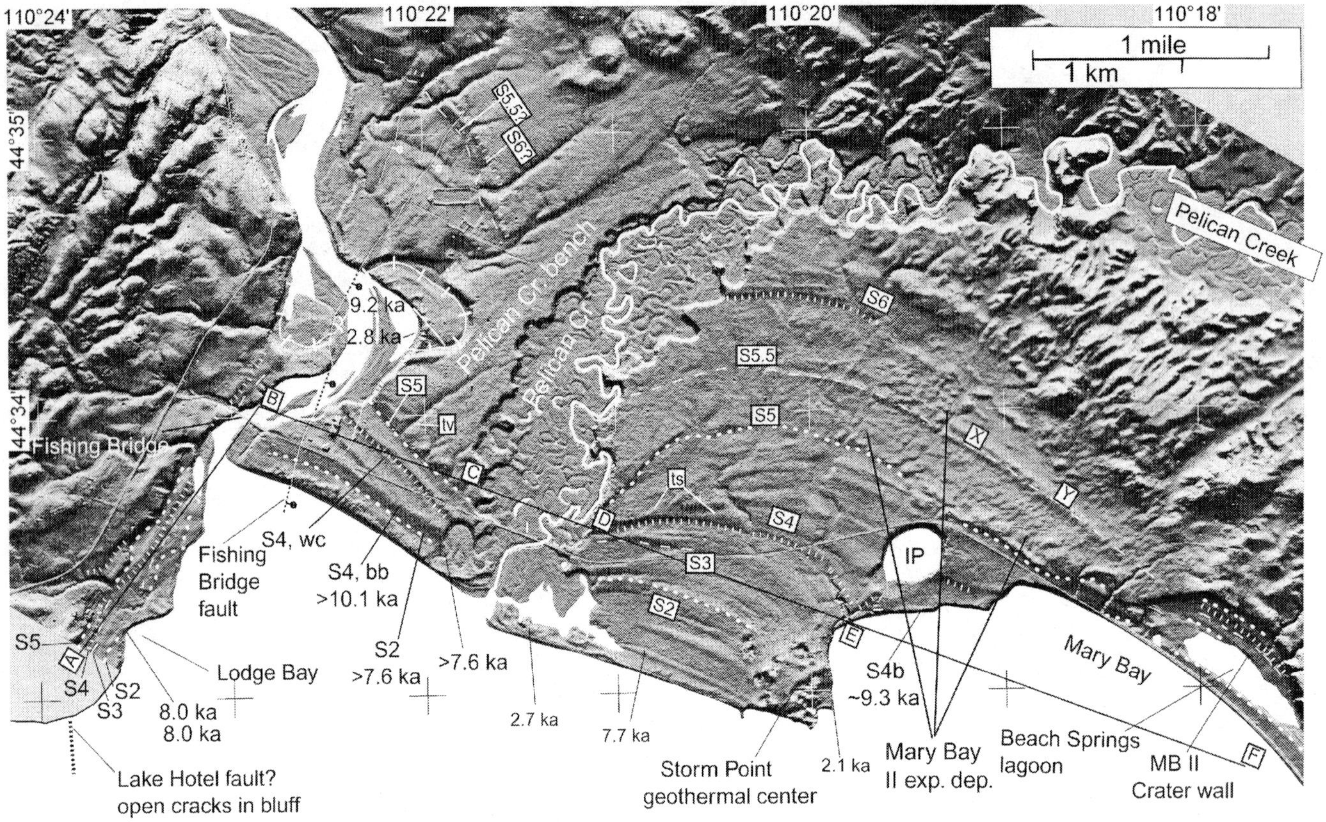

Figure 15. Lidar image of north shore area of Yellowstone Lake (from Pierce et al., 2002). Annotations show numbered shorelines, location of radiocarbon ages, Holocene faulting, and Mary Bay II explosion deposits. The wide channel of Yellowstone River is depositing sand bars and has been drowned by uplift downstream. The S–meander is shown by white line with crossing bar. IP– Indian Pond explosion crater, tv– truncated valley, ts– truncated shorelines.

DAY 4. LAKE VILLAGE TO JACKSON, WY

STOP 4–1. HIKE TO STORM POINT

About 3 cal ka B.P., a hydrothermal explosion ejected low–grade hydrothermal materials (weakly cemented lake sediments, sand, and gravel) and left the crater occupied by Indian Pond. The creek west of Indian Pond exposes platform sand, ~13–ka Mary Bay II hydrothermal explosion deposits, lake gravel, lake sediments with bryophytes, and varved lake sediments with Glacier Peak ash. Storm Point is a complex of hydrothermal craters as well as exposures of steeply dipping Pleistocene gravel. Uplift of Storm Point center has tilted shorelines westward at 6 m/km (Pierce et al., 2002).

YELLOWSTONE LAKE BOTTOM

Discoveries from multi–beam sonar mapping and seismic–reflection surveys of the northern, central, and West Thumb basins of Yellowstone Lake provide new insight into the extent of post–collapse volcanism and active hydrothermal processes occurring in a large lake above a large magma chamber (Morgan et al., 2003). Yellowstone Lake has an irregular bottom covered with dozens of features directly related to hydrothermal, tectonic, volcanic, and sedimentary processes (Fig. 16). Detailed bathymetric, seismic reflection, and magnetic evidence reveals that rhyolitic lava flows underlie much of Yellowstone Lake and exert fundamental control on lake bathymetry and localization of hydrothermal activity. Many previously unknown features have been identified and include over 250 hydrothermal vents, several >1,500 ft (500 m) diameter, hydrothermal explosion craters, many ~300–500 ft (100–200 m) diameter, hydrothermal vent craters, domed lacustrine sediments related to hydrothermal activity, elongate fissures cutting post–glacial sediments, hydrothermal siliceous spires, sub–lacustrine landslide deposits, submerged former shorelines, and a recently active graben. Sampling and observations with a submersible, remotely operated vehicle (ROV) confirm and extend our understanding of the identified features. Faults, fissures, hydrothermally inflated, domal structures, hydrothermal explosion craters, and sublacustrine landslides constitute potentially significant geologic hazards. Toxic elements derived from

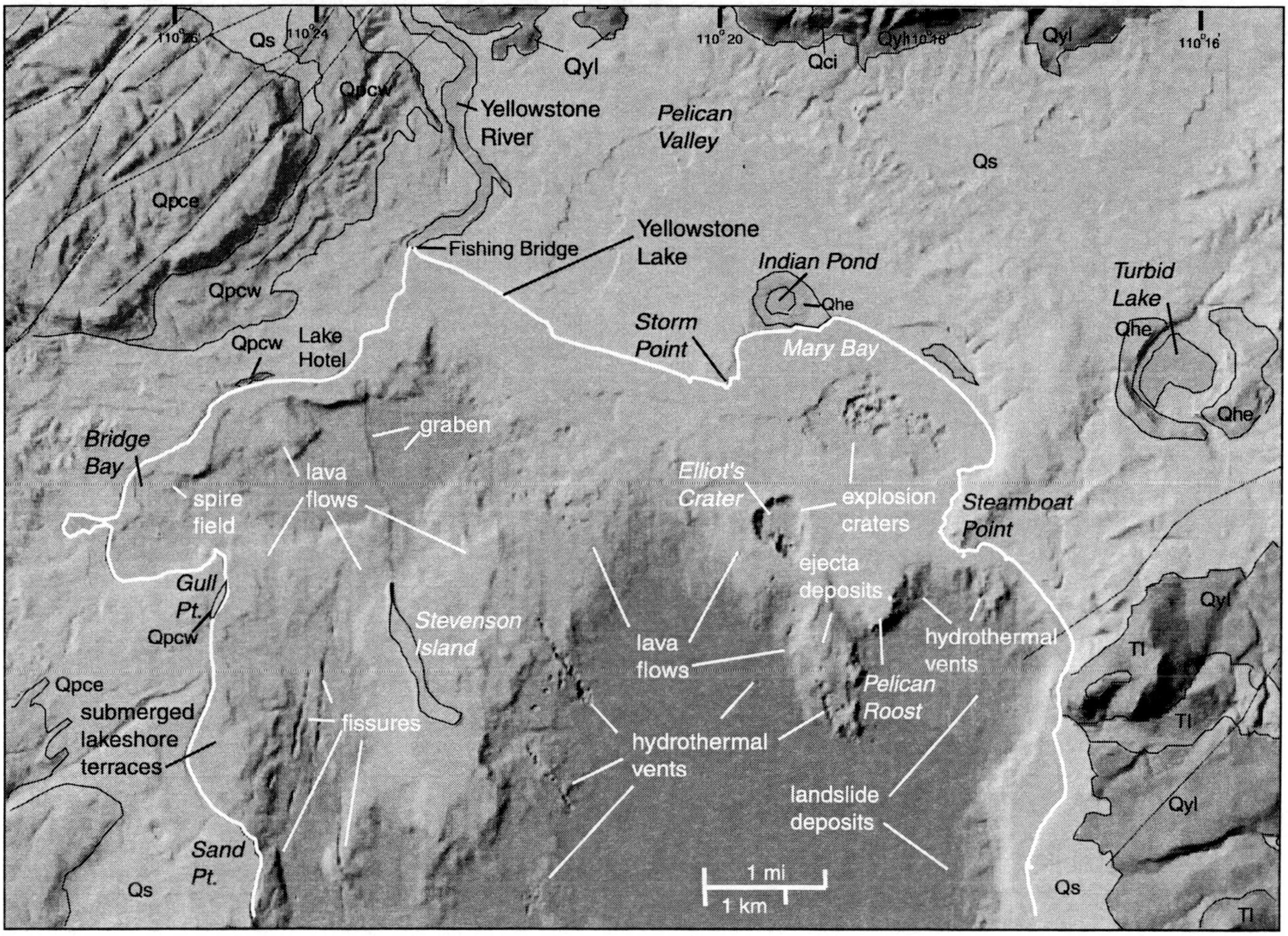

Figure 16. Shaded relief map showing underwater features on the floor of northern Yellowstone Lake (from Lisa Morgan, 2003). Onshore symbols: Qyl– Lava Creek Tuff, Qpcw– West Thumb flow, Qpce– Elephant Back flow, Qs– Quaternary sediments, and Qhe– hydrothermal explosion deposits.

hydrothermal processes may also significantly affect the Yellowstone ecosystem.

STOP 4–2. LAKE BUTTE VIEWPOINT

Absaroka volcanic and intrusive rocks here form the rim of the 640–ka Yellowstone caldera. During full–glacial time, the Yellowstone Plateau icecap formed a dome over Yellowstone Lake with ice flowing south from this icecap into Jackson Hole, west toward West Yellowstone, north over the top of the Washburn Range, and east through Sylvan Pass. The glacier flowing eastward through Sylvan Pass was so high that it flowed *up* the southern tributary valleys of Middle Creek and deposited lateral moraines 1200 ft (400 m) above the stream that slope *up* the tributary valley. During deglaciation, the Plateau icecap downwasted, and a succession of kame deposits was built from Sylvan Pass westward to Yellowstone Lake as well as around most of the lake basin. In a valley tributary to the southeast arm of Yellowstone Lake, peat dated at 15.5–16 cal ka B.P. overlies kame deposits and underlies gravel considered late Pinedale outwash from small glaciers upstream.

About halfway between here and Cody, Wyoming is Mummy Cave, which has an unusually complete archeologically record and serves as a major reference for archeological changes through time. (Husted and Edgar, 2002; Hughes, 1998).

The origin of Yellowstone Lake is complex. Excepting south arm, southeast arm, and the ~160–ka West Thumb caldera, the lake is in the southeast sector of the moat of the 0.64–Ma caldera. As concluded by Morgan et al. (2003), rhyolite flows are present on the west and south part of the lake, and a flow or intrusion is completely buried near the northeastern margin of the lake, but these flows are not as high as the flows that completely fill the "moat" in the north, west, and south parts of the caldera. The 300 ft (100 m) deep east side of the lake and its continuation up the southeast arm are a fiord–like trough apparently excavated by a glacier flowing northward from the high upper Yellowstone terrain during the early part of a glacial cycle.

STOP 4–3. WEST THUMB THERMAL AREA

West Thumb is located where a NNW–trending band of young intra–caldera vents intersect the ring–fracture zone of the Yellowstone caldera (Christiansen, 2001). Eruption of the tuff of Bluff Point 162 ± 2 ka formed the West Thumb caldera. Some sinter cones protrude above the lake and similar sinter cones are entirely underwater and were formed when the lake was at least 14–18 ft (4–5 m) lower. One of these is the "cutthroat Jacuzzi", often populated by circling trout. Along the south shore of Yellowstone Lake shorelines S2 and S4 are dated and of similar height as along the north shore. Although inflation and deflation of the caldera has occurred, the horizontality of shorelines suggests no net buildup of magmatic material at depth associated with the inflation–deflation cycles. This is compatible with inflation deflation cycles driven by geothermal fluid buildup and release.

Pull over to view the valleys of Thumb Creek and Little Thumb Creeks drowned by rising lake levels. The area burned here was part of the 1988 Shoshone fire. Further ahead in the Lewis River Canyon, the steep margin of the Yellowstone Plateau is deeply incising.

STOP 4–4. JACKSON LAKE PICNIC AREA

This stop (Fig. 17) is on the margin of the Snake River delta with its abundant archeology. Across the reservoir from here, the bench about 1,500 ft (500 m) above the lake is mantled by hummocky moraines of the last glaciation, deposited by the Snake River lobe from Yellowstone. The terminus of this glacial lobe was 13 mi (20 km) to the south beyond the south shore of Jackson Lake, and much of the terrain in between has drumlin–like glacial morphology. Glacial scour excavated Jackson Lake, which is 400 ft (120 m) deep with as much as 400 additional feet of sedimentary fill (Smith et al., 1993).

About 100 years ago, a dam was built to raise the lake 35 ft (10m). Before damming, the Snake River had built a delta southward past this stop. Post–glacial scarps on the Teton fault are locally apparent near the base of the Teton Range. Because the outlet of Jackson Lake is farther away (east) of the Teton fault than any other part of the lake, fault offsets would result in submergence of shorelines. Events here in the natural delta area are considered to represent the last two offsets of the Teton fault: (1) transgression of the lake up onto the riverine delta and construction of beaches on top of the scroll–work pattern of the riverine delta; this can be explained by ~1 m downfaulting of the delta relative to the lake outlet dated by habitation of the beaches ~1.6 cal ka B.P., and (2) formation of ridge–and–basin terrain apparently by shaking during fault offset, followed by Native American occupation of these ridges about 4 ka B.P..

Overview of Jackson Hole archeology

Jackson Lake has been the focus of intense archeological investigations (Reeve et al., 1979), including work associated with reconstruction of the Jackson Lake dam (Connor 1998). Although archeological sites experienced up to 1 m reservoir wave erosion, Boeka (2002) found only minimal horizontal movement of artifacts, thereby allowing human behavior to be confidently interpreted from the spatial patterning of artifacts

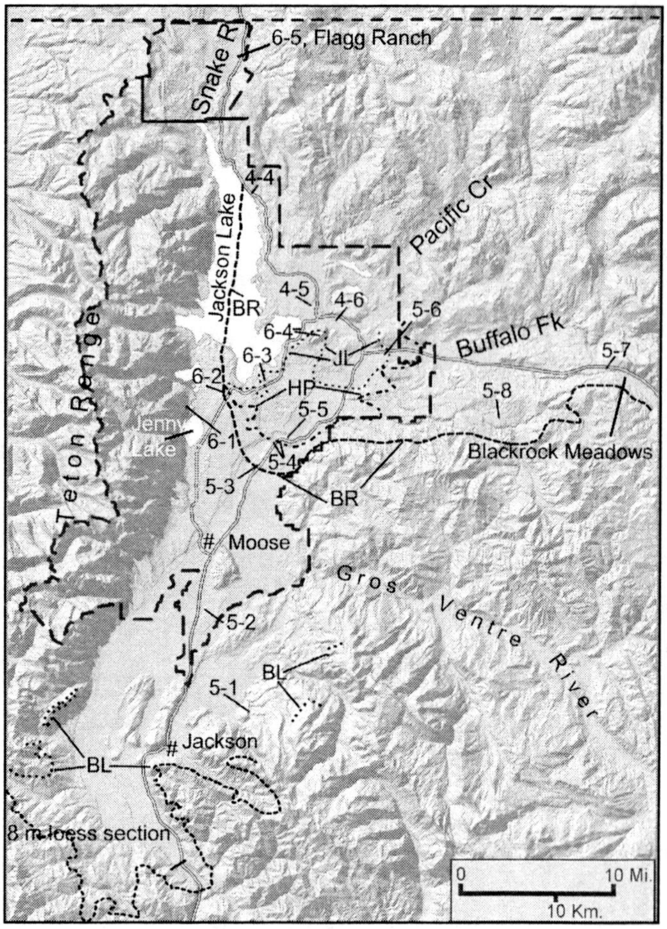

Figure 17. Shaded relief map of Jackson Hole area showing location of stops on Day 4, 5, and 6. Locations of boundaries of Bull Lake and Pinedale glaciations transferred by eye and not accurate; see Pierce and Good (1992) for better rendition. Ice margins labeled as follows: BL– Bull lake, BR– Burned Ridge, HP– Hedrick Pond, JL– Jackson Lake.

The interdisciplinary nature of the Jackson Lake investigations provided for the development of a model of landform evolution and subsequent human occupation (Pierce et al., 1998). Earliest occupation dates from the early Paleoindian Period (14.0 to 10.3 cal ka B.P.). Later Paleoindian complexes (10.3–7.6 cal ka B.P.), which are characterized by various lanceolate projectile point styles, are common in both Jackson Hole (Connor 1998) and the Yellowstone Plateau (Cannon and Hughes 1997), but few stratified sites have been found. A broad–based subsistence economy is suggested for this time period.

The earliest radiocarbon date (6.7 cal ka B.P., 5850 ± 50 ^{14}C yrs. B.P.) from the shores of Jackson Lake coincides with Archaic cultures of the mid–Holocene (7.6 to 1.9 cal ka B.P.) characterized by large side–notched and corner–notched projectile points, fired–rock earth ovens, and possible increased reliance on plant resources. Two subsistence–settlement patterns have been suggested for this period—one focused on bison which is common in the intermountain basins and on the plains, the other a mountain–oriented pattern for which mule deer and mountain sheep were the primary prey (Frison 1992). Connor (1998) suggests a pattern of change in subsistence–settlement systems occurred around 6 cal ka B.P.

The late Holocene of Jackson Hole (2.8 to 1.9 cal ka B.P.) mirrors a trend of increased site quantity and density seen throughout the Central Rocky Mts. (Connor 1998) and Northwestern Plains (Frison, 1992). Corner–notched points, such as Pelican Lake and Elko, are common during this period, as are smaller side– and corner–notched points representing bow–and–arrow technology. Subsistence patterns in the mountains appear to be a continuation from earlier times with mule deer and mountain sheep the primary prey. Fire pits, stone circles, and grinding stones are also common. However, Cannon (2001) has suggested that subsistence patterns in Jackson Hole may be different with bison being a more common prey. During the last 200 years, population seems to have decreased (Connor 1998). While Wright (1984) suggests collapse of the system around AD 1600, Conner (1998) suggests decreased temperatures associated with the Little Ice Age may have reduced seasonal use.

STOP 4–5. WILLOW FLATS VIEWPOINT

Willow Flats is part of a huge, post–glacial fan of Pilgrim Creek where natural sub–irrigation of the lower part of the fan fosters a willow habitat ideal for moose. On and near Signal Mt., the dip of dated volcanic units suggests that tilting into the Teton fault started about 5 Ma and has been relatively constant since then (Fig. 18). The 4.45–Ma Kilgore tuff flowed into this area from its source caldera on the west side of the Teton Range (Lisa Morgan, oral communication, 2002) indicating the Teton Range was no barrier then. The high, steep, imposing front of the Teton Range is another reflection of this young, ongoing formation of the range.

This stop is within the Jackson Lake phase of the Pinedale glaciation (Fig. 19). East of here is a "two–sided" kame terrace between Christian Pond on the west and Emma Matilda Lake on the east that formed between the Snake River glacial lobe on the west and the Pacific Creek glacial lobe on the east and defines the exact the relation between the two lobes at a given time.

Earlier in the Pinedale history during Burned Ridge time (Fig. 19), the Pacific Creek lobe flowed westward through this area and excavated a deep basin that now extends from Pacific Creek westward up the Snake River, beneath the present dam, and under the lake north of Signal Mountain. Drill holes beneath the dike section of the dam encountered 600 ft of silty lake sediments now filling this glacial trough. During an earthquake on the Teton fault, the fill of this trough, particularly the unconsolidated sand beneath the dam and dike, might liquefy and result in failure of the dam. The dam and dike were recently rebuilt and liquefiable material firmed up by both dynamic compaction and in–situ manufacture of concrete columns by injection of cement into long, rotating augers.

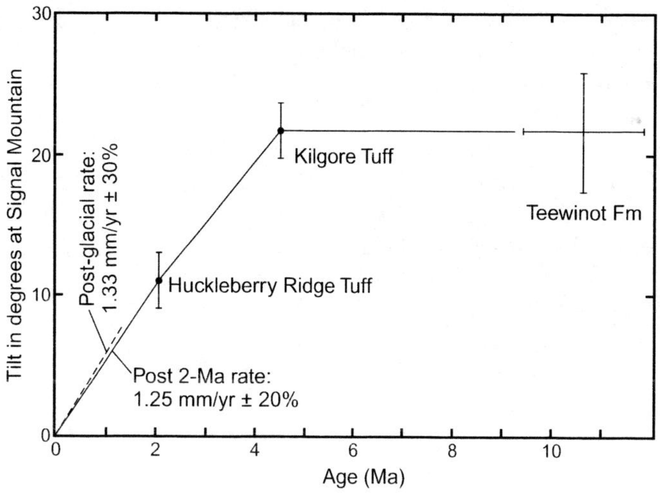

Figure 18. Tilting of late Cenozoic units in the Signal Mountain area westward into the Teton fault showing uniform high rate of tilting after about 5Ma. Dashed line is tilt rate reconstructed from post–glacial offset.

Pinedale Glacial Phase	NW Snake River glacial lobe	Pacific Creek glacial lobe	SE Buffalo Fork glacial lobe
Burned Ridge (oldest)	not even conflent	smaller	largest
Hedrick Pond	smaller	largest	not even conflent
Jackson Lake (youngest)	large	large	not even conflont

Figure 19. Changes in the relative size of the three glacial lobes flowing into northern Jackson Hole during the three phases of the Pinedale glaciation. Arrows indicate migration of dominant lobes from southeast to northwest towards direction of moisture supply.

STOP 4–6. SNAKE RIVER OXBOWS

This is a popular viewing area of waterfowl, moose, otter, and raptors. At the edge of the meadows east of here is a barrier beach of a late glacial lake at an altitude of 6800 ft (2072 m), 65 ft (20 m) above the natural level of Jackson Lake. The lush, flat meadows below this barrier beach are on the bottom of the associated lake. The water in the Oxbows reaches 15 ft (5 m) in depth.

This deep, tranquil stretch of the Snake River may result from westward tilting on the Teton fault. We are above a deep bedrock trough excavated in Burned Ridge time. The Snake River established its present course by following the lowest post–glacial terrain from this Pacific scour basin into the Triangle X scour basin. Continue south to Jackson.

DAY 5 JACKSON HOLE

STOP 5–1. BULL LAKE GLACIAL TERRAIN

Drive east from Jackson through the Elk Refuge to the viewpoint at the Curtis Canyon Campground. This viewpoint is in glacial terrain of Bull Lake age and overlooks Jackson Hole, which was filled by glacial ice during the Bull Lake glaciation. The strong, southwesterly grain of the local topography (Fig. 20) is the result of both the southwest flow of Bull Lake ice, whose upper limit is shown by till and erratics 1,200 ft (360 m) vertically upslope from this stop, and multiple ice–marginal, meltwater channels 30–300 ft (10–100 m) deep. The upper limit of Bull Lake deposits descends southward and at Jackson is at the top of the Snow King chair lift. Bull Lake terminal moraines and erratics drape the middle and lower slopes of the mountains at the southern end of Jackson Hole, 30 miles (50 km) south of the Pinedale terminal moraines.

The view from here includes: (1) Burned Ridge moraines and outwash, overlapped downvalley by successively younger Hedrick Pond outwash on antelope Flats and Jackson Lake(?) outwash at the airport, (2) Blacktail Butte and the hills of the National Elk Refuge overridden by Bull Lake ice, with the loess–mantled Bull Lake fan of the Gros Ventre River southwest of the Elk Refuge hills, (3) below that, the large, Pinedale, outwash fan of Flat Creek, (4) Pinedale moraines at the mouths of seven canyons along the front of the Teton Range, (5) the Teton fault, forming a scarp on the lower part of the range and offsetting the lateral moraines, (6) Gros Ventre Buttes streamlined by overriding Bull Lake glacier, and (7) Phillips Ridge, above these buttes on the far side of Jackson Hole, where the Bull Lake glacial limit is at 8,100 ft (2,470 m).

If the large glaciation filling all of Jackson Hole was pre–Bull Lake, the local valley glaciers would be expected to display Bull Lake and Pinedale moraines. Only Pinedale moraines are recognized for the seven Teton valley glaciers or for the several valley glaciers from the Gros Ventre Ranges, such as Sheep Creek just east of this stop (Fig. 20). As we will see later in the trip, the Bull Lake glaciation near West Yellowstone is dated 140 ka by combined obsidian–hydration and K–Ar dating techniques. For the greater Yellowstone glacial system, Bull Lake moraines are far beyond Pinedale ones to the west (West Yellowstone), southwest (Ashton), and south (Jackson Hole), but not recognized to the north and east.

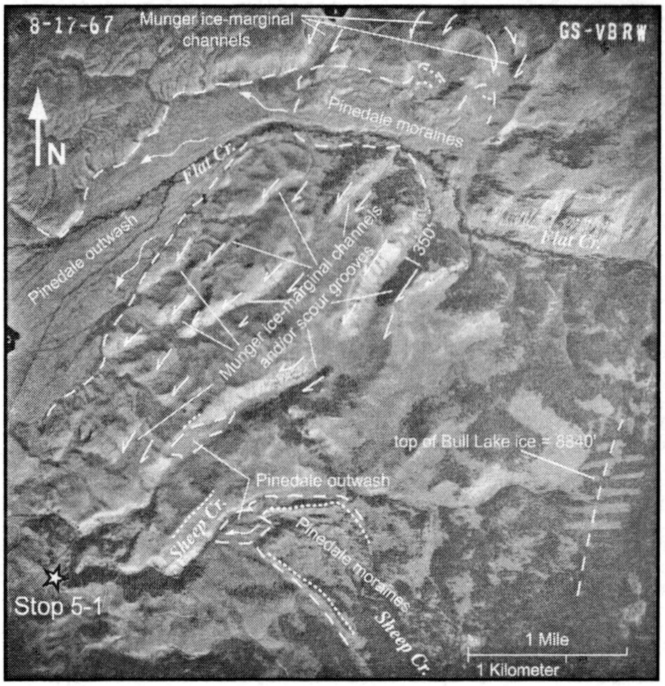

Figure 20. Moraines and meltwater channels on the eastern side of Jackson Hole from the Bull Lake glaciation that filled Jackson Hole. Also shown are Pinedale moraines and outwash from local mountain–valley glaciers.

Where slopes are gentle, Bull Lake deposits are mantled by several meters of loess with a well–developed, buried soil near its base. For a section 10 miles (15 km) south of Jackson, Figure 21 shows loess, buried soils, and ages on a Bull Lake outwash terrace 400 ft (120 m) above the Snake River and about 2.5 mi (4 km) inside the Bull Lake terminus. Return to Jackson and head north.

STOP 5–2. TERRACES NEAR THE AIRPORT

Park on the road into an old gravel pit (now a NPS firing range) and climb up onto loess–mantled terrace (Fig. 22). This terrace is a Gros Ventre River deposit and a candidate for an early Wisconsin age. A soil pit 11 ft (3.3 m) deep encountered only a reddened zone near the base of the loess and thin carbonate coats in the underlying gravel, indicating only a weakly developed, buried soil.

A graben offsets this terrace 1 mi. (2 km) northeast of here. The scarp at this stop has also been mapped as a postglacial fault (Love et al., 1992), but Pierce considers it a fluvial scarp because (1) alluvial fans deposited across the scarp show no evidence of offsets, whereas multiple offset events less than 10 ft (3 m) would be expected to create this 60 ft (20m) high scarp, (2) the scarp is scalloped by channels at the margin of the lower terrace, perhaps by floods, and (3) the scarp is more linear, gently arcuate, and constant than typically seen in normal fault scarps and truncates abruptly to the south.

Beneath the terrace of Jackson Lake (?) age near the airport, well logs note clays near 100 ft (30 m) and resistivity surveys confirm a highly conductive material. These may be lake sediments deposited in a lake scoured out during the Bull Lake glaciation.

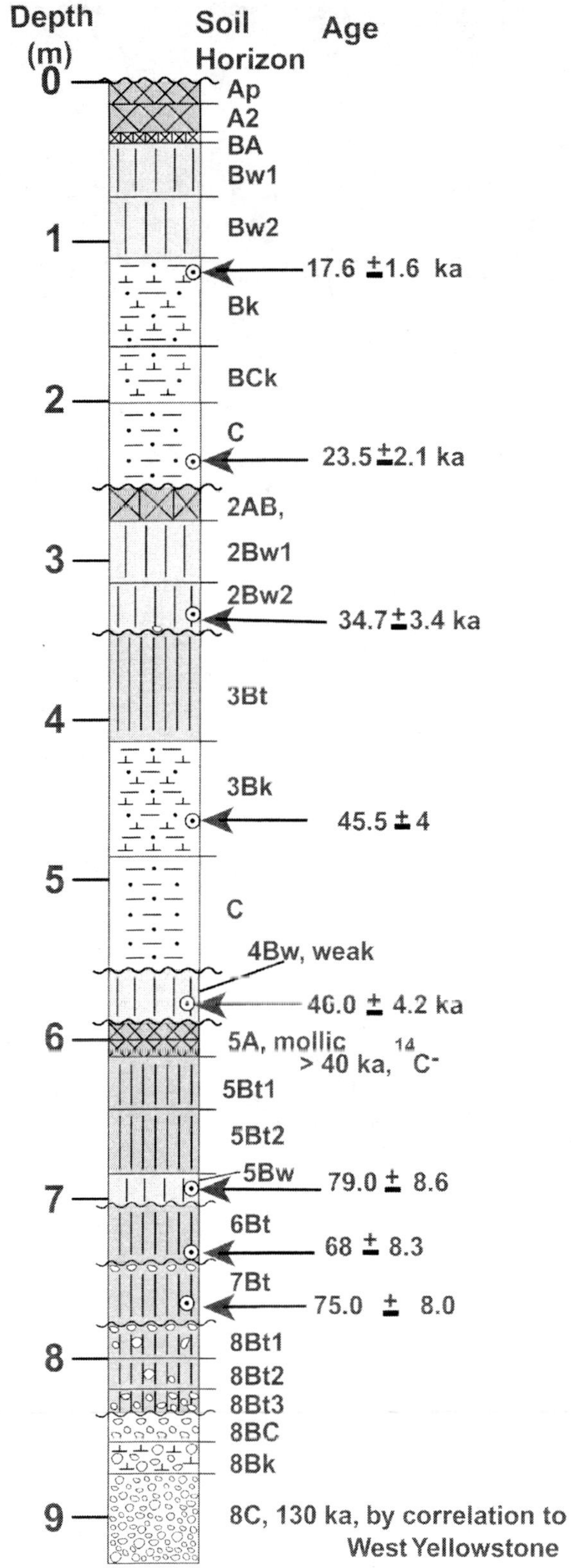

Figure 21. Loess, soil, and provisional TL ages in Porcupine Creek section, southern Jackson Hole. Symbols as follows: stronger development of B–horizons– darker shading and closer spaced vertical lines, A–horizons– cross hatch and shaded, carbonate– horizontal dash with vertical tic; and unweathered loess– dash dot pattern and no shading

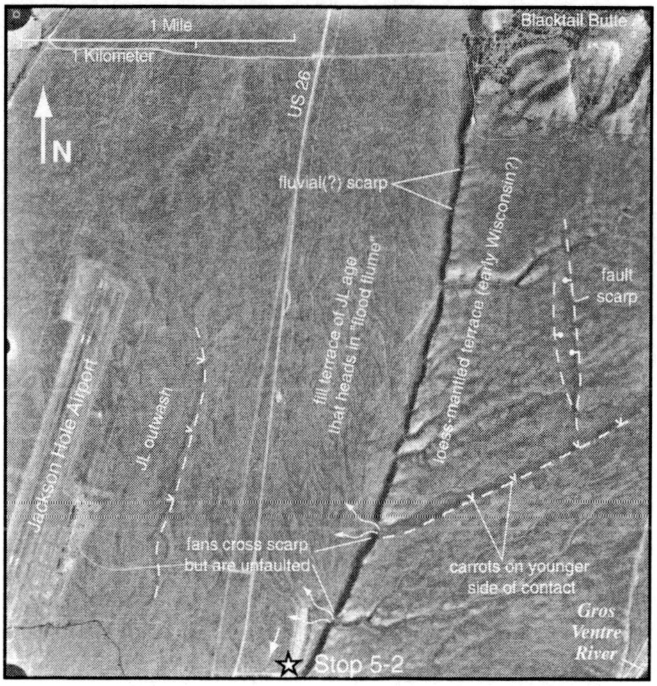

Figure 22. Snake River outwash terrace of Jackson Lake age around the Jackson Hole Airport undercutting loess–mantled Gros Ventre River terrace of early Wisconsin? age.

STOP 5–3. TETON POINT TURNOUT

The highest peak is Grand Teton, altitude 13,771 ft (4197 m), south of which are Middle Teton, South Teton, and Buck Mountain. North of Grand Teton are Mt Owen and Teewinot. The Snake River floodplain is down in the cottonwoods and is characterized by branching and continuously changing channels.

For locations that follow, the clock position of 12:00 is the upvalley trend of the bluff on which we are standing. This stop is near the onlap edge of the Spaulding Bay outwash fan (11:00) of Hedrick Pond age onto the Antelope flats outwash fan (1:00) of Burned Ridge age (Fig. 23). The Snake River follows the approximate boundary between these two large outwash fans of different ages and discharge points. The inset terrace is of Jackson Lake age. We call it a flood flume, for it displays large–scale bedforms produced by flood waters 30–45 ft. (10–15 m) deep and ~1 mi. (1.8 km) wide, and flood bars occur up to the top of the fill terrace. Looking downvalley along the length of the Snake River, this Jackson Lake terrace converges with the Hedrick Pond terrace.

Burned Ridge moraines occur at the head of outwash between 12:00 and 1:00 and represent the youngest (BR–3) of three sets of Burned Ridge moraines. The BR–2 moraines protrude through the outwash at 1:30. Kettles where the grassy ridge at 2:00 comes down to the outwash level represent the BR–1 limit. Along the Teton front at 8:00, moraines of Pinedale valley glaciers form large mounds that encompass Bradley and Taggart Lakes. The Teton fault scarp is along the front above these moraines.

Along the southern margin of the greater Yellowstone glacial system in Pinedale time, three glacial lobes poured into northern Jackson Hole (Fig. 19). The dominance of these lobes shifted through time as follows: Buffalo Fork lobe (1:00–2:00) from the Absaroka Range, Pacific Creek lobe (12:30) from the Two Ocean Plateau, and Snake River lobe (11:00), from the Yellowstone Plateau. This northwestward migration is towards the storm conduit of the Snake River Plain.

STOP 5–4A. BURNED RIDGE MORAINES

Turn right on Lost Creek road and depart busses at next sharp right turn. Hike north past a kettle and up onto the youngest of three Burned Ridge moraines (BR3; Fig. 24). Erratics are mostly Pinyon–type quartzite, with minor Tensleep sandstone, Paleozoic limestone, Absaroka volcanic rocks, Mesozoic sandstone, rhyolite tuff, and Precambrian crystalline rock.

To the north is a drop–off 100–200 ft (30–60 m) high that represents the cast of Burned Ridge ice, the front of which was largely buried by outwash. The middle Burned Ridge stand (BR2) is represented by moraines mostly buried by outwash 1.5 mi (2.5 km) south of here (Fig. 24), and the oldest stand (BR–1) is represented by kettles in outwash 2.5 mi (3.8 km) south of here. The Buffalo Fork lobe deposited these moraines and outwash. The outwash terrace rises to the northeast (Fig. 24) and heads at the ice–marginal position now occupied by Spread Creek. Outwash channel patterns (Fig. 24) show stream flow was to the west. The original depositional slope is an order of magnitude greater than the estimated tectonic tilting into the Teton fault.

The glacial deposits along the southern margin of the greater Yellowstone glacial system are predominantly outwash gravel of Pinyon–type quartzite. The great thickness of outwash gravel results from glacial erosion of weakly indurated quartzite conglomerate of upper Cretaceous to Paleocene age. Great quantities of these quartzite clasts were glacially transported to the head of outwash where they overloaded the outwash system and deposited outwash 200 ft (60 m) or more thick. These thick outwash accumulations bury much of the associated glacial moraines. Walk from moraine to the highway and carefully cross to Snake River overlook (Fig. 24) to where vehicles are parked.

STOP 5–4B. SNAKE RIVER OVERLOOK

This stop is on west–sloping, Burned Ridge outwash (Fig. 24). Across the Snake River, the forested landform is Burned Ridge, a moraine mantled or modified by Hedrick Pond deposits. South of Burned Ridge is outwash of Hedrick Pond age of the Spaulding Bay outwash fan. It has distinctly fresher channel morphology on aerial photographs than Burned Ridge outwash.

About 100 ft (30 m) lower than the Hedrick Pond outwash and to the north of Burned Ridge is outwash of Jackson Lake age of the Potholes channelway, where ice blocks of Hedrick Pond age were buried by Jackson Lake outwash. Jackson Lake out-

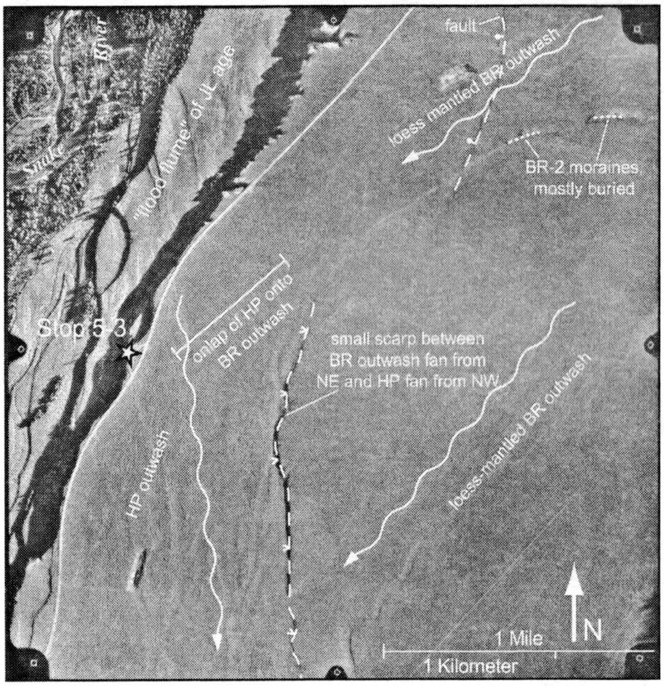

Figure 23. Onlap of outwash fan of Hedrick Pond age from north onto outwash fan of Burned Ridge age from northeast near Teton Point turnout.

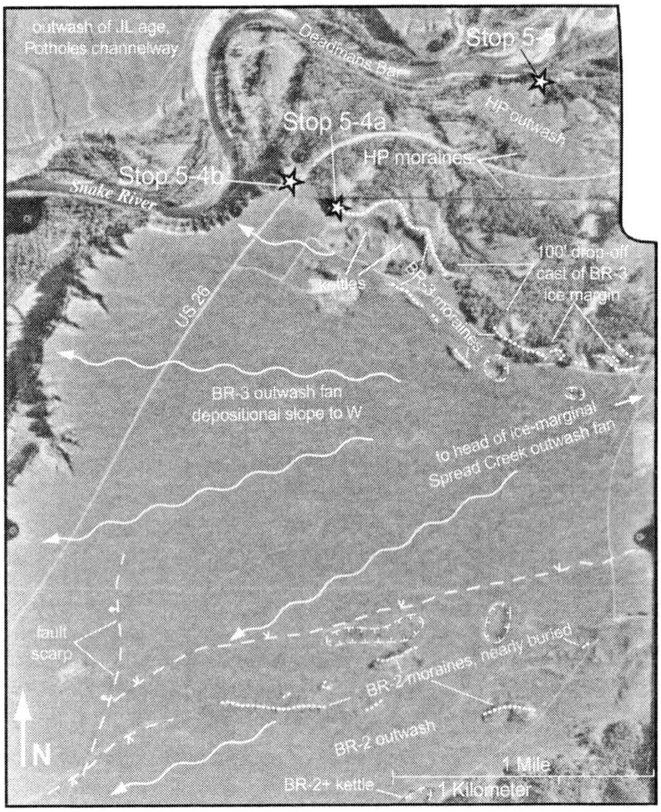

Figure 24. Outwash and moraines and of Burned Ridge and Hedrick Pond age near Snake River Overlook.

wash prograded southeast into the Triangle X2 lake. The outlet to this lake was here at the head of the "flood flume." Repeated flooding lowered the outlet from the ~6840 ft (2085 m) Hedrick Pond spill point to 6720 ft (2048 m), the present inset terrace level. The floods may be from release of glacially dammed lakes along the near STOP 5–6. The eroding bluffs from the Snake River up expose: gray gravel mantled by loess (?), yellow gravel (from the northeast?), gray gravel of Jackson Lake age.

STOP 5–5. HEDRICK POND SECTION AND PALEOECOLOGY

Turn left on the road to Deadman's Bar and park in the first parking area. Hike north across Hedrick Pond moraines, intersect the Snake River Bluffs on Hedrick Pond outwash, and walk up the Snake River to where it impinges on the bluff, creating a fresh, high exposure (Fig. 24). The following section is exposed (Pierce and Good, 1992):

Ft	(m)	Description
16	(5)	Gravel with scattered lenses of flow till. Bedded, but beds discontinuous; surface has kettles
6	(1.8)	Diamicton, flow till
11	(3.4)	Gravel
11	(3.4)	Sand, cross – bedded
2.5	(0.8)	Gravel
80	(24.1)	Diamicton, muddy sand matrix with scattered pebbles, locally wispy silt beds. The diamicton is rich in fines and poor in gravel and was probably eroded from lake sediments that accumulated in the Triangle X–1 lake that extended from the Snake River overlook to Moran Junction and up the Buffalo Fork valley from there. Diamicton deposited near ice as a lacustral till or flow till.

Kettles occur at this stop, as well as across the highway at the Hedrick–Pond core site (Fig. 25). An open forest of lodgepole pine and Douglas–fir and sagebrush surrounds Hedrick Pond. A 22.0–ft–(6.71 m)–long core yielded a basal date of ~20.3 cal ka B.P. (17,160 ± 210 ^{14}C yrs. B.P.). The pollen record indicates a period of alpine meadow and shrub communities prior to 13.5 cal ka B.P. (11,500 ^{14}C yrs. B.P.). Juniper (probably *Juniperus communis*) was the first conifer to become widespread in the deglaciated region. Spruce parkland developed at ca. 17.3 cal ka B.P. (14,750 ^{14}C yrs. B.P.), suggesting that spruce was growing close to the ice margin. Fir and pine (probably *whitebark pine*) were present with spruce in a mixed conifer forest between 15.3 and 10.5 cal ka B.P. (12,900 and 9500 ^{14}C yrs. B.P.). Lodgepole pine and Douglas–fir were abundant in the early Holocene ca. 10.5–6.3 cal ka B.P. (9500–5500 ^{14}C yrs. B.P.), implying that conditions were warmer than at present. In the last 6,000 years, spruce and fir reestablished near the site, compatible with the present mesophytic communities and a concomitant decrease in fire frequency (Whitlock, 1993). Figure 26 shows the five stages of the postglacial vegetation history in southern Yellowstone and Grand Teton National Parks.

Pull over before reaching the Triangle X Ranch turnoff. The white sediments are recycled from the Teewinot Formation and deposited as deltas in the Triangle X–2 lake. Deltas from 6,800 to 6720 ft (2,073–2,048 m) altitude represent lowering of lake outlet. Farther north on highway, Spread Creek is actively depositing gravel resulting from erosion along this ice marginal drainage. At Moran Junction, bear right on US 285 and where the highway leaves the flood plain and climbs up onto a moraine, turn right and take the two–track road to fence.

STOP 5–6. ADVANCES UP BUFFALO FORK

During Burned Ridge time, the Buffalo Fork lobe covered nearly all the terrain visible up Buffalo Fork and was more than 1,500 ft (500 m) thick here. It excavated a scour basin probably at least as deep as 100 ft (30 m) drill holes in unconsolidated material. This stop is on a moraine of the Pacific Creek lobe that advanced *up* the valley of Buffalo Fork in Jackson Lake time. Ridges and depressions in the moraine arc up the valley of Buffalo Fork and lake sediments occur on the upvalley side from these moraines (Fig. 27). Across the valley of Buffalo Fork and about 500 feet (150 m) higher, Hedrick Pond moraines also descend upvalley, revealing that the Pacific Creek lobe pushed into the deglaciated Buffalo Fork valley. On this side of the valley, the Pacific Creek lobe scoured south into the ice–free lower Buffalo Fork (Fig. 27). Similarly in the next drainage to the south on the north slope of "Spread Creek hill," Hedrick Pond moraines slope eastward towards the source of the Buffalo Fork lobe. Sudden releases of glacially dammed lakes from Buffalo Fork valley are the favored source for the floods down the "flood flume". Return to highway and continue east.

Pull Over as you descend from these moraines to view lake sediments overlying kame gravels in roadcut on north side of road.

STOP 5–7. THE BUFFALO FORK LOBE

Park in the large area north of the road, carefully cross the road, and climb up onto the moraines. We are on kettled moraine built across the valley of Blackrock Creek and that rises to the north across the highway. The age of this backfill moraine is Hedrick Pond, forming after the Buffalo Fork lobe had receded from its maximal size in Burned Ridge time. About 0.3 miles (1/2 km) south of here, Blackrock Creek exposes lake sediments, sand, and gravel more than 100 ft (30 m) thick.

During full–glacial, Burned Ridge time, the Buffalo Fork lobe headed both north of here in three large valleys draining the Absaroka Range as well as in the Blackrock Meadows just east of here. Evidence for a Pinedale glacier filling Blackrock Meadows consists of sharp–crested Pinedale moraines 500 ft (150 m) above the Meadows and fresh glacial striations with rat tails demonstrating flow westward from Blackrock Meadows. This ice margin connects along the trend of the ice marginal Spread Creek with the Burned Ridge moraines on the floor of Jackson Hole. Soil profiles are weakly developed, morainal topography is rela-

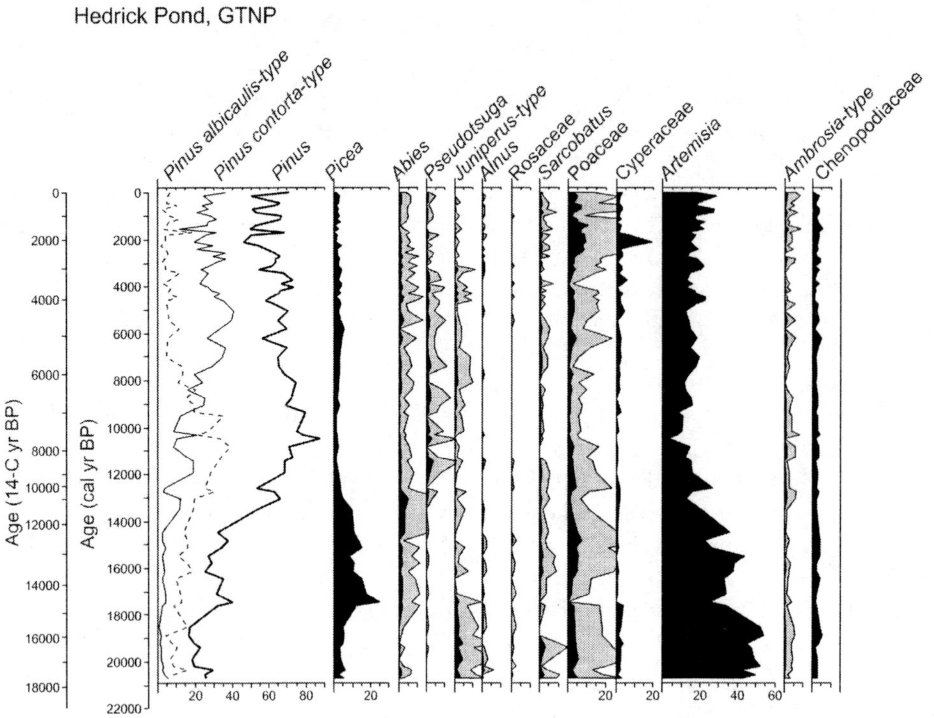

Figure 25. Pollen percentage diagram for Hedrick Pond. Shaded curves are 5X exaggeration (from Whitlock, 1993).

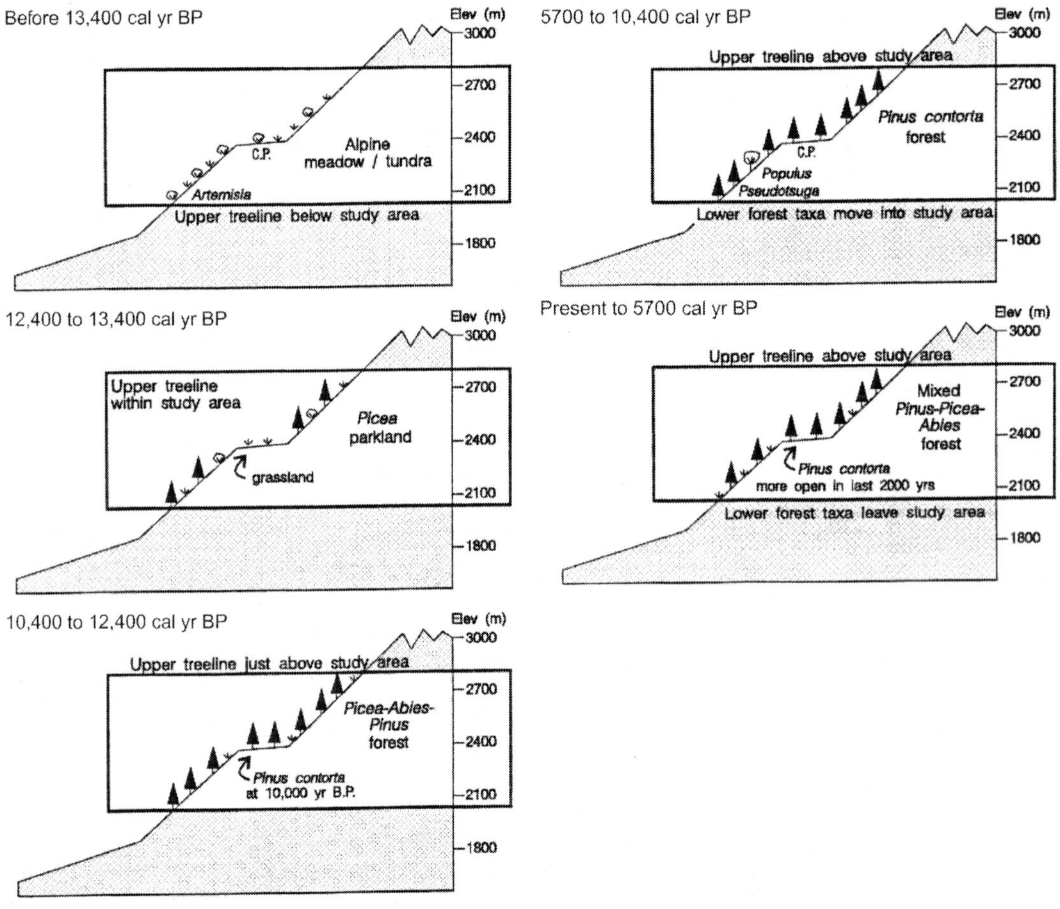

Figure 26. Vegetation history of the montane forest zones in the Yellowstone–Grand Teton region (based on Whitlock, 1993).

tively fresh, and weathering rinds are thin (0.1 mm), consistent with a Pinedale age.

Buffalo Fork flows in a deep valley 2 miles (3 km) north and 1,500 ft (460 m) below here. Blackwelder (1915) attributed this 1,500 ft high bench to the Blackrock erosion cycle and the deposits on it to the "Buffalo" Glaciation, inferring the height above drainage indicated a pre–Bull Lake age. But this height above the drainage here simply reflects the thickness of the Buffalo Fork lobe.

The Teton Range is about 30 miles (45 km) to the west and part of the reason for this stop is to show that the major glaciers did not come from the Teton Range. One mile (1.7 km) west of here, the Togowotee Lodge fault dams a sag pond on its east side (Love and Love, 1997, p. 15). The post–glacial scarp is 30 ft (10 m) high, 40° steep, and part of a set of fault scarps that are recognized for 2 miles (3 km) to the south, including a related fault that offsets Bull Lake moraines by 70 ft (20 m). This new faulting is part of an overall pattern of active faulting associated with northeast migration of the Yellowstone hotspot (Pierce and Morgan, 1992).

About 70 miles (110 km) farther east on the highway is Blackwelder's (1915) type Bull Lake Glaciation. Cosmogenic ages by Phillips et al. (1997, primarily ^{36}Cl) are >130 ka for older Bull Lake moraines, and 95–120 ka for younger Bull Lake moraines. But new Uranium series ages for carbonate coats on pebbles are 167 ± 6.4 ka and 150 ± 8.6 ka for older (WR4) and younger (WR3) Bull Lake terraces (Sharp et al., in press). In the same area, a high terrace dated by the 0.64 Ma Lava Creek ash increases in altitude 100 feet (30 m) more toward Yellowstone than does the 150 ka Bull Lake "Circle terrace." Pierce and Morgan (1992) interpret this 100 ft increase to represent uplift and tilting away from the Yellowstone hotspot. Continue east but turn right on Flagstaff Road.

Pull over and view Blackrock Meadows where Burned Ridge glacier 500 ft (152 m) feet thick formed in Burned Ridge time. This area was snowier in time than in Hedrick Pond time, apparently due to the progressive northwestward buildup of ice towards the Yellowstone Plateau. Continue down gravel road to Lily Lake along the general trend of the ice–marginal Spread Creek

STOP 5–8. LILY LAKE PALEOECOLOGY, BURNED RIDGE ICE MARGIN, FAULTING

Lily Lake is one of several kettles in the sediment–filled valley of Spread Creek. Spread Creek has incised a canyon 400 feet (120 m) deep to the south of here. South of Spread Creek, north–flowing drainages were blocked along the southern margin of Buffalo Fork lobe and inwash kame gravel hundreds of feet thick accumulated. To the north, Burned Ridge moraines drape around Baldy Mountain. A north–south fault offsets these moraines 15 ft (5 m), which is at least half the total offset on this fault. This "new" fault is on the leading margin of the northeastward progression of faulting associated with the Yellowstone hotspot.

Figure 27. Lower Buffalo Fork area showing backfill and damming by Pacific Creek lobe in Jackson Lake (JL) and Hedrick Pond time.

A sediment core from the fen adjacent to Lily Lake has a basal date of ~18.9 cal ka B.P. (16,040 ±220 ^{14}C yrs. B.P.). The pollen record resembles that of Hedrick Pond, with an early period of alpine meadow, juniper, and birch (*Betula*), followed by the development of spruce (*Picea*) parkland and then subalpine forest with *Picea*, *Abies*, and *Pinus albicaulis* in the late–glacial period. The early Holocene featured an expansion of lodgepole pine, Douglas–fir and quaking aspen during the summer insolation maximum (Whitlock, 1993). Return to Jackson and possible talk in evening on Yellowstone hotspot.

DAY 6. JACKSON TO OLD FAITHFUL

STOP 6–1. TETON FAULT SCARP

Drive through Moose to the String Lake area and park near the rest room facility. Walk to the trailhead at footbridge across String Lake outlet (Fig. 28). Note the bouldery, outer Jenny Lake moraine that descends to the outlet and is buried east of the outlet by outwash from the Spaulding Bay terminus. From the footbridge, hike 0.3 miles (0.5 km) along the Jenny Lake moraine to where a trail comes in from the right, turn right (north) on the trail toward Indian Paintbrush Canyon walking over forested Jenny Lake moraines. Enter an open, avalanche area where in 1986, snow and trees from a large avalanche littered this area and extended out into String Lake. The steep slope above the trail was formed by a cascade of morainal debris coming to rest at its angle of repose after being delivered to the snout of a Pinedale glacier in the Laurel Lake basin. The trail enters the forest and then comes into a meadow formed on a steep alluvial fan. Near the far side of the fan, look upslope to see the oversteepened fault scarp to the right of the gully at the fan head and leave trail and hike up the fan to the fault scarp. *(If you hike on the trail past the large, isolated Douglas–firs, you have gone too far).*

At the base of the fault scarp is a partially filled graben. The fault scarp is 113 ft (34 m) high, as steep as 39º, and has surface offset of 63 ft (19 m) (Gilbert et al., 1983). Pinedale moraines of steep slab–glaciers occur just above the fault scarp and the scarp offsets debris from this slab glacier estimated to have receded about 14 ka. This age and offset yield a high rate of 1.4 mm/yr.

The lakes along the Teton front are, from south to north, Jenny, String, Leigh, and Jackson. String Lake is a shallow, pan–like lake only about 5 ft (2 m) deep, which on sunny summer days is the warmest lake around for swimming. In Jackson Lake time, a large outwash stream headed at the upper end of String Lake. The outlet is 0.2 mi (0.6 km) farther away from the Teton fault than is the west lake shore. Tilting of Jackson Hole into the Teton fault combined with additional strands of faulting noted east of the main fault would submerge the western part of the lake area more than its outlet, thus backflooding this glacial stream course and converting it into a lake.

Given the clock direction to Mt. Leidy at 12:00, some key features are: 1:30–the southern margin of Snake River–Teton lobe at the narrows of String Lake; 11:00–the head of Spaulding Bay outwash fan of Hedrick Pond age with incised channels of Jackson Lake age, and on the far side of this fan, forested landform of Burned Ridge age draped with moraines of Hedrick Pond age; 12:00–across the Snake River at the far side of Antelope Flats meadows, the Burned Ridge outwash fan heading on the ice–marginal Spread Creek. The source areas for three glacial lobes and interlobe areas are: 10:00–12:30–Buffalo Fork lobe heading in the Absaroka Range; 10:00–11:00–highlands between Buffalo Fork and Pacific Creek lobes including Gravel Mountain and Mt Randolph; 10:00–Pacific Creek lobe; 9:00–10:00–Pilgrim Creek highlands that were invaded and surrounded by ice, but not traversed by ice, and 8:00–the Snake River lobe. The Gros Ventre slide of 1925 is at 1:30, and Sheep Mt., with the upper limit of Bull Lake ice near the upper treeline, is at 2:00. Return to busses parked near toilet facility at String Lake.

OPTIONAL STOP 6–1.5 CATHEDRAL GROUP OVERLOOK

The Teton fault scarp offsets moraines and debris of Pinedale slab–glaciers. Tilting of this surface into the Teton fault is much less than the original depositional slope. To the north are forested moraines of Jackson Lake (?) age containing large boulders from the Teton Range (Fig. 28). Quartzite cobbles are common in the outwash in front of the moraine, but are uncommon in the moraine itself, indicating glacial flow from the Tetons was fronted by outwash from the Spaulding Bay terminus.

Figure 28. String Lake area showing outwash from greater Yellowstone glaciation and local moraines enclosing Jenny Lake.

STOP 6–2. HEAD OF SPAULDING BAY OUTWASH FAN

Turn into off–road parking area. We are on the Hedrick Pond level that heads at the Spaulding Bay terminus. Note the same level to the east across the channelway incised in Jackson Lake time. The outwash fan extends northward into "thin air" and to understand its origin, one needs to visualize the snout of a large glacier in the vacant space extending hundreds of feet above Spaulding Bay. This outwash fan extends eastward almost to the Snake River Overlook and southward to Moose. Much of the gravel that forms this fan was excavated from the deep, glacial trough of Jackson Lake. The head of this fan was incised in Jackson Lake time and both the original fan and its incised channels became relict features when the Snake River lobe pulled back from the Spaulding Bay terminus and outwash drained farther east. Across the channelway at the head of outwash are forested moraines of Hedrick Pond age that trend easterly. These moraines have a tilted–slab–like form and may be a glacially thrust mass.

STOP 6–3. POTHOLES CHANNELWAY VIEWPOINT

Park in Mount Moran scenic turnout (or pullout 0.4 miles to the east) and hike north across outwash flats to the top of the bare slope (Fig. 29). After the Burned Ridge recession and prior to Hedrick Pond time, the Triangle X–1 lake occupied the glacial scour basin that extended 20 miles (32 km) from Deadman's bar up the Snake–Buffalo Fork drainage system. We are looking down the Potholes channelway, named for the abundant glacial

Figure Figure 29. Head of Potholes channelway where ice–blocks of Hedrick Pond age were buried by outwash of Jackson Lake age.

kettles that represent ice blocks of Hedrick Pond age buried by outwash of Jackson Lake age (Fig. 29). The kettles form alignments that indicate several ice–margin positions of the Snake River lobe of Hedrick Pond age. The highest outwash terrace along the left side of the Potholes channel is of Jackson Lake age and has linear trends of kettles that suggest the west margin of buried blocks from the Pacific Creek lobe of Hedrick Pond age.

Looking down The Potholes channelway and across the Snake River, the higher alluvial bench is Burned Ridge (BR–3) outwash and moraines, which climb to the left where the ice–marginal Spread Creek followed the southern margin of the Buffalo Fork lobe. The Hedrick Pond bench is 100–200 feet (30–60 m) below the Burned Ridge bench and climbs more gently to the left. This gentler, ice–margin gradient may result from the Hedrick Pond advance moving across Triangle X–1 lake sediments with basal sliding aided by shearing in the lake sediments. The biggest difference in the extent of ice between Hedrick Pond and Jackson Lake time is that the Triangle X lake basin was ice filled in Hedrick Pond time, but in Jackson Lake time only its northeast margin was occupied by the Pacific Creek lobe and only its north margin was occupied by the Snake River lobe. The escarpment at this stop is one of a set of kame moraines that descend to the east. The distal slope of the moraine is a gently sloping (5–8°) bench rather than a ridge.

The glacial geology of the Burned Ridge landform is puzzling. The northern half of Burned Ridge has Hedrick Pond moraines of the Snake River lobe draped on its east side and part of its west side. Hedrick Pond ice advanced through a breach in Burned Ridge through a gap from 0.9 and 1.5 mi (1.4–2.3 km) this side of the Snake River. The margins of this breach are marked by a loop of aligned kettles on the outside of this breach and by west–sloping small moraines in the breach through Burned Ridge.

Pull over into parking area on north side of South Landing Channelway. This outwash fan of Jackson Lake age has remarkably well–preserved channel morphology with downstream branching channels. It has much fewer kettles than the Potholes Channelway.

OPTIONAL STOP 6–4A. TOP OF SIGNAL MT.

Signal Mt. was ice covered in Burned Ridge time. Pinedale moraines drape around it with the highest moraines draped as consistent with draping from a northeastern source and the lower moraines consistent for a northern source, compatible with the change through time in dominance from the Buffalo Fork lobe to the Snake River lobe (Fig. 19). The view from the top of Signal Mt. shows many aspects of the glacial geology of the floor of Jackson Hole. Also visible are the sources of the three glacial lobes: Buffalo Fork lobe and its scoured basin to the east, the Pacific Creek lobe and the scour basin to the northeast, Snake River lobe, and the Jackson Lake scour basin to the north.

STOP 6–4B. SIGNAL MT. LODGE

Between Donoho Point (northwest across the lake) and us is a deep scour basin excavated by the Pacific Creek lobe in Burned Ridge time. If offset were to occur today on the Teton fault, the western shoreline of Jackson Lake would move down more than the natural outlet at Jackson Lake dam, and when the water in the lake rose to spill at the outlet, shorelines on the west side of the lake would be submerged. During rebuilding of the Jackson Lake dam in the late 1980's, the lake was at times lowered to its natural level. From a canoe in calm water, we could see deeply enough to distinguish two submerged shorelines at depths of 3.5–4 feet (1–1.2 m) and 8–9 feet (2.4–2.7 m). Combined side–scan sonar and fathometer records gathered in 1986 by Steve Colman and Ken Pierce show shorelines east of Bearpaw Bay about 2 km east of the Teton fault at depths below the natural shoreline of about 3, 8, 15, 20, 24, 27, 31, and 36 feet (1, 2.4, 4.6, 6.1, 7.3, 8.2, 9.4, and 11 m). These shorelines are inferred to be submergence events related to downfaulting of Jackson Hole, and if so, suggest at least 8 faulting events in post–glacial time.

STOP 6–5. FLAGG RANCH

The ridge to the east is Huckleberry Mt., which is the type section of the 2.1–Ma Huckleberry Ridge Tuff (Christiansen, 2001). The tuff forms a dip slope inclined about 9° to the west. On the western skyline, the same tuff caps ridges at the northern end of the Tetons and also dips west. East dipping normal faults are responsible for these tilts and offsets (Christiansen, 2001). Thirteen miles (20 km) northeast of here on the east side of Mt

Sheridan, post–glacial scarps as high as 70 ft (20 m) represent the main postglacial faulting near the southern Yellowstone National Park boundary.

On Huckleberry Ridge, a well–defined Pinedale lateral moraine occurs at an altitude of 9,100 ft (2774 m) and defines the maximum of the Snake River lobe in Hedrick Pond time. Southward 9 miles (14 km), the ice surface descends about 600 feet (180 m) to the high bench across the lake from Stop 4–6. Continue driving to just past West Thumb.

STOP 6–6. DUCK LAKE AND CENTRAL PLATEAU

Duck Lake is a hydrothermal explosion crater of Holocene age. Another explosion crater beneath Yellowstone Lake northeast of here has lurked unnoticed on the floor of Yellowstone Lake until recently labeled Duck Lake's "evil" twin by Lisa Morgan. Duck Lake has been the site of core studies related to recent fire history and patterns of charcoal accumulation following the 1988 fires in Yellowstone (Millspaugh and Whitlock, 1995; Whitlock and Millspaugh, 1996).

Between here and Old Faithful, the road traverses rhyolite flows mostly in the 150 ka range. Evidence of glaciation, such as erratics, striated stones, and glacial till, is not easy to recognize in this rhyolite terrain. The most recognizable deposits of glacial origin are recessional kame and outwash gravel generally in low areas between rhyolite flows. A sector of the plateau icecap flowed southwest from this area to beyond the southwest corner of Yellowstone National Park. Richmond mapped about eight places on the Central Plateau as nunataks (U.S. Geological Survey, 1972a). Pierce found Richmond's upper limit of Pinedale ice resulted in basal shear stress values in the range of 0.1–0.01 bars, unreasonably low values above permeable bedrock. Pierce also found obsidian pebbles with hydration the same as Pinedale end moraines (7 μm) on the one nunatak examined.

Drive to lodging at Old Faithful and in the evening watch Old Faithful erupt and stroll through Upper Geyser Basin.

DAY 7. OLD FAITHFUL TO BOZEMAN

STOP 7–1. MIDWAY GEYSER BASIN

Cross the Firehole River on the footbridge and view the steamy Excelsior Geyser Crater and colorful Grand Prismatic Spring, which are two of the largest geothermal pools in Yellowstone. Southwest along the skyline is the faulted profile of the Mallard Lake resurgent dome, the southwest of the two resurgent domes of the 0.64 Ma caldera. These faults strike NW, trend towards the Midway Geyser Basin, and are likely to serve as conduits for thermal fluids.

STOP 7–2. MADISON JUNCTION (Rest stop)

Madison Junction is along 0.64–Ma caldera margin. National Park Mt. to the southwest is the northeast margin of the large West Yellowstone rhyolite flow that was emplaced between the Bull Lake and Pinedale glaciations. Pinedale ice almost reached the top of the mountain to the north. The road through the Madison Canyon to West Yellowstone obliquely crosses the caldera margin.

STOP 7–3. PINEDALE TERMINAL MORAINE

Park in off–road parking area on north (right) side of the road and walk up onto the Pinedale moraine of an outlet glacier that flowed down the Madison Canyon from the Yellowstone Plateau icecap. This moraine consists mostly of rhyolite from the Madison Canyon. The soils are weakly developed, except that the porous till has gray, clay lamellae translocated as deep as 7 ft. These moraines probably correlate with the Eightmile moraines north of Yellowstone and the Hedrick Pond moraines in Jackson Hole because the Yellowstone Plateau icecap was a key source in the glacial system that deposited these three end moraines. Across the Madison River from here, the end moraine complex is well expressed with more than 5 morainal positions. Outwash channels from the inner moraines trend westward through the outer moraines. The Madison Canyon moraines are at the head of the West Yellowstone obsidian sand–plain, a low–gradient, outwash fan rich in granule–sized obsidian that floors most of the West Yellowstone basin. The obsidian hydration thickness on glacial cracks is about 7 μm, yielding an age of 25 ka (Pierce et al., 1976). The hydration on Deckard Flats deposits is about 5 μm. The thickness of weathering rinds on basalt is about 0.4 mm, whereas that on Bull Lake deposits is 0.8 mm (Colman and Pierce, 1981). Continue into West Yellowstone and drive north on US 191.

STOP 7–3. OBSIDIAN SAND–PLAIN AND ITS ECOLOGY

Cross the Madison River, climb to the top of the obsidian sand–plain, and turn west onto a two–track road leading to a bluff above the Madison River. The obsidian sand plain heads in the Pinedale moraines seen at the last stop. Glacial–pressure cracks on obsidian pebbles from here have hydration rinds 7 μm thick, the same as that on the end moraines. The soil with only a color B–horizon is consistent with a Pinedale age. The bedding of the obsidian sand deposit is sub–horizontal, with primary beds about 0.5 feet (15 cm) thick, capped by a silty zone a few millimeters thick. The sand plain has a remarkably low gradient of about 8 ft/mile (1.5 m/km). These and other features suggest that the obsidian sand plain was built by large sheet flow discharges carrying sand and granules in traction across a wide extent of the surface. The western part of the obsidian sand plain may be deltaic and deposited in a lake dammed by the Beaver Creek moraines just beyond the west margin of the West Yellowstone Basin. Such large discharge might result from release of water in the geyser basins, held either as glacially dammed lakes or as subglacial water. Coring for bridge abutments in the Lower Geyser Basin revealed at least three cycles of lake sediments capped with gravel, suggesting filling and sudden draining of glacially dammed lakes. The upper surface of the sand plain has

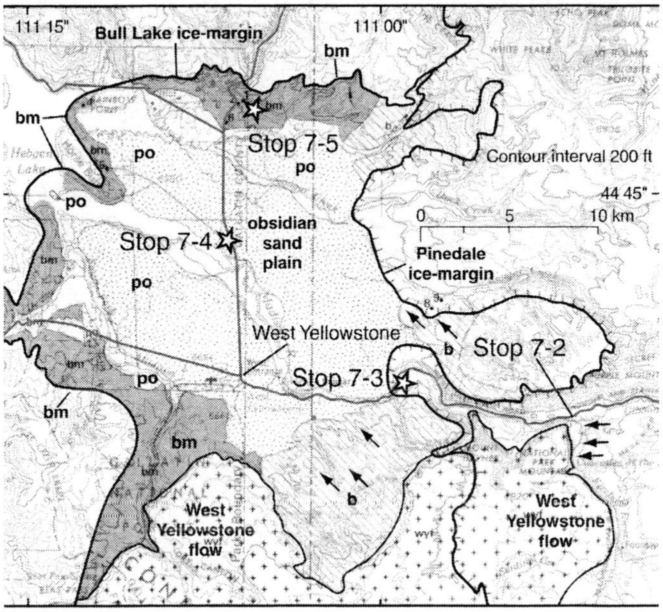

Figure 30. Map of Pinedale and Bull Lake glacial extents in West Yellowstone Basin (after Pierce, 1979, Fig. 8). bm– Bull Lake moraines, b– area glaciated by Bull Lake ice, po– Pinedale outwash forming obsidian sand plain. The West Yellowstone flow is ~120 ka and post–dates the Bull Lake moraines.

a thin topsoil of silty sand, and is excessively well drained, with an even lower water–holding capacity than rhyolite flows. The obsidian sand has the chemistry of rhyolite and thus is also low in nutrients. Such conditions allow lodgepole pine to form a climax community with Idaho fescue and antelope bitter brush (*Persia tridentata*) on the forest floor. The trees don't get very large here, even though most of them are more than 300 yrs. old.

STOP 7–4. BULL LAKE MORAINES

Continue north on US 191 past the Duck Creek Y and turn right into the old cemetery. Park and walk south onto a broad moraine crest in the open meadow. These Bull Lake moraines consist of glacially transported debris from the southern Gallatin Range and the Yellowstone Plateau. The moraines are broad, subdued, and capped with loess, commonly about 1.5–3 ft. (1/2–1 m) thick. A moderately developed, clayey–to–loamy soil has a high water–holding capacity. The roadcut near the turnoff shows a complex colluvial history with oxidized soil material filling concavities in the moraines. Although boulders protruding from the moraines are uncommon, boulders nearly flush with the surface are readily found. In soil pits, contortions and obliteration of soil horizons suggest solifluction has locally obliterated textural B–horizons and submerged boulders into the moraines.

The West Yellowstone Basin was nearly filled by Bull Lake ice from the Yellowstone Plateau. Bull Lake moraines drape around Horse Butte in the western part of the basin. Two grassy moraines descend westward to The Narrows of Hebgen Lake and connect with the moraines near this stop. The Bull Lake terminal moraines are 14 miles (22 km) beyond the Pinedale terminal moraines of STOP 7–3 but this large separation is much less than the 30 miles (50 km) in Jackson Hole. The Bull Lake moraines are not significantly tilted northward into the Hebgen–Red Canyon fault, suggesting that few offsets of 15 ft (5 m) have occurred in the last 140 ka.

On the southern skyline is the West Yellowstone rhyolite flow that overlaps and post–dates the Bull Lake moraines. When the Bull Lake Glaciation occurred, the West Yellowstone and several more flows south of it were not yet emplaced. Hydration rinds on obsidian pebbles from moraines in the West Yellowstone Basin averaged 14.7 μm thick. Hydration was measured in thin–sections across pebble corners where pressure cracks are produced by glacial abrasion. Based on 12.3 μm hydration on the 123 ± 3 ka West Yellowstone flow and 16.2 μm on the 183 ± 3 ka Obsidian Cliff flow, the 14.7 hydration yields an age of about 140 ka for glacial abrasion prior to deposition of obsidian clasts in the Bull Lake moraines (K–Ar ages from Obradovich, 1992, processed by Christiansen, 2001 and age of West Yellowstone flow is from Obradovich, oral communication, 2002 with temperature and rate adjustments of hydration in Pierce et al., 1976.).

The Bull Lake moraines support a productive rangeland for bison and cattle, including mountain big sage, Idaho fescue, sticky geranium, cinquefoils, stickseeds (*Lappula redowskii*), and little sunflower. After absorbing water from spring snowmelt, these moraines are covered with lush, knee–high grasses and forbs with a carpet of blooming wildflowers.

STOP 7–5. HEBGEN FAULT AT CABIN CREEK

Return to the Duck Creek Y and turn right on U.S. 287 to Cabin Creek. The magnitude 7.5 Hebgen Lake earthquake (Doser, 1985) occurred on August 18, 1959. The Cabin Creek stop is near the west end of the 1959 fault scarp and offsets two terrace levels. The lower terrace is about 1–3 ka and is offset 10.2 ft (3.1 m); the upper terrace is about 11–17 ka and is offset 17.4 ft (5.3 m) (cosmogenic ages from Jerome Van Der Woerd, written communication, 2001). A trench across fault on the upper terrace exposed: (1) the 1959 colluvial wedge with a height of 8.9 ft (2.7 m), equivalent to ~17 ft (5.2 m) vertical throw on the main fault (Tm) and (2) a penultimate colluvial wedge with a height 3.9 ft (1.2 m) equivalent to 7.9 ft (2.4 m) Tm, totaling 25 ft (7.6 m) Tm. At Cabin Creek, only one pre–1959 Holocene event is documented, which occurred about 3 ka, based on radiocarbon ages and cosmogenic ages.

Continue west driving past the Beaver Creek moraines of Pinedale age that may have dammed a lake into which the obsidian sand plain was deposited. Farther west is the Madison Canyon slide triggered by the 1959 Hebgen Lake earthquake. Note the trees blown down and pointing away from the slide, indicating an air blast accompanied the slide.

Pull over to view Missouri Flats terrace sequence near highway 87 turnoff. Scott Lundstrom (1986) studied sediments, soils, obsidian hydration, and range front faulting. He concluded this terrace 30 m above the Madison River was of Pinedale age, whereas the terrace 60–70 m on the far side of the river was of Bull Lake age.

STOP 7–6. INDIAN CREEK MORAINES AND FAULTING

A high lateral moraine is built of both Pinedale and Bull Lake deposits. The Pinedale moraines with lumpy morphology spill out beyond the high lateral. Smooth morainal ridges on the outside of the lateral appear to represent the Bull Lake. Across Indian Creek, Pinedale fan deposits are offset 13–20 ft (4–6 m) (Ruleman, 2002).

Continue north on US 287 through Ennis, and at Norris Hot Springs, take 289 to Bozeman. Arrive in Bozeman and end of field trip by 5:30 PM at City Center Motel. People are responsible for their own lodging.

REFERENCES

Arthur, G.W., 1962, The Emigrant Bison drives of Paradise Valley, Montana: *in*, Malouf, C., and Connor, S. eds., Symposium on Buffalo Jumps: Montana Archaeological Society Memoir No. 1, p. 16–27.

Arthur, G.W., 1966a, An Archaeological Survey of the Upper Yellowstone River Drainage, Montana, Montana Agricultural Experiment Station, Agricultural Economics Research Report No. 26, 111 p.

Arthur, G.W., 1966b, The Eagle Creek Site, 24PA301: Archaeology in Montana, v. 7(2), p. 11.

Berger, A.L., 1978, Long–term variations of caloric insolation resulting from the Earth's orbital elements: Quaternary Research, v.9, p. 139–167.

Blackwelder, Eliot, 1915, Post–Cretaceous history of the mountains of central western Wyoming: Journal of Geology, v. 23, p. 97–117, 193–217, 307–340.

Boeka, M.S., 2002, Spatial Analysis of Cultural and Natural Distributions of Artifacts at the Lawrence Site, Jackson Lake, Wyoming: Investigating Inundated Deposits: MA thesis, Department of Anthropology, University of Wyoming, Laramie, 59 p.

Bond, G., Showers, W., Cheseby, M., Lotti, R., Almasi, P., deMenocal, P., Priore, P., Cullen, H., Hajdas, I., and Bonani, G., 1997, A pervasive millennial–scale cycle in North Atlantic Holocene and glacial climates: Science, v. 278, p. 1257–1266.

Brackett, W.S., 1893, Indian Remains on the Upper Yellowstone: Annual Report of the Board of Regents of the Smithsonian Institution to July, 1892: Government Printing Office, Washington, D.C., p. 677–580.

Brown, B., 1932, The Buffalo Drive: Natural History, v 32(1), p. 75–82.

Christiansen, R.L., 2001, The Quaternary and Pliocene Yellowstone Plateau Volcanic Field of Wyoming, Idaho, and Montana: U.S. Geological Survey Professional Paper 729–G, 145 p.

Cannon, K.P., 2001, What the past can provide: contribution of prehistoric bison studies to modern bison management: Great Plains Research, v. 11, p. 145–174.

Cannon, G.M., Crothers, G.M., and Pierce, K.L., 1994, Archeological investigations along the Fishing Bridge Peninsula, Yellowstone National Park, Wyoming—the archeology, geology, and paleoenvironment: Manuscript on file at the Midwest Archeological Center, Lincoln, Nebraska, 200 p.

Cannon, K.P., and Hughes, R.E., 1993, Obsidian source characterization of Paleoindian projectile points from Yellowstone National Park: Current Research in the Pleistocene, v 10, p. 54–56.

Cannon, K.P., and Hughes, R.E., 1997, Provenance analysis of obsidian Paleoindian projectile points from Yellowstone National Park, Wyoming: Current Research in the Pleistocene, v 14, p. 101–104.

Connor, M.A., 1998, Final Report on the Jackson Lake Archeological Project, Grand Teton National Park, Wyoming: Midwest Archeological Center, Technical Report No. 46, Lincoln, Nebraska, 278 p.

Colman, S.M., and Pierce, K.L., 1981, Weathering rinds on andesitic and basaltic stones as a Quaternary age indicator, western United States: U.S. Geological Survey Professional Paper 1210, 56 p.

Despain D.G. 1990. Yellowstone Vegetation: Consequences of environment and history in a natural setting: Boulder, Colorado, Roberts Rinehart, Inc., 239 p.

Doser, D.I., 1985, Source parameters and faulting processes of the 1959 Hebgen Lake, Montana, earthquake sequence: Journal of Geophysical Research, v. 90, p. 4537–4555.

Dzurisin, Daniel and Yamashita, K.M., 1987, Vertical surface displacements at Yellowstone Caldera, Wyoming, 1976–1986: Journal of Geophysical Research, v. 92, no. B13, p. 13,753–13,766.

Dzurisin, Daniel, Yamashita, K.M., and Kleiman, J.W., 1994, Mechanisms of crustal uplift and subsidence at the Yellowstone caldera, Wyoming: Bulletin of Volcanology, v. 56, p. 261–270.

Fournier, R.O., Christiansen, R.L., Hutchinson, R.A., and Pierce, K.L., 1994, A field–trip guide to Yellowstone National Park, Wyoming, Montana, and Idaho—Volcanic, hydrothermal, and glacial activity in the region: U.S. Geological Survey Bulletin 2099, 46 p.

Frison, G.C., 1992, The foothills–mountains and the open plains: The dichotomy in Paleoindian subsistence strategies between two ecosystems: *in* Stanford, D.J. and Day, J.S., eds., Ice Age Hunters of the Rockies: University Press of Colorado, Niwot, CO, p. 323–342.

Gennett, J.A., and Baker, R.G., 1986, A late Quaternary pollen sequence from Blacktail Pond, Yellowstone National Park, Wyoming, U.S.A.: Palynology, v. 10, p. 61–71.

Gilbert, J.D., Ostenaa, Dean, and Wood, Christopher, 1983, Seismotectonic study, Jackson Lake Dam and Reservoir, Minidoka Project, Idaho–Wyoming: U.S. Bureau of Reclamation Seismotectonic Report 83–8, 122 p.

Good, J.D., 1964, Prehistoric landslide in Yankee Jim Canyon, Park County, Montana: Geological Society of America Special Paper 82, p. 327–328.

Good, J.D., and Pierce, K.L., 1996, Interpreting the landscapes of Grand Teton and Yellowstone national parks, recent and ongoing geology: Grand Teton National History Association, Moose, Wyoming, 58 p.

Graumlich, L.J., Pisaric, M.F.J, Waggoner, L.A., Littell, J.S., and King, J.C., in press, Upper Yellowstone River flow and teleconnections with Pacific basin climate variability during the past three centuries: Climatic Change.

Graumlich, L.J., and Littell, J., 2001, Multi–decadal variability of precipitation in the Greater Yellowstone region as inferred from tree rings: Proceedings, 18th Annual Pacific Climate Workshop on Climate Variability of the Eastern North Pacific and Western North America.

Haines, A.L., 1966, The Rigler Bluff hearth site, 24PA401: Archaeology in Montana, v. 7(2), p. 5.

Horberg, L., 1940, Geomorphic problems and glacial geology of the Yellowstone Valley, Park County, Montana: Journal of Geology, v. 48, p. 275–303.

Houston, D.B., 1973, Wild fires in Yellowstone National Park: Ecology, v. 54, p. 1111–1117.

Hughes, S.S., 1998, Getting to the point: Evolutionary change in Prehistoric weaponry: Journal of Archaeological Method and Theory, v. 5(4), p. 345–408.

Husted, W.M., and Edgar, R., 2002, The archeology of Mummy Cave, Wyoming: An introduction to Shoshonean prehistory: Midwest Archeological Center Special Report No. 4, Lincoln, Nebraska, 232 p.

Licciardi, J.M., Clark, P.U., Brook, E.J., Pierce, K.L., Kurz, M.D., Elmore, D., and Sharma, P., 2001, Cosmogenic ^{3}He and ^{10}Be chronologies of the northern outlet glacier of the Yellowstone ice cap, Montana, USA: Quaternary Research, v. 29, no 12, p. 1095–1098.

Littell, J. S. 2002. Determinants of fire regime variability in lower elevation forests of the northern Greater Yellowstone Ecosystem: MS thesis, Montana State University, Bozeman, MT, 122 p.

Locke, W.W., Clarke, W.D., Elliott, J.E., Lageson, D.R., Mokt, D.W., Montagne, J., Schmidt, J.G., and Smith, Michael, 1995, The middle Yellowstone valley from Livingston to Gardiner, Montana: A microcosm of northern Rocky Mountain geology: Northwest Geology, v. 24, p. 1–65.

Locke, W.W., and Meyer, G.A., 1994, A 12,000–year record of vertical deformation across the Yellowstone caldera margin: The shorelines of Yellowstone Lake: Journal of Geophysical Research, v. 99, No. B10, p. 20,079–20,094.

Love, J.D., and Love, J.M., 1997, Road log, Jackson to Dinwoody and return: Geological Survey of Wyoming Public Information Circular 20, 37 p.

Love, J.D., Reed, J.C. Jr., and Christiansen, A.C., 1992, Geologic map of Grand Teton National Park: U.S. Geological Survey Miscellaneous Investigations Series Map I–2031.

Lundstrom, S.C., 1986, Soil stratigraphy and scarp morphology studies applied to the Quaternary geology of the southern Madison Valley, Montana: MS thesis, Humboldt State University, Arcata, CA, 53 p.

Meyer, G.A., and Locke, W.W., 1986, Origin and deformation of Holocene shoreline terraces, Yellowstone Lake, Wyoming: Geology, v. 14, p. 699–702.

Meyer, G.A., Wells, S.G., and Jull, A.J.T., 1995, Fire and alluvial chronology in Yellowstone National Park: Climatic and intrinsic controls on Holocene geomorphic processes: Geological Society of America Bulletin, v. 107, p. 1211–1230.

Millspaugh, S.H., and Whitlock, C., 1995, A 750–year fire history based on lake sediment records in central Yellowstone National Park, USA: The Holocene, v. 5, p. 283–292.

Millspaugh, S.H. and Whitlock C., in press, Postglacial fire, vegetation, and climate history of the Yellowstone–Lamar and Central Plateau provinces, Yellowstone National Park: *in* After the Fires, Wallace, L., ed., The ecology of change in Yellowstone National Park: Yale University Press, New Haven, Conn.

Millspaugh, S. H., Whitlock, C., and Bartlein, P. J., 2000, A 17,000–year history of fire for the Central Plateau of Yellowstone National Park: Geology, v. 28, p. 211–214.

Montagne, John, and Locke, W.W., 1989, Trip 7 road log, Cenozoic history of Yellowstone Valley between Livingston and Gardiner, Montana: *in* French, D.E., and Grabb, R.F., eds., Geologic resources of Montana, Guidebook 1989 Field Conference: Montana Geological Society, v. II, p. 502–521

Morgan, L.A., Shanks, W.C. III, Lovalvo, D.A., Johnson, S.Y., Stephenson, W.J., Pierce, K.L., Harlan, S.S., Finn, C.A., Lee, G., Webring, M., Schulze, B., Dühn, J., Sweeney, R., and Balistrieri, L., 2003, Exploration and discovery in Yellowstone Lake: Results from high–resolution sonar imaging, seismic reflection profiling, and submersible studies: Journal of Volcanology and Geothermal Research, p. 221–242.

Norris, P.W., 1880, Prehistoric remains in Montana between Fort Ellis and the Yellowstone River: Annual report of the Board of Regents of the Smithsonian Institution for the year 1879: Government Printing Office, Washington, D.C., p. 327–328.

Obradovich, J.D., 1992, Geochronology of the late Cenozoic volcanism of Yellowstone National Park and adjoining areas, Wyoming and Idaho: U.S. Geological Survey Open–File Report 92–408, 45 p.

Obradovich, J.D., and Izett, G.A., 1991, ^{40}Ar/^{39}Ar ages of upper Cenozoic Yellowstone Group tuffs: Geological Society of America, Abstracts with Programs, v. 23, No. 2, p. 84.

Pelton, J.R., and Smith, R.B., 1979, Recent crustal uplift in Yellowstone National Park: Science, v. 206, p. 1179–1182.

Phillips, F.M., Zreda, M.G., Gosse, J.C., Klein, Jeffrey, Evenson, E.B., Hall, R.D., Chadwick, O.A., and Sharma, Pankaj, 1997, Cosmogenic ^{36}Cl and ^{10}Be ages of Quaternary glacial and fluvial deposits of the Wind River Range, Wyoming: Geological Society of America Bulletin, v. 109, p. 1453–1463.

Pierce, K.L., 2003, Pleistocene glaciations of the Rocky Mountains: *in* Gillespie, A..R., Porter, S.C., and Atwater, B.F., eds., Elsevier Science, in press.

Pierce, K.L., 1979, History and dynamics of glaciation in the northern Yellowstone National Park area: U.S. Geological Survey Professional Paper 729 F, 91 p.

Pierce, K.L., Adams, K.D., and Sturchio, N.C., 1991, Geologic setting of the Corwin Springs Known Geothermal Resource area–Mammoth Hot Springs area in and adjacent to Yellowstone National Park: *in* Sorey, M.L., ed., Effects of potential geothermal development in the Corwin Springs known geothermal resource area, Montana, on the thermal features of Yellowstone National Park: U.S. Geological Survey Water Resources Investigations Report 91–4052, p. C1–C37.

Pierce, K.L., Cannon, K.P., Meyer, G.A., Trebesch, M.J., and Watts, Raymond, 2002, Post–glacial inflation–deflation cycles, tilting, and faulting in the Yellowstone caldera based on Yellowstone Lake shorelines: U.S. Geological Survey Open–File Report 02–0142, 30 p.

Pierce, K.L., and Good, J.D., 1992, Field guide to the Quaternary geology of Jackson Hole, Wyoming: U.S. Geological Survey Open–File Report 92–504, 49 p.

Pierce, K.L., Lundstrom, Scott, and Good, John, 1998, Geologic setting of archeological sites in the Jackson Lake area, Wyoming: *in* Connor, Melissa, Final Report on the Jackson Lake Archeological Project, Grand Teton National Park, Wyoming, Midwest Archeological Center, National Park Service, Lincoln, Neb., Technical Report No. 46, p. 29–48, p. 222–242.

Pierce, K.L., and Morgan, L.A., 1992, The track of the Yellowstone hot spot: volcanism, faulting and uplift: *in* Link, P.K., Kuntz, M.A., and Platt, L.W., eds., Regional geology of eastern Idaho and western Wyoming: Geological Society of America Memoir 179, p. 1–53.

Pierce, K.L., Obradovich, J.D., and Friedman, Irving, 1976, Obsidian hydration dating and correlation of Bull Lake and Pinedale glaciations near West Yellowstone, Montana: Geological Society of America Bulletin, v. 87, p. 703–710.

Richmond, G.M., 1977, Surficial geologic map of the Canyon Village quadrangle, Yellowstone National Park, Wyoming: U.S. Geological Survey Miscellaneous Geological Investigations Map I–652, scale 1:62,500.

Richmond, G.M., 1976, Surficial geologic history of the Canyon Village quadrangle, Yellowstone National Park, Wyoming, for use with Map I–652: U.S. Geological Survey Bulletin 1427, 35 p.

Reeve, S.A.,. Wright, G.A., Marceau, T.E., and Meecham, P., 1979, Archeological Investigations at the Lawrence Site (48TE509), Grand Teton National Park, Wyoming: Manuscript on file at the Midwest Archeological Center, Lincoln, Nebraska. 111 p.

Ruleman, C.A. III, 2002, Quaternary tectonic activity within the northern arm of the Yellowstone tectonic parabola and associated seismic hazards, southwest Montana: MS thesis, Montana State University, Bozeman, MT, 157 p.

Shafer, S.L., Bartlein, P.J., and Whitlock, C. in press, Understanding the spatial heterogeneity of global environmental change in mountain regions: *in* Huber, U., Reasoner, M., and Bugmann, H., eds., Global Change and Mountain Regions, Kluwer, Dordrecht, The Netherlands.

Sharp, Warren, Ludwig, K.R., Chadwick, O.A., Amundson, Ronald, and Glaser, L.L., 2003, Dating fluvial terraces by ^{230}Th/U on pedogenic carbonate, Wind River Basin, Wyoming: Quaternary Research, in press.

Smith, M., Lageson, D., Heatherington, A., and Harlan, S., 1995, Geochronology, geochemistry, and isotopic systematics of the basalt of Hepburns Mesa, Yellowstone River Valley, Montana: Geological Society of America Abstracts with Programs, v. 27, no. 4, p. 56.

Smith, R.B., Pierce, K.L., and Wold, R.J., 1993, Seismic surveys and Quaternary history of Jackson Lake, Wyoming: *in* Snoke, A.W., Steidtmann, J.R., and Roberts, S.M., eds., Geology of Wyoming: Wyoming Geological Survey Memoir No. 5, p. 668–693.

Smith, R.B., and Siegel, L.J., 2000, Windows into the Earth, the Geologic Story of Yellowstone and Grand Teton National Parks: Oxford University Press, 242 p.

Stuiver, Minze, and Reimer, P.J. 1993, Extended ^{14}C data base and revised CALIB 3.0 ^{14}C age calibration program: Radiocarbon, v. 35, p. 215–230.

Sturchio, N.C., Pierce, K.L., Morrell, M.T., and Sorey, M.L., 1994, Uranium–series ages of travertines and timing of the last glaciation in the northern Yellowstone area, Wyoming–Montana: Quaternary Research, v. 41, p. 265–277.

U.S. Geological Survey, 1972a, Surficial geologic map of Yellowstone National Park: U.S. Geological Survey Miscellaneous Investigations Map 710.

U.S. Geological Survey, 1972b, Bedrock geologic map of Yellowstone National Park: U.S. Geological Survey Miscellaneous Geologic Investigations Map I–711.

Whitlock, C., 1993, Postglacial vegetation and climate of Grand Teton and southern Yellowstone National Parks: Ecological Monographs, v. 63, p. 173–198.

Whitlock, C., and Bartlein, P.J., 1993, Spatial variations of Holocene climatic change in the Yellowstone region: Quaternary Research, v. 39, p. 231–238.

Whitlock, C., and Millspaugh, S.H., 1996, Testing the assumptions of fire–history studies: An examination of modern charcoal accumulation in Yellowstone National Park, USA: The Holocene, v. 6, p. 7–15.

Wicks, Charles Jr., Thatcher, Wayne, and Dzurisin, Daniel, 1998, Migration of fluids beneath Yellowstone caldera inferred from satellite radar interferometry: Science, v. 282, p. 458–462.

Wright, G.A., 1984, People of the high country: Jackson Hole before the settlers: Peter Lang, NY, 197 p.

The Rocky Mountain glacial model: The Wind River Range, Wyoming

Authors and Leaders: **D. E. Dahms,** *Department of Geography, University of Northern Iowa, Cedar Falls, IA 50614*
R.D. Hall, *Department of Geology, Indiana-Purdue University at Indianapolis, 723 W. Mich. St., Indianapolis, IN 46202*
R.R. Shroba, *U.S. Geological Survey, MS 913, Box 25046, Federal Center, Denver, CO 80225*
C.J. Sorenson, *Department of Geography, University of Kansas, Lawrence, KS 66045*
E.A. Lynch, *Biology Department, Luther College, Decorah, IA 52101*
M.T. Applegarth, *Department of Geography and Earth Science, Shippensburg University, Shippensburg, PA 17257*

INTRODUCTION

The largest Middle Rocky Mt. Pleistocene glaciers outside of Yellowstone National Park formed in the Wind River Range, many advancing as far as 25 mi (40 km). The Green River glacier on the northwest flank of the Wind River Range advanced to ~2250 m (7400 ft.) during the Bull Lake Glaciation (although the ice stream may have been enlarged by ice flowing out of the Gros Ventre Range). Glaciers also advanced to elevations near 2200 m (7250 ft.) on the southwest flank of the range, while many glaciers on the northeast flank advanced to 1800 m (5950 ft.) or lower because of the higher gradient of the valleys. This difference is caused by the asymmetrical anticlinal form of the range, so that valley elevations on the west are higher than on the east. Unlike the west slope, glaciers did not often merge on the east flank of the range and most remained as isolated valley glaciers. At Bull Lake, valley glaciers advanced 35-40 km to ~1740 m (~5750 ft.) during the Bull Lake and Pinedale Glaciations (Richmond and Murphy, 1965). In 1975, the Wind River Range still held 63 active glaciers (Denton 1975).

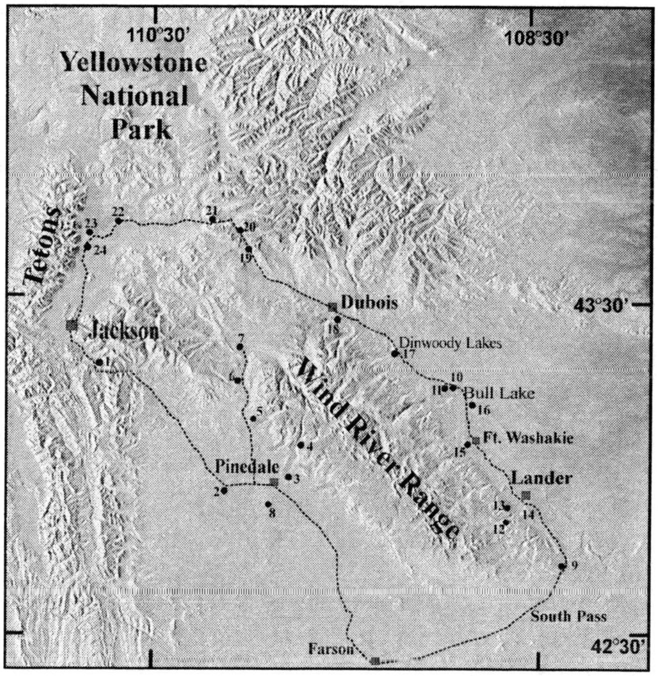

Figure 1. Fieldtrip route itinerary with fieldtrip stops numbered in consecutive order. Background graphic by Chalkbutte, Inc.

THE WIND RIVER RANGE GLACIAL MODEL

The Wind River Range contains the type areas for deposits that represent all of the Pleistocene glaciations identified in the U.S. Rocky Mountains (Blackwelder, 1915; Mears, 1974; Richmond, 1965, 1986, 1987; Dahms, 2003). Blackwelder laid the foundation for the Rocky Mt. glacial model from glacial deposits in the Wind River, Bighorn, and Teton Mts. in Wyoming and the Uinta Mts. in Utah. Blackwelder identified deposits of three glacial advances in the Wind River Range and named them the Pinedale, Bull Lake, and Buffalo. He correlated the Pinedale Glaciation with the late Wisconsin Glaciation of the midwestern U.S., the Bull Lake Glaciation with the early Wisconsin Glaciation, and the Buffalo with the Illinoian Glaciation. Deposits of the Buffalo Glaciation had no type locality and Blackwelder assumed that they represented a pre-canyon glaciation (Blackwelder, 1915; Mears, 1974).

Richmond subsequently refined Blackwelder's model and provided the outline for our present understanding of glacial events in the Wind River Range (Richmond 1964, 1965, 1983, 1986; Richmond and Murphy, 1989; Hall and Shroba, 1995; Hall and Jaworowski, 1999). Richmond subdivided the Buffalo Glaciation into three, separate, pre-Bull Lake glaciations, the Sacagawea Ridge (middle-late Pleistocene), Cedar Ridge, and Washakie Point. (Richmond, 1964; Richmond and Murphy, 1965). The type Sacagawea Ridge deposit is a diamicton on the east side of Dinwoody Lakes (Richmond and Murphy, 1989). The Cedar Ridge and Washakie Point glaciations were known only from Cedar Ridge, a locality on the north shore of Bull Lake where deposits representing these two events were buried under Bull Lake and Sacagawea Ridge deposits (Richmond, 1964).

Hall and Shroba (1995) identified an early Wisconsin moraine in the Wind River Range. The moraine (Richmond's 'moraine V', 1987) is located between the terminal Pinedale moraine and the

Dahms, D.E., Hall, R.D., Shroba, R.R., Sorenson, C.J., Lynch, E.A., Applegarth, M.T., and Dillon, J., 2003, The Rocky Mountain glacial model: The Wind River Range, Wyoming: *in* Easterbrook, D.J., ed., Quaternary Geology of the United States, INQUA 2003 Field Guide Volume, Desert Research Institute, Reno, NV, p. 345–364.

inner Bull Lake moraine at Fremont Lake. Moraine morphology, weathering characteristics, and soil development on this moraine are intermediate between those on the adjacent Bull Lake and Pinedale moraines. To date, however, few early Wisconsin glacial deposits are reported from the western United States (Bursik and Gillespie, 1993; Colman and Pierce, 1986, 1992; Gillespie and Molnar, 1995; Richmond, 1986; Easterbrook, 1992).

Recent studies have refined the Wind River Range glacial succession with badly needed numeric ages from the Pinedale and Bull Lake type localities and from correlative deposits. ^{10}Be exposure ages from boulders on moraines at the Pinedale type locality at Fremont Lake indicate an age range of from 22.5 to 15.5 ka for the Pinedale maxima there (Gosse et al., 1995b). Exposure ages calculated from ^{10}Be, ^{10}Be/^{36}Cl, and ^{36}Cl/ ^{10}Be measurements from boulders on moraines at the type Bull Lake locality range from >130 ka to 95 ka (Chadwick et al., 1997; Phillips, et al., 1997). The age of Sacagawea Ridge till at the type locality at Dinwoody Lakes is estimated to be >600 ka. Richmond identified Pearlette "O" tephra (now identified as Lava Creek B; Izett et al., 1992; Gansecki et al., 1998) at Cedar Ridge and in the gravel of two Wind River terrace remnants. He correlated the terrace remnants with an outwash deposit adjacent to the younger of two Sacagawea Ridge lateral moraines mapped at Dinwoody lakes (Richmond and Murphy, 1989).

Hall and Jaworowski (1999) suggest that the terms Cedar Ridge and Washakie Point now should be abandoned because they find no evidence of pre-Sacagawea Ridge diamicts at Richmond's Cedar Ridge section. Richmond identified the Cedar Ridge and Washakie Point glaciations based on lithologically distinct tills, deposits interpreted as glacial lake beds, and the presence of buried soils in the Cedar Ridge till and throughout the Washakie Point diamicton/lake beds (Richmond, 1964; Hall and Jaworowski, 1999). Richmond (1986) correlated the Washakie Point glaciation with marine oxygen isotope stage 20 and the Cedar Ridge glaciation with marine isotope stages 18-16, based on purported correlations with outwash terraces over 10 km from the Cedar Ridge section.

Hall and Jaworowski report that no evidence exists at Cedar Ridge for the Cedar Ridge or Washakie Point glaciations (1999). They questioned the designation of the upper Washakie Point deposit as till, based on lithologic differences between it and typical till identified in the area, and they could not identify a buried soil in this deposit. Most importantly, Hall and Jaworowski reported that all of Richmond's glacial deposits are normally polarized and can be no older than ~770 ka. They interpret these data to mean that no pre-Sacagawea Ridge glacial deposits are present at Cedar Ridge.

DAY 1. JACKSON TO PINEDALE, WYOMING

Drive south from the Jackson city limits on U.S. Highway 26 for 10.5 mi (16.9 km) to Hoback Junction. For the next ~30 mi (48 km) the road traverses the eastern edge of the Sevier thrust belt (thin-skinned thrust structures) where it meets the western margin of thick-skinned, east-to-west-reverse-faulted Laramide structures. For the first ~15 mi (24 km), the route generally parallels the Hoback normal fault. The Hoback fault is hinged on the west and tilted eastward, so movement along the fault is opposite to that of the Teton fault that we will see on DAY 5. From the first bridge across the Snake River south of Jackson, we can see to the east what J.D. Love called the "man-made" Squaw Creek landslide. Initial highway construction activated the slide until slides became so common that the highway was relocated to the west side of the river (Love, 1989).

At Alpine Junction, the route turns east onto U.S. Highway 187-189 and continues upstream along the Hoback River past gray cliffs (to the east) of the basal conglomerate of the Miocene Camp Davis Fm. This conglomerate is derived from the Gros Ventre Range and now is part of the eastward-dipping block formed by the Hoback normal fault (Basin and Range). Estimated displacement here may be as much as 10,000 feet (Love, 1989).

STOP 1-1. VIEW OF BULL LAKE ICE LIMIT

Turn off on the right at 14.9 mi (24 km) at the entrance to Camp Davis, the University of Michigan's geology camp. At this point we are just beyond the southern margin of Bull Lake ice from the Yellowstone Ice Cap (Fig. 2).

Continue through Hoback Canyon into Hoback Basin and past Bondurant. The road crosses Hoback Rim and enters the northern Green River Basin about 14 mi (22.5 km) past Bondurant (43.5 total mi, 70 km). Drainage down the Hoback River flows into the Snake River while south of Hoback Rim drainage becomes part of the Green-Colorado River system.

By the 46-mile mark, the road passes over deposits identified by G.M. Richmond (unpublished maps) as Bull Lake moraines. These moraines were deposited by ice that advanced down the Green River as far as five mi (8 km) into the basin from the Wind River Range-Gros Ventre Range ice field. At mile 58 (93.3 km), the road crosses the southernmost margin of deposits interpreted to be "old till" by Richmond (unpublished maps).

Routes 189-191 divide at mile 62 (99.8 km). Turn east on Highway 191 and continue toward Pinedale for 4.8 mi (7.7 km). Turn south (right) at the Junction of Highway 191 and Wyoming 352 onto a dirt road. Drive 0.1 mi (0.16 km) to the brown sign; turn right onto the short access road to Trapper's Point.

STOP 1-2. TRAPPER'S POINT: VIEW OF THE WIND RIVER RANGE FROM THE SITE OF THE 1833-1839 GREEN RIVER RENDEZVOUS

To the west of this point is the Green River Valley and the mountains of the Overthrust Belt. Charles Love and his students at Western Wyoming College mapped as many as fifteen terraces along the Green River between here and the lower Green River

Figure 2. Mapped extent of Bull Lake ice from the Yellowstone ice cap through Jackson Hole and into Hoback Canyon (K.L Pierce, personal communication, 2002).

Basin. To the east lies the Wind River Range, which is typical of the Laramide structures formed in the Middle Rocky Mountains. It is a broad-backed, asymmetrical, doubly-plunging anticline. The western margin of the range is bounded by the Wind River fault, a ~30° reverse fault that extends to a depth of ~30 km and strikes roughly NW-SE (Fig. 3). The northern margin of the range appears just south of the Absaroka mountains in the north and trends southeast for about 90 mi (145 km). The range's southern margin is located near South Pass (Fig. 1).

The highest peaks of the Wind River Range are located in the northern region and many are visible from Trapper's Point. Among the most notable of these are Gannett Peak (Wyoming's highest point, 13,804 ft) and Fremont Peak (13,745 ft). Along nearly the entire west side of the range below the high peaks is a broad bedrock platform at about 9,400 ft. Known as the Fremont surface, this feature is thought to be a Miocene-Pliocene erosion surface (pediment) that represents the level to which the mountains were buried by sediments during Tertiary deposition in the Green River Basin (Love, 1960). Subsequent uplift of the northern Wind River massif and associated southward tilting of the Green River Basin caused the Green River to excavate as much as 3000 ft (~900 m) of Tertiary sediments and create the present relief.

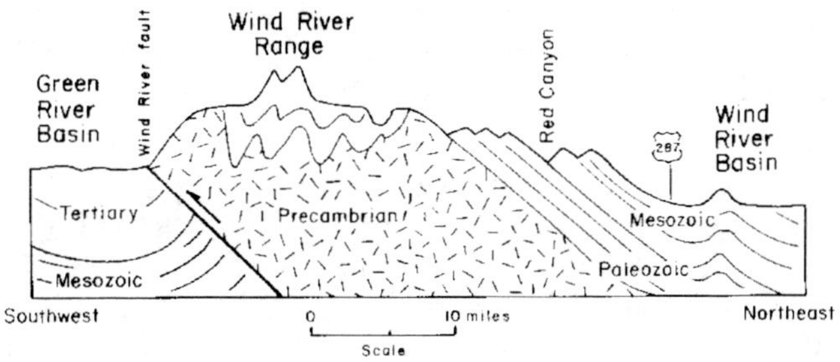

Figure 3. Generalized cross section of the Wind River Range showing the different structural relations, east vs. west. Note that the hanging wall of the Wind River Thrust buries the sedimentary rocks of the Green River Basin, whereas the sedimentary beds of the Wind River Basin are bent up into hogback formations against the Wind River massif. (From Mears et al., 1986, Courtesy of Wyoming Geological Survey)

At the base of the Fremont surface lie long, deep, glacially-carved lakes such as Fremont, New Fork, Boulder, and Willow lakes. Pleistocene till deposited by coalescent piedmont glaciers now covers a large part of the Fremont surface's lower escarpment south from Willow Lake to the East Fork drainage southeast of Boulder.

Return to U.S. Highway 191 and drive east 4.8 mi (7.7 km) to Pinedale. At 6 mi (9.7 km), as the highway swings to the right, turn left onto Fremont Lake Road. From this turn-off, the road immediately climbs onto a Bull Lake outwash terrace. At the northeast corner of the cemetery, the road climbs onto Bull Lake terminal moraines, finally reaching the Pinedale terminal moraines 2 mi (3.2 km) farther ahead. Drive 3 mi (4.8 km) past the Fremont Lake overlook (on left). Turn left at the sign for Lakeside Lodge. Turn left again at 3.1 mi (5 km) at the second turnoff to Lakeside Lodge. Continue 3.7 mi (6 km) and turn right at the entrance to the lodge. Check into the lodge and have lunch.

Leave Lakeside Lodge and return to Fremont Lake Road. Turn south (right) and return 0.9 mi (1.4 km).

STOP 1-3. FREMONT LAKE OVERLOOK

Fremont Lake is approximately 12.5 mi (20 km) long and is 641 ft deep at its deepest point. The lake is dammed at its southern end by Bull Lake and Pinedale moraines. Note the high peaks and the Fremont surface beyond the lake. The prominent flat-topped ridge to the east is Half Moon Mt. It represents the upper limit of Tertiary deposits that once filled the Green River Basin. Richmond identified a pre-Bull-Lake, glacial diamicton at the top of Half Moon Mt. (G.M. Richmond, personal communication, 1995).

The moraines rimming the lake here at the type locality are part of the Pinedale terminal moraine complex. The adjacent ridge to the south is formed by Bull Lake moraines (Fig. 4). A small moraine that Richmond identified as "Moraine V" (1973, 1986) lies between the two complexes. Once assigned to the Bull Lake moraine complex, it is now thought to have been deposited by an early Wisconsin glaciation (Hall and Shroba, 1995). ^{10}Be exposure ages from erratics on the Pinedale morainal complex constrain its age here to ~21.7-16 ka (Gosse et al., 1995a,b; Gosse, this volume). A single ^{10}Be exposure age from a Bull Lake moraine (~138 ka) corresponds to the oldest ages from Bull Lake deposits at the Bull Lake type area (Phillips et al., 1997).

Contrasting Soils of the Pinedale and Bull Lake Moraines, Fremont Lake area (Sorenson)

The soils of the Fremont Lake South quadrangle, first mapped by Richmond (1973), were remapped by Sorenson (1987). Soils developed in glacial deposits surrounding Fremont Lake formed from tills rich in igneous and metamorphic detritus eroded from the Wind River Range massif.

The Fremont Lake South quadrangle is famous as the type locality for Pinedale glacial deposits. It also contains remnants of earlier glacial sediments, including those correlated to the Bull Lake, and older Pleistocene stages of Rocky Mt. glaciation. Holocene glacial sediments are restricted to the cirques and high glacial valleys of the Wind River Mountains.

Richmond used soil characteristics as one among several means to identify individual units on his geologic map of the quadrangle. On his 1965 INQUA field trip, he repeatedly asserted that "soils could be used as a first approximation to separate Pinedale and Bull Lake moraines." Richmond's observations were the basis for the challenge to map soils in the quadrangle and examine the degree of correspondence between his geologic map and the area's soils. His observation that Pinedale and Bull Lake soils could easily be distinguished was born out early in the soil mapping. However, substages of Pinedale or Bull Lake glaciations could not be readily distinguished in the soil record. So, what one sees in the soils exposed for this field trip are the gross differences that Richmond first noted in soils originating from tills of Pinedale and Bull Lake glaciations. More specifically, the Pinedale equivalent soils are A/C profiles that are noncalcareous, with many of them displaying a Bw horizon. Bw horizons are most commonly identified where glacial action incorporated Tertiary sediments in tills and provided a readily available clay source for subsequent translocation or on landscape positions that have been especially stable over post-glacial time. In contrast, the Bull Lake soils commonly include a Bt and/or a Bk horizon, with $CaCO_3$ distributed below the solum into the uppermost parent material. The soils formed on these older Pleistocene deposits are located on surfaces of generally lower relief than their Pinedale counterparts. They appear to contain more Tertiary sediments, which may account for their generally higher clay content and perhaps some of their carbonates. Soils formed on piedmont tills of the Pinedale and Bull Lake glaciations typically have mollic epipedons and have been

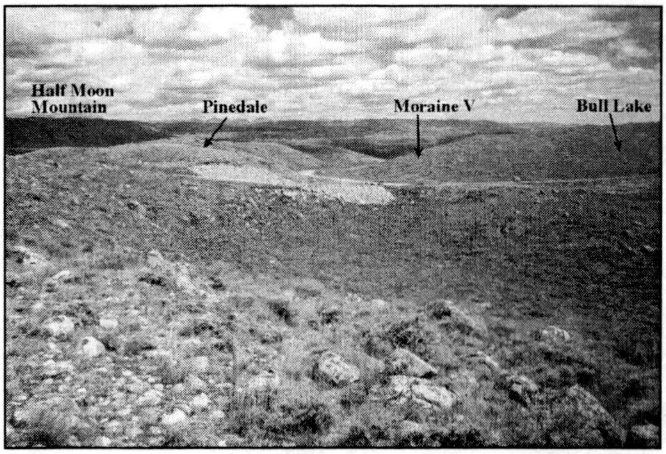

Figure 4. View to east of the Pinedale-Early Wisconsin-Bull Lake moraine sequence at the south end of Fremont Lake. Note the bouldery nature of the younger (Pinedale) till, whereas the older tills are progressively muted. A Sacagawea Ridge diamicton is reported from the top of Half Moon Mountain (upper left) by Gosse and Evenson (this volume). (D.E. Dahms photo)

enriched by the addition of loess to the profile. Eolian addition of carbonates from the Green River Basin also may account for some of the contrast between the generally carbonate rich Bull Lake soils and carbonate-free soils on Pinedale moraines.

Geomorphic processes, acting subsequent to deposition of the glacial drift of various ages, have had a profound effect on the variability in soils found on surfaces of different ages, as well as those located in different landscape positions on surfaces that are presumably of the same or similar age. For example, intermorainal soils typically have pachic epipedons owing to erosion and accumulation of sediments from steep adjacent moraine slopes. Evidence of mass wastage and buried soils is identified beneath debris shed from adjacent moraines. In one instance, a buried soil in a swale between two Pinedale lateral moraines is dated at 6070 ± 100 ^{14}C yrs. B.P., indicating early postglacial soil formation followed by burial from active Holocene slopewash and subsequent soil development in the colluvial sediment. Elsewhere, a basal date of 9305 ± 75 ^{14}C yrs. B.P. from peat in a former lake bed gives a minimum date for the development of Histosols in the map area.

The cosmogenic exposure ages for the Pinedale moraines in the quadrangle (Gosse et al, 1995a,b; Gosse, this volume) establish a maximum period for soil development here (15 ka to 20 ka). Since the Pinedale soils (A/C profiles) are weakly developed, we should examine processes that result in post depositional alteration or inhibition of their development.

Dahms (1991) offered compelling mineralogical evidence that eolian sediments have had a substantial effect on the soils of piedmont and alpine tills of the region wherein soil horizon development is retarded by the combination of pedoturbation and high amounts of eolian sedimentation (Fig. 5). Add wind and water erosion to the mix (see Hall, 1999) and the setting is con-

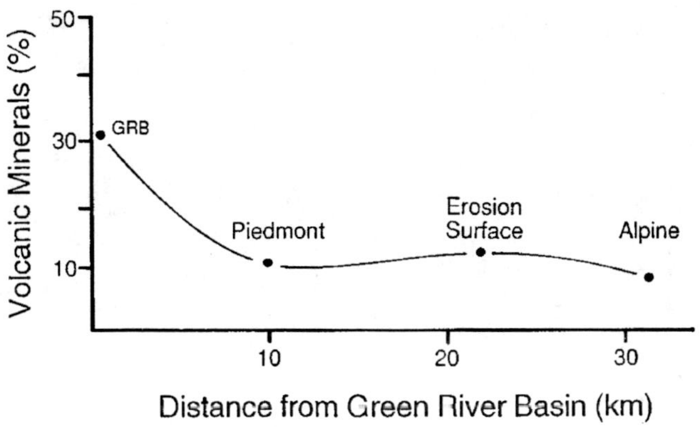

Figure 5. Plot of the average percentages of volcanic hornblende and zircon species identified in the eolian (20-50 μm) fraction of moraine soil B-horizons along transects through the three main physiographic areas of the WRR. Distance was calculated from a reference position in the Green River Basin. (Modified from Dahms, 1991)

sistent with a low level of soil development in this region during late glacial and postglacial time. The discovery of numerous buried soils at low catenal positions on the landscape provides ample evidence of the erosion of sediment from moraine crests and slopes and its redeposition in kettles and depressions throughout the region (Sorenson, 1987, Dahms, 1993, 1994; Fig. 6). Thus, while Bull Lake soils are more well-developed than Pinedale soils, late-glacial and post-glacial erosion has significantly weakened the morphological characteristics of both Pinedale and Bull Lake soils.

Turn north (left) from Fremont Lake overlook, back onto Fremont Lake Road. Continue past the Lakeside Lodge turnoff (toward Elkhart Park) for ~9.5 mi (14.6 km) to an overlook into the core of the range.

STOP 1-4. VIEW OF THE PEAKS OF THE NORTHERN WIND RIVER RANGE; POST-PINEDALE GLACIAL ACTIVITY

View the highest peaks of the range, including Gannett Peak (13,804 ft, 4208 m), Fremont Peak (13,745 ft, 4190 m), and Jackson Peak (13,517 ft, 4120 m). Behind this line of peaks lie the most extensive glaciers of the range (Gannett, Dinwoody, Sacagawea, Fremont, Bull Lake, and Knifepoint glaciers).

Four, post-Pinedale, glacial advances have now been described in at least 17 cirque valleys in the Wind River Range (e.g., Miller and Birkeland, 1974; Mahaney, 1978, 1984a,b; Dahms, 2002; Dahms and Birkeland, 2000; Dahms et al., in press; Gosse, this volume). The Temple Lake is the earliest post-Pinedale glaciation reported in the Wind River Range. The type locality was originally described by Hack (1943) and Moss (1951a,b) in the Temple Lake valley near the southern end of the range. The Temple Lake Glaciation is estimated to correspond to the Younger Dryas (Gosse et al., 1995a). Later (neoglacial) glacial episodes in the Wind River Range are called the Alice Lake, Black Joe, and Gannett Peak (LIA) glaciations (Dahms and Birkeland, 2000; Dahms et al., in press). Adequate numeric ages are not available for the neoglacial deposits, but relative-age data suggest the following ages: the Temple Lake is coeval with the Younger Dryas climate event, 12,800 ± 200 to 11,500 ± 300 calendar years ago (Alley et al., 1993; Gosse et al., 1995a); Alice Lake, less than 6000 cal. yrs. and Black Joe, 2000-1000 cal. yrs. (Dahms, 2002; Dahms et al., in press). Type localities for the Alice Lake and Black Joe glaciations are in valleys adjacent to Temple Lake valley. The Gannett Peak glaciation corresponds to the Little Ice Age; its type locality is below Gannett Peak, near the north end of the range (Richmond, 1957).

Return to Lakeside Lodge for the evening. Evening presentations are scheduled after dinner.

DAY 2. GREEN RIVER LOBE AND THE MESA

Return to Pinedale and turn west (right) on U.S. Highway 191. Continue through Pinedale and continue west from the Pinedale city limit for 4.5 mi (7.2 km). Turn north on WY 352.

At ~7.6 mi (12 km) the route rises slightly from Tertiary basin deposits onto a terrace just below moraines mapped by Richmond as Bull Lake. Note that these deposits appear different from the till and outwash deposits seen at Fremont Lake. A significant amount of ice that occupied this valley came from the Gros Ventre Mountains and may have only been augmented by ice from the Wind River Range (Green River lobe). Because of this, the diamicts here are significantly more fine-grained than those derived entirely from the crystalline core of the Wind River Range.

Continue north to 13.8 mi (22.2 km), turn east (right) onto New Fork Lakes road.

STOP 2-1. NEW FORK LAKES PINEDALE MORAINES

New Fork Lakes are twin-lakes that occupy a deeply incised, glaciated canyon. The outer lake is separated from the inner lake by a Pinedale recessional moraine.

Return to WY 352 and turn north for 0.2-0.3 mi (0.3-0.5 km). Turn right onto a gravel road leading into a small housing development (we call 'em Ranchettes out here). Drive north of the last house on a 2-track dirt road across. Stop at the end of the road on the crest of a moraine. Here, an outwash plain separates coarse-grained till of the New Fork Lakes moraines from the fine-grained till of the Green River lobe.

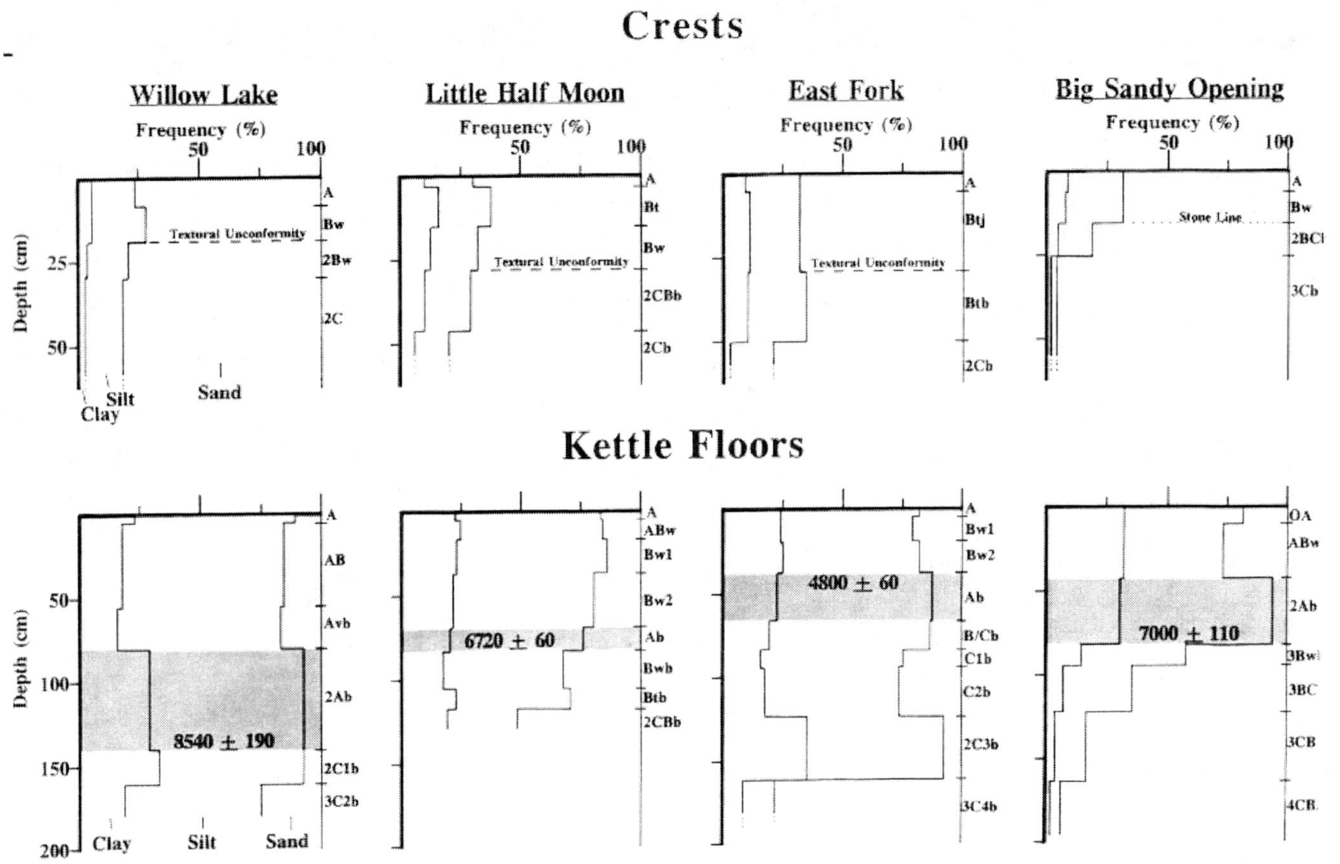

Figure 6. Particle-size distributions and profile-horizon designations of soils on crests and kettle floors of four Pinedale-age moraine catenas along the west slope of the Wind River Range. Samples for the ^{14}C ages indicated for each buried A-horizon were taken from the position shown. It appears that slope erosion was generally more active during the early middle Holocene at these locations. (Modified from Dahms, 1991, 1994)

Return to WY 352 and turn north (right). Continue for 7-8 mi (12 km) to Dew Lake.

STOP 2-2. DEW LAKE

Dew Lake is visible just north (downslope and ~500 m west) of the highway as it passes between Black Butte and Middle Butte. If time permits, turn left just past "The Place" onto a gravel road that serves as access to cabin sites surrounding Dew Lake. The gravel road to the lake is just past Boulder Basin Road (on the right) that goes to Buffalo Springs Estate (these may not be marked). Access to the lake is across private land, so park at the best view before reaching the lake.

Numeric ages and pollen stratigraphy at Dew Lake (Lynch)

This small lake, located at the steppe-forest border on the west flank of the northern Wind River Range (Sec 24, T. 37 N., R 110 W), was cored in 1960 by Bob Bright and Herb Wright. Bright prepared pollen diagrams showing three zones: a relatively warm period at the base of the core, followed by cold conditions, then warmer, post-glacial conditions. Radiocarbon dates on bulk marl sediments suggested that the cold period lasted from ~21,000 to ~11,000 ^{14}C yrs. B.P., and the warmer signal preceding it reflected warmer conditions before 21,000 ^{14}C yrs. B.P. Full glacial records from the Rocky Mountains are rare, and are potentially of great value in understanding the behavior of the glacial dynamics, paleoclimate, and vegetation patterns during this period of rapid change. AMS ^{14}C dates and pollen from a new 10-m-long sediment core obtained by Lynch from Dew Lake establish a chronology of full-glacial to post-glacial vegetation changes.

Six new AMS dates on wood and charcoal indicate that the chronology established by Bright was too old, possibly due to carbonate error resulting from bulk dates on marl. However, the oldest pollen zone described by Bright was not present in the Lynch core. The pollen and general sediment stratigraphy from the Lynch core match the upper 8.5 m of Bright's core. The lower zone of the new core (Zone 1) has very low tree pollen percentages and high grass and *Artemisia* percentages. This zone extends from at least 12,700 to ~10,270 ^{14}C yrs. B.P. The lack of tundra indicators in the pollen assemblage suggests that the vegetation surrounding the lake at this time was *Artemisia* steppe. The pine pollen percentages (10-30%) are much lower than those in modern tundra and are more similar to those found in surface samples from steppe areas tens of km away from montane forest (McAndrews and Wright, 1969; Fall, 1994), suggesting that forests did not grow within 40-50 km of the lake during the late-glacial. If trees grew at the base of the Wind River Range during the Pinedale glaciation, they must not have formed a continuous band of forest around the lower elevations of the range but instead may have been restricted to patchy refugia. In addition, the ratio of *Pinus* subgenus *Pinus* pollen (from *P. contorta*) to *Pinus* subgenus *Strobus* pollen (from *Pinus flexilis* and *P. albicaulis*) is generally lower than in Zone 2, suggesting that lodgepole pine populations were displaced to a greater extent than were limber and/or whitebark pine during the late-glacial.

Spruce (*Picea*) pollen percentages increase to 3-8% in the upper part of Zone 1, suggesting that spruce may have grown near the lake from ~11,000 to ~10,270 ^{14}C yrs. B.P. These may represent trees expanded from glacial refugia elsewhere, or possibly down-slope migration during the Temple Lake (Younger Dryas) glacial advance (Gosse, this volume; Dahms, 2002).

Zone 2 (~10,270-present) has a pollen assemblage typical of modern steppe vegetation near the lower forest border (Fall, 1994). Pine pollen percentages are 40-50% and *Artemisia* percentages are 20-35% throughout this zone. Most of the pine that could be differentiated was from *Pinus* subgenus *Pinus*, probably from the extensive lodgepole pine forests at higher elevation. Although the pollen profile in this zone doesn't show much change, but may be a hint of cooler and/or moister conditions after about 3000-4000 ^{14}C yrs. B.P. Percentages of spruce and fir (*Abies*) increased slightly at this time, perhaps representing the expansion of these types in the forests above the site. The increases in Cyperaceae and willow (*Salix*) pollen at this time probably represent the establishment of the sedge-mat and willow fringe in the shallow portion of the lake basin that had by this time become filled with marl.

Return to WY 352 and turn northeast (left). Continue along the south Green river valley past Kendall Guard Station to Union Pass Road. Turn north on Union Pass Road (dirt road) and continue to a turnoff just prior to Mosquito Lake (see Mosquito Lake 7.5' quadrangle).

Turn left past Mosquito Lake.

STOP 2-3. WAGON CREEK MORAINE COMPLEX

Along this road is a spectacular view of the northern end of the Wind River Range and the Green River drainage. The headwaters of Wagon Creek represent a small lobe of Pinedale–ice from the Green River valley that, after advancing into the eastern Gros Ventre range, stagnated below Bacon Ridge. This represents a problem that has not been resolved in this area. Moraines of the Green river lobe are composed of much finer-grained material than those found along the flanks of the Wind River Range. Thus, how much ice did the Gros Ventre Range add to the Green River lobe, and how did this addition affect the ice dynamics of the lobe? This area is not well-mapped and so we have no answers to these questions at present.

Fine-grained till is exposed in an open soil pit at this locality. Soil profiles here are deeper and show better B horizon development than those on the west slope of the range. We attribute this to a combination of greater available moisture and the fine-grained nature of the till (more clay in the parent materials).

Retrace the route back along Union Pass road to WY 352. Turn west (right) and go back past Cora to U.S. Highway 191. Turn east to Pinedale. Go through Pinedale and turn south (right) onto Tyler Avenue. Continue south on Tyler Avenue/The Mesa Road. By 5.4 mi (8.7 km) the road climbs south into a low topographic saddle; turn left (east) for 0.4 mi (0.6 km). Pull off to the left and stop below a prominent knob (Stewart Point).

STOP 2-4. Eroded soils on The Mesa

Return to Lakeside Lodge for the evening.

DAY 3. RED CANYON AND BULL LAKE

Return to Pinedale on Fremont Lake Road and turn south (left) onto U.S. Highway 191. From the south city limit, the road traverses Pinedale outwash deposits for about 10 mi (16 km) along Pine Creek and the New Fork River. The road follows below the escarpment of a Bull Lake terrace prior to crossing Boulder Creek. The road then rises onto deposits of the lower Eocene Wasatch Fm. for about 30 mi (48 km), then onto deposits of the middle Eocene Green River Fm. for the next 20 mi (32 km) to Farson.

Turn east (left) at Farson onto U.S. Highway 28. [Farson Merchantile: convenience store and rest stop-best ice cream in Wyoming.] Over the next ~35 mi (56 km) the road rises toward the southern end of the Wind River Range and crosses the Continental Divide a few miles to the north of South Pass. A mile or two past the second access road to Atlantic City the road passes the Atlantic City Mine. This taconite mine was operated by U.S. Steel from 1962 to 1983. Mine tailings here include spectacular examples of an Archean banded-iron member of the Goldman Meadows Fm..

STOP 3-1. RED CANYON OVERLOOK

Turn north (left) into the scenic-area parking lot just before the road swings north and downward along the east edge of Red Canyon. This view shows how the overlying Paleozoic and Mesozoic formations were upwarped during the Laramide uplift to form spectacular hogbacks along the flank of the range (Fig. 7). The most notable bedrock units in Red Canyon include the

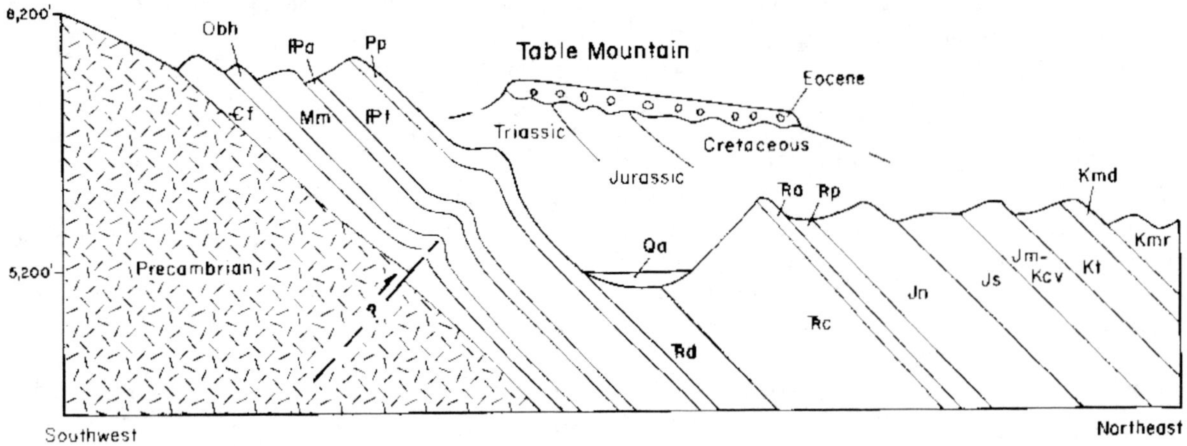

Figure 7. Schematic cross-section of the southeast flank of the Wind River Range and western margin of the Wind River Basin, looking north from Red Canyon overlook. The important stratigraphic units for this fieldtrip include the PreCambrian and Paleozoic sections, as well as the Triassic and Jurassic portions of the Mesozoic section. This section is ca. 10 miles across; vertical relief is ~3000 ft. (Modified from Mears et al., 1986, courtesy of the Wyoming Geological Survey.)

grass/sagebrush covered dipslope of the Phosphoria Fm. (Permian), the soft, eroded shale of the Dinwoody Fm. (Triassic) that underlie the main strike valley (Red Creek), the eastern escarpment of red sandstone and shale of the Triassic Chugwater Group, and the overlying pink-orange, eolian crossbeds of the Nugget Sandstone.

The nearly horizontal surface of Table Mt. appears in the distance (Fig. 7). The Tertiary upper beds on Table Mt. truncate the underlying Mesozoic formations. The unconformable relationship between flat-lying Tertiary sediments and tilted Mesozoic beds originally suggested the late Cretaceous age for the Laramide Orogeny. Table Mt. is now an erosional remnant whose surface is underlain by pre-glacial sediments eroded from the Wind River Range and deposited at the edge of the Wind River Basin during the middle-to-late Tertiary.

Continue north on Highway 28 to Lander. [Convenience stores and rest stop.] Cross the intersection of WY 789 and continue down the hill on Main Street and continue north through Lander.

Lander to Bull Lake - From the Lander city limit, the route continues 13.6 mi (21.8 km) to Ft. Washakie. Continue past Ft. Washakie north on Highway 287 to the Junction with U.S. Highway 26. Turn north onto Highway 26 for 2.3 mi (3.7 km) and turn west (left) into the Bull Lake rest area.

STOP 3-2. MORAINE MORPHOLOGY, AGES, AND SOILS AT THE BULL LAKE TYPE AREA

The moraines and stratigraphic sequence at Bull Lake have been studied in detail since Blackwelder (1915) first defined identified the type locality of the Bull Lake Glaciation (Richmond, 1957, 1962, 1964; Richmond and Murphy, 1965; Murphy and Richmond, 1965, Hall, 1999; Hall and Jaworowski, 1999). The Bull Lake rest area is on a complex of Pinedale moraines at the east end of the type area (Fig. 8). Pinedale lateral moraines and the recessional moraines that partly dam the

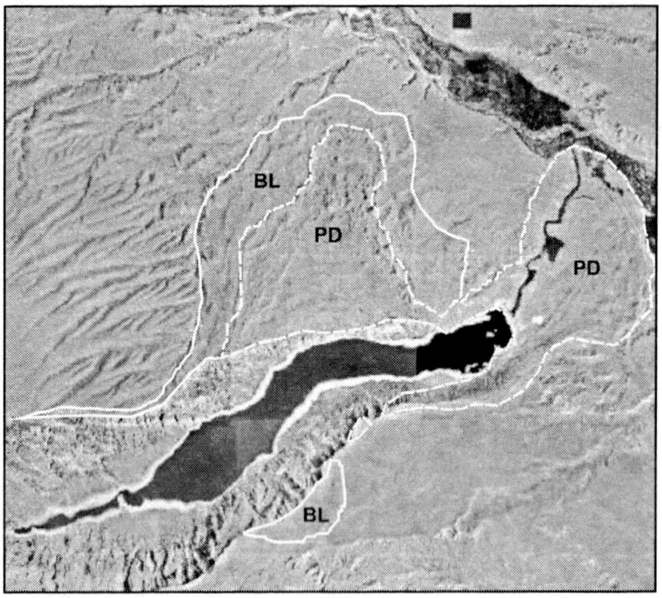

Figure 8. Digital image of the Bull Lake Type Area, Wyoming showing the generalized extents of the Pinedale (dashed outline) and Bull Lake (solid outline) moraines, as mapped by Richmond and Murphy (1965) and Murphy and Richmond (1965). Samples for the cosmogenic exposure ages reported by Chadwick et al. (1997) and Phillips et al. (1997) were taken from the moraine complexes on the north side of the lake. Image compiled by assembling USGS digital orthophoto quarter-quadrangles of the Bull Lake East, Bull Lake West, Crowheart, and Crowheart Butte 7.5' DOQQ's.

valley now occupied by Bull Lake (not visible from the parking area) can be seen from the rest area. The Pinedale moraines, like those at Fremont Lake, display well-developed knob-and-kettle topography with many fresh, relatively unweathered boulders at the surface. Pinedale ice breached the Bull Lake moraines here at the southeast end of the lake and forced the Wind River to the far (east) side of its floodplain. Bull Lake moraines have rounded crests and are more subdued than the Pinedale moraines. Integration of drainages has breached nearly all kettles on the Bull Lake moraines. Boulders on the Bull Lake moraines have thicker weathering rinds, pitted surfaces, and deep frost (lightning?) cracks.

Bull Lake moraines are best represented north of the lake (Fig. 8). Pinedale ice did not breach the Bull Lake moraines here so the Bull Lake moraines surround the Pinedale moraines on the west, north, and east. The stratigraphic sequence involves several nested moraine complexes. Soil profiles on the northern moraine complexes together with profiles at Dinwoody Lakes and Torrey Lake show significant differences between soils of the western and eastern slopes of the Wind River Range (Shroba, 1989; Hall, 1999; Applegarth and Dahms, 2001).

Characteristics of soils in the Bull Lake and Dinwoody Lakes areas compared with those in the Fremont Lake area. (Hall and Shroba)

Amounts of pedogenic carbonate and carbonate morphology in soil Bk and K horizons are the most useful soil criteria for distinguishing deposits of the Pinedale and Bull Lake Glaciations near Bull Lake (Hall, 1999; Shroba, 1989; Tables 1, 2; Fig. 9). B-horizon properties are less useful due to erosion of near-surface horizons (Hall, 1999; Applegarth and Dahms, 2001).

Bull Lake soils

Morphologic characteristics indicate that soils on glacial deposits in the area along the east slope of the Wind River Range are generally less-developed than are those near Fremont Lake (Table 1). Average thicknesses of the solum and B-horizons, average value of the structural index, and indices for texture and profile development (PDI) (Harden, 1982) clearly indicate greater development of soils at Fremont Lake.

Sufficient illuvial clay to form Bt horizons occurs in the soils in both areas. Some of the soils near Fremont Lake contain both Bw and Bt horizons, whereas those in the Bull Lake type area have Bt and Btk horizons. The depth of clay accumulation in soils is greater near Fremont Lake, whereas the maximum clay content occurs just below the ground surface in at Bull Lake. Although the average maximum and average profile weighted mean clay contents are similar in the two areas, the average clay accumulation index is about four times greater for the Fremont Lake soils.

Soils in both areas have Bk horizons, which commonly have stage II carbonate morphology. However, the soils at Bull Lake have an average carbonate content and an average carbonate accumulation index that are about 2-3 times greater than those of the Fremont Lake soils (Table 2).

Pinedale soils

As in the case of soils in Bull Lake moraines, most morphologic characteristics indicate that soils in Pinedale moraines at Fremont Lake are better developed than those in the Bull Lake area (Table 1). The average thickness of the B-horizons is about 3 times greater and the average index values for structure and rubification are 3 and 1.5 times greater, respectively, for

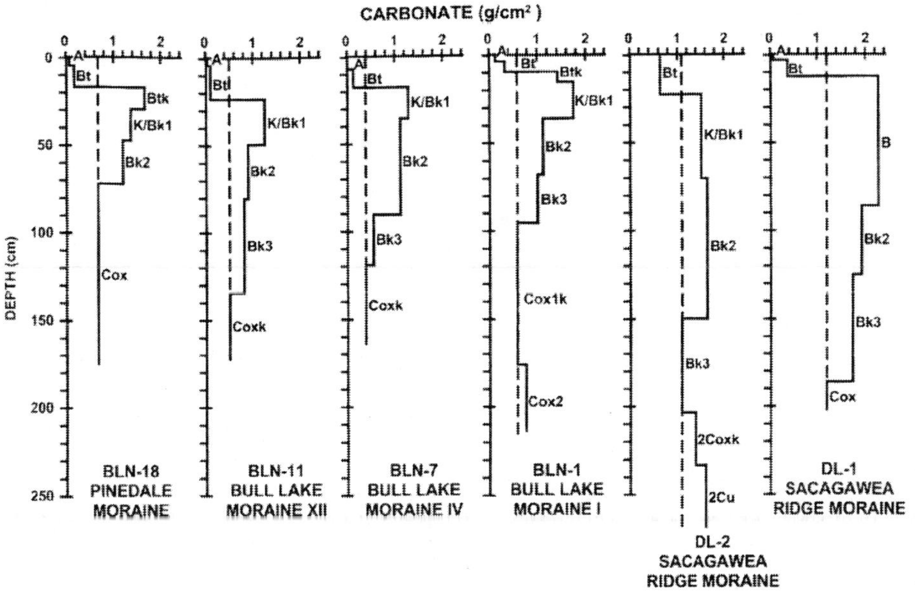

Figure 9. Carbonate distribution with depth for selected soil profiles. Dashed lines indicate the amounts of primary carbonate in the parent tills. (From Hall, 1999, Figure 7.)

TABLE 1. COMPARISON OF AVERAGE VALUES FOR SOIL CHARACTERISTICS FOR BULL LAKE AND PINEDALE SOILS IN THE FREMONT LAKE AND BULL LAKE TYPE AREAS*

Characteristic	Fremont Lake soils		Bull Ridge soils		Ratios: Fremont Lake vs. Bull Ridge	
	Pd	BL	Pd	BL	Pd soils	BL soils
Solum thickness (cm)	38	>125	86	75	0.4	>1.7
B-horizon thickness (cm)	28	>78	9	29	3.1	>2.7
Rubification index	12	15	8	14	1.5	1.1
Texture index	36	65	63	72	0.6	0.9
Structure index	32	54	11	31	2.9	1.7
PDI	25	39	20	26	1.3	1.5
Max. clay %	10	19	18	18	0.6	1.1
PWM clay %	6.1	14.6	9.3	11.8	0.7	1.2
Clay accumulation index	122	1452	115	344	1.1	4.2

* Modified from Hall, 1999.

TABLE 2. CHARACTERISTICS OF CaCO3 CONTENT OF SOILS AT BULL LAKE AND DINWOODY LAKES*

Age†	Moraines	Site	Carbonate stage index (cm—%)§	Lower boundary of visible $CaCO_3$	PWM# pedogenic carbonate² (%)
SR	SR	DL	80.7 ± 11.2	219 ± 28	5.5 ± 5.2
BL-A	I thru III	BLN	39.6 ± 14.2	155 ± 35	2.1 ± 0.8
BL-B	IV and V	BLN	34.0 ± 9.9	150 ± 13	2.7 ± 0.8
BL-C	IX thru XI	BLN	73.0 ± 43.9	125 ± 14.9	1.9 ± 0.4
BL-D	XII and XIV	BLN	32	160 ± 0	2.2 ± 2.6
PD	Bull Ridge PD	BLN	23.6 ± 4.9	114 ± 37	1.3 ± 0.8

* Modified from Hall, 1999.
†SR—Sacagawea Ridge; BL-A through BL-D, oldest to youngest Bull Lake groups; PD—Pinedale.
§Terminology discussed in Hall, 1999.
#PWM, Profile-weighted mean.

Fremont Lake soils. Surprisingly, the average solum thickness is about twice as great for soils at Bull Lake. These large values appear to be due chiefly to the thick Bk horizons in these soils. Average PDI values clearly indicate greater soil development near Fremont Lake, but the difference in values is less than for the Bull Lake soils in these areas.

In contrast with the Bull Lake soils, clay accumulation is more evident in the Pinedale soils at Bull Lake than in soils near Fremont Lake. Most of the Fremont Lake Pinedale soils have Bw horizons, whereas Bt horizons have formed in the soils at Bull Lake. The average maximum and average PWM clay contents of the Pinedale soils at Bull Lake are about twice those of the Fremont Lake Pinedale soils. This is probably due, in part, to the abundant calcium carbonate in the Bull Lake soils, which would inhibit deep translocation of clay. However, the average value of the clay accumulation index for the Pinedale soils in both areas is similar.

The amount of calcium carbonate that has accumulated in the Pinedale soils at Bull Lake is much greater than that in the Fremont Lake Pinedale soils (Table 2). All soils at Bull Lake have Bk horizons. However, these horizons are not present in the Fremont Lake Pinedale soils (which only contain negligible amounts of calcium carbonate). The main reason for this discrepancy apparently is the presence of Paleozoic limestone and dolomite in the canyons of the eastern slope of the Wind River Range (see Fig. 3), through which valley glaciers advanced. The presence of carbonate-rich rock and sediments resulted in significant amounts of primary carbonate in the tills of the eastern slope, but not on the western slope of the range. For the soils at Bull Lake, the average maximum calcium carbonate content is about 28 times greater, the average PWM carbonate content is about eight times greater, and the average accumulation index value is 46 times greater than for soils of similar age near Fremont Lake.

STOP 11 CEDAR RIDGE STRATIGRAPHIC SECTION
(If accessible)

The Cedar Ridge stratigraphic section exposed in the escarpment north of Bull Lake is one of the most significant section of Quaternary sediments in the Rocky Mountains. This section contains thick Bull Lake till, but its greater significance is as the reference section for deposits of three, purported, pre-Bull Lake glaciations: Washakie Point (oldest), Cedar Ridge, and Sacagawea Ridge (reference section, type section at Dinwoody Lakes). However, after reexamining the tills, lake sediments, and buried soils in this section, Hall and Jaworowski (1999) concluded that the stratigraphic record is best interpreted as the product of only two glaciations (Bull lake and Sacagawea Ridge; Fig. 10), not the four glaciations of previous investigations (Richmond, 1957, 1962, 1964). Hall and Jaworowski interpret deposits formerly called Cedar Ridge to represent an early stade of the Sacagawea Ridge glaciation and identify no deposits that correspond to the Washakie Point glaciation. The absence of buried soils previously described by Richmond (1964) and different interpretations of Richmond's "interglacial" and "interstadial" soils form the bases of the new interpretation. Additionally, they present paleomagnetic data that reveal all portions of the section above the Eocene deposits have normal polarity and thus must be younger than 770 ka (Brunhes-Matuyama boundary).

Return on Highway 287 south through Ft. Washakie to Lander. In Lander turn west (left) onto Fifth Street; at the end of Fifth Street turn north (right) on WY 131 (Fremont Street) which

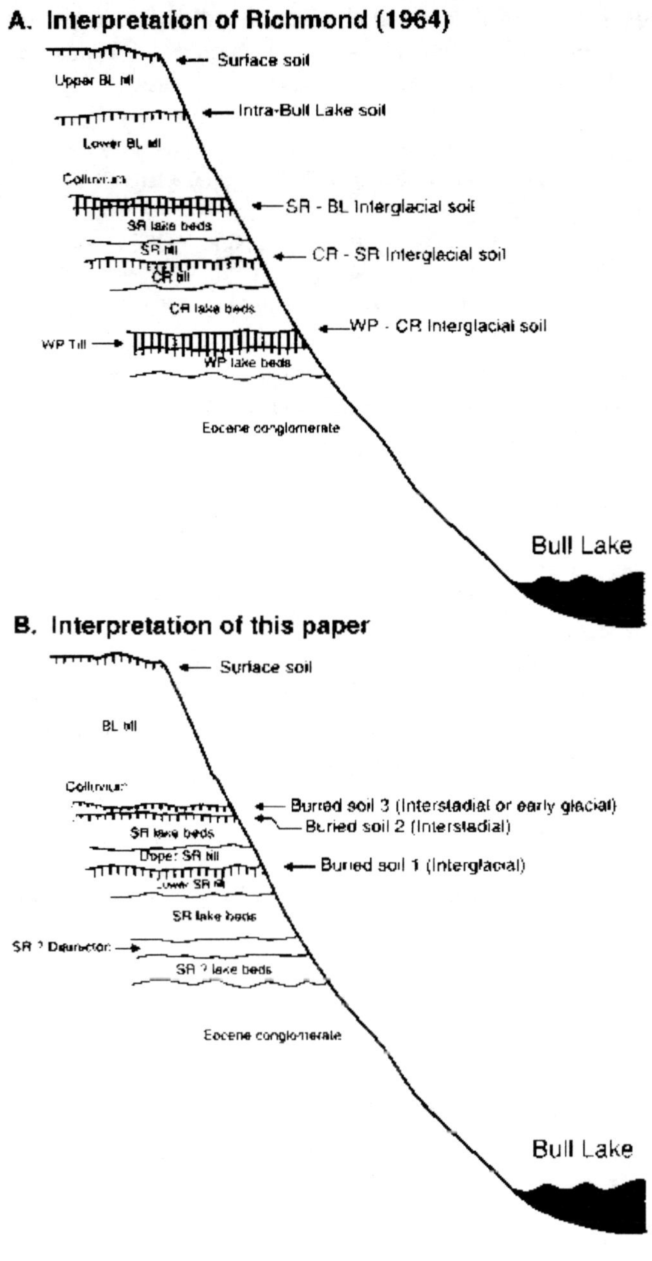

Figure 10. Interpretations of the Cedar Ridge section as discussed by Richmond, 1964 (A) and Hall and Jaworowski, 1999 (B). (Modified from Hall and Jaworowski, 1999, Fig. 2.)

turns into Sinks Canyon Road. Continue SW on Sinks Canyon Road for 9.5 mi (15.3 km) past the mouth of Sinks Canyon and through Sinks Canyon State Park to Camp Branson.

DAY 4. SINKS CANYON AND TABLE MT.

Deposits in Sinks Canyon represent the complete late-Pleistocene glacial succession in the Wind River Range (Pinedale, Bull Lake, Early Wisconsin, and Sacagawea Ridge). We will discuss the relative and numeric age-data that support the correlation of the deposits to the type localities at Fremont Lake, Bull Lake, and Dinwoody Lakes. The deposits in Sinks Canyon represent the second reported occurrence of early Wisconsin deposits in the Wind River Range.

We will also observe deposits adjacent to Sinks Canyon that appear to represent two pre-Sacagawea Ridge glaciations and discuss the relative and numeric-age evidence that suggest the oldest of these was deposited prior to ~800 ka. If the deposits mapped here as pre-Sacagawea Ridge represent older glacial activity, then Sinks Canyon contains the most complete record of glaciation reported from the Wind River Range and the only moraine record of Pleistocene glaciation prior to oxygen isotope stage 18 in the Rocky Mts.

Sinks Canyon

Sinks Canyon, 15 km southwest of Lander, Wyoming, is the most southerly of four canyons from which piedmont glaciers emerged along the eastern front of the Wind River Range (Blackwelder, 1915; Richmond, 1986; Murphy and Richmond, 1965; Richmond and Murphy, 1965, 1989; Shroba, 1989; Chadwick et al., 1997; Phillips et al., 1997; Applegarth and Dahms, 2001). The canyon was the single outlet for ice that accumulated in the middle Popo Agie Basin. Sinks Canyon is the only one of the four from which the glacial sequence is derived predominantly from bedrock and unconsolidated deposits within the canyon.

Sinks Canyon is incised into a nearly complete section of Paleozoic strata in this portion of Wyoming (see Fig. 7). The upper section of the canyon formed in Precambrian granite of the Louis Lake Fm. while the walls of the lower canyon consist of Paleozoic limestone, dolomite, and sandstone.

The canyon derives its name from "The Sinks" where the Middle Popo Agie River disappears into fissures eroded in the Madison Limestone, flows underground for about ? mile (?), and reemerges at the "Rise." Where the river goes is uncertain, but USGS dye tests indicate that the water that flows into the Sinks spends over two hours underground before it re-appears at the Rise.

Turn SW (left) from the Camp Branson road onto WY 131 for 0.75 mi (1.2 km) to the parking lot at Bruce's Campground.

STOP 4-1. BRUCE'S CAMPGROUND

Here the road begins to traverse the south canyon wall in a series of switchbacks for ~3 mi (4.8 km). The switchbacks cross the Pinedale, early Wisconsin, Bull Lake, and Sacagawea Ridge lateral moraines on the south canyon wall. Just past the last hairpin switchback, (~3 mi, 4.8 km) pull off to the right at a small dirt pullout before the road begins to swing south behind the southwest shoulder of Fossil Hill. From this overlook we see the relations among the moraines in the upper reaches of Sinks Canyon.

Glacial Deposits of Sinks Canyon (Dahms)

Pinedale. Five terminal-recessional-lateral moraines represent the limits of Pinedale ice in the upper 9 km of Sinks Canyon. The complex extends from a terminal moraine (unit P1) at 2037 m (6700 ft) just upstream from The Sinks to recessional moraine unit P5 at 2462 m (8100 ft), roughly 1 km southwest of Popo Agie Falls (Fig. 11). End moraines here are steep and sharp-crested. Kettles are not common on moraines in Sinks Canyon, probably because of the limited amounts of buried ice and thin till. Lateral moraines are steeper and sharper crested than the end moraines. The lateral moraines were deposited against steep valley walls and have undergone considerable erosion and mass wasting since deposition. Boulders on Pinedale moraines are unweathered, fresh, and unpitted, except where older till has been incorporated locally into the moraines (Table 3). Biotite-rich boulders show no evidence of grussification.

Early Wisconsin. Early Wisconsin deposits are identified at three locations. Two lateral moraines occur on the north and south walls of the canyon above the Missouri geology field camp. The crests of both moraines have elevations of approximately 2220 m (7300 ft). A third, early Wisconsin deposit lies north of The Sinks, adjacent to and downvalley from the Pinedale terminal moraine (P1). Early Wisconsin moraines exhibit relative age characteristics similar to the Pinedale moraines. Boulders show little post-depositional burial by loess. Small-diameter, biotite-rich boulders show only slight grussification in soil exposures.

Bull Lake. Bull Lake deposits occur just outside of (on the valley wall) or down valley from the Pinedale and early Wisconsin moraines. The highest Bull Lake deposits are identified on the north and south canyon walls at 2432 m (8000 ft) and 2402 m (7900 ft), respectively. Two adjacent lateral moraines are mapped at each of these localities. Bull Lake moraines exhibit more subdued morainal topography than early Wisconsin and Pinedale moraines. Locally, the innermost of the Bull Lake lateral moraines are partially buried or overtopped by the early Wisconsin or Pinedale deposits. The early Wisconsin and Pinedale deposits apparently incorporated older till (Bull Lake) because early Wisconsin and Pinedale soils in this part of the canyon have slightly redder hues and contain slightly more clay than those in other parts of the canyon. Bull Lake deposits also are identified farther downvalley on the south canyon wall on the slopes between talus cones. Bull Lake deposits are not exposed on the north wall of the canyon downvalley from the Missouri geology field camp.

Deposits representing the downvalley limit of Bull Lake ice are mapped just below the mouth of Sinks Canyon. Bull Lake ice apparently stagnated here and deposited both till and outwash from 1885 m (6200 ft) to 1760 m (5800 ft) on both sides of the Middle Popo Agie River. These deposits exhibit subtle knob-and-kettle topography, but the kettle walls have been breached (or the

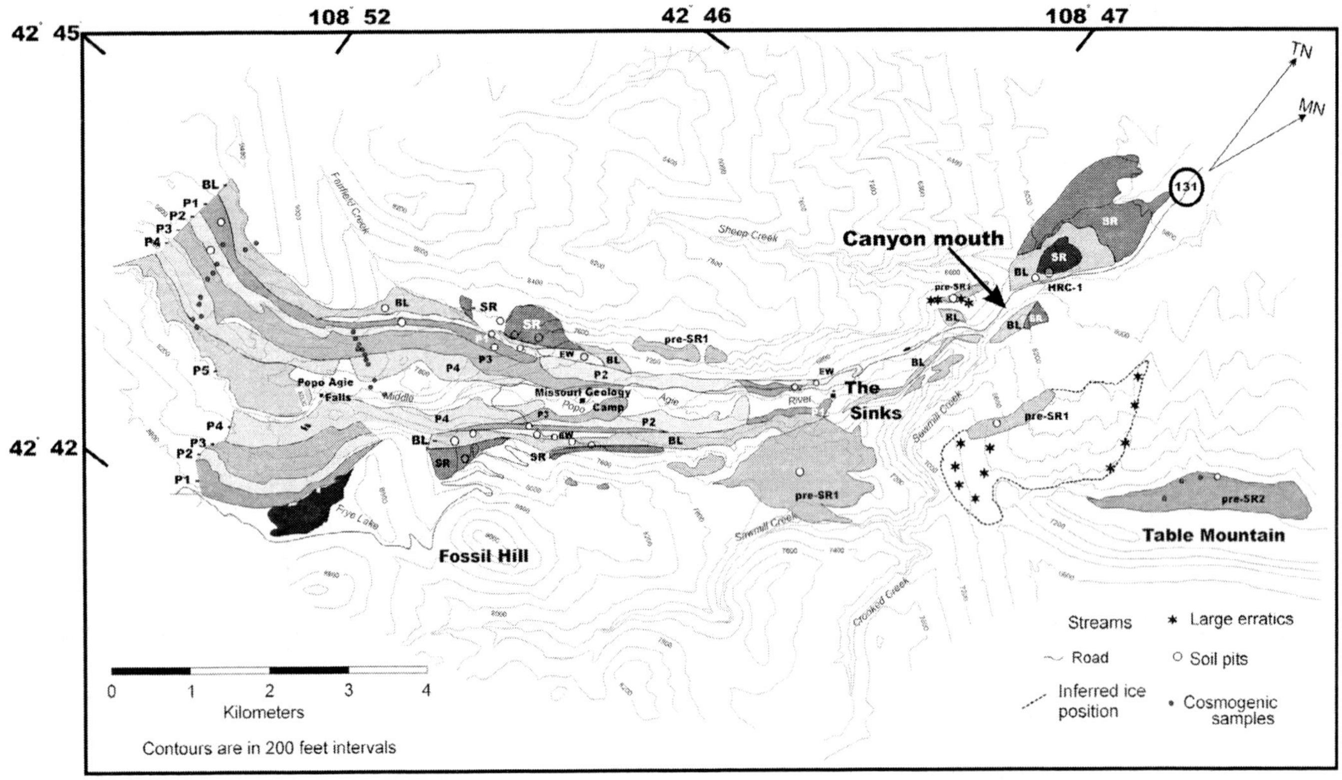

Figure 11. Map of the Pleistocene glacial deposits identified in Sinks Canyon. Major natural and cultural features are noted for reference. Elevations of contours are given in feet to correspond to the USGS 7.5' quadrangles (Wyoming) from which the contours are generalized: Cony Mountain, Fossil Hill, Mt. Arter SE, Lander, Wolf Point.

TABLE 3. RELATIVE WEATHERING CHARACTERISTICS FOR SINKS CANYON GLACIAL DEPOSITS

Map Unit	Moraine morphology	Pit Depths (>100; >1 m dia.)	Percent split boulders	Boulder frequency (per 10^4 m^{-2})
Pinedale (south)	Sharp-crested	<2	<5	268-299 (north); 195-267
Early Wisconsin	Sharp-crested	2-3	5-10	231-242 (n); 175-194 (s)
Bull Lake	Rounded	3-4.5	30-40	252-264 (n); 87-98 (s)
Sacagawea Ridge	Faint morainal topography	5-10	95-100	4-20 (n/s)
Younger pre-Sacagawea Ridge	Faint morainal topography; ~1-2 m of erosion	n/a	100	4
Older pre-Sacagawea Ridge	Diamicton; ~2-4 m of erosion	n/a	100	2

kettles filled). Drainages between depressions are integrated, so the original topography is now extremely muted. The stagnant ice deposit is deposited adjacent to and behind material deposited in a similar manner by an older (Sacagawea Ridge) ice advance.

Boulders are more weathered on Bull Lake deposits than on the early Wisconsin and Pinedale deposits. They are often imbedded in a ca. -50-cm-thick deposit of loess mixed with till that is characteristic of Bull Lake moraines in this region (Shroba and Birkeland, 1983). Weathering pits are deeper on biotite-rich surface boulders and approximately 50 percent of buried boulders are disintegrated.

Sacagawea Ridge. Sacagawea Ridge deposits are the oldest deposited here by ice that was constrained entirely within Sinks Canyon. Sacagawea Ridge deposits are identified on the north and south walls of the canyon above the Missouri geology field camp and below the canyon mouth. Sacagawea Ridge deposits exhibit only faint morainal topography. In the upper part of the canyon, Sacagawea Ridge deposits appear to form lateral moraines that are indistinct down valley where they are buried or destroyed by Bull Lake till and younger talus. Sacagawea Ridge deposits beyond the canyon mouth also are indistinct diamicts. They consist of stagnant-ice deposits (till and outwash) that lie outside and above adjacent Bull Lake deposits.

Boulder weathering is more pronounced and weathering-pit depths are much deeper on Sacagawea Ridge deposits than on any of the younger tills. Weathering pit size is not a reliable measure of boulder weathering on the older deposits here because weathering rinds tend to spall off the boulders.

Younger and older pre-Sacagawea Ridge. Younger and older pre-Sacagawea Ridge deposits are preserved only on or outside the rims of Sinks Canyon and are associated with ice that overflowed Sinks Canyon. Few boulders protrude above the ground surface due to a combination of boulder weathering and surface erosion. Most boulders that are level to the present ground surface are characteristically weathered to red to orange-yellow Munsell colors (10R 5-6/8 and 7.5YR5-6/8 colors). Larger (whaleback) boulders often are weathered into 'mushroom rock' shapes that are severely undercut as much as ~1 m above ground level. These weathering forms emphasize the importance of the deflation of the older moraine surfaces here. The intensity of boulder weathering and evidence of soil truncation indicates that erosion has removed as much as 4 m of material from the pre-Sacagawea Ridge deposits.

Only preliminary numeric age data are available to constrain the age of these deposits. Preliminary analyses of ^{10}Be concentrations from three boulders on the older pre-Sacagawea Ridge deposit mapped on Table Mt. produce maximum ages ranging from ~135 to 783 ka using Stone's (2000) production rate of 5.1 ^{10}Be atoms $g^{-1}a^{-1}$ and an erosion rate of 2 mm/kyr.

The ^{10}Be ages from Table Mt. suggest this glacial diamict was deposited prior to ~1 Ma if we assume the boulders were initially buried and subsequently exhumed. We assume considerable erosion of the till based on boulder age estimates, the presence of meso-channeling on the deposits, a ubiquitous cover of rock fragments on the soil surface, and soil erosion estimates (above) based on differences in soil thickness on adjacent (younger and older) moraines.

Glacial Erosion of Sinks Canyon

The aim of this study was to test the hypothesis that variations in cosmogenic radionuclide concentrations in samples collected across a glaciated valley could be used to determine the pattern of bedrock loss due to erosion during the last glaciation (Fabel et al., in press). Two cross-valley transects were sampled across the southwestextent of Sinks Canyon (see map). Samples of weathered bedrock, erratic boulders, and polished and/or striated bedrock were collected along each transect at roughly 20 m intervals from the canyon floor to the side-slope areas above the limit of the Pinedale glaciation.

Transect A consists of eight samples over an elevation range of ~310 m with exposure ages ranging from ~14 ka on the valley floor to ~111 ka on the highest side-slope sample (Table 4). Transect B is 2 km northeast of profile A and consists of nine samples with exposure ages ranging from ~17 ka on the valley floor to ~106 ka on the highest side-slope sample. The six lower samples of profile A have statistically identical exposure ages with a mean of roughly 15 ka that are statistically identical, whereas the four samples at the bottom of profile B have statistically identical exposure ages with a mean of 18.0 ± 1.2 ka. Some samples along the upper transect positions that were taken from positions *below* the Pinedale ice limit yield significantly older exposure ages (48 to 87 ka). Moreover, a threshold exists at a bedrock step between transects A and B. Above the step, glacial erosion has deeply scoured the valley and removed the cosmogenic depth profile (transect A). Below the step (transect B), the exposure ages indicate a significant amount of inheritance 120 m above the valley floor. Several results of this study include:

- Ages on the bedrock valley floor are statistically identical to Gosse's (1995b) exposure ages on the till of the Pinedale glaciation at its type area at Fremont Lake;
- Ages along the bedrock valley floor provide an estimate for deglaciation of Sinks Canyon (~18 ka) that appears to have occurred earlier than deglaciation began at Fremont Lake;
- Variations in cosmogenic inheritance in samples on opposite sides of the bedrock step between transects A and B indicate that variations in ice dynamics (velocities, basal pressures) over small distances can cause large variations in erosion characteristics. Thus, three-dimensional aspects of ice flow need to be considered in glacial erosion studies, not simply the position within a valley cross-section;
- The older exposure ages of samples high on the side-slopes below the limit of Pinedale ice are consistent with modeled predictions that most erosion by temperate valley glaciers should occur where the ice is thickest (valley floor/axis), whereas less erosion should occur at the ice-margins (Harbor, 1995; Harbor and Wheeler, 1992);
- The maximum calculated erosion rate along the valley axis was about 4-5 cm/ka^{-1};
- Transect A shows Pinedale ice scoured an average of about 1.9 m of bedrock from across the entire section. Closest to the valley axis the maximum depth of rock removed was ~4.9 m. Thus the amount of bedrock removed at the location of maximum ice thickness at each section is less than 2% of the current valley depth. Therefore, many glacial cycles appear to be needed to significantly impact valley scale topography.

Return back down the switchbacks; observe variations in soils at open soil pits on Sacagawea Ridge, Bull Lake, early Wisconsin, and Pinedale lateral moraines (see map; Table 5).

Continue NE on WY 131 through Sinks Canyon, past the canyon mouth 0.2 mi (0.3 km) and turn north (left) onto Homestead Park Road. Continue on this gravel road for 0.8 mi (1.3 km) to a locked cattle gate. We will drive up a set of steep switchbacks just north of Sheep Canyon to a view of Sinks Canyon from Homestead Park on the north rim of the canyon (Fossil Hill 7.5′ quadrangle, NW 1/4, Section 18, T32N, R100W).

STOP 4-2. HOMESTEAD PARK

View the relations among Pinedale, early Wisconsin, Bull Lake, and Sacagawea Ridge inner canyon deposits and the pre-Sacagawea Ridge deposits on the canyon rims and beyond to Table Mt..

TABLE 4. RELATIVE DEVELOPMENT CHARACTERISTICS FOR SINKS CANYON SOILS

Stratigraphic Unit	Solum thickness (cm)*#@	Depth to Cox/C boundary (cm)	AI[1]	Clay % Data Clay Max (B)	PWM[1]	Pedogenic Clay[2] (Profile g/cm^{-2})	Rubification index[1] (PWM)	Texture index[1] (PWM)	PDI[3]
Pinedale	*40	76	389	10.5	1.9	3.9	5.9	9.2	14.7
	#(20–52)	(50–120)	(22–1256)	(7–20.5)	(1.3–4.0)	(0–13.3)	(2.2-10.5)	(2.6-16.7)	(10.5-16.3)
	+17.7	29	450	5.1	1.04	5.0	3.7	6.1	2.2
Early Wisc.	59	111	833	17.5	3.8	8.9	24.2	23.9	27.1
	(45–80)	(65–140)	(228–2509)	(7–20.3)	(2.1–6.3)	(0.9–30.1)	(10.2-37.7)	(8.9-57.4)	(19.8-33.3)
	15.2	34	892	14.5	1.78	11.5	9.7	18.1	5.1
Bull Lake	94	118	1153	18.1	4	13.1	36.3	47.5	44.7
	(85–130)	(100–170)	(115–1652)	(12–52)	(3.3–5.3)	(0.5–20.5)	(24.4-48.1)	(23.3-59.3)	(31.4-58.7)
	18	34	702	6	0.94	9.3	8.7	14.1	9.9
Sacagawea Ridge	193	330	3766	17.7	13	46.6	111.9	130.7	111.6
	(100–500)	(210–500)	(824–6944)	(14–21)	(8.4–15.8)	(9.1–86.6)	(47.2-161.1)	(38.9-238.9)	(82.8-159.7)
	136	191	3066	3.3	4.02	43.7	58.5	101	41.9
Younger pre-Sacagawea Ridge	115	145	7557	18.8	16.9	63.1	184.6	293.3	148
	(80–150+)	(140–150)	(111–-9281)	(15–24)	(11.7–20.3)	(0.4–105.7)	(139.5-229.6)	(148.2-438.3)	(34.5-257.5)
	50	7.1	6484	6.6	6	74.0	64	205	158
Older pre-Sacagawea Ridge	300	900	5832	17.6	17.8	84.0	0	5.6	152

* Mean; # Variation; + Standard deviation

[1] Procedures follow those used by Hall and Shroba (1993, 1995) and Hall (1999).

[2] Procedures follow McCalpin and Berry (1994) and Dahms (2002).

[3] Procedures follow Harden (1982) as modified by Birkeland et al. (1991).

TABLE 5. SINKS CANYON EROSION TRANSECTS - ^{10}BE AND ^{26}AL DATA WITH FULL UNCERTAINTIES

Sample	Sample type	Elevation (m)	Geometric correction	Magnetic correction	Thickness (cm) [correction]	^{10}Be† x10^5 atom g^{-1}	Apparent ^{10}Be Exposure Age ka**	Native Al ppm	^{26}Al † x10^5 atom g^{-1}	Apparent ^{26}Al Exposure Age ka**	^{26}Al/^{10}Be
Sinks Canyon Profile A											
97-108	weathered bedrock	2830	0.49	0.88	1 [0.992]	23.8 ± 0.8	110.8 ± 15.6	16.7 ± 1.7	127.7 ± 8.5	100.6 ± 13.6	5.4 ± 0.4
97-109	erratic	2780	0.98	0.89	2 [0.983]	21.1 ± 0.7	50.7 ± 7.0	12.7 ± 1.3	118.6 ± 7.3	47.6 ± 6.3	5.6 ± 0.4
97-110	erratic	2720	1.00	0.99	2 [0.983]	5.7 ± 0.3	15.5 ± 2.3	11.7 ± 1.2	33.3 ± 2.5	15.0 ± 2.2	5.8 ± 0.5
97-111	striated bedrock	2680	1.00	0.98	1 [0.992]	7.1 ± 0 ± 0.4	19.5 ± 2.9	16.1 ± 1.6	35.3 ± 3.3	16.0 ± 2.5	5.0 ± 0.5
97-113	striated bedrock	2650	0.98	0.98	1 [0.992]	5.6 ± 0.3	15.8 ± 2.4	19.8 ± 2.0	34.6 ± 2.5	16.4 ± 2.3	6.2 ± 0.6
97-117	striated bedrock	2590	1.00	0.98	1 [0.992]			9.2 ± 0.9	32.6 ± 2.6	15.7 ± 2.3	
97-118	striated bedrock	2560	0.97	0.99	1 [0.992]			12.6 ± 1.3	27.3 ± 2.0	14.0 ± 2.0	
97-120	striated bedrock	2520	0.99	0.99	1 [0.992]			9.5 ± 0.6	27.8 ± 1.8	14.3 ± 2.0	
Sinks Canyon Profile B											
97-42	weathered bedrock	2560	1.00	0.89	2 [0.983]			25.7 ± 2.6	119.6 ± 7.0	54.8 ± 6.96	
97-43	polished bedrock	2530	0.53	0.88	1 [0.992]	20.1 ± 0.6	106.2 ± 14.8				
97-44	polished bedrock	2500	0.51	0.89	5 [0.959]	12.7 ± 0.3	71.6 ± 9.8				
97-45	polished bedrock	2480	0.99	0.89	2 [0.983]	29.1 ± 1.1	84.7 ± 11.9	21.8 ± 2.2	175.0 ± 9.8	85.2 ± 10.8	6.0 ± 0.4
97-46	polished bedrock	2460	0.99	0.89	1 [0.992]	29.9 ± 1.1	87.5 ± 12.3				
97-47	polished bedrock	2440	0.99	0.96	1 [0.992]	6.1 ± 0.4	19.5 ± 2.9				
97-48	striated bedrock	2420	0.98	0.96	3 [0.975]	5.9 ± 0.4	19.4 ± 3.0				
97-49	polished bedrock	2410	0.99	0.97	1 [0.992]	5.5 ± 0.4	17.8 ± 2.8				
97-85	striated bedrock	2350	0.99	0.98	1 [0.992]	5.1 ± 0.4	17.2 ± 2.8	20.0 ± 2.0	30.0 ± 2.1	17.0 ± 2.4	5.9 ± 0.6

* From Fabel et al., in press.
All latitudes are 42.73°N.
** Exposure ages calculated assuming sea-level, high latitude spallogenic and muon production rates of ^{10}Be = 4.8 ± 0.4 at g^{-1} yr^{-1}, ^{26}Al = 28.8 ± 1.3 at g^{-1} yr^{-1}, and ^{10}Be = 0.11 ± 0.02 at g^{-1} yr^{-1}, ^{26}Al = 0.81 ± 0.14 at g^{-1} yr^{-1} respectively. Uncertainties represent one standard error measurement uncertainty and maximum quoted magnetic palaeointensity uncertainty (Frank et.al., 1997). Systematic uncertainties in production rates (10%), and radioactive decay rates are added in quadrature.
† ^{26}Al/^{27}Al and ^{10}Be/^{9}Be measured by accelerator mass spectrometry at FRIME Lab. Quartz [Al] measured by flame atomic absorption spectrophotometry and assigned 8% uncertainty. ^{10}Be/^{9}Be measured against standards derived from NIST SRM 4325, with an assigned ratio of 3.05 x 10^{-11}.

STOP 4-3. TABLE MT.

Time and weather permitting, we will take a late afternoon or early evening hike up Table Mt. to observe the characteristics of the purported till and associated boulders and soils. The hike begins at the end of a private access road at the north base of Table Mt. and takes approximately 4-5 hours (round trip).

Evening at Camp Branson.

DAY 5. LANDER TO JACKSON; JACKSON HOLE GLACIATION

Turn NE onto WY 131 from Camp Branson and return to Lander. Turn east (left) on Fifth St. to Main St. (Highway 287). Turn north (left) on Highway 287 and continue through Lander and on to Ft. Washakie (13.6 mi, 21.9 km).

STOP 5-1. SACAGAWEA CEMETERY

Sacagawea Cemetery is just west of Ft. Washakie. The Shoshone tribe has erected a monument to Sacagawea here and buried her two sons beside it. The monument states that the tribe is thought to have buried Sacagawea "somewhere west of here in the Wind River Range."

Return back through Ft. Washakie to Highway 287. Continue north (left) for 13.4 mi (21.6 km). Pull off onto the right shoulder at the north edge of a high terrace just before the road descends steeply toward the Wind River valley.

STOP 5-2. PLEISTOCENE TERRACES OF THE WIND RIVER

The major outwash terraces that have been described along the Wind River in this area of the basin can be seen from here (Phillips et al., 1997).

Terraces of the Wind River (This discussion is modified from Chadwick et al., 1994, p. 5-8)

Some of the best-preserved alluvial terraces in the Middle Rocky Mt. region occur along the Wind River (Figs. 12, 13). Young terraces are still well preserved and, although the oldest terraces are mostly represented by scattered remnants, many middle to middle-late Pleistocene terraces are still easily recognized along 140 km of the Wind River from near Dubois to Riverton, Wyoming. Many older terraces in this area are identified on buttes and mesas from coarse-grained stream clasts that cap straths on the Wind River Fm.. The coarse clasts now armor the terraces against wind erosion and headward migration of gullies. Terrace caps have average thicknesses of about 10 m between Bull Lake and Dinwoody Lakes, thining downstream to about 3 m at Riverton. This thickening of bedload upstream suggests that it is glacially-derived for the most part, although local variations are present.

Fifteen terraces are mapped between Dubois and Riverton. Longitudinal profiles can be obtained for all terraces except WR11 and WR15 (although WR5, WR6, WR10, and WR 14 are poorly constrained because only a few remnants are preserved). Thus, terraces WR0, WR3, WR7, and WR9 are the most completely preserved (Fig.13). Estimated terrace ages are given in the caption to Figure 12.

Mapping indicates that earlier courses of the Wind River flowed nearer to the mountain front than at present. Glaciers issuing from the canyons progressively forced the river northward and eastward until about the time of the WR7 terrace. Since that time, the river has incised close to its previous bed. In some cases, this led to poor preservation of the younger terraces. Terraces older than WR9 have excellent preservation at only two localities, downbasin from Bull Lake and northwest of Riverton.

Continue north on Highway 287 to the Junction of U.S. Highway 26; turn north. From the Junction continue on Highway 26 for 13 mi (21 km) to Crowheart Store. [Convenience store and restroom]. North of Bull Lake Creek, the highway traverses younger and older Bull Lake outwash that grades to WR3 and WR4 terraces, respectively.

STOP 5-3. DINWOODY LAKES MORAINES

Continue on Highway 26 another 11 mi (17.7 km) to Dinwoody Lakes access road. All vans meet here and continue to the stop overlooking the type Sacagawea Ridge moraine(s).

Continue north on Highway 26 to Dubois [convenience stores, restrooms, and lots of cowboy arts and crafts]. From the north Dubois city limit, continue north on Highway 26 for ~5.2 mi (8.4 km). Here we drive around and below a Bull Lake (?) terrace below the confluence of ice from the Wind River and DuNoir valleys.

At 6.7 mi (10.8 km) the route passes the confluence of the Wind and DuNoir rivers. The DuNoir River drains the southern margin of the Absaroka Range. Lateral moraines of both the Pinedale and Bull Lake glaciations can be seen on either side of the DuNoir valley here. Highway 26 traverses the west side of the DuNoir right lateral moraine. Notice the fine-grained nature of the till here. This characteristic continues nearly into Jackson Hole, because the bedrock in this area is composed mainly of easily weathered Tertiary (Eocene) volcaniclastics (especially lahar deposits) associated with the Absaroka volcanic region.

Turn left into the pullout(s) at 21.2-21.7 mi (34.1-34.9 km).

STOP 5-4. WHISKEY BASIN

Continue north from Dinwoody Lakes Access Road for 13.3 mi (21.4 km) to Whiskey Basin Access Rd. Turn left across the

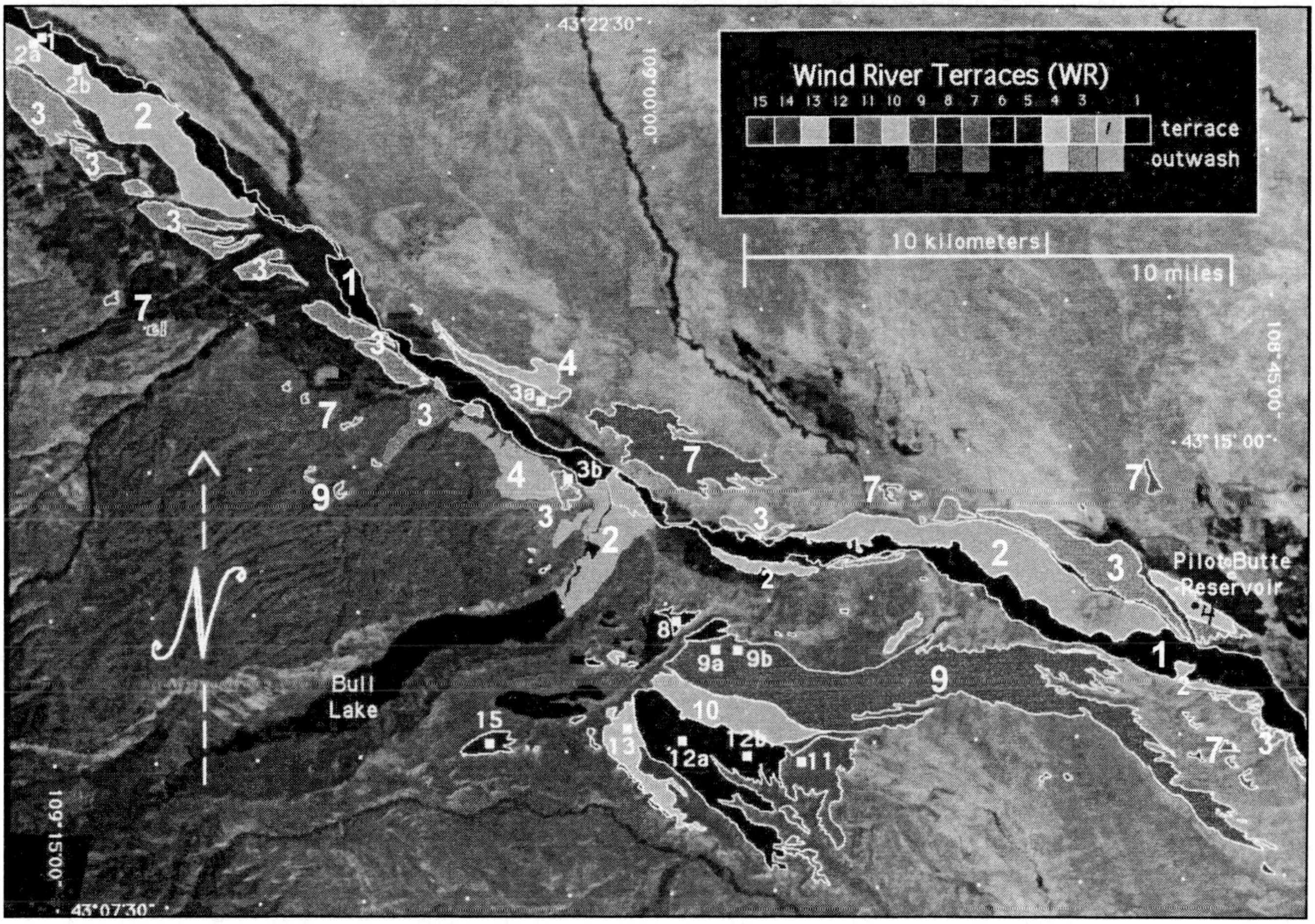

Figure 12. The suite of Pleistocene terraces identified in the Bull Lake area (Chadwick, et al., 1994, 1997). The most extensive terraces in this area are WR1 and WR2 (Pinedale), WR3 and 4 (upper and lower Bull Lake), WR7 (younger Sacagawea Ridge), and WR9 (lower Sacagawea Ridge), and WR10-11-12-13 (all pre-Sacagawea Ridge). Numeric-Arabic notations (9a, etc.) refer to terrace soil-sampling sites of Peacock et al., 1994.

cattle guard then immediately left again and continue 5.2 mi (8.4 km) to Whiskey Basin.

Soils and Glacial Deposits of Whiskey Basin (Applegarth)

Glacial deposits mapped in Whiskey Basin indicate that two discrete valley glaciers originated from the main Wind River Range massif and flowed into the area during the Pinedale and Bull Lake glaciations. One lobe of ice flowed south and east of Whiskey Mt. in the Torrey Creek valley and incorporated sediments from the Tensleep (calcareous sandstone), Madison, and Bighorn Formations (Boyd, 1993) into associated tills. The other lobe flowed north of Whiskey Mt. in the Jakey's Fork valley and incorporated sediments from the Dinwoody and Chugwater Formations (calcareous sandstones and shales) into associated tills. The Chugwater sandstone and shale in the Jakey's Fork valley is strongly oxidized and gives the tills a redder color (5YR, 7.5YR) than corresponding tills of the Torrey Creek valley (10YR, 2.5Y) (Picard, 1993).

Soil catenas on the Whiskey Basin moraines have gravelly sandy loam textures and structures range from granular in A horizons to subangular blocky Btk and some Ck horizons (Applegarth and Dahms, 2001). Solum colors mostly are 7.5YR and 5YR with occasional 10YR and 2.5YR hues. Soils along Pinedale catenas typically are Aridisols with generally A-Bk-Ck or Ak-Btk-Ck profiles (Soil Survey Staff, 1997). Soils along Bull Lake catenas also tend to be Aridisols, though Mollisols can occur in lower slope positions. Most Bull Lake soils have A-Btk-Ck and Ak-Btk1-Btk2-Ck profiles. Profiles are thicker and pedogenic clay and carbonate contents are greater in Bull Lake soils than Pinedale soils. Weighted Mean Profile Development Index values (Birkeland et al., 1991) generally are larger in Bull Lake profiles. Differences between Pinedale and Bull Lake catenas are attributed mainly to their difference in age and erosion history.

The soils of the Whiskey Basin catenas are similar to those described in tills at the Bull Lake type area and the Dinwoody Lakes area (Shroba, 1989; Hall, 1999). All the soils developed in till parent materials contain significant quantities of primary carbonates derived from Paleozoic and Mesozoic sedimentary rocks.

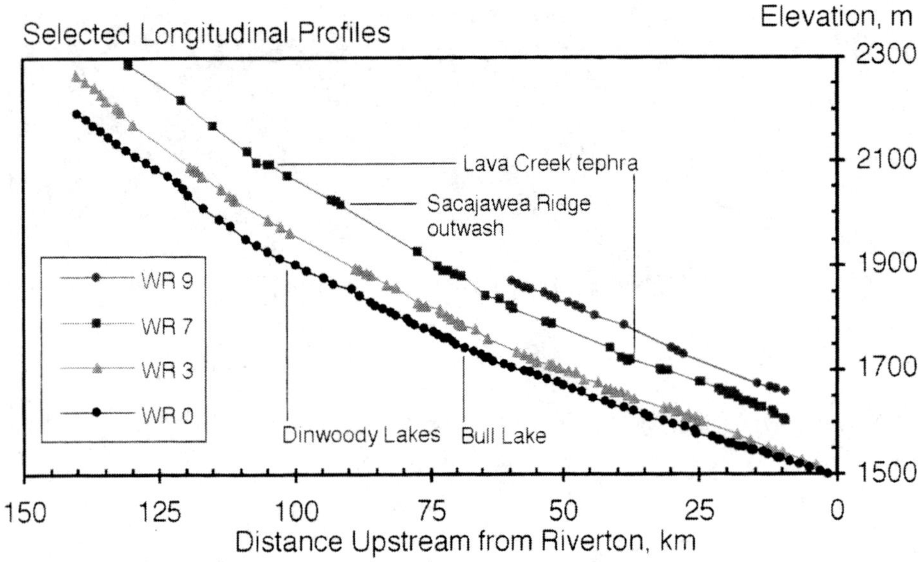

Figure 13. Selected longitudinal profiles for the most continuous remnants of terraces along the Wind River between Dubois and Riverton, Wyoming. (From Chadwick et al., 1994, Figure 5).

Evidence from soil morphology, clay content, and the carbonate morphology and content of soils developed in Bull Lake tills suggest soil erosion was quite severe on tills of the east slope (Hall, 1999; Applegarth and Dahms, 2001). Evidence for erosion takes the form of (1) thin surface horizons (often without A horizons), (2) lack of rubification in near-surface horizons, (3) weak relationships between the amount of pedogenic carbonate in the upper carbonate horizon and soil age, (4) apparent redistribution of pedogenic carbonate into Cox and Cu horizons, (5) similar clay contents of A horizons and B horizons or abrupt increases in clay content from thin A horizons to B horizons. Thus, near-surface horizons apparently are significantly younger than the rest of the soils' profiles. Near-surface soil horizons possibly have formed since the last deglaciation and probably as a result of wind and periglacial erosion processes that occur during glacier advances (Hall, 1999).

STOP 5-5. VIEW OF PINNACLE BUTTES

Turn left into the pullout(s) at 21.2-21.7 mi (34.1-34.9 km). This stop offers a spectacular view of the sequence(s) of lahars that form the majority of the Absaroka volcanic field.

STOP 5-6. ABSAROKA LAHARS

Pull off to the right around and behind a road cut exposure at 25.4 mi (40.9 km). This is a good chance to see details of the Absaroka lahars.

STOP 5-7. TOGWOTEE OVERLOOK.

Turn right into a viewpoint and rest area at 35.3 mi (56.8 km). This is a spectacular view of the Grand Tetons and Jackson Hole and gives us a chance to overview the interaction between alpine glaciers and the Yellowstone Ice Cap. Turn west back onto Highway 26 for 0.2 mi (0.3 km) to Cowboy Village [gas, convenience store, restrooms].

Continue west on Highway 26 past the east boundary of Grand Teton National Park (49.9 mi, 80.3 km). Turn north (right) onto Highway 89/191/287 at Moran Junction (51.9 mi, 83.5 km) to the park's east entrance station (52.1 mi, 83.8 km). Continue on Highway 89/191/287 and turn left (SW) at 55.8 mi (89.8 km) onto Teton Park Road at the Jackson Lake Junction.

STOP 5-8. SIGNAL MT.

Continue 3.9 mi (6.3 km) from Jackson Lake Junction; turn left at Signal Mt. turnoff. Drive to summit parking lot 5 mi (8 km) from the turnoff.

Signal Mt. provides a view of all the sources of ice that entered Jackson Hole during the Pleistocene. To the east and northeast we see the Buffalo and Pacific Creek valleys down which alpine glaciers flowed during the initial (Burned Ridge and Hedrick Pond) phases of the Pinedale glaciation (Pierce and Good, 1992; Table 6). To the north the region was occupied by the southern margin of Yellowstone ice cap, which was responsible for most of the ice that flowed into Jackson Hole during the height of the Pinedale Glaciation (Jackson Lake phase). To the west, the steep canyons provided little of the ice that filled Jackson Hole, but channeled the valley glaciers from the Teton Range that formed the terminal moraines now damming Jenny Lake and similar, small, scenic lakes along the mountain front.

STOP 5-9. TETON FAULT SCARP

Return to the Signal Mt. turnoff; turn left back onto Teton Park Road and go 6.5 mi (10.5 km) to the Cathedral Group

Turnout to view the Teton fault scarp. Here, normal displacement for about 30 km along the fault marks the eastern margin of the uplift and results in local relief of as much as 6000 ft (1828 m) between the peaks and the floor of Jackson Hole. Although the uplift of the Teton Range may have had a long and complicated history that goes back to the Eocene, the present range front is apparently between 4-2 Ma old (Fig. 14). Gilbert et al. (1983) estimated the uplift rate to be about 2.2 mm/yr^{-1} during the Quaternary.

STOP 5-10. JENNY LAKE

Continue on Jenny Lake Loop Road (one way) for 2.1 mi (3.4 km) to the Jenny Lake parking area (on the right). The overlook is directly across Jenny Lake from Cascade Canyon, which is a classic U-shaped glacial valley down which Pinedale ice flowed to deposit the moraine, which now encloses the lake.

TABLE 6. RELATIVE SIZES OF THE GLACIERS THAT FED ICE INTO JACKSON HOLE DURING THE PINEDALE GLACIATION*

Glacial phase	Snake River lobe	Pacific Creek lobe	Buffalo Fork lobe
Burned Ridge (*oldest*)	---	X	XXX
Hedrick Pond	X	XXX	---
Jackson Lake (*youngest*)	XX	XX	---

* From Pierce and Good, 1992, fig. 2, p. 7.
--- = lobe not confluent with other lobes at terminus.
X = lobe confluent but slightly smaller than other lobes.
XX = lobe confluent and very similar in size to the other lobe at terminus.
XXX = lobe confluent slightly larger than other lobe at terminus.

Figure 14. View of the Teton fault scarp near the north end of the Cathedral Group, Grand Teton National Park. This young (<14,000 yrs) fault scarp offsets Pinedale slope deposits and is about 35 m (120 ft) high (Pierce and Good, 1992). View is to west from the Cathedral Group Turnout on the north Jenny Lake Loop. (D.E. Dahms photo)

From Jenny Lake, continue for 1 mi (1.6 km) to the junction of Teton Park Road; turn right and continue another 7.6 mi (12.2 km) to the Moose Visitor Center (restrooms, snacks, bookstore).Continue 0.6 mi (1 km) to Moose Junction. Turn south (right) on Highway 26/89/191 and return to Jackson.

REFERENCES

Alley, R.B., Meese, D.A., Shuman, C.A., Gow, A.J., Taylor, K.C., Grootes, P.M., White, J.W.C., Ram, M., Waddington, E.D., Mayewski, P.A. and Zielinski, G.A., 1993, Abrupt increase in Greenland snow accumulation at the end of the Younger Dryas event: Nature, v. 362, p. 527-529.

Applegarth, M.T., and Dahms, D.E., 2001, Soil catenas of calcareous tills, Whiskey Basin, Wyoming, USA: Catena, v. 42, p. 17-38.

Birkeland, P.W., Machette, M.N., Haller, K.M., 1991. Soils as a tool for Quaternary geology: Miscellaneous Publication 91-3, Utah Geological and Mineral Survey, Salt Lake City, 63 p.

Blackwelder, E., 1915, Post-Cretaceous history of the mountains of western Wyoming: Journal of Geology, v. 23, p. 302-340.

Boyd, D.W., 1993. Paleozoic history of Wyoming: *in* Snoke, H., Steidtmann, J.R., Roberts, S.M., eds., Geology of Wyoming: Geological survey of Wyoming, Memoir #5, p. 164-187.

Bursik, M.I., and Gillespie, A.R., 1993, Late Pleistocene glaciation of Mono Basin, California: Quaternary Research, v. 39, p. 24-35.

Chadwick, O.A., Hall, R.D., Kelly, G., Amundson, R., Gosse, J., Phillips, F., and Jaworowski, C., eds., 1994, Quaternary Geology of the Wind River Basin, Wyoming: Guidebook to Friends of the Pleistocene-Rocky Mt. Cell fieldtrip.

Chadwick, O.A., Hall, R.D., Phillips, F.M., 1997, Chronology of Pleistocene glacial advances in the central Rocky Mountains: Geological Society of America Bulletin, v. 109, p. 1443-1452.

Colman, S., and Pierce, K.L., 1986, Glacial sequence near McCall, Idaho: Weathering rinds, soil development, morphology, and other relative-age criteria: Quaternary Research, v. 25, p. 25-42.

Colman, S., and Pierce, K.L., 1992, Varied records of early Wisconsinan alpine glaciation in the western United States derived from weathering rind thicknesses: *in* Clark, P.U., and Lea, P.D., eds., The Last Interglacial-Glacial Transition in North America, Geological Society of America Special Paper 270, p. 269 278.

Dahms, D.E., 1991, Eolian sedimentation and soil development on moraine catenas of the Wind River Mountains, West-Central Wyoming: PhD thesis, University of Kansas, 340 p.

Dahms, D.E., 1993, Mineralogical evidence for eolian sediments in soils on late Quaternary moraines, Wind River Mountains, Wyoming: Geoderma, v. 59, p. 175-196.

Dahms, D.E., 1994, Mid-Holocene erosion of soil catenas on moraines near the type Pinedale till, Wind River Range, Wyoming: Quaternary Research, v. 42, p. 41-48.

Dahms, D.E., 2002, Glacial stratigraphy of Stough Creek Basin, Wind River Range, Wyoming: Geomorphology, v. 42, p. 59-83.

Dahms, D.E., 2003, Glacial Limits in the Middle and Southern Rocky Mountains, U.S.A., south of the Yellowstone Ice Cap: *in* Gibbard, P.L. and Ehlers, J., eds., World-Wide Extent and Chronology of Glaciations: International Quaternary Association (INQUA), Commission on Glaciations, Workgroup 5.

Dahms, D.E., and Birkeland, P.W., 2000, Glacial stratigraphy of Stough Creek Basin and a latest Quaternary glacial and periglacial stratigraphy for the Wind River Range, Wyoming: Geological Society of America Abstracts with Programs, v. 32, no. 7, A16.

Dahms, D.E., Birkeland, P.W., Miller, C.D., Shroba, R.R., and Kihl, R., in press, Latest Quaternary glacial and periglacial stratigraphy, Wind River Range, Wyoming: Wyoming Geological Survey, Report of Investigations series.

Denton, G.H., 1975, Glaciers of the American Rocky Mountains. in Field, W.O., ed., Mountain glaciers of the Northern Hemisphere, Vol. 1: Cold Regions Research and Engineering Laboratory, Hanover, New Hampshire. p. 509-542.

Easterbrook, D.J., 1992, Advance and retreat of Cordilleran ice sheets in Washington, U.S.A.: Géographie physique et Quaternaire, v. 46, p. 51-68.

Fabel, D., Harbor, J., Dahms, D.E., James, A., Elmore, D., Horn, L., Daley, K., and Steele, C., in press, Spatial patterns of glacial erosion at a valley scale derived using cosmogenic radionuclide concentrations in rock: Annals of the Association of American Geographers.

Fall, P.L., 1994, Modern pollen spectra and vegetation in the Wind River Range, Wyoming: Arctic and Alpine Research, v. 26, p.383-392.

Gansecki, C.A., Mahood, G.A., and McWilliams, M., 1998, New ages for the climactic eruptions at Yellowstone: single-crystal $^{40}Ar/^{39}Ar$ dating identifies contamination: Geology, v. 26, p. 343-346.

Gilbert, J.D., Ostenna, D., and Wood, C., 1983, Seismotectonic study, Jackson Lake Dam and Reservoir: Minidoka Project, Idaho-Wyoming: U.S. Bureau of Reclamation Seismotectonic Report 83-8.

Gillespie, A.R., and Molnar, P., 1995, Asynchronous maximum advances of mountain and continental glaciers: Reviews in Geophysics, v. 33, p. 311-364.

Gosse, J.C., Evenson, E.D., Klein, J., Lawn, B., Middleton, R., 1995a, Precise cosmogenic ^{10}Be measurements in western North America: Support for a global Younger Dryas cooling event: Geology, v. 23, p. 877-880.

Gosse, J.C., Klein, J., Evenson, E.B., Lawn, B., Middleton, R., 1995b, Beryllium-10 dating of the duration and retreat of the last Pinedale glacial sequence: Science, v. 268, p. 1329-1333.

Hack, J.T., 1943, Antiquity of the Finley site: American Antiquity, v. 8, p. 235-241.

Hall, R.D., 1999, Effects of Climate change on soils in glacial deposits, Wind River Basin, Wyoming: Quaternary Research, v. 51, p. 248-261.

Hall, R.D., and Shroba, R.R., 1993, Soils developed in the glacial deposits of the type areas of the Pinedale and Bull Lake Glaciations, Wind River Range, Wyoming, U.S.A.: Arctic and Alpine Research, v. 25, p. 368-373.

Hall, R.D., and Jaworowski, C., 1999, Reinterpretation of the Cedar Ridge section, Wind River Range, Wyoming: Implications for the glacial chronology of the Rocky Mountains: Geological Society of America Bulletin, v. 111, p. 1233-1249.

Hall, R.D., and Shroba, R.R., 1995, Soil evidence for a glaciation intermediate between the Bull Lake and Pinedale glaciations at Fremont Lake, Wind River Range, Wyoming: Arctic and Alpine Research, v. 27, p. 89-98.

Harbor, J., and Wheeler, D.A., 1992, On the mathematical description of glaciated valley cross-sections: Earth surface Processes and Landforms, v. 17, p. 477-485.

Harbor, J., 1995, Development of glacial-valley cross sections under conditions of spatially variable resistance to erosion: Geomorphology, v. 14, p. 99-107.

Harden, J.W., 1982, A Quantitative index of soil development from field descriptions: Examples from a chronosequence in central California: Geoderma, v. 28, p. 1-28.

Izett, G.A., Pierce, K.L., Naeser, N.D., and Jaworowski, C., 1992, Isotopic Dating of Lava Creek B tephra in terrace deposits along the Wind River, Wyoming: Implications for post-0.6 Ma uplift of the Yellowstone hotspot: Geological Society of America Abstracts with Programs, v. 24, p. 102.

Love, J.D., 1960, Cenozoic sedimentation and crustal movement in Wyoming: American Journal of Science, v. 258A (Bradley volume), p. 204-213.

Love, J.D. (Leader), 1989, Yellowstone and Grand Teton National Parks and the Middle Rocky Mountains, Field Trip Guidebook: American Geophysical Union, Washington, D.C., 93 p.

Mahaney, W.C., 1978, late Quaternary Stratigraphy and soils in the Wind River Mountains, Western Wyoming: *in* Mahaney, W.C., ed., Quaternary Soils, GeoAbstracts, Norwich, UK, p. 223-264.

Mahaney, W.C., 1984a, Indian Basin Advance in Western and North-Central Wyoming: Northwest Science, v. 58, p. 94-102.

Mahaney, W.C., 1984b, superposed neoglacial and late Pinedale (Wisconsinan) tills, Titcomb Basin, Wind River Mountains, western Wyoming: Palaeogeography, Palaeoclimatology, Palaeoecology, v. 45, p. 149-163.

McAndrews, J. H., and Wright, H.E., Jr., 1969, Modern pollen rain across the Wyoming basins and the northern Great Plains (U.S.A.): Review of Palaeobotany and Palynology, v. 9, p. 17-43.

Mears, B., 1974, The evolution of the Rocky Mountain glacial model: *in* Coates, D.R., ed., Glacial Geomorphology: SUNY-Binghamton, NY, p. 11-40.

Miller, C.D., and Birkeland, P.W., 1974. Probable pre-Neoglacial age of the type Temple lake moraine, Wyoming: Discussion and additional relative-age data: Arctic and Alpine Research, v. 6, p. 301-306.

Moss, J.H., 1951a, Late Glacial Advances in the southern Wind River Mountains, Wyoming: American Journal of Science, v. 249, p. 865-883.

Moss, J.H., 1951b, Glaciation in the Wind River Mountains and its relation to early Man in the Eden Valley: *in* Eiseley, L.C., ed., Early Man in the Eden Valley, Museum Monograph No. 6, University of Pennsylvania, Philadelphia, PA, p. 9-94.

Murphy, J.F., and Richmond, G.M., 1965, Geologic map of Bull Lake West, Wyoming: U.S. Geological Survey Geologic Quadrangle Map GQ-432.

Picard, M.D., 1993, The early Mesozoic history of Wyoming, in Snoke, H., Steidtmann, J.R., Roberts, S. M., eds., Geology of Wyoming: Geological survey of Wyoming, Memoir #5, p. 210-248.

Phillips, F.M., Zreda, M.G., Gosse, J.C., Klein, J., Evenson, E.B., Hall, R.D., Chadwick, O.A., and Sharma, P., 1997, Cosmogenic ^{36}Cl and ^{10}Be ages of Quaternary glacial and fluvial deposits of the Wind River Range, Wyoming: Geological Society of America Bulletin, v. 109, p. 1453-1463.

Pierce, K.L, and Good, J.D., 1992, Field Guide to the Quaternary Geology of Jackson Hole, Wyoming: U.S. Geological survey, Open File Report 92-504, 54 p.

Richmond, G.M., 1957, Three pre-Wisconsinan glacial stages in the Rocky Mountain region: Geological Society of America Bulletin, v. 68, p. 239-62.

Richmond, G.M., 1962, Three pre-Bull Lake tills in the Wind River Mountains, Wyoming: U.S. Geological Survey Professional Paper #450-D, U.S. Government Printing Office, Washington D.C., p. D132-D136.

Richmond, G.M., 1964, Three pre-Bull Lake tills in the Wind River Mountains, Wyoming: A reinterpretation: U.S. Geological Survey Professional Paper 501-D, p. 104-109.

Richmond, G.M., 1965, Glaciation of the Rocky Mountains: *in* Wright, H.E., Jr., and Frey, D.G., eds., The Quaternary of the United States, Princeton University Press, NJ, p. 217-230.

Richmond, G.M., 1973, Geologic Map of the Fremont lake South quadrangle, Sublette County, Wyoming: U.S. Geological Survey, Geologic Quadrangle Map GQ-1138.

Richmond, G.M., 1983, Modification of glacial sequence along Big Sandy River, southern Wind River Range, Wyoming: Geological Society of America Abstracts, v. 15, no. 5, p. 431.

Richmond, G.M., 1986, Stratigraphy and correlation of glacial deposits of the Rocky Mountains, the Colorado Plateau, and the ranges of the Great Basin: *in* Sibrava, V., Bowen, D.Q., and Richmond, G.M., eds., Quaternary Glaciations of the Northern Hemisphere: Pergamon Press, Oxford, UK, p. 99-127.

Richmond, G.M., 1987, Type Pinedale Till in the Fremont lake area, Wind River Range, Wyoming: Geological Society of America Centennial Field Guide - Rocky Mountain Section: p. 201-204, Geological Society of America, Boulder, Colorado.

Richmond, G.M., and Murphy, J.F., 1965, Geologic map of Bull Lake East, Wyoming: U.S. Geological Survey, Geologic Quadrangle Map GQ-431, scale 1:24,000.

Richmond, G.M., and Murphy, J.F., 1989, Preliminary Quaternary geologic map of the Dinwoody Lake area, Fremont County, Wyoming: U.S. Geological Survey Open-File Report 89-435, 2 p.

Shroba, R.R., 1989, Physical Properties and laboratory data for soils formed in Pleistocene tills at Bull Lake, Dinwoody lakes, and Fremont lake, Fremont and Sublette counties, Wyoming: U.S. Geological Survey, Open-File Report 89-370, 14 p.

Shroba, R.R. and Birkeland, P.W., 1983, Trends in Late Quaternary Soil Development in the Rocky Mountains and Sierra Nevada of the western United States: *in* Wright, H.E., Jr., ed., Late Quaternary Environments of the United States, Volume I. The Late Pleistocene: University of Minnesota Press, Minneapolis, p. 145-156.

Soil Survey Staff, 1997, Keys to soil taxonomy (7th ed.): Pocahontas Press, Blacksburg, Va.

Sorenson, C.J., 1987, Soils Map of the Fremont Lake South Quadrangle, Sublette, County, Wyoming: U.S. Geological Survey, Miscellaneous Investigations Map I-1800.

Stone, J.O., 2000, Air pressure and cosmogenic isotope production: Journal of Geophysical Research, v. 105-B10, p. 23,753-23,760.

Quaternary and geomorphic processes and landforms along a traverse across northern New England

Authors and Leaders: **P.T. Davis,** *Dept. Natural Sciences, Bentley College, Waltham, MA 02452-4705*
P.R. Bierman, *Dept. Geology, Univ. of Vermont, Burlington, VT 05405-0122*
J. Brigham-Grette, *Dept. Geosciences, Univ. of Massachusetts, Amherst, MA 01003*
D.M. Fitzgerald, *Dept. Earth Sciences, Boston Univ., Boston, MA 02215*
B.K. Fowler, *North American Reserve, 67 Water St., Suite 207, Laconia, NH 03246-3300*
M.J. Retelle, *Dept. Geology, Bates College, Lewiston, ME 04240-6084*
J.C. Ridge, *Dept. Geology, Tufts Univ., Medford, MA 02155*
W.B. Thompson, *Maine Geological Survey, 22 State House Station, Augusta, ME 04333-0022*
T.K. Weddle, *Maine Geological Survey, 22 State House Station, Augusta, ME 04333-0022*

INTRODUCTION

This 7-day trip through northern New England begins and ends at Logan International Airport in Boston. We will examine raised glaciomarine deltas and related landforms, including moraines along the Maine coast, eskers in western Maine, landforms of alpine glaciers and continental ice sheets, along with mass-wasting deposits, in the White Mts. of New Hampshire and the Green Mts. of Vermont, and the famous varves of glacial Lake Hitchcock in the Connecticut Valley from northern New Hampshire and Vermont to central Massachusetts. We will debate the most recent chronologies derived from ^{14}C dating, varve counting, and cosmogenic exposure dating. We will also demonstrate GPR and useful field mapping techniques, and discuss various interpretations of glacial recession in New England, such as the morpho-sequence concept.

Because of the long lead-time required in the writing of this guidebook, some of the planned stops may no longer be available, so we may need to substitute some new sites. We are most grateful to landowners, too numerous to mention here, for permission to access their property to examine exposures of surficial deposits.

Some ages provided in this field trip guidebook are given in ^{14}C yrs. B.P. and others in calendar yrs. B.P.

DAY 1. DEGLACIAL AND MARINE DEPOSITS OF THE CASCO BAY LOWLAND - LOWER ANDROSCOGGIN VALLEY (Weddle and Retelle)

INTRODUCTION

The recession of the late Wisconsin, Casco Bay sublobe of the Laurentide Ice Sheet in southwestern Maine is represented by glaciomarine sediments deposited in an ice-marginal sea from a marine-based ice sheet, along with nearshore deposits associated with sea-level and postglacial isostatic rebound. Timing of late-glacial events and ice retreat is constrained by detailed mapping and stratigraphically-controlled, ^{14}C ages related to ice-marginal positions and marine-emergence deposits (Weddle et al., 1994; Weddle and Retelle, 1995, 1998). The altitudes of ice-marginal deltas and marine limit indicators, such as raised beaches deposited during ice retreat, provide a record of relative sea-level fluctuations and characterize the nature of postglacial uplift and emergence of the region (Thompson et al., 1989; Koteff et al., 1993; Barnhardt et al., 1995).

On DAY 1 (Fig. 1), we will examine deposits of late Wisconsin ice retreat and marine transgression/regression. Revised deglaciation and sea-level chronologies are based on recent detailed surficial mapping and stratigraphic studies that allow comparison with offshore records in Maine (Barnhardt et al., 1997; Schnitker et al., 2001) and onshore/offshore records of marine-based ice retreat in eastern Maine and adjacent Nova Scotia (Dorion et al., 2001; Stea et al., 2001). In addition, the Casco Bay chronology provides a link for comparison with terrestrial-based ice retreat through central New England and the White Mts. (Ridge et al., 1999, 2001; Thompson et al., 1999).

LATE WISCONSIN GLACIATION, DEGLACIATION, AND RELATIVE SEA-LEVEL

The southwestern margin of the Laurentide Ice Sheet in New England had a lobate geometry controlled by subglacial topography. The western Gulf of Maine (Fig. 1) was inundated by ice from the Charles-Merrimack lobe in southwestern Maine (whose southern extent formed the Cape Cod Bay lobe) and the South Channel lobe (Stone and Borns, 1986).

The Laurentide Ice Sheet reached its late Wisconsin maximum in the Gulf of Maine shortly after 22,000 ^{14}C yrs. B.P. (Stone and Borns, 1986; King, 1996). A regional deglaciation model proposes the existence of an ice stream in the Gulf of Maine, fed by inland ice that occupied lowland valleys (Hughes et al, 1985; Hughes, 1987). At its maximum extent, the ice stream fed a floating ice tongue over several of the deeper basins in the gulf. Based on seismic data and microfossil evidence in sediment cores, Belknap and Shipp (1991) proposed

Davis, P.T., Bierman, P.R., Brigham-Grette, J., Fitzgerald, D.M., Fowler, B.K., Retelle, M.J., Ridge, J.C., Thompson W.B., and Weddle, T.K., 2003, Quaternary and geomorphic processes and landforms along a traverse across northern New England, U.S.A.: *in* Easterbrook, D.J., ed., Quaternary Geology of the United States, INQUA 2003 Field Guide Volume, Desert Research Institute, Reno, NV, p. 365–398.

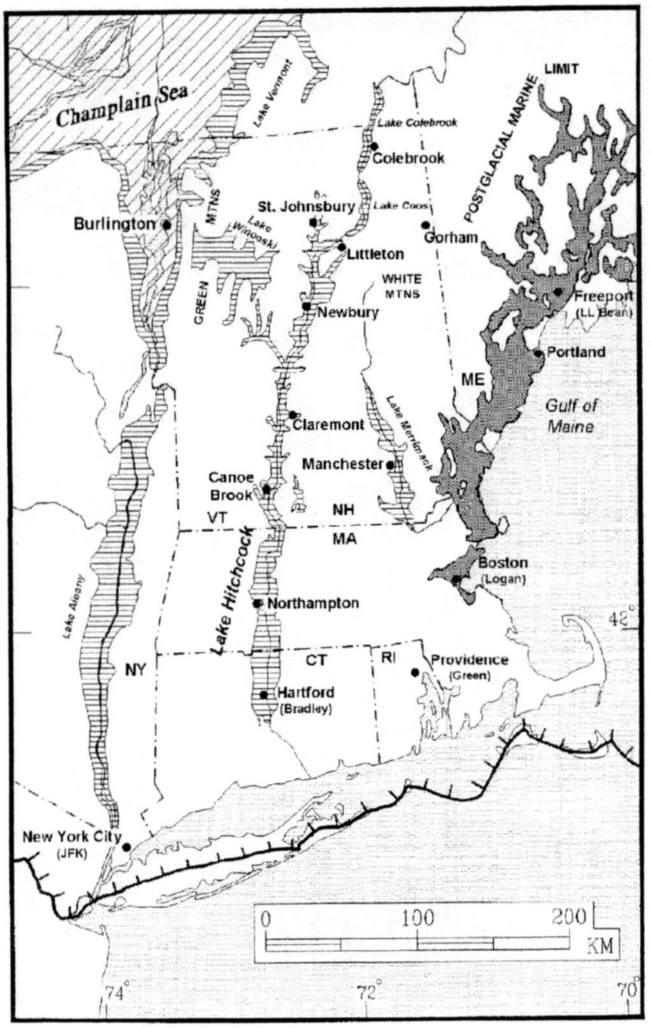

Figure 1. General location map of northern New England showing locations of larger regional airports, nightly accommodations and numbered stops for field trip, former glacial lakes (hachured), and areas that underwent postglacial marine submergence (shaded). Heavy hachured line shows LGM limit of Laurentide Ice Sheet (modified from Ridge, 2001).

that during initial deglaciation of the outer continental shelf, the ice shelf was pinned in several places on bedrock topographic highs in the Gulf of Maine.

Schnitker (1986) estimates that initial deglaciation in Wilkinson Basin in the central Gulf of Maine occurred prior to 17,600 ^{14}C yrs. B.P. To the northeast, Gipp and Piper (1989) and King (1996) place the deglaciation of the central Emerald Basin south of Nova Scotia between 17,700 and 18,000 ^{14}C yrs. B.P., although sediments interpreted as ice-proximal are found on the margin of the basin as late as 15,000 ^{14}C yrs. B.P.

Hughes et al. (1985) and Hughes (1987) proposed that a calving bay in the Gulf of Maine replaced ice streams and floating ice shelves by 14,000 ^{14}C yrs. B.P. On the inner continental shelf, Belknap and Shipp (1991) and Bacchus (1993) describe a glaciomarine unit that drapes over bedrock and is overlain by a horizontally-bedded glaciomarine mud. They interpret this stratigraphy to represent the shift from a floating ice margin to a calving embayment during deglaciation. As ice retreat progressed, calving along the floating margin eventually gave way to a grounded ice margin with calving along a tidewater cliff, similar to the conditions reported from glaciomarine tidewater settings in southeastern Alaska (Molnia, 1983; Phillips et al., 1991; Powell 1991).

Alternatively, Oldale (1989) suggests that ice was grounded at its maximum position in the western gulf and that the glaciomarine drift was deposited from point sources at the grounded tidewater terminus of the icesheet. Rapid deglaciation of the South Channel and Cape Cod Bay lobes may have by a calving front over Wilkinson Basin in the western Gulf of Maine (Stone and Peper, 1982).

Regardless of the mode of ice retreat on the continental shelf, numerous studies of glacial deposits on land in Maine and adjacent regions have demonstrated that moraines and grounding line fans (cf. Powell, 1991) are interbedded with glaciomarine sediments, inferring that deglaciation of the coastal lowland by a grounded tidewater glacier was contemporaneous with marine submergence of the region (Borns and Hagar, 1965; Stuiver and Borns, 1975; Thompson, 1982; Smith, 1985, Retelle and Bither, 1989, Ashley et al., 1991). Lowell and Borns (1994) have suggested that the effectiveness of a calving embayment in the major valleys across Maine was likely impeded by topography and shallow water depths of the transgressing sea, and that deglaciation may have been more influenced by climate at that time. By 14,000 to 12,000 ^{14}C yrs. B.P., the ice waned, and smaller sublobes of the retreating icesheet occupied the lowlands. We use the name Casco Bay sublobe for the lobate ice mass that occupied the Casco Bay Lowland.

Shoreline features, such as deltas and beaches, which define the local marine limit increase in elevation from the coast to the interior of Maine in response to the former increased isostatic load of late Wisconsin ice (Thompson et al., 1989). In the field area, the marine limit increases from about 75 m along the outer coast to over 100 m in the upper Androscoggin Valley (Thompson et al., 1989). The marine limit was established synchronously with ice retreat, so the oldest marine sediments transgressed the lowland from 14,000 to 13,000 ^{14}C yrs. B.P.

The marine transgression was followed quickly by regression of the synglacial sea as relative sea level dropped due to rapid glacio-isostatic uplift (Bloom, 1960, 1963; Borns and Hagar, 1965; Schnitker, 1974; Stuiver and Borns, 1975; Belknap et al., 1987; Crossen, 1984, 1991; Kelley et al., 1992; Barnhardt et al., 1995). Koteff et al. (1993) estimated that emergence in coastal central New England began sometime between 14,000 and 13,300 ^{14}C yrs. B.P. Smith (1985). Smith and Hunter (1989) estimate maximum marine inundation of the coastal lowland by 12,600 ^{14}C yrs. B.P. and emergence of the present landmass of southwestern Maine by 11,500 ^{14}C yrs. B.P. The postglacial low stand of the sea extended onto the inner continental shelf to a depth of -55 m by 10,500 ^{14}C yrs. B.P. (Barnhardt et al., 1997).

STOP 1-1. BRUNSWICK SAND PLAIN, BRUNSWICK, MAINE

At this stop, we will examine a section of nearshore bioturbated sediments overlain by fluvial deposits. These deposits are the upper part of the Brunswick sand plain, a regressive coastal delta (Retelle and Weddle, 2001; Weddle 2001).

The surficial geology in the Brunswick quadrangle records the advance and retreat of the Laurentide Ice Sheet, and subsequent Holocene events and deposits. The Brunswick sand plain is the most prominent deposit associated with this record (Fig. 2), with ^{14}C ages on marine fossils dating the formation of the sand plain to a time of worldwide rapid rise of sea level.

As ice sheets retreated at the end of the late Wisconsin and the Laurentide ice margin reached the present-day coast of Maine, the synglacial sea flooded the Gulf of Maine, making contact with the ice front. This transgression was in response to isostatic downwarping. As the ice melted, the depressed crust did not respond immediately to the release of weight from the ice, and as a result the sea flooded inland, up river valleys and across lowlands. In the Brunswick quadrangle, the marine limit is ~84 m above present sea level. During ice retreat and sea-level rise, meltwater discharging from the ice deposited sediment in the sea. Later, isostatic emergence of the land resulted in nearshore deposits associated with dropping, relative sea level.

The elevation of the sand plain is as much as 53 m below the highest elevation of older marine deposits, indicating that the sand plain is younger than the marine sediments. Exposures in the surface of the plain show trough-cross-bedded sand, typical of braided streams. Boring logs and geophysical data record the transition of late-glacial, isostatic emergence from marine to nearshore conditions and deposition of the Brunswick sand plain. Surficial sand of the plain overlies a sandy-silt zone that includes discrete sand units that dip gently eastward. Beneath the sandy deposits, thick glaciomarine mud overlies sand and till lying on bedrock.

Marine fossils just above the highest surface of the sand plain yielded an age of 12,800 ^{14}C yrs. B.P. Marine fossils from an offshore, seismically-identified sequence provide the youngest ages for the sand plain (12,200 ^{14}C yrs. B.P.; Oakley, 2001). The seismically identified units are likely the distal equivalent of the sand plain.

Marine fossil ^{14}C ages must be corrected for the ^{14}C marine reservoir effect (lag time for atmospheric ^{14}C to mix with sea water). For the Pleistocene Gulf of Maine, the correction is estimated at ~-600 to -700 years (Dorion et al., 2001). Thus, using 12,200 and 11,600 ^{14}C yrs. B.P. as corrected bracketing ages for the time of formation of the Brunswick sand plain, the time overlaps the period of worldwide sea level during meltwater pulse-1A, which occurred between 12,600 and 11,700 ^{14}C yrs. B.P. (Bard et al., 1990; Adkins et al., 1998). The data support the interpretation that the sand plain is a coastal delta formed as emergence continued, but during a period of relative sea-level stability, when the rate of local isostatic emergence kept pace with eustatic rise of sea level.

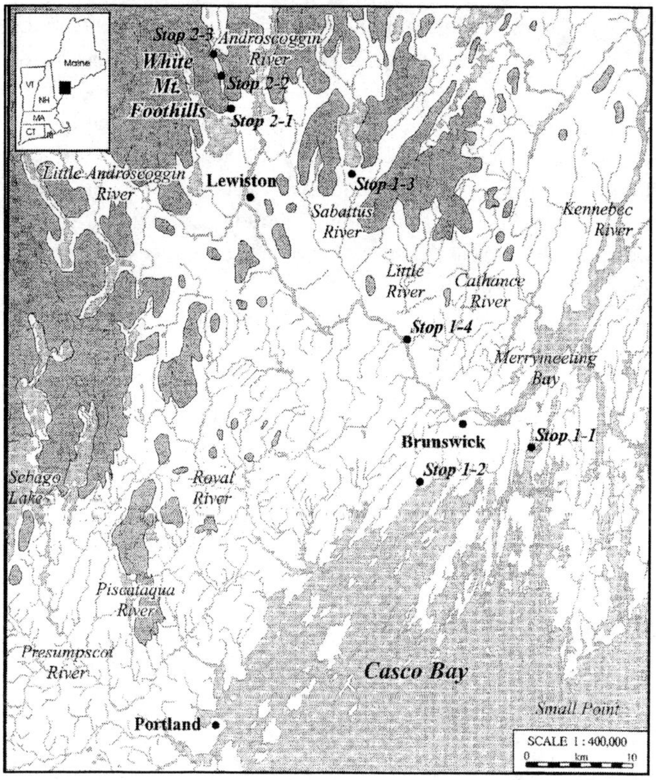

Figure 2. Field trip area for Days 1 and 2 (Stops 1.1 to 1.4 and 2-1 to 2-3). Gray shaded area is land above 100 m elevation above sea level (approximate elevation of local marine limit).

STOP 1-2. BUNGANUC BLUFF, BRUNSWICK, MAINE

This stop can be very muddy! Several large slump exposures along the coast reveal a section of glacial, glaciomarine, and nearshore deposits that represent changing environments in southwestern Maine (Fig. 2). Subglacial and ice-marginal sediments are present at the west end of the bluff, overlain by a thick, muddy, medial-to-distal glaciomarine fan or submarine plain in the central and eastern end of the section. In turn, nearshore deposits and an eolian cap overlie these. Belknap et al. (1987) and Belknap and Shipp (1991) described the section as an example of glaciomarine mud deposited in part under an ice shelf or at least near the ice-grounding line. Graded, rhythmically-bedded, fine sand, silt, and clay exposed here and identified elsewhere in subsurface boring data represent distal deposition with cyclical sorting of fine sediment. The lower deposits are most likely glaciomarine sediments, whereas the upper deposits are more likely marine and nearshore sediments. One question to consider at this stop is the influence of cyclical process on discharge, such as daily, seasonal, and tidal processes as described by Cowan and Powell (1990) and Phillips et al. (1991). A better understanding of the depositional environment at this exposure would allow comparisons with modern glaciomarine environments.

At the west end of this site, a very-poorly-exposed section consists of a thin veneer of sandy till (represented by numerous boulders at the base), massive sand, and glaciomarine mud, which overlie striated (170°) bedrock (Weddle, 1994). Eastward, near the top of the section is 2-3 m of medium-to-fine-grained sand, which contains rip-up mud clasts. The sand is found in erosional channels cut into underlying syn-depositionally-deformed, stratified sand, silt, and clay. In gradational contact with the upper sandy portion of the bluff, lower muddy sediments are best exposed in slumps at the eastern end of the section. These lower deposits consist of thin, well-stratified, laterally-continuous, silt and clay beds with silty, fine-sand laminations, grading upward into more massive mud. Dropstones are rare. At the east end of the section, striated (195° to 201°) bedrock is exposed at the waterline. Southwest-oriented striae are found along many of the peninsulas and islands in this part of Casco Bay and reflect the strong topographic control on late Wisconsin ice by the northeast trend of bedrock structure in the area. Seismic surveys show that depth to bedrock in the bay troughs east of the bluff is over 300 feet in places, the deep northeast-trending troughs preferentially controlling ice flow.

LUNCH STOP: Bradbury Mt. State Park, Pownal, Maine

STOP 1-3. DELTAS, FANS, AND MORAINES, SABATTUS, MAINE

Along the east side of Sabattus Lake (Fig. 2), we will examine fine examples of glaciomarine fans and deltas, and minor moraines (cf Ashley et al., 1991; Powell, 1991). The collective assemblage forms a beaded morphology (Sollid and Carlson, 1984; Caldwell et al., 1985) between two larger Gilbert-type deltas along the east side of Sabattus Pond. The two deltas and a series of smaller, elongate, lobe-shaped fans represent distinct ice-marginal positions where submarine outwash emanated from the ice-tunnel system during retreat of the ice. As many as four or five small moraines flank each of the beads along the east side of the valley. The moraine ridges have a general east-west alignment, are convex down-valley, and typically cluster in the valley, pinned to higher hillocks on the adjacent valley upland to the east. As many as 40 minor moraine ridges, with an average spacing of 50 to 100 m, are mapped in the 2-km-long valley east of Sabattus Pond (Bernotavicz, 1994). These moraines, which usually occur in clusters, are 3 to 10 m high, 30 to 50 m wide, and hundreds of meters long, but their internal composition is unknown.

The two deltas in the valley have different morphologies but share some noteworthy characteristics. The relatively flat surface elevation of each delta is about 100 m. Both have an elongate northern end, most prominent on the Marr Point delta. A ridge, traceable along Pleasant Hill Road, enters the north end of the Pleasant Hill delta just east of Round Pond (a kettle). These features represent the position of an ice-tunnel at or near the ice margin where sediment emanated from the tunnel to build the deposits. The Pleasant Hill delta is kidney-shaped, more lobate than the Marr Point delta, which is more wedge-shaped. The Pleasant Hill delta likely represents laterally migrating and coalescing delta or fan lobes along the ice margin. The Marr Point delta was apparently deposited with little or no lateral migration of a tunnel feeder source. Bernotavicz (1994) noted a change in orientation of end moraines near the Marr Point delta front and suggested that the shift from a roughly east-west trend (parallel to the ice margin) to a northwest-southeast trend reflects opening and enlargement of the ice tunnel, allowing lateral expansion of the Marr Point delta. During deglaciation, as the ice margin retreated in the Sabattus Pond lowland, the tunnel was pinned to the east at Sabattus Mt..

STOP 1-4. FAN DEPOSIT, MAINE DOT PIT, TOPSHAM, MAINE

The purpose of this stop is to see a Pleistocene, nearshore deposit associated with regression of the sea after uplift was underway and to discuss ^{14}C chronology in southwestern coastal Maine (Fig. 2). The deposit overlies glaciomarine mud, which rests on distal fan deposits of interbedded silt and fine sand. The larger exposure in the core of the fan immediately adjacent to the north shows the coarse nature of the proximal fan deposits associated with the ice tunnel from which the fan deposits originated. The flanks of the fan are mantled by massive glaciomarine mud from which the pit operator claims fossil shells were found. Bloom (1960) reported a varied assemblage of fossil shells from this pit, including *Hiatella arctica, Macoma calcarea, Musculus substriata, Mya arenaria, Mytilus edulis, Nuculana jacksoni, Serripes groenlandicus, Natica clausa, Neptunea decemcostata*, and *Balanus*, representing mixed intertidal and deep-water affinities. Molds of pelecypods have been found in the mud under the nearshore deposit, but fossils have not been observed recently in the pit. Retelle and Weddle (2001) report an age of 13,240 ± 190 ^{14}C yrs. B.P. (GX1632) on *Mesodesma arctatum* from the adjacent gravel pit to the north. In context with other earlier reported ages in the region (Attig, 1975; Stuiver and Borns, 1975), these ages bracket the marine regression between about 13,300 to 11,500 ^{14}C yrs. B.P.

DAY 2. DEGLACIATION FROM THE MARINE LIMIT TO THE WHITE MOUNTAINS, WESTERN MAINE AND EASTERN NEW HAMPSHIRE (Thompson)

On Day 2, we will start near Lewiston, Maine, close to the inland limit of late-glacial marine submergence and proceed northwestward across the foothills of the White Mts. (Fig. 2). This part of the trip will examine glaciomarine deltas, eskers, eolian deposits, and a group of moraines on the Maine-New Hampshire border.

Seaward from the marine limit, esker systems are punctuated by deltas and submarine fans where sediments washed into the sea at the mouths of ice tunnels. These deposits resemble the "DeGeer eskers" in Scandinavia. The contacts between topset and foreset beds in the marine deltas record sea level during deglaciation. Inland from the zone of marine submergence, many esker systems consist of long, variably segmented ridges that are locally associated with glacial lake deposits.

STOP 2-1. AUBURN PLAINS DELTA, AUBURN, MAINE

The gravel pit at STOP 2-1 (Fig. 2) is located in one of the many ice-contact glaciomarine deltas that formed as the Laurentide Ice Sheet receded from Maine's coastal lowland (Thompson et al., 1989). The Auburn Plains delta is associated with a segmented esker system that extends from Lewiston north to Sumner (Thompson and Borns, 1985).

Glaciomarine deltas in Maine are the Gilbert type, consisting of gravelly topset beds (fluvial delta-plain deposits) overlying sandy to gravelly foreset beds that accumulated on the prograding delta front. The contact between the topset and foreset beds marks the sea level to which the deltas were graded. Exposures at STOP 2-1 have shown up to ~2 m of poorly-sorted topset gravel with local cut-and-fill fluvial structures indicating generally southward flow. The foresets consist of well-stratified sand and gravelly sand beds that dip between northeast and south-southwest. The topset/foreset contact, surveyed at three locations in the pit, ranges in elevation from 106.4 to 107.0 m (349-351 ft). Coarse gravel, deep in the deposit as reported by the pit operator, may be a buried ice-tunnel facies.

Thompson et al. (1989) surveyed the elevations of topset/foreset contacts throughout southern Maine to determine the gradient of isostatic crustal tilt. Thompson and Koteff (2002) refined these elevation data and found that the deltas north of Sanford lie on a plane that slopes SSE at 0.50 m/km. This is a minimum value for tilt because the deltas were deposited diachronously as the ice withdrew and relative sea level fell. Fig. 3 shows the elevation profile of both the unmodified high-stand deltas (including Auburn Plains) and those that were eroded during marine regression. The lowest erosion surfaces (Fig. 3) define a stillstand position of relative sea level, which has been subsequently tilted with a gradient similar to that of the marine-limit plane.

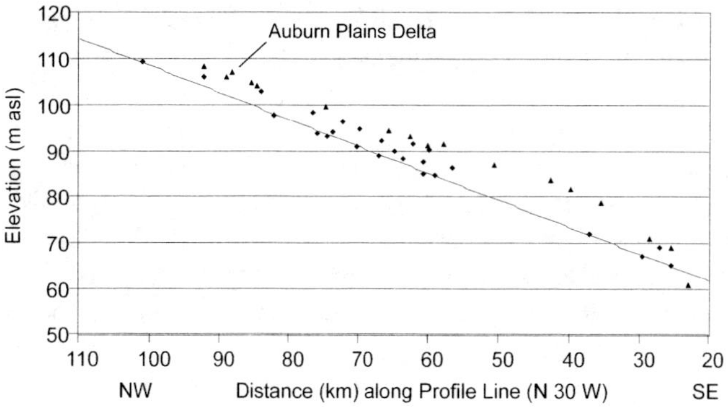

Figure 3. Glaciomarine delta elevations in southwestern Maine. Triangles indicate topset/foreset contacts of marine-limit (highstand) deltas. Diamonds show lower relative sea levels (RSL) marked by fluvial downcutting and shoreline erosion of delta surfaces. Linear regression line ($r^2 = 0.98$) indicates a strandline formed during stillstand of RSL (see Thompson and Koteff, 2002).

STOP 2-2. WOOD STREET DELTA, TURNER, MAINE

The Wood Street pit (Fig. 2) is located in the same cluster of marine deltas as STOP 2-1. Contours on the Lake Auburn West quadrangle suggest that the top of the Wood Street delta has an elevation similar to Auburn Plains. This locality is in a narrow delta splay that extends westward from the adjacent esker system. The foresets consist of well-stratified sand and pebbly sand beds dipping between west-southwest and northwest. In late 2002, the upper part of the pit face showed several meters of foreset beds (10-100 cm) with small-scale cross bedding. At the base of this sequence, convex-upward beds formed a climbing dune. These structures, as well as high-energy current ripples elsewhere in the delta foresets, indicate traction currents that are unusual in the glaciomarine deltas (where flow separation occurred as fresh water spread over the denser underlying sea water). They suggest estuarine conditions with a high input of glacial meltwater in the vicinity of the ice tunnel mouth. The delta foresets in the northern part of the pit are locally overlain by ~2 m of thin planar sand beds, which are probably eolian.

STOP 2-3. ESKER, TURNER, MAINE

The Turner town gravel pit will be visited if fresh exposures are present at the time of the field trip (Fig. 2). The working face is located in the north end of the long pit that follows the esker. Exposures reveal an east-west cross section through the esker ridge, which is about 12 m high and reaches an elevation of about 110 m (360 ft, essentially the local marine limit). Material in the pit face appears to be mostly pebble-boulder gravel, but a better exposure would likely show that the ridge consists of a coarser tunnel facies overlain by fan deposits, as is common in other eskers in the glaciomarine environment.

LUNCH STOP: Snow Falls Gorge, Little Androscoggin River, West Paris, Maine

STOP 2-4. ESKER, BETHEL, MAINE

The Hilltop pit is located in one of the major esker systems in Maine, named the Portland system by Stone (1899). This system extends discontinuously from the Quebec border southward through several large lake basins, then through a gap in the high mountains of western Maine, across the foothills, and down the Little Androscoggin Valley to New Gloucester (Thompson and Borns, 1985). A series of glaciomarine deltas formed in association with the esker as the ice margin retreated from New Gloucester to the imarine limit at South Paris (Fig. 2).

Much has been speculated about the origin of Maine's eskers and the glacial drainage networks in which they formed. Several investigators have proposed that the ice tunnels existed continuously along their entire lengths, rather than developing in shorter segments as the ice margin receded. Arguments in favor of simultaneous, full-length tunnels include (1) the well-integrated, dendritic patterns of the esker systems over their entire lengths, (2) their tendency to cut across valleys and climb over hills in response to the gradient of the ice surface, and (3) the great volumes of sediment in marine deltas at the ends of some of the systems (Shreve, 1985; Ashley et al., 1991; Boothroyd, 1995; Hooke, 2000). However, some authors propose that the deposits comprising the eskers formed near the ice margin in a south-to-north progression as the tunnel mouths receded (e.g. Ashley et al., 1991; Weddle et al., 1994). Sediment probably was delivered to the tunnels by convergent flow of adjacent glacial ice toward the tunnel axes, though field evidence of this process remains poorly documented in Maine.

The esker seen here is a narrow and nearly continuous ridge of sand and gravel along Route 232. It leaves the Androscoggin Valley and ascends the gently north-sloping valley of Barkers Brook. Stone (1899) showed a view of the esker from a nearby hill (Fig. 4). A parallel esker occurs near here on the west side of the Barkers Brook valley, and also a short distance south on Route 232. These sections of the system are examples of what have been called "multiple-crested eskers" (Shreve, 1985) or "esker nets" (Hooke, 2000). Shreve noted that they occur "in gently ascending reaches of esker paths," which is the case here, while Hooke relates esker nets to "places where the ice is so thin that tunnel closure rates cannot keep up with melt rates...".

STOP 2-5. SAND DUNE, BETHEL, MAINE

This stop is on East Bethel Road, on the south side of the Androscoggin Valley. A small barrow pit at the base of the hillslope exposes ~5 m of well-stratified sand with thin planar beds. This deposit is interpreted as eolian on the basis of its morphology, good sorting, lack of stones, and absence of sedimentary structures characteristic of waterlaid deposits.

Figure 4. Illustration from Stone (1899), looking south along the esker in Barkers Brook valley from viewpoint near STOP 2-4.

Dune deposits are common on the downwind (south and east) sides of major valleys in southwestern Maine. The dunes vary in age from late-glacial to those that have been reactivated in historic time. Dunes may be parabolic or longitudinal ridges, but commonly are irregular mounds and sheets that range in thickness from 1 to 10 m or more. Till-stone ventifacts are found at the contacts of the eolian sand with underlying till or gravel. In some places, pebbles and cobbles are found within the dunes, but how they got there is uncertain.

STOP 2-6. ANDROSCOGGIN MORAINE COMPLEX, SHELBURNE, N.H.

The Androscoggin moraine complex is located on the Maine -New Hampshire border in the Androscoggin Valley (Figs. 5, 6). Stone (1880, 1899) first recognized these moraines and striations indicating local eastward ice flow. Thompson and Fowler (1984) discovered additional moraine segments on both sides of the valley. The ridges are sharp-crested, very bouldery, and up to 30 m high. The latter investigators dug backhoe test pits at the five locations (Fig. 5). Four pits east of Stock Farm Mt. showed glacial diamictons containing lenses of sand, silt, and gravel. Pit 5, near the valley axis in the lowest part of the moraine complex, revealed a waterlaid diamicton with interbedded silt, fine sand, and flowtill(?).

Stone counts from the test pits indicated that local rock types are predominant. However, basaltic and rhyolitic dike rocks totaled 16 percent on the north side of the valley (Pit 5), while samples from the south side contained only 0-2 percent of these lithologies. Bedrock mapping by Billings and Fowler-Billings (1975) shows dike swarms on the north side of the Androscoggin Valley at Gorham and farther north in the Berlin area (Fig. 6). Thompson and Fowler (1989) concluded that the northern side of the Androscoggin ice tongue in the Shelburne area would have incorporated more material from these dikes, thus explaining the variation in stone counts.

The age of the moraine complex has not been determined. Borns et al. (2003) suggest an age of about 12,800 ^{14}C yrs. B.P. Whether the moraines represent a readvance down the valley or just a stillstand during progressive ice retreat is uncertain. The moraines may have formed in response to funneling of ice flow along the valley after the late Wisconsin icesheet thinned over the crest of the adjacent northeast-trending mountain chain.

DAY 3. GLACIAL GEOLOGY OF MT. WASHINGTON AND SURROUNDING "NOTCHES," NORTHERN WHITES MOUNTAINS (Davis and Fowler)

Glacial geology of the Presidential Range

The name Goldthwait is synonymous with the glacial history of the Presidential Range (Figs. 6, 7). J.W. Goldthwait (1913, 1916a) was the first to carry out an extensive study of glaciation in the range, where he reached three major conclusions: (1) the

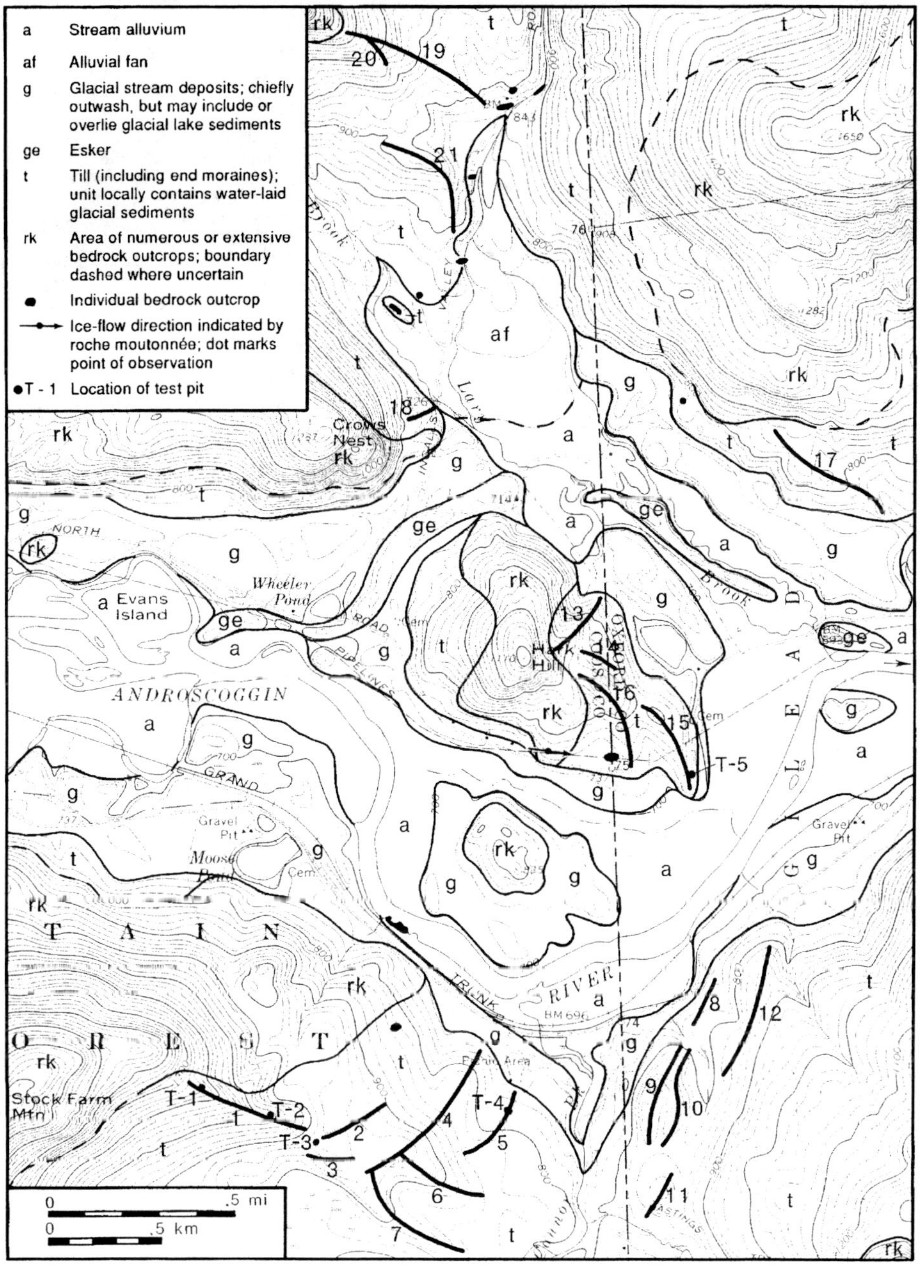

Figure 5. The Androscoggin moraine system (from Thompson and Fowler, 1989).

uplands above the cirques were eroded by both fluvial and glacial activity, (2) the cirques were carved by alpine glaciers, as opposed to continental ice, stream erosion, or frost action, and (3) continental glaciation followed the last cirque glacier activity on the range (Davis, 1999). His evidence that cirque glaciers were not active following continental glaciation included: (1) the lack of looped end moraines on cirque floors, (2) till of a northern provenance on cirque floors, and (3) asymmetric cirque cross-valley profiles. Goldthwait (1913) did not support the concept that local glaciers extended far down valleys from an icecap centered on the Presidential Range, as proposed by Packard (1867), Vose (1868), Agassiz (1870), and Hitchcock (1878).

Over the next two decades, two workers disputed the conclusions of J.W. Goldthwait concerning the timing of continental and cirque glaciation in the Presidential Range. Johnson (1933) suggested the lack of end moraines in cirques is not sufficient evidence to conclude that continental ice post-dated cirque glacier activity in the Presidential Range, as he noted other alpine areas in the world that have never undergone continental glaciation but have cirques that lack moraines. Antevs (1932) sided with Johnson, concluding that late Wisconsin cirque glaciers existed in the Presidential Range; however, neither author provided a convincing explanation for the till of northern provenance on the cirque floors.

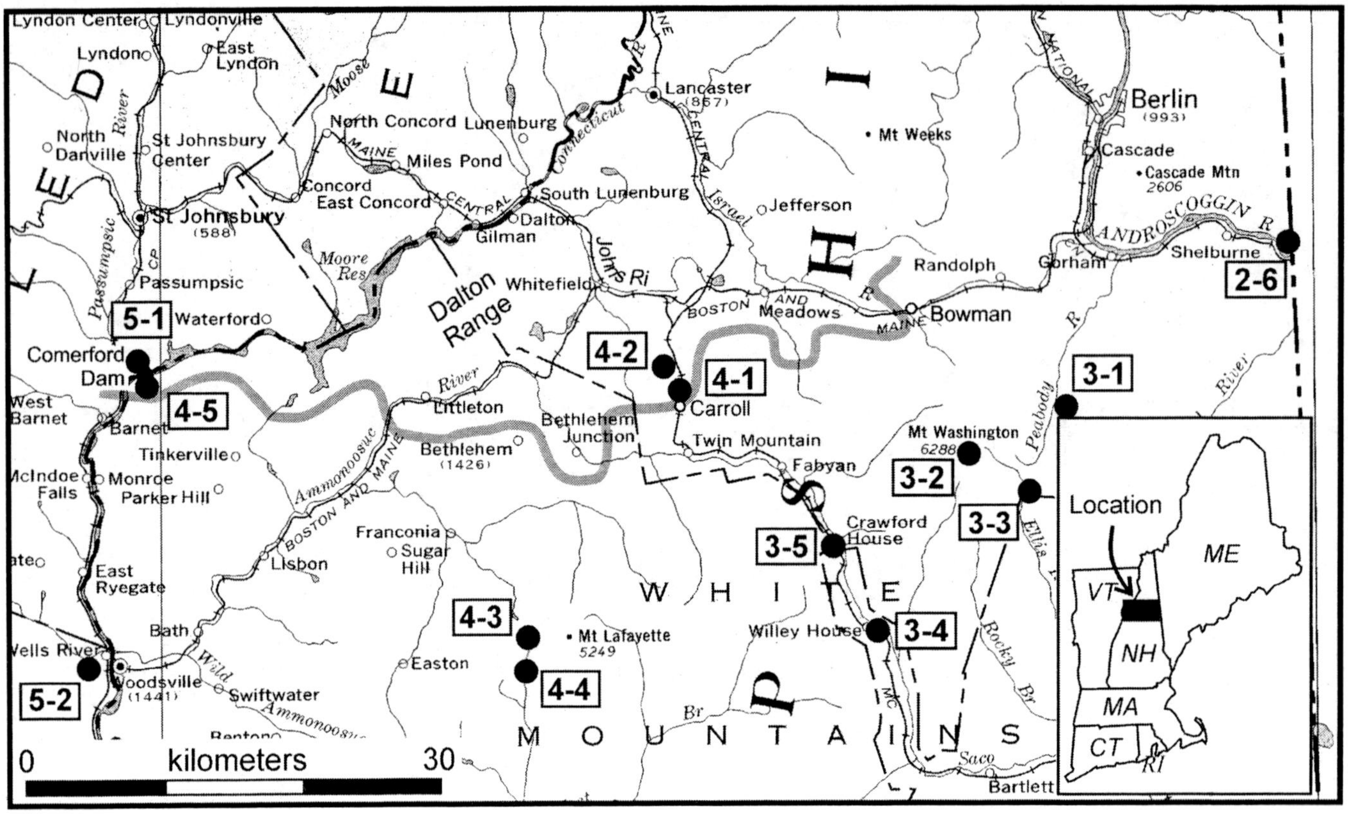

Figure 6. White Mt. area showing many of the stops for Days 2 through 5.

R.P. Goldthwait (1938, 1939, 1940, 1970a,b) carried on his father's interest in the glacial history of the Presidential Range. In his 1939 and 1940 publications, he not only noted the observations of his father in support of cirque glacier activity only preceding the last overriding by continental ice, but he also observed roche moutonnees on cirque floors along with striae and grooves on cirque headwalls, which he believed could only have been formed by continental ice (Davis, 1999). Also, in his latter paper, he presented morphometric data on cirques and altitudinal estimates of firn lines for the former cirque glaciers in the Presidential Range (Fig. 7). From these data, he calculated that, depending on the amount of winter precipitation, a 5 to 10°C mean summer temperature lowering would be necessary to support cirque glaciers in the Presidential Range today.

During the late 1950's, W.F. Thompson (1960, 1961) analyzed aerial photographs of the Presidential Range and refuted the Goldthwaits' view by arguing that the steep headwalls and sharp arêtes were indicative of active cirque glaciers following continental ice sheet deglaciation. Although Thompson (1960, 1961) did not present field data to support his view, he believed that moraines of cirque glaciers had been obliterated by post-glacial mass wasting processes (Davis, 1999). Work in Tuckerman Ravine by D.J. Thompson (1999) suggests that a deposit consisting of large blocks believed to be a moraine by Antevs (1932) is a relict, tongue-shaped, rock glacier unrelated to cirque glacier activity (Fig. 7).

Bradley (1981) challenged the Goldthwaits' view of the timing for cirque glaciation in the Presidential Range by noting that large boulders and diamicts at the mouths of north-facing cirques were composed of lithologies derived from bedrock in the cirques. However, Gerath and Fowler (1982), Fowler (1984), Gerath et al. (1985), and Waitt and Davis (1988) examined the diamicts at the cirque mouths and concluded that the sediments are not till, but rather debris flow deposits.

Opponents of post-ice sheet local ice support their views that continental ice was last with the following evidence: (1) roches moutonnees indicating upvalley ice flow on cirque floors, (2) striae trending obliquely across cirque headwalls, (3) ice sheet erratics in cirque–floor drift, and (4) apparent absence of moraines in or downvalley from cirques. Proponents counter that cirque glaciers did not completely remove ice sheet drift and that moraines in cirques are obscure because they are small, subdued by post-glacial mass wasting, or thickly forested. An important theoretical argument against post-ice sheet local ice in the Presidential Range is that, based on oxygen isotope and other proxies of late and immediate postglacial climate, such as pollen records, equilibrium-line altitudes rose rapidly to elevations well above cirque floors. Thus, cirque glaciers could not likely have been developed or sustained in the cirques following recession of continental ice.

Opportunities for developing a radiocarbon chronology for the deglaciation of cirques are limited because of the small num-

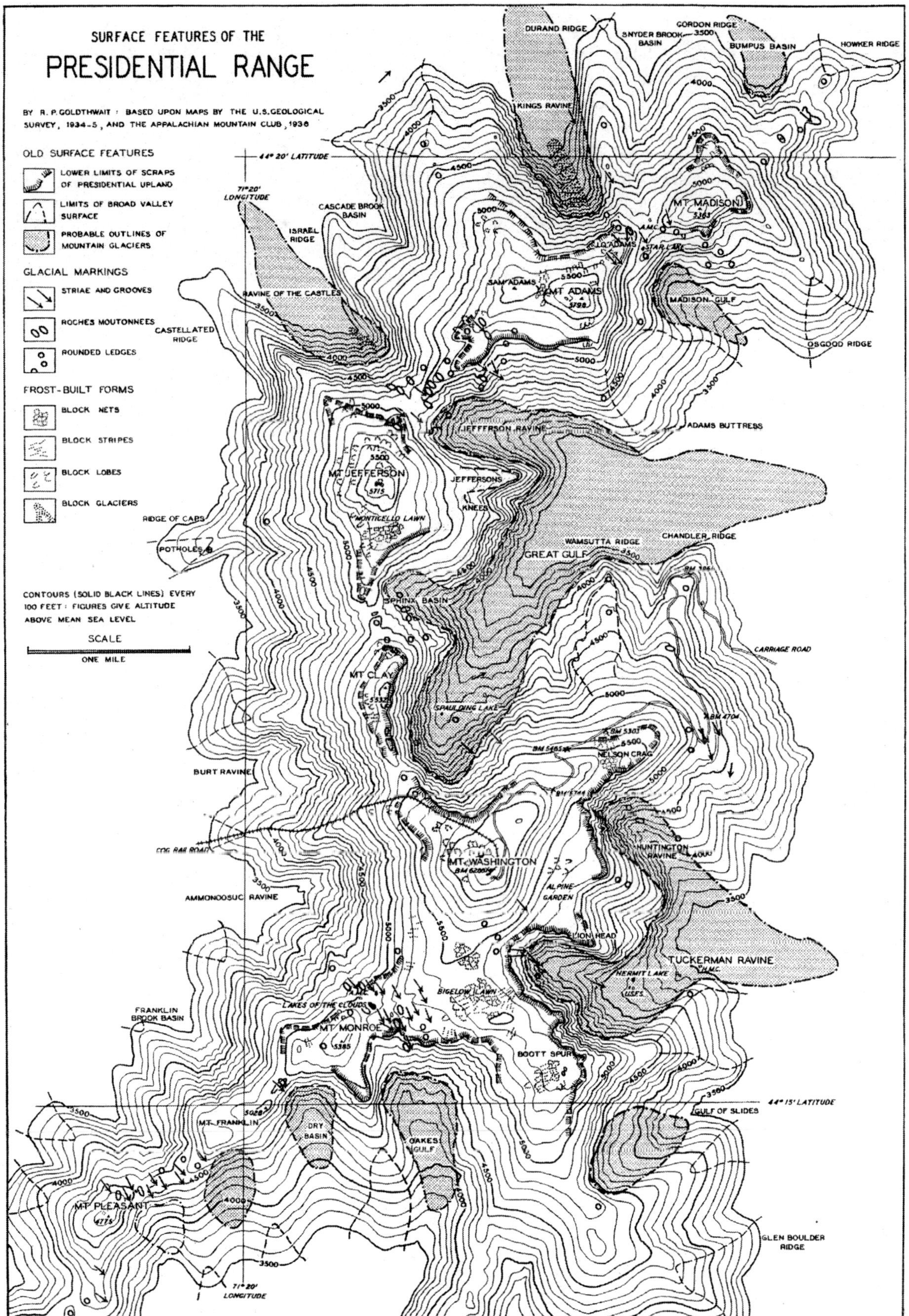

Figure 7. Surface features of the Presidential Range, including outlines of former cirque glaciers (from Goldthwait, 1940).

ber of tarns distributed across the range. Spaulding Lake in the Great Gulf and Hermit Lake in Tuckerman Ravine, although shallow, may provide useful continuous postglacial records of sediment accumulation and should be cored (Fig. 7).

Ponds near the cirques in the Presidential Range have provided minimum ^{14}C ages for continental ice retreat (Davis et al., 1980; Spear, 1989; Spear et al., 1994). Organic material from sediment at the base of a core retrieved from Lost Pond at an elevation 650 m in Pinkham Notch on the east side of the Presidential Range provided an age of 12,870 ± 370 ^{14}C yrs. B.P. (QL-985) (Spear et al., 1994). Organic material from sediments near the base of cores taken from the lower of the two Lakes of the Clouds at an elevation of 1542 m in the alpine zone between Mts. Monroe and Washington yielded an age of 11,530 ± 420 ^{14}C yrs. B.P. (I-10684) (Spear, 1989). Pollen data from sediments below the ^{14}C-dated level in the lower Lakes of the Clouds site correlate with the tundra pollen zone from Deer Lake Bog at an elevation of 1300 m on Mt. Moosilauke, which provides an age of 13,000 ± 400 ^{14}C yrs. B.P. (QL-1133) (Davis et al., 1980).

Given the model that continental ice thinned, separated, and retreated northward from the mountains of northern New England during late Wisconsin deglaciation (Goldthwait and Mickelson, 1982; Hughes et al., 1985; Stone and Borns, 1986; Borns, 1987; Davis and Jacobson, 1985; Thompson and Fowler, 1989), this entire process appears to have been very rapid. If these ^{14}C ages are taken at face value, they require almost 900 m of continental ice thinning in less than a few hundred years, a circumstance requiring very rapid warming of the local climate, warming too rapid to support local cirque glaciers. These ideas on the surficial geology of the White Mts. have been discussed on previous society field trips in the area (Davis et al., 1988; Davis et al., 1993; Davis et al., 1996; Allen et al., 2001).

Current work designed to refine the deglaciation chronology for the Presidential Range uses cosmogenic radionuclides ^{10}Be and ^{26}Al produced in quartz from boulders and bedrock. These exposure dating techniques (Bierman, 1994; Gosse and Phillips, 2001) may not provide the temporal resolution of AMS ^{14}C dating, but the method directly dates moraines and allows samples to be collected from sites where ^{14}C-datable materials are not available. As a test of the thinning continental ice model for deglaciation of the Presidential Range, a suite of bedrock and boulder samples with quartz veins were collected on an altitudinal transect from the summit of Mt. Washington to the floor of Pinkham Notch near Lost Pond for cosmogenic nuclide dating. The abundance of ^{10}Be and ^{26}Al in frost-riven bedrock samples from near the summit area of Mt. Washington is much higher (1.5 to 8 times) than expected had the peaks been covered by active, erosive ice during the late Wisconsin maximum (Bierman and Davis, 2000). Two summit samples provide ^{10}Be ages of 124,000 and 22,000 years exposure since deglaciation or the last active ice erosion. In contrast, a sample from one of the large boulders on the tongue-shaped rock glacier on the floor of Tuckerman Ravine (D.J. Thompson, 1999) appears to be late-glacial (Bierman and Davis, 2000). The mountain-top samples are consistent with two different scenarios, both of which have significant implications for understanding the spatial and temporal patterns of glaciation and glacial erosion in northern New England: (1) late Wisconsin continental ice was thinner than previously supposed leaving the tops of Mt. Washington's summit exposed since marine isotope stage 6, or (2) the summit was covered by glacial ice during the late Wisconsin, but the ice was thin enough to be frozen to its bed (<100 meters thick). Thus, the cold-based ice was unable to erode much rock, allowing cosmogenic nuclides to be inherited from prior periods of exposure. In either case, late Wisconsin continental ice in New England was thinner than previously believed, consistent with low, basal-shear stresses and/or the drawdown of continental ice by active ice streams flowing through the notches ("notch" is a local name for glacial U-shaped valley).

STOP 3-1. CIRQUES OF THE NORTHERN PRESIDENTIAL RANGE, PINKHAM NOTCH, N.H.

Mt. Washington (elev. 1950 m; 6450 ft.) is the highest peak in the Presidential Range and America's highest peak east of the Mississippi River and north of the Carolinas (Fig. 7). Mt. Washington also boasts the world's worst weather, its famous observatory having recorded the world's highest wind gust of 231 mph in 1934, along with other extremes of cloudiness, temperature, and precipitation. Pinkham Notch is located at ~650 m along the eastern base of Mt. Washington. Weather cooperating, the base of the Auto Road provides a superb panoramic view of the large east-facing cirques of the Presidential Range (Fig. 7). Glacially-eroded, bedrock forms that appear to be roches moutonnees occur nearby, but what do they indicate about ice-flow direction at this location?

Pinkham, Crawford, and Franconia Notches, from east to west, are the largest examples of glacial troughs in the White Mts., and all three are accessible from highways. Other smaller, but no less spectacular, notches in the White Mts. include, from east to west, Grafton, Mahoosuc, Carter, Carrigain, Zealand, and Kinsman Notches, some of which are only accessible by hiking trails. Each of these notches was a low-elevation mountain pass prior to the start of the multiple episodes of Pleistocene continental glaciation. The three notches that we will visit on this trip are substantially deeper than other higher elevation notches in the White Mts. because glacial ice moved through them for a longer period of time and/or was more erosive during each glacial episode. Also, the notches were likely locations of ice streams within ice sheets when the White Mts. were completely inundated by ice.

Looking northwest of the White Mts. highland area, continental glacial ice had few topographic obstacles interfering with its movement southeastward until it got to the steep northwesterly slopes of the White Mts. Because continental ice sought the path of least resistance through the mountains, it converged first into the deeper pre-existing passes through the mountains and continued to flow through them for the entire duration of each glacial episode. As continental ice thickened during each glacial episode, it began to move through the higher–elevation passes and eventually covered all the peaks at least once.

We know that all of the peaks here were over-ridden at least once by continental ice because lodgment till has been found near the summit of Mt. Washington at an elevation of 1,900 m (6270 ft.). Most workers agree that the last pulse of the late-Wisconsin Laurentide Ice Sheet arrived here about 25,000 ^{14}C yrs. B.P., reached its peak about 18,000 ^{14}C yrs. B.P., and had fully retreated from lowlands in the area by about 12,000 ^{14}C yrs. B.P. Thompson et al. (1999) have provided a detailed review of ^{14}C ages for the region north and northwest of the notches. However, because many of these ^{14}C ages are on bulk disseminated organic material, Ridge et al. (2001) question their accuracy, and based on an AMS ^{14}C-controlled varve chronology propose that deglaciation in the White Mts. was hundreds of years younger than the oldest ages summarized by Thompson et al. (1999).

On face value the ^{14}C ages from bog and pond bottoms in the White Mts. might suggest that deglaciation and organic sediment accumulation occurred earlier at lower elevations (ex. Lost Pond) than at higher elevations (ex. Lakes of the Clouds). However, palynological data (Davis et al., 1980; Spear, 1989; Spear et al., 1994; Cwynar and Spear, 2001) suggest that postglacial revegetation and organic sediment accumulation in bogs and ponds occurred at nearly all elevations at roughly the same time. Hence, the pollen data suggest that regional climate warmed rapidly during the early post-glacial, which supports the argument that glacial equilibrium-line altitudes rose too quickly to allow cirque glaciers to exist during the post-glacial.

STOP 3-2. SUMMIT, MT. WASHINGTON, N.H., LUNCH STOP

We will drive the 19 km (7.6 mi) Auto Road to the summit of Mt. Washington, making one or two stops on the way up to afford views of the Great Gulf cirque complex on the eastern side of the Northern Presidential Range (Fig. 7). On the summit, we will tour the Mt. Washington Observatory, the longest continuously operating mountain summit observatory on Earth. We will also examine sites where bedrock and frost-rived blocks sampled for cosmogenic nuclide exposure dating suggest that the summit was not overridden by erosive glacial ice during the last glacial maximum (LGM).

STOP 3-3. GLEN ELLIS FALLS, PINKHAM NOTCH, N.H.

A plaque above the falls titled "Geology of Glen Ellis Falls" reads "The Ellis River which had flowed uninterrupted during pre-glacial times was forced by violence and the struggle of the landmasses to plunge over the headwall of a glacial cirque." Not true! No cirque occurs here. The Cutler River, which feeds the falls, only recently adopted its present course (Fig. 6). Glen Ellis Falls is here because a major silicified zone, likely a Mesozoic fault with only a few meters of offset, juxtaposes the calc-silicate Wildcat granite or Rangeley Fm. migmatite against the Smalls Falls Fm. (D. Eusden, oral communicatioin, 2002). The 3-m-wide zone of massive quartz with many cross-cutting quartz veins is quite resistant and forms the escarpment over which the falls plunge. The silicified zone can be followed for about a mile in total, up and downstream from the falls.

STOP 3-4. WILLEY HOUSE SITE, CRAWFORD NOTCH, N.H.

Both walls of Crawford Notch (Mt. Webster on the east side and the Willey Range on the west side) are heavily scarred by landslide tracks (Figs. 6, 8). The largest landslide scar on the west side of the notch occurs on Mt. Willey, the site of the infamous Willey slide that occurred at 3 a.m. on Monday, 28 August 1826. As the story goes, upon hearing the roar of the slide, all family members ran out of their house, only to be obliterated while the house survived. Apparently a large boulder sitting directly upslope split the slide into two paths around the house. Following a hot and dry summer, landslides occurred all over the White Mts. during this two–day storm in 1926 (Silliman et al., 1829; Ramsey, 1988).

STOP 3-5. CRAWFORD HOUSE SITE AND "THE GATEWAY," HEAD OF CRAWFORD NOTCH, N.H.

We will walk east along the railroad tracks to a spectacular view down Crawford Notch. The steep slope at the head of the notch is not a cirque headwall at this low altitude, just as the precipice at Glen Ellis Falls in Pinkham Notch is not a cirque headwall (Figs. 6, 8). This valley's nick point probably predates Quaternary glaciation, and its presence here may result form harder bedrock types, such as the hornfels of Silurian Rangeley gneiss in the contact aureole of an intrusion of the Jurassic Conway granite. The darker crest of the ridge on Mt. Webster on the east side of the valley is also composed of the Rangeley and marks the roof of the Conway granite intrusion. This contact is exposed at road level on US 302 about 0.9 mi. east from the head of the Notch.

The head of Crawford Notch also was a spillway for glacial Lake Crawford to the west, which was contemporaneous with glacial Lake Pigwacket in the Conway area south of Crawford Notch (Thompson, 1997). The voluminous glacial meltwater that flowed down Crawford Notch during late-glacial time was orders of magnitude greater than postglacial runoff, and was responsible for the smaller V-shaped incision superposed on the classic U-shaped trough.

DAY 4. GLACIAL GEOLOGY OF THE LOWLANDS WEST OF MT. WASHINGTON AND FRANCONIA NOTCH, N.H. (Thompson, Davis, Fowler, and Ridge)

BETHLEHEM MORAINE COMPLEX AND ICE RECESSION IN THE NORTHWESTERN WHITE MTS., N.H. AND VT.

As originally mapped by J.W. Goldthwait (1916a), the Bethlehem moraine consists of dozens of hummocks and ridges in the Ammonoosuc River basin from Littleton east to Bethlehem (Fig. 6). The moraine ridges trend east to northeast, range in height from 3 m to over 30 m, and are up to 1300 m long. Few

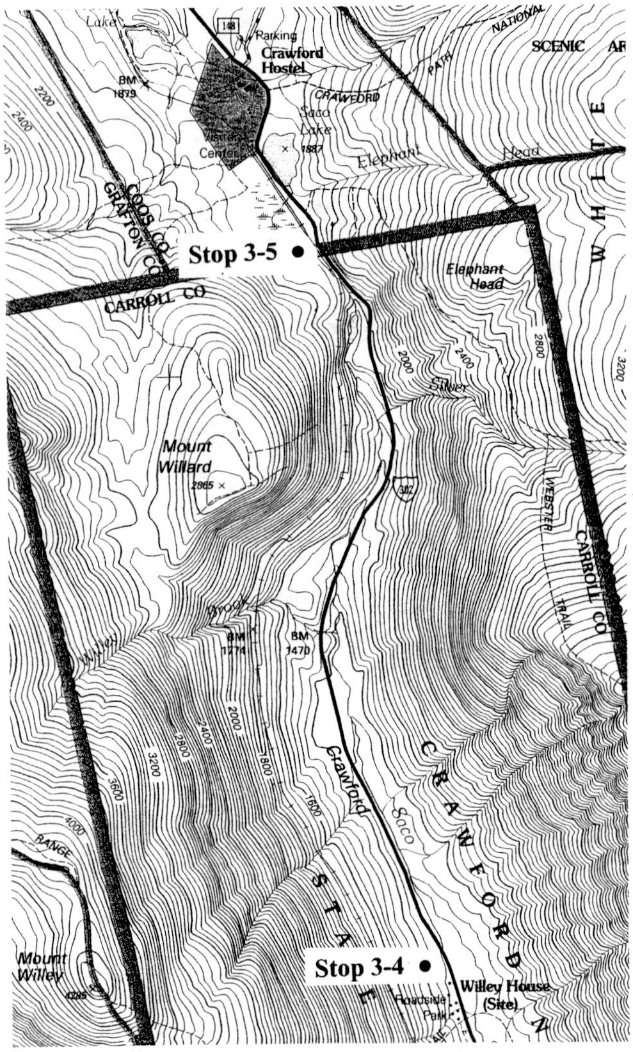

Figure 8. Part of the Crawford Notch, N.H., 1:24,000 quadrangle, showing locations for Stops 3-4 and 3-5 and contours indicating classic U-shape valley. Crawford Notch State Park boundary shown with shaded line.

bedrock exposures occur within the Bethlehem moraine complex, but striated outcrops immediately to the north and south indicate ice flow directions of 175-185° (Goldthwait, 1916a; W.B. Thompson, unpublished data). Surficial deposits in the moraine complex are generally thick (up to at least 50 m). The moraine ridges are composed of silty, sandy till and minor waterlaid sediments. Large boulders of granite and gneissic granite are abundant on the moraine surfaces. The elevations of the ridges suggest that they were deposited in glacial Lake Ammonoosuc, an interpretation supported by deformed glaciolacustrine sediments within the moraines.

The moraines of the Bethlehem area have been the focus of major controversies involving the modes of glaciation and ice retreat in the White Mts. (Thompson, 1999). Early workers accepted them as moraines deposited by active ice, but in the mid-1900s, ice-stagnation proponents opposed this concept. Thompson et al. (1999) re-examined the Bethlehem moraine complex and correlative deposits to the east. They concluded that the numerous subparallel till ridges are in fact end moraines deposited by oscillatory northward retreat of the ice margin during late Wisconsin deglaciation.

Thompson et al. (1996, 1999, 2002) compiled ^{14}C ages from ponds in the White Mts. and adjacent areas. Most ages are from the lower or basal portions of pond sediment cores and thus limit the time of deglaciation of each site. Pond of Safety, in the mountains just north of the Presidential Range, is favorably located to provide a minimum age for deposition of the Androscoggin moraine to the east and a maximum age for the Bethlehem moraine complex to the west. The basal age from Pond of Safety is 12,450 ± 60 ^{14}C yrs. B.P. (OS-7125), which is compatible with other limiting ages from northern N.H., as well as an age of 12,250 ± 55 ^{14}C yrs. B.P. (OS-7119) from Surplus Pond in western Maine. The cores from both ponds show sedimentological evidence of a climatic reversal that probably occurred in the Younger Dryas (Thompson et al., 1996).

Additional ^{14}C ages were recently obtained from cores taken in other ponds in the northern White Mts. (Boisvert et al., 2002). Together with previous results, the new data suggest that the Littleton-Bethlehem Readvance (and deposition of the Bethlehem moraine complex) occurred ~ 12,000 ^{14}C yrs. B.P. (~ 14,000 cal. yrs. B.P.). A short distance west of Littleton, Ridge et al. (1996, 1999, this volume) found stratigraphic evidence supporting the hypothesis of previous authors that a glacial readvance occurred in the vicinity of Comerford dam on the Connecticut River (Antevs, 1922; Crosby, 1934a; Lougee, 1935). The latter authors equated this readvance with the one that deposited the Bethlehem moraine complex. Lougee (1935) referred to this event as the "readvance at Littleton," and Thompson et al. (1999) named it the "Littleton-Bethlehem Readvance" to emphasize the correlation with the deposits historically known as the Bethlehem moraine.

Based on recent work on the glacial Lake Hitchcock varve sequence in the Connecticut Valley, Ridge et al. (1999) determined an age of 11,900-11,800 ^{14}C yrs. B.P. for the Littleton-Bethlehem Readvance (see also STOPS 5-1 to 5-3). Moreover, Larsen (2001) obtained an age of 11,900 ± 50 ^{14}C yrs. B.P. for the Middlesex Readvance near Montpelier, Vt., which he correlated with the readvance in the Comerford Dam area (and presumably with the Littleton-Bethlehem Readvance in general). These findings suggest that the Littleton-Bethlehem Readvance and its correlatives to the east and west were a major glacial event across this part of northern New England. Thompson (1998) proposed that the Littleton-Bethlehem Readvance occurred during the Older Dryas. This cold interval began ~ 12,200-12,000 ^{14}C yrs. B.P. and lasted only about 200 years (Donner, 1995; Wohlfarth, 1996). The GISP2 ice core from Greenland likewise indicates a brief cold period ~ 12,000 ^{14}C yrs. B.P. that is equated with the Older Dryas (Stuiver et al., 1995). Ridge's work in the Connecticut Valley varve sequence (STOP 5-3) further supports the Older Dryas age for the Littleton-Bethlehem readvance. If deposition of the Bethlehem moraine complex were limited to such a brief interval, the ice sheet must have responded quickly in northern N.H.

STOP 4-1. TWIN MT. SAND & GRAVEL PIT, CARROLL, N.H. (Thompson)

Glacial lake deposits are among the most useful features for determining the mode and sequence of deglaciation in the northern White Mts. Ice-dammed lakes existed in north-and west-draining valleys in this area, and most of the Bethlehem moraine complex was deposited in a lake that occupied the Ammonoosuc Valley. Goldthwait (1916a) named this water body "Lake Ammonoosuc," which resulted from damming of the valley when the late Wisconsin ice margin stood in the Littleton-Bethlehem area. As the margin receded, successively lower spillways for the lake were uncovered and the lake level fell. The earliest stage of the lake (Crawford Stage) drained eastward through Crawford Notch, where the spillway has an elevation of about 573 m (1880 ft).

Following the Crawford Stage, glacial Lake Ammonoosuc drained southwestward through five progressively lower spillways into Gale River valley in the Franconia area (Fig. 9). The spillway for the Gale River 2 stage is a prominent channel that will be seen during this trip along Route 3 southwest of Twin Mt. Later spillways north of Bethlehem village drained the Bethlehem and Wing Road Stages of Lake Ammonoosuc into Indian Brook and finally into the lower Ammonoosuc River (Lougee, n.d.; Thompson et al., 1999).

The large pit at this stop is located in the ice-contact "Carroll delta" (Lougee, 1940), which built southward into the Gale River 2 stage of glacial Lake Ammonoosuc. This deposit formed when the ice margin was pinned against Beech Hill to the west and Cherry Mt. to the east. Figures 9 and 10 show the relationship of the Carroll delta to meltwater channels and ice-margin positions. A short esker segment along the railroad track north of the pit marks a feeder channel for the delta. Additional meltwater and sediment came from channels on the hillside to the northeast. The lake level dropped when recession of the ice tongue blocking the lower Ammonoosuc Valley opened up the next lower spillway (G3 on Fig. 9). The resulting incision of the Carroll delta formed the prominent channel seen along the railroad track. The water

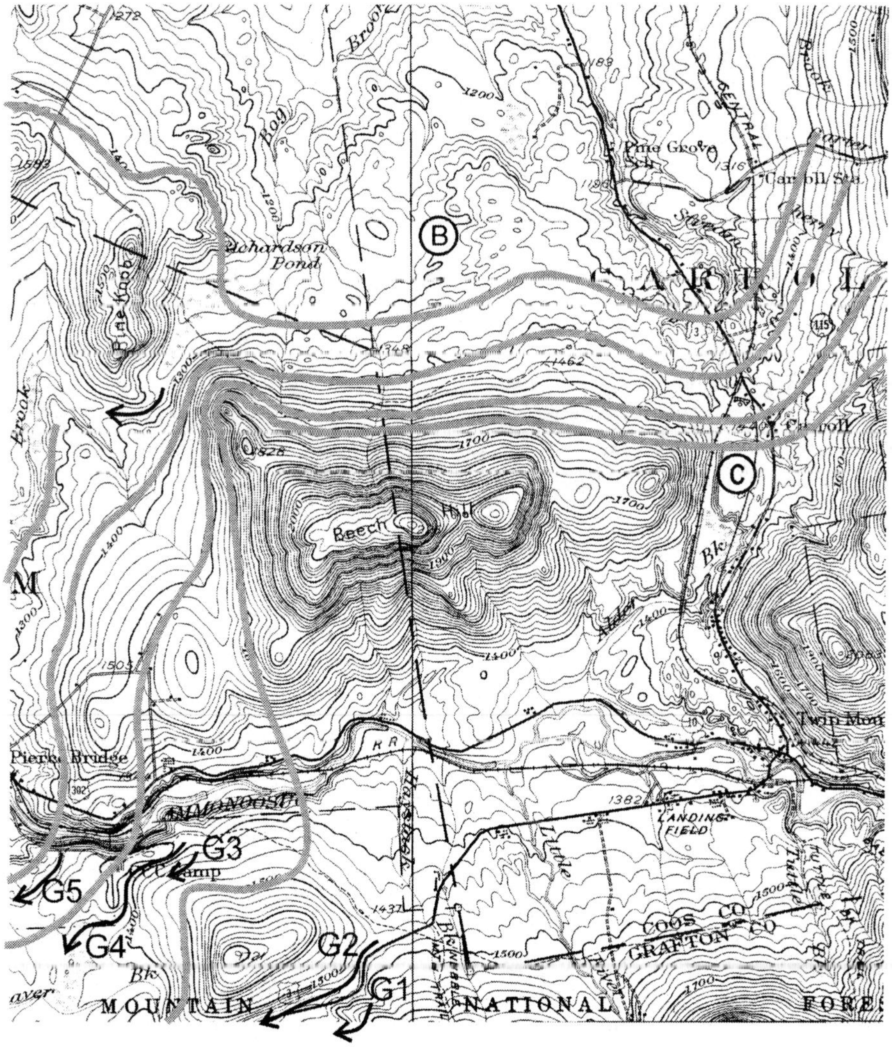

Figure 9. Part of the Whitefield, N.H., 1:62.500 quadrangle, showing inferred ice-margin positions (gray lines) and meltwater spillway channels (arrows). G1-G5 are spillways for successive Gale River stages of glacial Lake Ammonoosuc. B: Beech Hill moraines (STOP 4-2). C: Carroll delta (STOP 4-1). (From Thompson et al., 1999).

that cut the channel may have come initially from the more northerly of ice-margin positions shown in Figure 10, and later from glacial Lake Carroll to the north.

Remnants of coarse gravel forming the delta topset beds can be seen along the upper east wall of the pit. In 2000, the foreset beds were best exposed in a newly deepened area of the pit just east of the processing plant. The southern part of this opening showed many thrust faults in the delta foresets, while the west side (closer to former ice margin) revealed bouldery, ice-contact gravel and till. The faulting indicates ice shove against the backside of the delta, which is further evidence of late-glacial ice activity in the region.

STOP 4-2. BEECH HILL MORAINES, CARROLL, N.H.
(Thompson)

The Beech Hill moraines (Fig. 9), at elevations of 396-427 m (1300-1400 ft), are till ridges that are 4-14 m high, up to 700 m long, and trend ENE-WSW. A small borrow pit in one of the ridges shows sandy, light olive-gray till. Large (1-3 m) boulders of coarse white granite are extremely abundant on the surfaces of the moraines. The boulders were derived from the local bedrock, which is "moderately to weakly foliated biotite granite" of the Oliverian plutonic suite (Lyons et al., 1997). Many of the granite boulders are weathered or disintegrated. A great variety of rock types in the pit include igneous and metamorphic lithologies.

Meltwater channels occur on the distal sides of the Beech Hill moraines. The largest channel lies at the southern margin of this moraine cluster and hosts a wetland and small pond seen on the newest Bethlehem quadrangle map. The channels developed sequentially in the swales between moraine ridges and thus are parallel to the moraines. Meltwater flowed southwest along the ice margin as it retreated from Beech Hill and drained through the 387-m gap between the northwest corner of Beech Hill and Pine Knob (Fig. 9).

Thompson et al. (1999) correlated the Beech Hill moraines with the Bethlehem moraine complex to the west and moraines in the upper Israel River valley to the northeast. They attributed all of these moraines to the Littleton-Bethlehem readvance. Many other moraines occur in the area between this stop and Whitefield village.

STOP 4-3A. VIEW OF OLD MAN OF THE MT., FRANCONIA NOTCH, N.H., LUNCH STOP
(Davis and Fowler)

Exit I-93 at the "Old Man Viewing" ramp sign in Franconia Notch State Park parking lot and walk ~ 500 meters to the viewpoint on the north shore of Profile Lake (Fig. 6).

The profile is made up of six fortuitously-broken, plate-shaped blocks of bedrock positioned above each other in such a way as to create the distinctly "human profile" best seen from the shore of Profile Lake, 550 meters below (Fig. 11). The Profile is about 21 meters high and is estimated to weigh between 10,000 and 13,000 tons. The six blocks or plates comprising the Profile represent, from top to bottom: (1) forehead, (2) eyebrow, (3) nose, (4) base of nose and upper lip, (5) lower lip and chin, portion separated from the cliff, and (6) low lip and chin, portion behind No. 5 and not separated from the cliff. These blocks are vertically separated from one another along sub-horizontal joints (N25°W, 23°NE) that are part of a single set of joints. The Profile itself has been created by selective breakage at the edge of these blocks along nearly vertical joints (S65°W, 73°SE), the predominant set forming Cannon Cliff; N20°E, 75°SE; N35°E, 80°SE; N60°W, 60°SW; N25°E, vertical; N20°W, 85°NE)(Fowler, 1982, 1997).

The information presented here is from a study conducted in 1976 as a part of the environmental impact statement analyzing alternatives for Interstate-93 through Franconia Notch (Fowler, 1982). The study concluded that limited blasting could be undertaken if closely monitored, and it went on to draw the following conclusions about the stability of the Profile. (1) The principal mechanism responsible for the Profile's stability is the compressive cantilevering created by the dead weight load of most of the blocks comprising the Profile acting at the location of their combined center of gravity behind the cliff-face, (2) The blocks in the rock mass that are the most stable are those from the nose upward to the forehead because their large intact masses and the location of the individual and combined centers of gravity well behind the cliff face act to enhance the cantilevering mechanism, (3) The blocks in the rock mass that are the least stable, and which pose the most important stability problems, are those from the upper-lip downward to the chin because their comparatively small intact masses and the location of their individual centers of gravity very near to or (in the case of the chin) in front of the cliff-face are not part of the overall cantilevering mechanism and they obtain no support from the blocks that comprise the rest of the profile. (4) Because of the positions and weights of the blocks in the profile, the active mechanical reinforcement of the various portions of the rock mass to increase their security appears possible, but additional structural and mechanical analyses are needed to design a system that can be installed without endangering the profile in the process.

STOP 4-3B. PROFILE LAKE, LANDSLIDES AND ROCKFALLS, FRANCONIA NOTCH, N.H.
(Davis and Rogers)

Walk about 100 m south of the Old Man viewpoint to a point overlooking Profile Lake and a view of the spectacular slopes of Franconia Notch to the south (Fig. 11). The broad expanse of Cannon Cliff dominates the western slope and the bulky shoulder of Mt. Lafayette looms over the eastern slope of the notch. Mt. Lafayette is composed of porphyritic quartz syenite. Cannon Cliff is composed of Conway granite, which exhibits textbook examples of exfoliation slabs, especially striking at the north end. A deep recess at the south end of the cliff owes its origin to a thick diabase dike, which trends roughly NNE through the higher notch separating Eagle Cliff and Mt. Lafayette on the eastern

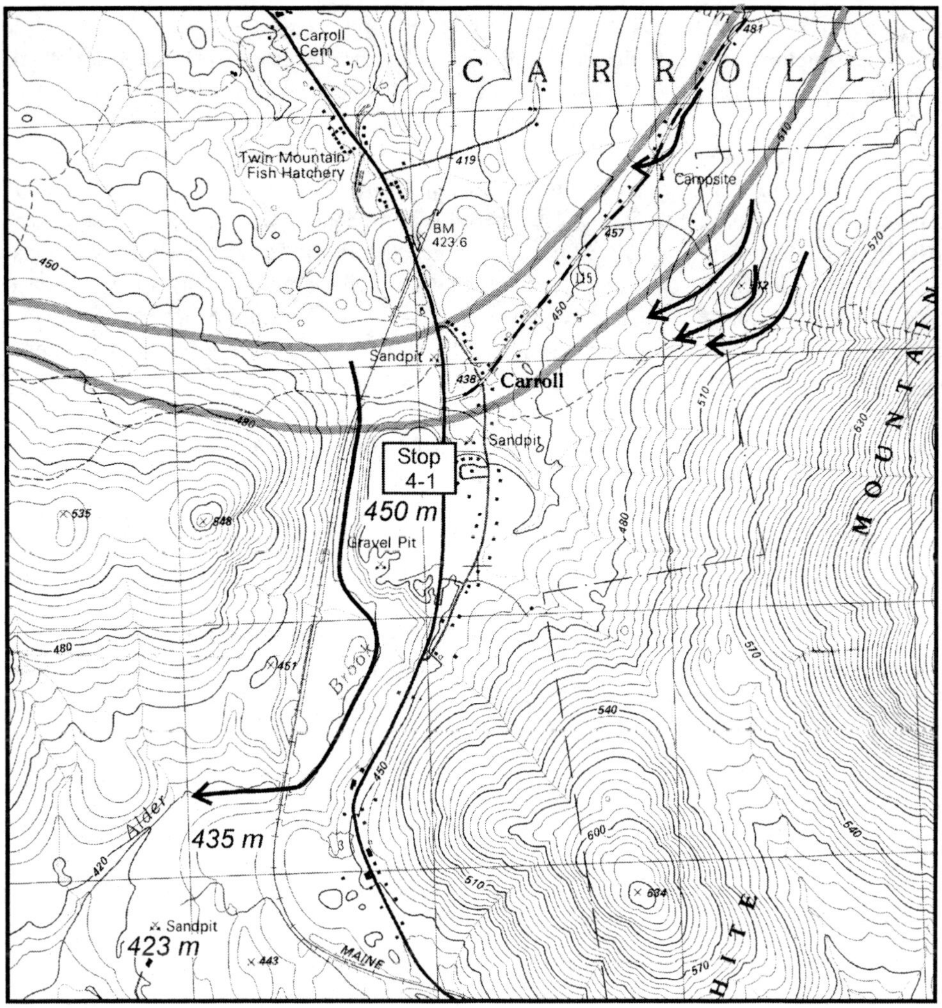

Figure 10. Part of the Bethlehem, N.H., 1:25,000 metric quadrangle, showing the location of the ice-contact Carroll delta at elevation of 450 m (STOP 4-1); the 435- and 423-m surfaces graded to lower stages of glacial Lake Ammonoosuc; inferred ice-margin positions (gray lines); and meltwater channels (arrows).

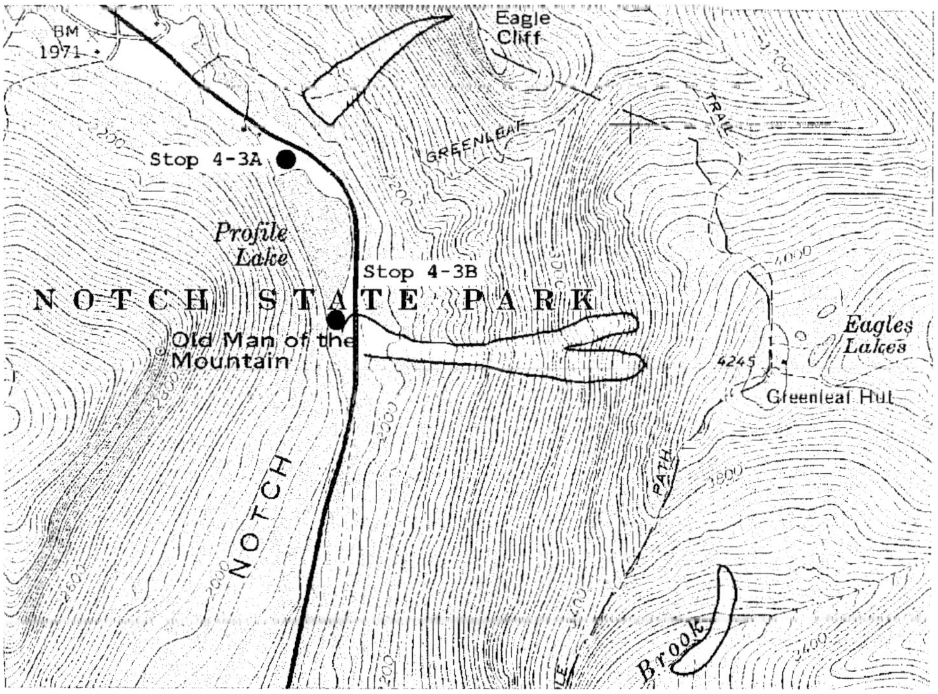

Figure 11. Part of Franconia Notch, N.H., 1:24,000 quadrangle, showing locations for STOPS 4-3a and 4-3b. Solid line is I-93 parkway. Note landslide scars on the east side of valley adjacent to Profile Lake. STOP 4-4 is located about 5 km south of Profile Lake on the west side of parkway.

slope of the Notch. Cannon Cliff is one of the East's premier destinations for rock and ice climbing. The south end of Profile Lake is dammed by landslide debris delivered from the west-facing slope of Mt. Lafayette in the area of the prominent landslide scar (Fig. 11). This form of mass wasting has been a very important mechanism in the modification of the landscape in Franconia Notch and other notches in the White Mts. since deglaciation.

As the regional climate rapidly warmed when continental ice was receding to the north, the newly-exposed bedrock surfaces of the mountains and valleys were attacked by the harshly variable climate that existed close to the retreating glacier. This climate was characterized by large and frequent differences in diurnal temperatures, which permitted intense freezing and thawing of water in cracks and crevices in the bedrock surfaces. This alternating freezing and thawing shattered the bedrock and created large volumes of blocky talus on the mountain peaks and slopes. As this talus continued to accumulate, it either slid or fell to the bases of the cliffs and valley walls, forming the Notch's ubiquitous talus slopes. This process lessened in intensity as the local climate continued to warm and stabilize, so the amount of shattering and rockfall has decreased but not stopped altogether.

The large talus slope at the base of the Cannon Cliff (Fig. 11) is an excellent example of rockfall debris accumulation and although rockfall is not as intense today as it was 10,000 to 12,000 years ago, it is still continuing and poses a considerable safety threat to rock climbers on the cliff and even those scrambling up to the base of the cliff. The now-sharply-pointed peaks of Franconia ridge (from north to south, Mts. Lafayette, Lincoln, Little Haystack, Liberty, and Flume) are the result of their glacially rounded summits being shattered and sharpened by the freeze-thaw process. Many of the alternative construction designs for Rt. I-93 were abandoned because of the expense and difficulty associated with construction in such unconsolidated materials (e.g. cut-and-cover tunnel). Finally, in the early 1980s, I-93 was completed as a parkway, and is the only non-four-lane section of the interstate highway system in the U.S.A.

Highway crews frequently clean up landslide debris from the bottom of the many slide tracks that scar all three major Notches visited on this trip. However, Franconia Notch has been particularly problematic. Numerous historic landslides have occurred on Eagle cliff and the western flank of Mt. Lafayette (Flaccus, 1958), with major events occurring in 1826, 1850, 1883, 1915, 1947, 1948, 1959, and 1974. The landslides of 1948 and 1959 covered old US 3 with so much debris at this location that many days were required to re-open the highway. Many of these landslides were caused by extreme rainfall events.

During the winters of 2000 and 2001, a research project was initiated to recover sediment cores from Profile Lake in order to reconstruct a pre-historic record of large landslides in Franconia Notch (Rogers et al, 2001; Rogers, 2003). Samples dated from near the base of the longest cores from the middle of the lake suggest that sediment accumulation began over 10,000 ^{14}C yrs. ago. From near the tops of sediment cores at the south end of Profile Lake, layers of coarse debris that fine upwards (based on detailed grain size analysis by Sedigraph) (Rogers, 2003), thicken toward the east shore of the lake. Magnetic susceptibility (MS) and loss-on-ignition (LOI) measurements allow correlation of coarse layers in near-shore, sediment cores with clastic lenses of finer grain-size in cores from the middle of the lake. Measurement of ^{210}Pb and ^{137}Cs profiles in the long sediment core tops from the middle of the lake suggest that the uppermost clastic lens reflects the 1959 landslide that covered the highway and extended into the south end of the lake. This analog for a major landslide triggered by an intense rainstorm (ex. Sharpe, 1938) is used to reconstruct a pre-historic record of landslides, which may indicate extreme rainfall events since deglaciation over 10,000 years ago. The sediment record from Profile Lake supports the work of Bierman et al. (1997), whose study of alluvial fans and pond sediments suggested that hillslope erosion in northern Vermont was caused by severe storms during the early and late Holocene. However, the Profile Lake landslide record does not support millennial-scale storminess variability as recorded by the compilation of sediment core data from 13 lakes in Vermont and eastern New York (Noren et al., 2002).

Cannon Mt. has one of the highest relief cliffs in northeastern United States (~300 m). At about 10 a.m. on June, 19, 1997, a large rockfall originated from near the top of the cliff's north end (elevation ~975m). The release involved a large section of steeply inclined exfoliation slabs of Conway granite, which became airborne over the lower half of the cliff. The rockfall was apparently not triggered by seismic, sonic, or meteorological events, as observers did not note an earthquake, sonic boom, or quarry blast, and the previous four days were oddly precipitation-free. Both the cliff scar and the impact swath on the talus slope below the cliff remain clearly visible from the valley floor.

Cannon Cliff's talus slope, with a relief of about 250 m, is composed of angular blocks generally <5 m in length. Many parts of the talus slope are vegetated with trees and shrubs, but in open areas most talus blocks are covered with gray to black lichens, in contrast to the green lichens on the cliff face. Because of the altitude of the cliff, frost action is severe and rockfalls are common, especially during the spring.

The June 1997 rockfall not only flattened all vegetation, including shrub birch and spruce, but mobilized most blocks for a width of about 100 m on the upper part of the talus slope, leaving immense exposures of fines (less than pebble size), which have begun to erode since June 1997. The talus slope absorbed most of the energy of impact, but many rockfall blocks traveled up to 425 m, cutting a 10–20 m wide, 5–10 m deep trough into the lower part of the slope. Beech and maple trees up to 45 cm DBH were splintered or sheared. One large rockfall block (5.1 m x 3.5 m x 3.3 m; ~3.5 tons) reached the valley floor, long axis oriented upslope, adjacent to a paved bicycle path ~380 m south of Profile Lake. This block originated from the cliff face, as part of the top surface remains mostly unscathed and covered with green lichens. The arrangement of blocks comprising the Old Man of the Mountains is susceptible to the same mass wasting processes demonstrated by the 1997 event, mechanisms that no doubt will be responsible for the Old Man's ultimate collapse, given its delicate state of stability.

STOP 4-4. THE BASIN POTHOLES, FRANCONIA NOTCH, N.H. (Davis)

Although not the largest and deepest in the White Mts., the potholes at the Basin are the most accessible. Conventional wisdom suggests that potholes are created by the abrasive action of sand and gravel in swirling eddies of meltwater streams. However, cases may be made for an entirely subglacial origin of potholes, as they have been described at the top of a bedrock ridge in Ontario (Gilbert, 2000) that requires overlying ice to provide water flow in such a topographic setting (i.e., via glacial moulins). The same case may be made to explain potholes at an altitude of over 1200 m along the Caps Ridge trail on the western flank of Mt. Jefferson in the north Presidential Range.

STOP 4-5. MILL BROOK SECTION, MONROE, N.H. (Ridge)

Three tills separated by sand, gravel, and bedded diamictons occur along Mill Brook at the south side of the Comerford dam (Fig. 6; Ridge et al., 1996; Thompson et al., 1999). The lower two tills were first described by Crosby (1934b) as lower and upper tills on the west side of the brook. The present exposure, farther south and on the brook's east bank, has an additional till at its top overlain by varved silt and clay. Crosby (1934a) interpreted his upper till (our middle till) to represent a late Wisconsin readvance correlative with the Littleton-Bethlehem moraines farther east (cf. Thompson et al., 1999). Our uppermost till appears to be a better candidate for this readvance. The entire section may be late Wisconsin. Alternatively, the lowermost till may be pre-late Wisconsin, but this is not supported by evidence of weathering, as at other places in New England (Koteff and Pessl, 1985; Newman et al., 1990).

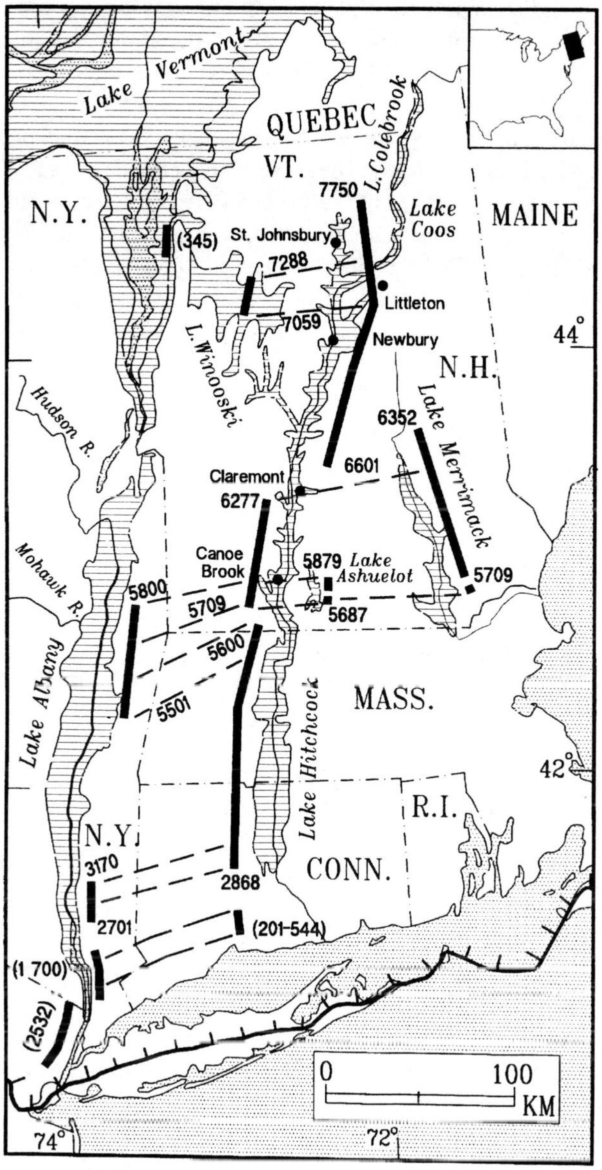

Figure 12. Important varve chronologies in the northeastern United States, including, from south to north, varve sequences from glacial Lake Hackensack in northern N.J. (2532 yr; Reeds, 1926; Antevs, 1928), correlated sequences from lake Quinnipiac near New Haven, Conn., and Lake Albany at Haverstraw, N.Y. (Antevs, 1928), correlated sequences (Antevs, 1922, 1928) of the lower (NE varves 2701-6352) and upper Connecticut varves (NE varves 6601-7750), and the Essex Junction, Vt., sequence (345 yr; Antevs, 1928).

DAY 5. CONNECTICUT RIVER LOWLAND, N.H.-VT. AND GREEN MTNS, VT.; THE NEW ENGLAND VARVE CHRONOLOGY (Ridge)

In the 1920's, Gerard De Geer and three of his students from the University of Stockholm, Ragnor Lidén, Ebba Hult de Geer, and Ernst Antevs, began work on varves in the northeastern United States to formulate a varve chronology similar to the Swedish varve chronology. However, Ernst Antevs (1922, 1928) almost single handedly formulated the two main sequences of the New England (NE) varve chronology (Fig. 12), in addition to sequences in the Hudson Valley, and N.J. with Chester Reeds (1926). Antevs began his work in southern Conn., with a varve sequence that arbitrarily starts at varve 2701, to accommodate older varves, and extends to varve 6352 (3652 varves). This varve sequence, the lower Connecticut varves, is mostly based on measurements in the Connecticut Valley from Hartford, CT., north to Claremont, N.H., and includes sequences from the Hudson Valley of N.Y., and Ashuelot and Merrimack Valleys of N.H. These additional sequences were critical to filling gaps and extending the chronology.

Antevs (1922, 1928) also compiled a younger sequence north of Claremont (Fig. 12), based on measurements from the upper Connecticut and Passumpsic Valleys of northern Vt. and

N.H., and the Winooski Valley of western Vt. This sequence is known as the upper Connecticut varves. It starts with varve 6601 and extends to varve 7750 (1150 varves) and does not have a clear overlap with the lower Connecticut varves. Antevs separated the sequences by what has become known as the Claremont Gap, which he interpreted to be the result of delayed ice recession that allowed deposition of the lower Connecticut varves while areas to the north remained ice covered. A crude similarity between the ends of the varve sequences, supported by similar paleomagnetic records and consistent ^{14}C ages, suggested an overlap of the varve sequences (Ridge et al., 1996, 1999), thus eliminating the Claremont Gap and implying continuous rapid deglaciation of the Claremont area.

An important step in using the varves as a chronologic tool has been the calibration of the NE varve chronology with ^{14}C ages so that both ^{14}C and calibrated (U-Th) time scales can be applied to the varves. This work began when ^{14}C ages were obtained from varve 6150 at Canoe Brook, Vt. (Ridge and Larsen, 1990). The calibration of the lower Connecticut varves appeared to be secure when consistent ^{14}C ages were also obtained from the upper Connecticut varves (Ridge and Toll, 1999; Ridge et al., 1999). However, the original ^{14}C ages from Canoe Brook were from the time of a plateau in the ^{14}C time scale (200 ^{14}C yr = 900 cal yr; 12.6-12.4 ^{14}C ka B.P.) that did not allow for a precise calibration of the lower Connecticut varves. A new ^{14}C age at Canoe Brook from varve 5858, (12,660±50, GX-25735), and from outside the ^{14}C plateau, has provided an improved calibration of the lower Connecticut varves. This new calibration makes the lower Connecticut varves older and based on ^{14}C ages they no longer appear to overlap the upper Connecticut varves (Ridge, 2003). Recent detailed mapping (Ridge, 1999, 2001) has also revealed three sets of end moraines associated with readvances in the Claremont area, indicating slowed ice recession and supporting a gap between the varve sequences (Ridge, 2003).

Along with the calibration of the NE varve chronology has been the refinement of paleomagnetic records from the varve sequence, especially a record of declination. Declination records were originally constructed by McNish and Johnson (1938), Johnson et al. (1948), and Verosub (1979a, 1979b). Recent work has focused on completing the declination record of the lower Connecticut varves and refining the upper Connecticut record with new laboratory techniques (Ridge et al., 1996, 1999, 2001). Correlation of the declination records from varves in New England and glacial lacustrine deposits in central N.Y. (Brennan et al., 1984; Ridge et al., 1990; Pair et al., 1994) has provided a means of testing ^{14}C ages and inferring correlations across the northeastern U.S. (Ridge et al., 1990, 1999, 2001; Ridge, 2003).

STOP 5-1. COMERFORD DAM SECTION, BARNET, VT.

At the lower end of the retaining wall embankment along the spillway for the Comerford dam is a gully exposure of varves resting on till (Fig. 6). Bedrock in the spillway channel is striated at 182°. The varve section is very close to a varve section on the spillway embankment studied by Lougee (1935) during the final phases of construction of the Comerford dam. He described a compact and deformed upper part of the varve section that was overlain by a "material resembling till." Lougee interpreted this till to represent the Littleton-Bethlehem Readvance. Lougee measured the varve exposure and matched the varve record to the New England varve chronology. The varve section at the present exposure has been measured (Ridge et al., 1996, 1999) and matches Lougee's measurements and the NE varve chronology (varves 7036-7154). However, the top of the varve section, with Lougee's overlying till unit, was truncated by completion of the retaining embankment. The basal varves resting on till allow a determination of the age of initial ice recession at this position in the Connecticut Valley. Further, if Lougee's (1935) observation of till at the top of the section is correct, the varve correlation, along with other nearby sections measured by Antevs (1922), precisely brackets the age of the Littleton-Bethlehem readvance.

STOP 5-2. VARVE SECTION, WELLS RIVER VALLEY, VT.

This exposure along the north side of the Wells River valley is a sand and gravel pit that exposes a thick section of varves from an embayment of Lake Hitchcock (Fig. 6). The sand and gravel is an esker that was buried beneath the varves. Several outcrops in the pit show the base of the varve section where it rests on sand and gravel, as well as the upper part of the varve sequence that grades upward into lacustrine sand. Many horizons in the pit show spectacular soft sediment contortions probably related to the instability of clayey sediment that was deposited on the flanks of the esker. The varves in the lower part of the section also contain concretions. Attempts to match varves in the upper stratigraphy with the NE varve chronology have failed and lower parts of the stratigraphy are presently being processed. Antevs (1922) also had trouble matching varve sections in this tributary valley. Local post-glacial depositional processes, related to meteoric or non-glacial runoff, may have masked annual weather patterns in the varves or may have recorded weather patterns differently than the glacially derived varves along the main axis of the Connecticut Valley to the east.

STOP 5-3. VARVE SECTION AND LUNCH STOP, NEWBURY, VT.

This section is along a small, unnamed stream that joins the Connecticut River at the outside of a meander just east of Rt. 5, about 15 km south of STOP 5-2. Antevs (1922) measured and matched 327 varves of the section to the NE varve chronology. He counted an additional 1184 varves above the measured section for a total of 1511 varves (varves 6990-8500 of Antevs, 1922, 1928). Most of the upper 1184 varves were too thin (some <1 mm) to measure realistically in the field with a ruler. In the 1940's, the Newbury section was again counted as part of one of the first studies in sedimentary paleomagnetism that recorded the secular variation of geomagnetic declination (Johnson et al., 1948). In the last 10 years, the entire section has been collected in duplicate PVC cores and measured with the aid of magnified video images and a computer measurement program (Ridge and Toll, 1999; Ridge et al., 1999). This recent analysis of the section has shown that it has 1735 varves spanning varves 6944-8679(+35/-20), after accounting for uncertainties in the interpretation of annual layers.

The lower Newbury section (varves 6944-7470) records deglaciation in the upper Connecticut Valley (Fig. 13). This part of the section has thick varves (mostly >1 cm) that are dominated by sediment from glacial runoff. The varves contain ostracods and the sinusoidal trails of nematodes. Scattered thick varves (more than an order of magnitude thicker than surrounding varves) are interpreted to be flood deposits formed by the release of water from tributary lakes impounded by the receding glacier. Most notable are the exceedingly thick couplets (up to 15 cm in varves 7200-7220) produced by the catastrophic release of water from lakes in the upper Ammonoosuc Valley about 50 km to the northeast (see STOP 4-5). These events do not appear in correlative records of the Winooski Valley. The implied source of the floods is based on the correlation of ice front positions in the Ammonoosuc and Connecticut Valleys at the Comerford dam, which is bracketed by the varve chronology (see STOP 5-1). The flood events mark the earliest recession of ice of the Littleton-Bethlehem Readvance. The period of readvance itself (varves 7115-7200), based on the Comerford dam correlation, is represented by relatively thin (< 1 cm) glacial varves that do not have erratic jumps in thickness representing floods. The period of ice advance appears to have been a time of reduced glacial meltwater production and the impounding, rather than catastrophic release, of tributary lakes by advancing ice.

At varve 7470, the varves abruptly become very thin (<5 mm), which represents recession of the glacial meltwater source and the start of non-glacial deposition in a paraglacial and periglacial environment. This non-glacial lake persisted in the Connecticut Valley for at least 1200 years after glacial runoff no longer contributed a significant volume of sediment to the lake. The non-glacial varves contain abundant fossil plants, ostracods of several species, and nematode trails. Two types of bivalves have been recovered including many *Sphaerid* clams and a single mussel specimen, *Pyganodon fragilis*, which is a cold-water species no longer found in New England (Smith and Ridge, 2001). Additional fossils include an as-yet-unidentified gastropod and sinusoidal, fish-swimming trails.

The plant fossils from the Newbury section, especially wood and woody twigs from terrestrial species, have yielded ^{14}C ages that provide a calibration of the upper Connecticut varves (Ridge et al., 1999; Ridge, 2003). Six ^{14}C ages from Newbury and an additional age from the Passumpsic River valley to the north, ranging 11.5-10.1 ka, are consistent with their positions in the varves and provide a calibration of the sequence to ^{14}C and calibrated (U-Th) time scales (Fig. 14). This calibration has allowed a comparison of the Newbury varve record with oxygen isotope records from Greenland ice cores (Ridge and Toll, 1999; Fig. 5-3a). The comparison indicates that the Littleton-Bethlehem Readvance corresponds to the Older Dryas. In addition, the Killarney Oscillation in North America (Intra-Allerød Cold Phase in Europe) appears to correspond to an interval of thicker varves that have a conspicuous 22-yr periodicity. Near the top of the non-glacial sequence the onset of the Younger Dryas also appears to be represented by a sudden thickening of varves. If these correlations are correct, periods of ice advances apparently were times of reduced sediment input, whereas cold intervals following the recession of ice from the basin

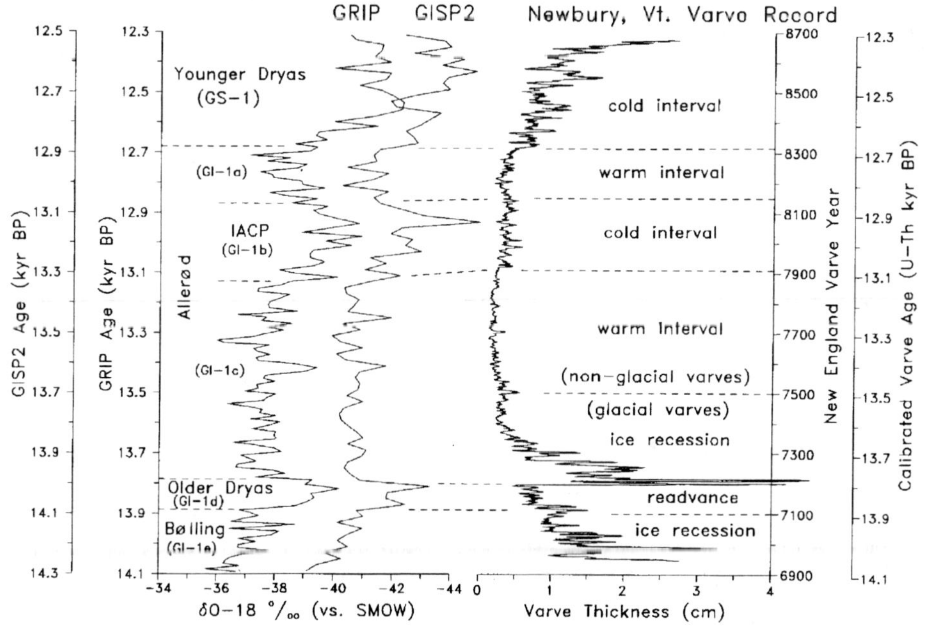

Figure 13. Comparison of oxygen isotope records from Greenland Summit ice cores (Dansgaard et al., 1998; Stuiver et al., 1995) and a 5-yr moving average of varve thickness at Newbury, Vt. Proposed event stratigraphy (Björck et al., 1998) shown in parentheses. Position of ice core time scales based on a correlation of ice core records. The Newbury record is lined up so that its calibrated time scale matches the GRIP time scale.

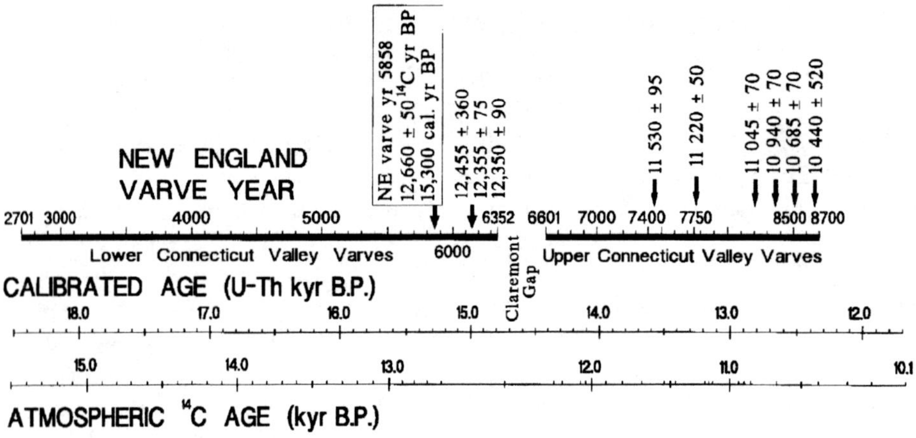

Figure 14. Atmospheric ^{14}C and calibrated (U-Th) time scales applied to the New England varve chronology (Ridge, 2003), based on the CALIB 4.3 computer program and INTCAL 98 calibration data (Stuiver and Reimer, 1993; Stuiver et al., 1998). Some individual ^{14}C ages are plotted for comparison.

were times of increased lacustrine sediment accumulation, possibly related to accelerated slope erosion.

LUNCH STOP: Quechee Gorge, Quechee, Vt.

Glacial geology of northwestern Vermont (Bierman)

The landscape of northwestern Vermont is one of dramatic contrasts. To the west are the Champlain Lowland and Lake Champlain, draining north to the Gulf of St. Lawrence over a bedrock-controlled spillway in the Richelieu River. To the east are the Green Mts., oriented north-south, following the strike of the dominant foliation, the structural grain of the schists and phyllites that dominate the range. The Champlain Lowland, underlain predominately by sedimentary rocks, is mantled in many places by sorted, glacial sediments, deposited directly off ice or in glacial lakes bordering the ice margin. The uplands of the Green Mts., in places where bedrock does not crop out, are covered primarily with varying thickness of till. In the uplands, sorted glacial and post-glacial sediments are rarely present outside river valleys, with the exception of isolated ice-marginal deposits (Stewart and MacClintock, 1969).

As ice melted and the ice margin retreated, what began as isolated ice-marginal lakes coalesced into glacial Lake Vermont, which drained into the Hudson Valley through a spillway near the southern end of Lake Champlain (Chapman, 1937). Glacial Lake Vermont at its lowest stage held about 240 km^3 of water, about ten times the volume of present-day Lake Champlain. Lake Vermont ended when ice-margin retreat allowed marine waters of the Champlain Sea into the isostatically-depressed Champlain Lowland. Over the next 1000 to 1500 years, isostatic rebound increased the elevation of the Richelieu River sill at a rate greater than eustatic sea-level rise, finally isolating Lake Champlain from marine water and freshening the lake by about 10,000 ^{14}C yrs. B.P. (Parent and Occhietti, 1988).

Human activity has significantly affected the Vermont landscape. Vermont was first settled in the late 1700s. By 1850, much of Vermont had been cleared for agriculture. Initial clearance was for cropland. Later clearance was for sheep grazing. In the 1870s, land at higher elevations was cleared for timber harvesting. Landscape response to this clearance is well preserved in the geologic record. Hillslopes became unstable and sediment yield appears to have increased as documented by aggrading alluvial fans and floodplains (Bierman et al., 1997).

STOP 5-4. MT. MANSFIELD, VT.

Mt. Mansfield, at 1340 m (4393 ft) is the highest point in Vermont (Fig. 15). Mt. Mansfield's rocky summit exposes multiply-deformed schist of the Underhill Fm. (Christman and Secor, 1961), covered in places by a thin mantle of till. Although the rock has been eroded into streamlined forms, >12,000 ^{14}C yrs. of post-glacial weathering and erosion have removed striations, except in areas recently exposed by human activity. Glacial grooves, striations, the orientation of streamlined features, and erratics derived from the Champlain Lowland can be used to show that ice flowed roughly NNW to SSE over Mt. Mansfield's summit.

On a clear day the summit provides views extending over 100 km. To the east is the Stowe Valley and Worcester Range in the foreground, with the White Mts. of N.H. in the background. To the south is the spine of the Green Mts., including Camels Hump. To the west is the Champlain Lowland and beyond, the Adirondacks. To the north are the lowlands of southern Canada and the Richelieu River, the outlet of Lake Champlain.

STOP 5-5. SMUGGLER'S NOTCH, VT.

Smuggler's Notch is a deep cleft that cuts across the main range of the Green Mts. just north of Mt. Mansfield (Fig. 15). The extreme topographic relief, recent landslide scars, and truck-sized blocks of rock that have fallen from the cliffs high overhead make this a much-visited site. Slope stability history and hazards in Smugglers Notch are discussed in Lee et al. (1994). Baskerville et al. (1988) described in some detail the 1983 slope failure. The rocks exposed in the cliffs above Smugglers Notch are all schist.

Rockfalls and debris flows are relatively frequent events in Smugglers Notch, many of which have been documented in the last 150 years (Lee et al., 1994). We will observe the debris slide that occurred during July 13, 1983, described by Baskerville et al. (1988). The landslide began at about 7 a.m. when a large block of rock (~10.4 x 10^6 kg), cantilevered over the valley, broke loose and fell onto the talus slope at the base of the cliff. The fall initiated a debris slide along the talus slope and material moved as far as the road (Baskerville et al., 1988). The rockfall occurred on a clear, sunny, midsummer morning and no rain had fallen for several days. Baskerville et al. (1988) suggest that the rock failure was most likely due to thermal expansion of the rock along a crack that had previously been extended by frost wedging.

A debris flow deposit visible along Rt. 108 near the Cambridge/Stowe town line is one of several that occurred during the night of May 22, 1986 and are described by Lee et al. (1994). This particular flow originated in the gully extending up the east side of the valley below Spruce Peak and incorporated colluvium as well as trees and soil (~250,000 m^3 of material; Lee et al., 1994). An intense rainfall apparently precipitated this and other debris flows that evening, loosening colluvium and organic debris that had accumulated in the chute. At present, this debris flow chute is almost barren of colluvium and will take some time to once-again accumulate sufficient debris to present a hazard.

STOP 5-6. LAKE MANSFIELD AND MILLER BROOK DEPOSITS, VT.

For almost 100 years, workers have debated the presence of late Pleistocene alpine ice in the mountains of Vermont and elsewhere in New England (e.g. Goldthwaitt, 1916b; see reviews by Davis, 1999, and Waitt and Davis, 1988). Two lines of evidence have been used to argue for post-Laurentide alpine glaciation: the presence of cirque-like features and the existence of linear ridges of sediment, interpreted as moraines. Both of these landforms are well-exhibited near Lake Mansfield in the Miller Brook valley, south of Stowe, Vermont (Fig. 15).

Beginning at Nebraska Notch, Miller Brook flows through three steep-sided, bowl-shaped segments of the valley, interpreted by some as cirques (Wagner, 1971). Lake Mansfield occupies the lowest of these and is artificially dammed. Downstream from the Lake Mansfield dam, the valley contains several distinct ridges (with relief up to 30 m) lying within or adjacent to fluvial landforms and older lacustrine terraces. Wagner (1970) first identified these ridges and interpreted them as moraines produced by alpine ice retreating up Miller Brook valley. Wagner's paper stimulated much discussion and raised again the question of whether or not alpine glaciers existed in New England following retreat of the Laurentide Ice Sheet.

Based on both their map pattern and the fluvial sediments within them, we interpret these ridges to be segments of an esker, and not moraines. The poorly-sorted veneer of sediment overlying the gravel likely originated from the low discharge of the stream occupying the esker tunnel towards the end of its history or from supraglacial debris dumped when the ice melted away. One ridge contains coarse diamict, including a few faceted, striated cobbles. This diamict likely melted out of the ice shortly before the ice tunnel was abandoned, so these clasts were probably never subjected to significant abrasion.

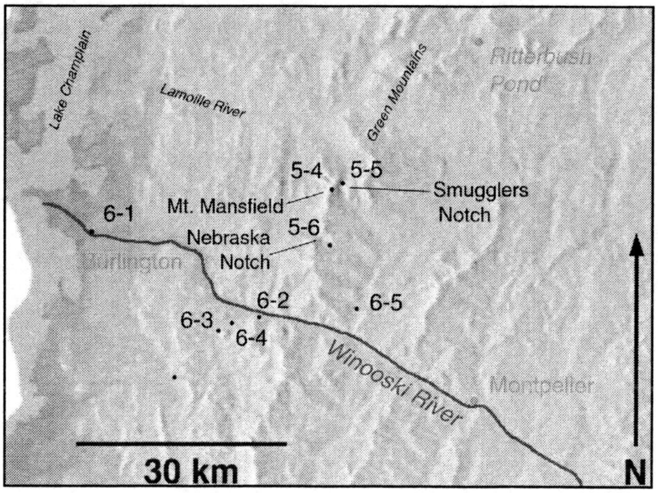

Figure 15. Digital terrain map of north central Vermont showing the location of Stops 5-4 through 5-6 and 6-1 through 6-5.

Nebraska Notch is the lowest gap through the Green Mts. between the Lamoille Valley to the north and the Winooski Valley to the south. Therefore, the Miller Brook valley must have contained Laurentide ice well after ice had melted from the eastern flanks of the Green Mts. immediately to the north and south. Once the ice elevation dropped below the elevation of Nebraska Notch, ice supply was shut off and the ice in the Miller Brook valley stagnated soon thereafter. The esker probably formed immediately before and during this stagnant ice stage. Loso et al. (1998) suggest climate could not have been cool enough in postglacial times to support a valley glacier.

DAY 6. POST-GLACIAL MASS WASTING PROCESSES, NORTHERN GREEN MTS., VT., AND SOUTHERN CONNECTICUT VALLEY (Bierman and Ridge)

STOP 6-1. TOWN LINE BROOK, WINOOSKI AND COLCHESTER, VT.

Town Line Brook is a small, deeply-incised tributary of the Winooski River. Its valley walls expose up to several meters of fluvial gravel cut into tan, well-sorted fine to very fine sand that overlies gray silt with interbedded fine sand. Underlying these sediments are thin silt/clay couplets (rhythmites) near the base of the exposure. The rhythmites were deposited in the quiet water of glacial Lake Vermont, whereas the gray silt and sand were deposited in the Champlain Sea. The Champlain Sea sediments can be distinguished by the absence of rhythmites and the presence of small, white bivalves (*Macoma Baltica*), which can occasionally be found on this outcrop. The sand signals the encroachment of the paleo-Winooski River delta to the area during the Champlain Sea. The gravel fills channels in the fine sand represent distributary channels of the paleo-Winooski River delta.

The valley of Town Line Brook has been widened by landsliding and deepened by fluvial incision, controlled to a great extent by the distribution of groundwater. The fine-grained sediments contain sandy interbeds along which groundwater preferentially flows. The interbeds wash out, causing small-scale slumps and toppling failures of the more cohesive, overhanging fine-grained sediments, which liquefies easily (try stamping on some failed material). The liquefaction is important because it allows failed material to be evacuated easily from the valley by rather modest stream flows. Once the fine-grained deposits have failed, the overlying, non-cohesive, and permeable deltaic sand and gravel also fail by translation and toppling. Such failures are particularly common during wet periods when the water table rises farther into the sand.

Today, only a few slides are active, but slide scars are prevalent along the watercourse. Ring counting of tree cores shows that most of the oldest trees within the currently inactive slides are about 100 years old (Baldwin et al., 1995). The age of these trees suggests that Town Line Brook hillslopes began to stabilize in the late 1800s, coincident with the reforestation of northwestern Vermont (Fig. 16). According to local residents, the major landslide complex became active within the past 20 years. Pin-line measurements suggest that the scarp has been retreating episodically over the past three years at rates of several cm to >1 m yr^{-1} (Fig. 17). Using the geometry of the slide, one can estimate that this slide alone provides 150 to 250 m^3 yr^{-1} of sediment to the Winooski River. A long-term average rate of sediment export from Town Line Brook (10 - 15 m^3 yr^{-1}) can be calculated using valley volume (about 100,000 - 150,000 m^3) and assuming that the paleo-Winooski River delta was abandoned ~ 10,000 ^{14}C years ago when the Champlain Sea drained.

STOP 6-2. STRIATED ROCK AND ICE RETREAT, JONESVILLE, VT.

This outcrop of Underhill mica schist preserves striations and groves indicating that the last ice flowing over this outcrop moved from N70°W (Fig. 18). This flow direction is parallel to the orientation of the Winooski Valley and indicates that ice flow was channeled, presumably during retreat, by this major topographic feature. Measurements of striae as a function of elevation above the valley bottom show that striae become oriented toward regional flow directions (~NNW-SSE) at higher elevations on Camel's Hump.

In a class study, students measured lichen size as a function of underlying tombstone age in the nearby Richmond cemetery and established that lichen growth rates over the past century are linear and on the order of 1 mm yr^{-1}. This rate is similar to that determined for lichens on tombstones less than 50 years old in

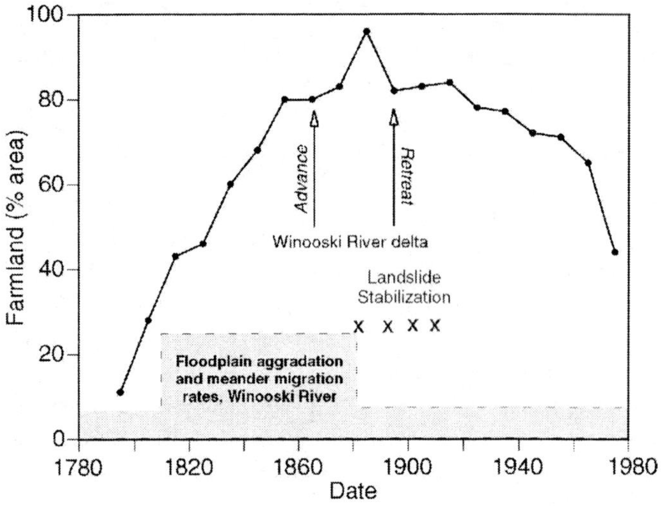

Figure 16. Historic landscape response, Chittenden County, northwestern Vermont. Open circles represent percentage farmland. Major expansion and contraction of Winooski River delta in Lake Champlain as deduced from historic maps are marked by arrows (Severson, 1991). Maximum age of trees growing in fossil landslide scars on tributary of the Winooski River (Town Line Brook), indicating when slides stabilized, are shown by triangles. Period of increased meander migration and flood plain aggradation in lower Winooski River flood plain shown schematically (Thomas, 1985).

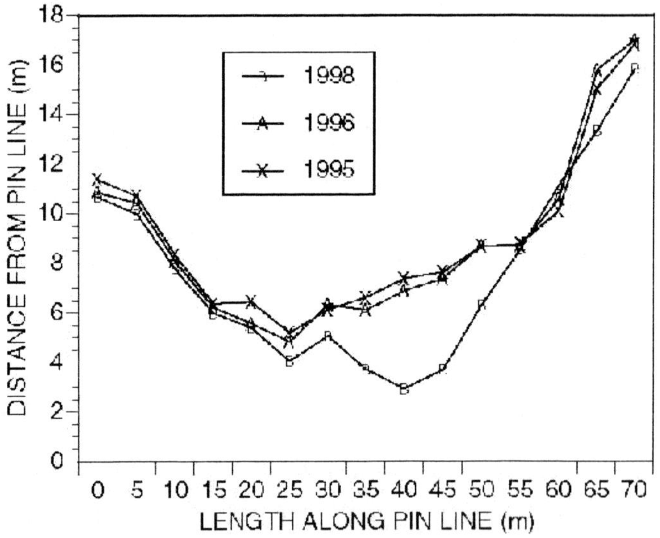

Figure 17. Pin line data for the main Town Line Brook landslide showing retreat of landslide scarp over the past three years. Data gathered by successive UVM Geohydrology classes from 1995 until 1998.

the Champlain Lowland. Using the Richmond calibration, lichen diameters on bare, striated bedrock outcrops similar to and near the Jonesville Rock, suggest exposure within the last century or two. Such recent exposure is consistent with the excellent preservation of striae on this relatively easily weathered rock. Exposure

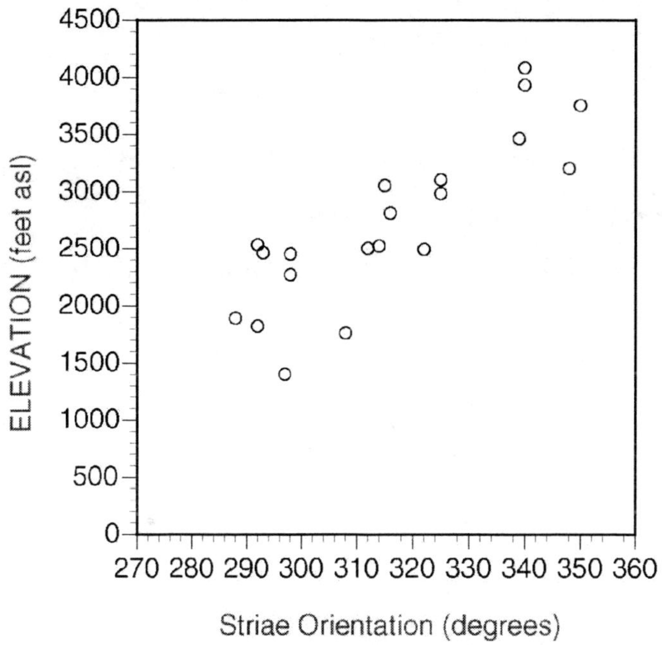

Figure 18. Orientation of striae as a function of elevation, transect from near Jonesville Rock to the summit of Camels Hump.

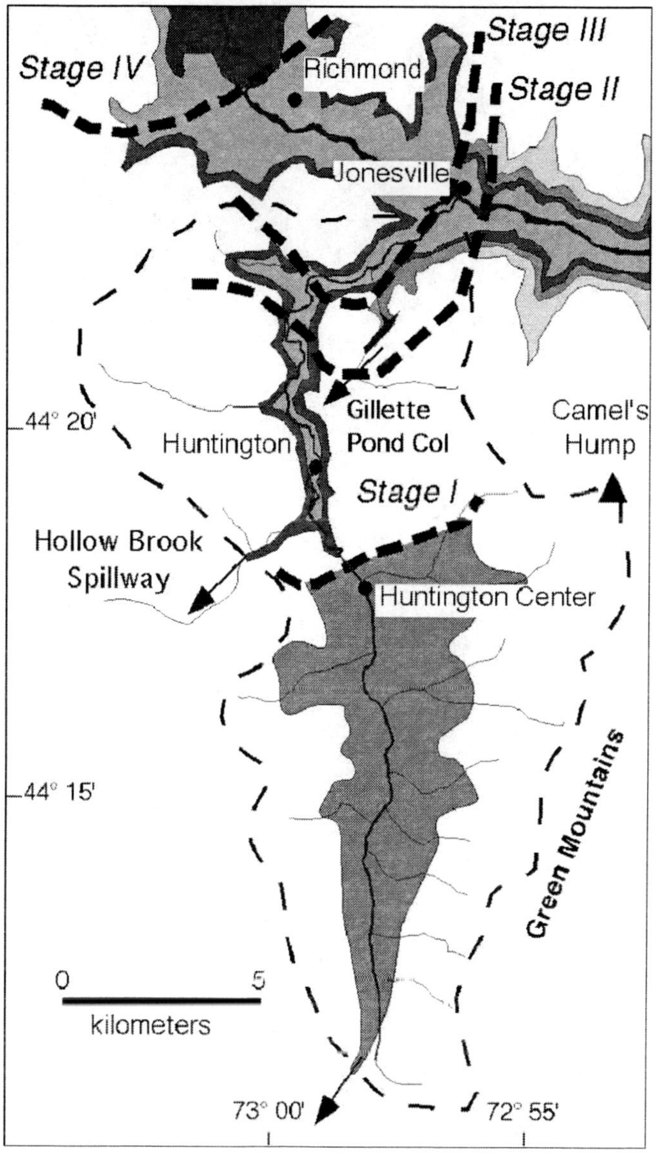

Figure 19. Map of ice margins and proglacial lake extents in the Huntington Valley of northwestern Vermont. Hollow Brook, Gillett Pond, and southern Huntington River valley thresholds are shown with arrows indicating flow directions. Ice margins are shown schematically and labeled by stage. Stage I is the earliest stage. Lake levels impounded by ice are shown in varying shades of gray. From Figure 6 of Wright et al. (1997).

of the bare rock surface was most likely the result of land clearance for farming and grazing during the 1800's.

STOP 6-3. HUNTINGTON RIVER TERRACES AND ALLUVIAL FANS, HUNTINGTON, VT.

Ten terrace levels in Huntington Valley range in age from >12,500 ^{14}C years to historic. A general history of valley evolution is illustrated in Figure 19. Initially, the river flowed over and augmented the glaciolacustrine valley fill (Fig. 19). These deposits are generally a coarsening-upward sequence of rhythmically bedded silt, grading into sand (some of which may locally exhibit foresets and are therefore deltaic in origin). The fill terraces are capped by fluvial deposits. Although these fill terraces have been mapped as deltas based on their morphology (i.e., Wagner 1972), many are located far from incoming streams.

Incision and gradient changes were a response to base-level drops in the lower reaches of the Winooski Basin. The amount of incision varied along the Huntington Valley and depended on two factors: the magnitude of base-level change and the distance from the base-level change. Base-level changes began with the lowering of proglacial lakes and continued with the decline of the Champlain Sea. Each drop in water level initiated a wave of incision that propagated upstream until a bedrock knick point was encountered or until the wave of incision diffused due to lack of stream power.

Backhoe trenches and shovel pits opened on nearly every terrace here reveal three different terrace stratigraphies (Fig. 20). Common to all terraces is the presence of a 25 cm deep plow zone capping the fluvial deposits. The upper meter of the glaciolacustrine fill terrace, T8 (Fig. 20), is composed of cross-bedded sand overlain by silty fine sand, which suggests deposition by a meandering fluvial system. The stratigraphy of the pits and trenches from T7 and T6 (represented by the T6 log in Fig. 20) is consistent with a braided fluvial system. Overbank deposits are thin or absent on these terraces. Terrace deposits from T5 to T1 all have a basal unit of imbricated gravel, deposited in the channel, overlain by laterally-accreted sand that is capped by over-

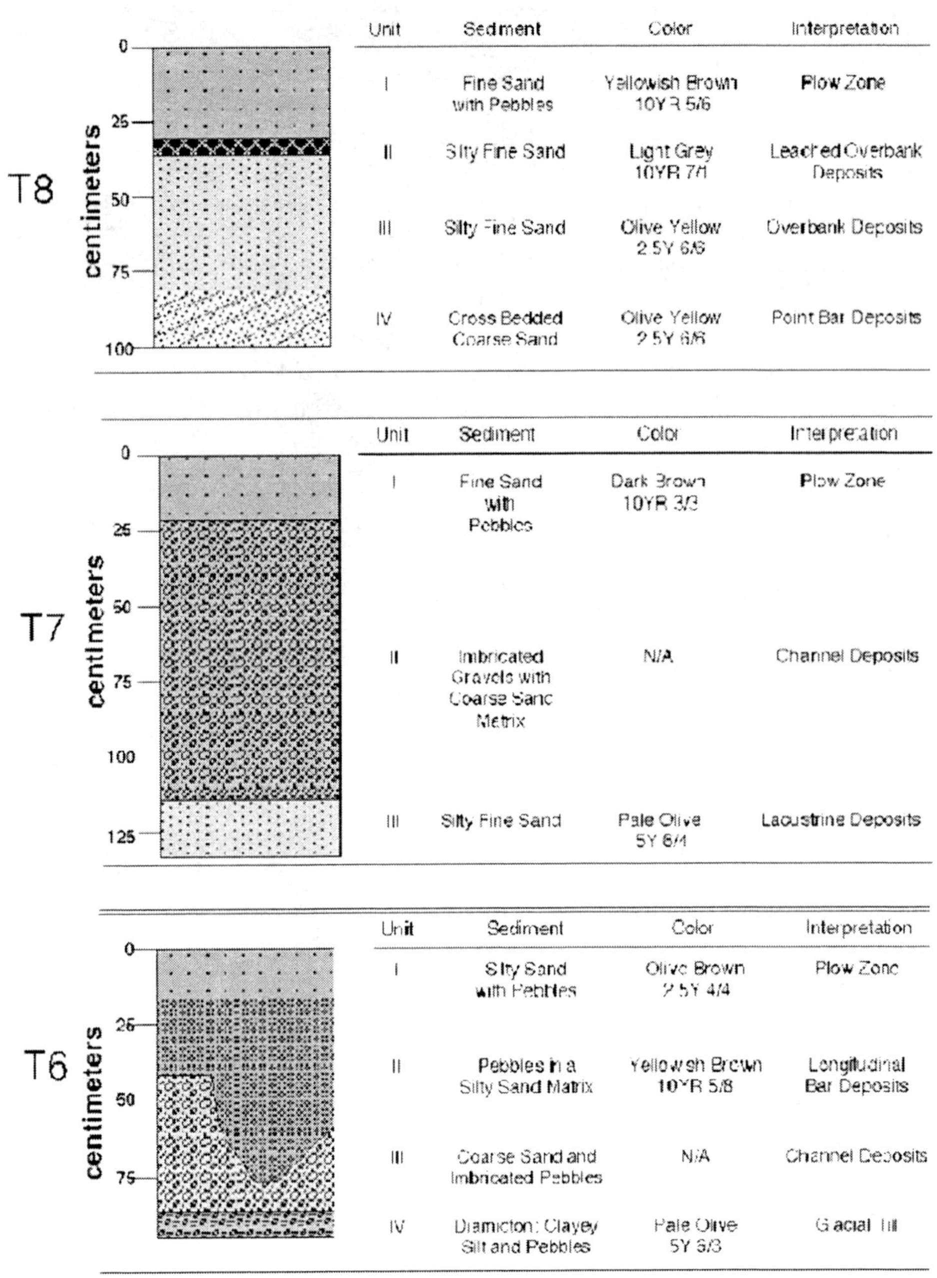

Figure 20. Examples of sedimentological difference in Huntington River terrace deposits. Oldest terraces have highest numbers. (From Whalen, 1998)

bank deposits of silty fine sand. This stratigraphy implies that the fluvial system behaved similarly to today's meandering river since the time of T5.

Alluvial fans are seldom-studied in the Northeast. We have investigated 22 alluvial fans in the Huntington River basin (Bierman et al., 1997; Zehfuss and Bierman, 1996), clustered in the northern part of the river valley where Huntington River terraces provide a platform for fan sedimentation. We have intensively studied three sites (Audubon, Moultroup, and Aldrich) and have used these sites to infer the regional history of hillslope stability during the Holocene. Recent work by Jennings et al. (2003) has extended this type of study throughout Vermont.

Most fan sediments are poorly sorted, although occasional thin (<10 cm) beds of well-sorted, clast-supported gravel occur, as well as black laminae that may represent decomposed leaf mats or concentrations of finely disseminated charcoal. Sediment deposited in the fans originates from distinct drainage basins located either on the surrounding mountainsides or on slopes created by the risers of higher Huntington River terraces. We ^{14}C-dated 14 samples of wood and charcoal from five Huntington valley alluvial fans in order to determine the timing of aggradation. These data show that two fans began to aggrade in the early Holocene (between 8530 and 8060 ^{14}C yrs. B.P., and between 7835 and 7360 ^{14}C yrs. B.P.) and two in the late Holocene (2500 and 1900 ^{14}C yrs. B.P.). One fan aggraded over 4 m during historic time (<100 ^{14}C yrs. B.P.).

Each fan contains a well-preserved, buried-soil profile, generally within the uppermost meter. But, in one fan, the soil is buried by four meters of debris, where ^{14}C ages confirm that the sediment above the buried soils was deposited coincident with historic destabilization of hillslopes. Without stability provided by root systems, the soils and sediments were more easily eroded from drainage basins, increasing sedimentation rates on the fan surfaces. Removal of vegetation interception of precipitation is also a factor. After initial deforestation, sediment continued to be deposited on the fans at a rapid rate. The recovery of forest growth on the hillsides above Vermont fans has returned some, but not complete, stability to slopes.

STOP 6-4. HUNTINGTON GORGE, HUNTINGTON CENTER, VT.

Huntington Gorge is cut through schist of the Underhill Fm. and displays well-developed potholes and plunge pools. The gorge appears to exploit joint sets trending N82°E, N75°W, and N51°W (Christman and Secor, 1961). Until recently, when the gorge formed was not known, although common speculations include incision immediately after deglaciation when poorly-vegetated slopes generated large amounts of sediment-charged runoff or catastrophic draining of a glacial lake. However, recent work by Whalen (1998), who surveyed longitudinal profiles of the Huntington River terraces, ^{14}C-dated terrace sediments, and correlated terraces to changing base-levels, constrains the age of the present Huntington Gorge to <11,700 ^{14}C yrs. B.P. The gorge may have been first exposed as late as 8500 ^{14}C yrs. B.P., the oldest age for charcoal pulled from the overbank sediments of terrace T5, the first terrace showing a gradient increase in the area of the gorge. These ages lead to the important conclusion that the gorge was formed by fluvial erosion at least 1000 ^{14}C years after local deglaciation, not by the catastrophic draining of any proglacial lake.

STOP 6-5. BEN AND JERRY'S FACTORY, WATERBURY, VT., LUNCH STOP

The Ben and Jerry's factory is currently the most popular tourist attraction in Vermont. The factory is located at an elevation of about 180 m, an area that would have been on the bottom of high-level lakes in the Winooski Valley, but near the shoreline as the ice sheet retreated, establishing glacial Lake Vermont (Figs. 1, 12). By Champlain Sea time, the factory would have stood on dry land.

STOP 6-6. CANOE BROOK SECTION, DUMMERSTON, VT. (Ridge)

The Canoe Brook section has been important in establishing the chronology of glacial Lake Hitchcock and calibration of the lower Connecticut varves (Figs. 1, 12). Ridge and Larsen (1990) measured the varves and matched the sequence to varves 5685-6229 of the lower Connecticut varves. Thick ice-proximal varves at the base of the section also give an estimate of the age of deglaciation in southern Vt. More important, several terrestrial plant fossils recovered from high in the section (varve 6150) yielded the first ^{14}C ages (average 12.3-12.5 ka ^{14}C yrs. B.P. or 14.3-14.8 cal. ka) from glacial Lake Hitchcock sediment. This allowed the first calibration and application of both ^{14}C and calibrated (U-Th) time scales to the lower Connecticut varves (Ridge and Toll, 1999; Ridge et al. 1996, 1999). Unfortunately, these ^{14}C ages are from the later stages of a plateau in the ^{14}C time scale (12.6-12.4 ka ^{14}C yrs. B.P.) when about 200 ^{14}C years span about 900 calibrated (U-Th) years. This situation led to uncertainty in the calibration of the varve chronology. Recently, another ^{14}C age at Canoe Brook from lower in the section that pre-dates the ^{14}C plateau (varve 5858, 12,660 $\pm$ 50 ^{14}C yrs. B.P. or 15,300 +227/-105 cal. yrs.) has provided a more precise calibration of the lower Connecticut varves (Ridge, 2003; Fig. 14).

DAY 7. GLACIAL LAKE HITCHCOCK, CONNECTICUT VALLEY, WESTERN MASS. (Brigham-Grette and Rittenour)

The Connecticut Valley in western Mass. contains a rich history of Wisconsin deglaciation and the drainage history of glacial Lake Hitchcock (Fig. 21). Studies in this region go back more than 150 years, but most surficial geologic maps for the areas we will examine were completed by Jahns, who was instrumental in developing the morphosequence concept (Jahns and Willard, 1942). The text that follows was compiled by a number of individuals for a recent field conference (Brigham-Grette et al., 2000). Earlier work of Koteff and Larsen (1989) suggested that glacial Lake Hitchcock drained at once from end to end due to erosion of the sediment dam at Rocky Hill, Conn. More recent work demonstrates that drainage of individual basins was sequential within the lake system, a process that took several thousand years as the Laurentide Ice Sheet retreated (Ridge et al., 1999; K. Curran and T.M. Rittenour, unpublished data). Direct dating of the varve sequence in the lake (Ridge et al., 1999) and sedimentological study of the varves indicate that they record an El Nino / Southern Oscillation-like (ENSO) signal for the period from about 17,000 to 13,000 cal. yrs. B.P. (Rittenour et al., 2000). This record is the first ever to demonstrate a teleconnection to western-Pacific, ENSO-like activity during the LGM.

STOP 7-1. BARTONS COVE KNICKPOINT AND GLACIAL LAKE HITCHCOCK DRAINAGE, TURNERS FALLS, GILL, MASS.

During the stable phase of glacial Lake Hitchcock, the Montague delta (fed by the Millers River) extended completely across the lake basin (Figs. 21, 22), evident from the delta foreset directions along the west side of Canada Hill and north of the Mineral Hills. The Montague delta separates the narrow, shallow-lake

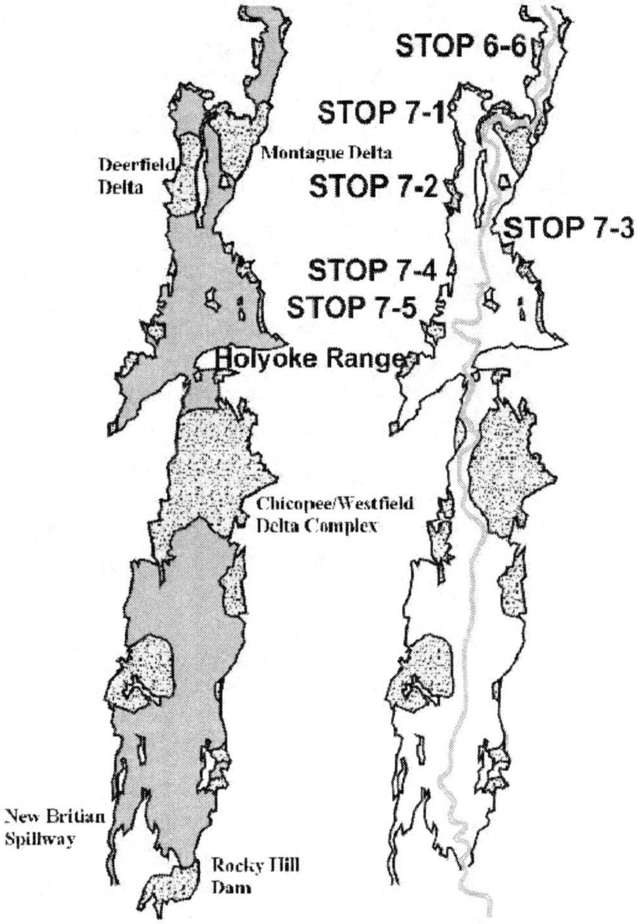

Figure 21. Schematic map showing stops for DAY 7. The sketch outlines glacial Lake Hitchcock in Mass. and Conn.; left sketch is lake filled; right sketch shows course of the modern Connecticut River. Stippled areas are prominent deltas graded to the highest "stable" lake level.

basin to the north (in the Northfield, Mass., area) from the deep, broad, Hadley basin to the south. The Hadley basin was separated from the remaining portions of glacial Lake Hitchcock to the south by the large Chicopee-Westfield delta complex, which also grew to extend across the lake basin.

Lake levels north of the Holyoke Range were controlled by incision into the Rocky Hill, Conn., sediment dam and subsequent entrenchment of the proto-Connecticut River into the Chicopee-Westfield Delta complex. The first lake-level drop in the Hadley basin is known as Lake Hadley (Fig. 23) and is recorded in topset / foreset contacts in the southernmost stretch of the Montague delta, 3-5 m below the stable lake level (Emerson, 1898; Jahns and Willard, 1942; Jahns and Lattman, 1962; Jahns, 1967; Koteff and Larsen, 1989). This drop in lake level may correlate with the terrace cut into the Rocky Hill sediment dam area at 2-5 m below the stable lake level and other deltas south of the Holyoke Range built at this level (Koteff and others, 1988; Koteff and Larsen, 1989; Stone and Ashley, 1992).

During initial lake drainage, a proto-Connecticut River channel was developed on the west side of Canada Hill (Fig. 22). This channel, called the White Ash Swamp channel, has a base at 84 m (270 ft) (~20 m below the stable lake level here) and is the oldest surface occupied by the Connecticut River during incision into the Montague delta. The White Ash Swamp channel was abandoned during further river incision and development of a more favorable route, similar to that of the modern Connecticut River, on the east side of the Canada Hill bedrock ridge.

Complete entrenchment through the Chicopee-Westfield delta complex eventually drained the Hadley basin. At this point, early strath terraces, cut by the flow of water across the lake bottom, were formed. Within a short period of time, channel flow was initiated in the proto-Connecticut River, and these early subtle terraces were abandoned. The exposed, lake-bottom sand quickly became subject to eolian transport, generating dunes and sand sheets. To the north, the drop in base level from the drainage of Lake Hadley caused the Connecticut River to entrench deeper into the Montague delta (Fig. 23). During incision the river became perched on a bedrock ledge near Turners Falls, Mass. (Fig. 22). Erosion continued south of this knick point but was suspended to the north. This situation created two waterfalls over the sandstone ridge, with associated plunge pools at the base (Fig. 22). Prior to the flooding of Barton's Cove by construction of the Turners Falls dam more than 100 years ago, these plunge pools were known as Lily Ponds and so the bedrock ridge has become known as the Lily Pond barrier (Emerson, 1898). Deep channels were cut into the barrier and prominent terraces formed upstream as the Connecticut River flowed over this knick point. Geomorphic and archeological evidence suggest this knick point may have been active for thousands of years, possibly having been abandoned before 10,200 cal. yrs. B.P. (9,000 ^{14}C yrs. B.P.) (Curran, 1999).

STOP 7-2. CONNECTICUT VALLEY VARVES, RIVER ROAD SITE, SUNDERLAND, MASS.

The River Road varve site, located just north of Sunderland (Fig. 21), lies on the west side of the Connecticut River and on the eastern flank of the Pocumtuck Range (Mt. Sugarloaf). The site lies midway between the Montague Plain (topset/foreset contact = 102.7 m) and Long Plain delta (89.9 m) and is one of the most studied exposures in this part of the valley. The inferred lake level at this site is 95.1 m. Two small slumps along this small creek expose over 200 years of sedimentation in glacial Lake Hitchcock. The two exposures have been correlated to each other (Levy, 1998) and to the Antev's master curve (varve #s 4980 to 5200) using patterns of varve thickness (Thomas, 1984). Using Ridge et al.'s (1999) calibrated varve curve, the age range for the site is inferred to be ca. 13,988 - 13,146 cal.yrs. B.P.

Carbonate concretions (40-45 wt. %) found at the River Road exposures range in size from 1 to 5 cm (long axis) and their morphologies appear to be closely related to the microstratigraphy of the host sediment (Levy, 1998). Based on carbon isotope data, the source of the concretion carbonate is

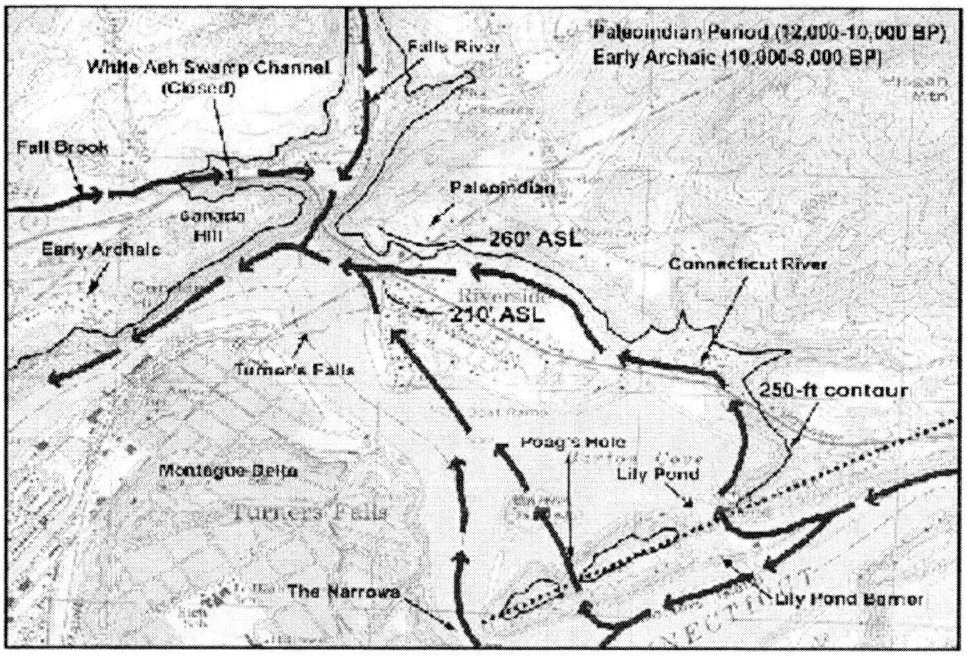

Figure 22. Part of Greenfield, Mass., 1:25,000 quadrangle, patterned area represents the once-continuous surface of the Montague delta connected by arrows. White Ash Swamp channel and Lily Pond barrier are labeled.

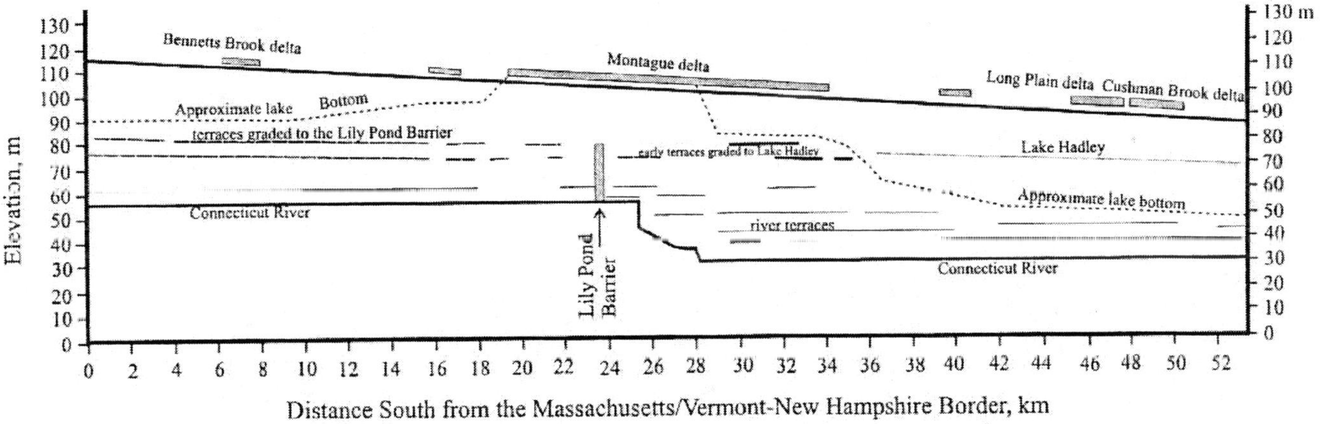

Figure 23. Longitudinal profile of highest Hitchcock terraces in the Hadley and Northfield basins, Mass. (from Rittenour, 1999). Dotted line is the ancient lake bottom and thick solid line the elevation of the modern river system. The Lilly Pond barrier at Barton's Cove in Turner's Falls, Mass., is shown schematically as a small vertical wall.

thought to be carbon originating from terrestrial organic matter deposited with the lake sediment. Individual concretion bands have been AMS ^{14}C-dated to evaluate their utility for reliable age control. Concretion centers date up to 2,500 years too old and the outer concretion samples too young (relative to the inferred age of the site), questioning the reliability of this approach. Recent work by Bosiljka Glumac (Smith College) indicates oxygen isotope ratios of –8.5 to –13.0 %δ (SMOW), values consistent with interglacial meteoric (local) water.

STOP 7-3. SUNDERLAND DELTA, MORPHOSEQUENCES AND ECHO DUNES, SUNDERLAND, MASS.

The Sunderland delta (or Long Plain delta, Fig. 24) is a classic example of the ice-contact / meteoric delta, graded to the stable level of glacial Lake Hitchcock. The topset / foreset contact occurs here at 90 m (295 ft). Once the ice margin had retreated northward and bifurcated around Mt. Toby, rain and meltwater continued to prograde the delta to lake level.

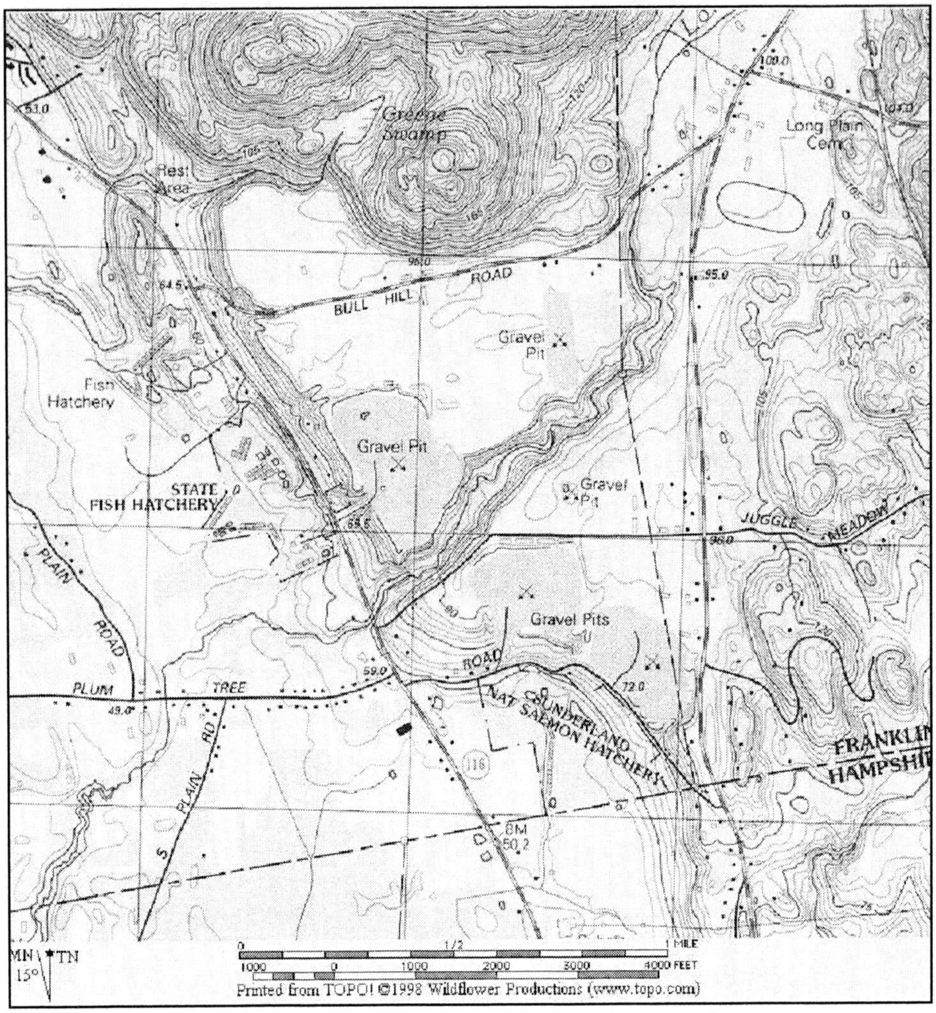

Figure 24. Part of Mt. Toby, Mass., 1:25,000 quadrangle, showing the Sunderland delta. Note the narrow canyon cut by the Long Plain River as glacial Lake Hitchcock drained and the strategic location of the fish hatcheries tapping freshwater springs from the base of the delta. Route 63 cuts through the echo dune at the base of the delta.

Following drainage of glacial Lake Hitchcock from the Hadley basin, Long Plain Brook quickly adjusted to a lower base level by incising a small canyon through the delta (Fig. 24). This downcutting produced an alluvial fan out over the floor of the former lake. Today, groundwater maintains a continuous flow through the delta, producing natural springs from sand and gravel beds confined by lake clay and silt at the foot of the delta slope. The State of Massachusetts maintains two fish hatcheries for trout at the base of the delta.

Periglacial conditions persisted in the valley for some time following lake drainage. Here, in front of the Sunderland delta, winds blowing across the exposed sediments of glacial Lake Hitchcock picked up sediments and created an echo dune. Such dunes are formed due to air turbulence as wind rises over an obstacle, commonly a steep bedrock outcrop, drumlin, or delta front on the eastern side of the lake basin.

LUNCH STOP: Mt. Sugarloaf State Park, Sunderland, Mass. (alternative site Mass. State Fish Hatchery or Yankee Candle Company picnic area)

STOP 7-4. UNIVERSITY OF MASSACHUSETTS-AMHERST NATIONAL GEOTECHNICAL EXPERIMENTATION SITE; ENSO-LIKE SIGNALS IN GLACIAL LAKE HITCHCOCK VARVES, AMHERST, MA

In the fall of 1997, two 10-cm-diameter cores were drilled at this site (Fig. 21), 1.5 km from the shore of glacial Lake Hitchcock, on a 43 m, lake-bottom surface in the broad, deep Hadley basin. In this area, the lake was about 20 km wide and the stable lake level was at 90 m. Water depth at the core site was 77.5 m at the initiation of varve deposition. Two cores were drilled to study the classic glacial varves in an area of the valley that lacked natural exposures and represented a large geographic gap in Antev's original varve record. One core was drilled to bedrock, and a second shorter core was taken to ensure complete recovery of the upper portion of the varve sequence (7.6 m). The U. Mass. core covers varves 4638-6027, capturing at one site a total of 1,389 varves, or nearly one-third of the entire lake history. The 32-m-long core drilled to bedrock provided a continuous sequence of varves ranging from thick (55 cm) ice-proximal varves at the

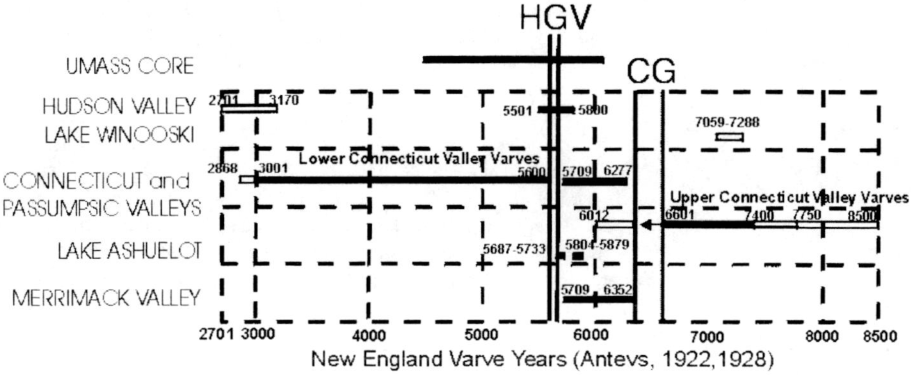

Figure 25. Time span in varve years (younger to the right) of overlapping sequences of the NE varve chronology (NE 2701 – 8500) plus the UMass core (4638 – 6027). Hudson Valley gap (corrected by Rittenour, 1999) and Claremont gap (corrected by Ridge et al., 1999)

base, through varves of average thickness for the region (1–4 cm), and into thin varves with relatively coarse-grained summer layers and thinner winter clay layers.

A ^{14}C age from small plant fragments picked from varves in a contorted zone between varves 5761-5768 produced a δ ^{13}C-corrected AMS age of 12,370 ± 120 ^{14}C yrs. B.P. (14.3 +1.2/-0.4 cal. kyr B.P.). AMS ages from the top and bottom of the core indicate that it records varve deposition from 15.4 to 14.0 cal. kyr B.P. (12.8 to 12.0 ka ^{14}C yrs. B.P., coincident with a large ^{14}C plateau at this time where ~1400 cal. yrs. equals only 800 ^{14}C years). The varves were correlated to normal curves from glacial Lake Hitchcock (datasets MASS 4-13, NH 13, VT 14-15, and VT-NH 16-17; Antevs, 1922) and the Canoe Brook section of Ridge and Larsen (1990). A 300-year section of the U. Mass. core was also matched with a normal curve from the Hudson Valley (dataset NY 13-14; Antevs, 1922) and used to bridge a critical gap in the Connecticut Valley sequence (Rittenour, 1999). This gap and the Claremont Gap in Vt. filled by Ridge et al. (1999) are the only two major adjustments that have been made to the Antevs' varve chronology. Other varve thickness measurements and correlations in the over 4000-year varve chronology have otherwise withstood vigorous testing (Fig. 25).

The chronology and varve counting from this site suggest that glacial Lake Hitchcock completely drained in the Hadley basin sometime after varve 6027 was deposited at the top of the core) at 14.0 ka cal. yrs. B.P. (12.0 ka ^{14}C yrs. B.P.). Counting to the base of the section, Laurentide ice had retreated from the Amherst, Mass. area as the first, ice-proximal varve was deposited (varve 4638) at 15.4 ka cal. B.P. (12.8 ka ^{14}C yrs. B.P.).

Spectral analyses were conducted on the New England varve chronology (varves 2868-6900) to test for climate signals in the varve-thickness record (Rittenour et al., 2000). Statistically significant climate signals occur at multi-decadal (>40 year) timescales, within the conventional 2.5-5 year and 7-9 year cycles associated with ENSO. The 22-year period signal also seen here has the same frequency as the Hale solar magnetic cycle, which has been correlated with modern Northern Hemisphere temperature variations. The absence of the lower-frequency component of variability during the latter part of the interval studied here (i.e., the latest Pleistocene) is consistent with the theoretical prediction of fewer large ENSO events in the early Holocene. Rittenour et al.'s (2000) results suggest that the ENSO system was operational during the late Pleistocene, when the Laurentide Ice Sheet was near its maximum extent and climatic boundary conditions were different than today.

STOP 7-5. GEOCHRONOLOGY OF DUNES ON THE FLOOR OF GLACIAL LAKE HITCHCOCK, HADLEY, MASS.

A variety of fossil dunes can be found throughout the Connecticut Valley and most are thought to have formed under cold, dry conditions just after the drainage of glacial Lake Hitchcock. Dunes are found on a variety of surfaces but of critical importance are those that formed on abandoned deltas and the floor of glacial Lake Hitchcock (Rittenour, 1999). Exposed at this site are sediments from a sand dune that formed on exposed lake bottom sediments in the Hadley basin (Fig. 21). It is one of many roughly north-south-trending, transverse dunes and is composed of shallowly-dipping beds that mimic the gentle lee slope of the dune.

Optically-stimulated luminescence (OSL) ages were obtained from parabolic and transverse sand dunes located on a number of genetic surfaces within the Connecticut Valley, Mass. (Rittenour, 1999; OSL analysis by Steve Forman, Univ. Illinois, Chicago). The ages from these dunes indicate that most formed between 14.0 ± 1.0 and 14.4 ± 1.0 ka cal. yrs. B.P. (12.0 to 12.4 ka ^{14}C yrs. B.P). An OSL age on one dune suggests that the Hadley basin completely drained prior to the deposition of the sand dune at 14.3 ± 1.6 ka cal. yrs. B.P. (12.3 ka ^{14}C yrs. B.P.). Also, ^{14}C ages from the U. Mass. core indicate that varve deposition continued within the Hadley basin until 14.0 ka cal. yrs. B.P. (12.0 ka ^{14}C yrs. B.P.). These ages constrain the complete drainage of glacial Lake Hitchcock in Massachusetts to between 14.0 and 14.3 ka cal. yrs. B.P.

End of field trip! Proceed to Logan International Airport, Boston, Mass.

ACKNOWLEDGMENTS

We thank Sheldon Eyster-Smith, Vivek Bishnoi, and Anna Tary, Bentley College, for assistance in preparing the field trip manuscript and figures. We dedicate this field trip to Hal Borns, Dick Goldthwait, Joe Hartshorn, and Ernie Mueller, four of our mentors in the glacial geology of the northeastern United States.

REFERENCES

Adkins, J.F., Cheng, H., Boyle, E.A., Druffel, E.R.M., and Edwards, R.L., 1998, Deep-sea coral evidence for rapid change in ventilation of the deep North Atlantic 15,400 years ago: Science, v. 280, p. 725-728.

Agassiz, L., 1870, On the former existence of local glaciers in the White Mountains: American Naturalist, v. 4, p. 550-558.

Allen, T.T., Creasy, J.W., Davis, P.T., Eusden, J.D., Fowler, B.K., and Thompson, W.B., 2001, The Notches: Bedrock and surficial geology of New Hampshire's White Mountains: *in* West, D.P., Jr., and Bailey, R.H., eds., Guidebook for Geological Field Trips in New England, 2001 Annual Meeting of the Geological Society of America, Boston, Mass., p. C1-33.

Antevs, E., 1922, The recession of the last ice sheet in New England: American Geographical Society, Research Series No. 11, 120 p.

Antevs, E., 1928, The last glaciation with special reference to the ice sheet in North America: American Geographical Society Research Series no. 17, 292 p.

Antevs, E., 1932, Alpine zone of the Mt. Washington Range: Auburn, Maine, Merrill and Webber, 118 p.

Ashley, G.M., Boothroyd, J.C., and Borns, H.W., Jr., 1991, Sedimentology of late Pleistocene (Laurentide) deglacial-phase deposits, eastern Maine; An example of a temperate marine grounded ice-sheet margin: *in* Anderson, J.B., and Ashley, G.M., eds., Glacial marine sedimentation; Paleoclimatic significance: Geological Society of America, Special Paper 261, p. 107-125.

Attig, J.W., Jr., 1975, Quaternary stratigraphy and history of the Androscoggin River valley, Maine: MS thesis, University of Maine, Orono, ME, 56 p.

Bacchus, T.S., 1993, Late Quaternary stratigraphy and evolution of the eastern Gulf of Maine: PhD thesis, , University of Maine, Orono, ME, 347 p.

Baldwin, L., Bierman, P., Schwartz, A., Church, A. and Larsen, P., 1995, The effects of colonial disturbance and subsequent reforestation on the Vermont landscape: Geological Society of America, Abstracts with Programs, v 27, no. 1, p. A28.

Bard, E., Hamelin, B., Fairbanks, R.G., and Zindler, A., 1990, Calibration of the ^{14}C timescale over the past 30,000 years using mass spectrometric U-Th ages from Barbados corals: Nature, v. 345, p. 405-410.

Barnhardt, W.A., Gehrels, W.R., Belknap, D.F., and Kelley J.T., 1995, Late Quaternary relative sea-level change in the western Gulf of Maine: Evidence for a migrating glacial forebulge: Geology, v. 23, p.317-320.

Barnhardt, W.A., Belknap, D.F., and Kelley, J.T., 1997, Stratigraphic evolution of the inner continental shelf in response to late Quaternary relative sea-level change, northwestern Gulf of Maine: Geological Society of America, v. 109, p. 612-630.

Baskerville, C.A., Ratte, C.A., and Lee, F.T., 1988, A rock fall and debris slide at Smugglers Notch, Mt. Mansfield, Cambridge, Vermont: Vermont Geological Survey, Studies in Vermont Geology, no. 4, 10 p.

Belknap, D.F., Andersen, B.G., Anderson, R.S., Anderson, W.A., Borns, H.W., Jr., Jacobson, G.L., Kelley, J.T., Shipp, R.C., Smith, D.C., Stuckenrath, R., Jr., Thompson, W.B., and Tyler, D.A., 1987, Late Quaternary sea-level changes in Maine: *in* Nummedal, N. Pilkey, O.R. and Howard, J.D., eds., Sea-level fluctuation and coastal evolution, Society of Economic Paleontologists and Mineralogists Special Publication 41, p. 71-85.

Belknap, D.F., and Shipp, R.C., 1991, Seismic stratigraphy of the glacial marine units, Maine inner shelf: *in* Anderson, J. B., and Ashley, G. M., eds., Glacial marine sedimentation; paleoclimatic significance, Geological Society of America, Special Paper 261, p. 137-157.

Bernotavicz, A.A., 1994, Glacial and postglacial history of the Sabattus valley, Sabattus, Maine: BA thesis, , Bates College, Lewiston, ID, 141p.

Bierman, P., 1994, Using in situ cosmogenic isotopes to estimate rates of landscape evolution: A review from the geomorphic perspective: Geophysical Research, v. 99 (B-7), p. 13,885-13,896.

Bierman, P.R and Davis, P.T., 2000, Old surfaces on New England summits imply thin Laurentide ice: Geological Society of America, Abstracts with Programs, v. 32, no. 7, p. A-330.

Bierman, P.R., Lini, A., Zehfuss, P., Church, A., Davis, P.T., Southon, J., and Baldwin, L., 1997, Postglacial ponds and alluvial fans: Recorders of Holocene landscape history: GSA Today, v. 7, no. 10, p.1-8.

Billings, M.P., and Fowler-Billings, K., 1975, Geology of the Gorham quadrangle, New Hampshire-Maine: New Hampshire Department of Resources and Economic Development, Bulletin 6, 120 p.

Björck, S., Walker, M.J.C., Cwynar, L.C., Johnsen, S., Knudsen, K.-L., Lowe, J.J., Wohlfarth, B., and INTIMATE members, 1998, An event stratigraphy for the last termination in the North Atlantic region based on the Greenland ice-core record: a proposal by the INTIMATE group: Journal of Quaternary Science, v. 13, p. 283-292.

Bloom, A.L., 1960, Late Pleistocene changes of sea level in southwestern Maine: Department of Economic Development, Maine Geological Survey, 143 p.

Bloom, A.L., 1963, Late Pleistocene fluctuations of sea level and postglacial crustal rebound in coastal Maine: American Journal of Science, v. 261, p. 862-879.

Boisvert, R.A., Dorion, C.C., and Thompson, W.B., 2002. Decline of the Ice Age and arrival of Paleoindians in the northern White Mountains-Part 3: Windswept (Mt. Washington Observatory Bulletin), v. 43, no. 1, p. 40-57.

Boothroyd, J.C., 1995, The Pine River and other esker systems of southwestern Maine and south-central New Hampshire: implications for regional deglaciation: Geological Society of America, Abstracts with programs, v. 27, no. 1, p. 31.

Borns, H.W., Jr., Dorion, C.C., Jacobson, G.L., Jr., Kreutz, K.J., Thompson, W.B., Weddle, T.K., Doner, L.A., Kaplan, M.R., and Lowell, T.V., in press, The deglaciation of Maine: *in* Ehlers, J., and Gibbard, P.L., eds., Quaternary glaciations-Extent and Chronology, Part II: North America, Elsevier, Amsterdam,.

Borns, H.W., Jr., and Hagar, D.J., 1965, Late-glacial stratigraphy of a northern part of the Kennebec River valley, western Maine: Geological Society of America, Bulletin, v. 76, p. 1233-1250.

Borns, H.W., and Hughes, T.J., 1977, The implications of the Pineo Ridge readvance in Maine: Geographie Physique et Quaternaire, v. 31, p. 203-206.

Bradley, D.C., 1981, Late Wisconsin mountain glaciation in the northern Presidential Range, New Hampshire: Arctic and Alpine Research, v. 13, no. 3, p. 319-327.

Brennan, W.J., Hamilton, M., Kilbury, R., Reeves, R.L., and Covert, L., 1984, Late Quaternary secular variation of geomagnetic declination in western New York: Earth and Planetary Science Letters, v. 70, p. 363-372.

Brigham-Grette, J., Rittenour, T., Stone, J., Ridge, J., Werner, A., Levy, L., Dincauze, D., Curren, K., Klekowski, E., and Little, R., 2000, A new drainage history for Glacial Lake Hitchcock: Varves, landforms, and stratigraphy, Northeastern Friends of the Pleistocene Field Conference, June 2-4, 2000, Amherst, University of Massachusetts Department of Geosciences Contribution No. 73, 125 p.

Caldwell, D.W., Hanson, L.S., and Thompson, W.B., 1985, Styles of deglaciation in central Maine: *in* Borns, H.W., Jr., LaSalle, P., and Thompson, W.B., eds., Late Pleistocene history of Northeastern New England and adjacent Quebec: Geological Society of America Special Paper 197, p. 45-57.

Chapman, D.H., 1937, Late-glacial and post glacial history of the Champlain valley: American Journal of Science, 5th series, v. 34, p. 89-124.

Christman, R.A. and Secor, D.T., 1961, Geology of the Camels Hump quadrangle, Vermont: Vermont Geological Survey, Bulletin 15, 70 p.

Cowan, E.A., and Powell, R.D., 1990, Suspended sediment transport and deposition of cyclically interlaminated sediment in a temperate glacial fjord, Alaska, U.S.A: *in* Dowdeswell, J.A., and Scourse, J.D., eds., 1990,

Glacimarine environments: processes and sediments: Geological Society Special Publication Number 53, The Geological Society, London, p. 75-89.

Crosby, I.B., 1934a, Extension of the Bethlehem, New Hampshire, moraine: Journal of Geology, v. 42, p. 411-421.

Crosby, I.B., 1934b, Geology of the Fifteen Mile Falls development: Civil Engineering, v. 4, p. 21-24.

Crossen, K.J., 1984, Glaciomarine deltas in southwestern Maine: formation and vertical movements: MS thesis, , University of Maine, Orono, ME, 121 p.

Crossen, K.J., 1991, Structural control of deposition by Pleistocene tidewater glaciers, Gulf of Maine: *in* Anderson, J. B., and Ashley, G. M., eds., Glacial marine sedimentation; paleoclimatic significance, Geological Society of America, Special Paper 261, p. 127-135.

Curran, K., 1999, Landscape evolution revealed by archeological excavations at Peskomskut: MS thesis, , University of Massachusetts, Amherst, MA.

Cwynar, L.C. and Spear, R.W., 2001, Lateglacial climate change in the White Mountains of New Hampshire: Quaternary Science Reviews, v. 20, no. 11, p. 1265-1274.

Dansgaard, W., White, J.W.C., and Johnsen, S.J., 1989, The abrupt termination of the Younger Dryas climate event: Nature, v. 339, p. 532-534.

Dansgaard, W., Johnsen, S.J., Clausen, H.B., Dahl-Jensen, D., Gundestrup, N.S., Hammer, C.U., Hvidberg, C.S., Steffensen, J.P., Sveinbjörnsdottir, A.E., Jouzel, J., and Bond, G., 1998, Evidence for general instability of past climate from a 250-kyr ice-core record: Nature, v. 364, p. 218-220.

Davis, M.B., Spear, R.W., and Shane, L.C.K., 1980, Holocene climate of New England: Quaternary Research, v. 14, p. 240-250.

Davis, P.T., 1999, Cirques of the Presidential Range, New Hampshire, and surrounding alpine areas in northeastern United States: Geographie physique et Quaternaire, v. 53, no. 1, p. 25-45.

Davis, P.T., Fowler, B.K., Thompson, D.J., and Thompson, W.B., 1996, Continental and alpine glacial sequence and mass wasting on Mt. Washington, northern New Hampshire: *in* Van Baalen, M.R., ed., Guidebook to field trips in northern New Hampshire and adjacent regions of Maine and Vermont, New England Intercollegiate Geological Conference, 88th annual meeting, Mt. Washington, New Hampshire, p. A4, 1-38.

Davis, P.T., Thompson, W.B. et al., 1988, Late Quaternary glacial and vegetational history of the White Mountains, New Hampshire, Guidebook for Trip A-1: 10th biennial meeting, American Quaternary Association, University of Massachusetts, Amherst, MA, 68p.

Davis, P.T., Thompson, W.B., Stone, B.D., Newton, R.M., and Fowler, B.K., 1993, Multiple glaciations and deglaciation along a transect from Boston, MA to the White Mountains, NH: Geological Society of America, Field Trip Guidebook–New England Section Mtg., v. 2, p. EE-1 to EE-27.

Davis, R.B., and Jacobson, G.L., Jr., 1985, Late-glacial and early Holocene landscapes in northern New England and adjacent areas of Canada: Quaternary Research, v. 23, p. 341-368.

Donner, J., 1995, The Quaternary history of Scandinavia: , Cambridge University Press, Cambridge, 200 p.

Dorion, C.C., Balco, G., Kaplan, M., Kreutz, K., Wright, J., and Borns, H.W., 2001, Stratigraphy, paleoceanography, chronology, and environment during deglaciation of Eastern Maine: *in* Weddle, T.K., and Retelle, M.J., eds., Deglacial history and relative sea-level changes, northern New England and adjacent Canada: Geological Society of America Special Paper 351, p. 193-216.

Emerson, B.K., 1898, Geology of old Hampshire county, Massachusetts, comprising Franklin, Hampshire and Hampden counties: U.S. Geological Survey Monograph 29, 790 p.

Flaccus, E., 1958, White Mountain landslides: Appalachia, v. 24, p. 175-191.

Fowler, B.K., 1982, Rock mass stability studies, Old Man of the Mountains, Franconia Notch, NH: *in* Farquhar, O.C., ed., University of Massachusetts, Conference Proceedings–Geotechnology in Massachusetts, p. 141-148.

Fowler, B.K., 1984, Late-Wisconsin cirque glacier activity in King ravine, Northern Presidential Range, NH: Arctic and Alpine Research, v. 16, no. 4, p. 431-437.

Fowler, B.K., 1997, Franconia Notch and The Old Man: Mt. Washington Observatory Bulletin, v. 38, no. 4, p. 8-15.

Gerath, R.F. and Fowler, B.K., 1982, Discussion of "Late Wisconsin mountain glaciation in the northern Presidential Range, New Hampshire" by D.C. Bradley: Arctic and Alpine Research, v. 14, no. 4, p. 369-370.

Gerath, R.F., Fowler, B.K., and Haselton, G.M., 1985, The deglaciation of the northern White Mountains of New Hampshire: *in* Borns, H.W., Jr.; LaSalle, P., and Thompson, W.B., eds., Late Pleistocene history of northeastern New England and adjacent Quebec: Geological Society of America Special Paper 197, p. 21-28.

Gilbert, R., 2000, The Devil Lake Pothole (Ontario): Evidence of subglacial fluvial processes: Geographie physique et Quaternaire, v. 54, no. 2, p. 245-250.

Gipp, M.R., and Piper, D.J.W., 1989, Chronology of late Wisconsin glaciation, Emerald Basin, Scotian Shelf: Canadian Journal of Earth Sciences, v. 26, p. 333-335.

Goldthwait, J.W., 1913, Glacial cirques near Mt. Washington: American Journal of Science, v. 235, p. 21-28.

Goldthwait, J.W., 1916a, Glaciation in the White Mountains of New Hampshire: Geological Society of America Bulletin, v. 27, p. 263-294.

Goldthwait, J.W., 1916b, Evidence for and against the former existence of local glaciers in Vermont: Vermont State Geologist, 10th Report, p. 42-73.

Goldthwait, R.P., 1938, The uncovering of New Hampshire by the last ice sheet: American Journal of Science, 5th Series, v. 36, p. 345-372.

Goldthwait, R.P., 1939, Mt. Washington in the great ice age: New England Naturalist, no. 5, p. 12-19.

Goldthwait, R.P., 1940, Geology of the Presidential Range: New Hampshire Academy of Science, Bulletin no. 1, 43 p.

Goldthwait, R.P., 1970a, Guidebook for the Mt. Washington area: Privately printed for 33rd Annual Reunion, Northeastern Friends of the Pleistocene, 12 p.

Goldthwait, R.P., 1970b, Mountain glaciers of the Presidential Range: Arctic and Alpine Research, v. 2, no. 2, p. 85-102.

Goldthwait R.P. and Mickelson, D.M., 1982, Glacier Bay: A model for the deglaciation of the White Mountains in New Hampshire: *in* Larson, G.J. and Stone B.D., eds., Late Wisconsin glaciation of New England, Kendall/Hunt, Dubuque, Iowa, p. 167-181.

Gosse, J.C. and Phillips, F.M., 2001, Terrestrial in situ cosmogenic nuclides: Theory and application: Quaternary Science Reviews, 20, p. 1475-1560.

Hitchcock, C.H., 1878, Glacial Drift, The Geology of New Hampshire, III: p. 177-340.

Hooke, R LeB., 2000, The Katahdin esker from Lincoln to Sunkhaze Stream: *in* Yates, M.G., Lux, D.R., and Kelley, J.T., eds., Guidebook for field trips in coastal and east-central Maine, 92nd annual meeting of the New England Intercollegiate Geological Conference, University of Maine, Orono, p. 82-93.

Hughes, T., Borns, H.W., Jr., Fastook, J.L., Hyland, M.R., Kite, J.S., and Lowell, T.V., 1985, Models of glacial reconstruction and deglaciation applied to Maritime Canada and New England: *in* Borns, H.W. Jr., LaSalle, P., and Thompson, W.B., eds., Late Pleistocene history of northeastern New England and adjacent Quebec: Geological Society of America Special Paper 197, p. 139-150.

Hughes, T., 1987, Ice dynamics and deglaciation models when ice sheets collapsed: *in* Ruddiman, W.F., and Wright, H.E., Jr., eds., North America and adjacent oceans during the last deglaciation: Boulder, Colorado, Geological Society of America, The Geology of North America, v. K-3, p. 183-220.

Jahns, R.H., 1967, Late Pleistocene of the Connecticut valley in northern Massachusetts: NEIGC Guide book, 59th Annual Meeting, Amherst College, Mt. Holyoke College, Smith College, and University of Massachusetts, Amherst, MA, p. 167-193.

Jahns, R.H., and Lattman, L.H., 1962, Sequence of late-glacial lacustrine sedimentation in the Hadley basin, Massachusetts: Abstracts with Programs, GSA Annual Meeting, p. 177.

Jahns, R.H., and Willard, M.E., 1942, Late Pleistocene and recent deposits in the Connecticut valley, Massachusetts, Parts I and II: American Journal of Science, v. 240, p. 161-191 and p. 265-287.

Jennings, K., Bierman, P., and Southon, J., 2003, Timing and style of deposition on humid-temperate fans, Vermont, U.S.A.. Geological Society of America Bulletin.

Johnson, D.W., 1933, Date of local glaciation in the White Mountains: American Journal of Science, v. 225, p. 399-405.

Johnson, E.A., Murphy, T., and Torreson, O.W., 1948, Pre-history of the earth's magnetic field: Terrestrial Magnetism and Atmospheric Electricity (now Journal of Geophysical Research), v. 53, p. 349-372.

Kelley, J. T., Dickson, S.M., Belknap, D.F., and Stuckenrath, R., Jr., 1992, Sea-level change and the introduction of late Quaternary sediment to the southern Maine inner continental shelf: *in* Fletcher, C.H., and Wehmiller, J.F., eds., Quaternary coasts of the United States: marine and lacustrine systems: Society of Economic Paleontologists and Mineralogists Special Publication No. 48, p. 23-34.

King, L.H., 1996, Late Wisconsin ice retreat from the Scotian Shelf: Geological Society of America Bulletin, v. 108, p. 1056-1067.

Koteff, C. and Pessl, F., Jr., 1985, Till stratigraphy in New Hampshire: Correlations with adjacent New England and Quebec: *in* Borns, H.W. Jr., LaSalle, P. and Thompson, W.B., eds., Late Pleistocene history of northeastern New England and adjacent Quebec. Geological Society of America Special Paper 197, p. 1-12.

Koteff, C., and Larsen, F.D., 1989, Postglacial uplift in western New England: Geologic evidence for delayed rebound: *in* Gregresen, S., and Bashman, P.W., eds., Earthquakes at North-Atlantic Passive Margins: Neotectonics and Postglacial Rebound, Kluwer Academic Publishers, Norwel, MA, p. 105-123.

Koteff, C., Robinson, G.R., Goldsmith, R., and Thompson, W.B., 1993, Delayed postglacial uplift and synglacial sea levels in coastal central New England: Quaternary Research, v. 40, p. 46-54.

Koteff, C., Stone, J.R., Larsen, F.D., Ashley, G.M., Boothroyd, J.C., and Dincauze, D.F., 1988, Glacial Lake Hitchcock, postglacial uplift, and post-lake archeology: *in* Brigham-Grette, J., ed., AMQUA 1988 Field Trip Guidebook, Contribution No. 63, Department of Geology and Geography, University of Massachusetts, Amherst, MA, 01003.

Larsen, F.D., 2001, The Middlesex Readvance of the late-Wisconsin ice sheet in central Vermont at 11,900 ^{14}C years BP: Geological Society of America Abstracts with Programs, v.33, no. 1, p. A-15.

Lee, F.T., Odum, J.K., and Lee, J.D., 1994, Rock falls and debris avalanches in the Smugglers Notch area, Vermont: U.S. Geological Survey Bulletin 2075, 33 p.

Levy, L.B., 1998, Interpreting the carbonate concretions of Glacial Lake Hitchcock: BA thesis, Mt. Holyoke College, South Hadley, Mass.

Loso, M., Schwartz, H., Wright, S., and Bierman, P., 1998, Morphology, composition, and genesis of a moraine-like feature in the Miller Brook valley, Vermont: Northeastern Geology and Environmental Sciences, v. 20 no. 1, p. 1-10.

Lougee, R.J., ca. 1930, The origin and occurrence of glacial washed deposits in the White Mountain region: unpublished manuscript based on field work for the New Hampshire Highway Department, 26 p.

Lougee, R.J., 1935, Time measurements of an ice readvance at Littleton, N.H.: Proceedings of the National Academy of Sciences, v. 21, no. 1, p. 36-41.

Lougee, R.J., 1940, Deglaciation of New England: Journal of Geomorphology, v. 3, p. 189-217

Lowell, T.V., and Borns, H.W., Jr., 1994, Maine's calving bay?: Geological Society of America, Abstracts with Programs, v. 26, no. 3, p. 57.

Lyons, J.B., Bothner, W.A., Moench, R.H., and Thompson, J.B., Jr., 1997, Bedrock geologic map of New Hampshire: U. S. Geological Survey, 1:250,000-scale map.

McNish, A.G., and Johnson, E.A., 1938, Magnetization of unmetamorphosed varves and marine sediments: Terrestrial Magnetism and Atmospheric Electricity (now Journal of Geophysical Research), v. 43, p. 401-407.

Molnia, B.F., 1983, Subarctic glacial-marine sedimentation; A model: *in* Molnia, B.F., ed., Glacial-marine sedimentation, Plenun Press, NY, p. 95-144.

Newman, W.A., Berg, R.C., Rosen, P.S., and Glass, H.D., 1990, Pleistocene stratigraphy of the Boston Harbor drumlins, Massachusetts: Quaternary Research, v. 34, p. 148-159.

Noren, A.J., Bierman, P.R., Steig, E.J., Lini, A., and Southon, J., 2002, Millennial-scale storminess variability in the northeastern United States during the Holocene epoch: Nature, v. 419, p. 821-824.

Oakley, A.J., 2001, The seismic and sedimentological characteristics of the upper New Meadows River, mid-coast Maine: BA thesis, , Bowdoin College, Brunswick, ME, 29 p.

Oldale, R.M., 1989, Timing and mechanisms for the deposition of the glaciomarine mud in and around the Gulf of Maine: a discussion of alternative models: *in* Tucker, R.D. and Marvinney, R.G., eds., Studies in Maine Geology, v. 5, Quaternary Studies, Geological Survey, Augusta, ME, p. 1-10.

Packard, A.S., 1867, Ice-marks and ancient glaciers in the White Mountains: American Naturalist, v. 1, p. 260-269.

Pair, D.L., Muller, E.H., and Plumley, P.W., 1994, Correlation of late Pleistocene glaciolacustrine and marine deposits by means of geomagnetic secular variation, with examples from northern New York and southern Ontario: Quaternary Research, v. 42, p. 277-287.

Parent, M. and Occhietti, S., 1988, Late Wisconsin deglaciation and Champlain Sea invasion in the St. Lawrence Valley, Quebec: Geographie physique et Quarternaire, v. 42, p. 215-246.

Phillips, A.C., Smith, N.D., and Powell, R.D., 1991 Laminated sediments in prodeltaic sediments, Glacier Bay Alaska: *in* Anderson, J.B., and Ashley, G.M., eds., Glacial marine sedimentation, Paleoclimatic Significance: Geological Society of America Special Paper 261, p. 51-60.

Powell, R.D., 1991, Grounding-line systems as to second-order controls on fluctuations of tidewater termini of temperate glaciers: *in* Anderson, J.B., and Ashley, G.M., eds., Glacial marine sedimentation; paleoclimatic significance, Geological Society of America, Special Paper 261, p. 75-93.

Ramsey, F.W., 1988, The Willey Slide; A new chronicle of the famous tragedy in ample detail and soberly recounted: Bond Cliff Books, Littleton, New Hampshire, 14 p.

Reeds, C.A., 1926, The varved clays at Little Ferry, New Jersey: American Museum Novitates no. 209, American Museum of Natural History, NY, p. 1-16.

Retelle, M.J., and Bither, K.M., 1989, Late Wisconsin glacial and glaciomarine sedimentary facies in the lower Androscoggin Valley, Topsham, Maine: *in* Tucker, R. D. and Marvinney, R. G., eds., Studies in Maine Geology, Quaternary Studies, Maine Geological Survey, Augusta ME, v. 6, p. 33-51.

Retelle, M.J., and Weddle, T.K., 2001, Deglaciation and relative sea-level chronology, Casco Bay Lowland and lower Androscoggin River valley, Maine: *in* Weddle, T.K., and Retelle, M.J. eds, Deglacial history and relative sea-level changes, northern New England and adjacent Canada, Geological Society of America, Special Paper 351, p. 193-216.

Ridge, J.C., 1999, Surficial Geologic Map of the Bellows Falls Quadrangle, Cheshire and Sullivan Counties, N.H. and Windham and Windsor Counties, Vt.: New Hampshire State Geological Survey Open-file Report, 2 map sheets.

Ridge, J.C., 2001, Surficial Geologic Map of part of the Springfield Quadrangle, Sullivan County, N.H. and Windsor County, Vt.: New Hampshire State Geological Survey Open-file Report, 3 map sheets.

Ridge, J.C., in press, The last deglaciation of the northeastern United States: A combined varve, paleomagnetic, and calibrated ^{14}C chronology: *in* Hart, J.P. and Cremeens D.L., eds., Geoarchaeology of Landscapes in the Glaciated Northeast U.S.: New York State Museum Bulletin 497.

Ridge, J.C., Besonen, M.R., Brochu, M., Brown, S.L., Callahan, J.W., Cook, G.J., Nicholson, R.S., and Toll, N.J., 1999, Varve, paleomagnetic, and ^{14}C chronologies for late Pleistocene events in New Hampshire and Vermont (U.S.A.): *in* Thompson, W. B., Fowler, B. K., and Davis, P. T., eds., Late Quaternary history of the White Mountains, New Hampshire and adjacent southeastern Québec, Géographie Physique et Quaternaire, v. 53, no. 1, p. 79-107.

Ridge, J.C., Brennan, W.J., and Muller, E.H., 1990, The use of paleomagnetic declination to test correlations of late Wisconsin glaciolacustrine sediments in central New York: Geological Society of America Bulletin, v. 102, p. 26-44.

Ridge, J.C., Canwell, B.A., Kelly, M.A., and Kelley, S.Z., 2001. Atmospheric ^{14}C chronology for late Wisconsin deglaciation and sea-level change in eastern New England using varve and paleomagnetic records: *in* Weddle, T. K., and Retelle, M. J., eds., Deglacial History and Relative Se-Level Changes, Northern New England and Adjacent Canada, Geological Society of American Special Paper 351, p. 171-189.

Ridge, J.C., and Larsen, F.D., 1990, Re-evaluation of Antevs' New England varve chronology and new radiocarbon dates of sediments from Glacial Lake Hitchcock: Geological Society of America Bulletin, v. 102, p. 889-899.

Ridge, J.C., Thompson, W.C., Brochu, M., Brown, S., and Fowler, B.K., 1996, Glacial Geology of the Upper Connecticut Valley in the Vicinity of the

Lower Ammonoosuc and Passumpsic Valleys of New Hampshire and Vermont: *in* Van Baalen, M.R., ed., Guidebook to Field Trips in Northern New Hampshire and Adjacent Regions of Maine and Vermont, 88th Annual Meeting, New England Intercollegiate Geologic Conference, Harvard University, Cambridge, MA, p. 309-339.

Ridge, J.C., and Toll, N.J., 1999, Are late-glacial climate oscillations recorded in varves of the upper Connecticut Valley, northeastern United States?: Geologiska Foreningens i Stockholm Forhandlingar, v.121, p.187-193.

Rittenour, T.M., 1999, Drainage history of Glacial Lake Hitchcock, Northeastern U.S.A.: MS thesis, University of Massachusetts, Amherst, MA, 178 p.

Rittenour, T.M., Brigham-Grette, J., and Mann, M.E., 2000, El Nino-like climate teleconnections in New England during the late Pleistocene: Science, v. 299, p. 1039-1042.

Rogers, J.N., 2003, Major Holocene debris flows of Franconia Notch (White Mountains, New Hampshire) recorded in Profile Lake: MS thesis, University of Massachusetts, Amherst, MA.

Rogers, J.N., McCoy, W.D., and Davis, P.T., 2001. A late Pleistocene and Holocene record of debris avalanches into Profile Lake, Franconia Notch, White Mountains, New Hampshire: Geological Society of America, Abstracts with Programs, v. 33, no. 7, p.

Schnitker, D., 1974, Postglacial emergence of the Gulf of Maine: Geological Society of America Bulletin, v. 85, p. 491-494.

Schnitker, D., 1986, Ocean basin with a past: Explorations, Journal of Research, University of Maine, Orono, ME, v. 3, p. 29-37.

Schnitker, D., Belknap, D.F., Bacchus, T.S., Friez, J.K., Lusardi, B.A. and Popek, D.M., 2001, Deglaciation of the Gulf of Maine: *in* Weddle, T. K., and Retelle, M. J., eds., Deglacial history and relative sea-level changes, northern New England and adjacent Canada: Geological Society of America, Special Paper 351, p. 193-216.

Severson, J. P., 1991, Patterns and causes of 19th and 20th century shoreline changes of the Winooski Delta: MS thesis, University of Vermont, Field Naturalist Program, 109 p.

Sharpe, C.F.S., 1938, Landslides and related phenomena: A study of mass-movements of soil and rock: Cooper Square Publishers, NY, 137 p.

Shreve, R.L., 1985, Esker characteristic in terms of glacier physics, Katahdin esker system, Maine: Geological Society of America Bulletin, v. 96, p. 639-646.

Silliman, B., Wilcox, C., and Baldwin, T., 1829, Miscellaneous notices on mountain scenery, and of slides and avalanches in the White and Green Mountains: American Journal of Science, 1st series, v. 15, p. 217-232.

Smith, D.G., and Ridge, J.C., 2001, *Pyganodon fragilis* (Lamarck, 1819) from late-glacial varves in northern New England: Abstract, Annual Meeting Freshwater Mussel Conservation Society, Pittsburgh, PA.

Smith, G.W., 1985, Chronology of Late Wisconsin deglaciation of coastal Maine: *in* Borns, H.W., Jr., LaSa1le, P., and Thompson, W.B., eds., Late Pleistocene history of Northeastern New England and adjacent Quebec, Geological Society of America Special Paper 197, p. 29-44.

Smith, G.W., and Hunter, L.E., 1989, Late Wisconsin deglaciation of coastal Maine: *in* Tucker, R.D. and Marvinney, R.G., eds., Studies in Maine Geology, Quaternary Studies, Maine Geological Survey, Augusta, v. 6, p. 13-32.

Sollid, J.L., and Carlson, A.B., 1984, DeGeer moraines and eskers in Parvik, north Norway: Striae, v. 20, p. 55- 61.

Spear, R.W., 1989, Late Quaternary history of high-elevation vegetation in the White Mountains of New Hampshire: Ecological Monographs, v. 59, no. 2, p. 125-151.

Spear, R.W., Davis, M.B., and Shane, L.C.M., 1994, Late Quaternary history of low and mid-elevation vegetation in the White Mountains of New Hampshire: Ecological Monographs, v. 64, p. 85-109.

Stea, R.R., Fader, G.B.J., Scott, D.B., and Wu, P., 2001, Glaciation and relative sea level change in Maritime Canada: *in* Weddle, T.K., and Retelle, M.J. , eds., Deglacial history and relative sea-level changes, northern New England and adjacent Canada, Geological Society of America Special Paper 351, p. 35-50.

Stewart, D.P. and MacClintock, P., 1969, The surficial geology and Pleistocene history of Vermont: Vermont Geological Survey Bulletin 31, 251p.

Stone, B.D., and Borns, H.W., Jr., 1986, Pleistocene glacial and interglacial stratigraphy of New England, Long Island, and adjacent Georges Bank and the Gulf of Maine: *in* Sibrava, V., Bowen, D.Q., and Richmond, G.M., eds., Quaternary glaciations in the Northern Hemisphere, Quaternary Science Reviews, v. 5, p. 39-52.

Stone, B.D., and Peper, J.D., 1982, Topographic controls of the deglaciation of eastern Massachusetts: Ice lobation and the marine incursion: *in* Larson, G.J. and Stone, B.D., eds., Late Wisconsin glaciation of New England, Kendall/Hunt, Dubuque, Iowa, p. 145-166.

Stone, J.R. and Ashley, G.M., 1992, Ice-wedge casts, pingo scars, and the drainage of Glacial Lake Hitchcock: *in* Robinson, P. and Brady, J.B., eds., Guidebook for Field Trips in the Connecticut Valley Region of Massachusetts and Adjacent States, v. 2.

Stone, G.H., 1880, Note on the Androscoggin glacier: American Naturalist: v. 14, p. 299-302.

Stone, G.H., 1899, The glacial gravels of Maine and their associated deposits: U.S. Geological Survey, Monograph 34, 489 p.

Stuiver, M., and Borns, H.W., Jr., 1975, Late Quaternary marine invasion in Maine; its chronology and associated crustal movement: Geological Society of America, Bulletin, v. 86, p. 99-104.

Stuiver, M., and Reimer, P.J., 1993, Extended ^{14}C database and revised CALIB 3.0 ^{14}C age calibration program: Radiocarbon, v. 35, p. 215-230.

Stuiver, M., Grootes, P.M., and Braziunas, T.F., 1995, The GISP2 $\delta^{18}O$ climate record of the past 16,500 years and the role of the sun, ocean, and volcanoes: Quaternary Research, v. 44, p. 341-354.

Stuiver, M., Reimer, P.J., Bard, E., Beck, J.W., Burr, G.S., Hughen, K.A., Kromer, B., McCormac, F.G., van der Plicht, J., and Spark, M., 1998, INTCAL98 radiocarbon age calibration 24,000–0 cal BP: Radiocarbon, v. 40, p. 1041-1083.

Thomas, G.M., 1984, A comparison of the paleomagnetic character of some varves and tills from the Connecticut Valley: MS thesis, University of Massachusetts, Amherst, MA. .

Thomas, P. A., 1985, Archeological and geomorphological evaluation; M5000 (3) northern connector material supply/disposal area, Howe Farm flood plain. Burlington, Vermont: University of Vermont, Department of Anthropology, no. 54, p. 41.

Thompson, D.J., 1999, Talus fabric in Tuckerman Ravine, New Hampshire: Evidence for a tongue-shaped rock glacier: Géographie physique et Quaternaire, v. 53, no. 1, p. 47-58.

Thompson, W.B., 1982, Recession of the late Wisconsin ice sheet in coastal Maine: *in* Larson, G.J., and Stone, B.D., eds., Late Wisconsin glaciation of New England, Kendall/Hunt, Dubuque, Iowa, p. 211-228.

Thompson, W.B., 1997, Surficial geology of the Fryeburg 7.5-minute quadrangle, Oxford County, Maine: Augusta, Maine Geological Survey, Open-File Report.

Thompson, W.B., 1998, Deglaciation of western Maine and the northern White Mountains: Geological Society of America, Abstracts with Programs, v. 30, no. 1, p. 79.

Thompson, W.B., 1999, History of research on glaciation in the White Mountains, New Hampshire: *in* Thompson, W.B., Fowler, B.K., and Davis, P.T., eds., Late Quaternary history of the White Mountains, New Hampshire and adjacent southeastern Québec: Géographie physique et Quaternaire, v. 53, no. 1, p. 7-24.

Thompson, W.B., and Borns, H.W., Jr., 1985, Surficial geologic map of Maine: Augusta, Maine Geological Survey.

Thompson, W.B., Crossen, K.J., Borns, H.W., Jr., and Anderson, B.G., 1989, Glaciomarine deltas of Maine and their relation to Late Pleistocene-Holocene crustal movements: *in* Anderson, W.A., and Borns, H.W., Jr., eds., Neotectonics of Maine-studies in seismicity, crustal warping, and sea-level change, Maine Geological Survey, Augusta, ME, p. 43-67.

Thompson, W.B., and Fowler, B.K., 1989, Deglaciation of the upper Androscoggin River valley and northeastern White Mountains, Maine and New Hampshire: *in* Tucker, R.D., and Marvinney, R.G., eds., Studies in Maine geology, Quaternary geology, Maine Geological Survey, Augusta, Maine, v. 6, p. 71-88.

Thompson, W.B., Fowler, B.K., and Dorion, C.C., 1999, Deglaciation of the northwestern White Mountains, New Hampshire: *in* Thompson, W.B., Fowler, B.K., and Davis, P.T., eds., Late Quaternary history of the White Mountains, New Hampshire and adjacent southeastern Québec, Géographie physique et Quaternaire, v. 53, no. 1, p. 59-77.

Thompson, W.B., Fowler, B.K., Flanagan, S.M., and Dorion, C.C., 1996, Recession of the Late Wisconsin ice sheet from the northwestern White Mountains, New Hampshire: *in* Van Baalen, M. R., ed., Guidebook to field trips in northern New Hampshire and adjacent regions of Maine and Vermont, Harvard University Department of Earth & Planetary Sciences, guidebook to 88th annual meeting of the New England Intercollegiate Geological Conference, Cambridge, MA, p. 203-234.

Thompson, W.B., Hildreth, C.T., Boisvert, R.A., Dorion, C.C., and Fowler, B.K., 2002, Glacial geology and archaeology of the northern White Mountains, New Hampshire: Guidebook for the 65th annual reunion of the Northeastern Friends of the Pleistocene, 41 p.

Thompson, W.B., and Koteff, C., 2002, Glaciomarine deltas in southwestern Maine: indicators of late-glacial relative sea level and isostatic uplift: Geological Society of America, Abstracts with Programs, v. 34, no. 1, p. A-26.

Thompson, W.F., 1960, The shape of New England mountains, Part I: Appalachia, v. 26, p. 145-159.

Thompson, W.F., 1961, The shape of New England mountains, Part II: Appalachia, v. 27 p. 316-335 and p. 457-478.

Verosub, K.L., 1979a, Paleomagnetism of varved sediments from western New England: secular variation: Geophysical Research Letters, v. 6, p. 245-248.

Verosub, K.L., 1979b, Paleomagnetism of varved sediments from western New England: variability of the recorder: Geophysical Research Letters, v. 6, p. 241-244.

Vose, G.L., 1868, Traces of ancient glaciers in the White Mountains of New Hampshire: American Naturalist, v. 2, p. 281-291.

Wagner, W.P., 1970, Pleistocene mountain glaciation: Northern Vermont Geological Society of America Bulletin, v. 81, p. 2465–2470.

Wagner, W.P., 1971, Pleistocene mountain glaciation: Northern Vermont, Reply: Geological Society of America Bulletin, v. 82, p. 1761–1762.

Wagner, W. P., 1972, Ice margins and water levels in northwester Vermont: *in* Doolan, B.L. and Stanley, R.S., eds., New England Intercollegiate Geologic Conference, Guidebook for Field Trips in Vermont, University of Vermont, Burlington, VT, p. 297-316..

Waitt, R.B. and Davis, P.T., 1988, No evidence for post-icesheet cirque glaciation in New England: American Journal of Science, v. 288, p. 495-533.

Weddle, T. K., 1994, Surficial geology of the Freeport quadrangle, Maine: Maine Geological Survey, Augusta, Open-file No. 94-5.

Weddle, T.K., 2001, Surficial geology of the Brunswick quadrangle, Maine: Maine Geological Survey, Augusta, Open-file No. 01-484.

Weddle, T.K., Lowell, T.V., and Dorion, C.C., 1994, Glacial geology of the Penobscot River basin between Millinocket and Medway: *in* Hanson, L. S., ed., Guidebook to field trips in north-central Maine, 85th annual meeting of the New England Intercollegiate Geological Conference W. C. Brown Publishers, Dubuque, Iowa, p. 193-212.

Weddle, T.K., and Retelle, M.J., 1995, Glaciomarine deposits of the Late Wisconsin Casco Bay Sublobe of the Laurentide Ice Sheet: *in* Hussey, A.M. II, and Johnston, R.J., eds., Guidebook to Field Trips in southern Maine and adjacent New Hampshire: New England Intercollegiate Geological Conference, 87th Annual Meeting, Brunswick, Maine, p.173-194.

Weddle, T.K., and Retelle, M.J., 1998, Deglacial style and relative sea-level chronology, Casco Bay lowlands to White Mountain foothills, southwestern Maine: Geological Society of America, Abstracts with Programs, v. 30, no. 1, p. 83.

Whalen, T., 1998, Interpretation of post-glacial fluvial terrace remnants in the Winooski River Basin, Vermont: MS thesis, University of Vermont, Burlington, VT, 186 p.

Wright, S. F., Whalen, T. N., Zehfuss, P. H., and Bierman, P. R., 1997, Late Pleistocene-Holocene history: Huntington River and Miller Brook valleys, northern Vermont: New England Intercollegiate Geologic Conference Guidebook.

Wohlfarth, B., 1996, The chronology of the last termination: a review of radiocarbon-dated, high-resolution terrestrial stratigraphies: Quaternary Science Reviews, v. 15, p. 267-284.

Zehfuss, P. and Bierman, P., 1996, Alluvial fans in Vermont as recorders of changes in sedimentation rates due to deforestation: Geological Society of America Abstracts with Programs, v. 28, p. 112.

Geo-Ubiquity: Chemical, hydrological and physical evolution of coastal terraces near Santa Cruz, California

Authors: **Marjorie Schulz,** *U.S. Geological Survey, 345 Middlefield Rd, Menlo Park, CA 94025*
Jennifer W. Harden, *U.S. Geological Survey, 345 Middlefield Rd, Menlo Park, CA 94025*
Gerald E. Weber, *University of California, Santa Cruz, Department of Earth Sciences*

Co-leaders: **Suzanne Anderson,** *CSIDE, Earth Sciences Dept., University of California, Santa Cruz, CA 95064*
Art White, *US Geological Survey, 345 Middlefield Road MS420 Menlo Park, 94025*
Lesley Perg, *University of Minnesota, Geology and Geophysics, 310 Pillsbury Dr. SE, Minneapolis, MN 55455*
Dave Stonestrom, *US Geological Survey, 345 Middlefield Road MS420 Menlo Park, 94025*
Dr. Paula Reimer, *Center for Accelerator Mass Spectrometry L-397, Lawrence Livermore National Laboratory, P.O. Box 808, Livermore, CA 94550*
Carrie Masiello, *University of California, Dept of Geography, Ellison Hall 3611, Santa Barbara, CA 93106*

INTRODUCTION

The term geo-ubiquity was created by Jennifer Harden to represent the multi-disciplinary research needed to fully describe a landscape. On this field trip, we plan to look at the diverse types of research necessary to understand and characterize the Santa Cruz marine terraces on Ben Lomond Mt. These tectonically uplifted marine terraces provide a natural laboratory to study pedologic, geologic, and geochemical processes over time. However, to understand how processes alter terraces, we must first understand the processes that created the terrace landscape.

The terraces north of Santa Cruz, California consist of bedrock platforms covered by sediment. Platforms were successively cut into the slowly rising coastline by wave action. Tectonic uplift and changing sea levels create emergent platforms that are blanketed with 1-5 ms of sediment. To fully describe the marine terrace landscape, understanding many interactive processes is necessary, including: formation of wave-cut platforms, sedimentation on the platforms (are sediments marine, fluvial or both?), sea level history, parent materials of the sediments, mechanisms of terrace uplift, chemical weathering of sediments, soil forming processes, carbon sequestration, unsaturated hydrology, watershed hydrology, erosion, cosmogenic isotope build up in sediments, U-series decay, and any land use changes which will have affected any of the previous factors. On this trip, we will discuss these processes and how they influenced each other in creating the current landscape.

On Day 1, we will investigate aspects of tectonic deformation along the coastline between Santa Cruz and Pigeon Point (Fig. 1) and explore the geomorphology of the terraces that marine processes and tectonic movement have created. Day 2 will be devoted to exploring the terraces in greater detail. For much of each day we will be walking and examining outcrops on State Parks land, where collecting is prohibited. Time permitting; we may visit some optional field trip stops which will provide additional background on the geology and cultural history of this stretch of coastline.

DAY 1.

Start at intersection of Gazos Creek Rd. and Cloverdale Rd., north on Cloverdale Rd.

1.0 mi 1.6 km Arroyo de Los Frijoles

Arroyo de Los Frijoles is a major drainage incised more than 180 m-(600 feet). However, the arroyo is presently an "underfit" stream whose headwaters were captured, probably by Gazos Creek, sometime during the late Pleistocene. The linearity of the Butano/Little Butano Creek valley that Cloverdale Road follows parallels the Arroyo de Los Frijoles valley, suggesting they are both fault controlled features. The extent to which these linear valleys are produced by active fault movement or by differential erosion along faults is not clear.

1.2 mi	*1.9 km*	*On right is the entrance to Butano State Park.*
5.6	*9.0*	*Intersection with Pescadero Road, Turn left (West) on Pescadero Road.*
6.0	*9.6*	*Pescadero is a 150 year old Portuguese fishing community.*
8.5	*13.6*	*Intersection with Hwy 1; turn left (South) onto Hwy 1.*
10.1	*1.6*	*Pebble Beach: the terrace surface is interrupted by a large paleosea-stack on the ancient wave-cut platform.*
10.9	*17.5*	*Bean Hollow State Beach is at the mouth of Arroyo de Los Frijoles.*
13.4	*21.6*	*Turn right on Pigeon Point Road.*

Schulz, M., W. Harden, J., and Weber, G.E., 2003, Geo-Ubiquity: Chemical, hydrological and physical evolution of coastal terraces near Santa Cruz, California: *in* Easterbrook, D.J., ed., Quaternary Geology of the United States, INQUA 2003 Field Guide Volume, Desert Research Institute, Reno, NV, p. 399–406.

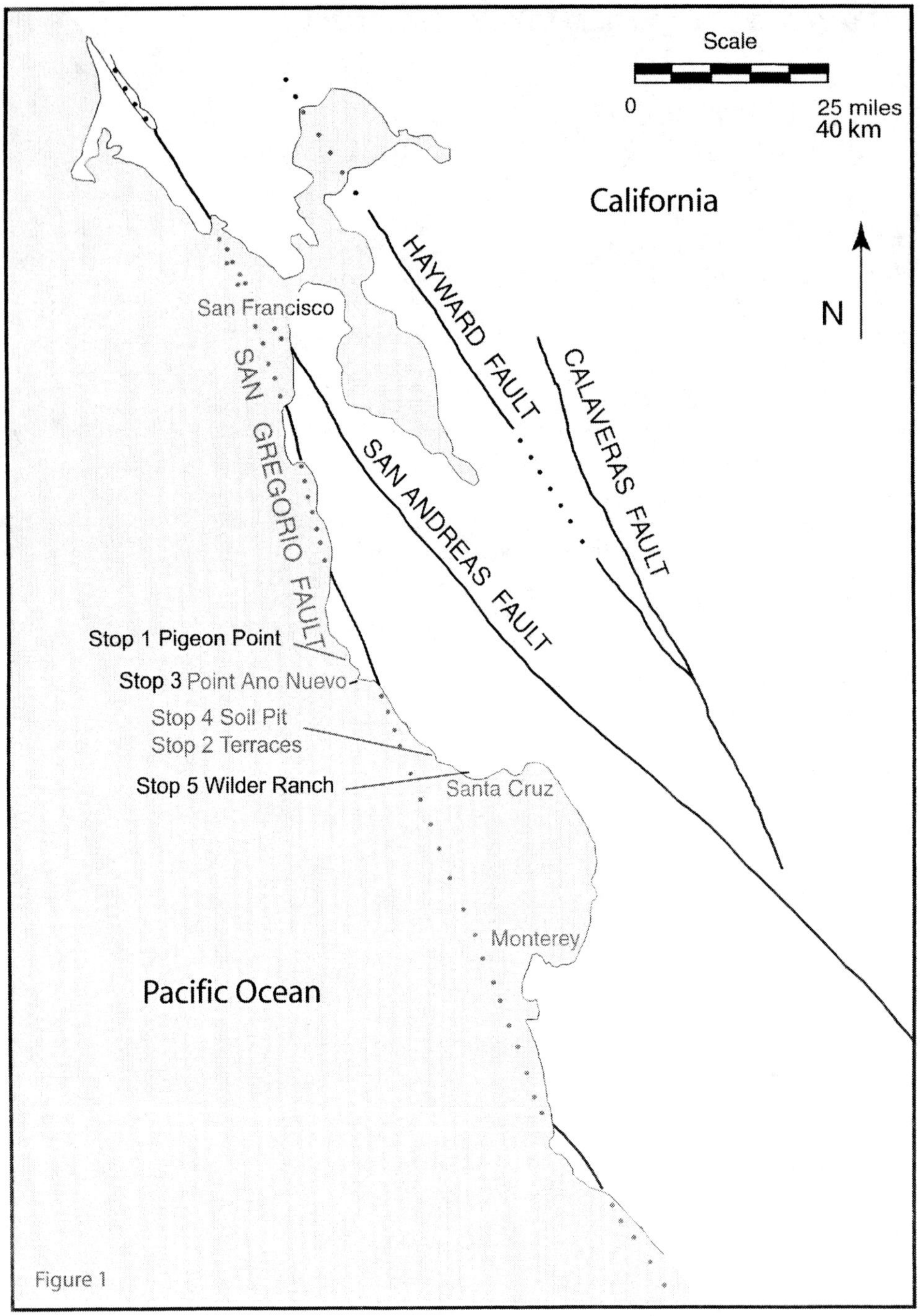

Figure 1. Field trip stops.

STOP 1-1. PIGEON POINT

Pigeon Point was named for the Yankee Clipper *Carrier Pigeon*, which sank off the point in 1853. The lighthouse was built in 1872 and at 35.05 m (115 ft) is the tallest lighthouse on the Pacific coast.

At Pigeon Point, the wave-cut platform of the first terrace formed on deposits of late Cretaceous age, known as the Pigeon Point Fm. The formation is comprised of 2500 m of immature mudstone, arkosic sandstone and conglomerate (Lowe, 1979). These sediments were deposited in a submarine canyon/slope depositional system. At Pigeon Point, the outcrop is composed of muddy conglomerate, sandy conglomerate, and pebbly mudstone. Individual beds are 0.3-1 m thick. Imbrication, lenticular bedding, and scour structures are common. We are here to see the contact of the Pigeon Point Fm. with the overlying terrace sediments, visible at the modern sea cliff. The irregular erosional surface that forms the contact between the Pigeon Point Fm. and the terrace deposits is an ancient wave-cut platform. Along this coastline the modern wave-cut platform is most obvious at low tide. Both the modern wave-cut platform and the ancient wave-cut platform are irregular surfaces. On both we observe sea-stacks, ridges of more resistant rocks and channels where weaker beds have eroded. Although we tend to conceptualize wave-cut platforms as smooth level surfaces, they typically have many topographic highs and lows.

Leaving Pigeon Point, turn south on Hwy 1.
15.4 mi 24.8 km Mouth of Gazos Creek.
19.6 31.6 Año Nuevo State Reserve.
22.0 35.4 Waddell Bluffs and Waddell Creek.

The Santa Cruz mudstone is the main bedrock formation forming outcrops in cliffs and roadcuts between Santa Cruz and Año Nuevo, where it is offset by the San Gregorio fault system. At the top of the bluffs is a poorly exposed, narrow remnant of a marine terrace, intermediate in elevation between the Santa Cruz (1st) and Western (2nd) terraces. This surface is probably the Cement terrace. A resistant bed of siliceous mudstone is exposed in the surf zone and forms a natural groin at this location. The result is a protected beach up-coast and active erosion down-coast. The highway below the cliffs is built entirely on artificial fill. Riprap was placed here in 1946 during construction of Highway 1. Because of its placement on a bedrock platform, this riprap has successfully protected the road for over 50 years. Prior to road construction, the high cliffs of Santa Cruz mudstone (Waddell Bluffs) were undercut by waves.

These bluffs formed a natural barrier to coastal travel in the 1800s when stagecoaches could pass the bluffs only during low tide on the wet beach. The southern tip of present-day San Mateo County was originally part of Santa Cruz County, but because access to the county seat in Santa Cruz was often impeded by this barrier, this land north of Waddell Bluffs was annexed by San Mateo County in 1868.

Debris from small rock falls and slides collects behind the cable netting on the eastern side of the road. The debris is periodically removed, stockpiled on the western side of Highway 1 and eventually dumped into the ocean to become part of the longshore drift of sediment to the south. Large rock falls are uncommon, probably because of the manner in which the Santa Cruz mudstone weathers—by the raveling of small blocks and chips typically less than 10 cm in dimension. Occasionally, blocks the size of a small car fall and bounce out onto Highway 1. About two decades ago, a passenger in a truck traveling north was killed by a rock that bounced through the window, and litigation against the state transportation agency ensued for improperly maintaining the debris trap on the eastern side of Highway 1 (then a simple trench and berm). To reduce potential liability, the state recently installed the Geobrugg® steel wire rope net barrier, which seems to be working satisfactorily.

26.6 mi 41.9 km Turn off for Quick View Stop

A Quick Stop View channel fill and Año Nuevo from a distance

Colluvium-filled gullies in the Santa Cruz mudstone: In approximately six locations between Scott Creek and Greyhound Rock, small V-shaped gullies filled with colluvium are present in roadcuts. Fluvial deposits are generally absent and the colluvial deposits are crudely stratified sub-parallel to the sides of the V-shaped drainage. These gullies apparently formed following the culmination of the sea-level rise that formed the sea-cliff. As erosion modified the original sea cliff a series of small gullies developed. Some gullies expanded into present day streams. The smaller gullies lost the battle for headwater drainage area, were abandoned and eventually filled with colluvium. Note that no surface evidence (geomorphic or vegetative) reveals the presence of these colluvium filled gullies beneath the level ground surface.
33.4 mi 53.8 km Majors Creek

The black-colored cliffs to the left (up Majors Creek) are composed of bitumen-saturated sandstone that was injected into the overlying Santa Cruz Mudstone in a fluid/plastic state. Numerous sandstone dikes and sills, most of which contain some bituminous material, are exposed in the modern sea cliff between Wilder Creek and Greyhound Rock. The Miocene Santa Margarita sandstone was the source of these bitumen sand intrusions. The Santa Margarita sandstone contains varying amounts of bitumen throughout its outcrop area from Santa Cruz to the vicinity of Big Basin Redwoods State Park. The hydrocarbons are believed to have migrated into the Santa Margarita sandstone from the underlying Monterey Formation.

The bituminous sandstone in this area has been mined since the late 1880s for paving material. The asphaltic content of the sand ranges from about 4% to 18% by weight. These oil-impregnated deposits range in thickness from 0.30 to 12.2 m and vary in quality from dry and brittle to soft and gummy. In some outcrops tar will drip or flow out of the bituminous sands when sufficiently warmed by the sun. San Francisco streets were reportedly paved in the 1890s with bituminous sandstone mined near Majors Creek and transported to San Francisco by boat. An estimated 614,000 tons of asphaltic paving material, worth approximately $2,360,000, was produced from this area between 1888 and 1914 (Page and Holmes, 1945). Production was intermittent after the 1920s, with the last of the quarries (Calrock Quarry) ceasing operations in the 1940s.

33.7 mi 54.3 km Pull off Highway 1.

STOP 1-2. BEN LOMOND MOUNTAIN MARINE TERRACES (HIKE AND LUNCH)

Walk up the road beyond the gate to examine the Western (2nd), Wilder (3rd), and Blackrock (4th) terraces. Upon reaching the Wilder terrace, take the right fork in the road and note the intricate flow structures in the asphalt pavement. This is one of the oldest paved roads in the county, utilizing locally quarried bituminous sandstone of the Santa Margarita Fm.

Since the initial study of Rode (1930), the exceptionally well-preserved Ben Lomond Mountain marine terrace sequence has been the subject of numerous studies and reinterpretations. These include Bradley (1957, 1958), Bradley and Griggs (1976), Lajoie et al. (1979), Hanks et al. (1984), Lajoie (1986), Weber (1990a, b), Anderson (1990, 1994), Anderson and Menking (1994), Anderson and Weber (1990), Lajoie et al. (1991), Weber et al. (1995), Perg et al. (2001), Weber et al. (2001), and Muhs et al. (2002).

Terrace names, elevations, estimated ages and estimated uplift rates for the Davenport area are shown in Table 1.

Bradley and Griggs (1976) describe the lowest emergent terrace (Santa Cruz terrace) as containing three separate wave-cut platforms and shoreline angles named, from youngest to oldest, the Davenport, Highway 1, and Greyhound Rock wave-cut platforms. Although not specifically stated, Bradley and Griggs (1976) imply that each of these platforms was formed by surf erosion during separate sea level stands. The shoreline angles and paleo-sea-cliffs between these platforms have been buried by fluvial and colluvial deposits. Consequently, there is only one terrace landform, but three abrasional surfaces.

Some question exists around the existence of the Greyhound Rock platform. Weber and others (1995) have concluded that the Greyhound Rock erosional surface is a localized shore platform associated with the Highway 1 wave-cut platform. Shore platforms are topographic irregularities at the inner edge (shoreline) of a wavecut platform. However, Muhs (2002, personnel communication) points out that the shoreline angle elevation difference is 8 to 11 m as reported by Bradley and Griggs (1976), which seems too high to be a shore platform,

The areal extent of the terraces on the flank of Ben Lomond Mountain is shown in Figure 2. The Santa Cruz terrace (Terrace 1) forms the broad, extensively cultivated bench closest to the ocean. Highway 1 is built on this terrace. The older terraces are not continuous along the entire coastline and show successively greater erosional modification and dissection with age and elevation. Note that the Wilder terrace (Terrace 3) is not preserved north of Laguna Creek, and the Cement terrace is restricted to the immediate vicinity of Davenport.

At the Wilder terrace (Terrace 3), we will examine exposures of the wave-cut platform for a positive indication of its formation as an ancient marine surface. The clusters of small holes in the ancient platform were created by rock-boring mollusks (pholadid bivalves) living in the ancient sea. Such pholad borings are considered strong evidence for a marine origin of an erosional surface. The platforms here are all cut in the Upper Miocene Santa Cruz Mudstone. Terrace sediment remnants on platforms cut into the Santa Cruz Mudstone in this area tend to be less than 2 m thick. We will stop to view and discuss a road cut exposure of a soil profile developed in the terrace sediments of the Wilder terrace (Terrace 3).

North on Highway 1
50.3 mi 81.0 km Stop 3 turn off on left.

STOP 1-3. QUATERNARY FAULTING AT POINT AÑO NUEVO

Leave the bus and hike down the dirt road to the beach. Hike north about 1? mi (2.4 km) along the beach to examine the evidence for late Pleistocene and Holocene faulting. Please note that the road is on private property. If you are taking this trip on your own, you must obtain a permit from Coastways Ranch to enter the property. Once we enter Año Nuevo State Reserve, collecting is prohibited.

The San Gregorio fault zone (SGfz) is the principal fault west of the San Andreas fault zone (SAfz) in central California and is part of the larger San Andreas fault system that represents the active tectonic boundary between the Pacific and North

TABLE 1. COMPARISON OF TERRACE AGES IN KA AND UPLIFT RATES IN M/KA.

		Bradley and Griggs 1976 (uplift rate)	Hanks and others 1984 (SAGT)	Lajoie and others, 1991 (SAGT)	Weber and others, 1995 (SAGT)	Perg and others, 2001 (CRN)	Muhs and others 2002 (U-series)
Average Uplift Rate		0.186m/ka	0.35 m/ka	0.21m/ka	0.44 m/ka	1.1 m/ka	
Santa Cruz aka Terrace 1	Davenport platform		83	81	80	65*	71.5 to 81
	Hwy 1 platform		104	124	103	65*	nd
	Greyhound Rock platform	160				65*	nd
Cement Terrace		260	120	320	125	nd	nd
Western aka Terrace 2		450	230	430	213	90	nd
Wilder aka Terrace 3		700	370	800	320	137	nd
Blackrock aka Terrace 4		900	450	1.0my	430	139	nd
Quarry aka Terrace 5		1.2my	650	1.3my	545	226	nd

Uplift Rate= ages assigned on assumed uplift rate and elevation of platforms
SAGT: Shoreline angle graphic technique.
CRN: Cosmogenic Radionuclide dating. * Davenport and Hwy 1 platforms were not differentiated.
U-series: Uranium series ages of solitary corals.
nd=not determined

American lithospheric plates. Here at Point Año Nuevo, the SGfz is 2-3 km wide and includes seven fault strands with evidence of late Quaternary activity. The Coastways fault, which crosses Highway 1 at the small dip in the road just before Coastways Ranch has long been considered the primary trace of the SGfz because of the obvious bedrock offset across a small re-entrant in the coastline. We will hike down this re-entrant to the beach. From Año Nuevo, the SGfz has been traced northward to its juncture with the SAfz near Bolinas Lagoon in Marin County and southward across Monterey Bay, a cumulative length approaching 200 km (125 miles). Based on its dimensions and late Quaternary activity, this fault zone appears to be a potential source of significant earthquakes and has been assigned a 7.3 maximum magnitude (Mmax) (Petersen and others, 1996).

At the beach, examine the rocks on either side of the re-entrant in the sea cliff. West of the re-entrant, fine-grained silty sandstone and sandy siltstone of the Pliocene Purisima Fm. are exposed in near-vertical sea cliffs. East of the re-entrant are poorly exposed outcrops of the Miocene Santa Cruz mudstone. Bedding, although difficult to find, strikes about N50°E and dips about 40°NW. This discordant juxtaposition of two units differing greatly in age, lithology, and structure can only be explained by the presence of a major, Neogene-active fault that extends up the brush- and-colluvium-covered drainage (Figure 3). Walk northward along the beach. From here to the mouth of Año Nuevo Creek, the sea cliff is cut in the Pliocene Purisima Fm. In this area, away from the Coastways fault, bedding strikes roughly N-S and dips gently 4° to 7° to the east. We will also examine several other faults as we hike toward point Año Nuevo. For more information on the San Andreas Fault system and marine terraces see Muhs et al. (2003, this volume).

DAY 2.

Start Intersection Gazos Creek Road with Hwy 1, go south
17.7 mi 28.5 km Back Ranch Road

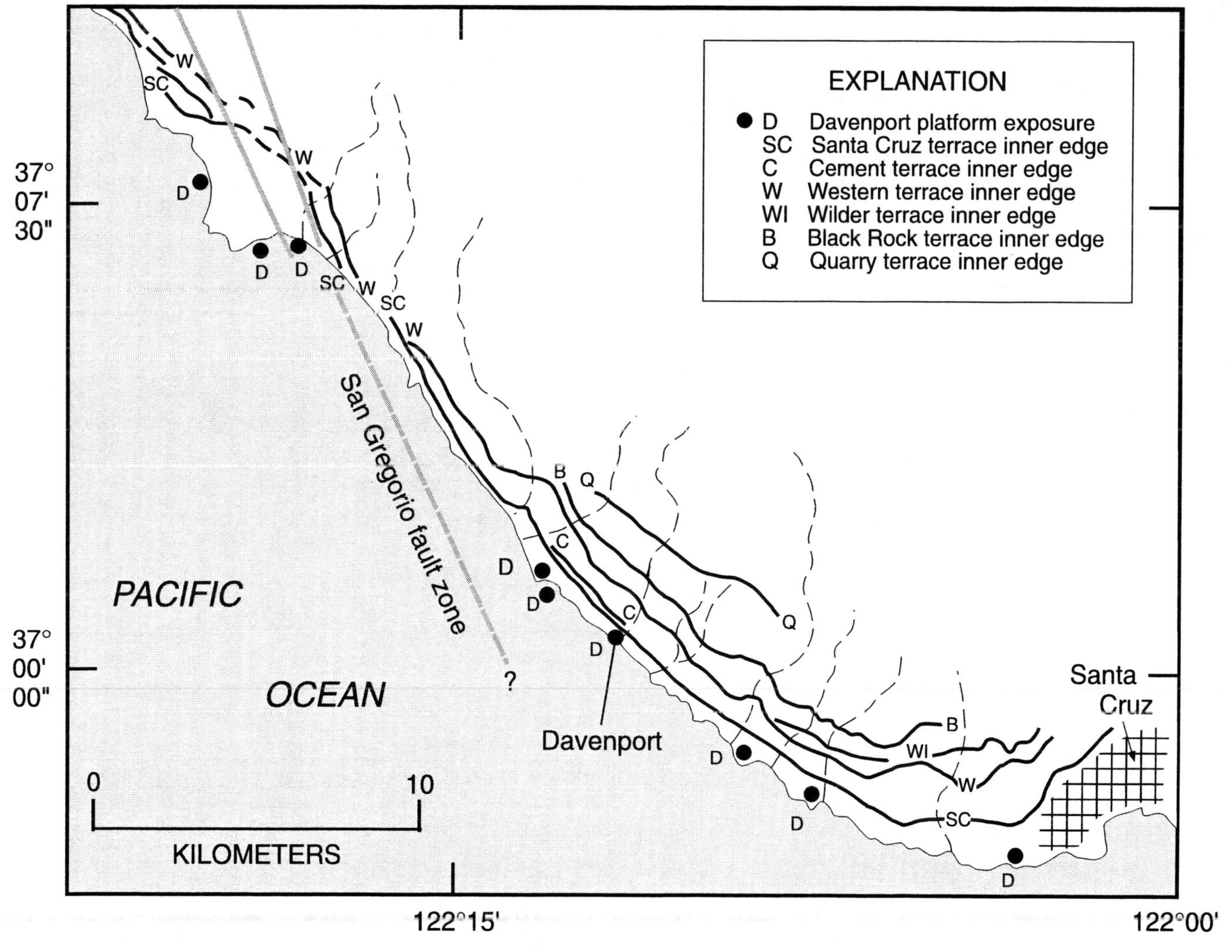

Figure 2. Marine Terraces on the flank of Ben Lomond Mountain, modified from Bradley and Griggs (1976).

STOP 2-1. SOIL PIT IN QUARRY TERRACE (TERRACE 5)

We will shuttle in small vehicles to the soil-pit as the road will not accommodate a bus. Unfortunately, the soil pit (located on private land) will be filled, and if you repeat this trip on your own at a later time, it will be no longer visible.

Terrace 5 (Quarry Terrace) has a well developed soil profile, leached in its upper horizons, with pedogenic clay and significant iron oxide accumulations in its lower horizons. Leaching of upper soil horizons is evident in the "bleached" appearance of sand grains that rest on top of the columns of clayey subsoil. "Eluvial" (meaning removed by leaching) horizons are depleted of exchangeable bases and even iron oxides. Deeper in the subsoil, this leached sand is also found along faces of the columnar clay structure, indicating that cracking of the soil occurs during summer months, providing pathways for sand grain descent.

Iron oxide accumulation, indicated by red and yellow soil colors, is significant in this soil. The accumulation zone of pedogenic minerals, clay and Fe-oxides extends down to 180 cm. This soil is a good illustration of the effects of seasonal saturation and periodic reducing conditions in parts of the soil. The color contrast between outside and inside of the columnar structure is significant with ped interiors of blue-grey clay surrounded by bright orange surfaces. These colors reflect the oxidation state of Fe with hematite and goethite on the outside of the peds but not on the inside. In a study of iron mineralogy of the Santa Cruz Terraces, Aniku and Singer (1990), found that geothite was abundant in all terraces, but that hematite also occurs in the older terraces (terraces 3 thru 5).

We will see in this pit the high Fe content of the terrace 5 soil and discuss how soil development may be influencing carbon cycling. The turnover time of soil organic matter varies in these soils and is controlled by mineral weathering, plant communities, climate, topography and land management. To understand the role of soils in carbon cycling, understanding the mechanisms that control the soil carbon system is necessary. Based on the work of Massiello (not yet published) and others on terraces in the Mattole River area in northern California, iron-oxides likely bind with soil organic matter thus stabilizing the carbon. These terraces provide a unique opportunity to study this hypothesis in detail.

Go back along Ranch Road and Hwy 1 go south
21.7 mi 34.9 km Wilder Ranch State Park Entrance, turn right into State Park

STOP 2-1. TERRACE WEATHERING, HYDROLOGY AND WATERSHED HYDROLOGY

Here we will hike to the Western terrace (2nd terrace), and will stop on the way at Wilder Creek and Peasley Gluch to discuss aspects of watershed hydrology. We will examine how terrace and watershed hydrology and chemistry evolve. Soils from five terraces have been sampled and instrumented with suction water samplers, tensiometers, and soil gas samplers. Physical characterization of the sediments includes bulk chemistry, grain size, and specific surface area. Chemical analyses of the sediments were done by X-ray fluorescence.

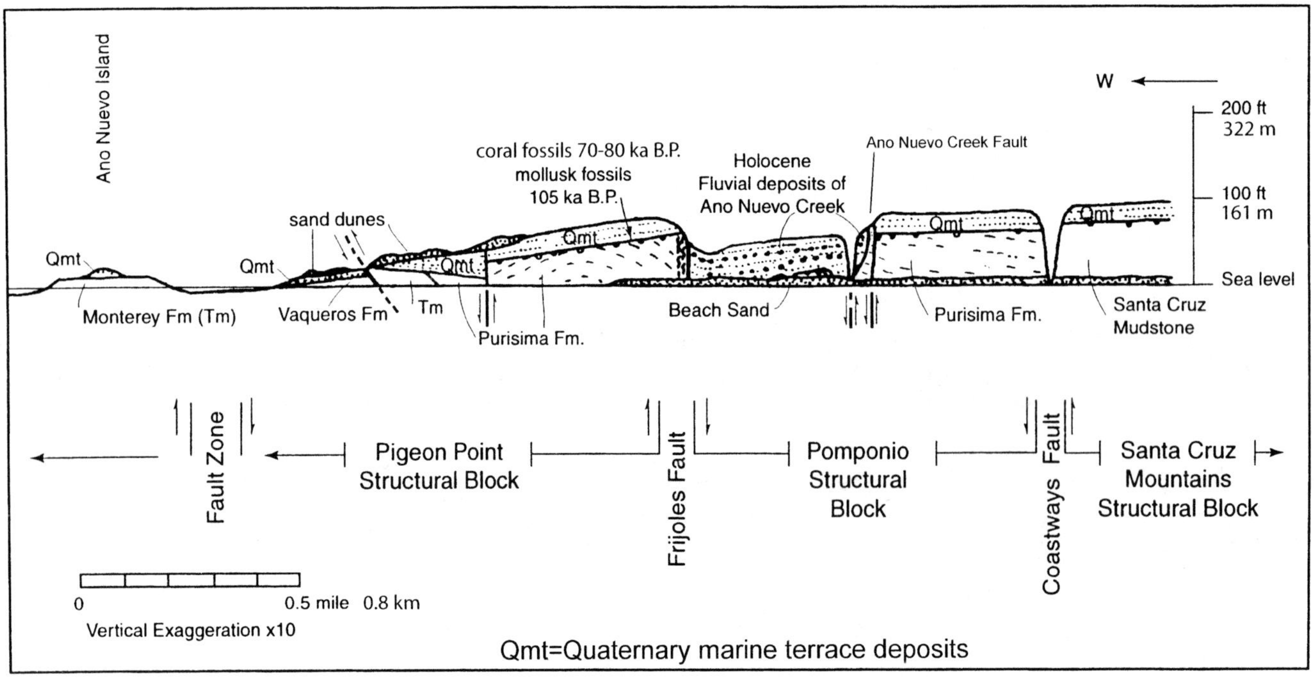

Figure 3. Sea cliff geology along the south shore of Point Año Nuevo; modified from Lajoie and others (1979).

Mass changes were calculated to indicate how much of a chemical component has been lost due to weathering (Fig. 4). SiO_2 was used as a preliminary initial proxy for a conservative component based on a high content of detrital quartz. Final calculations will been done using quartz abundances. Iron and aluminum were lost from the topsoils (<0.5 m) and gained in the subsoils (0.5m to 2m) relative to Si. The argillic B horizons contain significant amounts of kaolinite and extractable iron oxides at roughly the same depths where net gains of Al and Fe occur. The progressive development of these argillic horizons with time has a strong influence on pore water hydrology and solute fluxes

In contrast to Al and Fe, base cations become progressively depleted in the terrace soils with age. A correlation between Na and Ca loss indicates plagioclase weathering. Mg and K (not shown in figure 4) exhibit less consistency in weathering trends corresponding to partial retention of mica in the weathering profiles.

Micromorphologic and larger-scale changes in pore geometry associated with pedogenesis and mineral weathering can have profound effects on hydraulic properties including soil-moisture retentivity and hydraulic conductivity (e.g., Stonestrom et al., 1998). The purpose of the present study is to relate chemically induced changes in the parametric functions that control subsurface flow to resulting changes in the hydraulic responses of terraces deposits. To do this, terraces have been instrumented with tensiometers, time-domain reflectometers, thermocouples, and soil-water samplers. Hydraulic properties are being determined with standard soil-physics techniques. The strong seasonality of precipitation associated with the prevailing Mediterranean climate provides annual-scale forcing at the land surface that is ideally suited for examining weathering and other pedogenic changes in the hydraulic responses of the Santa Cruz profiles.

All of the processes we have touched upon in the last two days, tectonic, sedimentary, hydrologic, and chemical have acted in concert to create the complex landscape of coastal California. We hope that this trip has provided an increased appreciation for the interplay of all factors in generating a given terrain.

Drive To Monterey Bay Aquarium for Banquet. Thank you for joining us.

ACKNOWLEDGEMENTS

The field trip leaders acknowledge that we rely on the work of many others. These kind and curious people are also helping to guide sections of the field trip and/or providing evening seminars: Suzanne Anderson, Bob Anderson, Roberto Anima, Carrie Massiello, Lesley Perg, Paula Reimer, Stephanie Ross, Andy Stevenson, Dave Stonestrum, and Art White. We thank them all. We would like to thank Pescadero Consevation alliance and Larry Miller for accommodations. Dan Landino not only allowed access to his property, but also enthusiastically dug a pit on it for

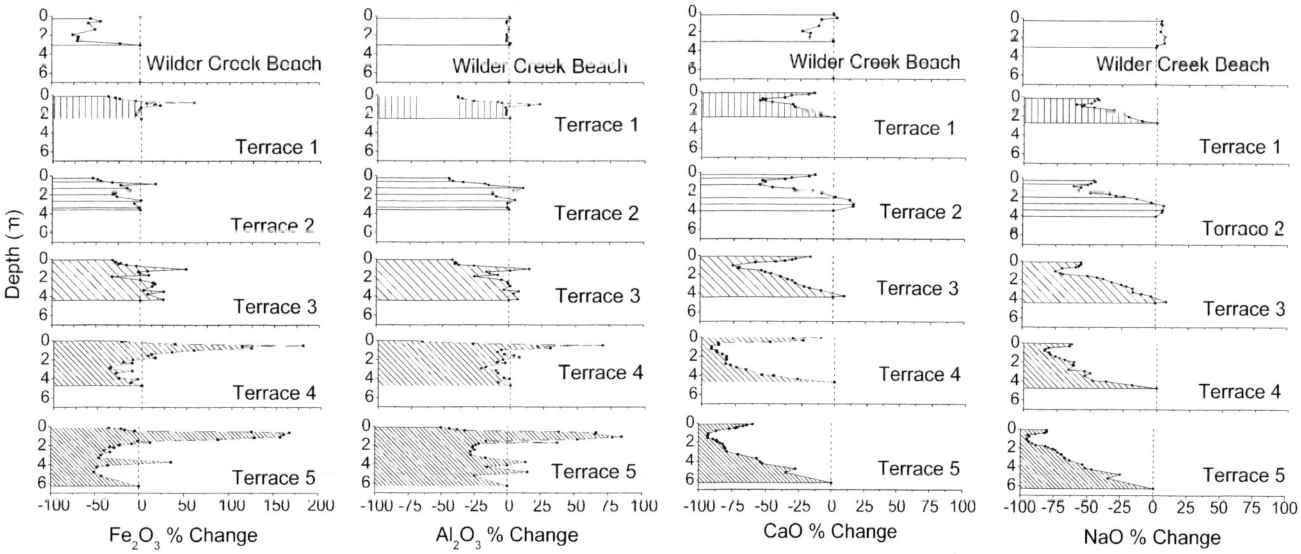

Figure 4. Solid phase mass changes during weathering; the vertical axis is depth below the ground surface and the horizontal axis is % mass change of the element in question from its abundance in a parent material (the unweathered sediment). The assumed parent material was generally the deepest sample taken. These preliminary calculations were made using Si as a conservative element Final calculations will use quartz as the conservative component. The translocation of Fe and Al from upper to lower horizons is graphically illustrated by increases in these elements at roughly the same depths where the argillic (B) horizon occurs. Unlike Al and Fe, Na and Ca become progressively depleted with age. Plagioclase feldspar is the major source of these elements. We interpret the loss of Na and Ca as an indication of plagioclase weathering. The dissolution of plagioclase shows we are also losing Si; therefore, Si is not a good conservative element to use in these calculations. The retention of Na and Ca through biotic retention and recycling appears as a spike of these elements at the top of the older profiles.

us. Wilder Ranch State Park and Lighthouse Field State Beach have allowed access and long-term site installation. Mills Ranch allowed access and long term site installation. We would like to thank Bob Anderson, Jane Reid and Alan Allwardt for informative discussions and for assistance providing electronic versions of the figures. Jennie Munster has been an invaluable assistant in the field and in the lab. We would like to thank Dan Muhs for providing an updated computer version of Figure 2. David Harwood, Dan Muhs and Phil Stoffer have provided helpful reviews of a previous version of this manuscript.

REFERENCES

Anderson, R. S., 1990, Evolution of the Northern Santa Cruz Mountains by advection of crust past a San Andreas Fault bend: Science, v. 249, p. 397–401.

Anderson, R. S., 1994, Evolution of the Santa Cruz Mountains, California through tectonic growth and geomorphic decay: Journal of Geophysical Research, v. 99, no. B7, p. 14013–14029.

Anderson, R. S., and Menking, K. M., 1994, The Quaternary marine terraces of Santa Cruz, California: Evidence for coseismic uplift on two faults: Geological Society of America Bulletin, v. 106, p. 649–664.

Anderson, R. S., and Weber, G. E., 1990, Marine terraces and dating of the Santa Cruz terrace sequence: *in* Schwartz, D. P., and Ponti, D. J., eds., Field Guide to Neotectonics of the San Andreas Fault System, Santa Cruz Mountains, in light of the 1989 Loma Prieta Earthquake, U.S.G.S. Open File Report, p. 8–11.

Aniku, J. R. F.,and Singer, M.J., 1990, Pedogenic iron oxide in a marrine terrace chronosequence: Soil Science Society of America Journal, v. 54, p. 147–152.

Bradley, W. C., 1957, Origin of marine-terrace deposits in the Santa Cruz Area: California: Geological Society of America Bulletin, v. 68, p. 421–444.

Bradley, W.C., 1958, Submarine abrasion and wave-cut platforms: Geological Society of America Bulletin, v. 69, p. 967–974.

Bradley, W. C., and Griggs, G. B., 1976, Form, genesis, and deformation of central California wave-cut platfoms: Geological Society of America Bulletin, v. 87, p. 433–449.

Clark, J. C., 1981, Stratigraphy, paleontology, and geology of the central Santa Cruz Mountains, California Coast Ranges U.S. Geological Survey Professional Paper 1168, 51 p.

Hanks, T. C., Buckman, K. R., Lajoie, K. R., and Wallace, R. E., 1984, Modification of wave-cut and faulting controlled landforms: Journal of Geophysical Research, v. 89, p. 5771–5790.

Lajoie, K. R., 1986, Coastal Tectonics, Tectonics: Studies in Geophysics: Washington D.C., National Academy Press, p. 95–124.

Lajoie, K. R., Ponti, D. J., Powell, C. L., Mathieson, S. A., and Sarna-Wojcicki, A. M., 1991, Emergent marine strandlines and associated sediments, coastal California; a record of Quarternary sea-level fluctuations, vertical tectonic movements, climatic changes, and coastal processes, Quaternary nonglacial geology: conterminous U.S.: The Geology of North America, DNAG, p. 190–213.

Lajoie, K. R., Weber, G. E., Mathieson, S. A., and Wallace, J., 1979, Quaternary tectonics of coastal Santa Cruz and San Mateo Counties, California, as indicated by deformed marine terraces and alluvial deposits: Coastal tectonics and coastal geologic hazards in Santa Cruz and San Mateo Counties, California: Geological Society of America, Cordilleran Section, Fieldtrip guidebook, p. 61–80.

Lowe, D. R., 1979, Stratigraphy and sedimentology of the Pigeon Point formation, San Mateo County, California: *in* Nilsen, T. H., and Brabb, E. E., eds, Geology of the Santa Cruz Mountains, California: Fieldtrip guidebook for the Geological Society of America Cordilleran Section. p. 17–26.

Muhs, D.R., Prentice, C., and Merritts, D.J., 2003, Marine terraces, sea level history and Quarternary tectonics of the San Andreas fault on the coast of California. This volume.

Muhs, D. R., Simmons, K. R., Kennedy, G. L., Ludwig, K. R., and Groves, L. T., 2002, A cool eastern Pacific Ocean at the close of the last interglacial complex ~80,00 yr B.P.: Geological Society of American Abstracts with Program, v. 34, No. 6, p. 130.

Page, B. M., and Holmes, C. N., 1945, Bituminous sandstone deposits near Santa Cruz County, California: U.S. Geological Survey, Preliminary Map 27

Perg, L. A., Anderson, R. S., and Finkle, R. C., 2001, Use of a new ^{10}Be and ^{26}Al inventory method to date marine terraces, Santa Cruz, California, USA: Geology, v. 29, no. 10, p. 879–882.

Petersen, M. D., Bryant, W. A., Cramer, C. H., T., C., Reichle, M. S., Frankel, A. D., Lienkaemper, J. J., McCrory, P. A., and Schwartz, D. P., 1996, Probabilistic seismic hazard assessment for the State of California: U.S. Geological Survey. 96–706.

Rode, K., 1930, Geomorphogenie des Ben Lomad (Kalifornien), eine Studie uber Terrassenbildung duch marine Abarsion: Zeitschrift fur Geomorphologie, v. 3, p. 16–78.

Weber, G. E., 1990, Pleistocene marine terraces and neotectonics of the San Gregorio fault zone *in* Griggs, G. B., and Weber, G. E., eds., Field Trip Guide: Coastal geologic hazards and coastal tectonics, northern Monterey Bay and Santa Cruz and San Mateo County coastlines: Association of Engineering Geologists, p. 146.

Weber, G. E., 1990, Late Pleistocene slip rates on the San Gregorio fault zone at Point Año Nuevo, San Mateo County, California: *in* Garrison, R. E., Greene, H. G., Hicks, K. R., Weber, G. E., and Wright, T. L., eds., Geology and tectonics of the central California coast region, San Francisco to Monterey, volume and guidebook, Pacific Section: Bakersfield, American Association of Petroleum Geologists, p. 193–204.

Weber, G. E., Nolan, J. M., and Zinn, E. N., 1995, Determination of Late Pleistocene- Holocene slip rates along the San Gregorio fault zone, San Mateo County, California: National Earthquake Hazard Reduction Program. Final Technical Report NEHRP Contract No. 1434-93-6-2336, 70 p.

Weber, G.E., and Allwardt, A.O., 2001 The Geology from Santa Cruz to Point Año Nuevo—The San Gregorio Fault Zone and Pleistocene Marine Terraces *in* Stoffer, P.W., and Gordon, L.C., eds. Geology and Natural History of the San Francisco Bay Area. U.S. Geological Survey Bulletin 2188, p. 1–32.

Pleistocene and Holocene geomorphology of Marble and Grand Canyons, canyon cutting to adaptive management

Authors: **Joel L. Pederson,** *Department of Geology, Utah State University, Logan, Utah*
John C. Schmidt, *Department of Aquatic, Watershed, and Earth Resources, Utah State University, Logan, Utah*
Matt D. Anders, *Department of Geology, Utah State University, Logan, Utah*

Organizers and Leaders: **Joel L. Pederson and John C. Schmidt**

INTRODUCTION

Grand Canyon is famous for its display of the bedrock stratigraphy of the Colorado Plateau, and it is also one of the premier geomorphic features of Earth (Fig. 1). This landscape has been a birthplace of many fundamental geomorphic ideas and it remains a focal point of ongoing research. This guide emphasizes the geomorphology of this spectacular environment, including the history of canyon cutting, the effects of climate change on the landscape, organization of the modern river, and the effects of Glen Canyon Dam on the river's flow, sediment transport, and alluvial deposits. Field stops are accessed by river expedition.

A field trip on the Colorado River is a unique scientific and outdoor experience. The trip is an unparalleled opportunity to examine the geomorphology and geology of Grand Canyon amidst its sublime beauty and the experience of running rapids and camping on sandy beaches. Nearly 25,000 persons raft the river each year, and the waiting time to receive a permit for a self–guided river trip exceeds 10 years.

Figure 1. Downstream view of the Colorado River from the mouth of Nankoweap Creek (RM 52.5). The vertical distance from the Bright Angel shale at river level to the Kaibab limestone at the canyon rim is ~1200 m.

This guide describes field stops between Lees Ferry and Whitmore Wash that cover a distance of 310 river km (189 mi.) in Marble and Grand Canyons. Although the entire area downstream from Lees Ferry is sometimes referred to as Grand Canyon, the first 100 km are properly called Marble Canyon, which extends to the Little Colorado River confluence (Fig. 2). Downstream from this point is Grand Canyon.

Field stops in this guide concern the Pleistocene and Holocene geomorphology that can be examined from the river and focus attention on research of the past decade. Previous field guides of the geology and geomorphology of the Colorado River corridor include Hamblin and Rigby (1968, 1969), Simons and Gaskill (1969), and Billingsley and Elston (1989). Graf et al.'s (1989) field guide focused on the hydraulic features of the river. Specific locations along the river are referenced to River Mile (RM) designations first established by Birdseye (1924) and used in modern river guidebooks (Belknap, 1969, Stevens, 1983).

THE GRAND CANYON LANDSCAPE

The Grand Canyon region is a landscape of broad plateaus into which the canyons of the Colorado River system have been cut. The region has an arid to semi-arid climate, and the Colorado River itself is mostly composed of water derived from the distant Rocky Mountains. Grand Canyon creates an east-west transect through the southwestern Colorado Plateau, exposing metamorphosed, Precambrian, basement rocks, a thick, tilted and faulted section of Meso-and–Neoproterozoic sedimentary and volcanic rocks, a relatively complete and undeformed Paleozoic sedimentary section above the Great Unconformity, Quaternary basalt flows in western Grand Canyon, and Quaternary surficial deposits that are well exposed where the valley is relatively wide.

Three types of surficial deposits: alluvial, colluvial, and eolian, are widespread in Grand Canyon. Alluvium includes fine-grained Holocene fill terraces and floodplain deposits of the Colorado River, Pleistocene gravelly fill terraces of the Colorado River, and fill terraces, debris-flow, and fluvial deposits of tributaries. Colluvium includes landslides, talus, and debris-flow deposits, commonly associated with large remnant sediment man tles on the slope-forming units or at the toe of canyon walls. Eolian deposits consist primarily of dunes derived from sands blown from channel bars when the Colorado River is at low flow stages.

Hillslopes in Grand Canyon are dominated by the compound escarpments of alternating cliffs and slopes of bedrock that create the classic Grand Canyon landforms (Fig. 1). Under present cli-

Pederson, J.L., Schmidt, J.C., and Anders, M.D., 2003, Pleistocene and Holocene geomorphology of Marble and Grand Canyons, canyon cutting to adaptive management: in Easterbrook, D.J., ed., Quaternary Geology of the United States, INQUA 2003 Field Guide Volume, Desert Research Institute, Reno, NV, p. 407–438.

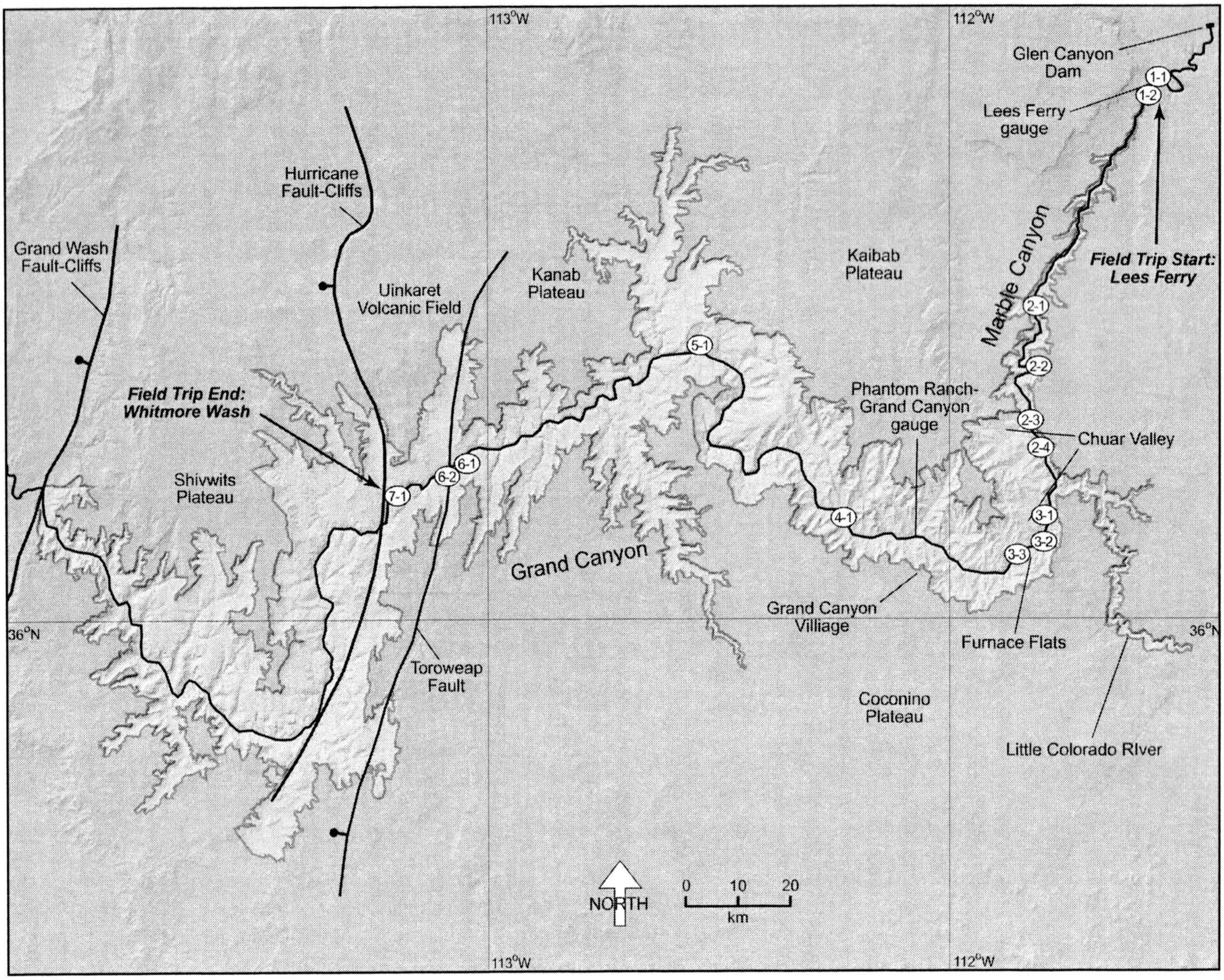

Figure 2. Map of the Grand Canyon region with the location of field trip stops and major physiographic, geologic, and cultural features labeled.

matic conditions, hillslopes are predominantly weathering-limited; that is, most slopes have bedrock exposed and sediment transport is limited by weathering rates. But abundant evidence exists that slope-forming components of hillslopes and escarpment toe slopes were buried beneath mantles of regolith and talus during past climatic episodes. Slope angle for a given geologic formation exposed in the canyon walls is likely a function of rock strength, and much of the excavation of Grand Canyon has been the result of cliff retreat and canyon widening rather than stream incision. The erosional potential expressed in the total relief of Grand Canyon, however, is controlled by the incision of the Colorado River and its tributaries.

LANDSCAPE EVOLUTION OF THE COLORADO PLATEAU

The timing of uplift and erosional exhumation of the Colorado Plateau has been long debated. The region was at sea level in late Cretaceous time, its landscape was relatively low in relief through much of middle Cenozoic time, and now the plateau is a deeply incised landscape at an average elevation of ~2 km. The mechanisms and detailed evolution of the landscape are much debated.

An average of 2117 m of rock uplift of the Colorado Plateau since late Cretaceous coastal sandstone deposited must be accounted for by some combination of: (1) Laramide orogenic uplift in the early Cenozoic, (2) middle–late Cenozoic epeirogenic uplift, and (3) isostatic rebound (Pederson et al., 2002b). This is less uplift than these three processes can supply, thereby suggesting that Laramide uplift of the plateau was significantly less than the neighboring Rocky Mountains or that little or no post-Laramide uplift beyond erosional isostasy occurred. Pederson et al. (2002b) estimate an average of about 850 m of erosional exhumation across the plateau since ~30 Ma, which itself can likely account for nearly a third of this rock uplift by isostatic processes. In addition to this, paleobotanical and fission-track data from the larger region suggest the Laramide orogeny alone should easily account for all of the rock uplift (e.g. Wolfe et al., 1998; Dimitru et al, 1994). Lastly, geophysical studies suggest changes in mantle buoyancy or "dynamic" asthenosphere can provide significant, late-Cenozoic, epeirogenic uplift (e.g., Lowry et al., 2000; Humphreys and Dueker, 1994; Morgan and Swanberg, 1985), and this may be supported by recent paleo–altimetry research along the edges of the plateau (Sahagian et al., 2002).

Grand Canyon ends where the Colorado Plateau meets the Basin and Range physiographic province at the Grand Wash Cliffs (Fig. 2). Basin and Range extension and normal faulting in the area adjacent to the plateau began in the Oligocene and was highly active in the Miocene, though little evidence exists for younger fault offset along the Grand Wash Cliffs (e.g., Young and Brennan, 1974; Faulds et al., 1997). The younger Hurricane and Toroweap fault zones are still active and cut across the canyon landscape, influencing the path the Colorado River between RM 200 and 225 (Fig. 2).

INCISION AND EVOLUTION OF GRAND CANYON

John Wesley Powell (1875) was the first geologist to complete a river expedition through Grand Canyon in 1869. He introduced the word "antecedent" to describe the path of the Colorado River as it crosses the uplifts in the Colorado Plateau, including the Kaibab uplift that forms the North Rim of Grand Canyon. Powell believed the Colorado River must be relatively old and that younger uplifts were raised across its path. Dutton (1882) concurred and suggested that the East Kaibab monocline at the eastern edge of the Kaibab uplift was Pliocene in age, coincident with the uplift of the entire Plateau and the incision of Grand Canyon. However, Gilbert (1876) admitted early on that the degree of antecedence versus superposition of regional drainages was unknown. That the monoclines of the Colorado Plateau are associated with the early Cenozoic Laramide orogeny has subsequently been established.

William Morris Davis visited the region at the turn of the 19th century and asserted that the river was superimposed, not antecedent, across the uplifts of the region. He envisioned drainages at one time flowing toward the northeast but then reversing direction due to block faulting and flowing westward (Davis, 1901). In Davis' model, later uplift rejuvenated the stream and caused an episode of incision wherein the river became superimposed across older uplifts. Workers in the early part of the 20th century interpreted the escarpments and plateaus of the region as remnants of several iterations of Davis' geographic cycle of peneplanation, uplift, and denudation (Breed, 1969). Eventually this interpretation was abandoned, because of lack of correlation of the many plateaus and the availability of alternative explanations for the broadly terraced landscape.

The ideas of Powell, Dutton, and Davis were supplemented by Charles Hunt (1969), Edwin McKee (1972), Chester Longwell (1946), Ivo Lucchitta (1972) and others, and the general idea that the paleo Colorado River is superimposed onto the plateau landscape has remained. The major working hypothesis for the past few decades has been that Grand Canyon was formed mostly in the Pliocene (McKee and McKee, 1972; Machette and Rosholt, 1991; Lucchitta, 1972, 2003) before Quaternary basalt flowed into and preserved the base of the paleo-canyon (Hamblin, 1994). However, the story is quite complicated and continues to be revised by new research.

The Grand Canyon that we see today is the product of a large-scale reversal in drainage direction that happened over the late Cenozoic. Laramide uplift of the ancestral Mogollon highlands to the south of the present-day Colorado Plateau created consequent, deeply incised, drainages that flowed to the north and northeast across the Grand Canyon region (Young and McKee, 1978; Potochnik and Faulds, 1998). These highlands collapsed during Basin–and–Range extension in the early-to-middle Miocene along the southern and western margins of the plateau (e.g. Young and Brennan, 1974; Jones et al., 1996; Faulds et al., 1997). This topographic inversion from highlands to extensional basins disrupted drainages and generated relief that would eventually cause a very significant drop in effective base level, steepening streams on the Colorado Plateau once they became integrated and reversed direction towards the opening Gulf of California (Lucchitta, 2003). Analysis of the sedimentary and erosional record of the Grand Wash Trough, where the Colorado River exits the Plateau and enters the Basin and Range, reveals a switch from internal-basin deposition in late-Miocene time to subsequent incision by a through-going river (Fig. 2). This indicates that drainage integration of the upper and lower Colorado Rivers was complete just after 6 Ma (Longwell, 1946; Lucchitta, 1966, 2003).

Most Grand Canyon incision has been thought to have happened between 6 and 1.2 Ma based upon K-Ar dates on basalts that flowed into the Grand Canyon from the Uinkaret volcanic field (Hamblin, 1994; Lucchitta, 2003). However, the basalt flows are substantially younger than previously thought, and we now have a better understanding of down to the west slip rates along the active Hurricane-Toroweap fault zone in this same area (Dalrymple and Hamblin, 1998; Fenton et al., 2001; Pederson et al., 2002a). Neotectonic slip on these normal faults has affected Quaternary incision rates of the Colorado River in Grand Canyon, though exactly how is debated (cf. Fenton et al., 2001; Pederson et al., 2002a).

In summary, the overall incision of Grand Canyon has been a response to a combination of Cenozoic uplift (the timing of which is unresolved), relief-generation at the edge of the Colorado Plateau in late Oligocene and Miocene time by extensional faulting, and drainage integration off the Grand Wash escarpment and effective base-level fall in the early Pliocene. Significant excavation of Grand Canyon has been accomplished in the Quaternary and continues. Stream incision rates in the Quaternary have varied along the length of Grand Canyon due to neotectonism, and erosion–aggradation activity has been strongly controlled by glacial-interglacial climate changes.

PLEISTOCENE STREAM TERRACES

Coarse-grained alluvium of the paleo-Colorado River is preserved in two areas of Grand Canyon where the valley is relatively wide: the "Furnace Flats" or "Big Bend" reach of eastern Grand Canyon and the Whitmore-Granite Park reach of western Grand Canyon in the vicinity of the Hurricane fault. Recent research described in the field trip stops indicate that inset fill terraces along the mainstem and tributaries record in detail landscape responses to climate changes, but are superimposed on the larger-scale signal of tectonic events.

Machette and Rosholt (1991) identified seven Colorado River terraces in eastern Grand Canyon formed episodically in the last 700 ka, based on uranium-trend dating. They interpreted incision rates as having increased, seemingly, towards the present day. In the same area, Lucchitta et al. (1995) identified seven mainstem terraces, and employed preliminary cosmogenic surface-exposure dating to obtain ages. The pertinence of the chronologic data of these studies is unclear because the uranium-trend dating method is now abandoned, and Lucchitta et al. (1995) did not utilize amalgamated sampling or correct for inheritance in their cosmogenic dates. In western Grand Canyon, Lucchitta et al. (2000) used Ar-Ar-dated basalt flows that cap stream gravel to calculate incision rates. Through comparison to preliminary data from eastern Grand Canyon, they recognized that incision rates in western Grand Canyon are significantly lower than to the east.

Lucchitta et al. (1995) concluded that two younger Pleistocene aggradation episodes in eastern Grand Canyon occurred as a result of sedimentologic and hydrologic changes in the headwaters of the Colorado River system during the Pinedale (OIS 2) and Bull Lake (OIS 6) glaciations, whereas a mid-Holocene deposit resulted from cannibalization of stored alluvium and colluvium in eastern Grand Canyon. Elston (1989) suggested that the sequence of inset fill terraces are instead a single deposit resulting from regional Miocene-Pliocene aggradation. Hamblin (1994) proposed instead that they were deposited in a series of lakes that formed behind lava dams in western Grand Canyon.

HOLOCENE TERRACES

The Holocene alluvial stratigraphy of the mainstem corridor, though not a focus of this field trip guide, has been well studied and provides an important record of fluvial history as well as prehistoric examples of features of the modern river. Hereford et al. (1996) presented the stratigraphy as a suite of inset sandy deposits and terraces formed during aggradation by the Colorado River or deposition during notable floods. They used ^{14}C dating, stratigraphic relations to archeological material, and carbonate clast dissolution pitting for age control (Fig. 3). Hereford et al's (1996) "striped alluvium" and especially the "alluvium of Pueblo II age" are associated with archeological sites of ancestral puebloan affiliation and generally were not inundated by historic pre-dam flows of the Colorado River (Fig. 3). Hereford (2002) correlated the "upper mesquite terrace" to the Naha Fm. of Hack (1942), which was deposited in many of the valleys of the southwestern U.S. during the Little Ice Age (ca. 1400-1880 A.D.).

Pre-Glen Canyon Dam and post-Glen Canyon Dam deposits and surfaces are inset into these Holocene terraces. Schmidt and Rubin (1995), Hereford (1993), and Hereford et al. (1996, 1998a, 1998b) subdivided post-dam deposits into flood sand of the summer 1983, the high-flow sand (1984-1986) deposited by floods whose maximum discharge was approximately 1270 m^3/s, sand deposited by the post-1986 flows that did not exceed 880 m^3/s, and 1996 controlled-flood deposits. The hydrology that gave rise to these deposits is described below.

Fine-grained alluvium can also be distinguished by depositional environment. Pre-dam deposits occur as narrow, linear bands along the river and are never inundated by the post-dam river. Lower elevation bars, formed and maintained by the post-dam river, are more extensive than the pre-dam deposits in most of Grand Canyon. A few of these bars occur in relatively straight reaches as alternate or mid-channel bars, but most bars occur in the lateral, recirculating flow of eddies (Schmidt, 1990).

MODERN HILLSLOPE AND TRIBUTARY PROCESSES

Rock falls, exfoliation–sheeting, and rock avalanches are active on the cliffs of Grand Canyon, and infiltration–excess overland flow is important for sediment transport on slopes, toeslopes of escarpments, and stream terraces. However, debris flows are probably the most important transport process delivering sediment from hillslopes to drainages in Grand Canyon (Webb et al., 1988, 1989; Webb 1996; Griffiths et al., 1996; Melis, 1997). In this setting, debris flows are initiated when intense or prolonged rainfall combines with antecedent soil moisture, triggering failure and mass movement. Typical of Grand Canyon are cases

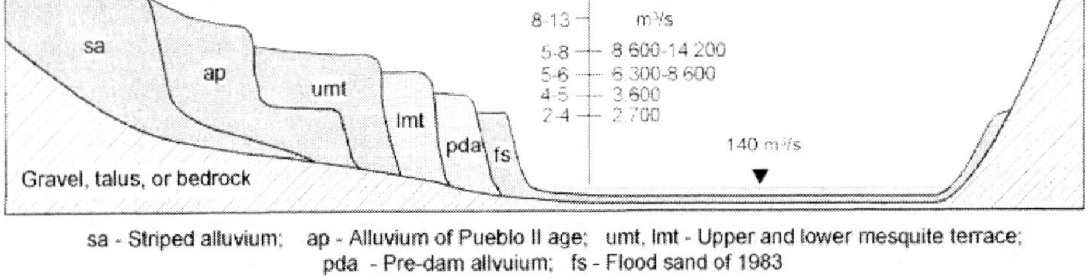

Figure 3. Holocene alluvial stratigraphy along the mainstem Colorado River corridor. Illustration provided by Richard Hereford, USGS.

where regolith stored in relict colluvial mantles are mobilized by a "firehose effect" of water pouring off cliffs and impacting slopes below (Griffiths, et al., 1996). A majority of modern debris flows are associated with summer monsoon thunderstorms, but less frequent, large events are triggered during low intensity winter precipitation (Cooley, et al., 1977; Griffiths, et al., 1996).

Griffiths et al. (1996) suggest that shale bedrock is critical to debris flow initiation. Shale bedrock fails readily and forms many of the slopes that store regolith. It also provides clay that increases flow mobility as well as competence by increasing the effective buoyancy of clasts. Griffiths et al. (1996) also state that debris-flow probability is higher in eastern versus western Grand Canyon because of greater precipitation, the southwest trend of drainages parallel to the direction of storm tracks, and optimum heights of the Permian Hermit Shale above the Colorado River. Hereford et al. (1998a) dated flows using the relief of carbonate dissolution pitting on clasts, calibrated with radiocarbon dating. He inferred from this that debris–flow activity, and supposedly precipitation, periodically increased at a recurrence interval of ~800 yrs during the late Holocene. Cerling et al. (1999) state that this scale of variability in debris flow frequency cannot be resolved with existing data. An average of two debris flows occurred each year between 1984 and 1997 in Grand Canyon (Webb et al., 1999).

CLIMATE AND PALEOCLIMATE OF GRAND CANYON

The present-day climate of Grand Canyon has a bimodal moisture distribution with about half of the annual precipitation in the form of high-intensity, localized, summer monsoonal storms and half as low-intensity precipitation or snow associated with frontal systems in the winter. The spring months are typically the driest of the year. Interannual variability in moisture in the southwestern U.S. has been linked to the El Nino–Southern Oscillation and, to a lesser degree to the Pacific Decadal Oscillation. Precipitation, temperature, and vegetation zones vary strongly with elevation. Mean annual precipitation is 647 mm and mean annual temperature is 6.3°C (43° F) on the North Rim (~2500 m), whereas at the bottom of the canyon (~600 m) they are 213 mm and 20.4°C (69° F), respectively. Desert scrub communities that include species such as sagebrush (*Artemisia tridentata*), mormon tea (*Ephedra nevadensis*), catclaw (*Acacia greggii*), and black-brush (*Coleogyne ramosissima*) occupy elevations less than 1,500 m (5,000 ft); pinyon-juniper shrub woodland (*Pinus edulis* and *Juniperus osteosperma*) dominates elevations between 1,500 and 2,200 m (5,000–7,200 ft), and fir (*Pseudotsuga menziesii*), pine (*Pinus ponderosa*), spruce (*Picea engelmannii*), and aspen (*Populus tremuloides*) are present above 2,200 m (7200 ft) (Cole, 1990; Weng and Jackson, 1999).

Regional paleoclimate records from pollen and plant macrofossils indicate that from 50-25 ka, prior to the last glacial maximum, mean annual temperature was 2.9-4.3°C cooler than the present, precipitation was greater than present, a winter precipitation regime dominated, and the altitudinal zones of plant communities were depressed 460-900 m relative to modern conditions (Anderson, 2000; Coats, 1997). During full-glacial conditions, 25-15 ka, pollen, macrofossils, and the isotopic composition of herbivore teeth indicate annual precipitation was 87 mm higher (13% increase), mean temperature was 6.7°C lower, and some plant species were depressed approximately 1000 m (e.g. Cole 1990; Mead and Phillips, 1981; Connin, et al., 1998). The glacial-interglacial transition began by ~15 ka and was characterized by a major vegetation disturbance as plant communities migrated upslope (Cole 1990; Weng and Jackson, 1999). By 8 ka, temperature was roughly similar to the present, but precipitation was still somewhat greater than present due partly to a stronger summer monsoon. The middle Holocene (8-4 ka) was more arid as annual temperature was ~1°C warmer and precipitation was slightly less than now (Cole 1990; Weng and Jackson, 1999).

THE MODERN RIVER

Bedrock exerts a strong control on width of the alluvial valley and ultimately on width of the active channel. The valley between Glen Canyon Dam and Lake Mead reservoir is narrow and width varies in relation to the lithology of the bedrock that occurs near river level (Howard and Dolan, 1981). The valley is widest where river level bedrock is shale or interbedded shale, sandstone, and limestone (Howard and Dolan, 1981; Schmidt and Graf, 1990) and is typically less than 20% wider than the channel itself. Thus, fine–grained deposits comprise a relatively small proportion of the entire valley because there is little space for these deposits. Reach–average channel width varies between 55 and 120 m (Table 1).

The longitudinal profile of the Colorado River through Grand Canyon has segments that are relatively steep and segments that are relatively flat (Fig. 4). The difference in segment steepness varies over a range between 0.001 and 0.0023 m/m (Table 1). Two views of the significance of these differences in steepness exist. Some argue that the couplet of a relatively flat segment, followed by steep segment, defines a convexity in the river's profile that may be related to regional tectonism. The other perspective is that relative segment steepness is determined by the coarse material brought by debris fans, which requires steepening of the gradient. Debris fan location in turn is established by the regional joint and fault system that defines the frequency of tributary canyons, since every tributary canyon has a debris fan at its mouth (Dolan et al., 1978).

Resolution of the differing perspectives on control of the longitudinal profile partly depends on whether the Colorado River generally flows on bedrock, in which case the river profile is likely to be determined by rock strength and tectonics, or if today's river flows within a fill of colluvium and alluvium. A systematic picture of the distribution of bedrock on the bed of the river has yet to emerge. Data about the thickness of the unconsolidated fill of the Colorado River has become available slowly, because of the absence of bore hole data, the difficulty of geo-

TABLE 1 REACH ATTRIBUTES (from Schmidt and Graf, 1990)

River miles that define the reach	Reach name	Major geologic formations that occur at river level	Average ratio of top width to mean depth	Average channel width, in meters	Reach slope
0-11.3	Permian section	Kaibab Limestone Toroweap Fm. Coconino Sandstone Hermit Shale	11.7	85	0.00099
11.4-22.5	Supai Gorge	Supai Group	7.7	64	.0014
22.6-39.9	Redwall Gorge	Redwall Limestone	9.0	67	.0015
40.0-61.5	Lower Marble Canyon	Muav Limestone Bright Angel Shale Tapeats Sandstone	19.1	107	.0010
61.6-77.4	Furnace Flats	Tapeats Sandstone Unkar Group	26.6	119	.0021
77.5-117.8	Upper Granite Gorge	Zoroaster Plutonic Complex Trinity and Elves Chasm Gneisses Vishnu Schist	7	58	.0023
117.9-125.5	Aisles	Tapeats Sandstone Vishnu Schist	11	70	.0017
125.6-139.9	Middle Granite Gorge	Unkar Group Vishnu Schist	8.2	64	.0020
140-159.9	Muav Gorge	Muav Limestone	7.9	55	.0012
160-213.8	Lower Canyon	Quaternary basalt Muav Limestone Bright Angel Shale	16.1	94	.0013

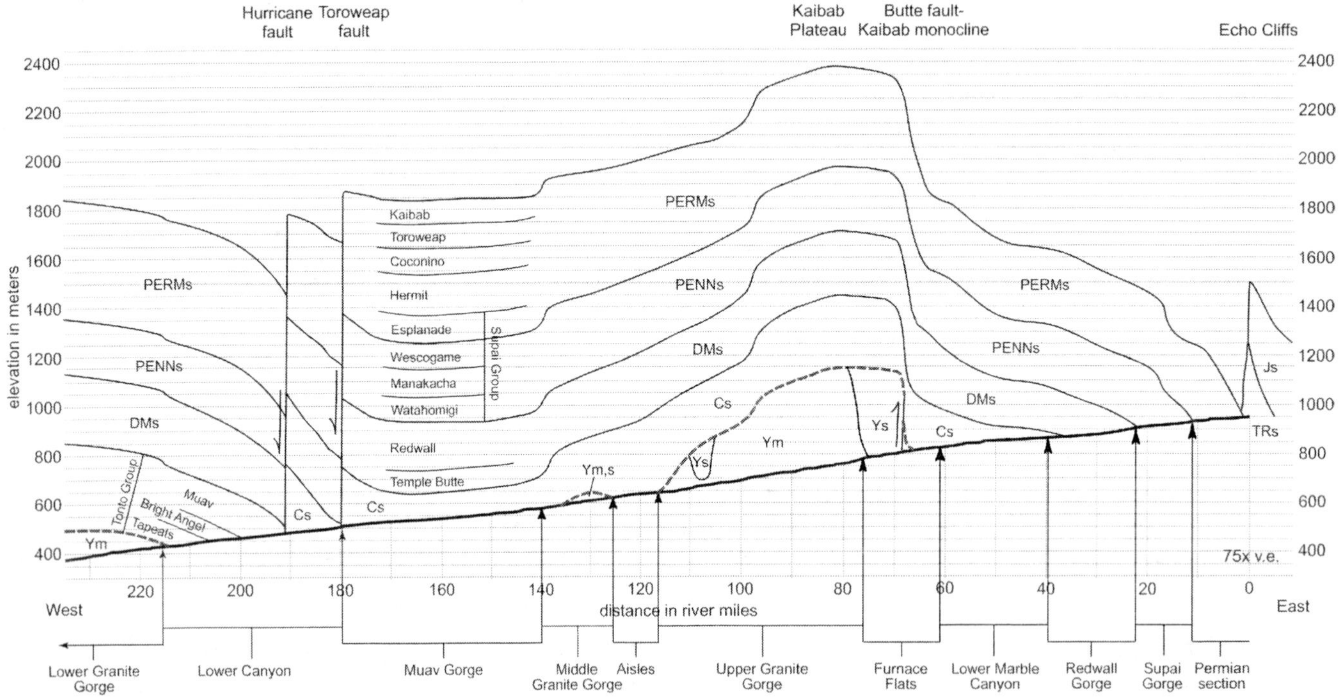

Figure 4. Longitudinal profile of the Colorado River through Grand Canyon. Water surface elevations are those surveyed by Birdseye (1924). Reaches are those of similar geomorphic characteristics (Table 1), following (Schmidt and Graf, 1990). Individual geologic formations labeled, and schematic groupings of bedrock units exposed in the canyon walls are as follows: Ym = Proterozoic metamorphic rocks, Ys = Proterozoic sedimentary and volcanic rocks, Cs = Cambrian Tonto Group sedimentary rocks, DMs = Devonian and Mississippian carbonates, PENNs = Pennsylvanian sandstone and mudstone, PERMs = Permian sedimentary rocks, TRs = Triassic sedimentary rocks, Js = Jurassic sandstone. The Great Unconformity is thicker and dashed. Generalized bedrock stratigraphy of canyon walls adapted from Hamblin and Rigby (1968, 1969).

physical data collection, and the absence of canyon–scale bed imaging. Excavation of Glen Canyon Dam revealed bedrock 39 m below the channel bed. Seismic refraction measurements of the depth to bedrock beneath 10 alluvial bars shows showed depths as great as 45 m (Rubin et al., 1994). However, the channel flows on bedrock elsewhere, because bedrock islands occur at some rapids (e.g. Walthenberg Rapid RM 112; Bedrock Rapid, RM 130).

Nevertheless, debris fans are the dominant control on the river's planform and profile at a small scale. The channel narrows and shallows as it crosses around each debris fan, and the small–scale profile of the river is that of a series of stair steps, with long, nearly flat treads and short steep steps where hydraulics are violent or exhilarating, depending upon one's perspective (Leopold, 1969; Kieffer, 1985). Schmidt and Rubin (1995) described the repeating series of geomorphic features centered on these debris fans as fan–eddy complexes, defined as a reach composed of those channel units whose hydraulics and depositional environments are determined by the existence of a debris fan (Fig. 5). The upstream end of a fan–eddy complex may be several kilometers upstream from the fan, because the constricted channel at the fan acts as a control and flow upstream is a hydraulic backwater. Downstream from each fan, the channel widens and depth may increase into a scour hole at the base of the rapid before resuming ambient conditions (Fig. 6). Recirculating eddies occur in the lee of these fans, and these eddies may fill half the river width and extend downstream for distances equivalent to several channel widths. These eddies are effective traps of the fine suspended load of the modern river. Downstream from the eddies, gravel bars occur in the middle of the channel and flow shallows. Webb et al. (1999) demonstrated that some of the gravel in these bars is comprised of reworked boulders from the upstream debris fan.

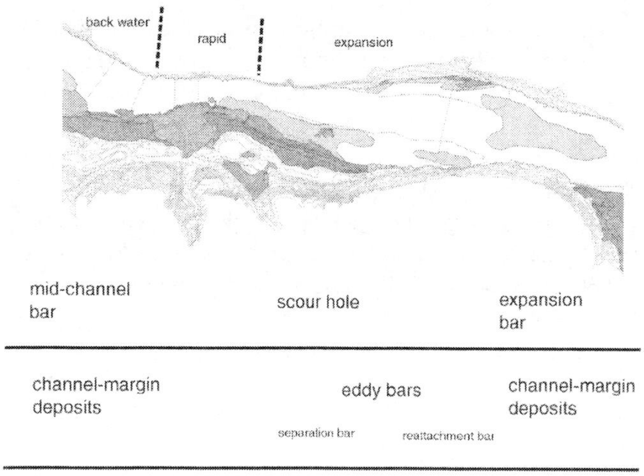

Figure 5. Distribution of modern alluvium and colluvium within a typical fan–eddy complex. River flows from left to right. The mapped site is Paria Riffle, located at RM 1.0, the downstream half of which is shown at the top of Figure 6. Features of the bed and the banks are schematically listed below the figure.

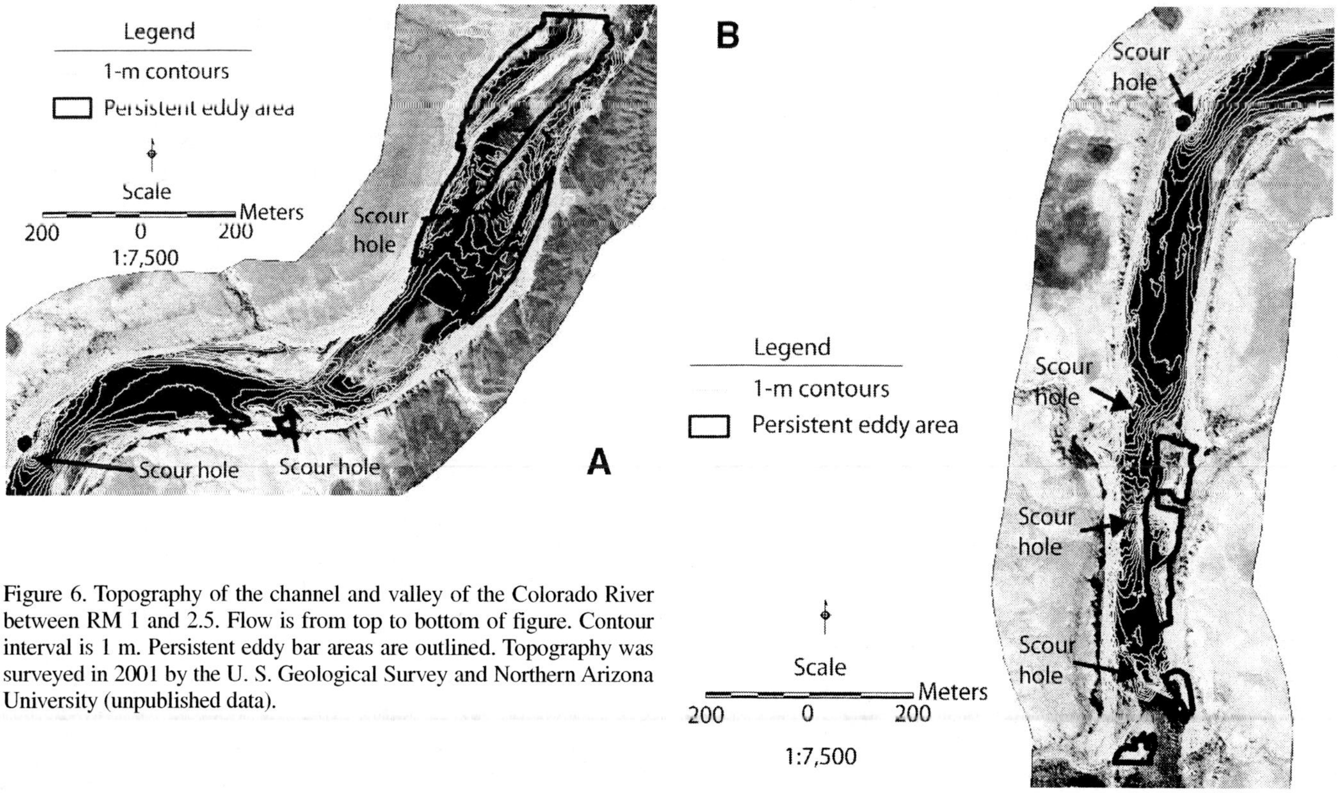

Figure 6. Topography of the channel and valley of the Colorado River between RM 1 and 2.5. Flow is from top to bottom of figure. Contour interval is 1 m. Persistent eddy bar areas are outlined. Topography was surveyed in 2001 by the U. S. Geological Survey and Northern Arizona University (unpublished data).

The distribution of fine–sediment deposits in 80 km of the Colorado River between Glen Canyon Dam and RM 72 have been mapped (Schmidt et al., 1999, 2002; Sondossi and Schmidt, 2001; Grams et al., 2002). These maps have been made from historical aerial photography for various years between 1935 and 2002. Schmidt et al. (1999) devised a scheme to delineate eddy bars and their changes through time and defined persistent eddy bar areas as the largest area in which emergent eddy bars have been photographed in all the years of mapping (Fig. 7). The frequency of those eddy bar areas larger than 1000 m^2 varies between 2 and 5 per kilometer in Grand Canyon.

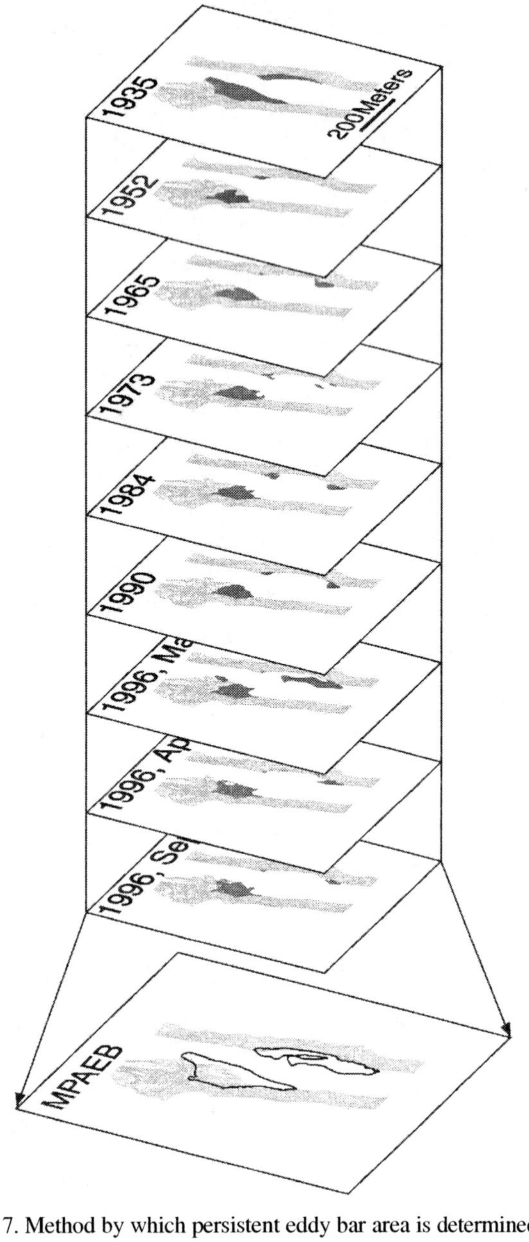

Figure 7. Method by which persistent eddy bar area is determined from aerial photograph information within a geographic information system. Each layer in this diagram represents a geographic information system database and the outline of the eddy bar in each year is shown in black. The persistent eddy bar area is the sum of all the areas in all of the years.

THE NATURAL HYDROLOGY AND SEDIMENT FLUX OF THE COLORADO RIVER

The natural flow regime of the Colorado River through Grand Canyon is dominated by pulses of snowmelt derived from the distant Rocky Mountains and High Plateaus of Colorado, Wyoming, and Utah (Fig. 8). The main snowmelt pulse typically reached its peak in early June. The median discharge of the Colorado River at the Lees Ferry gage was 227 m^3/s for the pre–dam period between May 8, 1921, and March 12, 1963, and the range of flows during the year was large. The 10% exceedence flow was 1359 m^3/s, and the 90% exceedence flow of 127 m^3/s was more than an order of magnitude less (Topping et al., in press).

Prior to completion of the dam, most of the fine sediment transported through Grand Canyon came from the upper Colorado River basin, rather then from tributaries that enter the river within Marble and Grand Canyons (Fig. 9). The total annual load of fine sediment was 57 $\pm$ 3 million metric tons at the Lees Ferry gage and 83 $\pm$ 4 million metric tons at the Grand Canyon gage (Fig. 2), based on measurements made after 1944 (Topping et al., 2000). Approximately 40% and 35% of the annual fine sediment load passing Lees Ferry and the Grand Canyon gage, respectively, was sand.

The concentration of sand in suspension and the discharge of water were well correlated at the Lees Ferry gage prior to closure of Glen Canyon Dam, but this correlation was poorer at the Grand Canyon gage at RM 87, where the concentration of sand in suspension varied as a function of the seasonal supply of sand (Topping et al. 2000). When flows were less than about 250 m^3/s, the concentration of sand in suspension was greater at the Lees Ferry gage than at the Grand Canyon gage, indicating deposition of sand in the reach between these gages. During periods of flow higher than about 500 m^3/s, the sand stored during lower flows in the reach between the gages was eroded, as indicated by the initially higher concentration of suspended sand at the Grand Canyon gage. As this stored sand was eroded, the concentration of sand measured in suspended samplers at the Grand Canyon gage decreased to approximately equal the concentration of sand in suspension at the Lees Ferry gage.

This longitudinal variation in fine–sediment transport led to seasonal accumulation between the Lees Ferry and Grand Canyon gages between July and the following March, when discharges were mostly less than about 250 m^3/s (Topping et al., 2000). The difference between fine–sediment delivery to, and export from, the river segment between these gages during this period was between 1 and 13 million metric tons, annually. During the spring snowmelt flood between April and June, the amount of sand exported past the Grand Canyon gage was approximately equal to the amount transported past the Lees Ferry gage plus the amount of fine sediment that had accumulated in the reach since the previous July. These calculations imply that fine sediment would have accumulated on the channel

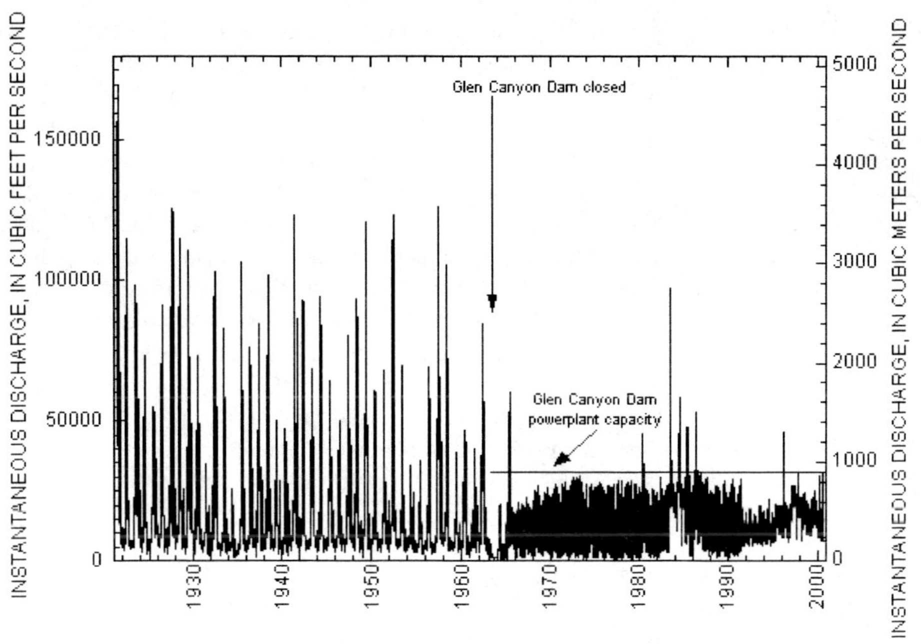

Figure 8. Hydrograph of the Colorado River at Lees Ferry based on instantaneous discharge, as determined from continuous stage records (Topping et al., in press).

bed and in eddies only during the 9 months of the year when the flows were typically low (July-March), and that the volume of fine sediment stored in the Colorado River during each year would have been least at the beginning of July.

The sedimentology of eddy bars is closely linked with the suspended–sediment flux, because eddy bars are effective traps of the suspended load. Topping et al. (2000) demonstrated that changes in the sizes of the suspended load during discrete floods are matched by vertical changes in the sizes of sediment in the eddy bars. Topping et al (2000) demonstrated that individual pre–dam and post–dam flood deposits in Glen, Marble, and Grand Canyons decrease in the proportion of silt and clay in an upward direction in about 80 % of all sampled deposits. However, they found that sand sizes typically coarsen upward in pre–dam flood deposits in Marble and Grand Canyons, but not in Glen Canyon. They concluded that pre–dam were, and post–dam floods are, supply limited in regards to sand in Marble and Grand Canyons.

Sedimentary structures in eddy bars are climbing ripples and dunes (Fig. 10). These structures are well preserved and provide evidence of the recirculating nature of eddy flow (Rubin et al., 1990). During the field trip, frequent opportunities will occur to examine sedimentologic evidence of flow stagnation points, migration of flow reattachment points, grain-size control on sedimentary structures, infilling of eddies, and the differences between separation and reattachment bars. A wide variety of well–preserved sedimentary structures in Grand Canyon alluvial deposits provide text–book quality examples (Rubin, 1987)

THE REGULATED HYDROLOGY AND SEDIMENT FLUX OF THE COLORADO RIVER

Glen Canyon Dam greatly reduced the magnitude of floods, increased the magnitude of base flows, and greatly reduced the amount of fine sediment annually delivered to the river. The median discharge of the Colorado River for the period between March 14, 1963, and September 30, 2000, was 74% higher than during the pre–dam period. The seasonal variation in flows was much less (Topping et al., in press), with the 10% exceedence flow for the post-dam period ~ 708 m^3/s and the 90% exceedence flow ~ 125 m^3/s.

The sediment supply to the Colorado River in the post–dam era was reduced more than was the capacity of the river to transport that load (Topping et al., 2000, in press). Though the dam has greatly reduced the floods that exported large quantities of fine sediment from Grand Canyon, the dam has also eliminated the naturally–occurring, lower flows that allowed temporary sediment storage during the 9–month accumulation season. Sediment delivery to Marble Canyon past the Lees Ferry gage decreased by about 99.5-99.6% after completion of Glen Canyon dam (Fig. 9). Transport past the Grand Canyon gage decreased a lesser amount, between 81 and 85%, because some fine sediment continues to enter the Colorado River from tributaries entering downstream from the dam (Topping et al., 2000).

Flow regimes of the post–dam river have changed since 1963, and management of Lake Powell reservoir and of water releases from Glen Canyon Dam can be divided into 3 time periods (Fig. 8). Between 1963 and 1980, the primary objective was to fill the reservoir and generate hydropower while still meeting the legally–defined downstream needs established by the Colorado River Compact of 1922. After the reservoir filled for the first time in 1980, dam operations were guided by traditional constraints of hydroelectric power production, downstream water needs, and dam safety. This period included a 4–year period between 1983 and 1986 when inflow was unusually large, and dam releases exceeded the capacity of the power plant. The highest post–dam discharge, 2750 m^3/s at the Lees Ferry gage, occurred on June 29, 1983.

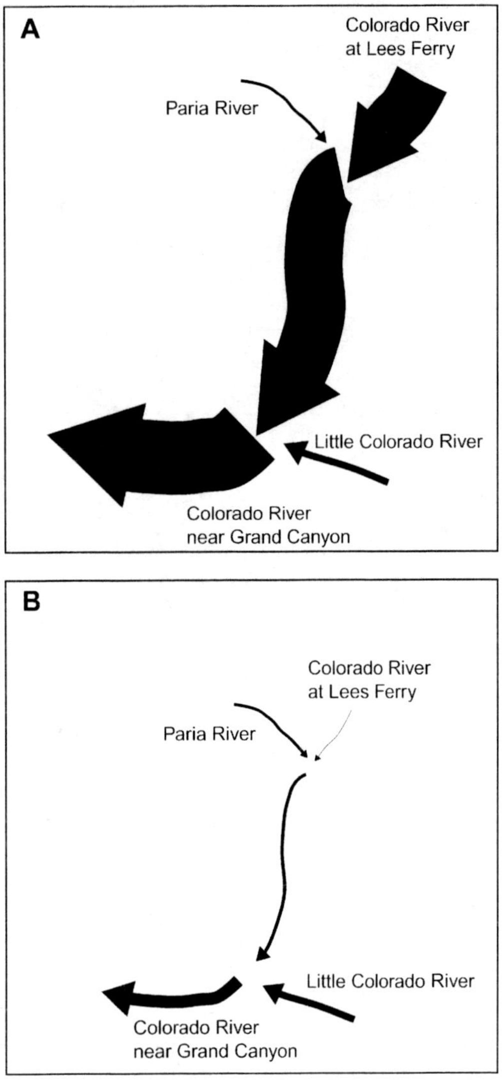

Figure 9. Flux of sediment transported through Grand Canyon (A) before and (B) after completion of Glen Canyon Dam

The daily range of reservoir releases were administratively constrained beginning in 1991, and flows less than the pre–dam median flow of 227 m^3/s now occur rarely. The post–1990 period included an 18–month period in 1990 and 1991 when the dam was operated to facilitate river–scale experiments and a 7–day experimental release of 1274 m^3/s that began on March 26, 1996. This event is generally referred to as the 1996 Controlled Flood. Three short–duration experimental releases of 878 m^3/s flows occurred in November 1997, May 2000, and September 2000.

RESEARCH AND MONITORING OF MODERN RIVER PROCESSES

The Colorado River in Grand Canyon is probably one of the most extensively studied rivers in the United States, and these studies have been funded since 1984 by the U.S. Bureau of Reclamation's Glen Canyon [Dam] Environmental Studies (GCES) and the U. S. Geological Survey. The GCES lasted until 1996 and focused on measuring and describing the relationship between dam operations and the environmental attributes of the Colorado River in the 400 km downstream to Lake Mead reservoir. This program provided the scientific background for an environmental impact statement concerning dam operations (U.S. Department of Interior 1995) and the decision of the Secretary of the Interior to revise those operations. As part of this decision, the Secretary created the Grand Canyon Monitoring and Research Center (GCMRC) and the Grand Canyon Adaptive Management Program. This program includes representatives of a wide range of interests affected by operations of the dam, including hydroelectric power users and professional whitewater river guides.

DAY 1. LEES FERRY AND ENTRANCE INTO MARBLE CANYON–DOWNSTREAM EFFECTS OF GLEN CANYON DAM ON FLUVIAL SYSTEM

STOP 1-1. LEES FERRY (RIVER MILE 0)

The river trip begins at Lees Ferry, the historic site of a ferry crossing of the Colorado River that was first operated by the Mormon Church in 1872 and then the State of Arizona until 1928 (Reilly, 1999). In that year, the ferry was abandoned, because Navajo Bridge, 7 km downstream, was completed.

Lees Ferry is located in a reentrant of the Vermillion and Echo Cliffs that mark the boundaries of the Paria and Kaibito Plateaus. These plateaus are located to the west, north, and east and are underlain by Triassic and Jurassic sedimentary rocks that are capped by the 400–m–thick, late–Triassic (?)–and–Jurassic, Navajo sandstone. The assemblage of Mesozoic rocks that now underlie the Paria and Kaibito Plateaus are more than 600 m thick and once covered the entire Grand Canyon region (Billingsley, 1989). Beneath the Navajo Sandstone are Triassic formations that are relatively erodible, thereby creating an open valley where the river banks can be accessed by vehicles.

About 3 km upstream from the boat ramp at Lees Ferry, the Colorado River emerges from Glen Canyon, a 100–200–m–deep canyon. Glen Canyon dam is 25 river km upstream from Lees Ferry, creates Lake Powell reservoir, and inundates most of Glen Canyon.

An important political boundary in the management of the Colorado River basin exists about 2 km downstream from the boat ramp. This point, downstream from the Colorado River's confluence with the Paria River, is called Lee Ferry (distinct from Lees Ferry) and was defined by the Colorado River Compact of 1922 as the division point between the upper Colorado River basin and the lower basin. By terms of this compact and subsequent treaties and administrative interpretations, 1.02 x 10^6 ha–m (8.25 x 10^6 acre-ft) of water must pass Lee Ferry. In order to measure this volume, the federal government established a

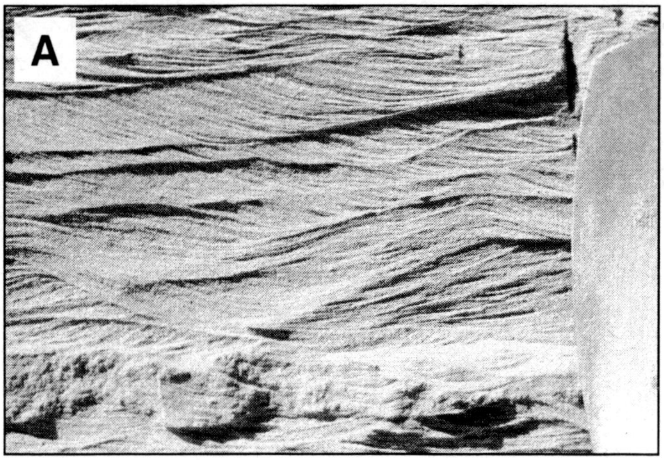

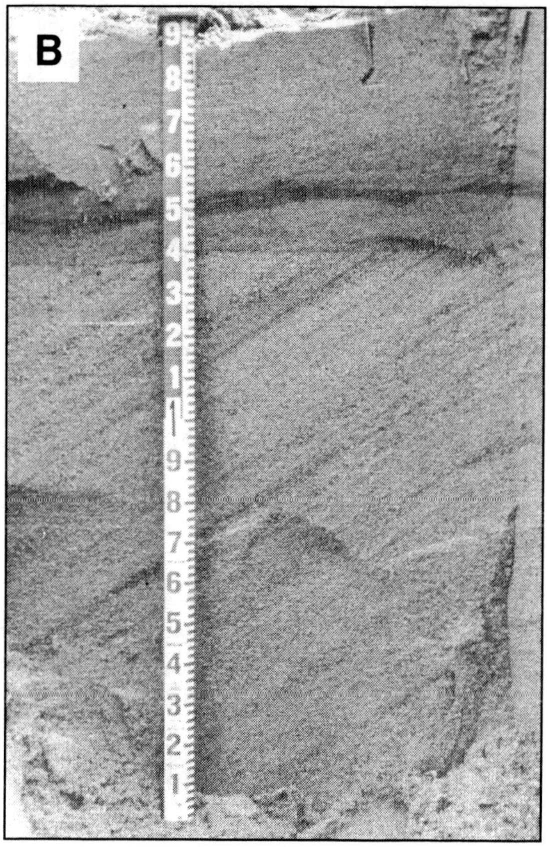

Figure 10. Sedimentary structures in eddy bars. A. Climbing ripples deposited near in the reattachment zone where upstream and downstream flow both occur. B. Tabular cross-beds overlain by climbing ripples in a reattachment bar.

stream–flow gaging site on the Colorado River near the site of the ferry in 1921 (U.S. Geological Survey gaging station 09380000) and on the Paria River in 1923 (U.S. Geological Survey gaging station 09382000).

The volume of water transferred from the upper to the lower basin is equivalent to a mean annual flow of 323 m^3/s, which was equaled or exceeded about 35 % of the time prior to completion of Glen Canyon dam. The dam was authorized in 1956 and completed in 1963 in order to reduce the risk that insufficient stream flow would be available to meet compact requirements during extended droughts and increased diversions in the upper basin. The project was funded by the federal government and is repaid to the U. S. Treasury by revenues from the generation of hydroelectric power. Thus, flows released from the dam traditionally follow the seasonal and daily needs for electricity in the region.

Water releases from the dam, completely absent of sediment, have caused large-scale bed degradation in Glen Canyon (Fig. 11). Bed degradation within 10 km of the dam began when the cofferdam was installed in 1959. The greatest degradation occurred in 1965 when high flows were released in a series of

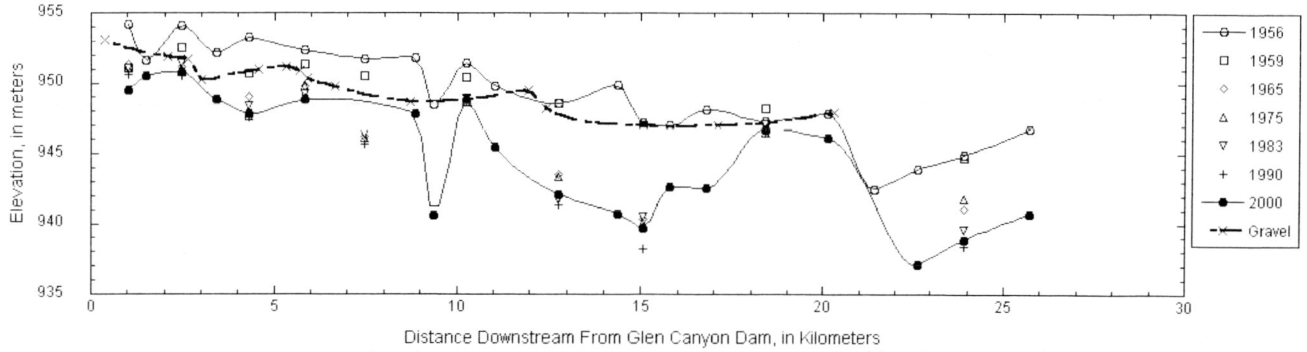

Figure 11. Longitudinal profile of the Colorado River between Glen Canyon Dam and Lees Ferry, thickness of sand on the bed at the time the dam was completed, and changes in the bed profile between 1956 and 2000. (Adaped from Grams et al., 2002)

pulses that achieved their intention of lowering average bed elevation in the 25-km tailwater (Fig. 12). Bed erosion occurred in the downstream part of Glen Canyon in the high releases of the mid-1980s, but the channel bed is much the same as it was 15 yrs. ago (Grams et al., 2002).

Degradation in Glen Canyon has been about 6 m in pools and 2–3 m in riffles, and degradation has entirely removed the former sand bed and transformed the bed to gravel (Grams et al., 2002). The aquatic ecosystem has been changed, and today's blue–ribbon trout fishery is dependent on a gravel–bed habitat created by bed scour. Degradation has also led to abandonment of alluvial surfaces that were once the river's active floodplain.

STOP 1-2. LEES FERRY REACH (RIVER MILES 0-8)

Immediately after launching, the boats are swept past the large alluvial fan at the mouth of the Paria River and enter Paria Riffle, a minor constriction and the most upstream, fan–eddy complex encountered on the river trip (Fig. 5, 6).

Mapping of the 14 km between the boat ramp and Badger Creek Rapids, at RM 8.0, indicates that the width of the alluvial valley here is between 100 and 200 m and the average channel top width is 110 m (Fig. 13). Eddies occur in the lee of debris fans and are discrete zones of potential sedimentation. In map view, these eddies are typically separated by reaches where flow is uniformally downstream. In the reaches with downstream flow, the banks may be constrained by talus or bedrock, or narrow floodplains and terraces may occur. The frequency of large eddy occurrence is directly tied to the frequency with which tributaries enter the Colorado River and form debris fans. The frequency that fan–eddy complexes occur in the Lees Ferry reach is less than elsewhere in Grand Canyon, presumably because jointing and faulting in this part of the river is less than downstream. The frequency of persistent eddy bar areas larger than 1000 m^2 is 2.2/km here, which is less than elsewhere in Marble or Grand Canyons.

Badger Creek rapids is the first major rapid encountered on the trip. It is formed where large debris fans at opposite tributary canyon mouths narrow the channel, which is choked by boulders. The bed has not degraded here, as it has in Glen Canyon, presumably because the coarse substrate of the rapid cannot be entrained. Thus, the stage–discharge relation has not changed and alluvial surfaces are inundated by the same discharge as prior to the dam.

However, the fine sediment deposits of the channel margins and eddies have been diminished. A long record of sand storage in the eddy on river left below Badger Creek Rapids has been reconstructed from historic ground and aerial photographs. These photographs show that the elevation of the separation bar on river left is now about 2 m lower than it was prior to completion of Glen Canyon dam and that the reattachment bar located farther downstream has been nearly eliminated (Fig. 14). The surface of the separation bar fluctuated by about 0.5 m during the pre–dam period and the denundation that has occurred since completion of the dam has not been reversed (Fig. 15). The timing of erosional and depositional periods indicates that eolian erosion is a primary

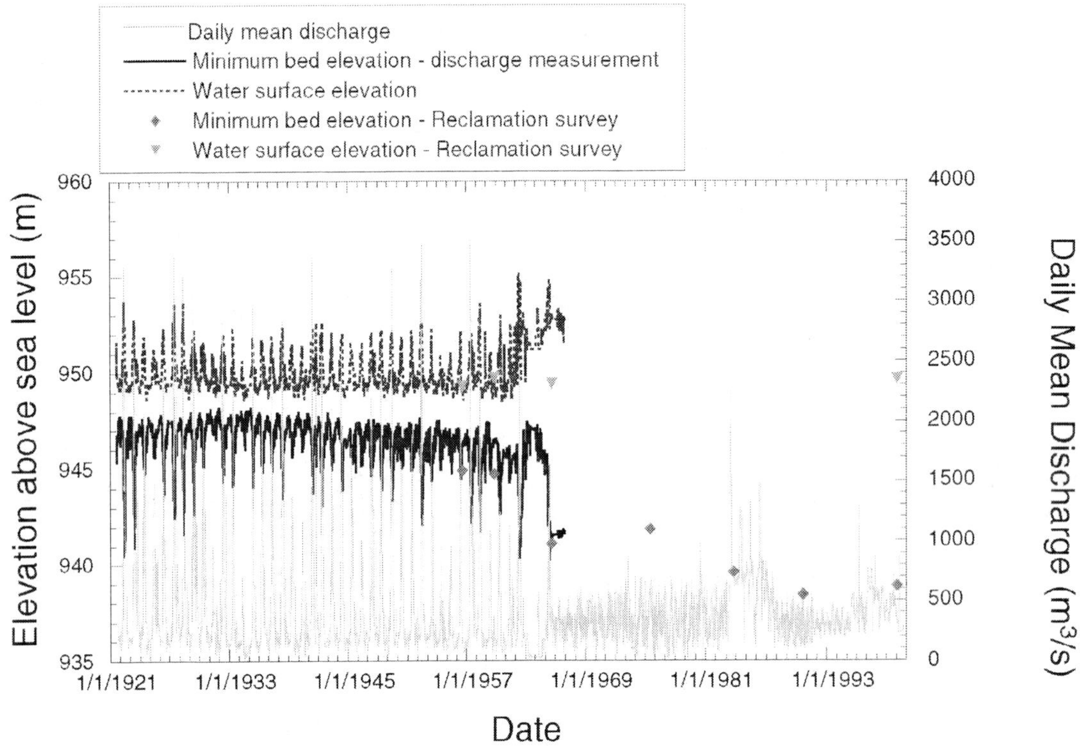

Figure 12. Elevation of the thalweg of the Colorado River at the upper cableway site at the Lees Ferry gage that was abandoned in 1965, water surface elevation at the time of each measurement, and subsequent resurveys of this site.

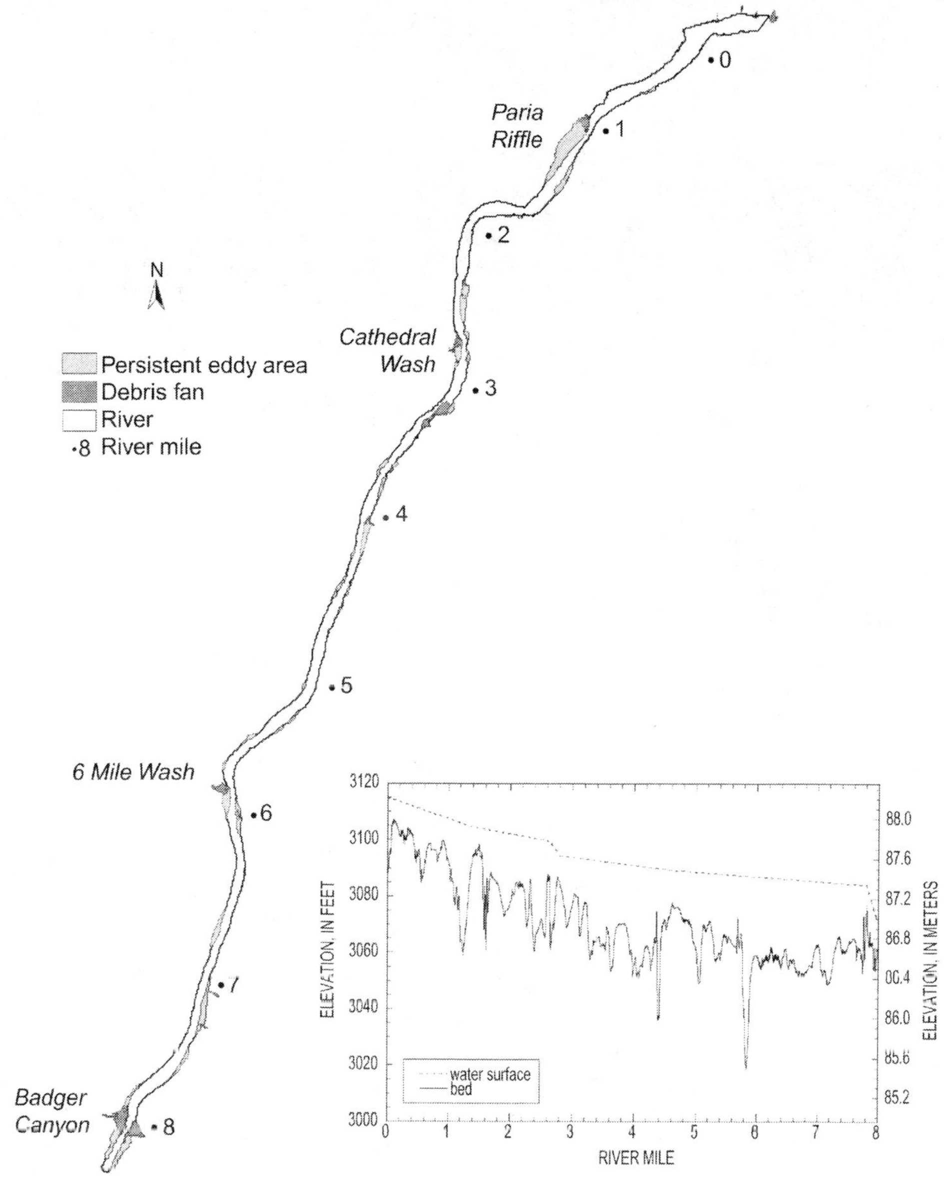

Figure 13. Lees Ferry reach and graph of the longitudinal profile, showing bed topography in 1984. River mile locations are indicated.

mechanism by which emergent sand bars are depleted. During the post-dam era, this material is not replaced during floods, because flood sediment concentrations are very low.

DAY 2. MARBLE CANYON—FAN–EDDY COMPLEXES, PLEISTOCENE STRATIGRAPHY AND THE POTENTIAL DAMMING OF THE COLORADO RIVER BY MASS–MOVEMENT

STOP 2-1. REDWALL GORGE REACH, STANTON'S CAVE AND REDWALL CAVERN (RIVER MILE 29 TO 35)

The Colorado River is an average of 74 m wide in the 10 km of this reach, and the canyon bottom that the river flows in is between 55 and 150 m wide. The river flows through the Mississippian Redwall Limestone and reach gradient is 0.0011, which is about the same as in the Lees Ferry reach. More geomorphically-significant tributaries enter the Colorado River here than in Stop 2, and the frequency of persistent eddy bar areas larger than 1000 m^2 is 3.3/km, which is about average for Marble and Grand Canyons. Springs issue from the canyon walls in places, and bedrock karst features are common. The channel is much narrower than in the Lees Ferry reach, and the area of persistent eddy bars, as defined above, is smaller.

Stanton's cave, the mouth of which lies ~45 m higher than the Colorado River at river mile 31.7R, is one of many caves associated with the Mississippian Redwall Limestone in Grand Canyon. Archeological excavation of the floor of the cave was undertaken in 1969 and 1970, and associated geomorphological and paleomagnetic research was conducted in 1976 and 1982

Figure 14. Downstream view of the separation and reattachment bars in the Badger Creek Rapids fan-eddy complex. A. Photograph taken in 1956; note boats for scale on river left at Jackass Creek beach. B. Photograph taken in 1989. Note deflation of the surface of the separation bar and large reduction in size of the reattachment bars.

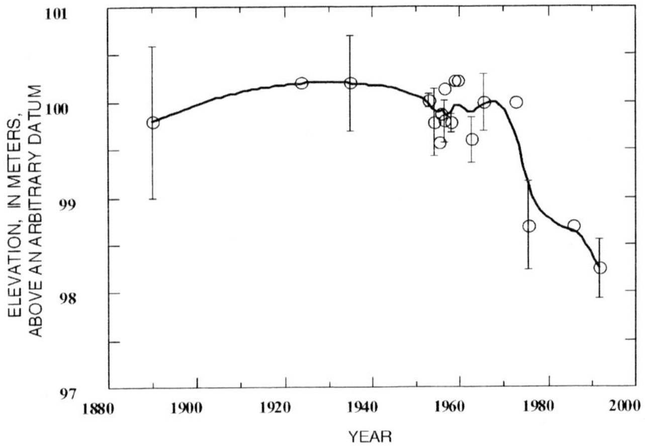

Figure 15. Decrease in elevation of the sand-bar surface at Jackass Creek camp (37 km downstream from Glen Canyon Dam), following construction of the dam. Elevations were determined by examination of oblique and aerial photographs of the site and field survey of the elevation and the former sand surface at its contact with large talus blocks. This graph shows the elevations near one prominent talus block that was inundated by the pre-dam mean annual flood but has been only infrequently inundated since the dam was completed (J.C. Schmidt and R.H. Webb, unpubl. data).

(Euler, 1984). In addition to discovering now famous split-twig figurines and other artifacts from the Archaic Period (7500-1900 cal. yrs. B.P.), the stratigraphy in the cave contained logs from cottonwood and Douglas fir trees. Based on weathering characteristics, the large size of logs in the cave, and tree-ring analyses, the wood debris was interpreted as driftwood that had entered the cave, mostly in a single event, by the damming or flooding of the Colorado River (Euler, 1984). Radiocarbon dating of this wood in the mid 1980's resulted in an age determination of 43,700 ± 1800 ^{14}C yrs. B.P., although this has been interpreted as an infinite age considering the upper age limit of the method at the time (Machette and Rosholt, 1991).

How the wood was transported into the cave by the Colorado River has been a conundrum, considering the entrance is well above any reasonable flood stage for the present-day channel (Cooley estimated 283,000 m^3/s (see Hereford, 1984)). Hereford (1984) suggested at least some of the wood was deposited at a time when the Colorado River was dammed by a large rock avalanche deposit 32 km downstream at the mouth of Nankoweap Canyon (see discussion at STOP 2–3). The original top of this rock avalanche is not preserved, and the reworked top (900 m) lies ~30 m below the elevation of the wood in Stanton's Cave. An alternative hypothesis, first noted by Machette and Rosholt (1991), is that the Colorado River has undergone large-amplitude, climate-driven cycles of aggradation and incision during the late Quaternary. Thus, wood could be deposited when the riverbed was higher than today. The channel bed throughout Marble Canyon was likely at a grade 35-40 m higher than present mean flow stage about 70-40 ka (Anders, 2003). Thus, changes in the form and grade of the Colorado River during the late Quaternary may account for the deposition of driftwood in Stanton's Cave without having to call upon large natural dams and extreme flood events.

STOP 2-2. POINT HANSBROUGH REACH AND EMINENCE BREAK (RIVER MILE 42 TO 48)

The 11 km of the Colorado River near Point Hansbrough have been studied in detail (Fig. 16). The distance across the canyon in this reach, as measured from rim to rim, is between 3100 and 4300 m, and the canyon is approximately 950 m deep. Along the

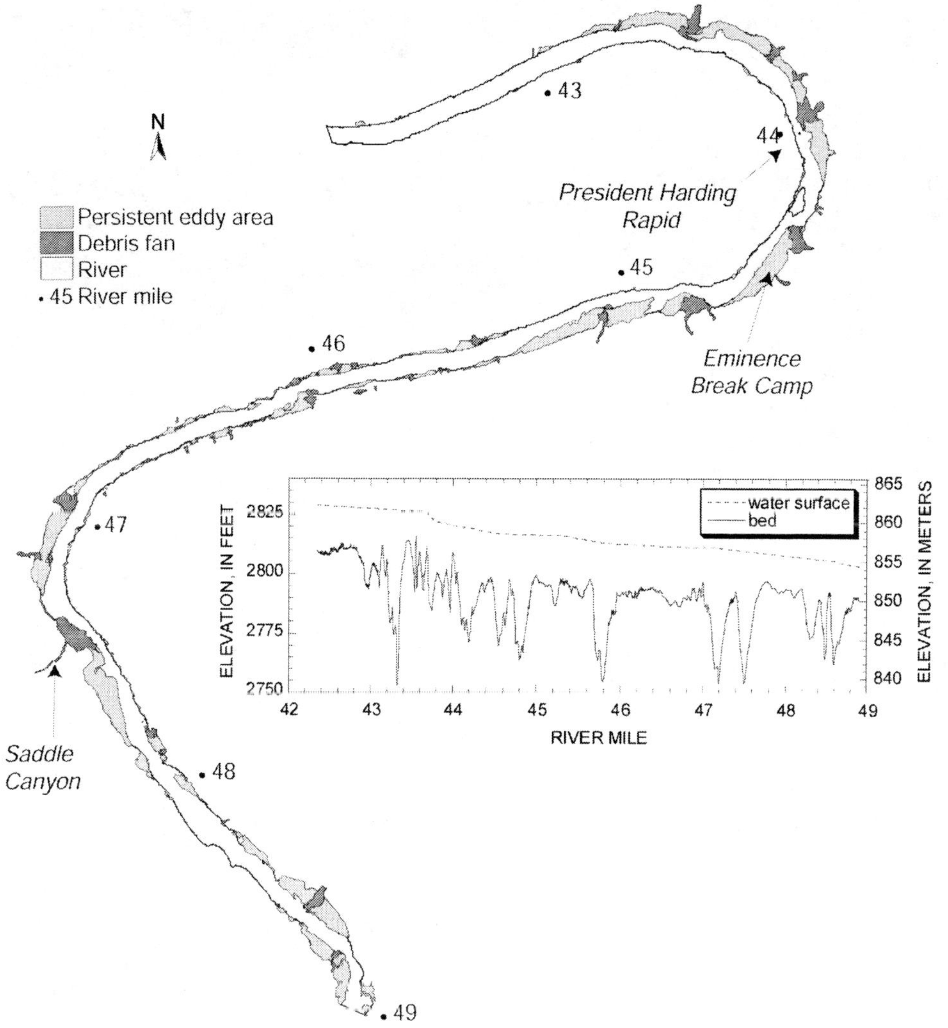

Figure 16. Point Hansbrough reach and graph of the longitudinal profile of the river and its bed.

river, the cross–sectional distance between bedrock outcrops is between 150 and 300 m, and the average channel width is about 100 m. About 3.5 persistent eddy bars larger than 1000 m^2 per river km occur in this reach, and most flood sediments are deposited in eddies. Comparison with the map of the Lees Ferry reach (Fig. 13) shows the greater size and frequency of eddies in this reach.

Bedrock at river level is the Cambrian Muav Limestone. Channel width–to–depth ratio in this reach, as measured at a discharge of 679 m^3/s, was between 16 and 27, and channel depths are between 4.1 and 6.3 m (U.S. Geological Survey - WRD, Tucson, written communication). Reach–average gradient is 0.0007, which is slightly flatter than the Lees Ferry reach.

The Eminence Break fan–eddy complex has been studied in detail since 1985. The site is not unique, but is typical of other fan–eddy complexes. The site is discussed here because we expect to camp at this site and a large amount of historical and process data are available. The dominance of eddy sedimentation as the primary mechanism of temporary storage of the suspended load and the recreational and ecological importance of these environments has focused research attention on eddies.

The Eminence Break debris fan constricts the Colorado River to a slightly greater extent than average for the Colorado River, but only a minor riffle occurs here. The ratio of the channel top width at the constriction to the average upstream channel width is 0.42 at 142 m^3/s, which is smaller than the average ratio of 0.49 for other constrictions formed by large fans in Marble and Grand Canyons (Schmidt and Graf, 1990). The constriction ratio widens at higher discharge, as do other constrictions in debris-fan-dominated canyons (Kieffer, 1985; Schmidt, 1990). Despite the narrow constriction, only a short length upstream is ponded flow, because another debris fan occurs a short distance upstream. The water surface drops between 0.12 and 0.15 m in the riffle, and a large gravel bar upstream from the riffle is emergent in the right half of the channel at base flow.

The eddy immediately downstream from Eminence Break debris fan is one of the largest in Marble or Grand Canyon (Schmidt, 1990), and this eddy persists at a wide range of flows. Aerial photographs demonstrate that this eddy has occurred at this site since at least since the mid-1930s. Riparian shrubs, especially saltcedar (*Tamarix* spp.) have expanded here, as they have throughout Marble and Grand Canyons. The largest documented area of exposed sand in the persistent eddy was in the mid–1930s when the entire sand bed of the eddy was exposed (Fig. 17).

In the post–dam era, the eddy has been partly filled by a reattachment bar. The primary features of a reattachment bar are that (1) they project upstream and offshore, (2) sedimentary structures in these bars reflect the circulation patterns of eddies and of the reattachment zone, and (3) evidence for migration of reattachment zones associated with shrinkage and swelling of eddies at different discharges is preserved in sediment structures.

Post–dam flows exceeding 890 m^3/s typically deposit fine sediment in nearshore areas to elevations that approximate the water surface. Deposition on the separation bar is typically more extensive than on the reattachment bar. The available sedimentological and historical data concerning the reattachment bar indicate that fill is preceded by little or no scour at the nearshore parts of the bar (Schmidt et al., 2002). However, offshore erosion may be extensive during the same flow event that causes near shore deposition. The subaerially-exposed parts of the separation and reattachment bars are extensively reworked by canyon winds during periods when flows do not exceed 890 m^3/s.

STOP 2-3. LARGE ROCK AVALANCHE ON NANKOWEAP DEBRIS FAN (RIVER MILE 52)

An unusually coarse and texturally immature diamicton underlies much of the prominent high ridge/terrace along the axis of the debris fan at the mouth of Nankoweap Creek (Fig 18). In particular, the high ridge on the debris fan south of Nankoweap Creek has a terrace-like surface that is 20-30 m above and imitates the slope of Nankoweap Creek as it crosses the debris fan from west to east. This ridge is underlain by two distinct deposits: (1) A texturally-mature, alluvial/debris-flow gravel, which stratigraphically underlies the younger, inset diamicton, and (2) an underlying deposit of medium-to- thickly-bedded, pebble-cobble gravel with a few boulders. Clasts are moderately to well rounded and composed of Proterozoic and Paleozoic (especially Redwall Limestone) sedimentary rocks, indicating that the source of this gravel is mostly the Nankoweap side drainage. The base of this older deposit is not exposed. Considering this, it's landscape position, and soil–profile development, it is tentatively correlated to side fill deposit 3 (S3) defined during mapping of the tributaries and mainstem canyon from River Mile 52–74 (Anders, 2003). The S3 deposit has been dated by optically stimulated luminescence (OSL) to 50-34 ka in tributary canyons, but can be as old as 65 ka at tributary mouths where it is influenced by the earlier aggradation of the mainstem river (Anders, 2003).

The inset diamicton appears to fill in erosional topography developed in the underlying gravel, and the entire thickness of the exposed ridge at its distal end comprises this deposit. It's texture and composition differ markedly from the underlying gravel: massive, poorly sorted, subangular to angular, pebble to boulder gravel, with a notable abundance of large angular boulders of chert-bearing carbonates of the Kaibab Fm. Hereford et al.

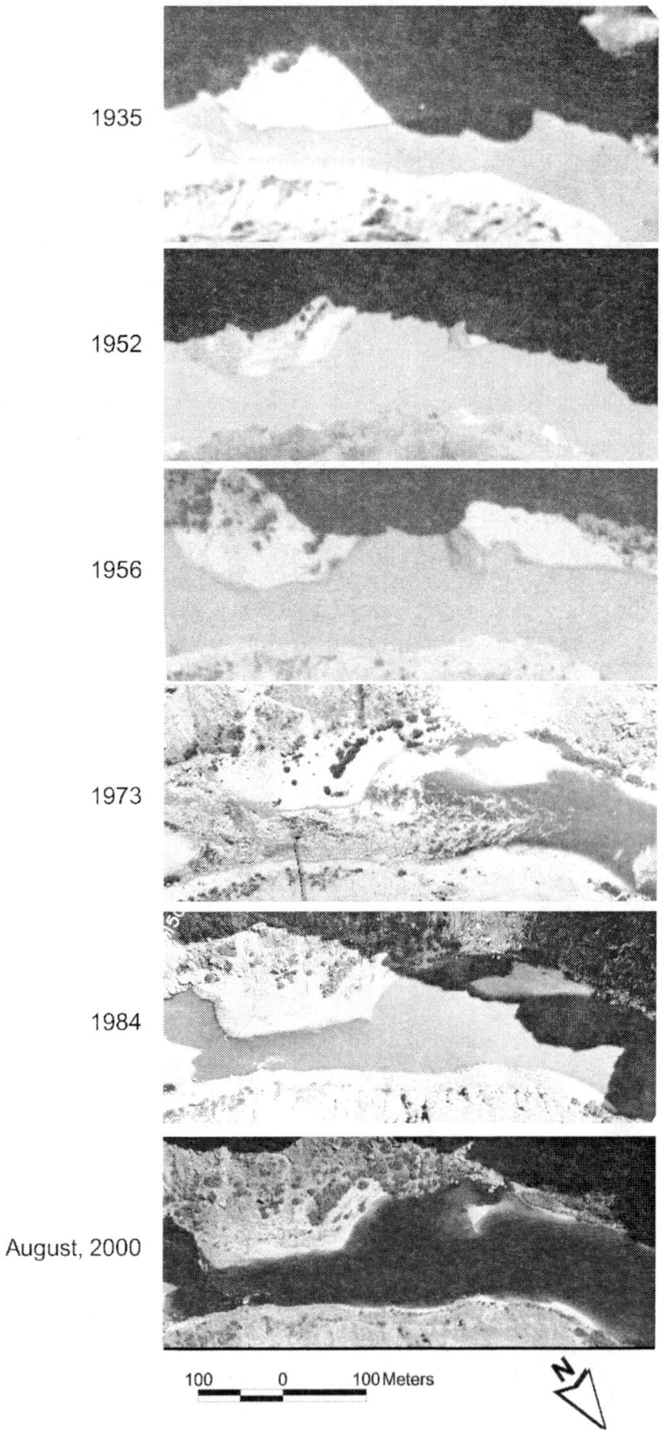

Figure 17. Eminence Break camp in different years.

(1998b) also note that more-rounded clasts of Nankoweap Creek provenance, like those of the underlying gravel, can be found in places near the top of the deposit, and that scattered mainstem Colorado River clasts and other evidence indicates that the surface of the ridge has been extensively reworked by the Colorado

Figure 18. View looking downstream at Pleistocene rock avalanche deposit on Nankoweap Fan (highlighted in sunlight in lower right) and escarpment across river from fan.

River. Machette and Rosholt (1991) interpreted the stratigraphy of the main ridge as having inset strath terraces at its eastern end, although this does not match Hereford et al's (1998b) extensive mapping of the debris fan. Machette and Rosholt (1991) used U-trend dating to obtain an age of 40 ± 24 ka for one of these terraces, but an age of 210 ± 25 ka for the rock avalanche deposit itself at the upper end of the ridge. The validity of these ages is unclear. Alternatively, the rock avalanche deposit's tentative overlying relation to the S3 deposit (<65 ka) and the presence of Holocene deposits inset into it on the Nankoweap debris fan suggest the diamicton is 65–10 ky old.

Hereford (1984) and Hereford et al. (1998b), based on their mapping of the Nankoweap debris fan, suggested the diamicton is the eroded remnant of a large rock avalanche or rock fall originating on the upper part of the escarpment on the left bank of the river. They argue that the deposit would have covered the entire canyon floor at this location, and Hereford (1984) proposed that the resultant damming of the river is responsible for the deposition of the wood in Stanton's Cave (see STOP 2–1 above). The presence of very large, angular, Kaibab Fm. blocks in the deposit is key to the rock–avalanche interpretation. The nearest outcrops of Kaibab Fm. are 1 km above the river on the east wall of Marble Canyon, whereas the nearest outcrops up the west–side drainages (Nankoweap and Little Nankoweap) are several kilometers away, and the typical combination of debris flow and streamflow processes transporting sediment in tributaries of the Colorado River form deposits that are somewhat better sorted with better rounded clasts. Other workers have suggested it could be a very large debris-flow deposit that issued from Nankoweap or Little Nankoweap canyons to the west, but this hypothesis is difficult to reconcile with the texture and composition of the diamicton (see Hereford et al., 1998a for a more complete discussion).

Certainly the Colorado River must have been affected, and to a degree, dammed by this rock avalanche. That the original top of the avalanche could have been everywhere ~30 m higher than the highest remnant, covering the entire canyon floor, cannot be demonstrated or ruled out. On the other hand, the mainstem Colorado River clearly has flowed at a higher grade episodically in the late Pleistocene, very likely along its entire profile through eastern Grand Canyon (see stops 2–4 and 3–2). The Colorado River was depositing gravel and sand ~40 m higher than present river stages at 70-40 ka (Anders, 2003), potentially just before the rock avalanche. Thus, the river was at a grade high enough to place driftwood in Stanton's Cave for an extended period of time during the last glacial epoch, providing an alternative hypothesis to damming by the Nankoweap rock avalanche.

STOP 2–4. KWAGUNT OUTCROPS, PLEISTOCENE MAINSTEM STRATIGRAPHY (RM 56 TO 61.5)

Spring travertine is particularly well preserved at the toe of the eastern canyon escarpment in the final reach of Marble Canyon from Kwagunt rapid (RM 56) to the confluence of Little Colorado River (RM 61.5). Travertine is present in several locations throughout Grand Canyon, and it commonly either interfingers with or cements stream gravel and colluvium. Previous work has established the ability to U-series date these carbonates and lead to the inference that they are preferentially formed during wetter full-glacial climates (Szabo, 1990). U-series dating has recently been applied by Warren Sharp, Berkeley Geochronology Center, to samples of flowstone that interfinger with stream gravel to provide age control for research on the Pleistocene stratigraphy. This has been particularly valuable in this lower reach of Marble Canyon where deposits in key exposures are cemented and middle

Pleisotocene in age. Samples were slabbed and polished to search for primary depositional textures, and duplicate or triplicate subsamples of 30-40 mm^3 were analyzed using thermal ionization mass spectrometry (TIMS). Results were discarded if samples were found to have excess 230-Th suggesting open-system behavior or if dates were not consistent among subsamples.

Two key samples of travertine from an outcrop across from the Kwagunt debris fan have been dated and provide ages that constrain the timing of a partial cycle of incision, then aggradation, of the Colorado River (Fig 19). A sample from travertine flowstone directly overlying the bedrock paleocanyon bottom and stratigraphically below a mainstem gravel deposit (M4) has an age of 151 ± 4.5 ka. This is interpreted to postdate an episode of bedrock incision, prior to an episode of hillslope and stream aggradation. A sample of travertine interfingering with mainstem river gravel within the M4 deposit above this bedrock strath has been dated at 118.3 ± 1.5 ka.

Two samples of travertine flowstone from the 57L outcrop (Fig 20), and another from an outcrop downstream provide ages for part of an older Colorado River aggradation–incision cycle. In these cases, travertine interfingers with gravelly alluvium indicating a paleoenvironment where spring deposits were precipitating at the canyon bottom proximal to the river. A 343 ± 28 ka date resulted from a sample 2.2 m above the bedrock strath of the M5 fill and a second date of 322 ± 9.7 ka is higher in this same deposit at an analogous outcrop at the confluence of the Little Colorado River. A third date of 280 ± 8.9 ka is from travertine interfingering with alluvium inset into the 343-322 ka M5 fill. This deposit lies 0.5 m above a beveled, gently-sloping, bedrock surface, and is interpreted to record a time of overall stream incision and lateral migration based on stratigraphic relations at the outcrop. The 342 ka and 280 ka samples provide minimum ages for when the paleo-river had incised to levels 33.6 and 22.0 m, respectively, above a reference river stage of 284 m^3/s (10,000 ft^3/s). Combining these data with a minimum estimate of the depth to bedrock under the modern river (the reach-averaged pool depth) of 15.7 ± 4.5 m yields bedrock incision rates of 144 ± 18 and 135 ± 17 m/m.y. (Pederson et al., 2002a). Likewise, the same calculation for the <151 ka age of the basal strath at the 56.3L outcrop yields a bedrock incision rate of 100-130 m/m.y., depending upon the stratigraphic assumptions made. Our incision rates in eastern Grand Canyon are generally lower and more precise than estimates by previous workers who reported that rates increase from 100 to 700 m/m.y. with progressively younger terraces (Machette and Rosholt, 1991). We consider our results more reliable because we avoid the bias of shorter time-interval measurements towards higher rates (Gardner, 1987), because the previous rates were based on the uranium-trend dating technique, and because we estimate true bedrock incision rates by comparing past bedrock surfaces to minimum estimates of present bedrock levels rather than terrace treads to the present river shoreline. We will discuss the timing of these sedimentary cycles relative to glacial-interglacial climate changes and other controlling factors at STOP 3–3.

DAY 3. FURNACE FLATS—PLEISTOCENE STRATIGRAPHY AND LANDSCAPE RESPONSES TO GLACIAL–INTERGLACIAL CLIMATE CHANGE

STOP 3–1. CARBON-LAVA CHUAR HIKE

A hike up Carbon Canyon, across the Butte fault, and into the southern part of the Chuar Valley provides an overview of the backcountry of eastern Grand Canyon (Fig 21). Moderate-sized tributary catchments like Carbon and Lava Chuar canyons have been a focus of recent study because they hold a spectacular and

Figure 19. View looking across Colorado River from Kwagunt camp. U-series dating of travertine provides data on the timing of the mainstem river's aggradation/incision cycles.

Figure 20. View looking across Colorado River at an outcrop just downstream from Kwagunt camp. U-series dating of travertine provides data on the timing of the mainstem river's aggradation/ incision cycles.

well-exposed Quaternary stratigraphy that records catchment processes and responses to climatic forcing. The Chuar Valley and Furnace Flats tributaries are all third or fourth-order, coarse-bedload-dominated, mostly- alluvial streams that have relatively low base discharges or are ephemeral. Five or more distinct alluvial-fill deposits have been recognized in these tributary canyons, and they are associated with at least eleven distinct terrace surfaces (Figs. 22,23,24). These can be correlated confidently between the study drainages by landscape position and surface and soil-profile characteristics. Hillslope colluvial remnants of the last major sedimentary hillslope and valley-fill episode can be traced directly to related tributary fill terraces (Fig. 21), and this alluvium can be traced downstream to the Colorado River corridor. At the confluences of side systems and the mainstem river, the very base of tributary fills interfinger with Colorado River deposits, but the majority of their thickness progrades out over and is younger than, related Colorado River terraces.

Tributary terrace deposits consist of thick lenticular beds of immature pebble-to-boulder gravel with a sandy matrix. Two distinct, interbedded facies are present in tributary stream deposits in varying proportions: (1) A predominance of clast-supported gravel with local clast imbrication, and (2) generally subordinate, matrix-supported gravel with randomly oriented clasts in a sandy matrix. The matrix-supported facies is interpreted to be debris flow deposits, whereas the clast- supported facies is interpreted to be alluvium derived in large part from the reworking of debris-flow deposits. In general, the lowest suite of tributary deposits and terraces (S_1) are heavily vegetated, have clear bar-and-swale topography, and have weakly developed desert soils. Older S1 terraces are physically correlative to Holocene deposits mapped along the Colorado River corridor that are middle to late Holocene in age (Lucchitta et al., 1995; Hereford et al., 1996), whereas some S1 deposits are the result of ~century old debris flows and flood events based on radiocarbon dating (Pederson, unpublished data). Pleistocene terraces are vegetated by desert scrub, grass, and cacti, have desert pavement in places, and are associated with greater soil profile development (calcic horizon stage and Bt horizon development) with increasing terrace height (Anders, 2003). Valley long-profiles illustrate that tributary terrace treads often converge downstream in lower reaches as they approach the Colorado River (Anders, 2003). Fills thicken locally at confluences of lower order tributary drainages, and the gradient of the master stream increases in the reach below such confluences, presumably due to local sediment loading and coarsening of bedload.

Figure 21. Overlooking part of the Carbon Canyon tributary drainage. A massive remnant colluvial apron that once covered entire toeslope of escarpment grades laterally to become the dominant alluvial valley fill. A secondary, yet significant, inset deposit along the axis of tributaries represents the initial pulse of reworking this major fill.

Figure 22. View northwest into the upper part of the Lava Chuar tributary drainage with the major terraces labeled.

At STOP 2-4, we reviewed some of the U-series dating of Pleistocene deposits in eastern Grand Canyon, but most age control, especially on younger fill terraces (1-3) is from optically stimulated luminescence (OSL) dating conducted by Tammy Rittenhour and Ron Goble at the University of Nebraska-Lincoln (Anders, 2003). OSL age summaries are given on figures 24 and 25. An advantage of both U-series and OSL dating is that they provides the timing of deposition within the stratigraphy rather than a difficult-to-decipher age of the terrace tread, as in the case of cosmogenic dating. The most recent major fill episode affecting the hillslopes and tributary streams of Grand Canyon dates from 50-34 ka, correlating to oxygen–isotope stage (OIS) 3 (Fig 21). A subordinate, yet significant, inset alluvial deposit in tributaries dates to 12-7 ka (Fig 24), and this fill terrace is important in that field relations indicate it is derived from incision and reworking of sediment stored in the thick S3 hills-

Figure 23. Example of fill terrace relations in Lava Chuar side canyon. Photograph looking downstream to southeast.

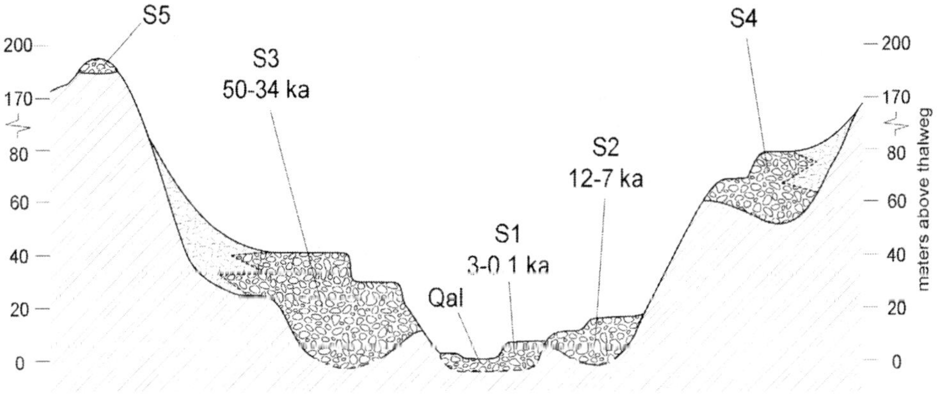

Figure 24. Tributary drainage stratigraphy with age ranges from OSL dating. Most recent major hillslope and stream aggradation was during the extended glacial epoch of OIS 4-2, and cannibalization of this deposit in the early Holocene resulted in a subsidiary stream gravel (S2) preserved in tributaries.

lope remnants. We will discuss the timing and meaning of these impressive deposits at STOP 3–3.

STOP 3–2. TANNER BAR

Recent mapping along the mainstem Colorado River has focused on Pleistocene deposits, where geomorphic and stratigraphic relations between stream terraces and side canyon and hillslope landscape components have been investigated. Five Pleistocene and one Holocene fill deposit of the Colorado River have been identified along the mainstem corridor, and each can have multiple terrace levels developed on it (Fig. 25). As in the tributary drainages, these are thick fill terraces (up to 64 m) with irregular basal straths representing buried valley topography.

The sandy, M1, middle late Holocene deposits have been studied by Hereford and others (e.g. 1996), as described in the introduction above. Distinct from these are the unconsolidated - to-strongly-cemented, clast-supported, pebble-cobble, gravel-to-gravelly-sand deposits M7-M3. Based on samples from M3, clast compositions are ~85% proximally derived limestone, sandstone, dolomite, and chert, and ~15% quartzite, volcanic porphyry, and granite clasts from far upstream in the Colorado River system, whereas the modern gravel bars in Furnace Flats are slightly more dominated by local rock types (Anders, 2003). Two facies in these deposits are: (1) clast- supported, imbricated, cross-bedded, pebble-cobble gravel; and (2) cross-bedded silty sand, and bedding is thin to medium- scale and has tabular to lenticular geometry (Fig. 26). M3 is the best preserved mainstem terrace in eastern Grand Canyon and was deposited during an episode that began prior to 71 ka and proceeded until after 64 ka based on OSL dating (Fig. 26). The M3 strath is below present-day river level and treads are up to 38 m high. M4 was deposited after 151 ka and aggradation proceeded until some time after 118 ka. Its strath is at river level, and

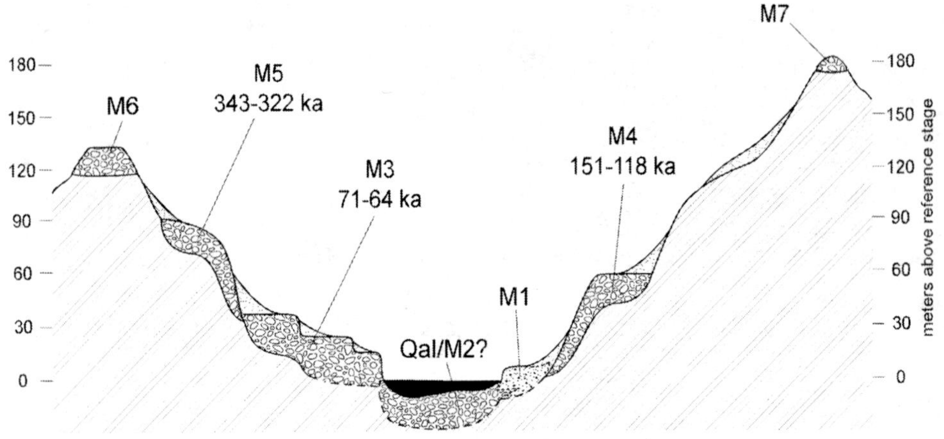

Figure 25. Mainstem Colorado River stratigraphy with numerical age ranges from OSL dating. Local colluvium and alluvium generally progrades over mainstem alluvium. Aggradation of mainstem river M3 was essentially complete up to 15 ky prior to the main pulse of tributary stream aggradation began (M3).

although the tread on this deposit is often obscured by colluvium or poorly preserved, we estimate it to be 60 m high. M5 was deposited during an episode that began prior to 343 ka and proceeded until some time after 322 ka. The strath is 28 m above river level and we estimate the poorly preserved tread to be 94 m above present river stage. M6 and M7 deposits are the highest and oldest preserved, but have not been dated. M6 is ~60 m thick and its strath is 117 m above river level. We will visit an example at STOP 3–3. M7 is ~ 50 m thick with a strath 133 m above the river. The true straths of these older fill terraces may have been lower and the fills were likely thicker than we observe in the present day.

Notably, no exposed deposits have been identified that date to the last glacial maximum (OIS2, 25-15 ka). However, the ubiquity of terraces elsewhere in the Colorado River drainage basin associated with glacially increased sediment load prompts us to suggest that a similar age deposit probably existed in eastern Grand Canyon (e.g. Phillips, et al., 1997; Repka, et al., 1997; Reheis et al., 1991). A possible location for such a deposit in the study area is under the present-day channel. Data on the composition of the Colorado River's channel bottom and dam-site exploration drilling records indicate significant thicknesses of sub-channel sediment, at least in some reaches of the river through Grand Canyon.

An interesting sedimentological disconnect has existed between the Pleistocene mainstem system and local catchments in eastern Grand Canyon. Local catchment alluvium and colluvium of major Pleistocene aggradation valley fill episodes commonly interfinger at their base with deposits of the Colorado River along the mainstem. But the bulk of tributary deposits overlie and prograde over mainstem fill terraces (Figs. 25, 27). For example, age control and stratigraphy indicate that M3 deposition preceeded S3 tributary aggradation by ~20 ky, and may have begun incising during the later stages of tributary aggradation (Anders, 2003). In addition, the Colorado River does not appear to have a deposit correlating to the transitional late Pleistocene-early Holocene S2 deposit of local catchments.

STOP 3–3. CARDENAS–HILLTOP RUIN

Hiking up Cardenas Canyon to Hilltop Ruin atop a M6 deposit gives a viewpoint with a commanding vista of the Furnace Flats region and the Quaternary stratigraphy. What are the driving mechanisms for the process changes that form this record? In terms of linkages between landscape components, why does a temporal disconnect exist between sedimentation in the local catchments versus the Colorado River? First, tectonism and drainage integration are the long-term mechanisms controlling incision of Grand Canyon but climate variability drives the short-term aggradation and incision episodes recorded in the thick Quaternary deposits of eastern Grand Canyon. The effects of glacial-interglacial climate changes rather than base-level or tectonic factors are responsible for the hydrology and sediment yield changes evident in the hillslope and stream records of local catchments. This is based upon evidence such as: (1) cyclic incision and aggradation of thick fill deposits with a period that matches glacial-interglacial timescales, (2) contemporaneous fill terraces in several local drainages, (3) thick colluvial mantles that grade into valley-fill alluvium, suggesting climate-controlled changes in hillslope weathering, hydrology, and sediment transport, (4) terrace long-profile characteristics, (5) the lack of Quaternary faulting affecting eastern Grand Canyon side drainages or evidence for knickpoints, and (6) high stream gradients and low discharges that reduce the streams ability to convey base-level effects (e.g. Leopold and Bull, 1979; Bull and Kneupfer, 1987; Merritts et al., 1994).

All evidence argues against the idea that proposed lava-dams in western Grand Canyon caused aggradation of the Colorado River and it's tributaries in eastern Grand Canyon (Hamblin, 1994). Topsets, foresets, and bottomsets indicative of prograding, coarse-grained, 'Gilbert-style' deltas are absent and deposits are significantly younger than most of the lava flows of western Grand Canyon (Dalrymple and Hamblin, 1998; Fenton, et al., 2001; Pederson, et al., 2002a). Elston (1989) proposed that the deposits are the result of regional Miocene-Pliocene

Figure 26. Mainstem Colorado River fill terrace 3 (M3) with OSL dates obtained from this locality. Basalt strath of the fill is below present river surface, note planar–tabular crossbedding interpreted as gravel bar foresets.

Figure 27. Overview of Colorado River corridor in the Furnace Flats reach of eastern Grand Canyon from Hilltop Ruin. Note that colluvium and tributary alluvium progrades over terrace gravel of the Colorado River.

aggradation. This is impossible because of the complicated, inset stratigraphy with deposits being older rather than younger with increasing height.

Overall incision of Grand Canyon. At a broader scale, incision along the Colorado River since the proposed initiation of incision at 6 Ma, using our minimum calculation of ~140 m/m.y., could carve less than half of the ~1800 m deep gorge of eastern Grand Canyon. This suggests that (1) rates have slowed through time, consistent with an attenuating wave of incision initiated by a relatively sudden, large, base-level fall, and/or (2) some canyon cutting, at least in eastern Grand Canyon, happened prior to the Colorado River taking its full present path at 6 Ma and creating the Grand Canyon we see today (Young and McKee, 1978; Elston and Young, 1991; Pederson et al., 2002a).

Landscape Response to Climate Change in Eastern Grand Canyon. We can summarize the Grand Canyon landscape responses to climate change based especially on the most recent, best preserved, and best dated set of Pleistocene to early Holocene deposits (Fig. 28). Mainstem Colorado River aggradation appears to be linked to glacial advances in its distant headwaters, which could have increased sediment yield by an order of magnitude (Hallet et al., 1996). Correlative deposits to M3 are found in the Wind River Mountains (WR2 terrace, minimum tread age = 55 ka) (Sharp et al., in press), and the Fremont River (FR2 terrace, cosmogenic date of tread = 60 ka) (Repka et al., 1997). According to the Devil's Hole paleoclimate record (Winograd et al., 1992), M3 aggradation in eastern Grand Canyon began roughly at the peak of a glacial advance and continued into the transition to warmer climate conditions. The same timing relative to glacial-interglacial climate changes holds true for both the M4 and M5 fill terraces, indicating a consistent response of the mainstem to full-glacial and deglacial sediment loading of the stream.

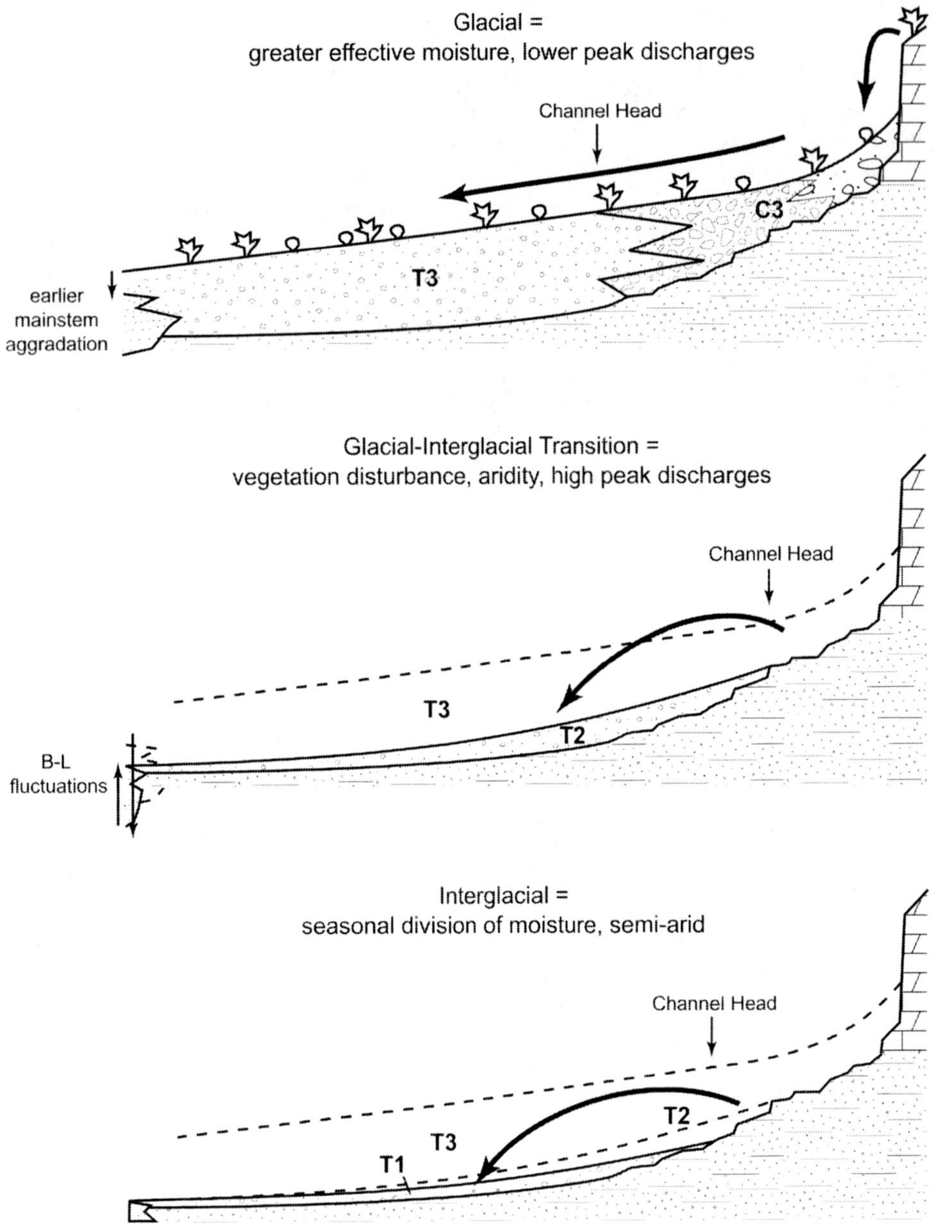

Figure 28. Conceptual model of Grand Canyon landscape–sedimentary responses to glacial-interglacial climate changes.

In contrast, tributary sedimentologic responses to glacial-interglacial climate changes appear to be delayed by 10^3-10^4 yrs, with tributary deposit S3 occurring late in the extended cold period of OIS 4 and OIS 3. Stratigraphic relations clearly show that tributary stream aggradation was caused by greatly increased sediment production and storage on local hillslopes that eventually translated downslope to channels. Relatively high effective moisture glacial conditions, through a series of positive feedbacks, created greatly enhanced bedrock-weathering rates, mantling lower slopes with colluvium, as seen with other dryland hillslopes (e.g. Bull, 1991; Harvey and Wells, 1994; Pederson et al. 2000). Combining this with decreased peak discharges due to lower precipitation intensity eventually caused decreased drainage density, lower stream concavity, and valley filling. The delayed tributary aggradation can be attributed to the lag time associated with generating and overwhelming hillslopes with colluvium.

Deposition of the tributary S2 deposit during the most recent glacial-interglacial transition the record in the Mojave and Sonoran deserts, and is consistent with the conceptual model of dryland piedmont aggradation during vegetation disturbance and intense summer monsoon precipitation (Bull, 1991). Increased overland flow and decreased regolith cohesion would cause increased channel erosivity, with channel heads extending up hillslopes and mobilizing sediment stored in colluvial mantles (Ritter and Gardner, 1993; Tucker and Slingerland, 1997). This cannibalization of the valley fill may have been aided by base-level fall due to Colorado River incision (Fig. 28), and S2-may be a time-transgressive deposit, representing the temporary storage of sediment during the first pulse of Holocene incision. Again, evidence indicates this sequence of responses to climate change during the glacial-interglacial transition is not evident in the record of the exotic Colorado River and may be unique to drylands. Sediment supply is key in the weathering-limited landscape of eastern Grand Canyon,, and hydrologic changes alone cannot account for the response of the landscape to climate change.

DAY 4. BOATING DAY—THE INNER GORGE AND RAPIDS

STOP 4–1. INNER GRANITE GORGE AND CRYSTAL RAPIDS (RM 99)

The segment of the Colorado River between Hance Rapid (RM 77) and Crystal Rapid (RM 99) includes some of the largest and most difficult river navigation of the entire river trip. Hance and Crystal are generally recognized as the most difficult boating opportunities in Grand Canyon except for Lava Falls. Other difficult navigation is confronted at Sockdolager Rapid (RM 79), Grapevine Rapid (RM 82), Horn Creek Rapid (RM 90.5), Granite Rapid (RM 93), and Hermit Rapid (RM 95).

Each of these rapids occurs opposite a debris fan at a tributary mouth, and this circumstance is no different than elsewhere in Grand Canyon. The succession of unusually difficult rapids is associated with the steep relief of the tributary basins, because the elevation between the river and the surrounding plateaus is large, the distance between the high country and the river is small, and the Inner Gorge of the Colorado River is very narrow.

Most river runners stop and scout Crystal Rapids, so we will have an opportunity to survey the scene. Until December 1966, this rapid was a relatively minor impediment to boaters. A series of storm cells brought large amounts of warm rain in short, high–intensity bursts to southern Utah and northern Arizona. Thirty sites of hillslope failure of weathered bedrock and colluvium were subsequently identified and these provided a large supply of fine sediment, gravel, and boulders to the streamflow, thereby creating a debris flow. Webb estimated that the discharge in Crystal Creek was about 280 m^3/s, and the concentration of sediment in the flow was about 75 % (Webb, 1996).

The debris flow caused the Colorado River channel to narrow from about 121 m in width to about 30 m. Farther downstream, an island of cobbles and small boulders formed, composed of debris that had originated in Crystal Creek. Thereafter, the rapid was considered very dangerous, requiring expert boating skills. The rapid remained that way until June 1983 when dam releases were the largest they had been in the post–dam era. At this very large discharge, a 17–m high, standing wave of a hydraulic jump established itself immediately at the downstream exit point from the narrowest part of the constriction (Kieffer, 1985). Motorized rafts as long as 37 ft flipped in this wave and lives were lost. Substantial widening of the channel occurred during this flood and boulders were transported from the debris fan edge (Webb, 1996).

DAY 5. BOATING DAY—THE ISLES AND MUAV GORGE

STOP 5–1. LANDSLIDES NEAR DEER CREEK FALLS

The exhumation and drainage history of the canyon is heavily influenced by large-scale landsliding from Tapeats Creek (RM 134) to Fishtail Canyon (RM 139), including the Surprise Canyon area between the river and the north rim. The most prominent example of this is on the right side of the river at RM 135, where a large, landslide-filled paleochannel of the Colorado River with it base ~70 m up the slope is displayed (Fig. 29; Huntoon, 2003). The upper Cambrian through the Pennsylvanian sedimentary section (Muav through Esplanade Formations) typically slides and rotates as massive blocks on a concave slip surface of the Cambrian Bright Angel Shale. Widening of Grand Canyon is thought to be largely due to landsliding in areas where streams incise the toe of relatively weak shale slopes. The load of the overlying, more-competent rock eventually overcomes cohesion and friction in the shale, leading to landslide failures. As a result, a new slope profile, with a lower-gradient, "buttressing" shale slope is formed (Huntoon, 2003). Another tempting idea is that this unusual concentration of landslides must also be influenced by the notable concentration of very large karst springs in this area (Thunder River, upper Deer Creek) and the higher pore pressure this could provide to facilitate bedrock failure.

Savage and others (2002) have identified at least six episodes of landsliding in this area, with each causing the diversion of tributary or mainstem drainages. They derive relative ages based on the height of buried channels above modern grade, and each of these episodes has caused the diversion of tributary or mainstem drainages. The spectacular incision of lower Deer Creek through Tapeats Sandstone is the product of landsliding in that it has occupied valleys both east and west of its present mouth but has been diverted by relatively recent landsliding to follow its present path through bedrock (Huntoon, 2003). Considering the heights of buried paleochannels in this area, and using a bedrock incision rate of ~140 m/my calculated both upstream and downstream from this point, preserved landslides may range in age from at least 2 Ma (near Tapeats Creek, ~290 m above grade) to possibly as young as Holocene in the case of the large landslide along the right bank of the Colorado River just downstream from Deer Creek.

DAY 6. FIRE AND ICE—FAULTING, VOLCANISM, AND LAVA FALLS

STOP 6–1. DIFFERENTIAL STREAM INCISION AND FAULT SLIP ON THE TOROWEAP AND HURRICANE FAULTS

Nearing the end of the fieldtrip, we enter what may be the most spectacular place in Grand Canyon geomorphically. Canyon cutting and highly active sediment transport processes here directly interact with Quaternary volcanism and active normal faulting, creating a complicated landscape. The Hurricane/Toroweap fault system is a 15-km-wide zone of active, normal faults that cut across the Grand Canyon, and the Uinkaret volcanic field lies between the Hurricane fault to the west and the Toroweap fault to the east (Fig 2). The Toroweap fault passes through Prospect Canyon and crosses the Colorado River at Lava Falls (Fig. 30). The basalts that have erupted in the canyon or flowed down into it from the rims have been studied in terms of hypothesized lava dams and large upstream lakes that supposedly drowned Grand Canyon several times in the Quaternary (Hamblin, 1994).

The offset of basalt flows also provides Quaternary slip rates along these faults. Neotectonic studies by Jackson (1990), Wenrich et al. (1997) and Pearthree (in Stenner et al., 1999) along with minimum ages provided by cosmogenic ^{3}He exposure dates of volcanic and sedimentary deposit surfaces give overall down-to-the-west Quaternary slip rates of 70-180 m/my for the Toroweap fault and 70-170 m/my for the Hurricane fault (Fenton et al., 2001). An Ar-Ar date of 509 ± 27 ka on a flow offset 48 m by the Toroweap fault gives a more precise slip rate of 94 ± 6 m/my at Grand Canyon (Pederson et al., 2002a). The precise rate of slip on the Hurricane fault where it crosses the river is not yet clear.

What the effect of these active normal faults is on the incision of Grand Canyon is not a simple question to answer. Lucchitta et al. (2000) recognized that incision rates at Granite Park, in the proximal hangingwall of the Hurricane fault (RM 208) are significantly less than estimates in eastern Grand Canyon. Pederson et al. (2002a), using Ar-Ar ages of basalt flows, compared the incision rates in Granite Park of <72-92 m/my to the rates from the outcrop at this stop (RM 177.5), just 1-2 river miles upstream from the Toroweap fault, as well as the incision rates discussed at STOP 2–4 in eastern Grand Canyon. Eastern Grand Canyon and the site just above the Toroweap fault have essentially the same bedrock-incision rate of ~140 m/my over the past ~350 ky. These estimates for bedrock incision rely upon comparing the bedrock strath of Pleistocene fill terraces preserved beneath basalt flows (Fig. 31) to the mean pool depth of the nearby reach of the Colorado River. The deeper pools along the bed of the river are used as a minimum estimate of present depth to bedrock under the river because the present river bottom flows on alluvium in most places (Roberto Anima, USGS, personal communication, 2001). In reaches where a significant amount of alluvium occurs everywhere between the channel bottom and the bedrock interface below, bedrock incision rates may be <50 m/my greater.

The slip rate of the Toroweap fault (94 m/m.y.) is approximately equal to the difference between upstream (~140 m/m.y.) and downstream (<72-92 m/m.y.) incision rates, considering that the range of values for Granite Park is a maximum estimate and upstream rates are minimum estimates. Subsidence in the hanging wall of the Toroweap fault may account for this differential incision by effectively counteracting river incision and reducing incision rates west of each fault. Conversely, geometry dictates that slip on these faults should not drive upstream incision or control effective base level for the Colorado River. Knickpoint regression is not a viable process in this setting because upstream incision rates are greater than individual fault-slip rates, and thus time–averaged vertical incision of the stream along its profile is greater than rates of scarp/knickpoint generation. Also evident is incision (albeit slower) downstream from the fault zone, rather than aggradation, which would be expected if faulting was faster than regional incision (cf. Hamblin et al., 1981). Instead of relatively slow faulting, surface drainage integration off the Grand Wash escarpment at the terminus of Grand Canyon ~6 Ma would have lowered effective base level by perhaps 1000 m, ultimately driving a large fraction of stream incision on the Colorado Plateau. Given the context of (1) regional incision driven by other factors such as drainage development, (2) faulting that is slower than regional incision; and (3) tectonics consistent with mostly subsidence of the hanging–wall near the faults, Pederson et al. (2002a) conclude that local faulting does not drive upstream incision, but rather, locally reduces incision in Grand Canyon.

Figure 29. Large Colorado River paleochannel (marked by dashed line) overlain by landslide (LS), RM 135. (Photograph by Peter Huntoon)

STOP 6–2. PROSPECT CREEK FAN AND LAVA FALLS

Lava Falls is the largest rapid in Grand Canyon, and is the standard against which the difficulty in navigating all other rapids is measured. Virtually every private and commercial boating party stops here to scout the rapid, and the site in a good one for geologic observation. At the overlook on the right side of the river, one can look across the river to the Prospect Creek debris fan, which is the second largest debris fan in Grand Canyon (Fig. 32). Webb (1996) described the sequence of debris flows that have occurred here since 1939 and the subsequent changes to the rapid that occurred each time a very large Colorado River flood encountered a newly–deposited debris-flow deposit.

The most recent, large, debris flow to occur here happened during the night of March 5-6, 1995, when Webb and colleagues

Figure 30. Mouth of Prospect Canyon from Tuweep on the north rim of the canyon, looking downstream to west-southwest. Note the Prospect debris fan, Lava Falls rapid, and the cinder cones and basalt flows of the Uinkaret volcanic field cascading down the canyon wall.

Figure 31. RM 177.5 basalt flow with pillows at its base indicating subaqueous component to flow. Height of bedrock strath and Ar-Ar age of basalt used to calculate minimum bedrock incision rate immediately upstream of the Toroweap Fault. (Photograph by Karl Karlstrom, Laura Crossey at bottom–center for scale)

were camped immediately upstream. Webb (1996) reported that the river was initially constricted to about 50% of the average upstream width and the rapid itself was unrunnable. The party remained at the site for 3 days, during which time bank erosion of the debris flow deposit widened the rapid to only 33% of the average upstream width, making navigation easier. Webb et al. (1999) and Pizzuto et al (1999) measured further reworking of the 1995 debris flow deposit that occurred during the 1996 controlled flood. The magnitude of this flood was about half the magnitude of the pre-dam, mean-annual flood, and other workers have speculated that only very large floods can significantly rework debris fan deposits. Pizzuto et al. (1999) found that the rapid widened an additional 10 % during this 7-day flood, and found that widening occurred mostly during the first days of the flood. Boulders as large as 2 m were transported downstream and entrainment of large boulders primarily occurred by slab failures of bank materials during the initial rise of the flood. Reworking ceased about 4 hrs after the flood had begun. Radio transmitters embedded in some boulders showed that other particles were entrained from the bed near the debris fan and that these particles were transported

Figure 32. Prospect debris fan and Lava Falls from river right.

into the pool immediately downstream from the rapid or onto the gravel bar located about 400 m farther downstream. The average travel distance of these boulders was about 230 m.

Cerling et al. (1999) used ^3He cosmogenic exposure dating, radiocarbon, and repeat photography to investigate the long-term history of debris-flow activity on the Prospect fan. They distinguish between large–scale "fan–forming" events that occur with a recurrence of hundreds to thousands of years and smaller flows, like that of 1995, that have recurrence intervals on the scale of years or decades. The smaller flows become inset within the channel of the larger deposits. Suprisingly, Cerling et al. (1999) found the entire fan was constructed in the late Holocene. The oldest surface/deposit is ~3 ka and large enough to have originally reached the northern bank of the river and raise the pool elevation ~30 m upstream for a short time. Lava Falls is a testament to the high activity of mass-movement processes in this steep, faulted side canyon formed in fractured Pleistocene basalt flows and tephra.

DAY 7. TAKEOUT, LAVA DAMS

STOP 7–1. LAVA FLOW AGES AND POSSIBLE LAVA DAMS OF WESTERN GRAND CANYON

Most of the Uinkaret volcanic field consists of Quaternary basalt, and it is still active. Some of these basalts have erupted or flowed into the canyon, extending several kms downstream and partially filling it in (Figs. 30, 33). John Wesley Powell (1875) was the first to propose the existence of high lava dams and extensive lakes in the geologic past, and Hamblin (1994) created a detailed story of flows that built 13 major dams and related lakes, the largest of which drowned the entire length of Grand Canyon. Reconstruction of this lava-dam history requires a detailed understanding of the age, duration, correlation, and stratigraphy of basalt flows within the canyon. Researchers are still working toward this goal. The stratigraphy of flows is extremely complicated, and recent Ar-Ar dating of flows indicates Hamblin's famous work will probably be revised (McIntosh et al., 2002).

Hamblin (1994) also concluded that incision of Grand Canyon was essentially complete by 1 Ma based on K-Ar ages of some basalts that nearly reach the present-day Colorado River. Dalrymple and Hamblin (1998) reported a more comprehensive K-Ar analyses of their samples with a handful of ~1 Ma dates, but these were interpreted as unreliable because results from the same flow were highly inconsistent, perhaps due to excess Ar. New Ar-Ar ages on these flows tend to be several hundred thousand years younger and significantly more precise (McIntosh et al., 2002). All newer $^{40}Ar/^{39}Ar$ ages thus far are younger than 720 ka, and the flows seem to cluster into three age groups. Older samples from units in the vicinity of Prospect Canyon–Toroweap fault and from sparse flow remnants below river mile 207 range in age from 720-480 ka. This includes the flows Lucchitta et al. (1999) dated in the Granite Park reach. A middle-aged set of flows dates from 355-296 ka, and the youngest basalts dated are 100-200 ka flows near Whitmore Canyon, suggesting that the flows along the Hurricane fault are relatively young compared to those in the Prospect Canyon area along the Toroweap fault (McIntosh et al., 2002; Karl Karlstrom, personal communication, 2002).

The existence of significant, Quaternary, lava-dam lakes in Grand Canyon has been questioned by those studying Hamblin's (1994) proposed lake deposits upstream (see STOP 3–3 above; Kaufman et al., 2002). Recent research efforts on the basalt flows may concur. As an example, the Prospect Canyon flow sequence was hypothesized by Hamblin (1994) to be the oldest (1.8 Ma) and highest of the lava dams, extending across the gorge and filling the Prospect and Toroweap side canyons. Ongoing research indicates these represent a composite of eruptions that may have lasted from >650 to 500 ka on both sides and within the canyon instead of a single, large dam edifice (Karl Karlstrom, personal communication). Other recent research supports a revised story of lava dams based on study of basalt-dominated, bouldery gravel in the reach below the proposed lava dams, which are interpreted as outburst-flood deposits from failed lava dams (Fenton et al., 2002;

Figure 33. View across Colorado River at Whitmore Canyon takeout. Thick basalt flow occupying canyon bottom is capped by coarse, immature boulder gravel interpreted by Fenton and others (2002) as a lava-dam failure flood deposit.

Fig. 33). Using correlations based on cosmogenic dating and basalt chemistry, Fenton and others (2002) identify two distinct flood deposits that thin and fine strongly downstream, probably both 160 ka old or older. In summary, a model of more dynamic basalt-river interactions, with the river eroding and circumventing lava flows and fewer, smaller, and leakier dams may be more valid (Fenton et al, 2002; Karl Karlstrom, personal communication).

REFERENCES

Anders, M.D., 2003, Quaternary geology and landscape evolution of eastern Grand Canyon, Arizona: MS thesis, Utah State University, Logan, UT, 147 p.

Anderson, R.S., Betancourt, J.L., Mead, J.I., Hevly, R.H., Adam, D.P., 2000, Middle and Late Wisconsin paleobotanical and paleoclimatic records from the southern Colorado Plateau, USA: Palaeogeography, Palaeoclimatology, Palaeoecology, v. 155, p. 31-57.

Belknap, B., 1969, Powell centennial Grand Canyon river guide:, Westwater Books, Boulder City, NV.

Billingsley, G.H., 1989, Chapter 16: Mesozoic strata at Lees Ferry, Arizona: *in* Elston, D. P., Billingsley, G. H., and Young, R. A., eds., Geology of Grand Canyon, Northern Arizona (with Colorado River Guides), American Geophysical Union, 28th International Geological Congress Field Trip Guidebook T115/315, p. 137-138.

Billingsley, G.H. and Elston, D.P. 1989, Chapter 1: Geologic log of the Colorado River from Lees Ferry to Temple Bar, Lake Mead, Arizona: *in* Elston, D. P., Billingsley, G.H., and Young, R.A., eds., Geology of Grand Canyon, Northern Arizona (with Colorado River Guides), American Geophysical Union, 28th International Geological Congress Field Trip Guidebook T115/315, p. 1-36.

Birdseye, C.H., 1924, Plan and profile of the Colorado River from Lees Ferry, Arizona, to Black Canyon, Arizona-Nevada, and the Virgin River, Nevada: U. S. Geological Survey Map Publication, scale 1:36,680, 21 sheets (A-U).

Breed, C.S., 1969, A century of conjecture on the Colorado River in Grand Canyon: *in* Geology and Natural History of the Grand Canyon Region; 5th Field Conference, Four Corners Geological Society Guidebook, p. 63-67.

Bull, W.B., 1991, Geomorphic responses to climate change: Oxford University Press, NY, 326 p.

Bull, W.B. and Knuepfer, P.L.K., 1987, Adjustments by the Charwell River, New Zealand to uplift and climatic changes: Geomorphology, v. 1, p. 15-32.

Cerling, T.E., Webb, R.H., Poreda, R.J., Rigby, A.D., and Melis, T.S., 1999, Cosmogenic 3-He ages and frequency of late Holocene debris flows from Prospect Canyon, Grand Canyon, USA: Geomorphology, v. 27, p. 93-111.

Coats, L.L., 1997, Middle to late Wisconsin vegetation change at Little Nankoweap, Grand Canyon National Park, Arizona: MS thesis, Northern Arizona University, Flagstaff, AZ, 139 p.

Cole, K.L., 1990, Reconstruction of past desert vegetation Along the Colorado River using packrat middens: Palaeogeography, Palaeoclimatology, Palaeoecology, v. 76, pgs. 349-366.

Connin, S.L., Betancourt, J., Quade, J., 1998, Late Pleistocene C_4 plant dominance and summer rainfall in the southwestern United States from isotopic study of herbivore teeth: Quaternary Research, v. 50, p. 179-193.

Dalrymple, G.B. and Hamblin, W.K., 1998, K-Ar ages of Pleistocene lava dams in the Grand Canyon in Arizona: Proceedings, National Academy of Science, v. 95, p. 9744-9749.

Davis, W.M., 1901, An excursion to the Grand Canyon of the Colorado: Bulletin of the Museum of Comparative Zoology, Harvard College, v. 38, Geological Series, v.4, no. 4, p. 107-201.

Dumitru, T.A., Duddy, I.R., and Green, P.F., 1994, Mesozoic-Cenozoic burial, uplift, and erosion history of the west-central Colorado Plateau: Geology, v. 22, p. 499-502.

Dolan, R., Howard, A., and Trimble, D., 1978, Structural control of the rapids and pools of the Colorado River in the Grand Canyon: Science, v. 202, p. 629-631.

Dutton, C.E., 1882, The Tertiary history of the Grand Canyon district: U. S. Geological Survey Monograph 2, 264 p.

Elston, D.P., 1989, Pre-Pleistocene deposits of aggradation, Lees Ferry to western Grand Canyon, Arizona: *in* Elston, D. P., Billingsley, G.H., and Young,

R.A., eds., Geology of Grand Canyon, Northern Arizona (with Colorado River Guides), American Geophysical Union, 28th International Geological Congress Field Trip Guidebook T115/315, p. 175-185.

Elston, D.P. and Young, R.A., 1991, Cretaceous-Eocene (Laramide) landscape development and Oligocene-Pliocene drainage reorganization of the Transition Zone and Colorado Plateau, Arizona: Journal of Geophysical Research, v. 96, p. 12,389-12,406.

Euler, R.C., 1984, The archeology, geology, and paleobiology of Staton's cave, Grand Canyon National Park, Arizona: Grand Canyon Natural History Association Monograph 6, 141p.

Faulds, J. E., Schreiber, B. C., Reynolds, S. J., Gonzalez, L. A., and Okaya, D., 1997, Origin and paleogeography of an immense, nonmarine Miocene salt deposit in the Basin and Range (western USA): Journal of Geology, v. 105, p. 19-36.

Fenton, C.R., Webb, R.H., Pearthree, P.A., Cerling, T.E., Poreda, R.J., 2001, Displacement rates on the Toroweap and Hurricane faults: implications for Quaternary downcutting in Grand Canyon, Arizona: Geology, v. 29, p. 1035-1038.

Fenton, C.R., Cerling, T.E., Nash, B.P., Webb, R.H., and Poreda, R.J., 2002, Cosmogenic 3-He ages and geochemical discrimination of lava-dam outburst-flood deposits in western Grand Canyon, Arizona: *in* House, P.K., ed., Ancient Floods, Modern Hazards, Principles and Applications of Paleoflood Hydrology, American Geophysical Union, Water Science and Application, v. 5, p. 191-215.

Gardner, T.W., Jorgensen, D.W., Shuman, C., and Lemieux, C.R., 1987, Geomorphic and tectonic process rates: Effects of measured time interval: Geology, v. 15, p. 259-261.

Gilbert, G.K., 1876, The Colorado Plateau Province as a field for geological study: American Journal of Science, v. 12, p. 16-24.

Graf, J. B., Schmidt, J. C., and Kieffer, S. W., 1989, Chapter 2: hydraulic log of the Colorado River from Lees Ferry to Diamond Creek, Arizona: *in* Elston, D. P., Billingsley, G. H., and Young, R. A., eds., Geology of Grand Canyon, Northern Arizona (with Colorado River Guides), American Geophysical Union, 28th International Geological Congress Field Trip Guidebook T115/315, p. 37-47.

Grams, P.E., Schmidt, J.C., Topping, D.J., and Goeking, S., 2002, The degraded reach: rate and pattern of bed and bank adjustment of the Colorado River in the 25 km immediately downstream from Glen Canyon Dam: report to the Grand Canyon Monitoring and Research Center, 83 p.

Griffiths, P.G., Webb, R.H., Melis, T.S., 1996, Initiation and frequency of debris flows in Grand Canyon, Arizona: U.S. Geological Survey Open-File Report 96-491, 35 p.

Hack, J.T., 1942, Erosion and sedimentation in the Jeddito Valley and other valleys of the western Navajo country: Cambridge, Papers of the Peabody Museum of American Archaeology and Ethnology, n. 35, p. 45-69.

Hallet, B, Hunter, L., Bogen, J., 1996, Rates of erosion and sediment evacuation by glaciers: A review of field data and Their Implications: Global and Planetary Change, v. 12, p. 213-235.

Hamblin, W.K., 1994, Late Cenozoic lava dams in the western Grand Canyon: Geological Society of America Memoir 183, 139 p.

Hamblin, W.K., Damon, P.E., and Bull, W.B., 1981, Estimates of vertical crustal strain rates along the western margins of the Colorado Plateau: Geology, v. 9, p. 293-298.

Hamblin, W. K. and Rigby, J. K., 1968, Guidebook to the Colorado River, Part 1: Lee's Ferry to Phantom Ranch in Grand Canyon National Park, Studies for Students No. 4: Brigham Young University Geology Series Provo, UT, v. 15, Part 5, 84 p.

Hamblin, W. K. and Rigby, J. K., 1969, Guidebook to the Colorado River, Part 2: Phantom Ranch in Grand Canyon National Park to Lake Mead, Arizona-Nevada, Studies for Students No. 5: Brigham Young University Geology Series, Provo, UT, v. 16, Part 2, 126 p.

Harvey, A.M. and Wells, S.G., 1994, Late Pleistocene and Holocene changes in hillslope sediment supply to alluvial fan systems: *in* Millington, A.C., Pye, K., (Eds.), Environmental Change in Drylands, Biogeographical and Geomorphological Perspectives, John Wiley and Sons, NY, p. 67-84.

Hereford, R., 1984, Driftwood in Stanton's Cave—The case for temporary damming of the Colorado River at Nankoweap Creek in Marble Canyon, Grand Canyon National Park, Arizona: *in* Euler, R.C., ed., The Archeology, Geology, and Paleobiology of Staton's Cave, Grand Canyon National Park, Arizona: Grand Canyon Natural History Association Monograph 6, p. 99-106.

Hereford, R., 1993, Map showing surficial geology and geomorphology of the Palisades Creek archeologic area, Grand Canyon National Park, Arizona: U.S. Geological Survey Open–File Report 94–563, 21 p.

Hereford, R., 2002, Valley-fill alluviation during the Little Ice Age (ca. A.D. 1400-1880), Paria River basin and southern Colorado Plateau, United States: Geological Society of America Bulletin, v. 114, p. 1550-1563.

Hereford, R., Thompson, K.S., Burke, K.J. and Fairley, H.C., 1996, Tributary debris fans and the late Holocene alluvial chronology of the Colorado River, eastern Grand Canyon, Arizona: Geological Society of America Bulletin, v. 108, p. 3-19.

Hereford, R. Thompson, K.S., and Burke, K.J., 1998a, Numerical ages of Holocene tributary debris fans inferred from dissolution pitting on carbonate boulders in the Grand Canyon of Arizona: Quaternary Research, v. 50, p. 139-147.

Hereford, R., Burke, K.J., and Thompson, K.S., 1998b, Quaternary geology and geomorphology of the Nankoweap rapids area, Marble Canyon, Arizona: U. S. Geological Survey Investigation Series Map I-2608.

Howard, A. and Dolan, R., 1981, Geomorphology of the Colorado River in Grand Canyon: Journal of Geology, v. 89, p. 269-297.

Humphreys, E.D. and Dueker, K.G., 1994, Western U. S. upper mantle structure: Journal of Geophysical Research, v. 99, p. 9615-9634.

Hunt, C. B., 1969, Geologic history of the Colorado River: *in* The Colorado River Region and John Wesley Powell: U.S. Geological Survey Professional Paper 669-C, p. 59-130.

Huntoon, P.W., 2003, Post-Precambrian tectonism in the Grand Canyon region: *in* Beus, S.S. and Morales, M., eds. Grand Canyon Geology: Oxford University Press, NY, p. 222-259.

Jackson, G., 1990, Tectonic geomorphology of the Toroweap fault, western Grand Canyon, Arizona: Implications for transgression of faulting on the Colorado Plateau: Arizona Geological Survey Open–File Report 90–4, 67 p.

Jones, C. H., Unruh, J. R., Sonder, L. J., 1996, The role of gravitational potential energy in active deformation in the Southwestern United States: Nature v. 381, p. 37-41.

Kaufman, D.S., O'Brien, G., Mead, J.I., Bright, J., and Umhoefer, P., 2002, Late Quaternary spring-fed deposits of the Grand Canyon and their implication for deep lava-dammed lakes: Quaternary Research, v. 58, p. 329–340.

Kieffer, S.W., 1985, The 1983 hydraulic jump in Crystal rapid: implications for river-running and geomorphic evolution in the Grand Canyon: Journal of Geology 93: 385-406.

Leopold, L.B., 1969, The rapids and the pools–Grand Canyon: *in* The Colorado River region and John Wesley Powell: U. S. Geological Survey Professional Paper 669, p. 131-145.

Leopold, L.B. and Bull, W.B., 1979, Base level, aggradation, and grade: Proceedings of the American Philosophical Society, v. 123, p. 168-202.

Longwell, C.R., 1946, How old is the Colorado River?: American Journal of Science, v. 244, p. 817-835.

Lowry, A.R., Rive, N.M., and Smith, R.B., 2000, Dynamic elevation of the Cordillera, western United States: Journal of Geophysical Research, v. 105, p. 23,371-23,390.

Lucchitta, I., 1966, Cenozoic geology of the upper Lake Mead area adjacent to the Grand Wash Cliffs, Arizona: PhD thesis, Pennsylvania State University, 218 p.

Lucchitta, I., 1972, Early history of the Colorado River in the basin and range province: Geological Society of America Bulletin, v. 83, p. 1933–1947.

Lucchitta, I., 2003, History of the Grand Canyon and of the Colorado River in Arizona: *in* Beus, S.S. and Morales, M., eds., Grand Canyon Geology, Oxford University Press, NY, p. 260-274.

Lucchitta, I, Curtis, G.H., Davis, M.E., Davis, S.W., and Turrin, B., 2000, Cyclic aggradation and downcutting, fluvial response to volcanic activity, and calibration of soil-carbonate stages in western Grand Canyon: Quaternary Research, v. 53, p. 23-33.

Lucchitta, I., Dehler, C.M., Davis, M.E., Basdekas, P.G., Burke, K.J., 1995, Quaternary geologic map of the Palisades Creek-Comanche Creek area, eastern Grand Canyon, Arizona: U.S. Geological Survey Open-File Report 95-832, p. 39.

Machette, M.N. and Rosholt, J.N., 1991, Quaternary geology of the Grand Canyon: *in* Morrison, R.B., ed., The Geology of North America Volume K-2, Quaternary Nonglacial Geology: Conterminous U.S.: Geological Society of America, Denver, p.397-401.

McIntosh, W.C., Peters, L., Karlstrom, K.E., and Pederson, J.L., 2002, New 40Ar-39Ar dates on basalts in Grand Canyon: Constraints on rates of Quaternary river incision and slip on the Toroweap fault, and implications for lava dams: Geological Society of America Abstracts with Programs, v. 22, no. 4, p 61.

McKee, E. D. and McKee, E. H., 1972, Pliocene uplift of the Grand Canyon region; time of drainage adjustment: Geological Society of America Bulletin, v. 83, p. 1923-1932.

Mead, J.I., Phillips, A.M. III, 1981, The late Pleistocene and Holocene fauna and flora of Vulture cave, Grand Canyon, Arizona: The Southwestern Naturalist, v. 26(3), p. 257-288.

Melis, T.S., 1997, Geomorphology of debris flows and alluvial fans in Grand Canyon National Park and their influence on the Colorado River below Glen Canyon dam, Arizona: PhD thesis, University of Arizona, Tucson, AZ, 490 p.

Merritts, D.J., Vincent, K.R., Wohl, E.E., 1994, Long river profiles, tectonism, and eustasy: A guide to interpreting fluvial terraces: Journal of Geophysical Research, v. 99, no. B7, p. 14,031-14,050.

Morgan, P. and Swanberg, C.A., 1985, On the Cenozoic uplift and tectonic stability of the Colorado Plateau: Journal of Geodynamics, v. 3, p. 39-63.

Pederson, J., Karlstrom. K., Sharp, W., McIntosh, W., 2002a, Differential incision of the Grand Canyon related to Quaternary faulting-Constraints from U-series and Ar/Ar Dating: Geology, v. 30, p. 739-742.

Pederson, J.L., Mackley, R.D., and Eddleman, J.L., 2002b, Colorado Plateau uplift and erosion—amounts and causes evaluated with GIS: GSA Today, v. 12, no. 8, p. 4-10.

Pederson, J.L., Pazzaglia, F., Smith, G., 2000, Ancient hillslope deposits: Missing links in the study of climate controls on sedimentation: Geology, v. 28, p. 27-30.

Phillips, F.M., Zreda, M.G., Gosse, J.C., Klein, J, Evanson, E.B., Hall, R.D., Chadwick, O.A., Sharma, P., 1997, Cosmogenic ^{36}Cl and ^{10}Be ages of Quaternary glacial fluvial deposits in the Wind River Range, Wyoming: Geological Society of America Bulletin, v. 128, p. 1453-1463.

Pizzuto, J.E., Webb, R.H., Griffiths, P.G., Elliott, J.G., Melis, T.S., 1999, Entrainment and transport of cobbles and boulders from debris fans: *in* Webb, R.H., Schmidt, J.C., Marzolf, G.R., and Valdez, R.A., eds., The controlled flood in Grand Canyon, American Geophysical Union Geophysical Monograph 110, p 53–70.

Potochnik, A.R. and Faulds, J.E., 1998, A tale of two rivers: Tertiary structural inversion and drainage reversal across the southern boundary of the Colorado Plateau: *in* Duebendorfer, E. M., ed., Geologic Excursions in Northern and Central Arizona, Geological Society of America, Rocky Mountain Section Field Trip Guidebook, p. 149-174.

Powell, J.W., 1875, Exploration of the Colorado River of the west and its tributaries: Smithsonian Institute Annual Report, 291 p.

Reheis, M.C., Palmquist, R.C., Agard, S.S., Jaworowski, C., Mears, B., Jr., Madole, R.F., Nelson, A.R., Osborn, G.D., 1991, Quaternary history of some of southern and central Rocky Mountain basins: *in* Morrison, R.B., ed., Quaternary Nonglacial Geology: Conterminous U.S., Geology of North America Volume K-2: Geological Society of America, Boulder, CO, p. 407-440.

Reilly, P. T., 1999, Lees Ferry, from Mormon crossing to national park: Utah State University Press, Logan, UT, 542 p.

Repka, J.L., Anderson, R.S., Dick, G.S., Finkel, R.C., 1997, Quaternary geology and geomorphology, northern Henry Mountains region; dating the Fremont River terraces: Part 7: *in* Link, P.K., Kowallis, B.J., eds., Mesozoic to Recent Geology of Utah, Brigham Young University Geology Studies, v. 42, p. 398-404.

Ritter, J.B., Gardner, T.W., 1993, Hydrologic evolution of drainage basins disturbed by surface mining, central Pennsylvania: Geological Society of America Bulletin, v. 105, p. 101-115.

Rubin, D.M., 1987, Cross-bedding, bedforms, and paleocurrents: Society of Economic Paleontologists and Mineralogists Concepts in Sedimentology and Paleontology, Volume 1, 187 p.

Rubin, D.M., Anima, R.A., and Sanders, R., 1994, Measurements of sand thickness in Grand Canyon, Arizona, and a conceptual model for characterizing changes in sand-bar volume through time and space: U. S. Geological Survey Open-File Report 94-597, 9 p.

Rubin, D.M., Schmidt, J.C., and Moore, J.N., 1990, Origin, structure, and evolution of a reattachment bar, Colorado River, Grand Canyon, Arizona: Journal of Sedimentary Petrology, v. 60, p. 982-991.

Sahagian, D., Proussevitch, A., Carlson, W., 2002, Timing of Colorado Plateau uplift: Initial constraints from vesicular basalt-derived paleoelevations: Geology, v. 30, p. 807-810.

Savage, J.E., Huntoon, P.W., and Warme, J.E., 2002, Effects of large-scale landsliding in central Grand Canyon National Park, Arizona: Geological Society of America Abstracts with Programs, v. 22, no. 4, p. 61.

Schmidt, J.C., 1990, Recirculating flow and sedimentation in the Colorado River in Grand Canyon: Journal of Geology, v. 98. p. 709-724.

Schmidt, J.C. and Graf, J.B., 1990, Aggradation and degradation of alluvial sand deposits, 1965-1986, Colorado River, Grand Canyon National Park, Arizona: U. S. Geological Survey Professional Paper 1493, 74 p.

Schmidt, J.C., Grams, P.E., and Leschin, M.F., 1999, Variation in the magnitude and style of deposition and erosion in three long (8-12 km) reaches as determined by photographic analysis: *in* Webb, R. H., Schmidt, J. C., Marzolf, G. R., and Valdez, R. A., eds., The Controlled Flood in Grand Canyon: American Geophysical Union Geophysical Monograph 110, p. 185-203.

Schmidt, J.C., and Rubin, D.M., 1995, Regulated streamflow, fine-grained deposits, and effective discharge in canyons with abundant debris fans: *in* Costa, J. E., Miller, A. J., Potter, K. W., and Wilcock, P. `R., eds., :Natural and Anthropogenic Influences in Fluvial Geomorphology, AGU Geophysical Monograph 89, p. 177-195.

Schmidt, J.C., Topping, D.J., Goeking, S., Sondossi, H., Hazel, Jr., J.E., and Grams, P.E., 2002, System-wide changes in the distribution of fine-grained alluvium in the Colorado River corridor between Lees Ferry and Bright Angel Creek, Arizona, 1980s to 2001: Report to the Grand Canyon Monitoring and Research Center, 86 p.

Sharp, W.D., Ludwig, K.R., Chadwick, O.A., Amundson, R.A., and Glaser, L.L., in press, Dating fluvial terraces by 230-Th/U on pedogenic carbonate, Wind River Basin, Wyoming: Quaternary Research.

Simmons, G.C. and Gaskill, D.L., 1969, River runners' guide to the Canyons of the Green and Colorado Rivers with emphasis on geologic features, volume III, Marble Gorge and Grand Canyon: Northland Press, Flagstaff, AZ, 132 p.

Sondossi, H.A., and Schmidt, J.C., 2001. Sand bars of the Colorado River near Lees Ferry, Arizona, and their changes with time: Utah State University report to the Grand Canyon Monitoring and Research Center, Logan, UT, contract 1425-98-FC-40-22640, 54 p.

Stenner, H.D., Lund, W.R., Pearthree, P.A., and Everitt, B.L., 1999, Paleoseismologic investigations of the Hurricane Fault in northwestern Arizona and southwestern Utah: Arizona Geological Survey Open-File Report 99-8, 136 p.

Stevens, Larry, 1983, The Colorado River in Grand Canyon–a comprehensive guide to its natural and human history: Red Lake Books, Flagstaff, AZ, 107 p.

Szabo, B.J., 1990, Ages of travertine deposits in eastern Grand Canyon National Park, Arizona: Quaternary Research, v. 34, p. 24-32.

Topping, D.J., Schmidt, J.C., and Vierra, L.E., Jr., in press, Discharge of the Colorado River at Lees Ferry, Arizona, during the 1884 flood and between May 8, 1921, and September 30, 2000: construction and analysis of a con tinuous record of instantaneous discharge: U. S. Geological Survey Professional Paper.

Topping, D.J., Rubin, D.M., and Vierra, Jr., L.E., 2000, Colorado River sediment transport 1. Natural sediment supply limitation and the influence of Glen Canyon Dam: Water Resources Research, v. 36, p. 515-542.

Tucker, G.E., Slingerland, R., 1997, Drainage basin responses to climate change: Water Resources Research, v. 33, p. 2031-2047.

U.S. Department of the Interior, 1995, Operation of Glen Canyon dam, fFinal environmental impact statement: Bureau of Reclamation, Salt Lake City, UT, 337 p.

Webb, R. H., 1996, Grand Canyon, a century of change: The University of Arizona Press, Tucson, AZ, 290 p.

Webb, R.H., Melis, T.S., Griffiths, P.G., and Elliott, J.G., 1999, Reworking of aggraded debris fans: *in* Webb, R.H., Schmidt, J.C., Marzolf, G.R., and Valdez, R.A., eds., The Controlled Flood in Grand Canyon, American Geophysical Union Geophysical Monograph 110, 37-51.

Webb, R.H., Pringle, P.T., Reneau, S.L., and Rink, G.R., 1988, Monument Creek debris flow, 1984, implications for formations of rapids on the Colorado River in Grand Canyon National Park: Geology, v. 16, p. 50-54.

Webb, R.H., Pringle, P.T. and Rink, G.R., 1989, Debris Flows from tributaries of the Colorado River, Grand Canyon National Park, Arizona: U.S. Geological Survey Professional Paper 1492, 39 p.

Weng, C., Jackson, S.T., 1999, Late Glacial and Holocene Vegetation History and Paleoclimate of the Kaibab Plateau, Arizona: Paleogeography, Paleoclimatology, Paleoecology, v. 153, p. 179-201.

Wenrich, K.J., Billingsley, G.H., and Huntoon, P.W., 1997, Breccia-pipe and geologic map of the northeastern part of the Hualapai Indian reservation and vicinity, Arizona: U.S. Geological Survey Map I-2440.

Winograd, I.J., Coplen, T.B., Landwehr, J.M., Riggs, A.C., Ludwig, K.R., Szabo, B.J., Kolesar, P.T., Revesz, K.M., 1992, Continuous 500,000-year climate records from vein calcite in Devils Hole, Nevada: Science, v. 258, p. 255-260.

Wolfe, J.A., Forest, C.E. and Molnar, P., 1998, Paleobotanical evidence for Eocene and Oligocene paleoaltitudes in midlatitude western North America: Geological Society of America Bulletin, v. 110, p. 664-678.

Young, R.A. and Brennan, W.J., 1974, Peach Springs Tuff: Its bearing on structural evolution of the Colorado Plateau and development of Cenozoic drainage in Mohave County, Arizona: Geological Society of America Bulletin, v. 85, p. 83-90.

Young, R.A. and McKee, E.H., 1978, Early and middle Cenozoic drainage and erosion in west-central Arizona: Geological Society of America Bulletin, v. 89, p. 1745-1750.